C000133425

1 MONTH OF
FREE
READING

at

www.ForgottenBooks.com

By purchasing this book you are eligible for one month membership to ForgottenBooks.com, giving you unlimited access to our entire collection of over 1,000,000 titles via our web site and mobile apps.

To claim your free month visit:

www.forgottenbooks.com/free1236207

* Offer is valid for 45 days from date of purchase. Terms and conditions apply.

ISBN 978-0-332-73780-5
PIBN 11236207

This book is a reproduction of an important historical work. Forgotten Books uses
state-of-the-art technology to digitally reconstruct the work, preserving the original format
whilst repairing imperfections present in the aged copy. In rare cases, an imperfection in
the original, such as a blemish or missing page, may be replicated in our edition. We do,
however, repair the vast majority of imperfections successfully; any imperfections that
remain are intentionally left to preserve the state of such historical works.

Forgotten Books is a registered trademark of FB &c Ltd.
Copyright © 2018 FB &c Ltd.
FB &c Ltd, Dalton House, 60 Windsor Avenue, London, SW19 2RR.
Company number 08720141. Registered in England and Wales.

For support please visit www.forgottenbooks.com

CARCINOGENESIS ABSTRACTS

'A monthly publication of the

National Cancer Institute

Editor
George P. Studzinski, M.D.

College of Medicine and Dentistry

of New Jersey, Newark

Associate Editor
Jussi J. Saukkonen, M.D.

Jefferson Medical College, Philadelphia

NCI Staff Consultants

Elizabeth Weisburger, Ph.D.

Joan W. Chase, M.S.

Literature Selected, Abstracted, and Indexed

by

The Franklin Institute Research Laboratories
Science Information Services
Biomedical Section

Bruce H. Kleinstein, Ph.D., J.D., Group Manager,

Biomedical Projects

Ruthann E. Auchinleck
Production Editor

Contract Number N01-CP-43293

Public Health Service, USDHEW

DHEW Publication No. (NIH) 76-301

616.99405
CAR
v.13
no. 5-8

Carcinogenesis Abstracts is a publication of the National Cancer Institute. The journal serves as a vehicle through which current documentation of carcinogenesis research highlights are compiled, condensed, and disseminated on a regular basis. It represents an integral part of the Institute's program of fostering and supporting coordinated research into cancer etiology. Issues of *Carcinogenesis Abstracts* normally contain three-hundred abstracts and three-hundred citations (unaccompanied by corresponding abstracts). Abstracts and citations refer to the current scientific literature that describes the most significant carcinogenesis research carried on at the National Cancer Institute, other governmental agencies, and private institutions. *Carcinogenesis Abstracts* is intended to be a highly useful current awareness tool for scientists engaged in carcinogenesis research or related areas. The great number and diversity of publications relevant to carcinogenesis make imperative the availability of this service to investigators whose work requires that they keep abreast with current developments in the field.

Carcinogenesis Abstracts is normally published monthly. Volume XIII covers the scientific literature published from Jan 1975 through Dec 1975. To increase the usefulness of *Carcinogenesis Abstracts*, Volume XIII, a Wiswesser Line Notation index and a Chemical Abstracts Service Registry Number index have been provided. These indexes reference compounds described in abstracted articles. A cumulative subject, author, CAS Registry Number, and Wiswesser Line Notation index for Volume XIII will be published shortly after the final regular issue.

Carcinogenesis Abstracts is available free of charge to libraries and to individuals who have a professional interest in carcinogenesis. Requests for *Carcinogenesis Abstracts* from qualified individuals should include statements of their relationship to carcinogenesis research. All correspondence should be addressed as follows.

> *Carcinogenesis Abstracts*
> Room C-325
> Landow Building
> National Cancer Institute
> National Institutes of Health
> Bethesda, Maryland 20014

The Secretary of Health, Education and Welfare has determined that the publication of this periodical is necessary in the transaction of the public business required by law of this Department. Use of funds for printing this periodical has been approved by the Director of the Office of Management and Budget through November 30, 1977.

DEPOSITORY,

OCT 14 1976

UNIV. OF ILL. LIBRARY
AT URBANA-CHAMPAIGN

NOTE

Journal names are abbreviated according to the list of abbreviations used by *Index Medicus*. For s not covered by *Index Medicus*, the abbreviations found in *Chemical Abstracts Service Source Index*, 74 Cumulative, are used. New journals are verified in *New Serial Titles* and abbreviated according rmational Standard ISO 833. An asterisk indicates the author to address (other than the primary) in ing reprints.

LANGUAGE ABBREVIATIONS

Afr.	Afrikaans	Ind.	Indonesian
Ara.	Arabic	Ita.	Italian
Bul.	Bulgarian	Jpn.	Japanese
Chi.	Chinese	Kor.	Korean
Cro.	Croation	Lav.	Latvian
Cze.	Czech	Lit.	Lithuanian
Dan.	Danish	Nor.	Norwegian
Dut.	Dutch	Pol.	Polish
Eng.	English	Por.	Portuguese
Est.	Estonian	Rum.	Rumanian
Fin.	Finnish	Rus.	Russian
Fle.	Flemish	Ser.	Serbo-Croatian
Fre.	French	Slo.	Slovak
Geo.	Georgian	Spa.	Spanish
Ger.	German	Swe.	Swedish
Gre.	Greek	Tha.	Thai
Heb.	Hebrew	Tur.	Turkish
Hun.	Hungarian	Ukr.	Ukrainian
Ice.	Icelandic	Vie.	Vietnamese

ABBREVIATIONS USED IN ABSTRACTS

	angstrom(s)	M	molar
	adrenocorticotropic hormone	mM	millimolar
	adenosine diphosphate	µM	micromolar
	adenosine monophosphate	mOsm	milliosmolar
	adenosine triphosphate	mEq	milliequivalents
	Bacillus Calmette Guerin	min	minute(s)
	twice daily'	mo	month(s)
	degrees centigrade	MTD	maximum tolerated dose
	calorie(s)	N	normal concentration
al	kilocalorie(s)	NAD	nicotinamide adenine dinucleotide
	cubic centimeter(s)	NADH	reduced nicotinamide adenine dinucleotide
	curie(s)	NADP	nicotinamide adenine dinucleotidephosphate
i	millicurie(s)	NADPH	reduced nicotinamide adenine dinucleotide-
i	microcurie(s)		phosphate
	centimeter(s)	ng	nanogram(s) (10^{-9})
	central nervous system	od	once daily
	counts per minute	Pa	ambient pressure
	deciliter(s)	PAS	periodic acid-Schiff
	milliliter(s)	pg	picogram(s) (10^{-12})
	microliter(s)	pgEq	picogram equivalent
	deoxyribonucleic acid	po	orally
	median effective dose	ppb	parts per billion
	ethylenediamine tetraacetic acid	ppm	parts per million
	erythrocyte sedimentation rate	qid	four times daily
	gram(s)	qod	every other day
	kilogram(s)	QO_2	oxygen quotient
	milligram(s)	R	roentgen(s)
	microgram(s)	RBC	red blood cells (erythrocytes)
	hemoglobin	RNA	ribonucleic acid
	hour(s)	sc	subcutaneous
	intra-arterial	sec	second(s)
	intracerebral	SGOT	serum glutamic-oxalacetic transaminase
	intracavitary	SGPT	serum glutamic-pyruvic transaminase
	intradermal ·	SRBS	sheep red blood cells
	increased life span	TCD	tissue culture dose
	intramuscular	TCD_{50}	median tissue culture dose
	intraperitoneal	tid	three times daily
	intrapleural	U	unit(s)
	intratumorous	mU	milliunit(s)
	International Unit	UV	ultraviolet
	intravenous	vol	volume
	Michaelis constant	WBC	white blood cells (leukocytes)
	lethal dose	wk	week(s)

CONTENTS

	Cross Reference Abbreviations	Abstracts, Citations	Page
REVIEW..............................(Rev)..........2401-2451			427
CHEMICAL CARCINOGENESIS..............(Chem).........2452-2560			436
PHYSICAL CARCINOGENESIS..............(Phys).........2561-2575			454
VIRAL CARCINOGENESIS.................(Viral)........2576-2663			457
IMMUNOLOGY...........................(Immun)........2664-2761			477
PATHOGENESIS.........................(Path).........2762-2879			493
EPIDEMIOLOGY AND BIOMETRY............(Epid-Biom).....2880-2911			507
MISCELLANEOUS........................(Misc).........2912-3000			513

AUTHOR INDEX...

SUBJECT INDEX..

CHEMICAL ABSTRACT SERVICES REGISTRY NUMBER INDEX.......................

WISWESSER LINE NOTATION INDEX............................

minobiphenyl. This compound and the corresponding deacetylated derivative have been detected in rat RNA treated with N-hydroxy-acetylaminobiphenyl esters of other carcinogenic arylhydroxamic acids reacted similarly with guanosine. Esters of N-hydroxy-2-acetylaminophenanthrene react equally well with adenosine and guanosine. In model experiments with N-acetoxy-AAF and nucleic acids *in vitro*, no N-acetyl groups have been lost in aqueous solutions. Once the N-acetyl group is removed, the resulting N-(guanine-8-yl)-2-AF moiety becomes unstable. This reaction may represent a mechanism by which release of the bound carcinogen leads to an altered nucleic acid. In addition, the role of guanine in nucleic acid structure is discussed. The authors suggest the possibility that the proton released by substitution of the 8-carbon of guanine would account for the reactivity of DNA to N-acetoxy-AAF and other arylhydroxamic acid esters. (21 references)

2403   FUNCTION AND HOMEOSTASIS OF COPPER AND ZINC IN MAMMALS. (Eng.) Fisher, G. L. (Radiobiology Lab., Univ. California, Davis, Calif. 95616). *Sci. Total Environ.* 4(4):373-412; 1975.

The biochemical and physiological functions of the micronutrients copper and zinc are reviewed; special attention is given to the maintenance of homeostatic levels, the physiologic changes associated with deficiency states, and the physiochemical properties of copper and zinc biomolecules. Trace elements are defined, and the general roles of essential, and non-essential trace elements are discussed. Hazardous trace elements are also noted. The characteristics and properties of metalloenzymes and metal ion activated enzymes are summarized, and direct and indirect interactions with enzyme systems are noted. A discussion of copper as a micronutrient notes the source and functions of the major copper metalloenzymes, and tabulates the levels of copper in tissues and fluids of numerous species; serum copper levels as a function of age in humans are graphically represented. Numerous effects of primary and secondary copper deficiencies are described in detail. Studies of the isolation, characterization, distribution, and function of the serum copper enzyme ceruloplasmin are also detailed, and a scheme for ceruloplasmin-iron interaction is proposed. Disease and environmental factors affecting serum copper levels in humans are summarized. In addition to the circadian pattern noted, serum copper levels are also found to reflect the activity of efficiency of treatment of leukemia, malignant lymphoma, osteosarcoma, and mammary cancers. Attempts to evaluate the physiologic and biochemical function of zinc also note its nonuniform distribution in biologic tissues and fluid of various species. The role of zinc in metalloenzymes, as enzyme activators, and associated with polypeptides, nucleotides, and other macromolecules is also acknowledged. The major zinc metalloenzymes of carbonic anhydrase, carboxypeptidase, alkaline phosphatase, and liver alcohol dehydrogenase are especially noted. The characteristics of zinc deficiency in higher animals are described, and the associated biochemical changes are discussed. In addi-

427

tion to noting the *in vitro* requirements of zinc, the zinc requirements of neoplastic growth are discussed. Changes in serum and plasma zinc levels in human tissues and fluid are reported, and the effects of radiation on zinc levels, uptake, and metabolism are described. However, the exact mechanisms of many of the effects noted are not yet determined. (176 references)

2404   THE CARCINOGENICITY OF DIELDRIN. PART
        II. (Eng.) Epstein, S. S. (Medical Sch.,
Case Western Reserve Univ., Cleveland, Ohio). *Sci. Total Environ.* 4(3):205-217; 1975.

Further results of investigations on the carcinogenicity of Dieldrin are reported. A standard definition of chemical carcinogens, i.e. agents that increase the probability of induction of benign or malignant tumors, is presented. Attempts are made to discount the significance of the Dieldrin-induced carcinogenicity data in mice, due to a high variability in their spontaneous hepatic tumors. However, numerous independent studies reveal an enhanced frequency of "type a" tumors after Dieldrin treatment; an increase from 0-47% of "type b" tumors in female mice has also been found. Despite claims that human babies are at no greater, or even at a lesser risk to Dieldrin carcinogenesis than the general population, newborn animals have been repeatedly shown more sensitive to carcinogenic effects than adult animals. Enhanced sensitivity to aflatoxin B_1, 3-methylcholanthrene, urethane, dimethylnitrosamine, condensed cigarette smoke, and organic extracts of particulate atmospheric pollutants has been demonstrated in infant animals. Much experimental data has indicated that the enhanced sensitivity to carcinogenicity relates to lipid-soluble carcinogens, such as Dieldrin. The carcinogenic effects of Dieldrin are both dose-dependent and time-dependent. In the assessment of policies regarding the regulation of Dieldrin use, the carcinogenicity demonstrated in several strains of mice, the close parallelism of cancers seen in humans and animals, and the involuntary nature of human exposure sbuld all be taken into account. Although no "safe" levels of carcinogens can be determined, continued Dieldrin exposure is an imminent hazard. (19 references)

2405   ORAL CONTRACEPTIVES: RELATION TO MAMMARY
        CANCER, BENIGN BREAST LESIONS, AND CERVICAL
CANCER. (Eng.) Drill, V. A. (Coll. Medicine, Univ. Illinois, Medical Center, Chicago, Ill. 60680). *Annu. Rev. Pharmacol.* 15:367-385; 1975.

The relationship of oral contraceptives to mammary cancer, benign breast lesions, and cervical cancer is reviewed. The type of response obtained depends largely on the species and strain of animal that is employed. A variety of clinical studies has failed to demonstrate that estrogen can cause mammary cancer. Similar relationships exist for oral contraceptives, and the clinical data show good agreement in demonstrating that the contraceptive steroids do not have a tumorigenic effect on the human mammary gland. Estrogen can increase the occurrence of cervical cancer

Current concepts of plutonium (Pu) metabolism are reviewed. Both the amounts and rates of Pu absorption
from the gastrointestinal tract, wound sites, or the
lung decrease in the order: soluble complexes > hy-
drolyzable salts > insoluble compounds. Absorption
of Pu from the gastrointestinal tract is greater in
very young animals than in adults. Deposition of Pu
in bone is greater in growing animals than in young
adults. When introduced as a metabolizable complex
or simple salt, Pu is transported in plasma chiefly
by transferrin. Circulating as the Pu-transferrin
complex, Pu is deposited as single atoms in liver parenchymal cells and at the cell-mineral interface of
those bone surfaces in closest proximity to erythropoietic marrow. Liver deposition is favored when Pu
is introduced directly into the body by parenteral
injection. Skeletal deposition is favored when Pu has
been transported across cell barriers as in the case
of ingestion or inhalation. In the dog, Pu in hepatic
cells is initially associated with soluble ferritin,
and later with subcellular structures. Upon the death
of hepatic cells, Pu-laden debris accumulates in liver
reticuloendotheial (RE) cells. Some of the phagocy-
tized Pu is eventually released from the liver. Concentrations of Pu on bone surfaces decrease in the
order: endosteal > periosteal > Haversian. Plutonium
concentrations on endosteal surfaces decrease in the
order: resorbing > resting > growing. Bone surfaces
labeled with Pu may remain unchanged, be buried by
formation of new bone, or be reabsorbed by osteoclasts.
Some resorbed Pu is dissolved and recirculated, and
may be redeposited in the skeleton. Apparently some
of the Pu in resorbed bone cannot be dissolved and accumulates in osteoclasts in the marrow. Following
death of the osteoclasts, the Pu is taken up by marrow
RE cells, which slowly disappear from the marrow. Insoluble Pu compounds initially deposited in lung or beneath the skin may gradually enter the body without
having been dissolved, and pass by way of RE cells into
the lymph and local lymph nodes, and ultimately into RE
cells in liver, spleen, and bone marrow. (2
references)

2410   SKELETAL AND TISSUE LESIONS RESULTING FROM
        EXPOSURE TO RADIUM AND FISSION PRODUCTS.
(Eng.) Warren, S. (N. Engl. Deaconess Hosp., Boston,
Mass.). *Ann. Clin. Lab. Sci.* 5(2):75-81; 1975.

The effects of absorbed radionuclides and external
radiation on bone and tissue are reviewed. The phys-
ical-chemical nature of the bone-seeking radioactive
isotopes largely determines their ultimate effects.
Those in soluble form are initially widely diffused
in the body; they are then partially excreted and
partially concentrated. Those in particulate form
experience localization according to their means of
access into the body, then redistribution by phagocytosis or slow solution; liver and lungs may be
exposed in addition to bone. ^{226}Ra tends to be
deposited locally more in cortical bone than
trabecular and may be localized in several adjacent
Haversian systems. It is about equal in patchy
distribution near periosteum and endosteum. Thorium
deposits mainly in the endosteum of both trabecular
and cortical bone. It involves some Haversian
systems, as does plutonium. Americum, less effec-

429

tive in inducing bone tumors than plutonium, is widely and equally distributed in endosteal and periosteal bone. The chief effects of absorbed radionuclides and external radiation are radiation osteitis, characterized by pain and evidence of rarefaction of bone; disturbed bone growth; spontaneous fractures; myelofibrosis; and bone tumors, including osteogenic sarcoma near the epiphyses, fibrosarcomas, and plasma cell myeloma. Another result of radium poisoning is cancer of the mucous membrane of the sinuses. In a survey of radiation-induced osteogenic sarcomas, a practical threshold is observed of 1,111 rads, implying a permissible burden of 0.1 mCi of ^{226}Ra. (14 references)

2411   SUNLIGHT, SKIN CANCER, AND SUNSCREENS.
         (Eng.) Kaminester, L. H. (Suite 201B,
The Gentry, 860 US Highway 1, North Palm Beach,
Fla. 33408). *J.A.M.A.* 232(13):1373-1374; 1975.

Effects of acute or prolonged exposure to UV light and measures to prevent or minimize such effects are discussed. On acute exposure, the erythrogenic spectrum causes not only sunburn but may also induce phototoxic and photoallergic reactions and precipitate changes such as xeroderma pigmentosum, solar urticaria, and systemic lupus erythematosus. A single exposure to erythrogenic UV light has been shown to inhibit DNA, RNA, and protein synthesis in human skin. Prolonged exposure to UV light may lead to the development of premalignant and malignant skin lesions, primarily actinic keratosis, basal cell carcinomas, and squamous cell carcinomas. The actinic or solar keratosis is found predominantly in sunny climates, on white-skinned persons who often have a poor ability to tan and may progress to frank squamous cell carcinoma. The basal cell carcinoma is usually a round or oval lesion with central depression, superficial telangiectasis and a gray or translucent pearly border. Some lesions also contain melanin. The squamous cell carcinoma has a variable clinical picture, from a small, firm, red nodule to a red, verrucose or scaly growth. Squamous cell carcinomas arising in sun-damaged areas have a far lower incidence of metastasis than do squamous cell carcinomas arising in scars or *de novo*. The most important factor in prevention of these unpleasant skin disorders is to avoid excessive sun exposure. There are two types of sunscreens, physical and chemical blockers. The physical blockers deflect and scatter light, providing a barrier to sunlight penetration. They have the general advantage of giving more inclusive coverage over a wider range of the UV spectrum, but are not cosmetically acceptable. The chemical sunscreens absorb light of a particular wavelength. While they are usually more cosmetically acceptable, they have several disadvantages, including sensitization or primary irritation and nonuniform coverage of the complete UV spectrum. The absorption characteristics and clinical efficacy studies performed with aminobenzoic acid indicate that it is one of the best agents for protection within the erythrogenic spectrum. Systemic photoprotective agents have not found widespread acceptance. Drugs such as indomethacin and aspirin, which cause a dramatic decrease in the erythema and elevated skin temperature of sunburn, may actually be deleterious in the long run by masking the acute changes of sun exposure while doing nothing to prevent the long-term sequelae. (12 references)

2412   NEWER KNOWLEDGE IN COMPARATIVE VIROLOGY--
         ITS CONTRIBUTION TO HUMAN HEALTH RESEARCH.
(Eng.) Cabasso, V. J. (Cutter Labs. Inc., Berkeley,
California). *Vet. Rec.* 96(26):563-566; 1975.

Progress in the field of comparative virology is reviewed. The study of Marek's disease found application in that of Burkitt's lymphoma, and may lead to a possible vaccine against the human disease. Equally useful information came from the study of canine distemper in the development of a chorio-allantoic membrane attenuated measles vaccine, and in present knowledge of subacute sclerosing panencephalitis (SSPE) of humans; from the study of reovirus-like agents of infant mice and neonatal calves in that of an acute nonbacterial gastroenteritis of infants and young children; and from that of the cancer-producing viruses of chickens, cats, and dogs to a better under standing of some human neoplasias. Finally, Aleutian mink disease may be an excellent natural model for the study of the collagen diseases of man, and scrapi of sheep one for that of a human chronic degenerative disease of the central nervous system of humans such as Kuru. It is concluded that the comparative study of viruses has allowed more rapid progress in the knowledge and control of human pathogenic viruses, and that this approach to the study of viral diseases should be continued. (46 references)

2413   EPSTEIN-BARR-VIRUS-(EBV)-ANTIGENS. (Eng.)
         Diehl, V. (Hämatologisch-onkologische
Abteilung, Medizinische Hochschule Hannover, West
Germany). *Proc. Int. Cancer Congr. 11th.* Vol. 2
(Chemical and Viral Oncogenesis). Florence, Italy,
October 20-26, 1974. Edited by Bucalossi, P.;
Veronesi, U.; Cascinelli, N. New York, American
Elsevier, 1975, pp. 182-187.

Epstein-Barr virus (EBV) macromolecules can be detected by immunological and biochemical means *in vitro* and *in vivo*. There exist at least four different EBV-associated antigens. The EBV-nuclear antigen (EBNA) may reflect a T antigen. Its appearance correlates in 100% to the presence of the EBV DNA. *In vivo*, restriction of the virus genome permits only the expression of EBNA and membrane antigen. The mechanisms of this restriction are still unknown. Virus particles have not yet been detected in biopsy material by electron microscopy. EBV-specific macromolecules have been discovered *in vivo* in two different cell types, the lymphoblastoid cells (B-cells) and the epithelial cells in nasopharyngeal carcinoma. Both cell types represent transformed cells. Thus far, the permissive *in vivo* cell for EBV has not been detected. *In vitro*, EBV is found exclusively in B-lymphocytes. T-cells do exclude infection by EBV and show no production of EBV associated products. Three types of B-lymphoblastoid cell lines exist with regard to EBV-macromolecule synthesis: (a) EBV-negative

2415   NEUROBLASTOMA, IMMUNODEFICIENCY, AND
         CATECHOLAMINES. (Eng.) Wood, L. (1188
North Euclid, Anaheim, Calif. 92801). *Lancet* 1
(7915):1091-1092; 1975.

The potential of the immune system in the prevention
and the response to treatment of malignancy is discussed. It is acknowledged that patients with congenital immunodeficiency have an increased incidence
of neoplasia, that many human neoplasms have tumor-
specific antigens, and that cancer patients with
immunological deficiency have a poor prognosis; the
importance of immunocompetence in cancer patients
is suggested. Adrenaline is noted to modify a wide
variety of immune processes by increasing leukocyte
cyclic-AMP levels, thus inducing production or secretion of antibody. Patients with neuroblastoma are
in double jeopardy--from a malignant tumor and from
the substances it may elaborate which specifically
impair their immune responses. Although urinary
adrenaline is generally found within the normal
range, various catecholamine precursors and metabolites are increased in neuroblastoma patients.
Whereas adrenaline may be of trivial importance, the
possibility of such compounds of similar structure
stimulating B-adrenergic receptors prompts the suggestion that adjuvant treatment with B-adrenergic
antagonists may improve the prognosis in patients
with neuroblastoma. (6 references)

2416   CURRENT STATUS OF CARCINOEMBRYONIC ANTIGEN
         ASSAY. (Eng.) Chu, T. M. (Roswell Park
Memorial Inst. and State Univ. New York at Buffalo,
Buffalo, N.Y.). *Semin. Nucl. Med.* 5(3):255-262; 1975.

The usefulness of carcinoembryonic antigen (CEA) in
the detection and diagnosis of cancer was evaluated.
CEA is a glycoprotein that can be measured by radioimmunoassay and other immunologic techniques. The
CEA reagent is currently commercially available in
kit form and has been found to be satisfactory for
laboratory use. The highest percentage of elevated
plasma CEA and the highest CEA titer have been found
in patients with entodermally derived tumors. Current CEA reagents do not assist in the differential
diagnosis of tumors, however. CEA has also been detected in patients with other tumors and in patients
with nonneoplastic disease, as well as in heavy cigarette smokers. The present CEA assay cannot be used to
screen for cancer in the general population. The
greatest clinical usefulness of current CEA assay is
in assessing prognosis, in the detection of residual
tumors and recurrent disease, and in monitoring chemotherapy and radiotherapy. The use of radioactive
anti-CEA antiserum as a tumor-localizing agent may be
of potential value in the future. (35 references)

2417   MITOGENS AND T-CELL HETEROGENEITY. (Eng.)
         Dutton, R. W. (Dep. Biol., Univ. California,
San Diego). *Ann. N.Y. Acad. Sci.* 249:43-46; 1975.

The results of a study on the effect of concanavalin
A (Con A) on the *in vitro* immune response of normal
mouse spleen cell suspensions are reviewed. Con
A added to spleen cell suspensions stimulated thy-

midine incorporation and inhibited primary and
secondary IgM response to SRBC or burro RBC (BRBC)
or the anti-trinitrophenyl (TNP) response to TNP-
SRBC. Optimal response occurred at 2 µg/ml Con A.
No inhibition was observed when Con A was added
24 hr after the culture initiation. The inhibitory
effects of Con A were also abolished by treatment
with goat anti-mouse brain antisera followed by
complement or by X-irradiation. In the absence
of inhibitory effects, Con A induced a stimulatory
effect on the immune response of heterologous RBC.
The cell-mediating stimulatory effect was present
after treatment with goat anti-mouse brain and
complement, in AT x BM mice, and persisted in adult
thymectomized mouse spleens long after thymectomy.
It was present in lymph node cells, but not in
thymus, and was also not found in the spleens of
nu/nu mice. Spleen cell suspensions preincubated
with Con A for 24 hr inhibited the humoral response
of fresh syngeneic spleen cells in the absence of
Con A. The author stresses the need to remember
that the data reported here represent net effects,
which result from combined inhibitory and stimulatory effects. He concludes that because activity
was absent from nu/nu mouse spleens, both the inhibitory and stimulatory activities are mediated by
T cells and proposes that the Con A-induced activities are mediated by two separate T cell types.
(6 references)

2418   IMMUNITY AND TUMORS OF THE NERVOUS SYSTEM.
(Eng.) Brooks, W. H. (VA Hosp., Lexing-
ton, Ky., 40506); Netsky, M. G.; Levine, J. E. *Surg.
Neurol.* 3(4):184-186; 1975.

The immunological privileged site of the brain, with
respect to the lack of immune response to antigens
within the brain, appears only relative. Animals
bearing a methylcholanthrene-induced ependymoma
transplant, either intracranial or sc, and subsequently cured by irradiation, show a 37% rejection
of a second cerebral implant in the first group and
64% in the second; this indicates that somewhat
greater immunity was induced by sc implants than by
implants within the brain. Patients with primary
tumors of the CNS show a cell-mediated response a-
gainst their own tumor cells in culture. Blocking
antibodies against glial tumor cells are demonstrable, but are lost postoperatively. Sera of 17 of
26 patients with astrocytomas have been shown to be
cytotoxic to tumor cells grown in culture. Prior
immunization of mice with brain tumor cells brings
about significant inhibition and rejection of challenge inoculations of the tumor. A scientific approach to immunotherapy against tumors of the nervous system in man requires precise antigenicity of
the tumor tissues, because surrounding normal brain
may have related antigens and may be subject to destruction by autoimmune mechanisms. The immunother-
apeutic approach thus appears highly speculative.
This approach may be highly dangerous because of the
nonspecificity of anti-glioma antibodies, the evidence of shared antigens between brain and lymphoid
tissue, and the danger of the production of enhancing antibodies augmenting tumor growth. (36 references).

mechanism, none can function as isolated proteins. Further investigations into the DNA growing point problem suggest two alternative solutions: existence of a 5'-3' DNA polymerase, or net elongation *via* discontinuous synthesis. Results support an optional discontinuous synthesis of both daughter strands. A knife and fork model proposed for the formation of the DNA fragment falters due to the disagreement with experimental predictions. However, using the single-stranded DNA phages φX and M13 as an experimental system, a rolling circle intermediate capable of forming both double-stranded and single-stranded progeny DNA is supported. Studies on initiating DNA strands *de novo* note the use of preformed oligonucleotides and of RNA oligonucleotides synthesized *in situ* as primers. The uniform size of the RNA moiety, the presence of initiating triphosphate end groups, and the copurification of the RNA with DNA fragments support the RNA primer hypothesis. A postulated replisome is discussed, and a teleological concept that initiating a DNA fragment with RNA facilitates the recognition, excision, and repair of any probable mistakes is presented. (223 references).

2423 A CRITICAL REASSESSMENT OF THE EVIDENCE BEARING ON SMOKING AS THE CAUSE OF LUNG CANCER. (Eng.) Sterling, T. D. (Dep. Appl. Math. Comput. Sci., Washington Univ., St. Louis, Mo.). *Am. J. Public Health* 65(9):939-953; 1975. (116 references)

2424 SACCHARIN AND BLADDER CANCER. (Eng.) Anonymous. *Br. Med. J.* 3(5984):610;1975. (6 references)

2425 ENCYCLOPEDIA OF FOOD CHEMICALS. (Eng.) Anonymous. *Food Can.* 35(4):33-56; 1975. (1 reference)

2426 CARCINOGENIC AGENT? (Eng.) Anonymous. *Nature* 257(5529):735; 1975. (No references)

2427 BACKGROUND INFORMATION ON TRICHLOROETHYLENE. (Eng.) Lloyd, J. W. (Office Occupational Health Surveillance and Biometrics, NIOSH, 5600 Fishers Lane, Rockville, Md. 20852); Moore, R. M., Jr.; Breslin, P. *J. Occup. Med.* 17(9):603-605; 1975. (20 references)

2428 BENZPYRENE--CARCINOGENIC RISKS. (Swe.) Sandberg, C. G. (Hagfors Jarnverk, S-683 01 Hagfors 1, Sweden). *Lakartidningen* 72(33):3081-3083; 1975. (32 references)

2429 METABOLIC ACTIVATION AND REACTIVITY OF CHEMICAL CARCINOGENS. (Eng.) Miller, J. A. (Univ. Wisconsin Medical Center, Madison, Wis. 53706); Miller, E. C. *Mutat. Res.* 33(1):25-26; 1975. (4 references)

2430 CARCINOGENS ARE MUTAGENS: A SIMPLE TEST SYSTEM. (Eng.) Ames, B. N. (Dept. Bio-

chemistry, Univ. California, Berkeley, Calif. 94720);
McCann, J.; Yamasaki, E. *Mutat. Res.* 33(1):27-28;
1975. (9 references)

2431   CANCER AT WORK. (Eng.) Anonymous. *Na-
        ture* 257(5523):170-171; 1975. (No refer-
ences)

2432   VINYL CHLORIDE--AN EXAMPLE OF NEGLECTED
        RISK IN THE WORKING ENVIRONMENT. (Swe.)
Olin, R. (Health Center, Technical Univ., Stockholm,
Sweden). *Lakartidningen* 72(7):540-541; 1975. (26
references)

2433   PRELIMINARY ASSESSMENT OF THE ENVIRONMENTAL
        PROBLEMS ASSOCIATED WITH VINYL CHLORIDE
AND POLYVINYL CHLORIDE. (Eng.) Anonymous (Environmental Protection Agency, Washington, D.C. Office
of Toxic Substances). 106 pp., 1974. [available
through National Technical Information Services,
Washington, D.C. Document No. PB-239 110/OWP].

2434   CHARACTERIZATION AND CONTROL OF ASBESTOS
        EMISSIONS FROM OPEN SOURCES. (Eng.) Har-
wood, C. F. (IIT Res. Inst., Chicago, Ill.); Blas-
zak, T. P. 204 pp., 1974. [available through
National Technical Information Services, Washington
D.C. Document No. PB-238 925/2GA].

2435   BIOLOGY OF AVIAN RNA RUMOUR VIRUSES.
        (Eng.) Hlozanek, I. (Inst. Experimental
Biology and Genetics, Czechoslovak Acad. of Sciences,
Prague, Czechoslovakia). In: *Reviews in Leukaemia
and Lymphoma, 1--Advances in Acute Leukaemia* edited
by Cleton, F. J.; Crowther, D.; Malpas, J. S. Amsterdam, North Holland Publishing Co., 1974, pp. 229-
301. (460 references)

2436   C-TYPE VIRUSES IN CHIMPANZEE (*PAN* SP.)
        PLACENTAS. (Eng.) Kalter, S. S. (South-
west Foundation for Res. and Education, P.O. Box
28147, San Antonio, Tex. 78284); Heberling, R. L.;
Smith, G. C.; Helmke, R. J. *J. Natl. Cancer Inst.*
55(3):735; 1975. (8 references)

2437   CELLULAR RECEPTORS OF THE IMMUNE SYSTEM.
        (Rus.) Galaktionov, V. G. (N. I. Pirogov
Second Moscow State Medical Inst., U.S.S.R.). *Usp.
Sovrem. Biol.* 80(1):84-101; 1975. (126 references)

2438   HOW CANCER AND THE FETUS ELUDE THE IMMUNE
        DEFENSE. (Fre.) Gallois, J.-M. (No af-
filiation given). *Recherche* 6(54):268-269; 1975.
(10 references)

2439   DIFFUSE NESIDIOBLASTIC HYPERPLASIA OF THE
        PANCREATIC ISLETS IN INFANTILE HYPOGLYCEMIA
SYNDROME. (Ger.) Kloppel, G. (Pathologisches Institut der Universitat, 2 Hamburg 20, Martinistr.
52, West Germany); Seifert, G. *Dtsch. Med. Wochenschr.* 100(38):1906-1910; 1975. (43 references)

ESOPHAGEAL CANCER IN CHINA. (Eng.) A-
onymous. *Br. Med. J.* 3(5975):61, 1975.
ces)

2450   CELL SURFACE PROTEIN: NO SIMPLE CANCER
        MECHANISMS. (Eng.) Kolata, G. B. (No
affiliation given). *Science* 190(4209):39-40; 1975.
(No references)

XCESS BLADDER CANCER IN BEAUTICIANS [let-
er to editor]. (Eng.) Menkart, J.
 Blachley Road, Stamford, Conn. 06902).
(4210):96-98; 1975. (No references)

2451   ARE LYSOSOMAL ENZYMES INVOLVED IN REGU-
        LATION OF CELL PROLIFERATION? (Ger.)
Seelich, F. (Institut fur Krebsforschung der Universitat Wien, Wien, Austria). *Oesterr. Z. Onkol.* 2
(2/3):31-37; 1975.

2452   CHEMICAL INDUCTION OF SPERM ABNORMALITIES
        IN MICE. (Eng.) Wyrobek, A. J. (Ontario
Cancer Inst., 500 Sherbourne St., Toronto, Ontario
M4X 1K9, Canada); Bruce, W. R. *Proc. Natl. Acad.
Sci. USA* 72(11):4425-4429; 1975.

The sperm of (C57BL x C3H)F_1 mice were examined 1,
4, and 10 wk after a subacute treatment (five consecutive days, daily ip injections) with one of 25
chemicals at two or more dose levels. The fraction
of sperm that were abnormal in shape was elevated
above control values of 1.2-3.4% for methyl methanesulfonate, ethyl methanesulfonate, griseofulvin, benzo[a]pyrene, tris(2-methyl-1-aziridinyl)phosphine
oxide, tris(1-aziridinyl)phosphine sulfide, mitomycin C, myleran, vinblastine sulphate, hydroxyurea,
3-methylcholanthrene, colchicine, actinomycin D, imu-
ran, cyclophosphamide, 5-iododeoxyuridine, dichlorvos, aminopterin, and trimethylphosphate. Dimethylnitrosamine, urethane, DDT [1,1,1-trichloro-2,2-bis
(p-chlorophenyl)ethane], 1,1-dimethyl-hydrazine, caffeine, and calcium cyclamate did not induce elevated
levels of sperm abnormalities. The results suggest
that sperm abnormalities might provide a rapid inexpensive mammalian screen for agents that lead to errors in the differentiation of spermatogenic stem
cells *in vivo* and thus indicate agents that might
prove to be mutagenic, teratogenic, or carcinogenic.

2453   MUTATIONS TO AZAGUANINE RESISTANCE IN-
        DUCED IN CULTURED DIPLOID HUMAN FIBRO-
BLASTS BY THE CARCINOGEN, *N*-ACETOXY-2-ACETYLAMINO-
FLUORENE. (Eng.) Maher, V. M. (Michigan Cancer
Found., Detroit); Wessel, J. E. *Mutat. Res.* 28(2):
277-284; 1975.

The ability of *N*-acetoxy-2-acetylaminofluorene (N-
AcO-AAF) to induce mutations to azaguanine resistance in diploid human fibroblasts was quantitatively
investigated. Male diploid cells in thymidine-hy-
poxanthine-aminopterin culture medium were plated
in dishes, allowed to attach, and exposed to 1-10
mM N-AcO-AAF for four hours. Following exposure to
the carcinogen, the medium was replaced with fresh
culture medium and sufficient time allowed for three
cell divisions to occur in order to overcome phenotypic lag of cellular expression of azaguanine resistance by diluting out of hypoxanthine(guanine)
phosphoribosyltransferase. A parallel series of cytotoxicity experiments was set up to determine the
number of viable cells per dish for each dose of
N-AcO-AAF. The cytotoxic and mutagenic effect of
exposure to N-AcO-AAF was clearly dose-dependent.
The killing action followed single hit kinetics
and the mutagenic action was directly proportional
to the number of mean lethal events per cell. Mutant clones resistant to azaguanine (2.10^{-5}M) retained normal sensitivity to N-AcO-AAF and also retained the mutant phenotype over several generations
in the absence of azaguanine. These results strength-
en the correlation between carcinogencity and muta-
gencity of N-AcO-AAF and imply that somatic mutations by such a carcinogen could ultimately be involved causally in the carcinogenetic process.

2458   REDUCTION IN BINDING OF [^{14}C] AFLATOXIN B$_1$
        TO RAT LIVER MACROMOLECULES BY PHENOBARBI-
TONE PRETREATMENT. (Eng.) Garner, R. C. (Dept. Exp.
Pathol. Cancer Res., Univ. Leeds, England) *Biochem.
Pharmacol.* 24(16):1553-1556; 1975.

The binding of [^{14}C]aflatoxin B to rat liver
macromolecules was investigated in control and
phenobarbitone-treated animals in an attempt to
clarify seemingly contradictory findings from previous
studies in some of which phenobarbitone exerted a protective effect against aflatoxin B$_1$, and in others increased aflatoxin B$_1$ metabolism. Male Wistar rats were
injected ip with 40 μg/100 g [^{14}C]aflatoxin B, dissolved in dimethylsulfoxide. Phenobarbitone-pre-
treated animals received 1 mg/ml sodium phenobarbitone
in the drinking water for seven days prior to carcinogen administration. Animals were killed six hours after aflatoxin administration, and the liver and kidney
radioactivities were determined. The percentage of
[^{14}C]aflatoxin B$_1$ remaining in the liver and kidneys
after carcinogen injection was the same for control
and phenobarbitone-treated rats. In control animals,
binding of carcinogen to liver nucleic acids was
higher than binding to protein. Phenobarbitone treatment decreased carcinogen binding to nucleic acids so
that on a per milligram basis binding to all liver
macromolecules was similar. Differences were also
seen in kidney macromolecular binding for the two
groups of animals; there was less nucleic acid-bound
carcinogen in enzyme-induced rats. Protein binding
in the liver and kidney was unaffected by prior phenobarbitone administration. The decreased binding of
aflatoxin B$_1$ to nucleic acids in phenobarbitone-in-
duced animals suggests that induction reduces the
amount of reactive carcinogen formed during metabolism
by liver mixed function oxidases. This reduction in
nucleic acid binding may account for the protective
effect of phenobarbitone on aflatoxin B$_1$ toxicity and
carcinogenicity. These data as well as previously
published reports indicate an apparent discrepancy
between *in vivo* and *in vitro* aflatoxin B$_1$ activation.
This difference may be due to a fundamental change in
the pharmokinetics of aflatoxin B$_1$ (more metabolism
proceeding *via* detoxification pathways in the animal
than in the artificial microsomal system).

2459   INDUCTION OF TUMORS OF THE URINARY BLADDER
        IN FEMALE MICE FOLLOWING SURGICAL IMPLANTA-
TION OF GLASS BEADS AND FEEDING OF BRACKEN FERN.
(Eng.) Miyakawa, M. (Faculty of Medicine, Kyoto
Univ., 53 Shogoin-kawaramachi, Sakyo-ku, Kyoto 606,
Japan); Yoshida, O. *Gann* 66(4):437-439; 1975.

The carcinogenic effect of bracken fern (*Pteris aqui-
line*) to the urinary bladder of mice was studied.
Powdered bracken fern mixed with a basic commercial
diet (1:4 by weight) was fed to a group of 40 ICR
strain mice for 20 wk after the implantation of a
glass bead into the bladder. The animals were observed for a total of 70 wk. Fifteen mice survived
40 or more wk and four mice (27%) developed
urinary bladder tumors. No urinary bladder
tumor was found in a group of mice fed a similar
bracken fern diet mixture without a glass bead in
their bladder and in a control group fed a normal
diet after the implantation of a glass bead. The

results may indicate that there is a metabolite(s) in the urine of bracken-fed mice carcinogenic to the bladder.

2460   ABSENCE OF TOXIC AND CARCINOGENIC EFFECTS
        AFTER ADMINISTRATION OF HIGH DOSES OF
CHROMIC OXIDE PIGMENT IN SUBACUTE AND LONG-TERM FEED-
ING EXPERIMENTS IN RATS. (Eng.) Ivankovic, S. (In-
stitut für Toxicologie und Chemotherapie, Deutsches
Krebsforschungszentrum, D-69 Heidelberg 1, Bundes-
republik Deutschland); Preussmann, R. *Food Cosmet.
Toxicol.* 13(3):347-351; 1975.

Cr_2O_3, which as chromium oxide green (C-green 9) or
pigment green 17 is used in cosmetics, was tested
in BD rats for subacute toxicity (90-day test) and
for long-term toxicity and carcinogenicity (2-yr
study). Administration of 2 or 5% in the feed for
90 days produced no signs of toxic effect. The only
striking observation was a dose-dependent reduction
in the liver and spleen wt in the 90-day study. No
significant histological or hematological changes
were found. Fertility of the animals was normal during treatment, and the young showed no malformations.
Feeding of 1, 2 or 5% Cr_2O_3 in the feed for 2 yr was
well tolerated. The animals were observed throughout
life and there was no reduction in the average life
expectancy of the experimental animals. At the end
of the 2-yr study the rats were fed control diet
until they were dead or moribund. All important organs including the CNS were studied histologically.
Even with the very high oral doses of Cr_2O_3 given,
no carcinogenic action was detected.

2461   CARCINOMA OF THE URINARY BLADDER IN PA-
        TIENTS RECEIVING CYCLOPHOSPHAMIDE. (Eng.)
Wall, R. L. (Univ. Hosp., 410 W. 10th Aven., Columbus,
Ohio 43210); Clausen, K. P. *N. Engl. J. Med.* 293(6):
271-273; 1975.

The occurrence of interstitial hemorrhagic cystitis
and carcinomas of the urinary bladder in five patients receiving large cumulative doses of cyclophosphamide is reported. A 47-yr-old man with multiple
myeloma received 295 g total cyclophosphamide over
five years; squamous-cell carcinoma of the urinary
bladder developed, and death due to urinary-tract
obstruction occurred at about 17 mo after the first
episode of hematuria. A similar course occurred in
a 63-yr-old man with IgG myeloma who had received a·
total dose of 114 g cyclophosphamide over three
years. A 57-yr-old man with IgG myeloma had received
252 g of cyclophosphamide over five years and ten
months; death was due to invasive squamous cell carcinoma of the urinary bladder. A 60-yr-old man with
IgG myeloma, receiving 192 g over four years, died of
undifferentiated carcinoma of the urinary bladder.
The only survivor was a man in whom Hodgkin's had
been diagnosed at age 27 yr. He had received 147 g
of cyclophosphamide over three years, which was discontinued because of hemorrhagic cystitis. For the
next $10\frac{1}{2}$ yr, he received a total dose of 23 g of
chlorambucil. He was then found to have epidermoid
carcinoma of the urinary bladder and associated extensive *in situ* carcinoma. The following cautions

terone. Duration of synthesis was 20 hr in mammary
gland, in HAN about 18 hr, and in mammary tumors a-
bout 10 hr. The low labeling index of HAN suggests
that HAN contains nonproliferating cell populations.
The results show that HAN cell populations differ
from mammary gland and mammary tumor cells by
their lack of response to hormonal stimulation for
growth and lack of specific estradiol receptors.

2465   TEMPERATURE AND BLOOD FLOW MEASUREMENTS
        IN AND AROUND 7,12-DIMETHYLBENZ(a)ANTHRA-
CENE-INDUCED TUMORS AND WALKER 256 CARCINOSARCOMAS
IN RATS. (Eng.) Moller, U. (Finsen Lab., Finsen
Inst., Strandboulevarden 49, Copenhagen, Denmark);
Bojsen, J. *Cancer Res.* 35(11/Part 1):3116-3121;
1975.

In order to determine the validity of using temperature as a parameter of blood flow in tumors, temperature and blood flow measurements were compared in
7,12-dimethylbenz(a)anthracene-induced tumors and in
Walker 256 carcinosarcomas in female Sprague-Dawley
rats. Mammary tumors were induced by administration
of 20 mg 7,12-dimethylbenz(a)anthracene po. After
3-5 mo, the tumors were 9-18 cm^3 and were mainly
adenocarcinomas. Other rats were inoculated sc with
10^6 Walker tumor cells; 10-14 days after inoculation
the tumors were 8-28 cm^3. Rats bearing Walker 256
carcinosarcomas died from metastases 4-6 wk after
inoculation. Neither tumor was adherent to the skin,
and there was no vascular connection between tumor
and skin. In most of the tumor measurements, temperature transmitters with two thermistors and a magnetic
microswitch were implanted. Temperature measurements
were carried out in rats bearing 7,12-dimethylbenz(a)-
anthracene-induced tumors continuously for one month,
and rats bearing Walker 256 carcinosarcoma were
measured for 2-4 wk. Blood flow was measured using
the ^{133}Xe washout technique. A circadian rhythm of
temperature was found both in 7,12-dimethylbenz(a)-
anthracene-induced tumors (38.0-39.4 C, light and
dark periods, respectively) and in Walker 256 carcinosarcomas (37.6-38.3 C, light and dark periods,
respectively), with no significant difference in the
temperatures of the two tumors. The temperature was
lowest in periods of rest, and the temperature difference between light and dark periods was about 1 C.
Both tumors had a higher temperature than the subcutis.
External temperature measurements of the skin covering
the 7,12-dimethylbenz(a)anthracene-induced tumors by
thermistor probe and by thermography showed temperatures 1-2 C below the temperatures of surrounding
skin areas. In the 7,12-dimethylbenz(a)anthracene-
induced tumors, the blood flow was low (2.5 ml/100
g/min), which should correspond to a relatively small
heat production, although the temperature was
relatively high. Blood flow in skin overlying the
tumor was high, presumably a perifocal hyperemic
reaction, although the temperature was relatively
low in this area. It is concluded that there is no
correlation between tissue temperature and blood
flow, and that it is not possible by means of the
circadian temperature rhythm to distinguish between
the hormone-dependent 7,12-dimethylbenz(a)anthracene-
induced tumor and the hormone-independent Walker
256 carcinosarcoma.

2466   ON THE MECHANISM OF PROLACTIN AND ESTRO-
        GEN ACTION IN 7,12 DIMETHYLBENZ(A)ANTHRA-
CENE-INDUCED MAMMARY CARCINOMA IN THE RAT. II. *IN
VIVO* TUMOR RESPONSES AND ESTROGEN RECEPTOR. (Eng.)
Leung, B. S. (Dept. Surgery, Univ. Oregon Health
Sciences Center, 3181 S.W. Sam Jackson Park Road,
Portland, Oreg. 97201); Sasaki, G. H. *Endocrino-
logy* 97(3):564-572; 1975.

In order to test the *in vivo* effect of prolactin on
estrogen receptor (ER) binding capacity in tumors
induced by 7,12 dimethylbenz(a)anthracene (DMBA-
tumor), growth of the tumors from changes in prolactin and estrogen levels was compared retrospectively
with cytoplasmic ER levels. The tumors were induced
by intragastric feeding of 16 mg DMBA to 50-day-old
Sprague-Dawley rats. ER values were determined only
from adenocarcinomas. ER was measured using Scat-
chard plot analysis and/or by a technique involving
the estimation of bound radioactive estradiol after
0.2 nM $[2,3,6,7-^3H]$ estradiol (105 Ci/mM) had been
added to aliquots of 40,000 x g post-mitochondria
tumor supernatant. Some tumors required prolactin
(2mg), some needed prolactin-estrogen (estradiol
0.01 µg; prolactin 2.mg) during their growth period
and a small number were not influenced by hormonal
milieu. ER was present in hormonally dependent tumors but was low or absent in hormonally-independent
tumors. Deletion of hormones by endocrine ablation
in the host rat resulted in tumor regression and loss
of ER. Replenishment of ER and subsequent tumor
growth were accomplished by injection of prolactin
or prolactin-estrogen in endocrine-ablated rats but
were not achieved in rats bearing tumors exposed to
prolactin-nafoxidine (0.1 mg, daily). The results
demonstrate that both estrogen and prolactin are essential for growth of hormonally dependent DMBA tumors. Tumor growth was also prevented when cytoplasmic ER was not replenished, indicating that ER
may be an indispensable prerequisite for growth.
Prolactin, independently of or cooperatively with
estrogen, stimulated ER binding capacity. These
results support the hypothesis that there may exist
a prolactin regulatory mechanism of estrogen action
at the tumor site. The interactions of estrogen and
prolactin *in situ* in modulating hormonal receptor
binding capacities may contribute to the overall
stimulatory effect of these two hormones on DMBA
tumors.

2467   INHIBITORY EFFECT OF ACTINOMYCIN D ON THE
        INDUCTION OF RAT MAMMARY TUMORS BY *IN VITRO*
EXPOSURE TO 7,12-DIMETHYLBENZ(a)ANTHRACENE. (Eng.)
Tominaga, T. (Res. Inst. Microbial Diseases, Osaka
Univ., Yamada-kami, Suita, Osaka, Japan); Tei, N.;
Taguchi, T.; Takeda, Y. *Cancer Res.* 35(7):1698-1701;
1975.

The inhibitory effect of actinomycin D on the induction of tumors by *in vitro* exposure of mammary glands
to 7,12-dimethylbenz(a)anthracene (DMBA) was studied.
Mammary tumors were induced by DMBA (25 mg/kg, iv)
in female Sprague Dawley rats. Transplantation was
to the back of the donor rat. When actinomycin D
was given ip 24 hr before excision of the mammary
glands, the number of tumors per number of grafts

free E_2, resp., and DHT. The percentage of binding
inhibition was calculated with reference to the
inhibition obtained with nafoxidine in a molar concentration ratio of 1000 for E_2 binding, and with
cyproterone acetate in a molar concentration ratio
of 10,000 for DHT. Receptors for both E_2 and DHT
were present in all 15 samples of myometrium tested.
Of 19 samples of mammary carcinoma tissue, one had
no binding activity, three bound E_2, five bound DHT,
and ten showed binding for both steroids. A 50%
inhibition of E binding in myometrial and tumor
tissue required a molar concentration ratio of 40
for Adiol, >2000 for T and DHT, and about 20,000 for
DHEA. No significant inhibiting activity occurred
with A up to a molar concentration ratio of 10,000
or with DHEA-S up to 40,000. With regard to DHT
binding, Adiol was more active than E_2 and less active than T. Of the substances tested, Adiol was the
only one which exerted a significant inhibiting
influence at a molar ratio not far beyond the physiological range. This signifies that Adiol might
interfere at the receptor level in the estrogenic
stimulation of mammary cancer cells.

2472   GONADOTROPIN-INDUCED PROLIFERATION OF
        ENDOPLASMIC RETICULUM IN AN ANDROGENIC
TUMOR AND ITS RELATION TO ELEVATED PLASMA TESTOSTER-
ONE LEVELS. (Eng.) Neaves, W. B. (Southwestern
Medical Sch., Dallas, Tex. 75235). *Cancer Res.*
35(10):2663-2669; 1975.

The development of gonadotropin-induced structural
change in a testosterone-secreting Leydig cell tumor
and its relationship to plasma testosterone levels
were studied in castrate, tumor-bearing C57Bl/6J
mice two hours after daily pc injections of human
chorionic gonadotropin (HCG, 50 IU). Tumor cells
from control animals were small, averaging less than
1.0×10^{-9} ml in volume, and were poorly differentiated, having very little smooth endoplasmic reticulum (SER) in their cytoplasm. Average plasma
testosterone levels in these mice were near 1.3 ng/
ml. Although a five-fold rise in plasma testosterone
was measured two hours after the first HCG injection,
no changes in the endoplasmic reticulum were detected
at this time. After the second injection, plasma
testosterone rose to only twice control levels,
and cell ultrastructure was largely unchanged.
Modest amounts of SER had appeared in many tumor cells after the third injection, and plasma
testosterone showed a six-fold rise. Between the
third and fourth injections, cell volume increased
by about 70% as large accumulations of SER appeared
in the cytoplasm of most cells. Plasma testosterone
again rose to five times control levels. Increased
cell volume and abundant SER were maintained by
continuing daily injections of HCG, while the two-
hour plasma testosterone response persisted at 5-6
times the control level. These findings show that a
maximal elevation of plasma testosterone can occur
prior to changes in the endoplasmic reticulum of
Leydig tumor cells. However, they also suggest that
daily repetition of the maximal functional response
requires that the cells acquire large quantities of
newly produced SER.

2473   CORTICAL IMPLANTATION OF HYDROCARBONS IN
        RATS AND MICE. (Eng.) Gershbein, L. L.
(Northwest Inst. for Medical Res., Chicago, Ill.
60634); Benuck, I. *Oncology* 31(2):76-82; 1975.

Early brain changes following the cortical implantation of fragments of several carcinogenic hydrocarbons
and azulene were studied in adult male BDF$_1$ mice and
Holtzman rats. Pellets composed of crystals of azulene, 3-methylcholanthrene (MCA), 1,2,5,6-dibenzan-
thracene (DBA), or 9,10-dimethyl-1,2-benzanthracene
(DMBA) were implanted cortically into groups of 14
rats; the hydrocarbons (1.0 mg/ml) were implanted ic
in saline by trocan in groups of 12 mice. After intervals of 56 (rats) or 140 (mice) days, the animals were
perfused with formalin, and the brains were examined
microscopically. In some experiments, the rats were
examined after intervals of 128-146 days. DMBA produced space-occupying lesions in 6 of 14 rats. The
inflammatory lesions were filled with necrotic tissue
and surrounded by an edematous border; several crystals of hydrocarbon were found within each lesion.
The remaining rats injected with DMBA displayed non-
space-occupying lesions, as did all animals treated
with DBA, MCA, and azulene. The latter produced no
edema or atypical cells. In the rats studied after
128-146 days, the treatment appeared to be somewhat
better tolerated; 3 of the 9 animals treated with
DMBA developed hemorrhagic space-occupying lesions,
all other changes being non-space-occupying lesions.
In contrast to the rats, the mice underwent few
changes in brain anatomy after hydrocarbon treatment.
Neoplasms as such were found in neither rats nor mice.

2474   ACTIVATION OF CARCINOGENIC POLYCYCLIC
        HYDROCARBONS IN POLYOMA-VIRUS-TRANSFORMED
CELLS AS A PREREQUISITE FOR POLYOMA VIRUS INDUCTION.
(Eng.) Huberman, E. (Dept. of Genetics, Weizmann
Inst. of Science, Rehovoth, Israel); Fogel, M. *Int.
J. Cancer* 15(1):91-98; 1975.

Studies were undertaken to determine whether or not
carcinogenic polycyclic hydrocarbons can induce the
synthesis of polyoma virus (PV) in inducible PV-
transformed rat cell clones, and to determine if this
induction requires metabolic activation of the carcinogens. In cultures of such clones, which metabolize
benzo(a)pyrene (BP) to a level of 30-60% of that of
normal cells, up to 10.4% of the cells were induced
for PV synthesis by BP (1 µg/ml), 20-methylcholan-
threne (MCA 0.1-10 µg/ml) and 7,12-dimethylbenz(a)-
anthracene (DMBA 0.1-10 µg/ml). No PV induction was
observed with the noncarcinogenic polycyclic hydrocarbons pyrene, chrysene, and benz(a)anthracene. A
proportion of subclones, isolated from a PV-inducible
clone, which metabolized 0.1 µg or less BP per 10^6
cells, proved noninducible by any of the carcinogenic
polycyclic hydrocarbons BP, MCA and DMBA. However,
subclones that metabolized more than 0.3 µg BP per
10^6 cells were all inducible for PV synthesis by
these carcinogens. Subclones isolated from an inducible clone pretreated with BP were shown to metabolize
less than 0.1 µg BP per 10^6 cells and were resistant
to virus induction by the carcinogenic polycyclic
hydrocarbons. Benzoflavone (3 µg/ml) which inhibits
the metabolism of BP in clones metabolizing high
levels of this carcinogen, prevented the induction of

chromatography, but type I changed into type II during storage. This suggests that type II is a modified product of type I. Type I and type II were also produced in ethanol solutions (33% and 100% ethanol) of DNA irradiated for six hours by the visible light of a high-pressure mercury lamp (200 V, 1.5 A), or treated with iodine (5 mM) or hydrogen peroxide. The type II spectrum of mouse DNA resembled that produced by hydrogen peroxide, rather than that produced by iodine. Treatment of *Escherichia coli* with benzo[a]pyrene (10 mg/10 ml dimethyl sulfoxide) produced only type I fluorescence. Although the relationship between types I and II remain to be established, the emission spectrum of mouse DNA indicated that the DNA-bound benzo[a]pyrene possessed the intact, conjugated ring structure of the parent compound. This suggests that hydroxylation by the substitution reaction at the 6,3,1 region of benzo[a]pyrene is responsible for the binding of benzo[a]pyrene to DNA. Although a cation radical cannot be excluded, it is concluded that the proximate (active) form of benzo[a]pyrene is probably an hydroxylated product including an oxy radical.

2479   FACTORS AFFECTING THE QUANTITATION OF DOSE-RESPONSIVE CURVES FOR MUTATION INDUCTION IN V_{79} CHINESE HAMSTER CELLS AFTER EXPOSURE TO CHEMICAL AND PHYSICAL MUTAGENS. (Eng.) Fox, M. (Christie Hosp., Manchester M20 9BX, Great Britain). *Mutat. Res.* 29(3):449-466; 1975.

Using four common mutagens; ethyl methanesulfonate (EMS), methyl methanesulfonate (MMS), UV-irradiation, and X irradiation; the relationship between dose of mutagen, cellular lethality, and frequency of 8-azaguanine-resistant (azg^r) colonies in V79 Chinese hamster cells was studied. After exposure to mutagen, the cells were incubated for up to 120 hr before the addition of 8-azaguanine (30 µg/ml) for determination of expression times. Reconstruction experiments were performed to determine the overall cell density that interferes with mutant recovery, and determinations were made of the incorporation of $[^{14}C]$hypoxanthine into cultures at different cell densities. Induced mutant frequences were assayed by two methods *in situ* and after replating. After exposure to X-rays, MMS, and UV, a significantly higher frequency of mutants was observed in the replated cultures as compared with the *in situ* assays; similar observations were made at all survival levels. With EMS, an increment on replating was observed only at high survival levels. The replating data suggest that two types of azg^r colonies are produced: those containing only azg^r cells; and those that, due to damage segregation, contain a mixture of azg^r and 8-azaguanine-sensitive cells. These mixed colonies appear to be lost by metabolic cooperation when mutation frequencies are assayed *in situ*. The proportion of mixed colonies to homogenous colonies differed with different mutagens. Taking into account such factors, EMS and UV irradiation were similarly mutagenic at a given survival level, but at equitoxic doses, fewer mutants were recovered after exposure of V79 cells to MMS and X-rays. In the case of EMS, there appears to be a component of "mutational" damage that is unrelated to cell lethality and therefore probably unrelated to chromosome damage.

2480 CARCINOGEN-INDUCED DNA REPAIR IN NUCLEO-
TIDE-PERMEABLE *ESCHERICHIA COLI* CELLS:
INDUCTION OF DNA REPAIR BY THE CARCINOGENS METHYL
AND ETHYL NITROSOUREA AND METHYL METHANESULFONATE.
(Eng.) Thielmann, H. W. (Institut fur Biochemie,
Deutsches Krebsforschungszentrum Heidelberg, D-
6900 Heidelberg 1, Im Neuenheimer Feld 280, Federal
Republic of Germany); Vosberg, H.-P; Reygers, U.
Eur. J. Biochem. 56(2):433-447; 1975.

DNA excision repair of ether-permeabilized (nucleo-
tide-permeable) cells of *Escherichia coli* was investigated after exposure to N-methyl-N-nitrosourea
(MeNOUr), N-ethyl-N-nitrosourea (EtNOUr) and methyl
methanesulfonate (MeSO$_2$OMe), which are known to bind
covalently to DNA. These cells, which had been pretreated with 2 ml diethylether, exhibited this repair after being treated with a 15-fold molar excess
(over DNA nucleotides) of one of the carcinogens.
Defect mutations in genes *uvrA, uvrB, uvrC, recA,
recB, recC* and *rep* did not. Enzymic activities involved in this repair were identified by measuring
size reduction of DNA, DNA degradation to acid-soluble nucleotides and repair polymerization. In per-
meabilized cells, methyl and ethyl nitrosourea induced endonucleolytic cleavage of endogenous DNA,
as determined by size reduction of denatured DNA in
neutral and alkaline sucrose gradients. An enzymic
activity from *E. coli* K-12 cell extracts was purified (greater than 2,000-fold) and was found to
cleave preferentially methyl nitrosourea-treated
DNA and to convert the methylated supercoiled DNA
duplex (RF I) of phage ⏀X174 into the nicked circular form. Degradation of alkylated cellular DNA
to acid solubility was diminished in a mutant lacking the 5'→3'exonucleolytic activity of DNA polymerase I, but was not affected in a mutant that
lacked the DNA polymerizing but retained the 5'→3'-
exonucleolytic activity of DNA polymerase I. An
easily measurable effect is carcinogen-induced repair polymerization, making it suitable for the detection of covalent binding of carcinogens and potentially carcinogenic compounds.

2481 INDUCTION OF MOUSE LUNG ADENOMAS BY
AMINES OR UREAS PLUS NITRITE AND BY
N-NITROSO COMPOUNDS: EFFECT OF ASCORBATE, GALLIC
ACID, THIOCYANATE, AND CAFFEINE. (Eng.) Mirvish,
S. S. (Univ. Nebraska Medical Center, Omaha, Nebr.
68105); Cardesa, A.; Wallcave, L.; Shubik, P. *J.
Natl. Cancer Inst.* 55(3):633-636; 1975.

Preliminary studies investigating the usefulness of
sodium ascorbate (NaASC) and gallic acid for blocking *in vivo* nitrosation are presented. The effects
of varying the concentrations of three N-nitroso
compounds and NaNO$_2$ concentration in morpholine plus
NaNO$_2$ and methylurea plus NaNO$_2$ systems were also
studied. Groups of 40 male strain A mice were given
N-nitroso compounds in drinking water; all other
chemicals were added to food. Beginning at ten
weeks of age, the mice were treated five days a week
for 20 wk, except the groups that received methylnitrosourea, which were treated for ten weeks.
Sodium ascorbate (NaASC) at the highest level tested
(11.5 or 23 g/kg food) gave 89-98% inhibition of

pepsinogen isozymes (Pg 1, 3, 4) normally present in
the pyloric mucosa had decreased or disappeared. Its
decrease was observed from one wk after the beginning
of MNNG treatment to at least three mos after the end
of the seven month MNNG administration. Remarkable
histopathologic changes were found from 8 months after
MNNG was given, and rats showing such unusual histopathologic alterations also had changes in their
pepsinogen isozyme pattern. In four of 27 rats, two
(Pg 1, 2) of the four isozymes of pepsinogen (Pg 1-4)
in the fundic mucosa decreased or disappeared from
three mo after the beginning of MNNG treatment to at
least 2 months after the end of its seven month administration. Histopathologic changes induced by MNNG
were not as remarkable in the fundic mucosa as in the
pyloric mucosa. The close relationship between histopathologic changes in the pyloric mucosa and changes
in pepsinogen isoenzyme patterns suggests that pep-
sinogens may mediate the presence of a nonreversible
phenotypic change related to the carcinogenic process.

2486   A STUDY OF TOBACCO CARCINOGENESIS. XIV.
        EFFECTS OF N'-NITROSONORNICOTINE AND N'-
NITROSONANABASINE IN RATS. (Eng.) Hoffmann, D.
(American Health Foundation, Valhalla, N.Y. 10595);
Raineri, R.; Hecht, S. S.; Maronpot, R.; Wynder,
E. L. *J. Natl. Cancer Inst.* 55(4):977-981; 1975.

N'-Nitrosonornicotine (NNN) and N'-nitrosoanabasine
(NAB) were synthesized and administered in drinking
water to male Fischer rats for 30 wk (total dose,
630 mg). 1,4-Dinitrosopiperazine (DNPI) served as
the positive control. By 11 mo, all surviving rats
given NNN developed esophageal tumors (12/20); one
had a pharyngeal tumor and three had invasive carcinomas originating in the nasal cavity. During
the same time, only 1 of 20 rats given NAB developed esophageal tumors. Compared to the strong
esophageal carcinogen DNPI, NNN was a moderately
active carcinogen, and NAB was a weak carcinogen.
Because NNN is highly concentrated in tobacco (0.3-
90 ppm) these findings should be applied to the
etiology of various cancers related to tobacco
chewing and smoking.

2487   METABOLISM OF NITROSAMINES *IN VIVO* IV.
        ISOLATION OF 3-HYDROXY-1-NITROSOPYRROLI-
DINE FROM RAT URINE AFTER APPLICATION OF 1-NITROSO-
PYRROLIDINE. (Eng.) Krüger, F. W. (Institut für
Toxicologie und Chemotherapie am Deutschen Krebs-
forschungszentrum D-6900 Heidelberg Im Neuenheimer
Feld 280 Federal Republic of Germany); Bertram, B.
Z. Krebsforsch. 83(3):255-260; 1975.

The urine of nitrosopyrrolidine-treated rats was
analyzed in order to elucidate the metabolic degradation of this substance. Male Sprague-Dawley rats
(160-250 g) were injected ip with unlabeled (400
mg/kg) nitrosopyrrolidine or 2,5- and 3,4-^{14}C-nitro-
sopyrrolidine (6 mg/30 µCi/kg). The urine from 96
rats was collected for 16 hr, and the urinary metabolites were analyzed by thin-layer chromatography,
gas chromatography, and mass spectrometry. The
analysis showed 7% of the injected radioactivity.
Some nitrosopyrrolidine was excreted unchanged, and

some (0.9%) of the radioactivity corresponded to synthetic 3-hydroxy-1-nitrosopyrrolidine. The $^{14}CO_2$ excretion was analyzed in a CO_2 exhalation apparatus in six rats. About one-third of the injected radioactivity was excreted in all forms, with over 20% in $^{14}CO_2$ form. The results show that β-hydroxylation of nitrosopyrrolidine does not occur as a minor pathway in its metabolism.

2488   INDUCTION OF NEUROGENIC TUMORS BY NITRO-
        SOTRIALKYLUREAS IN RATS. (Eng.) Lijin-
sky, W. (Oak Ridge Natl. Lab., P.O. Box Y, Oak Ridge,
Tenn. 37830); Taylor, H. W. *Z. Krebsforsch.* 83(4):
315-321; 1975.

Four nitrosotrialkylureas were each fed to 30 male and female Sprague-Dawley rats for 50 wk in drinking water at the same molar concentration to correlate reactions *in vivo* of nitrosamides with tumor induction. Tumors of nervous origin arose after treatment with nitrosotrimethylurea (3/30), nitrosotriethylurea (7/30), nitrosomethyldiethylurea (23/30), and nitrosoethyldimethylurea (8/30). A comparison of the relative stabilities of the four nitrosoureas in aqueous solution at various pH's showed no correlation with the tumorigenicities of the compounds. It is concluded that with the exception of nitrosotrimethylurea, the nitrosotrialkylureas have a pronounced tumorigenic action in rats.

2489   INTESTINAL TUMORS IN MICE TREATED WITH A
        SINGLE INJECTION OF *N*-NITROSO-*N*-BUTYLUREA.
(Eng.) Ward, J. M.'(Natl. Cancer Inst., Bethesda,
Md. 20014); Weisburger, E. K. *Cancer Res.* 35(8):
1938-1943; 1975.

The induction of intestinal tumors by *N*-nitroso-*N*-butylurea (BNU) was studied in 80 male C57Bl/6 mice, divided into four groups of 20 each. Two groups consisted of 3-wk-old mice, and the other two of 6-wk-old mice. In each age group, one group was given 75 mg/kg of BNU and the other 150 mg/kg in a single ip injection. The control group received either vehicle injection or were untreated. All mice dying after 15 wk were autopsied, and all survivors were sacrificed and autopsied at 67 wk. Intestinal tumors developed in 14 mice receiving 75 mg/kg BNU at six weeks of age, and in 17 of those receiving 150 mg/kg at six weeks. Of those receiving BNU at three weeks of age, seven of those treated with 75 mg/kg and ten of those given 150 mg/kg developed intestinal tumors. One of the untreated and four of the vehicle-treated controls also developed intestinal tumors. The tumors were seen primarily at the pyloroduodenal junction and in the anterior portion of the small intestine; a few occurred in the cecum, colon and rectum. The tumors at the junction were not very invasive and were frequently polypoid. Tumors at the other locations were invasive adenocarcinomas, which did not metastasize. Colorectal tumors were adenomas and adenocarcinomas. BNU also induced tumors of the stomach, liver, lung, and hematopoietic system. It is concluded that older mice are more susceptible to BNU-induced tumor production and that this effect is more profound at higher doses of BNU.

from medical records. The prevalence of recorded
rauwolfia use among the controls was 20%, and that
of other drug use was correspondingly high. The risk
ratio for rauwolfia use was estimated to be 1.2 (95%
confidence interval, 0.7-2.2). These results do not
support the hypothesis that reserpine causes breast
cancer.

2495   EVALUATION OF ORAL, PHARYNGEAL, LARYNGEAL
         AND ESOPHAGEAL CANCER RISK IN REVERSE
SMOKERS OF CHUTTAS. (Eng.) Reddy, C. R. R. M.
(Andhra Med. Coll., Visakhapatnam, India); Kameswari,
V. R.; Chandramouli, K. B.; Prahlad, D.; Ramulu, C.
Int. Surg. 60(5):266-269; 1975.

The effect of pure tobacco smoking, especially reverse smoking, on the development of oral, pharyngeal, laryngeal and esophageal cancers was investigated. Five hundred and twenty cancer patients were
questioned about their smoking habits, chewing habits, diet, hygiene and religion. Controls were
matched with patients by sex, economic status, literacy, religion, occupation and locality; 48% of the
patients could be exactly age-matched. The relative
risks of cancer in each site was calculated from
each smoking habit. The hard palate was the most
common cancer site, comprising 197 of the 520 cases;
133 of these were women. Oral cavity and oropharyngeal cancers comprised 400 cases and pharyngeal,
laryngeal and esophageal cancer comprised 120 cases.
Reverse smoking of chuttas gave the highest risk of
hard palate cancer. This was 19.5 times the relative risk, with floor of the mouth cancer 4.8 times,
gum cancer 4.6 times, and tongue cancer 4.4 times
the relative risk. Cancers of the larynx and hypo-
larynx gave the lowest risk (0.3 times the relative
risk). The hypopharynx was the only site for which
smoking of chuttas in the usual way had any significance (5.7 times relative risk). There was no
significance attributed to cigarette or bidi smoking
for developing cancer, but these habits were rare
in areas studied. When the relative risk was expressed for all smokers, smoking was significant
for oral cancers of hard palate. Dental hygiene
was poor in 87% of the patients and 77% of controls,
and 70% of both were in the lowest socioeconomic
class. This study reemphasizes the strong association between reverse smoking and cancer of the hard
palate. The conventional chutta smoker runs a slight
risk of oral cancer.

2496   ASBESTOS, SMOKING, AND LARYNGEAL CARCIN-
         OMA. (Eng.) Shettigara, P. T. (Dept.
Preventive Medicine, Univ. Toronto, McMurrich
Building, Toronto, Ontario, Canada M5S 1A8); Morgan,
R. W.* *Arch. Environ. Health* 30(10):517-519; 1975.

The relationship between exposure to asbestos and
laryngeal cancer was investigated. A retrospective
study of 43 pairs of patients with laryngeal cancer
and their matched controls examined a number of
variables, including smoking, exposure to asbestos,
and other occupational factors. Patients with
laryngeal cancer and the controls were matched for
age, sex, and place of residence. The data indicate
a substantial association between asbestos exposure

and laryngeal cancer. Of the 43 laryngeal cancer
patients, ten (23.3%) had previous exposure to asbestos; none of the control patients reported exposure. There was also a significantly higher proportion of smokers among the patients with laryngeal
cancer (40, or 93.02%) than among controls (31, or
72.09%). Exposure to uranium, chromium, nickel, cobalt, arsenic, x-rays, and alcohol did not appear
related to later development of carcinoma of the
larynx. It is concluded that exposure to asbestos
and cigarette smoking are potent factors in the
development of this disease.

2497   METABOLISM OF CIGARETTE SMOKE CONDENSATES
        BY HUMAN AND RAT HOMOGENATES TO FORM MUTA-
GENS DETECTABLE BY *SALMONELLA TYPHIMURIUM* TA1538.
(Eng.) Hutton, J. J. (Veterans Admin. Hosp., Lexington, Ky. 40507); Hackney, C. *Cancer Res.* 35(9):
2461-2468; 1975.

Nineteen fractions of whole condensate of smoke from
the University of Kentucky Reference Cigarette IRI
were tested for mutagenicity *in vitro* using a bacterial indicator system. As little as 25 μg of the
active fractions were mutagenic toward histidine-
requiring *Salmonella typhimurium* TA1538, if the condensates were incubated in the presence of ARS/
Sprague-Dawley rat or human liver homogenates during treatment of the bacteria. Homogenates of lung
were relatively inactive. Homogenates from livers
of rats that were treated with 3-methylcholanthrene
(100 mg/kg, ip) converted condensates to mutagens
more efficiently than did liver homogenates from
man or from normal or phenobarbital-treated (0.1%
in drinking water for one week prior to sacrifice)
rats. Use of homogenates from animals treated with
3-methylcholanthrene gave much more reproducible
results in smoke fraction assays because larger numbers of revertants were obtained, and dose-response
curves were linear over the range 25-250 μg condensate. The linear dose-response curves permitted
quantitative comparison of the various fractions.
The mutagenicity per milligram of basic fractions
of whole smoke condensate was very high and that of
neutral polycyclic hydrocarbons was very low. Because of the preferential sensitivity of the TA1538
test system to polycyclic amines and insensitivity
to alkyl polycyclics, there was a poor quantitative
correlation between mutagenicity and carcinogenicity, as measured by skin painting or *in vitro* cell
transformation. If further development improves
the sensitivity of the bacterial testing system to
mutagenic derivatives of alkyl polycyclic and hetero-
polycyclic hydrocarbons, it may provide a convenient,
rapid, quantitative, and inexpensive bioassay for
the detection of potentially carcinogenic substances
in tobacco smoke condensates.

2498   MOUSE SKIN TUMORIGENESIS AND INDUCTION OF
        ARYL HYDROCARBON HYDROXYLASE BY TOBACCO
SMOKE FRACTIONS. (Eng.) Akin, F. J. (Richard B.
Russell Agric. Res. Cent., Agric. Res. Serv., U.S.
Dept. Agric., Athens, Ga.); Chamberlain, W. J.;
Chortyk, O. T. *J. Natl. Cancer Inst.* 54(4):907-912;
1975.

:ette
)carbon
as studic
>olynucic
)rigenic
basic, a
d silici
stermined
d recon-
55-day-
ipplicati
he concer
(ali phat
i spectro
l 522 m
ie (BP)
to deter
iators.
ing 0.0.
he F22
ial to
account.
n a 5-
inducti
and F24
reat as
activit
ather th
sults to
etected
f a gen
apabilit
ENZO(a)
MMALIA
Y P, P
AMINO-
nat. Sci
, L. J.

p⁰,
cAMP
3 af
-dit
at)
e ac
for
(0.
activit
and pro
he cell
[³H]les
esse,
zophyll
)':5'-
;cAMP,
decreas
d war
14 hr.
e did
.ould
)antiw
actin
s D.
)antim
antim

hylline, and benz(a)anthracene converted benzo(a)-
-yrene to similar alkali-extractable metabolites with
fluorescence spectra similar to that of 3-hydroxy-
enzo(a)pyrene. These induced enzyme activities also
howed a similar heat stability. Induction by Bt_2-
AMP and aminophylline, like induction by benz(a)-
nthracene, required continued protein synthesis and
only an initial period of RNA synthesis. Compared to
the benz(a)anthracene-induced hydroxylase with a K_m
of 4.3 μM, the hydroxylase induced by Bt_2cAMP and
aminophylline showed a K_m of 0.14 μM, and was 100-fold
more sensitive to inhibition by 7,8-benzoflavone
(10^{-4}-10^7 M). Increasing the serum concentration in
the culture medium stimulated the induction by aminophylline but did not stimulate induction by benz(a)-
anthracene. The results indicate that aryl hydrocarbon (benz(a)pyrene) hydroxylase can be induced by
compounds that increase the level of AMP, and that
this induction and induced enzyme activity differs
from that caused by benz(a)anthracene.

2500   INDUCTION OF ARYL HYDROCARBON HYDROXYLASE
        ACTIVITY IN VARIOUS CELL CULTURES BY 2,3-
7,8-TETRACHLORODIBENZO-p-DIOXIN. (Eng.) Niwa, A.
(Natl. Inst. Child Health and Human Development,
Bethesda, Md. 20014); Kumaki, K.; Nebert, D. W.
Mol. Pharmacol. 11(4):399-408; 1975.

Hydroxylase induction by 2,3,7,8-tetrachlorodibenzo-
p-dioxin (TCDD) was characterized in a number of established cell lines and primary cell cultures from
various species. The established cell lines included
TRL-2, ERL-2, NRKE, Chang liver, VERO, HTC, LB82, H-
4-II-E, Hepa-1, MA, and E-3. Primary cell cultures
were derived from fetal hamsters, rats, chickens, rabbits, and four inbred strains of mice. Cultured human
lymphocytes were also studied. TCDD and/or 3-methyl-
cholanthrene (MC) was added in varying amounts to the
cultures 24 hr after plating; 24 hr later, the cultures were assayed for hydroxylase activity and protein content. The kinetics of aryl hydrocarbon
(benzo(α)pyrene) hydroxylase induction by TCDD among
the established lines and primary cultures was similar
to the time course of hydroxylase induction by MC.
The TCDD-inducible process was sensitive to actinomycin D (40 nM) and cycloheximide (350 nM or 3.5 μM).
The induced enzyme activity in the cells treated with
TCDD (0-100 nM) plus MC (0-10 μM) was not greater than
that in cells exposed to either agent alone. There
was no relationship between cytotoxicity by TCDD
(1.5, 300, or 30 nM) and the level of inducible hydroxylase activity in culture. The estimated ED_{50}
values for the hydroxylase induction by TCDD ranged
from about 0.12 nM in C57BL/6N mouse cultures and
0.23 nM in the H-4-II-E cell line to more than 100 nM
in the VERO and HTC cell lines. No hydroxylase activity was detectable in control, MC-, or TCDD-treated
LB82 cells. In several cell lines and in the primary
cultures, the responsiveness to TCDD was 250-900 times
greater than that to MC. The responsiveness to TCDD
in C57BL/6N mouse-derived cultures was about 16 times
greater than that to TCDD in DBA/2N-derived cultures;
this difference is similar to that previously observed
in these two mouse strains in vivo. A bioassay with
H-4-II-E cells is suggested for the detection of minute (10^{-14} M) levels of TCDD.

2501   FOREIGN-BODY TUMORIGENESIS BY VINYL CHLO
        RIDE VINYL ACETATE COPOLYMER: NO EVIDEN
FOR CHEMICAL COCARCINOGENESIS. (Eng.) Brand, K.
(Univ. of Minnesota Medical Sch., Minneapolis, Min
55455); Buoen, L. C.; Brand, I. J. Natl. Cancer
Inst. 54(5):1259-1262; 1975.

Vinyl chloride monomers, released from implants of
vinyl chloride vinyl acetate copolymer (VCA), were
investigated in order to determine if they had a
cocarcinogenic effect in foreign body (FB) tumorigenesis. One of the following substances was implanted into CBA/H and CBA/H-T6 mice: VCA films (2
mm thick); VCA powder equal by weight to two films
and glass films (#2 and #3 thickness). Pulverized
FB was implanted ip or sc with a nebulizer dispen-
ser. The VCA powder was implanted in 76 6-wk-old
mice. Autopsy of the mice, most of which died aft
30 mo, revealed no tumors that could have been unequivocally caused by VCA. Roughened VCA films we
implanted in 26 mice and smooth VCA films were implanted in 53 mice. There was no increase in the
rate of tumorigenesis by virtue of the increased
surface area in the rough versus smooth implant
groups. VCA films were implanted in ten female mi
and for ten months, after which the implants were
removed and a second VCA film was implanted in the
opposite flank. There was no significant shortening of tumor latency in these animals. VCA powder
equal to four VCA films, was implanted sc into 11
mice and ip into ten mice. At eight months, singl
smooth VCA films were implanted in the same mice.
There was no significant acceleration of tumor appearance in these animals. The results may be indicative of tumor production by FB tumorigenesis
alone, and not FB/VCA cocarcinogenesis. The study
also disproves the contention that tumorigenesis
induced by VCA implants is due to both FB and chem
ical tumorigenesis. It is concluded that VCA implants are not suitable for studying this type of
cocarcinogenesis.

2502   BIOLOGICAL REACTIVITY OF PVC DUST. (Eng
        Richards, R. J. (Dep. Biochem., Univ. Co
lege, Cardiff, Wales); Desai, R.; Hext, P. M.;
Rose, F. A. Nature 256(5519):664-665; 1975.

The biological reactivity of polyvinyl chloride
(PVC) dust was studied by a hemolysis technique an
compared with the hemolytic potential of reactive
chrysotile asbestos A; the effect of polyvinyl chl
ride on lung fibroblast cultures was also investigated. Two samples of PVC were tested (KM 1 and K
2) both of which were obtained as finely-divided
dried powders intended for use in fabrication work
The first sample (KM 1) was highly hemolytic at re
latively low concentrations, 100% hemolysis (over
50 min) being achieved by 7.5-10 mg dust. The sec
ond sample (KM 2) was practically nonhemolytic.
The sample of 100 mg of KM 2 PVC gave an equivalen
hemolytic effect to 1 mg of sample KM 1. When the
hemolytic activity with time of a sample of KM 1 P
(7.5 mg) was compared with an equivalent mass of
chrysotile asbestos A, the asbestos was a faster a
gent, although total lysis was achieved by PVC aft
one hour. Samples of KM 1 PVC (7.5 mg) were washe

in 3 ml veronal buffer followed by centrifugation (2,000 revolutions/min for 20 min), and the hemo-lytic potency of the washed dust samples and the supernatant were determined. After a single wash, the hemolytic potency was reduced by over 60%, and subsequent washes reduced the activity of the dust even further. The effect of KM 1 PVC was studied on the levels of cell mat DNA, RNA, protein and hy-droxyproline in lung fibroblast cultures maintained *in vitro* for 24 days. Before addition of different concentrations of KM 1 PVC (50-200 µg/ml culture medium or 0.5-2.0 mg/culture) the dust sample was first washed in a balanced salt solution containing antibiotics. Fibroblast cultures treated with the dust all had lower levels of cell mat hydroxypro-line after 24 days. There was some fluctuation in DNA levels, but little change occurred in the level of total protein or RNA. It is concluded that cer-tain forms of PVC dusts exhibit a high hemolytic potential because of the presence of a readily sol-uble, surface-associated agent.

2503   EFFECTS OF VINYL CHLORIDE EXPOSURES TO
        RATS PRETREATED WITH PHENOBARBITAL. (Eng.)
Drew, R. T. (Natl. Inst. Environmental Health Sci-ences, P.O. Box 12233, Res. Triangle Park, N.C. 27709); Harper, C.; Gupta, B. N.; Talley, F. A. *Environ. Health Perspect.* 11:235-242; 1975.

Male Charles River CD-1 rats were exposed to ten consecutive days, six hours/day, to vinyl chloride vapors at an average concentration of 13,500 ppm. The exposed rats were divided into three groups of eight rats each: one group was pretreated with 3-methylcholanthrene; (15 mg/kg, ip for two days prior to first exposure); one group was pretreated with phenobarbital (1 mg/ml in drinking water for three days prior to exposure and during exposure); and the third group received no treatment. Half the animals in each group were sacrificed 18 hr after the last exposure and half were sacrificed four days later. In a second experiment, four rats pretreated with phenobarbital were exposed to vinyl chloride vapors at a concentration of 17,300 ppm for two days and sacrificed on the third day. In both experiments, control animals, also treated with phenobarbital or 3-methylcholanthrene, were exposed to air only. At the time of sacrifice, lungs, kidneys, spleen, heart, and a small piece of liver from each animal were preserved for histological examination. The remain-der of the liver was processed for assay of micro-somal enzyme activity. The following parameters were investigated: growth rate, organ weights, mor-phological changes, and both benzphetamine-*N*-de-methylase activity and cytochrome P-450 content of microsomes prepared from the livers. In both ex-periments the only marked difference noted in any group was a decrease in the growth rate of the ani-mals exposed to vinyl chloride and treated with phenobarbital. This decreased growth rate was par-ticularly apparent on the third day of the vinyl chloride exposures. Occasional morphological changes were also seen in the livers of the animals treated with phenobarbital and exposed to vinyl chloride. Although changes resulting from exposure to VC and phenobarbital have been demonstrated, these changes have not been shown to be preneoplastic.

2504   UNIQUE PROPERTIES OF INBRED SYRIAN HAM-
        STERS FOR CARCINOGENESIS STUDIES [abstract]
(Eng.) Homburger, F. (Bio-Res. Inst., Cambridge, Mass.); Handler, A.; Laird, C. *Lab. Invest.* 32(3): 427; 1975.

2505   THE IMMUNOLOGICAL EFFECTS OF THYMECTOMY
        AND BCG CELL WALL SKELETON ON PRODUCTION
OF LUNG CANCER IN RABBITS INDUCED BY CHEMICAL CAR-CINOGENS [abstract]. (Jpn.) Hirao, F. (Osaka Univ. Medical Sch., Osaka, Japan); Nishikawa, H.; Taniyama, T.; Azuma, I.; Ogura, T.; Yamamura, Y. *Gann, Proc. Jpn. Cancer Assoc., 34th Annual Meeting,* October, 1975, p. 47.

2506   TUMOR INDUCTION BY A SINGLE SUBCUTANEOUS
        ADMINISTRATION OF STERIGMATOCYSTIN IN
NEWBORN MICE [abstract]. (Jpn.) Fujii, K. (Sch. Med Univ. Tsukuba, Ibakari, Japan); Kurata, H.; Oda-shima, S. *Gann, Proc. Jpn. Cancer Assoc., 34th Annual Meeting,* October, 1975, p. 45.

2507   SPECIES AND STRAIN DIFFERENCES IN SUS-
        CEPTIBILITY TO THE CARCINOGENICITY OF
FLOWER STALKS OF *PETASITES JAPONICUS* MAXIM [ab-stract]. (Jpn.) Fushimi, K. (Gifu Univ., Sch. Medicine, Gifu, Japan); Kato, K.; Kato, T.; Mat-subara, N.; Hironi, I. *Gann, Proc. Jpn. Cancer Assoc., 34th Annual Meeting,* October, 1975, p. 45.

2508   MALONATE AS A PRECURSOR IN THE BIOSYN-
        THESIS OF AFLATOXINS. (Eng.) Gupta,
S. R. (Vallabhbhai Patel Chest Inst., Univ. Delhi, Delhi-7, India); Prasanna, H. R.; Viswanathan, L.; Venkitasubramanian, I. A. *J. Gen. Microbiol.* 88 (2):317-320; 1975.

2509   ACTION OF WEAK BASES UPON AFLATOXIN B_1
        IN CONTACT WITH MACROMOLECULAR REACTANTS.
(Eng.) Beckwith, A. C. (Agric. Res. Serv., Peoria, Ill.); Vesonder, R. F.; Ciegler, A. *J. Agric. Food Chem.* 23(3):582-587; 1975.

2510   AMMONIATION OF AFLATOXIN B_1. MASS SPEC-
        TRAL ANALYSIS OF COMPOUNDS SEPARATED BY
MICROSUBLIMATION. (Eng.) Stanley, J. B. (South Reg. Res. Cent. Agric. Res. Serv., New Orleans, La.); Lee, L. S.; Cucullu, A. F.; deGruy, I. V. *J. Agric. Food Chem.* 23(3):447-449; 1975.

2511   IMPROVEMENTS IN THE THIN-LAYER CHROMA-
        TOGRAPHY OF NATURAL PRODUCTS. I. THIN-
LAYER CHROMATOGRAPHY OF THE AFLATOXINS. (Eng.) Heathcote, J. G. (Univ. Salford, England); Hibbert, J. R. *J. Chromatogr.* 108(1):131-140; 1975.

2512   MYCOTOXINS IN HOT SPOTS IN GRAINS. I.
        AFLATOXIN AND ZEARALENONE OCCURRENCE IN
STORED CORN. (Eng.) Shotwell, O. L. (Northern Regional Res. Lab., Agricultural Res. Service, U.S.

2521   CADMIUM POTENTIATION OF DRUG RESPONSE--
        ROLE OF THE LIVER. (Eng.) Johnston, R.
E. (Sch. Pharm. Pharm. Sci., Purdue Univ., West
Lafayette, Indiana); Miya, T. S.; Schnell, R. C.
Biochem. Pharmacol. 24(8):877-881; 1975.

2522   PROLIFERATION KINETICS OF BILE DUCT EPI-
        THELIA IN THE REGENERATING MOUSE LIVER
AFTER CCl_4-POISONING. (Ger.) Gerhard, H. (Insti-
tut fur Medizinische, Strahlenkunde der Universi-
tat, D8700 Wurzburg, Versbacher Landstrasse, West
Germany); Schultze, B.; Maurer, W. *Virchows Arch.*
[*Zellpathol.*] 17(3):213-227; 1975.

2523   STUDIES OF BRONCHOGENIC CARCINOMA AMONG
        CHROMATE WORKERS. (Jpn.) Abe, S. (Sch.
Medicine, Hokkaido Univ., Hokkaido, Japna); Tsuneta,
I.; Honma, Y.; Osaki, Y.; Murao, M. *Lung Cancer*
15(2):125-131; 1975.

2524   DIET AND WESTERN DISEASE. (Eng.) Taylor,
        G. (The Old Rectory, Cricket Malherbie,
Ilminster, Somerset, England). *Lancet* 1(7907):644;
1975.

2525   EFFECT OF PRENATAL EXPOSURE OF MICE TO
        DIETHYLSTILBESTROL ON REPRODUCTIVE TRACT
FUNCTION IN THE OFFSPRING [abstract]. (Eng.)
McLachlan, J. A. (Natl. Inst. Environmental Health
Sciences, Res. Triangle Park, N.C.); Shah, H. C.;
Newbold, R. R.; Bullock, B. C. *Toxicol. Appl. Phar-
macol.* 33(1):190; 1975.

2526   *N*-HYDROXY-2-FLUORENYLACETAMIDE, AN ACTIVE
        INTERMEDIATE OF THE MAMMARY CARCINOGEN
N-HYDROXY-2-FLUORENYLBENZENESULFONAMIDE. (Eng.)
Malejka-Giganti, D. (Veterans Administration Hosp.,
Minneapolis, Minn. 55417); Gutmann, H. R. *Proc.
Soc. Exp. Biol. Med.* 150(1):92-97; 1975.

2527   ALTERATION OF PLASMA PROTEIN SYNTHESIS
        AND REAPPEARANCE OF α-FETOPROTEIN (α FP)
IN ORGAN CULTURED EMBRYONIC RAT LIVER INDUCED BY
N-2-FLUORENYLACETAMIDE (2FAA) [abstract]. (Eng.)
Parsa, I. (Downstate Med. Cent., Brooklyn, N.Y.).
Fed. Proc. 34(3):828; 1975.

2528   ALTERATION OF HEPATOCYTES BY SUBCARCINO-
        GENIC EXPOSURE TO *N*-2-FLUORENYLACETAMIDE.
(Eng.) Becker, F. F. (New York Univ. Sch. of Medicine, New York, N.Y. 10016). *Cancer Res.* 35(7):
1734-1736; 1975.

2529   DEHYDRORETRONECINE INDUCED RHABDOMYOSAR-
        COMAS IN RATS [abstract]. (Eng.) Allen,
J. R. (Univ. Wisconsin, Madison); Hsu, I. C.; Lalich,
J. J. *Fed. Proc.* 34(3):828; 1975.

2530   STRUCTURAL SPECIFICITY OF NITROSAMINE
        DEALKYLASES [abstract]. (Eng.) Bryant,
G. M. (Tulane Med. Sch., New Orleans, La.); Argus,
M. F.; Arcos, J. C. *Fed. Proc.* 34(3):756; 1975.

2531   DETECTION OF MUTAGENIC ACTIVITY OF METRONI-
        DAZOLE AND NIRIDAZOLE IN BODY FLUIDS OF
HUMANS AND MICE. (Eng.) Legator, M. S. (Roger
Williams Gen. Hosp., Providence, R.I.); Connor, T.
H.; Stoeckel, M. *Science* 188(4193):1118-1119; 1975.

2532   THE REACTION OF PLANT SEEDLINGS TO THE
        EFFECT OF IMINO-DI-(PARA-DIMETHYL AMINO
PHENYL)-METHANE-AURAMINE-OO. (Rus.) Gabaraeva,
N. I. (Botanic Inst. U.S.S.R. Acad. Sciences, Leningrad, U.S.S.R.); Slepian, E. I. *Vopr. Onkol.*
21(8):73-80; 1975.

2533   GAS CONTAMINANT INHIBITION OF MAMMARY
        GLAND DIFFERENTIATION *IN VITRO.* (Eng.)
Warner, M. R. (Baylor Coll. of Medicine, Houston,
Tex. 77025); Medina, D. *J. Natl. Cancer Inst.* 54
(5):1257-1258; 1975.

2534   CARCINOGENICITY EXAMINATION OF MILLIAMINE
        A [abstract]. (Jpn.) Fushimi, K. (Gifu
Univ., Sch. Medicine, Gifu, Japan); Kato, K.; Hiro-
no, I. *Gann, Proceedings Jpn. Cancer Assoc., 34th
Annual Meeting,* October, 1975, p. 45.

2535   DETERMINATION OF FOUR-AND FIVE-RING CON-
        DENSED HYDROCARBONS. II. ANALYSIS OF
POLYNUCLEAR AROMATIC COMPOUNDS IN *n*-PARAFFIN FEED
OIL FOR YEAST FERMENTATION. (Eng.) McGinnis, E.
L. (Gulf Res. Dev. Co., Pittsburgh, Pa.). *J. Agric.
Food Chem.* 23(2):226-229; 1975.

2536   METABOLISM OF POLYCYCLIC HYDROCARBONS IN
        THE LUNG [abstract]. (Eng.) Jakobsson,
S. (Stockholm, Sweden). *Forensic Sci.* 5(2):138-139;
1975.

2537   DETERMINATION OF FOUR-AND FIVE-RING CON-
        DENSED HYDROCARBONS. I. ANALYSIS OF
POLYNUCLEAR AROMATIC HYDROCARBONS IN YEAST PRO-
DUCED ON BOTH *n*-HYDROCARBON AND DEXTROSE FEEDS.
(Eng.) McGinnis, E. L. (Gulf Res. Dev. Co., Pitts-
burgh, Pa.); Norris, M. S. *J. Agric. Food Chem.*
23(2):221-225; 1975.

2538   ON THE CHARACTERIZATION OF THE h-PROTEIN,
        A PRIMARY TARGET OF CARCINOGENIC HYDRO-
CARBONS [abstract]. (Eng.) Sarrif, A. M. (McArdle
Lab., Univ. Wisconsin, Madison); Ketterer, B.; Woodman, P. W.; Heidelberger, C. *Fed. Proc.* 34(3):811;
1975.

2539   REVERSAL OF ISCHEMIC CELL INJURY AND LONG
        TERM STORAGE OF HUMAN BRONCHUS (HB) AND
HAMSTER TRACHEA (HT) [abstract]. (Eng.) Barrett,

2554   CHEMICALLY INDUCED HEPATIC NEOPLASMS IN
        THE FISCHER RAT (F-344) [abstract].
(Eng.) Garner, F. M. (Litton Bionetics, Kensington,
Md.); Cockrell, B. Y. *Toxicol. Appl. Pharmacol.*
33(1):172; 1975.

2555   DRUG-RELATED THYROID TUMORS IN THE
        FISCHER RAT (F-344) [abstract]. (Eng.)
Cockrell, B. Y. (Litton Bionetics, Kensington,
Md.); Garner, F. M. *Toxicol. Appl. Pharmacol.* 33
(1):172; 1975.

2556   TERATOGENICITY STUDY IN RATS GIVEN HIGH
        DOSES OF PYRIDOXINE (VITAMIN B_6) DURING
ORGANOGENESIS. (Eng.) Khera, K. S. (Health Protection Branch, Natl. Health Welfare, Tunney's Pas-
ture, Ottawa, Canada). *Experientia* 31(4):469-470;
1975.

2557   TUMOR INDUCTION BY SUBCUTANEOUS ADMINIS-
        TRATIONS OF SOME METABOLITES OF TRYPTOPHAN
IN INFANT MICE [abstract]. (Jpn.) Fujii, K. (Sch.
Medicine, Univ. Tsukuba, Ibaraki, Japan); Watanabe,
M. *Gann, Proc. Jpn. Cancer Assoc., 34th Annual
Meeting,* October, 1975, p. 46.

2558   AN AUTORADIOGRAPHIC AND MORPHOLOGICAL
        STUDY OF MOUSE BONE MARROW LITTORAL CELLS
DURING AND AFTER TREATMENT WITH URETHANE. (Eng.)
Langdon, H. L. (Univ. Miami Sch. Med., Fla.); Ber-
man, I. *Cell Tissue Kinet.* 8(3):285-292; 1975.

2559   VINYL CHLORIDE DETECTION USING CARBON
        MONOXIDE AND CARBON DIOXIDE INFRARED
LASERS. (Eng.) Freund, S. M. (Opt. Phys. Div.,
Natl. Bur. Stand., Wash., D.C.); Sweger, D. M.
Anal. Chem. 47(6):930-932; 1975.

2560   UPTAKE AND PHARMACODYNAMICS OF VINYL
        CHLORIDE ADMINISTERED TO RATS BY DIFFER-
ENT ROUTES [abstract]. (Eng.) Withey, R. J.
(Branch of Chemical Safety, Health and Welfare,
Canada, Ottawa, Ontario, Canada). *Toxicol. Appl.
Pharmacol.* 33(1):146; 1975.

See also:

* (Rev):   2401, 2402, 2403, 2404, 2405, 2406, 2407,
           2408, 2423, 2424, 2425, 2526, 2427, 2428,
           2429, 2430, 2431, 2432, 2433, 2434, 2449
* (Phys): 2563, 2569, 2570
* (Immun): 2666, 2683, 2714, 2727
* (Path): 2769, 2770, 2792, 2834, 2858
* (Epid): 2893, 2905, 2910

2561   CHROMOSOMAL ABERRATIONS IN BONE MARROW OF
        CONTINUOUSLY IRRADIATED RATS. (Eng.)
Chlebovsky, O. (Faculty of Natural Sciences, P. J.
Safarik Univ., Moyzesova 11., 04 167 Kosice, Czecho-
slovakia); Praslicka, M.; Chlebovska, K. *Biologia*
(Bratislava) 30(6):427-433; 1975.

The effects of continuous Co^{60} irradiation (0.5-
82.5 R/day) on chromosomal aberrations in the bone
marrow were studied in Wistar rats. An increase
in chromosomal aberrations was seen on the fifth
day in animals receiving 0.5 R/day. This initial
increase in chromosome aberrations was also seen
at other dose levels. Up until the 30th day of
treatment, the number of chromosome aberrations
varied considerably, and there was no increase in
aberrations despite increasing total exposure. After the 30th day, the increased aberration rates
stabilized, the absolute increase being 20% at 0.5-
2.5 R/day, 30% at 5-10 R/day, 35% at 20-36 R/day,
45% at 53 R/day, and 55% at 82.5 R/day. There was
no appreciable increase in the number of chromosomal aberrations between the 30th and 18th days of
treatment. These and other previously reported data
suggest that relatively low doses of continuous radiation cause fewer chromosomal aberrations than do
single larger doses, and/or that damaged chromosomes
are able to recover from radiation damage.

2562   *UMBRA LIMI:* A MODEL FOR THE STUDY OF
        CHROMOSOME ABERRATIONS IN FISHES. (Eng.)
Kligerman, A. D. (Dept. Poultry Science, Cornell
Univ., Ithaca, N.Y. 14853); Bloom, S. E.; Howell,
W. M. *Mutat. Res.* 31(4):225-233; 1975.

Due to the lack of information available on the
effects of various clastogenic agents on the chromosomes of fishes, an *in vivo* cytogenetics model
system was developed. The central mudminnow, *Umbra*
limi, was chosen for this study because of its ideal
karyotype consisting of 22 large meta- and submeta-
centric chromosomes. Intestine, stomach, kidney,
and gill tissues were found to be the most suitable
for clastogenic studies. Phase contrast observations were made on the chromosomes of control mudminnows and mudminnows exposed to 325 R of X-radiation. The control rate of spontaneous chromosome
aberrations was found to be low (about 0.03%). In
contrast, fish exposed to 325 R of X-rays had aberrations in approximately 30% of the metaphases per
fish examined. An apparent increase in clumping
and a decrease in the mitotic index were also noted.
It is concluded that the chromosomes of *Umbra limi*
display typical responses to low level radiation
exposure, and that this fish would be an ideal cytogenetics model for the study of induced chromosome
aberrations in fishes.

2563   EFFECTS OF X-RAY IRRADIATION ON THE SUB-
        SEQUENT GONADOTROPIN SECRETION IN NORMAL
AND NEONATALLY ESTROGENIZED FEMALE RATS. (Eng.)
Matsumoto, A. (Juntendo Univ. Sch. Medicine, Hongo,
Tokyo 113, Japan); Asai, T.; Wakabayashi, K. *Endo-*
crinol. Jpn. 22(3):233-241; 1975.

for males were about 25% lower than those for females. It is postulated that estrogen might promote
the transport of yttrium to its ultimate site of
deposition in the bones.

2567   LATE RESULTS OF IRRADIATION OF NEWBORN
        MICE WITH GAMMA RAYS. IV. SUMMARY OF
INVESTIGATIONS AND THE EFFECT OF SEX AND DOSE OF
RADIATION ON THE EXTENT OF RADIATION-INDUCED IN-
JURY. (Pol.) Gajewski, A. K. (00-791 Warszawa,
ul. Chocimska 24, Poland); Slowikowska, M. G.;
Chomiczewski, K.; Szram, S. *Rocz. Panstw. Zakl.
Hig.* 26(2):263-277; 1975.

2568   STUDY, BY A SEDIMENTATION METHOD, OF THE
        REPAIR OF SINGLE STRAND BREAKS IN DNA OF
THE REGENERATING LIVER OF γ-IRRADIATED RATS. (Rus.)
Zakrzhevskaia, D. T. (Inst. Biological Physics,
Acad. Sciences, Pushchino, U.S.S.R.); Gaziev, A.
I. *Radiobiologiia* 15(4):593-596; 1975.

2569   TUMOR INDUCTION BY X-RADIATION AND URE-
        THANE IN RATS [abstract]. (Eng.) Myers,
D. K. (Biol. Branch, Atomic Energy Canada Ltd.,
Chalk River, Ontario, Canada). *Radiat. Res.* 62(3):
597; 1975.

2570   MULTICARCINOGENESIS BY URETHAN AND X-
        IRRADIATION [abstract]. (Eng.) Vessel-
inovitch, S. D. (Dep. Radiol., Med., Pathol., Univ.
Chicago, Ill.); Simmons, E.; Mihailovich, N. *Radiat.
Res.* 62(3):597-598; 1975.

2571   HISTOPATHOLOGIC CHANGES INDUCED IN RATS
        BY LOCALIZED X-IRRADIATION OF AN EXTER-
IORIZED SEGMENT OF THE SMALL INTESTINE. (Eng.)
Sebes, J. I. (Univ. Tennessee Center for Health
Sciences, 865 Jefferson Ave., Memphis, Tenn. 38163);
Zaldivar, R.; Vogel, H. H., Jr. *Strahlentherapie*
150(4):403-410; 1975.

2572   MOLECULAR BASIS FOR THE MUTAGENIC AND
        LETHAL EFFECTS OF ULTRAVIOLET IRRADIATION,
PROGRESS REPORT, 4 OCT. 1973 - 29 MAY 1974. (Eng.)
Grossman, L. (Brandeis Univ., Waltham, Mass.). 11
pp., 1974. [available through National Technical
Information Services, Washington, D.C. Document No.
COO-3232-3]

2573   O.E.R. AND R.B.E. FOR 10 MeV ELECTRONS,
        AND THE TWO-COMPONENT THEORY OF RADIATION
ACTION. (Eng.) Winston, B. M. (Christie Hosp. and
Holt Radium Inst., Manchester M20 9BX, England);
Oliver, R.; Berry, R. J. *Int. J. Radiat. Biol.*
28(2):187-190; 1975.

2574   COMPARISON OF THE CARCINOGENICITY OF ^{131}I
        AND ^{125}I [abstract]. (Eng.) de Ruiter,
J. (Radiobiological Inst. TNO, 151 Lange Kleiweg,
Rijswijk, Netherlands); van Putten, L. M.; Boorman,
G. A.; Hollander, C. F. *Int. J. Radiat. Biol.* 27
(6):593; 1975.

2575   THE EFFECTS OF MILLING ON DIAMETERS AND
        LENGTHS OF FIBROUS GLASS AND CHRYSOTILE
ASBESTOS FIBERS. (Eng.) Assuncao, J. (Graduate

Sch. Public Health, Univ. Pittsburgh, Pittsburgh,
Pa. 15261); Corn, M. *Am. Ind. Hyg. Assoc. J.* 36
(11):811-819; 1975.

See also:

* (Rev):   2409, 2410, 2411
* (Chem):   2459, 2479
* (Immun): 2707, 2718
* (Epid):   2897, 2902

level measured in the presence of 10% calf serum.
After an initial lag, uptake was linear and reached a
plateau at approximately two hours after inoculation,
when about 3-9% of the viral DNA were adsorbed. After 2-6 hr, part of the DNA initially adsorbed was
shed into the adsorption medium and later partly read-
sorbed. Viral DNA reached the nucleus beginning at
0.5-2 hr after infection; at 24 hr, 70% of the DNA
taken up reached the nucleus. After 24 hr, the amount
of label associated with cells exceeded that at any
earlier time period. The results were similar when
intact type 2 virions were adsorbed to KB cells.
Viral DNA was not detected in lysosomes or in cytoplasmic vesicles. Adenovirus-2 DNA was not infectious
on HeLa and human embryo kidney cells. Treatment of
DNA with pancreatic DNase abolished the activity, and
pretreatment with pronase markedly reduced the infectivity. Incubation of type 2 virus DNA with pronase
did not alter the sedimentation profiles of viral DNA
in neutral and alkaline sucrose gradients. Endonucleolytic fragmentation of viral DNA occurred in the
medium. It is suggested that the parental DNA associated with cellular DNA is integrated into the host
genome.

2579   A STUDY OF THE ONCOGENICITY OF ADENOVIRUS
        TYPE 2 TRANSFORMED RAT EMBRYO CELLS.
(Eng.) Harwood, L. M. J. (Medical Sch., Univ. of
Birmingham, Birmingham B15 2TJ, England); Gallimore,
P. H. Int. J. Cancer 16(3):498-508; 1975.

The oncogenicity of several adenovirus type 2 (AD-
2)-transformed rat cell lines and of several tumor
cell lines derived from the transformed cells was
studied. Secondary cultures of whole rat embryo
fibroblasts (REF), rat embryo muscle (REM), and
rat embryo brain (REB) cells from 18-day embryos
were transformed with Ad-2. Newborn Hooded Lister
(HL) rats and specific pathogen-free Wistar White
rats were inoculated ip or sc with 2×10^6 to $3 \times
10^6$ transformed cells from in vitro passages 5-15.
In some cases, five daily doses of antithymocyte
serum (ATS) (0.2 ml) were administered beginning
eight days after virus inoculation. Tumors were
transplanted by ip or sc inoculation into newborn
18-day-old rats; no immunosuppression was used.
All transformed lines and cultured tumor lines
were examined for Ad-2 T-antigen by indirect immunofluorescence. The amount of T-antigen and its
position within the cell varied considerably between different cell lines, although there was some
similarity between the T-antigen staining of a
transformed line and the tumors derived from it.
Of 48 AD-2-transformed cell lines, only 14 were
oncogenic in newborn rats immunosuppressed with
ATS, and only one line was oncogenic in nonimmuno-
suppressed rats. The invasiveness of the tumors
resulting from the inoculation of Ad-2-transformed
cells appeared to depend to some extent on the
route of administration and the transforming virus
dose. All tumors resulting from the ip inoculation
of lines transformed at 50 and 20 plaque forming
units (PFU)/cell invaded and metastasized, while
less than 50% of the tumors resulting from the ip
inoculation of lines transformed at 10 PFU/cell
were invasive; ip inoculation produced more invasive

tumors than sc inoculation. All tumors were histologically identified as undifferentiated round-
or spindle-cell sarcomas. The number of multinucleate cells and the proportion of round and
spindle cells varied between tumors arising from
different transformed lines. Cell lines transformed
by adenovirus type 12 were more oncogenic than those
transformed by Ad-2. The transforming virus dose
had no significant effect on the transplantability
of tumors derived from the transformed cells, although the tumor incidence resulting from transplantation was reduced in older rats. The tumor
cell lines were not homogeneous in terms of oncogenicity. The results indicate that the most
important factor influencing the oncogenicity of
Ad-2-transformed cells is the viral dose used for
transformation.

2580   COMPLEMENTATION OF HUMAN ADENOVIRUS TYPE 5
          ts MUTANTS BY HUMAN ADENOVIRUS TYPE 12.
(Eng.) Williams, J. (Inst. Virol., Glasgow, Scotland); Young, H.; Austin, P. J. Virol. 15(3):675-
678; 1975.

To assess the degree to which human adenovirus A and
C are functionally related, complementation tests
were carried out between type 5 (group A) temperature-
sensitive (ts) mutants of 15 different complementation groups and wild type adenovirus type 12 strain
1131. Confluent HeLa cell monolayers were infected
either singly at an input multiplicity of 10 plaque-
forming U/cell, or doubly with mutant or wild type 12,
each at an input multiplicity of 5 plaque-forming
U/cell. In a yield from mixed-infected cells, only
type 5 will form plaques and only at 32.5 C, unless
recombination has taken place. Type 5 mutants included eight enhanced by type 12; hexon-deficient
mutants (ts 17 and ts 20); mutants with abnormal
hexon transport (ts 1, 2, 3, and 4); and late mutants
whose defects are not known (ts 18 and ts 24). In
crosses with these mutants, type 12 products functioned efficiently to replace defective type 5 functions and enhanced the yield of infectious virus.
The other seven mutants that complemented poorly or
not at all by type 12 included: DNA-negative mutants
(ts 36 and ts 125); fiber-deficient mutants (ts 5, 9,
and 22); and two mutants whose defects are not known
(ts 19 and ts 31). In these crosses, type 12 products did not substitute for the defective type 5
functions, and yields of infectious virus were enhanced less than 10-fold. Hamster embryo cells were
infected with the non-type 12-complemented type 5
mutants singly or doubly as described, and no mutants
were enhanced. Coinfection of HeLa cells with type
12 and type 5 had little or no effect on the replication of the wild type adenovirus 5 in these cells,
and similar results were obtained with type 5 replication in hamster cells. Monospecific antisera
against types 5 and 12 were neutralized by type 5
antiserum and not by type 12; similar results were
obtained with the yield from cells coinfected by
wild type adenovirus type 5 and strain 1131. Progeny clones from the antigenically-mixed yields were
isolated, and all were serotype 5, plaqueing on HeLa
cells only at 32.5 C. No evidence was found for viable wild-type recombinants in any cross.

2581   HEXON PEPTIDES OF TYPE 2,3, AND 5 ADENO-
          VIRUSES AND THEIR RELATIONSHIP TO HEXON
STRUCTURE. (Eng.) Stinski, M. F. (Sch. Med. Univ.
Pennsylvania, Philadelphia); Ginsberg, H. S. J. Vi-
rol. 15(4):898-905; 1975.

The chemical analysis of three different adenovirus
serotypes belonging either to subgroup I (type 3
adenovirus) or III (type 2 and 5 adenoviruses) was
carried out. The prototype strains of these viruses,
which had been plaque-purified three times, were used
After infecting monolayers of KB cells, virus was
quantitated by plaque or fluorescent focus assay.
Peptides of hexons from type 2, 5 and 3 adenoviruses
were produced by treatment with cyanogen bromide and
separated by isoelectric focusing on polyacrylamide
gels containing 8 M urea. Peptides with identical
isoelectric points, but from different hexon types,
were considered to have structural similarities.
Accordingly, approximately two-thirds of the type 2
and 5 hexon peptides were considered similar. The
pI was indicative of the net charge of the peptide,
which was influenced primarily by its amino acid composition. These data implied that the similar peptides originated from a common segment(s) of the
polypeptide chain, and that the peptides with distinct pI values were from the variable segment(s) of
the polypeptide. When the CNBr peptides of type 2
and 3 hexons were compared, the inconstancy of the
^3H:^{14}C ratios of the peptides that electrophoresed
suggested that some of these may not be identical.
The amount of variability was much greater between
2 and 3 hexons than between the hexons of type 2 and
5. When virion hexons were iodinated by the lactoperoxidase method, one set of peptides with pI values
greater than 6.8 was iodinated, and this set contained unique as well as some common peptides.
Hexons assembled in virions and those free in solution were iodinated differently. These results suggest that, immunologically, the hexons in viral capsids react differently from unassembled hexons because the polypeptide chains assume different folding configurations in the hexon forms and expose different regions of the protein to antibodies.

2582   THE BINDING OF POLYNUCLEOTIDES TO THE DNA
          POLYMERASE OF AVIAN MYELOBLASTOSIS VIRUS.
(Erickson, R. J. (Mol. Biol. Dept., Miles Lab., Inc.,
Elkhart, Ind.). Arch. Biochem. Biophys. 167(1):238-
246; 1975.

The relationship between the structure of synthetic
nucleic acids and their ability to interact with avian
myeloblastosis virus DNA polymerase was studied. This
interaction was studied using an adaptation of the
membrane filter binding technique. Isotopically labeled Bacillus subtilis DNA was used as a substrate.
The presence of detergent deoxynucleoside triphosphate or deoxyribonuclease all greatly diminished the
retention of the [^3H]DNA on the membrane filter, where
as denaturation of the B. subtilis DNA significantly
increased filter retention. The binding kinetics of
the [^3H]DNA to the partially purified enzyme preparation was a function of both nucleic acid and protein
concentration. Binding of [^3H]DNA to methylated bovine serum albumin was compared to the binding in the

Localization of the avian myeloblastosis viral genome of the RNA sequences complementary to rapidly reassociating cellular DNA was attempted. The poly(A) at the 3' terminus of the viral genome was used as a reference to localize such sequences in the viral RNA. Fragmented avian myeloblastosis virus (AMV) RNA was adsorbed to poly(U)sepharose and the eluted poly(A)-containing RNA species were fractionated into different sizes. RNA fragments of specific sizes were hybridized with excess leukemic cell DNA. Approximately 80% of the 35S AMV RNA hybridized to leukemic DNA by a C_0t of 10,000 (M/sec/l) in a biphasic curve. In the initial transition, 38% of the RNA hybridized; in the second, an additional 42% hybridized. The C_0t 1/2 of the chicken 28S RNA monophasic hybridization curve was used to estimate the reiteration frequency of the cellular DNA sequences that hybridized to AMV RNA. The midpoint of the initial transition occurred at a C_0t of 85 M/sec/l, and the second was half-completed at 4,000 M/sec/l. These were thus complementary to cellular DNA sequences reiterated about 100 times per haploid genome, and 2-3 times per haploid genome, respectively. Fragments of AMV RNA used as probes in liquid hybridization experiments were prepared. The 35S, 27S, 15.5S, 12.5S and 8S poly(A)-containing fragments were hybridized to chicken DNA. These hybridizations showed rapidly hybridizing sequences in all size classes of AMV RNA. These data indicate that the RNA sequences that hybridize rapidly with excess of cellular DNA are not restricted to any one region of the AMV 35S RNA. It is suggested that they may be randomly distributed over the entire 35S AMV molecule, with some positioned within 200 nucleotides of the poly(A) tract at the 3' end of the RNA.

2585   INTEGRATION OF AVIAN SARCOMA VIRUS SPECI-
FIC DNA IN MAMMALIAN CHROMATIN. (Eng.)
de la Maza, L. M. (Dept. Lab. Medicine and Pathology, Mayo Memorial Building, Univ. Minnesota, Minneapolis, Minn. 55455); Faras, A.; Varmus, H.; Vogt, P. K.; Yunis, J. J. *Exp. Cell. Res.* 93(2): 484-487; 1975.

Constitutive heterochromatin and euchromatin fractions from normal and avian sarcoma virus (ASV)-transformed cells of *Mus musculus* and *Microtus agretis* were isolated in order to characterize the sites of integration of the viral-specific DNA sequences. The cells used were BALB/c 3T3 cells transformed with the B77 strain of ASV, *M. agrestis* fibroblasts transformed with the Schmidt-Ruppin strain of Rous sarcoma virus (UMMA-RSV-21), and a revertant subclone of UMMA-RSV-21. The number of viral copies per diploid genome was studied by the technique of DNA-DNA hybridization in solution; this involved determining the rate of reassociation of ^{3}H-labeled viral DNA with total DNA or DNA from euchromatin, intermediate chromatin, and heterochromatin fractions. Total DNA and euchromatin DNA increased the rate of reassociation approximately two-fold when the ratio of cellular DNA to ^{3}H-labeled viral DNA was 1.3×10^6. This increase indicated the presence of 1-2 viral copies per diploid mammalian genome. The ASV-specific DNA sequences were

found in the euchromatin fraction in all cases.
Constitutive heterochromatin and satellite DNA are
apparently not favored sites for integration.

2586   MAMMALIAN TROPISM OF B77(RBI) VIRUS. EX-
        PRESSION OF VIRUS GENOME IN THE HAMSTER SAR-
COMA CELL CLONES. (Eng.) Svec, J. (Cancer Res. Inst.,
Slovak Acad. Sciences, 880 32 Bratislava, Czechoslo-
vakia); Hlubinova, K.; Lizonova, A.; Thurzo, V.
Neoplasma 22(2):133-145; 1975.

The interaction of the B77 virus and its rat variant,
B77(RBI), with newborn Syrian hamster cells was studied in vivo and in vitro. In contrast to B77, the
B77(RBI) continuously and efficiently induced progressively growing tumors in hamsters and transformed hamster embryo cells in vitro. Both induced tumors and
transformed cell cultures showed production of infectious virus and presence of the avian gs-antigen.
Clonal analysis of the established hamster sarcoma
cell line RBH$_{tc}$ revealed that despite the quantitative
differences in the virus production and gs-antigen
content among various clonal populations, full expression of the B77(RBI) virus genome had occurred in each
cell of the virus-induced hamster sarcoma. The permissive relationship of the B77(RBI) virus with rat and
hamster cells, as well as changed antigenic composition, implies that a genetic change of the virus occurred which is reflected in its increased mammalian
tropism and its genetically stable virus-productive
interaction with mammalian cells. It is proposed that
either the rat cells select for the virus variant, or
the variant carries part of the rat cell genome.

2587   ANALYSIS OF ANTIGENIC DETERMINANTS OF
        STRUCTURAL POLYPEPTIDES OF AVIAN TYPE C
TUMOR VIRUSES. (Eng.) Stephenson, J. R. (Animal
Physiology and Genetics Inst., Beltsville, Md.
20705); Smith, E. J.; Crittenden, L. B.; Aaronson,
S. A. J. Virol. 16(1):27-33; 1975.

Competition immunoassays were developed and used in
comparing the reactivities of representatives of
major type C avian virus subgroups. Purified viral
polypeptides were labeled with ^{135}I. A double antibody radioimmunoassay was employed for 27,000 molecular weight polypeptides, while competition immunoassays were developed for the 19,000, 15,000, and
12,000 molecular weight polypeptides of avian myeloblastosis virus (AMV) and for the 19,000 and 12,000
polypeptides of RAV-0, a subgroup E avian tumor virus.
Reactivity studies indicated a lack of significant
type-specific differences between the 127 polypeptides of five major subgroups, including Prague
strain Rous sarcoma virus (RSV) of subgroup A (Pr-
RSV-A) and C (Pr-RSV-C); Bryan high-titer RSV of
subgroups B, BH-RSV(RAV-1) and BH-RSV(RAV-2); Carr-
Zilber RSV of subgroup D; and RAV-0. Agarose gel
filtration revealed that all three low molecular
weight ^{125}I-labeled polypeptides migrated as single
peaks, relative to standards; the molecular weights
were 19,000, 15,000, and 12,000. Unlabeled p27
failed to compete in homologous AMV p19, p15, or p12
assays. Testing for immunological type specificity
of p19, p15, and p12 revealed no detectable differ-

murine viruses also contained a second discrete set of antigenic determinants related to those in infectious primate viruses and endogenous porcine viruses, but not detected in the feline leukemia virus group. The p30 proteins of endogenous viruses of baboons and domestic cats shared a third set of cross-reactive determinants not detected in type C viruses isolated from other species of animals. Enzyme inhibition studies performed with antisera raised toward the reverse transcriptases of these same groups of type C viruses showed the same patterns of immunological cross-reactions as observed with p30 proteins. The results suggest that animals of one species may be infected by type C viruses derived from a second, distantly related group of animals, leading to the incorporation of new virogenes and the eventual perpetuation of these genes in the germ line.

2590   BIOLOGIC AND IMMUNOLOGIC PROPERTIES OF
        PORCINE TYPE C VIRUSES. (Eng.) Lieber,
M. M. (Viral Leukemia and Lymphoma Branch, Natl.
Cancer Inst., Bethesda, Md. 20014); Sherr, C. J.;
Benveniste, R. E.; Todaro, G. J. *Virology* 66(2):
616-619; 1975.

Six type C virus-producing porcine cell lines [MPK, ST-Mo, PORC, PK(15), 38A-1, and PFT] were compared in immunologic and infectivity studies. Reverse transcriptase inhibition studies were performed using immune IgG directed against the partially purified polymerase from the PK(15) virus, and additional experiments were performed to determine if the porcine type C viruses could rescue the Kirsten murine sarcoma virus (Ki-MSV) from mink, rat, and mouse nonproducer cell lines after cocultivation. Attempts were also made to transmit the six porcine type C viruses to other mammalian host cell lines. All six porcine cell lines spontaneously began to release type C viruses after long-term propagation *in vitro*. The reverse transcriptases of all viral isolates were inhibited by immune IgG directed against the PK(15) polymerase, whereas the enzymes from feline leukemia virus, the endogenous baboon virus M7, and the endogenous feline virus RD-114 were not significantly inhibited. Transmission of infection to other host cell lines was detected only with MPK and 38A-1, and then only in the porcine ST-Iowa host cell line. 38A-1 virus also rescued focus-forming Ki-MSV in low titer when cocultivated with the mink and rat nonproducer cells, while the MPK rescued the Ki-MSV genome from the rat nonproducer cells and the PORC rescued Ki-MSV from the mouse nonproducer cell line. The other three porcine viruses were unable to rescue the Ki-MSV genome and produce foci of transformation on any indicator host cell line tested. Since endogenous porcine type C viruses are present as virogens in all pig cells, and since certain of these viruses can replicate in pig cells, these viruses may play an important role in porcine pathophysiology.

2591   STUDIES OF EPSTEIN-BARR VIRUS (EBV)-ASSOCI-
        ATED NUCLEAR ANTIGEN. I. ASSAY IN HUMAN
LYMPHOBLASTOID CELL LINES BY DIRECT AND INDIRECT DE-

TERMINATION OF ^{125}I-IgG BINDING. (Eng.) Brown, T. D. K. (Karolinska Inst., Stockholm, Sweden); Ernberg I.; Klein, G. *Int. J. Cancer* 15(4):606-616; 1975.

A quantitative assay for Epstein-Barr virus (EBV)-associated nuclear antigen (EBNA) in human lymphoblast cells was developed. Cells were grown in RPMI 1640 dium supplemented with 10% fetal bovine serum and a biotics. Nuclei were isolated by harvesting and washing cells in 0.01 M N-2-hydroxyethylpiperazine-N'-2 ethane sulfonic acid and agitated with 22.5% Ficoll and 0.1 mM dithiothreitol, and then centrifuged at 1,000 x g for 15 min at 4C. Isolated IgG antigen w labeled with ^{125}I by a modified chloramine T method The assay used employs EBV-associated nuclear antig positive or -negative $[^{125}$I]IgG preparations, and c be used in a direct or indirect manner. $[^{125}$I]IgG binding is measured by incubation in antigen-contai ing medium for 18 hr at 4C, and the cells are filte in 24 mm glass fiber filters and washed in buffer. Indirect assay is performed by a similar incubation followed by pelleting at 3,500 x g for 10 min; the supernatant is then tested for residual binding to target cells or nuclei using the filter. The assay is sensitive, and detected antigen in 10^6 cells. T indirect assay offers a rudimentary test of antigen relationships among nonproducer cell lines; however it is a more vigorous test of specificity than the direct assay, once the basic specificity of the bin ing to the reference cell line has been established Both antigens of purified nuclear and frozen-thawed cells were efficient in removing $[^{125}$I]IgG binding against the alternative antigen. Most of the cell lines tested contained the EBV genome, and the expressed EBNA contained an antigen and cross-reacted with Raji cells. Fixatives had an adverse effect on antigen activity. The measurement of blocking of EBV-specific $[^{125}$I]IgG binding by human sera has demonstrated the potential of this method for the detection of anti-EBNA antibodies.

2592   EPSTEIN-BARR VIRUS-INDUCED TRANSFORMATION
        OF HUMAN LEUKOCYTES AFTER CELL FRACTIONA-
TION. (Eng.) Schneider, U. (Institut für Klinisc Virologie der Universität Erlangen-Nürnberg, 852 Erlangen, Loschgestr. 7, West Germany); zur Hausen, H. *Int. J. Cancer* 15(1):59-66; 1975.

The efficiency of transformation of human lymphocyt after infection with Epstein-Barr virus (EBV) was determined in fractionated and non-fractionated pre parations derived from 16 human cord blood samples and two blood samples from adult donors. Macrophag depleted leukocytes had a consistently lower transformation efficiency than non-fractionated leukocyt Additional depletion of B-cells resulted in a further decrease. Reduction of T-cells did not signi ficantly change the transformation rate. In nonfra tionated leukocyte cultures, as well as in macrophage-depleted and B-cell enriched cultures, coloni of transformed cells were regularly observed within the first week of cultivation. All cell lines esta lished after EBV-infection revealed membrane-bound immunoglobulin. Reconstitution of macrophage-deple ted, B-cell enriched or B-cell depleted cultures with autologous macrophages or additions of heterologous human embryonic lung fibroblasts increased the transformation efficiency up to the values of nonfractionated leukocyte preparations. The results support the interpretation that EBV transforms only bone-marrow derived cells of the hematopoietic system. The transformation efficiency is increased by cocultivation of lymphocytes with macrophages and heterologous human fibroblasts which seem to exert a feeder-layer effect by enhancing survival of lymphocytes *in vitro*.

2593   ANTIBODIES TO EPSTEIN-BARR VIRUS-ASSOCI-
         ATED NUCLEAR ANTIGEN AND TO OTHER VIRAL
AND NON-VIRAL ANTIGENS IN HODGKINS DISEASE. (Eng.)
Rocchi, G. (Clinica Medica Generale, III Cattedra,
Univ. Rome, Rome, Italy); Tosato, G.; Papa, G.; Ra-
gona, G. *Int. J. Cancer* 16(2):323-328; 1975.

Antibody reactivity to Epstein-Barr virus (EBV)-asso-
ciated nuclear antigen (EBNA) was investigated with
the anticomplement immunofluorescence technique on
sera from 100 patients with Hodgkin's disease (HD)
and from appropriate controls. Antibody levels to
other EBV-determined antigens [i.e., viral capsid
(VCA) and early antigens (EA)], to measles and rubella viruses, to cytomegalovirus (CMV), and to *Toxo-
plasma gondii* were also measured. Results of anti-
EBV antibody titrations demonstrated that anti-VCA,
anti-EA, and anti-EBNA reactivity was significantly
higher in HD patients than in healthy subjects. There
was no significant difference between the distribution
of high rubella and measles antibody titers in HD and
control sera. The geometric mean antibody titer and
the incidence of high titer anti-CMV and *T. gondii*
antibodies were greater in HD patients than in controls. Analysis of the data according to histological
subtypes showed that the condition of lymphocyte depletion in HD patients was associated with the highest anti-EBNA antibody levels and the lymphocyte predominance with the lowest. This pattern seemed to be
peculiar for anti-EBV reactivity, since anti-CMV and
anti-*T. gondii* antibody levels in the lymphocyte-de-
pleted group of patients did not significantly differ from those of controls. No correlation was found
between anti-VCA and anti-EBNA in individual sera of
HD patients. This observation suggests that different
mechanisms are probably responsible in HD for the release of EBV-related antigens from infected cells.

2594   VIRAL AND CELLULAR DNA SYNTHESIS IN NUCLEI
         FROM HUMAN LYMPHOCYTES TRANSFORMED BY EP-
STEIN-BARR VIRUS. (Eng.) Benz, W. C. (Biological
Lab., Harvard Univ., Cambridge, Mass. 02138); Strominger, J. L. *Proc. Natl. Acad. Sci. USA* 72(6):2413-
2417; 1975.

A DNA-synthesizing system *in vitro* was developed
using nuclei prepared by treatment of human lymphocytes with the detergent Brij 58. Lymphocytes from
freshly drawn blood are prepared by Ficoll-hypaque
density gradient centrifugation and washed in dex-
trose/gelatin/veronal buffer. Cultured logarithmic-
phase, phytohemagglutinin-stimulated lymphocytes and
fresh peripheral lymphocytes are sedimented out of
the culture medium, washed, and then resuspended at

pH 8.0 The cells are then chilled and diluted with
an equal volume of a chilled solution of 0.15 M sucrose, 5 mM CaCl$_2$, 25 mM Hepes (pH 8.0) and 0.5% Brij
58. The suspension is gently agitated on ice for 5
min, diluted 20- to 30-fold with chilled 0.25 M sucro
buffer containing 2% Dextran type C 100, pelleted,
and resuspended to a final concentration of 2 x 10^7
cells/ml. The resultant nuclei were used immediately
for determining DNA synthesis using an incorporation
mixture containing dATP, dGTP, dCTP, and [^{3}H]dTTP.
Radioactivity was determined by trichloracetic acid
precipitation. Nuclei from cultured lymphocytes synthesized DNA for as long as five hours (cells in
early logarithmic phase); nuclei prepared from cells
in midlogarithmic phase continued incorporation for
at least 3.5 hr. In nuclei from a parially synchronized line of cultured lymphocytes carrying several
hundred copies of the Epstein-Barr viral genome, synthesis *in vitro* was predominantly viral in the early
S phase and cellular in late S phase. The synthesis
was strongly ATP-dependent, suggesting that the *in
vitro* system incorporates radioactivity primarily by
a replicative process. This is substantiated by the
fact that repair synthesis stimulated by addition of
DNase was ATP-independent.

2595   THE SYNTHESIS OF POLYADENYLATED MESSENGER
         RNA IN HERPES SIMPLEX TYPE I VIRUS IN-
FECTED BHK CELLS. (Eng.) Harris, T. J. R. (Animal
Virus Res. Inst., Pirbright, Woking, Surrey, U.K.);
Wildy, P. *J. Gen. Virol.* 28(3):299-312; 1975.

The pattern of polyadenylated messenger RNA (mRNA)
synthesis in baby hamster kidney cell monolayers, infected under defined conditions with herpes simplex
type I virus (40 plaque-forming U/cell) was investigated by polyacrylamide gel electrophoresis of pulse-
labeled RNA isolated by oligo dT-cellulose chromatography. Two classes of mRNA molecules were synthesized in infected cells; these were not detected in
uninfected cells. The rate of synthesis of the
larger, 18 to 30S, RNA class reached a maximum soon
after infection and then declined, whereas the rate
of synthesis of the 7 to 11S RNA class did not reach
a maximum until much later and did not decline. In
the presence of cytosine arabinoside (50 μg/ml), the
rate of mRNA synthesis in infected cells was reduced
but the electrophoretic pattern remained the same.
The latter observation is consistent with the view
that there are no classical "early" and "late" controls on transcription but rather purely time-dependent controls.

2596   PRESENCE OF HERPES SIMPLEX VIRUS-RELATED
         ANTIGENS IN TRANSFORMED L CELLS. (Eng.)
Chadha, K. C. (Dep. Med. Viral Oncol., Roswell Park
Mem. Inst., Buffalo, N. Y.); Munyon, W. *J. Virol.*
15(6):1475-1486; 1975.

Antiserum prepared against herpes simplex virus type
1 (HSV-1)-infected L cells, i.e., lytic antiserum,
was shown by an indirect immunofluorescence test to
stain 90 to 95% of HSV-transformed L or HeLa cells.
Immunofluorescence in these cells was always most

Cancer 15(5):786-798; 1975.

The induction, isolation, and preliminary characterization of six cell lines transformed by UV-irradiated
herpes simplex virus type 2 (HSV-2) and a temperature
sensitive mutant (ts mutant) of HSV-2 are described.
Irradiated virus was adsorbed to 2-day-old hamster
embryo fibroblasts for two hours at a multiplicity of
400 UV-inactivated particles per cell. Rabbit antiserum to HSV-2 was added to half the infected cultures before incubation. Foci of transformed cells
were picked 40 days after infection and grown to confluency. In additional experiments, hamster embryo
fibroblast monolayers were pretreated with 5-bromode-
oxyuridine or UV irradiation prior to infection and
cell foci were picked 34 days after infection. To
obtain transformation with ts mutants of HSV, hamster
embryo fibroblast monolayers were infected at a multiplicity of 5-10 plaque forming units per cell, and the
virus was adsorbed for two hours at 37 C. Infected
monolayers were incubated at the nonpermissive temperature, and cell foci were picked 30 days after infection. Cultures were tested for oncogenicity by
injecting 0.1 ml (5 x 10^5 cells) of each cell suspension sc into newborn hamsters which were observed
for six months for tumor development. Two of 25 foci
picked 40 days after infection could be subcultured
indefinitely. Three cell lines were established from
foci picked from pretreated infected cultures: two
from 18 foci derived from UV-treated cultures, and
one from 29 foci picked from bromodeoxyuridine-pre-
treated cultures. One of eight foci picked from cultures infected with a HSV-2 DNA⁻ ts mutant survived
more than 70 passages. Two additional mutant-trans-
formed cell lines were established from 130 foci
picked from cultures infected with a DNA⁺ HSV-1 ts
mutant and a DNA⁺ HSV-2 ts mutant. All six transformed cell lines produced undifferentiated fibrosarcomas in newborn hamsters after, but not before,
33 passages. Four of six lines produced metastatic
tumors in the lungs. Sera from tumor-bearing hamsters contained neutralizing antibody to HSV-2. HSV-
specific cell-surface antigen was detected by immunofluorescence in all six of the lines established from
primary tumors. Antibody to an early, HSV-specific,
nonstructural polypeptide reacted by immunofluorescence with antigen(s) in fixed preparations of 3 of 6
tumor cell lines. No type-C virus group-specific-3
antigen was detected in either transformed or tumor-
cell lines. It is suggested that the absence of such
markers is presumptive evidence that type-C virus
played no role in the transforming event, and that
transformation was induced by HSV-2.

2599   EXPERIMENTAL INFECTION WITH HERPES SIMPLEX
        VIRUS TYPE 2 IN NEWBORN RATS: EFFECTS OF
TREATMENT WITH IODODEOXYURIDINE AND CYTOSINE ARABINO-
SIDE. (Eng.) Percy, D. H. (Health Sciences Centre,
Univ. Western Ontario, London 72, Ontario, Canada);
Hatch, L. A. *J. Infect. Dis.* 132(3):256-261; 1975.

Sprague-Dawley rats were inoculated ip, within 36 hr
of birth, with 0.05 ml of either of two strains of
herpes simplex virus type 2. The effects of treatment for five days with either 5 iododeoxyuridine or
cytosine arabinoside (12.5 or 25 mg/kg/day) were

studied. Evaluation of treatment based on rate of
survival, incidence of lesions, and isolation of
virus from the CNS (100 or 200 mg/kg/day) showed no
appreciable difference between treated and untreated
littermates. Drug-related effects of retardation of
growth and defective development of the cerebellum
and retina were found. Areas of retinal hyperplasia
were present in the majority of rats that survived
for more than ten days after treatment with the
higher 5-iododeoxyuridine dose. Cerebellar hypoplasia and marked retinal dysplasia were present in
animals treated with either dose of cytosine arabinoside, particularly those killed at 14 days. The
virus content of brain and skin suggested that neither
drug had been effective. The validity of therapy
with either of these drugs in generalized herpetic
infections is questioned because activity was minimal against the infecting agent even when the dosage
of the drug was sufficient to produce serious
defects in certain developing tissues.

**2600   HERPESVIRUS ANTIBODIES AND ANTIGENS IN
PATIENTS WITH CERVICAL ANAPLASIA AND IN
CONTROLS.** (Eng.) Pacsa, A. S. (University Medical
Sch., H-7643 Pecs, Hungary); Kummerlander, L.; Pejt-
sik, B.; Pali, K. *J. Natl. Cancer Inst.* 55(4):775-
781; 1975.

Antibody activity to herpesvirus hominis type 2
(HVH-2) in 151 women cured of cervical carcinoma (60
in situ, 91 invasive) and in 106 controls was investigated. The two control groups consisted of women with
or without cervical atypia; controls of both groups
were matched to women with cervical atypia by age
(10-yr intervals), social status, marital status, number of pregnancies, and number of children. The antibody activity to HVH-2 differed significantly, especially between the *in situ* (73%) and control (17%)
groups. Sera of 57 patients with cervical atypia
showed an increased antibody to HVH-2 (58%), compared
with that of the 57 matched controls (23%). Antibodies to cytomegalovirus (CMV) were detected more
frequently in sera of women with atypia (61%) than in
sera of women with cervical disorders other than
atypia (42%) or in sera of healthy controls (33%).
HVH antigens were present in cervical cells from patients with atypia and from matched controls. Not
only exfoliated (imprints) but also cultured and cocultured cervical cells contained HVH antigens; there
was no correlation bewteen antigen positivity and
antibody activity to HVH-2. The data support the
association between HVH-2 and cervical anaplasia and
indicate that CMV may also be implicated in its etiology.

**2601   SEGREGATION OF RD-114 FeLV-RELATED SE-
QUENCES IN CROSSES BETWEEN DOMESTIC CAT
AND LEOPARD CAT.** (Eng.) Benveniste, R. E. (Natl.
Cancer Inst., Bethesda, Md. 20014); Todaro, G. J.
Nature 257(5527):506-508; 1975.

Male leopard cats (*Felis bengalensis*) lacking RD-
114 and feline leukemia virus (FeLV)-related DNA
sequences (RD⁻, FL⁻) were mated to domestic cats
(*F. catus*, RD⁺, FL⁺). The segregation of both sets

of virogenes was studied in F₁ hybrids and in the
progeny of a backcross to the RD⁻, FL⁻ parent. The
cellular DNA of the F₁ hybrids contained half the
number of each set of sequences present in the domestic cat parent. One of two kittens obtained
from an F₁ hybrid female backcrossed to the leopard
cat contained all the RD-114-related information
but, like the F₁ parent, only half the number of
virogene copies. This kitten also had the same
number of FeLV-related virogene copies of F₁ parent.
The second kitten lacked both RD-114 and FeLV-related DNA sequences. In the backcrossed animals,
the RD and FL virogenes segregated together in a
manner consistent with their localization at a
single chromosomal site. The data rule out both
the possibility that each of the multiple copies of
RD-114 or FeLV-related virogenes occurs on a different linkage group, and models of nonchromosomal
inheritance of the multiple viral copies.

**2602   FRIEND LEUKEMIA AND ENDOGENOUS VIRUSES.
I. PRODUCTION OF FRIEND AND ENDOGENOUS
C-TYPE PARTICLES BY MOUSE CELLS CULTURED *IN VITRO*.**
(Eng.) Stewart, M. L. (Albert Einstein Coll. Med.,
Bronx, N.Y.); Maizel, J. V., Jr. *Virology* 65(1):55-
66; 1975.

The production of Friend leukemia virus (FLV) and
other particles produced by BALB/c mouse cells was
followed in serial cultures after infection *in vitro*.
BALB/c mouse cells, free of C-type particles as
shown by electron microscopy, were seeded at
5 x 10⁵/75 cm² and infected with 4 ml of mouse spleen
extract containing 2 x 10⁶ plaque forming units FLV.
Cells were maintained in exponential phase of growth
by transfer at 90% confluency. RNA in cells was
radiolabeled with [³H]uridine (50 μCi/ml added to
the medium for 16 hr. After removal of cells from
the medium by centrifugation, the supernatant medium
was centrifuged to remove subcellular debris before
the addition of concanavalin A for 2.5 hr. Concanavalin A-agglutinated virus was fractioned in sucrose equilibrium gradients. Virus production was
assayed by analysis of 60-70S RNA in sodium dodecyl
sulfate-solubilized Concanavalin A aggregates.
Newly infected cells released infectious leukemia
virus (density 1.16) together with a heterogeneous
population of other particles of densities 1.18,
1.22, and 1.24. A temporal relationship existed
between the kinds of particles produced by the cells
and the extent of cell passage after infection.
Particles of a density greater than 1.16 were released in gradually diminishing amounts during successive cell passages. Virus was continually released by the cells during these serial passages,
and it was consistently and quantitatively isolated
by agglutination. Three days after infection, 65S
RNA was detectable in particles released into the
culture medium. The amount of 65S RNA increased
significantly through the second cell passage and
remained constant in subsequent passages. Infected
and uninfected eighth passage cells were examined
in thin sections in the electron microscope for C-
type particles. Extracellular virus particles 100-
150 nm in diameter having typical C-type morphology
were visible in sections from both cultures. These

day 0. Irradiated mice were inoculated ip on day 5
and nonirradiated mice on days 7-9 with 10^{-7} M of
5-fluoro-2'-deoxyuridine to depress endogenous thymidine synthesis; 30-60 min later they were inoculated with 0.2-0.5 µCi of ^{125}5-iodo-2'-deoxyuridine
(^{125}IUR) to monitor leukemic cell multiplication.
The uptake of ^{125}IUR in the spleen increased in
inoculated mice with Friend leukemic cells as compared with uninoculated mice. This uptake was inhibited when treated with interferon. The ^{125}IUR
uptake in leukemic-infected mice was much greater
than those given by frozen and thawed cells representing a difference in the proliferation of injected
viable cells. Daily treatment of mice with interferon inhibited this leukemia-induced ^{125}IUR uptake.
Treatment of mice with mouse cell culture interferon
or with Newcastle disease virus (NDV) inhibited the
^{125}IUR uptake, whereas treatment with either rabbit
interferon or control cell cultures had no effect.
These results show that daily interferon treatment
can inhibit the multiplication of transplanted Friend
leukemic cells in the spleen. The influence of interferon may be a direct or indirect (host-mediated)
interference of cell multiplication.

2605   ERYTHROID COLONY INDUCTION WITHOUT ERY-
         THROPOIETIN BY FRIEND LEUKEMIA VIRUS *IN*
VITRO. (Eng.) Clarke, B. J. (Faculty of Medicine,
Univ. Toronto, Toronto, Ontario, M5S 1A8, Canada);
Axelrad*, A. A.; Shreeve, M. M.; McLeod, D. L.
Proc. Natl. Acad. Sci. USA 72(9):3556-3560; 1975.

The induction of erythroid colony formation without erythropoietin (Epo) by bone marrow cells in-
feeted *in vitro* with Friend leukemia virus (FV) is
reported. Bone marrow cells from C3Hf/BiUt mice
were treated with media from IS cell cultures,
which continuously produce FV in culture. The
cells were then incubated for two days in plasma
culture, and the number of erythroid colonies were
observed. The effects of centrifuging the IS medium on the induction of erythroid colonies was
determined, as were the effects of subjecting it
to elevated temperatures (56 C). In addition,
determinations were made of the effect of using
bone marrow cells from different mouse strains
(SIM and C57BL/6), the fetal calf serum concentration (0-20%), the presence of XC plaque-forming
virus in the FV-containing medium, and the dose
of FV-containing IS culture medium. Compared with
control medium-treated cultures, the number of erythroid colonies was 2- to 3-fold higher in the IS
medium-treated cultures. The activity in the IS
medium could be removed by centrifugation and recovered from the resuspended pellet in concentrated
pellets. The number of erythroid colonies produced *in vitro* was related to the Friend spleen
focus-forming virus (SFFV) titer *in vivo*. Similarly, both the capacity of IS culture medium to
induce erythroid colonies and its Friend SFFV titer
were lost after heating. While Epo induced the
formation of erythroid colonies by bone marrow
cells independently of their strain of origin, the
IS medium induced erythroid colony formation only
when mixed with C3H or SIM bone marrow cells, but
not with those of the B6 strain. Reduction of the

fetal calf serum concentration lowered the number
of erythroid colonies in control cultures, but had
much less effect on the IS medium-treated cultures.
The presence of Friend SFFV activity was necessary
for erythroid colony induction by FV *in vitro*, and
the number of erythroid colonies produced in the IS
medium-treated cultures was approximately proportional to the concentration of IS medium. Also,
the capacity of IS medium to induce the formation
of erythroid colonies *in vitro* was related to its
capacity to induce spleen foci *in vivo*. This system
should prove useful for the *in vitro* investigation
of Friend virus-host interactions which lead to
Epo-independent erythropoiesis.

2606   ADAPTATION OF N-TROPIC FRIEND LEUKAEMIA
        VIRUS AND ITS MURINE SARCOMA VIRUS PSEU-
DOTYPE TO NON-PERMISSIVE B-TYPE C57BL/6 MOUSE CELL
LINE. (Eng.) Yoshikura, H. (Inst. Medical Science,
Univ. Tokyo, Japan). *J. Gen. Virol.* 29(1):1-9;
1975.

The adaptation of N-tropic Friend leukemia virus
(FLV) and a murine sarcoma virus (MuSV) pseudotype
to nonpermissive B-type cells was studied. The
MuSV pseudotype was obtained by infecting S+L- cells
with the FLV. N-tropic FLV was serially passaged
in the B-type C57BL/6 mouse cell line, YH-7, as was
MuSV(FLV). After a single passage, the infectious
efficiency of FLV in the restrictive YH-7 cells was
significantly increased. This adapted character of
FLV could be reversed by a single passage in permissive N-type MLg cells. The true host range conversion from N to NB was accomplished after 11-12 passages in YH-7 cells; no reversion to N-tropism was
observed with the NB-tropic-converted twelfth passage FLV in YH-7 cells. When the NB-tropic Rauscher
leukemia virus (RLV) was serially diluted in the
medium or in FLV and used to infect the YH-7 cells,
the virus in the harvest was predominantly NB-
tropic. This was true even if the N-tropic FLV
existed in a large excess. The N to NB conversion
of FLV was much slower than that of MuSV(FLV); in
the latter, the host range conversion from N to NB
was accomplished in two passages, the converted
virus remaining NB-tropic even after propagation
back into the N-type MLg cells. Thus, the N-tropic
FLV was readily converted to NB-tropic virus in the
presence of the MuSV genome, but the conversion required many more passages in the absence of MuSV.

2607   ECOTROPIC LEUKEMIA VIRUSES IN CONGENIC
        C57BL MICE: NATURAL DISSEMINATION BY
MILK-BORNE INFECTION. (Eng.) Melief, C. J. M.
(Tufts Univ. Sch. Medicine, 171 Harrison Ave., Bos-
ton, Mass. 02111); Louie, S.; Schwartz, R. S.*   *J.
Natl. Cancer Inst.* 55(3):691-698; 1975.

The incidence of infectious murine leukemia virus
(MuLV) was studied in ten lines of congenic-resistant
C57BL mice. No constant relationship between the
incidence of MuLV and the H-2 complex was apparent.
Two lines, B10 and B10.A, were examined in detail because the incidence of MuLV in B10 was low and the
virus appeared relatively late in life, whereas B10.A

animals had a relatively high incidence of infection
by MuLV early in life. Further studies of B10.A mice
revealed an almost universal concordance between the
virologic status of the mother and her offspring.
This was particularly evident when (B10 x B10.A)F₁
animals were compared with (B10.A x B10)F₁ mice. Although genetically identical, the incidence of MuLV
in the latter was high, whereas in the former it was
low. Transmission of MuLV by milk was proved by fos-
ter-nursing experiments; when the infants of MuLV-
positive B10.A mothers were suckled on MuLV-negative
B10.A mothers, they were free of MuLV. Milk-borne
infection may account for the natural dissemination
of MuLV among some inbred lines of mice.

2608   HYBRIDIZATION OF MOUSE LEUKEMIA VIRUS c-
        DNA TO MOUSE REPEATED DNA SEQUENCES.
(Eng.) Ginelli, E. (Centro di Patologia Molecolare,
Via Pace 15, I-20122 Milano, Italy); Gianni, A. M.;
Corneo, G.; Polli, E. *Acta Haematol. (Basel)* 54(4):
221-226; 1975.

Hybridization experiments between mouse leukemia virus synthetic-³H-DNA probe and mouse main band and
satellite DNAs were carried out to determine the percentage of viral sequences in satellite DNA. Transformed mouse cell line 1798 was obtained by infection
of a normal 3T6 mouse with Harvey strain Murine sarcoma virus. Mouse satellite DNA was isolated from
main band DNA by silver-cesium or cesium chloride
centrifugation. Viral DNA probe was hybridized to
isolated satellite and main band DNAs at 60 C in
0.48 M phosphate buffer. The amount of radioactive
viral complementary DNA hybridized was evaluated by
batch elution at 60 C. No higher concentrations of
viral sequences were observed in the satellite DNA;
the hybridization was not due to highly reiterated
sequences present in the viral genome. There was
no significant difference between main band and satellite and main band DNAs at 60 C in 0.48 M phosphate
buffer. The amount of radioactive viral complementary DNA hybridized was evaluated by batch elution
at 60 C. No higher concentrations of viral sequences
were observed in the satellite DNA; the hybridization
was not due to highly reiterated sequences present in
the viral genome. There was no significant difference between main band and satellite DNAs; however,
a difference in the thermal stability in the hybrids
obtained with normal and transformed DNA was evident.
The viral sequences appeared to be enriched in the
fast renaturing intermediate main band DNA.

2609   IDENTIFICATION OF THE VIRIONS IN THE *IN
        VITRO* L1210(V) LEUKEMIA CELL LINES BY MOR-
PHOLOGICAL, VIROLOGICAL, AND IMMUNOLOGICAL TECHNIQUES.
(Eng.) Chen, P. L. (Inst. Med. Res., Camden, N.J.);
Hutchison, D. J.; Sarkar, N. H.; Kramarsky, B.; Moore,
D. H. *Cancer Res.* 35(3):718-728; 1975.

Studies of the structural and biological features of
virus in two cell lines, L1210(V) Glu⁻ and L1210(V)-
Asn⁻, are reported. Sera used in immunological tests
included: normal rabbit, normal rat, rabbit anti-mam-
mary tumor virus of RIII milk, rat anti-Gross leukemia
virus, rat anti-Rauscher leukemia virus proteins, goat

466

anti-rabbit α-globulin, and anti-rat α-globulin. The cultures were subjected to immunoelectron microscopic examinations and immunodiffusion tests. The cell lines from *in vitro* cultures were inoculated ip into four strains of mice: C57Bl, Af, RIIIf, and BALB/c, having mammary tumor incidences of 0, 47, 10, and 27%, respectively. A fifth mouse strain, BD2F₁, was used in initial infectivity experiments. The investigation involved thin-section, whole-cell mount, and negative-strain electron microscopy. About 20-30% of the cells of both lines bore two conspicuous surface extensions, i.e., stout protrusions of varying shape and size, and microvilli. Cells cultured *in vitro* gradually lost their efficiency in virion production, and required passage *in vivo* to replenish their productivity. Intracytoplasmic A-particles exhibited three general distribution patterns: aggregated, scattered, and near the cell margin. Five modes of particle morphogenesis were observed: (1) spiked virions with nucleoids in the process of assembly, (2) spiked virions with pre-formed nucleoids, (3) spikeless particles with nucleoids in the process of assembly, (4) intracisternal A-particles, and (5) spikeless particles with pre-formed nucleoids. Only two morphologically distinct types of extracellular particles were observed: virions covered by uniformly spaced surface spikes, and virions with spikeless envelopes. Density gradient separation of B- and C-type particles was unsuccessful; the majority of the budding virions of L1210 cells prepared for immuno-electron microscopy had performed nucleoids. Generalized leukemia occurred in 100% of the BD2F₁ and 47% of the BALB/c mice within 8-10 days after inoculation. However, generalized leukemia did not occur in any of the C57BL, Af, and RIIIf strains tested; the purified virions were also ineffective. The L1210(V) cell suspension lines, differing in one nutritional requirement, nonetheless produced the same types of viral particles. That the cells also gave cytoplasmic immunofluorescence when tested against the six protein antigens of Rauscher virus and demonstrated the strongest reaction to the 30,000 molecular weight protein, suggests that at least some of the viral antigens are related to Rauscher leukemia virus.

2610   *IN VITRO* INFECTION OF LYMPHOID CELLS BY
          THYMOTROPIC RADIATION LEUKEMIA VIRUS
GROWN *IN VITRO*. (Eng.) Haas, M. (Dept. Cell Biology, Weizmann Inst. Science, Rehovot, Israel); Hilgers, J. *Proc. Natl. Acad. Sci. USA* 72(9):3546-3550; 1975.

Some of the biological properties of a cell-culture-derived lymphatic leukemia virus (LLV) of the mouse are described. Murine lymphoid cells were infected *in vitro* with purified leukemogenic radiation leukemia virus (RadLV) produced by virus-induced lymphoblast cell lines (BALB/3T3, NIH/3T3, normal rat kidney, BL-5, M-6, and SC-1). Using indirect immunofluorescence and xc plaque assays, thymocytes were shown to be highly susceptible to infection by the virus, whereas murine or other fibroblasts were refractory to it. Murine bone marrow and spleen cells were shown to be much less sensitive to infection by this thymotropic RadLV. By comparison, a B-tropic RadLV isolate (RadLV*), propagated on a mouse fibroblast cell line, was noninfectious for lymphoid

cells but infected fibroblasts. A correlation was shown to exist between *in vitro* infection of thymocytes, as assayed by immunofluorescence, and *in vivo* leukemogenicity of the thymotropic RadLV. This constitutes a rapid *in vitro* test for *in vivo* leukemogenicity of a natural lymphatic leukemia virus.

2611   HEMOLYTIC ANEMIA INDUCED BY MURINE ERY-
          ROBLASTOSIS VIRUS: POSSIBLE MECHANISMS
OF HEMOLYSIS AND EFFECTS OF AN INTERFERON INDUCER
(Eng.) Slamon, D. J. (Dept. Anatomy, Univ. Chicago, 1025 E. 57th St., Chicago, Ill. 60637). *J. Natl. Cancer Inst.* 55(2):329-338; 1975.

Viral, morphologic, and biochemical differences between the murine erythroblastosis virus (MuEV, also called murine leukemia virus-Kirsten)-infected and MuEV-infected, poly I·poly C treated W/Fu rats were studied to gain insight into the nature of the hemolytic anemia seen in erythroblastosis and the possible mechanisms by which poly I·poly C prevents the development of this disorder. The ultrastructure of the virus and of the viral reproductive process were also studied. MuEV infection in 1-2 wk-old rats induced a fulminant hemolytic anemia accompanied by a marked splenomegaly and the characteristic splenic histology seen in MuEV-induced erythroblastosis. Animals treated with poly I·poly C within three days after MuEV infection did not develop splenomegaly or the histologic changes, although those beginning treatment after day 3 post-infection showed a pathologic response pattern similar to that in the untreated animals. In comparison with the untreated rats, animals receiving poly I· poly C also had markedly reduced levels of virus reproduction as measured by bioassay and electron microscopic examination of the erythroid cells. The proliferation of erythroblasts after MuEV infection in the animals not receiving poly I·poly C appeared to be an erythropoietin-dependent compensatory response to hemolysis. The hemolysis itself seemed to require virus reproduction in the affected cell types. Morphologically, the virus itself was indistinguishable from any other C-type RNA viruses in the murine leukemia-sarcoma group. The budding process during virus replication began as a dense linear thickening immediately subjacent to the plasma membrane. As the assembly process continued the nucleoid grew in size and budded from the cell taking with it the overlying plasma membrane and underlying cytoplasm. The results suggest that poly I·poly C treatment in MuEV-infected rats inhibits virus reproduction and thus may circumvent the hemolytic disease syndrome.

2612   VIRAL-RELATED INFORMATION IN ONCORNAVIRUS-
          LIKE PARTICLES ISOLATED FROM CULTURES OF
MARROW CELLS FROM LEUKEMIC PATIENTS IN RELAPSE AND
REMISSION. (Eng.) Mak, T. W. (Dept. Med. Biophys., Univ. Toronto, Canada); Kurtz, S.; Manaster, J.; Housman, D. *Proc. Natl. Acad. Sci.* 72(2):623-627; 1975.

The nucleic acid component of particles found in tissue from patients with leukemia was characterized. Twenty-nine marrow specimens were obtained from 26

patients with leukemia, and nine specimens were obtained from patients without leukemia. The size distribution of RNA contained in particles released by
leukemic cells was analyzed by velocity sedimentation
in sucrose gradients containing sodium dodecyl sulfate. As a control, a tissue culture cell line actively producing a murine oncornavirus (Friend murine
leukemia virus) was incubated under conditions identical to those used for the marrow samples. Seventeen marrow samples obtained from leukemic patients
in relapse or in a chronic phase of the disease
yielded particles containing high-molecular-weight
RNA with a sedimentation velocity (about 70S) similar
to that obtained for murine oncornavirus RNA. These
particles were released even after a short period (16-
18 hr) from cultures of cells from patients with the
following leukemias: acute myeloblastic, acute
lymphoblastic, acute myelomonocytic, chronic lymphocytic (in relapse or chronic phase), and chronic myelocytic (in chronic phase). Eight of nine marrow
samples from nonleukemic patients did not yield detectable high-molecular-weight RNA. Among patients
in firm hematological remission, 3 of 3 samples from
patients with acute lymphoblastic leukemia and 3 of
9 samples from patients with acute myeloblastic leukemia were positive for high-molecular-weight RNA.
The base sequence of the RNA in particles was characterized by synthesizing complementary [^{3}H]DNA in an
endogenous reaction and hybridizing to excess RNA
from known oncornaviruses. Hybridization of 40-60%
of input complementary DNA to simian sarcoma virus
RNA was detected. No homology was detected with an
avian oncornavirus (Rous sarcoma virus), while an
intermediate level of homology (10-30%) was found
in hybridization to murine sarcoma virus (Kirsten) and
murine leukemia viruses (Raucher, Moloney, and Gross).
The results provide additional evidence that the virus-related particles produced by marrow from leukemia
patients share many properties with viruses known to
be oncogenic in other species.

2613   BIOLOGICAL AND PHYSICAL MODIFICATIONS OF A
        MURINE ONCORNAVIRUS BY 2-DEOXY-D-GLUCOSE.
(Eng.) Prochownik, E. V. (Pritzker Sch. Med., Univ.
Chicago, Ill.); Panem, S.; Kirsten, W. H. J. Virol.
15(6):1323-1331, 1975.

The effect of 2-deoxy-D-glucose (2-DG) on the release
of Kirsten murine sarcoma-leukemia virus [KiMSV-
(KiMuLV)] from transformed rat kidney (NRK-K) cells
was investigated. ^{3}Uridine-labeled virus was pellete.'
and banded in sucrose. The total number of ^{3}H counts
per minute in the density region of the viral peak
(∼1.14-1.18 g/cm^3) was counted. Aliquots (50 μl) of
each gradient fraction were assayed for protein.
At 30 mM 2-DG, RNA synthesis in NRK-K cells was inhibited by approximately 30%, and protein synthesis
was inhibited by as much as 80% of control levels.
RNA synthesis was not inhibited in nontransformed
normal rat kidney (NRK) cells, although protein synthesis was equally suppressed in NRK and NRK-K cells.
After treatment with 2-DG, the release of physical
particles of KiMSV(KiMuLV) from NRK-K cells was not
reduced as determined by equilibrium density gradient
centrifugation and assays for RNA-dependent DNA polymerase of culture fluids. The ability to detect

equivalent in length to a complete transcript of the
35S virion RNA. The second of the proviral DNA, as
detected by complementarity to ^{3}H-complementary DNA
(^{3}H-cDNA)-sedimented molecules of 5 x 10^5 daltons;
the largest had a molecular wt of about 9 x 10^5.
The proviral DNA derived from the upper band was not
derived from the super-coiled closed-circular DNA.
Boiling before hybridization increased the number of
sites available to hybridize to ^{125}I-RNA by a factor
of 1.5; prior denaturation increased the hybridization of proviral DNA to ^{3}H-cDNA. Treatment with S_1
nuclease confirmed the observation that the first-
strand DNA contained sites available for hybridization by virtue of their single-strandedness. Further
experiments indicated that about half of the first-
strand DNA was initially found in a single-stranded
state, while the second strand DNA was virtually all
in a double-stranded configuration. Further confirmation was obtained by isopycnic centrifugation of
the complex on sodium iodide gradients, indicating
the presence of large amounts of single-stranded DNA
in the proviral DNA.

2616   GENETIC CONTROL OF ONCOGENESIS BY MURINE
          SARCOMA VIRUS MOLONEY PSEUDOTYPE. I.
GENETICS OF RESISTANCE IN AKR MICE. (Eng.) Colombatti, A. (Inst. of Pathological Anatomy, Univ. of
Padova, Italy); Collavo, D.; Biasi, G.; Chieco-Bianchi, L. *Int. J. Cancer* 16(3):427-434; 1975.

The differential susceptibility to M-MSV in hybrids
of AKR mice was studied. Injection (im, thigh) of
Moloney mouse sarcoma virus (M-MSV) in adult mice of
several inbred strains revealed that all strains except AKR are highly susceptible (80-100%) to M-MSV
tumor development. F1 hybrids between AKR and CBA,
DBA/2 or NIH mice are as resistant (93%) as the
parental AKR strain, which indicates that resistance
is transmitted as a dominant character. First backcross mice to the susceptible parent show a 3:1
ratio of resistant to susceptible mice. This is the
expected ratio for two segregating loci which independently confer resistance. The incidence of resistant F2 mice is somewhat lower than expected.
Further support for the two-gene hypothesis was obtained in second backcross mice. The major genes
affecting murine leukemia virus infection (Fv-2 and
H-2) do not seem to play any role in this system and
no linkage was found with Thy.1 and albino, dilute,
agouti and brown markers.

2617   GENETIC CONTROL OF ONCOGENESIS BY MURINE
          SARCOMA VIRUS MOLONEY PSEUDOTYPE. II.
A DOMINANT EPISTATIC SUSCEPTIBILITY GENE. (Eng.)
Colombatti, A. (Inst. of Pathological Anatomy, Univ.
of Padova, Italy); Collavo, D.; Biasi, G.; Chieco-
Bianchi, L. *Int. J. Cancer* 16(3):435-441; 1975.

The tumor response pattern of F_1, F_2, and backcross
hybrids between resistant AKR and susceptible
strains (C57Bl/6, BALB/c, and BIOBR) of mice following injection with the Moloney mouse sarcoma virus
(M-MSV) was studied. Fv-1 typing was determined
by progeny testing using the Friend virus spleen
focus assay, and a humoral cytotoxic assay on peri-

pheral leukocytes was performed for the detection
of H-2^k haplotype. Among adult hybrids between AK
and B6, BALB, or BIOBR, 86% developed tumors withi
ten days after M-MSV injection; this was nearly th
same percentage as was observed in the susceptible
parents. Segregation was observed in the first
backcross (Bc1) and F_2 mice, and the segregation
ratios up to Bc3 mice fit a one-gene model. In
tests in which susceptible F_1 hybrids were mated
with a third nonparental strain, 48% of the progen
were susceptible, indicating that in some strains
and their hybrids a dominant epistatic gene was
masking the resistant phenotype. Progeny testing
for Fv-1 showed no close linkage between susceptibility and H-2, although a slight suggestion of
linkage with albino phenotype was found in F_2, but
not Bc1, mice. The data conclusively indicate tha
M-MSV tumor induction is under host genetic contro
At least three genes appear to govern the tumor re
sponse in the offspring of crosses with AKR mice:
two dominant autosomal loci independently confer
resistance to M-MSV tumor development in AKR; and
dominant autosomal epistatic gene masks the resistant phenotype in some individuals. There is also
some evidence that additional factors may exist
which exert a minimal influence on tumor response.

2618   EXPRESSION OF MOUSE MAMMARY TUMOR VIRAL
          POLYPEPTIDES IN MILKS AND TISSUES. (Eng
Noon, M. C. (Natl. Cancer Inst., Bethesda, Md.); W
ford, R. G.; Parks, W. P. *J. Immunol.* 115(3):653-
658; 1975.

A 14,000-dalton polypeptide (p14) from RIII murine
mammary tumor virus (MMTV) was isolated by column
chromatography in 6 M GuHCl. Antiserum prepared i
rabbits specifically precipitated ^{125}I-labeled p14
in double antibody competition, radioimmunoassays
performed with limiting amounts of antibody, both
purified p14 and disrupted MMTV, competed specifically with labeled antigen. The expression of thi
MMTV type B virus antigen could be measured by com
petition radioimmunoassays in milks, mammary gland
tumors, and tissue culture cells. MMTV expression
measured by p14 immunoassay correlated well with t
spontaneous incidence of mammary adenocarcinomas i
different murine strains but not with type C MuLV
antigen expression. Levels of MMTV gp52, the majo
type-B viral glycoprotein, corresponded to p14 lev
suggesting that their control is comparably regula
Evidence that this low molecular weight polypeptide
is present in feral and inbred strains of widely d
fering geographic origin and in MMTV with apparent
different biologic properties suggests surprising
conservation of MMTV protein homology.

2619   MOUSE MAMMARY TUMOR VIRUS POLYPEPTIDE
          PRECURSORS IN INTRACYTOPLASMIC A PAR-
TICLES. (Eng.) Smith, G. H. (Natl. Cancer Inst.,
Bethesda, Md. 20014); Lee, B. K. *J. Natl. Cancer
Inst.* 55(2):593-496; 1975.

Intracytoplasmic A particles were analyzed by immunodiffusion and sodium dodecyl sulfate-polyacryl
amide gel electrophoresis (SDS-PAGE) before and

after enzymatic cleavage with trypsin. The particles were purified from transplantable Leydig cell tumors. Mouse mammary tumor virus (MMTV) was isolated from C3H-A^{vy}/He mammary tumors by sequential rate zone sedimentation in two linear gradients. A common antigen in A particles was detected by antisera prepared against this virus and against a purified MMTV core polypeptide (p28) and purified intracytoplasmic A particles. Despite this correlation, no SDS-polyacrylamide band migrating at p28 was observed in purified intracytoplasmic A particles. However, after incubation with trypsin (1.25 µg/100 µg A particle protein, 30 min, 37 C), A particles subjected to SDS-PAGE produced only two polypeptide bands. They were observed at p28 and p15-10. Ouchterlony analysis of the trypsin-cleaved A particles revealed no alteration in the antigenicity of the particles. These results suggest that some structural components of intracytoplasmic A particles are polypeptide precursors of MMTV core proteins.

2620   PROPERTIES OF A CHICKEN LYMPHOBLASTOID
        CELL LINE FROM MAREK'S DISEASE TUMOR.
(Eng.) Nazerian, K. (Reg. Poult. Res. Lab., Agric. Res. Serv., East Lansing, Mich.); Witter, R. L. *J. Natl. Cancer Inst.* 54(2):453-458; 1975.

Properties of a chicken lymphoblastoid cell line (MSB-1) from a Marek's disease tumor (MD) are described. MBS-1 cells were cultured at 41 C and 37 C; subcultures were made every 48 hr. MSB-1 cells died after the sixth subculture at 37 C, but were growing rapidly eight months after culturing at 41 C. They measured 8-10 µ in diameter, had a smooth surface with round nuclei which stained pink with Wright's stain. Secondary monolayers of duck embryo fibroblasts (DEF) or chick embryo fibroblasts (CEF) exhibited microplaques typical of MD-virus after 6-7 days of co-cultivation with 1 x 10^3 to 2.5 x 10^5 MSB-1 cells. In virus-producing cells, the nuclear chromatin was marginated, nucleoli were fragmented and naked nucleocapsids appeared in the nucleus. Ten chicks (7$_2$ line) were inoculated ip with 24 plaque-forming units (PFU) of MSB-1 cells; all developed tumors in the pectoral muscle (3/10 died). Two of five untreated chicks kept in the same cage developed similar tumors. All of nine chicks, inoculated ip with 310 PFU of DEF (co-cultivated with MSB-1 cells), and two of five untreated chicks in the same cage developed visible pectoral muscle tumors. Controls failed to develop tumors. These results demonstrate that co-cultivation of MSB-1 with DEF or CEF results in the transfer of the disease agent to these cells.

2621   EFFECTS OF CYCLOHEXIMIDE ON VIRUS DNA
        REPLICATION IN AN INDUCIBLE LINE OF POLY-
OMA-TRANSFORMED RAT CELLS. (Eng.) Manor, H. (Dept. Biology Technion-Israel Inst. Technology Haifa, Israel); Neer, A. *Cell* 5(3):311-318; 1975.

The biochemical transformations causing induction of virus synthesis in an inducible polyoma-transformed rat cell line (LPT) was studied by analyzing the effect of cycloheximide (CH) on the replication of viral DNA in the repressed and activated states. At a concentration of 10 µg/ml, CH had two distinct effects on the replication of polyoma virus (PV). Exposure of LPT cells to CH caused up to an 8-fold increase in the cellular concentration of PV DNA determined by molecular hybridization. The treatment also inhibited cell division and chromosomal DNA replication, but did not affect the amount of chromosomal DNA per cell. In LPT cells treated with mitomycin C (1 µg/ml), PV DNA replication was enhanced after seven hours. Between 7-24 hr, the concentration of virus DNA increased at least 100-fold. CH added to the cells 0-7 hr after treatment with mitomycin C inhibited the replication of PV DNA by 90-100%. The inhibition was less effective in cells exposed to CH from 7 hr and on. The inhibitory effect was reversible: virus DNA synthesis was resumed after removal of CH from the growth medium. The data indicate that CH acts as an inducer of virus DNA synthesis in cells whose resident viral genome is repressed, but inhibits the autonomous replication of the activated genome following induction with MMC. At least two proteins (or groups of proteins) apparently control the replication of PV DNA; one directly or indirectly involved in maintaining the repressed state of the viral genome, and the other playing an essential role in the autonomous replication of PV DNA.

2622   TEMPERATURE-DEPENDENT PROPERTIES OF CELLS
        TRANSFORMED BY A THERMOSENSITIVE MUTANT
(TS-121) OF POLYOMA VIRUS. II. CHARACTERIZATION OF 121-6 CELLS. (Eng.) Okada, Y. S. (Res. Inst. for Microbial Diseases, Osaka Univ., Suita, Osaka, Japan); Hakura, A. *Int. J. Cancer* 16(3):394-403; 1975.

The growth and agglutinability properties of a cell line transformed by a temperature-sensitive mutant of polyoma virus (ts-121) in liquid media were studied. The cell lines used were: A31-714, a recloned cell line from BALB/3T3 cells; PV-4, a line of A31-714 cells transformed by wild-type polyoma virus; and 121-6-5, a subclone isolated from 121-6 (a transformed line induced by ts-121). Determinations were made of cell density and agglutinability by concanavalin A (Con A); autoradiography was performed after treatment with 5 µCi/ml of ^{3}H-thymidine. The cells were grown at 35 or 39 C. Both the morphology and saturation density of the 121-6 cells were affected by temperature; at 39 C, they grew to monolayer sheets and remained contact inhibited for at least 25 days, while at 35 C, they grew beyond the monolayer density, thus resembling the PV-4 cells. When the cells were incubated at 39 C and the temperature was shifted to 35 C before the cells became confluent, they continued to grow beyond the saturation density normally reached at 39 C. However, when they were downshifted after reaching confluence, they began to grow only after a 34 day latent period. When sparse cultures were moved from 35 to 39 C, they continued to grow to form a monolayer; when cultures which had grown beyond confluence were similarly upshifted, growth stopped and a gradual decrease in cell density was observed.

studied in clones of a Syrain hamster cell line (NIL)
and subclones of this line transformed by polyoma
virus (NIL-Py) or hamster sarcoma virus (NIL-HSV).
The synthesis of DNA-binding proteins in NIL and in
its virus-transformed derivatives, NIL-Py and NIL-
HSV, was very similar in exponentially growing cells,
but in dense culture there was a very significant
difference in the level of a protein (P8), which was
much higher in the transformed lines than in untransformed NIL. Experiments with synchronized cells indicated that the time of maximal P8 synthesis relative to cellular DNA synthesis in NIL-HSV precedes
that observed in NIL cells. P8 has a molecular
weight of 30,000 as determined by polyacrylamide gel
electrophoresis in the presence of sodium dodecyl
sulfate and is present in large amounts in the transformed cells in dense culture, where it makes up 0.5
to 1% of the total soluble protein. It is concluded
that P8 is greatly reduced in dense NIL cells, but
is prominent in dense transformed cells. The fact
that the high levels in the latter were observed in
all the clones of NIL-HSV indicates that this behavior is due to the transformed state, and not to for-
tuitous clonal selection.

2625   DETECTION OF TYPE A ONCORNAVIRUS-LIKE
        STRUCTURES IN CHICKEN EMBRYO CELLS AFTER
INFECTION WITH ROUS SARCOMA VIRUS. (Eng.) Maul,
G. G. (Wistar Inst. Anatomy and Biology, Philadelphia, Pa. 19104); Lewandowski, L. J.* *J. Virol.*
16(4):1071-1074; 1975.

Occasional aggregates of high electron-dense particles were observed within the cytoplasm of chicken
embryo cells transformed by Rous sarcoma virus (RSV).
A clone-purified stock of subgroup C of RSV was used
to infect C/E chicken embryo cells, which were incubated with and without 20 μM glucosamine/ml for 16
hr. Thin sections of cells were then stained with
uranyl acetate and studied with the electron microscope. The particles were present in approximately
10% of all cells and appeared to be composed of mem-
brane-free nucleoids resembling type A oncornavirus,
with diameters of about 60.0 nm, and electron-lucent
centers of about 25.0 nm surrounded by two dense
rings. The aggregates were not glycogen because the
uranyl acetate stain would have leached glycogen, and
they were not lysosomes because they were not surrounded by a membrane. Complete and incomplete core-
like structures with electron-lucent centers were also seen in unstained preparations. These particles
were not detected in noninfected cells, and they did
not accumulate in excess under conditions of glucosamine block. It is suggested that the particles may
be either newly synthesized viral cores or infecting
virus actively engaged in virus replication.

2626   ABSENCE OF SURFACE PROJECTIONS ON SOME NON-
        INFECTIOUS FORMS OF RSV. (Eng.) de Giuli,
C. (Rockefeller Univ., New York, N. Y. 10021); Kawai,
S.; Dales, S.; Hanafusa*, H. *Virology* 66(1):253-260;
1975.

Purified virus particles of two strains of Rous sarcoma virus that are defective in glycoprotein synthe-

sis were examined with the electron microscope to determine if there is a correlation between the absence of glycoproteins and the absence of surface projections. Chicken embryo cells either positive or negative for both the group-specific antigen and endogenous helper factor were used. Primary or secondary chicken cells, from either helper factor-negative or -positive embryos, were infected with the defective mutant of the Schmidt-Ruppin strain of Rous sarcoma virus or the Bryan high-titer strain of Rous sarcoma virus that had been propagated in helper factor-negative cells in the presence of 25 µg/ml of diethylaminoethyl dextran and 1,500 hemagglutinating U/ml of UV-inactivated Sendai virus. The cultures were transferred at 4- to 5-day intervals until the entire culture consisted of only transformed cells. The virus shed into the medium was harvested, purified by centrifugation through a sucrose gradient, and examined by electron microscopy following negative staining with 2% phosphotungstic acid. The viruses defective in formation of virion glycoproteins, were seen to be lacking in projections if they were propagated in helper factor-negative cells, but they had projections when they were replicated in helper factor-positive cells. The defective sarcoma viruses were noninfectious when grown in helper factor-negative cells. It is suggested that the glycoproteins residing on the projections are essential for the interaction with cellular receptors to permit initiation of the infectious process. Although virions without projections are incapable of penetrating the cells, they appear to become adsorbed to the host cell surface at almost the same rate as infectious types, indicating that the projection is not necessary for the adsorption step.

2627   COMPARISONS OF MAJOR CELL-SURFACE PROTEINS OF NORMAL AND TRANSFORMED CELLS. (Eng.) Robbins, P. W. (Cent. Cancer Res., Massachusetts Inst. Technol., Cambridge). *Am. J. Clin. Pathol.* 63 (5):671-676; 1975.

Transformation of the chick fibroblast surface was studied in cells infected with Schmidt-Ruppin Rous sarcoma virus and its temperature sensitive mutant, TS-68. Transformation caused a dramatic decrease in the levels of these cellular proteins: Z (molecular weight 230,000), Ω (206,000), and Δ (47,000). TS-68-infected cells were pulse-labeled with ^{35}S-methionine. When they were transferred from 41 C to 36 C they stopped synthesizing the Ω protein. The intracellular location of this protein is not known. The rate of synthesis of Δ protein dropped to 10-20% of the 41 C rate within 3-6 hr after the TS-68 infection culture was shifted to 36 C. This change in concentration was not due to increased proteolysis or turnover. When the TS-68-infected cells were maintained at the permissive temperature for an extended time and then shifted to 41 C, increased synthesis began promptly. The exact physiologic role of Δ, a plasma membrane protein, is unknown. It is suggested that it is a structural component that governs or restrains general plasma membrane protein mobility. There was a cessation or slowing of the synthesis of the large trypsin- and collaginase-sensitive protein, Z, but only after days of

morphological transformation by TS-68 virus. The concentration of collaginase necessary for 50% degradation of Z was three times higher than for trypsin, and was correlated to stimulated deoxyglucose transport. However, the breakdown of Z was not essential for activating fibroblasts, as the Vmax of thrombin was 10-fold lower to stimulate deoxyglucose uptake than that causing 50% breakdown of Z. The significance of Z in growth control is uncertain. Generally, the TS-68-infected cells differed by increased quantities of type-specific viral antigen in the membranes at 41 C, as compared to 36 C.

2628   PARTIAL PURIFICATION AND CHARACTERIZATION OF DNA POLYMERASES FROM A ROUS SARCOMA VIRUS-TRANSFORMED RAT CELL LINE. (Eng.) Bandyopadhyay, A. K. (N.C.I. Frederick Cancer Res. Center, Frederick, Md. 21701). *Biochim. Biophys. Acta* 407(1) 1-13; 1975.

A new DNA polymerase was partially purified from cell free extracts of a continuous rat cell line (XC). The XC cells had been transformed by the Prague strain of Rous sarcoma virus but did not produce infectious virus. The molecular wt of the DNA polymerase is 70,000, as estimated by glycerol gradient centrifugation and by Sephadex gel filtration. This enzyme can be distinguished from the other cellular DNA polymerases by its elution pattern on DNA-cellulose column chromatography, its molecular wt, and its primer-template specificity. The enzyme has some characteristics of the murine leukemia virus reverse transcriptase. It is partially inhibited by immunoglobulin G (IgG) purified from rabbit antiserum prepared against Rauscher leukemia virus reverse transcriptase, but it is not inhibited by IgG from rat antiserum prepared against avian myeloblastosis virus reverse transcriptase. However, the XC cell enzyme can be distinguished from the murine leukemia virus reverse transcriptase by its inefficiency in copying an oligo-$(dG)_{12} \cdot poly(rC)$ primer-template.

2629   CHARACTERIZATION OF DEFECTIVE SV40 ISOLATED FROM SV40-TRANSFORMED CELLS. (Eng.) Huebner, K. (Wistar Inst. Anat. Biol., Philadelphia, Pa.); Santoli, D.; Croce, C. M.; Koprowski, H. *Virology* 63(2):512-522; 1975.

Defective simian virus 40 (SV40) virions isolated from SV40-transformed monkey, human, and hamster cell lines that had failed to yield wild type virus after Sendai virus-mediated fusion with permissive African green monkey kidney TC7 cells were characterized. KBr-purified pools were processed in several ways: (1) A sample was inoculated into TC7 cell cultures, and coverslips were examined for SV40-induced T or V antigens; (2) An aliquot was centrifuged to equilibrium in CsCl solution, and fractions were assayed for density and total radioactivity; and (3) An aliquot of DNA was extracted and analyzed by ethidium bromide CsCl centrifugation and by velocity sedimentation in neutral CsCl for the presence of closed circular double-stranded DNA. The bouyant densities in CsCl of the defective viruses ranged between 1.32 and 1.33 g/cc.

The DNA isolated from defective viruses was heterogenous in size and contained some covalently closed
double-stranded circular molecules. These data indicate that defective virion preparations rescued from
the nonrescuable transformed cell lines contain DNA
of the size and conformation of SV40 component I DNA.
Although defective viruses isolated from nonrescuable
transformed monkey, human, and hamster cells were
capable of replication to a limited extent (as evidenced by the formation of microplaques on TC7 monolayers and by the incorporation of radioactive precursors into virion DNA and protein), the defective
viruses had a reduced ability to induce detectable
levels of viral antigens in permissive, semipermissive, and nonpermissive cells. T antigen-positive
mouse cell clones, established after infection with
a defective virus stock rescued from H50 cells, did
not have the characteristics of wild type SV40-
transformed cells (except that they were T antigen-
positive) and were presumably nontransformed. Recovery of defective viruses from these transformed
cell lines is consistent with the proposal that cells
that are permissive or semipermissive to SV40 lytic
infection and that survive SV40 lytic infection to
become transformed contain defective viral genomes.

2630   UV-INDUCED REVERSION OF A TEMPERATURE-SEN-
        SITIVE LATE MUTANT OF SIMIAN VIRUS 40 TO
A WILD-TYPE PHENOTYPE. (Eng.) Cleaver, J. E. (Lab.
Radiobiology, Univ. California, San Francisco, Calif.
94143); Weil, S. *J. Virol.* 16(1):214-216; 1975.

UV radiation (50 ergs/mm^2/sec of 254-nm light for
1-7 min) of a simian virus 40 temperature-sensitive
late mutant before infection of CV1 monkey cells
produced viruses that were phenotypically wild-type
revertants (calculated from the ratio of the number
of plaques formed on infected monolayers at 39 C)
was high, and increased approximately as the square
of the UV dose. The revertants grew at 39 C, but
were more sensitive than the temperature-sensitive
mutant to heating at 60 C. The revertants produced
by UV light must therefore be suppressor mutations
rather than reversions of the original temperature-
sensitive mutant; this may explain the relatively
high frequency of the UV-induced reversion. It is
suggested that suppression may involve second-site
mutations in the viral genome in a manner similar
to that previously seen in spontaneous revertants
isolated from a simian virus 40 temperature-sensitive mutant.

2631   PRODUCTION OF LYMPHOKINE-LIKE FACTORS
        (CYTOKINES) BY SIMIAN VIRUS 40-INFECTED
AND SIMIAN VIRUS 40-TRANSFORMED CELLS. (Eng.)
Bigazzi, P. E. (Sch. of Medicine, 327 Sherman Hall,
State Univ. of New York at Buffalo, Buffalo, N. Y.
14214); Yoshida, T.; Ward, P. A.; Cohen, S. *Am.
J. Pathol.* 80(1):69-78; 1975.

Macrophage migration inhibitory (MIF-like) activity
was demonstrated in the supernatant fluids from
primary cultures of African green monkey kidney
cells infected with simian virus 40 (SV40). Virus-
free kidney cell cultures had no MIF activity.

Supernatant fluids from continuous cultures of non
transformed and SV40-transformed human fibroblasts
contained MIF-like activity. Productive infection
with SV40 virus results in the production of a lym
okine-like factor. SV40 infection does not induce
chemotactic factors. The results reported here,
taken in conjunction with previous observations,
suggest that the production of lymphokine-like fac
tors (cytokines) may be a general biologic phenomenon, and that many, if not all, cell types may be
capable of such activity.

2632   QUANTITATIVE MEASUREMENT OF SV 40 T-ANT
        GEN PRODUCTION. (Eng.) Horan, M. (Los
Alamos Sci. Lab., Univ. California, N.M.); Horan,
P. K.; Williams, C. A. *Exp. Cell Res.* 91(2):247-
252; 1975.

African green monkey cells (CV-1) were infected wi
simian virus 40 (SV40) virus at high multiplicitie
of infection (MOI), and the production of T-antige
was studied. A new instrumentation, flow microflu
orometry, coupled with indirect immunofluorescence
permitted quantitative evaluation of this antigen.
Optimum conditions were determined for antibody ex
cess. Antigen production was not detected for the
first six hours post-infection. The value of this
technology is discussed in relation to quantitati
evaluation of expression of cellular antigens. By
permitting comparison of antigen production rates
among various strains of SV40 and the study of cor
tions that might alter these rates, the technology
could yield valuable information on the control of
virus infection.

2633   SIMIAN VIRUS 40 DNA DIRECTS SYNTHESIS
        AUTHENTIC VIRAL POLYPEPTIDES IN A LINKE
TRANSCRIPTION-TRANSLATION CELL-FREE SYSTEM. (Eng
Roberts, B. E. (Biol. Dep., Massachusetts Inst. Te
nol., Cambridge); Gorecki, M.; Mulligan, R. C.; Da
K. J.; Rozenblatt, S.; Rich, A. *Proc. Natl. Acad.
Sci. USA* 72(5):1922-1926; 1975.

The characteristics and requirements of a linked c
free system capable of transcribing and translatir
mammalian viral DNA are described. The system is
heterologous one utilizing *Escherichia coli* RNA po
erase to transcribe the DNA template and wheat germ
extracts to translate the RNA into protein. Simia
virus 40 (SV40) DNA Form I (supercoiled) directed
synthesis of discrete polypeptides up to 85,000 da
tons in size. One of these products was indistinguishable from authentic major virus capsid prote
VPI, as judged by motility on sodium dodecyl sulfa
polyacrylamide gels, antibody precipitation, and p
tide analyses. The cell-free products larger than
VPI comprised a number of polypeptides ranging in
molecular weight from 50,000 to 85,000. These pol
peptides demonstrated no immunological relationshi
to the structural protein VPI. However, two of th
products, along with one of approximately 25,000 d
tons, were precipitated with antiserum to SV40 tur
antigen. Linear SV40 DNA generated by the cleava
of Form I DNA with the restriction endonuclease Ec
was an efficient template in this system and also

directed the synthesis of a polypeptide migrating
with VPI on polyacrylamide gels. This cell-free system should facilitate the identification of exclusively virus-coded polypeptides, the direct mapping
of viral DNA, the study of control mechanisms regulating the flow of information from DNA into polypeptides, and, in combination with hybridization techniques, the purification of specific eukaryotic genes.
With SV40, the use of restriction enzyme DNA fragments
or DNA from homogeneous defective populations as templates will facilitate functional mapping of the viral
genome.

2634　　TOPOGRAPHY OF SUBSTRATE-ATTACHED GLYCOPRO-
　　　　TEINS FROM NORMAL AND VIRUS-TRANSFORMED
CELLS. (Eng.) Culp, L. A. (Sch. Medicine, Case
Western Reserve Univ., Cleveland, Ohio 44106).
Exp. Cell Res. 92(2):467-477; 1975.

The topographical distribution of substrate-attached
materials (SAM) which may mediate adhesion of BALB/c
3T3 and SV40-transformed 3T3 cells to the culture
substrate was studied. Autoradiographic detection
of these 'glycoproteins' on the substrate after incorporation of ^{14}C-glucosamine or ^{35}S-methionine
indicated: glucosamine-labeled SAM on the substrate
of confluent cultures was evenly distributed, the
methionine content of SAM from transformed cells was
deficient, after these materials were directly deposited onto the substrate at cell colony locations,
and not randomly on the substrate secretion into
the medium. Autoradiographic detection of these
'glycoproteins' subsequent to ^{3}H-glucosamine (or
^{3}H-leucine) incorporation and exposure to liquid
emulsion indicated that substrate-attached glycoproteins were present in focal pools on the underside
of the cell, and that these focal pools were distributed comparably, in terms of number of foci per
area of substrate, for confluent normal and transformed cells. The patterns of SAM deposited by
freshly-attaching and spreading 3T3 cells have also
been examined. The evidence suggests that cells
are adherent to the substrate at localized areas
which appear to be evenly distributed on the underside of normal or transformed cells.

2635　　MAPPING OF THE SV40 SPECIFIC SEQUENCES
　　　　TRANSCRIBED *IN VITRO* FRCM CHROMATIN OF
SV40 TRANSFORMED CELLS. (Eng.) Astrin, S. M. (Inst.
for Cancer Res., Fox Chase Cancer Center, Philadelphia, Pa. 19111). *Biochemistry* 14(12):2700-2704;
1975.

Chromatin was isolated from simian virus 40 (SV40)-
transformed mouse cells (SV3T3) and transcribed with
Escherichia coli RNA polymerase. The SV40 specific
transcripts were analyzed by annealing the RNA to the
minus strands of purified fragments of SV40 DNA produced by cleavage of the DNA with a restriction enzyme isolated from *Hemophilus aegyptius*. Quantitation
of the frequency of transcription from the regions
represented by the fragments showed that the early
region (fragments A and D) was transcribed 5-10 times
more frequently than the remaining regions. These
results are in good agreement with the transcription
pattern observed in the transformed cell. In contrast,

that the choice of specific fragments ultimately
transported to the cytoplasm occurs after transcription.

2639   PURIFICATION OF R-DNA POLYMERASE FROM AN
        ADENOVIRUS DNA REPLICATION COMPLEX [ab-
stract]. (Eng.) Arens, M. Q. (St. Louis Univ., Mo.);
Ito, K.; Yamashita, T.; Green, M. Fed. Proc. 34(3):
494; 1975.

2640   SULFATED COMPONENTS OF ENVELOPED VIRUSES.
        (Eng.) Pinter, A. (Memorial Sloan-Ketter-
ing Cancer Center, New York, N.Y. 10021); Compans,
R. W. J. Virol. 16(4):859-866; 1975.

2641   STUDIES OF EPSTEIN-BARR VIRUS NUCLEAR
        ANTIGEN AND ANTISERA USING A QUANTITATIVE
MICRO COMPLEMENT FIXATION ASSAY [abstract]. (Eng.)
Baron, D. (Harvard Univ., Cambridge, Mass.); Benz,
W. C.; Strominger, J. L. Fed. Proc. 34(3):527; 1975.

2642   PRESENCE OF AN AUTOCOMPLEMENTARY RNA WITH
        VIRAL SPECIFICITY IN HERPES VIRUS-INFECTED
CELLS. (Fre.) Bechet, J.-M. (Unite d'Oncologie
Virale, Institut Pasteur, 25, rue de Docteur-Roux,
75015 Paris, France); Montagnier, L. C.R. Acad.
Sci. [D] (Paris) 280(2):217-220; 1975.

2643   THE ROLE OF GENITAL HERPES VIRUS IN CAN-
        CER OF THE UTERINE CERVIX. (Fre.) Thiry,
L. (Departement de Virologie, Institut Pasteur du
Brabant, Bruxelles, Belgium); Sprecher-Goldberger,
S. Bull. Acad. Med. Belg. 129(1):109-141; 1975.

2644   DETECTION OF VIRAL CORES HAVING TOROID
        STRUCTURES IN EIGHT HERPESVIRUSES. (Eng.)
Nii, S. (Res. Inst. Microb. Dis., Osaka Univ., Japan); Yasuda, I. Biken J. 18(1):41-46; 1975.

2645   A HERPESVIRUS-TYPE AGENT ASSOCIATED WITH
        SKIN LESIONS OF GREEN SEA TURTLES IN
AQUACULTURE. (Eng.) Rebell, G. (Sch. Medicine,
Univ. of Miami, Miami, Fla. 33152); Rywlin, A.;
Haines*, H. Am. J. Vet. Res. 36(8):1221-1224; 1975.

2646   POLYADENYLATION OF HERPES SIMPLEX VIRUS
        RNA [abstract]. (Eng.) Swanstrom, R. I.
(Univ. California, Irvine); Pivo, K. R.; Wagner, E.
K. Fed. Proc. 34(3):526; 1975.

2647   TUBULAR STRUCTURES IN RABBIT KIDNEY CELLS
        INFECTED WITH HERPESVIRUS CUNICULI. (Eng.)
Nii, S. (Res. Inst. for Microbial Diseases, Osaka
Univ., Yamadi-kami, Suita, Osaka, Japan); Yasuda,
I. Biken J. 18(2):145-148; 1975.

2648   INFECTIOUS SPREAD AND CONTROL OF FELINE
        LEUKEMIA VIRUS. (Eng.) Hardy, W. D., Jr.
(Memorial Sloan-Kettering Cancer Center, New York,
N.Y.); McClelland, A. J. *Transplant. Proc.* 7(2):265-
267; 1975.

2649   DEMONSTRATION OF FELINE LEUKEMIA VIRUS
        ANTIBODY IN CATS BY PASSIVE HEMAGGLUTINA-
TION. (Eng.) Wang, J. T. (New York State Veterinary Coll., Cornell Univ., Ithaca, N.Y. 14850);
Lee, K. M. *Proc. Soc. Exp. Biol. Med.* 148(2):557-
561; 1975.

2650   EFFECT OF 1-β-D-RIBOFURANOSYL-1,2,4-
        TRIAZOLE-3-CARBOXAMIDE (RIBAVIRIN) ON
FRIEND LEUKEMIA VIRUS INFECTIONS IN MICE. (Eng.)
Sidwell, R. W. (Nucleic Acid Res. Inst., Irvine,
Calif. 92664); Allen, L. B.; Huffman, J. H.; Wit-
kowski, J. T.; Simon, L. N. *Proc. Soc. Exp. Biol.
Med.* 148(3):854-858; 1975.

2651   CONTINUOUS REPLICATION OF FRIEND VIRUS
        COMPLEX (SPLEEN FOCUS-FORMING VIRUS-
LYMPHATIC LEUKEMIA-INDUCING VIRUS) IN MOUSE EMBRYO
FIBROBLASTS: RETENTION OF LEUKEMOGENICITY AND LOSS
OF IMMUNOSUPPRESSIVE PROPERTIES. (Eng.) Eckner,
R. J. (Boston Univ. Sch. Medicine, Boston, Mass.
02118). *J. Exp. Med.* 142(4):936-948; 1975.

2652   EFFECT OF DIETHYLAMINOETHYL-DEXTRAN ON THE
        REPLICATION OF A MURINE SARCOMA (MOLONEY)-
LEUKEMIA VIRUS COMPLEX IN MOUSE EMBRYO CULTURES.
(Eng.) Hirschman, S. Z. (Mount Sinai Sch. Medicine
of City Univ. of New York, New York, N.Y. 10029);
Funke, V. *Proc. Soc. Exp. Biol. Med.* 148(2):527-
531; 1975.

2653   CHARACTERIZATION OF ENDOGENOUS RNA-DIRECTED
        DNA POLYMERASE ACTIVITY OF RETICULOENDO-
THELIOSIS VIRUSES. (Eng.) Kang, C.-Y. (Univ. of
Texas Southwestern Medical Sch., Dallas, Tex. 75235).
J. Virol. 16(4):880-886; 1975.

2654   BIOLOGY OF MOUSE RNA TUMOR VIRUSES. (Jpn.)
        Takano, T. (Keio Univ. Med. School, Tokyo,
Japan). *Tampakushitsu Kakusan Koso* 20(2):140-150;
1975.

2655   STUDY OF THE EFFECT OF STORAGE ON ROUS
        SARCOMA VIRUS. (Rus.) Zelenskii, V. P.
(All-Union Scientific Res. Inst. of Avian Diseases,
USSR); Laretskaia, L. I.; Rozhnova, L. I.; Rachkov-
skaia, L. A. *Veterinariia* (8):36; 1975.

2656   CONCENTRATION OF ROUS SARCOMA VIRUS FROM
        TISSUE CULTURE FLUIDS WITH POLYETHYLENE
GLYCOL. (Eng.) Bronson, D. L. (Univ. of Minnesota
Health Sciences, Minneapolis, Minn. 55455); Elliott,
A. Y.; Ritzi, D. *Appl. Microbiol.* 30(3):464-471;
1975.

2657   HUMAN PAPOVAVIRUS-BK DNA: CHARACTERIZA-
        TION AND POLYNUCLEOTIDE SEQUENCE HOMOLOGY
TO SV40 DNA [abstract]. (Eng.) Howley, P. M. (Natl.
Inst. Allergy Infect Dis., Bethesda, Md.); Khoury, G.;
Takemoto, K. K.; Mullarkey, M. F.; Martin, M. A.
Fed. Proc. 34(3):526; 1975.

2658   CLONING AND AMPLIFICATION OF SPECIFIC DE-
        FECTIVE GENOMES OF SV40 [abstract]. (Eng.)
Ganem, D. (Harvard Med. Sch., Boston, Mass.); Davoli,
D.; Fareed, G. C. *Fed. Proc.* 34(3):526; 1975.

2659   METHYLATED SV40-SPECIFIC RNA FROM NUCLEI
        AND CYTOPLASM OF INFECTED BSC-1 CELLS [ab-
stract]. Lavi, S. (Roche Inst. Mol. Biol., Nutley,
N.J.); Shatkin, A. J. *Fed. Proc.* 34(3):526; 1975.

2660   BIOLOGICAL ACTIVITIES OF CLONED EVOLU-
        TIONARY VARIANTS OF SIMIAN VIRUS 40 [ab-
stract]. (Eng.) Brockman, W. W. (Johns Hopkins Univ.
Sch. Med., Baltimore, Md.); Scott, W. A. *Fed. Proc.*
34(3):526; 1975.

2661   *IN VITRO* SYNTHESIS OF RNA IN ISOLATED NUCLEI
        FROM SIMIAN VIRUS 40 INFECTED CELLS [ab-
stract]. (Eng.) Qasba, P. K. (Natl. Cancer Inst.,
Bethesda, Md.); Yi, P.-N.; Finlayson, B. *Fed. Proc.*
34(3):526; 1975.

2662   CELL-MEDIATED CORRECTION OF MISMATCHED
        BASES IN HETERODUPLEX MOLECULES OF SIMIAN
VIRUS 40 DNA [abstract]. (Eng.) Lai, C.-J. (Johns
Hopkins Univ. Sch. Med., Baltimore, Md.); Nathans, D.
Fed. Proc. 34(3):515; 1975.

2663   ADENYLATE CYCLASE OF NORMAL AND TRANS-
        FORMED HUMAN FIBROBLASTS: INFLUENCE OF
HORMONES AND 2-CHLOROADENOSINE [abstract]. (Eng.)
Makman, M. H. (Albert Einstein Coll. Med., New York,
N.Y.); Keehn, E. *Fed. Proc.* 34(3):264; 1975.

See also:

*   (Rev):   2412, 2413, 2414, 2435, 2436
*   (Chem): 2470, 2474
*   (Immun): 2665, 2668, 2669, 2671, 2673, 2675, 2680,
             2687, 2688, 2690, 2701, 2703, 2705, 2709,
             2710, 2742, 2747, 2752,
*   (Path): 2783, 2807
*   (Epid): 2882, 2901, 2909

2666   EFFECT OF MULTIPLE INJECTIONS OF ALLOGEN-
         EIC SPLEEN CELLS ON METHYLCHOLANTHRENE
CARCINOGENESIS IN THE MOUSE. (Eng.) Oth, D. (Unite
INSERM, U 95, Plateau de Brabois, 54500 Vandoeuvre-
les-Nancy, France); Liegey, A. *Biomedicine* 23(1):
17-19; 1975.

Mice were pretreated with allogeneic transplantation
antigens in an attempt to diminish the establishment
of primary tumors induced by methylcholanthrene (MCA,
0.1 mg). The mice were treated with 10^6 spleen cells,
sc, weekly for seven weeks. In all experiments, allogeneic spleen cells (type 1, origin of spleen cells
changed every week) treatment diminished the number
of tumors in comparison with untreated controls (91/
123 or 73% and 121/135 or 90%, respectively). Mean
survival time of the tumor-bearing animals was not
influenced by the treatment. When the allogeneic
cell donors were always of the same strain (type 2
spleen cells), no protection was observed. These
data suggest that immunoprophylaxis occurs as a result of pretreatment of mice with allogeneic spleen
cells of multiple origin against MCA carcinogenesis.

2667   CELLULAR IMMUNOLOGIC RESPONSIVENESS TO
         EXTRACTS OF OVARIAN EPITHELIAL TUMORS.
(Eng.) Melnick, H. (Lenox Hill Hosp., New York,
N. Y.); Barber, H. R. K. *Gynecol. Oncol.* 3(1):77-
86; 1975.

The leukocyte migration inhibition assay was employed in an investigation of cellular immunological
reactivity to a solubilized extract of pooled, epi-
thelially derived ovarian carcinomas. Lymphocyte-
rich populations of leukocytes were prepared from
heparinized blood of patients having primary ovarian
cancers and of various control subjects. The epithelial extract was prepared from fresh surgical
specimens of serous and mucinous cystadenocarcinomas
of the ovary. In 85% of patients having adenocarcinoma of the ovary, leukocyte migration was 30-
77% inhibited in the presence of the ovarian tumor
extract in the assay. Leukocytes obtained from
patients having endometroid, mesometanephric, and
teratocarcinomas of the ovary were not inhibited
by incubation with the epithelial ovarian extract.
  similar absence of migration inhibition was observed in patients with squamous cell carcinoma of
the cervix, endometrial carcinoma, i.e., malignant
tumors other than seromucinous adenocarcinomas.
A lack of migration inhibition was also noted in
leukocytes obtained from patients with various benign ovarian neoplasms from normal or pregnant females, or from normal males. It is suggested that
the demonstrated cellular immunologic reactivity
to a specific ovarian epithelial tumor antigen may
serve as an approach to an early diagnostic test
for histologically specific types of ovarian cancer.

2668   MAREK'S DISEASE IN IMMUNOSUPPRESSED CHICK-
         ENS: GROWTH OF A TRANSPLANTABLE LYMPHOMA
AND DEVELOPMENT OF THE DISEASE BY NATURAL EXPOSURE.
(Eng.) Cotter, P. F. (Dep. Pathobiol., Univ.
Connecticut, Storrs); Jakowski, R. M.; Fredrickson,
T. N.; Schierman, L. W.; McBride, R. A. *J. Natl.
Cancer Inst.* 54(4):969-973; 1975.

The effect of treatment of syngeneic White Leghorn chickens with Cytoxan (CY) on immunocompetence and on susceptibility to Marek's disease (MD) was investigated. Chicks were immunosuppressed by intra-abdominal injections of CY shortly after hatching, or at two-three wks of age. Immunocompetence was tested by survival of allogeneic,full-thickness skin grafts and by humoral response to injected sheep red blood cells (SRBC) virus. Exposure to MD was effected either by s.c. inoculation of the left wing web or by forced-air contact with MDV-infected birds. The incidence of wing web lymphomas in chicks treated with CY at hatching was 100%, as in saline-treated controls; further, no anti-SRBC hemagglutinins were detected in 96% of the CY-treated chicks, while skin graft survival was not affected. In assessing the immunocompetence of chicks treated with CY at two-three wk, however, hemagglutinins were detected and titers were highest in chicks receiving 37 and 75 mg CY/kg weight. Only chicks receiving 150 mg CY/kg had a delayed rejection time to allogeneic skin grafts and significantly fewer bursa-dependent foci observable on midsaggital spleen sections. Thus CY treatment at hatching produced almost complete unresponsiveness to antigenic stimulation and affected primarily the humoral immune system. However, later CY immunosuppression resulted in a dose-dependent enhancement or suppression of anti-SRBC antibody production, and only a slight increase in skin graft survival time.

2669   EVALUATION OF THE CELL-MEDIATED IMMUNE
        RESPONSE TO MURINE SARCOMA VIRUS BY [^{125}I]-
IODODEOXYURIDINE ASSAY AND COMPARISON WITH CHROMIUM
51 AND MICROCYTOTOXICITY ASSAYS. (Eng.) Fossati, G.
(Natl. Cancer Inst., Bethesda, Md. 20014). Holden,
H. T.; Herberman, R. B. *Cancer Res.* 35(9):2600-2608;
1975.

The cell-mediated immune response of C57BL/6 mice to murine sarcoma virus (MSV) was examined by the [^{125}I]-iododeoxyuridine ([^{125}I]IUdR) release cytotoxicity assay using MSV-induced sarcoma tissue culture cell lines as target cells. Cellular cytotoxicity was detected as early as three days after virus inoculation. Most mice assayed between 12 and 17 days after MSV inoculation im gave positive results with maximum levels of activity present on days 13 and 14. Reactivity was frequently detected for up to 100 days after MSV inoculation, although at low levels (5 to 10%). Additional experiments comparing the kinetics of the cellular response as measured by different *in vitro* cytotoxicity assays were performed. The results showed a good direct correlation between the [^{125}I]-IUdR release assay and a ^{51}Cr release assay. A similar pattern of reactivity was also observed when the cellular response was measured by a visual microcytotoxicity assay, although reactivity dropped off more rapidly and became undetectable in most instances by 20 days after injection of MSV. Studies on effector cell type revealed that cytotoxicity in all three assays was T-cell dependent, being eliminated by treatment with anti-θ plus complement. Macrophages did not appear to play a role, since treatment with carbonyl iron and magnet had no effect. It is suggested that additional testing against other target cells is indicated.

2670   CELL-MEDIATED IMMUNITY TO MURINE TUMOR
        ALLOGRAFTS. INCREASE IN THE ACTIVITIES
OF ACTIVATED THYMUS-DERIVED CELLS FOLLOWING *IN
VITRO* INCUBATION. (Eng.) Gorczynski, R. M. (Ontario Cancer Inst., 500 Sherbourne St., Toronto, Ontario M4X 1K9, Canada); Tigelaar, R. E. *Cell. Immunol.* 18(1):121-143; 1975.

The immunity of C57Bl spleen cells to an allogeneic mastocytoma, P815, was studied with a direct cytotoxicity assay and an assay of macrophage migration inhibition. Spleen cells, obtained from C57Bl mice given P815 cells (10 x 10^6) cells 18 days before, were separated by velocity sedimentation into large and small cell fractions. Irradiated (850 rads) C57Bl mice, divided into groups of six mice each, were inoculated with these cells or unfractionated cells. At 24 hr postinoculation, spleens and peripheral lymph nodes were pooled within each group; aliquots were then tested for their activity in cytotoxic lymphocyte (CL) and migration inhibition factor (MIF) assays. The CL activity of unfractionated immune spleen cells showed a marked preference for migration to the spleens rather than lymph nodes of recipients, while the MIF cell activity showed no such preferential migration to the spleen. Both large CL and MIF cells showed a marked preference for migration to the spleen, while small CL and MIF cells gave rise to activity in both spleens and lymph nodes of recipients. Overnight incubation in the absence of deliberately added antigen substantially increased CL and MIF activity of immune spleen cells and resulted in the detection of MIF activity in small cells that were initially inactive. The activity in each assay was abolished by pretreatment of immune cells with rabbit antimouse brain-associated θ serum (anti-BAθ) and guinea pig complement, and no new activity was seen after overnight incubation in anti-BAθ serum. Thus, the increase in activity after incubation in the absence of antigen depended on the existence of T cells in the initial culture. Short-term exposure to trypsin (100 µg/ml) depressed the MIF activity of large T cells, but increased the activity of small cells. Increases in CL activity and the development of MIF activity from small cells could be blocked by incubation in the presence of serum from spleen cell donors. These results suggest that blocking factors exist on the surface of some small T lymphocytes taken from P815 immune C57Bl mice, and that these factors can suppress T cell activity in both of the assay systems used.

2671   VIRAL WARTS, HERPES SIMPLEX AND HERPES
        ZOSTER IN PATIENTS WITH SECONDARY IMMUNE
DEFICIENCIES AND NEOPLASMS. (Eng.) Morison, W. L.
(St. John's Hosp., Leicester Square, London, England).
Br. J. Dermatol. 92(6):625-630; 1975.

The incidence of viral warts, recurrent herpes simplex, and herpes zoster was assessed in 633 patients with possible secondary immune deficiency states and in 348 control subjects. The 633 patients included 78 with multiple myeloma, 159 with Hodgkin's disease, 109 with malignant lymphoma, 51 with chronic lymphatic leukemia (CLL), 156 with systemic malignancy, and 80 with cutaneous basal cell carcinoma. The pa-

PRESSION. Hirsch, M. S. (Mass. Gen. Hosp., Boston); Ellis, D. A.; Kelly, A. P.; Proffitt, M. R.; Black, P. H.; Monaco, A. P.; Wood, M. L. *Int. J. Cancer* 15(3):493-502; 1975.

Both mouse-tropic and zenotropic C-type RNA viruses were activated in BALB/c mice during rejection reactions against skin grafts from mice differing at the major histocompatibility locus (A/J) or sharing this locus (DBA/2). Animals that maintained their skin grafts following immunosuppression with antilymphocyte serum (ALS, 0.25 ml ip twice weekly for two wk beginning the day after grafting and 0.1 ml twice weekly thereafter) or cyclophosphamide (20 mg/kg/day ip from day of grafting; a total dose of 560 mg/kg) had a much higher incidence of viruspositivity than animals that did not receive ALS or cyclophosphamide, or that rejected their skin grafts despite immunosuppressive therapy. ALS itself activated type-C viruses from a small proportion of BALB/c mice whereas cyclophosphamide alone did not. Type-C viruses were detected in spleens and regional nodes draining skin-graft sites between one and two wk following transplantation; thereafter they reached maximum titers in spleens. Viruses were not detected at skin-graft sites, in tails, or in thymuses of grafted animals. Virus activation was associated with splenic histological changes consisting of germinal center formation, diffuse hyperplasia of reticulum cells, depletion of periarteriolar lymphocytes and hyperplasia of red pulp hematopoietic elements. These results suggest that detection of C type virus activation following immunostimulation reactions may depend on persistent immunosuppression following the induction event.

2674   EFFECT OF WATER-SOLUBLE ADJUVANTS ON
        IN VITRO LYMPHOCYTE IMMUNIZATION. (Eng.)
Sharma, B. (Sch. Med., Univ. California, Los Angeles); Kohashi, O.; Mickey, M. R.; Terasaki, P. I. *Cancer Res.* 35(3):666-669; 1975.

Evidence for the potentiation of lymphocyte transformation and *in vitro* immunization by water-soluble adjuvants (WSA) from *Mycobacterium smegmatis* and human mixed mycobacteria strains is presented. Lymphocytes purified over Ficoll-Hypaque were immunized *in vitro* by culturing with mitomycin C-treated tumor cells, and cytotoxic activity was tested using a ^{51}Cr release assay. Lymphocyte transformation by WSA did not occur in the range of 5-40 μg/ml; however, one-way MLCs were amplified by WSA. An increase in thymidine incorporation was generally observed as the WSA concentration was increased to 40 μg/ml; the average uptake increased significantly (72 ± 14%). *In vitro* immunization of lymphocytes to cultured human tumor cells revealed a significant amplification in response against breast, lung, Hep-2, and bladder tumor cell lines in the presence of WSA. Lymphocyte response to phytohemagglutinin (PHA) was depressed in the presence of WSA. The data suggest that WSA produce amplification of the MLC response in approximately the same range as produced by polyadenylic acid and polyuridylic acid. WSA may be useful in boosting immunization of lymphocytes *in vitro*.

2675   A KILLED VACCINE DERIVED FROM THE ONCOGENIC
        HERPESVIRUS ATELES. (Eng.) Laufs, R. (Hy-
giene-Institut der Universitat Gottingen, D-34 Gottingen, West Germany); Steinke*, H. *J. Natl. Cancer
Inst.* 55(3):649-651; 1975.

A killed herpesvirus vaccine was prepared by inactivation of the oncogenic Herpesvirus ateles (HVA) with
heat and formaldehyde. The vaccine was used safely
in marmosets (Saguinus oedipus) for 461 days. The
vaccinated marmosets developed high titers of serum
antibodies against HVA and were resistant to 316
lethal dose 50 (LD50) cell-free HVA. The nonvaccinated control monkeys died of malignant lymphoma.
The five challenged monkeys were clinically well and
were under observation for 362 days without signs of
HVA infection. A nonhuman primate species can be
successfully vaccinated against malignant lymphoma
induced by cell-free HVA. The prevention of malignant lymphoma with a killed HVA vaccine supports the
concept that HVA is the etiologic agent for the tumors observed in the nonvaccinated control monkeys.
The induction of a cellular immune response remains
to be determined. It seems possible to develop an
HVA vaccine free of viral nucleic acid similar to
that being used to protect against Marek's lymphomas
in chickens.

2676   VARIANTS OF A MOUSE MYELOMA CELL LINE THAT
        SYNTHESIZE IMMUNOGLOBULIN HEAVY CHAINS
HAVING AN ALTERED SEROTYPE. (Eng.) Preud'Homme,
J. L. (Albert Einstein Coll. Med., Bronx, N. Y.);
Birshtein, B. K.; Scharff, M. D. *Proc. Natl. Acad.
Sci. USA* 72(4):1427-1430; 1975.

A selected group of MPC-11 mouse myeloma cell
variants having serological, chemical, and assembly characteristics of a different subclass
from the parent was described. The cell lines
were derived from the MPC-11 mouse myeloma tumor,
which produces an IgG_{2b} immunoglobulin. These different lines were subjected to *in vitro* mutagenesis
with ICR-191 or Melphalan and subsequent cloning in
soft agar. The resulting clones were overlaid with
rabbit antibody against the Fc region of the MPC-11
H chain, and those not covered by a visible precipitate were selected for mass culture. The presence
of heavy chains was detected by Ouchterlony analysis
of cytoplasmic lysates and secreted immunoglobulins
using antisera to the MPH-11 chain and its fragments
and antisera directed against other mouse subclasses.
Molecular weights of H chains were determined by
electrophoresis of radiolabeled immunoglobulin molecules on acrylamide gels containing sodium dodecyl
sulfate. The variants were also subjected to peptide analysis by ion exchange chromatography. The
Ouchterlony analyses showed that where H chains of
normal size were produced by variants, such chains
were not recognized by antiserum against the parental IgG_{2b} serotype, but rather were recognized by
an antiserum against the IgG_{2a} subclass. Four variants synthesizing H chains of the γ2a subclass possessed chains of normal size (55,000 molecular weight)
and one synthesized a γ2a subclass H chain of 75,000
molecular weight. When the variants and the parent
were compared using antibody against the Fc region

of the MPC-11 protein, spurs of incomplete identity
were seen, indicating that the variants lacked antigenic determinants present in the parent chain. An
antiserum directed against the Fab region of the
MPC-11 protein failed to distinguish variants from
the parent. Further, the γ2a variants lacked many
of the tryptic/chymotryptic peptides found in the
parent, contained tryptic/chymotryptic peptides not
present in the parent but present in an unrelated
γ2a myeloma chain, and assembled with light chains
by a pathway typical of IgG_{2a} myelomas. A suggested
nonmutational mechanism was the turning off of one
gene and the turning on of another. Mutational
mechanisms suggested involved mitotic crossing over
or deletion involving parts of neighboring γ2b and
γ2a genes.

2677   IMMUNOGLOBULINS UNDER THE INFLUENCE OF
        NONSPECIFIC FACTORS. II. THE INFLUENCE
OF WORK-STRESS ON LEVELS OF IMMUNOGLOBULINS (IgG,
IgA, IgM) OF MINERS IN URANIUM MINES. (Eng.) Wagner, V. (Inst. Health Uranium Min., Pribram, Czechoslovakia); Andrlikova, J.; Palek, V. *Z. Immunitaetsforsch.* 148(4):356-365; 1975.

Serum IgG, IgA, and IgM levels were measured before
and after the workshift in two groups of 22 miners
each, exposed 5.4 and 10.8 yr, respectively, to the
environment of an underground uranium mine. The
average age was 30.5 yr in the first group and 36.95
yr in the second. Immunoglobulin levels were determined by radial immunodiffusion. After the workshift, miners with the longer exposure showed a significant decrease in all three immunoglobulin classes; in miners with the shorter exposure, only IgG
decreased significantly. Initial IgG levels were
considerably lower in miners exposed for ten years
compared with those with shorter exposure. Differences in the decreases of immunoglobulin levels between the two groups were significant only in the
case of IgA. This study shows that the stress of
a single workshift in uranium mines decreases immunoglobulin levels, and that this decrease differs
with the length of previous exposure and, apparently,
with age of the miners.

2678   DETECTION OF ANTIBODY AND COMPLEMENT COM-
        PLEXED *IN VIVO* ON MEMBRANES OF HUMAN CAN-
CER CELLS BY MIXED HEMADSORPTION TECHNIQUES. (Eng.)
Irie, K. (Cent. Health Sci., Univ. California, Los
Angeles); Irie, R. F.; Morton, D. L. *Cancer Res.*
35(5):1244-1248; 1975.

The mixed hemadsorption (MHA) techniques were used
to demonstrate antibody and complement fixed *in vivo*
to the surface of human cancer cells. Tumors from
twelve cancer patients and normal tissues from five
cancer patients and eight patients with cerebrovas-
cular or cardiac diseases were collected from biopsy
and autopsy for *in vitro* testing. Antiserum to human
whole immunoglobulins and antiserum to human complement 3 (C3) were used in the MHA techniques. Positive MHA patterns were demonstrated on the surface
of cancer cells by both methods. Positive reactions
ranged from 12 to 32% in mixed hemadsorption for

antibody detection and from 10 to 34% in mixed hemadsorption for C3 detection. Normal tissues obtained
from cancer patients or from patients who died of
causes other than cancer rarely exhibited distinct
MHA reactivity. Collectively, the data suggest that
most human cancers are antigenic in the autologous
host and that tumor-associated antigens of cancer
cells react *in vivo* with their humoral antibody to
fix complement.

2679   REACTIVITY OF SMOOTH-MUSCLE ANTIBODIES, SURFACE ULTRASTRUCTURE, AND MOBILITY IN CELLS OF HUMAN HEMATOPOIETIC CELL LINES. (Eng.)

Fagraeus, A. (Natl. Bacteriological Lab., Stockholm,
Sweden); Nilsson, K.; Lidman, K.; Norberg, R. *J.
Natl. Cancer Inst.* 55(4):783-789; 1975.

Seventeen human hematopoietic cell lines were tested
by indirect immunofluorescence (IF) for reactivity
with human serum containing smooth-muscle antibodies
(SMA). The correlation of the IF pattern to the cell
surface ultrastructure was revealed by scanning electron microscopy (SEM). Lymphoma cells, viewed by
SEM, had short villi over the entire cell surface,
but, by IF, showed membrane fluorescence. Cells of
lymphoblastoid lines had thin, long, surface villi,
sometimes asymmetric but most often distributed over
the whole cell surface. Myeloma and leukemia cells,
which had few membrane villi but a surface covered
by "blebs" revealed by SEM, demonstrated, by IF, only
a few stub-like projections extending from the surface. Time-lapse cinematography revealed a positive
correlation between long, thin, surface villi demonstrated by IF and SEM and the degree of motility. Indirect IF with human SMA-positive serum might be used
in the classification of cell lines derived from human hematopoietic tissue.

2680   CLONAL NATURE OF THE IMMUNE RESPONSE TO PHOSPHORYLCHOLINE (PC). V. CROSS-IDIO-TYPIC SPECIFICITY AMONG HEAVY CHAINS OF MURINE ANTI-PC ANTIBODIES AND PC-BINDING MYELOMA PROTEINS.

(Eng.) Claflin, J. L. (Washington Univ. Sch. Med.,
St. Louis, Mo.); Davie, J. M. *J. Exp. Med.* 141(5):
1073-1083; 1975.

To test the implication that a prerequisite for all
phosphorylcholine (PC)-combining activity in mice is
a heavy chain subtype, PC-binding myeloma proteins
and anti-PC antibodies from different mouse strains
were studied using a cross-idiotypic assay. Seven
mouse myeloma proteins with specificity for PC were
found to share a common antigenic determinant. This
group of proteins contained members which differed
in genetic origin, heavy chain class, κ-chain subgroup, individual antigenic determinants, and specificity for choline analogues. The cross-idiotypic
determinant, designated V_H-PC, was antigenically
similar in each of the proteins, and was associated
with the variable portion of the heavy chain in the
region of the antibody combining site. Further
studies showed that an indistinguishable determinant
was present on IgM anti-PC antibodies isolated from
all strains of mice tested, regardless of histocompatibility or heavy chain allotype. In view of the
finding that this cross-idiotypic determinant was

not found on antibodies or myeloma proteins that
lacked specificity for PC, the data suggest that a
particular heavy chain variable region is preserve
in all mouse antibodies with specificity for PC.

2681   EFFECT OF TRANSPLANTED SYNGENEIC MYELOMA ON THE ANTIBODY RESPONSE OF MICE. (Eng.)

Warr, G. W. (Univ. Med. Sch., Edinburgh, Scotland)
Willmott, N.; James, K. *Eur. J. Cancer* 11(5):351-
357; 1975.

Antibody responses to alum precipitated bovine ser
albumin (BSA) and sheep erythrocytes (SRBC) in mic
bearing various syngeneic myelomas were investigat
Inbred BALB/c mice of both sexes, age 10-12 weeks,
were used. Standard immunizing doses of 1 mg BSA
in alum or 3×10^8 SRBC in 0.3 ml were injected
i.p., and the antibody response was assessed by
antigen-binding capacity and relative binding affi
ity. The growth of all of the myelomas significantly depressed the antibody response to alum BSA
while the relative binding affinity of the antibod
produced was not reduced by the growth of any tumo
The splenic PFC response to SRBC in mice simultaneously transplanted with viable tumor varied, how
ever, as follows. During early tumor growth, mela
nomas MOPC 47A, MPC 25, and ADJPC 5 significantly
depressed the IgG1 response, while MOPC 47A also
suppressed the IgG2b response and ADJPC 5 also sup
pressed the IgG2a response. The IgM response was
unaffected by the early growth of melanomas MOPC
47A, MPC 25, and ADJPC 5, while it was enhanced by
melanoma MOPC 104EM. During late growth of all
of the tumors, the IgM response was unaffected,
all IgG subclass responses were unaffected by MPC
25 or MOPC 104EM, MOPC 47A enhanced all of IgG sub
class responses, and ADJPC 5 enhanced the IgG1 response. No indication of the influence of the cla
of Ig secreted by the melanomas on the effects of
the myelomas on the class of PFC produced were not
While the failure of established myelomas to suppress the primary IgG response to SRBC was thus
observed, the manner of their depressing effects
on humoral immune response to alum BSA was not cle

2682   THE MITOGENIC ACTIVITY OF POLYADENYLIC-POLYURIDYLIC ACID COMPLEXES. (Eng.)

Han, I. H. (Univ. Michigan Med. Sch., Ann Arbor);
Johnson, A. G. *Ann. N.Y. Acad. Sci.* 249:370-379;
1975.

The capacity of polyadenylic and polyuridylic acid
complexes (poly A:U) in increasing antibody synthesis to a number of different antigens was inves
tigated. BALB/Aj mice were inbred and used throug
out the study. Single cell suspensions were prepared from spleens, mesenteric lymph nodes, and
thymuses, and labeled with [3H]thymidine. Mice we
killed by cervical dislocation; the spleen was removed, homogenized, and cultured. By 24 hr, the
cells given antigen alone showed a 38% increase in
[3H]thymidine incorporation; poly A:U alone produc
a 24% increase, while combined stimulation of poly
A:U and antigen effected a 150% increase. Poly A:
alone was as mitogenic as SRBC. *In vivo* effects

were studied *via* administration of poly A:U. Two days after injection, 260% stimulation was observed, with similar values in animals receiving poly A:U plus SRBC. The elimination of cortisone-sensitive lymphocytes did not reduce or modify the mitogenic effects of poly A:U, indicating an effect other than T_0-cells. In studying the differences in homeostatic factors related to cortisone treatment, it was found that poly A:U increased the stimulatory effect and the level of DNA synthesis. The *in vitro* effect of poly A:U on bone marrow cells resulted in a 180% increase in [^{3}H]thymidine incorporation, irrespective of the dose level or presence or absence of antigen. The results indicate that mitogenic activity by poly A:U can be demonstrated in mouse spleen cells, that such stimulatory effects on [^{3}H]thymidine incorporation can be elicited in the absence of injected antigen, and that cortisone-resistant T-cells may be responsible for the enhanced incorporation of [^{3}H]thymidine.

2683   NEW CELL SURFACE ANTIGENS IN RAT DEFINED BY TUMORS OF THE NERVOUS SYSTEM. (Eng.) Fields, K. L. (Dep. Zool., Univ. Coll., London, England); Gosling, C.; Megson, M.; Stern, P. L. *Proc. Natl. Acad. Sci. USA* 72(4):1296-1300; 1975.

S.c. injection of N-ethyl-N-nitrosourea into newborn, inbred W/FU rats resulted in tumors of the brain, cord, lumbar cord and roots, and spinal roots. Several cell lines were developed from these tumors, 13 clonal cell lines from other alkylnitrosourea-induced tumors were also used, and antisera were prepared by i.p. injection of cultured cells of rat tumor line 33B into C34/He mice. The antisera were tested mainly by a two-stage dye exclusion cytotoxicity assay. They had high cytotoxic titers (1:4000) against all rat cells tested, tumor or normal. After extensive adsorption with normal liver, spleen and thymus, the serum was not cytotoxic for thymocytes, yet killed cells of rat tumor lines 21A and 33B equally well. In contrast, absorption with line 21A cells had very little effect on the cytotoxic titer for line 33B cells. These findings, together with further analysis, defined two antigens: a common antigen shared by brain, embryonic tissue, and neural tumors, and a restricted antigen present only on a subset of neural tumors. In screening for the restrictive antigen, no normal tissues showed definite adsorption, numerous clonal cell lines tested negative, and neither brain nor embryonic tissue showed clearly detectable amounts. The restricted antigen thus separated various tumor cells into two classes.

2684   INDUCTION OF THE IMMUNE RESPONSE TO CELL SURFACE ANTIGENS IN VITRO. (Eng.) Manson, L. A. (Wistar Inst., 36th and Spruce Streets, Philadelphia, Pa. 19104); Palmer, J. C. *In Vitro* 11(4):186-204; 1975.

Immune responses to cell surface antigens were demonstrated on two tissue culture incubation systems. In the one-way "mixed lymphocyte interaction" system, a specific stimulation of thymidine uptake was induced

by a particulate membrane antigen fraction, the microsomal lipoproteins (MLP), when low levels (0.01 to 0.001 µg per ml) were incubated with spleen or lymph node cells from nonsensitized mice. No stimulation was seen when allogeneic MLP was used at high levels, 10 µg per ml, nor at any level with syngeneic MLP. Specific effectors were demonstrated after 72-hr incubation with stimulatory levels of allogeneic MLP in three separate *in vitro* assays, a plaque-forming cell reduction assay, a tumor target assay, and an antigen-binding cell assay. In the latter assay, [^{125}I]MLP was used as the source of antigen. This system has limited potential as mouse spleen cells do not survive in it beyond the fourth day of culture. The second tissue culture system, the Marbrook system, has much greater possibilities because at least 25% of the inoculum is recovered seven days later. In this culture system, a cell-free sheep erythrocyte membrane preparation can induce plaque-forming cells in the absence of macrophages. Using a sensitive radioimmunoassay, free specific antibody was detected in culture supernatant fluids. With the same culture system, allogeneic lymphocytotoxic cells (killer) have been induced with spleen cells from unprimed mice in strains differing at the major histocompatibility locus (H-2). Allogeneic MLP induced very significant "killer" cell activity with spleen cells from primed mice. In a syngeneic tumor system, significant amounts of killer cell activity were induced with unprimed spleen cell inocula, and much larger amounts induced with spleen cells from immunized mice.

2685   BIOLOGIC EFFECTS OF SOLUBILIZED H-2 AND TUMOR-SPECIFIC ANTIGENS. (Eng.) Law, L. W. (Natl. Cancer Inst., Bethesda, Md.); Appella, E.; Chang, K. S. S. *Transplant. Proc.* 7(2):233-241; 1975.

The genetically well-defined histocompatibility antigens (H-antigens), specifically H-2 antigens, derived from dissociated spleen cells of A/J mice and the tumor-associated cell-surface antigens of the transplantation type (TSTA) of BALB/c mice were studied. The H-antigens and TSTA were solubilized by papain digestion and purified by G-150 Sephadex fractionation. All allogenic specificities (detected by cytotoxicity inhibition of a ^{51}Cr release assay) and transplantation antigens of H-2 were confined to fraction F2 following chromatography. All biological activity assayed (i.e., alloantigenicity, enhancing capacity, and immunogenicity) were confined to a single peak (fraction F2). Complete tolerance was not achieved; however, humoral tolerance was easily induced. No class I and II fragments were detected. In addition, β-2-like molecules were associated with alloantigenic activity. These results suggest that the H-2 antigen represents a complete molecule containing all the major determinants detected *in situ*. TSTA of neoplastic cells transformed by simian virus 40 (SV40) demonstrated immunologic activity concentrated mainly in the peak (fraction F2) that was positive for allogeneic activity. As little as 5.0 µg of crude soluble (CS) material produced protection against a challenge of 7×10^3 and 7×10^4 SV40-transformed neoplastic cells of BALB/c mice. This represents an immunogen yield of approximately 20% for CS preparations. TSTA on Rauscher-induced

peared in the blood of most mice with regressed
splenic tumors, suggesting the persistance of the
virus complex in the animals. Antibody responsiveness, as determined by the numbers of hemolytic
plaque-forming cells (PFC) after a single immunization with SRBC, was suppressed in leukemia progression and spontaneously recovered during regression
of leukemia. However, the PFC elicited by the SRBC
in both progressors and regressors expressed the
specific virus-induced membrane antigen, FVMA, as
demonstrated by inhibition of the PFC test with FVMA-
specific antiserum and complement. The recovery of
immunological responsiveness also included the spontaneous appearance of virus-neutralizing antibody to
FV. However, this was not paralleled by the appearance of antibody of FVMA. Traces of anti-FVMA antibody were occasionally detectable in serum of both
progressors and regressors and did not correlate
with virus neutralization. This may explain the observed susceptibility of regressors to secondary
relapse and to infection. It is suggested that the
general improvement of immune responsiveness during
regression would point to this phase of the disease
as a crucial stage for active immunotherapy.

2688   T ANTIGEN BINDS TO SIMIAN VIRUS 40 DNA
         AT THE ORIGIN OF REPLICATION. (Eng.)
Reed, S. I. (Stanford Univ. Sch. Med., Calif.);
Ferguson, J.; Davis, R. W.; Stark, G. R. *Proc. Natl.
Acad. Sci. USA* 72(4):1605-1609; 1975.

A technique employing ferritin-conjugated antibody
was developed to visualize specific protein-depen-
dent DNA complexes, and was employed in demonstrating the preferential location of binding of SV40
T antigen to SV40 DNA. SV40 DNA in the form of covalently closed supercoiled circles was obtained
from line Ma-134 African green monkey kidney cells
infected with strain WT-830 of the virus. T antigen was obtained in partially-purified form from
SV40-infected CV-1 or Cl3/SV28 lines of cells.
Anti-T sera were prepared by immunizing hamsters
with purified T antigen. After serial incubation
of the SV40 DNA with T antigen, anti-T γ-globulin,
and ferritin-conjugated anti-hamster γ-globulin,
followed by cleavage of the circular DNA to linear
form by treatment with *Eco*RI endonuclease, electron
dense ferritin cores could be seen bound to DNA.
To be able to map the position of these bound cores,
a marker for orientation, namely, *E. coli* DNA unwinding protein, for which SV40 has an affinity at
specific sites, was added. Purified *E. coli* DNA
unwinding protein incubated with SV40 DNA and fixed
with glutaraldehyde, followed by cleavage with
Hemophilus parainfluenzae restriction endonuclease
II, permitted assignment of two preferred sites for
the unwinding protein at 0.46 and 0.90 fractional
length relative to the *Eco* endonuclease cleavage
site. Ferritin binding mediated by T antigen could
then be mapped at 0.65 and 0.70 fractional length
from the *Eco*RI cleavage site. The affinity of
SV40 DNA for DNA unwinding protein thus served to
demonstrate the preferential binding of SV40 T
antigen at or near the origin of replication of SV40
DNA, namely, 0.67 fractional length clockwise from
the *Eco*RI cleavage site.

2689   SPECIFIC DEPRESSION OF THE ANTITUMOR ·
CELLULAR IMMUNE RESPONSE WITH AUTOLOGOUS
TUMOR HOMOGENATE. (Eng.) Paranjpe, M. S. (Natl.
Cancer Inst., Bethesda, Md.); Boone, C. W. *Cancer
Res.* 35(5):1205-1209; 1975.·

The source and characteristics of the factor responsible for the specific depression (eclipse) of antitumor cellular immune response were investigated in
simian virus 40 (SV40)-transformed tumor murine host
system of BALB/c AnN mice. A group of E_4 (a cell
line started from an explant of solid fibrosarcoma
in a BALB/c mouse produced by inoculating it with
SV40-transformed BALB /,T3 cells) tumor-immune mice
received, ip, 0.25 g of homogenized homologous tumor. The mice were· challenged 24 hr later with
10^6 E_4 tumor cells, and the cell-mediated immune
response was measured *in vitro* by radioisotopic
(^{51}Cr) footpad assay. Treated mice gave a lower
foot count ratio than untreated E_4-immune mice
(1.22 versus 2.83, respectively). No ·depression of
foot count ratio·was observed with an ip inoculate
of an equal amount of homogenate from an antigenically unrelated syngeneic methylcholanthrene-induced tumor. Inoculation of the tumor homogenate
to tuberculin-sensitized mice did not alter their
footpad reactivity to tuberculin. The fraction of
tumor homogenate that suppressed antitumor response
of immune mice could be extracted by 3 M KCl. The
authors conclude that the ip inoculation of E_4 tumor
homogenate does not cause general immunosuppression,
but rather produces specific eclipse of the antitumor cellular· immune response.

2690   DIFFERENTIAL INDUCTION OF TUMOUR ANTIGENS BY·
TRANSFORMATION-DEFECTIVE VIRUS MUTANTS.
(Eng.) Kurth, R. (Dept. Tumour Virology, Imperial
Cancer Res. Fund Lab,, P.O. Box 123, Lincoln's Inn
Fields, London, WC2A·3PX, England). *J. Gen. Virol.*
28(2):167-177; 1975.

Normal rat kidney cells infected by a variety of
transformation-defective temperature-sensitive avian
leukosis sarcoma virus mutants were tested for the
expression of transformation characteristics at permissive and restrictive temperature. Morphology,
growth behavior and agglutinability by concanavalin A
corresponded fully to the phenotype of the infected
cells; at permissive temperature the cells resembled
wild type virus transformed cells, whereas when grown
under restrictive conditions they became virtually
indistinguishable from normal cells. The quantitative expression of. allo- or xenogeneic cell surface
antigens was not significantly affected by the phenotype of the cells. Two out of the five tested mutants
induced tumor antigens in the expected temperature-
dependent manner, whereas the other three mutants were
able to induce tumor-specific cell surface antigens
even in the revertant cells cultured at the restrictive temperature. · These findings extend previous results about tumor antigen induction in mutant-infected
cells of the natural host, the chicken embryo fibroblasts. There may be value in the transformation-
defective tumor antigen-positive mutants for vaccination purposes.

2691   REACTIVITY TO TUMOUR-ASSOCIATED ANTIGENS
DETECTED IN MICE UNDERGOING LIVER REGEN-
FRATION. (Eng.) Hellstrom, I. (Univ. Washington
Med. Sch., Seattle); Hellstrom, K. E.; Nishioka, M.
Nature 253(5494):744-746; 1975.

The reactivity of lymph node cells (LNC) from mice
undergoing liver regeneration to tumor-associated
antigens was investigated. Partial hepatectomy was
performed on BALB/c and C3H mice, removing 65-75%
of the hepatic parenchyma; littermates underwent
sham-hepatectomy. Mice were bled from the retro-
orbital sinus; sera' were stored for testing of blocking activity; and the animals were killed and tested
for LNC-mediated reactivity. A BALB/c methylcholanthrene-induced sarcoma line, a BALB/c Moloney sarcoma
virus-transformed 3T3 line, BALB/c fibroblasts from
newborn mice, and nontransformed BALB/c 3T3 cells
were used as targets in microcytotoxicity assays.
LNC from hepatectomized mice were found to be cytotoxic to the two tumor lines used as targets, as
compared to LNC from sham-hepatectomized mice. However, no consistent cytotoxic effect against the untransformed, control cells was observed. The data
indicated that 'sera from the hepatectomized mice
could abrogate the cytotoxic effect of LNC from
multiparous and hepatectomized mice, as compared with
sera from sham-operated controls. The specificity
demonstrated suggests that an immune reaction to
some antigens present on tumor cells and regenerating
liver cells is responsible for these findings, rather
than a more nonspecific destruction of tumor cells.
It is further suggested that a normal tissue repair
process, involving rapid cell division, may lead to
a detectable immunological reaction against antigen
present on neoplastic cells.

2692   A FETUIN-LIKE ANTIGEN FROM HUMAN NEPHRO-
BLASTOMA. (Eng.) Wise, K. S. (Sch. Den-
tistry, Univ. Southern California, Los Angeles,
Calif. 90007); Allerton, S. E.; Trump, G.; Powars,
D.; Beierle, J. W. *Int. J. Cancer* 16(2):199-210;
1975.

The characterization of an antigen in human nephroblastomas is presented. The antigen was detected
in pooled tumor suspensions using antiserum prepared
in rabbits against an EDTA extract of the tumors.
This antigen was not found in normal human plasma
or kidney extracts, and was not related to the ABO
or Forssman blood groups. The antigen was detected
in extracts of cultured nephroblastoma cells, but
was not present in extracts of normal human fetal
kidney cell cultures. It is believed to be present
at the cell surface, as cell viability was not significantly lowered during the extraction procedure.
A reaction of complete identity was demonstrated by
Ouchterlony double diffusion experiments with this
antigen and purified bovine fetuin. The antigen was
not detected in extracts of human fetal spleen, thymus or kidney, nor in human fetal serum. Furthermore, it does not possess determinants in common
with the human alpha-fetoprotein of hepatomas, nor
was it detected in human renal clear cell carcinoma.
Initial characterization of the antigen showed it to
be nondialysable, not sedimentable at 100,000 x g for
two hours, stable to repeated freeze-thawing and to

tified as malignant or normal by their size. The mixed antiglobulin reaction showed that the malignant cells did not carry the surface Ig characteristic of B cells, whereas these malignant cells formed non-immune rosettes with guinea-pig RBC. Among lymph node cells, most surviving normal small lymphocytes, from outside the thymus-dependent areas, reacted as B cells. The morphologic evidence therefore corroborated the test results, which indicated that the formation of rosettes with guinea-pig RBC seems a reliable means for the demonstration of T cells in the cat.

2695   IDENTIFICATION OF LYMPHOID CELLS IN CULTURES OF MURINE LEUKOCYTES AND THYMUS.
(Eng.) Burke, T. R. (Springville Lab., N.Y.);
Moore, G. E.; Stobo, J. D. *Cancer Res.* 35(3):673-678; 1975.

Several culture conditions and media were studied in an effort to establish long-term cultures of murine lymphoid·cells from blood and thymus. Mouse strains C3H/St Ha MF+, C57BL/6Ha, and DBA/I were used for a source of thymus cells, while WBC were obtained from C3Hf /HeHa mice. Media including RPMI 1640, 1700, 1701, 1715, GEM1717, NCTC, fetal calf serum (FCS) and horse-serum supplements, and conditioned medium. Cells in the third passage were harvested for chromosomal analysis, and subcultures from thymus cell lines and cultured lymphoid cells were used in studies of mitogen reactivity in mixed lymphocyte cultures, respectively. The probability of thymus cell line-initiation seemed highest when thymuses from 1-mo-old mice were cultured with GEM 1717 medium supplemented with 20% heat-inactivated FCS. Use of supplemented RPMI medium yielded the longest survival times for WBC cultures, generally a few months. Light microscopy of thymus cells revealed many small, round to ovoid cells possessing small, dark, round nuclei which were apparently lymphoid; other pleomorphically shaped cells were also present, with a preponderant epithelioid cell type noted in a late passage. All chromosomes seen were normal acrocentries of $2n$ karyograms, with an abundance of binucleated cells. A fluorescence antibody assay of cultured thymic theta lymphoid cells revealed that they were negative for surface immunoglobulin and θ antigen. Mitogen reactivity indicated that subculture cells were capable of reacting to phytohemagglutin and concanavalin A by initiating DNA synthesis, and that there was a nondifferential response to the classical T-dependent mitogens. One-way MCL results indicated low but positive mixed lymphocyte reactivity function by the long-term cultured lymphoid cells. It is suggested that the cells represent a class of T cells that have all but lost θ antigen, possibly due to prolonged passage.

2696   *IN VIVO* STUDIES OF DIFFERENTIATION OF THYMUS-DERIVED LEUKEMIC CELLS. (Eng.)
Barker, A. D. (Battelle's Columbus Lab., Ohio);
Waksal, S. D. *Ann. N.Y. Acad. Sci.* 249:484-491; 1975.

Changes in differentiation patterns in neoplastic

thymus-derived lymphocytes (T-cells) subpopulations following second and third *in vivo* passage were investigated. The behavior of leukemic T-cell subpopulations in splenectomized and thymectomized preleukemic mice was also studied. In the original experiments, transplantation of 10^5 LTC_1, LSC_1, or $LLNC_1$ from AKR mice with spontaneous leukemia into preleukemic AKR mice resulted in three different disease pathologies. When the lymphoid cells from these mice were repassaged into preleukemic recipients, all mice exhibited the distinct disease pattern seen in animals originally implanted with $LLNC_1$. Preleukemic mice implanted with LTC_3, or $LLNC_3$ also displayed enlargement of lymphoid tissues reminiscent of mice given $LLNC_1$. These three groups of mice showed the largest percentage of animals dying before the 18 day termination date, all displaying thymic atrophy, gross splenomegaly, and moderate lymph node enlargements. The three different types of disease pathologies resulted from a sequential T-cell differentiation process. Passage of LTC_1 and LSC_1 leukemic T-cells into thymectomized or splenectomized mice caused splenomegaly and lymph node enlargement. The stage of differentiation of the subpopulations was apparently related to the lymphoid compartment of origin; the number of differentiated cells in each compartment was lymph node > spleen > thymus. Thymus removal did not appear to block the differentiation process, while control of the differentiation and release of thymocytes occurred in both intact and splenectomized animals. The data suggest that leukemic T-cells exist as subpopulations in various stages of differentiation, and retain many characteristics of normal T-cells.

2697   ADULT THYMECTOMY RESULTS IN LOSS OF T-
        DEPENDENT MITOGEN RESPONSE IN MOUSE SPLEEN
CELLS. (Eng.) Jacobs, D. M. (Dep. Biol., Univ. California, San Diego); Byrd, W. *Nature* 255(5504): 153-155; 1975.

Responses to phytohemagglutinin (PHA, 2.5 µg/ml) and concanavalin A (Con A, 10 µg/ml) were investigated in adult thymectomized mice to determine if responsiveness to these mitogens could be attributed to subpopulations of T cells. Female BDF_1 (C57-BL/6 female x DBA/2 male) mice were subjected to complete thymectomy or sham thymectomy at four weeks of age. Spleen cells obtained from mice (aged 8-20 wk) were cultured with the mitogens, and the cultures were pulsed with 50 µl of ^{125}I-5-iodo-2-deoxyuridine (1 µCi/ml). The cells were then counted in an automated gamma-counter. The results show that adult thymectomy initially caused an increase in the response of spleen cells to PHA and Con A. This was followed by a rapid and sustained decay of responsiveness to the same mitogens, whereas cells from sham-thymectomized mice continued to increase to their mature adult level of responsiveness to the two mitogens. This divergence of responsiveness between thymectomized and normal mice was not observed with a thymus-independent B-cell mitogen, lipopolysaccharide (20 µg/ml). The decay in responsiveness to PHA and Con A is consistent with the existence of two populations of T cells: one required for the bulk of mitogen response which decays completely by nine weeks after thymectomy, and the other a long-lived population that responds poorly to PHA and Con A.

2698   STIMULATION IN THE MIXED LEUKOCYTE CULTURE
        AND GENERATION OF EFFECTOR CELLS IN CELL
MEDIATED LYMPHOLYSIS BY A HUMAN T LYMPHOBLAST CELL LINE. (Eng.) Callewaert, D. M. (Wayne State Univ. Sch. Medicine, 540 E. Canfield, Detroit, Mich. 48201); Kaplan, J.; Peterson, W. D.; Jr.; Lightbody, J. J. *Cell Immunol.* 19(2):276-281; 1975.

A lymphoblast cell line possessing B cell surface markers and a line possessing T cell surface markers, each derived from the same patient, were tested for their ability to stimulate in the mixed leukocyte culture (MLC) reaction and their ability to generate effector cells in cell-mediated lympholysis (CML). Both the T and B cell lines were capable of stimulating human peripheral blood lymphocytes in the MLC reaction; however, the T cell line required four times as many cells for an equivalent response. Generation of effector cells in CML was achieved when both the T and B cells were used as stimulator cells, but the T cell line generated a much higher degree of ^{51}Cr release. These results are in agreement with a previous report indicating that human B cells possess strong stimulating capacity while T cells possess weak but significant MLC stimulating activity.

2699   REACTIVITY OF ANTIHUMAN THYMOCYTE WITH ACUTE
        LEUKEMIA BLASTS. (Eng.) Mills, B. (Milwaukee Children's Hosp., 1700 West Wisconsin Ave., Milwaukee, Wis. 53233); Sen, L.; Borella*, L. *J. Immunol.* 115(4):1038-1044; 1975.

Normal and leukemic lymphoid populations were examined to determine if the formation of spontaneous E rosettes correlated with the presence of thymus-associated cell surface antigens. Lymphoblasts from four of 12 children with acute lymphoblastic leukemia (ALL) formed spontaneous E rosettes at 4°C and 37°C(E^+) and bound rabbit antihuman thymocyte serum, as determined by indirect immunofluorescence and by an indirect radiolabeled antibody assay. The blasts from the other eight children did not form E rosettes at 4 or 37°C(E^-) and did not bind antithymocyte serum. Absorption of the antiserum with peripheral blood leukocytes removed all detectable reactivity with normal peripheral blood lymphocytes, normal bone marrow cells and E^- bone marrow cells, but did not remove all reactivity with thymus cells or E^+ leukemic blasts. Absorption of the antiserum with thymus from one donor removed all reactivity against thymus cells from another donor. After absorption with normal peripheral blood leukocytes, the antithymocyte serum reacted with E^+ leukemic blasts but not with remission lymphocytes from the same patients. Thus, there are at least two distinct thymus-associated antigens-one that is present on T lymphocytes from peripheral blood as well as thymus and another that is present only on thymus and on ALL blasts that are identified by their ability to form E rosettes at 37°C.

2700   E RECEPTORS ON BLASTS FROM UNTREATED ACUTE
        LYMPHOCYTIC LEUKEMIA (ALL): COMPARISON OF
TEMPERATURE DEPENDENCE OF E ROSETTES FORMED BY NOR-
MAL AND LEUKEMIC·LYMPHOID CELLS. (Eng.) Borella,
L. (St. Jude Child. Res. Hosp., Memphis, Tenn.) and
Sen, L. *J. Immunol.* 114(1):187-190; 1975.

The temperature sensitivity of SRBC rosettes formed
by normal human lymphocytes was compared with the
sensitivity of SRBC rosettes formed by the blasts
of 29 untreated children with acute lymphocytic leukemia (ALL). Rosette formation by normal thymic
cells, by bone marrow lymphocytes from ALL children,
by blood and bone marrow lymphocytes from children
in remission for a minimum of 4.5 yr were compared
at 4 C and 37 C. TC199 (0.2 ml) containing 1.5 x
10^6 lymphoid cells and an equal volume of 0.5% SRBC
was incubated at 37 C for five minutes, after which
the suspension was centrifuged and incubated at 4 C
or 37 C for one hour. The cells were resuspended,
and 1,000 cells were counted to determine the percentage of rosettes. The lymphocytes of six patients formed·rosettes (SRBC-positive cells). Of
these patients, four had a thymic mass and all had
WBC counts greater than 5 x $10^4/mm^3$. Two of 23
with SRBC-negative ALL had WBC counts greater than
5 x $10^4/mm^3$, and none of the 23 had a thymic mass.
Cells from children in ALL remission formed 11% rosettes at 4 C, and 1% rosettes at 37 C. Thymus
cells formed 83% rosettes at 4 C and 76% at 37 C,
while SRBC-positive ALL cells formed 57% at 4 C and
48% at 37 C. Rosette formation in the lymphocytes
of two children was 58% at 4 C and 38% at 37 C prior
to treatment; after six days of chemotherapy (vincristine, prednisone and asparaginase) this fell to
42% at 4 C and 0% at 37 C. The results suggest rosette formation in ALL is correlated with high WBC
counts and/or a thymic mass at diagnosis. ALL cells
and thymic cells form rosettes at 37 C, while normal
lymphocytes do not. SRBC rosette formation may serve
as another parameter to evaluate the differential ef-
feets of cytotoxic agents on normal and neoplastic
cells.

2701   DEPRESSION OF IN VITRO RESPONSIVENESS TO
        PHYTOHEMAGGLUTININ IN SPLEEN CELLS CUL-
TURED FROM CHICKENS WITH MAREK'S DISEASE. (Eng.)
Theis, G. A. (New York Medical Coll., Valhalla,
N.Y. 10595); McBride, R. A.; Schierman, L. W. *J.
Immunol.* 115(3):848-853; 1975.

Studies were undertaken to determine what effect
Marek's disease virus (MDV) infection and oncogenesis might exert on phytohemagglutinin (PHA) responsiveness of T lymphocytes in two sublines of Marek's
disease (MD)-susceptible inbred White Leghorn chickens (lines G-B1 and G-B2). Cultures of dispersed
spleen cells, prepared from chickens with MD visceral lymphomas that had been given an ip injection
of blood from MDV-infected chickens (0.5 ml 1:5 dilution), showed marked depression of responsiveness
to PHA (2 µl/ml), as measured by tritiated thymidine
(^{3}H-Tdr) incorporation in cells *in vitro*. When data
were expressed quantitatively in terms of cpm/10^5
viable cells, the functional depletion of PHA-re-
sponsive cells appeared to result from lower levels
of ^{3}H-Tdr incorporation in the PHA-stimulated spleen

cultures from chickens with acute MD symptoms, as
compared to similar cultures from uninfected iso-
lator-reared control chickens. It is suggested th
depression of PHA-induced blastogenesis in spleen
cell cultures from chickens with acute MD reflects
virus-related alterations in T lymphocytes.

2702   REDUCED LYMPHOCYTE TRANSFORMATION IN EAR
        CANCER OF THE BREAST. (Eng.) Knight, L
A. (King's Coll. Hosp. Medical Sch., London, Engla
Davidson, W. M. *J. Clin. Pathol.* 28(5):372-376; 1

The cell-mediated immune response was measured *in
vitro* by lymphocyte transformation in 53 patients
with malignant breast tumors and in 53 patients op
erated upon for benign tumors. Both groups were o
similar ages. Lymphocyte preparations from WBC-ri
plasma were sedimented by gravity or slight centri
fugation, placed in medium TE 199 supplemented wit
20% autologous or human serum, and treated with ph
hemagglutinin (PHA, 4 µg/ml), or with pokeweed mit
(0.04 ml/ml). After the cultures were incubated a
37 C for 48 hr, tritiated thymidine (1 µCi/ml) was
added, and after 68 hr incubation, Colcemid (0.6 µ
was added. After reincubation for four hours, the
uptake of the radioactive DNA precursor was measur
in portions of the cultures by scintillation proce
dures, and the percentage of cells incorporating t
label was determined by autoradiography. When the
medium contained standard human serum, the uptake
of tritiated thymidine was 26.40% by lymphocytes
from patients with malignant breast cancer, and
34.63% by lymphocytes from subjects with benign
tumors. The corresponding values for the percenta
incorporation were 47.60% and 60.79%, respectively
The differences were statistically significant. S
ilar results were found when the medium contained
autologous serum. Differences in the same directi
were observed with the use of pokeweed mitogen, bu
they were not statistically significant. Reduced
lymphocyte transformation was present preoperative
and even very early in the disease. The reduction
appeared to be even more marked in the advanced
cases. It appears significant that plasma from th
cancer patients contains a factor which reduces th
PHA transformation of lymphocytes from a healthy
donor, although contradictory findings have been
reported by other investigators.

2703   BIOLOGICAL DIFFERENCES BETWEEN EPSTEIN-
        BARR VIRUS (EBV) STRAINS WITH REGARD TO
LYMPHOCYTE TRANSFORMING ABILITY, SUPERINFECTION AN
ANTIGEN INDUCTION. (Eng.) Menezes, J. (Ste. Jus-
tine Hosp., Montreal, P.Q., Canada); Leibold, W.;
Klein, G. *Exp. Cell Res.* 92(2):478-484; 1975.

Epstein-Barr viruses (EBV) that inhibit growth of
an established EBV-carrying Raji line or transform
cord blood lymphocytes (CBL) were subjected to fou
tests in order to determine if transformation and
growth inhibition are due to the same type of viru
particle. Six different sources of EBV were teste
for CBL transformation, early antigen induction in
Raji cells, inhibition of Raji cell growth, and in
duction of EBV-determined nuclear antigen (EBNA) i

CBL. The EBV-producer lines used as virus donors were P3HR-1, B95-8, 833L, QIMR-WIL, F137, and cb-8-7. Lymphocytes prepared from human umbilical cord blood within 20 hr after collection were separated on a Ficoll-isopaque gradient and cultured in RPMI-HEPES buffer. Primary CBL cultures were infected with serially diluted EBV, and transformation was assessed in a microculture system with twice-weekly microscopic examination. EBV-determined antigens were detected by immunofluorescence. Growth inhibition was measured by a micromethod, the end-point being taken as the highest virus dilution that produced at least 75% inhibition in relation to uninfected controls. The B95-8 virus transformed and induced EBNA in CBL; the virus neither induced early antigen in Raji cells, nor inhibited their growth. P3HR-1 virus did not transform CBL or induce EBNA or EV in CBL, but did induce early antigen in Raji cells and inhibit their growth. EBV isolated from cell lines QIMR-WIL, 833L, F137 and cb-8-7 resembled the B95-8 virus. The findings indicate that the early antigen inducing and growth-inhibitory properties of the P3HR-1 virus, and the transforming activity of the B95-8 and other cell line viruses, are mutually exclusive viral functions.

2704   THE ROLE OF MACROPHAGES IN THE GENERATION OF T-HELPER CELLS: I. THE REQUIREMENT FOR MACROPHAGES IN HELPER CELL INDUCTION AND CHARACTERISTICS OF THE MACROPHAGE-T CELL INTERACTION. (Eng.) Erb, P. (Dept. Zoology, Univ. Coll., London, WC1E 6BT, England); Feldmann, M. *Cell Immunol.* 19(2):356-367; 1975.

The role of macrophages in the induction of helper cells *in vitro* was investigated. When either soluble or particulate antigens were used, macrophages were found to be essential. This was true regardless of the anatomical source of the T cells (spleen, lymph node or the cortisone resistant pool of the thymus), or of the method of macrophage depletion, (adherence to polystyrene or nylon wool, or by the use of carbonyl iron). There were some differences, however, depending on the physical nature of the antigen used. With soluble antigen, 2-mercaptoethanol or allogeneic macrophages would not overcome the macrophage deficit, whereas they did with particulate antigen. The nature of the interaction between macrophages and T cells was investigated using flasks with double chambers, separated by a nucleopore membrane with 0.2 μm pores. The results indicate that cell contact was not necessary; macrophages functioned normally when in either the upper or the lower compartments. Since physical contact was not required between T cells and macrophages for helper cell induction, macrophages release mediators which interact with T cells.

2705   SUSCEPTIBILITY AND RESISTANCE OF CHICKEN MACROPHAGES TO AVIAN RNA TUMOR VIRUSES. (Eng.) Gazzolo, L. (Unit Virology, I.N.S.E.R.M.,

Lyon, France); Moscovici, M. G.; Moscovici, C. *Virology* 67(2):553-565; 1975.

This study provides new information on the susceptibility of chicken macrophages to avian leukosis and sarcoma viruses of various subgroups. Cultures of embryonic macrophages were derived from chicken yolk sac; those of adult macrophages, from peripheral blood of chickens. Cultures of fibroblasts were derived from chick embryos. Cells for most experiments came from line 6 white Leghorn embryos, but fertile eggs of other lines and of various fibroblast phenotypes (relating to resistance to virus subgroups) were also used. Viruses studied were Rous-associated leukosis viruses, myeloblastic-associated leukosis viruses, transformation-defective B77 virus, ring-necked pheasant virus, and golden pheasant virus, together representing a variety of subgroups. Also studied were Bryan high-titer Rous sarcoma virus R(-), R(-) pseudotypes with leukosis viruses, R(-) pseudotype with chicken helper factor R(chf), Prague strain Rous sarcoma viruses, Schmidt-Ruppin Rous sarcoma viruses, and avian sarcoma virus B77, also representing a variety of subgroups. Established methods were used for assaying the different viruses. For studies involving isolation of avian sarcoma virus host range recombinants, Chf-negative C/E fibroblasts were infected with sarcoma virus PRA, PRB, PRC or B77 and, when completely transformed, they were superinfected with a leukosis virus of another envelope subgroup. The mixed harvests were plated at low multiplicities in an infectious center assay. The indicator cells in the assay were chosen to exclude the subgroup of the parental sarcoma virus, so that only those sarcoma viral infectious centers registered which released progeny belonging to the subgroup of the parental leukosis virus. Single foci produced under these conditions were recloned three times at high virus dilutions. The envelope subgroup of these recombinant clones was determined on selectively resistant chicken cells by interference tests with leukosis virus of known subgroup affiliation, and by antibody neutralization. Macrophages from several genetic lines of chickens were susceptible to avian leukosis and sarcoma viruses of subgroups B and C, but were resistant to viruses of subgroups A, D, E, F, and G, even when fibroblasts derived from the same embryos were susceptible. This suggested that the macrophage-specific resistance was epigenetic. If the fibroblasts of a chicken embryo were genetically resistant to subgroup B or C, this resistance was also maintained in macrophages. Although chicken macrophages showed active phagocytosis in cultures, they could not initiate infection by the defective Bryan high-titer strain of Rous sarcoma virus, R(-). Subgroup B and C viruses also activated the endogenous chicken helper factor when they replicated in macrophages derived from factor-positive embryos. However, compared to fibroblasts, the levels of exogenous as well as endogenous virus synthesized by macrophages were reduced. Studies with viral pseudotypes and with recombinants of viral envelope markers suggested that the avian tumor viral host range in macrophages was controlled by an envelope component of the virion. In the experiments with recombinants, several viral clones were found that showed host range characteristics of both parental viruses.

Albert-Thomas, F 69008 Lyon, France). *Bull. Cancer (Paris)* 62(3):265-276; 1975.

2710   GLOMERULAR IMMUNE COMPLEXES ASSOCIATED
          WITH CANCER [abstract]. (Eng.) Pascal,
R. R. (Columbia Univ. Coll. Physicians Surg., New
York, N.Y.); Rollwagen, F. M.; Koss, M. N.; Iannacco-
ne, P. M. *Lab. Invest.* 32(3):432; 1975.

2711   ASSAYS OF IMMUNE FUNCTION IN PATIENTS
          WITH COLON CANCER. (Eng.) Slade, M. S.
(Univ. Minnesota, Minneapolis); Greenberg, L. J.;
Goldberg, S. M.; Hallgren, H.; Simmons, R. L.;
Yunis, E. J. *Fed. Proc.* 34(3):823; 1975.

2712   CORRELATION OF *IN VITRO* IMMUNE RESPONSE
          WITH CLINICAL COURSE OF MALIGNANT NEO-
PLASIA IN DOGS. (Eng.) Brodey, R. S. (Sch. Vet.
Med., Univ. Pennsylvania, Philadelphia); Fidler,
I. J.; Bech-Nielsen, S. *Am. J. Vet. Res.* 36(1):75-80;
1975.

2713   IMMUNOLOGIC EVALUATION OF HUMAN BLADDER
          CANCER: *IN VITRO* STUDIES. (Eng.) Elhil-
ali, M. M. (Univ. California Los Angeles Med. Cent.);
Nayak, S. K. *Cancer* 35(2):419-431; 1975.

2714   OBSERVATIONS ON POSSIBLE AUTOIMMUNE
          THERAPEUTIC EFFECTS IN EXPERIMENTALLY
PRODUCED RAT BLADDER TUMORS. (Eng.) Fingerhut,
B. (Coll. Physicians Surg., Columbia Univ., New
York, N.Y.); Veenema, R. J. *Urol. Int.* 30(4):255-
265; 1975.

2715   ONCO-FOETAL CROSS REACTIVITY BETWEEN HU-
          MAN OSTEOGENIC SARCOMA AND FOETAL PERI-
OSTEAL FIBROBLASTS GROWN *IN VITRO*. (Eng.) Gangal,
S. G. (Biology Div., Cancer Res. Inst., Parel,
Bombay 400 012, India); Agashe, S. S.; Nair, P. N.
M.; Rao, R. S.; Ranadive, K. J. *Indian J. Med. Res.*
63(6):851-857; 1975.

2716   COMPARATIVE STUDIES OF TUMOUR REACTIVITY
          IN NEWBORN AND ADULT MICE. (Rus.) Ni-
konova, M. F. (Sci. Lab. Exp. Immunobiol., U.S.S.R.
Acad. Med. Sci., Moscow, U.S.S.R.); Maiskii, I. N.;
Pokrovskaia, T. A. *Bull. Exp. Biol. Med.* 59(6):85-
88; 1975.

2717   SPLENIC REGULATION OF THE CLINICAL AP-
          PEARANCE OF SMALL TUMORS. (Eng.) Nord-
lund, J. J. (Yale Univ. Med. Cent., New Haven,
Conn.); Gershon, R. K. *J. Immunol.* 114(5):1486-
1490; 1975.

2718   RESTING STAGES OF THE JB-1 ASCITES TU-
          MOUR IN THE IRRADIATED MOUSE. (Eng.)
Bichel, P. (The Inst. of Cancer Research, Radium-

stationen, Aarhus, Denmark); Dombernowsky, P.
Eur. J. Cancer 11(6):425-431; 1975.

2719   LYMPHORETICULAR RESPONSE TO A SYNGENEIC
        RAT TUMOUR: GRAVIMETRIC AND HISTOLOGI-
CAL STUDIES. (Eng.) Flannery, G. R. (Monash Univ.
Medical Sch., Melbourne, Victoria, Australia,
3181); Muller, H. K.; Nairn, R. C. *Br. J. Cancer*
31(6):614-619; 1975.

2720   MODULATION OF THE IMMUNE RESPONSE TO
        TRANSPLANTATION ANTIGENS. IV. ENHANCED
GROWTH OF TUMOUR ALLOGRAFTS IN MICE IMMUNISED WITH
PAPAIN-SOLUBILISED TRANSPLANTATION ANTIGEN. (Eng.)
Zola, H. (Dep. Protein Chem., Wellcome Res. Lab.,
Beckenham, England). *Ann. Immunol. (Paris)*
126C(1):51-62; 1975.

2721   IMMUNOLOGICAL RESPONSIVENESS AGAINST TWO
        PRIMARY ANTIGENS IN UNTREATED PATIENTS
WITH HODGKIN'S DISEASE. (Eng.) De Gast, G. C.
(Dep. Med., Univ. Groningen, Netherlands); Halie,
M. R.; Nieweg, H. O. *Eur. J. Cancer* 11(4):217-
224; 1975.

2722   STUDIES ON THE CONTACT SENSITIZATION OF
        MAN WITH SIMPLE CHEMICALS. IV. TIMING
OF SKIN REACTIVITY, LYMPHOKINE PRODUCTION, AND
BLASTOGENESIS FOLLOWING RECHALLENGE WITH DINITRO-
CHLOROBENZENE USING AN AUTOMATED MICROASSAY. (Eng.)
Powell, J. A. (Natl. Cancer Inst., Bethesda, Md.);
Whalen, J.; Levis, W. R. *J. Invest. Dermatol.*
64(5);357-363; 1975.

2723   STUDY BY BIOASSAY OF DISPOSAL OF LIVE
        LYMPHOMA CELLS BY RECIPIENT IMMUNE AND
CONTROL RATS [abstract]. (Eng.) McCreary, P. (Rush-
Presbyt.-St. Luke's Med. Cent., Chicago, Ill.); Laing,
G.; Hass, G. *Lab. Invest.* 32(3):430; 1975.

2724   EFFECT OF A PROTEIN-FREE DIET ON LYMPH
        NODE AND SPLEEN CELL RESPONSE *IN VIVO*
TO BLASTOGENIC STIMULANTS. (Eng.) Aschkenasy, A.
(Centre National de la Recherche Scientifique, Cen-
tre Marcel Delépine, Laboratoire d'Hematologie
nutritionnelle, 45045 Orléans Cedex, France).
Nature 254(5495):63-65; 1975.

2725   IMMUNO-PHYSIOLOGICAL MECHANISMS BY WHICH
        NEOPLASMS AVOID DESTRUCTION BY THE IMMUNE
SYSTEMS: ACTION OF BACCILUS CALMETTE AND GUERIN
(BCG) [abstract]. (Eng.) Hakim, A. A. (Univ. Illinois Med. Cent., Chicago); Grand, N. G. *Fed. Proc.*
34(3):382; 1975.

2726   MYCOBACTERIAL ADJUVANT AND ITS CARRIER.
        (Eng.) Hiu, I. J. (Institut de Cancer-
ologie et d'Immunogenetique, Hopital Paul-Brousse,
14, Avenue Paul-Vaillant-Couturier, F-94 Villejuif,
France). *Experientia* 31(8):983-985; 1975.

2727   RADIOIMMUNOASSAY FOR THE CHEMICAL CAR-
        CINOGEN 2-ACETYLAMINO-FLUORENE [abstract].
(Eng.) Cernosek, S. F., Jr. (Univ. Arkansas Sch.
Med., Little Rock); Gutierrez-Cernosek, R. M. *Fed.
Proc.* 34(3):513; 1975.

2728   LEUKOCYTE MIGRATION IN AGAROSE: A SPECI-
        FIC TEST FOR CELLULAR IMMUNITY TO SOLUBLE
TUMOR ANTIGENS IN PATIENTS WITH CANCER [abstract].
(Eng.) Boddie, A. W., Jr. (Div. Surg. Oncol., Univ.
California Los Angeles); Uriat, M. M.; Chee, D. O.;
Holmes, E. C.; Morton, D. L. *Fed. Proc.* 34(3):268;
1975.

2729   CELL-MEDIATED IMMUNE RESPONSES IN PA-
        TIENTS WITH WARTS. (Eng.) Morison, W.
L. (Massachusetts General Hosp., Boston, Mass.
02114). *Br. J. Dermatol.* 93(5):553-556; 1975.

2730   HISTOLOGIC EVIDENCE FOR IMMUNOLOGIC
        ENHANCEMENT AND SUPPRESSION OF HUMAN
BREAST CANCER [abstract]. (Eng.) Hunter, R. (Dep.
Pathol., Univ. Chicago, Ill.); Ferguson, D.; Copple-
son, W. *Fed. Proc.* 34(3):846; 1975.

2731   GROWTH OF HUMAN TUMOURS IN IMMUNE-
        SUPPRESSED MICE. (Eng.) Franks, C. R.
(Guy's Hosp., London, England); Perkins, F. T.;
Holmes, J. T. *Proc. R. Soc. Med.* 68(5):287-290;
1975.

2732   ZINC SUPPRESSION OF INITIATION OF SARCOMA
        180 GROWTH. (Eng.) Woster, A. D. (Dep.
Microbiol., Indiana Univ., Bloomington); Failla,
M. L.; Taylor, M. W.; Weinberg*, E. D. *J. Natl.
Cancer Inst.* 54(4):1001-1003; 1975.

2733   PRESENCE OF IMMUNOSUPPRESSIVE AGENTS
        WITH VARIOUS ACTIVITIES IN EHRLICH AS-
CITES FLUID. (Eng.) Nitta, K. (Dept. Antibiotics,
Natl. Inst. Health, Kamiosaki 2-10-35, Shinagawa-
ku, Tokyo 141, Japan); Umezawa, H. *Gann* 66(4):459-
460; 1975.

2734   *IN VITRO* SUPPRESSION OF IMMUNOCOMPETENT
        CELLS BY LYMPHOMAS FROM AGING MICE.
(Eng.) Jaroslow, B. N. (Div. Biological and Medical Res., Argonne Natl. Lab., Argonne, Ill.
60439); Suhrbier, K. M.; Fry, R. J. M.; Tyler,
S. A. *J. Natl. Cancer Inst.* 54(6):1427-1432; 1975.

2735   AN ABSOLUTE REQUIREMENT FOR SERUM MACRO-
        MOLECULES IN PHYTOHAEMAGGLUTININ-INDUCED
HUMAN LYMPHOCYTE DNA SYNTHESIS. (Eng.) Yachnin,
S. (Univ. Chicago, 950 East 59th St., Box 420,
Chicago, Ill. 60637); Raymond, J. *Clin. Exp.
Immunol.* 22(1):153-166; 1975.

2736   THE ANTITUMOUR ACTIVITY OF PHYTOHEMAG-

Choi, H.-S. (Veterans Adm. Hosp., Bronx, N.Y.);
Paronetto, F. *Fed. Proc.* 34(3):846; 1975.

2746   NATURALLY OCCURRING HUMAN ANTIBODY TO
        NEURAMINIDASE-TREATED HUMAN LYMPHOCYTES.
ANTIBODY LEVELS IN NORMAL SUBJECTS, CANCER PATIENTS,
AND SUBJECTS WITH IMMUNODEFICIENCY. (Eng.) Rogentine, G. N., Jr. (Natl. Cancer Inst., Bethesda, Md.
20014). *J. Natl. Cancer Inst.* 54(6):1307-1311;
1975.

2747   CELL-FREE SYNTHESIS OF HUMAN INTERFERON.
        (Eng.) Pestka, S. (Roche Inst. Molecular
Biology, Nutley, N.J. 07110); McInnes, J.; Havell,
E. A.; Vilcek, J. *Proc. Natl. Acad. Sci. USA*
72(10):3898-3901; 1975.

2748   THE BINDING OF CARCINOEMBRYONIC ANTIGEN
        BY ANTIBODY AND ITS FRAGMENTS. (Eng.)
Morris, J. E. (Batelle Pacific Northwest Lab.,
Richland, Wash. 99352); Egan, M. L.; Todd, C. W.
Cancer Res. 35(7):1804-1808; 1975.

2749   VARIATIONS IN CARCINOEMBRYONIC ANTIGEN
        LOCALIZATION IN TUMORS OF THE COLON.
(Eng.) Rogalsky, V. Y. (P. A. Herzen Res. Inst. of
Oncology, Moscow, USSR). *J. Natl. Cancer Inst.*
54(5):1061-1071; 1975.

2750   EPITHELIAL BLOOD GROUP ANTIGENS IN COLON
        POLYPS. I. MORPHOLOGIC DISTRIBUTION AND
RELATIONSHIP TO DIFFERENTIATION. (Eng.) Denk, H.
(Univ. Vienna Sch. Medicine, Spitalgasse 4, A-1090
Vienna, Austria); Holzner, J. H.; Obiditsch-Mayr,
I. *J. Natl. Cancer Inst.* 54(6):1313-1317; 1975.

2751   HUMAN LYMPHOCYTE ALLOANTIGEN(S) SIMILAR
        TO MURINE Ir REGION-ASSOCIATED (Ia)
ANTIGENS. (Eng.) Arbeit, R. D. (Immunology
Branch, Natl. Cancer Inst., Bethesda, Md. 20014);
Sachs, D. H.; Amos, D. B.; Dickler, H. B. *J.
Immunol.* 115(4):1173-1175; 1975.

2752   FELINE ONCORNAVIRUS-ASSOCIATED CELL MEM-
        BRANE ANTIGEN. II. ANTIBODY TITERS IN
HEALTHY CATS FROM HOUSEHOLD AND LABORATORY COLONY
ENVIRONMENTS. (Eng.) Essex, M. (Harvard Univ.
Sch. Public Health, Boston, Mass.); Cotter, S. M.;
Carpenter, J. L.; Hardy, W. D., Jr.; Hess, P.;
Jarrett, W.; Schaller, J.; Yohn, D. S. *J. Natl.
Cancer Inst.* 54(3):631-635; 1975.

2753   RAPID QUANTITATION OF MEMBRANE ANTIGENS.
        (Eng.) Welsh, K. I. (Queen Victoria
Hosp., East Grinstead, England); Dorval, G.; Wigzell, H. *Nature* 254(5495):67-69; 1975.

2754   CELL CYCLE DEPENDENCY OF TUMOR ANTIGEN

EXPRESSION IN HUMAN SARCOMA CELLS [abstract].
(Eng.) Burk, K. H. (M. D. Anderson Hosp. Tumor Inst.,
Houston, Tex.); Drewinko, B.; Lichtiger, B.; Tru-
jillo, J. M. *Am. J. Pathol.* 78(1):28a; 1975.

2755   AN IMMUNOENZYMATIC ASSAY OF TUMOR-SPECIFIC
        ANTIGEN [abstract]. (Eng.) Natali, P. G.
(Lab. Cell Biol. C.N.R., Rome, Italy); Radojkovic, J.;
Vasile, C.; Celada, F. *Fed. Proc.* 34(3):850; 1975.

2756   MEMBRANE DIFFERENCES IN PERIPHERAL BLOOD
        LYMPHOCYTES FROM PATIENTS WITH CHRONIC
LYMPHOCYTIC LEUKEMIA AND HODGKIN's DISEASE. (Eng.)
Mintz, U. (Dept. Genetics, Weizmann Inst. Science,
Rehovot, Israel); Sachs, L. *Proc. Natl. Acad. Sci.
USA* 72(6):2428-2432; 1975.

2757   SELECTIVE ADHERENCE OF IMMUNE LYMPHO-
        CYTES, SYNTHESIZING DNA TO CORRESPONDING
TARGET CELLS. (Rus.) Brondz, B. D. (N.F. Gamaleya
Inst. Epidemiology and Microbiology, Inst. Biologi-
cal Trial of Chemical Compounds, Moscow, U.S.S.R);
Kirkin, A. F.; Epikhina, S. O. *Biull. Eksp. Biol.
Med.* 53(7):68-71; 1975.

2758   TERMINAL DEOXYNUCLEOTIDYL TRANSFERASE AND
        ADENOSINE DEAMINASE IN HUMAN LYMPHOBLAS-
TOID CELL LINES. (Eng.) Coleman, M. S. (Dept.
Medicine, Univ. Kentucky, Lexington, Ky. 40506);
Hutton, J. J. *Exp. Cell Res.* 94(2):440-442; 1975.

2759   T-CELL ORIGIN OF ACID-PHOSPHATASE-POSITIVE
        LYMPHOBLASTS [letter to editor]. (Eng.)
Catovsky, D. (Royal Postgraduate Medical Sch., London W12 OHS, England). *Lancet* 2(7929):327-238; 1975.

2760   CONCANAVALIN A RECEPTORS ON THE SURFACE
        MEMBRANE OF LYMPHOCYTES FROM PATIENTS WITH
HODGKIN'S DISEASE AND OTHER MALIGNANT LYMPHOMAS.
(Eng.) Ben-Bassat, H. (Heb. Univ.-Hadassah Med.
Sch., Jerusalem, Israel); Goldblum, N. *Proc. Natl.
Acad. Sci. USA* 72(3):1046-1049; 1975.

2761   THE RELATIVE OCCURRENCE OF MACROPHAGES IN
        REGRESSING AND PROGRESSING MOLONEY SAR-
COMAS [abstract]. (Eng.) Russell, S. W. (Scripps
Clin. Res. Found., La Jolla, Calif.); Doe, W. F.;
Tozier, A. *Fed. Proc.* 34(3):268; 1975.

See also:

*   (Rev):   2415, 2416, 2417, 2418, 2437, 2438
*   (Chem): 2475, 2482, 2490, 2505
*   (Viral): 2583, 2587, 2590, 2600, 2611, 2632
*   (Path): 2770, 2800, 2807, 2858
*   (Epid): 2882

of the scapula and stretching movements of the upper extremities apparently play a main part in necrotic tissue change.

2764   MELANOTIC NEUROECTODERMAL TUMOR OF INFANCY.
        (Eng.) Brekke, J. H. (214 N. Fifth Ave.,
Virginia, Minn. 55792); Gorlin, R. J. *J. Oral Surg.*
33(11):858-865; 1975.

A case history is presented of a white male infant
with a melanotic neuroectodermal tumor of infancy,
the recurrence of which was clinically evident eight
wk after the original lesion had been surgically removed. The lesion was first noticed at five mo. The
3 x 2 x 1.5-cm firm tumescence of the anterior left
maxilla extended 3.0 cm posteriorly from the midline.
The lesion was hard in the posterosuperior regions,
and was slightly compressible anteriorly and infer-
iorly. The lesion was removed, and the postoperative
course was uncomplicated until the seventh wk, when
there was a swelling in the left maxilla posterior
to the original lesion. The second lesion was also
excised and the diagnosis of melanotic neuroectodermal tumor was confirmed histologically. The child
had an uncomplicated postoperative course, and was
without evidence of recurrence in his 17th postoperative month. The rapid, aggressive growth of this tumor and its propensity for recurrence make it a particularly serious and difficult lesion to manage. The
author suggests that these melanotic neuroectodermal
tumor cells may be induced by normal odontoblasts.
If this theory is valid, it would explain the recurrences of the tumor in the maxilla and mandible.

2765   SOME NEW OBSERVATIONS IN AN INTRACRANIAL
        GERMINOMA. (Eng.) Hirano, A. (Monte-
fiore Hosp. and Medical Center, 111 East 210 St.,
Bronx, N.Y. 10467); Llena, J. F.; Chung, H. D.
Acta Neuropathol. (Berl.) 32(2):103-113; 1975.

A case of an intracranial germinoma from the supra-
sellar region of a 9-yr-old girl was examined in the
electron microscope. The tumor consisted for the
most part, of both large polygonal and small lympho-
cyte-like elements. Annulate lamellae were common
in the epithelial cells. The small blood vessels
were fenestrated, and the endothelial cells contained
tubular bodies, membrane-bounded vacuoles containing
dense fluid and occasional tubules, arrays of tubules
within the nuclear envelope and rough endoplasmic
reticulum, and a markedly irregular luminal surface.
Dense, lamellated structure were present in the
widened, collagen-containing perivascular spaces.
The high frequency of annulate lamellae within the
large polygonal cell components (which have previously been confirmed to be normal components of
oocytes) supports the conclusion that intracranial
germinomas are probably unrelated to the pineal
gland and are, instead, identical to germinomas of
the ovary.

2766   FINE STRUCTURE OF INTERCELLULAR JUNC-
        TIONS AND BLOOD VESSELS IN MEDULLOBLAS-
TOMAS. (Eng.) Hassoun, J. (Laboratoire de Neuro-
pathologie, Faculte de Medecine. Bd. Jean-Moulin,

F-Marseille 13885, France); Hirano, A.; Zimmerman,
H. M. *Acta Neuropathol. (Berl.)* 33(1):67-78; 1975.

Specimens from six cerebellar medulloblastomas
(all in children) were studied by light and electron
microscopy. The specimens for light microscopy
were fixed in formaldehyde; for electron microscopy they were fixed in glutaraldehyde with phosphate buffer. Specimens were postfixed in osmium
tetroxide, dehydrated in ethanol, and embedded in
Epon or Araldite. Thin sections were stained with
uranyl acetate and lead citrate. The six tumors
all displayed the usual undifferentiated pattern
generally associated with medulloblastoma. Two
features were found which are suggested to be
constant and essential characteristics of medulloblastoma. First, cell junctions were abundant
between tumor cells; these were mostly desmosome-
like but other closer junctions were also seen.
Second, the capillary endothelia contained frequent
tubular bodies and other inclusions which may be
related to them. The function of the tubular
bodies is unknown, but their occurrence in brain
tumors is emphasized because tubular bodies are
normally present in brain capillaries only in much
more limited numbers.

2767   ELECTRON MICROSCOPIC AND ELECTRON HISTO-
        CHEMICAL STUDIES ON EMBRYONAL RHABDOMYO-
SARCOMA OF THE ORBIT. (Eng.) Amemiya, T. (Dept.
Ophthalmology, Kyoto Univ., Sakyo-ku, Kyoto-shi,
Japan); Tsuboi, M.; Uchida, S. *Z. Krebsforsch.*
83(4):305-314; 1975.

Two case reports of embryonal rhabdomyosarcoma of the
orbit are presented, and the results of electron
microscopic and electron histochemical studies of
the tumor are described. The tumors were removed
from a 9-yr-old Japanese boy and a 5-yr-old Japanese
boy. A small number of differentiated tumor cells
were found along with a great number of undifferentiated ones in both tumors. Abnormal dense granules
were found in the nuclei of undifferentiated tumor
cells, but not in differentiated ones. Nonspecific
filaments were seen in the cytoplasm of undifferentiated tumor cells. Differentiated tumor cells demonstrated various stages of myofibrillar structures
such as A, I, and Z bands, many glycogen granules,
mitochondria and a basement membrane. Polyglucose
particles synthesized from glucose-1-phosphate by
phosphorylase activity were located in the cytoplasmic matrix of undifferentiated and undifferentiated
tumor cells and in the karyolymph of undifferentiated
tumor cells only. Polyglucose particles increased
in number according to the degree of differentiation
of tumor cells. These results suggest that the
intranuclear dense granules found only in the undifferentiated tumor cells are glycogen granules or some
products in cell division, but not destructive products.

2768   INTRAVASCULAR ANGIOMATOSIS: DEVELOPMENT
        AND DISTINCTION FROM ANGIOSARCOMA.
(Eng.) Salyer, W. R. (Johns Hopkins Hosp., Baltimore, Md. 21205); Salyer, D. C. *Cancer* 36(3):995-
1001; 1975.

Two hundred and seventy-five arterial thromboemboli
and 140 venous thrombi were compared with 12 angiosarcomas in order to study the development of the
angiosarcoma-like pattern in organizing thrombi and
to note certain features that aid in their histologic differentiation from angiosarcoma. Three-
fourths of the venous thrombi were in hemorrhoidal
vessels; the remainder, and the arterial thromboemboli, were taken from autopsy and other surgical
cases. Features similar to those present in angiosarcomas were found in organizing thrombi. Within
the thrombi, freely anastomosing small channels,
often lined by one or more layers of prominent,
occasionally atypical, endothelial cells were noted.
Papillary-like projections of organizing thrombus
material lined by similar cells heightened the
similarity of the process to angiosarcomas. The
papillary-like structures appeared to develop due
to a combination of endothelialization of thrombus
fragments and of ingrowth of interlacing vessels.
The peculiar process of thrombus organization,
which has been called "intravascular angiomatosis,"
may be mistaken for true angiosarcoma and thus lead
to unnecessary irradiation or radical surgery. The
pseudoangiosarcoma differs from angiosarcoma in its
confinement entirely within large vascular lumens
and in its lack of mitoses, necrosis, and true
solid cellular areas devoid of vascular differentiation.

2769   STRUCTURAL AND FUNCTIONAL CHARACTERISTICS
        OF THE MICROCIRCULATION IN NEOPLASMS.
(Eng.) Papadimitriou, J. M. (Dept. Pathology, Univ.
Western Australia, Perth, Western Australia); Woods,
A. E. *J. Pathol.* 116(2):65-72; 1975.

The microcirculation in a transplantable rat fibrosarcoma was investigated. Methylcholanthrene-induced
fibrosarcoma suspension was injected sc into neonatal
Wistar rats. Venograms were performed by injecting
the large veins overlying the neoplasm, and arterio-
grams involved intracardiac puncture. Normal and
neoplastic tissues were observed by electron microscopy. Permeability studies were carried out by injecting a 2.5% trypan blue solution (iv, 25 mg/100 g)
into the dermis and into the neoplasm. Histamine
(0.01 mg at each injection site) was then administered, and the amount of exuded trypan blue was estimated after 30 min. The vasculature of the neoplasm
consisted of irregular channels lined by plump endothelium that displayed mainly pentalaminar junctions,
although a few heptalaminar junctions were also
found. Few pericytes surround the endothelial layer,
while the basement-membrane varied in thickness and
was often duplicated and triplicated. Challenge with
histamine resulted in increased permeability in comparison with normal connective tissues similarly
treated. The increased permeability was accompanied
by the formation of interendothelial gaps in these
irregular vascular channels. Carrageenan (0.2 ml of
1% solution, sc) induced a leukocytic exudate in the
neoplasm which, apart from an increase in the mononuclear concentration two weeks after injection,
varied little from normal connective tissues. It is
concluded that the majority of vascular channels in
this neoplasm bear some structural and functional
relationship to small venules.

2770   INDUCTION OF METASTASIZING CARCINOMA IN
        RATS AND THEIR BIOLOGICAL CHARACTERISTICS.
(Eng.) Harada, Y. (Shionogi Res. Lab., 2-47 Sagisu-
Kami, Fukushima-ku, Osaka, Japan). *Acta Pathol.*
Jpn. 25(4):451-461; 1975.

Induction of a spontaneously metastasizing carcinoma
in rats was attempted. Four-wk-old Sprague-Dawley
female rats were thymectomized and/or splenectomized
and fed 200 mg (20 mg x 10) of 3-methylcholanthrene
from seven wk of age twice weekly for five wk. The
early-appearing tumors were excised in order to select by isoimmunity the late-appearing ones that
were less antigenic. The latter were easily transplanted into normal syngeneic female rats with metastasis to remote organs. This metastasizing capacity became an inherent character in syngeneic
normal rats from generation to generation of transplantation. With one of these tumors (MRMT-1) many
cancer cells were histologically detected in circulating blood three days after tumor transplantation and were arrested in the capillary beds of
lungs. The spontaneous metastasis to lymph nodes
and lungs was macroscopically found within several
weeks after tumor transplantation. It is concluded
that the MRMT-1 carcinoma is useful as a model for
study of the mechanism of cancer metastasis, and
might be applied as a screening tool for anticancer
drugs that affect the initial phase of metastasis
after excision of the primary tumor.

2771   DISTRIBUTION OF ELECTROPHORETIC MOBILITIES
        OF MOUSE THYMOCYTE SUBPOPULATIONS IN THE
PRESENCE OF TUMOUR CELLS. (Eng.) Jenkins, R.
(Bristol Royal Infirmary, Horfield Road, Bristol
BS2 8ED England). *Immunology* 29(5):893-902; 1975.

Analysis of the electrophoretic mobility of mouse
thymus cells showed two main populations, with mean
mobility values of $0 \cdot 77 \pm 0 \cdot 023$ μm sec^{-1} $V^{-1}cm$ and
$0 \cdot 99 \pm 0 \cdot 015$ μm $sec^{-1}V^{-1}cm$; these absolute values
varied slightly from one strain to another. Implan-
tation of tumor cells caused the relative proportions
of these two populations to change dramatically
within 48 hr, when an increase in the fast-moving
'immunocompetent' thymocytes was observed. The ratio
of slow to fast cells changed from 9:1 in the normal
BALB/c mouse to 2:1 in the presence of the tumor
cells, and this 2:1 ratio persisted throughout the
remainder of the animal's life. However, inoculation
of histocompatible spleen cells from a normal individual evoked only a brief response in the host's
thymus. This change in ratio of slow to fast cells
in the thymus is interpreted as an increased production of immunocompetent cells in response to the
presence of the tumor cells.

2772   STUDIES OF A CELL LINE DERIVED FROM A HU-
        MAN MALIGNANT MELANOMA. (Eng.) Chen, T.
R. (Graduate Sch. Biomedical Sciences, Univ. Texas
at Houston, Houston, Tex. 77025); Shaw, M. W. *In*
Vitro 10(3/4):216-224; 1974.

Employing an *in vitro* method and new staining techniques, changes in karyotype of a heteroploid tumor

cell line derived from a pulmonary metastasis of a
malignant melanoma were studied over ten months of
continuous cultivation. A tissue specimen was obtained from a 69-yr-old man who had been diagnosed
as having a metastatic melanoma of the lung. The
following chromosome staining procedures were used
in most instances: (a) acid-saline-Giemsa G-banding
(3-hr incubation in a saline solution containing 0.
M NaCl and 0.06 M sodium citrate adjusted to pH 7);
(b) trypsin G-banding; (c) Q-banding; and (d) C-
banding using the Q--C procedure. The modal karyot
of tumor cells in a 3-wk-old culture was composed c
41 intact and 11 marker chromosomes. These markers
contained three paired (disomic) and five other
(monosomic) elements. The derivative cell populations in 5- to 10-mo-old cultures showed changes
merely in the paired markers from disomy to monosom
by loss of one element or by formation of a new
marker chromosome. Prolonged treatment in a define
medium completely devoid of serum resulted in a
selection of heteroploid cells and suppression of
diploid fibroblasts. The heteroploid cultures were
further enriched by collecting freely suspended cel
Growth of tumor and normal cells in mixed culture
had no apparent influence on either cell type.

2773   A HUMAN/MOUSE HYBRID MODEL FOR THE STUDY
        OF HUMAN GENETIC FACTORS INFLUENCING TUMC
CELL GROWTH. (Eng.) Koshman, R. W. (Dept. of Surgery and Surgical-Medical Res. Inst., Univ. of Alberta, Edmonton, Alberta, Canada); Koo, J.; Thursto
O. G. *J. Surg. Oncol.* 7(4):323-327; 1975.

Somatic cell hybridization (fusion) was carried out
between cells of the murine lymphoma L-5178Y(r) and
a human/mouse hybrid cell model. The L-5178Y(r) 1s
is deficient in hypoxanthine-guanine phosphorybosyl
transferase (HGPRT). Using a selective medium syst
it was possible to isolate hybrid cell clones havir
the complete complement of murine chromosomes as we
as the human X chromosome on which the gene for HGF
is located. This model system may be used to study
the effect of the human X chromosome on the phenoty
of the murine lymphoma. It is possible that the ef
fect of X-linked genes on behavior of the tumor *in*
vivo can be studied.

2774   HIGHLY MALIGNANT CELLS WITH NORMAL
        KARYOTYPE IN G-BANDING. (Eng.)
Mitelman, F. (University Hosp., S-221 85 Lund,
Sweden); Levan, G.; Brandt, L. *Hereditas* 80(2):
291-293; 1975.

The Giemsa banding pattern of diploid tumor cells
was studied during the transformation of Ph^1-
negative chronic myeloid leukemia (CML) into acute
leukemia, and during the progression of Rous sarcoma virus (RSV)-induced sarcomas in rats. Initial
chromosome analysis of a bone marrow preparation
from a 72-yr-old man with CML showed that 100% of
the metaphases had a completely normal banding
pattern. All 50 cells counted after blastic
transformation had the normal chromosome 2n = 46
and the 21 cells karyotyped in detail all displayed
a normal Giemsa banding karyotype. Five sarcomas

induced by RSV in inbred W/Fu rats were transplanted to syngeneic rats; each tumor had a normal diploid stem line. A total of 42 cells from the five transplanted tumors had the normal chromosome number 2n = 42 and a completely normal banding karyotype. These results confirm previous findings indicating that neither the transformation of a premalignant to a malignant condition nor the progression of advanced tumors is necessarily associated with visible chromosomal alterations.

2775   A NEW HEMATOLOGIC SYNDROME WITH A DIS-
      TINCT KARYOTYPE: THE 5q - CHROMOSOME.
(Eng.) Sokal, G. (Hematology -- U.C.L., Ave.
Chapelle aux Champs 4, 1200 Brussels, Belgium);
Michaux, J. L.; Van Den Berghe, H.; Cordier, A.;
Rodhain, J.; Ferrant, A.; Moriau, M.; De Bruyere,
M.; Sonnet, J. *Blood* 46(4):519-533; 1975.

Cytogenetic findings, clinical picture, and hematologic data obtained from five patients with refractory anemia and the same abnormal bone marrow karyotype, are described. The patients (four women and one man, age 32-80 yr, all Caucasian, had a partial deletion of the long arm of the No. 5 chromosome. The hematologic syndrome was practically the same in these five cases. Examination of the blood by light and electron microscopy revealed a moderate to severe, generally macrocytic anemia with slight leukopenia but normal or elevated platelet count. The bone marrow showed a depressed erythroid series and some abnormalities of the granulocytic series with an occasional excess of myeloblasts. Most of the megakaryocytes had a nonlobulated nucleus. Of the detailed hemostatic function tests carried out in four patients, the main abnormalities were low values for adhesiveness of platelets to glass beads, as well as insufficient release and poor aggregation in the presence of adrenalin (2 μg/ml) and collagen (40 μg/ml). No consistent pattern of immunoglobulin deficiency or quantitative immunoglobulin abnormality was observed. The relationship of this newly established syndrome to other hematologic diseases is discussed. The syndrome constitutes another example of the association between a specific abnormal chromosome and a distinct hematologic disorder.

2776   GENE EXPRESSION OF FOREIGN METAPHASE
       CHROMOSOMES INTRODUCED INTO CULTURED
MAMMALIAN CELLS? (Eng.) Sekiguchi, T. (Div.
Radiobiology, Natl. Cancer Center Res. Inst.,
Tukiji 5-Chome, Chu-O-Ku, Tokyo 104, Japan); Sekiguchi, F.; Tachibana, T.; Yamada, T.; Yoshida, M.
Exp. Cell Res. 94(2):327-338; 1975.

Transfer of genetic information from isolated hamster chromosomes to mouse cells is described. Metaphase chromosomes isolated from Chinese hamster diploid cells were incubated with mouse C1.1-d cells deficient in thymidine kinase activity. Two viable colonies appeared from the treated mouse cells after culturing in selective medium (HAT: 1×10^{-4} M hypoxanthine, 4×10^{-7} M amethopterine, 1.6×10^{-5} M thymidine, and 1×10^{-4} M glycine) with a frequency of about 10^{-8}. The first colony isolated (C1.1) failed to grow. The second colony

isolated (C1.2) grew well in HAT medium and was subcultured for more than 70 generations. C1.2 cell possessed an elevated tetrahydrofolate dehydrogenase activity of molecular species resembling that of Chinese hamster cells as shown by disc electrophoresis. The cell line also expressed surface antigen(s) specific to hamster species, as shown by mixed hemadsorption test and immune cell electrophoresis. This latter phenotype disappeared after prolonged cultivation (59 generations) of the cells in nonselective medium. The karyotype of C1.2 cells corresponded to that of the mouse species, and was quite different from that of hamster cells. Hamster chromosomes could not be identified in any of the cell clones by detailed analysis by the banding method (Q- and C-band). Not one revertant cell was obtained among 4.2×10^{8} C1.1-d cells in the control. The technique for direct transfer of genetic information from isolated metaphase chromosomes to cells of another species may provide a complementary method for cell hybridization for genetic mapping.

2777   CYTOGENETIC ANALYSIS OF ERYTHROLEUKAEMIA
       IN TWO CHILDREN: EVIDENCE OF NONMALIGNANT
NATURE OF ERYTHRON. (Eng.) Inoue, S. (Child Res.
Cent., Detroit, Michigan); Ravindranath, Y.; Zuelzer,
W. W. *Scand. J. Haematol.* 14(2):129-139; 1975.

Sequential cytogenic and cytological studies were conducted in cases of erythroleukemia in two 5-yr-old girls. Bone marrow aspirations were used to determine the differential counts, mitotic index, and classification of mitotic figures, and for chromosome analysis. In the first case the bone marrow and blood were studied on eight occasions over a period of nine months. Initial bone marrow showed no aneuplody, except occasional tetraploid cells. Four months later the marrow consisted almost entirely of pathologic blasts, and 45 of 55 cells showed the karyotype 47, +C. This aneuploid cell line persisted as a dominant clone in all subsequent specimens. In the second case the first bone marrow test showed five of 103 cells to be G trisomic; over the next three months the percentage of such cells increased to 50%. In both cases a correlation was observed between the percentage of aneuploid cells and blasts in mitosis; an inverse correlation between peripheral normoblast count and hemoglobin level was also observed, indicating that erythropoiesis was under physiological control. The authors conclude that in the case of erythroleukemia, erythroid cells do not have inherent malignant properties. Cytogenic evidence, although indirect, adds another parameter to the differentiation of this type of erythroleukaemia from other forms of DiGuglielmo's syndrome.

2778   MEGALOBLASTIC CHANGES AND CHROMOSOME AB-
       NORMALITIES OF ERYTHROPOIETIC CELLS IN
ACUTE MYELOID LEUKAEMIA. (Eng.) Brandt, L. (University Hosp., S-221 85, Lund, Sweden); Mitelman, F.;
Sjogren, U. *Acta Haematol. (Basel)* 54(5):280-283;
1975.

Using a Giemsa banding technique, bone marrow chromosomes were studied in nine patients with acute myeloid leukemia (AML). The age-range of the patients

(early polychromatic normoblasts) mainly in the G_1-phase. They were therefore largely comparable to the megaloblastoid erythroblasts in erythroleukemia. Erythroblasts in preleukemia with nuclear abnormalities occurred in a high percentage in the G_2-phase, or were unlabeled with a DNA content of between diploid and tetraploid value. They showed a similar proliferative behavior to megaloblasts in pernicious anemia. Early polychromatic erythroblasts arrested in G_2-phase could differentiate without mitosis into tetraploid mature erythroblasts (E_5). They could divide elsewhere endomitotically, produce binucleated E_5, or take up the DNA synthesis and become polyploid. E_4 with nuclear abnormalities did not proliferate and were mainly found in the premitotic phase.

2781   NODULAR LYMPHOMA: BONE MARROW AND BLOOD
        MANIFESTATIONS. (Eng.) McKenna, R. W.
(Box 198 Mayo, Univ. Minnesota Hosp., Minneapolis,
Minn. 55455); Bloomfield, C. D.; Brunning, R. D.
Cancer 36(2):428-440; 1975.

Morphological and clinical features of 39 patients with nodular lymphoma were studied with particular reference to bone marrow and blood involvement. To detect bone marrow involvement, trephine biopsies followed by aspiration biopsies were performed on the posterior superior iliac spines of all patients. The degree of blood involvement was graded on the basis of the number of lymphoma cells and the degree of WBC count elevation. Twenty-one patients (54%) had bone marrow involvement at the time of initial diagnosis. Thirteen (33%) also had some degree of peripheral blood involvement. A specific cell type, the small nodular lymphoma cell, was observed in the bone marrow smears of 19 of the 21 patients with bone marrow involvement and in the peripheral blood smears of all patients with blood involvement. When large cells predominated in the lymph node or bone marrow sections, the bone marrow smears and the imprints of the lymph node and trephine biopsy demonstrated these large cells to have morphological qualities of lymphocytes rather than histiocytes. This result may indicate that nodular histiocytic and nodular lymphocytic-histiocytic lymphomas are much less common than is generally accepted.

2782   CELL RECEPTOR STUDIES ON SEVEN CASES OF
        DIFFUSE HISTIOCYTIC MALIGNANT LYMPHOMA
(RETICULUM CELL SARCOMA). (Eng.) Habeshaw, J. A.
(Univ. Med. Sch., Edinburgh, Scotland); Stuart, A.
E. *J. Clin. Pathol.* 28(4):289-297; 1975.

A study was made of the cell surface receptors in seven cases of reticulum cell sarcoma in order to determine B or T lymphoid or histiocytic origin. Electron microscopy of five of the specimens confirmed the diagnosis, confirmed common features, and showed differences in the degree of cytoplasmic differentiation. The cytoplasm generally contained large numbers of ribosomes with variable amounts of rough-surfaced endoplasmic reticulum. All cases showed abundant mitochondria and prominent nucleoli. The neoplastic cells were not conspicuously fiber-forming. Ingestion of dead cells by macrophages was

noted, but there was no indication of overt phagocytosis. Phagocytic activity was assessed functionally by the ability to ingest neutral red. A "sand-
wich" technique was used to detect immunoglobulin-
bearing cells, and their morphology was assessed by
dark ground and phase contrast microscopy. Rosette
tests were conducted using washed SRBC, SRBC sensitized with IgG antibody (SRBCAIgG), and SRBC with
IgM antibody and complement (SRBCAC). Wide variability in receptor patterns was found for both the
cases studied and for control studies: SRBC rosettes
(T cells) 15-72%; SRBCAIgG rosettes 6-43%; neutral
red cells 1-60%; SRBCAC receptors 7-68%; and immunofluorescent cells (B cells) 6-39%. Of the seven
cases, one case indicated a histiocytic origin. The
cells were phagocytically active and expressed receptors for SRBCAIgG in culture, but lacked those for
SRBC and SRBCAC. In two cases cells expressed both
T and B lymph markers, indicating that either the
neoplastic population is derived from a subclass of
lymphoid cells bearing both markers or that the lymphocytes may acquire another marker during neoplastic
transformation. These cases, and two showing a normal pattern of expression, have a lymphoid origin.
Two additional cases appear to be of uncertain pathogenesis, having no detectable lymphocyte receptors
and showing no evidence of phagocytic function.
These results appear to confirm the view of both
histiocytic and lymphoid origins for reticulum cell
sarcoma.

2783   VIRUS-LIKE PARTICLES IN CYSTIC MAMMARY
            ADENOMA OF A SNOW LEOPARD. (Eng.)
Chandra, S. (Chicago Zoological Park, Brookfield,
Ill. 60513); Laughlin*, D. C. Cancer Res.
35(11/Part 1):3069-3074; 1975.

A new type of intracisternal virus-like particle
was observed in the giant cells of a cystic mammary adenoma of a snow leopard in captivity.
Masses noted in each of the four mammary glands
were surgically removed and a few pieces of tumor
tissues placed in formalin. Various areas of the
tumor were dissected and placed in 2.5% glutaraldehyde. The tissues were sliced in 1-mm cubes,
postfixed in chrome-osmium, dehydrated in ethanol,
and embedded in Epon-Araldite. Thin sections were
stained with uranyl acetate and lead citrate for
study with the electron microscope. Of the various
cell types seen in the cystic mammary adenoma,
virus-like particles were seen only in the giant
cells. These cells contained many relatively small
nuclei that were always quite far apart. The
nuclear surface was indented and the chromatin matter was marginated. The cytoplasm contained two
unusual structures, one consisting of lipoid
material that appeared to be disintegrating, and
the other consisting of membranous elements condensed into a large mass and enclosed by a single
membrane. Elements of rough endoplasmic reticulum
and mitochondria were scant. Virus-like particles,
115-125 nm in diameter, budded from the lamella of
endoplasmic reticulum and were studded on their
inner surface with dense granules (12 nm) giving
them their unique ultrastructural morphology. More
than two budding particles were never seen within

a cisterna. Such particles were neither seen
extracellularly nor in any other cell type. Type B
and Type C particles were not seen in the tumor
tissue. The fact that the intracisternal particles
in the snow leopard mammary tumor have a morphological similarity with other virus-like particles
during the budding process indicates that they are
not secretory in nature. Because particles of such
an ultrastructural morphology have not been describ
in any other species, they might be considered
characteristic to the snow leopard.

2784   PROGNOSIS OF THE SECOND BREAST CANCER.
            THE ROLE OF PREVIOUS EXPOSURE TO THE
FIRST PRIMARY. (Eng.) Khafagy, M. M. (Mem. Hosp.
Cancer Allied Dis., New York, N.Y.); Schottenfeld,
D.; Robbins, G. F. Cancer 35(3):596-599; 1975.

The prognosis of multiple metachronous tumors after
successful treatment of the first primary tumor was
investigated. Eighty-two patients with asynchronous
bilateral breast cancer were evaluated; the median
age for the first and second cancer was 46 and 57,
respectively. The majority of the first and second
breast cancers (70.0% and 74.4% respectively) were
infiltrating duct carcinomas; less frequently, they
were lobular, medullary, colloid, and papillary
types. The first and second mammary cancers were
of the same histologic type in 74.4%. Axillary
lymph node metastases were detected in 34.1% of
the patients with first breast cancer and in 40.2%
with the second breast cancer; no significant difference in the incidence of axillary lymph node
metastases from the second breast cancer was found
in relation to the presence or absence of positive
axillary lymph nodes from the first cancer. No
significant difference was found when the 5-yr survival rate of 79 patients with potentially curable
second breast cancer was compared with the total
number of mastectomy patients. Thus, no difference
in the course of the second cancer as compared with
the first has been found.

2785   SEX HORMONES IN BREAST DISEASE. (Eng.)
            England, P. C. (University Hosp. South
Manchester, Manchester, England); Skinner, L. G.;
Cottrell, K. M.; Sellwood, R. A. Br. J. Surg.
62(10):806-809; 1975.

Radioimmunoassay was used to measure serum concentrations of estradiol 17β and progesterone daily
throughout one menstrual cycle in 32 normal women
(aged 20-49 yr), 32 women with benign disease of
the breast (aged 20-49 yr) and ten women with
breast cancer (aged 40-49 yr). In normal women,
concentrations varied with age; women in the
fourth decade had mean estradiol concentrations
significantly higher during the luteal phase than
did women in the fifth and third decades, and the
mean concentration of progesterone in women in the
fifth decade was significantly lower than that in
those in the third or fourth decades. Women with
cysts had concentrations of estradiol 17β which
were significantly higher than normal (P < 0.01)
in both the follicular and luteal phases. In
women with cancer, the concentrations of estradiol

a patient with typical middle lobe syndrome suffering
from chronic glomerulonephritis. Tumorlets were discovered in the subpleural and peribronchiolar fibrous
areas. Areas containing tumorlets were comparatively
whitish with less anthracosis than those in other
inflammatory scars. Six out of eight foci were
found in the right middle lobe, one in the right
upper lobe, and one in the left lower lobe. The
tumorlets consisted of small clusters of epithelial
cells, which were uniform and round or spindle-
shaped and had hyperchromatic nuclei with very few
mitoses surrounded by fibrous tissue. Proliferating
cells of five foci from four tumorlets and the allied
lesion showed argyrophilia but none showed argentaf-
finity. Three foci from two cases were examined by
electron microscopy. Electron dense granules
1,200 A - 3,000 A in diameter, were found scattered
in the cytoplasm. These granules were similar to
those of Kultschitzky-like cells and also to those
of carcinoid tumor and oat-cell carcinoma. It is
suggested that the tumorlet develops more closely in
undergrowth areas of the lungs than in chronic
inflammatory areas, and that this type of epithelial
proliferation may have a hamartomatous character.

2788   THE FINE STRUCTURE OF SO-CALLED MINUTE
        PULMONARY CHEMODECTOMAS. (Eng.) Kuhn,
C., III. (Washington Univ. Sch. Medicine, St. Louis,
Mo. 63110); Askin, F. B. *Hum. Pathol.* 6(6):681-
691; 1975.

Electron microscopy was performed on several minute
tumors of the type called chemodectomas, all from
the lung of a 69-yr-old Negro woman. The cells
had a whorling pattern with extensive interdigitating cytoplasmic processes joined by desmosomes.
Except for tangles of cytoplasmic fibrils, the
tumor cells had few distinctive organelles. They
had no endocrine-like granules and were not associated with nerves or basement membranes. The tumors
had little resemblance to paragangliomas, but
displayed a puzzling similarity to meningiomas.
No definite conclusions were made as to the histogenesis of these lung tumors. Viewed in the light
of recent physiologic studies, they cast doubt on
the presence of special chemoreceptive paraganglia
in the lung.

2789   SMALL CELL EPIDERMOID CARCINOMA OF THE ESO-
        PHAGUS: AN OAT-CELL-LIKE CARCINOMA. (Eng.)
Rosen, Y. (State Univ. New York Downstate Medical
Center, 450 Clarkson Ave., Brooklyn, N.Y. 11203);
Moon, S.; Kim, B. *Cancer* 36(3):1042-1049; 1975.

An esophageal tumor in a 67-yr-old Puerto Rican woman is described. The tumor was found on autopsy
to consist mostly of sheets of small pleomorphic
anaplastic cells with markedly hyperchromatic nuclei and little cytoplasm. There were foci of squamous differentiation in the tumor and epidermoid
carcinoma adjacent to it. This tumor is considered
to be a small cell anaplastic variant of epidermoid
carcinoma. Metastases found in the liver, bones,
and lymph nodes exhibited only the anaplastic small
cell appearance. Electron microscopy showed advanced

postmortem autolysis. Demonstration of intracytoplasmic neurosecretory granules is necessary for the
certain identification of oat cell carcinoma. Four
similar cases in the literature could not be diagnosed with certainty. The occurrence of true oat
cell carcinoma of the esophagus is still in question.

2790   MUCOCELES OF THE APPENDIX: THEIR RELATION-
        SHIP TO HYPERPLASTIC POLYPS, MUCINOUS
CYSTADENOMAS, AND CYSTADENOCARCINOMAS. (Eng.) Qizil-
bash, A. H. (Henderson General Hosp., 711 Concession
St., Hamilton, Ontario, Canada L8V 1C3). *Arch.
Pathol.* 99(10):548-555; 1975.

Sixty-four cases of hyperplastic polyp, mucinous
cystadenoma, and cystadenocarcinoma of the appendix
were studied in relation to the development of mucocele and "pseudomyxoma peritonei." Thirty-three
cases were examples of hyperplastic polyps. In 11,
the appendix was transformed into a mucocele; eight
were associated with mucinous cystadenoma and one
with mucinous cystadenocarcinoma. The hyperplastic
polyp alone was the cause of mucocele formation in
two. Thirty-five cases represented examples of mucinous cystadenoma; 32 resulted in mucocele formation.
Rupture with localized pseudomyxoma peritonei was
found in six; generalized pseudomyxoma peritonei was
encountered only once. In four of the five cases of
mucinous cystadenocarcinoma, the appendix was grossly
transformed into mucoceles. The histological features
of mucinous cystadenoma are identical to villous adenoma of the large bowel and probably represent its
counterpart within the appendix.

2791   A REAPPRAISAL OF STAGING AND THERAPY FOR
        PATIENTS WITH CANCER OF THE RECTUM.
I. DEVELOPMENT OF TWO NEW SYSTEMS OF STAGING.
(Eng.) Feinstein, A. R. (333 Cedar St., New Haven,
Conn. 06516); Schimpff, C. R.; Hull, E. W. *Arch.
Intern. Med.* 135(11):1441-1453; 1975.

A taxonomy was prepared for unclassified medical
data. Two new systems of staging in a cohort of
318 patients were developed and tested. The first
system, which can be applied before treatment, is
divided into four composite stages that contain
elements of symptomatic, chronometric, co-morbid,
and para-morbid data, as well as information obtained from physical examination, sigmoidoscopy,
and roentgenography. The second system, applicable
to patients with resected tumors, is based on a
combination of pretherapeutic clinical information
and postsurgical anatomic evidence. The two systems
produce prognostic gradients that are clinically
distinctive and statistically efficacious.

2792   FUNCTIONAL AND STRUCTURAL ALTERATIONS OF
        LIVER ERGASTOPLASMIC MEMBRANES DURING
DL-ETHIONINE HEPATOCARCINOGENESIS. (Eng.) Gravela,
E. (Istituto di Patologia generale, Universita di
Torino, Corso Raffaello 30, 10125 Torino, Italy);
Feo, F.; Canuto, R. A.; Garcea, R.; Gabriel, L.
Cancer Res. 35(11/Part 1):3041-3047; 1975.

Different functional and structural properties of

tissue was serially sectioned and examined histologically. Of the seven patients, five had silent
but extensive intraductal prostatic involvement. In
three of these, the carcinoma *in situ* was associated
with microinvasion. The mean age at the time of
diagnosis was 68 yr. All had symptoms characteristic of carcinoma *in situ*, including hematuria, dysuria, and urgency. In three patients the prostatic
involvement was diagnosed on transurethral resection.
In two patients it was discovered only after radical
cystectomy. The prostatic involvement was neither
suspected clinically nor has it been previously emphasized. Although three patients were alive, apparently free of disease up to 15 mo postcystectomy,
two had died, one of metastatic disease. The remaining patient was alive, but had clinical symptoms indicative of persistent disease. The importance of
prostatic assessment in the evaluation of the patient with carcinoma in situ of the urinary bladder
is emphasized.

2796   ULTRASTRUCTURE OF OVARIAN TERATOMAS IN
          LT MICE. (Eng.) Damjanov, I. (Dept.
Pathology, UCONN Health Center, Farmington, Conn.
06032); Katic, V.; Stevens, L. C. Z. *Krebsforsch.*
83(4):261-267; 1975.

The ultrastructure of ovarian teratomas in LT mice
was investigated with emphasis on tumor stem cells.
These cells formed nests consisting of undifferentiated embryonic cells in the center with more
differentiated cells toward the periphery. The
stem cells contained intracisternal A-type viral
particles. Ultrastructurally the stem cells of
the ovarian teratomas did not differ from stem
cells of testicular or embryo-derived teratomas.
The cells were, however, distinct from the cleavage
stage embryonic cells and/or the unfertilized
ovum from which they arose. Both cytologically
and developmentally, the stem cells corresponded
to ectodermal embryonic cells from the egg-cylinder,
the most advanced development stage of parthenotes
previously observed in the ovary of LT mice.
The finding of A-type particles in the stem cells
indicated that they had passed the three to four
cell embryo stage of development.

2797   OCCURRENCE OF OTHER ENDOCRINE TUMOURS IN
          PRIMARY HYPERPARATHYROIDISM. (Eng.)
Taylor, S. (Royal Postgraduate Medical Sch. Hammer-
smith Hosp., London W12 OHS, England); Boey, J. H.;
Cooke, T. J. C.; Gilbert, J. M.; Sweeney, E. C.
Lancet 2(7939):781-784; 1975.

A series of 119 patients who underwent parathyroid-
ectomy between 1955 and 1975 were followed up in
order to determine the occurrence of multiple endocrine adenomatosis (MEA). The patients were deemed
to show evidence of an associated endocrinopathy
on the basis of the following criteria: clinically
overt disease (e.g., acromegaly), radiological enlargement of the pituitary, biochemical findings of
raised plasma-hormone concentration, and postmortem discovery of hyperplasia or adenoma of an
endocrine gland. Twenty-one patients (17.5%) were

found to have evidence of associated endocrine disease and were deemed to have MEA. The clinical pattern of hypercalcemia in no way distinguished these patients from other hyperparathyroid patients. MEA was most commonly found in patients with several diseased parathyroid glands. The results suggest that MEA is more common in hyperparathyroid patients than earlier reports have indicated.

2798   PITUITARY TUMOR WITH PRIMARY HYPOTHYROID-
         ISM; POSSIBLE ETIOLOGIC RELATIONSHIP.
(Eng.) Balsam, A. (Montefiore Hosp. and Medical Center, 111 East 210th St., Bronx, N.Y. 10467); Oppenheimer*, J. H. *N.Y. State J. Med.* 75(10):1737-1741; 1975.

The case of a 59-yr-old woman with a pituitary tumor, chiasmatic compression, and presumed secondary hypo-thyroidism was reevaluated. Lack of augmentation of the 24-hr radioiodine uptake following the adminis-tration of exogenous thyroid stimulating hormone (TSH) had previously been attributed to thyroidal atrophy due to prolonged absence of thyrotrophic stimulation. However, the new finding of elevated plasma levels of endogenous TSH (25 μU/ml) which were suppressed by thyroid hormone replacement 25 gr./day, established the diagnosis of primary hypothyroidism. These findings prompted a review of sella x-ray films of 40 cases of primary hypothyroidism; four patients had enlargement of the sella turcica. The develop-ment of reactive pituitary enlargement in primary hypothyroidism may be more frequent than was pre-viously suspected.

2799   SMALL CELL MALIGNANT TUMORS OF THE THYROID:
         A LIGHT AND ELECTRON MICROSCOPIC STUDY.
(Eng.) Cameron, R. G. (Pathological Inst., 3775 Uni-versity St., Montreal, Province of Quebec, Canada); Seemayer*, T. A.; Wang, N.-S.; Ahmed, M. N.; Tabah, E. J. *Hum. Pathol.* 6(6):731-740; 1975.

Observations obtained with the light and electron microscopes on three small cell malignant tumors in-volving the thyroid gland are presented. The case reports of the patients; an 86-yr-old woman, a 61-yr-old man, and a 57-yr-old woman; are also presented. In each tumor numerous similarities were present on light microscopic analysis rendering interpretation difficult. In the 86-yr-old woman the small cells were identified by electron microscopy as moderately well differentiated lymphocytes. This tumor occurred as a locally invasive thyroid tumor, subsequently involving distant sites, including the liver, spleen, lymph nodes, and soft tissue. The tumor repeatedly regressed following radiotherapy. The patient even-tually died with disseminated lymphocytic lymphoma. In the man, the thyroid tumor also locally invasive was composed principally of neoplastic epithelial cells, which were identified with the electron micro-scope. This patient responded poorly to radiation and died within a year after diagnosis. The third patient presented with an enlarging thyroid mass, which ultrastructurally was found to be composed principally of well differentiated lymphocytes. Sub-sequent clinical evaluation established a diagnosis

of chronic lymphocytic leukemia. Small cell maligna tumors of the thyroid represent a difficult diagnoa-tic problem for surgical pathologists. Ultrastruc-tural study is a useful adjunct in the differentia-tion of these tumors.

2800   VITILIGO AND MALIGNANT MELANOMA.
         (Eng.) Fodor, J. (Natl. Cancer Inst., Budapest, Hungary); Bodrogi, I. *Neoplasma* 22(4):445-448; 1975.

In 4 of 6 cases of vitiligo and malignant melanoma, the latter was detected a few years after onset of vitiligo (ages 44-73 yr). The authors conclude that old-age vitiligo may be a skin marker of malignant neoplasm.

2801   MULTIFOCAL SCLEROSING OSTEOID OSTEOMA
         (ONE CASE). (Rus.) Selivanov, V. P.
(Novokuznetsk Municipal Clinical Hosp. No. 1, U.S.S.R.); Fetinin, V. A. *Vopr. Onkol.* 21(9):94-95; 1975.

2802   METASTASES TO THE BONES OF THE HAND.
         (Eng.) Kumar, P. P. (Howard Univ. Coll. Medicine, Washington, D.C.). *J. Natl. Med. Assoc.* 67(4):275-276; 1975.

2803   BONE LESIONS OF MALIGNANT LYMPHOMA. (Jpn.)
         Watanabe, T. (Dept. of Radiology, Shin-shu Univ., Asahi 3-1-1, Matsumoto, Japan). *Nippon Acta Radiol.* 35(3):111-118; 1975.

2804   RHABDOMYOSARCOMA OF THE BRAIN: CASE RE-
         PORT. (Eng.) Matsukado, Y. (Kumamoto Univ. Medical Sch., Kumamoto, Japan 860); Yokota, A.; Marubayashi, T. *J. Neurosurg.* 43(2):215-221; 1975.

2805   MEDULLOBLASTOMA IN TWO BROTHERS. (Eng.)
         Yamashita, J. (Kyoto Univ. Medical Sch., Sakyoku, Kyoto, Japan 606); Handa, H.; Toyama, M. *Surg. Neurol.* 4(2):225-227; 1975.

2806   "MALIGNANT LYMPHOMA" OF THE BRAIN FOLLOW-
         ING RENAL TRANSPLANTATION. (Eng.) Ker-sting, G. (Institut fur Neuropathologie d. Univ., Annaberger Weg, 5800 Bonn-Venusberg, West Germany); Neumann*, J. *Acta Neuropathol. [Suppl.] (Berl.)* 6:131-133; 1975.

2807   CYTOMEGALOVIRUS AND LYMPHOMA IN A PEDIA-
         TRIC TRANSPLANT RECIPIENT [letter to editor]. (Eng.) Matas, A. J. (Mayo Memorial Bldg., Univ. Minnesota, Minneapolis, Minn. 55455); Sim-mons, R. L.; Kersey, J. H.; Kjellstrand, C. M.; Najarian, J. S. *J. Pediatr.* 87(3):494-495; 1975.

2808   PRIMARY LYMPHOMAS OF THE CENTRAL NERVOUS

2817   MICROGLIOMA AND/OR RETICULOSARCOMA OF
        THE NERVOUS SYSTEM. (Eng.) Polak, M.
(Registro Latino Americano de tumores del Systema
Nervioso, Terrada 1164 Buenos Aires, Argentina).
Acta Neuropathol. [Suppl.] (Berl.) 6:115-118; 1975.

2818   THE CLASSIFICATION OF MICROGLIOMATOSIS
        WITH PARTICULAR REFERENCE TO DIFFUSE
MICROGLIOMATOSIS. (Eng.) Adams, J. H. (Inst.
Neurological Sciences, Southern General Hosp.,
Glasgow G 51 4 TF, Scotland). *Acta Neuropathol.
[Suppl.] (Berl.)* 6:119-123; 1975.

2819   PRIMARY CEREBRAL RETICULOSIS AND PLASMA
        CELL DIFFERENTIATION. (Eng.) Vuia, O.
(Inst. Neuropathology, Justus Liebig Univ., Arndt-
strasse 16, D-63 Giessen, West Germany). *Acta
Neuropathol. [Suppl.] (Berl.)* 6:161-166; 1975.

2820   PATTERNS OF PROLIFERATION IN CEREBRAL
        LYMPHORETICULAR TUMOURS. (Eng.) Bar-
nard, R. O. (Maida Vale Hosp., London, W. 9 ITL,
England); Scott, T. *Acta Neuropathol. [Suppl.]
(Berl.)* 6:125-130; 1975.

2821   DIFFUSE RETICULOSIS WITH LEUKOMALACIA.
        (Eng.) Liss, L. (Neuropathology Div.,
Upham Hall, Ohio State Univ., Columbus, Ohio
43210); Gogate, S. A. *Acta Neuropathol. [Suppl.]
(Berl.)* 6:257-260; 1975.

2822   PRIMARY INTRACRANIAL OLFACTORY ESTHESIO-
        NEUROEPITHELIOMA. (Jpn.) Kinoshita, K.
(Kumamoto Univ. Medical Sch., Kumamoto, Japan);
Marubayashi, T.; Tokuda, H.; Matsukado, Y. *No To
Shinkei* 27(1):39-47; 1975.

2823   BREAST CANCER. (Ita.) Bianucci, P. (Ente
        Ospedaliero San Paolo Di Savona, Piazza
Liguria, 3/3 17012 Albissola Mare, Savona, Italy).
Cancro 27(6):345-351; 1974.

2824   A SARCOMATOUS METASTASIS OF A CARCINOMA
        OF THE BREAST. (Ger.) Korb, G. (Stad-
tisches Krankenhaus, Pathologisches Institut, D-8480
Weiden i.d. Oberpfalz, Bismarckstrasse 30, West
Germany); Weiss, R. *Z. Krebsforsch.* 83(3):251-253;
1975.

2825   ELECTRONMICROSCOPIC STUDIES ON SOLID
        MAMMARY CARCINOMAS IN BITCHES. (Ger.)
von Bomhard, D. (Institut fur Allgemeine Pathologie
und Pathologische Anatomie der Tierarztlichen
Fakultat der Universitat, D-8000 Munchen 22,
Veterinarstrasse 13, West Germany); Raddatz, R.
Raddatz, R. *Z. Krebsforsch.* 83(2):129-143; 1975.

2826   A CASE OF 'CROHN'S CARCINOMA'. (Eng.)
        Fleming, K. A. (Univ. Dep. Pathol., Glas-

gow R. Infirm., Scotland); Pollock, A. C. *Gut* 16 (7):533-537; 1975.

2827   CARCINOMA COMPLICATING CROHN'S DISEASE:
        REPORT OF SEVEN CASES AND REVIEW OF THE
LITERATURE. (Eng.) Lightdale, C. J. (Memorial
Sloan-Kettering Cancer Center, 1275 York Ave., New
York, N.Y. 10021); Sternberg, S. S.; Posner, G.;
Sherlock, P. *Am. J. Med.* 59(2):262-268; 1975.

2828   POLYPOSIS OF THE LARGE BOWEL AND CANCER:
        LONGITUDINAL STUDIES [abstract]. (Eng.)
Burbige, E. J. (Dept. of Med., The Johns Hopkins
Univ., Baltimore, Md.); Cohen, S. B.; Krush, A.
J.; Levin, L. S.; Wennstrom, J.; Murphy, E.; Mil-
ligan, F. D. *Gastroenterology* 68(4/Part 2):869;
1975.

2829   ADENOCARCINOMA OF THE JEJUNUM ASSOCIATED
        WITH NONTROPICAL SPRUE. (Eng.) Petre-
shock, E. P. (Long Island Jewish-Hillside Medical
Center, New Hyde Park, N.Y. 11040); Pessah, M.;
Menachemi, E. *Am. J. Dig. Dis.* 20(8):796-802; 1975.

2830   CANCER OF THE GASTRIC STUMP FOLLOWING
        OPERATIONS FOR BENIGN GASTRIC CANCER OR DUODENAL
ULCERS. (Eng.) Gazzola, L. M. (30 rue Jean Vio-
lette, 1205 Geneva, Switzerland); Saegesser, F.
J. Surg. Oncol. 7(4):293-298; 1975.

2831   EXTRAMAMMARY PAGET'S DISEASE AND ANAPLAS-
        TIC BASALOID SMALL-CELL CARCINOMA OF THE
ANUS: REPORT OF A CASE. (Eng.) Jackson, B. R.
(St. Vincent's Hosp., Los Angeles, Calif.). *Dis.
Colon Rectum* 18(4):339-345; 1975.

2832   DOUBLE AND MULTIPLE CARCINOMAS OF THE
        COLON [abstract]. (Dut.) Heystraten,
F. M. J. (Nijmegen, Netherlands); Schillings, P.
H. M.; Rosenbusch, G. *Ned. Tijdschr. Geneeskd.*
119(37):1440-1441; 1975.

2833   FAMILIAL CONONIC POLYPOSIS. (Pol.)
        Chrominski, J. (20-081 Lublin, ul. Stas-
zica 11, Klinika Chirurgii Dzieciecej Instytutu
Pediatrii AM., Poland); Wnuk-Katynska, U.; Drylska,
M. *Pol. Tyg. Lek.* 30(13):569-570; 1975.

2834   A CASE OF IDIOPATHIC THROMBOCYTOPENIC
        PURPURA ACCOMPANIED BY ESOPHAGEAL CAR-
CINOMA AND COLONIC POLYPS DURING NINE YEARS OF
PREDNISOLONE THERAPY. (Eng.) Ouchi, E. (Sch.
Med., Tohoku Univ., Sendai, Japan); Sato, J.; Wa-
tabe, S.; Seiji, K.; Nomura, N.; Yamagata, S.
Jpn. J. Clin. Oncol. 5(1):59-64; 1975.

2835   COLUMNAR-LINED OESOPHAGUS: SUPPORT FOR
        THE ACQUIRED THEORY IN A PATIENT WITH

Dermatol. 111(6):610-614; 1975.

2855   MULTIPLE PRIMARY CUTANEOUS MELANOMAS.
        (Eng.) Beardmore, G. L. (Queensland
Melanoma Proj., Brisbane, Australia); Davis, N.
C. *Arch. Dermatol.* 111(6):603-609; 1975.

2856   INTRA-EPIDERMAL CARCINOMA OF THE EYELID
        MARGIN. (Eng.) McCallum, D. I. (Raig-
more Hosp., Inverness, Scotland); Kinmont, P. D.
C.; Williams, D. W.; Cotton, R. E.; Wroughton, M.
A. *Br. J. Dermatol.* 93(3):239-252; 1975.

2857   BASAL-CELL CARCINOMA IN OFFICE PRACTICE.
        (Eng.) Biro, L. (7502 Ridge Boulevard,
Brooklyn, N.Y. 11209); Price, E.; MacWilliams, P.
N.Y. State J. Med. 75(9):1427-1433; 1975.

2858   EPITHELIOMAS IN TWO PATIENTS WITH PSORI-
        ASIS TREATED WITH IMMUNE-SUPPRESSIVE DRUG
THERAPY. (Fre.) Rimbaud, P. (No affiliation
given); Meynadier, J.; Guilhou, J. J.; Barneon, G.
Bull. Soc. Fr. Dermatol. Syphiligr. 82(2):215-217;
1975.

2859   MUCINOUS ADENOCARCINOMA OF THE RENAL
        PELVIS: DISCUSSION OF POSSIBLE PATHO-
GENESIS. (Eng.) Liwnicz, B. H. (Lincoln Hosp.
Albert Einstein Coll. Medicine, Bronx, N.Y.);
Lepow, H.; Schutte, H.; Fernandez, R.; Caberwal, D.
J. Urol. 114(2):306-310; 1975.

2860   BILATERAL DIFFUSE LYMPHOSARCOMA OF THE
        KIDNEYS WITH LEUKEMIC TRANSFORMATION.
(Ger.) Hüttig, G. (Universitäts-Kinderklinik 87,
Wurzburg, Josef Schneider-Strasse 2, West Germany);
Gekle, D. *Pad. Praxis* 15(1):55-60; 1975.

2861   ON PATHOGENETIC RELATIONSHIPS BETWEEN
        CARCINOMA OF THE URINARY BLADDER AND
DISEASES OF THE SPINAL CORD. (Ger.) Schnoy, N.
(Pathologisches Institut der FU im Klinikum West-
end, 1 Berlin 19, Spandauer Damm 130, West Germany);
Leistenschneider, W. *Med. Monatsschr.* 29(3):133-
135; 1975.

2862   MALIGNANT TRANSFORMATION OF DIVERTICULA
        OF THE BLADDER. (Ita.) Stigliani, V.
(Cattedra di Urologia, Universita di Modena, Viale
Medaglie d'oro, 22, Modena, Italy). *Cancro* 27(6):
301-304; 1974.

2863   SQUAMOUS CARCINOMA OF BLADDER WITH PSEUDO-
        SARCOMATOUS STROMA. (Eng.) Jao, W.
(Michael Reese Medical Center, 29th St. and Ellis
Ave., Chicago, Ill. 60616); Soto, J. M.; Gould, V.
E. *Arch. Pathol.* 99(9):461-466; 1975.

2864   OVARIAN GIANT CELL TUMOR WITH CYSTADENO-
         CARCINOMA. (Eng.) Veliath, A. J. (Jaw-
aharlal Inst. Postgraduate Medical Education and
Res., Pondicherry-6, India); Sankaran, V.; Aurora,
A. L. *Arch. Pathol.* 99(9):488-491; 1975.

2865   SIX FAMILIES PRONE TO OVARIAN CANCER.
         (Eng.) Fraumeni, J. F., Jr. (Natl. Can-
cer Inst., Bethesda, Md. 20014); Grundy, G. W.;
Creagan, E. T.; Everson, R. B. *Cancer* 36(2):364-
369; 1975.

2866   CLEAR CELL CARCINOMATA OF THE FEMALE GENI-
         TAL TRACT. (Fre.) Van Bogaert, L.-J.
(Laboratoire de Pathologie et Cytologie Tumorale,
Kapucijnenvoer 35, 3000 Louvain, Belgium); Malda-
gue, P. *J. Gynecol. Obstet. Biol. Reprod. (Paris)*
4(1):51-63; 1975.

2867   ADENOMYOMATOSIS OF ENDOMETRIUM AND ENDO-
         CERVIX -- A HAMARTOMA? (Eng.) Silver-
berg, S. G. (Univ. Colorado Sch. Med., Denver).
Am. J. Clin. Pathol. 64(2):192-199; 1975.

2868   DIFFERENT AGGLUTINABILITY OF FIBROBLASTS
         UNDERLYING VARIOUS PRECURSOR LESIONS OF
HUMAN UTERINE CERVICAL CARCINOMA. (Eng.) Chaud-
huri, S. (Temple Univ. Health Sciences Center,
Philadelphia, Pa. 19140); Koprowska, I.; Rowinski,
J. *Cancer Res.* 35(9):2350-2354; 1975.

2869   BENIGN PROSTATIC HYPERPLASIA, PROSTATIC
         CANCER, AND CARCINOGENESIS. (Eng.)
Rotkin, I. D. (Univ. Illinois Coll. Medicine, 835
South Wolcott Ave., Chicageo, Ill. 60612). *Lancet*
2(7930):359-360; 1975.

2870   MALIGNANT GONADAL TUMOUR FORMATION IN
         INTERSEXUAL STATES. (Eng.) Pigott, H.
W. S. (Westminster Hosp., London, England). *Post-
grad. Med. J.* 51(594):252-255; 1975.

2871   INTERSTITIAL CELL TUMOR OF THE TESTIS:
         TISSUE CULTURE AND ULTRASTRUCTURAL STUDIES
[abstract]. (Eng.) Kay, S. (Medical Coll. Virginia,
Richmond, Va. 23298); Fu, Y.-S.; Koontz, W. W.;
Chen, A. T. L. *Lab. Invest.* 32(3):449; 1975.

2872   PITUITARY ADENOMA PRODUCING AMYLOID-LIKE
         SUBSTANCE. (Eng.) Bilbao, J. M. (St.
Michael's Hosp., 30 Bond St., Toronto, Ontario,
Canada M5B 1W8); Horvath, E.; Hudson, A. R.; Kovacs,
K. *Arch. Pathol.* 99(8):411-415; 1975.

2873   FINE STRUCTURE AND ORIGIN OF AMYLOID
         DEPOSITS IN PITUITARY ADENOMA. (Eng.)
Schober, R. (Stanford Univ. Sch. Medicine, Stanford, Calif. 94305); Nelson, D. *Arch. Pathol.*
99(8):403-410; 1975.

2880   FAMILIAL LEUKAEMIA: A STUDY OF 909 FAMI-
        LIES. (Eng.) Gunz, F. W. (Sydney Hosp.,
Sydney, N.S.W. 2000, Australia); Gunz, J. P.; Veale,
A. M. O.; Chapman, C. J.; Houston, I. B. Scand. J.
Haematol. 15(2):117-131; 1975.

A family survey was conducted among 909 patients with
leukemia of all types, with the purpose of establishing the incidence of further cases of leukemia
among relatives. Among 41,807 relatives 8,349 were
deceased, and the cause of death was objectively
confirmed in 5,011. Seventy-two patients had one
or more relatives with leukemia. First degree relatives with leukemia were much more frequent in
families of patients with chronic lymphocytic leukemia than in those of patients with chronic granulocytic leukemia. The incidence of leukemia among
first degree relatives was established to be 2.8-
3.0 times, among more distant relatives about 2.3
times, and overall about 2.5 times that expected.
This excess is of the order of that observed in relatives of patients with solid tumors. Genetic factors may have accounted for much of the excess incidence in chronic lymphocytic and acute leukemia,
but there was little evidence for a genetic background in chronic granulocytic leukemia. With the
possible exception of one family with multiple cases,
a simple Mendelian mechanism did not appear to be
involved in the leukemia families investigated. It
appears more likely that a polygenic mechanism led
to a heightened susceptibility to the disease in
these families.

2881   CANCER BY COUNTY: NEW RESOURCE FOR ETIO-
        LOGIC CLUES. (Eng.) Hoover, R. (Epide-
miology Branch, Natl. Cancer Inst., Bethesda, Md.
20014); Mason, T. J.; McKay, F. W.; Fraumeni, J. F.,
Jr. Science 189(4207):1005-1007; 1975.

To provide leads to the causes of cancers, cancer-
mortality rates were collected for the 3056 counties
of the contiguous U. S. The age-standardized mortality rates for 35 cancer sites were calculated
from the age-, race-, and sex-specific rates for
1950-1969. The distribution of bladder cancer in
white men so obtained is clustered in areas with
heavy industry; this pattern was not duplicated in
females. To characterize the possible hazards, 64
areas were selected in which: bladder-cancer mortality for men was significantly higher than the
national rate, the male-to-female ratio of bladder
cancer mortality was higher than the national ratio,
and the lung-cancer rate among men did not differ
significantly from the national rate. These criteria selected areas where bladder cancer is more
likely to be related to industrial exposure than to
cigarette smoking. The patterns obtained indicate
that automobile and nonelectrical-machinery manufacturing are correlated with high incidences of
bladder cancer. The known association with chemical industry was confirmed. Mapping of stomach-
cancer mortality rates showed high rates in major
cities and areas characterized by low socioeconomic
class, but also in rural counties with notable populations of Russian, Austrian, Scandinavian, and
German descent.

2882   EPIDEMIOLOGICAL STUDIES OF α-FETOPROTEIN
        AND HEPATITIS B ANTIGEN IN TOMIE TOWN,
NAGASAKI, JAPAN. (Eng.) Koji, T. (Nagasaki Univ.
Sch. Medicine, Nagasaki, Japan); Munehisa, T.; Yam
guchi, K.; Kusumoto, Y.; Nakamura, S. Ann. N.Y.
Acad. Sci. 259:239-247; 1975.

Annual mass physical examinations were conducted o
1,582 individuals (613 male, 969 female, 30-84 yr
old) in Tomie Town, Japan in order to screen for
liver diseases and for the presence of α-fetoprote
hepatitis B antigen and hepatitis B antibody. The
clinical protocol involved a medical history, a
physical examination and five laboratory tests:
SGOT, glutamic pyruvic transaminase, thymol turbid
ity, zinc sulfate turbidity, and alkaline phosphatase. Further examinations were performed on indi
viduals in whom hepatomegaly was detected or who
showed abnormal clinical findings. Among 668 adul
subjects (246 male, 422 female), the hepatitis B
antigen and antibody levels were measured by immun
adherence hemagglutination and passive hemagglutin
tion. Serum α-fetoprotein levels were estimated b
hemagglutination. There were 484 cases (30.6%) of
hepatomegaly and 208 (13.1%) cases of abnormal liv
function. The number of individuals who had both
conditions was 101 (6.4%). Among the 81 subjects
who underwent liver biopsy, seven cases of liver
cirrhosis and one hepatoma were observed. Ten cas
of chronic hepatitis and 40 cases of nonspecific
liver fibrosis were also seen. The incidence of
positive hepatitis B antigen and hepatitis B antibody serum levels were 5.7% and 25.4%, respectivel
(the highest rates in Japan). Of five cases of
liver cirrhosis or hepatoma, four were positive fo
hepatitis B antigen. Two of 668 subjects were pos
tive for α-fetoprotein, showing radioimmunoassay
serum concentrations of 130 and 260 ng/ml, respectively. The first subject had clinically diagnose
liver cirrhosis with hepatoma but soon died; the
second is a case with previously unsuspected hepatoma who survived surgery, and for whom a brief ca
report is presented. In both hepatoma cases the
serum was positive for hepatitis B antigen. These
findings suggest, but do not prove, a possible rol
for hepatitis B antigen as one of the etiological
factors in liver disease. The fact that one clinically latent hepatoma patient was detected by the
α-fetoprotein assay and underwent a successful ope
ation verifies the validity of the α-fetoprotein
assay for the early diagnosis of hepatoma.

2883   MALIGNANT TUMORS IN AMERICAN BLACK AND
        NIGERIAN CHILDREN: A COMPARATIVE STUDY.
(Eng.) Olisa, E. G. (Dept. Pathology, Howard Univ
Washington, D.C. 20001); Chandra, R.; Jackson, M.
A.; Kennedy, J.; Williams, A. O. J. Natl. Cancer
Inst. 55(2):281-284; 1975.

Results of a study on the relative frequencies of
tumors in 162 American black and 1,325 Nigerian
children were compared with data from the Childhoo
Cancer Registries in Manchester, United Kingdom
(994 children), and Kampala, Uganda (776). The
American black child living in Washington, D.C. an
the Caucasian child living in Manchester had simi-

lar high frequencies for leukemia (27.8% and 29.5%, respectively) and glioma (11.2% and 17.0%, respectively), whereas the incidence of lymphoma (9.8% and 8.7%, respectively) and retinoblastoma (4.9% and 3.1%, respectively) was low. African children living in Nigeria or Uganda had the opposite frequency patterns. These were, for Uganda and Nigeria respectively, 7.1% and 4.5% leukemia; 49.2% and 59.0% lymphoma; 1.3% and 2.2% gliomas; and 7.4% and 7.4% retinoblastoma. These differences in frequencies of tumors between two ethnologically related population groups (American black and Nigerian) suggest the influence of environmental factors in the etiology of these tumors, even though exposure to environmental carcinogens was short. The rarity of Ewing's sarcoma and testicular tumors in American black (0.6% and 0%, respectively) and Nigerian children also suggest a genetic influence.

2884   MALIGNANT MELANOMA IN THE IGBOS OF NIGERIA.
(Eng.) Onuigbo, W. I. B. (General Hosp.,
Enugu, Nigeria). Br. J. Plast. Surg.28(2):114-117;
1975.

Of 21 cases of malignant melanoma occurring in the Igbos of Nigeria 17 were on the foot; 11 of these were on the sole. This result supports the conclusion that Negroes tend to have the disease in the non-pigmented parts. Wearing shoes might decrease the incidence of melanoma of the sole in undeveloped countries.

2885   PREVALENCE OF CARCINOMA OF THE PENIS WITH
SPECIAL REFERENCE TO INDIA. (Eng.) Reddy,
C. R. R. M. (D-8 Doctor's Colony, Visakhapatnam
530002, India); Raghavaiah, N. V.; Mouli, K. C. Int.
Surg. 60(9):474-476; 1975.

The frequency of carcinoma of the penis throughout the world is tabulated and compared with similar data from 32 hospitals in India. There is a wide variation in the frequency of penile carcinoma in India, the disease ranking first among male cancers in some areas and tenth or lower in other areas. In Visakhapatnam, penile carcinoma accounted for 12.5% of male cancers and 5.5% of all cancers. The highest incidence was in Kurnool, where carcinoma of the penis accounted for 13.8% of male cancers and 6.9% of all cancers. Hindus, who do not usually practice circumcision, predominate in India; however, in some regions with many Muslims (as in Kurnool) the frequency of penile carcinoma is high. In other Muslim areas, such as Hyderabad, the incidence is low. Other etiologic factors such as poor sexual hygiene may account for this variability in frequency. There was no relationship between the frequency of cervical and penile carcinomas.

2886   PROSTATIC MALIGNANCY IN INDIA. (Eng.)
Raghavaiah, N. V. (New Padma Buildings,
Maharanipet, Visakhapatnam 530002 Andhra Pradesh,
India); Singh, S. M. Int. Surg. 60(9):482-485; 1975.

Clinicopathologic data are presented for 120 cases

of carcinoma of the prostate diagnosed and treated at a New Delhi hospital between January 1, 1965 and June 30, 1972. The cases comprised 0.14% of all hospital admissions, 1.8% of all urological patients, and 22.5% of patients admitted with prostatic disease. The highest incidence of prostatic carcinoma was in patients aged 61-70 yr. The most common presenting symptom was prostatism followed by anemia. Hematuria was less frequent than in benign prostatic hypertrophy. In 80% of the cases, the prostate was hard and nodular and typical of malignancy; rectal palpation was the only investigation used to screen the majority of carcinoma cases. Histologically, 75% of the carcinoma were adenocarcinoma, 15.3% were anaplastic, and 0.7% were squamous cell carcinomas. Almost all the patients presented at a very late stage in the disease; hence, radical surgery was not feasible. Palliative treatment consisted of diethylstilbestrol therapy, bilateral orchidectomy, or estrogen administration (1-15 mg/day in divided doses). Results of treatment could not be assessed because 70-80% of the patients were not followed.

2887   EPIDEMIOLOGIC ASSOCIATION BETWEEN GONORRHEA
AND PROSTATIC CARCINOMA. (Eng.) Heshmat,
M. Y. (520 "W" St., N.W., Washington, D.C.); Kovi,
J.; Herson, J.; Jones, G. W.; Jackson, M. A. Urology
6(4):457-460; 1975.

The relationship between gonorrheal infection and later development of prostatic carcinoma was assessed. Two methods included: (a) an associative study of the correlation between gonorrhea incidence and prostatic cancer mortality rates in Denmark, and (b) a retrospective epidemiologic study of patients with prostatic cancer and matching controls. Figures for the incidence of gonorrhea (1891-1940) and for mortality from prostatic cancer (1941-1970) were obtained for Denmark. Curves for gonorrhea incidence rates per 100,000 men and for death rates from prostatic cancer per 100,000 men were subjected to a time series analysis using a computer program. All patients with prostatic cancer diagnosed since 1973 in five hospitals in Washington, D.C. will be interviewed for a past history of gonorrhea. The curves for death rates from prostatic cancer and gonorrhea incidence rates in Denmark, over a span of 30 yr, matched well with a lag period of 45 yr. The retrospective study of 75 cancer patients and 75 age-matched controls demonstrated a statistically significant (P < 0.05) association between gonorrheal infection and subsequent development of prostatic cancer. The findings could be explained by two hypotheses: the viral-venereal theory, and the chronic prostatitis theory. A recent increase in the incidence of prostatic cancer in the U.S. could be the beginning of an epidemic.

2888   SOME MARITAL-SEXUAL CONCOMITANTS OF CAR-
CINOMA OF THE CERVIX. (Eng.) Vincent, C.
E. (Boeman Gray Sch. Med., Winston-Salem, N.C.); Vincent, B.; Greiss, F. C.; Linton, E. B. South. Med. J.
68(5):552-558; 1975.

Fifty patients were given four interviews concerning

melanoma distribution has changed. According to a
Connecticut tumor registry (1935-1972), there was a
small increase in the incidence of tumors of the
head and neck, a striking increase in tumors of the
lower limbs of females, the trunk in men, and the
upper limbs in both sexes. The site-sex differences gave some clues to possible etiology; it appeared that the change was greatest in those sites
where exposure to the sun was increased as a result
of changes in clothing and recreational habits after
World War II. This was demonstrated by a study in
Queensland, Australia (1963-1967), where the incidences of melanoma by body site showed that areas
with significant excesses of melanomas were usually
exposed areas (face, leg, neck, and back in men).
The frequency of melanocytes varied with the region and the thickness of the skin. These data
suggest a dose-response relation between UV radiation, exposure, and the development of melanoma;
environmental changes that produce even small increases in UV radiation should be viewed with concern. The relation of skin cancer mortality rates
to latitude within North America is also described.
Melanoma mortality and latitude exhibited a linear
relation over the 48 states and ten Canadian provinces; mortality decreased (22 deaths per million
to seven deaths per million) as the degree of north
latitude increased (25-55°). The death rates were
greater for males than for females, suggesting the
action of a hormonal or other biochemical factors.

2891   BURKITT'S LYMPHOMA. (Eng.) Vianna, N. J.
        (Cancer Control Bureau, New York State
Dept. Health, Albany, N.Y.). In: *Lymphoreticular
Malignancies: Epidemiologic and Related Aspects*.
Baltimore, University Park Press, 1975, pp. 49-59.

Various etiologic and epidemiologic features of
Burkitt's lymphoma are presented. Contrasting incidence patterns were studied in relation to the
histologic similarity of Burkitt's tumor in endemic
and nonendemic areas. In tropical Africa, it is
the most frequently encountered childhood malignancy; it is endemic in Uganda and New Guinea. In
Canada, Great Britain, the U.S., and Brazil, it
occurs only sporadically. Burkitt's lymphoma demonstrates a climatic dependence. In Africa, it
occurs with greatest frequency in regions with a
mean temperature of 60 F during the coldest month.
Burkitt's lymphoma occurs around lakes, but not in
regions with an altitude over 5000 ft. Evidence
of time-space clustering has been variable; it was
demonstrated in the West Nile district and Bwamba,
but not in the North Mara district in Tanzania.
Additional evidence is presented for an epidemic
drift of this disorder, suggesting that some environmental agent had moved through the community.
Studies in several countries have indicated that
the prevalence of Epstein-Barr virus (EBV) antibody varies according to the level of socio-economic development, and that EBV is not solely responsible for Burkitt's lymphoma. One study in
Uganda suggests a close association between the
incidence of Burkitt's lymphoma and malaria; in
areas where malaria control has been instituted,
Burkitt's lymphoma rates have decreased. The re-

lation between Burkitt's lymphoma and acute lymphatic leukemia is also discussed. One hypothesis suggests that they are alternative forms of one another. This theory is based on a study which indicated that lymphoblastic leukemia is the major childhood lymphoproliferative malignancy in the northern hemisphere; in much of Africa and New Guinea, however, Burkitt's lymphoma predominates. Other studies suggest that while a deficiency in child lymphatic leukemia in certain African countries does exist, it is not inversely related to the incidence of Burkitt's lymphoma. These contrasting findings demonstrate the need for additional studies to elucidate the true nature of Burkitt's lymphoma.

2892   PROPORTIONAL MORTALITY AMONG ALCOHOLICS.
        (Eng.) Monson, R. R. (Harvard Sch. Public Health, 677 Huntington Ave., Boston, Mass. 02115); Lyon, J. L. *Cancer* 36(3):1077-1079; 1975.

To test the hypothesis that there is a positive association between chronic alcoholism and carcinoma of the pancreas, the mortality experience of 1,382 chronic alcoholics (1,139 men and 243 women) was studied. Analysis was limited to a comparison of observed and expected proportional mortality of different causes of death in the 894 Caucasians who were known to have died. For carcinoma of the pancreas, three deaths were observed and 5.2 were expected. The observed/expected ratios for other causes of death, including other sites of cancer, were in accordance with prior studies, and there does not appear to be an association between alcohol consumption and cancer of the pancreas.

2893   ORAL CANCER AND PRECANCEROUS LESIONS IN
        57,518 INDUSTRIAL WORKERS OF GUJARAT,
INDIA. (Eng.) Smith, L. W. (Government Dental Coll. and Hosp., Ahmedabad-380 016 India); Bhargava, K.; Mani, N. J.; Malaowalla, A. M.; Silverman, S., Jr. *Indian J. Cancer* 12(2):118-123; 1975.

The objectives and results of a study of oral cancer and precancerous lesions in a selected population of 57,518 cohorts are reported. The objectives were (a) to determine the prevalence of oral and oropharyngeal cancer and precancerous lesions, and (b) to describe the relationship of oral and oropharyngeal cancer and lesions. All subjects were 35 yr of age or older. An examination of lips, oral cavity, oral pharynx and lymph node palpability was made. A chart was kept for data on specific lesions, and a biopsy was performed when possible. Men made up 95% of the population ranging in age from 35-83 yr. As literacy increased there was a corresponding decrease in the prevalence of leukoplakia, nicotine, stomatitis, and leukoedema. Most participants were Hindus (86%). No association was made between the use of chillies in food preparation and oral lesions. Ninety-three percent of the cohorts cleaned their teeth once a day and used a twig chewed in the form of a brush known as a "Datan". Twenty-nine cases of cancer were diagnosed along with 69 cases of epithelial dysplasia. Of the total

study population, 28.3% had lesions. Of the lesions observed, 52.5% were nicotine stomatitis, and 32.1% were leukoplakia. The rate of oral cancer was 50.4 per 100,000. The high rate of oral cancer and precancerous lesions appears due to smoking and chewing habits of the study population. A close relationship existed between nicotine stomatitis and smoking; leukoedema also appeared to be related to a lesser extent.

2894   ADENOCARCINOMA OF THE NOSE AND PARANASAL
        SINUSES IN WOODWORKERS IN THE STATE OF
VICTORIA, AUSTRALIA. (Eng.) Ironside, P. (Cancer Inst., 481 Little Lonsdale St., Melbourne 3000, Australia); Matthews, J. *Cancer* 36(3):1115-1121; 1975.

In view of findings in other countries, an investigation was carried out in Australia to determine whether or not adenocarcinoma of the nose and sinuses is associated with occupational exposure to wood dust. The case index of the Cancer Institute of Victoria contained 19 cases of adenocarcinoma of the nose and paranasal sinuses. Eighteen cases occurred in men and one occurred in a woman. Routine questioning of these patients revealed an occupation involving woodworking in seven cases, whereas among 80 cases of other malignant tumors of the nose and sinuses there were only four who had been woodworkers. Among the patients with adenocarcinoma of the nose and sinuses, there was a significantly higher proportion of woodworkers than in the general population. No tumors of specific salivary type were associated with exposure to wood dust. The findings are consistent with European reports associating nasal adenocarcinoma with wood dust, but whereas the workers at risk in Europe are mainly in the furniture industry, some of the workers affected in Victoria have been sawmillers or carpenters.

2895   THE OCCUPATION OF FISHING AS A RISK FACTOR
        IN CANCER OF THE LIP. (Eng.) Spitzer, W. O. (Faculty Medicine, Memorial Univ. Newfoundland, St. John's, Newfoundland); Hill, G. B.; Chambers, L. W.; Helliwell, B. E.; Murphy, H. B. *N. Engl. J. Med.* 293(9):419-424; 1975.

A project that included a case-control study and a cohort analysis was undertaken in Newfoundland to study the role of commercial fishing and related factors in the development of lip cancer. The cases were 366 patients with the diagnosis of squamous cell carcinoma of the lip and three control groups: (a) 132 patients with squamous cell carcinoma of the oral cavity, (b) 81 patients with squamous cell carcinoma of the skin of the head and neck, and (c) 210 randomly selected men, matched for age and geographic location. Household survey data were linked with cancer-registry and census data. In comparison with other men, fishermen had a probability of development of lip cancer that was 1.5 times higher (by the case-control method, $P < 0.05$) or 4.4 times higher (by cohort analysis, $P < 0.001$). Despite the effect of pipe smoking, "outdoorness" and age on the development of lip cancer in general, the occupation of fishing was an additional, independent contribution to the risk. Unexpectedly,

radiation levels were on the surface of the various
product pumps. Propane reflux pumps exhibited the
highest level in all but one case. No external
radiation levels above 8 mR/hr were found. Measurements were made at equipment surfaces but, in
all cases, the equipment was in low occupancy areas
and there was· no way to hypothesize a whole body
exposure rate in excess of 2 mR/hr (the generally
acceptable level for nonrestricted areas). It is
concluded that no substantial external radiation
hazard exists in the plants surveyed. Since only
nine of the existing 805 plants in the U.S. were
surveyed, there may be some plants with whole body
radiation levels above 2 mR/hr. The potential hazard of long-lived radon-222 daughters built up on
the internal surfaces of processing equipment is
recognized but could not be evaluated because no
equipment was encountered in the disassembled condition.

2898   CELL KINETICS METHODS FOR CLINICAL USE:
        ASSESSMENT OF AUTORADIOGRAPHY AND ^{125}I-
IODO-DEOXYURIDINE WITH MOUSE EHRLICH TUMOUR. *(Eng.)*
Jenkinson, I. S. (St. Vincent's Hosp., Sydney, Australia); Shuter, B. J.; Wright, P. N. M. *Cell Tissue Kinet*. 8(4):307-320, 1975.

Superficial Ehrlich tumors in male Quackenbush mice
were used to assess how much information on cell
kinetics could be obtained from the simplest autoradiography techniques and the *in situ* monitoring of
^{125}I-iododeoxyuridine (^{125}I-IUdR). These techniques
were selected as being readily applicable to clinical situations. From 1-12 days after initiation,
id and im tumors were labeled with ^{3}H-thymidine or
^{125}I-IUdR by the ip injection of 20 μCi or the id
injection of the superficial tumors with 3-5 μCi; in
some cases, the tumors were produced by ascites
cells prelabeled with ^{125}I-IUdR. A hand-held scintillation counter was used for *in situ* monitoring
of tumor radioactivity, and measurements were made
of tumor growth and labeling indices. The tumors
labeled with ^{3}H-thymidine were excised one hour
later. The results indicated that the simple techniques of a single localized application of ^{3}H-
thymidine followed by a single biopsy, or a single
biopsy followed by *in vitro* labeling, were sufficient to assess the morphological and spatial distribution of labeling index values in the tumor.
From this the cell production and loss rates were
readily calculated, provided that the tumor's net
growth rate was measurable or could be inferred from
the tumor's classification in clinical applications.
A single localized injection of ^{125}I-IUdR followed
by daily monitoring for 1-2 wk was sufficient to
assess the rate of cell loss. In slow growing
tumors, this was very nearly equal to the cell production rate.

2899   NEOPLASMS IN CHILDREN AGED UP TO 14 YEARS
        IN THE PROVINCE OF WARSAW. (Pol.) Zaor-
ska, B. (Wojewodski Osrodek Matki i Dziecka, ul.
Zamieniecka 2/16, 04-158 Warszawa, Poland); Worowska-Rogowska, J. *Pediatr. Pol*. 50(1):75-82; 1975.

2900   INCIDENCE OF MALIGNANT NEOPLASMS IN
        GOIANIA. (Por.) Reboucas, M. A. (Faculd-
ada de Medicina da Universidade Federal de Goiás,
Brazil); Lopes, G. de M. P.; Cardoso, V. M. *Rev.
Goiana Med.* 20(3/4):243-247; 1975.

2901   PRIMARY LIVER CANCER COINCIDENT WITH
        SCHISTOSOMIASIS JAPONICA. A STUDY OF 24
NECROPSIES. (Eng.) Nakashima, T. (Chiba Univ.
Hosp., Chiba 280, Japan); Okuda*, K.; Kojiro, M.;
Sakamoto, K.; Kubo, Y.; Shimokawa, Y. *Cancer*
36(4):1483-1489; 1975.

2902   EPIDEMIOLOGICAL ASPECTS OF CARCINOMA OF
        THE ENDOMETRIUM. (Ger.) Lau, H.-U.
(Frauenklinik des Bereichs Medizin, Charite, DDR-
104 Berlin, Tucholskystr. 2, East Germany); Petschelt,
E.; Poehls, H.; Pollex, G.; Unger, H.-H.; Zegenhagen,
V. *Zentralbl. Gynaekol.* 97(17):1025-1036; 1975.

2903   A DIETARY METHOD FOR AN EPIDEMIOLOGIC
        STUDY OF GASTROINTESTINAL CANCER. (Eng.)
Hankin, J. H. (Sch. Public Health, 1960 East-West
Road, Honolulu, Hawaii); Rhoads, G. G.; Glober, G.
A. *Am. J. Clin. Nutr.* 28(9):1055-1061; 1975.

2904   LEUKOPLAKIA REVISITED. A CLINICOPATHO-
        LOGIC STUDY 3256 ORAL LEUKOPLAKIAS.
(Eng.) Waldron, C. A. (Emory Univ. Sch. Dentistry,
Atlanta, Ga. 30322); Shafer, W. G. *Cancer* 36(4):
1386-1392; 1975.

2905   EPIDEMIOLOGY AND AETIOLOGY OF CANCER OF
        THE UTERINE CERVIX: INCLUDING THE DETEC-
TION OF CARCINOGENIC N-NITROSAMINES IN THE HUMAN
VAGINAL VAULT. (Eng.) Harington, J. S. (Cancer
Res. Unit Natl. Cancer Assoc. South Africa, South
African Inst. Medical Res., Johannesburg). *S. Afr.
Med. J.* 49(12):443-446; 1975.

2906   EVALUATION OF THE ENVIRONMENTAL FACTORS
        IN WOMEN TREATED BECAUSE OF OVARIAN
TUMORS. (Pol.) Pilawski, Z. (70-111 Szczecin, Al.
Powstancow 72, Poland); Lazar, W.; Zoltowski, M.;
Sieja, K. *Ginekol. Pol.* 46(1):39-47; 1975.

2907   CERVICAL NEOPLASIA IN RESIDENTS OF A LOW-
        INCOME HOUSING PROJECT: AN EPIDEMIO-
LOGIC STUDY. (Eng.) Ory, H. W. (Center Disease
Control, Public Health Service, U.S. Dept. Health,
Education, Welfare, Atlanta, Ga.); Jenkins, R.;
Byrd, J. Y.; Jones, C. J.; Smith, J.; Tyler, C. W.,
Jr. *Am. J. Obstet. Gynecol.* 123(3):275-277; 1975.

2908   THE EPIDEMIOLOGY AND INTERRELATIONSHIP
        OF CERVICAL DYSPLASIA AND TYPE 2 HERPES-
VIRUS IN A LOW-INCOME HOUSING PROJECT. (Eng.) Ory,
H. W. (Center Disease Control, Public Health Service
U.S. Dept. Health, Education, Welfare, Atlanta,
Ga.); Jenkins, R.; Byrd, J. Y.; Nahmias, A. J.;
Tyler, C. W., Jr.; Allen, D. T.; Conger, S. B.
Am. J. Obstet. Gynecol. 123(3):269-274; 1975.

2909   INTERPERSONAL TRANSMISSION OF EB-VIRUS
        INFECTION [letter to editor]. (Eng.)
Chang, R. S. (Univ. of California Sch. Medicine,
Davis, Calif.). *N. Engl. J. Med.* 293(9):454-455;
1975.

2910   SCREENING METHOD FOR THE DETECTION OF
        AFLATOXINS IN MIXED FEEDS AND OTHER
AGRICULTURAL COMMODITIES WITH SUBSEQUENT CONFIRMA-
TION AND QUANTITATIVE MEASUREMENT OF AFLATOXINS
IN POSITIVE SAMPLES. (Eng.) Romer, T. R. (Ralston
Purina Co., St. Louis, Mo.). *J. Assoc. Off. Anal.
Chem.* 58(3):500-506; 1975.

2911   THE RADIOAUTOGRAPHIC TRANSFER FUNCTION
        AND ITS IMPLICATIONS FOR RADIOAUTOGRAPHIC
METHODOLOGY. (Eng.) Shackney, S. E. (Natl. Cancer
Inst., Bethesda, Md. 20014). *J. Natl. Cancer Inst.*
55(4):811-820; 1975.

See also:

*   (Rev): 2401, 2403, 2407, 2419, 2425, 2432, 2434,
            2447, 2448, 2449
*   (Chem): 2522, 2523, 2559
*   (Immun): 2752
*   (Path): 2771, 2791

they produced a benign tumor that regressed after three weeks. When exponentially growing cells were infected with the Moloney strain of murine sarcoma virus, a change in morphology occurred about four days after infection. Cells became spindle-shaped and re-fractile, and grew in a criss-cross fashion forming a multilayer; this was in contrast to the confluent monolayer formed by controls. These results indicate that hamster embryo cells develop into a permanent cell line by chance, with an incidence of roughly 10-30%; this incidence varies according to culture conditions.

2914   STUDIES ON THE HUMORAL REGULATION OF GRAN-
        ULOPOIESIS IN LEUKAEMIC RFM MICE. (Eng.)
Gordon, M. Y. (Inst. Cancer Res., Belmont, Sutton, Surrey, England); Lindop, P. J. *Br. J. Cancer* 32 (2):186-192; 1975.

Intraperitoneal diffusion chambers were used to in-vestigate changes in humoral factors during the de-velopment of myeloid leukemia in RFM mice. Normal mouse bone marrow cells form colonies of granulo-cytes and macrophages when cultured in semi-solid agar medium within intraperitoneal diffusion cham-bers. The use of mice bearing transplanted myeloid leukemia as agar diffusion chamber (ADC) hosts en-hanced colony formation from normal marrow. The humoral basis for this stimulation was shown by the colony stimulating activity of the fluid entering the diffusion chambers when assayed against normal mouse bone marrow cells in agar culture *in vitro*. The stimulus to colony growth in ADCs and the *in vitro* colony stimulating activity depended on the phase in the development of the leukemia investi-gated, and the stimulation was abolished by splenec-tomy. There was no apparent relationship between the growth of the leukemic cell population *in vivo* and the level of the stimulating factor detected in leukemic mice. The results show that the combined effects of factors that inhibit or stimulate granu-lopoiesis change radically during the development of transplantable leukemia in RFM mice. These data may be compared with the increased colony stimulating factor levels associated with leukemia in man, al-though evidence of an early phase characterized by low levels of stimulation has not been among pre-leukemia patients.

2915   GRANULOPOIETIC PROGENITORS IN SUSPENSION
        CULTURE: A COMPARISON OF STIMULATORY
CELLS AND CONDITIONED MEDIA. (Eng.) Niho, Y. (Ontario Cancer Inst., Toronto, Canada); Till, J. E.; McCulloch, E. A. *Blood* 45(6):811-821; 1975.

Kinetic studies were carried out to investigate the functional heterogeneity in populations of human mar-row or peripheral blood cells separated by velocity sedimentation. Peripheral blood was obtained from seven normal volunteers and a patient with hemochro-matosis; marrow was obtained from 18 patients with cancer or anemia. Cultures of these cells were as-sayed for granulopoietic progenitor cells and factors, or for cells that stimulate colony formation. The slowly-sedimenting cells were found to stimulate both

colony formation by granulopoietic progenitors and an
increase in numbers of granulopoietic progenitors in
suspension culture, while rapidly-sedimenting cell
stimulated only colony formation and not increased
progenitors in suspension cultures. Investigations of
the properties of media conditioned by these two subpopulations of cells revealed no clear differences
between them; both stimulated suspension cultures as
well as colony formation, and both lost the former
activity, but not the latter, after dialysis. The
results evidence that more than one process is
regulated in cultures of granulopoietic progenitor
cells.

2916   THE EFFECTS OF TOPOINHIBITION AND CYTO-
        CHALASIN B ON METABOLIC COOPERATION.
(Eng.) Stoker, M. (Imperial Cancer Res. Fund Laboratories, Lincoln's Inn Fields, London WC2, England). *Cell* 6(2):253-257; 1975.

The effects of topoinhibition, topostimulation, and
cytochalasin B on metabolic cooperation (the ability
to form intercellular junctions) of 3T3 cells were
examined in a stationary monolayer culture. A line
of polyoma-transformed BHK cells (TG1) lacking hypoxanthine guanine phosphoribosyl transferase was used
as an indicator. TG1 cells incorporated [^{3}H]hypo-
xanthine into nucleic acids when in prolonged (20 hr)
or brief (1 hr) contact with donor 3T3 cells, whether
the 3T3 cells were in a quiescent layer or in the
region of stimulation at the edge of the wound. Cytochalasin B (1 μg/ml) prevented the development of
metabolic cooperation between donor 3T3 cells and
recipient BHK cells and abolished most pre-existing
cooperation. Metabolic cooperation resumed three
hours after the removal of cytochalasin B. The proportion of residual cooperating cells at different
doses of the drug remained unaffected by the position
in the edge or in the layer, and was not correlated
with topostimulation. The results support the hypothesis that alterations in the capacity to form stable intercellular junctions is not a necessary feature of the topoinhibition phenomenon.

2917   OCCLUDING JUNCTIONS AND CELL BEHAVIOR
        IN PRIMARY CULTURES OF NORMAL AND NEO-
PLASTIC MAMMARY GLAND CELLS. (Eng.) Pickett, P. B.
(Dept. Zool., Univ. California, Berkeley); Pitelka,
D. R.; Hamamoto, S. T.; Misfeldt, D. S. *J. Cell
Biol.* 66(2):316-332; 1975.

Dome-forming primary cultures of cells dissociated
from normal prelactating mouse mammary glands or
from spontaneous mammary adenocarcinomas were examined with particular attention to morphological
polarity and cell junctions to determine to what
extent cultured cells resemble mammary epithelium
in structural organization. Spontaneous mammary
tumors from BALB/cfC3H/Crgl or from C3H/Crgl multiparous mice and normal mammary glands from 14- to
19-day pregnant females of the same strains or
BALB/cCrgl were minced and dissociated. The cell
suspension was washed, filtered and plated at high
density (5 x 10^{-5} cells/cm^2 of substrate). Tumorigenicity of cultured cells was tested by sc injec-

tion into isogenic mice of 10^5 cells from 14-day-
old primary normal and tumor cultures. Tumors
developed from all of six tumor cell implants and
from neither of two normal cell implants. The
cultures were examined microscopically alive and in
thick and thin sections and freeze-fracture replica
Pavement cells in all cases were polarized toward
the bulk medium as a lumen equivalent, with microvilli and continuous, well-developed occluding
junctions at this surface. Between the pavement
and the substrate were other cells, of parenchymal
or stromal origin, scattered or in loose piles;
these sequestered cells were relatively unpolarized
and never possessed occluding junctions. Small gap
junctions were seen in the pavement layer, and
desmosomes could be seen linking epithelial cells
in any location. Development of the epithelial
secretory apparatus was not demonstrable under the
culture conditions used; normal and neoplstic cells
did not differ consistently in any property examined. The roof of a dome was merely a raised part
of the epithelial pavement and did not differ from
the latter in either cell or junction structure.
Dome formation may demonstrate the persistance of
some transport functions and of the capacity to form
effective occluding junctions. These basic epithelial properties may survive both neoplastic
transformation and transition to culture.

2918   THE CONTROL OF ANAEROBIC GLYCOLYSIS BY
        GLUCOSE TRANSPORT AND OUABAIN IN SLICES
OF HEPATOMA 3924A. (Eng.) Van Rossum, G. D. V.
(Temple Univ. Sch. Med., Philadelphia, Pa. 19140);
Galeotti, T.; Palombini, G.; Morris, H. P. *Biochim.
Biophys. Acta* 394(2):267-280; 1975.

The activities of glycolysis and K$^+$ transport were
studied in slices of Morris hepatoma 3924A incubated
under anaerobic conditions in the presence of glucose
at 1-50 mM. The tumors were grown for 20-25 days in
ACI/T rats. Ouabain-sensitive net transport of K$^+$
was observed at all glucose concentrations greater
than 1 mM; ouabain reduced the rate of glycolysis by
about 25% at all glucose concentrations able to support ion transport. The net entry of glucose into
the intracellular phase was studied at varying glucose concentrations. The rate of net entry of glucose was similar to the rate of glucose utilization
by anaerobic glycolysis at concentrations of 10 mM
and less, but exceeded the rate of glycolysis at 20
mM and above. The glucose entry was not Na$^+$-depend-
ent and was not inhibited by ouabain. The results
suggest that the reduction in glycolytic activity
caused by ouabain is not due to an inhibition of glucose transport and that the glucose transport system
of this hepatoma has properties similar to that of
normal liver.

2919   THE EFFECT OF FLAVONOIDS ON AEROBIC GLYCOLY-
        SIS AND GROWTH OF TUMOR CELLS. (Eng.) Suo-
linna, E. M. (Section of Biochemistry, Molecular and
Cell Biology, Cornell Univ., Ithaca, N.Y. 14853);
Buchsbaum, R. N.; Racker, E. *Cancer Res.* 35(7):1865-
1872; 1975.

The effects of bioflavonoids on the aerobic glycolysis

and with dieldrin at both concentrations. Increasing
concentrations of carbaryl caused subsequent increases
in sphingomyelin content of the phospholipid fractions of exposed cells suggesting inhibition of
sphingomyelinase. Both carbaryl and dieldrin elicited decreases in the amounts of phosphatidylcholine,
and a slight increase in lysophosphatidylcholine was
noted in cells exposed to carbaryl. The results indicate that insecticides from different chemical
groups are nonspecific in their action upon mammalian
cells *in vitro*. The variations in phospholipid content of HeLa cells caused by carbaryl and dieldrin
suggest an alteration of the structure of cellular
membranes.

2922   REGULATION OF PROTEIN SYNTHESIS AT THE
        TRANSLATIONAL LEVEL IN NEUROBLASTOMA CELLS.
(Eng.) Zucco, F. (C.N.R., Laboratory of Molecular
Embryology, 80072 Arco Felice, Naples, Italy); Per-
sico, M.; Felsani, A.; Metafora, S.; Augusti-Tocco, G.
Proc. Natl. Acad. Sci. USA 72(6):2289-2293; 1975.

In vitro protein synthesis in neuroblastoma cells
placed in two different growth conditions was studied.
Neuroblastoma clone 41A3 was grown in monolayer and
suspension culture. Assays of standard cell-free
protein synthesis, total soluble activity, and total
transfer activity were performed. Polysome sucrose
density gradient analysis and acrylamide gel electrophoresis were also performed. The specific activity
of the lysate increased during growth of the monolayer cultures, yet showed little change in the suspension cultures. Both the rate and extent of protein synthesis were markedly higher in the lysate from
monolayer cells. The specific activity of monolayer
cell sap was twice as high as that of suspension.
Although the activities of elongation factors I and II
and of aminoacyl-transfer RNA synthetases were the
same in both cell saps, the suspension cells were two
times more active in elongating polypeptide chains.
Ribosome activity was found to be two times higher in
the monolayer cells, as was the polysomes/monosomes
ratio. Although the total ribosomal population in
monolayer cells was about twice that present in suspension cells, there was no significant change in the
polysome size among the two cell types. Cross experiments suggested that both ribosomes and some of the
soluble factors were involved in determining the
higher activity of the monolayer lysate. The overall
patterns of the proteins released from reconstituted
cell-free systems were similar, but quantitative differences were noted. The data indicate that the
transition of neuroblastoma cells from suspension to
monolayer growth conditions involve a marked increase
in the protein-synthesizing activity.

2923   PROPERTIES OF CHROMOSOMAL PROTEINS OF
        HUMAN LEUKEMIC CELLS. (Eng.) Desai, L.
S. (Harvard Medical Sch., Boston, Mass. 02115); Wulff,
U. C.; Foley, G. E. *Biochimie (Paris)* 57(3):315-323;
1975.

Purified chromatin isolated from lymphocytic cells
derived from patients with infectious mononucleosis
(lines CCRF-RKB and CCRF-SB) or acute lymphoblastic

leukemia (lines CCRF-H-SB and CCRF-CEM) was compared with chromatin isolated from normal human lymphocytic cells (lines CCRF-SLT and CCRF-TOH) by gel electrophoresis and differential gradient ultracentrifugation. Thermal denaturation studies showed higher melting temperature values for chromatin from leukemic cells (84.7 and 84.3 for lines CCRF-H-SB and CCRF-CEM, respectively) as compared to that of lymphocytic cells from normal donors (76.8 and 77.1 for CCRF-SLT and CCRF-TOH, respectively) from patients with infectious mononucleosis (81.4 and 82.2 for CCRF-RKB and CCRF-SB, respectively). This reflects the diverse complexity of these chromatins with respect to their varying chemical compositions. There were significant differences in the ratios of DNA:RNA:protein, as, well as in the ratios of chromatin-associated histone and nonhistone proteins; although chromatin-associated histones were more homogeneous than were the nonhistone proteins, as judged by amino acid analyses and acrylamide gel electrophoresis. These differences in chromatin structure may relate to the differences in gene expression characteristic of these lymphocytic cells. The chromosomal acidic proteins isolated from the purified chromatin of human leukemic cells greatly stimulated the template activity of the chromatin in *in vitro* RNA synthesis. The nonhistone proteins selectively interact with chromatins and influence the RNA polymerase reactions, indicating that there is selective tissue specificity of nonhistone proteins. These proteins may play a greater role in transcription mechanisms and are more loosely bound to chromatins than to histones.

2924   SULFHYDRYL GROUP QUANTITATION OF HEPATOMA
        AND LIVER MICROSOMAL FRACTIONS. (Eng.)
Stratman, F. W. (Inst. Enzyme Res., Univ. Wisconsin, Madison); Hochberg, A. A.; Zahlten, R. N.; Morris, H. P. *Cancer Res.* 35(6):1476-1484; 1975.

The relationship of sulfhydryl and disulfide groups to protein synthesis in normal and rapidly growing tissues was investigated by quantitation of sulfhydryl groups in endoplasmic reticulum and polyribosomes of livers from fed or 24-hr-fasterd Sprague-Dawley rats, normal Buffalo rats, or rats with Morris hepatoma. Stripping the smooth endoplasmic reticulum of normal liver with potassium chloride and EDTA reduced the sulfhydryl groups available for carboxamide methylation by iodoacetamide by 15%, while stripping the rough endoplasmic reticulum increased the available groups by 30%. This could reflect the removal of ribosomes from rough endoplasmic reticulum with the subsequent exposure of sulfhydryl groups. Exposed sulfhydryl groups of normal mature female rat liver smooth endoplasmic reticulum were decreased to a similar degree by the stripping procedure with EDTA and potassium chloride when quantitated by either iodoacetamide or 4,4'-dithiodipyridine. In immature male and female rats, however, the stripping procedure failed to decrease the exposed sulfhydryl groups of smooth endoplasmic reticulum. Stripping rough endoplasmic reticulum from the livers of mature and immature rats and using iodoacetamide showed an increased quantity of exposed sulfhydryl groups, but using 4,4'-dithiodipyridine resulted in no change. The negative correlation between exposed sulfhydryl

groups in the polyribosomes and the rate of growth of normal liver and of Morris hepatomas 6 and 38B suggests that the conformation of the free polyribosomal proteins could be a control factor for the rat of protein synthesis. Faster growing hepatomas also have greater quantities of sulfhydryls and disulfides

2925   LOCALIZATION AND PHOSPHORYLATION OF NUCLEAR
        NUCLEOLAR AND EXTRANUCLEOLAR NON-HISTONE
PROTEINS OF NOVIKOFF HEPATOMA ASCITES CELLS. (Eng.)
Olson, M. O. J. (Baylor Coll. Medicine, Houston, Tex. 77025); Ezrailson, E. G.; Guetzow, K.; Busch, H. *J. Mol. Biol.* 97(4):611-619; 1975.

Novikoff hepatoma ascites cells were studied to determine whether any non-histone proteins are localized in either the nucleolar or extranucleolar chromatin fraction, and whether any of the proteins in either fraction were specifically labeled with ^{32}P. Rats (male, Holtzman) bearing ascites cells were injected ip with 20 mCi of carrier-free [^{32}P]orthophosphate 2 hr before sacrifice. Chromatin was prepared from whole nuclei, nucleolar and extranucleolar fractions. After removal of histones by 0.4-NH_2SO_4 and digestion of the DNA by DNase I, the nonhistone proteins were subjected to two-dimensional electrophoresis and autoradiography. Although most non-histone proteins were common to both fractions, the nucleolar fraction was enriched in proteins C18, C21, Cg' and CB. Proteins CM, C6, CA, Cb, B24, B22 and BF were found in higher concentrations in the extranucleolar component. In the nucleolar fraction, only protein C18 was labeled significantly with ^{32}P. In the extranucleolar fraction, ^{32}P was incorporated into protein spots C18, Cg', CM, CN and C6. Analyses of ^{32}P-labeled spots for phosphoamino acids indicated that all spots analyzed contained ^{32}P-labeled phosphoserine. The results indicate that nucleolar chromatin contains specific proteins which have a unique ^{32}P labeling pattern.

2926   IDENTIFICATION OF A HIGH MOLECULAR WEIGHT
        TRANS-MEMBRANE PROTEIN IN MOUSE L CELLS.
(Eng.) Hunt, R. C. (Univ. Virginia Sch. Medicine, Charlottesville, Va. 22901); Brown, J. C. *J. Mol. Biol.* 97(4):413-422; 1975.

A new method for the identification of "trans-membrane" proteins in cultured animal cells is described. Proteins exposed on the outer cell surface are labeled with ^{125}I by the lactoperoxidase technique. "Inside-out" membrane vesicles are prepared from the labeled cells using the polystyrene latex bead procedure. These inside-out vesicles are then treated briefly with trypsin and analyzed for the presence of ^{125}I-labeled protein species which were degraded by proteolytic attack. Such proteins must be exposed on both the outer and inner membrane surfaces and, therefore, they must pass through the lipid barrier. The method showed a particular high molecular wt polypeptide chain spans the plasma membrane of mouse L cells. It is similar in its iodinatability and molecular wt to a protein recently found on the surface of normal, but not virus-transformed fi-

cusing not found in normal liver ferritin. Comparisons of ferritins purified from human liver cell carcinoma and normal liver showed that both ferritins consisted of a subunit species with an identical molecular weight of approximately 18,500. Two subunits were demonstrable in normal liver ferritin by means of acrylamide electrophoresis and 8 M urea in acid pH. The same two subunits were also demonstrable in ferritin isolated from human liver cell carcinoma along with a third subunit, intermediate in charge between the two normal liver subunits. This was demonstrable in different amounts in ferritins from two hepatomas. The isolated acidic isoferritin was immunologically identical to normal liver isoferritins. It is concluded that the multiple isoferritins of the human liver ferritin consist of two subunits, which are identical in molecular weight but which differ in net charge. Ferritin, isolated from two human liver carcinoma tissues, was composed of the same two subunits and a third unique subunit. Different amounts of these subunits may account for the several normal isoferritins and a unique tumor-specific acid isoferritin found in hepatoma.

2929   DNA SYNTHESIS IN TUMOR-BEARING RATS. (Eng.)
        Shirasaka, T. (Sch. Med., Tokushima Univ.,
Japan); Fujii, S. *Cancer Res.* 35(3):517-520; 1975.

The synthesis of DNA in tumor-bearing male Donryu rats was studied. Yoshida sarcoma and AH130 cells were transplanted ip. After five days, 5×10^7 ascites cells were transplanted subepidermally on the back of rats, to produce solid tumors of each type. Rats bearing either tumor were killed and their liver, spleen, and tumor tissues were used for DNA and enzyme assays. Solid tumor was used 1 or 2 wk after transplantation. Partial hepatectomy was performed on some rats, and the regenerated tissue was removed 48 hr after hepatectomy. Incorporation of tritiated thymidine (TdR) into DNA was measured and the activities of TdR kinase, thymidine monophosphate (TMP) kinase and DNA polymerase were also assessed. In the partially hepatectomized rats, TdR incorporation into DNA was increased in the liver but not in the spleen; however, the activities of TdR and TMP kinases and DNA polymerase were increased in liver and in spleen. High activities of these enzymes were also found in both Yoshida sarcoma and AH130 tumors, and TdR incorporation was also increased. Elevated levels of these enzymes were also found in the liver and spleen of animals bearing both cell types. Activity of the enzymes was shown to rise at the fourth day after transplantation in the liver of rats bearing Yoshida sarcoma. However, in the sera of these animals, only TdR kinase and DNA polymerase increased after four days. TdR incorporation and TMP kinase were not detectable in the sera of these animals. Diethylaminoethyl-cellulose column chromatography showed the serum TdR kinase of Yoshida sarcoma-bearing rats to consist of two fractions while liver TdR kinase had only one fraction that was identical to one of the serum fractions. Serum TdR kinase activity was shown to decrease after the removal of the tumor. It is suggested that the serum TdR kinase is derived from the Yoshida sarcoma, while that of the liver is derived from *de novo* synthesis and may be enhanced by some tumor substance.

2930   CORRELATION BETWEEN DNA SYNTHESIS AND INTRA-
        CELLULAR NAD IN CULTURED HUMAN LEUKEMIC LYM-
PHOCYTES. (Eng.) Chang, S. C. S. (Lab. Mol. Biol.,
Mayo Found., Rochester, Minn.); Bernofsky, C. *Bio-
chem. Biophys. Res. Commun.* 64(2):539-545; 1975.

The relationship between DNA synthesis and intracellular nicotinamide (NAD$^+$) was examined in lymphocytes
from an individual with acute monocytic leukemia,
which were adapted to tissue culture. The nicotinamide concentration of the culture medium was 7.0 μM.
The cells were maintained at a concentration of 2-4
x 10^6 cells/ml. The NAD$^+$ level of the freshly-iso-
lated leukemia lymphocytes was 67 nM/10^9 cells (56 nM
/10^9 cells normal). The NAD$^+$ content of the acute
monocytic leukemia cells increased steadily and reached a final level of about 350 nM/10^9 cells at about
the time the cells became established (four wk). When
the NAD$^+$ level of established cells was lowered by
means of a nicotinamide-poor medium or by the action
of l-methyl-l-nitrosourea, there was a concomitant
decrease in the rate of DNA synthesis. Upon incubation of the leukemic lymphocytes with 4 nM l-methyl-
l-nitrosourea, the NAD$^+$ level decreased rapidly,
reaching 50% of its initial value in 24 min.
Measurement of [^{3}H]thymidine incorporation into DNA
showed that the rate of DNA synthesis in l-methyl-l-
nitrosourea-treated cells immediately decreased to
7% of the rate in untreated cells. When established
cells were transferred to a nicotinamide-poor medium,
the level of NAD$^+$ and the rate of DNA synthesis decreased in a parallel fashion, and after four days
were 61% and 47% of their initial values, respectively. Cells treated comparably but with complete
medium maintained a NAD$^+$ level of 307 nM/10^9 cells
and incorporated [^{3}H]thymidine at a mean rate of 16.8
x 10^4 cpm. These results indicate that there is a
direct relationship between the level of NAD$^+$ and the
rate of DNA synthesis.

2931   ON THE CHARACTER OF THE NUCLEAR MEMBRANE-
        ASSOCIATED DNA OF HELA CELLS. (Eng.)
Cabradilla, C. D. (Hazelton Lab., 9200 Leeberg Pike,
Vienna, Va. 22180); Toliver, A. P. *Physiol. Chem.
Phys.* 7(2):153-166; 1975.

The nature of DNA-membrane complexes isolated from
synchronized HeLa cells was studied. Monolayer cultures of HeLa cells, strain S$_3$, were used throughout
the study. 5-Fluorodeoxyuridine was used as the syn-
chronizing agent. DNA isolated from either nuclei or
nuclear membranes by the chloroform:isoamyl alcohol
method was analyzed in either neutral or alkaline
CsCl equilibrium density gradients. Although most of
the DNA synthesis occurring in the nucleoplasm appeared to be the result of repair replication, virtually all of the membrane-associated DNA appeared to
be the result of semiconservative replication. Hy-
droxylapatite column chromatography suggested that the
nascent DNA associated with the membrane may exist as
a nicked duplex; also, 14-15% eluted as single-
stranded DNA. Buoyant density studies and the thermal melting profile suggested a guanine-cytosine base
content of 40%, and indicated that the base content of
the membrane-associated and bulk DNA were identical.
Stability studies revealed sensitivity (release) of

the membrane-DNA to DNase I, proteolytic enzymes,
NaCl, and deoxycholate and phospholipase C. The
results of treating nuclear membranes isolated on
sucrose-shelf gradients with such agents suggested
that the DNA is associated with a lipoprotein structure.

2932   DNA FROM EUKARYOTIC CELLS CONTAINS UNUSU-
        ALLY LONG PYRIMIDINE SEQUENCES. (Eng.)
Birnboim, H. C. (Biology and Health Physics Div.,
Atomic Energy Canada Ltd., Chalk River, Ontario,
Canada); Straus, N. A. *Can. J. Biochem.* 53(5):640-
643; 1975.

The presence of polypyrimidines in DNA from several
different higher organisms was investigated. Samples
of [^{3}H]thymidine-labeled DNA were subjected to acid
hydrolysis and gel electrophoresis; nucleosides were
separated by paper chromatography. In quantitating
polypyrimidines *via* formic acid-diphenylamine hydrolysis of the samples, long pyrimidine tracts were
separated by ethanol precipitation followed by polyacrylamide gel electrophoresis. Duplicate analyses
of a single sample of L-cell DNA resulted in values
agreeing within 10-30%. The highest content of polypyrimidines was found in mouse L-cell DNA (0.8%).
Human HeLa cells, rabbit cells, and Chinese hamster
cells had about one half the amount (0.3-0.5%). Small
amounts of polypyrimidines were also found in xenopus
and bovine DNA (0.07-0.08%), and lesser amounts were
detected in adenovirus type 12, polyoma, and *Escherichia coli* DNA (0.025, 0.023, and 0.009%, respectively). It is unlikely that such polypyrimidines
arose from culture contaminants. Thus, the data
indicate that polypyrimidines are not unique to human
DNA, but are found in DNA from several different organisms. This suggests that polypyrimidines represent a relatively old evolutionary component, possibly
with a chromosomal or regulatory function.

2933   IS FANCONI'S ANAEMIA DEFECTIVE IN A PRO-
        CESS ESSENTIAL TO THE REPAIR OF DNA CROSS
LINKS? (Eng.) Sasaki, M. S. (Tokyo Medical and
Dental Univ., Yushima, Tokyo 113, Japan). *Nature*
257(5527):501-503; 1975.

The hypothesis that lymphocytes from patients with
Fanconi's anemia (FA) are defective in DNA repair
was tested by determining the chromosome response
of lymphocytes from two male FA patients (aged 6
and 7 yr) to mitomycins in relation to the cell
cycle phase at time of treatment. At various times
before fixation, FA cells were treated for 30 min
with difunctional mitomycin C (MMC, 0.1 μg/ml) or
monofunctional decarbamoyl mitomycin C (DCMMC, 50
μg/ml) in medium containing [^{3}H]thymidine. After
a corresponding recovery time in medium free of test
compounds, the chromosome preparations were processed for autoradiography and analyzed for both their
metaphase labeling pattern and for the presence of
chromosome aberrations in labeled and unlabeled mitoses. Treatment with mitomycins reduced the rate
of progression through the cell cycle; the effect
was more marked in cells treated with MMC. Both
DCMMC and MMC induced significant chromosome aber-

poly(A)-poly(U) into cell nuclei was demonstrated
by autoradiography, by recovery of acid-precipitable
material from isolated nuclei, and by sucrose gradient centrifugation of nuclear lysates. The majority of poly(A)-poly(U) remained intact in the
nuclei for at least 2½ hr. This penetration increased 20-fold by pretreatment of the cells with
DEAE Dextran. In cells treated with DEAE Dextran.
DNA and RNA syntheses stimulated by poly(A)-poly(U)
from the time the polymer complex was added to at
least 2½ hr. The authors postulate that this early
increased nucleic acid synthesis might be related
to immune stimulation by poly(A)-poly(U).

2936   POLY ADENYLIC ACID SYNTHESIS *IN VITRO* IN
        ISOLATED HeLa CELL NUCLEI AND WHOLE CELL
HOMOGENATES. (Eng.) Jelinek, W. R. (Dep. Biol.
Sci., Columbia Univ., New York, N.Y.). *Cell* 2(3):
197-204; 1975.

The possibility that HeLa cell nuclei could synthesize poly(A) similar to that found in growing cells,
and whether *in vitro*-synthesized poly(A) was part of
heterogeneous nuclear RNA were studied in cell-free
homogenates. Synthesis of poly(A) by isolated nuclei
did not occur at low [^{3}H]ATP concentrations as demonstrated on poly(U)-sepharose column affinity chromatography. Nuclei from 2×10^7 to 3×10^7 cells were
purified and resuspended in 1 ml of the cytoplasmic
supernatant from the original cell homogenate, 0.1
mM each guanosine triphosphate and cytosine triphosphate, and 50 μCi [^{3}H]uridine. After 20 min incubation at 37 C, RNA was obtained by phenol extraction and poly(A)-containing molecules were selected
by chromatography. When ATP concentration in the
medium was increased to 0.10 or 0.20 mM, the total
AMP incorporated into poly(A) was stimulated 180- to
660-fold, respectively, whereas total RNA synthesis
as measured by total AMP incorporated was increased
only 2.5-3.5 times. In the presence of cytoplasmic
extract, total RNA synthesis was stimulated about
10-fold over that when a concentration of ATP equal
to that found in cytoplasm was present. Poly(A)
synthesized in both purified nuclei and in whole cell
homogenates was identical to that synthesized *in vivo*.
When undigested with T1 and pancreatic ribonuclease,
bound molecules of *in vitro*-synthesized RNA contained
poly(A) the size of *in vivo*-synthesized, poly(A)-con-
taining heterogeneous nuclear RNA. Polyacrylamide
gel electrophoresis showed *in vitro*-synthesized poly-
(A) comigrating with *in vivo*-synthesized poly(A). It
is concluded that any RNA labeled *in vitro* with [^{3}H]-
uridine triphosphate found to contain poly(A) must
have received poly(A) synthesized during the *in vitro*
incubation period.

2937   METHYLATED, BLOCKED 5' TERMINI IN HeLa
        CELL mRNA. (Eng.) Furuichi, Y. (Roche
Inst. Mol. Biol., Nutley, N.J.); Morgan, M.; Shatkin,
A. J.; Jelinek, W.; Salditt-Georgieff, M.; Darnell,
J. E. *Proc. Natl. Acad. Sci. USA* 72(5):1904-1908;
1975.

The presence of terminal m^{7}G(5') ppp(5')N^m-Np "caps"
on HeLa cell messenger RNA (mRNA) was studied. Sus-

pension cultures of HeLa cells were grown in Eagle's
medium and labeled with ^{32}P or $[methyl ^3H]$methionine.
Enzymatic digestion of mRNA was accomplished with
RNase T$_2$, RNase A, *Penicillium* nuclease (P$_1$), bacterial alkaline phosphate (BAP), and/or nucleotide
pyrophosphatase. Nucleotide and oligonucleotide
analysis included column chromatography on DEAE-
cellulose high voltage electrophoresis alone and
coupled with DEAE paper electrophoresis, and descend-
ing paper chromatography. HeLa cells exposed to
$[methyl-^3H]$methionine were found to contain radioactivity in poly(A)-terminated molecules the size of
mRNA. Further analysis on the labeled oligonucleotides eluted from DEAE revealed two "caps". "Cap 1"
consisted of m^{7}G and a second methylated nucleoside
of 2'-O-methylguanosine and/or 2'-O-methyluridine;
"cap 2" contained two adjacent 2'-O-methylated nucleosides, one of which was retained in the P$_1$-resistant
portion of the cap. Enzymatic digestion of mRNA followed by a "two-dimensional fingerprint" demonstrated
the presence of a variety of "capped" structures in
HeLa cell mRNA (designated 0.5-1 cap 1 mRNA molecule)
and the presence of N^{6}mA. Studies on the localization of the methylated nucleotides within HeLa cell
mRNA showed about 33% of the $[methyl-^3H]$methionine in
N^6-methyl-adenylic acid after RNase T$_2$ digestion,
indicating that it is not adjacent to the capped 5'
terminus. HeLa cell mRNA appeared to contain several
forms of 5'-terminal, "capped" oligonucleotide structures. All were terminated with m^{7}G linked *via* three
phosphates to either A^m, G^m, U^m, or C^m. The methylated cap structures were grouped into two general
types: m^{7}G pppN^m–N^p and m^{7}GpPpN^m–N^m–NP. The great
variabilities observed suggest important effects of
cell growth conditions of the composition of caps in
mRNA.

2938 CONTROL OF rRNA SYNTHESIS: EFFECT OF PRO-
    TEIN SYNTHESIS INHIBITION. (Eng.) Franze-
Fernández, M. T. (Departamento de Química Biológica,
Facultad de Farmacia y Bioquímica, Universidad de
Buenos Aires, Junín 956, Buenos Aires, Argentina);
Fontanive-Sanguesa, A. V. *FEBS Lett.* 54(1):26-29;
1975.

The time required to synthesize the 45S ribosomal RNA
(rRNA) precursor in control and protein synthesis-
inhibited Ehrlich ascites tumor cells was determined.
[^{3}H]uridine was added to cell cultures (1 x 10^6 cells
/ml) and the incubation was continued for periods ranging from 7-18 min. Other cells were incubated for 90
min, after which 0.25 µg/ml pactamycin was added and
the incubation was continued for 15 min; the cells
were then labeled for 8-33 min. The amount of labeled RNA was determined by polyacrylamide gel electrophoresis. At 14 and 18 min pulses of [^{3}H]uridine,
a small peak of radioactivity appeared at the 32S position in the gels from control cells. No 32S peak
appeared, even after 30 min of labeling, in the gels
from cells incubated with pactamycin. After an initial delay of about 5.5 min, radioactivity increased
linearly in cells incubated without pactamycin. This
delay reflected the time at which linear incorporation of the label into total RNA began, plus the time
required to transcribe a complete 45S RNA chain. Incorporation of [^{3}H]uridine into total RNA was linear

Different rates of purine nucleotide catabolism were produced by varying the concentration of 2-deoxyglucose in culture in order to investigate the regulation of enzyme activities in intact Ehrlich ascites tumor cells. Tumor cells were incubated with 5.5 mM glucose for 20 min; 100 μM [^{14}C]adenine was added, and incubation was continued for 30 min to synthesize [^{14}C]ATP. The same procedure was followed to prepare cells containing [^{14}C]GTP. Unutilized purine bases were removed and ATP or GTP catabolism was induced by resuspending the cells in medium containing 5.5 mM deoxyglucose or mixtures of glucose and deoxyglucose with a final total hexose concentration of 5.5 mM or with 50 or 100 μM 2,4-dinitrophenol in the absence of glucose. The concentrations of radioactive purine nucleotides, nucleosides, and bases were measured during the course of nucleotide catabolism. From these data, the apparent activities of enzymes of nucleotide catabolism were calculated. The linear relationship between the amount of adenylate deaminated and the amount formed suggested that the latter is the major factor in the determination of the rate of adenylate deaminase activity. The ATP concentrations did not markedly affect adenylate deaminase activity except as source of the substrate. A linear relationship was also seen between inosinate dephosphorylation and formation, suggesting that a major regulatory factor is the rate of inosinate formation. A few points, however, deviated from this linear relationship and these points corresponded to the experimental conditions that lead to the accumulation of radioactive inosinate to concentrations greater than 500 nM/g. A linear relationship was also seen between the rate of adenylate dephosphorylation and adenylate formation except under those conditions that led to the accumulation of radioactive adenylate to concentrations greater than 200 nM/g. Dehydrogenation of inosinate proceeded at a much slower rate than dephosphorylation, but a linear relationship was again seen between the amount of inosinate dehydrogenated and the amount formed, except under conditions that led to the accumulation of radioactive inosinate to concentrations greater than 150 nM/g. Dinitrophenol induced ATP catabolism by a different mechanism than deoxyglucose. The major products of dinitrophenol-induced ATP catabolism were mainly the nucleoside monophosphates, whereas these tended to be dephosphorylated when ATP catabolism was induced by deoxyglucose. Although the basis for this difference is not clear, it is concluded that substrate concentration is not highly important in the regulation of these processes.

2941   IMBALANCE OF PURINE METABOLISM IN HEPA-
          TOMAS OF DIFFERENT GROWTH RATES AS EX-
PRESSED IN BEHAVIOR OF GLUTAMINE-PHOSPHORIBOSYL-
PYROPHOSPHATE AMIDOTRANSFERASE (AMIDOPHOSPHORIBO-
SYLTRANSFERASE, EC 2.4.2.14). (Eng.) Prajda, N.
(Indiana Univ. Sch. Medicine, Indianapolis, Indiana
46202); Katunuma, N.; Morris, H. P.; Weber*, G.
Cancer Res. 35(11/Part 1):3061-3068; 1975.

The behavior of glutamine-phosphoribosylpyrophosphate amidotransferase was determined in normal,

differentiating, and regenerating liver and in a spectrum of hepatomas of widely different growth rates. The liver and tumor enzymes were measured in 100,000 x g supernatants prepared from 20% tiss homogenates containing 0.25 M sucrose and 1 mM MgCl$_2$. Kinetic studies were carried out on the amidotransferase in the crude supernatant from liv and rapidly growing hepatoma 3924A so that under optimum standard assay conditions only the enzyme amount would be the limiting factor. The liver ar hepatoma enzyme exhibited apparent Km values for: glutamine, 1.7 and 2.3 mM; MgCl$_2$, 0.7 and 1.1 mM, and phosphoribosylpyrophosphate. S$_{0.5}$ for 0.9 and 0.4 mM, respectively. The liver amidotransferase showed a sigmoid curve for phosphoribosylpyrophosphate, but under the same conditions the behavior of the hepatoma amidotransferase was similar to Michaelis-Menten kinetics. The liver and hepatoma enzymes exhibited a pH optimum at approximately 6.5-7.2. However, the plateau for liver was broac but for the tumor it was sharp. Neither enzyme had activity under pH 4.8. At and above pH 9.0 th hepatoma exhibited no amidotransferase activity; however, the liver enzyme retained approximately 50% of its maximal activity. A standard assay was developed for liver and hepatoma in which good proportionality was achieved over a 120-min incuba tion period and with various amounts of enzyme added. In a spectrum of hepatomas, where the grow rate of the various tumor lines was 12.4-0.5 mo, t amidotransferase activity in the average cell was significantly increased in all neoplasms to approx mately 175-256% of the values observed in the norm liver of control rats of the same strain, sex, age and weight. In the regenerating liver the amido-transferase activity in the average cell was signi ficantly increased at 24 hr (142%), 48 hr (181%), 72 hr (124%) after partial hepatectomy above the values observed in the sham-operated control liver The activity returned to normal 96 hr after operation. In postnatal differentiation the amidotrans ferase activity in the average liver cell at 1, 6, 18, and 25 days of age was 45, 46, 50, and 59%, respectively, of the activity observed in the live of the normal adult rat. Since the tumor amido-transferase was increased in activity, it exhibite more favorable kinetics for phosphoribosylpyrophos phate and was less inhibited by AMP; these changes in the activity and molecular properties of this key purine-synthesizing enzyme should provide an increase in the capacity of the purine-synthetic pathway. These alterations in gene expression, wh are manifested in the behavior of amidotransferase in the hepatoma, should confer selective advantage to the neoplastic cells.

2942   FOLATE METABOLISM IN THE RAT LIVER DUR
          REGENERATION AFTER PARTIAL HEPATECTOMY.
(Eng.) Barbiroli, B. (Istituto de Chimica Biolo-
gica e di Biochimica Applicata dell'Universita di
Bologna, Via Irnerio 48, 40126 Bologna, Italy);
Bovina, C.; Tolomelli, B.; Marchetti, M. Biochem.
J. 152(2):229-232; 1975.

Folate metabolism was studied during the early phases of liver regeneration after partial hepatec

tomy in seven-wk-old male Wistar rats accustomed
to eating during the first eight hours of a daily
12-hr dark period. The content of 5-methyltetra-
hydrofolate was drastically decreased during the
first hours of regeneration. The total HCO-H_4-
folate coenzymes showed a constant decrease during
the first three days of regeneration, and a continuous interconversion between 5-HCO-H_4folate and
10-formyltetrahydrofolate. 10-Formyltetrahydro-
folate synthetase, serine hydroxymethyltransferase,
and 5,10-methylenetetrahydrofolate dehydrogenase
activities were relatively low during the first
hours after the operation, and increased only
several hours later. The increase in enzyme activities showed a stepwise pattern, apparently due to
an interaction between the regeneration process
and the controlled feeding schedules.

2943   INCREASED PARTICULATE AND DECREASED SOLU-
        BLE GUANYLATE CYCLASE ACTIVITY IN REGENER-
ATING LIVER, FETAL LIVER, AND HEPATOMA. (Eng.) Ki-
mura, H. (Dept. Intern. Medicine, Univ. Virginia,
Charlottesville); Murad, F. *Proc. Natl. Acad. Sci.
USA* 72(5):1965-1969; 1975.

The activities of soluble and particulate guanylate
cyclase were determined in regenerating rat liver,
fetal and neonatal rat liver, and hepatoma. Male
Sprague-Dawley rats were partially hepatectomized;
immediately afterwards, some of these rats were
injected ip with 0.1 mg/100 g cycloheximide. Female
ACI rats were inoculated sc with Morris hepatoma
3924A 47 days before sacrifice. Guanylate cyclase
activity was determined by radioimmunoassay of
cyclic GMP formed in culture. These tissues
showed increased particulate and decreased soluble
enzyme activities in comparison to normal adult
rat liver. The particulate activity increased 12
hr after partial hepatectomy, reached maximal ac-
zyme activity decreased within eight hours and continued to decline. The activity of homogenates did
not change. Guanylate cyclase activity was increased
in plasma membrane and microsome fractions from regenerating liver. Cycloheximide prevented the increase in particulate guanylate cyclase activity and
the increased incorporation of L-^{3}H-leucine (10 μCi,
ip, 30 min prior to sacrifice) into protein after
the partial hepatectomy. Decreased soluble and increased particulate enzyme activities were found in
fetal liver. After birth, the soluble activity increased and the particulate activity decreased.
Seven to 14 days after birth, the activities of
soluble and particulate fractions were similar to
those of adult rat liver. In hepatoma 3924A, the
activity of particulate guanylate cyclase was 9-fold
greater and that of the soluble enzyme was 50% that
of normal liver. The results suggest that guanylate
cyclase activity and its subcellular distribution may
be related to liver growth through some unknown mechanism.

2944   ULTRASTRUCTURAL EVIDENCE OF ECTOGLYCO-
        SYLTRANSFERASE SYSTEMS. (Eng.) Porter,
C. W. (Grace Cancer Drug Center, Roswell Park Memo-
rial Inst., 666 Elm St., Buffalo, N.Y. 14263);
Bernacki, R. J. *Nature* 256(5519):648-650; 1975.

of 25 controls. Of 11 cases of predominantly monocytic leukemia, ten had elevated scores for the enzyme; of 29 cases of acute myelo-monocytic leukemia,
21 had high scores; and of 15 cases classified as
acute myeloblastic leukemia, all but four had low
scores. The results point to the possible usefulness of the enzyme as a monocyte marker.

2948   IDENTITY, RELEASE, AND BINDING OF MITOCHON-
        DRIAL-BOUND HEXOKINASES IN MAMMARY GLANDS
AND ADENOCARCINOMAS OF LACTATING MICE. Bartley, J. C.
(Children's Hosp. Medical Center of Northern Cali-
fornia, Oakland, Calif. 94609); Barber, S.; Abraham,
S. *Cancer Res.* 35(7):1649-1653; 1975.

The mitochondrial-bound hexokinases (ATP:D-hexose 6-
phosphotransferase) of mammary adenocarcinoma and of
normal gland were compared in lactating C3H/Blmrl
mice. Some of the tumors used arose spontaneously in
old mice, and some were from a transplanted line.
Treatment of mitochondria isolated from both the normal and neoplastic tissue with 0.5 M NaCl or 0.1 mM
glucose 6-phosphate effected the release of about 50%
of the bound hexokinase. In the presence of Mg^{+2} enzyme from either source attached to mitochondria from
either tissue and in all combinations to the same extent. Identification of the isoenzyme complement in
the mitochondrial extract by diethylaminoethylcellu-
lose chromatography revealed only types I and II. In
the tumor, the hexokinase activity in both the cytosol
and the fraction solubilized from mitochondria was
predominantly in the form of type I ($\sim$60%). In contrast, the activity released from mitochondria isolated from normal gland was predominately type II,
while the cytosol contained almost equivalent amounts
of types I and II. The difference in the complement
of released hexokinase may provide a useful means of
distinguishing between normal and neoplastic tissue.

2949   PROPORTIONAL ACTIVITIES OF GLYCEROL KIN-
        ASE AND GLYCEROL 3-PHOSPHATE DEHYDROGEN-
ASE IN RAT HEPATOMAS. (Eng.) Harding, J. W., Jr.
(Dept. Chem., Univ. Delaware, Newark); Pyeritz, E.
A.; Morris, H. P.; White, H. B., III. *Biochem. J.*
148(3):545-550; 1975.

The activities of glycerol 3-phosphate dehydrogenase,
glycerol kinase, lactate dehydrogenase, "malic" enzyme (L-malate-NADP$^+$ oxidoreductase), and the β-oxo-
acyl-(acyl-carrier protein) reductase component of
the fatty acid synthetase complex were measured in
nine hepatoma lines (eight in rats, one in mouse)
and in the livers of host animals. With the single
exception of Morris hepatoma 16, which had unusually
high glycerol 3-phosphate dehydrogenase activity, the
activities of glycerol 3-phosphate dehydrogenase and
glycerol kinase were highly correlated in normal
livers and hepatomas (r=0.97; P < 0.01). The lowest
activities for both enzymes were found in the fast-
growing Morris hepatoma 3924A and the highest activities were found in the slower growing rat hepatomas
7787, 9618B, and the mouse tumor BW7756. In general,
the activities of both enzymes increased with decreasing growth rate as is typical of gluconeogenic
enzymes. The activities of the two enzymes were not
strongly correlated with the activities of any of the

other three enzymes. The primary function of hepatic glycerol 3-phosphate dehydrogenase appears to be in gluconeogenesis from glycerol.

2950   ISOENZYMES OF ALANINE ARYLAMIDASE (AAP, EC 3.4.11.2) AND GAMMA-GLUTAMYLTRANSPEP-TIDASE (GGTP, EC 2.3.2.2) IN CHRONIC PANCREATITIS AND PANCREAS NEOPLASM. (Ger.) Bornschein, W. (II. Medizinische Klinik Rechts der Isar der Technischen Hochschule Munchen, Munchen, West Germany). *Clin. Chim. Acta* 61(3):325-333; 1975.

Gamma-glutamyltranspeptidase (GGTP), alanine arylamidase (AAP), and isoenzyme patterns were studied in sera from 22 patients with chronic pancreatitis and from 16 patients with neoplasms of the pancreas. Gel electrophoresis on agar revealed two fractions of AAP and GGTP isoenzymes: alpha-1 and alpha-2 (GGTP between alpha-1 and beta-globulin). The alpha-1 fraction of AAP and GGTP seens to be a specific liver isoenzyme. The slower fraction of both enzymes was also detected in chronic pancreatic diseases and cholestatic diseases as well as in neoplasms of the liver, pancreas, and biliary tract. However, the practical importance of the findings is diminished by the large variation coefficients of the results. A significantly low ratio of alpha-1 to alpha-2 fraction (or beta-globulin) of AAP and GGTP was found in the group with neoplasms of the pancreas, and especially in the head of the pancreas as compared with the group with intrahepatic cholestasis. A narrow fraction, designated as isoenzyme 3, occurred in a few cases of tumors in the pancreatobiliary region; it was not organ-specific.

2951   PURIFICATION OF RNA-INSTRUCTED DNA POLYMERASE FROM HUMAN LEUKEMIC SPLEENS. (Eng.) Witkin, S. S. (Memorial Sloan-Kettering Cancer Center, New York, N.Y. 10021); Ohno, T.; Spiegelman, S. *Proc. Natl. Acad. Sci. USA* 72(10): 4133-4136; 1975.

Isopycnic separation of virus particles and their conversion to cores by nonionic detergents provided a successful method for purifying a reverse transcriptase from the spleen of a patient with chronic lymphocytic leukemia. Spleen cell supernatants were layered on discontinuous sucrose gradients; the material present in the 25-50% sucrose interface was removed and treated with 30% solid $(NH_4)_2SO_4$. Viral cores were prepared by treating the $(NH_4)_2SO_4$ precipitates with 1% Nonidet P-40 and 0.1 M dithiothreitol and 30-65% sucrose gradient centrifugation. Particles possesssing a density of 1.16 g/ml and encapsulating a 70S RNA and a RNA-instructed DNA polymerase (reverse transcriptase) were isolated. These particles were converted to cores with a density of 1.26 g/ml and containing the enzyme-RNA complex, in complete analogy to the known RNA tumor viruses of avian and murine origin. The reverse transcriptase was purified from the cores by column chromatography to a stage showing a single major protein band of 70,000 daltons in gel electrophoresis. The enzyme was capable of transcribing heteropolymeric RNA into DNA complements as demonstrated

fetoprotein was 77%. The occurrence of this enzyme in sera of four patients, whose case histories are presented, with hepatoma was, accordingly, independent of the serum alpha-fetoprotein concentration, and also independent of the appearance of the Regan or the Nagao isoenzymes and of the serum alkaline phosphatase activity. Patients with the enzyme had a massive type of hepatocellular carcinoma with grade III differentiation by Edmondson's classification. The detection of this enzyme in serum may be of help in confirming the diagnosis of hepatoma.

2956   THE RELEASE OF HIGH-MOLECULAR-WEIGHT ALKALINE PHOSPHATASE AND LEUCINE AMINO-PEPTIDASE INTO THE MEDIA OF CULTURED HUMAN CELLS. (Eng.) Singer, R. M. (Tufts Cancer Res. Center, Boston, Mass. 02111); White, L. J.; Perry, J. E.; Doellgast, G. J. *Cancer Res.* 35(11/Part 1):3048-3050; 1975.

Using exclusion from Sepharose 4B as a criterion, a high-molecular-weight form of alkaline phosphatase and of leucine aminopeptidase were found which were released into the culture media by the FL amnion cell line. A low-molecular-weight form of leucine aminopeptidase was also found to contribute to the total levels of this enzyme in the media. The levels of these enzymes increased during the growth cycle of the culture, paralleling the increase in cell density; this suggests that the two events may be related. This phenomenon in culture suggests a possible explanation for the appearance of similar enzyme forms in patient serum and fluids originating from diseased tissue.

2957   NATURE OF THE INCREASE IN RENAL ORNITHINE DECARBOXYLASE ACTIVITY AFTER CYCLOHEXIMIDE ADMINISTRATION IN THE RAT. (Eng.) Levine, J. H. (Medical Univ. South Carolina, 80 Barre St., Charleston, S.C. 29407); Nicholson, W. E.; Orth, D. N. *Proc. Natl. Acad. Sci. USA* 72(6):2279-2283; 1975.

The increase in rat renal ornithine decarboxylase activity after cycloheximide administration was studied in Holtzman rats to determine whether it was mediated primarily by intracellular events within the kidney, or was secondary to the action of pituitary or adrenocortical hormones, whose release was stimulated by the antimetabolite. Within 30 min, 250 µg of cycloheximide essentially abolished renal protein synthesis. Renal ornithine decarboxylase activity was reduced approximately 70% one hour after ip administration of doses of cycloheximide that also inhibited renal protein synthesis by 68-95% within one hour. Protein synthesis began to recover by the second hour, accompanied by a rise in decarboxylase activity. Peak ornithine decarboxylase activity was directly proportional to cycloheximide doses up to 250 µg; larger doses were inhibitory. Plasma corticosterone rose rapidly after cycloheximide treatment peaking at two hours. Corticosterone response was also dose-dependent up to 250 µg, but larger doses were inhibitory. Hypophysectomy greatly reduced baseline renal decarboxylase activity within nine hours, and all but abolished the increase in enzyme

activity normally seen after cycloheximide administration to the intact rat. The hypophysectomized
animal exhibited apparent increased sensitivity to
cycloheximide. As protein synthesis was recovering
in the hypophysectomized animals, renal decarboxylase
activity responded adequately to the injection of a
crude pituitary extract. These data suggest that
renal ornithine decarboxylase turnover is rapid,
that the early fall in its activity after cycloheximide treatment reflects inhibition of new enzyme
synthesis, and that the subsequent rise in decarboxylase activity requires the action of pituitary
factors released in response to the stress of cycloheximide administration.

2958   SPERMINE INHIBITS INDUCTION OF ORNITHINE
        DECARBOXYLASE BY CYCLIC AMP BUT NOT BY
DEXAMETHASONE IN RAT HEPATOMA CELLS. (Eng.) Theoharides, T. C. (Yale Univ. Sch. Medicine, New Haven,
Conn. 06510); Canellakis, Z. N. *Nature* 255(5511):
733-734; 1975.

The induction of ornithine decarboxylase (ODC) activity in rat hepatoma cells by cyclic AMP (cAMP) and
dexamethasone was studied. Induction by cAMP reached
a maximal 4-fold increase after three hours and declined to basal levels after seven hours; dexamethasone induction reached a 6-fold increase after 12
hours, and continued for 20 hours. A comparative
study of the types of induction revealed that 10^{-5} M
spermine abolished the cAMP induction and depressed
the ODC activity below basal level; 6×10^{-5} M spermidine was equally effective. However, neither concentration affected the dexamethasone-induced ODC activity. Neither cycloheximide nor actinomycin D affected
the basal enzyme level. Cycloheximide abolished induction by both agents; actinomycin D completely inhibited induction by dexamethasone, but only partially
inhibited induction by cAMP. The low effective
concentration of spermine, its speed of action, plus its
selective inhibition of cAMP-induced activity suggests
a unique role for spermine and a post-transcriptional
site of action for cAMP.

2959   CYCLIC AMP-MEDIATED INDUCTION OF ORNI-
        THINE DECARBOXYLASE OF GLIOMA AND NEURO-
BLASTOMA CELLS. (Eng.) Bachrach, U. (Natl. Heart
and Lung Inst., Bethesda, Md. 20014). *Proc. Natl.
Acad. Sci. USA* 72(8):3087-3091; 1975.

When confluent cultures of C6-BU-1 glioma or N115
neuroblastoma cells were treated with norepinephrine
or isoproterenol, and prostaglandin E_1 or adenosine.
L-ornithine decarboxylase activity was increased
significantly in both cultures. Ornithine decarboxylase activity was elevated in confluent C6-BU-1
glioma cells treated with dibutyryl adenosine-3':5'-
cyclic monophosphate and theophylline, or after the
glioma cells were fed with a serum-depleted medium
in the presence of catecholamines and inhibitors of
cyclic nucleotide phosphodiesterase. The activity
of the enzyme increases 500- to 1000-fold, 2-6 hr
after stationary-phase N115 neuroblastoma cells were
fed with a serum-free medium, supplemented with phosphodiesterase inhibitors, plus adenosine, or plus

rated with cofactor. In vitamin-deficient animals,
only 6% of the tumor enzyme was saturated with the
cofactor. The percent saturation of host liver TAT
varied, with minimal values found in the vitamin-
deficient animals. Hepatic and tumor pyridoxal phosphate content of pair-fed animals was unusually high
(10 μg/g); in vitamin-deficient animals, only the
coenzyme content of hepatomas was high (7.0 μg/g).
The results indicate that the presence of the tumor
altered the (a) specific activity level of TAT and
tissue content of cofactor, (b) pattern of hormonal
induction of the enzyme, and (c) effects of the absence of dietary pyridoxine on TAT induction observed in animals without tumors.

2963   MELANOCYTE-STIMULATING HORMONE PROMOTES
        ACTIVATION OF PRE-EXISTING TYROSINASE
MOLECULES IN CLOUDMAN S91 MELANOMA CELLS. (Eng.)
Wong, G. (Yale Univ. Sch. of Medicine, New Haven,
Conn. 06510); Pawelek, J. *Nature* 255(5510):644-646;
1975.

The control of melanocyte-stimulating hormone (MSH)
by cyclic AMP was studied in Cloudman S91 melanoma
cells. Metaphase cells were synchronized by exposure
to colchicine and MSH, actinomycin D, or cycloheximide.
Tyrosinase activity was found to increase in response
to MSH during the G2 phase of the cell cycle; the increase was not suppressed by either actinomycin D or
cycloheximide. In nonsynchronized cells, MSH did not
alter the decay of tyrosinase activity. MSH also
apparently promoted the inactivation of an inhibitor
of tyrosinase functions. Partial purification of
tyrosinase affinity chromatography and subsequent activity studies confirmed the presence of tyrosinase
inhibitors in the cells. A further purification procedure indicated that the same number of enzyme molecules were present in cells regardless of pretreatment
with MSH. It is concluded that MSH does not promote
the synthesis of new molecules of the soluble form of
tyrosinase. The data suggest that the mode of action
of MSH is to partially relieve an inhibition of the
enzyme, and that cAMP control is exerted at a posttranslational level.

2964   THE ANDROGENIC REGULATION OF THE ACTIVI-
        TIES OF ENZYMES ENGAGED IN THE SYNTHESIS
OF DEOXYRIBONUCLEIC ACID IN RAT VENTRAL PROSTATE
GLAND. (Eng.) Rennie, P. S. (Imperial Cancer Res.
Fund, P.O. Box 123, Lincoln's Inn Fields, London
WC2A 3PX, U.K.); Symes, E. K.; Mainwaring, W. I. P.
Biochem. J. 152(1):1-16; 1975.

The restoration of mitosis and growth of the prostate gland of castrated Sprague-Dawley rats (270-
300 g) by androgens provides an experimental system
for studying the hormonal regulation of enzymes engaged in DNA replication. The most marked change
in the many DNA polymerase activities found in prostate after androgenic stimulation was a 9S form
using denatured DNA as a template. This enzyme increased 14-fold during four days of testosterone
administration. Androgenic stimulation of castrated
rats led to a 100-fold increase in prostate DNA
ligase activity which was not mimicked in spleen
extracts. Thymidine kinase provided a sensitive in-

dicator of the hormonal regulation of DNA replication. By electrophoretic criteria, one novel form
of the enzyme appeared precisely with the onset of
mitosis. Only one form (pI7.0) of the DNase activities in the prostate gland can be said to be active in DNA replication. Androgenic stimulation
of the prostate gland led to the appearance of a
component capable of denaturing or unwinding prostate DNA. This component was distinct from RNA
or DNA polymerase activities on the basis of several
physicochemical characteristics. The conspicuous
feature of all the changes in enzyme activities
evoked by androgens in the prostate gland is their
tissue- and steroid-specificity. Such changes could
not be mimicked in liver or spleen and the regulatory role of androgens could not be stimulated by
other classes of steroid hormones. On the basis of
studies with the anti-androgen cyproterone acetate,
it is concluded that the changes are initially mediated by the androgen-receptor system and the high-
affinity binding of 5α-dihydrotestosterone in the
prostate gland.

2965   STUDY OF GROWTH HORMONE SECRETION IN
        LUNG CARCINOMA. (Eng.) Claeys-de
Clercq, P. (Service de Médecine et d-Investigation
Clinique de l-Institut J. Bordet, Bruxelles, Belgium); Levin, S.; Borkowski, A. *J. Cancer* 11(8):
565-569; 1975.

The possibility of ectopic growth hormone (HGH) secretion in 26 patients with lung carcinoma was studied by measuring plasma HGH during an oral glucose
tolerance test (GTT). Only four patients had raised
basal HGH concentrations and these were only slightly
above normal, and did not result from an autonomous
secretion since they were modified by the GTT.
Five patients had a paradoxical response to glucose.
In the other 18 patients, basal HGH was normal and
remained within normal limits after glucose administration. No evidence for a significant ectopic HGH
secretion was found. A slight ectopic secretion
without increased plasma HGH concentrations cannot
be excluded, since during the GTT some plasma HGH
was measurable in most of our 26 patients. There
was no correlation between plasma HGH and any other
clinical or biological parameter, including clubbing.

2966   PRE-PROPARATHYROID HORMONE IDENTIFIED BY
        CELL-FREE TRANSLATION OF MESSENGER RNA
FROM HYPERPLASTIC HUMAN PARATHYROID TISSUE. (Eng.)
Habener, J. F. (Harvard Medical Sch., Boston, Mass.
02114); Kemper, B.; Potts, J. T., Jr.; Rich, A. *J.
Clin. Invest.* 56(5):1328-1333; 1975.

The earliest precursor of parathyroid hormone (PTH)
and the initial product of the gene for PTH was
identified by translating messenger RNA (mRNA) isolated from human parathyroid tissue (5 g) from a
patient with primary chief-cell hyperplasia. RNA
translation was performed in a heterologous cell-
free system derived from wheat germ. Three fractions of RNA were observable on a sucrose gradient;
one fraction (C) was significantly more active in
directing amino acid incorporation into protein.

2978   L-LACTATE TRANSPORT IN EHRLICH ASCITES TUMOR CELLS [abstract]. (Eng.) Spencer, T. L. (Johns Hopkins Univ. Sch. Med., Baltimore, Md.). *Fed. Proc.* 34(3):250; 1975.

2979   CHARACTERISTICS OF FIBRINOLYSIN SECRETED BY CULTURED RAT BREAST CARCINOMA CELLS [abstract]. (Eng.) Wu, M. C. (Univ. Miami Sch. Med., Fla.); Schultz, D. R.; Yunis, A. A. *Fed. Proc.* 34(3):529; 1975.

2980   RELATION OF INTEGRAL MEMBRANE PROTEINS AND PERIPHERAL BOVINE SERUM PROTEINS TO HeLa 71 GROWTH CONTROL [abstract]. (Eng.) Johnson, H. V. (Oklahoma Med. Res. Found., Oklahoma City); Killion, J. J.; Griffin, M. J. *Fed. Proc.* 34(3): 614; 1975.

2981   IDENTIFICATION OF A PROTEIN WHICH SELEC-TIVELY REDUCES THE APPARENT BINDING AF-FINITY FOR CYCLIC AMP IN HEPATOMA CELL EXTRACTS [abstract]. (Eng.) Stellwagen, R. H. (Univ. South. California Sch. Med., Los Angeles); Mackenzie, C. W., III. *Fed. Proc.* 34(3):543; 1975.

2982   GLYCOPROTEINS ISOLATED FROM THE PLASMA MEMBRANE OF LYMPHOCYTES MAINTAINED IN LONG TERM CULTURE [abstract]. (Eng.) O'Brien, K. J. (Med. Sch., Univ. Minnesota, Minneapolis); Edstrom, R. D. *Fed. Proc.* 34(3):615; 1975.

2983   CHEMICAL STRUCTURE OF PARAPROTEINS OF MYELOMAS. (Eng.) Roholt, O. A. (Dept. Immunology and Immunochemistry Res., Roswell Park Mem. Inst., Buffalo, N.Y.); Seon, B.-K.; Grossberg, A. L.; Pressman, D. *Transplant. Proc.* 7(2):215-218; 1975.

2984   AUTORADIOGRAPHIC DEMONSTRATION OF DNA REPLICATION IN ULTRATHIN SECTIONS OF PLASTIC-EMBEDDED TISSUES USING AN EXOGENEOUS DNA POLYMERASE. (Eng.) Geuskens, M. (Institut de Recherches Scientifiques sur le Cancer, Villejuif, France); de Recondo, A. M.; Chevaillier, P. *Chromosoma* 52(2):175-188; 1975.

2985   ANALYSIS OF THE DEOXYRIBONUCLEIC ACID CONTENT IN THE EPITHELIUM OF THE LARGE INTESTINE AND ITS APPLICATION TO THE STUDY OF CARCINOGENESIS. (Jpn.) Shindo, K. (Osaka Univ., Medical School, Japan). *Osaka Daigaku Igaku Zasshi* 26(9-12):313-329; 1974.

2986   DNA SYNTHESIS BY AMV REVERSE TRANS-CRIPTASE: TEMPLATE-SPECIFIC CATION EF-FECTS [abstract]. (Eng.) Marcus, S. L. (Mem. Sloan-Kettering Cancer Cent., New York, N.Y.); Modak, M. J. *Fed. Proc.* 34(3):609; 1975.

MISCELLANEOUS

2987   EXTENDED RNA SYNTHESIS BY ISOLATED NUCLEI
         FROM PITUITARY TUMOR CELLS [abstract].
(Eng.) Martin, T. F. J. (Harvard Sch. Dent. Med.,
Boston, Mass.); Biswas, D. K.; Tashjian, A. H., Jr.
Fed. Proc. 34(3):629; 1975.

2988   CHARACTERIZATION OF NOVIKOFF HEPATOMA
         MESSENGER RNA METHYLATION [abstract].
(Eng.) Desrosiers, R. (Dep. Biochem., Michigan
State Univ., East Lansing); Friderici, K.; Rottman,
F. *Fed. Proc.* 34(3):628; 1975.

2989   GROWTH DEPENDENT CHANGES OF GLYCOLIPID
         COMPOSITIONS OF MOUSE ASCITES TUMOR
CELLS. (Eng.) Egawa, K. (Inst. Medical Science,
Univ. of Tokyo, Shirokanedai, Minato-ku, Tokyo 108,
Japan); Choi, Y. S.; Tanino, T. *Jpn. J. Exp. Med.*
45(3):155-159; 1975.

2990   COMPARISON OF GLYCOSPHINGOLIPID METABOLISM
         IN MOUSE GLIAL TUMORS AND CULTURED CELL
STRAINS OF NEURAL ORIGIN [abstract]. (Eng.) Stool-
miller, A. C. (Dep. Pediatr., Univ. Chicago, Ill.);
Dawson, G.; Schachner, M. *Fed. Proc.* 34(3):634; 1975.

2991   DRUG METABOLISM BY MICROSOMES FROM EXTRA-
         HEPATIC ORGANS OF RAT AND RABBIT PREPARED
BY CALCIUM AGGREGATION. (Eng.) Litterst, C. L.
(Natl. Cancer Inst., Bethesda, Md. 20014); Mimnaugh,
E. G.; Reagan, R. L.; Gram, T. E. *Life Sci.* 17(5):
813-818; 1975.

2992   EFFECTS OF LYSOZYME ON THE PATTERN OF
         COLONY GROWTH OF CULTURED RAT HEPATOMA
CELLS [abstract]. (Eng.) Quincy, D. A. (Inst.
Cancer Res., Columbia Univ., New York, N.Y.);
Osserman, E. F. *Fed. Proc.* 34(3):269; 1975.

2993   MECHANISM OF ALKALINE PHOSPHATASE OF
         MURINE PLACENTA AND LYMPHOMA [abstract].
(Eng.) Davis, R. (Atlanta Univ., Ga.); Prioleau,
J. C., III; Lumb, J. R. *Fed. Proc.* 34(3):599; 1975.

2995   GUANYLATE CYCLASE, CYCLIC GMP PHOSPHO-
         DIESTERASE AND CYCLIC GMP IN CULTURED
FIBROBLASTIC CELLS [abstract]. (Eng.) Nesbitt, J.
(Natl. Inst. Health, Bethesda, Md.); Russell, T. R.;
Miller, Z.; Pastan, I. *Fed. Proc.* 34(3):616; 1975.

SERUM ALBUMIN, BOVINE
    LYSOSOMES, KIDNEY, LIVER, 2519*

ACETOHYDROXAMIC ACID, N-FLUOREN-2-YL-
    BENZENESULFONAMIDE, 2-FLUORENYL-N-
        HYDROXY-
            CARCINOGENIC METABOLITE, 2526*
    HEPATOMA
        ACETAMIDE, N-FLUCREN-2-YL-, 2528*
        FETAL GLOBULINS, 2528*
    RNA
        BINDING SITES, 2402

N-ACETOXY-N-2-ACETYLAMINOFLUORENE
    SEE ACETAMIDE, N-(ACETYLOXY)-N-9H-
        FLUOREN-2-YL-

2-ACETYLAMINOFLUORENE
    SEE ACETAMIDE, N-FLUCREN-2-YL

ACHLORHYDRIA
    NEOPLASMS
        ACID-BASE EQUILIBRIUM, 2967

ACID PHOSPHATASE
    LEUKEMIA, LYMPHOBLASTIC
        B-LYMPHOCYTES, 2759*
        T-LYMPHOCYTES, 2759*

ACRYLAMIDE, 2-(2-FURYL)-3-(5-NITRO-2-FURYL)-
    BONE MARROW CELLS
        CHROMOSOME ABERRATIONS, 2454

ACTINOMYCIN D
    GLIOMA
        CARBOXY-LYASES, 2959
    MAMMARY NEOPLASMS, EXPERIMENTAL
        BENZ(A)ANTHRACENE, 7,12-DIMETHYL-,
            2467
        PROTECTIVE EFFECT, MOUSE, 2467
    MELANOMA
        CELL DIVISION, 2977*
        PLASMINOGEN ACTIVATORS, 2977*
        TYROSINASE, 2963
    NEUROBLASTOMA
        CARBOXY-LYASES, 2959
    SPERMATOZOA
        ABNORMALITIES, MCUSE, 2452

ACUTE DISEASE
    LEUKEMIA
        ERYTHROPOIESIS, 2780

ADENOACANTHOMA
    SEE ADENOCARCINOMA

ADENOCARCINOMA
    ANTIGENIC DETERMINANTS
        IGG, 2748*
    BLADDER NEOPLASMS
        GENETICS, 2794
    BLOOD VOLUME DETERMINATIONS
        BENZ(A)ANTHRACENE, 7,12-DIMETHYL-,
            2465
    BREAST NEOPLASMS
        RADIATION, IONIZING, 2565
    BRONCHIAL NEOPLASMS
        EPIDEMIOLOGY, SWEDEN, 2889

CHOLANTHRENE, 3-METHYL-
    HISTOLOGICAL STUDY, RABBIT, 2542*
    NEOPLASM METASTASIS, 2770
COLONIC NEOPLASMS
    CARCINOEMBRYONIC ANTIGEN, 2748*
    ENTERITIS, REGIONAL, 2827*
    IMMUNOGLOBULINS, FAB, 2748*
    UREA, N-NITROSO-N-PROPYL-, 2491
ENTERITIS, REGIONAL
    CASE REPORT, REVIEW, 2826*
ESOPHAGEAL NEOPLASMS
    CASE REPORT, 2835*
INTESTINAL NEOPLASMS
    CELIAC DISEASE, 2829*
    ENTERITIS, REGIONAL, 2826*, 2827*
    UREA, 1-BUTYL-1-NITROSO-, 2489
    UREA, N-NITROSO-N-PROPYL-, 2491
KIDNEY NEOPLASMS
    IMMUNE SERUMS, 2713*
LUNG NEOPLASMS
    CHOLANTHRENE, 3-METHYL-, 2542*
MAMMARY NEOPLASMS, EXPERIMENTAL, 2725*
    BENZ(A)ANTHRACENE, 7,12-DIMETHYL-,
        2465
    ETHYNODIOL DIACETATE, 2405
    GROWTH, 2725*
    IMMUNITY, CELLULAR, 2712*
    NORETHYNODREL, 2405
    OCCLUDING JUNCTIONS, 2917
    TRANSPLANTATION, HOMOLOGOUS, 2948
    UREA, N-NITROSO-N-PROPYL-, 2491
    VIRUS, MURINE MAMMARY TUMOR, 2618
NOSE NEOPLASMS
    OCCUPATIONAL HAZARD, 2894
    WOOD, 2894
OVARIAN NEOPLASMS
    MIGRATION INHIBITORY FACTOR, 2667
PARANASAL SINUS NEOPLASMS
    OCCUPATIONAL HAZARD, 2894
    WOOD, 2894
PROSTATIC NEOPLASMS
    EPIDEMIOLOGY, INDIA, 2886
RADIATION, IONIZING
    CASE REPORT, 2565
RECTAL NEOPLASMS
    EPIDEMIOLOGY, 2791
    STAGING AND THERAPY, 2791
STOMACH NEOPLASMS
    UREA, N-NITROSO-N-PROPYL-, 2491
THYROID NEOPLASMS
    DNA, 2934
    IMMUNITY, CELLULAR, 2712*   ADE
    RADIATION, IONIZING, 2934
    URACIL, 6-METHYL-2-THIO-, 2934
TRANSPLANTATION, HOMOLOGOUS
    T-LYMPHOCYTES, 2771
UROGENITAL NEOPLASMS
    ULTRASTRUCTURAL STUDY, 2866*
UTERINE NEOPLASMS
    UREA, N-NITROSO-N-PROPYL-, 2491   ADE
VIRUS, MURINE MAMMARY TUMOR
    PEPTIDES, 2618

ADENOFIBROMA
    MAMMARY NEOPLASMS, EXPERIMENTAL   ADE
        UREA, N-NITROSO-N-PROPYL-, 2491

ADENOMA
    ADRENAL GLAND NEOPLASMS

CARCINOGENIC ACTIVITY, RAT, 2458
FOOD CONTAMINATION
PEANUT MEAL, DETECTION, 2910*
HEPATOMA
CARCINOGENIC POTENTIAL, REVIEW,
2404
LIVER NEOPLASMS
BARBITURIC ACID, 5-ETHYL-5-PHENYL-,
2458
RNA
ACETAMIDE, THIO-, 2457

AFLATOXIN B2
FOOD CONTAMINATION
PEANUT MEAL, DETECTION, 2910*

AFLATOXIN G1
FOOD CONTAMINATION
PEANUT MEAL, DETECTION, 2910*

AFLATOXIN G2
FOOD CONTAMINATION
PEANUT MEAL, DETECTION, 2910*

AGAMMAGLOBULINEMIA
LYMPHOCYTES
NEURAMINIDASE, 2746*

AGGLUTINATION
DIBUTYRYL CYCLIC AMP
CELLS, CULTURED, 2972*
HEPATOMA
CONCANAVALIN A, 2706
PLANT AGGLUTININS, 2706
VIRUS, AVIAN LEUKOSIS-SARCOMA
CONCANAVALIN A, 2690

AIR POLLUTANTS
ASBESTOS
CONTROL TECHNOLOGY, 2434*
INDUSTRIAL WASTE, 2434*
MINING, 2434*
ASBESTOSIS
PARTICLE SIZE, 2434*

ALBUMINS
BACILLUS SUBTILIS
DNA, BACTERIAL, 2582

ALCOHOLISM
BRAIN NEOPLASMS
EPIDEMIOLOGY, 2892
BREAST NEOPLASMS
EPIDEMIOLOGY, 2892
DIGESTIVE TRACT NEOPLASMS
EPIDEMIOLOGY, 2892
GYNECOLOGIC NEOPLASMS
EPIDEMIOLOGY, 2892
LEUKEMIA
EPIDEMIOLOGY, 2892
LIVER NEOPLASMS
EPIDEMIOLOGY, 2892
PANCREATIC NEOPLASMS
EPIDEMIOLOGY, 2892
RESPIRATORY TRACT NEOPLASMS
EPIDEMIOLOGY, 2892
UROGENITAL NEOPLASMS
EPIDEMIOLOGY, 2892

ALDRICH SYNDROME
  VIRUS, POLYOMA, BK
    HYBRIDIZATION, 2657*
    ISOLATION AND CHARACTERIZATION,
    2657*

ALKALINE PHOSPHATASE
  COLONIC NEOPLASMS
    SERUM LEVELS, 2955
  GROWTH
    CELL CYCLE KINETICS, 2956
  HEPATOMA
    CASE REPORT, 2955
    FETAL GLOBULINS, 2955
  LUNG NEOPLASMS
    SERUM LEVELS, 2955
  LYMPHOMA
    ISOENZYMES, 2993*
  PANCREATIC NEOPLASMS
    SERUM LEVELS, 2955
  PLACENTA
    ISOENZYMES, 2993*
  STOMACH NEOPLASMS
    SERUM LEVELS, 2955

ALKYLATING AGENTS
  HYDRAZINE, 1,2-DIMETHYL-
    CARCINOGENIC METABOLITE, RAT, 2463

ALPHA-FETOPROTEIN
  SEE FETAL GLOBULINS

ALTITUDE
  NEOPLASMS
    EPIDEMIOLOGY, 2967

AMERICIUM
  BONE NEOPLASMS
    REVIEW, 2410

AMINO ACID NAPHTHYLAMIDASES
  PANCREATIC NEOPLASMS
    ISOENZYMES, 2950
  PANCREATITIS
    ISOENZYMES, 2950

AMINOPTERIN
  SPERMATOZOA
    ABNORMALITIES, MOUSE, 2452

AMINOTRANSFERASES
  HEPATOMA
    GROWTH, 2941

AMMONIA
  AFLATOXIN B1
    FOOD CONTAMINATION, 2509*
    REACTION PRODUCTS, 2509*, 2510*

AMOSITE
  SEE ASBESTOS

AMYLOID
  PITUITARY NEOPLASMS
    ADENOMA, 2873*

AMYLOIDOSIS
  BENZ(A)ANTHRACENE, 7,12-DIMETHYL-

    CARCINOGENIC POTENTIAL, MOUSE, 24
  PITUITARY NEOPLASMS
    ADENOMA, 2872*, 2873*

ANABASINE, 1-NITROSO-
  NICOTINE, 1'-DEMETHYL-N'-NITROSO-
    COCARCINOGENIC EFFECT, RAT, 2486

ANDROGENS
  BREAST NEOPLASMS
    EPIDEMIOLOGY, REVIEW, 2420

ANDROST-5-ENE-3,17-DIOL
  BREAST NEOPLASMS
    ESTRADIOL, 2471
  ESTRADIOL
    SPECIFIC RECEPTOR BINDING, 2471
  URETERAL NEOPLASMS
    ESTRADIOL, 2471

5ALPHA-ANDROSTAN-3-ONE, 17BETA-HYDROXY-
  BREAST NEOPLASMS
    ESTRADIOL, 2471
  ESTRADIOL
    SPECIFIC RECEPTOR BINDING, 2471

ANEMIA
  VIRUS, FELINE LEUKEMIA
    HORIZONTAL TRANSMISSION, 2648*

ANEMIA, APLASTIC
  CHROMOSOME ABNORMALITIES
    ULTRASTRUCTURAL STUDY, 2775
  ERYTHROPOIESIS
    PRECANCEROUS CONDITIONS, 2780
  MITOMYCIN C
    CHROMOSOME ABERRATIONS, 2933
    DECARBAMOYL DERIVATIVES, 2933
    DNA REPAIR, 2933
  PRECANCEROUS CONDITIONS
    CASE REPORT, 2775
    IGA, 2775
    IGG, 2775
    IGM, 2775
    IMMUNOGLOBULINS, 2775

ANEMIA, FACONI'S
  SEE ANEMIA, APLASTIC

ANEMIA, HEMOLYTIC
  GRANULOPOIETIC PROGENITOR CELLS
    COLONY FORMATION, 2915
  VIRUS, KIRSTEN MURINE LEUKEMIA
    INTERFERON, 2611
    ULTRASTRUCTURAL STUDY, 2611
  VIRUS, MURINE ERYTHROBLASTOSIS
    INTERFERON, 2611
    ULTRASTRUCTURAL STUDY, 2611

ANEMIA, MACROCYTIC
  CONTRACEPTIVES, ORAL
    FOLIC ACID, 2406
    VITAMIN B12, 2406

ANESTHETICS
  BONE MARROW CELLS
    CARCINOGENIC EFFECT, 2407
  ETHANE, 2-BROMO-2-CHLORO-1,1,1-
    TRIFLUORO-

CARCINOMA, 2678
GANGLIONEUROMA
    IMMUNE SERUMS, 2678
HODGKIN'S DISEASE
    IMMUNE SERUMS, 2678
MELANOMA
    IMMUNE SERUMS, 2678
PANCREATIC NEOPLASMS
    CARCINOMA, 2678
SARCOMA, OSTEOGENIC
    IMMUNE SERUMS, 2678
SARCOMA, RETICULUM CELL
    IMMUNE SERUMS, 2678
SYNOVIOMA
    IMMUNE SERUMS, 2678

ANTIBODIES, VIRAL
    BURKITT'S LYMPHOMA
        EPSTEIN-BARR NUCLEAR ANTIGEN, 2641*
    CERVIX NEOPLASMS
        CARCINOMA, 2600
        CARCINOMA IN SITU, 2600
        VIRUS, HERPES SIMPLEX 1, 2600
        VIRUS, HERPES SIMPLEX 2, 2600
    HELA CELLS
        VIRUS, HERPES SIMPLEX 1, 2596
    INFECTIOUS MONONUCLEOSIS
        EPSTEIN-BARR NUCLEAR ANTIGEN, 2641*
    LEUKEMIA
        MAST CELLS, 2649*
        VIRUS, FELINE LEUKEMIA, 2649*
    PRECANCEROUS CONDITIONS
        VIRUS, HERPES SIMPLEX 2, 2908*
    VIRUS, EPSTEIN-BARR
        INTERPERSONAL TRANSMISSION, 2909*
    VIRUS, FELINE LEUKEMIA
        IMMUNE SERUMS, 2649*
        PASSIVE HEMAGGLUTINATION, 2649*
    VIRUS, FRIEND MURINE LEUKEMIA
        B-LYMPHOCYTES, 2687
        T-LYMPHOCYTES, 2687

ANTIGEN-ANTIBODY REACTIONS
    BRONCHIAL NEOPLASMS
        IGE, 2740*
        IGG, 2740*
    HODGKIN'S DISEASE
        ANTIGENS, 2721*
    VIRUS, EPSTEIN-BARR
        CELL LINE, 2641*

ANTIGENIC DETERMINANTS
    ADENOCARCINOMA
        IGG, 2748*
    HEPATOMA
        FERRITIN, 2928
    INTERFERON
        RNA, MESSENGER, 2747*
    T-LYMPHOCYTES
        GRAFT VS HOST REACTION, 2693
    ONCOGENIC VIRUSES
        VIRAL PROTEINS, 2589
    VIRUS, AVIAN MYELOBLASTOSIS
        VIRAL PROTEINS, 2583
    VIRUS, C-TYPE AVIAN TUMOR
        PEPTIDES, 2587
    VIRUS, C-TYPE PORCINE

```
        VIRUS, FELINE LEUKEMIA, 2590
        VIRUS, GIBBON LEUKEMIA, 2590
        VIRUS, M7 BABOON, 2590
        VIRUS, RAUSCHER MURINE LEUKEMIA,
          2590
        VIRUS, RD114, 2590
    VIRUS, FELINE LEUKEMIA
        REVERSE TRANSCRIPTASE, 2589
        VIRAL PROTEINS, 2589
    VIRUS, FELINE LEUKEMIA SARCOMA
        REVERSE TRANSCRIPTASE, 2589
        VIRAL PROTEINS, 2589
    VIRUS, GIBBON LEUKEMIA
        REVERSE TRANSCRIPTASE, 2589
        VIRAL PROTEINS, 2589
    VIRUS, MOLONEY MURINE SARCOMA
        IMMUNE RESPONSE, 2616
    VIRUS, RADIATION LEUKEMIA
        T-LYMPHOCYTES, 2610
    VIRUS, RAUSCHER MURINE LEUKEMIA
        REVERSE TRANSCRIPTASE, 2589
        VIRAL PROTEINS, 2589
    VIRUS, RD 114
        REVERSE TRANSCRIPTASE, 2589
        VIRAL PROTEINS, 2589
    VIRUS, ROUS SARCOMA
        PEPTIDES, 2587
        REVIEW, 2435*
    VIRUS, SIMIAN SARCOMA
        REVERSE TRANSCRIPTASE, 2589
        VIRAL PROTEINS, 2589

ANTIGENS
    HODGKIN'S DISEASE
        ANTIGEN-ANTIBODY REACTIONS, 2721*
    LEUKEMIA, LYMPHOBLASTIC
        T-LYMPHOCYTES, 2699
    T-LYMPHOCYTES
        CELL MEMBRANE, 2695
        IMMUNE RESPONSE, 2704
    LYMPHOKINES
        BENZENE, 1-CHLORO-2,4-DINITRO-,
          2722*
    NERVOUS SYSTEM NEOPLASMS
        UREA, ETHYL NITROSO-, 2683
    SARCOMA
        BENZO(A)PYRENE, 2475
        CHOLANTHRENE, 3-METHYL-, 2475
    VIRUS, AVIAN LEUKOSIS-SARCOMA
        CELL MEMBRANE, 2690

ANTIGENS, HETEROGENETIC
    T-LYMPHOCYTES
        CELL DIVISION, 2682

ANTIGENS, NEOPLASM
    BLADDER NEOPLASMS
        IMMUNITY, CELLULAR, 2713*
    BRAIN NEOPLASMS
        IMMUNITY, CELLULAR, 2418
    FIBROMA
        VIRUS, BOVINE PAPILLOMA, 2665
    HEPATOMA
        DIETHYLAMINE, N-NITROSO-, 2708*
        IMMUNITY, 2708*
        IMMUNITY, ACTIVE, 2716*
        IMMUNITY, CELLULAR, 2716*
    LUNG NEOPLASMS
```

IA

UTANTS, 2622
FMIA
IZATION,

IZATION,

IZATION,

RAT, 2482

-CIMETHYL-,

MOR

ARY

SMA, 2954

ETRACHLORO-.

CHOLANTHRENE, 3-METHYL-, 2500
   ENZYME INDUCTION, 2500
DIBUTYRYL-CYCLIC AMP
   ENZYME INDUCTION, 2499
DIBUTYRYL CYCLIC GMP
   ENZYME INDUCTION, 2499
ISOQUINOLINE, 6,7-DIMETHOXY-1-VERATRYL-

   ENZYME INDUCTION, 2499
SKIN NEOPLASMS
   SMOKING, 2498
THEOPHYLLINE
   ENZYME INDUCTION, 2499
THEOPHYLLINE, COMPOUND WITH
   ETHYLENEDIAMINE (2:1)
   ENZYME INDUCTION, 2499

ASBESTOS
   AIR POLLUTANTS
      CONTROL TECHNOLOGY, 2434*
      INDUSTRIAL WASTE, 2434*
      MINING, 2434*
   FIBERS
      OCCUPATIONAL HAZARD, 2575*
   LARYNGEAL NEOPLASMS
      CARCINOMA, 2496
   NEOPLASMS
      PARTICLE SIZE, 2434*
   SMOKING
      COCARCINOGENIC EFFECT, 2496

ASBESTOSIS
   AIR POLLUTANTS
      PARTICLE SIZE, 2434*

ASCORBIC ACID
   BLOOD VESSELS
      DIETARY FACTORS, 2524*
   COLLAGEN DISEASE
      DIETARY FACTORS, 2524*
   GALLIC ACID
      NITROSATION-BLOCKING AGENT, 2481
   MORPHOLINE
      NITROSAMINES, 2547*
      NITROUS ACID, SODIUM SALT, 2547*

ASPARAGINASE
   LEUKEMIA, LYMPHOBLASTIC
      T-LYMPHOCYTES, 2700
      ROSETTE FORMATION, 2700

ASPERGILLUS FLAVUS
   AFLATOXIN B1
      CORN, 2512*, 2513*, 2514*
   MYCOTOXINS
      FOOD CONTAMINATION, 2456
      ISOLATION AND CHARACTERIZATION,
         2511*
   ZEARALENONE
      CORN, 2512*, 2513*, 2514*

ASPERGILLUS NIGER
   MYCOTOXINS
      FOOD CONTAMINATION, 2456

ASPERGILLUS PARASITICUS
   MYCOTOXINS
      ACETATES, 2508*
      MALONATES, 2508*

* INDICATES A PLAIN CITATION WITHOUT ACCOMPANYING ABSTRACT

ASPERGILLUS REPENS
    MYCOTOXINS
        FOOD CONTAMINATION, 2456

ASTROCYTOMA
    BRAIN NEOPLASMS
        IMMUNITY, CELLULAR, 2418
        TRANSPLANTATION IMMUNOLOGY, 2418
        UREA, N-NITROSO-N-PROPYL-, 2491
    NERVOUS SYSTEM NEOPLASMS
        EPIDEMIOLOGY, RACIAL FACTORS,
            CHILD, 2883
        UREA, 1,1-DIETHYL-3-METHYL-3-
            NITROSO-, 2488
        UREA, 3-ETHYL-1,1-DIMETHYL-3-
            NITROSO-, 2488
        UREA, 1,1,3-TRIETHYL-3-NITROSO-,
            2488
        UREA, 1,1,3-TRIMETHYL-3-NITROSO-,
            2488

ATAXIA TELANGIECTASIA
    LYMPHOCYTES
        ANTIBODIES, 2746*
        NEURAMINIDASE, 2746*

AUTOIMMUNE DISEASES
    BLADDER NEOPLASMS
        TRANSPLANTATION IMMUNOLOGY, 2714*

AZATHIOPRINE
    SEE PURINE, 6-((1-METHYL-4-
        NITROIMIDAZOL-5-YL)THIC)-

AZULENE
    BRAIN NEOPLASMS
        PRECANCEROUS CONDITIONS, 2473

BACILLUS SUBTILIS
    DNA, BACTERIAL
        ALBUMINS, 2582

BARBITURIC ACID, 5-(1-CYCLOHEXEN-1-YL)-1,5-
    DIMETHYL-, SODIUM SALT
    CADMIUM
        POTENTIATION, LIVER, 2521*

BARBITURIC ACID, 5-ETHYL-5-PHENYL-
    ACETAMIDE, N-FLUOREN-2-YL-
        TISSUE RESIDUES, 2516*
    AFLATOXIN B1
        CARCINOGENIC ACTIVITY, RAT, 2458
    DIETHYLAMINE, N-NITROSO-
        CARCINOGENIC ACTIVITY, RAT, 2482
    ETHYLENE, CHLORO-
        COCARCINOGENIC EFFECT, RAT, 2503
    LIVER NEOPLASMS
        AFLATOXIN B1, 2458

BENZ(A)ANTHRACENE
    PETROLEUM
        FOOD ANALYSIS, 2535*
    YEAST
        FOOD ANALYSIS, 2537*

BENZ(A)ANTHRACENE, 7,12-DIMETHYL-
    ADENOCARCINOMA
        BLOOD VOLUME DETERMINATIONS, 2465

AMYLOIDOSIS
    CARCINOGENIC POTENTIAL, MOUSE,
        2468
    BRAIN NEOPLASMS
        PRECANCEROUS CONDITIONS, 2473
CARCINOMA, EPIDERMOID
    CARCINOGENIC POTENTIAL, MOUSE,
    CELL LINE
        CELL TRANSFORMATION, NEOPLASTI
        2913
GRAFT VS HOST REACTION
    CARCINOGENIC ACTIVITY, MOUSE,
HEPATOMA
    CARCINOGENIC POTENTIAL, MOUSE,
        2468
KERATOACANTHOMA
    CARCINOGENIC POTENTIAL, MOUSE,
        2468
LYMPHOMA
    ANTILYMPHOCYTE SERUM, 2469
MAMMARY NEOPLASMS, EXPERIMENTAL
    ACTINOMYCIN D, 2467
    ADENOCARCINOMA, 2465
    CARCINOGENIC ACTIVITY, RAT, 25
    ESTROGENS, 2466
    PORTACAVAL ANASTOMOSIS, 2540*
    PRECANCEROUS CONDITIONS, 2464
    PROLACTIN, 2466
MUTAGENS
    SALMONELLA/MICROSOME OXIDATION
        TEST SYSTEM, 2430*
PAPILLOMA
    CARCINOGENIC POTENTIAL, MOUSE,
        2468
SARCOMA
    CARCINOGENIC POTENTIAL, MOUSE,
        2468
SPLENECTOMY
    CARCINOGENIC ACTIVITY, MOUSE,
VIRUS, POLYOMA
    VIRUS REPLICATION, 2474

BENZ(A)ANTHRACENE, 7-METHYL-
    DNA
        ISOLATION AND CHARACTERIZATION
        2476

BENZENE
    HEMOPOIETIC SYSTEM
        TOXIC METABOLITES, 2541*
    LEUKEMIA
        REVIEW, 2447*

BENZENE, 1-CHLORO-2,4-DINITRO-
    HODGKIN'S DISEASE
        IMMUNITY, CELLULAR, 2721*
    LYMPHOCYTE TRANSFORMATION
        IMMUNITY, CELLULAR, 2722*
    LYMPHOCYTES
        IMMUNITY, CELLULAR, 2722*
    LYMPHOKINES
        ANTIGENS, 2722*
    SKIN NEOPLASMS
        BENZO(A)PYRENE, 2545*

BENZENE, 1-FLUORO-2,4-DINITRO-
    UROGENITAL NEOPLASMS
        HYPERSENSITIVITY, 2737*

SUBJECT   8

POTENTIATION, LIVER, 2521*

BENZYL ALCOHOL, 3,4-DIHYDROXY-ALPHA-(
  (METHYLAMINO)-METHYL)-
    SEE EPINEPHRINE

4-BIPHENYLAMINE
    BLADDER NEOPLASMS
        OCCUPATIONAL HAZARD, REVIEW, 2431*

BLADDER NEOPLASMS
    ADENOCARCINOMA
        GENETICS, 2794
    ADJUVANTS, IMMUNOLOGIC
        LYMPHOCYTE TRANSFORMATION, 2674
    ANTIGENS, NEOPLASM
        IMMUNITY, CELLULAR, 2713*
    BENZIDINE
        OCCUPATIONAL HAZARD, REVIEW, 2431*
    1,2-BENZISOTHIAZOLIN-3-ONE, 1,1-
        DIOXIDE
        DIABETES MELLITUS, 2424*
        EPIDEMIOLOGY, REVIEW, 2424*
    BENZO(A)PYRENE
        OCCUPATIONAL HAZARD, REVIEW, 2431*
    4-BIPHENYLAMINE
        OCCUPATIONAL HAZARD, REVIEW, 2431*
    CARCINOMA
        BRACKEN FERN, 2459
        CASE REPORT, 2861*, 2862*
        GLASS BEAD IMPLANTATION, 2459
    CARCINOMA, BASAL CELL
        GENETICS, 2794
    CARCINOMA, EPIDERMOID
        CYCLOPHOSPHAMIDE, 2461
        GENETICS, 2794
        PSEUDOSARCOMATUS STROMA, 2863*
        ULTRASTRUCTURAL STUDY, 2863*
    CARCINOMA IN SITU
        NEOPLASM METASTASIS, 2795
    CARCINOMA, TRANSITIONAL CELL
        ACETAMIDE, N-FLUOREN-2-YL-, 2714*
        COMPLEMENT, 2741*
        GENETICS, 2794
        HISTOCOMPATIBILITY ANTIGENS, 2794
        IMMUNITY, CELLULAR, 2713*
        B-LYMPHOCYTES, 2794
        MORRIS DIET, 2714*
        NEOPLASM METASTASIS, 2795
        TOBACCO, 2794
        TRYPTOPHAN, 2794
    DIABETES MELLITUS
        EPIDEMIOLOGY, REVIEW, 2424*
    DYES
        HAIRDRESSERS, 2449*
    HISTOCOMPATIBILITY ANTIGENS
        IMMUNITY, CELLULAR, 2713*
    LYMPHOCYTES
        IMMUNE RESPONSE, 2713*
    MILLIAMINE A
        CARCINOGENIC ACTIVITY, MOUSE,
            2534*
    2-NAPTHYLAMINE
        OCCUPATIONAL HAZARD, REVIEW, 2431*
    OCCUPATIONAL HAZARD
        EPIDEMIOLOGY, 2881
    TRANSPLANTATION IMMUNOLOGY
        AUTOIMMUNE DISEASES, 2714*

BLOOD PROTEINS
    HELA CELLS
        GROWTH, 2980*

BLOOD VESSELS
    ASCORBIC ACID
        DIETARY FACTORS, 2524*

BONE MARROW
    LEUKEMIA, MYELOBLASTIC
        CHROMOSOME ABNORMALITIES, 2778
        ERYTHROBLASTS, 2778

BONE MARROW CELLS
    ANESTHETICS
        CARCINOGENIC EFFECT, 2407
    CARBAMIC ACID, ETHYL ESTER
        HEMATOPOIETIC STEM CELLS, 2558*
    CHROMOSOME ABERRATIONS
        ACRYLAMIDE, 2-(2-FURYL)-3-(5-NITRO-
          2-FURYL)-, 2454
        FOOD PRESERVATIVES, 2454
    COBALT RADIOISOTOPES
        CHROMOSOME ABERRATIONS, 2561
    ERYTHROLEUKEMIA
        CASE REPORT, 2777
        CHROMOSOME ABERRATIONS, 2777
    LYMPHOMA
        HISTOLOGICAL STUDY, 2781
    LYMPHOSARCOMA
        HISTOLOGICAL STUDY, 2781
    SARCOMA, RETICULUM CELL
        HISTOLOGICAL STUDY, 2781

BONE NEOPLASMS
    AMERICIUM
        REVIEW, 2410
    CHORDOMA
        CASE REPORT, 2762
        CELL DIFFERENTATION, 2762
        ULTRASTRUCTURAL STUDY, TISSUE
          CULTURE, 2762
    HODGKIN'S DISEASE
        CLASSIFICATION, 2803*
    LYMPHOMA
        CLASSIFICATION, 2803*
    LYMPHOSARCOMA
        CLASSIFICATION, 2803*
    NEOPLASM METASTASIS
        CASE REPORT, 2832*
    PLUTONIUM
        REVIEW, 2410
    RADIATION
        REVIEW, 2410
    RADIUM
        REVIEW, 2410
    SARCOMA, RETICULUM CELL
        CLASSIFICATION, 2803*
    THORIUM
        REVIEW, 2410
    ZINC   BR
        REVIEW, 2403

BRACKEN FERN
    BLADDER NEOPLASMS
        CARCINOMA, 2459

BRAIN NEOPLASMS

EPIDEMIOLOGY, REVIEW, 2420
CARCINOEMBRYONIC ANTIGEN
  DIAGNOSIS AND PROGNOSIS, REVIEW,
    2416
CARCINOMA
  ANTIBODIES, NEOPLASM, 2678
  NEOPLASM METASTASIS, 2824*
  NEOPLASMS, MULTIPLE PRIMARY, 2784
CARCINOMA, BRONCHIOLAR
  NEOPLASMS, MULTIPLE PRIMARY, 2784
CARCINOMA, COLLOID
  NEOPLASMS, MULTIPLE PRIMARY, 2784
CARCINOMA, DUCTAL
  GROWTH, 2730*
  IMMUNOSUPPRESSION, 2730*
  NEOPLASMS, MULTIPLE PRIMARY, 2784
CARCINOMA, PAPILLARY
  NEOPLASMS, MULTIPLE PRIMARY, 2784
CONTRACEPTIVES, ORAL
  CARCINOGENIC POTENTIAL, REVIEW, 2405
COPPER
  REVIEW, 2403
DIET
  EPIDEMIOLOGY, REVIEW, 2420
ESTRADIOL
  ANDROST-5-ENE-3,17-DIOL, 2471
  5ALPHA-ANDROSTAN-3-ONE, 17BETA-
    HYDROXY-, 2471
  MENSTRUATION, 2785
  PRECANCEROUS CONDITIONS, 2785
ESTROGENS
  EPIDEMIOLOGY, REVIEW, 2420
  PROLACTIN, 2421
GENETICS
  EPIDEMIOLOGY, REVIEW, 2420
IMMUNITY, CELLULAR
  LYMPHOCYTE TRANSFORMATION, 2702
  PLANT AGGLUTININS, 2702
LYMPHOCYTES
  CONCANAVALIN A, 2760*
NEOPLASM METASTASIS
  ESTROGEN RECEPTORS, REVIEW, 2421
NEOPLASMS, MULTIPLE PRIMARY
  DIAGNOSIS AND PROGNOSIS, 2784
  HISTOLOGICAL STUDY, 2784
  NEOPLASM METASTASIS, 2784
OVARIAN NEOPLASMS
  GENETICS, 2865*
PRECANCEROUS CONDITIONS
  DIAGNOSIS AND PROGNOSIS, 2823*
PROGESTERONE
  ESTRADIOL, 2785
  MENSTRUATION, 2785
  PRECANCEROUS CONDITIONS, 2785
RADIATION, IONIZING
  CASE REPORT, 2565
RESERPINE
  EPIDEMIOLOGY, 2494
VIRUS, C-TYPE RNA TUMOR
  COMPARATIVE VIROLOGY, REVIEW, 2412

OMODEOXYURIDINE
  SEE URIDINE, 5-BROMO-2'-DEOXY'

BROMODEOXYURIDINE
  SEE URIDINE, 5-BROMO-2'-DEOXY'

ONCHIAL NEOPLASMS

ADENOCARCINOMA
  EPIDEMIOLOGY, SWEDEN, 2889
CARCINOMA
  EPIDEMIOLOGY, SWEDEN, 2889
CARCINOMA, EPIDERMOID
  EPIDEMIOLOGY, SWEDEN, 2889
CARCINOMA, OAT CELL
  EPIDEMIOLOGY, SWEDEN, 2889
CYLINDROMA
  EPIDEMIOLOGY, SWEDEN, 2889
IGE
  ANTIBODIES, 2740*
  ANTIGEN-ANTIBODY REACTIONS, 2740*
IGG
  ANTIGEN-ANTIBODY REACTIONS, 2740*

BURKITT'S LYMPHOMA
  ANTIBODIES, VIRAL
    EPSTEIN-BARR NUCLEAR ANTIGEN,
      2641*
  CEREBROSPINAL FLUID
    CASE REPORT, 2811*
  CHROMOSOME ABERRATIONS
    REVIEW, 2444*
  EPIDEMIOLOGY
    AFRICA, 2891
    REVIEW, 2419
  GENETICS
    REVIEW, 2444*
  LEUKEMIA, LYMPHOBLASTIC
    EPIDEMIOLOGY, AFRICA, 2891
  B-LYMPHOCYTES
    IMMUNE SERUMS, 2679
    IMMUNOGLOBULINS, 2679
  MALARIA
    EPIDEMIOLOGY, AFRICA, 2891
    REVIEW, 2444*
  VIRUS, EPSTEIN-BARR
    ANTIGENS, VIRAL, 2591
    COMPARATIVE VIROLOGY, REVIEW, 2412
    EPIDEMIOLOGY, AFRICA, 2891
    REVIEW, 2444*
    ULTRASTRUCTURAL STUDY, 2679
    VACCINES, 2412

1-BUTANAMINE, N-BUTYL-N-NITROSO-
  HEPATOMA
    CARCINOGENIC ACTIVITY, MOUSE, 2550*
  LUNG NEOPLASMS
    CARCINOGENIC ACTIVITY, MOUSE, 2550*

1-BUTANAMINE, N-(3-CARBOXYBUTYL)-N-NITROSO-
  HEPATOMA
    CARCINOGENIC ACTIVITY, MOUSE,
      2550*
  LUNG NEOPLASMS
    CARCINOGENIC ACTIVITY, MOUSE,
      2550*

1,4-BUTANEDIOL DIMETHYLSULFONATE
  SPERMATOZOA
    ABNORMALITIES, MOUSE, 2452

1-BUTANOL, 4-(BUTYLNITROSOAMINO)-
  HEPATOMA
    CARCINOGENIC ACTIVITY, MOUSE,
      2550*
  LUNG NEOPLASMS

        CARCINOGENIC ACTIVITY, MOUSE,
          2550*

BUTYRIC ACID, 2-AMINO-4-(ETHYLTHIO)-
    HEPATOMA
        AMINOPYRINE DEMETHYLASE, 2792
        CYTOCHROME P-450, 2752
        CYTOCHROME REDUCTASES, 2792
        MICROSOMES, LIVER, 2792
        PHOSPHOLIPIDS, 2792
        PRECANCEROUS CONDITICNS, 2792
        RIBOSOMES, 2792
        RNA, TRANSFER, METHYLTRANSFERASES,
          2952

C.I. ACID RED 27
    FRUCTOSE
        GLUCOSE, 2518*

CADMIUM
    BARBITURIC ACID, 5-(1-CYCLOHEXEN-1-YL)-
      1,5-DIMETHYL-, SODIUM SALT
        POTENTIATION, LIVER, 2521*
    BENZOXAZOLE, 2-AMINO-5-CHLORO-
        POTENTIATION, LIVER, 2521*
    METALLOPROTEINS
        RNA, MESSENGER, 2520*
        SYNTHESIS, LIVER, KICNEY, 2520*

CAFFEINE
    ADENOMA
        ANTINEOPLASTIC EFFECT, MOUSE, 2481

CALCIUM
    MICROSOMES
        ENZYMATIC ACTIVITY, 2991*

CANCER
    SEE NEOPLASMS

CAPSID ANTIGENS
    SEE VIRAL PROTEINS

CARBAMIC ACID, ETHYL ESTER
    BONE MARROW CELLS
        HEMATOPOIETIC STEM CELLS, 2558*
    LEUKEMIA
        CARCINOGENIC POTENTIAL, REVIEW,
          2404
        RADIATION, IONIZING, 2569*
    LIVER NEOPLASMS
        RADIATION, IONIZING, 2570*
    LUNG NEOPLASMS
        RADIATION, IONIZING, 2570*
    LYMPHATIC NEOPLASMS
        RADIATION, IONIZING, 2570*
    RADIATION, IONIZING
        COCARCINOGENIC ACTIVITY, 2570*
        COCARCINOGENIC ACTIVITY, RAT,
          2569*

CARBAMIC ACID, METHYL-, 1-NAPHTHYL ESTER
    HELA CELLS
        GROWTH, 2921
        LECITHINS, 2921
        PHOSPHOLIPIDS, 2921
        PROTEINS, 2921
        SPHINGOMYELINS, 2921

        LIVER CIRRHOSIS, 2851*
        MONOCROTALINE, 2529*
        NCI CO2OJ6, 2554*
        PRECANCEROUS CONDITIONS, 2851*
        RETRONECINE, 3,8-DIDEHYDRO-, 2529*
LUNG NEOPLASMS
    SOMATOTROPIN, 2965
MAMMARY NEOPLASMS, EXPERIMENTAL
    FIBRINOLSYIN, 2979*
    INCIDENCE, HAMSTER, 2971*
    ULTRASTRUCTURAL STUDY, DOG, 2825*
MOUTH NEOPLASMS
    EPIDEMIOLOGY, INDIA, 2893
NOSE NEOPLASMS
    OCCUPATIONAL HAZARD, 2894
PANCREATIC NEOPLASMS
    ANTIBODIES, NEOPLASM, 2678
    2-PROPANOL, 1,1'-(NITROSOIMINO)DI-,
      2483
PARANASAL SINUS NEOPLASMS
    OCCUPATIONAL HAZARD, 2894
PARATHYROID NEOPLASMS
    HYPERCALCEMIA, 2797
    HYPERPARATHYROIDISM, 2797
    PRECANCEROUS CONDITIONS, 2797
PENILE NEOPLASMS
    EPIDEMIOLOGY, INDIA, 2885
PHARYNGEAL NEOPLASMS
    EPIDEMIOLOGY, INDIA, 2893
    NICOTINE, 1'-DEMETHYL-N'-NITROSO-,
      2486
PRECANCEROUS CONDITIONS
    GASTROENTEROSTOMY, 2830*
THYROID NEOPLASMS
    CASE REPORT, 2799
    GENETICS, 2875*
    GOITER, 2876*
    NCI CO2186, 2555*
    PRECANCEROUS CONDITIONS, 2875*
    ULTRASTRUCTURAL STUDY, 2799
URETERAL NEOPLASMS
    COMPLEMENT, 2741*
UTERINE NEOPLASMS
    EPIDEMIOLOGY, GERMANY, 2902*
    HYDROXYSTEROID DEHYDROGENASES,
      2961

CARCINOMA, ANAPLASTIC
    RECTAL NEOPLASMS
        EPIDEMIOLOGY, 2791
        STAGING AND THERAPY, 2791

CARCINOMA, BASAL CELL
    BLADDER NEOPLASMS
        GENETICS, 2794
    EYELID NEOPLASMS
        CASE REPORT, 2856*
    SKIN NEOPLASMS
        EPIDEMIOLOGY, 2857*
        IMMUNITY, CELLULAR, 2671
        RADIATION, 2857*
        ULTRAVIOLET RAYS, 2411

CARCINOMA, BRONCHIOLAR
    BREAST NEOPLASMS
        NEOPLASMS, MULTIPLE PRIMARY, 2784

CARCINOMA, BRONCOGENIC
    LUNG NEOPLASMS
        CHROMATES, 2523*
        SOMATOTROPIN, 2965

CARCINOMA, COLLOID
    BREAST NEOPLASMS
        NEOPLASMS, MULTIPLE PRIMARY, 2784
    RECTAL NEOPLASMS
        EPIDEMIOLOGY, 2791
        STAGING AND THERAPY, 2791

CARCINOMA, DUCTAL
    BREAST NEOPLASMS
        GROWTH, 2730*
        IMMUNOSUPPRESSION, 2730*
        NEOPLASMS, MULTIPLE PRIMARY, 2784
    CHOLANTHRENE, 3-METHYL-
        NEOPLASM METASTASIS, 2770
    MESTRANOL
        NORETHYNODREL, 2405

CARCINOMA, EHRLICH TUMOR
    CELL MEMBRANE
        LACTATES, 2978*
    DNA REPLICATION
        THYMIDINE INCORPORATION, 2898
        URIDINE, 2'-DEOXY-5-IODO-, 2898
    FLAVONES
        ADENOSINE TRIPHOSPHATASE, 2919
        GLYCOLYSIS, 2919
        GROWTH, 2919
    GLYCOLIPIDS
        GROWTH, 2989*
        ISOLATION AND CHARACTERIZATION,
        2989*
    IMMUNOSUPPRESSIVE AGENTS
        ISOLATION AND CHARACTERIZATION,
        2733*
    NUCLEOTIDES
        CATABOLISM, 2940
    PLANT AGGLUTININS
        ANTI-NEOPLASTIC ACTIVITY, 2736*
    RNA, RIBOSOMAL
        PACTAMYCIN, 2938
        PROTEINS, 2938
    URIDINE, 2'-DEOXY-5-IODO
        CELL CYCLE KINETICS, 2898

CARCINOMA, EMBRYONAL
    SEE TERATOID TUMOR

CARCINOMA, EPIDERMOID
    BENZ(A)ANTHRACENE, 7,12-CIMETHYL-
        CARCINOGENIC POTENTIAL, MOUSE,
        2468
    BLADDER NEOPLASMS
        CYCLOPHOSPHAMIDE, 2461
        GENETICS, 2794
        PSEUDOSARCOMATUS STRCMA, 2863*
        ULTRASTRUCTURAL STUDY, 2863*
    BRONCHIAL NEOPLASMS
        EPIDEMIOLOGY, SWEDEN, 2889
    CHOLANTHRENE, 3-METHYL-
        HISTOLOGICAL STUDY, RABBIT, 2542*
    EAR NEOPLASMS
        UREA, N-NITROSO-N-PRCPYL-, 2491
    ESOPHAGEAL NEOPLASMS

        CASE REPORT, 2789
        NICOTINE, 1'-DEMETHYL-N'-NITROS
        2486
    EYELID NEOPLASMS
        CASE REPORT, 2856*
    HEAD AND NECK NEOPLASMS
        EPIDEMIOLOGY, FISH, 2895
    LIP NEOPLASMS
        EPIDEMIOLOGY, FISH, 2895
    LUNG NEOPLASMS
        CHOLANTHRENE, 3-METHYL-, 2542*,
        2543*
        CHROMATES, 2523*
        HISTOLOGICAL STUDY, DOG, 2543*
        INCIDENCE, HAMSTER, 2971*
        SOMATOTROPIN, 2965
    LYMPHATIC NEOPLASMS
        LYMPHORETICULAR RESPONSE, 2719*
    MOUTH NEOPLASMS
        EPIDEMIOLOGY, INDIA, 2893
    NOSE NEOPLASMS
        OCCUPATIONAL HAZARD, 2894
    ORAL CAVITY
        LEUKOPLAKIA, 2904*
    PARANASAL SINUS NEOPLASMS
        OCCUPATIONAL HAZARD, 2894
    PHARYNGEAL NEOPLASMS
        EPIDEMIOLOGY, INDIA, 2893
    PROSTATIC NEOPLASMS
        EPIDEMIOLOGY, INDIA, 2886
    PSEUDOSARCOMA STROMA
        ULTRASTRUCTURAL STUDY, 2863*
    RADIATION
        REVIEW, 2410
    SKIN NEOPLASMS
        EPIDEMIOLOGY, FISH, 2895
        INCIDENCE, HAMSTER, 2971*
        ULTRAVIOLET RAYS, 2411
    SPLENIC NEOPLASMS
        LYMPHORETICULAR RESPONSE, 2719*

CARCINOMA IN SITU
    BLADDER NEOPLASMS
        NEOPLASM METASTASIS, 2795
    CERVIX NEOPLASMS
        ANTIBODIES, VIRAL, 2600
        CONCANAVALIN A, 2868*
        PLANT AGGLUTININS, 2868*
    ORAL CAVITY
        LEUKOPLAKIA, 2904*

CARCINOMA, MUCINOUS
    COLONIC NEOPLASMS
        CARCINOEMBRYONIC ANTIGEN, 2749*
    KIDNEY NEOPLASMS
        CASE REPORT, 2859*

CARCINOMA, OAT CELL
    BRONCHIAL NEOPLASMS
        EPIDEMIOLOGY, SWEDEN, 2889
    ESOPHAGEAL NEOPLASMS
        CASE REPORT, 2789
    LUNG NEOPLASMS
        CELL LINE, 2973*
        CHROMATES, 2523*
        ULTRASTRUCTURAL STUDY, 2787

CARCINOMA, PAPILLARY
    BREAST NEOPLASMS

CELL TRANSFORMATION, NEOPLASTIC
LETS PROTEIN, REVIEW, 2450*

CELL DIFFERENTIATION
SEE ALSO GROWTH
BONE NEOPLASMS
CHORDOMA, 2762
BRAIN NEOPLASMS
SARCOMA, RETICULUM CELL, 2819*
LEUKEMIA
T-LYMPHOCYTES, 2696
LEUKEMIA, LYMPHOBLASTIC
CEREBROSPINAL FLUID, 2813*
LEUKEMIA, MYELOBLASTIC
CEREBROSPINAL FLUID, 2813*
MAMMARY NEOPLASMS, EXPERIMENTAL
INHIBITORY FACTCR, IN VITRO, 2533*
SARCOMA, RETICULUM CELL
CEREBROSPINAL FLUID, 2813*
UTERINE NEOPLASMS
HYDROXYSTEROID DEHYDROGENASES,
2961

CELL DIVISION
SEE ALSO GROWTH
CELL TRANSFORMATION, NEOPLASTIC
LETS PROTEIN, REVIEW, 2450*
LEUKEMIA
SPLENECTOMY, 2696
THYMECTOMY, 2696
T-LYMPHOCYTES
ANTIGENS, HETEROGENETIC, 2682
CORTISONE, 2682
POLY A-U, 2682
MELANOMA
ACTINOMYCIN D, 2977*
MITOMYCIN C, 2977*

CELL FUSION
MELANOMA
PLASMINOGEN ACTIVATORS, 2977*

CELL LINE
BENZ(A)ANTHRACENE, 7,12-DIMETHYL-
CELL TRANSFORMATION, NEOPLASTIC,
2913
FIBROSARCOMA
CHROMOSOMES, 2970*
VIRUS-LIKE PARTICLES, 2970*
GUANIDINE, 1-METHYL-3-NITRO-1-NITROSO-
CELL TRANSFORMATION, NEOPLASTIC,
2913
LUNG NEOPLASMS
MELANOMA, 2772
LYMPHOCYTES
VIRUS, EPSTEIN-BARR, 2744*
T-LYMPHOCYTES
CARCINOGENIC POTENTIAL, 2695
CHROMOSOMES, 2695
HISTOLOGICAL STUDY, MOUSE, 2695
MAREK'S DISEASE
ISOLATION AND CHARACTERIZATION,
2620
MULTIPLE MYELOMA
IGG, 2676
IMMUNOGLOBULINS, HEAVY CHAIN, 2676
NEUROBLASTOMA
TRANSPLANTATION, HETEROLOGOUS, 2912

SARCOMA
    CHROMOSOMES, 2970*
    VIRUS-LIKE PARTICLES, 2970*
VIRUS, EPSTEIN-BARR
    ANTIGEN-ANTIBODY REACTIONS, 2641*
    ANTIGENS, VIRAL, 2641*
    GROWTH, 2703
    LYMPHOCYTE TRANSFORMATION, 2703
VIRUS, MAREK'S DISEASE HERPES
    ISOLATION AND CHARACTERIZATION,
     2623
VIRUS, MOLONEY MURINE SARCOMA
    CELL TRANSFORMATION, NEOPLASTIC,
    2913

CELL MEMBRANE
  CARCINOMA, EHRLICH TUMOR
    LACTATES, 2978*
  CELLS, CULTURED
    GLYCOPEPTIDES, 2972*
  FIBROBLASTS
    CONCANAVALIN A, 2671
  HELA CELLS
    DNA, 2931
  HYBRID CELLS
    ISOANTIGENS, 2776
  L CELLS
    PEPTIDES, 2926
  LEUKEMIA L1210
    GLYCOPROTEINS, 2944
    SIALIC ACID INCORPORATION, 2944
    TRANSFERASES, 2944
  LYMPHOCYTES
    GLYCOPROTEINS, 2982*
  T-LYMPHOCYTES
    ANTIGENS, 2695
  TERATOID TUMOR
    GALACTOSIDASES, 2882
    HYALURONIC ACID, 2882
    HYALURONIDASE, 2882
    MONOSACCHARIDES, 2882
    NEURAMINIDASE, 2882
  VIRUS, AVIAN LEUKOSIS-SARCOMA
    ANTIGENS, 2690
    ANTIGENS, NEOPLASM, 2690
    ISOANTIGENS, 2690
  VIRUS, AVIAN RNA TUMOR
    ANTIGENS, VIRAL, 2414
    GLYCOPROTEINS, 2414
    LIPIDS, 2414
    PLANT AGGLUTININS, 2414
  VIRUS, FELINE LEUKEMIA
    ANTIGENS, VIRAL, 2752*
  VIRUS, FELINE SARCOMA
    ANTIGENS, VIRAL, 2752*
  VIRUS, HERPES SIMPLEX 1
    ANTIGENS, VIRAL, 2597
  VIRUS, HERPES SIMPLEX 2
    ANTIGENS, VIRAL, 2597
  VIRUS, ROUS SARCOMA
    PEPTIDES, 2627
  VIRUS, SV40
    CONCANAVALIN A, 2671
    GLYCOPROTEINS, 2634

CELL MOVEMENT
  CYTOCHALASIN B
    WOUNDS AND INJURIES, 2916

EPIDEMIOLOGY
    SOCIOECONOMIC FACTORS, 2907*, 2908*
NEOPLASM METASTASIS
    CASE REPORT, 2802*
PRECANCEROUS CONDITIONS
    CONCANAVALIN A, 2868*
    PLANT AGGLUTININS, 2868*
RADIATION, IONIZING
    EPIDEMIOLOGY, SOCIOECONOMIC
        FACTORS, 2888
TRICHOMONAS VAGINALIS
    DIETHYLAMINE, N-NITROSO-, 2905*
    DIMETHYLAMINE, N-NITROSO-, 2905*
    DIPROPYLAMINE, N-NITROSO-, 2905*
    EPIDEMIOLOGY, AFRICA, 2905*
VIRUS, CYTOMEGALO
    PRECANCEROUS CONDITIONS, 2600
VIRUS, HERPES SIMPLEX 1
    ANTIBODIES, VIRAL, 2600
    ANTIGENS, VIRAL, 2643*
    ISOLATION AND CHARACTERIZATION,
        2643*
    PRECANCEROUS CONDITIONS, 2600
VIRUS, HERPES SIMPLEX 2
    ANTIBODIES, VIRAL, 2600
VIRUS HERPES SIMPLEX 2
    ANTIGENS, VIRAL, 2643*
VIRUS, HERPES SIMPLEX 2
    EPIDEMIOLOGY, 2908*
    EPIDEMIOLOGY, AFRICA, 2905*
VIRUS HERPES SIMPLEX 2
    ISOLATION AND CHARACTERIZATION,
        2643*
VIRUS, HERPES SIMPLEX 2
    PRECANCEROUS CONDITIONS, 2600

CHEMODECTOMA
    SEE PARAGANGLIOMA, NONCHROMAFFIN

CHEMORECEPTORS
    LUNG NEOPLASMS
        PARAGANGLIOMA, NONCHROMAFFIN, 2788

CHICKENPOX
    LEUKEMIA, LYMPHOBLASTIC
        REVIEW, 2444*

CHLORAMPHENICOL
    LEUKEMIA
        REVIEW, 2447*

CHOLANTHRENE, 3-METHYL-
    ACETAMIDE, N-FLUOREN-2-YL-
        TISSUE RESIDUES, 2516*
    ADENOCARCINOMA
        HISTOLOGICAL STUDY, RABBIT, 2542*
        NEOPLASM METASTASIS, 2770
    ADENOMA
        NEOPLASM METASTASIS, 2770
    BRAIN NEOPLASMS
        PRECANCEROUS CONDITIONS, 2473
    CARCINOMA
        NEOPLASM METASTASIS, 2770
    CARCINOMA, DUCTAL
        NEOPLASM METASTASIS, 2770
    CARCINOMA, EPIDERMOID
        HISTOLOGICAL STUDY, RABBIT, 2542*
    CELL TRANSFORMATION, NEOPLASTIC

27

GUANOSINE CYCLIC 3',5'
    MONOPHOSPHATE, 2994*
GUANYL CYCLASE, 2994*
DIBENZO-P-DIOXIN, 2,3,7,8-TETRACHLORO-
    ARYL HYDROCARBON HYDROXYLASES, 2500
EPENDYMOMA
    IMMUNITY, CELLULAR, 2418
    TRANSPLANTATION IMMUNOLOGY, 2418
FIBROSARCOMA
    MICROCIRCULATION, PERMEABILITY,
    2769
    ULTRASTRUCTURAL STUDY, RAT, 2769
HISTOCOMPATIBILTY ANTIGENS
    IMMUNITY, CELLULAR, 2666
    TRANSPLANTATION IMMUNOLOGY, 2666
LUNG NEOPLASMS
    ADENOCARCINOMA, 2542*
    CARCINOGENIC POTENTIAL, REVIEW,
    2434
    CARCINOMA, EPIDERMOIC, 2542*,
    2543*
MAMMARY NEOPLASMS, EXPERIMENTAL
    SYRIAN HAMSTER, 2504*
NEOPLASM TRANSPLANTATION
    NEOPLASM METASTASIS, 2770
OCCUPATIONAL HAZARD
    CARCINOGENIC ACTIVITY, REVIEW,
    2428*
PAPILLOMA
    NEOPLASM METASTASIS, 2770
PETROLEUM
    FOOD ANALYSIS, 2535*
PROTEINS
    BINDING, 2538*
SARCOMA
    ANTIGENS, 2475
    ANTIGENS, NEOPLASM, 2691, 2755*
    GROWTH, 2976*
    HISTOLOGICAL STUDY, 2976*
SPERMATOZOA
    ABNORMALITIES, MOUSE, 2452
VIRUS, POLYOMA
    VIRUS REPLICATION, 2474
VIRUS, SV40
    IMMUNITY, CELLULAR, 2689
    TRANSPLANTATION IMMUNOLOGY, 2689

CHOLINE
GAMMA GLOBULINS
    BINDING SITES, ANTIBODY, 2680
IGA
    BINDING SITES, ANTIBODY, 2680
IGG
    BINDING SITES, ANTIBODY, 2680
IGM
    BINDING SITES, ANTIBODY, 2680
IMMUNOGLOBULIN FRAGMENTS
    BINDING SITES, ANTIBODY, 2680
IMMUNOGLOBULINS
    BINDING SITES, ANTIBODY, 2680
MYELOMA PROTEINS
    BINDING SITES, ANTIBODY, 2680
SERUM ALBUMIN
    BINDING SITES, ANTIBODY, 2680

CHOLINE, CHLORIDE, CARBAMATE
GLIOMA
    CARBOXY-LYASES, 2959

NEUROBLASTOMA
    CARBOXY-LYASES, 2959

CHONDROID SYRINGOMA
    SEE NEOPLASMS, EMBRYONAL AND MIXED

CHONDROMYXOID FIBROMA
    SEE CHONDROMA

CHONDROSARCOMA
    EPIDEMIOLOGY
        RACIAL FACTORS, CHILD, 2883
    GLYCOSYLTRANSFERASES
        ENZYMATIC ACTIVITY, 2945

CHORDOMA
    BONE NEOPLASMS
        CASE REPORT, 2762
        CELL DIFFERENTATION, 2762
        ULTRASTRUCTURAL STUDY, TISSUE
        CULTURE, 2762

CHORIOCARCINOMA
    ENZYMES
        ULTRASTRUCTURAL STUDY, 2999*

CHROMATES
    LUNG NEOPLASMS
        CARCINOMA, BRONCOGENIC, 2523*
        CARCINOMA, EPIDERMOID, 2523*
        CARCINOMA, OAT CELL, 2523*
        OCCUPATIONAL HAZARD, 2523*

CHROMATIN
    INFECTIOUS MONONUCLEOSIS
        ISOLATION AND CHARACTERIZATION,
        2923
        PROTEINS, 2923
    LEUKEMIA, LYMPHOBLASTIC
        ISOLATION AND CHARACTERIZATION, 292
        PROTEINS, 2923
    VIRUS, AVIAN SARCOMA
        DNA, SATELLITE, 2585
        DNA, VIRAL, 2585
    VIRUS, ROUS SARCOMA
        DNA, VIRAL, 2585
    VIRUS, SV40
        DNA-RNA HYBRIDIZATION, 2635
        DNA, VIRAL, 2635

CHROMIUM OXIDE
    CARCINOGENIC EFFECT
        TOXICITY, RAT, 2460

CHROMOSOME ABERRATIONS
    ANEMIA, APLASTIC
        MITOMYCIN C, 2933
    BONE MARROW CELLS
        ACRYLAMIDE, 2-(2-FURYL)-3-(5-NITRO-
        2-FURYL)-, 2454
        COBALT RADIOISOTOPES, 2561
        FOOD PRESERVATIVES, 2454
    BURKITT'S LYMPHOMA
        REVIEW, 2444*
    ERYTHROLEUKEMIA
        BONE MARROW CELLS, 2777
        CYTOGENIC ANALYSIS, 2777
    GUANIDINE, NITROSO-

COBALT RADIOISOTOPES
    BONE MARROW CELLS
        CHROMOSOME ABERRATIONS, 2561
    RADIATION, IONIZING
        CHROMOSOME ABERRATIONS, 2561

COENZYME A
    ACETAMIDE, N-FLUOREN-2-YL-
        DNA, 2515*

COLCHICINE
    MELANOMA
        TYROSINASE, 2963
    SARCOMA, MAST CELL
        IMMUNITY, CELLULAR, 2670
        MIGRATION INHIBITORY FACTOR, 2670
    SPERMATOZOA
        ABNORMALITIES, MOUSE, 2452

COLLAGEN DISEASE
    ASCORBIC ACID
        DIETARY FACTORS, 2524*

COLONIC NEOPLASMS
    ADENOCARCINOMA
        CARCINOEMBRYONIC ANTIGEN, 2748*
        ENTERITIS, REGIONAL, 2827*
        IMMUNOGLOBULINS, FAB, 2748*
        UREA, N-NITROSO-N-PROPYL-, 2491
    ADENOMA
        UREA, N-NITROSO-N-PROPYL-, 2491
    ALKALINE PHOSPHATASE
        SERUM LEVELS, 2955
    CARCINOEMBRYONIC ANTIGEN
        DIAGNOSIS AND PROGNOSIS, REVIEW,
            2416
    CARCINOMA
        CARCINOEMBRYONIC ANTIGEN, 2749*
        COMPLEMENT, 2741*
        POLYPS, 2832*
        PRECANCEROUS CONDITIONS, 2832*
    CARCINOMA, MUCINOUS
        CARCINOEMBRYONIC ANTIGEN, 2749*
    DNA
        EPITHELIAL CELLS, 2985*
    ENTERITIS, REGIONAL
        CASE REPORT, 2827*
    FIBROMA
        UREA, N-NITROSO-N-PROPYL-, 2491
    FIBROSARCOMA
        UREA, N-NITROSO-N-PROPYL-, 2491
    HEMANGIOENDOTHELIOMA
        UREA, N-NITROSO-N-PROPYL-, 2491
    HEMANGIOMA
        UREA, N-NITROSO-N-PROPYL-, 2491
    HISTOCOMPATIBILITY ANTIGENS
        GENETICS, 2686
    INTESTINAL POLYPS
        GENETICS, 2442*, 2828*
        PRECANCEROUS CONDITIONS, 2442*
        SCREENING, 2828*
    MYOSARCOMA
        UREA, N-NITROSO-N-PROPYL-, 2491
    PAPILLOMA
        ISOANTIGENS, 2750*
    PEUTZ-JEGHERS SYNDROME
        GENETICS, 2828*
        SCREENING, 2828*
    PLANT AGGLUTININS

IMMUNITY, CELLULAR, 2711*
POLYPS
  CASE REPORT, 2833*
  GENETICS, 2833*
  ISOANTIGENS, 2750*

COMPLEMENT
  BLADDER NEOPLASMS
    CARCINOMA, TRANSITIONAL CELL,
    2741*
  COLONIC NEOPLASMS
    CARCINOMA, 2741*
  KIDNEY NEOPLASMS
    CARCINOMA, TRANSITIONAL CELL,2741*
  URETERAL NEOPLASMS
    CARCINOMA, 2741*

CONCANAVALIN A
  BREAST NEOPLASMS
    LYMPHOCYTES, 2760*
  CELL TRANSFORMATION, NEOPLASTIC
    CELL LINE, VARIANT, 2974*
  CERVIX NEOPLASMS
    CARCINOMA IN SITU, 2868*
    PRECANCEROUS CONDITIONS, 2868*
  DIBUTYRYL CYCLIC AMP
    CELLS, CULTURED, 2972*
  FIBROBLASTS
    CELL MEMBRANE, 2671
  HEPATOMA
    AGGLUTINATION, 2706
    LECTIN BINDING, 2706
  HODGKIN'S DISEASE
    LYMPHOCYTES, 2756*, 2760*
  INFECTIOUS MONONUCLEOSIS
    LYMPHOCYTES, 2756*
  LEUKEMIA, LYMPHOCYTIC
    LYMPHOCYTES, 2756*
  T-LYMPHOCYTES
    IMMUNE RESPONSE, 2417, 2695
  LYMPHOMA
    LYMPHOCYTES, 2760*
  LYMPHOSARCOMA
    LYMPHOCYTES, 2760*
  SARCOMA, RETICULUM CELL
    LYMPHOCYTES, 2760*
  THYMECTOMY
    T-LYMPHOCYTES, 2697
  VIRUS, AVIAN LEUKOSIS-SARCOMA
    AGGLUTINATION, 2690
  VIRUS, FRIEND MURINE LEUKEMIA
    GLYCOPEPTIDES, 2603
    T-LYMPHOCYTES, 2651*
    PEPTIDES, 2603
    VIRAL PROTEINS, 2603
    VIRUS-LIKE PARTICLES, 2602
  VIRUS, FRIEND SPLEEN-FOCUS FORMING
    T-LYMPHOCYTES, 2651*
  VIRUS, SV40
    CELL MEMBRANE, 2671

CONTRACEPTIVES, ORAL
  ANEMIA, MACROCYTIC
    FOLIC ACID, 2406
    VITAMIN B12, 2406
  BREAST NEOPLASMS
    CARCINOGENIC POTENTIAL, REVIEW,
    2405

CERVIX NEOPLASMS
  CARCINOGENIC POTENTIAL, R
  2405
  FOLIC ACID, 2406
  PRECANCEROUS CONDITIONS,

COPPER
  BREAST NEOPLASMS
    REVIEW, 2403
  HODGKIN'S DISEASE
    REVIEW, 2403
  LEUKEMIA
    REVIEW, 2403
  LUNG NEOPLASMS
    REVIEW, 2403
  LYMPHOMA
    REVIEW, 2403
  NEOPLASMS
    REVIEW, 2403
  PROSTATIC NEOPLASMS
    REVIEW, 2403
  SARCOMA, OSTEOGENIC
    NEOPLASM METASTASIS, 2403
    REVIEW, 2403

CORTICOSTERONE
  CYCLOHEXIMIDE
    CARBOXY-LASES, 2957
  DNA REPLICATION
    PROSTATE, 2964
  THYMIDINE KINASE
    PROSTATE, 2964

CORTICOTROPIN
  CYCLOHEXIMIDE
    CARBOXY-LASES, 2957

CORTISOL
  HELA CELLS
    GROWTH, 2980*
  HEPATOMA
    TYROSINE AMINOTRANSFERASE,
    VITAMIN B DEFICIENCY, 2962

CORTISONE
  T-LYMPHOCYTES
    CELL DIVISION, 2682
    DNA REPLICATION, 2682

CROCIDOLITE
  SEE ASBESTOS

CYCLIC AMP
  SEE ADENOSINE CYCLIC 3',5'
  MONOPHOSPHATE

CYCLIC AMP PHOSPHODIESTERASE
  SEE PHOSPHODIESTERASES

CYCLOHEXIMIDE
  CARBOXY-LASES
    ADRENAL CORTEX HORMONES, 29
    CORTICOSTERONE, 2957
    CORTICOTROPIN, 2957
    ENZYMATIC ACTIVITY, 2957
  DNA REPLICATION
    MITOMYCIN C, 2621
  GLIOMA

CYTOCHALASIN B
    CELL MOVEMENT
        WOUNDS AND INJURIES, 2916
    VIRUS, POLYOMA
        CELL MOVEMENT, 2916
        WOUNDS AND INJURIES, 2916

CYTOCHROME P-450
    HEPATOMA
        BUTYRIC ACID, 2-AMINO-4-
        (ETHYLTHIO)-, 2792
    POLUCYLIC HYDROCARBONS
        COAL-TAR, 2536*

CYTOCHROME REDUCTASES
    HEPATOMA
        BUTYRIC ACID, 2-AMINO-4-
        (ETHYLTHIO)-, 2792

CYTOSINE, 1-BETA-D-ARABINOFURANOSYL-
    BRAIN NEOPLASMS
        VIRUS, HERPES SIMPLEX 2, 2599
    EYE NEOPLASMS
        VIRUS, HERPES SIMPLEX 2, 2599
    LIVER NEOPLASMS
        VIRUS, HERPES SIMPLEX 2, 2599
    SKIN NEOPLASMS
        VIRUS, HERPES SIMPLEX 2, 2599
    VIRUS, HERPES SIMPLEX 1
        RNA, MESSENGER, 2595
    VIRUS, HERPES SIMPLEX 2
        VIRUS REPLICATION, 2599

CYTOSOL
    MAMMARY NEOPLASMS, EXPERIMENTAL
        HEXOKINASE, 2948
        PHOSPHOTRANSFERASES, ATP, 2948

CYTOXAN
    SEE CYCLOPHOSPHAMIDE

2667   DACTINOMYCIN
    SEE ACTINOMYCIN D

DEOXY SUGARS
    VIRUS, ROUS SARCOMA
        CELL TRANSFORMATION, NEOPLASTIC,
2667   2470

DEOXYIODOURIDINE
2667   SEE URIDINE, 2'-DEOXY-5-IODO-

DEOXYRIBONUCLEASE
    DNA POLYMERASE
        DNA, BACTERIAL, 2582

DEOXYRIBONUCLEOSIDES
    DNA POLYMERASE
        DNA, BACTERIAL, 2582

DERMATOFIBROMA
    SEE FIBROMA

34   DERMOID CYST
    OVARIAN NEOPLASMS
        MIGRATION INHIBITORY FACTOR, 2667

DEXAMETHASONE

SPERMINE
    CARBOXY-LYASES, 2958

DEXTRAN
    VIRUS, MOLONEY MURINE LEUKEMIA
        VIRUS REPLICATION, 2652*
    VIRUS, MOLONEY MURINE SARCOMA-LEUKEMIA
        VIRUS REPLICATION, 2652*
    VIRUS, MURINE SARCOMA
        VIRUS REPLICATION, 2652*
    VIRUS, ROUS SARCOMA, 2656*

DIABETES MELLITUS
    BLADDER NEOPLASMS
        1,2-BENZISOTHIAZOLIN-3-ONE, 1,1-
            DIOXIDE, 2424*
        EPIDEMIOLOGY, REVIEW, 2424*
    LUNG NEOPLASMS
        EPIDEMIOLOGY, REVIEW, 2424*

DIBENZ(A,H)ANTHRACENE
    BRAIN NEOPLASMS
        PRECANCEROUS CONDITIONS, 2473
    PETROLEUM
        FOOD ANALYSIS, 2535*
    SKIN NEOPLASMS
        OCCUPATIONAL HAZARD, REVIEW, 2431*
    YEAST
        FOOD ANALYSIS, 2537*

DIBENZO-P-DIOXIN, 2,3,7,8-TETRACHLORO-
    ARYL HYDROCARBON HYDROXYLASES
        ENZYME INDUCTION, 2500
    CHOLANTHRENE, 3-METHYL-
        ARYL HYDROCARBON HYDROXYLASES, 2500

DIBUTYRYL CYCLIC AMP
    ARYL HYDROCARBON HYDROXYLASES
        ENZYME INDUCTION, 2499
    CELLS, CULTURED
        AGGLUTINATION, 2972*
        CONCANAVALIN A, 2972*
        PLANT AGGLUTININS, 2972*
    GLIOMA
        CARBOXY-LYASES, 2959
    NEUROBLASTOMA
        CARBOXY-LYASES, 2959

DIBUTYRYL CYCLIC GMP
    ARYL HYDROCARBON HYDROXYLASES
        ENZYME INDUCTION, 2499

DIELDRIN
    HELA CELLS
        GROWTH, 2921
        LECITHINS, 2921
        PHOSPHOLIPIDS, 2921
        PROTEINS, 2921
        SPHINGOMYELINS, 2921
    LIVER NEOPLASMS
        CARCINOGENIC POTENTIAL, REVIEW, 2404

DIET
    BREAST NEOPLASMS
        EPIDEMIOLOGY, REVIEW, 2420

DIETHYLAMINE, N-NITROSO-
    ANTILYMPHOCYTE SERUM

MICROSOMES, LIVER
    ACETAMIDE, N-FLUOREN-2-YL-, 2515*
THYMIDINE
    AUTORADIOGRAPHY, 2911*
    PHORBOL 12,13-DIMYRISTATE, 2462
THYROID NEOPLASMS
    ADENOCARCINOMA, 2934
    ADENOMA, 2934
    CARCINOMA, PAPILLARY, 2934
    CYSTADENOMA, PAPILLARY, 2934
    HYPERPLASIA, 2934
VIRUS, HARVEY MURINE SARCOMA-LEUKEMIA
    DNA, VIRAL, 2608

DNA, BACTERIAL
    BACILLUS SUBTILIS
        ALBUMINS, 2582
    DNA POLYMERASE
        DEOXYRIBONUCLEASE, 2582
        DEOXYRIBONUCLEOSIDES, 2582

DNA NUCLEOTIDYLTRANSFERASES
    SEE ALSO DNA POLYMERASE
    DNA REPLICATION
        SPERMATAZOA, 2984*
        TEMPLATE RESPONSE, 2984*
    SARCOMA, YOSHIDA
        DNA REPLICATION, 2929
    VIRUS, ADENO 2
        ISOLATION AND CHARACTERIZATION,
        2639*
    VIRUS, ROUS SARCOMA
        ISOLATION AND CHARACTERIZATION,
        2628

DNA POLYMERASE
    SEE ALSO DNA NUCLEOTIDYLTRANSFERASES
    DNA, BACTERIAL
        DEOXYRIBONUCLEASE, 2582
        DEOXYRIBONUCLEOSIDES, 2582
    DNA REPAIR
        REVIEW, 2422
    DNA, REPLICATION
        REVIEW, 2422
    POLYNUCLEOTIDES
        NUCLEIC ACIDS, 2582
    VIRUS, AVIAN MYELOBLASTOSIS
        NUCLEIC ACIDS, 2582
        POLYNUCLEOTIDES, 2582

DNA REPAIR
    ANEMIA, APLASTIC
        MITOMYCIN C, 2933
    DNA POLYMERASE
        REVIEW, 2422
    ESCHERICHIA COLI
        DIETHYLAMINE, N-NITROSO-, 2480
        DIMETHYLAMINE, N-NITROSO-, 2480
        METHANESULFONIC ACID, METHYL ESTER,
        2480
        ULTRAVIOLET RAYS, 2572*
        UREA, ETHYL NITROSO-, 2480
        UREA, METHYL NITROSO-, 2480
    KIDNEY NEOPLASMS
        QUINOLINE, 4-NITRO-, 1-OXIDE, 2493
    LIVER NEOPLASMS

QUINOLINE, 4-NITRO-, 1-OXIDE, 2493
RADIATION, IONIZING, 2568*
LUNG NEOPLASMS
QUINOLINE, 4-NITRO-, 1-OXIDE, 2493
MICROCOCCUS LUTEUS
ULTRAVIOLET RAYS, 2572*

DNA REPLICATION
CARCINOMA, EHRLICH TUMOR
THYMIDINE INCORPORATION, 2898
URIDINE, 2'-DEOXY-5-IODO-, 2898
CORTICOSTERONE
PROSTATE, 2964
DNA NUCLEOTIDYLTRANSFERASES
TEMPLATE RESPONSE, 2984*
DNA POLYMERASE
REVIEW, 2422
DNA, SINGLE STRANDED
REVIEW, 2422
DNA, VIRAL
REVIEW, 2422
DROSOPHILA
REVIEW, 2422
ESCHERICHIA COLI
REVIEW, 2422
ESTRADIOL
PROSTATE, 2964
LEUKEMIA, MONOCYTIC
NICOTINAMIDE, 2930
UREA, METHYL NITROSO-, 2930
LYMPHOCYTES
SERUM ROLE, 2735*
T-LYMPHOCYTES
CORTISONE, 2682
MITOMYCIN C
CYCLOHEXIMIDE, 2621
POLYNUCLEOTIDE SYNTHETASES
REVIEW, 2422
PROGESTERONE, 6-CHLORO-6-DEHYDRO-17-
ALPHA-HYDROXY-
1,2ALPHA-METHYLENE-, ACETATE, 2964
RNA
REVIEW, 2422
RNA POLYMERASE
REVIEW, 2422
SARCOMA, YOSHIDA
DNA NUCLEOTIDYLTRANSFERASES, 2929
THYMIDINE INCORPORATION, RAT, 2929
THYMIDINE KINASE, 2929
SPERMATAZOA
DNA NUCLEOTIDYLTRANSFERASES, 2984*
HYBRIDIZATION, 2984*
TESTOSTERONE, PHENYLPROPIONATE
PROSTATE, 2964
VIRUS, ADENO 2
DNA, VIRAL, 2578
VIRUS, AVIAN MYELOBLASTOSIS
REVERSE TRANSCRIPTASE, 2986*
VIRUS, POLYOMA
CYCLOHEXIMIDE, 2621
MITOMYCIN C, 2621
PROTEINS, 2624
VIRUS, RNA HAMSTER SARCOMA
PROTEINS, 2624
VIRUS, ROUS SARCOMA
REVIEW, 2435*
VIRUS, SV40
ANTIGENS, VIRAL, 2688

CLONED VARIANTS, 2658*, 2660*
VIRUS, SV40
DNA, VIRAL, 2637

DNA, RIBOSOMAL
DROSOPHILA MELANOGASTER
CARBAMIC ACID, N-METHYL-N-NITROS
ETHYL ESTER, 2492

DNA, SATELLITE
VIRUS, AVIAN SARCOMA
CHROMATIN, 2585

DNA, SINGLE STRANDED
DNA, REPLICATION
REVIEW, 2422
VIRUS, MOLONEY MURINE LEUKEMIA
DNA, VIRAL, 2615

DNA, VIRAL
DNA, REPLICATION
REVIEW, 2422
VIRUS, ADENO 2
CELL CYCLE KINETICS, 2578
DNA REPLICATION, 2578
VIRUS, AVIAN SARCOMA
CHROMATIN, 2585
VIRUS, EPSTEIN-BARR, 2594
VIRUS, FELINE LEUKEMIA
CROSSES, GENETIC, 2601
VIRUS, HARVEY MURINE SARCOMA-LEUKEMIA
DNA, 2608
VIRUS, MOLONEY MURINE LEUKEMIA
DNA, SINGLE STRANDED, 2615
VIRUS, POLYOMA
CYCLOHEXIMIDE, 2621
ENDONUCLEASES, 2623
VIRUS, POLYOMA, BK
HYBRIDIZATION, 2657*
ISOLATION AND CHARACTERIZATION,
2657*
VIRUS, RD114
CROSSES, GENETIC, 2601
VIRUS, ROUS SARCOMA
CHROMATIN, 2585
VIRUS, SV40
ANTIGENS, VIRAL, 2688
CHROMATIN, 2635
DEFECTIVE VIRUSES, 2629
VIRUS, SV40
DNA HYBRIDIZATION, 2637
DNA REPLICATION, 2637
VIRUS, SV40
ENDONUCLEASES, 2635
PEPTIDES, 2633
SUBCELLULAR FRACTIONS, 2633
TEMPERATURE SENSITIVE MUTANTS,
2662*
VIRAL PROTEINS, 2633
VIRUS, SV40
VIRUS REPLICATION, 2637

INTESTINAL NEOPLASMS
    ADENOCARCINOMA, 2826*, 2827*
    CASE REPORT, 2827*

ENVIRONMENTAL HAZARD
    ETHYLENE, CHLORO-
        ASSESSMENT, 2433*
    ETHYLENE, CHLORO- PCLYMER
        ASSESSMENT, 2433*

ENZYMES
    CHORIOCARCINOMA
        ULTRASTRUCTURAL STUDY, 2999*
    HYDATIDIFORM MOLE
        ULTRASTRUCTURAL STUDY, 2999*
    LYSOSOMES
        REVIEW, 2451*
    NEOPLASMS
        REVIEW, 2451*

EPENDYMOMA
    BRAIN NEOPLASMS
        ULTRASTRUCTURAL STUDY, 2809*
    CHOLANTHRENE, 3-METHYL-
        IMMUNITY, CELLULAR, 2418
        TRANSPLANTATION IMMUNOLOGY, 2418
    NERVOUS SYSTEM NEOPLASMS
        EPIDEMIOLOGY, RACIAL FACTORS,
            CHILD, 2883

EPIDEMIOLOGY
    LEUKEMIA
        REVIEW, 2447*

412   EPINEPHRINE
        NEOPLASMS
412   ADENOSINE CYCLIC 3',5'
                MONOPHOSPHATE, 2998*
        NEUROBLASTOMA
            ADENOSINE CYCLIC 3',5'
                MONOPHOSPHATE, 2415
            CATECHOLAMINES, 2415
        RECTAL NEOPLASMS
            METABOLISM, 3000*
        VIRUS, SV40
            ADENYL CYCLASE, 2663*
67
        EPITHELIAL CELLS
            COLONIC NEOPLASMS
                DNA, 2985*

        ERYTHROBLASTS
            LEUKEMIA, MYELOBLASTIC
                BONE MARROW, 2778

        ERYTHROLEUKEMIA
            BONE MARROW CELLS
                CASE REPORT, 2777
                CHROMOSOME ABERRATIONS, 2777
            CLASSIFICATION
                ULTRASTRUCTURAL STUDY, 2445*
            CYTOGENIC ANALYSIS
                CASE REPORT, 2777
                CHROMOSOME ABERRATIONS, 2777
            GENETICS
                EPIDEMIOLOGY, 2880
            UREA, N-NITROSO-N-PROPYL-

        ULTRASTRUCTURAL STUDY, RAT, 2491
    VIRUS, FRIEND SPLEEN-FOCLS FORMING
        VIRUS, FRIEND MURINE LEUKEMIA, 2651*

ERYTHROPOIESIS
    ANEMIA, APLASTIC
        PRECANCEROUS CONDITICNS, 2780
    LEUKEMIA
        ACUTE DISEASE, 2780
        PRECANCEROUS CONDITICNS, 2780
    LEUKEMIA, MYELOBLASTIC
        PRECANCEROUS CONDITICNS, 2780

ERYTHROPOIETIN
    VIRUS, FRIEND MURINE LEUKEMIA
        COLONY FORMATION, 26C5

ESCHERICHIA COLI
    DIETHYLAMINE, N-NITROSO-
        DNA REPAIR, 2480
    DIMETHYLAMINE
        DIMETHYLAMINE, N-NITROSO-, 2484
    DIMETHYLAMINE, N-NITROSO-
        DNA REPAIR, 2480
        MICROSOMES, LIVER, 2551*
        MUTAGENESIS, 2551*
    DNA REPAIR
        ULTRAVIOLET RAYS, 2572*
    DNA, REPLICATION
        REVIEW, 2422
    ENDONUCLEASES
        ULTRAVIOLET RAYS, 2572*
    2-FURANACRYLAMIDE, ALPHA-2-FURYL-5-
        NITRO-
        MUTAGENIC ACTIVITY, 2455
    METHANESULFONIC ACID, METHYL ESTER
        DNA REPAIR, 2480
    UREA, ETHYL NITROSO-
        DNA REPAIR, 2480
    UREA, METHYL NITROSO-
        DNA REPAIR, 2480

ESOPHAGEAL NEOPLASMS
    ADENOCARCINOMA
        CASE REPORT, 2835*
    CARCINOMA
        EPIDEMIOLOGY, CHINA, REVIEW, 2448*
        MOLYBDENUM, 2448*
        NITROSAMINES, 2448*
    CARCINOMA, EPIDERMOID
        CASE REPORT, 2789
        NICOTINE, 1'-DEMETHYL-N'-NITROSO-,
        2486
    CARCINOMA, OAT CELL
        CASE REPORT, 2789
    INTESTINAL POLYPS
        PREDNISOLONE, 2834*
    MELANOMA
        CASE REPORT, 2836*, 2837*, 2839*
    NEOPLASM METASTASIS
        CASE REPORT, 2802*
    NEOPLASMS, EMBRYONAL AND MIXED
        CASE REPORT, 2838*
    PAPILLOMA
        NICOTINE, 1'-DEMETHYL-N'-NITROSO-,
        2486
    PREDNISOLONE
        IMMUNITY, CELLULAR, 2834*

ETHYNODIOL DIACETATE
    MAMMARY NEOPLASMS, EXPERIMENTAL
        ADENOCARCINOMA, 2405

EXUDATES AND TRANSUDATES
    FIBROSARCOMA
        CARRAGEEN, 2769

EYE NEOPLASMS
    VIRUS, HERPES SIMPLEX 2
        CYTOSINE, 1-BETA-D-
            ARABINOFURANOSYL-, 2599
        URIDINE, 2'-DEOXY-5-IODO-, 2599

EYELID NEOPLASMS
    CARCINOMA, BASAL CELL
        CASE REPORT, 2856*
    CARCINOMA, EPIDERMOID
        CASE REPORT, 2856*

FD AND C RED NO. 40
    FRUCTOSE
        GLUCOSE, 2518*

FERRITIN
    HEPATOMA
        ANTIGENIC DETERMINANTS, 2928
        IRON(II) SULFATE, 2927
        ISOLATION AND CHARACTERIZATION,
            2928

FETAL GLOBULINS
    ACETAMIDE, N-FLUOREN-2-YL-
        EMBRYO, LIVER, 2527*
    HEPATOMA
        ACETOHYDROXAMIC ACID, N-FLUOREN-2-
            YL-, 2528*
        ALKALINE PHOSPHATASE, 2955
        DIMETHYLAMINE, N-NITROSO-, 2528*
    LIVER NEOPLASMS
        CARCINOMA, 2851*
    NEPHROBLASTOMA
        ISOLATION AND CHARACTERIZATION,
            2692
    TERATOID TUMOR
        CASE REPORT, 2841*

ALPHA-FETOPROTEIN
    SEE FETAL GLOBULINS

FIBERS
    ASBESTOS
        OCCUPATIONAL HAZARD, 2575*

FIBRINOLSYIN
    MAMMARY NEOPLASMS, EXPERIMENTAL
        CARCINOMA, 2979*
        ISOLATION AND CHARACTERIZATION,
            2979*

FIBROBLASTS
    CELL MEMBRANE
        CONCANAVALIN A, 2671
    LYMPHOCYTE TRANSFORMATION
        VIRUS, EPSTEIN-BARR, 2592
    PROSTATIC NEOPLASMS
        IMMUNE SERUMS, 2713*

SARCOMA, OSTEOGENIC
    CROSS REACTIONS, 2715*

FIBROMA
   COLONIC NEOPLASMS
      UREA, N-NITROSO-N-PROCPYL-, 2491
   NEOPLASMS, CONNECTIVE TISSUE
      EPIDEMIOLOGY, 2763
      HISTOLOGICAL STUDY, 2763
      PRECANCEROUS CONDITICNS, 2763
   OVARIAN NEOPLASMS
      MIGRATION INHIBITORY FACTOR, 2667
   VIRUS, BOVINE PAPILLOMA
      ANTIGENS, NEOPLASM, 2665
      IMMUNE SERUMS, 2665
      TRANSPLANTATION IMMUNOLOGY, 2665

FIBROSARCOMA
   CARRAGEEN
      EXUDATES AND TRANSUDATES, 2769
   CELL LINE
      CHROMOSOMES, 2970*
      VIRUS-LIKE PARTICLES, 2970*
   CHOLANTHRENE, 3-METHYL-
      MICROCIRCULATION, PERMEABILITY,
        2769
      ULTRASTRUCTURAL STUDY, RAT, 2769
   COLONIC NEOPLASMS
      UREA, N-NITROSO-N-PROCPYL-, 2491
   EPIDEMIOLOGY
      RACIAL FACTORS, CHILD, 2883
   LYMPHOCYTES
      IMMUNITY, CELLULAR, 2712*
   OVARIAN NEOPLASMS
      UREA, N-NITROSO-N-PROCPYL-, 2491
   PLANTS
      CARCINOGENIC ACTIVITY, MOUSE, 2537*

   SKIN NEOPLASMS
      UREA, N-NITROSO-N-PROCPYL-, 2491
   VIRUS, SV40
      IMMUNITY, CELLULAR, 2689
      IMMUNOSUPPRESSION, 2689
      TRANSPLANTATION IMMUNOLOGY, 2689

FLAVONES
   CARCINOMA, EHRLICH TUMOR
      ADENOSINE TRIPHOSPHATASE, 2919
      GLYCOLYSIS, 2919
      GROWTH, 2919
   LEUKEMIA
      ADENOSINE TRIPHOSPHATASE, 2919
      GLYCOLYSIS, 2919
      GROWTH, 2919
   VIRUS, KIRSTEN MURINE SARCOMA
      ADENOSINE TRIPHOSPHATASE, 2919
      GLYCOLYSIS, 2919
      GROWTH, 2919

FLUORENYL ACETAMIDE
   SEE ACETAMIDE, N-FLUOREN-2-YL-

FOLIC ACID
   ANEMIA, MACROCYTIC
      CONTRACEPTIVES, ORAL, 2406
   CERVIX NEOPLASMS
      CONTRACEPTIVES, ORAL, 2406
   HEPATECTOMY
      ENZYMATIC ACTIVITY, 2942

REVIEW, 2444*
LEUKEMIA
    REVIEW, 2444*
LEUKEMIA, LYMPHOBLASTIC
    EPIDEMIOLOGY, 2880
    REVIEW, 2444*
LEUKEMIA, LYMPHOCYCTIC
    REVIEW, 2444*
LEUKEMIA, LYMPHOCYTIC
    EPIDEMIOLOGY, 2880
LEUKEMIA, MONOBLASTIC
    EPIDEMIOLOGY, 2880
LEUKEMIA, MYELOCYTIC
    EPIDEMIOLOGY, 2880
LYMPHOMA
    HYBRID CELL MODEL, 2773
    REVIEW, 2444*
NASOPHARYNGEAL NEOPLASMS
    VIRUS, EPSTEIN-BARR, 2709*
OVARIAN NEOPLASMS
    BREAST NEOPLASMS, 2865*
    CYSTADENOCARCINOMA, 2865*
    PRECANCEROUS CONDITIONS, 2865*
RETINOBLASTOMA
    CLEFT PALATE, 2878*
    HEART DEFECTS, CONGENITAL, 2878*
SARCOMA, RETICULUM CELL
    REVIEW, 2444*
SKIN NEOPLASMS
    CYLINDROMA, 2854*
THYROID NEOPLASMS
    CARCINOMA, 2875*
    PHEOCHROMOCYTOMA, 2875*
UTERINE NEOPLASMS
    HISTOCOMPATIBILITY ANTIGENS, 2686
VIRUS, B-TROPIC MURINE LEUKEMIA
    VIRUS ACTIVATION, 2607

GIANT CELL TUMOR
    OVARIAN NEOPLASMS
        CYCTADENOMA, MUCINOUS, 2864*

GLIOMA
    ACTINOMYCIN D
        CARBOXY-LYASES, 2959
    ADENOSINE
        CARBOXY-LYASES, 2959
    ARTERENOL
        CARBOXY-LYASES, 2959
    CHOLINE, CHLORIDE, CARBAMATE
        CARBOXY-LYASES, 2959
    CYCLOHEXIMIDE
        CARBOXY-LYASES, 2959
    DIBUTYRYL CYCLIC AMP
        CARBOXY-LYASES, 2959
    2-IMIDAZOLIDONE, 4-(3-BUTOXY-4-
      METHOXYBENZYL)-
        CARBOXY-LYASES, 2959
    ISOPROTERENOL
        CARBOXY-LYASES, 2959
    NERVOUS SYSTEM NEOPLASMS
        UREA, 1,1-DIETHYL-3-METHYL-3-
          NITROSO-, 2488
        UREA, 3-ETHYL-1,1-DIMETHYL-3-NITRO-
          , 2488
        UREA, 1,1,3-TRIETHYL-3-NITROSO-,
          2488
        UREA, 1,1,3-TRIMETHYL-3-NITROSO-,
          2488

PROSTAGLANDINS E
    CARBOXY-LYASES, 2959
    XANTHINE, 3-ISOBUTYL-1-METHYL-
      ADENOSINE CYCLIC 3',5'
      MONOPHOSPHATE, 2997*
    CARBOXY-LYASES, 2959
    GUANOSINE CYCLIC 3',5'
      MONOPHOSPHATE, 2997*
    PHOSPHODIESTERASES, 2997*

GLOMERULONEPHRITIS
    LEUKEMIA
      IMMUNE COMPLEX DISEASE, 2710*
    LIVER NEOPLASMS
      SARCOIDOSIS, 2850*
    LUNG NEOPLASMS
      SARCOIDOSIS, 2850*
    LYMPHOMA
      IMMUNE COMPLEX DISEASE, 2710*
    MAMMARY NEOPLASMS, EXPERIMENTAL
      IMMUNE COMPLEX DISEASE, 2710*
    SPLENIC NEOPLASMS
      SARCOIDOSIS, 2850*
    SYNOVIOMA
      IMMUNE COMPLEX DISEASE, 2710*

GLUCONEOGENESIS
    HEPATOMA
      GLYCEROLPHOSPHATE DEHYDROGENASE,
      2949
      LACTATE DEHYDROGENASE, 2949
      NADH, NADPH OXIDOREDUCTASES, 2949
      OXIDOREDUCTASES, 2949
      PHOSPHOTRANSFERASES, ATP, 2949

GLUCOSAMINIDASE
    LEUKEMIA, MONOBLASTIC
      ISOLATION AND CHARACTERIZATION,
      2947
    LEUKEMIA, MYELOBLASTIC
      ISOLATION AND CHARACTERIZATION,
      2947
    LEUKOCYTES
      ISOLATION AND CHARACTERIZATION, 2947

GLUCOSE
    C.I. ACID RED 27
      FRUCTOSE, 2518*
    FD AND C RED NO. 40
      FRUCTOSE, 2518*

GLUCOSE, 2-DEOXY-
    VIRUS, KIRSTEN MURINE SARCOMA-LEUKEMIA
      PROTEINS, 2613
      RNA REPLICATION, 2613

GLUCOSIDASES
    HEPATOMA
      GLYCOGEN, 2946
      ISOENZYMES, 2946

GLUTAMYL TRANSPEPTASE
    PANCREATIC NEOPLASMS
      ISOENZYMES, 2950
    PANCREATITIS
      ISOENZYMES, 2950

GLYCEROLPHOSPHATE DEHYDROGENASE

HEPATOMA
    GLUCONEOGENESIS, 2949

GLYCOGEN
    HEPATOMA
      GLUCOSIDASES, 2946

GLYCOLIPIDS
    CARCINOMA, EHRLICH TUMOR
      GROWTH, 2989*
      ISOLATION AND CHARACTERIZATI
      2989*
    HEPATOMA
      GROWTH, 2989*
      ISOLATION AND CHARACTERIZATI
      2989*
    LEUKEMIA L1210
      GROWTH, 2989*
      ISOLATION AND CHARACTERIZATI
      2989*
    MAMMARY NEOPLASMS, EXPERIMENTAL
      GROWTH, 2989*
      ISOLATION AND CHARACTERIZATI
      2989*

GLYCOLYSIS
    CARCINOMA, EHRLICH TUMOR
      FLAVONES, 2919
    LEUKEMIA
      FLAVONES, 2919
    VIRUS, KIRSTEN MURINE SARCOMA
      FLAVONES, 2919

GLYCOPEPTIDES
    CELL MEMBRANE
      CELLS, CULTURED, 2972*
    VIRUS, FRIEND MURINE LEUKEMIA
      CONCANAVALIN A, 2603

GLYCOPROTEINS
    CELL MEMBRANE
      VIRUS, AVIAN RNA TUMOR, 2414
    LEUKEMIA L1210
      CELL MEMBRANE, 2944
    LEUKEMIA, MYELOCYTIC
      TRANSPLANTATION, HOMOLOGOUS,
    LYMPHOCYTES
      CELL MEMBRANE, 2982*
    VIRUS, ROUS SARCOMA
      SURFACE PROPERTIES, 2626
      ULTRASTRUCTURAL STUDY, 2626
    VIRUS, SV40
      CELL MEMBRANE, 2634

GLYCOSPHINGOLIPIDS
    NERVOUS SYSTEM NEOPLASMS
      METABOLISM, 2990*
    NEUROBLASTOMA
      METABOLISM, 2990*

GLYCOSYLTRANSFERASES
    CHONDROSARCOMA
      ENZYMATIC ACTIVITY, 2945

GOITER
    THYROID NEOPLASMS
      ADENOMA, 2876*
      CARCINOMA, 2876*

        CELL CYCLE KINETICS, 2956
LEUKEMIA
    FLAVONES, 2919
    RADIATION, 2720*
    ZINC, 2403
LEUKEMIA L1210
    GLYCOLIPIDS, 2989*
LYMPHOMA
    HYBRID CELL MODEL, 2773
MAMMARY NEOPLASMS, EXPERIMENTAL
    ADENOCARCINOMA, 2725*
    GLYCOLIPIDS, 2989*
NEOPLASMS
    IMMUNITY, CELLULAR, 2712*
    TRANSPLANTATION, HETEROLOGOUS,
      2975*
PLANTS
    ANLINE, 4,4'-(IMIDOCARBONYL)BIS(N,
      N-DIMETHYL)-, 2532*
SARCOMA, 2725*
    CHOLANTHRENE, 3-METHYL-, 2976*
    ZINC, 2403
SARCOMA 180, CROCKER
    ZINC, 2732*
SKIN NEOPLASMS
    IMMUNOSUPPRESSION, 2717*
    T-LYMPHOCYTES, 2717*
    MELANOMA, 2717*
VIRUS, EPSTEIN-BARR
    CELL LINE, 2703
VIRUS, KIRSTEN MURINE SARCOMA
    FLAVONES, 2919

GUANIDINE, 1-METHYL-3-NITRO-1-NITROSO-
CELL LINE
    CELL TRANSFORMATION, NEOPLASTIC,
      2913
PEPSINOGEN
    ISOENZYMES, 2485

GUANIDINE, NITROSO-
CHROMOSOME ABERRATIONS
    MUTATION, 2548*

GUANOSINE CYCLIC 3',5' MONOPHOSPHATE
CHOLANTHRENE, 3-METHYL-
    CELL TRANSFORMATION, NEOPLASTIC,
      2994*
GLIOMA
    XANTHINE, 3-ISOBUTYL-1-METHYL-,
      2997*
HEPATOMA
    GUANYL CYCLASE, 2943
NEUROBLASTOMA
    XANTHINE, 3-ISOBUTYL-1-METHYL-,
      2997*
VIRUS, HARVEY MURINE SARCOMA
    CELL TRANSFORMATION, NEOPLASTIC,
      2994*
VIRUS, KIRSTEN MURINE SARCOMA
    CELL TRANSFORMATION, NEOPLASTIC,
      2994*
VIRUS, MOLONEY MURINE SARCOMA
    CELL TRANSFORMATION, NEOPLASTIC,
      2994*
VIRUS, SV40
    CELL TRANSFORMATION, NEOPLASTIC,
      2994*

GUANYL CYCLASE
    CHOLANTHRENE, 3-METHYL-
        CELL TRANSFORMATION, NEOPLASTIC,
          2994*
    HEPATOMA
        CYCLOHEXIMIDE, 2943
        GUANOSINE CYCLIC 3',5'
          MONOPHOSPHATE, 2943
    VIRUS, HARVEY MURINE SARCOMA
        CELL TRANSFORMATION, NEOPLASTIC,
          2994*
    VIRUS, KIRSTEN MURINE SARCOMA
        CELL TRANSFORMATION, NEOPLASTIC,
          2994*
    VIRUS, MOLONEY MURINE SARCOMA
        CELL TRANSFORMATION, NEOPLASTIC,
          2994*
    VIRUS, SV40
        CELL TRANSFORMATION, NEOPLASTIC,
          2994*

GYNECOLOGIC NEOPLASMS
    ALCOHOLISM
        EPIDEMIOLOGY, 2892

HAMARTOMA
    KIDNEY NEOPLASMS
        CASE REPORT, CHILD, 2793
    UTERINE NEOPLASMS
        CASE REPORT, 2867*

HEAD AND NECK NEOPLASMS
    CARCINOEMBRYONIC ANTIGEN
        DIAGNOSIS AND PROGNOSIS, REVIEW,
          2416
    CARCINOMA
        IMMUNOGLOBULINS, 2782
        ROSETTE FORMATION, 2782
    CARCINOMA, EPIDERMOID
        EPIDEMIOLOGY, FISH, 2895
    NEUROEPITHELIOMA
        CASE REPORT, 2822*

HEART DEFECTS, CONGENITAL
    RETINOBLASTOMA
        GENETICS, 2878*

HEART NEOPLASMS
    HEMANGIOSARCOMA
        CASE REPORT, 2843*
        PERICARDIAL EFFUSION, 2843*
    NEUROFIBROMA
        UREA, 1,1-DIETHYL-3-METHYL-3-
          NITROSO-, 2488
        UREA, 3-ETHYL-1,1-DIMETHYL-3-
          NITROSO-, 2488
        UREA, 1,1,3-TRIETHYL-3-NITROSO-,
          2488
        UREA, 1,1,3-TRIMETHYL-3-NITROSO-,
          2488

HELA CELLS
    BLOOD PROTEINS
        GROWTH, 2980*
    CARBAMIC ACID, METHYL-, 1-NAPHTHYL
      ESTER
        GROWTH, 2921
        LECITHINS, 2921

TOSARCOMA
SEE ANGIOSARCOMA

CYANIN
HODGKIN'S DISEASE
    IMMUNITY, CELLULAR, 2721*
T-LYMPHOCYTES
    IMMUNE RESPONSE, 2704

LYSIS
ETHYLENE, CHLORO-
    BIOLOGICAL ACTIVITY, DUST, 2502
ETHYLENE, CHLORO- POLYMER
    BIOLOGICAL ACTIVITY, DUST, 2502

POIETIC SYSTEM
BENZENE
    TOXIC METABOLITES, 2541*

TECTOMY
FOLIC ACID
    ENZYMATIC ACTIVITY, 2942

TITIS
LIVER NEOPLASMS
    CARCINOMA, 2851*

TOCARCINOMA
SEE HEPATOMA

TOCELLULAR CARCINOMA
SEE HEPATOMA

TOMA
ACETOHYDROXAMIC ACID, N-FLUCREN-2-YL-
    ACETAMIDE, N-FLUOREN-2-YL-, 2528*
    FETAL GLOBULINS, 2528*
ADENOSINE CYCLIC 3',5' MCNOPHOSPHATE
    CARRIER PROTEINS, 2981*
AFLATOXIN B1
    CARCINOGENIC POTENTIAL, REVIEW, 2404
ALKALINE PHOSPHATASE
    CASE REPORT, 2955
    FETAL GLOBULINS, 2955
AMINOTRANSFERASES
    GROWTH, 2941
ANTIGENS, NEOPLASM
    IMMUNITY, ACTIVE, 2716*
    IMMUNITY, CELLULAR, 2716*
BENZ(A)ANTHRACENE, 7,12-CIMETHYL-
    CARCINOGENIC POTENTIAL, MOUSE, 2468
1-BUTANAMINE, N-BUTYL-N-NITROSO-
    CARCINOGENIC ACTIVITY, MCUSE, 2550*
1-BUTANAMINE, N-(3-CARBOXYBUTYL)-N-
NITROSO-
    CARCINOGENIC ACTIVITY, MOUSE, 2550*
1-BUTANOL, 4-(BUTYLNITROSOAMINO)-
    CARCINOGENIC ACTIVITY, MOUSE,
    2550*
BUTYRIC ACID, 2-AMINO-4-(ETHYLTHIO)-
    AMINOPYRINE DEMETHYLASE, 2792
    CYTOCHROME P-450, 2792
    CYTOCHROME REDUCTASES, 2792
    MICROSOMES, LIVER, 2792
    PHOSPHOLIPIDS, 2792
    PRECANCEROUS CONDITICNS, 2792
    RIBOSOMES, 2792
    RNA, TRANSFER, METHYLTRANSFERASES,

2952
CONCANAVALIN A
    AGGLUTINATION, 2706
    LECTIN BINDING, 2706
CORTISOL
    TYROSINE AMINOTRANSFERASE, 2962
    VITAMIN B DEFICIENCY, 2962
DIETHYLAMINE, N-NITROSO-
    ANTIGENS, NEOPLASM, 2708*
    CARCINOGENIC ACTIVITY, RAT, 2482
    IMMUNITY, 2708*
    NEOPLASM TRANSPLANTATION, 2549*
DIMETHYLAMINE, N-NITROSO-
    ACETAMIDE, N-FLUCREN-2-YL-, 2528*
    FETAL GLOBULINS, 2528*
    KARYOTYPING, 2528*
ENDOPLASMIC RETICULUM
    DISULFIDES, 2924
    SULFHYDRYL COMPCUNDS, 2924
EPIDEMIOLOGY
    RACIAL FACTORS, CHILD, 2883
ETHYLENE, TRICHLORO-
    CARCINOGENIC ACTIVITY, REVIEW, 2427*
    OCCUPATIONAL HAZARD, 2427*
FERRITIN
    ANTIGENIC DETERMINANTS, 2928
    ISOLATION AND CHARACTERIZATION,
    2928
GLUCOSIDASES
    GLYCOGEN, 2946
    ISOENZYMES, 2946
GLYCEROLPHOSPHATE DEHYDROGENASE
    GLUCONEOGENESIS, 2949
GLYCOLIPIDS
    GROWTH, 2989*
    ISOLATION AND CHARACTERIZATION,
    2989*
GUANYL CYCLASE
    CYCLOHEXIMIDE, 2943
    GUANOSINE CYCLIC 3',5'
    MONOPHOSPHATE, 2943
IMMUNITY
    ANTIGENS, NEOPLASM, 2708*
IRON(II) SULFATE
    FERRITIN, 2927
LACTATE DEHYDROGENASE
    GLUCONEOGENESIS, 2949
LIVER NEOPLASMS
    UREA, N-NITROSO-N-PROPYL-, 2491
MICROSOMES, LIVER
    DISULFIDES, 2924
    SULFHYDRYL COMPCUNDS, 2924
MURAMIDASE
    GROWTH, 2992*
    HISTOLOGY, 2992*
NADH, NADPH OXIDOREDUCTASES
    GLUCONEOGENESIS, 2949
OUABAIN
    GLYCOLYSIS, 2918
    ION TRANSPORT, 2918
OXIDOREDUCTASES
    GLUCONEOGENESIS, 2949
    GROWTH, 2960
    ISOLATION AND CHARACTERIZATION,
    2960
PHOSPHOTRANSFERASES, ATP
    GLUCONEOGENESIS, 2949
PLANT AGGLUTININS

AGGLUTINATION, 2706
POLYRIBOSOMES
    DISULFIDES, 2924
    SULFHYDRYL COMPOUNDS, 2924
PROTEINS
    LOCALIZATION AND PHOSPHORYLATION,
    2925
PYRIDOXAL, 5-(DIHYDROGEN PHOSPHATE)-
    TYROSINE AMINOTRANSFERASE, 2962
PYRIDOXOL
    VITAMIN B DEFICIENCY, 2962
RNA, MESSENGER
    METHYLATION, 2988*
SCHISTOSOMAISIS
    EPIDEMIOLOGY, 2901*
SPERMINE
    CARBOXY-LYASES, 2958
STERIGMATOCYSTIN
    CARCINOGENIC ACTIVITY, MOUSE,
    2506*
TRANSPLANTATION, HOMOLOGOUS
    OXIDOREDUCTASES, 2960

HEXOKINASE
    MAMMARY NEOPLASMS, EXPERIMENTAL
        CYTOSOL, 2948
        ISOLATION AND CHARACTERIZATION,
        2948
        MITOCHONDRIA, 2948

HISTIOCYTES
    SARCOMA, RETICULUM CELL
        ULTRASTRUCTURAL STUDY, 2782

HISTOCOMPATIBILITY ANTIGENS
    BLADDER NEOPLASMS
        CARCINOMA, TRANSITIONAL CELL, 2794
        IMMUNITY, CELLULAR, 2713*
    CHOLANTHRENE, 3-METHYL-
        IMMUNITY, CELLULAR, 2666
        TRANSPLANTATION IMMUNOLOGY, 2666
    COLONIC NEOPLASMS
        GENETICS, 2686
    GENETICS
        MIXED LYMPHOCYTE CULTURE, 2879*
    IMMUNOGLOBULINS, FC
        B-LYMPHOCYTES, 2751*
    LEUKEMIA
        TRANSPLANTATION, HOMOLOGOUS, 2720*
    LEUKEMIA, LYMPHOBLASTIC
        VIRUS, RAUSCHER MURINE LEUKEMIA,
        2685
    LEUKEMIA, LYMPHOCYTIC
        IMMUNITY, CELLULAR, 2664
    LYMPHOCYTES, 2684
    B-LYMPHOCYTES
        ISOLATION AND CHARACTERIZATION,
        2753*
    T-LYMPHOCYTES
        ISOLATION AND CHARACTERIZATION,
        2753*
    LYMPHOMA
        GRAFT REJECTION, 2684
    MELANOMA
        IMMUNITY, CELLULAR, 2664

    NEOPLASMS
        CANCER FAMILY SYNDROME, 2686

SARCOMA
    VIRUS, SV40, 2685
UTERINE NEOPLASMS
    GENETICS, 2686
VIRUS, B-TROPIC MURINE LEUKEMIA
    MILK-BORNE TRANSMISSION, 2607
    VIRUS ACTIVATION, 2607
VIRUS, MURINE LEUKEMIA
    TRANSPLANTATION IMMUNOLOGY, 26
VIRUS, RAUSCHER MURINE LEUKEMIA
    ISOLATION AND CHARACTERIZATION
    2685
    TRANSPLANTATION IMMUNOLOGY, 26
VIRUS, SV40
    ISOLATION AND CHARACTERIZATION
    2685
    TRANSPLANTATION IMMUNOLOGY, 26

HODGKIN'S DISEASE
    ANTIBODIES, NEOPLASM
        IMMUNE SERUMS, 2678
    ANTIGENS
        ANTIGEN-ANTIBODY REACTIONS, 2
    BENZENE, 1-CHLORO-2,4-DINITRO-
        IMMUNITY, CELLULAR, 2721*
    BONE NEOPLASMS
        CLASSIFICATION, 2803*
    BRAIN NEOPLASMS
        TRANSPLANTATION, HOMOLOGOUS, 2
        ULTRASTRUCTURAL STUDY, 2809*
    CHROMOSOME ABERRATIONS
        REVIEW, 2444*
    COPPER
        REVIEW, 2403
    EPIDEMIOLOGY
        RACIAL FACTORS, CHILD, 2883
    GASTROINTESTINAL NEOPLASMS
        REVIEW, 2443*
    GENETICS
        REVIEW, 2444*
    GRANULOPOIETIC PROGENITOR CELLS
        COLONY FORMATION, 2915
    HEMOCYANIN
        IMMUNITY, CELLULAR, 2721*
    IMMUNOGLOBULINS
        ROSETTE FORMATION, 2782
    INFLUENZA
        REVIEW, 2444*
    LYMPHOCYTES
        CONCANAVALIN A, 2756*, 2760*
    NERVOUS SYSTEM NEOPLASMS
        HISTOLOGICAL STUDY, 2812*
    TOXOPLASMA GONDII
        ANTIGENS, VIRAL, 2593
    ULTRASTRUCTURAL STUDY, 2877*
    VIRUS, CYTOMEGALO
        ANTIGENS, VIRAL, 2593
    VIRUS, EPSTEIN-BARR
        ANTIGENS, VIRAL, 2593
    VIRUS, HERPES SIMPLEX
        IMMUNITY, CELLULAR, 2671
    VIRUS, HERPES ZOSTER
        IMMUNITY, CELLULAR, 2671
    VIRUS, MEASLES
        ANTIGENS, VIRAL, 2593
    VIRUS, RUBELLA
        ANTIGENS, VIRAL, 2593

ONES
MELANOMA
  TYROSINASE, 2963

URONIC ACID
TERATOID TUMOR
  CELL MEMBRANE, 2882

URONIDASE
TERATOID TUMOR
  CELL MEMBRANE, 2882

ID CELLS
CELL MEMBRANE
  ISOANTIGENS, 2776
CHROMOSOMES
  ISOLATION AND CHARACTERIZATION,
  2776
HYPOXANTHINE
  THYMIDINE KINASE DEFICIENCY, 2776
METHOTREXATE
  THYMIDINE KINASE DEFICIENCY, 2776
TETRAHYDROFOLATE DEHYDROGENASE
  THYMIDINE KINASE DEFICIENCY, 2776
THYMIDINE
  THYMIDINE KINASE DEFICIENCY, 2776

NTOIN, 1-((5-NITROFURFURYLIDINE)AMINO)-

MICROSOMES, LIVER
  HYPERSENSITIVITY, 2530*
  METABOLITES, 2530*

TIDIFORM MOLE
ENZYMES
  ULTRASTRUCTURAL STUDY, 2999*

AZINE, 1,2-DIMETHYL-
ALKYLATING AGENTS
  CARCINOGENIC METABOLITE, RAT, 2463

OCORTISONE
SEE CORTISOL

DROXY-2-ACETYLAMINOFLUORENE
SEE ACETOHYDROXAMIC ACID, N-FLUOREN-2-
  YL-

OXYSTEROID DEHYDROGENASES
UTERINE NEOPLASMS
  CARCINOMA, 2961
  CELL DIFFERENTATION, 2961
  ENZYMATIC ACTIVITY, 2961

RCALCEMIA
ENDOCRINE GLAND NEOPLASMS
  HYPERPARATHYROIDISM, 2797
PARATHYROID NEOPLASMS
  ADENOMA, 2797
  CARCINOMA, 2797

RNEPHROID CARCINOMA
SEE ADENOCARCINOMA

RNEPHROMA
SEE ADENOCARCINOMA

RPARATHYROIDISM

ENDOCRINE GLAND NEOPLASMS
  HYPERCALCEMIA, 2797
PANCREATIC NEOPLASMS
  ADENOMA, 2797
  ISLET CELL TUMOR, 2797
  ZOLLINGER-ELLISON SYNDROME, 2797
PARATHYROID NEOPLASMS
  ADENOMA, 2797
  CARCINOMA, 2797

HYPERPLASIA
PARATHYROID HORMONE
  ISOLATION AND CHARACTERIZATION,
  PRECURSOR, 2966
PARATHYROID NEOPLASMS
  PARATHYROID HORMONE, 2966
  RNA, MESSENGER, 2966
THYROID NEOPLASMS
  DNA, 2934
  RADIATION, IONIZING, 2934
  URACIL, 6-METHYL-2-THIO-, 2934

HYPERSENSITIVITY
UROGENITAL NEOPLASMS
  BENZENE, 1-FLUORO-2,4-DINITRO-,
  2737*

HYPOGLYCEMIA
PANCREATIC NEOPLASMS
  PRECANCEROUS CONDITIONS, 2439*

HYPOTHYROIDISM
PITUITARY NEOPLASMS
  CASE REPORT, 2798

HYPOXANTHINE
HYBRID CELLS
  THYMIDINE KINASE DEFICIENCY, 2776

IGA
ANEMIA, APLASTIC
  PRECANCEROUS CONDITIONS, 2775
CHOLINE
  BINDING SITES, ANTIBODY, 2680
MYELOMA PROTEINS, 2983*
  LIGAND-BINDING, 2983*
PLASMACYTOMA
  LIGAND-BINDING, 2983*
RADIATION, IONIZING
  STRESS, 2677
STREPTOCOCCUS
  PROTEIN UPTAKE STUDY, MYELOMA, 2739*
URANIUM
  OCCUPATIONAL HAZARD, 2677

IGE
BRONCHIAL NEOPLASMS
  ANTIBODIES, 2740*
  ANTIGEN-ANTIBODY REACTIONS, 2740*

IGG
ANEMIA, APLASTIC
  PRECANCEROUS CONDITIONS, 2775
ANTIGENIC DETERMINANTS
  ADENOCARCINOMA, 2748*
BRONCHIAL NEOPLASMS
  ANTIGEN-ANTIBODY REACTIONS, 2740*
CHOLINE

        BINDING SITES, ANTIBODY, 2680
    LEUKEMIA, LYMPHOCYTIC
        ANTIBODIES, 2743*
    B-LYMPHOCYTES
        STAPHYLOCOCCUS AUREUS, 2753*
        VIRUS, EPSTEIN-BARR, 2592
    T-LYMPHOCYTES
        VIRUS, EPSTEIN-BARR, 2592
    MULTIPLE MYELOMA
        CELL LINE, 2676
    RADIATION, IONIZING
        STRESS, 2677
    SARCOMA, RETICULUM CELL
        B-LYMPHOCYTES, 2782
    STREPTOCOCCUS
        PROTEIN UPTAKE STUDY, MYELOMA,
            2739*
    URANIUM
        OCCUPATIONAL HAZARD, 2677
    VIRUS, EPSTEIN-BARR
        ANTIGENS, VIRAL, 2591, 2742*

IGM
    ANEMIA, APLASTIC
        PRECANCEROUS CONDITIONS, 2775
    CHOLINE
        BINDING SITES, ANTIBODY, 2680
    RADIATION, IONIZING
        STRESS, 2677
    SARCOMA, RETICULUM CELL
        B-LYMPHOCYTES, 2782
    URANIUM
        OCCUPATIONAL HAZARD, 2677
    VIRUS, EPSTEIN-BARR
        ANTIGENS, VIRAL, 2742*

IMIDAZOLE-1-ETHANOL, 2-METHYL-5-NITRO-
    SALMONELLA TYPHIMURIUM
        MUTAGENESIS, 2531*

2-IMIDAZOLIDINONE, 4-(3-BUTOXY-4-
  METHOXYBENZYL)-
    NEUROBLASTOMA
        CARBOXY-LYASES, 2959

2-IMIDAZOLIDINONE, 1-(5-NITRO-2-THIAZOLYL)-
    SALMONELLA TYPHIMURIUM
        MUTAGENESIS, 2531*

2-IMIDAZOLIDONE, 4-(3-BUTOXY-4-
  METHOXYBENZYL)-
    GLIOMA
        CARBOXY-LYASES, 2959

IMMUNE COMPLEX DISEASE
    LEUKEMIA
        GLOMERULONEPHRITIS, 2710*
    LYMPHOMA
        GLOMERULONEPHRITIS, 2710*
    MAMMARY NEOPLASMS, EXPERIMENTAL
        GLOMERULONEPHRITIS, 2710*
    SYNOVIOMA
        GLOMERULONEPHRITIS, 2710*

IMMUNE SERUMS
    ACETAMIDE, N-FLUOREN-2-YL-
        RADIOIMMUNOASSAY, 2727*
    BURKITT'S LYMPHOMA

                    CARCINOMA, BASAL CELL, 2671
            THYMOMA
                LYMPHOCYTES, 2712*
            THYROID NEOPLASMS
                ADENOCARCINOMA, 2712*
            VIRUS, MOLONEY MURINE SARCOMA
                CYTOTOXICITY, TESTS, IMMUNOLOGIC,
                    2669
                URIDINE, 2'-DEOXY-5-IODO-, 2669
            VIRUS, SV40
                CHOLANTHRENE, 3-METHYL-, 2689
                TUBERCULIN, 2689
            WARTS
                PLANT AGGLUTININS, 2729*

        IMMUNIZATION
            LYMPHOMA
                VIRUS, HERPES ATELES, 2675

        IMMUNOGLOBULIN FRAGMENTS
707*   CHOLINE
                BINDING SITES, ANTIBODY, 2680
664

        IMMUNOGLOBULINS
            SEE ALSO IGA/IGD/IGE/IGG/IGM
            ANEMIA, APLASTIC
                PRECANCEROUS CONDITIONS, 2775
            BURKITT'S LYMPHOMA
                B-LYMPHOCYTES, 2679
            CHOLINE
                BINDING SITES, ANTIBODY, 2680
            HEAD AND NECK NEOPLASMS
                CARCINOMA, 2782
            HODGKIN'S DISEASE
                ROSETTE FORMATION, 2782
            LEUKEMIA, LYMPHOCYTIC
                B-LYMPHOCYTES, 2679
            LYMPHATIC NEOPLASMS
                B-LYMPHOCYTES, 2679
2684   B-LYMPHOCYTES
                VIRUS, EPSTEIN-BARR, 2592
            T-LYMPHOCYTES
                VIRUS, EPSTEIN-BARR, 2592
            LYMPHOSARCOMA
                B-LYMPHOCYTES, 2679
664   MULTIPLE MYELOMA
                ANTIBODIES, 2681
                B-LYMPHOCYTES, 2679
            RADIATION, IONIZING
                STRESS, 2677
            SARCOMA, RETICULUM CELL
                B-LYMPHOCYTES, 2782
            URANIUM
                OCCUPATIONAL HAZARD, 2677
            VIRUS, EPSTEIN-BARR
                ANTIGENS, VIRAL, 2591, 2742*
2667
        IMMUNOGLOBULINS, FAB
            COLONIC NEOPLASMS
                ADENOCARCINOMA, 2748*

        IMMUNOGLOBULINS, FC
            B-LYMPHOCYTES
                BINDING SITES, 2751*
                HISTOCOMPATIBILITY ANTIGENS, 2751*

        IMMUNOGLOBULINS, HEAVY CHAIN
            MULTIPLE MYELOMA
                CELL LINE, 2676

IMMUNOGLOBULINS, SURFACE
    CARCINOMA 256, WALKER
        LYMPHOCYTES, 2738*
    LEUKEMIA, LYMPHOCYTIC
        LYMPHOCYTES, 2743*
    LYMPHOSARCOMA
        B-LYMPHOCYTES, 2694
        T-LYMPHOCYTES, 2694

IMMUNOLOGIC DEFICIENCY SYNDROMES
    LYMPHOCYTES
        ANTIBODIES, 2746*
        NEURAMINIDASE, 2746*

IMMUNOSUPPRESSION
    BREAST NEOPLASMS
        CARCINOMA, DUCTAL, 2730*
    CYCLOPHOSPHAMIDE
        TRANSPLANTATION, HOMOLOGOUS, 2668
    DIETHYLAMINE, N-NITROSO-
        CARCINOGENIC ACTIVITY, RAT, 2482
    FIBROSARCOMA
        VIRUS, SV40, 2689
    LARYNGEAL NEOPLASMS
        PSORIASIS, 2858*
    LYMPHOMA
        VIRUS, MAREK'S DISEASE HERPES,
            2731
    MAREK'S DISEASE
        T-LYMPHOCYTES, 2701
    NEOPLASM METASTASIS
        TRANSPLANTATION, HETEROLOGOUS,
            2731*
    NEOPLASMS
        IMMUNITY, CELLULAR, 2438*
    SKIN NEOPLASMS
        GROWTH, 2717*
    SPLENIC NEOPLASMS
        LYMPHOMA, 2734*
    VIRUS, C-TYPE MURINE RNA TUMOR
        ANTILYMPHOCYTE SERUM, 2673
        CYCLOPHOSPHAMIDE, 2673
        IMMUNE SERUMS, 2673
    VIRUS, FRIEND MURINE LEUKEMIA
        T-LYMPHOCYTES, 2651*
    VIRUS, MAREK'S DISEASE HERPES
        CYCLOPHOSPHAMIDE, 2668

IMMUNOSUPPRESSIVE AGENTS
    CARCINOMA, EHRLICH TUMOR
        ISOLATION AND CHARACTERIZATION,
            2733*

INDUSTRIAL WASTE
    ASBESTOS
        AIR POLLUTANTS, 2434*

INFECTIOUS MONONUCLEOSIS
    ANTIBODIES, VIRAL
        EPSTEIN-BARR NUCLEAR ANTIGEN,
            2641*
    CHROMATIN
        ISOLATION AND CHARACTERIZATION,
            2923
        PROTEINS, 2923
    LYMPHOCYTES
        CONCANAVALIN A, 2756*

INFLATION

NEOPLASMS
    IMMUNITY, CELLULAR, 2438*

INFLUENZA
    HODGKIN'S DISEASE
        REVIEW, 2444*

INSULIN
    MAMMARY NEOPLASMS, EXPERIMENTA
        PRECANCEROUS CONDITIONS, 2

INTERFERON
    RNA, MESSENGER
        ANTIGENIC DETERMINANTS, 27
    VIRUS, FRIEND MURINE LEUKEMIA
        URIDINE, 2'-DEOXY-5-FLUORO
        URIDINE, 2'-DEOXY-5-IODO-,
    VIRUS, KIRSTEN MURINE LEUKEMIA
        ANEMIA, HEMOLYTIC, 2611
    VIRUS, MURINE ERYTHROBLASTOSIS
        ANEMIA, HEMOLYTIC, 2611

INTESTINAL NEOPLASMS
    ADENOCARCINOMA
        CELIAC DISEASE, 2829*
        ENTERITIS, REGIONAL, 2826*
        UREA, 1-BUTYL-1-NITROSO-,
        UREA, N-NITROSO-N-PROPYL-,
    ENTERITIS, REGIONAL
        CASE REPORT, 2827*
    HEMANGIOENDOTHELIOMA
        UREA, N-NITROSO-N-PROPYL-,
    POLYPS
        UREA, 1-BUTYL-1-NITROSO-,
    RADIATION, IONIZING
        PRECANCEROUS CONDITIONS, 2
    SARCOMA, RETICULUM CELL
        UREA, 1-BUTYL-1-NITROSO-,

INTESTINAL POLYPS
    COLONIC NEOPLASMS
        GENETICS, 2442*, 2828*
        PRECANCEROUS CONDITIONS, 2
        SCREENING, 2828*
    ESOPHAGEAL NEOPLASMS
        PREDNISOLONE, 2834*

IODINE RADIOISOTOPES
    THYROID NEOPLASMS
        CARCINOGENIC ACTIVITY, RAT

IPON(II) SULFATE
    HEPATOMA
        FERRITIN, 2927

ISLET CELL TUMOR
    PANCREATIC NEOPLASMS
        HYPERPARATHYROIDISM, 2797

ISOANTIGENS
    COLONIC NEOPLASMS
        PAPILLOMA, 2750*
        POLYPS, 2750*
    HYBRID CELLS
        CELL MEMBRANE, 2776
    T-LYMPHOCYTES
        IMMUNE SERUMS, 2693
    VIRUS, AVIAN LEUKOSIS-SARCOMA
        CELL MEMBRANE, 2690

                IMMUNE RESPONSE, 2713*
        LYMPHOSARCOMA
            CASE REPORT, 2860*
            ULTRASTRUCTURAL STUDY, 2860*
        NEPHROBLASTOMA
            UREA, N-NITROSO-N-PROPYL-, 2491

    KLEBSIELLA PNEUMONIAE
        DIMETHYLAMINE
            DIMETHYLAMINE, N-NITROSO-, 2484

    L CELLS
        CELL MEMBRANE
            PEPTIDES, 2926
        DNA
            ISOLATION AND CHARACTERIZATION,
            2932

    LACTATE DEHYDROGENASE
        HEPATOMA
            GLUCONEOGENESIS, 2949

    LACTATES
        CARCINOMA, EHRLICH TUMOR
            CELL MEMBRANE, 2978*

    LACTOBACILLUS FERMENTI
        DIMETHYLAMINE
            DIMETHYLAMINE, N-NITROSO-, 2484

    LARYNGEAL NEOPLASMS
        CARCINOMA
            ASBESTOS, 2496
            ULTRASTRUCTURAL STUDY, 2840*
        PSORIASIS
            IMMUNOSUPPRESSION, 2858*
        SMOKING
            EPIDEMIOLOGY, INDIA, 2495
        TOBACCO
            EPIDEMIOLOGY, INDIA, 2495

    LECITHINS
        HELA CELLS
            CARBAMIC ACID, METHYL-, 1-NAPHTHYL
            ESTER, 2921
            DIELDRIN, 2921

    LEIOMYOMA
        UTERINE NEOPLASMS
            UREA, N-NITROSO-N-PROPYL-, 2491

    LETTERER-SIWE DISEASE
        NERVOUS SYSTEM NEOPLASMS
            HISTOLOGICAL STUDY, 2812*

    LEUCINE AMINOPEPTIDASE
        GROWTH
            CELL CYCLE KINETICS, 2956

    LEUKEMIA
        ACUTE DISEASE
            ERYTHROPOIESIS, 2780
        AGE FACTORS
            EPIDEMIOLOGY, CHILD, POLAND, 2899*
        ALCOHOLISM
            EPIDEMIOLOGY, 2892
        BENZENE
            REVIEW, 2447*
        CARBAMIC ACID, ETHYL ESTER

CARCINOGENIC POTENTIAL, REVIEW,
  2404
RADIATION, IONIZING, 2569*
CELL DIVISION
  SPLENECTOMY, 2696
  THYMECTOMY, 2696
CHLORAMPHENICOL
  REVIEW, 2447*
CHROMOSOME ABERRATIONS
  REVIEW, 2444*
COPPER
  REVIEW, 2403
EPIDEMIOLOGY
  RACIAL FACTORS, CHILD, 2883
  REVIEW, 2419, 2447*
ERYTHROPOIESIS
  PRECANCEROUS CONDITIONS, 2780
FLAVONES
  ADENOSINE TRIPHOSPHATASE, 2919
  GLYCOLYSIS, 2919
  GROWTH, 2919
GENETICS
  REVIEW, 2444*
IMMUNE COMPLEX DISEASE
  GLOMERULONEPHRITIS, 2710*
T-LYMPHOCYTES
  CELL DIFFERENTIATION, 2696
  PRECANCEROUS CONDITIONS, 2696
  TRANSPLANTATION, HOMOLOGOUS, 2696
MAST CELLS
  ANTIBODIES, VIRAL, 2649*
3,5-PYRAZOLIDINEDIONE, 4-BUTYL-1,2-
  DIPHENYL-
  REVIEW, 2447*
RADIATION
  GROWTH, 2720*
  REVIEW, 2447*
TRANSPLANTATION, HOMOLOGOUS
  GRAFT REJECTION, 2720*
  HISTOCOMPATIBILITY ANTIGENS, 2720*
TRANSPLANTATION IMMUNOLOGY
  IMMUNITY, CELLULAR, 2707*
  LYMPHOCYTES, 2707*
VIRUS, FELINE LEUKEMIA
  ANTIBODIES, VIRAL, 2649*
  REVIEW, 2444*
ZINC
  GROWTH, 2403

LEUKEMIA L1210
  CELL MEMBRANE
    GLYCOPROTEINS, 2944
    SIALIC ACID INCORPORATION, 2944
    TRANSFERASES, 2944
  GLYCOLIPIDS
    GROWTH, 2989*
    ISOLATION AND CHARACTERIZATION,
      2989*

VIRUS, GROSS MURINE LEUKEMIA
  ANTIGENS, VIRAL, 2609

VIRUS-LIKE PARTICLES
  ISOLATION AND CHARACTERIZATION,
    2609
VIRUS, MURINE MAMMARY TUMOR
  ANTIGENS, VIRAL, 2609
VIRUS, RAUSCHER MURINE LEUKEMIA
  ANTIGENS, VIRAL, 2609

LEUKEMIA, LYMPHOBLASTIC
  ASPARAGINASE
    T-LYMPHOCYTES, 2700
    ROSETTE FORMATION, 2700
  BURKITT'S LYMPHOMA
    EPIDEMIOLOGY, AFRICA, 2891
  CELL DIFFERENTIATION
    CEREBROSPINAL FLUID, 2813*
  CELL TRANSFORMATION, NEOPLASTIC
    ULTRASTRUCTURAL STUDY, 2813*
  CHICKENPOX
    REVIEW, 2444*
  CHROMATIN
    ISOLATION AND CHARACTERIZATION,
      2923
    PROTEINS, 2923
  CHROMOSOME ABERRATIONS
    REVIEW, 2444*
  CLASSIFICATION
    ULTRASTRUCTURAL STUDY, 2445*
  EPIDEMIOLOGY
    AGE FACTORS, 2446*
  GENETICS
    EPIDEMIOLOGY, 2880
    REVIEW, 2444*
  LEUROCRISTINE
    T-LYMPHOCYTES, 2700
    ROSETTE FORMATION, 2700
  B-LYMPHOCYTES
    ACID PHOSPHATASE, 2759*
    NUCLEOSIDE DEAMINASES, 2758*
    NUCLEOTIDYLTRANSFERASES, 2758*
  T-LYMPHOCYTES
    ACID PHOSPHATASE, 2759*
    ANTIGENS, 2699
    IMMUNE SERUMS, 2699
    NUCLEOSIDE DEAMINASES, 2758*
    NUCLEOTIDYLTRANSFERASES, 2758*
    ROSETTE FORMATION, 2699, 2700
  NEPHRITIS, INTERSTITIAL
    CASE REPORT, 2849*
    PRECANCEROUS CONDITIONS, 2849*
    ULTRASTRUCTURAL STUDY, 2849*
    URIC ACID, 2849*
  PREGNA-1,4-DIENE-3,11,20-TRIONE, 17,2
    DIHYDROXY-
    T-LYMPHOCYTES, 2700
    ROSETTE FORMATION, 2700
  UREA, N-NITROSO-N-PROPYL-
    ULTRASTRUCTURAL STUDY, RAT, 2491
  VIRUS, C-TYPE RNA TUMOR
    COMPARATIVE VIROLOGY, REVIEW, 241
    VIRUS-LIKE PARTICLES, 2612
  VIRUS, RAUSCHER MURINE LEUKEMIA
    HISTOCOMPATIBILITY ANTIGENS, 2685
  VIRUS, VARICELLA-ZOSTER
    IN UTERO EXPOSURE, 2446*

LEUKEMIA, ACUTE GRANULOCYTIC
  SEE LEUKEMIA, MYELOBLASTIC

LEUKEMIA, ACUTE LYMPHOCYTIC
  SEE LEUKEMIA, LYMPHOBLASTIC

LEUKEMIA, CHRONIC GRANULOCYTIC
  SEE LEUKEMIA, MYELOCYTIC

                    KARYOTYPING, 2779

          CHROMOSOMES
              G-BANDING, 2779
              KARYOTYPING, 2779
          CLASSIFICATION
              ULTRASTRUCTURAL STUDY, 2445*
          ERYTHROPOIESIS
              PRECANCEROUS CONDITIONS, 2780
          GLUCOSAMINIDASE
              ISOLATION AND CHARACTERIZATION,
                  2947
          UREA, N-NITROSO-N-PROPYL-
              ULTRASTRUCTURAL STUDY, RAT, 2491
          VIRUS, C-TYPE RNA TUMOR
              VIRUS-LIKE PARTICLES, 2612
          VIRUS, GROSS MURINE LEUKEMIA
              HYBRIDIZATION, 2612
          VIRUS, KIRSTEN MURINE SARCOMA
              HYBRIDIZATION, 2612
          VIRUS, MOLONEY MURINE LEUKEMIA
              HYBRIDIZATION, 2612
          VIRUS, SIMIAN SARCOMA
              HYBRIDIZATION, 2612

     LEUKEMIA, MYELOCYTIC
         CHROMOSOME ABERRATIONS
             DIAGNOSIS AND PROGNOSIS, 2968*
         GENETICS
             EPIDEMIOLOGY, 2880
         HEMATOLOGY
             CASE REPORT, DOG, 2845*
         KARYOTYPING
             CASE REPORT, 2774
         LEUKOCYTES
             GEL DIFFUSION TESTS, 2914
         LEUKOCYTOSIS
             CASE REPORT, 2846*
         RETRONECINE, 3,8-DIDEHYDRO-
             MONOCROTALINE, 2529*
         TRANSPLANTATION, HOMOLOGOUS
             GLYCOPROTEINS, 2914
         VIRUS, C-TYPE RNA TUMOR
             VIRUS-LIKE PARTICLES, 2612

     LEUKEMIA, MYELOGENOUS
         SEE LEUKEMIA, MYELOCYTIC

     LEUKEMIA, MYELOID
         SEE LEUKEMIA, MYELOCYTIC

     LEUKEMIA, PROMYELOCYTIC
         SEE LEUKEMIA, MYELOBLASTIC

     LEUKEMIA, SUBLEUKEMIC
         UREA, N-NITROSO-N-PROPYL-
             ULTRASTRUCTURAL STUDY, RAT, 2491

     LEUKOCYTES
         GLUCOSAMINIDASE
             ISOLATION AND CHARACTERIZATION,
                 2947
         LEUKEMIA, MYELOCYTIC
             GEL DIFFUSION TESTS, 2914
         OVARIAN NEOPLASMS
             MIGRATION INHIBITORY FACTOR, 2667

     LEUKOPLAKIA
         ORAL CAVITY
             CARCINOMA, EPIDERMOID, 2904*
             CARCINOMA IN SITU, 2904*

DYSPLASIA, 2904*
EPIDEMIOLOGY, 2904*

LEUROCRISTINE
    LEUKEMIA, LYMPHOBLASTIC
        T-LYMPHOCYTES, 2700
        ROSETTE FORMATION, 2700

LEYDIG CELL TUMOR
    GONADOTROPINS
        TESTOSTERONE, 2472
    TESTICULAR NEOPLASMS
        CASE REPORT, 2871*

LH
    ESTRONE
        RADIATION, IONIZING, 2563

LIP NEOPLASMS
    CARCINOMA, EPIDERMOID
        EPIDEMIOLOGY, FISH, 2895

LIPIDS
    CELL MEMBRANE
        VIRUS, AVIAN RNA TUMOR, 2414
    LUNG NEOPLASMS
        CERIUM RADIOISOTOPES, 2564
    MYCOBACTERIUM BOVIS
        ADJUVANTS, IMMUNOLOGIC, 2726*

LIPOPOLYSACCHARIDES
    T-LYMPHOCYTES
        PROTEINS, 2724*
    THYMECTOMY
        B-LYMPHOCYTES, 2697

LIPOPROTEINS
    LYMPHOMA
        GRAFT REJECTION, 2684

LIPOSARCOMA
    GRANULOPOIETIC PROGENITOR CELLS
        COLONY FORMATION, 2915

LIVER CELL CARCINOMA
    SEE HEPATOMA

LIVER CIRRHOSIS
    LIVER NEOPLASMS
        CARCINOMA, 2851*

LIVER NEOPLASMS
    ADENOMA
        NCI C02006, 2554*
    ALCOHOLISM
        EPIDEMIOLOGY, 2892
    ANGIOSARCOMA
        EPIDEMIOLOGY, 2896
        ETHYLENE, CHLORO-, 2896
        ETHYLENE, CHLORO- POLYMER, 2896
        HISTOGENESIS, 2852*
    BARBITURIC ACID, 5-ETHYL-5-PHENYL-
        AFLATOXIN B1, 2458
    CARBAMIC ACID, ETHYL ESTER
        RADIATION, IONIZING, 2570*
    CARBON TETRACHLORIDE
        BILE DUCT EPITHELIA PROLIFERATION,
        2522*
        CELL CYCLE KINETICS, 2522*

RHABDOMYOSARCOMA
    CASE REPORT, 2842*
SARCOIDOSIS
    GLOMERULONEPHRITIS, 2850*
    PRECANCEROUS CONDITIONS, 2850*
SARCOMA
    CASE REPORT, 2842*
STERIGMATOCYSTIN
    CARCINOGENIC ACTIVITY, MOUSE,
    2506*
SYNDROME, SCHWARTZ-BARTTER
    CELL LINE, 2973*
TUMORLET
    PRECURSOR CELL, 2787
    ULTRASTRUCTURAL STUDY, 2787

LYMPHATIC NEOPLASMS
    ANESTHETICS
        CARCINOGENIC EFFECT, 2407
    BRAIN NEOPLASMS
        HISTOLOGICAL STUDY, 2820*
    CARBAMIC ACID, ETHYL ESTER
        RADIATION, IONIZING, 2570*
    CARCINOMA, EPIDERMOID
        LYMPHORETICULAR RESPONSE, 2719*
    B-LYMPHOCYTES
        IMMUNE SERUMS, 2679
        IMMUNOGLOBULINS, 2679
        ULTRASTRUCTURAL STUDY, 2679
    VIRUS, EPSTEIN-BARR
        ULTRASTRUCTURAL STUDY, 2679

LYMPHOCYTE TRANSFORMATION
    BENZENE, 1-CHLORO-2,4-DINITRO-
        IMMUNITY, CELLULAR, 2722*
    BLADDER NEOPLASMS
        ADJUVANTS, IMMUNOLOGIC, 2674
    BREAST NEOPLASMS
        ADJUVANTS, IMMUNOLOGIC, 2674
        IMMUNITY, CELLULAR, 2702
    CARCINOMA
        ADJUVANTS, IMMUNOLOGIC, 2674
    LUNG NEOPLASMS
        ADJUVANTS, IMMUNOLOGIC, 2674
    LYMPHOMA
        MITOGENS, 2847*
        PLANT AGGLUTININS, 2847*
    VIRUS, EPSTEIN-BARR
        CELL LINE, 2703
        FIBROBLASTS, 2592
        MACROPHAGES, 2592

LYMPHOCYTES
    AGAMMAGLOBULINEMIA
        NEURAMINIDASE, 2746*
    ATAXIA TELANGIECSTASIA
        ANTIBODIES, 2746*
        NEURAMINIDASE, 2746*
    BENZENE, 1-CHLORO-2,4-DINITRO-
        IMMUNITY, CELLULAR, 2722*
    BLADDER NEOPLASMS
        IMMUNE RESPONSE, 2713*
    BREAST NEOPLASMS
        CONCANAVALIN A, 2760*
    CARCINOMA 256, WALKER
        IMMUNOGLOBULINS, SURFACE, 2738*

CELL LINE
    VIRUS, EPSTEIN-BARR, 2744*
CELL MEMBRANE
    GLYCOPROTEINS, 2982*
DNA REPLICATION
    SERUM ROLE, 2735*
FIBROSARCOMA
    IMMUNITY, CELLULAR, 2712*
HEMANGIOPERICYTOMA
    IMMUNITY, CELLULAR, 2712*
HISTOCOMPATIBILITY ANTIGENS, 2684
HODGKIN'S DISEASE
    CONCANAVALIN A, 2756*, 2760*
IMMUNOLOGIC DEFICIENCY SYNDROMES
    ANTIBODIES, 2746*
    NEURAMINIDASE, 2746*
INFECTIOUS MONONUCLEOSIS
    CONCANAVALIN A, 2756*
KIDNEY NEOPLASMS
    IMMUNE RESPONSE, 2713*
LEUKEMIA
    TRANSPLANTATION IMMUNOLOGY, 2707*
LEUKEMIA, LYMPHOCYTIC
    CONCANAVALIN A, 2756*
    IMMUNOGLOBULINS, SURFACE, 2743*
LYMPHOMA
    CONCANAVALIN A, 2760*
LYMPHOSARCOMA
    CONCANAVALIN A, 2760*
MELANOMA
    IMMUNITY, CELLULAR, 2712*
NEOPLASMS
    ANTIBODIES, 2746*
    NEURAMINIDASE, 2746*
PLASMACYTOMA
    NEOPLASM TRANSPLANTATION, 2718*
POLY A-U
    NUCLEIC ACIDS, 2935
PROSTATIC NEOPLASMS
    IMMUNE RESPONSE, 2713*
SARCOMA
    IMMUNITY, CELLULAR, 2712*
SARCOMA, OSTEOGENIC
    IMMUNITY, CELLULAR, 2712*
SARCOMA, RETICULUM CELL
    CONCANAVALIN A, 2760*
THYMOMA
    IMMUNITY, CELLULAR, 2712*
VIRUS, EPSTEIN-BARR, 2594
VIRUS, MOLONEY MURINE SARCOMA
    DNA, 2672
    IMMUNE RESPONSE, 2672

B-LYMPHOCYTES
    BLADDER NEOPLASMS
        CARCINOMA, TRANSITIONAL CELL, 2794
    BURKITT'S LYMPHOMA
        IMMUNE SERUMS, 2679
        IMMUNOGLOBULINS, 2679
    HISTOCOMPATIBILITY ANTIGENS
        ISOLATION AND CHARACTERIZATION,
        2753*
    IGG
        STAPHYLOCOCCUS AUREUS, 2753*
    IMMUNE RESPONSE
        REVIEW, 2437*
    IMMUNOGLOBULINS, FC
        BINDING SITES, 2751*
        HISTOCOMPATIBILITY ANTIGENS, 2751*

LEUKEMIA, LYMPHOBLASTIC
    ACID PHOSPHATASE, 2759*
    NUCLEOSIDE DEAMINASES, 2758*
    NUCLEOTIDYLTRANSFERASES, 2758*
LEUKEMIA, LYMPHOCYTIC
    IMMUNOGLOBULINS, 2679
    ULTRASTRUCTURAL STUDY, 2679
LYMPHATIC NEOPLASMS
    IMMUNE SERUMS, 2679
    IMMUNOGLOBULINS, 2679
    ULTRASTRUCTURAL STUDY, 2679
LYMPHOMA
    IMMUNITY, CELLULAR, 2698
    LEUKOCYTE CULTURE TEST, MIXED,
    2698
LYMPHOSARCOMA
    IMMUNE SERUMS, 2679
    IMMUNOGLOBULINS, 2679
    IMMUNOGLOBULINS, SURFACE, 2694
    ROSETTE FORMATION, 2694
    ULTRASTRUCTURAL STUDY, 2679
MULTIPLE MYELOMA
    IMMUNE SERUMS, 2679
    IMMUNOGLOBULINS, 2679
    ULTRASTRUCTURAL STUDY, 2679
PROTEINS
    PLANT AGGLUTININS, 2724*
SARCOMA, RETICULUM CELL
    IGG, 2782
    IGM, 2782
    IMMUNOGLOBULINS, 2782
    ULTRASTRUCTURAL STUDY, 2782
THYMECTOMY
    LIPOPOLYSACCHARIDES, 2697
VIRUS, EPSTEIN-BARR
    IGG, 2592
    IMMUNOGLOBULINS, 2592
    REVIEW, 2413
VIRUS, FRIEND MURINE LEUKEMIA
    ANTIBODIES, VIRAL, 2687
    ANTIGENS, VIRAL, 2687
    IMMUNE RESPONSE, 2687

T-LYMPHOCYTES
    ADENOCARCINOMA
        TRANSPLANTATION, HOMOLOGOUS, 277
    ANTIGENIC DETERMINANTS
        GRAFT VS HOST REACTION, 2693
    ANTIGENS
        IMMUNE RESPONSE, 2704
    ANTIGENS, HETEROGENETIC
        CELL DIVISION, 2682
    CELL LINE
        CARCINOGENIC POTENTIAL, 2695
        CHROMOSOMES, 2695
        HISTOLOGICAL STUDY, MOUSE, 2695
    CELL MEMBRANE
        ANTIGENS, 2695
    CONCANAVALIN A
        IMMUNE RESPONSE, 2417, 2695
    CORTISONE
        CELL DIVISION, 2682
    DNA REPLICATION
        CORTISONE, 2682
    HEMOCYANIN
        IMMUNE RESPONSE, 2704
    HISTOCOMPATIBILITY ANTIGENS
        ISOLATION AND CHARACTERIZATION,
        2753*

IGG, 2592
IMMUNOGLOBULINS, 2592
VIRUS, FRIEND MURINE LEUKEMIA
ANTIBODIES, VIRAL, 2687
ANTIGENS, VIRAL, 2687
CONCANAVALIN A, 2651*
IMMUNE RESPONSE, 2687
IMMUNOSUPPRESSION, 2651*
VIRUS, FRIEND SPLEEN-FOCUS FORMING
CONCANAVALIN A, 2651*

VIRUS, MAREK'S DISEASE HERPES
CYCLOPHOSPHAMIDE, 2668
VIRUS, RADIATION LEUKEMIA
ANTIGENIC DETERMINANTS, 2613
VIRUS REPLICATION, 2610

LYMPHOKINES
BENZENE, 1-CHLORO-2,4-DINITRO-
ANTIGENS, 2722*
VIRUS, SV40
MIGRATION INHIBITORY FACTOR, 2631

LYMPHOMA (GENERAL AND UNSPECIFIED)
SEE ALSO HODGKIN'S DISEASE/BURKITT'S
LYMPHOMA/SARCOMA, RETICULUM CELL
LYMPHOSARCOMA

ALKALINE PHOSPHATASE
ISOENZYMES, 2993*
BENZ(A)ANTHRACENE, 7,12-DIMETHYL-
ANTILYMPHOCYTE SERUM, 2469
BONE MARROW CELLS
HISTOLOGICAL STUDY, 2781
BONE NEOPLASMS
CLASSIFICATION, 2803*
CHROMOSOME ABERRATIONS
REVIEW, 2444*
COPPER
REVIEW, 2403
EPIDEMIOLOGY
REVIEW, 2419
GENETICS
HYBRID CELL MODEL, 2773
REVIEW, 2444*
GRANULOPOIETIC PROGENITOR CELLS
COLONY FORMATION, 2915
GROWTH
HYBRID CELL MODEL, 2773
HISTOCOMPATIBILITY ANTIGENS
GRAFT REJECTION, 2684
IMMUNE COMPLEX DISEASE
GLOMERULONEPHRITIS, 2710*
LIPOPROTEINS
GRAFT REJECTION, 2684
LYMPHOCYTE TRANSFORMATION
MITOGENS, 2847*
PLANT AGGLUTININS, 2847*
LYMPHOCYTES
CONCANAVALIN A, 2760*
B-LYMPHOCYTES
IMMUNITY, CELLULAR, 2698
LEUKOCYTE CULTURE TEST, MIXED,
2698
T-LYMPHOCYTES
GRAFT REJECTION, 2684
IMMUNITY, CELLULAR, 2698
LEUKOCYTE CULTURE TEST, MIXED,
2698
MAGNESIUM

DIETARY FACTORS, 2723*
MONOCYTES
    GLUCOSE METABOLISM, 2847*
NERVOUS SYSTEM NEOPLASMS
    HISTOLOGICAL STUDY, TISSUE
        CULTURES, 2808*
    HISTOLOGY, REVIEW, 2441*
    VIRUS, MAREK'S DISEASE HERPES,
        2441*
    VIRUS, RNA TUMOR, 2441*
NOSE NEOPLASMS
    OCCUPATIONAL HAZARD, 2894
PARANASAL SINUS NEOPLASMS
    OCCUPATIONAL HAZARD, 2894
SPLENIC NEOPLASMS
    ANTIBODY PRODUCING CELLS, 2734*
    IMMUNOSUPPRESSION, 2734*
THYMUS NEOPLASMS
    UREA, 1-BUTYL-1-NITROSO-, 2489
THYROID NEOPLASMS
    CASE REPORT, 2799
    ULTRASTRUCTURAL STUDY, 2799
TRANSPLANTATION, HOMOLOGOUS
    GRAFT REJECTION, 2684
    IMMUNITY, CELLULAR, 2684
TRANSPLANTATION IMMUNOLOGY
    MAGNESIUM DEFICIENCY, 2723*
VIRUS, C-TYPE RNA TUMOR
    COMPARATIVE VIROLOGY, REVIEW, 2412
VIRUS, FELINE LEUKEMIA
    REVIEW, 2444*
VIRUS, HERPES ATELES
    IMMUNIZATION, 2675
    VIRAL VACCINES, 2675
VIRUS, HERPES SIMPLEX
    IMMUNITY, CELLULAR, 2671
VIRUS, HERPES ZOSTER
    IMMUNITY, CELLULAR, 2671
VIRUS, MAREK'S DISEASE HERPES
    CYCLOPHOSPHAMIDE, 2668
    IMMUNOSUPPRESSION, 2701
    T-LYMPHOCYTES, 2701
VIRUS, MOLONEY MURINE SARCOMA
    DNA, 2672
    IMMUNE RESPONSE, 2672
LYMPHOMA, GIANT FOLLICULAR
GASTROINTESTINAL NEOPLASMS
    REVIEW, 2443*

LYMPHOSARCOMA
    BONE MARROW CELLS
        HISTOLOGICAL STUDY, 2781
    BONE NEOPLASMS
        CLASSIFICATION, 2803*
    BRAIN NEOPLASMS
        HISTOLOGICAL STUDY, 2810*
    EPIDEMIOLOGY
        RACIAL FACTORS, CHILD, 2883
        REVIEW, 2419
    GASTROINTESTINAL NEOPLASMS
        REVIEW, 2443*
    KIDNEY NEOPLASMS
        CASE REPORT, 2860*
        ULTRASTRUCTURAL STUDY, 2860*
    LYMPHOCYTES
        CONCANAVALIN A, 2760*
    B-LYMPHOCYTES
        IMMUNE SERUMS, 2679

IMMUNOGLOBULINS, 2679
    IMMUNOGLOBULINS, SURFACE, 2694
    ROSETTE FORMATION, 2694
    ULTRASTRUCTURAL STUDY, 2679
T-LYMPHOCYTES
    IMMUNOGLOBULINS, SURFACE, 2694
    ROSETTE FORMATION, 2694
NERVOUS SYSTEM NEOPLASMS
    HISTOLOGICAL STUDY, 2812*
THYMUS NEOPLASMS
    UREA, N-NITROSO-N-PROPYL-, 2491
VIRUS, EPSTEIN-BARR
    ULTRASTRUCTURAL STUDY, 2679
VIRUS, FELINE LEUKEMIA
    HORIZONTAL TRANSMISSION, 2648*
    REVIEW, 2648*
    VACCINATION, 2648*

LYSOSOMES
    ACETIC ACID, CADMIUM SALT
        BOVINE SERUM ALBUMIN, KIDNEY,
            LIVER, 2519*
    ENZYMES
        REVIEW, 2451*

MACROPHAGES
    LYMPHOCYTE TRANSFORMATION
        VIRUS, EPSTEIN-BARR, 2592
    T-LYMPHOCYTES
        IMMUNE RESPONSE, 2704
    VIRUS, MOLONEY MURINE SARCOMA
        REGRESSION, 2761*

MAGNESIUM
    ACETAMIDE, N-FLUOREN-2-YL-
        DNA, 2515*
    LYMPHOMA
        DIETARY FACTORS, 2723*

MALARIA
    BURKITT'S LYMPHOMA
        EPIDEMIOLOGY, AFRICA, 2891
        REVIEW, 2444*

MALONATES
    ASPERGILLUS PARASITICUS
        MYCOTOXINS, 2508*

MAMMARY NEOPLASMS, EXPERIMENTAL
    ACTINOMYCIN D
        PROTECTIVE EFFECT, MOUSE, 2467
    ADENOCARCINOMA, 2725*
        BENZ(A)ANTHRACENE, 7,12-DIMETHYL-,
            2465
        ETHYNODIOL DIACETATE, 2405
        GROWTH, 2725*
        IMMUNITY, CELLULAR, 2712*
        NORETHYNODREL, 2405
        OCCLUDING JUNCTIONS, 2917
        TRANSPLANTATION, HOMOLOGOUS, 2948
        UREA, N-NITROSO-N-PROPYL-, 2491
        VIRUS, MURINE MAMMARY TUMOR, 2618
    ADENOFIBROMA
        UREA, N-NITROSO-N-PROPYL-, 2491
    ADENOMA
        VIRUS-LIKE PARTICLES, 2783
    ANILINE COMPOUNDS
        SYRIAN HAMSTER, 2504*

BENZ(A)ANTHRACENE, 7,12-DIMETHYL-
    ACTINOMYCIN D, 2467
    CARCINOGENIC ACTIVITY, RAT, 2540*
    ESTROGENS, 2466
    PORTACAVAL ANASTOMOSIS, 2540*
    PRECANCEROUS CONDITICNS, 2464
    PROLACTIN, 2466
BENZENESULFONAMIDE, 2-FLUORENYL-N-
    HYDROXY-
    MECHANISM OF ACTION, 2526*
CARCINOMA
    FIBRINOLYSIN, 2979*
    INCIDENCE, HAMSTER, 2971*
    ULTRASTRUCTURAL STUDY, DOG, 2825*
CELL DIFFERENTIATION
    INHIBITORY FACTOR, IN VITRO, 2533*
CHOLANTHRENE, 3-METHYL-
    SYRIAN HAMSTER, 2504*
CYTOSOL
    HEXOKINASE, 2948
    PHOSPHOTRANSFERASES, ATP, 2948
ESTRADIOL
    PRECANCEROUS CONDITICNS, 2464
    PROGESTERONE, 2533*
ESTRADIOL, 17-ETHINYL-
    CONTRACEPTIVES, ORAL, REVIEW, 2405
FIBRINOLYSIN
    ISOLATION AND CHARACTERIZATION,
    2979*
GLYCOLIPIDS
    GROWTH, 2989*
    ISOLATION AND CHARACTERIZATION,
    2989*
HEXOKINASE
    ISOLATION AND CHARACTERIZATION,
    2948
IMMUNE COMPLEX DISEASE
    GLOMERULONEPHRITIS, 2710*
INSULIN
    PRECANCEROUS CONDITICNS, 2464
MESTRANOL
    CONTRACEPTIVES, ORAL, REVIEW, 2405
MITOCHONDRIA
    HEXOKINASE, 2948
    PHOSPHOTRANSFERASES, ATP, 2948
NITROSO COMPOUNDS
    SYRIAN HAMSTER, 2504*
PHOSPHODIESTERASES
    KINETICS, 2996*
PHOSPHOTRANSFERASES, ATP
    ISOLATION AND CHARACTERIZATION, 2948
POLYCYCLIC HYDROCARBONS
    SYRIAN HAMSTER, 2504*
PROGESTERONE
    CONTRACEPTIVES, ORAL, REVIEW, 2405
    PRECANCEROUS CONDITICNS, 2464
PROLACTIN
    PRECANCEROUS CONDITICNS, 2464
VIRUS-LIKE PARTICLES
    ULTRASTRUCTURAL STUDY, SNOW
    LEOPARD, 2783

REK'S DISEASE
CELL LINE
    ISOLATION AND CHARACTERIZATION, 2620
T-LYMPHOCYTES
    IMMUNOSUPPRESSION, 2701
PLANT AGGLUTININS

T-LYMPHOCYTES, 2701
VACCINES
    COMPARATIVE VIROLOGY, REVIEW, 2412

MAST CELLS
LEUKEMIA
    ANTIBODIES, VIRAL, 2649*

MAXILLARY NEOPLASMS
    MELANOTIC NEUROECTODERMAL TUMOR
    CASE REPORT, 2764
    NEOPLASM RECURRENCE, LOCAL
    VANILMANDELIC ACID, 2764

MEDULLOBLASTOMA
    BRAIN NEOPLASMS
    CASE REPORT, 2805*
    GENETICS, 2805*
    ULTRASTRUCTURAL STUDY, 2809*
    CEREBELLAR NEOPLASMS
    CELLULAR INCLUSICNS, 2766
    ULTRASTRUCTURAL STUDY, 2766
    NERVOUS SYSTEM NEOPLASMS
    EPIDEMIOLOGY, RACIAL FACTORS,
    CHILD, 2883

MEDULLOEPITHELIOMA
    SEE NEUROEPITHELIOMA

MELANOMA
    ACTINOMYCIN D
    CELL DIVISION, 2977*
    PLASMINOGEN ACTIVATORS, 2977*
    TYROSINASE, 2963
    ANTIBODIES, NEOPLASM
    IMMUNE SERUMS, 2678
    ANTIGENS, NEOPLASM
    IMMUNITY, CELLULAR, 2728*
    CELL FUSION
    PLASMINOGEN ACTIVATORS, 2977*
    COLCHICINE
    TYROSINASE, 2963
    CYCLOHEXIMIDE
    TYROSINASE, 2963
    ESOPHAGEAL NEOPLASMS
    CASE REPORT, 2836*, 2837*, 2839*
    HORMONES
    TYROSINASE, 2963
    IMMUNITY, CELLULAR
    HISTOCOMPATIBILTY ANTIGENS, 2664
    LUNG NEOPLASMS
    CELL LINE, 2772
    CHROMOSOMES, 2772
    LYMPHOCYTES
    IMMUNITY, CELLULAR, 2712*
    T-LYMPHOCYTES
    ROSETTE FORMATION, 2664
    MITOMYCIN C
    CELL DIVISION, 2977*
    PLASMINOGEN ACTIVATORS, 2977*
    NOSE NEOPLASMS
    OCCUPATIONAL HAZARD, 2894
    PARANASAL SINUS NEOPLASMS
    OCCUPATIONAL HAZARD, 2894
    SKIN NEOPLASMS
    EPIDEMIOLOGY, 2855*, 2890
    EPIDEMIOLOGY, NIGERIA, 2884
    GROWTH, 2717*

PRECANCEROUS CONDITIONS, 2800
VITILIGO, 2800
TRANSFERASES
ENZYMATIC ACTIVITY, PLASMA, 2954
ULTRAVIOLET RAYS
EPIDEMIOLOGY, 2890
URIDINE, 5-BROMO-2'-DEOXY-
PLASMINOGEN ACTIVATORS, 2977*

MENINGIOMA
BRAIN NEOPLASMS
ANTIBODIES, 2745*

MENSTRUATION
BREAST NEOPLASMS
ESTRADIOL, 2785
PROGESTERONE, 2785

MESONEPHROMA
OVARIAN NEOPLASMS
MIGRATION INHIBITORY FACTOR, 2667

MESTRANOL
CARCINOMA, DUCTAL
NORETHYNODREL, 2405
MAMMARY NEOPLASMS, EXPERIMENTAL
CONTRACEPTIVES, ORAL, REVIEW, 2405

METALLOPROTEINS
CADMIUM
RNA, MESSENGER, 2520*
SYNTHESIS, LIVER, KIDNEY, 2520*

METHANE, SULFINYLBIS-
ACETAMIDE, N-FLUOREN-2-YL-
EMBRYO, LIVER, 2527*

METHANESULFONIC ACID, ETHYL ESTER
MUTAGENS
AZAGUANINE RESISTANCE, HAMSTER, 2479
DOSE RESPONSE, HAMSTER, 2479
SPERMATOZOA
ABNORMALITIES, 2452

METHANESULFONIC ACID, METHYL ESTER
ESCHERICHIA COLI
DNA REPAIR, 2480
MUTAGENS
AZAGUANINE RESISTANCE, HAMSTER,
2479
DOSE RESPONSE, HAMSTER, 2479
SPERMATOZOA
ABNORMALITIES, 2452

METHOTREXATE
HYBRID CELLS
THYMIDINE KINASE DEFICIENCY, 2776

N-METHYL-N'-NITRO-N-NITROSOGUANIDINE
SEE GUANIDINE, 1-METHYL-3-NITRO-1-
NITROSO-

N-METHYL-N-NITROSOUREA
SEE UREA, METHYL NITROSO-

1-METHYL-1-NITROSOUREA
SEE UREA, METHYL NITROSO-

METHYL SULFOXIDE
SEE METHANE, SULFINYLBIS-
METHYLATION
HEPATOMA
RNA, MESSENGER, 2988*

3-METHYLCHOLANTHRENE
SEE CHOLANTHRENE, 3-METHYL-

20-METHYLCHOLANTHRENE
SEE CHOLANTHRENE, 3-METHYL-

METHYLNITROSOUREA
SEE UREA, METHYL NITROSO-

MICROCOCCUS LUTEUS
DNA REPAIR
ULTRAVIOLET RAYS, 2572*
ENDONUCLEASES
ULTRAVIOLET RAYS, 2572*

MICROSOMES
CALCIUM
ENZYMATIC ACTIVITY, 2991*

MICROSOMES, LIVER
ACETAMIDE, N-FLUOREN-2-YL-
DNA, 2515*
RNA, 2515*
BENZO(A)PYREN-3-OL
NADH, NADPH OXIDOREDUCTASES, 2477
BENZO(A)PYREN-6-OL
NADH, NADPH OXIDOREDUCTASES, 2477
BENZO(A)PYREN-9-OL
NADH, NADPH OXIDOREDUCTASES, 2477
ESCHERICHIA COLI
DIMETHYLAMINE, N-NITROSO-, 2551*
HEPATOMA
BUTYRIC ACID, 2-AMINO-4-(ETHYLTHIO
-, 2792
DISULFIDES, 2924
SULFHYDRYL COMPOUNDS, 2924
HYDANTOIN, 1-((5-NITROFURFURYLIDINE)
AMINO)-
HYPERSENSITIVITY, 2530*
METABOLITES, 2530*

MIGRATION INHIBITORY FACTOR
OVARIAN NEOPLASMS
ADENOCARCINOMA, 2667
CYSTADENOCARCINOMA, 2667
CYSTADENOCARCINOMA, MUCINOUS, 2667
CYSTADENOMA, 2667
DERMOID CYST, 2667
ENDOMETRIOSIS, 2667
FIBROMA, 2667
IMMUNE RESPONSE, 2667
IMMUNITY, CELLULAR, 2667
LEUKOCYTES, 2667
MESONEPHROMA, 2667
TERATOID TUMOR, 2667
SARCOMA, MAST CELL
COLCHICINE, 2670
IMMUNE SERUMS, 2670
T-LYMPHOCYTES, 2670
VIRUS, SV40
LYMPHOKINES, 2631

NITROSAMINES, 2547*
NITROUS ACID, SODIUM SALT, 2547*

MORPHOLINE, 1-NITROSO-
    LUNG NEOPLASMS
        ADENOMA, 2481
    NITROUS ACID, SODIUM SALT
        COCARCINOGENIC EFFECT, MOUSE, 2481

MOUTH NEOPLASMS
    CARCINOMA
        EPIDEMIOLOGY, INDIA, 2893
    CARCINOMA, EPIDERMOIC
        EPIDEMIOLOGY, INDIA, 2893
    PRECANCEROUS CONDITIONS
        EPIDEMIOLOGY, INDIA, 2893
    SMOKING
        EPIDEMIOLOGY, INCIA, 2495, 2893
    TOBACCO
        EPIDEMIOLOGY, INDIA, 2495, 2893

MUCOCELE
    APPENDICEAL NEOPLASMS
        CYSTADENOMA, MUCINOUS, 2790
        POLYPS, 2790

MUCOPOLYSACCHARIDES
    VIRUS, SENDAI
        ISOLATION AND CHARACTERIZATION,
            2640*
    VIRUS, SINDBIS
        ISOLATION AND CHARACTERIZATION,
            2640*
    VIRUS, SV5
        ISOLATION AND CHARACTERIZATION,
            2640*

MULTIPLE MYELOMA
    CELL LINE
        IGG, 2676
        IMMUNOGLOBULINS, HEAVY CHAIN, 2676
    IMMUNOGLOBULINS
        ANTIBODIES, 2681
    B-LYMPHOCYTES
        IMMUNE SERUMS, 2679
        IMMUNOGLOBULINS, 2679
        ULTRASTRUCTURAL STUDY, 2679
    SERUM ALBUMIN, BOVINE
        ANTIBODIES, 2681
    VIRUS, HERPES SIMPLEX
        IMMUNITY, CELLULAR, 2671
    VIRUS, HERPES ZOSTER
        IMMUNITY, CELLULAR, 2671
    ZINC
        REVIEW, 2403

MURAMIDASE
    HEPATOMA
        GROWTH, 2992*
        HISTOLOGY, 2992*

MUTAGENS
    ACETAMIDE, N-(ACETYLOXY)-N-9H-FLUOREN-
    2-YL-
        AZAGUANINE RESISTANCE, FIBROBLASTS,
            2453
    BENZ(A)ANTHRACENE, 7,12-DIMETHYL-
        SALMONELLA/MICROSOME OXIDATION

TEST SYSTEM, 2430*
BENZO(A)PYRENE
    SALMONELLA/MICROSOME OXIDATION
        TEST SYSTEM, 2430*
CELL TRANSFORMATION, NEOPLASTIC
    ELECTROPHILIC FORMS, 2429*
2-FURANACRYLAMIDE, ALPHA-2-FURYL-5-
    NITRO-
        AZAGUANINE RESISTANCE, HAMSTER, 2455
METHANESULFONIC ACID, ETHYL ESTER
    AZAGUANINE RESISTANCE, HAMSTER, 2479
    DOSE RESPONSE, HAMSTER, 2479
METHANESULFONIC ACID, METHYL ESTER
    AZAGUANINE RESISTANCE, HAMSTER, 2479
    DOSE RESPONSE, HAMSTER, 2479
RADIATION
    AZAGUANINE RESISTANCE, HAMSTER, 2479
    DOSE RESPONSE, HAMSTER, 2479
ULTRAVIOLET RAYS
    AZAGUANINE RESISTANCE, HAMSTER, 2479
    DOSE RESPONSE, HAMSTER, 2479

MUTATION
    GUANIDINE, NITROSO-
        CHROMOSOME ABERRATIONS, 2548*
    VIRUS, SV40
        ULTRAVIOLET RAYS, 2630

MYCOBACTERIUM
    NEOPLASMS
        IMMUNE RESPONSE, 2674

MYCOBACTERIUM BOVIS
    LIPIDS
        ADJUVANTS, IMMUNOLOGIC, 2726*
    LUNG NEOPLASMS
        IMMUNITY, CELLULAR, 2505*

MYCOBACTERIUM SMEGMATIS
    NEOPLASMS
        IMMUNE RESPONSE, 2674

MYCOSIS FUNGOIDES
    NERVOUS SYSTEM NEOPLASMS
        HISTOLOGICAL STUDY, 2812*

MYCOTOXINS
    ASPERGILLUS FLAVUS
        FOOD CONTAMINATION, 2456
        ISOLATION AND CHARACTERIZATION,
            2511*
    ASPERGILLUS NIGER
        FOOD CONTAMINATION, 2456
    ASPERGILLUS PARASITICUS
        ACETATES, 2508*
        MALONATES, 2508*
    ASPERGILLUS REPENS
        FOOD CONTAMINATION, 2456
    MONILIALES
        FOOD CONTAMINATION, 2456
    PENICILLIN
        FOOD CONTAMINATION, 2456
    RHIZOPUS
        FOOD CONTAMINATION, 2456

MYELOMA
    SEE MULTIPLE MYELOMA

MYELOMA PROTEINS
    CHOLINE
        BINDING SITES, ANTIBODY, 2680
    IGA, 2983*
        LIGAND-BINDING, 2983*
    PLASMACYTOMA
        IMMUNE SERUMS, 2680
    STREPTOCOCCUS
        UPTAKE, 2739*

MYOSARCOMA
    COLONIC NEOPLASMS
        UREA, N-NITROSO-N-PROPYL-, 2491
    SKIN NEOPLASMS
        UREA, N-NITROSO-N-PROPYL-, 2491

NADH, NADPH OXIDOREDUCTASES
    HEPATOMA
        GLUCONEOGENESIS, 2949
    MICROSOMES, LIVER
        BENZO(A)PYREN-3-OL, 2477
        BENZO(A)PYREN-6-OL, 2477
        BENZO(A)PYREN-9-OL, 2477

ALPHA-NAPHTHOFLAVONE
    SEE 7,8-BENZOFLAVONE

BETA-NAPHTHOFLAVONE
    SEE 5,6-BENZOFLAVONE

2-NAPTHYLAMINE
    BLADDER NEOPLASMS
        OCCUPATIONAL HAZARD, REVIEW, 2431*

NASOPHARYNGEAL NEOPLASMS
    VIRUS, EPSTEIN-BARR
        ANTIGENS, VIRAL, 2591, 2709*
        GENETICS, 2709*

NCI CO2006
    LIVER NEOPLASMS
        ADENOMA, 2554*
        CARCINOMA, 2554*
        PRECANCEROUS CONDITIONS, 2554*

NCI CO2186
    THYROID NEOPLASMS
        ADENOMA, 2555*
        CARCINOMA, 2555*
        PRECANCEROUS CONDITIONS, 2555*

NEOPLASM METASTASIS
    ADENOCARCINOMA
        CHOLANTHRENE, 3-METHYL-, 2770
    ADENOMA
        CHOLANTHRENE, 3-METHYL-, 2770
    BLADDER NEOPLASMS
        CARCINOMA IN SITU, 2795
        CARCINOMA, TRANSITIONAL CELL, 2795
    BONE NEOPLASMS
        CASE REPORT, 2802*
    BRAIN NEOPLASMS
        SARCOMA, RETICULUM CELL, 2821*
    BREAST NEOPLASMS
        CARCINOMA, 2824*
        ESTROGEN RECEPTORS, REVIEW, 2421
        NEOPLASMS, MULTIPLE PRIMARY, 2784
    CARCINOMA

CHOLANTHRENE, 3-METHYL-, 2770
CARCINOMA, DUCTAL
    CHOLANTHRENE, 3-METHYL-, 2770
CERVIX NEOPLASMS
    CASE REPORT, 2802*
ESOPHAGEAL NEOPLASMS
    CASE REPORT, 2802*
HELA CELLS
    TRANSPLANTATION, HETEROLOGOUS, 2731*
IMMUNOSUPPRESSION
    TRANSPLANTATION, HETEROLOGOUS, 2731*
NEOPLASM TRANSPLANTATION
    CHOLANTHRENE, 3-METHYL-, 2770
NOSE NEOPLASMS
    NICOTINE, 1'-DEMETHYL-N'-NITROSO-,
        2486
PAPILLOMA
    CHOLANTHRENE, 3-METHYL-, 2770
PROSTATIC NEOPLASMS
    HISTOLOGICAL STUDY, 2795
RECTAL NEOPLASMS
    EPIDEMIOLOGY, 2791
    STAGING AND THERAPY, 2791
SARCOMA, OSTEOGENIC
    COPPER, 2403
TRANSFERASES
    ENZYMATIC ACTIVITY, PLASMA, 2954

OPLASM RECURRENCE, LOCAL
MAXILLARY NEOPLASMS
    VANILMANDELIC ACID, 2764

OPLASM TRANSPLANTATION
CHOLANTHRENE, 3-METHYL-
    NEOPLASM METASTASIS, 2770
HEPATOMA
    DIETHYLAMINE, N-NITROSO-, 2549*
T-LYMPHOCYTES
    ELECTROPHORETIC MOBILITY, 2771
    IMMUNE RESPONSE, 2771
    ISOLATION AND CHARACTERIZATION, 2771
NEURILEMMOMA
    UREA, ETHYL NITROSO-, 2490
PLASMACYTOMA
    LYMPHOCYTES, 2718*

OPLASMS (GENERAL AND UNSPECIFIED)
SEE ALSO UNDER PARTICULAR SITE

ACETAMIDE, N-(5-SULFAMOYL-1,3,4-
    THIADIAZOL-2-YL)-
    ACID-BASE EQUILIBRIUM, 2967
ACHLORHYDRIA
    ACID-BASE EQUILIBRIUM, 2967
ADENOSINE CYCLIC 3',5' MONOPHOSPHATE
    EPINEPHRINE, 2998*
    ESTRADIOL, 2998*
    ISOPROTERENOL, 2998*
    KINETICS, 2998*
    PROGESTERONE, 2998*
AGE FACTORS
    EPIDEMIOLOGY, CHILD, POLAND, 2899*
ALTITUDE
    EPIDEMIOLOGY, 2967
ASBESTOS
    PARTICLE SIZE, 2434*
COPPER
    REVIEW, 2403

ENZYMES
    REVIEW, 2451*
EPIDEMIOLOGY
    BRAZIL, 2900*
GROWTH
    IMMUNITY, CELLULAR, 2712*
HISTOCOMPATIBILITY ANTIGENS
    CANCER FAMILY SYNDROME, 2686
IMMUNITY, CELLULAR
    IMMUNOSUPPRESSION, 2438*
    INFLATION, 2438*
LYMPHOCYTES
    ANTIBODIES, 2746*
    NEURAMINIDASE, 2746*
MYCOBACTERIUM
    IMMUNE RESPONSE, 2674
MYCOBACTERIUM SMEGMATIS
    IMMUNE RESPONSE, 2674
PHOSPHODIESTERASES
    ISOPROTERENOL, 2998*
PULMONARY EMPHYSEMA
    ACID-BASE EQUILIBRIUM, 2967
SCROTUM
    LUBRICATING OIL, 2431*
    OCCUPATIONAL HAZARD, REVIEW, 2431*
    SOOT, 2431*
TRANSFERASES
    ENZYMATIC ACTIVITY, PLASMA, 2954
TRANSPLANTATION, HETEROLOGOUS
    ATHYMIC NUDE MICE, 2975*
    GROWTH, 2975*
ZINC
    REVIEW, 2403

NEOPLASMS, CONNECTIVE TISSUE
FIBROMA
    EPIDEMIOLOGY, 2763
    HISTOLOGICAL STUDY, 2763
    PRECANCEROUS CONDITIONS, 2763

NEOPLASMS, EMBRYONAL AND MIXED
ESOPHAGEAL NEOPLASMS
    CASE REPORT, 2838*

NEOPLASMS, EXPERIMENTAL
POLY A-U
    NUCLEIC ACIDS, 2935

NEOPLASMS, MULTIPLE PRIMARY
BREAST NEOPLASMS
    CARCINOMA, 2784
    CARCINOMA, BRONCHIOLAR, 2784
    CARCINOMA, COLLOID, 2784
    CARCINOMA, DUCTAL, 2784
    CARCINOMA, PAPILLARY, 2784
    DIAGNOSIS AND PROGNOSIS, 2784
    HISTOLOGICAL STUDY, 2784
    NEOPLASM METASTASIS, 2784

NEPHRITIS, INTERSTITIAL
LEUKEMIA, LYMPHOBLASTIC
    CASE REPORT, 2849*
    PRECANCEROUS CONDITIONS, 2849*
    ULTRASTRUCTURAL STUDY, 2849*
    URIC ACID, 2849*

NEPHROBLASTOMA
ANTIGENS, NEOPLASM

ISOLATION AND CHARACTERIZATION,
2692
EPIDEMIOLOGY
RACIAL FACTORS, CHILD, 2883
FETAL GLOBULINS
ISOLATION AND CHARACTERIZATION,
2692
KIDNEY NEOPLASMS
UREA, N-NITROSO-N-PROPYL-, 2491

NERVOUS SYSTEM NEOPLASMS
ASTROCYTOMA
EPIDEMIOLOGY, RACIAL FACTORS,
CHILD, 2883
UREA, 1,1-DIETHYL-3-METHYL-3-
NITROSO-, 2488
UREA, 3-ETHYL-1,1-DIMETHYL-3-
NITROSO-, 2488
UREA, 1,1,3-TRIETHYL-3-NITROSO-,
2488
UREA, 1,1,3-TRIMETHYL-3-NITROSO-,
2488
EPENDYMOMA
EPIDEMIOLOGY, RACIAL FACTORS,
CHILD, 2883
GANGLIONEUROMA
EPIDEMIOLOGY, RACIAL FACTORS,
CHILD, 2883
GLIOMA
UREA, 1,1-DIETHYL-3-METHYL-3-
NITROSO-, 2488
UREA, 3-ETHYL-1,1-DIMETHYL-3-NITRO-
, 2488
UREA, 1,1,3-TRIETHYL-3-NITROSO-,
2488
UREA, 1,1,3-TRIMETHYL-3-NITROSO-,
2488
GLYCOSPHINGOLIPIDS
METABOLISM, 2990*
HODGKIN'S DISEASE
HISTOLOGICAL STUDY, 2812*
INFLAMMATORY LYMPHORETICULOSES
HISTOLOGY, REVIEW, 2441*
LETTERER-SIWE DISEASE
HISTOLOGICAL STUDY, 2812*
LYMPHOMA
HISTOLOGICAL STUDY, TISSUE
CULTURES, 2808*
HISTOLOGY, REVIEW, 2441*
VIRUS, MAREK'S DISEASE HERPES,
2441*
VIRUS, RNA TUMOR, 2441*
LYMPHOSARCOMA
HISTOLOGICAL STUDY, 2812*
MEDULLOBLASTOMA
EPIDEMIOLOGY, RACIAL FACTORS,
CHILD, 2883
MYCOSIS FUNGOIDES
HISTOLOGICAL STUDY, 2812*
NEUROBLASTOMA
EPIDEMIOLOGY, RACIAL FACTORS,
CHILD, 2883
PLASMACYTOMA
HISTOLOGICAL STUDY, 2812*
SARCOMA, RETICULUM CELL
CASE REPORT, 2815*, 2816*
CLINICOPATHOLOGIC STUDY, 2816*
HISTOLOGICAL STUDY, 2812*
HISTOLOGY, REVIEW, 2441*

ULTRASTRUCTURAL STUDY, 2814*,
2817*
UREA, ETHYL NITROSO-, 2814*
UREA, ETHYL NITROSO-
ANTIGENS, 2683
IMMUNE SERUMS, 2683

NEURAMINIDASE
AGAMMAGLOBULINEMIA
LYMPHOCYTES, 2746*
ATAXIA TELANGIECTASIA
LYMPHOCYTES, 2746*
IMMUNOLOGIC DEFICIENCY SYNDROMES
LYMPHOCYTES, 2746*
NEOPLASMS
LYMPHOCYTES, 2746*
TERATOID TUMOR
CELL MEMBRANE, 2882

NEURILEMMOMA
UREA, ETHYL NITROSO-
NEOPLASM TRANSPLANTATION, 2490
ULTRASTRUCTURAL STUDY, 2490

NEUROBLASTOMA
ACTINOMYCIN D
CARBOXY-LYASES, 2959
ADENOSINE
CARBOXY-LYASES, 2959
ANTIGENS, NEOPLASM
IMMUNE RESPONSE, 2415
CELL LINE
TRANSPLANTATION, HETEROLOGOUS,
2912
CHOLINE, CHLORIDE, CARBAMATE
CARBOXY-LYASES, 2959
CYCLOHEXIMIDE
CARBOXY-LYASES, 2959
DIBUTYRYL CYCLIC AMP
CARBOXY-LYASES, 2959
EPINEPHRINE
ADENOSINE CYCLIC 3',5'
MONOPHOSPHATE, 2415
CATECHOLAMINES, 2415
GLYCOSPHINGOLIPIDS
METABOLISM, 2990*
2-IMIDAZOLIDINONE, 4-(3-BUTOXY-4-
METHOXYBENZYL)-
CARBOXY-LYASES, 2959
ISOPROTERENOL
CARBOXY-LYASES, 2959
NERVOUS SYSTEM NEOPLASMS
EPIDEMIOLOGY, RACIAL FACTORS,
CHILD, 2883
NOREPINEPHRINE
CARBOXY-LYASES, 2959
NOSE NEOPLASMS
OCCUPATIONAL HAZARD, 2894
PARANASAL SINUS NEOPLASMS
OCCUPATIONAL HAZARD, 2894
PROSTAGLANDINS E
CARBOXY-LYASES, 2959
PROTEINS
REGULATION, 2922
XANTHINE, 3-ISOBUTYL-1-METHYL-
ADENOSINE CYCLIC 3',5'
MONOPHOSPHATE, 2997*
CARBOXY-LYASES, 2959
GUANOSINE CYCLIC 3',5'

SEE UREA, METHYL NITROSO-

N-NITROSOMETHYLUREA
    SEE UREA, METHYL NITROSO-

NITROUS ACID, SODIUM SALT
    FOOD PRESERVATION
        DIMETHYLAMINE, N-NITROSO-, 2552*
    MORPHOLINE
        ASCORBIC ACID, 2547*
    MORPHOLINE, 1-NITROSO-
        COCARCINOGENIC EFFECT, MOUSE, 2481
    PIPERAZINE, 1-NITROSO-
        COCARCINOGENIC EFFECT, MOUSE, 2481
    UREA, METHYL-
        COCARCINOGENIC EFFECT, MOUSE, 2481
    UREA, METHYL NITROSO-
        COCARCINOGENIC EFFECT, MOUSE, 2481

NOREPINEPHRINE
    NEUROBLASTOMA
        CARBOXY-LYASES, 2959
    RECTAL NEOPLASMS
        METABOLISM, 3000*

NORETHYNODREL
    CARCINOMA, DUCTAL
        MESTRANOL, 2405
    MAMMARY NEOPLASMS, EXPERIMENTAL
        ADENOCARCINOMA, 2405

NORNICOTINE, N-NITROSO-
    TOBACCO
        CARCINOGENIC ACTIVITY, 2553*

NOSE NEOPLASMS
    ADENOCARCINOMA
        OCCUPATIONAL HAZARD, 2894
        WOOD, 2894
    CARCINOMA
        OCCUPATIONAL HAZARD, 2894
    CARCINOMA, EPIDERMOID
        OCCUPATIONAL HAZARD, 2894
    CARCINOMA, TRANSITIONAL CELL
        OCCUPATIONAL HAZARD, 2894
    CYLINDROMA
        OCCUPATIONAL HAZARD, 2894
    LYMPHOMA
        OCCUPATIONAL HAZARD, 2894
    MELANOMA
        OCCUPATIONAL HAZARD, 2894
    MIXED SALIVARY GLAND TUMOR
        OCCUPATIONAL HAZARD, 2894
    NEOPLASM METASTASIS
        NICOTINE, 1'-DEMETHYL-N'-NITROSO-,
            2486
    NEUROBLASTOMA
        OCCUPATIONAL HAZARD, 2894
    PHOSPHORIC TRIAMIDE, HEXAMETHYL-
        CARCINOGENIC POTENTIAL, RAT, 2426*
    PLASMACYTOMA
        OCCUPATIONAL HAZARD, 2894

NUCLEIC ACIDS
    SEE ALSO DNA/RNA
    DNA POLYMERASE
        POLYNUCLEOTIDES, 2582
    LYMPHOCYTES
        POLY A-U, 2935

NEOPLASMS, EXPERIMENTAL
   POLY A-U, 2935
VIRUS, AVIAN MYELOBLASTOSIS
   DNA POLYMERASE, 2582

NUCLEOSIDE DEAMINASES
   LEUKEMIA, LYMPHOBLASTIC
      B-LYMPHOCYTES, 2758*
      T-LYMPHOCYTES, 2758*

NUCLEOTIDES
   CARCINOMA, EHRLICH TUMOR
      CATABOLISM, 2943

NUCLEOTIDYLTRANSFERASES
   LEUKEMIA, LYMPHOBLASTIC
      B-LYMPHOCYTES, 2758*
      T-LYMPHOCYTES, 2758*

NUTRITION
   SEE DIET

OCCUPATIONAL HAZARD
   ASBESTOS
      FIBERS, 2575*
   BLADDER NEOPLASMS
      EPIDEMIOLOGY, 2881
   CHOLANTHRENE, 3-METHYL-
      CARCINOGENIC ACTIVITY, REVIEW,
        2428*
   ETHANE, 2-BROMO-2-CHLORO-1,1,1-
      TRIFLUORO-
      ANESTHETICS, 2408
   ETHER, 1-CHLORO-2,2,2-TRIFLUOROETHYL
      DIFLUOROMETHYL
      ANESTHETICS, 2408
   ETHER, 2,2-DICHLORO-1,1-
      DIFLUOROETHYLMETHYL-
      ANESTHETICS, 2408
   ETHYLENE, CHLORO-
      CARCINOGEN, ENVIRONMENTAL, 2432*
      DETECTION DEVICE, 2559*
   ETHYLENE, TRICHLORO-
      ANESTHETICS, 2408
   HEPATOMA
      ETHYLENE, TRICHLORO-, 2427*
   IGA
      URANIUM, 2677
   IGG
      URANIUM, 2677
   IGM
      URANIUM, 2677
   IMMUNOGLOBULINS
      URANIUM, 2677
   LIVER NEOPLASMS
      ETHYLENE, CHLORO-, 2896
      ETHYLENE, CHLORO- POLYMER, 2896
      ETHYLENE, TRICHLORO-, 2427*
   LUNG NEOPLASMS
      CHROMATES, 2523*
      EPIDEMIOLOGY, 2881
   NOSE NEOPLASMS
      ADENOCARCINOMA, 2894
      CARCINOMA, 2894
      CARCINOMA, EPIDERMOID, 2894
      CARCINOMA, TRANSITIONAL CELL, 2894
      CYLINDROMA, 2894
      LYMPHOMA, 2894
      MELANOMA, 2894

      MIXED SALIVARY GLAND TUMOR, 2894
      NEUROBLASTOMA, 2894
      PLASMACYTOMA, 2894
PARANASAL SINUS NEOPLASMS
   ADENOCARCINOMA, 2894
   CARCINOMA, 2894
   CARCINOMA, EPIDERMOID, 2894
   CARCINOMA, TRANSITIONAL CELL, 289
   CYLINDROMA, 2894
   LYMPHOMA, 2894
   MELANOMA, 2894
   MIXED SALIVARY GLAND TUMOR, 2894
   NEUROBLASTOMA, 2894
   PLASMACYTOMA, 2894
   SARCOMA, 2894
RADON
   RADIATION EXPOSURE, 2897

OLIGODENDROGLIOMA
   BRAIN NEOPLASMS
      ULTRASTRUCTURAL STUDY, 2809*

ONCOGENIC VIRUSES
   VIRAL PROTEINS
      ANTIGENIC DETERMINANTS, 2589

ORAL CAVITY
   LEUKOPLAKIA
      CARCINOMA, EPIDERMOID, 2904*
      CARCINOMA IN SITU, 2904*
      DYSPLASIA, 2904*
      EPIDEMIOLOGY, 2904*

ORBITAL NEOPLASMS
   RHABDOMYOSARCOMA
      ULTRASTRUCTURAL STUDY, 2767

OSTEOMA, OSTEOID
   ULTRASTRUCTURAL STUDY
      CASE REPORT, 2801*

OUABAIN
   HEPATOMA
      GLYCOLYSIS, 2918
      ION TRANSPORT, 2918

OVARIAN NEOPLASMS
   ADENOCARCINOMA
      MIGRATION INHIBITORY FACTOR, 2667
   BREAST NEOPLASMS
      GENETICS, 2865*
   CARCINOEMBRYONIC ANTIGEN
      DIAGNOSIS AND PROGNOSIS, REVIEW,
        2416
   CYSTADENOMA, MUCINOUS
      GIANT CELL TUMOR, 2864*
   CYSTADENOCARCINOMA
      GENETICS, 2865*
      MIGRATION INHIBITORY FACTOR, 2667
   CYSTADENOCARCINOMA, MUCINOUS
      MIGRATION INHIBITORY FACTOR, 2667
   CYSTADENOMA
      MIGRATION INHIBITORY FACTOR, 2667
   DERMOID CYST
      MIGRATION INHIBITORY FACTOR, 2667
   ENDOMETRIOSIS
      MIGRATION INHIBITORY FACTOR, 2667
   FIBROMA
      MIGRATION INHIBITORY FACTOR, 2667

GLUTAMYL TRANSPEPTASE
    ISOENZYMES, 2950

PAPILLOMA
    BENZ(A)ANTHRACENE, 7,12-DIMETHYL-
        CARCINOGENIC POTENTIAL, MOUSE,
            2468
    CHOLANTHRENE, 3-METHYL-
        NEOPLASM METASTASIS, 2770
    COLONIC NEOPLASMS
        ISOANTIGENS, 2753*
    EAR NEOPLASMS
        UREA, N-NITROSO-N-PROPYL-, 2491
    ESOPHAGEAL NEOPLASMS
        NICOTINE, 1'-DEMETHYL-N'-NITROSO-,
            2486
    STOMACH NEOPLASMS
        UREA, 1-BUTYL-1-NITROSO-, 2489

PARAGANGLIOMA, NONCHROMAFFIN
    LUNG NEOPLASMS
        CHEMORECEPTORS, 2788
        ULTRASTRUCTURAL STUDY, 2788

PARANASAL SINUS NEOPLASMS
    ADENOCARCINOMA
        OCCUPATIONAL HAZARD, 2894
        WOOD, 2894
    CARCINOMA
        OCCUPATIONAL HAZARD, 2894
    CARCINOMA, EPIDERMOID
        OCCUPATIONAL HAZARD, 2894
    CARCINOMA, TRANSITIONAL CELL
        OCCUPATIONAL HAZARD, 2894
    CYLINDROMA
        OCCUPATIONAL HAZARD, 2894
    LYMPHOMA
        OCCUPATIONAL HAZARD, 2894
    MELANOMA
        OCCUPATIONAL HAZARD, 2894
    MIXED SALIVARY GLAND TUMOR
        OCCUPATIONAL HAZARD, 2894
    NEUROBLASTOMA
        OCCUPATIONAL HAZARD, 2894
    PLASMACYTOMA
        OCCUPATIONAL HAZARD, 2894
    SARCOMA
        OCCUPATIONAL HAZARD, 2894

PARATHYROID HORMONE
    HYPERPLASIA
        ISOLATION AND CHARACTERIZATION,
            PRECURSOR, 2966
    PARATHYROID NEOPLASMS
        HYPERPLASIA, 2966

PARATHYROID NEOPLASMS
    ADENOMA
        HYPERCALCEMIA, 2797
        HYPERPARATHYROIDISM, 2797
        PRECANCEROUS CONDITIONS, 2797
    CARCINOMA
        HYPERCALCEMIA, 2797
        HYPERPARATHYROIDISM, 2797
        PRECANCEROUS CONDITIONS, 2797
    PARATHYROID HORMONE
        HYPERPLASIA, 2966
    RNA, MESSENGER
        HYPERPLASIA, 2966

PENICILLIN
MYCOTOXINS
· FOOD CONTAMINATION, 2456

PENILE NEOPLASMS
CARCINOMA
EPIDEMIOLOGY, INDIA, 2885

PEPSINOGEN
GUANIDINE, 1-METHYL-3-NITRO-1-NITROSO-
ISOENZYMES, 2485

PEPTIDES
ADENOCARCINOMA
VIRUS, MURINE MAMMARY TUMOR, 2618
L CELLS
CELL MEMBRANE, 2926
MILK
VIRUS, MURINE MAMMARY TUMOR, 2618
VIRUS, ADENO 2
HEXON STRUCTURE, 2581
VIRUS, ADENO 3
HEXON STRUCTURE, 2581
VIRUS, ADENO 5
HEXON STRUCTURE, 2581
VIRUS, C-TYPE AVIAN TUMOR
ANTIGENIC DETERMINANTS, 2587
ISOLATION AND CHARACTERIZATION,
2587
VIRUS, FRIEND MURINE LEUKEMIA
CONCANAVALIN A, 2603
VIRUS, MURINE MAMMARY TUMOR
ANTIGENS, VIRAL, 2619
ISOLATION AND CHARACTERIZATION,
2618
VIRUS-LIKE PARTICLES, 2619
VIRUS, ROUS SARCOMA
ANTIGENIC DETERMINANTS, 2587
CELL MEMBRANE, 2627
ISOLATION AND CHARACTERIZATION,
2587
VIRUS, SV40
DNA, VIRAL, 2633

PERICARDIAL EFFUSION
HEART NEOPLASMS
HEMANGIOSARCOMA, 2843*
SKIN NEOPLASMS
HEMANGIOMA, 2843*

PETROLEUM
BENZ(A)ANTHRACENE
FOOD ANALYSIS, 2535*
BENZO(A)PYRENE
FOOD ANALYSIS, 2535*
CHOLANTHRENE, 3-METHYL-
FOOD ANALYSIS, 2535*
DIBENZ(A,H)ANTHRACENE
FOOD ANALYSIS, 2535*
POLYCYCLIC HYDROCARBONS
FOOD ANALYSIS, 2535*

PEUTZ-JEGHERS SYNDROME
COLONIC NEOPLASMS
GENETICS, 2828*
SCREENING, 2828*

PHAGOCYTOSIS
BRAIN NEOPLASMS

SARCOMA, RETICULUM CELL, 2440*

PHARYNGEAL NEOPLASMS
CARCINOMA
EPIDEMIOLOGY, INDIA, 2893
NICOTINE, 1'-DEMETHYL-N'-NITROS
2486
CARCINOMA, EPIDERMOID
EPIDEMIOLOGY, INDIA, 2893
PRECANCEROUS CONDITIONS
· EPIDEMIOLOGY, INDIA, 2893
SMOKING
EPIDEMIOLOGY, INDIA, 2495, 2893
TOBACCO
EPIDEMIOLOGY, INDIA, 2495, 2893

PHENOBARBITAL
SEE BARBITURIC ACID, 5-ETHYL-5-PHENY

PHENOBARBITONE
SEE BARBITURIC ACID, 5-ETHYL-5-PHENY

PHEOCHROMOCYTOMA
THYROID NEOPLASMS
GENETICS, 2875*
PRECANCEROUS CONDITIONS, 2875*

PHORBOL 12,13-DIMYRISTATE
DNA
THYMIDINE, 2462
RNA
THYMIDINE, 2462
SKIN NEOPLASMS
PRECANCEROUS CONDITIONS, 2462

PHORBOL MYRISTATE ACETATE
SEE 12-O-TETRADECANOYLPHORBOL-13-
ACETATE

PHOSPHINE OXIDE, TRIS(2-METHYL-1-
AZIRIDINYL)-
SPERMATOZOA
ABNORMALITIES, 2452

PHOSPHINE SULFIDE, TRIS(1-AZIRIDINYL)-
SPERMATOZOA
ABNORMALITIES, 2452

PHOSPHODIESTERASES
CELL TRANSFORMATION, NEOPLASTIC
PLASMINOGEN ACTIVATORS, 2995*
GLIOMA
XANTHINE, 3-ISOBUTYL-1-METHYL-,
2997*
MAMMARY NEOPLASMS, EXPERIMENTAL
KINETICS, 2996*
NEOPLASMS
ISOPROTERENOL, 2998*
NEUROBLASTOMA
XANTHINE, 3-ISOBUTYL-1-METHYL-,
2997*
VIRUS, HARVEY MURINE SARCOMA
CELL TRANSFORMATION, NEOPLASTIC,
2994*
VIRUS, KIRSTEN MURINE SARCOMA
CELL TRANSFORMATION, NEOPLASTIC,
2994*
VIRUS, MOLONEY MURINE SARCOMA
CELL TRANSFORMATION, NEOPLASTIC,

ISOENZYMES, 2993*

PLANT AGGLUTININS
    BREAST NEOPLASMS
        IMMUNITY, CELLULAR, 2702
    CARCINOMA, EHRLICH TUMOR
        ANTI-NEOPLASTIC ACTIVITY, 2736*
    CELL MEMBRANE
        VIRUS, AVIAN RNA TUMOR, 2414
    CERVIX NEOPLASMS
        CARCINOMA IN SITU, 2868*
        PRECANCEROUS CONDITIONS, 2868*
    COLONIC NEOPLASMS
        IMMUNITY, CELLULAR, 2711*
    DIBUTYRYL CYCLIC AMP
        CELLS, CULTURED, 2972*
    HEPATOMA
        AGGLUTINATION, 2706
    B-LYMPHOCYTES
        PROTEINS, 2724*
    T-LYMPHOCYTES
        HISTOLOGICAL STUDY, MOUSE, 2695
        PROTEINS, 2724*
    LYMPHOMA
        LYMPHOCYTE TRANSFORMATION, 2847*
    MAREK'S DISEASE
        T-LYMPHOCYTES, 2701
    THYMECTOMY
        T-LYMPHOCYTES, 2697
    WARTS
        IMMUNITY, CELLULAR, 2729*

PLANTS
    ANLINE, 4,4'-(IMIDOCARBONYL)BIS(N,N-
        DIMETHYL)-
        GROWTH, 2532*
    FIBROSARCOMA
        CARCINOGENIC ACTIVITY, MOUSE,
            2507*

PLASMACYTOMA
    IGA
        LIGAND-BINDING, 2983*
    LYMPHOCYTES
        NEOPLASM TRANSPLANTATION, 2718*
    MYELOMA PROTEINS
        IMMUNE SERUMS, 2680
    NERVOUS SYSTEM NEOPLASMS
        HISTOLOGICAL STUDY, 2812*
    NOSE NEOPLASMS
        OCCUPATIONAL HAZARD, 2894
    PARANASAL SINUS NEOPLASMS
        OCCUPATIONAL HAZARD, 2894
    TRANSPLANTATION, HOMOLOGOUS
        T-LYMPHOCYTES, 2771

PLUTONIUM
    BONE NEOPLASMS
        REVIEW, 2410
    METABOLISM
        REVIEW, 2409

PLUTONIUM OXIDE
    METABOLISM
        REVIEW, 2409

POLY A
    HELA CELLS
        ADENOSINE TRIPHOSPHATE, 2936

        RNA, 2936
    VIRUS, HERPES SIMPLEX 1
        RNA, VIRAL, 2646*

POLY A-U
    LYMPHOCYTES
        NUCLEIC ACIDS, 2935
    T-LYMPHOCYTES
        CELL DIVISION, 2682
    NEOPLASMS, EXPERIMENTAL
        NUCLEIC ACIDS, 2935

POLYCYCLIC HYDROCARBONS
    CELL TRANSFORMATION, NEOPLASTIC
        COAL-TAR, 2536*
    CYTOCHROME P-450
        COAL-TAR, 2536*
    MAMMARY NEOPLASMS, EXPERIMENTAL
        SYRIAN HAMSTER, 2504*
    PETROLEUM
        FOOD ANALYSIS, 2535*

POLYCYTHEMIA VERA
    LEUKOCYTOSIS
        CASE REPORT, 2846*

POLYETHYLENE GLYCOLS
    VIRUS, ROUS SARCOMA
        VIRUS CONCENTRATION, 2656*

POLYNUCLEOTIDE SYNTHETASES
    DNA, REPLICATION
        REVIEW, 2422

POLYNUCLEOTIDES
    DNA POLYMERASE
        NUCLEIC ACIDS, 2582
    VIRUS, AVIAN MYELOBLASTOSIS
        DNA POLYMERASE, 2582

POLYPS
    APPENDICEAL NEOPLASMS
        MUCOCELE, 2790
    COLONIC NEOPLASMS
        CARCINOMA, 2832*
        CASE REPORT, 2833*
        GENETICS, 2833*
        ISOANTIGENS, 2750*
    INTESTINAL NEOPLASMS
        UREA, 1-BUTYL-1-NITROSO-, 2489
    RECTAL NEOPLASMS
        EPIDEMIOLOGY, 2791
        STAGING AND THERAPY, 2791

POLYRIBOSOMES
    HEPATOMA
        DISULFIDES, 2924
        SULFHYDRYL COMPOUNDS, 2924

POLYVINYL CHLORIDE
    SEE ETHYLENE, CHLORO- POLYMER

PORTACAVAL ANASTOMOSIS
    MAMMARY NEOPLASMS, EXPERIMENTAL
        BENZ(A)ANTHRACENE, 7,12-DIMETHYL-,
            2540*

PRECANCEROUS CONDITIONS

```
                    PRECANCEROUS CONDITIONS, 2464
        PROLINE
            SODIUM CHLORIDE
                NITROSAMINES, 2546*

        PRONASE
            VIRUS, ROUS SARCOMA
                VIRUS CONCENTRATION, 2656*

        2-PROPANOL, 1,1'-(NITROSOIMINO)DI-
            PANCREATIC NEOPLASMS
                ADENOMA, 2483
                CARCINOMA, 2483

        PROSTAGLANDINS E
            GLIOMA
                CARBOXY-LYASES, 2959
            NEUROBLASTOMA
                CARBOXY-LYASES, 2959

        PROSTAGLANDINS F
            VIRUS, SV40
                ADENYL CYCLASE, 2663*

        PROSTATIC NEOPLASMS
            ADENOCARCINOMA
                EPIDEMIOLOGY, INDIA, 2886
            CARCINOEMBRYONIC ANTIGEN
                DIAGNOSIS AND PROGNOSIS, REVIEW,
                2416
            CARCINOMA, EPIDERMOID
                EPIDEMIOLOGY, INDIA, 2886
            COPPER
                REVIEW, 2403
            FIBROBLASTS
                IMMUNE SERUMS, 2713*
            GONORRHEA
                EPIDEMIOLOGY, 2887
            GRANULOPOIETIC PROGENITOR CELLS
                COLONY FORMATION, 2915
            LYMPHOCYTES
                IMMUNE RESPONSE, 2713*
            NEOPLASM METASTASIS
                HISTOLOGICAL STUDY, 2795
            PRECANCEROUS CONDITIONS, 2869*
            PROSTATITIS
                EPIDEMIOLOGY, 2887
            URETHRITIS
                EPIDEMIOLOGY, 2887
            ZINC
                REVIEW, 2403

        PROSTATITIS
            PROSTATIC NEOPLASMS
                EPIDEMIOLOGY, 2887

        PROTEINS
            CARCINOMA, EHRLICH TUMOR
                RNA, RIBOSOMAL, 2938
            CHOLANTHRENE, 3-METHYL-
                BINDING, 2538*
            HELA CELLS
                CARBAMIC ACID, METHYL-, 1-NAPHTHYL
                ESTER, 2921
                DIELDRIN, 2921
            HEPATOMA
                LOCALIZATION AND PHOSPHORYLATION,
                2925
```

INFECTIOUS MONONUCLEOSIS
    CHROMATIN, 2923
LEUKEMIA, LYMPHOBLASTIC
    CHROMATIN, 2923
B-LYMPHOCYTES
    PLANT AGGLUTININS, 2724*
T-LYMPHOCYTES
    LIPOPOLYSACCHARIDES, 2724*
    PLANT AGGLUTININS, 2724*
NEUROBLASTOMA
    REGULATION, 2922
VIRUS, KIRSTEN MURINE SARCOMA-LEUKEMIA
    GLUCOSE, 2-DEOXY-, 2613
VIRUS, POLYOMA
    DNA REPLICATION, 2624
VIRUS, RNA HAMSTER SARCOMA
    DNA REPLICATION, 2624

PSEUDOTUMORS, INFLAMMATORY
    SEE FIBROMA

PSORIASIS
    LARYNGEAL NEOPLASMS
        IMMUNOSUPPRESSION, 2858*

PULMONARY EMPHYSEMA
    NEOPLASMS
        ACID-BASE EQUILIBRIUM, 2967

PULMONARY NEOPLASMS
    SMOKING
        EPIDEMIOLOGY, REVIEW, 2423*

PURPURA, THROMBOPENIC
    ESOPHAGEAL NEOPLASMS
        PREDNISOLONE, 2834*

3,5-PYRANZOLIDINEDIONE, 4-BUTYL-1,2-
DIPHENYL-
    LEUKEMIA
        REVIEW, 2447*

PYRIDOXAL, 5-(DIHYDROGEN PHOSPHATE)-
HEPATOMA
        TYROSINE AMINOTRANSFERASE, 2962

PYRIDOXOL
    HEPATOMA
        VITAMIN B DEFICIENCY, 2962
    TERATOGENS
        DEPRESSION, 2556*

PYRROLIDINE, 1-NITROSO-
3-PYRROLIDINOL, 1-NITROSO-
        METABOLIC DEGRADATION, RAT, 2487

3-PYRROLIDINOL, 1-NITROSO-
    PYRROLIDINE, 1-NITROSO-
        METABOLIC DEGRADATION, RAT, 2487

QUINOLINE, 4-NITRO-, 1-OXIDE
    KIDNEY NEOPLASMS
        DNA REPAIR, 2493
    LIVER NEOPLASMS
        DNA REPAIR, 2493
    LUNG NEOPLASMS
        DNA REPAIR, 2493

RADIATION
    BONE NEOPLASMS
        REVIEW, 2410
    CARCINOMA, EPIDERMOID
        REVIEW, 2410
    LEUKEMIA
        GROWTH, 2720*
        REVIEW, 2447*
    MUTAGENS
        AZAGUANINE RESISTANCE, HAMSTER,
            2479
        DOSE RESPONSE, HAMSTER, 2479
    RADIOBIOLOGICAL PREDICTIONS
        TWO-COMPONENT MODEL, 2573*
    SARCOMA, OSTEOGENIC
        REVIEW, 2410
    SKIN NEOPLASMS
        CARCINOMA, BASAL CELL, 2857*

RADIATION INJURIES, EXPERIMENTAL
    CHROMOSOME ABERRATIONS
        RADIATION, IONIZING, MOUSE, 2567

RADIATION, IONIZING
    ADENOCARCINOMA
        CASE REPORT, 2565
    BREAST NEOPLASMS
        ADENOCARCINOMA, 2565
        CASE REPORT, 2565
    CARBAMIC ACID, ETHYL ESTER
        COCARCINOGENIC ACTIVITY, 2570*
        COCARCINOGENIC ACTIVITY, RAT,
            2569*
    CERVIX NEOPLASMS
        EPIDEMIOLOGY, SOCIOECONOMIC
            FACTORS, 2888
    CHROMOSOME ABERRATIONS
        CLASTOGENIC MODEL, FISH, 2562
        RADIATION INJURIES, MOUSE, 2567*
    COBALT RADIOISOTOPES
        CHROMOSOME ABERRATIONS, 2561
    ESTRONE
        FSH, 2563
        LH, 2563
    IGA
        STRESS, 2677
    IGG
        STRESS, 2677
    IGM
        STRESS, 2677
    IMMUNOGLOBULINS
        STRESS, 2677
    INTESTINAL NEOPLASMS
        PRECANCEROUS CONDITIONS, 2571*
    LEUKEMIA
        CARBAMIC ACID, ETHYL ESTER, 2569*
    LIVER NEOPLASMS
        CARBAMIC ACID, ETHYL ESTER, 2570*
        DNA REPAIR, 2568*
    LUNG NEOPLASMS
        CARBAMIC ACID, ETHYL ESTER, 2570*
    LYMPHATIC NEOPLASMS
        CARBAMIC ACID, ETHYL ESTER, 2570*
    THYROID NEOPLASMS
        ADENOCARCINOMA, 2934
        ADENOMA, 2934
        CARCINOMA, PAPILLARY, 2934
        CYSTADENOMA, PAPILLARY, 2934

,2598

ISOLATION AND CHARACTERIZATION,2951
VIRUS-LIKE PARTICLES, 2951
VIRUS, AVIAN INFECTIOUS ANEMIA
ISOLATION AND CHARACTEPIZATION,
2653*
VIRUS, AVIAN MYELOBLASTOSIS
DNA REPLICATION, 2986*
VIRUS, AVIAN SPLEEN NECROSIS
ISOLATION AND CHARACTERIZATION,
2653*
VIRUS, C-TYPE PORCINE
VIRUS REPLICATION, 2590
VIRUS, FELINE LEUKEMIA
ANTIGENIC DETERMINANTS, 2589
VIRUS, FELINE LEUKEMIA SARCOMA
ANTIGENIC DETERMINANTS, 2589
VIRUS, GIBBON LEUKEMIA
ANTIGENIC DETERMINANTS, 2589
VIRUS, RAUSCHER MURINE LEUKEMIA
ANTIGENIC DETERMINANTS, 2589
VIRUS, RD 114
ANTIGENIC DETERMINANTS, 2589
VIRUS, RETICULOENDOTHELIOSIS
ISOLATION AND CHARACTERIZATION,
2653*
VIRUS, ROUS SARCOMA
REVIEW, 2435*
VIRUS, SIMIAN SARCOMA
ANTIGENIC DETERMINANTS, 2589

RHABDOMYOSARCOMA
BRAIN NEOPLASMS
CASE REPORT, 2804*
PRECANCEROUS CONDITIONS, 2804*
EPIDEMIOLOGY
RACIAL FACTORS, CHILD, 2883
LIVER NEOPLASMS
MONOCROTALINE, 2529*
RETRONECINE, 3,8-DIDEHYDRO-, 2529*
LUNG NEOPLASMS
CASE REPORT, 2842*
ORBITAL NEOPLASMS
ULTRASTRUCTURAL STUDY, 2767

RHIZOPUS
MYCOTOXINS
FOOD CONTAMINATION, 2456

RIBONUCLEOSIDES
VIRUS, FRIEND MURINE LEUKEMIA
CELL TRANSFORMATION, NEOPLASTIC,
2650*
SPLENOMEGALY, 2650*

RIBOSOMES
HEPATOMA
BUTYRIC ACID, 2-AMINO-4-(ETHYLTHIO)
-, 2792

RNA
ACETAMIDE, N-(ACETYLOXY)-N-9H-FLUOREN-
2-YL-
BINDING SITES, 2402
ACETAMIDE, N-FLUOREN-2-YL-
BINDING SITES, 2402
ACETOHYDROXAMIC ACID, N-FLUOREN-2-YL-
BINDING SITES, 2402
AFLATOXIN B1

ACETAMIDE, THIO-, 2457
DNA, REPLICATION
  REVIEW, 2422
HELA CELLS
  POLY A, 2936
MICROSOMES, LIVER
  ACETAMIDE, N-FLUOREN-2-YL-, 2515*
PITUITARY NEOPLASMS
  SUBCELLULAR FRACTURES, 2987*
THYMIDINE
  PHORBOL 12,13-DIMYRISTATE, 2462

RNA, MESSENGER
CADMIUM
  METALLOPROTEINS, 2520*
HELA CELLS
  METHYLATION, 2937
HEPATOMA
  METHYLATION, 2988*
INTERFERON
  ANTIGENIC DETERMINANTS, 2747*
PARATHYROID NEOPLASMS
  HYPERPLASIA, 2966
VIRUS, HERPES SIMPLEX 1
  CYTOSINE, 1-BETA-D-
    ARABINOFURANOSYL-, 2595
  ISOLATION AND CHARACTERIZATION,
    HAMSTER, 2595

RNA POLYMERASE
DNA, REPLICATION
  REVIEW, 2422

RNA REPLICATION
VIRUS, KIRSTEN MURINE SARCOMA-LEUKEMIA
  GLUCOSE, 2-DEOXY-, 2613

RNA, RIBOSOMAL
CARCINOMA, EHRLICH TUMOR
  PACTAMYCIN, 2938
  PROTEINS, 2938

RNA, TRANSFER, METHYLTRANSFERASES
HEPATOMA
  BUTYRIC ACID, 2-AMINO-4-(ETHYLTHIO)
    -, 2952

RNA, VIRAL
VIRUS, ADENO 2
  DNA-RNA HYBRIDIZATION, 2576, 2577
  ISOLATION AND CHARACTERIZATION,
    2576, 2577
  VIRUS REPLICATION, 2577
VIRUS, AVIAN MYELOBLASTOSIS
  DNA-RNA HYBRIDIZATION, CHICKEN,
    2584
VIRUS, FRIEND MURINE LEUKEMIA
  VIRUS-LIKE PARTICLES, 2602
VIRUS, GRAFFI HAMSTER LEUKEMIA
  ISOLATION AND CHARACTERIZATION,
    HAMSTER, 2588
VIRUS, HARVEY MURINE SARCOMA
  ISOLATION AND CHARACTERIZATION,
    HAMSTER, 2588
VIRUS, HERPES SIMPLEX 1
  ISOLATION AND CHARACTERIZATION,
    2642*
  POLY A, 2646*

VIRUS, KIRSTEN MURINE SARCOMA
  ISOLATION AND CHARACTERIZATIO,
    MOUSE, 2614
  ISOLATION AND CHARACTERIZATION,
    HAMSTER, 2588
  VIRUS, MURINE ERYTHROBLASTOSIS,
    2614
VIRUS, MOLONEY MURINE SARCOMA
  ISOLATION AND CHARACTERIZATION,
    HAMSTER, 2588
VIRUS, MURINE ERYTHROBLASTOSIS
  ISOLATION AND CHARACTERIZATIO,
    MOUSE, 2614
VIRUS, SV40
  DNA-RNA HYBRIDIZATION, 2638
  METHYLATION, 2659*
  TEMPERATURE-DEPENDENT DEGRADATIO
    2661*
  VIRUS REPLICATION, 2638

SALMONELLA TYPHIMURIUM
IMIDAZOLE-1-ETHANOL, 2-METHYL-5-NITR
  MUTAGENESIS, 2531*
2-IMIDAZOLIDINONE, 1-(5-NITRO-2-
  THIAZOLYL)-
  MUTAGENESIS, 2531*
SMOKING
  MUTAGENIC ACTIVITY, 2497

SARCOIDOSIS
LIVER NEOPLASMS
  GLOMERULONEPHRITIS, 2850*
  PRECANCEROUS CONDITIONS, 2850*
LUNG NEOPLASMS
  GLOMERULONEPHRITIS, 2850*
  PRECANCEROUS CONDITIONS, 2850*
SPLENIC NEOPLASMS
  GLOMERULONEPHRITIS, 2850*
  PRECANCEROUS CONDITIONS, 2850*

SARCOMA
ANTIGENS, NEOPLASM
  IMMUNITY, CELLULAR, 2728*
  LIVER REGENERATION, 2691
BENZ(A)ANTHRACENE, 7,12-DIMETHYL-
  CARCINOGENIC POTENTIAL, MOUSE,
    2468
BENZO(A)PYRENE
  ANTIGENS, 2475
CELL LINE
  CHROMOSOMES, 2970*
  VIRUS-LIKE PARTICLES, 2970*
CHOLANTHRENE, 3-METHYL-
  ANTIGENS, 2475
  ANTIGENS, NEOPLASM, 2691, 2755*
  GROWTH, 2976*
  HISTOLOGICAL STUDY, 2976*
GROWTH, 2725*
IMMUNITY, CELLULAR
  LIVER REGENERATION, 2691
LUNG NEOPLASMS
  CASE REPORT, 2842*
LYMPHOCYTES
  IMMUNITY, CELLULAR, 2712*
PARANASAL SINUS NEOPLASMS
  OCCUPATIONAL HAZARD, 2894
SPLENIC NEOPLASMS
  UREA, N-NITROSO-N-PROPYL-, 2491

CLINICAL ATTRIBUTES, REVIEW, 2440*
HISTOLOGICAL STUDY, 2820*
NEOPLASM METASTASIS, 2821*
PHAGOCYTOSIS, 2440*
ULTRASTRUCTURAL STUDY, 2818*
CELL DIFFERENTIATION
CEREBROSPINAL FLUID, 2813*
CELL TRANSFORMATION, NEOPLASTIC
ULTRASTRUCTURAL STUDY, 2813*
CHROMOSOME ABERRATIONS
REVIEW, 2444*
EPIDEMIOLOGY
RACIAL FACTORS, CHILD, 2883
REVIEW, 2419
GASTROINTESTINAL NECPLASMS
REVIEW, 2443*
GENETICS
REVIEW, 2444*
HISTIOCYTES
ULTRASTRUCTURAL STUDY, 2782
INTESTINAL NEOPLASMS
UREA, 1-BUTYL-1-NITROSO-, 2489
LYMPHOCYTES
CONCANAVALIN A, 2760*
B-LYMPHOCYTES
IGG, 2782
IGM, 2782
IMMUNOGLOBULINS, 2782
ULTRASTRUCTURAL STUDY, 2782
T-LYMPHOCYTES
ROSETTE FORMATION, 2782
ULTRASTRUCTURAL STUDY, 2782
NERVOUS SYSTEM NEOPLASMS
CASE REPORT, 2815*, 2816*
CLINICOPATHOLOGIC STUDY, 2816*
HISTOLOGICAL STUDY, 2812*
HISTOLOGY, REVIEW, 2441*
ULTRASTRUCTURAL STUDY, 2814*, 2817*
UREA, ETHYL NITROSO-, 2814*
PRECANCEROUS CONDITIONS
GASTROENTEROSTOMY, 2830*
TRANSPLANTATION, HOMOLOGOUS
CASE REPORT, 2807*
VIRUS, CYTOMEGALO
TRANSPLANTATION, HOMOLOGOUS, 2807*

SARCOMA, YOSHIDA
DNA REPLICATION
DNA NUCLEOTIDYLTRANSFERASES, 2929
THYMIDINE INCORPORATION, RAT, 2929
THYMIDINE KINASE, 2929

SCHISTOSOMAISIS
HEPATOMA
EPIDEMIOLOGY, 2901*

SCROTUM
NEOPLASMS
LUBRICATING OIL, 2431*
OCCUPATIONAL HAZARD, REVIEW, 2431*
SOOT, 2431*

SEMINOMA
SEE DISGERMINOMA

SERUM ALBUMIN
CHOLINE
119*   BINDING SITES, ANTIBODY, 2680

SERUM ALBUMIN, BOVINE
  ACETIC ACID, CADMIUM SALT
    LYSOSOMES, KIDNEY, LIVER, 2519*
  MULTIPLE MYELOMA
    ANTIBODIES, 2681

SKIN NEOPLASMS
  BENZO(A)PYRENE
    BENZENE, 1-CHLORO-2,4-DINITRO-,
      2545*
  CARBAMIC ACID, ETHYL ESTER
    RADIATION, IONIZING, 2569*
  CARCINOMA, BASAL CELL
    EPIDEMIOLOGY, 2857*
    IMMUNITY, CELLULAR, 2671
    RADIATION, 2857*
    ULTRAVIOLET RAYS, 2411
  CARCINOMA, EPIDERMOID
    EPIDEMIOLOGY, FISH, 2895
    INCIDENCE, HAMSTER, 2971*
    ULTRAVIOLET RAYS, 2411
  CYLINDROMA
    GENETICS, 2854*
  DIBENZ(A,H)ANTHRACENE
    OCCUPATIONAL HAZARD, REVIEW, 2431*
  FIBROSARCOMA
    UREA, N-NITROSO-N-PROPYL-, 2491
  HEMANGIOMA
    CASE REPORT, 2843*
    PERICARDIAL EFFUSION, 2843*
  IMMUNOSUPPRESSION
    GROWTH, 2717*
  T-LYMPHOCYTES
    GROWTH, 2717*
  MELANOMA
    EPIDEMIOLOGY, 2855*, 2890
    EPIDEMIOLOGY, NIGERIA, 2884
    GROWTH, 2717*
    PRECANCEROUS CONDITIONS, 2800
    VITILIGO, 2800
  MYOSARCOMA
    UREA, N-NITROSO-N-PROPYL-, 2491
  PRECANCEROUS CONDITIONS
    ACETIC ACID, 2462
    CASE REPORT, 2853*
    PHORBOL 12,13-DIMYRISTATE, 2462
  SMOKING
    ARYL HYDROCARBON HYDROXYLASES, 2498
  VIRUS, HERPES
    ULTRASTRUCTURAL STUDY, 2645*
    VIRUS-LIKE PARTICLES, 2645*
  VIRUS, HERPES SIMPLEX 2
    CYTOSINE, 1-BETA-D-
      ARABINOFURANOSYL-, 2599
    URIDINE, 2'-DEOXY-5-IODO-, 2599

SMOKING
  ASBESTOS
    COCARCINOGENIC EFFECT, 2496
  ESOPHAGEAL NEOPLASMS
    EPIDEMIOLOGY, INDIA, 2495
  LARYNGEAL NEOPLASMS
    EPIDEMIOLOGY, INDIA, 2495
  MOUTH NEOPLASMS
    EPIDEMIOLOGY, INDIA, 2495, 2893
  PHARYNGEAL NEOPLASMS
    EPIDEMIOLOGY, INDIA, 2495, 2893

PULMONARY NEOPLASMS
  EPIDEMIOLOGY, REVIEW, 2423*
SALMONELLA TYPHIMURIUM
  MUTAGENIC ACTIVITY, 2497
SKIN NEOPLASMS
  ARYL HYDROCARBON HYDROXYLASES,
    2498

SODIUM CHLORIDE
  PROLINE
    NITROSAMINES, 2546*
  VIRUS, ROUS SARCOMA
    VIRUS CONCENTRATION, 2656*

SODIUM FLUORIDE
  VIRUS, SV40
    ADENYL CYCLASE, 2663*

SOMATOTROPIN
  LUNG NEOPLASMS
    CARCINOMA, 2965
    CARCINOMA, BRONCHOGENIC, 2965
    CARCINOMA, EPIDERMOID, 2965

SPERMATOZOA
  ACTINOMYCIN D
    ABNORMALITIES, MOUSE, 2452
  AMINOPTERIN
    ABNORMALITIES, MOUSE, 2452
  BENZO(A)PYRENE
    ABNORMALITIES, 2452
  1,4-BUTANEDIOL DIMETHYLSULFONATE
    ABNORMALITIES, MOUSE, 2452
  CHOLANTHRENE, 3-METHYL-
    ABNORMALITIES, MOUSE, 2452
  COLCHICINE
    ABNORMALITIES, MOUSE, 2452
  CYCLOPHOSPHAMIDE
    ABNORMALITIES, MOUSE, 2452
  DNA REPLICATION
    DNA NUCLEOTIDYTRANSFERASES, 2984*
    HYBRIDIZATION, 2984*
  GRISEOFULVIN
    ABNORMALITIES, 2452
  METHANESULFONIC ACID, ETHYL ESTER
    ABNORMALITIES, 2452
  METHANESULFONIC ACID, METHYL ESTER
    ABNORMALITIES, 2452
  MITOMYCIN C
    ABNORMALITIES, 2452
  PHOSPHINE OXIDE, TRIS(2-METHYL-1-
    AZIRIDINYL)-
    ABNORMALITIES, 2452
  PHOSPHINE SULFIDE, TRIS(1-AZIRIDINYL)-
    ABNORMALITIES, 2452
  PHOSPHORIC ACID, TRIMETHYL ESTER
    ABNORMALITIES, MOUSE, 2452
  UREA, HYDROXY-
    ABNORMALITIES, MOUSE, 2452
  VINCALEUKOBLASTINE, SULFATE
    ABNORMALITIES, MOUSE, 2452

SPERMINE
  ADENOSINE CYCLIC 3',5' MONOPHOSPHATE
    CARBOXY-LYASES, 2958
  DEXAMETHASONE
    CARBOXY-LYASES, 2958

8

YL-, 1-NAPHTHYL

-METHYL-3-

IMETHYL-3-NITRO-

L-3-NITROSO-,

YL-3-NITROSC-,

PCNSE, 2719*

RCPYL-, 2491

CELLS, 2734*
2734*

2850*
ICNS, 2850*

RCPYL-, 2491

UKEMIA
50*

TY, MOUSE,

TY, MOUSE,.

PHA'-CIETHYL-

REPRODUCTIVE

REPRODUCTIVE

ALPHA, ALPHA'-

UREA, N-NITROSO-N-PRCPYL-, 2491
ADENOMA
UREA, N-NITROSO-N-PRCPYL-, 2491
ALKALINE PHOSPHATASE
SERUM LEVELS, 2955
CARCINOEMBRYONIC ANTIGEN
DIAGNOSIS AND PROGNOSIS, REVIEW,
2416
CARCINOMA, SCIRRHOUS
CELL CYCLE KINETICS, 2969*
HEMANGIOENDOTHELIOMA
UREA, N-NITROSC-N-PROPYL-, 2491
PAPILLOMA
UREA, 1-BUTYL-1-NITROSO-, 2489

STREPTOCOCCUS
IGA
PROTEIN UPTAKE STUDY, MYELOMA, 2739*
IGG
PROTEIN UPTAKE STUDY, MYELOMA, 2739*
MYELOMA PROTEINS
UPTAKE, 2739*

STREPTOCOCCUS FAECALIS
DIMETHYLAMINE
DIMETHYLAMINE, N-NITROSO-, 2484

STRESS
IGA
RADIATION, IONIZING, 2677
IGG
RADIATION, IONIZING, 2677
IGM
RADIATION, IONIZING, 2677
IMMUNOGLOBULINS
RADIATION, IONIZING, 2677

STRONTIUM RADIOISOTOPES
YTTRIUM RADIOISOTOPES
ACTIVITY RATIO, AGE, SEX FACTORS,
2566

SUCCINAMIC ACID, 3-AMINC-N-(ALPHA-
CARBOXYPHENETHYL)-
SWEETENING AGENTS
CARCINOGENIC ACTIVITY, MOUSE, 2517*

SULFHYDRYL COMPOUNDS
4H-FURO(3,2-C)PYRAN-2(6H)-ONE, 4-
HYDROXY-
CARCINOGENIC POTENTIAL, REVIEW, 2401
HEPATOMA
ENDOPLASMIC RETICULUM, 2924
MICROSOMES, LIVER, 2924
POLYRIBOSOMES, 2924

SWEAT GLAND NEOPLASMS
ADENOMA
CASE REPORT, 2874*

SWEETENING AGENTS
SUCCINAMIC ACID, 3-AMINO-N-(ALPHA-
CARBOXYPHENETHYL)-
CARCINOGENIC ACTIVITY, MOUSE, 2517*

SYNDROME, SCHWARTZ-BARTTER
LUNG NEOPLASMS
CELL LINE, 2973*

SYNOVIOMA
  ANTIBODIES, NEOPLASM
    IMMUNE SERUMS, 2678
  IMMUNE COMPLEX DISEASE
    GLOMERULONEPHRITIS, 2710*
TERATOGENS
  PYRIDOXOL
    DEPRESSION, 2556*

TERATOID TUMOR
  EPIDEMIOLOGY
    RACIAL FACTORS, CHILD, 2883
  FETAL GLOBULINS
    CASE REPORT, 2841*
  GALACTOSIDASES
    CELL MEMBRANE, 2882
  HYALURONIC ACID
    CELL MEMBRANE, 2882
  HYALURONIDASE
    CELL MEMBRANE, 2882
  MONOSACCHARIDES
    CELL MEMBRANE, 2882
  NEURAMINIDASE
    CELL MEMBRANE, 2882
  OVARIAN NEOPLASMS
    CASE REPORT, 2864*
    HEMATOPOIETIC STEM CELLS, 2796
    MIGRATION INHIBITORY FACTOR, 2667
    ULTRASTRUCTURAL STUDY, MOUSE, 2796
    VIRUS-LIKE PARTICLES, 2796
  UROGENITAL NEOPLASMS
    HERMOPHRODITISM, 2870*

TERATOMA, EMBRYONAL
  SEE TERATOID TUMOR

TESTICULAR NEOPLASMS
  LEYDIG CELL TUMOR
    CASE REPORT, 2871*

TESTOSTERONE
  LEYDIG CELL TUMOR
    GONADOTROPINS, 2472

TESTOSTERONE, PHENYLPROPIONATE
  DNA REPLICATION
    PROSTATE, 2964
  THYMIDINE KINASE
    PROSTATE, 2964

TETRAHYDROFOLATE DEHYDROGENASE
  HYBRID CELLS
    THYMIDINE KINASE DEFICIENCY, 2776

THEOPHYLLINE
  ARYL HYDROCARBON HYDROXYLASES
    ENZYME INDUCTION, 2499
  VIRUS, SV40
    ADENYL CYCLASE, 2663*

THEOPHYLLINE, COMPOUND WITH
  ETHYLENEDIAMINE (2:1)
  ARYL HYDROCARBON HYDROXYLASES
    ENZYME INDUCTION, 2499

THIOCYANIC ACID, SODIUM SALT
  ADENOMA
    ANTINEOPLASTIC EFFECT, MOUSE, 2481

THORIUM
  BONE NEOPLASMS
    REVIEW, 2410

THROMBOSIS
  ANGIOSARCOMA
    ANGIOMATOSIS, 2768

THYMECTOMY
  LEUKEMIA
    CELL DIVISION, 2696
  B-LYMPHOCYTES
    LIPOPOLYSACCHARIDES, 2697
  T-LYMPHOCYTES
    CONCANAVALIN A, 2697
    PLANT AGGLUTININS, 2697

THYMIDINE
  DNA
    AUTORADIOGRAPHY, 2911*
    PHORBOL 12,13-DIMYRISTATE, 2462
  HYBRID CELLS
    THYMIDINE KINASE DEFICIENCY, 2776
  RNA
    PHORBOL 12,13-DIMYRISTATE, 2462

THYMIDINE KINASE
  CORTICOSTERONE
    PROSTATE, 2964
  ESTRADIOL
    PROSTATE, 2964
  PROGESTERONE, 6-CHLORO-6-DEHYDRO-17-
    ALPHA-HYDROXY-
    1,2ALPHA-METHYLENE-, ACETATE, 296
  SARCOMA, YOSHIDA
    DNA REPLICATION, 2929
  TESTOSTERONE, PHENYLPROPIONATE
    PROSTATE, 2964

THYMOMA
  LYMPHOCYTES
    IMMUNITY, CELLULAR, 2712*
  THYMUS NEOPLASMS
    TRANSPLANTATION, HOMOLOGOUS, 2696
  ULTRASTRUCTURAL STUDY, 2877*

THYMUS NEOPLASMS
  LYMPHOMA
    UREA, 1-BUTYL-1-NITROSO-, 2489
  LYMPHOSARCOMA
    UREA, N-NITROSO-N-PROPYL-, 2491
  PRECANCEROUS CONDITIONS
    TRANSPLANTATION, HOMOLOGOUS, 2696
  THYMOMA
    TRANSPLANTATION, HOMOLOGOUS, 2696

THYROID NEOPLASMS
  ADENOCARCINOMA
    DNA, 2934
    IMMUNITY, CELLULAR, 2712*
    RADIATION, IONIZING, 2934
    URACIL, 6-METHYL-2-THIO-, 2934
  ADENOMA
    DNA, 2934
    GOITER, 2876*
    NCI CO2186, 2555*
    RADIATION, IONIZING, 2934
    URACIL, 6-METHYL-2-THIO-, 2934
    UREA, N-NITROSO-N-PROPYL-, 2491

                                    MELANOMA
                                        ENZYMATIC ACTIVITY, PLASMA, 2954
                                    NEOPLASM METASTASIS
                                        ENZYMATIC ACTIVITY, FLASMA, 2954
                                    NEOPLASMS
CNS, 2875*   ENZYMATIC ACTIVITY, FLASMA, 2954
Y, 2799

                                TRANSPLANTATION ANTIGENS
                                    SEE HISTOCOMPATIBILITY ANTIGENS

 2934
HIO-, 2934   TRANSPLANTATION, HETEROLOGOUS
                                    HELA CELLS
                                        NEOPLASM METASTASIS, 2731*
 2934   IMMUNOSUPPRESSION
HIO-, 2934   NEOPLASM METASTASIS, 2731*
                                    NEOPLASMS
                                        ATHYMIC NUDE MICE, 2975*
 2934   GROWTH, 2975*
HIO-, 2934   NEUROBLASTOMA
                                        CELL LINE, 2912
Y, RAT, 2574*   VIRUS, SV40
                                        ATHYMIC NUDE MICE, 2975*

Y, 2799   TRANSPLANTATION, HOMOLOGOUS
                                    ADENOCARCINOMA
CNS, 2555*   T-LYMPHOCYTES, 2771
                                    BRAIN NEOPLASMS
                                        GRANULOMA, 2806*
CNS, 2875*   HODGKIN'S DISEASE, 2806*
                                    CYCLOPHOSPHAMIDE
                                        IMMUNOSUPPRESSION, 2668
                                    HEPATOMA
NAL CELL, 2794   OXIDOREDUCTASES, 2960
                                    LEUKEMIA
 2495   GRAFT REJECTION, 2720*
                                        HISTOCOMPATIBILITY ANTIGENS, 2720*
 2495, 2893   T-LYMPHOCYTES, 2696
                                    LEUKEMIA, MYELOCYTIC
Y, 2553*   GLYCOPROTEINS, 2914
                                    LYMPHOMA
 2495, 2893   GRAFT REJECTION, 2684
                                        IMMUNITY, CELLULAR, 2684
                                    MAMMARY NEOPLASMS, EXPERIMENTAL
                                        ADENOCARCINOMA, 2948
                                    PLASMACYTOMA
                                        T-LYMPHOCYTES, 2771
                                    SARCOMA
                                        T-LYMPHOCYTES, 2771
                                    SARCOMA, MAST CELL
                                        IMMUNE RESPONSE, 2670
                                    SARCOMA, RETICULUM CELL
 3   CASE REPORT, 2807*
                                        VIRUS, CYTOMEGALO, 2807*
                                    THYMUS NEOPLASMS
                                        PRECANCEROUS CONDITIONS, 2696
                                        THYMOMA, 2696
                                    VIRUS, SV40
                                        ATHYMIC NUDE MICE, 2975*

CNS, 2539*   TRANSPLANTATION IMMUNOLOGY
                                    BLADDER NEOPLASMS
                                        AUTOIMMUNE DISEASES, 2714*
                                    BRAIN NEOPLASMS
PLASMA, 2954   ASTROCYTOMA, 2418
                                    CHOLANTHRENE, 3-METHYL-
                                        HISTOCOMPATIBILTY ANTIGENS, 2666
                                    EPENDYMOMA
PLASMA, 2954   CHOLANTHRENE, 3-METHYL-, 2418

FIBROMA
    VIRUS, BOVINE PAPILLOMA, 2665
FIBROSARCOMA
    VIRUS, SV40, 2689
LEUKEMIA
    IMMUNITY, CELLULAR, 2707*
    LYMPHOCYTES, 2707*
LYMPHOMA
    MAGNESIUM DEFICIENCY, 2723*
VIRUS, C-TYPE MURINE RNA TUMOR
    GRAFT VS HOST REACTION, 2673
VIRUS, MURINE LEUKEMIA
    HISTOCOMPATIBILITY ANTIGENS, 2685
VIRUS, RAUSCHER MURINE LEUKEMIA
    HISTOCOMPATIBILITY ANTIGENS, 2685
VIRUS, SV40
    CHOLANTHRENE, 3-METHYL-, 2689
    HISTOCOMPATIBILITY ANTIGENS, 2685
    TUBERCULIN, 2689

TRICHOMONAS VAGINALIS
    CERVIX NEOPLASMS
        DIETHYLAMINE, N-NITROSO-, 2905*
        DIMETHYLAMINE, N-NITROSO-, 2905*
        DIPROPYLAMINE, N-NITROSO-, 2905*
        EPIDEMIOLOGY, AFRICA, 2905*

TRYPTOPHAN
    BLADDER NEOPLASMS
        CARCINOMA, TRANSITIONAL CELL, 2794
    CARCINOGENIC METABOLITE
        MOUSE, 2557*

TUBERCULIN
    VIRUS, SV40
        IMMUNITY, CELLULAR, 2689
        TRANSPLANTATION IMMUNOLOGY, 2689

TUMORLET
    LUNG NEOPLASMS
        PRECURSOR CELL, 2787
        ULTRASTRUCTURAL STUDY, 2787

TYROSINASE
    MELANOMA
        ACTINOMYCIN D, 2963
        COLCHICINE, 2963
        CYCLOHEXIMIDE, 2963
        HORMONES, 2963

TYROSINE AMINOTRANSFERASE
    HEPATOMA
        CORTISOL, 2962
        PYRIDOXAL, 5-(DIHYDROGEN PHOSPHATE)-, 2962

ULTRAVIOLET RAYS
    ESCHERICHIA COLI
        DNA REPAIR, 2572*
        ENDONUCLEASES, 2572*
    MELANOMA
        EPIDEMIOLOGY, 2890
    MICROCOCCUS LUTEUS
        DNA REPAIR, 2572*
        ENDONUCLEASES, 2572*
    MUTAGENS
        AZAGUANINE RESISTANCE, HAMSTER, 2479
        DOSE RESPONSE, HAMSTER, 2479

SKIN NEOPLASMS
    CARCINOMA, BASAL CELL, 2411
    CARCINOMA, EPIDERMOID, 2411
    VIRUS, SV40
        MUTATION, 2630
        TEMPERATURE SENSITIVE MUTANTS, 26

URACIL, 6-METHYL-2-THIO-
    THYROID NEOPLASMS
        ADENOCARCINOMA, 2934
        ADENOMA, 2934
        CARCINOMA, PAPILLARY, 2934
        CYSTADENOMA, PAPILLARY, 2934
        HYPERPLASIA, 2934

URANIUM
    IGA
        OCCUPATIONAL HAZARD, 2677
    IGG
        OCCUPATIONAL HAZARD, 2677
    IGM
        OCCUPATIONAL HAZARD, 2677
    IMMUNOGLOBULINS
        OCCUPATIONAL HAZARD, 2677

UREA, 1-BUTYL-1-NITROSO-
    INTESTINAL NEOPLASMS
        ADENOCARCINOMA, 2489
        POLYPS, 2489
        SARCOMA, RETICULUM CELL, 2489
    LUNG NEOPLASMS
        ADENOMA, 2489
    STOMACH NEOPLASMS
        PAPILLOMA, 2489
    THYMUS NEOPLASMS
        LYMPHOMA, 2489

UREA, 1,1-DIETHYL-3-METHYL-3-NITROSO-
    HEART NEOPLASMS
        NEUROFIBROMA, 2488
    NERVOUS SYSTEM NEOPLASMS
        ASTROCYTOMA, 2488
        GLIOMA, 2488
    SPINAL CORD NEOPLASMS
        NEUROFIBROMA, 2488

UREA, 3-ETHYL-1,1-DIMETHYL-3-NITRO-
    NERVOUS SYSTEM NEOPLASMS
        GLIOMA, 2488
    SPINAL CORD NEOPLASMS
        NEUROFIBROMA, 2488

UREA, 3-ETHYL-1,1-DIMETHYL-3-NITROSO-
    HEART NEOPLASMS
        NEUROFIBROMA, 2488
    NERVOUS SYSTEM NEOPLASMS
        ASTROCYTOMA, 2488

UREA, ETHYL NITROSO-
    ESCHERICHIA COLI
        DNA REPAIR, 2480
    NERVOUS SYSTEM NEOPLASMS
        ANTIGENS, 2683
        IMMUNE SERUMS, 2683
        SARCOMA, RETICULUM. CELL, 2814*
    NEURILEMMOMA
        NEOPLASM TRANSPLANTATION, 2490
        ULTRASTRUCTURAL STUDY, 2490

A, HYDROXY-
DIETHYLAMINE, N-NITROSO-
  CARCINOGENIC ACTIVITY, RAT, 2482
SPERMATOZOA
  -ABNORMALITIES, MOUSE, 2452

A, METHYL-
LUNG NEOPLASMS
  ADENOMA, 2481
NITROUS ACID, SODIUM SALT
  COCARCINOGENIC EFFECT, MOUSE, 2481

A, METHYL NITROSO-
ESCHERICHIA COLI
  DNA REPAIR, 2480
LEUKEMIA, MONOCYTIC
  DNA REPLICATION, 2930
  NICOTINAMIDE, 2930
LUNG NEOPLASMS
  ADENOMA, 2481
NITROUS ACID, SODIUM SALT
  COCARCINOGENIC EFFECT, MOUSE, 2481

A, N-NITROSO-N-PROPYL-
ADRENAL GLAND NEOPLASMS
  ADENOMA, 2491
BRAIN NEOPLASMS
  ASTROCYTOMA, 2491
COLONIC NEOPLASMS
  ADENOCARCINOMA, 2491
  ADENOMA, 2491
  FIBROMA, 2491
  FIBROSARCOMA, 2491
  HEMANGIOENDOTHELIOMA, 2491
  HEMANGIOMA, 2491
  MYOSARCOMA, 2491
EAR NEOPLASMS
  CARCINOMA, EPIDERMOIC, 2491
  PAPILLOMA, 2491
ERYTHROLEUKEMIA
  ULTRASTRUCTURAL STUDY, RAT, 2491
INTESTINAL NEOPLASMS
  ADENOCARCINOMA, 2491
  HEMANGIOENDOTHELIOMA, 2491
KIDNEY NEOPLASMS
  NEPHROBLASTOMA, 2491
LEUKEMIA, LYMPHOBLASTIC
  ULTRASTRUCTURAL STUDY, RAT, 2491
LEUKEMIA, MYELOBLASTIC
  ULTRASTRUCTURAL STUDY, RAT, 2491
LEUKEMIA, SUBLEUKEMIC
  ULTRASTRUCTURAL STUDY, RAT, 2491
LIVER NEOPLASMS
  HEMANGIOENDOTHELIOMA, 2491
  HEPATOMA, 2491
LUNG NEOPLASMS
  ADENOMA, 2491
MAMMARY NEOPLASMS, EXPERIMENTAL
  ADENOCARCINOMA, 2491
  ADENOFIBROMA, 2491
OVARIAN NEOPLASMS
  FIBROSARCOMA, 2491
PANCREATIC NEOPLASMS
  ADENOMA, 2491
SKIN NEOPLASMS
  FIBROSARCOMA, 2491
  MYOSARCOMA, 2491
SPLENIC NEOPLASMS

HEMANGIOENDOTHELIOMA, 2491
  SARCOMA, 2491
STOMACH NEOPLASMS
  ADENOCARCINOMA, 2491
  ADENOMA, 2491
  HEMANGIOENDOTHELIOMA, 2491
THYMUS NEOPLASMS
  LYMPHOSARCOMA, 2491
THYROID NEOPLASMS
  ADENOMA, 2491
UTERINE NEOPLASMS
  ADENOCARCINOMA, 2491
  LEIOMYOMA, 2491

UREA, 1,1,3-TRIETHYL-3-NITROSO-
HEART NEOPLASMS
  NEUROFIBROMA, 2488
NERVOUS SYSTEM NEOPLASMS
  ASTROCYTOMA, 2488
  GLIOMA, 2488
SPINAL CORD NEOPLASMS
  NEUROFIBROMA, 2488

UREA, 1,1,3-TRIMETHYL-3-NITROSO-
HEART NEOPLASMS
  NEUROFIBROMA, 2488
NERVOUS SYSTEM NEOPLASMS
  ASTROCYTOMA, 2488
  GLIOMA, 2488
SPINAL CORD NEOPLASMS
  NEUROFIBROMA, 2488

URETERAL NEOPLASMS
CARCINOMA
  COMPLEMENT, 2741*
ESTRADIOL
  ANDROST-5-ENE-3,17-DIOL, 2471
  SPECIFIC RECEPTOR BINDING, 2471

URETHANE
SEE CARBAMIC ACID, ETHYL ESTER

URETHRITIS
PROSTATIC NEOPLASMS
  EPIDEMIOLOGY, 2687

URIC ACID
LEUKEMIA, LYMPHOBLASTIC
  NEPHRITIS, INTERSTITIAL, 2849*

URIDINE, 5-BROMO-2'-DEOXY-
MELANOMA
  PLASMINOGEN ACTIVATORS, 2977*

URIDINE, 2'-DEOXY-5-FLUORO-
VIRUS, FRIEND MURINE LEUKEMIA
  INTERFERON, 2604

URIDINE, 2'-DEOXY-5-IODO-
BRAIN NEOPLASMS
  VIRUS, HERPES SIMPLEX 2, 2599
CARCINOMA, EHRLICH TUMOR
  CELL CYCLE KINETICS, 2898
CARCINOMA, EHRLICH TUMOR
  DNA REPLICATION, 2898
EYE NEOPLASMS
  VIRUS, HERPES SIMPLEX 2, 2599
LIVER NEOPLASMS

VIRUS, HERPES SIMPLEX 2, 2599   VIR
  SKIN NEOPLASMS
    VIRUS, HERPES SIMPLEX 2, 2599
  VIRUS, FRIEND MURINE LEUKEMIA
    INTERFERON, 2604
  VIRUS, HERPES SIMPLEX 2
    VIRUS REPLICATION, 2599
  VIRUS, MOLONEY MURINE SARCOMA
    IMMUNITY, CELLULAR, 2669

URIDINE KINASE
  SEE PHOSPHOTRANSFERASES, ATP

UROGENITAL NEOPLASMS
  ADENOCARCINOMA
    ULTRASTRUCTURAL STUDY, 2866*
  ALCOHOLISM
    EPIDEMIOLOGY, 2892
  BENZENE, 1-FLUORO-2,4-DINITRO-
    HYPERSENSITIVITY, 2737*
  CYSTADENOCARCINOMA
    HERMOPHRODITISM, 287C*
  TERATOID TUMOR
    HERMOPHRODITISM, 287C*

UTERINE NEOPLASMS
  ADENOCARCINOMA
    UREA, N-NITROSO-N-PROPYL-, 2491
  CARCINOMA
    EPIDEMIOLOGY, GERMANY, 2902*
    HYDROXYSTEROID DEHYDROGENASES, 2961
  HAMARTOMA
    CASE REPORT, 2867*
  HISTOCOMPATIBILITY ANTIGENS
    GENETICS, 2686
  HYDROXYSTEROID DEHYDROGENASES
    CELL DIFFERENTATION, 2961
    ENZYMATIC ACTIVITY, 2961
  LEIOMYOMA
    UREA, N-NITROSO-N-PROPYL-, 2491
  VIRUS, HERPES SIMPLEX 1   VIR
    ANTIGENS, VIRAL, 2643*
    ISOLATION AND CHARACTERIZATION,
      2643*
  VIRUS HERPES SIMPLEX 2
    ANTIGENS, VIRAL, 2643*
    ISOLATION AND CHARACTERIZATION,   VIR
      2643*

VACCINES
  BURKITT'S LYMPHOMA
    VIRUS, EPSTEIN-BARR, 2412
  MAREK'S DISEASE
    COMPARATIVE VIROLOGY, REVIEW, 2412

VANILMANDELIC ACID
  MAXILLARY NEOPLASMS
    NEOPLASM RECURRENCE, LOCAL, 2764

VINCALEUKOBLASTINE, SULFATE
  SPERMATOZOA
    ABNORMALITIES, MOUSE, 2452

VINCRISTINE
  SEE LEUROCRISTINE

VINYL CHLORIDE   VIR
  SEE ETHYLENE, CHLORO-

VIRUS, AVIAN SYNCYTIAL
    VIRUS, RETICULOENDOTHELIOSIS
        HYBRIDIZATION, 2653*

VIRUS, B-TROPIC MURINE LEUKEMIA
    GENETICS
        VIRUS ACTIVATION, 2637
    HISTOCOMPATIBILITY ANTIGENS
        MILK-BORNE TRANSMISSION, 2607
        VIRUS ACTIVATION, 2637

VIRUS, BOVINE PAPILLOMA
    FIBROMA
        ANTIGENS, NEOPLASM, 2665
        IMMUNE SERUMS, 2665
        TRANSPLANTATION IMMUNOLOGY, 2665

VIRUS, C-TYPE AVIAN TUMOR
    PEPTIDES
        ANTIGENIC DETERMINANTS, 2587
        ISOLATION AND CHARACTERIZATION,
            2587

VIRUS, C-TYPE MURINE RNA TUMOR
    ANTILYMPHOCYTE SERUM
        IMMUNOSUPPRESSION, 2673
    CYCLOPHOSPHAMIDE
        IMMUNOSUPPRESSION, 2673
    GRAFT VS HOST REACTION
        TRANSPLANTATION IMMUNOLOGY, 2673
    IMMUNE SERUMS
        IMMUNOSUPPRESSION, 2673

VIRUS, C-TYPE PORCINE
    VIRUS, FELINE LEUKEMIA
        ANTIGENIC DETERMINANTS, 2590
    VIRUS, GIBBON LEUKEMIA
        ANTIGENIC DETERMINANTS, 2590
    VIRUS, KIRSTEN MURINE SARCOMA
        VIRUS, RESCUE, 2590
    VIRUS, M7 BABOON
        ANTIGENIC DETERMINANTS, 2590
    VIRUS, RAUSCHER MURINE LEUKEMIA
        ANTIGENIC DETERMINANTS, 2590
    VIRUS, RD114
        ANTIGENIC DETERMINANTS, 2590
    VIRUS REPLICATION
        REVERSE TRANSCRIPTASE, 2590

VIRUS, C-TYPE RNA TUMOR
    BREAST NEOPLASMS
        COMPARATIVE VIROLOGY, REVIEW, 2412
    LEUKEMIA, LYMPHOBLASTIC
        COMPARATIVE VIROLOGY, REVIEW, 2412
        VIRUS-LIKE PARTICLES, 2612
    LEUKEMIA, LYMPHOCYTIC
        VIRUS-LIKE PARTICLES, 2612
    LEUKEMIA, MYELOBLASTIC
        VIRUS-LIKE PARTICLES, 2612
    LEUKEMIA, MYELOCYTIC
        VIRUS-LIKE PARTICLES, 2612
    LYMPHOMA
        COMPARATIVE VIROLOGY, REVIEW, 2412
    SARCOMA
        COMPARATIVE VIROLOGY, REVIEW, 2412

VIRUS, CANINE DISTEMPER
2653* .   ENCEPHALITIS

COMPARATIVE VIROLOGY, REVIEW, 2412

VIRUS, CYTOMEGALO
  CERVIX NEOPLASMS
    PRECANCEROUS CONDITIONS, 2600
  HODGKIN'S DISEASE
    ANTIGENS, VIRAL, 2593
  SARCOMA, RETICULUM CELL
    TRANSPLANTATION, HOMOLOGOUS, 2807*
  VIRAL PROTEINS
    ULTRASTRUCTURAL STUDY, 2644*

VIRUS, EPSTEIN-BARR
  ANTIBODIES, VIRAL
    INTERPERSONAL TRANSMISSION, 2909*
  ANTIGENS, VIRAL
    IGG, 2591, 2742*
    IGM, 2742*
    IMMUNOGLOBULINS, 2591, 2742*
    ISOLATION AND CHARACTERIZATION, 2742*
    REVIEW, 2413
  BURKITT'S LYMPHOMA
    ANTIGENS, VIRAL, 2591
    COMPARATIVE VIROLOGY, REVIEW, 2412
    EPIDEMIOLOGY, AFRICA, 2891
    REVIEW, 2444*
    ULTRASTRUCTURAL STUDY, 2679
    VACCINES, 2412
  CELL LINE
    ANTIGEN-ANTIBODY REACTIONS, 2641*
    ANTIGENS, VIRAL, 2641*
    GROWTH, 2703
    LYMPHOCYTE TRANSFORMATION, 2703
  DNA, VIRAL, 2594
  HODGKIN'S DISEASE
    ANTIGENS, VIRAL, 2593
  LEUKEMIA, LYMPHOCYTIC
    ULTRASTRUCTURAL STUDY, 2679
  LYMPHATIC NEOPLASMS
    ULTRASTRUCTURAL STUDY, 2679
  LYMPHOCYTE TRANSFORMATION
    FIBROBLASTS, 2592
    MACROPHAGES, 2592
  LYMPHOCYTES, 2594
    CELL LINE, 2744*
  B-LYMPHOCYTES
    IGG, 2592
    IMMUNOGLOBULINS, 2592
    REVIEW, 2413
  T-LYMPHOCYTES
    IGG, 2592
    IMMUNOGLOBULINS, 2592
  LYMPHOSARCOMA
    ULTRASTRUCTURAL STUDY, 2679
  NASOPHARYNGEAL NEOPLASMS
    ANTIGENS, VIRAL, 2591, 2709*
    GENETICS, 2709*

VIRUS, FELINE LEUKEMIA
  ANEMIA
    HORIZONTAL TRANSMISSION, 2648*
  ANTIBODIES, VIRAL
    IMMUNE SERUMS, 2649*
    PASSIVE HEMAGGLUTINATION, 2649*
  ANTIGENS, VIRAL
    CELL MEMBRANE, 2752*
  CHROMOSOMES
    CROSSES, GENETIC, 2601

DNA, VIRAL
  CROSSES, GENETIC, 2601
  LEUKEMIA
    ANTIBODIES, VIRAL, 2649*
    REVIEW, 2444*
  LYMPHOMA
    REVIEW, 2444*
  LYMPHOSARCOMA
    HORIZONTAL TRANSMISSION, 2648*
    REVIEW, 2648*
    VACCINATION, 2648*
  REVERSE TRANSCRIPTASE
    ANTIGENIC DETERMINANTS, 2589
  VIRAL PROTEINS
    ANTIGENIC DETERMINANTS, 2589
  VIRUS, C-TYPE PORCINE
    ANTIGENIC DETERMINANTS, 2590
  VIRUS, RD114
    CROSSES, GENETIC, 2601

VIRUS, FELINE LEUKEMIA SARCOMA
  REVERSE TRANSCRIPTASE
    ANTIGENIC DETERMINANTS, 2589
  VIRAL PROTEINS
    ANTIGENIC DETERMINANTS, 2589

VIRUS, FELINE SARCOMA
  ANTIGENS, VIRAL
    CELL MEMBRANE, 2752*

VIRUS, FRIEND MURINE LEUKEMIA
  CELLS, CULTURED
    VIRUS REPLICATION, 2651*
  CONCANAVALIN A
    GLYCOPEPTIDES, 2603
    PEPTIDES, 2603
    VIRAL PROTEINS, 2603
  ERYTHROLEUKEMIA
    VIRUS, FRIEND SPLEEN-FOCUS FORMI
    2651*
  ERYTHROPOIETIN
    COLONY FORMATION, 2605
  HISTOCOMPATIBILITY ANTIGENS
    TRANSPLANTATION IMMUNOLOGY, 2685
  LEUKEMIA, LYMPHOCYTIC
    VIRUS, FRIEND SPLEEN-FOCUS FORMI
    2651*
  B-LYMPHOCYTES
    ANTIBODIES, VIRAL, 2687
    ANTIGENS, VIRAL, 2687
    IMMUNE RESPONSE, 2687
  T-LYMPHOCYTES
    ANTIBODIES, VIRAL, 2687
    ANTIGENS, VIRAL, 2687
    CONCANAVALIN A, 2651*
    IMMUNE RESPONSE, 2687
    IMMUNOSUPPRESSION, 2651*
  RIBONUCLEOSIDES
    CELL TRANSFORMATION, NEOPLASTIC,
    2650*
    SPLENOMEGALY, 2650*
  URIDINE, 2'-DEOXY-5-FLUORO-
    INTERFERON, 2604
  URIDINE, 2'-DEOXY-5-IODO-
    INTERFERON, 2604
  VIRUS-LIKE PARTICLES
    CONCANAVALIN A, 2602

IMMUNIZATION, 2675
VIRAL VACCINES, 2675
VIRAL VACCINES
IMMUNE RESPONSE, 2675

VIRUS, HERPES CANICULI
ULTRASTRUCTURAL STUDY
TUBULAR STRUCTURES, KIDNEY, 2647*
VIRAL PROTEINS
ULTRASTRUCTURAL STUDY, 2644*

VIRUS, HERPES SIMPLEX
HODGKIN'S DISEASE
IMMUNITY, CELLULAR, 2671
LEUKEMIA, LYMPHOCYTIC
IMMUNITY, CELLULAR, 2671
LYMPHOMA
IMMUNITY, CELLULAR, 2671
MULTIPLE MYELOMA
IMMUNITY, CELLULAR, 2671

VIRUS, HERPES SIMPLEX 1
ANTIGENS, VIRAL
CELL MEMBRANE, 2597
ISOLATION AND CHARACTERIZATION,
HAMSTER, 2597
CERVIX NEOPLASMS
ANTIBODIES, VIRAL, 2600
ANTIGENS, VIRAL, 2643*
ISOLATION AND CHARACTERIZATION,
2643*
PRECANCEROUS CONDITIONS, 2600
CYTOSINE, 1-BETA-D-ARABINOFURANOSYL-
RNA, MESSENGER, 2595
HELA CELLS
ANTIBODIES, VIRAL, 2596
ANTIGENS, VIRAL, 2596
IMMUNE SERUMS, 2596
RNA, MESSENGER
ISOLATION AND CHARACTERIZATION,
HAMSTER, 2595
RNA, VIRAL
ISOLATION AND CHARACTERIZATION,
2642*
POLY A, 2646*
UTERINE NEOPLASMS
ANTIGENS, VIRAL, 2643*
ISOLATION AND CHARACTERIZATION,
2643*
VIRAL PROTEINS
ULTRASTRUCTURAL STUDY, 2644*

VIRUS, HERPES SIMPLEX 2
ANTIGENS, VIRAL
CELL MEMBRANE, 2597
ISOLATION AND CHARACTERIZATION,
HAMSTER, 2597
BRAIN NEOPLASMS
CYTOSINE, 1-BETA-D-
ARABINOFURANOSYL-, 2599
URIDINE, 2'-DEOXY-5-IODO-, 2599
CELL TRANSFORMATION, NEOPLASTIC
RADIATION, IONIZING, 2598

TEMPERATURE-SENSITIVE MUTANTS,
HAMSTER, 2598
CERVIX NEOPLASMS
ANTIBODIES, VIRAL, 2600

ANTIGENS, VIRAL, 2643*
  EPIDEMIOLOGY, 2908*
  EPIDEMIOLOGY, AFRICA, 2905*
  ISOLATION AND CHARACTERIZATION,
    2643*
  PRECANCEROUS CONDITIONS, 2600
CYTOSINE, 1-BETA-D-ARABINOFURANOSYL-
  VIRUS REPLICATION, 2599
EYE NEOPLASMS
  CYTOSINE, 1-BETA-D-
    ARABINOFURANOSYL-, 2599
  URIDINE, 2'-DEOXY-5-IODO-, 2599
LIVER NEOPLASMS
  CYTOSINE, 1-BETA-D-
    ARABINOFURANOSYL-, 2599
  URIDINE, 2'-DEOXY-5-IODO-, 2599
PRECANCEROUS CONDITIONS
  ANTIBODIES, VIRAL, 2908*
SKIN NEOPLASMS
  CYTOSINE, 1-BETA-D-
    ARABINOFURANOSYL-, 2599
  URIDINE, 2'-DEOXY-5-IODO-, 2599
URIDINE, 2'-DEOXY-5-IODO-
  VIRUS REPLICATION, 2599
UTERINE NEOPLASMS
  ANTIGENS, VIRAL, 2643*
  ISOLATION AND CHARACTERIZATION,
    2643*
VIRAL PROTEINS
  ULTRASTRUCTURAL STUDY, 2644*

VIRUS, HERPES ZOSTER
  HODGKIN'S DISEASE
    IMMUNITY, CELLULAR, 2671
  LEUKEMIA, LYMPHOCYTIC
    IMMUNITY, CELLULAR, 2671
  LYMPHOMA
    IMMUNITY, CELLULAR, 2671
  MULTIPLE MYELOMA
    IMMUNITY, CELLULAR, 2671

VIRUS, KIRSTEN MURINE LEUKEMIA
  ANEMIA, HEMOLYTIC
    INTERFERON, 2611
    ULTRASTRUCTURAL STUDY, 2611

VIRUS, KIRSTEN MURINE SARCOMA
  CELL TRANSFORMATION, NEOPLASTIC
    GUANOSINE CYCLIC 3',5'
      MONOPHOSPHATE, 2994*
    GUANYL CYCLASE, 2994*
    PHOSPHODIESTERASES, 2994*
  FLAVONES
    ADENOSINE TRIPHOSPHATASE, 2919
    GLYCOLYSIS, 2919
    GROWTH, 2919
  LEUKEMIA, MYELOBLASTIC
    HYBRIDIZATION, 2612
  RNA, VIRAL
    ISOLATION AND CHARACTERIZATIO,
      MOUSE, 2614
    ISOLATION AND CHARACTERIZATION,
      HAMSTER, 2588
  VIRUS, C-TYPE PORCINE
    VIRUS, RESCUE, 2590
  VIRUS, MURINE ERYTHROBLASTOSIS
    DNA-RNA HYBRIDIZATION, 2614
    RNA, VIRAL, 2614

VIRUS, KIRSTEN MURINE SARCOMA-LEUKEMIA
  GLUCOSE, 2-DEOXY-
    PROTEINS, 2613
    RNA REPLICATION, 2613

VIRUS-LIKE PARTICLES
  CHIMPANZEE PLACENTA
    C-TYPE, 2436*
  FIBROSARCOMA
    CELL LINE, 2970*
  LEUKEMIA L1210
    ISOLATION AND CHARACTERIZATION,
      2639
  LEUKEMIA, LYMPHOBLASTIC
    VIRUS, C-TYPE RNA TUMOR, 2612
  LEUKEMIA, LYMPHOCYTIC
    REVERSE TRANSCRIPTASE, 2951
    VIRUS, C-TYPE RNA TUMOR, 2612
  LEUKEMIA, MYELOBLASTIC
    VIRUS, C-TYPE RNA TUMOR, 2612
  LEUKEMIA, MYELOCYTIC
    VIRUS, C-TYPE RNA TUMOR, 2612
  MAMMARY NEOPLASMS, EXPERIMENTAL
    ADENOMA, 2783
    ULTRASTRUCTURAL STUDY, SNOW
      LEOPARD, 2783
  OVARIAN NEOPLASMS
    TERATOID TUMOR, 2796
  SARCOMA
    CELL LINE, 2970*
  SKIN NEOPLASMS
    VIRUS, HERPES, 2645*
VIRUS, FRIEND MURINE LEUKEMIA
  CONCANAVALIN A, 2602
  RNA, VIRAL, 2602
VIRUS, MURINE MAMMARY TUMOR
  PEPTIDES, 2619
VIRUS, ROUS SARCOMA
  ISOLATION AND CHARACTERIZATION,
    2625
  ULTRASTRUCTURAL STUDY, CHICKEN,
    2625

VIRUS, M7 BABOON
  VIRUS, C-TYPE PORCINE
    ANTIGENIC DETERMINANTS, 2590

VIRUS, MAREK'S DISEASE HERPES
  CELL LINE
    ISOLATION AND CHARACTERIZATION,
      2620
  CYCLOPHOSPHAMIDE
    IMMUNOSUPPRESSION, 2668
    T-LYMPHOCYTES, 2668
  LYMPHOMA
    CYCLOPHOSPHAMIDE, 2668
    IMMUNOSUPPRESSION, 2701
    T-LYMPHOCYTES, 2701
  NERVOUS SYSTEM NEOPLASMS
    LYMPHOMA, 2441*
  VIRAL PROTEINS
    ULTRASTRUCTURAL STUDY, 2644*

VIRUS, MEASLES
  ENCEPHALITIS
    COMPARATIVE VIROLOGY, REVIEW, 241
  HODGKIN'S DISEASE
    ANTIGENS, VIRAL, 2593

PEPTIDES
    ANTIGENS, VIRAL, 2619
    ISOLATION AND CHARACTERIZATION,
        2618
    VIRUS-LIKE PARTICLES, 2619

VIRUS, MURINE RNA TUMOR
    ENZYMATIC ACTIVITY
        REVIEW, 2654*
    ISOLATION AND CHARACTERIZATION
        REVIEW, 2654*

VIRUS, MURINE SARCOMA
    DEXTRAN
        VIRUS REPLICATION, 2652*

VIRUS, N-TROPIC
    VIRUS, FRIEND MURINE LEUKEMIA
        HOST RANGE CONVERSION, 2606
    VIRUS, FRIEND MURINE SARCOMA
        HOST RANGE CONVERSION, 2606

VIRUS, POLYOMA
    ANTIGENS, VIRAL
        TEMPERATURE SENSITIVE MUTANTS,
            2622
    BENZ(A)ANTHRACENE, 7,12-DIMETHYL-
        VIRUS REPLICATION, 2474
    CHOLANTHRENE, 3-METHYL-
        VIRUS REPLICATION, 2474
    CYCLOHEXIMIDE
        DNA REPLICATION, 2621
        DNA, VIRAL, 2621
    CYTOCHALASIN B
        CELL MOVEMENT, 2916
        WOUNDS AND INJURIES, 2916
    DNA REPLICATION
        PROTEINS, 2624
    DNA, VIRAL
        ENDONUCLEASES, 2623
    MITOMYCIN C
        DNA REPLICATION, 2621
    TEMPERATURE SENSITIVE MUTANTS
        ISOLATION AND CHARACTERIZATION,
            2622

VIRUS, POLYOMA, BK
    ALDRICH SYNDROME
        HYBRIDIZATION, 2657*
        ISOLATION AND CHARACTERIZATION,
            2657*
    DNA, VIRAL
        HYBRIDIZATION, 2657*
        ISOLATION AND CHARACTERIZATION,
            2657*

VIRUS, RADIATION LEUKEMIA
    LEUKEMIA, LYMPHOCYTIC
        VIRUS, THYMOTROPIC, 2610
    T-LYMPHOCYTES
        ANTIGENIC DETERMINANTS, 2610
        VIRUS REPLICATION, 2610

VIRUS, RAUSCHER MURINE LEUKEMIA, 2640*
    ANTIGENS, VIRAL
        ISOLATION AND CHARACTERIZATION,
            2685
    HISTOCOMPATIBILITY ANTIGENS

ISOLATION AND CHARACTERIZATION,
  2685
TRANSPLANTATION IMMUNOLOGY, 2685
LEUKEMIA L1210
  ANTIGENS, VIRAL, 2639
LEUKEMIA, LYMPHOBLASTIC
  HISTOCOMPATIBILITY ANTIGENS, 2685
PHOSPHOLIPIDS
  CONCENTRATION, SPLEEN, 2939
REVERSE TRANSCRIPTASE
  ANTIGENIC DETERMINANTS, 2589
VIRAL PROTEINS
  ANTIGENIC DETERMINANTS, 2589
VIRUS, C-TYPE PORCINE
  ANTIGENIC DETERMINANTS, 2590

VIRUS, PD114
  CHROMOSOMES
    CROSSES, GENETIC, 2601
  DNA, VIRAL
    CROSSES, GENETIC, 2601
  REVERSE TRANSCRIPTASE
    ANTIGENIC DETERMINANTS, 2589
  VIRAL PROTEINS
    ANTIGENIC DETERMINANTS, 2589
  VIRUS, C-TYPE PORCINE
    ANTIGENIC DETERMINANTS, 2590
  VIRUS, FELINE LEUKEMIA
    CROSSES, GENETIC, 2601

VIRUS, REO
  GASTROENTERITIS
    COMPARATIVE VIROLOGY, REVIEW, 2412

VIRUS REPLICATION
  CYTOSINE, 1-BETA-D-ARABINOFURANOSYL-
    VIRUS, HERPES SIMPLEX 2, 2599
  URIDINE, 2'-DEOXY-5-IODO-
    VIRUS, HERPES SIMPLEX 2, 2599
  VIRUS, ADENO 2
    RNA, VIRAL, 2577
  VIRUS, AVIAN MYELOBLASTOSIS
    ANTIGENS, VIRAL, 2705
  VIRUS, AVIAN SARCOMA
    ANTIGENS, VIRAL, 2705
  VIRUS, AVIAN SARCOMA B77
    ANTIGENS, VIRAL, 2586
  VIRUS, C-TYPE PORCINE
    REVERSE TRANSCRIPTASE, 2590
  VIRUS, FRIEND MURINE LEUKEMIA
    CELLS, CULTURED, 2651*
  VIRUS, FRIEND SPLEEN-FOCUS FORMING
    CELLS, CULTURED, 2651*
  VIRUS, GOLDEN PHEASANT
    ANTIGENS, VIRAL, 2705
  VIRUS, MOLONEY MURINE LEUKEMIA
    DEXTRAN, 2652*
  VIRUS, MOLONEY MURINE SARCOMA-LEUKEMIA
    DEXTRAN, 2652*
  VIRUS, MURINE SARCOMA
    DEXTRAN, 2652*
  VIRUS, POLYOMA
    BENZ(A)ANTHRACENE, 7,12-DIMETHYL-,
      2474
    CHOLANTHRENE, 3-METHYL-, 2474
  VIRUS, RADIATION LEUKEMIA
    T-LYMPHOCYTES, 2610
  VIRUS, RINGNECKED PHEASANT

ANTIGENS, VIRAL, 2705
VIRUS, ROUS AVIAN LEUKOSIS
  ANTIGENS, VIRAL, 2705
VIRUS, ROUS SARCOMA
  ANTIGENS, VIRAL, 2705
  ULTRASTRUCTURAL STUDY, CHICKEN,
    2625
VIRUS, SV40
  DEFECTIVE VIRUSES, 2629
VIRUS, SV40
  DNA, VIRAL, 2637
VIRUS, SV40
  RNA, VIRAL, 2638

VIRUS, RESCUE
  VIRUS, C-TYPE PORCINE
    VIRUS, KIRSTEN MURINE SARCOMA,
      2590

VIRUS, RETICULOENDOTHELIOSIS
  REVERSE TRANSCRIPTASE
    ISOLATION AND CHARACTERIZATION,
      2653*
  VIRUS, AVIAN SYNCYTIAL
    HYBRIDIZATION, 2653*
  VIRUS, ROUS SARCOMA-ROUS ASSOCIATED
    HYBRIDIZATION, 2653*

VIRUS, RINGNECKED PHEASANT
  ANTIGENS, VIRAL
    VIRAL REPLICATION, 2705

VIRUS, RNA HAMSTER SARCOMA
  DNA REPLICATION
    PROTEINS, 2624

VIRUS, RNA TUMOR
  NERVOUS SYSTEM NEOPLASMS
    LYMPHOMA, 2441*

VIRUS, ROUS AVIAN LEUKOSIS
  ANTIGENS, VIRAL
    VIRAL REPLICATION, 2705

VIRUS, ROUS SARCOMA
  ANTIGENIC DETERMINANTS
    REVIEW, 2435*
  ANTIGENS, VIRAL
    VIRAL REPLICATION, 2705
  BIOLOGICAL PROPERTIES
    REVIEW, 2435*
  CELL MEMBRANE
    PEPTIDES, 2627
  CELL TRANSFORMATION, NEOPLASTIC
    DEOXY SUGARS, 2470
    ETHIDIUM BROMIDE, 2470
    KARYOTYPING, 2774
    REVIEW, 2435*
    TEMPERATURE SENSITIVE MUTANTS, 26?
  CHROMATIN
    DNA, VIRAL, 2585
  DEXTRAN, 2656*
  DNA NUCLEOTIDYLTRANSFERASES
    ISOLATION AND CHARACTERIZATION,
      2628
  DNA REPLICATION
    REVIEW, 2435*
  GLYCOPROTEINS

ADENYL CYCLASE, 2663*
ANTIGENS, VIRAL
  DNA REPLICATION, 2688
  DNA, VIRAL, 2688
  IMMUNE SERUMS, 2632
  ISOLATION AND CHARACTERIZATION,
    2632
BASE SEQUENCE
  TEMPERATURE SENSITIVE MUTANTS,
    2662*
CELL MEMBRANE
  CONCANAVALIN A, 2671
  GLYCOPROTEINS, 2634
CELL TRANSFORMATION, NEOPLASTIC
  ADENYL CYCLASE, 2663*
  GUANOSINE CYCLIC 3',5'
    MONOPHOSPHATE, 2994*
  GUANYL CYCLASE, 2994*
  PHOSPHODIESTERASES, 2994*
CHOLANTHRENE, 3-METHYL-
  IMMUNITY, CELLULAR, 2689
  TRANSPLANTATION IMMUNOLOGY, 2689
CHROMATIN
  DNA-RNA HYBRIDIZATION, 2635
  DNA, VIRAL, 2635
CHROMOSOMES, HUMAN, 6-12
  HYBRID CELLS, 2636
DEFECTIVE VIRUSES
  ANTIGENS, VIRAL, 2629
  DNA, VIRAL, 2629
  ISOLATION AND CHARACTERIZATION,
    2629
  VIRUS REPLICATION, 2629
DNA REPLICATION
  CLONED VARIANTS, 2658*, 2663*
DNA, VIRAL
  DNA HYBRIDIZATION, 2637
  DNA REPLICATION, 2637
  PEPTIDES, 2633
  SUBCELLULAR FRACTIONS, 2633
  TEMPERATURE SENSITIVE MUTANTS,
    2662*
  VIRAL PROTEINS, 2633
  VIRUS REPLICATION, 2637
ENDONUCLEASES
  DNA-RNA HYBRIDIZATION, 2635
  DNA, VIRAL, 2635
EPINEPHRINE
  ADENYL CYCLASE, 2663*
FIBROSARCOMA
  IMMUNITY, CELLULAR, 2689
  IMMUNOSUPPRESSION, 2689
  TRANSPLANTATION IMMUNOLOGY, 2689
HISTOCOMPATIBILITY ANTIGENS
  ISOLATION AND CHARACTERIZATION, 2685
  TRANSPLANTATION IMMUNOLOGY, 2685
ISOQUINOLINE, 6,7-DIMETHOXY-1-VERATRYL-

  ADENYL CYCLASE, 2663*
MIGRATION INHIBITORY FACTOR
  LYMPHOKINES, 2631
PROSTAGLANDINS F
  ADENYL CYCLASE, 2663*
RNA, VIRAL
  DNA-RNA HYBRIDIZATION, 2638
  METHYLATION, 2659*
  TEMPERATURE-DEPENDENT DEGRADATION,
    2661*

VIRUS REPLICATION, 2638
SARCOMA
    HISTOCOMPATIBILITY ANTIGENS, 2685
SODIUM FLUORIDE
    ADENYL CYCLASE, 2663*
THEOPHYLLINE
    ADENYL CYCLASE, 2663*
TRANSPLANTATION, HETEROLOGOUS
    ATHYMIC NUDE MICE, 2975*
TRANSPLANTATION, HOMOLOGOUS
    ATHYMIC NUDE MICE, 2975*
TUBERCULIN
    IMMUNITY, CELLULAR, 2689
    TRANSPLANTATION IMMUNOLOGY, 2689
ULTRAVIOLET RAYS
    MUTATION, 2630
    TEMPERATURE SENSITIVE MUTANTS,
        2630

VIRUS, TURKEY HERPES
    VIRAL PROTEINS
        ULTRASTRUCTURAL STUDY, 2644*

VIRUS, VARICELLA-ZOSTER
    LEUKEMIA, LYMPHOBLASTIC
        IN UTERO EXPOSURE, 2446*
    VIRAL PROTEINS
        ULTRASTRUCTURAL STUDY, 2644*

VITAMIN A
    SEE RETINOL

VITAMIN B12
    ANEMIA, MACROCYTIC
        CONTRACEPTIVES, ORAL, 2436

VITAMIN C
    SEE ASCORBIC ACID

VITILIGO
    SKIN NEOPLASMS
        MELANOMA, 2800

WARTS
    IMMUNITY, CELLULAR
        PLANT AGGLUTININS, 2729*

WILM'S TUMOR
    SEE NEPHROBLASTOMA

WOOD
    NOSE NEOPLASMS
        ADENOCARCINOMA, 2894
    PARANASAL SINUS NEOPLASMS
        ADENOCARCINOMA, 2894

WOUNDS AND INJURIES
    CYTOCHALASIN B
        CELL MOVEMENT, 2916
    VIRUS, POLYOMA
        CYTOCHALASIN B, 2916

XANTHINE, 3-ISOBUTYL-1-METHYL-
    GLIOMA
        ADENOSINE CYCLIC 3',5'
            MONOPHOSPHATE, 2997*
        CARBOXY-LYASES, 2959
        GUANOSINE CYCLIC 3',5'

MONOPHOSPHATE, 2997*
        PHOSPHODIESTERASES, 2997*
NEUROBLASTOMA
    ADENOSINE CYCLIC 3',5'
        MONOPHOSPHATE, 2997*
    CARBOXY-LYASES, 2959
    GUANOSINE CYCLIC 3',5'
        MONOPHOSPHATE, 2997*
    PHOSPHODIESTERASES, 2997*

YEAST
    BENZ(A)ANTHRACENE
        FOOD ANALYSIS, 2537*
    BENZO(A)PYRENE
        FOOD ANALYSIS, 2537*
    BENZO(E)PYRENE
        FOOD ANALYSIS, 2537*
    BENZO(G,H,I,)PERYLENE
        FOOD ANALYSIS, 2537*
    DIBENZ(A,H)ANTHRACENE
        FOOD ANALYSIS, 2537*

YTTRIUM RADIOISOTOPES
    STRONTIUM RADIOISOTOPES
        ACTIVITY RATIO, AGE, SEX FACTORS
            2566

ZEARALENONE
    ASPERGILLUS FLAVUS
        CORN, 2512*, 2513*, 2514*

ZINC
    BONE NEOPLASMS
        REVIEW, 2403
    CARCINOMA 256, WALKER
        GROWTH, 2403
    LEUKEMIA
        GROWTH, 2403
    LIVER NEOPLASMS
        REVIEW, 2403
    MULTIPLE MYELOMA
        REVIEW, 2403
    NEOPLASMS
        REVIEW, 2403
    PROSTATIC NEOPLASMS
        REVIEW, 2403
    SARCOMA
        GROWTH, 2403
    SARCOMA 180, CROCKER
        DIETARY FACTORS, 2732*
        GROWTH, 2732*
    SARCOMA, OSTEOGENIC
        REVIEW, 2403

ZOLLINGER-ELLISON SYNDROME
    PANCREATIC NEOPLASMS
        HYPERPARATHYROIDISM, 2797

ARTHUR, M.
2494
ASAI, T.
2563
ASCHKENASY, A.
2724*
ASKIN, F.B.
2788
ASSUNCAO, J.
2575*
ASTRIN, S.M.
2635
AUGUST, J.T.
2583
AUGUSTI-TOCCO, G.
2922
AUKCRA, A.L.
2864*
AUSTIN, P.
2580
AXELRUD, A.A.
2635
AXEMO, P.
2442*
AXLER, D.A.
2970*
AZUMA, I.
2505*
BACHRACH, U.
2959
BAGNARA, A.S.
2940
BALCERZAK, S.P.
2847*
BALSAM, A.
2798
BALUDA, M.A.
2584
BANDYOPADHYAY, A.K.
2628
BAPTISTA, J.
2494
BARBER, H.R.K.
2667
BARBER, S.
2948
BARBIROLI, B.
2942
BARKER, A.D.
2696, 2970*
BARNARD, R.O.
2820*
BARNEON, G.
2858*
BARON, D.
2641*
BARONI, C.D.
2469
BARRETT, L.A.
2539*
BARRIERE, H.
2844*
BARTHOLD, S.W.
2665
BARTLEY, J.C.
2948
BASSET, F.
2789

BAUER, H.
2414
BAUMGARTNER, G.
2743*
BEARDMORE, G.L.
2855*
BECHET, J.-M.
2642*
BECKER, F.F.
2528*
BECKWITH, A.C.
2509*
BEIERLE, J.W.
2692
BEN-BASSAT, H.
2760*
BENJAMIN, S.A.
2564
BENNETT, J.M.
2947
BENUCK, I.
2473
BENVENISTE, R.E.
2589, 2590, 2601
BENZ, W.C.
2594, 2641*
BERGE, T.
2889
BERMAN, I.
2558*
BERN, M.M.
2741*
BERNACKI, R.J.
2944
BERNARD, W.
2837*
BERNAUDIN, J.F.
2786
BERNOFSKY, C.
2930
BERRY, R.J.
2573*
BERTRAM, B.
2487
BHARADWAJ, V.P.
2553*
BHARGAVA, K.
2893
BIANUCCI, P.
2823*
BIASI, G.
2616, 2617
BICHEL, P.
2718*
BIGAZZI, P.E.
2631
BILBAO, J.M.
2872*
BING, D.H.
2741*
BINZ, H.
2693
BIRNBOIM, H.C.
2932
BIRO, L.
2857*
BIRSHTEIN, B.K.
2676

SWAS, D.K.
2987*
ACK, P.H.
2673
ANCHON, P.
2839*
ASZAK, T.P.
2434*
EVINS, R.D.
2921
OOM, S.E.
2562
OOMFIELD, C.D.
2781
ODIE, A.W., JR.
2728*
OROGI, I
2800*
ECKER, B.B.
2564
EY, J.H.
2797
JSEN, J.
2465
NA, C.
2744*
NAITI-PELLIE, C.
2878*
ONE, C.W.
2689
ORMAN, G.A.
2574*
QUOI, E.
2961
RELLA, L.
2699, 2700
RKOWSKI, A.
2965
RNSCHEIN, W.
2950
THAST, R.J.
2512*
UTWELL, R.K.
2462
VINA, C.
2942
WDEN, G.T.
2462
YER, J.F.
2999*
AND, I.
2501
AND, K.G.
2531
ANDT, L.
2774, 2778
EKKE, J.H.
2764
EMNER, C.G.
2835*
ESLIN, P.
2427*
IARD-GUILLEMOT, M.L.
2878*
OCKMAN, W.W.
2660*
ODEY, R.S.
2712*

BRONDZ, B.D.
2757*
BRONSON, D.L.
2656*
BROOKES, P.
2476
BROOKS, W.H.
2418
BROWN, D.T.
2578
BROWN, J.C.
2926
BROWN, T.D.K.
2591
BRUCE, W.R.
2452
BRUNNING, R.D.
2781
BRYANT, G.M.
2530*
BUCHANAN, J.W., JR
2896
BUCHSBAUM, R.N.
2919
BULLERMAN, L.B.
2401
BULLOCK, B.C.
2525*
BUOEN, L.C.
2501
BUPBIGE, E.J.
2828*
BUREAU, B.
2844*
BURK, K.H.
2754*
BURKE, T.R.
2695
BURNETT, J.W.
2853*
BURTON, A.C.
2967
BUSCH, H.
2925
BYRD, J.Y.
2907*, 2908*
BYRD, W.
2697
CABASSO, V.J.
2412
CABERWAL, D.
2859*
CABRADILLA, C.D.
2931
CAIN, J.A.
2519*
CALLEWAERT, D.M.
2698
CALTON, G.J.
2853*
CAMERON, R.G.
2799
CAMPBELL, D.R.
2896
CANELLAKIS, Z.N.
2958
CANUTO, R.A.
2792

CARDESA, A.
2481
CARDOSO, V.M.
2900*
CARPENTER, J.L.
2752*
CATOVSKY, D.
2759*
CELADA, F.
2755*
CERNOSEK, S.F.
2727*
CERNY, J.
2687
CERVI-SKINNER, S.
2875*
CHADHA, K.C.
2596
CHAMBERLAIN, W.J.
2498
CHAMBERS, L.W.
2895
CHAN, P.C.
2920, 2996*
CHANDRA, R.
2883
CHANDRA, S.
2783
CHANDRAMOULI, K.B
2495
CHANG, K.S.S.
2685
CHANG, R.S.
2909*
CHANG, S.C.S.
2930
CHAPERON, E.
2686
CHAPMAN, C.J.
2880
CHAUDHURI, S.
2868*
CHEE, D.O.
2728*
CHEN, A.T.L.
2871*
CHEN, P.L.
2609
CHEN, T.R.
2772
CHEVAILLIER, P.
2984*
CHIECO-BIANCHI, L.
2616, 2617
CHIRIKJIAN, J.G.
2953
CHLEBOVSKA, K.
2561
CHLEBOVSKY, O.
2561
CHOI, H.-S.
2745*
CHOI, Y.S.
2989*
CHOMICZEWSKI, K.
2567*
CHORTYK, O.T.
2498

CORDIER, A.
2775
CORN, M.
2575*
CORNEO, G.
2638
COTTER, P.F.
2668
COTTER, S.M.
2752*
COTTON, R.E.
2856*
COTTRELL, K.M.
2785
COX, R.
2493
CRAVIOTO, H.
2814*
CREAGAN, E.T.
2865*
CRITTENDEN, L.B.
2587
P.
CROCE, C.M.
2629, 2636
CUCULLU, A.F.
2510*
CULP, L.A.
2634
DAHLE, H.K.
2552*
CALES, S.
2626
DAMJANOV, I.
2796
DANNA, K.J.
2633
CAO, T.L.
2464
DARLE, N.
2442*
CARNELL, J.E.
2937
DAS, S.K.
2912
DAVIDSON, W.M.
2702
DAVIE, J.M.
2680
DAVIOLI, D.
2658*
DAVIS, N.C.
2855*
DAVIS, R.
2993*
DAVIS, R.W.
2688
DAWSON, G.
2990*
DAY, N.K.
2879*
DE BRUYERE, M.
2775
DE GAST, G.C.
2721*
DE GIULI, C.
2626
DE LA MAZA, L.M.
2585

DE RECONDO, A.M.
2984*
DE RUITER, J.
2574*
DE-THE, G.
2709*
DE WAARD, F.
2420
DECENZO, J.M.
2737*
DEGRUY, I.V.
2510*
DELACROIX, R.
2838*
DELARUE, J.
2838*
DELELLIS, R.
2875*
DENK, H.
2753*
DENLINGER, R.
2490
DENT, D.M.
2876*
DESAI, L.S.
2923
DESAI, R.
2502
DESMOND, W.
2975*
DESROSIERS, R.
2988*
DICKENS, J.W.
2456
DICKLER, H.B.
2751*
DIEHL, V.
2413, 2664
DOE, W.F.
2761*
DOELLGAST, G.J.
2956
DOERFLER, W.
2578
DOLL, A.H.
2896
DOMBERNOWSKY, P.
2718*
DORFMAN, A.
2945
DORVAL, G.
2753*
DOYON, D.
2839*
DRESSLER, D.
2422
DREW, R.T.
2503
DREWINKO, B.
2754*
DRILL, V.A.
2405
DROHAN, W.N.
2584
DRYLSKA, M.
2833*
DUNN, W.C., JR.
2921

ONT, A.
2838*
ONT, B.
2879*
BIN, P.W.
2409
TON, R.W.
2417
ERT, G.
2664
NER, R.J.
2651*
TROM, R.D.
2982*
N, M.L.
2748*
WA, K.
2989*
ILALI, M.M.
2713*
IOTT, A.Y.
2656*
IS, D.A.
2673
OOD, J.M.
2890
O, Y.
2841*
LAND, P.C.
2785
KHINA, S.O.
2757*
TEIN, S.S.
2404
, P.
2704
CKSON, R.J.
2582
BERG, I.
2591
EX, M.
2687, 2752*
RSON, R.B.
2865*
AILSON, E.G.
2925
SH, P.F.
2995*
RAEUS, A.
2679
MY, M.J.
2492
MY, O.G.
2492
LLA, M.L.
2732*
, T.-Y.
2547*
AS, A.
2585
EED, G.C.
2658*
CHER, J.-N.
2810*
LKNER, C.S.
2973*
ELE, L.A.
2589

FEINGOLD, J.
2878*
FEINSTEIN, A.R.
2791
FELDMAN, M.
2704
FELDMAN, Z.
2875*
FELSANI, A.
2922
FENSTER, E.D.
2935
FEO, F.
2792
FERGUSON, D.
2730*
FERGUSON, J.
2688
FERNANDEZ, R.
2859*
FERPANT, A.
2775
FERRELL, J.F.
2517*
FETININ, V.A.
2831*
FIDLER, I.J.
2712*
FIELDS, K.L.
2683
FINGERHUT, B.
2714*
FINLAYSON, B.
2661*
FISCHER, M.
2780
FISHER, G.L.
2403
FLANNERY, G.R.
2719*
FLANNERY, V.L.
2598
FLEMING, K.A.
2826*
FODOR, J.
2800*
FOGEL, M.
2474
FOLEY, G.E.
2923
FONCIN, J.-F.
2810*
FONTANIVE-SANGUESA, A.V.
2938
FORD, J.H.
2779
FORNACIARI, G.
2836*
FOSSATI, G.
2669
FOTINO, M.
2879*
FOX, C.F.
2974*
FOX, M.
2479
FRANK, A.
2539*

FRANKS, C.R.
2731*
FRANZE-FERNANDEZ, M.T.
2938
FRAUMENI, J.F., JR.
2865*, 2881
FRAYSSINET, C.
2457
FREDRICKSON, T.N.
2668
FREUND, S.M.
2559*
FREZAL, J.
2878*
FRIDERICI, K.
2988*
FRIED, M.
2623
FRITZE, D.
2475
FRY, R.J.M.
2734*
FU, S.M.
2879*
FU, Y.S.
2762
FU, Y.-S.
2871*
FUCHS, H.B.
2811*
FUJII, K.
2506*, 2550*, 2557*
FUJII, S.
2929
FUNKE, V.
2652*
FURIHATA, C.
2485
FURUICHI, Y
2937
FUSHIMA, K.
2507*, 2534*
GABARAEVA, N.I.
2532*
GABRIEL, L.
2792
GAGEL, R.F.
2875*
GAJEWSKI, A.K.
2567*
GALAKTINOV, V.G.
2437*
GALEOTTI, T.
2918
GALLIMORE, P.H.
2579
GALLOIS, J.-M.
2438*
GANEM, D.
2658*
GANGAL, S.G.
2715*
GARCEA, R.
2792
GARELLICK, G.
2442*
GARNER, F.M.
2554*, 2555*

GOULD, V.
2863*
GOULDEN, M.L.
2512*, 2514*
GRAM, T.E.
2991*
GRAND, N.G.
2725*
GRAVELA, E.
2792
GREEN, M.
2639*
GREEN, M.H.L.
2551*
GREENBERG, L.J.
2711*
GREISS, F.C.
2888
GRESSER, I.
2604
GRIFFIN, B.E.
2623
GRIFFIN, M.J.
2980*
GRONEBERG, J.
2578
GROSSBERG, A.L.
2983*
GROSSMAN, L.
2572*
GRUNDY, G.W.
2865*
GUETZOW, K.
2925
GUILHOU, J.J.
2858*
GUIRGIS, H.A.
2686
GUNZ, F.W.
2779, 2880
GUNZ, J.P.
2880
GUPTA, B.N.
2503
GUPTA, P.
2583
GUPTA, S.R.
2508*
GUSEO, A.
2848*
GUTIERREZ-CERNOSEK, R.M.
2727*
GUTMANN, H.R.
2526*
HAAS, M.
2610
HABENER, J.F.
2966
HABESHAW, J.A.
2782
HACKNEY, C.
2497
HADA, T.
2955
HAINES, H.
2645*
HAJDU, S.I.
2912

HAKIM, A.A.
2725*
HAKURA, A.
2622
HALIE, M.R.
2721*
HALLGREN, H.
2711*
HAMAMOTO, S.T.
2917
HAN, I.H.
2682
HANAFUSA, H.
2626
HANDA, H.
2805*
HANDLER, A.
2504*
HANKIN, J.H.
2903*
HANSEN, J.A.
2879*
HARADA, Y.
2770
HARDING, J.W., JR.
2949
HARDY, A.S.
2708*
HARDY, W.D., JR.
2648*, 2687, 2752*
HAREL, J.
2935
HARINGTON, J.S.
2905*
HARPER, C.
2503
HARRIS, C.C.
2539*
HARRIS, T.J.R.
2595
HARWOOD, C.F.
2434*
HARWOOD, L.M.J.
2579
HASEGAWA, H.
2841*
HASHIMOTO, S.
2484
HASHINOTSUME, M.
2955
HASHIZUME, K.
2841*
HASS, G.
2723*
HASSOUN, J.
2766
HATANAKA, M.
2588
HATCH, L.A.
2599
HAVELL, E.A.
2747*
HAYASHI, J.
2549*
HAYASHI, N.
2851*
HAYATA, Y.
2542*, 2543*

M.

YATSU, H.
2706
ATHCOTE, J.G.
2511*
BERLING, R.L.
2436*
CHT, S.S.
2486
IDELBERGER, C.
2538*
LLIWELL, B.E.
2895
LLSTROM, I
2691
LLSTROM, K.E.
2691
LMKE, R.J.
2436*
LSON, L.
2912
NDEKSON, B.E.
2494
NDERSON, J.F.
2940
NNEKFUSER, H.H.
2813*
RBERMAN, R.B.
2669, 2672
RSON, J.
2887
SHMAT, M.Y.
2887
SS, P.
2752*
SSELTINE, C.W.
2512*, 2513*, 2514*
XT, P.M.
2502
YSTRATEN, F.M.J.
2832*
BBERT, J.R.
2511*
GASHINO, K.
2955
LDRUM, K.I.
2546*
LGERS, J
2610
LL, G.B.
2895
RANO, A.
2765, 2766, 2809*
RAO, F.
2505*
RONI, I.
2507*
RONO, I.
2534*
RSCH, M.S.
2673
RSCHMAN, S.Z.
2652*
SAMITSU, T.
2851*
U, I.J.
2726*
OZANEK, I.
2435*

HLUBINOVA, K.
2586
HOCHBERG, A.A.
2924
HOCHSTRASSER, D.L.
2896
HOFFMAN, D.
2486
HOLDEN, H.T.
2669, 2672
HOLLANDER, C.F.
2574*
HOLLER, A.
2839*
HOLMES, E.C.
2728*
HOLMES, J.T.
2731*
HOLZNER, J.H.
2750*
HOMBURGER, F.
2504*
HONMA, Y.
2523*
HOOVER, R.
2881
HORAN, M.
2632
HORAN, P.K.
2632
HORVATH, E.
2872*
HOURI, M.
2740*
HOUSMAN, D.
2612
HOUSTON, I.B.
2880
HOWELL, W.M.
2562
HOWLEY, P.M.
2657*
HSU, I.C.
2529*
HSU, K.C.
2850*
HUBERMAN, E.
2474, 2499
HUBNER, K.
2780
HUDSON, A.R.
2872*
HUEBNER, K.
2629
HUFFMAN, J.H.
2650*
HULL, E.W.
2791
HUNT, R.C.
2926
HUNTER, R.
2730*
HUTCHISON, D.J.
2609
HUTH, F.
2852*
HUTTIG, G.
2860*

HUTTON, J.J.
2497, 2758*
IANNACCONE, P.M.
2710*
IKNAYAN, H.F.
2565
INOSE, T.
2816*
INOUE, S.
2777
IRIE, K.
2678
IRIE, R.F.
2678
IRONSIDE, P.
2894
IRVING, C.C.
2493
ISHIDA, M.
2841*
ISHIDATE, M.
2548*
ITO, K.
2639*
IVANKOVIC, S.
2460
IWATA, K.K.
2974*
JACKSON, B.R.
2831*
JACKSON, M.A.
2883, 2887
JACKSON, R.C.
2960
JACOBS, D.M.
2697
JAKOBSSON, S.
2536*
JAKOWSKI, R.M.
2668
JAMES, K.
2681
JANISCH, W.
2812*
JAO, W.
2863*
JAROSLOW, B.N.
2734*
JARRETT, O.
2694
JARRETT, W.
2694, 2752*
JARVI, O.H.
2763
JELINEK, W.
2937
JELINEK, W.R.
2936
JENKINS, R.
2771, 2907*, 2908*
JENKINSON, I.S.
2898
JEPSON, A.M.
2514*
JERSILD, C.
2879*
JESKE, R.
2961

KAWAUCHI, T.
  2542*, 2543*
KAY, S.
  2871*
KAZAMA, S.
  2485
KEEHN, E.
  2663*
KELLY, A.P.
  2673
KEMPER, B.
  2966
KENNDEY, J.
  2883
KERN, D.H.
  2475
KERR, S.J.
  2952
KERSEY, J.H.
  2807*
KERSTING, G.
  2806*
KESSEL, D.
  2954
KETTERER, B.
  2538*
KHAFAGY, M.M.
  2784
KHERA, K.S.
  2556*
KHOURY, G.
  2657*
KILLION, J.J.
  2980*
KIM, B.
  2789
KIMURA, H.
  2943
KIMURA, S.
  2598
KING, E.J.
  2850*
KING, G.W.
  2847*
KINMONT, P.D.C.
  2856*
KINOSHITA, J.
  2816*
KINOSHITA, K.
  2822*
KIRCHNER, H.
  2672
KIRKIN, A.F.
  2757*
KIRSTEN, W.H.
  2613
KISHIKAWA, H.
  2842*
KISS, J.
  2939
KJELLSTRAND, C.M.
  2807*
KLEIN, G.
  2591, 2703
KLEMENT, V.
  2614
KLIGERMAN, A.D.
  2562

KLOPPEL, G.
  2439*
KNAACK, J.
  2795
KNAPP, W.
  2743*
KNIGHT, L.A.
  2702
KOCSIS, J.
  2541*
KODAMA, M.
  2478
KOESTNER, A.
  2441*
KOGURE, K.
  2485
KOHASHI, O.
  2674
KOJI, T.
  2882
KOJIMA, H.
  2946
KOJIRO, M.
  2901*
KOLATA, G.B.
  2450*
KOO, J.
  2773
KOONTZ, W.W.
  2871*
KOPROWSKA, I.
  2868*
KOPROWSKI, H.
  2629, 2636
KORB, G.
  2824*
KORENMAN, S.G.
  2421
KORMAN, D.B.
  2969*
KOSELNIK-GLUGLA, B.
  2840*
KOSHMAN, R.W.
  2773
KOSS, M.N.
  2710*
KOTCHEN, J.M.
  2896
KOTODA, K.
  2851*
KOURILSKY, F.M.
  2742*
KOVACS, K.
  2872*
KOVI, J.
  2887
KRAFT, C.
  2686
KRAMARSKY, B.
  2609
KREUTZER, G.
  2961
KRIEK, E.
  2402
KRUGER, F.
  2483
KRUGER, F.W.
  2437

KRUSH, A.J.
2826*
KRUTOVA, T.V.
2969*
KRYGIER-STOJALOWSKA, A.
2840*
KUBO, Y.
2851*, 2901*
KUDO, S.
2955
KUHN, C., III
2788
KUMAKI, K.
2500
KUMAR, P.P.
2802*
KUMMERLANDER, L.
2600
KUNKEL, H.G.
2879*
KURATA, H.
2506*
KUROKI, T.
2913
KURTENOV, O.A.
2545*
KURTH, R.
2414, 2690
KURTZ, S.
2612
KUSHKO, L.IA.
2976*
KUSUMOTO, Y
2882
KWOLEK, W.F.
2513*, 2514*
KYNER, D.S.
2977*
LACOUR, F.
2935
LAFARGE-FRAYSSINET, C.
2457
LAFFARGUE, F.
2999*
LAI, C.-J.
2662*
LAING, G.
2723*
LAIRD, C.
2534*
LALICH, J.J.
2529*
LAMM, D.L.
2794
LANDON, J.
2743*
LANGDON, H.L.
2558*
LANGELIER, Y.
2637
LANSIMIES, P.H.
2763
LARETSKAIA, L.I.
2655*
LARSSON, A.
2442*
LAU, H.-U.
2902*

LAUB, O.
2638
LAUFS, R.
2675
LAUGHLIN, D.C.
2783
LAVI, S.
2659*
LAW, L.W.
2685
LAZAR, W.
2906*
LEADBETTER, G.W., JR.
2737*
LECKY, J.H.
2437
LEE, B.K.
2619
LEE, E.W.
2541*
LEE, J.A.H.
2890
LEE, J.C.K.
2927
LEE, K.M.
2649*
LEE, L.S.
2510*
LEE, S.S.C.
2927
LEESMAN, A.E.
2516*
LEFEBRE, J.-C.
2744*
LEGATOR, M.S.
2531*
LEIBOLD, W.
2703
LEISTENSCHNEIDER, W.
2861*
LEONARD, E.J.
2708*
LEPOW, H.
2859*
LEUNG, B.S.
2466
LEVAN, G.
2774
LEVASEUR, P.
2839*
LEVIN, L.S.
2828*
LEVIN, S.
2465
LEVINE, J.E.
2418
LEVINE, J.H.
2957
LEVIS, W.R.
2722*
LEVY, B.
2598
LEVY, R.B.
2970*
LEWANDOWSKI, L.J.
2625
LIABEUF, A.
2742*

LICHTIGER, B.
2754*
LIDMAN, K.
2679
LIEBER, M.M.
2590
LIEGEY, A.
2666
LIGHTBODY, J.J.
2698
LIGHTDALE, C.J.
2827*
LIJINSKY, W.
2488
LIKHACHEV, A.J.
2463
LILLEHOJ, E.B.
2513*
LIN, C.S.
2953
LIN, E.
2542*, 2543*
LINDENBAUM, J.
2436
LINDOP, P.J.
2914
LINTON, E.B.
2888
LIPCHINA, L.P.
2969*
LISS, L.
2821*
LITOUX, P.
2844*
LITTERST, C.L.
2991*
LIWNICZ, B.H.
2859*
LIZONOVA, A.
2586
LLENA, J.F.
2765
LLOYD, J.W.
2427*
LOBUGLIO, A.R.
2847*
LOMAX, C.A.
2940
LOPES, G. DE M.P.
2900*
LOUIE, S.
2637
LOUW, J.H.
2876*
LUBBERT, H.
2961
LUKASEWYCZ, O.A.
2707*
LUMB, J.R.
2993*
LUTSIK, M.D.
2736*
LYNCH, H.T.
2686
LYNCH, J.
2686
LYON, J.L.
2892

MATAS, A.J.
2807*
MATSUBARA, N.
2507*
II
MATSUKADO, Y.
2804*, 2822*
MATSUMOTO, A.
2563
MATTHEWS, J.
2894
MATUCHANSKY, C.
2443*
MAUDERLY, J.L.
2564
MAUL, G.G.
2625
MAURER, L.H.
2973*
MAURER, W.
2522*
MCBRIDE, R.A.
2668, 2701
MCCALLUM, D.I.
2856*
MCCANN, J.
2430*
MCCLELLAN, R.O.
2564
MCCLELLAND, A.J.
2643*
MCCONNELL, R.G.
2517*
D.
MCCREARY, P.
2723*
MCCULLOCH, E.A.
2915
MCCULLOUGH, D.L.
2794
MCDANIEL, M.N.
2516*
MCDOWELL, E.M.
2539*
MCGINNIS, E.L.
2535*, 2537*
MCINNES, J.
2747*
MCKAY, F.W.
2881
MCKENNA, R.W.
2781
MCLACHLAN, J.A.
2525*
MCLAUGHLIN, A.P., III
2794
MCLEOD, D.L.
2605
MEDINA, D.
2533*
MEGO, J.L.
2519*
MEGSON, M.
2683
MELERO, J.A.
2624
MELIEF, C.J.M.
2607
MELNICK, H.
2667

MELTZER, M.S.
2708*
MELVIN, K.E.W.
2875*
MENACHEMI, E.
2829*
MENEZES, J.
2703
MENKART, J.
2449*
MERRETT, T.G.
2740*
METAFORA, S.
2922
MEYNADIER, J.
2858*
MICHAUX, J.L.
2775
MICKEY, M.R.
2674
MIHAILOVICH, N.
2570*
MIKUZ, G.
2837*
MILLER, A.
2876*
MILLER, E.C.
2429*
MILLER, J.A.
2429*
MILLER, Z.
2994*
MILLIGAN, F.D.
2828*
MILLS, B.
2699
MIMNAUGH, E.G.
2991*
MINTZ, U.
2756*
MIRME, H.J.
2545*
MIRVISH, S.S.
2481
MISFELDT, D.S.
2917
MITELMAN, F.
2774, 2778
MITROU, P.S.
2780
MIYA, T.S.
2521*
MIYAKAWA, M.
2459
MIYASHITA, S.Y.
2913
MOBIUS, W.
2813*
MOCOVICI, C.
2705
MODAK, M.J.
2986*
MOHR, U.
2483
MOLLER, U.
2465
MONACO, A.P.
2673

MONSON, R.R.
2892
MONTAGNIER, L.
2642*
MOON, S.
2789
MOORE, D.H.
2609
MOORE, G.E.
2695
MOORE, R.M., JR.
2427*
MORGAN, M.
2937
MORGAN, R.W.
2496
MORIAU, M.
2775
MORISON, W.L.
2671, 2729*
MORRIS, H.P.
2918, 2924, 2941, 2949,
2962
MORRIS, J.E.
2748*
MORTON, D.L.
2678, 2728*
MOSCOVICI, M.G.
2705
MOTT, D.M.
2995*
MOULI, K.C.
2885
MOYER, G.H.
2515*
MUCHMORE, A.V.
2672
MULLARKEY, M.F.
2657*
MULLER, H.K.
2719*
MULLIGAN, R.C.
2633
MUNEHISA, T.
2882
MUNYON, W.
2596
MURAD, F.
2943
MURAO, M.
2523*
MURIEL, W.J.
2551*
MURPHY, E.
2828*
MURPHY, H.B.
2895
MURPHY, W.H.
2707*
MUTAI, M.
2484
MYERS, D.K.
2569*
NAGATA, C.
2478
NAGATA, E.
2851*
NAHMIAS, A.J.

2908*
NAIR, P.N.M.
2715*
NAIFN, R.C.
2719*
NAJAPIAN, J.S.
2607*
NAKADATE, M.
2491, 2548*
NAKAJIMA, H.
2542*, 2543*
NAKAMURA, S.
2882
NAKASHIMA, T.
2901*
NASS, M.M.K.
2470
NATALI, P.G.
2755*
NATHANS, D.
2662*
NAYAK, S.K.
2713*
NAZERIAN, K.
2620
NEAVES, W.B.
2472
NEBERT, D.W.
2500
NEER, A.
2621
NELSON, D.
2873*
NELSON, R.A.
2742*
NESBITT, J.
2994*
NETSKY, M.G.
2418
NEUFELD, G.R.
2407
NEUMANN, J.
2806*
NEWBOLD, R.R.
2525*
NEWCOMB, E.
2977*
NICHOLSON, W.E.
2957
NICOLAS, R.
2844*
NIEWEG, H.O.
2721*
NIHO, Y.
2915
NII, S.
2644*, 2647*
NIKONOVA, M.F.
2716*
NILSSON, K.
2679
NILSSON, L.
2442*
NISHIKAWA, H.
2505*
NISHIOKA, M.
2691
NITTA, K.

2733*
NIWA, A.
2500
NOMURA, N.
2834*
NOON, M.C.
2618
NOONAN, K.D.
2972*
NORBERG, R.
2679
NORDLUND, J.J.
2717*
NORRIS, M.S.
2537*
NOVAL, J.J.
2540*
OBATA, H.
2851*
OBIDITSCH-MAYR, I.
2750*
O'BRIEN, K.J.
2982*
ODASHIMA, S.
2491, 2506*, 2550
OGIU, T.
2491
OGURA, T.
2505*
OHE, K.
2577
OHKOCHI, T.
2955
OHMI, K.
2841*
OHNO, T.
2951
OHO, K.
2542*, 2543*
OHTANI, R.
2955
OKADA, M.
2550*
OKADA, Y.S.
2622
OKUDA, K.
2851*, 2901*
OLIN, R.
2432*
OLISA, E.G.
2883
OLIVER, R.
2573*
OLSON, M.O.J.
2925
ONUIGBO, W.I.B.
2884
OPPENHEIMER, J.H.
2798
ORTH, D.N.
2957
ORY, H.W.
2907*, 2908*
OSAKI, Y.
2523*
OSSERMAN, E.F.
2992*
OTH, D.

2564
PHILIPSON, L.
2576
PICKETT, P.B.
2917
PICKRELL, J.A.
2564
PIGOTT, H.W.S.
2870*
PIKE, M.C.
2494
PILAWSKI, Z.
2906*
PILCH, Y.H.
2475
PINTER, A.
2640*
PITELKA, D.R.
2917
PITTMAN, S.M.
2779
PIVO, K.R.   M.
2646*
PLISS, G.B.
2544*
POEHLS, H.
2902*
POGOSIANZ, E.E.
2971*
POKROVSKAIA, T.A.
2716*
POLAK, M.
2817*
POLLEX, G.
2902*
POLLI, E.
2608
POLLOCK, A.C.
2826*
POLLOW, B.
2961
POLLOW, K.
2961
POON, T.P.
2850*
POORTMAN, J.
2471
PORTER, C.W.
2944
POSNER, G.
2827*
POTTS, J.T., JR.
2966
POUR, P.
2483
POWARS, D.   JR.
2692
POWELL, J.A.
2722*
POZHARISSKI, K.M.
2463
PRAHLAD, D.
2495
PRAJDA, N.
2941
PRASANNA, H.R.
2538*
PRASLICKA, M.

2561
PRENEN, J.A.C.
2471
PRESSMAN, D.
2983*
PREUD'HOMME, J.L.
2676
PREUSSMANN, R.
2460
PRICE, E.
2857*
PRIGOZHINA, E.L.
2971*
PRIOLEAU, J.C., III
2993*
PRITCHETT, P.S.
2762
PROCHOWNIK, E.V.
2613
PROFFITT, M.R.
2673
PUFFER, H.W.
2998*
PYERITZ, E.A.
2949
QASBA, P.K.
2661*
QIZILBASH, A.H.
2790
QUINCY, D.A.
2992*
RACHKOVSKAIA, L.A.
2655*
RACKER, E.
2919
RADDATZ, R.
2825*
RADOJKOVIC, J.
2755*
RAGHAVAIAH, N.V.
2885, 2886
RAGUNA, G.
2593
RAINERI, R.
2486
RAMBAUD, J.C.
2443*
RAMULU, C.
2495
RANADIVE, K.J.
2715*
RAO, K.S.
2517*
RAO, R.S.
2715*
RAPP, F.
2597
RAPP, H.J.
2708*
RASMUSSEN, J.E.
2854*
RAVINDRANATH, Y.
2777
RAYMOND, J.
2735*
REAGAN, R.L.
2991*
REBELL, G.

2645*
BOUCAS, M.A.
2900*
DAI, I.
2939
DDY, C.R.R.M.
2495, 2885
DMAN, H.C.
2564
ED, C.E.
2947
ED, C.L.
2597
ED, S.I.
2688
ICHLE, F.A.
2540*
ICHLE, R.M.
2543*
IN, F.R.
2852*
ISEGGER, W.
2837*
MUS, I.
2812*
NNIE, P.S.
2964
NO, F.E.
2517*
YGERS, U.
2480
WNER, F.
2436
YNIER, M.O.
2457
ZNIK, M.
2815*
DADS, G.G.
2903*
CH, A.
2633, 2966
CH, M.A.
2687
CHARDS, R.J.
2502
CHTER, G.W.
2927
ECHLIN, S.
2875*
MBAUD, P.
2858*
TTENHOUSE, H.G.
2974*
TZI, D.
2656*
BBINS, G.F.
2784
BBINS, P.W.
2627
BERTS, B.E.
2633
BERTS, R.M.
2972*
CCHI, G.
2593
OHAIN, J.
2775
GALSKY, V.Y.

2749*
ROGENTINE, G.N., JR.
2746*
ROHOLT, O.A.
2983*
ROLLWAGEN, F.M.
2710*
ROMER, T.R.
2910*
ROSAN, R.C.
2811*
ROSE, F.A.
2502
ROSEN, Y.
2789
ROSENBUSCH, G.
2832*
ROSS, K.D.
2518*
ROSSI, G.B.
2604
ROTKIN, I.D.
2869*
ROTTMAN, F.
2988*
ROWINSKI, J.
2868*
ROY-BURMAN, P.
2614
ROZENBLATT, S.
2633
ROZHNOVA, L.I.
2655*
RUCO, L.
2469
RUF, H.
2999*
RUSSEL, T.R.
2994*
RUSSELL, S.W.
2761*
RYWLIN, A.
2645*
SACHS, D.H.
2751*
SACHS, L.
2499, 2756*
SAEGESSER, F.
2830*
SAGONE, A.L.
2847*
SAIER, M.H.
2975*
SAITO, S.
2841*
SAKAMOTO, K.
2901*
SAKATO, K.
2842*
SALAS, J.
2624
SALAS, M.L.
2624
SALDITT-GEORGIEFF, M.
2937
SALOMON, M.I.
2850*
SALYER, D.C.

2768
SALYER, W.R.
2768
SANDBERG, C.G.
2428*
SANDRITTER, W.
2934
SANKARAN, V.
2864*
SANTOLI, D.
2629
SARKAR, N.H.
2609
SARRIF, A.M.
2538*
SASAJIMA, K.
2485
SASAKI, G.H.
2466
SASAKI, M.S.
2933
SATO, G.
2975*
SATO, H.
2549*
SATO, J.
2834*
SATO, T.
2549*
SAUS, F.L.
2962
SCANLAN, R.A.
2546*
SCHACHNER, M.
2990*
SCHAEFFER, L.D., JR.
2998*
SCHAFFER, P.A.
2598
SCHALLER, J.
2752*
SCHARFETTER, H.
2837*
SCHARFF, M.D.
2676
SCHARMER, B.A.
2516*
SCHIERMAN, L.W.
2668, 2701
SCHILLINGS, P.H.M.
2832*
SCHIMPFF, C.R.
2791
SCHLESINGER, K.J.
2927
SCHMAHL, D.
2483
SCHNEIDER, U.
2592
SCHNELL, R.C.
2521*
SCHNOY, N.
2861*
SCHOBER, R.
2873*
SCHOTTENFELD, D.
2784
SCHULTZ, D.R.

2827*
SHERR, C.J.
2589, 2590
SHETTIGARA, P.T.
2496
SHIMOKAWA, Y.
2851*, 2901*
SHINDO, K.
2985*
SHIFASAKA, T.
2929
SHOTWELL, O.L.
2512*, 2513*, 2514*
SHOYAB, M.
2584
SHREEVE, M.M.
2605
SHUBICK, P.
2481
SHUTER, B.J.
2898
SIDWELL, R.W.
2650*
SIEJA, K.
2936*
SILAGI, S.
2977*
SILVERBERG, S.G.
2867*
SILVERMAN, S., JR.
2893
SILVERSTEIN, S.C.
2977*
SIMARD, R.
2637
SIMMONS, E.
2570*
SIMMONS, R.L.
2711*, 2807*
SIMON, L.N.
2650*
R.A.   SINGER, R.M.
2936
SINGH, S.
2779
SINGH, S.M.
2886
SINHA, D.
2464
SJOGREN, U.
2778
SKAAPE, J.U.
2552*
SKINNER, L.G.
2785
SLADE, M.S.
2711*
SLAGA, T.J.
2462
D.   SLAMON, D.J.
2611
SLAUGHTER, J.C.
2896
SLEPIAN, E.I.
2532*
SLOWIK, F.
2808*
SLOWIKOWSKA, M.G.

2567*
SMITH, E.J.
2587
SMITH, G.C.
2436*
SMITH, G.H.
2619
SMITH, J.
2907*
SMITH, J.C.
2520*
SMITH, L.W.
2893
SMULSON, M.
2953
SNYDER, R.
2541*
SOKAL, G.
2775
SOKOVA, O.I.
2971*
SOLER, P.
2786
SONNET, J.
2775
SORAVITO, G.
2469
SORENSON, G.D.
2973*
SOROF, S.
2995*
SOSLAU, G.
2470
SOTO, J.M.
2863*
SPENCER, T.L.
2978*
SPIEGELMAN, S.
2951
SPITZER, W.O.
2895
SPRECHER-GOLDBERGER, S.
2643*
SPURLOCK, S.E.
2741*
ST. PIERRE, R.L.
2970*
STANG-VOSS, C.
2813*
STANLEY, J.B.
2510*
STARK, G.R.
2688
STEINKE, H.
2675
STELLWAGEN, R.H.
2981*
STEPHENSON, J.R.
2587
STERLING, T.D.
2423*
STERN, P.L.
2683
STERN, R.
2879*
STERNBERG, S.S.
2827*
STEVENS, L.C.

2796
STEVENSON, A.F.G.
2566
STEWART, M.L.
2602, 2603
STIGLIANI, V.
2862*
STILES, C.D.
2975*
STINSKI, M.F.
2581
STOBO, J.D.
2695
STOECKEL, M.
2531*
STOKER, M.
2916
STOOLMILLER, A.C.
2945, 2990*
STOTT, W.T.
2401
STRAKA, D.
2996*
STRATMAN, F.W.
2924
STRAUS, N.A.
2932
STROMINGER, J.L.
2594, 2641*
STUART, A.E.
2782
SUGIMURA, T.
2485
SUGIYAMA, T.
2454
SUHRBIER, K.M.
2734*
SULTAN, C.
2968*
SUOLINNA, E.M.
2919
SVEC, J.
2586
SWANSTROM, R.I.
2646*
SWEENEY, E.C.
2797
SWEGER, D.M.
2559*
SWENBERG, J.A.
2490
SYMES, E.K.
2964
SZRAM, S.
2568*
TABAH, E.J.
2799
TACHIBANA, M.
2542*, 2543*
TACHIBANA, T.
2776
TAGUCHI, T.
2467
TAKAHASHI, M.
2485
TAKAHASHI, Y.
2955
TAKANO, T.

2654*
TAKAYAMA, S.
2553*
TAKEDA, Y.
2467
TAKEMOTO, K.K.
2657*
TAMIYA, M.
2851*
TANAKA, K.
2842*
TANIMURA, A.
2553*
TANINO, T.
2989*
TANIYAMA, T.
2505*
TANNENBAUM, S.R.
2547*
TARUTINOV, V.I.
3000*
TASHJIAN, A.H., JR.
2987*
TASHJIAN, A., JR.
2875*
TATEMATSU, M.
2485
TAWARA, M.
2542*, 2543*
TAYLOR, G.
2524*
TAYLOR, H.W.
2488
TAYLOR, M.W.
2732*
TAYLOR, S.
2797
TCHERTKOFF, V.
2850*
TEI, N.
2467
TERASAKI, P.I.
2674, 2686
THAMBIAH, A.
2874*
THEIS, G.A.
2701
THELMO, W.L.
2795
THEOHARIDES, T.C.
2958
THIELMANN, H.W.
2480
THIJSSEN, J.H.H.
2471
THIPY, L.
2043*
THOMAS, C.
2934
THOMAS, R.J.
2686
THURSTON, O.G.
2773
THURZO, V.
2586
TIGELAAR, R.E.
2673
TILL, J.E.

2915
TING, A.
2686
TODARO, G.J.
2589, 2590, 260
TODD, C.W.
2748*
TOKUDA, H.
2822*
TOLIVER, A.P.
2931
TOLOMELLI, B.
2942
TOMINAGA, T.
2467
TOREMALM, N.G.
2889
TORIKATA, C.
2787
TOSATO, G.
2593
TOYAMA, M.
2805*
TOZIER, A.
2761*
TRUJILLO, J.M.
2754*
TRUMP, B.F.
2539*
TRUMP, G.
2692
TRUTTER, J.A.
2517*
TRYFIATES, G.P.
2962
TSUBOI, M.
2767
TSUCHIDA, N.
2588
TSUCHIDA, Y.
2841*
TSUNETA, I.
2523*
TURSZ, T.
2443*
TYLER, C.W., JR.
2907*, 2908*
TYLER, S.A.
2734*
TYLER, T.R.
2516*
UCCINI, S.
2469
UCHIDA, S.
2767
UENAKA, H.
2454
UKITA, T.
2706
UMANSKII, IU.A.
2976*
UMEZAWA, H.
2733*
UNGER, H.-H.
2902*
URANO, Y.
2841*
URBANO, U.

2904*
WALL, R.
2584
WALL, R.L.
2461
WALLCAVE, L.
2481
WANG, J.T.
2649*
WANG, N.S.
2795
WANG, N.-S.
2799
WARD, J.M.
2482, 2489
WARD, P.A.
2631
WARNER, M.R.
2533*
WARNER, N.E.
2998*
WARR, G.W.
2681
WARREN, S.
2410
WASS, E.J.
2779
WATABE, S.
2834*
WATANABE, M.
2557*
WATANABE, T.
2803*
WATSON, A.D.J.
2845*
WEBER, G.
2941, 2960
WEIL, S.
2630
WEINBERG, E.D.
2732*
WEINBERG, R.A.
2615
WEISBURGER, E.K.
2482, 2489
WEISBURGER, J.H.
2482
WEISMAN, M.
2547*
WEISS, R.
2824*
WELIN, J.
2844*
WELSH, K.I.
2753*
WELTY, R.E.
2456
WERNSTROM, J.
2828*
WESSEL, J.E.
2453
WHALEN, J.J.
2722*
WHITE, D.R.
2846*
WHITE, H.B., III
2949
WHITE, L.J.

2956
WHITEHEAD, N.
2406
WIEBEL, F.J.
2477
WIGGER, H.J.
2793
WIGZELL, H.
2693, 2753*
WILD, D.
2455
WILDY, P.
2595
WILLIAMS, A.O.
2883
WILLIAMS, C.A.
2632
WILLIAMS, C.W.
2856*
WILLIAMS, J.
2580
WILLIAMS, J.L.
2546*
WILLIAMS, R.E.
2974*
WILLMOTT, N.
2681
WILSON, L.
2694
WINSTON, B.M.
2573*
WISE, K.S.
2692
WITHEY, R.J.
2560*
WITKIN, S.S.
2951
WITKOWSKI, J.T.
2650*
WITMER, C.
2541*
WITTER, R.L.
2620
WNUK-KATYNSKA, U.
2833*
WOLFE, H.
2875*
WOLFORD, R.G.
2618
WONG, G.
2963
WOOD, L.
2415
WOOD, M.L.
2673
WOODMAN, P.W.
2538*
WOODS, A.E.
2769
WOROWSKA-ROGOWSKA, J.
2899*
WOSTER, A.D.
2732*
WRIGHT, D.H.
2419
WRIGHT, P.N.M.
2898
WROUGHTON, M.A.

, I.A.

D.

891

2856*
WU, M.C.
  2979*
WULFF, U.C.
  2923
WURSTER-HILL, D.H.
  2973*
WYATT, P.H.
  - 2896
WYNDER, E.L.
  2486
WYROBEK, A.J.
  2452
YACHNIN, S.
  2735*
YAKUMARU, K.
  2787
YAKUSHIJI, F.
  2851*
YAMADA, T.
  2553*, 2776
YAMAGATA, S.
  2834*
YAMAGUCHI, K.
  2882
YAMAMURA, Y.
  2505*, 2955
YAMASAKI, E.
  2430*
YAMASAKI, H.
  2499
YAMASHITA, J.
  2805*
YAMASHITA, T.
  2639*
YASUDA, I.
  2644*, 2647*
YESUDIAN, P.
  2874*
YI, P.-N.
  2561*
YOHN, D.S.
  2752*
YOKOTA, A.
  2804*
YOSHIDA, M.
  2776
YOSHIDA, O.
  2459
YOSHIDA, T.
  2631
YOSHIKAWA, K.
  2548*
YOSHIKURA, H.
  2606
YOUNG, H.
  2580
YOUNG, H.F.
  2762
YUASA, Y.
  2913
YUNIS, A.A.
  2979*
YUNIS, E.J.
  2711*
YUNIS, J.J.
  2585
ZAHLTEN, R.N.

  2924
ZAKRZHEVSKAIA, D.T.
  2568*
ZALDIVAR, R.
  2571*
ZAORSKA, B.
  2899*
ZEGENHAGEN, V.
  2902*
ZELENSKII, V.P.
  2655*
ZIMMERMAN, E.A.
  2973*
ZIMMERMAN, H.M.
  2766
ZOLA, H.
  2720*
ZOLTOWSKI, M.
  2906*
ZUCCO, F.
  2922
ZUELZER, W.W.
  2777
ZUR HAUSEN, H.
  2592

CAS REGISTRY NUMBER	ABSTRACT NUMBER
53065............................	2682
53167............................	2563
53703............................	2473
53952............................	2402
53963............................	2402
54126............................	2794
54422............................	2599, 2604, 2669, 2898
54477............................	2962
54626............................	2452
55174............................	2952, 2792
55185............................	2480, 2482
55981............................	2452
56042............................	2934
56495............................	2404, 2418, 2452, 2473, 2474, 2475, 2500, 2666, 2689, 2691, 2769, 2770
56575............................	2493
57227............................	2700
57396............................	2452
57636............................	2405
57830............................	2405, 2464, 2785
57976............................	2464, 2465, 2466, 2467, 2468, 2469, 2473, 2474, 2913, 2920

CHEMICAL ABSTRACT SERVICES REGISTRY NUMBER INDEX

CAS REGISTRY NUMBER	ABSTRACT NUMBER	CAS REGISTRY NUMBER	ABSTRACT N
58082...............	2481	68235...............	2405
58220...............	2472	68940...............	2776
58559...............	2499	70257...............	2485, 2913
58617...............	2959	71443...............	2958
58742...............	2499	72333...............	2405
59052...............	2776	73223...............	2794
59303...............	2406, 2942	75014...............	2501, 2502, 2896
59665...............	2967	76380...............	2408
59892...............	2481	79016...............	2408
60571...............	2404, 2921	83443...............	2367
62497...............	2680	98920...............	2930
62500...............	2452, 2479	124403...............	2484
62555...............	2457	126078...............	2452
62759...............	2404, 2480, 2484	127071...............	2452, 2482
63252...............	2921	143679...............	2452
64197...............	2462	147944...............	2595, 2599
64868...............	2452, 2670, 2963	149291...............	2401
65236...............	2962	149917...............	2481
66273...............	2452, 2479, 2480	151677...............	2408
66819...............	2621, 2943, 2957, 2959, 2963	154176...............	2613
67210...............	2952, 2792	275514...............	2473
68199...............	2406	297767...............	2405

RACT SERVICES REGISTRY NUMBER INDEX

ER	CAS REGISTRY NUMBER	ABSTRACT NUMBER
	1746016	2500
	2541697	2476
	3475636	2488
	3688537	2454, 2455
	6098448	2402, 2453
	7440075	2409, 2410
	7440246	2566
	7440291	2410
	7440359	2410
	7440451	2564
2930	7440484	2561
2683	7440611	2677
	7440655	2566
	7441044	2410
	7720787	2927
	7757791	2484
	8015950	2769
2458	8063943	2501, 2502, 2896
	9002624	2421, 2464, 2466
	9002862	2501, 2502, 2896
	9004108	2464
	9005792	2946

CAS 3

CAS REGISTRY NUMBER	ABSTRACT NUMBER	CAS REGISTRY NUMBER	ABSTRACT N
9007732	2927, 2928		
9008111	2604, 2611		
10043922	2897		
11078925	2409		
12001284	2496		
12001295	2496		
12059959	2409		
12172735	2496		
13345216	2477		
13823272	2409		
15158119	2403		
16543558	2486		
17068789	2496		
17573216	2477		
23713497	2403		
26675467	2408		
28822584	2959		
29925175	2959		
33953730	2477		
37558193	2462		
53609646	2483		

ER	WLN	ABSTRACT NUMBER

L E5 B666TTT&J E FQ F1UU1 OQ 2405

L E5 B666TTT&J E FQ OQ 2464, 2471, 2785, 2964

L E6 B6656 1A T&&&T&J R 2404, 2418, 2452, 2473, 2474, 2475, 2500, 2666, 2689, 2691, 2769, 2770

L G6 D6 B666J 2473

L66J BOVM1 2921

ONN1&VN1&1 2488

2503, ONN1&VO2 2492

ONN1&1 .... 2404, 2480, 2484

ONN2&2 .... 2480, 2482

QR BQ DYQ1MY -L 2959
2963

QR BQ DYQ1M1 -LQR BQ DYQ1M1 2415
2466,
2469,
2913, QVR CQ DQ EQ 2481

QVYZ2S2 .... 2792, 2952

QVYZ2S2 -DL 2792, 2952
2476,

QV1 2462

Q2K &Q 2680

T C6 B5665 2AB S BX IN QN NU JH&&TTTJ FO1
    IVH KVO1 KQ LOV1 M2 E- NT F6 E596 A BA
    LM&&TJ NVO1RQ R2................... 2700

T C666 BO EV INJ D FZ N G- K-/VM- OT5-16
    AN FVN IVN LVO PVM SVTJ G J KY N RY 2..
2785   2452, 2467, 2959, 2963

T D3 B556 BN EM JV MVTTT&J GO1 H1OVZ KZ L
                                    2452, 2621, 2933

VLN	ABSTRACT NUMBER	WLN	ABSTRACT NUM
T E3 D5 C555 A D- FO KUTJ AG AG BG JG KG LG	2404	T6NTJ ANO B- CT6NJ	2486
T F5 C6 B655 DOV GV OO QO RUT&&TTJ LO1	2404, 2457, 2458	T6NVMVJ EF A- ET5OTJ B1Q CQ	2604
T F6 D5 C666 EM ON&&TTTJ HO1 SOVR CO1 DO1 EO1& TO1 UVO1..................	2494	T6NVMVJ EI A- ET5OTJ B1Q CQ -A&C	2599, 2604, 2898
T3NTJ A- 3PST3NTJ A- 3PS	2452	T6NVNJ DZ A- BT5OTJ CQ DQ E1Q	2595, 2599
T3NTJ B A- 3PO	2452	T6VMVMV FHJ F2 FR	2458, 2482,
T5NN DSJ CSZW EMV1	2967	T6VMVTJ E1YQ- BL6VTJ D F	2621, 2943, 2959, 2963
T5NTJ ANO	2487	T66 BN DN GN JNJ CZ EQ H1MR DVMYVQ2VQ .	2406, 2942
T5OJ BYVZU1- BT5OJ ENW	2454, 2455	T66 BN DN GN JNJ CZ EZ HIMR DVMYVQ2VQ .	2452
T5OV EHJ CQ DQ EYQ1Q	2481	T66 BN DN GN JNJ CZ EZ H1N1&R DVMYVQ2VQ	2776
T56 BM DN FNVNVJ F H	2499	T66 BNJ BO ENW	2493
T56 BN DM FVM INJ	2776	T66 CNJ B1R CO1 DO1& HO1 IO1	2499
T56 BN DN FN HNJ IZ D- BT5OTJ CQ DQ E1Q -A&CD	2959	VH1YQYQYQ1Q -BAA -D	2613
T56 BN DN FNVNVJ B F H	2481	:NMYUM&N1&NO	2485, 2913
T56 BOV GO IU&TJ FQ	2401	WS1&O1	2452, 2479,
T56 BOXVJ FO1 HO1 IG C-& DL6V DX BUTJ CO1 E	2452	WS1&O2	2452, 2479
T6M DNTJ DNO	2481	WS1&O2 2U -C	2452
T6MPOTJ BO BN2G2G	2452, 2461, 2668, 2673	ZVM1	2481
T6MYMVJ BUS F	2934	ZVN1&NO	2480, 2481,
T6N DOTJ ANO	2481	ZVN2&NO	2480, 2490,
T6NJ B CQ D1Q E1Q	2962	ZVN4&NO	2489
T6NJ CVZ	2930	ZVO2	2404

SSER LINE NOTATION INDEX

NUMBER

Volume XIII, No. 6
June, 1975

Abstract and Citation Nos. 3001 - 3600

CARCINOGENESIS ABSTRACTS

A monthly publication of the

National Cancer Institute

Editor

George P. Studzinski, M.D.

College of Medicine and Dentistry

of New Jersey, Newark

Associate Editor

Jussi J. Saukkonen, M.D.

Jefferson Medical College, Philadelphia

NCI Staff Consultants

Elizabeth Weisburger, Ph.D.

Joan W. Chase, M.S.

DEPOSITORY

OCT 2 0 1976

UNIV. OF ILL. LIBRARY
AT URBANA-CHAMPAIGN

Literature Selected, Abstracted, and Indexed

by

The Franklin Institute Research Laboratories
Science Information Services
Biomedical Section

Bruce H. Kleinstein, Ph.D., J.D., Group Manager,
Biomedical Projects

Ruthann E. Auchinleck
Production Editor

Contract Number N01-CP-43293

Public Health Service, USDHEW

DHEW Publication No. (NIH) 76-301

PREFACE

Carcinogenesis Abstracts is a publication of the
National Cancer Institute. The journal serves as a
vehicle through which current documentation of carcinogenesis research highlights are compiled, condensed, and disseminated on a regular basis. It
represents an integral part of the Institute's program of fostering and supporting coordinated research
into cancer etiology. Issues of *Carcinogenesis Abstracts* normally contain three-hundred abstracts and
three-hundred citations (unaccompanied by corresponding abstracts). Abstracts and citations refer to
the current scientific literature that describes the
most significant carcinogenesis research carried on
at the National Cancer Institute, other governmental
agencies, and private institutions. *Carcinogenesis
Abstracts* is intended to be a highly useful current
awareness tool for scientists engaged in carcinogenesis research or related areas. The great number and
diversity of publications relevant to carcinogenesis
make imperative the availability of this service to
investigators whose work requires that they keep
abreast with current developments in the field.

Carcinogenesis Abstracts is normally published monthly. Volume XIII covers the scientific literature published from Jan 1975 through Dec 1975. To increase
the usefulness of *Carcinogenesis Abstracts,* Volume
XIII, a Wiswesser Line Notation index and a Chemical
Abstracts Service Registry Number index have been
provided. These indexes reference compounds described in abstracted articles. A cumulative subject,
author, CAS Registry Number, and Wiswesser Line Notation index for Volume XIII will be published shortly
after the final regular issue.

Carcinogenesis Abstracts is available free of charge
to libraries and to individuals who have a professional interest in carcinogenesis. Requests for
Carcinogenesis Abstracts from qualified individuals
should include statements of their relationship to
carcinogenesis research. All correspondence should
be addressed as follows:

> *Carcinogenesis Abstracts*
> Room C-325
> Landow Building
> National Cancer Institute
> National Institutes of Health
> Bethesda, Maryland 20014

The Secretary of Health, Education and Welfare has
determined that the publication of this periodical
is necessary in the transaction of the public business required by law of this Department. Use of
funds for printing this periodical has been approved
by the Director of the Office of Management and Budget through November 30, 1977.

NOTE

Journal names are abbreviated according to the list of abbreviations used by *Index Medicus*. For ʝurnals not covered by *Index Medicus*, the abbreviations found in *Chemical Abstracts Service Source Index*, ʝ07-1974 Cumulative, are used. New journals are verified in *New Serial Titles* and abbreviated according ʝ *International Standard ISO 833*. An asterisk indicates the author to address (other than the primary) in ʝquesting reprints.

LANGUAGE ABBREVIATIONS

Afr.	Afrikaans	Ind.	Indonesian
Ara.	Arabic	Ita.	Italian
Bul.	Bulgarian	Jpn.	Japanese
Chi.	Chinese	Kor.	Korean
Cro.	Croation	Lav.	Latvian
Cze.	Czech	Lit.	Lithuanian
Dan.	Danish	Nor.	Norwegian
Dut.	Dutch	Pol.	Polish
Eng.	English	Por.	Portuguese
Est.	Estonian	Rum.	Rumanian
Fin.	Finnish	Rus.	Russian
Fle.	Flemish	Ser.	Serbo-Croatian
Fre.	French	Slo.	Slovak
Geo.	Georgian	Spa.	Spanish
Ger.	German	Swe.	Swedish
Gre.	Greek	Tha.	Thai
Heb.	Hebrew	Tur.	Turkish
Hun.	Hungarian	Ukr.	Ukrainian
Ice.	Icelandic	Vie.	Vietnamese

ABBREVIATIONS USED IN ABSTRACTS

A	angstrom(s)	M	molar
ACTH	adrenocorticotropic hormone	mM	millimolar
ADP	adenosine diphosphate	µM	micromolar
AMP	adenosine monophosphate	mOsm	milliosmolar
ATP	adenosine triphosphate	mEq	milliequivalents
BCG	Bacillus Calmette Guerin	min	minute(s)
bid	twice daily	mo	month(s)
C	degrees centigrade	MTD	maximum tolerated dose
cal	calorie(s)	N	normal concentration
kcal	kilocalorie(s)	NAD	nicotinamide adenine dinucleotide
cc	cubic centimeter(s)	NADH	reduced nicotinamide adenine dinucleotide
Ci	curie(s)	NADP	nicotinamide adenine dinucleotidephosphate
mCi	millicurie(s)	NADPH	reduced nicotinamide adenine dinucleotide-
µCi	microcurie(s)		phosphate
cm	centimeter(s)	ng	nanogram(s) (10^{-9})
CNS	central nervous system	od	once daily
cpm	counts per minute	Pa	ambient pressure
dl	deciliter(s)	PAS.	periodic acid-Schiff
ml	milliliter(s)	pg	picogram(s) (10^{-12})
µl	microliter(s)	pgEq	picogram equivalent
DNA	deoxyribonucleic acid	po	orally
ED$_{50}$	median effective dose	ppb	parts per billion
EDTA	ethylenediamine tetraacetic acid	ppm	parts per million
ESR	erythrocyte sedimentation rate	qid	four times daily
g	gram(s)	qod	every other day
kg	kilogram(s)	QO$_2$	oxygen quotient
mg	milligram(s)	R	roentgen(s)
µg	microgram(s)	RBC	red blood cells (erythrocytes)
Hb	hemoglobin	RNA	ribonucleic acid
hr	hour(s)	sc	subcutaneous
ia	intra-arterial	sec	second(s)
ic	intracerebral	SGOT	serum glutamic-oxalacetic transaminase
icav	intracavitary	SGPT	serum glutamic-pyruvic transaminase
id	intradermal	SRBS	sheep red blood cells
ILS	increased life span	TCD	tissue culture dose
im	intramuscular	TCD$_{50}$	median tissue culture dose
ip	intraperitoneal	tid	three times daily
ipl	intrapleural	U	unit(s)
it	intratumorous	mU	milliunit(s)
IU	International Unit	UV	ultraviolet
iv	intravenous	vol	volume
K$_m$	Michaelis constant	WBC	white blood cells (leukocytes)
LD	lethal dose	wk	week(s)
LD$_{50}$	median lethal dose	wt	weight
m	meter(s)	x	times
mm	millimeter(s)	yr	year(s)

CONTENTS

	Cross Reference Abbreviations	Abstracts, Citations	Page
REVIEW............................(Rev)..........3001-3075			531
CHEMICAL CARCINOGENESIS.............(Chem).........3075-3177			545
PHYSICAL CARCINOGENESIS.............(Phys).........3178-3195			563
VIRAL CARCINOGENESIS................(Viral)........3196-3297			566
IMMUNOLOGY..........................(Immun)........3298-3396			587
PATHOGENESIS........................(Path).........3397-3502			608
EPIDEMIOLOGY AND BIOMETRY...........(Epid-Biom).....3503-3535			621
MISCELLANEOUS.......................(Misc).........3536-3600			627

AUTHOR INDEX...

SUBJECT INDEX..

CHEMICAL ABSTRACT SERVICES REGISTRY NUMBER INDEX.......................

WISWESSER LINE NOTATION INDEX..

the surface cell loss, probably consists of a system of tissue-specific chemical regulatory substances. The S-factor (G_1-chalone) and the M-factor (G_2-chalone) are known to inhibit the progress of cells through the cell cycle. It is noted that most of the experimental evidence on epidermal cell kinetics and early carcinogenesis is based on situations employing one or few large applications of chemical carcinogens and irradiation. One single application of a strong hydrocarbon carcinogen primarily leads to a toxic "shock" phase which is followed by a proliferative phase of increased DNA synthesis and mitotic rate, then consequent transient hyperplasia. The initial blocking of DNA synthesis seems to follow all types of carcinogenic influences; alterations induced by promoting substances in two-stage carcinogenesis models are very similar to those induced by high doses of complete carcinogens. Repeated doses of carcinogens also have kinetic consequences, and regularly lead to a situation where both cell proliferation is moderately increased and cell maturation is disturbed. Many studies have indicated that cells may be more sensitive to carcinogens when they are in DNA synthesis, or that tissues with a high rate of cell turnover, e.g. rapidly dividing epidermal basal cells, are most sensitive to carcinogens. It is concluded that carcinogens provoke obvious changes in almost all parameters of the epidermal cell kinetics; however, it is unknown whether such changes are specific, or toxic and nonspecific. (23 references)

3003 BREAST CANCER AND ORAL CONTRACEPTION. (Eng.) Taber, B. Z. (Palo Alto, Calif. 94304). *J. Reprod. Med.* 15(3):97-99; 1975.

A review of human retrospective and prospective studies, and studies on dogs and monkeys, indicates no association between oral contraceptive intake and benign or malignant breast disease. (13 references)

3004 IMMUNOSUPPRESSIVE ANTICANCER DRUGS IN MAN: THEIR ONCOGENIC POTENTIAL. (Eng.) Harris, C. C. (Building 37, Room 3C03, Natl. Inst. Health, Bethesda, Md. 20014). *Radiology* 114(1):163-166; 1975.

Case reports of new cancers in patients treated with anticancer drugs have recently appeared. Animal studies indicate that alkylating drugs (nitrogen mustard), those that bind DNA (actinomycin D), and those that form electrophilic reactants, may be oncogens. Antimetabolites (methotrexate) may be co-oncogens. Radiation therapy closely followed by intensive chemotherapy may increase the risk of a second neoplasm in Hodgkin's disease. The use of these drugs to treat nonfatal diseases indicates the need for studies to evaluate the hazard in humans. (72 references)

3005 OCCUPATIONAL DISEASES OF THE LUNGS. PART II. INHALATION DISEASES DUE TO INORGANIC DUST. (Eng.) Wolf, A. F. (Scott and White Clinic, Temple, Tex. 76501). *Ann. Allergy* 35(2):87-92; 1975.

Occupational diseases of the lungs caused by inhalation of inorganic dusts are described. In general, most dusts are capable of irritating or fibrotic effects, depending on the particle size, concentration in the inhaled air, duration of exposure, and individual susceptibility. Silicosis, due to exposure to silicon dioxide (quartz dust), occurs in gold miners, granite and tombstone workers, sand blasters, foundry workers, and those exposed to potter's clay and diatomaceous earth. The importance of particle size, characteristic nodular and fibrotic lesion's, roentgenographic patterns, and clinical evaluation is discussed in detail. Coal miners' pneumoconiosis, also referred to as anthracosis or anthracosilicosis, is due to the inhalation of a mixture of carbon, silica, and other dusts. The morphological changes due primarily to carbon deposition, the deposition around bronchioles and arterioles, and the disabling focal emphysema are described. The classification and symptomology of simple and progressive coal dust pneumoconiosis are discussed. Pneumoconioses due to graphite dust or iron dust exposure are also described. Asbestosis is also an important peumoconiosis, occurring chiefly in asbestos miners, but also in persons directly and indirectly exposed to the fiber. The significance of fiber length, the roentgenographic changes of fibrosis and focal emphysema, and the frequent lack of clinical symptoms are discussed. Complications due to bronchogenic carcinoma, bronchoalveolar carcinoma, and mesothelioma have been often reported in cases of asbestosis. Talcosis, resulting from exposure to magnesium silicate (tremolite talc) dust, affects miners, millers, and rubber workers; fibrosis and disabling symptoms are described. Nonpathological pneumoconiosis can also be due to radiopaque dusts of iron, tin, barium, antimony, and emery. Although a marked decrease of berylliosis has been observed, both chronic and acute forms are being found. Despite prior predictions, no pulmonary diseases have yet been ascribed to fiberglass exposure. Exposure to inorganic dusts may result in some roentgenographic similarities, but a wide range of symptoms of susceptibilities does occur. (No references)

3006   GENETIC DIFFERENCES IN SUSCEPTIBILITY TO
        CANCER AND TOXICITY CAUSED BY ENVIRONMENTAL CHEMICALS. (Eng.) Nebert, D. W. (Natl. Inst. Child Health and Human Development, Bethesda, Md.); Felton, J. S.; Robinson, J. R. *Proc. Eur. Soc. Study Drug Toxic.* 16:82-95; 1975.

Experimental data are presented showing that genetic differences in aromatic hydrocarbon responsiveness exist among various mouse strains, including inbred C57BL/6 (B6) and DBA/2 (D2) mice in progeny from various back- and inter-crosses. A responsive animal has higher inducible hydroxylase activities than a nonresponsive animal, not only in liver but in numerous nonhepatic tissues. About 11 inducible monooxygenase activities are known to co-segregate with increased hepatic cytochrome P_1-450 in aromatic hydrocarbon-treated responsive mice. Recent data indicate that hydroxylase induction by aromatic hydrocarbons is regulated by at least two genetic loci, instead of a single gene difference. By comparing responsive and nonresponsive siblings in 14 inbred mouse strains, it was possible to evaluate the sus-

ceptibility of each animal to various carcinogenic or toxic environmental agents. Tumorigenesis caused by the sc administration of 150 μg 3-methylcholanthrene, but not similar treatment with benzo(a)pyrene or with 7,12-dimethylbenz(a)anthracene, is highly correlated wih aromatic hydrocarbon responsiveness. Benzo(a)pyrene or 7,12-dimethylbenz(a) anthracene is more cytotoxic in cell cultures derived from responsive mice. Benzo(a)pyrene, 7,12-dimethylbenz(a)anthracene, or ß-naphthoflavone given in large ip doses cause responsive mice to die earlier than nonresponsive mice; these data are in agreement with those found in cell cultures and suggest that the activation of these chemicals to toxic metabolites is associated with increased aromatic hydrocarbon hydroxylase activities. However, nonresponsive mice receiving nonresponsive, 7,12-dimethylbenz(a)anthracene, 3-methylcholanthrene, or ß-naphthoflavone die sooner than responsive mice. On the other hand, responsive mice receiving polychlorinated biphenyls po die sooner than nonresponsive mice. When given lethal ip doses of the insecticide lindane (300 mg/kg), 3-methylcholanthrene-treated nonresponsive mice died much earlier than 3-methylcholanthrene-treated responsive mice. These studies demonstrate the importance of genetic predisposition in determining the extent to which a foreign compound may be carcinogeneic or toxic to a particular individual. (52 references)

3007   RETROSPECTIVE SURVEY OF THE ALLEGED
        DISEASES ASSOCIATED WITH VINYL-CHLORIDE IN THE FEDERAL REPUBLIC OF GERMANY. (Eng.) Thiess, A. M. (6700 Ludwigshafen am Rhein, W. Germany); Frentzel-Beyme, R. *J. Occup. Med.* 17(7):430-432; 1975.

Various aspects and problems of conducting epidemiological studies of disease associated with vinyl chloride/polyvinyl chloride (VC/PVC) exposure in the Federal Republic of Germany are reviewed. According to a survey made in 1974, a total of 180 cases of vinyl chloride-related diseases has been recorded. Of that number, 132 cases were workers employed in VC/PVC production plants and 48 were employed in PVC processing factories. Only 57 of 180 original cases were recognized as being occupational illnesses according to expert opinion from the Occupational Medical Officer. Five cases of hemangioendotheliosarcoma have been established in Germany among workers exposed to vinyl chloride. Four of these cases, all men exposed to VC/PVC for 11-17 yr, have died. The only living subject (49 yr old) had been employed as an autoclave cleaner for 11.75 yr. Three research projects are currently being carried out on the health effects of vinyl chloride: (1) an investigation into metabolism and pharmacokinetics of vinyl chloride; (2) a study of the documentation, methods of assessment and analysis of vinyl chloride; and (3) an epidemiological investigation of health consequences due to vinyl chloride. Preliminary results from the epidemiological investigation indicate that there have been 201 deaths of vinyl chloride-exposed workers, 49 of which have been caused by cancer (25%). In an epidemiological investigation being conducted by BASF, only 46% of those workers who have ever been exposed to

is deposited on bone surfaces of the trabecular matrices adjacent to the thin layer of endosteal tissue, which is the most critical tissue in this case. The use of a uniform distribution value also raises the question of the "hot particle" problem and the adequacy of a value of body burden or organ burden for ^{239}Pu based on the lung as the critical organ. The most reliable values of body burden based on the bone as the critical tissue can be obtained by comparing the incidence of bone carcinoma and sarcoma in dogs that have been injected with known amounts of ^{226}Ra and ^{239}Pu. Such studies suggest that the present maximum permissible body burden of ^{239}Pu should be reduced by at least a factor of 200. (23 references)

3010   CELL INJURY WITH VIRUSES. (Eng.) Tamm, I. (Rockefeller Univ., New York, N.Y. 10021). *Am. J. Pathol.* 81(1):163-178; 1975.

A review of literature concerning biochemical alterations that occur in cytocidal virus-cell interactions is presented. The inhibitions of cellular protein and RNA synthesis in picornavirus-infected cells are early events requiring only limited synthesis of virus-specific proteins. The inhibition of cellular protein synthesis appears to be due to blocking of the association of ribosomes with host messenger RNA. Inhibition of cellular RNA synthesis involves inactivation of the template-enzyme complex and affects ribosomal RNA synthesis before messenger RNA synthesis. Inhibition of cellular DNA synthesis also occurs early and may be secondary to inhibition of cellular protein synthesis. Early chromatid breaks may be related to inhibitions of cellular protein and nucleic acid synthesis. Stimulation of phospholipid synthesis in picornavirus-infected cells also requires only limited synthesis of virus-specific proteins. In contrast, release of lysosomal enzymes, proliferation of smooth cytoplasmic membranes, cellular vacuolization, retraction, and rounding, and diffuse chromosomal changes related to karyorrhexis and pyknosis are all dependent on considerable synthesis of virus-specific proteins during the middle part of the picornavirus growth cycle. The early virus-induced alterations in cellular biosynthetic processes are not the direct or sole cause of the subsequent marked pathologic changes in the membranes and chromosomes of the cell. (62 references)

3011   PART XV: TUMORS: STUDIES ON EPSTEIN-BARR VIRUS-ASSOCIATED ANTIGENS. (Eng.) Ernberg, I. (Karolinska Inst., S-10401 Stockholm 60, Sweden). *Ann. N.Y. Acad. Sci.* 254:516-522; 1975.

The Epstein-Barr virus (EBV) is discussed in association with three diseases in man, i.e., Burkitt's lymphoma, nasopharyngeal carcinoma, and infectious mononucleosis. The association of these diseases to EBV is based on the altered serological patterns, raised antibody titers against EBV-associated antigens, the presence of virus-associated antigens, and the discovery of viral genetic material. Studies of patient antibody reactivity to virus-associated antigens involve membrane antigens, early antigens, viral cap-

sid antigen, and Epstein-Barr viral nuclear antigen
(EBNA). *In vitro* studies on lymphoblastoid cell
lines, surface characteristics of B cells have been
assigned to EBV-infected cells, and comparatively
few EBNA sites per cell have been found. The cell
lines containing EBV genetic material are classified
into two groups, i.e., nonproducer and producer cell
lines, depending upon their ability to synthesize
viral DNA and late viral products. Such studies have
demonstrated an additional membrane antigen, the late
membrane antigen. The EBV genome in certain nonproducer cell lines can be induced to synthesize viral
antigens *via* addition of iododeoxyuridine or bromodeoxyuridine. Antigen synthesis is also induced by
superinfection of cell lines with purified EBV. The
varying degrees of spontaneous expression of viral
antigens suggests mechanisms of cellular control.
The studies thus far have indicated that early membrane antigens, early antigens, viral capsid antigen,
and late membrane antigens are associated with the
virus-producing cycle. Other experiments have suggested that EBNA is still present in the cell when
early antigen is synthesized. (70 references)

3012   EPSTEIN-BARR VIRUS (EBV). (Eng.) Klein,
          G. (Dept. Tumor Biology, Karolinska Insti-
tutet, S 104 01 Stockholm 60, Sweden). *Proc. Int.
Cancer Congr. 11th.* Vol. 2 *(Chemical and Viral
Oncogenesis)*. Florence, Italy, October 20-26,
1974. Edited by Bucalossi, P.; Veronesi, U.;
Cascinelli, N. New York, American Elsevier, 1975,
pp. 282-297.

Research on the Epstein-Barr virus (EBV) is reviewed
and the role of EBV in human malignancy is discussed.
Six major findings are especially noted: (1) normal
human lymphocytes do not grow as established lines
in vitro, unless and until EBV is added; established
lymphoblastoid lines then carry multiple viral
genome equivalents/cell. (2) primary infection
of young children with EBV leads to seroconversion
but no clinically recognizable disease; however,
primary infection of teenagers or young adults
leads to acute infectious mononucleosis in 50% of
the cases. (3) antibody titers against EBV-determined antigens are increased in various disease
states; only two human neoplastic diseases, i.e.,
nasopharyngeal carcinoma and African Burkitt lymphoma cases tested, by nucleic acid hybridization
and/or EBNA tests, contain the viral genome. (5)
EBV can induce malignant lymphoproliferative disease
in marmosets and owl monkeys; the resulting tumors
carry the EBV genome. (6) poorly differentiated
or anaplastic nasopharyngeal carcinoma cells also
contain the EBV genome. There is thus no doubt
that the EBV can transform normal lymphocytes into
potentially malignant cells, can induce tumors in
some nonhuman primates, and is capable of causing
a self-limited lymphoproliferative disease. Several
hypotheses of the role of EBV in human malignancy
are then presented. The "passenger hypothesis,"
i.e., the notion that the ubiquitous EBV is a frequent contaminant of lymphoid tissue, is deemed
untenable because the viral genome or its antigenic
"footstep" (EBNA) is not found in the neoplastic
cells of any human lymphoproliferative disease
except African Burkitt's lymphoma. It is then

postulated that primary infection and neoplastic
transformation coincide closely in time, or, alter-
natively, that the neoplastic transformation responsible for African Burkitt lymphoma preferentially
affects an EBV-genome carrying cell. Two main
alternatives are presented; the "immunological
hypothesis" suggests that the EBV-converted lymphoid
cell requires additional change prior to attaining
full malignant potential. It is also suggested
that the EBV-genome negative non-African Burkitt
lymphomas and the few (3%) genome negative African
cases probably have a different, entirely EBV-
unrelated origin. Speculations on the role of
EBV in nasopharyngeal carcinoma also note the
association of the viral genome and suggest a major
genetic influence. Considering also the possible
existence of different EBV variants of different
oncogenic properties, the exact role of EBV in
the neoplastic process is not yet resolved.
(44 references)

3013   HUMAN HERPESVIRUSES. I: A MODEL FOR
          MOLECULAR ORGANIZATION OF HERPESVIRUS
VIRIONS AND THEIR DNA. (Eng.) Roizman, B. (Dept.
Microbiology, Univ. Chicago, Chicago, Ill.); Bayward,
G.; Jacob, R.; Wadsworth, S.; Honess, R. W. *Proc.
Int. Cancer Congr. 11th.* Vol. 2 *(Chemical and
Viral Oncogenesis)*. Florence, Italy, October 20-26,
1974. Edited by Bucalossi, P.; Veronesi, U.;
Cascinelli, N. New York, American Elsevier, 1975,
pp. 188-198.

Current knowledge concerning the molecular organization of Herpes simplex virus (human herpesvirus-
1, HSV-1) virion and its DNA is reviewed. DNA,
proteins, glycoproteins, polyamines, and lipids
are currently recognized as the major components
of the virion. The protein:DNA ratio for purified
HSV-1 is 10.70 ± 0.97, the molecular wt of viral
DNA is 10^8 and the wt of virion proteins is $19.4
\times 10^{-6}$ gm/particle. Polyacrylamide gel electro-
phoresis of purified HSV-1 virions reveals the
presence of at least 33 species of polypeptides
greater than 25,000 in molecular wt; the major
differences in the electrophoretic profiles of
various strains of HSV-1 DNA at $99 + 5 \times 10^8$. Background studies in the repetitive regions in the DNA
are summarized; recent studies on the nature and
organization of the repetitive regions are discussed
in detail. Examination of DNA denatured with formamide and allowed to renature yields two kinds of
structures, i.e., DNAs with single small loops with
short double-stranded regions and DNAs with barbell
structures. Precise localization of the observed
cyclodome and its relationship to the termini is
inferred from the partial denaturation maps presented and discussed. Analysis of viral DNA with
Hin III and Eco RI restriction enzymes note that
all of the HSV-1 and HSV-2 DNAs tested to date
yield two sets of fragments, i.e., major and minor
fragments. The virus strain specific gel electrophoretic profiles of the fragments are also presented
and discussed. Implications arising from the presence of cyclodomes in the DNA structure are considered, and a model of the molecular organization of
the DNA is constructed. A detailed analysis of the
molecular weight and molar ratios of the HinIII

534

Clinical, cytological, and epidemiological studies
of feline viral leukemia are reviewed. A 5-yr survey revealed that 33% of feline malignant diseases
are hematopoietic, covering a broad spectrum. The
great majority are lymphosarcoma. Although the presenting signs are vague and nonspecific, the major
tumor predilection sites are the kidney, anterior
thymus, intestine, and mesenteric lymph nodes.
Granulocytic leukemia, eosinophilic leukemia, erythremic myelosis, and erythroleukemia are described
in the cat. Transmission of feline leukemias into
neonatal kittens *via* inoculations of cell free tumor extract, purified virus, cellular tumor extract,
and homogenized tumor tissue have been reported.
Detailed ultrastructural studies have revealed a
wide distribution of C-type particles in infected
cats. Further studies have indicated that feline
leukemia virus (FeLV) is capable of inducing malignant change in all cell types of the hematopoietic
system. The antigenic structure of FeLV corresponds
to that of other C-type leukoviruses. Type- or
strain-specific antigenic determinants are present
on the viral envelope, group-specific antigenic components, and an RNA-directed DNA polymerase have also been isolated. Cells infected with FeLV develop
an antigenic determinant on their surface membrane.
Antibodies to the envelope and group-specific antigens and to the viral polymerase have also been detected in sera of normal and leukemia cats. A correlation between antibody titer against the cell
membrane antigens and resistance to the virus-induced
neoplasms has been suggested as having a role of depressed cellular immune responses. Epidemiological
and serological surveys have supported a horizontal
mode of viral transmission, as the virus has been
isolated from urine, saliva, nasal, and tracheal
epithelium of infected cats. Preliminary findings
have also indicated vertical transmission. However,
no correlation between feline viral leukemia and human leukemia is found. (56 references)

3016   MAREK'S DISEASE AND AVIAN LEUKOSIS. (Eng.)
        Biggs, P. M. (Houghton Poultry Res. Sta-
tion, Houghton, Cambs. PE17 2DA, England). *Proc.
Int. Cancer Congr. 11th*. Vol. 2 *(Chemical and Viral
Oncogenesis)*. Florence, Italy, October 20-26,
1974. Edited by Bucalossi, P.; Veronesi, U.;
Cascinelli, N. New York, American Elsevier, 1975,
pp. 166-175.

The pathogenesis, virology, natural history, and
control of Marek's disease (MD) and of avian leukosis are reviewed. Although the herpesvirus of MD
is ubiquitous among poultry populations, mortality
varies from 0 to 80% of a flock. Two forms of MD
are recognized; the classical form is characterized by neural involvement, while the main manifestation of the acute form is lymphoid neoplasia.
Pathogenesis studies have described three phases
of the development of lesions of MD: cytolytic
infection, proliferative changes, and chronic inflammation. The development of the pathological
changes in the disease is summarized, and observations note that infection is persistent and non-
productive in many cells. Various theories are
proposed concerning the cause of the widespread
proliferation of lymphoid tissue in MD; intrinsic

theories suggest that the stimulus comes from within the cells while extrinsic theories propose a stimulus outside the cell. Virology studies divide MD viruses into three groups on the basis of their pathogenicity, i.e., those that produce the acute form of MD, those producing the classical form, and those that are apathogenic. While virus is always horizontally transmitted, usually early in life, such infection does not necessarily result in clinical MD. Factors affecting the development of the disease may include: the genetic constitution of the chicken, the sex, the age at infection, the immune status of the host, and the strain(s) of the virus. Although the mechanism of the protection is unknown, an attenuated pathogenic strain and an apathogenic strain of MD virus are shown to immunize chickens against later challenge. The leukoses, caused by a group of RNA tumor viruses called the avian leukosis/sarcoma group, consist of several clearly recognizable pathological entities; lymphoid leukosis is especially noted. The avian leukosis/-sarcoma group of viruses is defined by the presence of group-specific antigens present in the core; however, the viruses are usually classified into subgroups on the basis of properties of the viral envelope: (1) intraspecies host range, (2) interference, (3) antigenic characteristics in neutralization tests. Resistance and susceptibility to the viruses are determined by properties of the host cell membrane. Both vertical and horizontal transmission of leukosis viruses occurs, and a cycle of infection is postulated. The chicken's phenotype, the presence of maternally-derived antibody, plus other undetermined factors may affect the infection and development of the disease. Thus, although the DNA Marek's disease herpesvirus and the RNA leukosis viruses are ubiquitous infections in poultry populations, the development of disease is the exception rather than the rule. (64 references)

3017   STUDIES OF VIRAL GENE PRODUCTS IN CELLS
        INFECTED OR TRANSFORMED WITH PAPOVA VIRUS.
(Eng.) Cuzin, F. (Centre de Biochimie, Universite de Nice, Nice, France). *Proc. Int. Cancer Congr. 11th.* Vol. 2 *(Chemical and Viral Oncogenesis).* Florence, Italy, October 20-26, 1974. Edited by Bucalossi, P.; Veronesi, U.; Cascinelli, N. New York, American Elsevier, 1975, pp. 199-206.

The purification and functional identification of the virus gene products of cells transformed or lytically infected by polyoma virus and simian virus 40 (SV40) were reviewed. Viral coded proteins belong to two main classes: the early gene products are the first to be synthesized during the virus cycle, independently of viral DNA synthesis; the late gene products are synthesized only after initiation of viral DNA synthesis and primarily include the structural proteins of the virions. Methodologies employed in current studies include: the purification of the early viral T antigen, the *in vitro* synthesis of viral peptides in viral DNA or mRNA-directed protein synthesizing systems, and the purely analytical identification of the new peptides present in a virus-infected cell. Infection of either permissive or non-permissive cells

with polyoma or SV40 leads to derepression of the synthesis of a number of cellular enzymes; therefore, a major problem is that of distinguishing the actual virus coded proteins from the induced cellular ones. The gel electrophoresis analysis of new peptides made in SV40 or polyoma virus-infected cells is then described. Immunological methods offer specific probes for the determination of new products or structures present in virus-infected or transformed cells; a whole series of such changes are designated, i.e., as T, U, S, TSTA, and V antigens. Studies on the nature of the "Tumor" (T) antigen are especially discussed. This antigen has been evidenced *via* complement fixation and immunofluorescence, and is present in the nuclei of transformed or lytically-infected permissive cells; further observations lead to the hypothesis that an early virus coded peptide, product of the temperature sensitive (ts) a gene, carries the T antigenic state. Although T antigen purification by direct immunoadsorption leads to important experimental difficulties, use of a complement fixation assay yields a partial purification of SV40 and polyoma T antigen. The partially purified T antigen has a molecular wt of 60-100,000 daltons, can strongly bind double stranded DNA, and is found in cell extracts under several different forms. The existence of two independent early gene products, in polyoma or SV40, is suggested but not proven. Meanwhile, several hypotheses regarding the function of the early viral protein(s) are presented; these models are derived from *in vivo* studies of ts mutants. The products of the A gene of SV40 and of the ts-a gene of polyoma are suggested to play the role of initiator for the viral replicon. A study of the phenotype of the polyoma ts-a mutant led to the hypothesis that the corresponding gene product is necessary for the integration of viral DNA into the cell genome. It is also suggested that the T antigen is necessary for the maintenance of the transformed state. Analysis of the structural components of the virus particle reveals similar patterns for polyoma and SV40. (45 references)

3018   GENE FUNCTIONS OF SIMIAN VIRUS 40. (Eng.)
        Kimura, G. (Tottori Univ. Sch. Medicine, Yonago 683, Japan). *Proc. Int. Cancer Congr. 11th.* Vol. 2 *(Chemical and Viral Oncogenesis).* Florence, Italy, October 20-26, 1974. Edited by Bucalossi, P.; Veronesi, U.; Cascinelli, N. New York, American Elsevier, 1975, pp. 214-219.

A physical map of the simian virus 40 (SV40) genome is described, the behavior of temperature-sensitive (ts) mutants of SV40 in permissive cells is reviewed, and data on the transforming functions of SV40 is summarized. More than 200 mutants of SV40 thermosensitive for viral replication in permissive cells have been isolated and characterized. These are divided into three complementation groups, i.e., groups A, B, and C or groups I, II, and III; a non-complementing D group and an abnormally-complementing BC group are also known. By using bacterial restrictive endonucleases, a physical cleavage map of the SV40 chromosome, the ts and deletion mutants, was

alcoholic pancreatitis. In serial pre- and postoperative determinations, negative CEA levels do not preclude primary, metastatic, or recurrent colon cancer. Serial determinations are useful in monitoring the response of patients with metastatic breast cancer to chemotherapy; the CEA assay may also be a valuable prognostic marker in lung cancer patients treated by resection and/or radio- and chemotherapy. Serum AFP levels may exceed 1,000 ng/ml in patients with primary liver carcinoma, and serial determinations may help to assess prognosis and recurrence in these patients. The role of AFP as a monitor of chemotherapy has not been adequately studied. Other tumor markers discussed include α_2-H fetoprotein, found in 60% of patients with primary liver carcinoma, 50% of patients with various malignant diseases, and 20-30% of those with benign diseases; β-S fetoprotein, present in 48% of patients with primary liver carcinoma; and fetal sulfoglycoprotein, found in gastric juice of 75 of 78 patients with gastric carcinoma. Two other markers under investigation are carcinoplacental alkaline phosphatase, a nonspecific marker, and an oncofetal antigen that has been found in the sera of patients with carcinoma of the pancreas. (68 references).

3020   PATHOLOGICAL CONSIDERATIONS IN PITUITARY
        TUMORS. (Eng.) Bergland, R. M. (M. S.
Hershey Medical Center, Pennsylvania State Univ.,
Hershey, Pa. 17033). *Prog. Neurol. Surg.* 6:62-94;
1975.

The histological, immunohistochemical, and light and electron microscopic observations of numerous pituitary tumors and pituitary capsules are reviewed. The early descriptions of pituitary cells and the early histological classifications by acidic-basic stains are presented. Problems of histochemical classifications of pituitary cell types are discussed; these problems include the misleading nature of target organ ablation, the biochemical species specificity of the pituitary trophic hormones, various technical problems, and the tangled nomenclature employed. A functional classification designating pituitary cells according to the hormones elaborated is then presented. Peculiar histological features of the human pituitary include the lack of the pars intermedia, the follicular cells, the posterior pituitary basophils, and the controversial "amphophils." The employment of electron microscopy in the study of pituitary cell types has revealed the vascular relationships, has demonstrated the morphology and packaging of pituitary hormones, and has revealed the lysosomes of prolactin cells and the presence of follicular cells. The use of horseradish peroxidase as a histochemical marker in electron microscopy is described, and a schematic illustration of the Nahane horseradish peroxidase immunohistochemical staining technique is presented. The early history of the use of acidic-basic stains in the classification of pituitary tumors has reported the histological descriptions of chromophobic tumors, acromegaly, and Cushing's disease. Subsequent operative histological observations, and problems in the evaluation of pituitary tumor histology, are discussed in detail. The Young series of 44 consecutive pituitary tumors associated with acromegaly, the histological results of the Schelin series of ten acromegalic patients, and the Ray series analyzing exenterated sellae in 55 acromegalic patients are described. Discrepancies between autopsy studies and operative studies are noted. Further electron microscopic observations, radioimmune assays of pituitary hormones, and immunohistochemical analyses of pituitary tumors are also discussed. Following a description of the normal microvascular structure of the pituitary, the tumor-pituitary interface (pseudocapsule), tumor-sellar interface (pseudo-capsule) and tumor-brain interface (capsule) are described and diagrammatically and photographically represented. Studies of the vascular supply of pituitary tumors note the structure and importance of the normal pituitary portal system, illustrate the blood supply of experimental tumors, and suggest that human pituitary tumors may likewise derive their blood supply wholly *via* the pituitary portal system. (32 references)

3021   PROGRESS IN THE CLASSIFICATION OF LYMPHOID
        AND/OR MONOCYTOID LEUKEMIAS AND OF LYMPHO
AND RETICULOSARCOMAS (NON-HODGKIN'S LYMPHOMAS).
(Eng.) Mathé, G. (Institut de Cancerologie et d'Immunogénétique (INSERM et Association Claude Bernard), Hôpital Paul-Brousse, 14-16, avenue Paul-Vaillant Couturier, 94800 Villejuif, France); Belpomme, D.; Dantchev, D.; Pouillart, P. *Biomede-cine* 22(3):177-185; 1975.

The construction of a new classification of lymphoid and monocytoid leukemias and hematosarcomas and new analytical methods are reviewed. A discussion of classical and new morphological and immunological methods emphasizes the value of immunological markers, ultrastructural analysis by conventional and scanning electron microscopy, and Giemsa smears. Non-thymic-dependent (B) lymphocytes have been found in chronic lymphoid leukemia, primary macroglobulinemia, and chronic macroglobulinemic proplasmocytic leukemia. The B nature of myleoma plasma cells is also evident in plasma-cell leukemia and heavy chain diseases. The thymic-dependent (T) cells of Sézary's disease and traditional acute lymphoid leukemia are discussed. A fifth type of T or B cells have been noted in immunoblastic acute lymphoid leukemia and leukemic lymphosarcoma. Difficulties in the classification of malignant histiocytosis and in the identification of chronic or subacute leukemias as lymphoid *versus* monocytoid are also acknowledged. The classification of hematosarcomas notes the B lymphoid cells of the nodular lymphosarcoma and the T or B lymphoid cells of the diffuse lymphosarcoma. Nodular lymphosarcomas are recognized as prolymphocytic or lymphoblastic, while diffuse lymphosarcomas may be prolymphocytic, lymphoblastic, or immunoblastic. True (African) Burkitt's lymphosarcomas are distinguished from pseudo (non-African) Burkitt's lymphosarcomas on the basis of macrophages and Epstein Barr virus genome. The mediastinal form of lymphosarcoma is composed of T immunoblasts, and the cells of mycosis fungoides are considered as T lymphocytes. The diagnosis of reticulosarcoma is based on the pre-

sence of many reticulin fibers at histiological examination and on the cytological aspect of the cells on smears. All the prognostic parameters in the four types of common acute lymphoid leukemia are seen to be related to cytological type, and a revision of the nomenclature and classification of the disease is considered justified. (62 references)

3022   CURRENT AND FUTURE CONCEPTS OF LUNG CAN-
        CER. (Eng.) Benfield, J. R. (Harbor
General Hosp., 1000 Carson St., Torrance, Calif. 90509); Pilch, Y. H.; Rigler, L. G.; Selecky, P. *Ann. Intern. Med.* 83(1):93-106; 1975.

Diagnosis and therapeutic advances in the treatment of lung cancer, and the objectives of future research are discussed. An understanding of the natural history of lung carcinoma is important; several detailed case studies are presented. Serial roentgenograms, have been useful in many retrospective diagnoses. Survival is a function of many variables; these include cell type, anatomic location, tumor size, and patient performance status. As an aid to reporting and discerning the prognosis, a staging system is described; such staging also contributes to determining surgical resectability. Diagnostic methods including bronchoscopy, mediastinoscopy, scalene node biopsy, percutaneous needle biopsy, aspiration needle biopsy, and bronchial brush biopsy are evaluated. Various screening procedures for the detection of lung cancer are reviewed. Such screenings involve chest roentgenograms, sputum cytology, and combined roentgenograms and cytology. Concepts of operative care are discussed, with particular attention to lobectomy and pneumonectomy. The role of radiotherapy in the management of lung carcinoma, including combined chemotherapy-radiation therapy approaches, is discussed and followed by a discussion of potential immunotherapy. The prognostic significance of delayed cutaneous hypersensitivity reactions is considered, and previous studies of immunotherapy are evaluated. (80 references)

3023   SARCOMAS OF THE LARYNX. (Eng.) Fried-
        mann, I. (Northwick Park Hosp., Watford
Road, Harrow, Middlesex, England, HA1 3UJ). *Can. J. Otolaryngol.* 4(2):297-302; 1975.

The epidemiology and histological features of benign mesenchymatous tumors, leiomyosarcomas, cartilaginous tumors, rhabdomyosarcomas, Kaposi's hemangiosarcomas, and giant cell sarcomas of the larnyx are reviewed. The benign tumors of the larynx (e.g., polyps, granulomas, and cysts) are not true neoplasms and arise in the stroma of the vocal cord. Diagnostic difficulties may arise when the stroma shows enhanced cellularity, and the surface epithelium is hyperplastic. Sarcomas of the larynx occur once to every 100 cases of carcinoma, more than 50% of these tumors being fibrosarcomas. Most fibrosarcomas occur in men over 50 yr of age, although they may occur at any age and in either sex. Histologically, fibrosarcomas in the larynx resemble fibrosarcomas in other parts of the body and are divided into well differentiated and poorly

either in negative findings or in weak positive results. It is suggested that the issue of clustering in leukemia remains open, not because there are a sufficient number of positive results, but because the phenomenon has only been studied superficially. It is concluded that there is no significant evidence that any human cancer has the characteristics of a communicable disease. (22 references)

3025   EPIDEMIOLOGY OF HODGKIN'S DISEASE: GEO-
        GRAPHIC VARIATIONS IN INCIDENCE AND HIS-
TOLOGIC SUBTYPES. (Eng.) Sacks, M. I. (Hadassah Univ. Hosp., Jerusalem, Israel). *Proc. Int. Cancer Congr. 11th.* Vol. 6 (*Tumors of Specific Sites*). Florence, Italy, October 20-26, 1974. Edited by Bucalossi, P.; Veronesi, U.; Cascinelli, N. New York, American Elsevier, 1975, pp. 345-350.

Recent studies in which pathologists from different parts of the world have reviewed pathological material from their countries have confirmed raw cancer registry data indicating that there are considerable geographic variations in the incidence and age distribution of Hodgkin's disease (HD). There are marked geographic differences in the relative frequency of the nodular sclerosis subtype HD and it seems most unlikely that these are merely due to the use of different histologic criteria by pathologists in different countries. A major determinant of the geographic differences in the epidemiologic and histologic patterns of HD seems to be the socioeconomic level of the community studied. Poor socioeconomic conditions seem to lead to HD at an earlier age, and to a predominance of the histologic types associated with a poor prognosis. In poorer countries, nodular sclerosis is uncommon at all ages, which may be a reflection of host resistance. These findings are consistent with the view that the pattern of HD seen in a particular population is the result of an interaction between the etiologic agent(s) of the disease and host susceptibility. According to this view, the different epidemiologic and histologic patterns of HD reflect different types of host response in different environments rather than multiple disease entities. While host susceptibility is probably largely dependent on environmental influences, the possibility that genetic factors may also play a role cannot be excluded and it has recently been suggested that the histologic picture of HD may be influenced by the HL-A phenotype of the host. Having accepted the evidence that environmental factors are important in the etiology of HD, the great challenge is to further elucidate the possible role of infective agents in the causation of the disease. (34 references)

3026   STATISTICS AND CANCER. (Eng.) Mould, R.
        F. (Westminster Hosp., Page Street Wing, London SW1, England). *Nurs. Times* 71(29):1122-1133; 1975.

The statistical analyses of cancer incidences, survival times, mortality rates, and cure rates are discussed. Survival time from treatment to death can be divided into two periods: apparent disease-free time (remission) and terminal disease time (time from recurrence to death). The distribution of terminal disease time for 854 cervical carcinoma patients is graphically represented. The prognosis differs with the clinical stage, but the terminal disease time is approximately independent of clinical stage. The survival time distribution of patients who eventually die with primary and/or metastatic stage three breast carcinoma, stage two cervix carcinoma, and pharyngeal carcinoma is also graphically represented; similar trends are noted in all three sections. Based on a series of 99 recorded patient histories, the random site distribution of second primaries subsequent to cervical cancer is illustrated. The necessary parameters required for a cancer registry data bank are tabulated. Wide variations have been found in the national patterns of cancer incidence throughout the world. The annual incidence rate and annual mortality rate are defined and cancer incidence rates are graphed. The unusual incidence and national registration patterns of cervical carcinoma *in situ* are discussed. Examples of both occupational and nonoccupational environmental links with cancer are tabulated. The recognition of and precaution for the occupational hazard associated with X-rays and radium is noted. Wide variations in national cancer mortality rates are also noted, and the annual mortality rates from 1851 to 1969 are discussed. National patterns of cancer mortality, and 1969 mortality rates for cancer and other diseases are tabulated; different rates in men and women are revealed. Computations and methods of calculating five-year survival rates are described, and five-year cancer survival rates are also tabulated. A discussion of the concept of "cure rate" indicates that the five-year survival rate will not always be a good indication of cure. Crude and age-corrected survival rates are analyzed, and estimates of the proportion cured are made using statistical prediction models. It is concluded that the definitive year is usually about the 10th yr after treatment. (No references)

3027   GROWTH OF PERMANENT LYMPHOID CELL CULTURES FROM HUMAN SOURCE: TENTH ANNIVERSARY. (Eng.) Wang, C. -H. (Univ. Texas System Cancer Center, Houston, Tex. 77025); Sinkovics*, J. G.; Kay, H. D.; Gyorkey, F.; Shullenberger, C. C. *Tex. Rep. Biol. Med.* 33(1):213-250; 1975.

The development and peculiar features of human lymphoid cell cultures are reviewed, and previously unpublished observations on 52 permanent lymphoid cell cultures are presented. The techniques of growing permanent cultures of human lymphoid cells from WBC, bone marrow explants, splenic tissue, and lymph node tissue are described. The events of lymphoblastoid transformation in tissue cultures of lymph nodes or spleen derived from malignant lymphomas of Hodgkin's disease are tabulated; the duration of elongated cells in cultures undergoing lymphoblastoid transformation is compared with cultures failing to produce lymphoid cells. Established cultures of lymphoid cells show morphological

features displaying a wide range of developmental stages (i.e., reticulum cells, monocytes or macrophages, lymphoblasts, plasmablasts, and large and small lymphocytes). Generally stable diploid, hyperdiploid, or hypodiploid chromosome ranges have been found, although abnormal karyotypes have also been observed. The kinetics of growth and cloning are discussed and the production of well-defined antigens, immune globulins, and various mediators of the delayed hypersensitivity reaction is documented. The inhibition of the growth of established target cells by permanent cell lines of lymphoid cells is graphically represented. Other studies have been concerned with the roles of the Epstein Barr virus in the establishment of permanent human lymphoid cell lines; herpesvirus morphology has been found in lymphoid cells established from explants of breast carcinoma, angiosarcoma, Kaposi's sarcoma, malignant melanoma, and thyroid carcinoma. The pathogenicity of human lymphoid cell lines in animals is discussed, and their apparent infinite growth potential *in vitro* has been noted. Lymphoid cell cultures are also obtained from normal healthy individuals and from patients with numerous pathological conditions, including malignant lymphomas, leukemias, nasopharyngeal carcinoma, Chediak-Higashi syndrome, and multiple myeloma. Exceptional cell lines developed from patients with lymphoproliferative diseases and exceptionally uniform suspension cultures of lymphoblastoid cells are discussed in detail. New applications of lymphoid cell cultures in immunology are then described. The reaction of human sera with established lymphoid cell lines may indicate an expression of antibody activity directed to the Epstein Barr virus or its antigens; established lymphoid cell lines express cytotoxicity indiscriminately to batteries of target cells, and suspension cultures of human lymphoid cells can be used as target cells in assays measuring migration inhibitory factors. The major paradox thus appears to be the morphological and functional heterogeneity of the cultures. (150 references)

3028   CARCINOEMBRYONIC ANTIGEN (CEA): MOLECULAR BIOLOGY AND CLINICAL SIGNIFICANCE. (Eng.) Fuks, A. (Montreal Gen. Hosp., Canada); Banjo, C.; Shuster, J.; Freedman, S.O.; Gold, P. *Biochim. Biophys. Acta* 417(2):123-152; 1975.

Information on the carcinoembryonic antigen (CEA) of the human digestive system is discussed. Tumor-specific antigens in experimentally-induced animal tumors, as well as the induction of experimental animal tumors by chemical carcinogenic agents, DNA or RNA oncogenic viruses, and human tumor antigens, are reviewed. Technique for the demonstration of CEA, including the use of antisera and immunochemical procedures, are noted. This is followed by a description of the known cellular localization of the CEA, the cellular and human responses to CEA, and observations on the circulation and metabolism of CEA. The initial experimental work on CEA summarized establishes the existance of a material associated with carcinomata originating from the entodermally-derived portion of the gastrointestinal

ing-Passey melanoma consisting of RNA and protein, and is augmented by hydrocortisone. The "antichalones" may have reactive sites separate from mitotic stimulators, but the actual sites are unknown. In melanocyte and lymphocyte G1 chalones, a sialic acid component is necessary for conserving activity. There is no positive correlation between cyclic AMP levels and chalone action. Chalones do act by restricting the nucleoside pool in cultured lymphocytes, with an inhibition of nucleoside kinase activity. Melanocyte G1 chalone depresses both protein and DNA synthesis. Chalones may be useful in cancer treatment as adjuncts to chemotherapy. However, the immunological responses have not been distinguished from the mitotic activity, and they may play only a part of the complex of tissue-stimulators and inhibitors. (45 references)

3030   THE ROLE OF CHOLESTERYL 14-METHYLHEXADE-
        CANOATE IN GENE EXPRESSION AND ITS SIGNI-
FICANCE FOR CANCER. (Eng.) Hradec, J. (Dept. Biochemistry, Oncological Inst., 18000 Prague 8, Czechoslovakia). *Prog. Biochem. Pharmacol.* 10:197-226; 1975.

The role of cholesteryl 14-methylhexadecanoate (CMH) in enzyme activation, gene expression, and malignant growth is discussed. Experimental results indicate that lipids are involved in the synthesis of the aminoacyl-tRNA complex. CMH, the only lipid found to be active in that respect, forms part of the molecule of amino acid-tRNA-ligase required for normal enzymatic activity. Experiments suggesting a close relationship between the chemical structure and biological activity of CMH indicate the presence of a fatty acid with a definite chain-length, and anteiso-type branching is essential for the activity in protein synthesis. CMH administration also enhances the rate of DNA-dependent RNA polymerase activity, and increases the specific activity of the ribosome finding enzyme (transferase I). The presence of CMH has thus been found to be essential for the normal function of amino acid-tRNA ligases, transferase I, transferase II, and peptidyl transferase. Studies on ribosomes indicate that CMH affects the function of the A site and the peptidyl transferase site. Preliminary isolation experiments have revealed higher quantities of CMH in tumor tissues than in other animal tissues. Despite a relatively large variability, the duration of individual periods of decreased or increased CMH levels are strictly dependent on the tumor growth rate. Further results indicate that tumor growth is accompanied by significant changes in CMH metabolism, although no qualitative differences in mechanisms of gene expression for normal and tumor cells have yet been found. While it is assumed that all tissues demand a constant supply of CMH, it is suggested that CMH affects the regulation of protein synthesis and the growth of malignant tissues. (71 references).

3031   THE ROLE OF DNA REPAIR AND SOMATIC MUTA-
        TION IN CARCINOGENESIS. (Eng.) Trosko,
J. E. (Dept. Human Development, Michigan State Univ., East Lansing, Mich.); Chu, E. H. Y. *Adv. Cancer Res.* 21:391-425; 1975.

Information on the role of DNA repair and somatic mutation in carcinogenesis is reviewed. A highly speculative and heuristic scheme that has been developed to account for a wide range of observations that make it appear that the carcinogenic process has several heritable components is described. Based on the assumptions (a) that germ-line mutations on several levels can influence carcinogenesis, and (b) that DNA repair and somatic mutagenesis are related in mammalian systems in a manner similar to bacteriological systems, a model linking genetic and environmental factors influencing DNA repair with observations on the initiation and promotion of tumors is presented. Within this model, genes that influence (a) the level of genetic damage, (b) the level and efficiency of the repair of DNA damage, and (c) the proliferation of the altered cells, are those that are subject to wide variation in expression owing to individual genetic or environmental predispositions. At this time, only circumstantial evidence links DNA repair to carcinogenesis in human beings. A critical component of the somatic mutation theory of cancer will be an elaboration of the role of the repair of carcinogen-induced DNA damage to mutagenesis in eukaryotic systems. Within the framework of the model presented, it appears that the goal of a total prevention of cancer and the development of a universal therapy for the cure of cancer will be extremely difficult, if not impossible, to attain. Even if the amount of potential carcinogens (initiators) in the human environment could be prevented from increasing, it could never be reduced to zero levels. There are many steps at which genetic variability could lead to the genetic predisposition to cancer, as well as many levels by which certain environmental factors can enhance the carcinogenic and aging processes. Genetic instability, due to faulty DNA repair (or replication) on the gene or chromosomal level in the germ cells, has selective advantage because it can contribute to sources of variation needed for the evolution of the species. At the same time, it is hypothesized that gene and chromosome instability in the soma can manifest themselves in both the carcinogenic and aging process. In essence, a cybernetic-feedback type of model has been advanced that links germ-line mutations, which favor the initiation and promotion of somatic mutations, to those somatic mutations that alter a cell's ability to respond to cell division regulators. (212 references)

3032   A CRITICAL EVALUATION OF THE METHODS USED FOR DETERMINING CARCINOGENICITY. (Eng.) Weisburger, E. K. (Natl. Cancer Inst., Bethesda, Md.). J. Clin. Pharmacol. 15(1):5-15; 1975.

In the selection of a bioassay method to determine the carcinogenicity or long-term toxicity of a drug, consideration should be given to the chemical structure of the test compound and its physical and chemical properties. Other factors that influence test results are: route of administration, dosage and frequency of exposure, the age of the animal at the start of the test, the species, strain, sex, diet, test duration, immune status, the possible presence of synergists or inhibitors and even the solvent or vehicle used for administration of the test substance. Guidelines for selecting the route of administration are summarized together with the bioassay procedure used at the National Cancer Institute and results of tests with chemotherapeutic agents. Present NCI protocols call for 50 animals of each sex and species to be tested at each of at least two dose levels. At the end of the experimental and/or holding period, preferably at least two yr for rats and 18-20 months for mice, experimental animals and controls are killed for histopathologic study. Cyclophosphamide (three i.p. injections/week for six months) had a weak but definite carcinogenic effect on the lung and bladder in Swiss-Webster mice and the breast in Sprague-Dawley rats. Female mice given azathioprine had a significant incidence of lymphosarcomas, while the number of lung tumors in male mice and s.c. tumors in male and female rats was also higher than normal. Methotrexate resulted in a slight increase in lymphosarcomas in male and female mice. It is concluded that these drugs should be prescribed only when the disease cannot be controlled by less toxic drugs. (76 references)

3033   AFLATOXIN RESEARCH: CURRENT STATUS AND OUTLOOK. (Eng.) Kensler, C. J. (Arthur D. Little, Inc.). Manuf. Confect. 55(1):25-27; 1975. (No references)

3034   SOCIAL AND ETHICAL IMPLICATIONS OF CLAIMS FOR CANCER HAZARDS [abstract]. (Eng.) Weisburger, J. H. (Naylor Dana Inst., Am. Health Found., New York, N.Y.). Proc. Am. Assoc. Cancer Res. 16:206; 1975. (No references)

3035   RESERPINE AND BREAST CANCER: A PERSPECTIVE [editorial]. (Eng.) Jick, H. (Boston Univ. Medical Center, Boston, Mass). J.A.M.A. 233(8):896-897; 1975. (6 references)

3036   ULTRAVIOLET RADIATION CARCINOGENESIS IN MICE AND MEN [abstract]. (Eng.) Blum, H. F. (Temple Univ. Health Sci. Cent., Philadelphia, Pa.). Proc. Am. Assoc. Cancer Res. 16:92; 1975. (No references)

3037   WASHINGTON SEEN. (Eng.) Norman, C. (No affiliation given). Nature 257(5523): 173; 1975. (No references)

3038   THE VCM -- PVC SYNDROME. (Eng.) Seymour, R. B. (Dep. Chem., Univ. Houston, Tex.). Aust. Plast. Rubber 25/26(12/1):9-11; 1974/1975. (No references)

3049   SOME PROBLEMS OF SPONTANEOUS AND INDUCED
        MUTAGENESIS IN MAMMALS AND MAN. (Eng.)
Kerkis, J. (Inst. Cytol. Genet., Novosibirsk, USSR).
Mutat. Res. 29(2):271-277; 1975. (19 references)

3050   GENETIC HAZARDS OF IRRADIATION OF PREG-
        NANT WOMEN. (Fre.) Feremans, W. P.
(Faculte de Medecine, Universite Libre de Bruxelles,
Bruxelles, Belgium). *J. Belge Radiol.* 58(1):47-
51; 1975. (No references)

3051   CARCINOMA OF THE THYROID AFTER RADIATION
        TO THE NECK [editorial]. (Eng.) Crile,
G., Jr. (Cleveland, Ohio). *Surg. Gynecol. Obstet.*
141(4):602; 1975. (No references)

08;

3052   ONCOGENIC VIRUSES OF NONHUMAN PRIMATES:
        A REVIEW. (Eng.) Raflo, C. P. (Edgewood
Arsenal, Aberdeen Proving Ground, Md.) 15 pp.,
1975. [available through National Technical Information Services, Washington, D.C. Document No.
AD/A-005 160/7GA]

3053   RELATIONSHIP BETWEEN THE IDIOPLASM OF
        CELLS AND TUMOR VIRUS. (Swe.) Ponten, J.
(Uppsala Universitet, Sweden). *Lakartidningen*
72(44):4264-4267; 1975. (No references)

3054   ONCORNAVIRUS AND HUMAN LEUKEMIA. (Fre.)
        Boiron, M. (Laboratoire d'Hématologie
expérimentale, Institut de Recherches sur les
Leucemies, Hopital Saint-Louis, 2, place du D
Alfred-Fournier, F 75475 Paris Cedex 10., France).
Bull. Cancer (Paris) 62(2):205-212; 1975. (30
references)

3055   TUMOUR-ASSOCIATED ANTIGENS IN ACUTE LEU-
        KAEMIA. (Eng.) Powles, R. L. (St.
Bartholomew's Hosp., London, E.C. 1, England). In:
*Reviews in Leukaemia and Lymphoma, 1--Advances in
Acute Leukaemia,* edited by Cleton, F. J.; Crowther,
D.; Malpas, J. S. Amsterdam, North Holland Pub-
lishing Co., 1974, pp. 115-141. (100 references)

3056   IMMUNOFLUORESCENCE IN CANCER INVESTIGA-
        TION AND RESEARCH. (Eng.) Nairn, R. C.
(Dept. of Pathology, Monash Univ., Melbourne, Australia); Rolland, J. M.; Ward, H. A.; Matthews, N.;
Chalmers, P. J. *Ann. N.Y. Acad. Sci.* 254:523-527;
1975. (8 references)

3057   SOME FUNCTIONAL AND BIOCHEMICAL CHARACT-
        ERISTICS OF LYMPHOCYTES IN NORMAL SUB-
JECTS AND IN CHRONIC LYMPHATIC LEUKEMIA (A REVIEW).
(Rus.) Luganova, I. S. (Lab. Biochem., Leningrad
Sci. Res. Inst. Hematol. Blood Transf., USSR);
Blinov, M. N. *Lab. Delo* (2):67-71; 1975. (77
references)

3058   LYMPHOCYTES IN SPLEEN IN HODGKIN'S DISEASE
        [letter to editor]. (Eng.) Grifoni, V.
(Cattedra di Clinica Medica RR, Universita di Genova,
Italy); Del Giacco, G. S.; Manconi, P. E.; Tognella,
S.; Mantovani, G. *Lancet* 1(7902):332-333; 1975. (6
references)

3059   HYPOTHESIS: DIFFERENTIATION OF THE HUMAN
        LYMPHOID SYSTEM BASED ON CELL SURFACE
MARKERS. (Eng.) Davis, S. (Natl. Cancer Inst.,
Bethesda, Md.). *Blood* 45(6):871-880; 1975. (41
references)

3060   CIRCULATING IgE LEVELS IN PATIENTS WITH
        CANCER [letter to editor]. (Eng.) Assem,
E. S. K. (Dept. Pharmacology, Univ. Coll. London,
Gower St., London WC1, England). *Lancet* 2(7923):
34-35; 1975. (4 references)

3061   ONCOLOGY AND IMMUNOLOGY: CROSS REACTIONS
        BETWEEN DEVELOPING SCIENCES. (Eng.)
Halliday, W. J. (Dept. Microbiology, Univ. Queensland, Brisbane, Australia). *Immunochemistry* 12(6/7):
573-575; 1975. (42 references)

3062   CLASSIFICATION OF MALIGNANT LYMPHOMAS AND
        THE RELATIONSHIP BETWEEN HISTOPATHOLOGY
AND CLINICAL COURSE. (Dut.) van Unnik, J. A. M.
(St. Elisabeth-ziekenhuis Tilburg, consulent af-
deling pathologie Antoni van Leeuwenhoek Ziekenhuis,
Amsterdam, The Netherlands); Somers, R. *Ned.
Tijdschr. Geneeskd.* 119(25):981-987; 1975. (42
references)

3063   *IN VITRO* STUDIES IN THE MYELOID LEUKAEMI-
        AS. (Eng.) Moore, M. A. S. (Royal Mel-
bourne Hosp., Victoria, Australia). In: *Reviews
in Leukaemia and Lymphoma, 1--Advances in Acute
Leukaemia* edited by Cleton, F. J.; Crowther, D.;
Malpas, J. S. Amsterdam, North Holland Publishing
Co., 1974, pp. 161-277. (154 references)

3064   EMBRYOLOGICAL PERSPECTIVES ON THE FINE
        STRUCTURE OF ORBITAL TUMORS. (Eng.)
Jakobiec, F. A. (No affiliation given); Tannenbaum,
M. *Int. Ophthalmol. Clin.* 15(1):85-110; 1975. (44
references)

3065   PREMALIGNANT LESIONS IN THE MOUTH. (Eng.)
        Cawson, R. A. (Guy's Hosp., London, Eng-
land). *Br. Med. Bull.* 31(2):164-168; 1975. (18
references)   .

3066   ADENOCARCINOMA OF THE LARYNX. (Eng.)
        Fechner, R. E. (Baylor Coll. Medicine,
1200 Moursund, Houston, Tex. 77025). *Can. J.
Otolaryngol.* 4(2):284-289; 1975. (18 references)

3067   GENETIC PROBLEMS IN TUMOURS OF THE GASTRO-
        INTESTINAL TRACT. (Eng.) Harper, P. S.
(Univ. Hosp. Wales, Cardiff). *Schweiz. Med. Wochenschr.* 105(18):564-569; 1975. (47 references)

3068   NOTES ON THE MODERN CRITERIA FOR THE
        HISTOGENETIC CLASSIFICATION OF OVARY
TUMOURS. (Ita.) Croce, C. (Vittorio Emanuele III
Hosp., Carate Brianza, Milan, Italy); Giordano, G.;
Di Lucrezia, F. *Arch. Sci. Med.* 131(4):226-233;
1974. (78 references)

3069   ONCOGENS - VIROGENS - PATHOGENS. (Ger.)
        Horing, F. O. (D-1000 Berlin, Kurfursten-
damm 139, West Germany). *Infection* 3(1):37-39;
1975. (9 references)

3070   PHACOMATOSES: OPTHALMOLOGICAL AND GENETIC
        ASPECTS. (Fre.) Cuendet, J. F. (31,
avenue de Rumine, 1005 Lausanne, Switzerland). *J.
Gent. Hum.* 23(Suppl.):193-209; 1975. (34 references)

3071   EPIDEMIOLOGICAL ASPECTS OF CHORIOCARCINO-
        MA. (Eng.) Anonymous. *Br. Med. J.*
3(5984):606-607; 1975. (15 references)

3072   STUDIES OF LEUKEMIA CELL PROLIFERATION.
        (Eng.) Mauer, A. M. (Children's Hosp.,
Cincinnati, Ohio 45229); Lampkin, B. C. In: *Re-
views in Leukaemia and Lymphoma, 1--Advances in
Acute Leukaemia* edited by Cleton, F. J.; Crowther,
D.; Malpas, J. S. Amsterdam, North Holland Publish-
ing Co., 1974, pp. 69-94. (119 references)

3073   CANCER: A SECOND OPINION. (Eng.) Is-
        sels, J. (No affiliation given). Hodder
and Stoughton, 1975, 216 pp.   .

3074   CHEMICAL STUDY CENTER FOR PHARMACEUTICAL
        DRUGS AND BIOLOGICALLY ACTIVE PRODUCTS,
PADOVA: SCIENTIFIC ACTIVITY IN THE YEAR 1973.
(Ita.) Rodighiero, G. (Centre di studio sulla
chimica del farmaco e dei prodotti biologicamente
attivi, Padova, Italy). *Ric. Sci.* 44(4):584-588;
1974. (23 references)

number of binding sites was doubled in the presence of calf thymus histones. The mechanism of the interaction remains unknown, but presumably intercalation is not involved.

3077   BINDING OF [^{14}C]AFLATOXIN B$_1$ TO CELLULAR
        MACROMOLECULES IN THE RAT AND HAMSTER.
(Eng.) Garner, R. C. (Dept. of Experimental Pathology and Cancer Res., Univ. of Leeds, 171 Woodhouse Lane, Leeds LS2 3AR, England); Wright, C. M. *Chem. Biol. Interact.* 11(2):123-131; 1975.

The uptake and binding of ring-labeled [^{14}C]aflatoxin B$_1$ (AFB$_1$) by rat and hamster liver and kidney was studied. Male Wistar rats or male Golden Syrian hamsters were injected ip with 40 µg/100 g [^{14}C]AFB$_1$ in dimethylsulphoxide (40 µg AFB$_1$/0.1 ml). Three animals were used to measure uptake and binding of the compound for each time period. DNA was estimated by the diphenylamine reaction using calf-thymus DNA as standard; rRNA was estimated by the orcinol reaction using yeast tRNA as standard and protein was determined by the Lowry-Folin procedure, using bovine serum albumin as standard. Binding of the carcinogen to nucleic acids was far greater than that to protein. Rat liver DNA bound ten times and ribosomal (rRNA) 20 times more carcinogen than protein. There were also differences in the amount of carcinogen bound to rat liver nucleic acids compared to those of the hamster, the latter species binding lower amounts of the carcinogen. Rat liver DNA bound four times and rRNA ten times as much AFB$_1$ six hr after carcinogen administration, whereas liver protein-bound AFB$_1$ was similar for the two species. Whereas rat liver nucleic acid-bound carcinogen decayed with time, no such fall was seen in the hamster; this remained at a low level throughout the 48-hr period. In contrast, reaction of the carcinogen with kidney macromolecules was similar for the two species. The much higher binding of AFB$_1$ to nucleic acids than to protein might account for the potent carcinogenicity of this compound in the rat, particularly since liver binding does not differ between a susceptible and a resistant species. Another factor in carcinogenic sensitivity may be the removal of nucleic acid-bound radioactivity with time, a possible repair process.

3078   EFFECT OF LIFETIME EXPOSURE TO AFLATOXIN
        B$_1$ IN RATS. (Eng.) Ward, J. M. (Natl. Cancer Inst., Bethesda, Md.); Sontag, J. M.; Weisburger, E. K.; Brown, C. A. *J. Natl. Cancer Inst.* 55(1):107-113; 1975.

Studies were conducted to determine the effects of exposure of rats to transplacental and milk metabolites of aflatoxin B$_1$ (AFB$_1$) followed by lifetime dietary exposure. AFB$_1$ was fed at 2 ppm to a group of pregnant F344 rats from the time of conception; it was then fed to their offspring until death. For comparison, the same diet was administered to another group of rats beginning at age 6-7 wk. The survival time of male rats was significantly shorter than that of the female rats of both groups. However, the survival times of rats of the same sex in both groups did not differ significantly. The major causes of death were hepatic neoplasms with metastases, al-

though some early deaths occurred before neoplasms
developed. Most deaths were from malignant hemorrhagic liver tumors, histologically diagnosed as
hemangiosarcomas, which caused rupture and hemorrhage into the peritoneal cavity or metastases to
the lungs. These hemangiosarcomas were readily
transplantable and did not produce α-fetoprotein.
Ultrastructurally, they were composed of poorly
differentiated cells resembling endothelial cells.
Nodules of hyperplasia induced by aflatoxin B_1 sometimes grew large (>1.5 cm), and two were transplanted. Colon tumors developed in more than 20% of
rats exposed *in utero* as well as in those given AFB_1
from six or seven weeks of age. A few rats had tumors of the kidney, oral cavity, and hematopoietic
system. These findings indicate that AFB_1 should be
considered a possible etiologic agent for colon cancer. Inhibition of the hepatocarcinogenic effect
of AFB_1 may induce colon and/or renal tumors.

3079   A POSSIBLE PROTECTIVE ROLE FOR REDUCED
        GLUTATHIONE IN AFLATOXIN B_1 TOXICITY: EF-
FECT OF PRETREATMENT OF RATS WITH PHENOBARBITAL AND
3-METHYLCHOLANTHRENE ON AFLATOXIN TOXICITY. (Eng.)
Mgbodile, M. U. K. (Vanderbilt Univ. Sch. Medicine,
Nashville, Tenn. 37232); Holscher, M.; Neal*, R. A.
Toxicol. Appl. Pharmacol. 34(1):128-142; 1975.

The possibility of a protective role for reduced glutathione (GSH) in aflatoxin B_1-induced hepatotoxicity
in male Sprague-Dawley rats was investigated. A
200-mg/kg dose of cysteine (a gluthathione precursor), given ip prior to the administration of aflatoxin B_1 (3 mg/kg), reduced the hepatic necrosis
and partially reversed the decrease in the activity
of the hepatic drug metabolizing enzyme system seen
on administration of the same dose of aflatoxin B_1
alone. In contrast, diethyl maleate (0.2 or 0.6
ml/kg), given ip two hours prior to the aflatoxin,
enhanced the aflatoxin-induced hepatotoxicity. In
contrast to untreated rats, pretreatment with phenobarbital (80 mg/kg, ip, daily for three days) appeared to block the liver necrosis and the decrease
in mixed function oxidase activity seen on administration of aflatoxin B_1. On the other hand, pretreatment of the rats with 3-methylcholanthrene
(20 mg/kg, ip, daily for three days; or one 20 mg/
kg dose 72 hr before aflatoxin) did not protect
against the hepatic damage or the decrease in the
activity of benzphetamine *N*-demethylase seen on
administration of aflatoxin B_1. The results suggest that glutathione may be involved in reducing
the toxicity of aflatoxin B_1 to the liver.

3080   BIOCHEMICAL AND MORPHOLOGICAL CHANGES
        IN HEPATIC NUCLEAR MEMBRANES PRODUCED BY
N-HYDROXY-2-ACETYLAMINOFLUORENE. (Eng.) Glazer,
R. I. (Dept. Pharmacology, Emory Univ., Atlanta,
Ga. 30322); La Via, M. F. *Cancer Res.* 35(9):2511-
2519; 1975.

The effect of *N*-hydroxy-2-acetylaminofluorene (N-
OH-AAF) on the ultrastructure and synthesis of
hepatic nuclear membranes was evaluated in partially
hepatectomized (male Sprague Dawley) rats. The

binding of BP metabolites to DNA as compared to that
of controls. The AHH activity of mice fed BHA was
much more sensitive to *in vitro* inhibition by α-
naphthoflavone (0.16, 0.32, or 0.64 μM) than that of
controls. The amount of cytochrome P450 was increased
per unit weight of microsomal protein and liver in
mice fed BHA. The ethyl isocyanide binding spectra
were measured to see if alterations of cytochrome
P450 might be produced by BHA feeding. The maximum
at 430 nm was the same in control and BHA-fed mice.
However, the maximum at 455 nm was lower in BHA-fed
mice than in controls, which indicated that BHA had
caused some change. The data show that BHA feeding
results in altered properties of liver microsomes,
including a decrease in BP metabolite binding to
DNA.

3084   INHIBITION OF BENZO(a)PYRENE METABOLISM
        CATALYZED BY MOUSE AND HAMSTER LUNG MICRO-
SOMES. (Eng.) Hill, D. L. (Kettering-Meyer Lab.,
Southern Res. Inst., Birmingham, Ala. 35205); Shih,
T.-W. *Cancer Res.* 35(10):2717-2723; 1975.

Microsomal enzymes that metabolize benzo(a)pyrene
in mouse and hamster lungs were characterized, and
their inhibition by various agents was studied.
Benzo(a)pyrene hydroxylase activity was induced in
female DBA/2 mice and Syrian golden hamsters by
giving them injections of 3 or 9 mg benz(a)anthra-
cene, respectively, 16 hr before they were sacrificed. Induced and constitutive microsomal enzymes
of mouse and hamster lungs catalyzed both the hydroxylation of benzo(a)pyrene and reactions that
led to its irreversible binding to macromolecules.
For mouse and hamster, the induced lung hydroxylases had K_m values of 1.10 and 0.52 μM, respectively. The induced hydroxylases were strongly
inhibited by 7,8-benzoflavone (0.25 and 50 μM) and
were stimulated by cyclohexene oxide (2 mM). Formation of the macromolecular product by the induced
"binding" enzyme followed Michaelis-Menten kinetics,
except for substrate inhibition, and had K_m values
of 0.52 and 0.25 μM for lung microsomes from mouse
and hamster, respectively. These reactions were
also inhibited by 7,8-benzoflavone (50 μM). The
reaction catalyzed by the constitutive hydroxylase
of mouse lungs was characterized by a brief lag
period, but it proceeded in a linear fashion after
the lag. The enzyme required 60 μM benzo(a)pyrene
to achieve maximum reaction velocity. Above this
concentration, strong substrate inhibition was observed; accurate values for V_{max} and K_m could not
be derived. The constitutive hydroxylases were
moderately inhibited by butylated hydroxytoluene
(50 μM), retinol (50 μM), cyclohexene oxide (2 mM),
and 7,8-benzoflavone (50 μM). The product of the
constitutive "binding" enzyme formed in a reaction
that followed Michaelis-Menten kinetics. The K_m
value for enzymes from mouse and hamster lungs were
11.8 and 4.9 μM, respectively. Formation of this
product was strongly inhibited by butylated hydroxytoluene and by retinol but not strongly by 7,8-
benzoflavone or cyclohexene oxide. Since other
evidence indicates that a constitutive enzyme may
be involved in carcinogenesis by benzo(a)pyrene,
and since this reaction is inhibited by two known

anticarcinogens, it is suggested that it may be
involved in this process.

3085   POSITION-SPECIFIC OXYGENATION OF BENZO[a]-
          PYRENE BY DIFFERENT FORMS OF PURIFIED CYTO-
CHROME P-450 FROM RABBIT LIVER. (Eng.) Wiebel, F.
J. (Natl. Cancer Inst., Bethesda, Md. 20014); Sel-
kirk, J. K.; Gelboin, H. V.; Haugen, D. A.; van der
Hoeven, T. A.; Coon, M. J. *Proc. Natl. Acad. Sci,
USA* 72(10):3917-3920; 1975.

Oxygenated products of benzo[a]pyrene formed in a
reconstituted microsomal mixed-function oxidase
system containing cytochrome P-450 (P-450LM), phospholipid, and NADPH-cytochrome P-450 reductase
(NADPH:ferricytochrome oxidoreductase) were detected
by high-pressure liquid chromatography. Three cytochrome fractions purified from hepatic microsomes
from phenobarbital-treated rabbits, were studied;
the various forms of the cytochrome are designated
by their relative electrophoretic mobilities. The
total benzo[a]pyrene oxygenation rate was greatest
for P-450LM$_{1,7}$, intermediate for P-450LM$_2$, and least
for P-450LM$_4$. The phenolic products were eluted in
two peaks, A and B, that contained primarily 9-hy-
droxy- and 3-hydroxybenzo[a]pyrene, respectively.
The ratio of peak A to peak B phenols was 0.11 for
P-450LM$_2$ and 0.45 for P-450LM$_4$. Thus, the relative
amounts of the various phenols formed by these two
cytochrome fractions differ markedly. The positional specificity of the hydroxylation is also indicated by large differences in the fluorescence
spectra of the phenolic products formed by the two
cytochromes. P-450LM$_2$ and P-450LM$_4$ did not form
benzo[a]pyrene dihydrodiols, thereby showing that
benzo[a]pyrene oxide hydratase activity was absent
from these purified preparations. Ninety percent
of the phenols formed by P-450LM$_{1,7}$ were eluted in
peak B; the metabolites produced by this preparation also included dihydrodiols, thus indicating
the presence of hydratase activity. The positional
specificities of different forms of cytochrome P-450
may channel polycyclic aromatic hydrocarbon metabolism into the various activation and detoxification
pathways and thereby help determine the cytotoxic
and carcinogenic activity of these compounds.

3086   EFFECT OF *CAPSELLA BURSA-PASTORIS* ON LIVER
          CATALASE ACTIVITY IN RATS FED 3'-METHYL-4-
(DIMETHYLAMINO)AZOBENZENE. (Eng.) Kuroda, K. (Res.
Inst. for Chemobiodynamics, Chiba Univ., Izumi-cho
3-9-1, Narashino 275, Japan); Akao, M. *Gann* 66(4):
461-462; 1975.

The effect of *Capsella bursa-pastoris* extract on
the liver catalase activity in male Donryu rats
treated with an azo dye hepatocarcinogen was studied. Catalase activity was determined by the ti-
tanyl sulfate method in three groups of rats: rats
fed a diet containing 0.06% 3'-methyl-4-(dimethylamino)azobenzene (3'-Me-DAB, 0.5 g) and then given
water containing 0.2% of *Capsella bursa-pastoris*
extract; rats given 3'-Me-DAB and untreated drinking water; and rats given neither 3'-Me-DAB nor the
herb extract. Catalase activity was expressed by

ing the modifying role of age, sex, duration of treatment, dose of carcinogen, and nonspecific toxicity of
the agent when selecting a bioassay system for evaluation of carcinogenicity.

3090   *IN VITRO* TRANSFORMATION OF BHK21 CELLS
        GROWN IN THE PRESENCE OF CALCIUM CHROMATE.
(Eng.) Fradkin, A. (Health Sci. Cent., State Univ.
New York Stony Brook); Janoff, A.; Lane, B. P.;
Kuschner, M. *Cancer Res.* 35(4):1058-1063; 1975.

BHK21 cells were grown in the presence of calcium
chromate ($CaCrO_4 \cdot 2H_2O$) to detect any *in vitro* trans-
formational changes in morphology and growth behavior
due to carcinogenic stimuli. BHK21 cells treated with
chromium were cultured in Dulbecco's medium with 10%
calf serum, penicillin and streptomycin. Plastic
tissue cultures were rinsed with phosphate-buffered
saline, sodium phosphate buffer and NaCl, followed by
trypsinization in 0.1% trypsin. Control (C) and
chromate-treated (CT) groups were maintained in tetra-
plicate to observe reproducibility in changes of
morphology and growth induced by chromate ions. CT
subgroups were grown in chromate salt (CT-H) followed
by introduction of chromate-containing medium added to
the culture. Random C cells and CT cells were grown
in methocel according to the method of Stoker with
modified amounts of chromate added to the CT cells
growing in methocel. *In vitro* transformation of BHK21
cells grown in the presence of $CaCrO_4$ revealed alterations in morphology and cell growth when exposed to
medium containing 0.25 and 0.5 µg/ml of $CaCrO_4$.
Treated cells were shortened and randomly oriented
and had enlarged nuclei and granular cytoplasm. Cells
grown in methocel treated with chromate underwent many
cell divisions to form clusters while normal cells
reproduced more slowly under the same growth conditions. These alterations were found to be irreversible and persisted for long periods of time. Transformation due to $CaCrO_4$ was found to be directly
related to the concentration of the chromium salt.
The results suggest that $CaCrO_4$ does cause *in vitro*
transformation in BHK21 cells. Further investigation
into *in vivo* transformation by $CaCrO_4$ and other
ions is indicated.

3091   ENHANCED HEPATOTOXICITY OF CARBON TETRA-
        CHLORIDE, THIOACETAMIDE, AND DIMETHYLNI-
TROSAMINE BY PRETREATMENT OF RATS WITH ETHANOL AND
SOME COMPARISONS WITH POTENTIATION BY ISOPROPANOL.
(Eng.) Maling, H. M. (Natl. Heart and Lung Inst.,
Bethesda, Md. 20014); Stripp, B.; Sipes, I. G.;
Highman, B.; Saul, W.; Williams, M. A. *Toxicol.
Appl. Pharmacol.* 33(2):291-308; 1975.

Elevated plasma glutamic-pyruvic transaminase (GPT)
activity induced by carbon tetrachloride (CCl_4),
thioacetamide, or dimethylnitrosamine in male
Sprague-Dawley rats was increased by pretreatment
with four doses (each 5 ml/kg) of ethanol administered po 48, 42, 24, and 18 hr before the hepatotoxic agent. Blood ethanol concentrations were
less than 5 mg/100 ml when a hepatotoxic agent was
injected ip. Pretreatment with ethanol did not
affect the hepatic concentrations of CCl_4 or its
metabolite, chloroform ($CHCl_3$), at one hr after ad-

ministration of CCl_4. The CCl_4-induced diene con-
jugation tended to increase after ethanol pretreatment and was significantly potentiated by pretreatment with isopropanol or pyrazole and a single dose
of ethanol. In rats pretreated with ethanol, covalent binding of $^{14}CCl_4$ to liver protein and lipid
in vivo was significantly greater at six and 24 hr
than in control rats. The *in vitro* binding of
$^{14}CCl_4$, $^{14}CHCl_3$, and $^{14}CBrCl_3$ to hepatic microsomal
protein was increased by ethanol pretreatment. Etha-
nol pretreatment also doubled the *in vitro* rate of
demethylation of dimethylnitrosamine by liver microsomes. The similarities in microsomal effects of
pretreatment with isopropanol and with ethanol suggest that similar mechanisms are involved in their
potentiation of CCl_4-induced hepatotoxicity. The
potentiation by pretreatment with ethanol, but not
with isopropanol, of the hepatotoxicity of thioacetamide and dimethylnitrosamine suggests that ethanol
pretreatment also activates some additional mechanisms.

3092 · IS THERE ANY ASSOCIATION BETWEEN ELE-
VATED SERUM PROLACTIN LEVEL AND MAMMARY
ADENOCARCINOMA INDUCED BY 7,12-DIMETHYLBENZ(a)-
ANTHRACENE? (Eng.) Feuer, G. (Dept. Clinical Bio-
chemistry, Banting Inst., Room 521, Univ. Toronto,
Toronto, Canada M5G 1L5); Kellen, J. A.; Kovacs, K.
Res. Commun. Chem. Pathol. Pharmacol. 11(3):435-444;
1975.

The association between mammary carcinogenesis induced by 7,12-dimethylbenz(a)anthracene (DMBA) in
the rat, the influence by manipulations of its
hepatic metabolism, and the secretion of prolactin
was investigated. Mammary tumors were induced in
female Wistar rats,' and the appearance of palpable
tumors was assessed twice weekly after the administration of various test compounds. The ability of
the liver to metabolize drugs was measured by assay
of coumarin 3-hydroxylase activity. Serum prolactin
levels were determined using a modified radioimmunoassay. Administration of the following compounds
seven days before DMBA treatment and continued for
14 days resulted in elevated serum prolactin: Coumarin (1 mM/kg, po), 4-methylcoumarin, phenobarbital
(0.2 mM/kg, ip), and CCl_4 (2.5 mM/kg, two doses
only). Only coumarin and 4-methylcoumarin reduced
tumor incidence. These observations do not support
the assumption that the suppression of DMBA-induced
breast adenocarcinoma by coumarin and 4-methylcou-
marin is mediated *via* prolactin.

3093 SPECIFIC ESTROGEN BINDING IN RAT MAMMARY
TUMORS INDUCED BY 7,12-DIMETHYLBENZ(A)
ANTHRACENE. (Eng.) Boylan, E. S. (Univ. Rochester
Cancer Cent., N.Y.); Wittliff, J. L. *Cancer Res.*
35(3):506-511; 1975.

The binding of 17β-[^{3}H]estradiol to ovarian-dependent and ovarian-independent 7,12-dimethylbenz(a)-
anthracene (DMBA)-induced mammary tumors in Sprague-
Dawley rats was studied to determine whether the two
classes of tumors differed in this respect. Seven-

week-old mice were given DMBA by gastric intubation
(25 mg/wk for five weeks), and bilateral ovariectomy
was performed within 2-3 wk of the appearance of tumors. Tumors were classified as ovarian-dependent
if the tumor mass decreased for a minimum of ten
days after ovariectomy, and as ovarian-independent
if growth continued for more than ten days after
ovariectomy. While all tumors classified as ovarian-dependent had substantial cytoplasmic estrogen-
binding capacity, 10 of 13 ovarian-independent tumors had a total binding capacity, a specificity
for estrogens, an extent of saturation *in vivo*, and
an ability for translocation which were similar to
these parameters of the ovarian-dependent tumors.
The capacity to bind estrogens in DMBA-induced mammary tumors may indicate that tumor growth may be
affected by ovariectomy, but it does not preclude
the possibility that some of these tumors will be
unaffected by endocrine ablation. The results are
similar to those observed among human patients with
breast tumors.

3094 INHIBITION OF THE ACUTE TOXICITY AND ADRE-
NOCORTICOLYTIC EFFECT OF 7,12-DIMETHYLBENZ
(a)ANTHRACENE BY ISOPROPYLVALERAMIDE AND ALLYLISO-
PROPYLACETAMIDE IN THE RAT. (Eng.) Somogyi, A.
(Univ. Nebraska Medical Center, Omaha, Nebr. 68105);
Levin, W.; Banerjee, S.; Kuntzman, R.; Conney, A. H.
Cancer Res. 35(9):2500-2505; 1975.

The effects of allylisopropylacetamide (AIA, 12.5-40
mg/kg, ip) and isopropylvaleramide (IVA, 2.50-25.00
mg/kg, ip) on the hemorrhagic adrenocortical necrosis and mortality induced by 7,12-dimethylbenz(a)-
anthracene (DMBA, 12 mg/kg, iv, one hour after AIA
or IVA administration) were studied in female Spragu
Dawley rats. The effects of secobarbital, aprobar-
bital, phenobarbital, and pentobarbital (37-46 mg/kg
ip or iv) on DMBA damage were also determined, as
was the influence of IVA on the clearance of [^{3}H]-
DMBA and its metabolites from the adrenals, liver,
and blood and on the *in vitro* metabolism of DMBA and
benzo(a)pyrene. In the latter experiments, the a-
mounts of fluorescent phenolic metabolites of benzo-
(a)pyrene and DMBA formed by liver homogenates from
IVA-treated (10-25 mg/kg) rats were determined. IVA
and AIA inhibited the hemorrhagic adrenocortical necrosis caused by DMBA, the anti-DMBA activity of
these compounds not being dependent upon the presence of the reactive allyl group in the molecule.
Similarly, the four barbiturates tested did not protect against DMBA damage, regardless of whether or
not they contained an allyl group. Thus, the protective action of IVA and AIA could not be attributed to the destruction of the microsomal enzyme
system responsible for the activation of DMBA. The
toxicity of dimethylnitrosamine, another carcinogen
that requires metabolic activation by microsomal
enzymes, was not influenced by either IVA and AIA.
IVA, which counteracted the adrenocorticolytic action of DMBA when given prior to, with, or after the
carcinogen, had no discernible effect on hydrocarbon
metabolism *in vitro* or *in vivo*. Although the mechanism of action of IVA remains unclear, it might be
valuable as a tool in studies of the nonmetabolic
aspects of DMBA carcinogenesis.

dine (10-20 mg/kg/wk, 10 wk, intragastrically), *N*-[4-(5-nitro-2-furyl)-2-thiazolyl]formamide (0.188% in diet, 18 wk) aflatoxin G_1 (25 µg/day, 41 days, intragastrically) or ethionine (50-100 mg/kg/wk, 22 wk, intragastrically). *N*-2-Fluorenylacetamide induced hepatocarcinomas more rapidly and in higher incidence (41%) in deficient rats than in control rats (19%). 3,3-Diphenyl-3-dimethylcarbamoyl-1-propyne induced a higher incidence of hepatocarcinomas but not gastric tumors in deficient rats. Aflatoxin B_1, (15 µg/day, 25 days, intragastrically), included as a positive control, was significantly more hepatocarcinogenic in deficient rats (87% incidence in deficient rats *vs* 11% in nondeficient rats). Gastric tumor induction by *N*-methyl-*N*-nitroso-*N'*-nitroguanidine and induction of tumors of the urinary bladder by *N*-[4-(5-nitro-2-furyl)-2-thiazolyl]formamide were not influenced by diet. Aflatoxin G_1 and ethionine were toxic to 82% of deficient rats, and carcinogenic doses could not be administered. These results may indicate that deficient dietary levels of the lipotropes and the metabolic load imposed by the high dietary level of fat are responsible for the enhancement of hepatocarcinogenesis.

3097   EFFECTS OF 5,5-DIPHENYLHYDANTOIN AND RELATED CARBAMATES AND HYDROXYLAMINES ON NUCLEIC ACID AND PROTEIN SYNTHESIS OF PROLIFERATING HUMAN LYMPHOCYTES. (Eng.) MacKinney, A. A. (Veteran's Admin. Hosp., Madison, Wis. 53706); Vyas, R. *Toxicol. Appl. Pharmacol.* 33(1):38-45; 1975.

In order to explore the mode of action of 5,5-diphenylhydantoin (DPH), the inhibitory effect of compounds with related hydroxylamine or carbamate structures were studied in the DNA and protein synthesis of normal proliferating human lymphocytes. DNA synthesis was estimated by treating phytohemagglutinin-stimulated lymphocytes with [methyl-^{3}H]-thymidine (0.12 mg/mCi) during the final two hours of incubation. Protein synthesis was estimated by incubating 3-day-old cultures with L[4,5^{3}H]leucine (4.3 mg/mCi). A series of 3-hydroxyhydantoins was weaker than the parent DPH, indicating that the N3 hydroxyl group did not activate the compounds, whereas substitutions at the C5 position generally diminished activity. Parahydroxylation of the benzene ring enhanced inhibition of DNA and protein synthesis. Acetoxylation of the N3 position of diphenylhydantoin greatly increased its potency; acetoxylation of 2-acetylaminofluorene had a similar effect. 1-Acetyl-3-acetoxy-5,5-dephenylhydantoin inhibited DNA synthesis to approximately the same degree as *N*-acetoxy-2-acetyl aminofluorene, xylyl phenyl carbamate, hydroxyurea, and hydroxylamine. This derivative of DPH illustrates the central role of the N3 position in metabolic reactions with lymphocytes. The contribution of this derivative to lymphomagenesis is not known.

3098   ASSOCIATION OF EXOGENOUS ESTROGEN AND ENDOMETRIAL CARCINOMA. (Eng.) Smith, D. C. (Mason Clinic, 1118 9th Ave., Seattle, Wash. 98101); Prentice, R.; Thompson, D. J.; Herrmann, W. L. *N. Engl. J. Med.* 293(23):1164-1167; 1975.

To determine the association between the incidence of endometrial cancer and the use of estrogen in menopausal and post-menopausal women, 317 patients with adenocarcinoma of the endometrium were compared with an equal number of matched controls having other gynecologic neoplasms. One-hundred fifty-two of the patients used estrogen, as compared to 54 of 317 controls. Thus, the risk of endometrial cancer was 4.5 times greater among women exposed to estrogen therapy. When estrogen use was adjusted for concomitant variables such as obesity, hypertension, diabetes, parity, referral pattern, age at diagnosis, year of diagnosis, and other gynecologic neoplasms, the magnitude of the increased relative risk was associated with several of these variables. The risk was highest in patients without obesity and hypertension. Exogenous estrogen therapy is associated with an increased risk of endometrial carcinoma, but this increased relative risk is less apparent in patients with physiologic characteristics previously associated with an increased risk.

3099   HORMONAL STATUS OF BREAST CANCER. II. AB-
        NORMAL URINARY STEROID EXCRETION. (Eng.)
Kodama, M. (Aichi Cancer Center Res. Inst., Chikusa-
ku, Nagoya, Japan); Kodama, T.; Yoshida, M.; Totania,
R.; Aoki, K. *J. Natl. Cancer Inst.* 54(6):1275-1282;
1975.

The urinary excretion of 14 neutral steroids was measured by gas-liquid chromatography in women with early and advanced breast cancer, in women with early uterine cancer, and in healthy women. For statistical analysis, measurements with a skewed distribution were normalized by the square root or logarithmic transformation. Urine samples from patients with neoplasia were taken 3-7 days before and 10-14 days after surgery; samples from controls were taken on the 20th days of the menstrual cycle. Measurements of age-dependent steroids were adjusted to a value of 35 yr of age for the premenopausal group and to 60 yr of age for postmenopausal group. The premenopausal patients with early breast cancer excreted subnormal amounts of five steroids (11-hydroxyandrosterone, 11-hydroxyethiocholanolone, prenanediol, pregnanetriol, and tetrahydrocorticosterone) and increased amounts of tetrahydrocortisol, compared to controls. Postmenopausal breast cancer patients excreted greater amounts of five steroids (one steroid from 17-ketosteroids and four from 17-hydroxycorticoids) that the controls. The discrepancy between premenopausal and postmenopausal breast cancer was tentatively related to ovarian-adrenal dysfunction in the course of aging. Oophorectomy induced a long-lasting tumor regression only in patients with a high value for the ratio of 11-deoxy-17-ketosteroid/17-hydroxycorticosteroid in urine taken before surgery; the ratio in the responsive patients decreased remarkably after surgery. A constitutional change in 17-ketosteroids, as observed in a postmenopausal breast-cancer patient and a premenopausal healthy women of urban origin, favored the geographic importance in the genesis of breast malignancy. 11-Deoxy-17-ketosteroid excretion at the premenopausal stage and the pregnanetriol/tetrahydrocortisone ratio at the postmenopausal stage differentiated breast can-

frequency of mutations. The polycyclic hydrocarbons, pyrene and benz(a)anthracene, which are not carcinogenic, were also not mutagenic. This assay demonstrates a relationship between the carcinogenicity of polycyclic hydrocarbons and their mutagenicity in mammalian cells, without having to isolate their reactive metabolic intermediates.

3103   THE USE OF THE FLUOROCHROME BIS-BENZIMID-
        AZOL DERIVATIVE (HOECHST 33258) IN THE
STUDY OF SPONTANEOUS AND INDUCED CHROMOSOME ABERRA-
TIONS. (Eng.) Raposa, T. (Semmelweis Univ. Medical
Sch., 1430 Budapest 81. POB. 21. Hungary); Natarajan,
A. T. *Cytobiologic Z. Exp. Zellforsch.* 11(2):230-
239; 1975.

The value of a new fluorochrome bis-benzimidazole derivative in the identification of chromosomal rearrangements in mammalian chromosomes was investigated. Human chromosome preparations were made from phytohemagglutinin-stimulated lymphocytes or fibroblasts, and normal mouse chromosome preparations were made from cultures derived from 12-14-day-old embryos. Chromosome aberrations were induced *in vivo* by treating tumor-bearing mice with mitomycin C (4 μg/ml ascites fluid), and aberrations were produced in normal human peripheral blood chromosomes by *in vitro* treatment with mitomycin C (0.6 μg/ml). The preparations were stained with fluorochrome at 0.1 μg/ml for ten minutes at 37 C; in some cases, the cells were pretreated with DNase (100 μg/ml) for five minutes at 37 C. With the normal human chromosomes, the fluorochrome produced a distinct banding pattern not quite similar to Q-banding. A pattern very similar to Q-banding was also produced with the mouse chromosomes, although, again, there were some differences. With the fluorochrome, the constitutive heterochromatin regions showed bright fluorescence, while the euchromatic regions exhibited a distinct banding pattern, allowing an easy and accurate evaluation. Mild DNase treatment increased the differential affinity of the fluorochrome for the eu- and heterochromatic regions. Mitomycin C-induced aberrations in both human and mouse chromosomes were therefore easily studied. In addition, identification of the Ph' chromosome and the association translocation in chromosome 9 was easy. The value of the fluorochrome in studying the distribution of constitutive heterochromatin regions in mouse tumors was also demonstrated. The new fluorochrome provides a valuable new banding technique. The mechanism of binding of the fluorochrome is unclear.

3104   A RAPID AND SENSITIVE LIQUID CHROMATO-
        GRAPHIC ASSAY FOR EPOXIDE HYDRASE. (Eng.)
Nesnow, S. (McArdle Lab. for Cancer Res., Univ.
Wisconsin, Madison, Wis. 53706); Heidelberger, C.
Anal. Biochem. 67(2):525-530; 1975.

A rapid method for the assay of epoxide hydrase activity is described. 3-Methylcholanthrene-11,12-oxide is the substrate. High-speed liquid chromatography is used to separate and quantitate *trans*-11,12-dihydro-11,12-dihydroxy-3-methylcholanthrene (product) formation. The product can be determined at picomole levels. Epoxide hydrase was determined in lower homogenate and microsomes of rats pretreated with 3-methylcholanthrene or phenobarbital. The enzyme levels obtained were comparable to those obtained with the styrene oxide method. An assay requires 35 min with this new method.

3105   RELATIONSHIP OF TUMOR IMMUNOGENICITY TO
        CONCENTRATION OF THE ONCOGEN. (Eng.)
Prehn, R. T. (Inst. for Cancer Res., Fox Chase Cancer Center, 7701 Burholme Ave., Philadelphia, Pa. 19111). *J. Natl. Cancer Inst.* 55(1):189-190; 1975.

Studies were conducted to determine if the immunogenicity of 3-methylcholanthrene (MCA)-induced tumors is related to the concentration of the chemical. Tumors were induced sc in (BALB/c x DBA/2)F_1 female mice by various concentrations of MCA in paraffin pellets. There was an inverse relationship between MCA concentration and tumor latency (interval between MCA implantation and detection of gross tumor). The tumors were transplanted into syngeneic recipients, and material from this first transplant generation was used to immunize a series of syngeneic mice; any resulting growth was excised. Nonimmunized mice were controls. Immunized and control mice were irradiated and given an sc inoculation of a near-threshold number of tumor cells (usually 10^5). Tumor growth from that inoculation was measured weekly in both groups and the antigenicity ratio (mean tumor size in controls/mean tumor size in immunized mice) was calculated. In a series of tumors with similar latencies, the only ones with high antigenicity ratios were those resulting from the high MCA concentration (5%). The results suggest that tumors induced by low levels of oncogen may be good models of spontaneous neoplasia. The data strengthen the hypothesis that "spontaneous" tumors may actually result from low levels of oncogen, and indicate that neoplastic transformation and the development of immunogenicity are, at least in chemically induced tumors, independent changes that may be produced in the same cell when the concentration of oncogen is sufficient.

3106   EFFECT OF 3-METHYLCHOLANTHRENE (NSC-
        21970) ON HUMAN PROSTATE IN ORGAN CUL-
TURE. (Eng.) Noyes, W. F. (Dept. Biological Resources, Roswell Park Memorial Inst., 666 Elm St., Buffalo, N.Y. 14203). *Cancer Chemother. Rep. (Part 1)* 59(1):67-71; 1975.

Long-term organ culture of normal human prostatic tissue with demonstration of functional and morphologic maintenance of the glandular epithelium is described. The tissues were obtained from 50- to 70-yr-old patients undergoing cystectomy. They were maintained in McCoy's 5a plus 20% fetal bovine serum and antibiotics, and were maintained as rocker cultures; the rocking motion provided a thin nutrient film and allowed adequate gas exchange (95% O_2, 5% CO_2, pH 7.2). 3-Methylcholanthrene (4 μg/ml) in the culture fluid induced morphologic changes that were not observed in untreated cultures. The changes observed after exposure to the

degradation of RNA to the total toxicity. Breaks in the RNA chain result from the hydrolysis of phosphotriesters, and were thus used as a measure of the extent of O-alkylation and of the S_N1-type mechanism of the reaction. In experiments with methyl methanesulphonate, no evidence of degradation was observed at up to 19 times the mean lethal dose (620 methylations/RNA molecule). Breaks in the RNA chain accounted for 1 in 10 of the lethal lesions with β-hydroxyethyl methanesulphonate, 1 in 60 with bis-(2-chloromethyl)methylamine (nitrogen mustard, HN2), less than 1 in 125 with 2,2-dichlorvinyl dimethyl phosphate (dichlorvos, DDVP), and 1 in 200 with propylene oxide. The hydrolysis rate of bis-(2-chloroethyl)ether was too slow for any reaction to be detected. In reactions with the carcinogen bis-(2-chloromethyl)ether, the toxicity observed could be accounted for by the formaldehyde produced on hydrolysis. Cross-linking of the bacteriophage components by formaldehyde reduced the survival range over which the physical state of the RNA could be studied. No evidence of RNA degradation was observed. Reaction of the formaldehyde led to a progressive loss of biological activity over 24 hr, a loss which was partially reversed by dialysis. The results suggest that the existence of O-alkylation as visualized by degradation following phosphotriester formation correlates with the known mutagenicity of some compounds.

3112   NEURO-ONCOGENIC ALKYLATING AGENTS: PERI-
        NATAL CARCINOGENESIS VERSUS APPLICATION
IN ADULT ANIMALS. (Eng.) Kleihues, P. (Max-Planck-
Institut für Hirnforschung, Abteilung Allgemeine
Neurologie, 5 Köln-Merheim, West Germany); Margison,
G. P. *Proc. Int. Cancer Congr. 11th.* Vol. 2 *(Chem-
ical and Viral Oncogenesis).* Florence, Italy,
October 20-26, 1974. Edited by Bucalossi, P.;
Veronesi, U.; Cascinelli, N. New York, American
Elsevier, 1975, pp. 146-149.

The relative amounts of methylguanine found in cerebral DNA hydrolysates of animals injected with methylnitrosourea or methyl methanesulphonate were measured chromatographically. Alkylguanine is an indicator of alterations in the DNA molecule because: (1) it is capable of miscoding (2) potent carcinogenic agents produce the base to a greater extent than weaker carcinogenic agents. Methyl methanesulfonate is an example of the group of potent carcinogens which alkylate DNA via a free alkyl carbonium ion released in chemical or enzymatic degradation of the compound. The agent was injected (50 mg/kg, sc) into newborn BD-IX rats, killed two hours later in order to measure the DNA cerebral hydrolysates. Methylnitrosourea (80 mg/kg, iv) was administered to adult BD-IX rats, killed six hours later. A weaker carcinogen, methylnitrosourea is an example of agents which react with DNA by direct transfer of an alkyl group. The relative amounts of methylguanine produced by equimolar doses of these agents differed by a factor of about 20 with methyl methanesulfonate injection resulting in the higher amount. Susceptibility to neuroonco-genic agents is maximal in the fetal and newborn rat and decreased to the adult level about one month post-natally. The difference in methylguanidine

levels in the DNA cerebral hydrolysates may be
attributable to the ages of the animals as well
as the carcinogenic action of the compounds.

3113   AMELOBLASTIC ODONTOMA IN RATS INDUCED BY
         N-BUTYLNITROSOUREA. (Eng.) Wang, H.-H.
(Kagoshima Univ. Sch. Medicine, Usuki-cho, Kagoshima
890, Japan); Terashi, S. I.; Fukunishi, R. *Gann*
66(3):319-321; 1975.

Four cases of ameloblastic odontomas were found in
68 Long-Evans rats treated with po administration of
300 mg/kg N-butylnitrosourea four times biweekly.
The tumor was found as an expansive enlargement over
the mandible and was roentgen-opaque. It was histologically characterized by various elements such as
odontogenic epithelium, stellated reticulum tissue,
enamel, and/or dentine, and was classified as ameloblastic odontoma. It is concluded that, although
the present incidence is not high enough, N-butylnitrosourea and the Long-Evans strain may be a good
combination for the induction of odontogenic tumors
in rats.

3114   MACROPHAGES IN BRAIN TUMOURS INDUCED
         TRANSPLACENTALLY BY N-ETHYL-N-NITROSOUREA
IN RATS: AN ELECTRON-MICROSCOPE STUDY. (Eng.) Lan-
tos, P. L. (Middlesex Hosp. Medical Sch., London,
England). *J. Pathol.* 116(2):107-115; 1975.

The fine structure of macrophages was studied in experimental brain tumors induced transplacentally in
BD-IX rats by a single iv injection of 30 mg of N-
ethyl-N-nitrosourea/kg on the 15th day of gestation.
The tumors, depending on their localization and size,
caused various lesions in the brain, namely axonal
degeneration, loss of myelin, edema, hemorrhage and
cell necrosis. The tumors and the resulting alterations elicited a strong reaction by macrophages;
activation of microglial cells *in situ* and infiltration of the brain by leucocytes, chiefly monocytes.
Since both microglial cells and monocytes underwent
morphological changes, it was difficult, or impossible, to establish the origin of these reacting cells.
In a few cases, however, microglial cells and monocytes could be distinguished, indicating that microglial cells were still being activated and leucocytes
were still entering the brain. The number of lysosomes and cytoplasmic inclusions was thought to indicate activation, phagocytosis and repletion. Activation was characterized by an increase of lysosomes
and by frequent cell divisions. Phagocytic activity
was accomplished by the appearance of inclusions
which varied in different lesions: protein-like material in edema, remnants of erythrocytes in hemor-
rhages and myelinlamellae with lipid droplets in de-
myelination. These various inclusions were frequently present in the same cell, since the different lesions not uncommonly occurred at the same time. In
the stage of repletion macrophages contained mainly
lipid droplets and unidentifiable debris in their
abundant cytoplasm and thus corresponded to the compound granular corpuscles.

The hyperplastic phase was followed by a phase of
reparation, leading to the development of non-
ciliated epithelium without observable clear cells,
and areas of uniform multi-layered epithelium.
After 11 wk, new changes analogous to those found
at the beginning of the experiment were observed.

3120   INHIBITORS OF POLYAMINE BIOSYNTHESIS. 3.
         (±)-5-AMINO-2-HYDRAZINO-2-METHYLPENTANOIC
ACID, AN INHIBITOR OF ORNITHINE DECARBOXYLASE. (Eng.)
Abdel-Monem, M. M. (Coll. Pharmacy, Univ. Minnesota,
Minneapolis, Minn. 55455); Newton, N. E.; Weeks, C.
E. *J. Med. Chem.* 18(9):945-948; 1975.

(±)-5-Amino-2-hydrazino-2-methylpentanoic acid (α-
hydrazino-α-methyl-(±)-ornithine) was synthesized
and evaluated as an inhibitor of ornithine decarboxylase *in vitro* and in transforming lymphocytes. The
title compound was obtained from 1-phthalimidopentan-
4-one by treatment with hydrazine and KCN followed by
acid hydrolysis. The title compound was found *in
vitro* to be a potent competitive inhibitor of ornithine decarboxylase obtained from the prostate glands
of rats. It was approximately 10 times more potent
than alpha-methyl-(±)-ornithine and equipotent to
alpha-hydrazino-L-ornithine *in vitro*. This inhibition was abolished at high concentrations of pyridoxal phosphate. The title compound also blocked the
increase in putrescine levels normally observed in
bovine lymphocytes transformed by conconavalin A.

3121   CYTOLOGICAL STUDIES ON ONION ROOT-TIP CELLS
         TREATED WITH WATER-SOLUBLE EXTRACT OF TO-
BACCO SMOKE CONDENSATE FROM COMMERCIAL CIGARETTES.
(Eng.) Sabharwal, P. S. (T. H. Morgan Sch. Bio-
logical Sciences, Univ. Kentucky, Lexington, Ky.);
Gulati, D. K.; Bhalla, P. R. *Mutat. Res.* 31(4):
217-224; 1975.

The effect of water-soluble extract of tobacco
smoke condensate (TSC) from two commercially available cigarettes with different types of filters was
studied on the cytology of root-tip cells of onion
(Allium cepa). One of the cigarettes had a 2-cm
cellulose acetate filter, and the other had a filter comprised of 1 cm of cellulose acetate and 1 cm
of activated charcoal. A total of 840 cigarettes
were smoked on a smoking machine, and the smoke was
collected in a flask containing 100 ml of water
cooled to 2 C. The flask was rinsed with water to
give a final concentration of 50 mg/ml of nonvola-
tile residue. This TSC was subsequently diluted
with distilled water to make 0.01, 0.025, 0.05, 0.1,
and 0.25% test concentrations. The treatment was
given by placing the onion bulbs with roots dipping
in small vials containing 15 ml of TSC solution for
24 hr. TSC induced mitotic abnormalities; with increasing concentrations (0.01% to 0.1%) of TSC from
cigarettes with filters and defiltered, percent mitotic abnormalities increased. These abnormalities
included scattering, stickiness, lagging, condensation, and breaking of chromosomes during metaphase. Bridging and lagging of chromosomes were
observed during anaphase. To investigate whether
these two commercial filters could retain cigarette
smoke component(s) responsible for mitotic irregu-

larities, the cigarettes were defiltered, and TSC was prepared and tested on the young roots of onion. The cytological effect of TSC from defiltered cigarettes was not significantly different from the effect of TSC from cigarettes with filters. Thus, the filters utilized in these cigarettes do not retain compound(s) responsible for mitotic irregularities in the root-tip cells of onion.

3122   INHALATION EXPERIMENTS WITH ^{14}C-LABELLED CIGARETTE SMOKE. I. DETERMINATION OF THE EFFECTIVENESS OF TWO DIFFERENT SMOKING SYSTEMS WITH LABELLED CIGARETTES. (Eng.) Reznik, G. (Medizinische Hochschule Hannover, 3000 Hannover-Kleefeld, Karl-Wiechert-Allee 9, West Germany); Kmoch, N.; Mohr, U. *Toxicology* 4(3):363-371; 1975.

Two different types of smoking systems (type RM 20/68 and type Hamburg II) were tested by exposing Syrian golden hamsters to the smoke from 60 mm cigarettes labeled with [^{14}C]dotriacontane-16,17 (DOT, 55 mCi/M). Two experiments were performed using the first type of smoking system, where individual animals were exposed to the smoke from four cigarettes (12 μCi [^{14}C]DOT/cigarette in the first experiment, and 12.8 μCi [^{14}C]DOT/cigarette in the second experiment). The third experiment utilizing the type Hamburg II system involved ten hamsters exposed simultaneously to smoke from 30 cigarettes (10.6 μCi [^{14}C]DOT/cigarette). In each experiment, the radioactivity of mainstream and sidestream total particulate matter (TPM), and of butts and ash was determined. In the system involving individual animal exposure, more than 4% of the radioactivity was found in the animals, whereas with group exposure only 1% of the total radioactivity passed into the animals. Therefore, the type RM 20/68 smoking system seemed to be preferable for obtaining high quantities of mainstream TPM for inhalation studies on induction of a pulmonary carcinoma.

3123   CLONAL ISOLATION OF EPITHELIAL CELLS FROM MOUSE LUNG ADENOMA. (Eng.) Stoner, G. D. (Sch. of Medicine, Univ. of California at San Diego, La Jolla, Calif. 92037); Kikkawa, Y.; Kniazeff, A. J.; Miyai, K.; Wagner, R. M. *Cancer Res.* 35(8):2177-2185; 1975.

The clonal isolation and characterization of an epithelial cell line derived from an A/He mouse lung adenoma (LA-4) induced by a 20-mg, ip injection of urethan are described. The clone LA-4 cells were characterized with respect to morphology, karyotype, growth, tumorigenicity, mycoplasma contamination, and response to an 8-mo treatment of hydrocortisone (HC, 1.4×10^{-5} M). The LA-4 cells and the tumors produced in athymic, nude mice by HC-treated LA-4 cells were examined microscopically. The parental adenoma cells appeared to be composed of a mixture of epithelial and fibroblastic cells; lamellar inclusions were observed in the epithelial-like, but not the fibroblast-like, cells. The clonal LA-4 cells were also epithelial in nature and contained inclusion bodies with a dense, concentric lamellar structure. The LA-4 cells, like their parental pre-

3129   THE EFFECTS OF SEVERAL CROTON OIL CON-
        STITUENTS ON DNA REPAIR AND CYCLIC NUCLE-
OTIDE LEVELS IN CULTURED MAMMALIAN CELLS [abstract].
(Eng.) Yager, J. D. Jr. (McArdle Lab., Univ. Wisconsin, Madison, Wis. 53706); Trosko, J. E.; Butch-
er, F. R. *Proc. Am. Assoc. Cancer Res.* 16:10; 1975.

3130   9-AMINOACRIDINE MUTAGENESIS OF BACTERIO-
        PHAGE T4 INTRACELLULAR DNA. (Eng.) Alt-
man, S. (Dept. Biology, Yale Univ., New Haven,
Conn.); Warner, V. *Mol. Gen. Genet.* 138(4):333-
343; 1975.

3131   THE EFFECT OF CYCLIC AMP ON TRITIATED
        THYMIDINE INCORPORATION IN RAT MAMMARY
TUMOR EXPLANTS [abstract]. (Eng.) Seidman, I.
(New York Univ. Medical Cent., New York, N.Y.);
Palekar, L. *Am. J. Pathol.* 78(1):29a-30a; 1975.

3132   BIOSYNTHESIS OF AVERUFIN FROM ACETATE
        BY *ASPERGILLUS PARASITICUS.* (Eng.)
Fitzell, D. L. (Dep. Environ. Toxicol., Univ. California, Davis); Hsieh, D. P. H.; Yao, R. C.; La
Mar, G. N. *J. Agric. Food Chem.* 23(3):442-444;
1975.

3133   THE EFFECT OF SELENIUM ON THE GROWTH OF
        CARCINOGENIC FUNGI AND CYTOTOXIC ACTION
OF AFLATOXIN B_1 IN CELL CULTURES OF LYMPHOCYTES AND
ON THE EMBRYONAL DEVELOPMENT OF *XENOPUS LAEVIS.*
(Pol.) Aleksandrowicz, J. (Klinika Hematologiczna
AM. w Krakowie ul. Kopernika 17, Poland); Dobrowol-
ski, J.; Lisiewicz, J.; Smyk, B.; Skotnicki, A.
Pol. Arch. Med. Wewn. 53(3):209-218; 1975.

3134   INFLUENCE OF L-DOPA ON SPONTANEOUS TUMOR
        DEVELOPMENT IN MALE MARSH MICE [abstract].
(Eng.) Bryson, G. (Cottage Hosp. Res. Inst., Santa
Barbara, Calif. 93102); Bischoff, F. *Proc. Am.
Assoc. Cancer Res.* 16:2; 1975.

3135   STUDY OF THE DEGRADATION REACTION OF
        PARANITROSODIMETHYLANILINE BY A YEAST C-
NITROSOREDUCTASE. (Fre.) Huynh, C. -H. (Labora-
toire de Microbiologie, U.E.R. des Sciences de la
Vie et du Comportement. Université de Caen, 14032
Caen Cédex, France); *Biochimie* 57(5):529-537; 1975.

3136   THE METABOLIC ACTIVATION OF N-METHYL-4-
        AMINOAZOBENZENE BY RAT LIVER [abstract].
(Eng.) Kadlubar, F. F. (Univ. Wisconsin Center
Health Sciences, Madison, Wis. 53706); Miller, J.
A.; Miller, E. C. *Proc. Am. Assoc. Cancer Res.*
16:14; 1975.

3137   ASBESTOS, SMOKING, AND LARYNGEAL CARCINO-
        MA. (Eng.) Shettigara, P. T. (Dept.
Preventive Medicine, Univ. of Toronto, McMurrich
Building, Toronto, Ontario, Canada M5S 1A8); Morgan,
R. W. *Arch. Environ. Health* 30(10):517-519; 1975.

3138   CHRYSOTILE AT THE INTRATHORACIC AND SUB-
        CUTANEOUS RODENT SITES [abstract]. (Eng.)
Bischoff, F. (Cottage Hosp. Res. Inst., Santa Bar-
bara, Calif. 93102); Bryson, G. *Proc. Am. Assoc.
Cancer Res.* 16:4; 1975.

3139   CHEMICALLY INDUCED CARCINOGENESIS IN HAM-
        STER EMBRYO CELLS IN MASS CULTURE. (Fre.)
Markovits, P. (Fondation Curie-Institut du Radium,
26, rue d'Ulm, F 75231 Paris Cedex 05); Daudel, P.;
Papadopoulo, D.; Mazabraud, A.; Hubert-Habart, M.
Bull. Cancer (Paris) 62(1):59-72, 1975.

3140   IMMUNOGENICITY OF GUINEA PIG CELLS TRANS-
        FORMED BY CHEMICAL CARCINOGENS IN CULTURE
[abstract]. (Eng.) Evans, C. H. (Natl. Inst.
Health, Bethesda, Md.); Ohanian, S. H.; DiPaolo,
J. A. *Fed. Proc.* 34(3):828; 1975.

3141   INVESTIGATIONS ON THE CARCINOGENIC BURDEN
        BY AIR POLLUTION IN MAN. XI. ABOUT THE
EFFECT OF ALUMINIUMHYDROXIDE UPON THE BENZO(a)PYRENE
CARCINOGENESIS. (Eng.) Pfeiffer, E. H. (Institut
f. Hygiene d. Universitat, D-65 Mainz, Hochhaus am
Augustusplatz). *Zentralbl. Bakteriol. [Orig. B.]*
160(2):99-107; 1975.

3142   ENZYMIC FORMATION OF 6-OXO-BENZO(a)PYRENE
        RADICAL FROM CARCINOGENIC BENZO(a)PYRENE
[abstract]. (Eng.) Lesko, S. (Johns Hopkins Univ.,
Baltimore, Md. 21205); Caspary, W.; Lorentzen, R.;
Ts'o, P. *Biophys. J.* 15(2/Part 2):324a; 1975.

3143   α-NAPHTHOFLAVONE ACTIVATION OF HYDROXY-
        METHYL-BENZO(α)PYRENE SYNTHETASE [ab-
stract]. (Eng.) Sloane, N. H. (Univ. Tennessee
Center for Health Sciences, Memphis, Tenn. 38163).
Proc. Am. Assoc. Cancer Res. 16:34; 1975.

3144   CORRELATION BETWEEN ARYL HYDROCARBON
        HYDROXYLASE (AHH) ACTIVITY AND THE ACTIV-
ITY FOR THE FORMATION OF DNA-BINDING METABOLITE(S)
FROM BENZO(α)PYRENE (BP) IN VARIOUS INBRED STRAINS
OF MICE [abstract]. (Eng.) Gurtoo, H. L. (Grace
Cancer Drug Cent., Roswell Park Mem. Inst., Buffalo,
N.Y.); Sebring, C. *Fed. Proc.* 34(3):755; 1975.

3145   TISSUE-MEDIATED MUTAGENICITY OF VINYL-
        IDENE CHLORIDE AND 2-CHLOROBUTADIENE IN
SALMONELLA TYPHIMURIUM. (Eng.) Bartsch, H. (International Agency Res. Cancer, 150, Cours Albert
Thomas, 69008 Lyon, France); Malaveille, C.; Montesano, R.; Tomatis, L. *Nature* 255(5510):641-643;
1975.

3146   DNA REPAIR SYNTHESIS FOLLOWING *IN VITRO*
        EXPOSURE OF GUINEA PIG PANCREATIC SLICES
TO NITROSOMETHYLURETHANE [abstract]. (Eng.) Iq-
bal, Z. M. (Sch. Medicine, Case Western Reserve
Univ., Cleveland, Ohio 44106); Hasumi, K.; Epstein,
S. S. *Proc. Am. Assoc. Cancer Res.* 16:204; 1975.

3164   MALONALDEHYDE IS A CARCINOGEN [abstract].
        (Eng.) Shamberger, R. J. (Cleveland
Clin., Ohio); Tytko, S. A.; Willis, C. E. *Fed Proc*
34(3):827; 1975.

3165   EXAMINATIONS OF THE POSSIBILITIES TO
        NITROSATE NICOTINE AND NORNICOTINE AND
OF THE FORMATION OF *N*-NITROSONORNICOTINE IN THE
STOMACH OF SMOKERS. (Ger.) Schweinsberg, F.
(Hygiene-Institut der Universitat, D-7400 Tubin-
gen, West Germany); Sander, J.; Schweinsberg, E.;
Kollat, P. *Z. Krebsforsch.* 84(1):81-87; 1975.

3166   ON ERRORS IN DNA REPLICATION AND CAR-
        CINOGENESIS [abstract]. (Eng.) Loeb,
L. A. (Fox Chase Cancer Center, Philadelphia, Pa.
19111); Sirover, M. A.; Weymouth, L.; Battula, N.
Proc. Am. Assoc. Cancer Res. 16:93; 1975.

3167   ALKYLATION OF PHENOLS BY PHENANTHRENE
        9,10-OXIDE. (Eng.) Okamoto, T. (Fac.
Pharm. Sci., Univ. Tokyo, Japan); Shudo, K.; Nagata,
S. *Chem. Pharm. Bull. (Tokyo)* 23(3):687-689; 1975.

3168   EFFECT OF VITAMIN A ON LUNG TUMOR INDUC-
        TION IN RATS [abstract]. (Eng.) Net-
tesheim, P. (Carcinogenesis Program, Biology Div.,
Oak Ridge Natl. Lab., Oak Ridge, Tenn. 37830);
Synder, C.; Williams, M. L.; Cone, M. V.; Kim,
J. C. S. *Proc. Am. Assoc. Cancer Res.* 16:54;
1975.

3169   TAMOXIFEN (ICI 46, 474) AND THE HUMAN
        CARCINOMA 8S OESTROGEN RECEPTOR. (Eng.)
Jordan, V. (Worcester Found. Exp. Biol., Shrewsbury,
Mass. 01545); Koerner, S. *Eur. J. Cancer* 11(3):
205-206; 1975.

3170   THE EFFECT OF ETHYL NITROSO UREA (ENU),
        NERVE GROWTH FACTOR (NGF), AND ANTI-NGF
ON NEURAL DIFFERENTIATION OF RAT PRIMARY CULTURES
[abstract]. (Eng.) Vinores, S. (Dept. Zoology,
Univ. Texas, Austin, Tex. 78712); Perez-Polo, J. R.
Biophys. J. 15(2):46a; 1975.

3171   THE TUMOROUS NATURE OF AN EXPERIMENTALLY
        INDUCED HYPERPLASIA. (Fre.) Aaron-da
Cunha, M. I. (Universite Pierre et Marie Curie,
Tour 53, Paris, France); Kurkdjian, A.; Le Goff,
L. *C. R. Soc. Biol. (Paris)* 169(3/Suppl.):755-
759; 1975.

3172   EFFECT OF INTRAGASTRIC INFUSION OF TOBAC-
        CO POWDER ON DNA CONTENT OF GASTRIC
ASPIRATE. (Eng.) Desai, H. G. (B. Y. L. Nair
Charitable Hosp., Bombay, India); Venugopalan, K.;
Antia, F.P. *Am. J. Dig. Dis.* 20(5):450-453; 1975.

3173   A CLASSIFICATION OF TRANSPLANTABLE TUMORS
        IN Nb RATS CONTROLLED BY ESTROGEN FROM
DORMANCY TO AUTONOMY. (Eng.) Noble, R. L. (Cancer
Res. Centre, Univ. British Columbia, Vancouver,
British Columbia, Canada V6T 1W5); Hoover, L. *Cancer Res.* 35(11/Part 1):2935-2941; 1975.

3174   THE EFFECT OF STEROID HORMONES ON THE
        GROWTH PATTERN AND RNA SYNTHESIS IN
HUMAN BENIGN PROSTATIC HYPERPLASIA IN ORGAN CUL-
TURE. (Eng.) Lasnitzki, I. (Strangeways Res.
Lab., Cambridge, England); Whitaker, R. H.; Withy-
combe, J. F. R. *Br. J. Cancer* 32(2):168-178; 1975.

3175   THE CHEMICAL MECHANISMS OF CARCINOGENESIS
        -- A HYPOTHESIS. (Eng.) Love, D. L.
(Naval Ordnance Lab., White Oak, Md.). 22 pp.,
1974. [available through National Technical Information Services, Washington, D.C. Document
No. AD/A-005 398/3GA].

3176   HAIR DYES [letter to editor]. (Eng.)
        MacPhee, D. G. (Dept. Genetics and Human
Variation, La Trobe Univ., Bundoora, Vic. 3083,
Australia); Podger, D. M. *Med. J. Aust.* 2(1):32-
33; 1975.

3177   ESTIMATING THE MAXIMUM TEST DOSE IN CAR-
        CINOGENESIS BIOASSAYS USING COMPUTERIZED
DATA HANDLING AND ANALYSIS [abstract]. (Eng.)
Gordon, E. (NCI Frederick Cancer Res. Center,
Frederick, Md.); Riggs, C.; Bostian, P.; Ulland,
B. M.; Callahan, L. *Toxicol. Appl. Pharmacol.*
33(1):164; 1975.

See also:

* (Rev):   3001, 3002, 3003, 3004, 3005, 3006,
            3007, 3008, 3033, 3034, 3035, 3037,
            3039, 3040, 3041, 3042, 3043, 3044,
            3045, 3046, 3047, 3048, 3049
* (Phys):   3101, 3103
* (Viral): 3217, 3229, 3249, 3268, 3270, 3275,
            3280
* (Immun): 3299, 3303, 3307, 3309, 3342, 3372
* (Path):   3431, 3432
* (Epid-Biom): 3510, 3517, 3518, 3519, 3520,
            3529, 3530

mitted from the surfaces were measured using silicon
surface barrier alpha particle detectors. In auto-
radiographs of transverse sections of diaphyseal
bone, the endosteal surfaces showed intense labeling.
In the case of plutonium-239, there was no evidence
of burial of the surface deposit by appositional bone
growth. The thickness of the initial deposit of plu-
tonium-239 was $\leq$ 0.2 μm; for radium-226, the thickness was $\leq$ 6.5 μm. Equations are presented for dose
calculation in which a deposit of infinitesimal thickness is assumed. When using these equations, the
error introduced for plutonium dose is small; in the
case of radium, the greater thickness of the deposit
would result in a somewhat greater error. If pluto-
nium-239 were deposited in uniform concentration to
a depth of 0.2 μm, then the rate of escape of α-par-
ticle energy from the surface would be 98% of that
from a plane surface source of infinitesimal thickness. For radium-226 uniformly deposited to a depth
of 5 μm, the energy escape would be 76-81% of that
from a surface source of infinitesimal thickness,
depending on the assumption made about retention of
daughter products.

3180   COMPARATIVE METABOLISM OF RADIONUCLIDES
         IN MAMMALS—IX. RETENTION OF ^{75}Se IN THE
MOUSE, RAT, MONKEY AND DOG. (Eng.) Furchner, J. E.
(Los Alamos Scientific Lab., Univ. California, Los
Alamos, N.M. 87544); London, J. E.; Wilson, J. S.
Health Phys. 29(5):641-648; 1975.

Whole-body retention of ^{75}Se-selenite in mice, rats,
dogs and monkeys was determined with 4π liquid scintillation counting after oral and iv administration
in all species, and after ip injection in mice and
rats. Oral administrations in rats and mice were
made by gastric intubation, and an additional group
of mice were given ^{75}Se in their drinking water.
The selenious acid was diluted with 0.01 N HNO immediately before injection. Mice and rats were injected with 0.1 cm[3] and dogs and monkeys with 0.5 cm[3]
of solution. The group of 40 mice given ^{75}Se in the
drinking water received a daily dose of 65 nCi for
up to 112 days. Whole-body activities were measured
periodically before and at death, when the following
tissues were removed and assayed for ^{75}Se activity:
pelt, brain, heart, lung, gut (esophagus to rectum,
with contents), spleen, liver and kidney. Blood,
fat, mesenteries and urogenital organs were pooled
and assayed. The residual carcass was assayed as a
unit. More than 90% was absorbed from the gut in
all species but monkeys, where much less was absorbed.
Retention was described by the sum of three exponen-
tials in all species, whatever the route of administration. Effective half-times of the long components
increased with body weight. The kidney was the most
important emunctory for ^{75}Se after the first three
days in all species and for all routes of administration. Radiation protection guides (indicating the
concentration level in water deemed to produce no
detectable effects) were estimated to be 1.5 nCi/cm[3],
which is half the value given as the radiation protection guide for continuous exposure. Figures giv-
en in previously published reports are discussed and
compared.

3181   INVESTIGATION OF IMMUNOGLOBULIN (IgG, IgA,
        IgM) LEVELS IN THE BLOOD SERUM OF URANIUM
MINERS AFTER HIGHER AND LOWER EXPOSURE TO IONIZING
RADIATION. (Eng.) Andrlíková, J. (Res. Dep. Health
Inst. Uranium Ind., Pribram, Czechoslovakia); Wag-
ner, V.; Pálek, V. *Strahlentherapie* 149(2):212-
218; 1975.

Serum immunoglobulin (1g) G, A, and M levels were
measured in 35 uranium miners (average age 29.9 yr)
in a mining district where an increase in exposure
to ionizing radiation occurred after a geological
disturbance. Serum samples were obtained before
the men starting working in the mine and after 1
and 2 yr of work in the mine. The second sampling
followed the geological disturbance, which increased
the mean exposure to ^{222}Rn daughters to 7.35 Working
Level Months (WLM) for about three months. During
the second year of work, the mean exposure was as
low as 0.13 WLM. IgG and IgM levels decreased significantly after one year of exposure in the mine
environment, while the IgA level changed only
slightly. After the second year, the IgG and IgM
levels rose significantly; IgG exceeded the initial
levels to a degree that indicated the occurrence
of a regenerative hypercompensation. The IgA level
was decreased at this time. There was no correlation between IgA and IgM and the radiation dose rate
and only a slight correlation in the case of the
decreased IgG levels. In some miners the reduction
in IgG reached hypogammaglobulinemic levels after
the first year and was followed by a propensity to
respiratory tract infections. It is suggested that
radiation may damage Ig-producing lymphoid cells
through an indirect mechanism and that regenerative
mechanisms are not the same for all three Ig classes.
The heterogeneity of Ig responses in the miners is
in agreement with the hypothesis that different
immunosurveillance mechanisms function in the resistance against the growth of emerging lung tumor
cells.

3182   AGE-DEPENDENCE OF THE X-RAY-INDUCED DEFI-
        CIENCY IN DNA SYNTHESIS IN HeLa S3 CELLS
DURING GENERATION 1. (Eng.) Griffiths, T. D. (Wash-
ington Univ. Sch. Medicine, St. Louis, Mo. 63110);
Tolmach, L. J. *Radiat. Res.* 63(3):501-520; 1975.

The radiation-induced deficiency in DNA synthesis in
Generation 1 was studied as a function of the age of
HeLa S3 cells at the time of exposure to 220 kV X-
rays in the previous generation (Generation 0). The
amount of DNA synthesized was dependent on the stage
in the generation cycle at which cells were irradiated. The smallest deficiency (20-35% after a dose
of 500 rad) was observed in cells irradiated in early
G1 or early G2, while the greatest deficiency (55-70%
after 500 rad) was found in cells irradiated at mitosis or at the G1/S transition. The high sensitivity
of cells at G1/S was also manifested by a steeper
dose-response curve. Cells irradiated in late G2,
past the point where their progression was temporarily
blocked by X-rays, synthesized a normal amount of DNA
in Generation 1, while cells held up in the G2 block
exhibited deficient synthesis in the next generation.
The extent of the deficiency in early G1 cells could

be enhanced by treatment with 1 mM hydroxyurea for
several hours immediately following irradiation. It
is concluded that the amount of DNA synthesized in
generation 1 depends on the stage in the cycle at
which generation 0 cells are irradiated; cells irrad
ated at mitosis or the G1/S transition showing the
greatest deficiency in synthesis, and those irradiat
in early G1 or late S/early G2, the least.

3183   INDUCTION OF CHROMOSOMAL ABERRATIONS IN
        HUMAN LYPMHOCYTES BY X RAYS AND FISSION
NEUTRONS: DEPENDENCE ON CELL CYCLE STAGE. (Eng.)
Carrano, A. V. (Lawrence Livermore Lab. Univ. California, Livermore, Calif. 94550). *Radiat. Res.*
63(3):403-421; 1975.

The induction of chromosomal aberrations after irradiation in the different stages of the cell cycle
was studied, and relative biological effectiveness
values for specific aberration types as a function o
dosage and cell cycle stage were determined. Human
lymphocytes were irradiated in culture with several
doses of 250 kVp X rays or fission neutrons (0.85 Me
average energy) either in the unstimulated state (G_0
at 17 hr after phytohemagglutinin stimulation (G_1) o
at approximately 48 hr after stimulation. In the
last case, the cells were labeled with tritiated thy
midine prior to irradiation to distinguish S and G_2
cells. All cultures were harvested at two successiv
intervals between 48 and 58 hr. The relative biolog
ical effectiveness for total aberration production w
highest in G_2, lowest in S, and intermediate in G_1.
The frequencies of neutron-induced chromatid dele-
tions in G_2 cells differed significantly among harvest times and definite trends were observed for
other aberrations. Dose-response curves for chromosome type aberrations followed the classical aberration production model; chromatid aberrations exhibit
more complex kinetics. The differential cell cycle
response to aberration induction varied with radiation, quality, dose, and aberration type. The relative biological effectiveness of fission neutrons fo
aberration production also varied with aberration
type, cell cycle stage, and neutron dose but remaine
principally between 2 and 6.

3184   DIFFERENTIAL REGENERATION OF INTESTINAL
        PROLIFERATIVE CELLS AND CRYPTOGENIC CELLS
AFTER IRRADIATION. (Eng.) Potten, C. S. (Christie
Hosp., Manchester M20 9BX, England); Hendry, J. H.
Int. J. Radiat. Biol. 27(5):413-424; 1975.

3185   QUANTITATIVE DOSE-RESPONSE OF GROWTH AND
        DEVELOPMENT IN *ARABIDOPSIS THALIANA* EX-
POSED TO CHRONIC GAMMA-RADIATION. (Eng.) Daly, K.
(Dept. Biology, California State Univ., Northridge,
Calif. 91324); Thompson, K. H. *Int. J. Radiat.
Biol.* 28(1):61-66; 1975.

3186   CONSEQUENCES OF THYROID RADIATION IN
        CHILDREN. (Eng.) Braverman, L. E. (St.
Elizabeth's Hosp., Boston, Mass.). *N. Engl. J. Med.*
292(4):204-205; 1975.

3192   ACUTE GRANULOCYTIC LEUKEMIA IN HODGKIN's
        DISEASE. (Eng.) Raich, P. C. (Univ. of
Wisconsin Medical Center, 1300 University Ave.,
Madison, Wis. 53706); Carr, R. M.; Meisner, L. F.;
Korst, D. R. *Am. J. Med. Sci.* 269(2):237-241; 1975.

3193   ENVIRONMENTAL MUTAGENS AND ENVIRONMENTAL
        FACTORS THAT CAN MODIFY THEIR ACTION.
(Eng.) de Marco, A. (Centro Genetica Evoluzionis-
tica del CNR, Istituto di Genetica, Rome (Italy));
Belloni, M. P.; Cozzi, R.; Olivieri, G. *Mutat. Res.*
29(2):253; 1975.

3194   DOSIMETRIC SIGNIFICANCE OF CYTOGENIC
        TESTS IN ACCIDENTALLY IRRADIATED SUBJECTS.
(Fre.) Doloy-Biola, M. T. (Département de protec-
tion, CEN de Fontenay-aux-Roses., FRANCE); Le Go,
R.; Ducatez, G.; Lepetit, J.; Bourguignon, M. *Bull*
Inf. Sci. Techn. Comm. Energ. Atom. (203):17-25;
1975.

3195   RADIATION-INDUCED INTESTINAL TUMORS IN
        RATS: CORRELATION BETWEEN THE AMOUNT OF
THE CIRCULATING TUMOR-RELATED ANTIGEN(S) AND TUMOR
SIZES [abstract]. (Eng.) Cheng, H. F. (Radiat. Res.
Lab. Univ. Iowa, Iowa City); Cantarero, L.; Osborne,
J. W. *Radiat. Res.* 62(3):604; 1975.

See also:

*   (Rev):   3005,   3009,   3036,   3048,   3049,   3050,
            3051
*   (Viral):   3229,   3249
*   (Path):   3456

3196   THE LOCATION OF THE GENES CODING FOR HEXON
        AND FIBER PROTEINS IN ADENOVIRUS DNA.
(Eng.) Mautner, V. (Natl. Inst. Med. Res., London,
England); Williams, J.; Sambrook, J.; Sharp, P. A.;
Grodzicker, T. *Cell* 5(1):93-99; 1975.

A serological analysis was performed of the structural proteins of 17 adenovirus recombinants that
permitted the genes coding for two of the chief components of the virus capsid (bexon and fiber) to be
assigned to precise locations on the genetic and
physical maps of adenovirus DNA. High titer stocks
of recombinants and the parental, wild-type viruses
were prepared in monolayers of HeLa cells. For complement-fixation assays, 10^7 cells were removed
three days after infection, resuspended in 1 ml of
infection medium, and sonicated. The aqueous layer
after centrifugation contained approximately 10^9
plaque-forming U/ml, and was used at a dilution of
1:20 in complement fixation tests. For gel diffusion experiments, 10^6 infected cells were resuspended in 0.2 ml of phosphate-buffered saline and disrupted in an ultrasonic bath for two minutes. Hexon
and fiber of AD2 and AD5 were prepared from soluble
cell extracts of infected HeLa cells and purified by
DEAE cellulose chromatography and crystallization.
Antisera to these antigens were raised in guinea
pigs. The phenotype of the recombinants was determined by the serological analysis. By correlating
this data with the genetic and physical maps of the
adenovirus genome, obtained by recombination and
restriction endonuclease analysis, the genes coding
for the hexon and fiber were assigned to specific
locations on the adenovirus DNA. It is concluded
that the region of adenovirus DNA that codes for
fiber protein lies towards the right hand end of the
genome between positions 0.85 and 0.98; this corresponds to a DNA fragment of about 2.7×10^6 daltons,
with the coding potential for a protein of 99,000
daltons. It is also concluded that the segment of
DNA that codes for bexon lies between positions 0.44
and 0.59, corresponding to a DNA fragment of $3.0 \times
10^6$ daltons, with the coding potential for a protein
of 130,000 daltons.

3197   ADENOVIRUS-2 mRNA IS TRANSCRIBED AS PART
        OF A HIGH-MOLECULAR WEIGHT PRECURSOR RNA.
(Eng.) Bachenheimer, S. (Rockefeller Univ., New
York, N.Y. 10021); Darnell, J. E. *Proc. Natl.
Acad. Sci. USA* 72(11):4445-4449; 1975.

The order of transcription and the length of nascent
RNA transcripts from adenovirus-2 (Ad-2) DNA in the
nucleus of infected cells were investigated. The
growing RNA chains were labeled with [^{3}H]uridine for
1 or 2 min. The RNA was separated on the basis of
size, and to the ordered *Eco*RI restriction endonuclease fragments derived from Ad-2 DNA. The majority of the virus-specific RNA molecules were synthesized as very high-molecular-weight units beginning
at a common point at least 25-30,000 base pairs from
one end of the Ad-2 DNA. These molecules could be
reduced in size without further RNA synthesis. The
experiments indicate the obligatory origin of Ad-2
messenger RNA from a high-molecular-weight precursor
molecule.

creased by the 18th day, while that of LDH_2 showed a sharp decrease by the 24th day. The motility of LDH_3 increased by the 18th day. The MDH_1 activity was significantly increased on the 5th and 18th day, the MDH_2 activities were increased on the 3rd, 5th, and 8th days, and the MDH_3 activity was increased on the third and 24th days. The electrophoretic motility of LDH_3 was significantly increased on the 5th and 8th days. The findings indicate that the partial morphological transformation, starting on about the 18th day, was accompanied by an increased LDH_1 and LDH_2 activity, followed by a drop in the LDH_1 activity, and by sharp changes in the MDH_3 activity. The changes observed in the isoenzyme synthesis are probably due to the epigenomic effect of oncovirus and may be used as indicators of the malignant transformation of tissue cultures.

3202   CONTROL OF NUCLEAR DIVISION IN SV40 AND
        ADENOVIRUS TYPE 12 TRANSFORMED MOUSE 3T3
CELLS. (Eng.) O'Neill, F. J. (Univ. Utah Med. Cent., Salt Lake City). *Int. J. Cancer* 15(5):715-723; 1975.

The control of nuclear division was investigated in simian virus 40 (SV40) and adenovirus type 12 (Ad-12)-transformed cells. The cells included BALB/c 3T3 cells, Swiss 3T3 cells (3T34E), BALB/c mouse embryo fibroblasts (MEF), and SVT2 cells (SV40-transformed Swiss-Webster mouse cells). The formation of multinuclear cells and premature chromosome condensation (PCC) was studied following incubation for seven days with 1.5 µg/ml cytochalasin B (CB). MEF and both BALB/c 3T3 and 3T34E cells became binucleated in the presence of CB and cells with three or more nuclei were very rare or undetectable. Transformation of MEF by SV40 produced a dramatic change in response to CB. These cells, which contain SV40 T-antigen, became highly multinucleated in the presence of CB. More than 20% of the cells had at least seven nuclei. PCC were frequent and occurred in at least 10% of mitoses. SV40 MEF were diploid or tetraploid. Transformation of 3T3 by SV40 or adenovirus type 12 did not result in a marked change in the response to CB. Although some trinucleate and tetranucleate cells were formed, cells with more nuclei were undetectable or rare. PCC were also rare. These cells showed chromosome numbers somewhat lower than their untransformed parents and in one line the chromosome number appeared to decrease with the passage of the cells. This failure to undergo a marked change in responsiveness to CB following transformation was not a characteristic of all transformed 3T3 cells. SVT2, a line of 3T3 which was transformed by SV40 prior to its establishment as a continuous line, responded to CB with a high degree of multinucleation. Because 3T3, 3T34E and their transformed derivatives were aneuploid or subtetraploid, an association must exist between chromosomal constitution and control of nuclear division. The transformed MEF and the SVT2 cells were diploid or near diploid. The results suggest that 3T3 and 3T34E may be constitutive for control of nuclear division because transformation does not result in loss of controlled nuclear division.

3203   DEMONSTRATION OF INFECTIOUS DNA IN TRANS-
FORMED CELLS: III. CORRELATION OF DETEC-
TION OF INFECTIOUS DNA-PROTEIN COMPLEXES WITH
PERSISTENCE OF VIRUS IN SIMIAN ADENOVIRUS SA7-INDUCED
TUMOR CELLS. (Eng.) Butel, J. S. (Baylor Coll.
Medicine, Houston, Tex. 77025); Talas, M.; Ugur,
J.; Melnick, J. L. *Intervirology* 5(1/2):43-56;
1975.

Primary tumors induced in newborn hamsters by simian
adenovirus SA7 were investigated in transfection
experiments. The animals were injected sc with
SA7 virus (5 x 10^7 plaque-forming units/0.1 ml)
within 24 hr after birth. Infectious DNA-protein
complexes were readily detected in both the supernatant and pellet fractions obtained by a modified
Hirt extraction procedure; DEAE-dextran was required
for infectivity to become manifest. Infectivity
could be abolished by exposure to DNase (100 µg/ml),
but it was unaffected by SA7-specific antiserum
or RNase (50 µg/ml), and only partially inactivated
by trypsin (300 µg/ml) treatment. SA7 tumor cells
serially passaged either in tissue culture or in
hamsters yielded infectious DNA complexes much
less frequently and appeared to evolve into non-
yielder cell lines. When large numbers of cells
were lysed, intact virus could be recovered from
all the primary tumors and from some of the subcultured cell lines. There was a correlation between
the persistence of complete virus and the presence
of infectious DNA-protein complexes. When the tumor
cells carried very small amounts of intact virus,
infectious DNA complexes could still be detected;
when virus could no longer be detected in the tumor
cells, infectious DNA complexes could no longer be
found. The results suggest that a portion of the
infectious DNA moieties exists as viral DNA-protein
complexes in the tumor cells.

3204   DIFFERENT HEMATOLOGICAL DISEASES INDUCED
BY TYPE C VIRUSES CHEMICALLY ACTIVATED
FROM EMBRYO CELLS OF DIFFERENT MOUSE STRAINS. (Eng.)
Greenberger, J. S. (Natl. Cancer Inst., Bethesda,
Md.); Stephenson, J. R.; Moloney, W. C.; Aaronson,
S. A. *Cancer Res.* 35(1):245-252; 1975.

The biological activities of two endogenous type C
viruses chemically induced from C58-derived and
BALB/c mouse embryo cells in culture were compared in
NIH Swiss mice. Preparations of BALB:virus-1 and
murine leukemia virus C58 (C58-MuLV) were tested for
the major mouse type C virion polypeptide (p30) and
for infectivity for NIH/3T3 and BALB/3T3 cells by
the polymerase induction assay. Each virus was more
infectious for NIH/3T3, and C58-MuLV was about 10-
fold more infectious per nanogram p30 than the BALB/
virus-1. Infectious type C virus recovery from
spleens of NIH Swiss mice inoculated at birth with
10^5 polymerase-inducing units of either virus or
serum-free medium were assayed after incubation for
14 days at 37 C. At 15 mo postinoculation, all mice
inoculated with BALB:virus-1 and all of those inoculated with C58-MuLV were virus-positive. A quantitative comparison was made of virus production by
mouse spleens ten months after inoculation. The
level of virus produced by spleens of C58-MuLV-

inoculated mice was much higher than that of spleens
releasing BALB:virus-1. Virus inoculation led to
the development of diseases associated with splenomegaly in 19 of 135 C58-MuLV-inoculated mice and 16
of 100 BALB:virus-1 inoculated mice. Lymphoadeno-
pathy also occurred in virus-inoculated mice. Histopathological diagnosis in the C58-MuLV group was
lymphoblastic leukemia, compared to the myeloid metaplasia or myelogenous leukemia found in the BALB:
virus-1 group. These data indicate that two distinct diseases were induced *in vivo* by each endogenous type C virus. It is also suggested that the
N-tropic endogenous C58 virus is the causative agent
in naturally occurring leukemia in this strain.

3205   PRIMATE TYPE C VIRUS p30 ANTIGEN IN CELLS
FROM HUMANS WITH ACUTE LEUKEMIA. (Eng.)
Sherr, C. J. (Natl. Cancer Inst., Bethesda, Md.);
Todaro, G. J. *Science* 187(4179):855-857; 1975.

The peripheral blood WBC of patients with acute leukemia were studied for the presence of type C virus
p30 antigen. The WBC were obtained from five patients by leukophoresis, and were extracted and partially purified by gel filtration. The extracts
were used as competing antigens in radioimmunoassays
for the p30 proteins of wooly monkey type C virus
(SSAV), gibbon ape leukemia virus (GALV), feline
leukemia virus (FeLV), Rauscher murine leukemia virus (MuLV), and an endogenous baboon type C virus
(M7/M28 group). The assay showed little or no difference between SSAV and GALV p30 proteins, and did
not detect those of other type C viral groups. Of
the five extracts from the patients tested, all competed for ^{125}I-labeled p30 protein of SSAV. None
competed for the p30 protein of MuLV or FeLV. Control studies with other assays for p30 proteins of
SSAV and GALV showed that the competition reactions
of the SSAV assay are independent of the source or
manner of purification of the labeled test antigens,
the antisera used, or the sensitivities of the assays. All five assays showed comparable antigen levels (2-20 ng/mg tissue). It is concluded that viruses of this group may be associated with acute leu
kemia in man.

3206   INDUCTION OF MOUSE TYPE-C VIRUS BY LIPO-
POLYSACCHARIDE. (Eng.) Greenberger, J. S
(Natl. Cancer Inst., Bethesda, Md.); Phillips, S. M.
Stephenson, J. R.; Aaronson, S. A. *J Immunol.* 115
(1):317-320; 1975.

The effects of lymphocyte mitogens and halogenated
pyrimidines on the induction of specific endogenous
viruses from BALB/c mouse spleen cells were compared
A cocultivation technique was used to quantitate
type-C virus release from BALB/c embryo fibroblasts
and spleen cells. NIH/3T3 and NRK cell lines, non-
productively transformed by the Kirsten strain of
murine sarcoma virus, were used. KIMSV nonproducer
lines of NIH/3T3 and NRK (K-NIH and K-NRK, respectively) were plated at 3 x 10^6 cells per Petri dish.
Sarcoma virus release by either K-NIH or K-NRK cells
cocultivated with untreated BALB/c spleen cells was
below the level of detection. Spleen cells exposed

DNA to the S-1 nuclease from *Aspergillus oryzea*. Single-stranded DNA transcripts of the rat type-C viruses hybridized specifically to DNA of several rat cell cultures with no obvious qualitative or quantitative differences. Similar products prepared from a pseudo-type sarcoma virus with contributions from rat and mouse type-C viruses hybridized to both rat and mouse cellular DNA; mouse viral transcripts did not hybridize to rat cell DNA. Viral RNA was detected in all rat cells by means of the rat viral DNA transcripts, with some differences between untreated low-passage cells and sister cultures treated with bromodeoxyuridine or bromodeoxyuridine plus methylcholanthrene. Cells treated with both compounds were previously shown to be transformed and tumorigenic, and these were distinguishable by kinetic analysis form the control cells. It is concluded that the genome of rat type-C viruses is represented, at least in part, in the DNA of normal rat cells.

3208   SEROEPIDEMIOLOGICAL EVIDENCE FOR HORIZONTAL TRANSMISSION OF BOVINE C-TYPE VIRUS. (Eng.) Piper, C. E. (Sch. Veterinary Medicine, Univ. Pennsylvania, Kennett Square, Pa. 19348); Abt, D. A.; Ferrer, J. F.; Marshak, R. R. *Cancer Res.* 35(10): 2714-2716; 1975.

Studies were performed to determine the natural mode of transmission of bovine C-type virus (BLV). Thirty colostrum-deprived calves from leukemia-free herds were foster nursed for ten weeks on cows infected with BLV from multiple-case herds or on cows from leukemia-free herds. After weaning, the calves were raised in continuous contact with BLV-infected animals of approximately the same ages. Sera collected at 6-18 and 43-48 mo of age were examined for the presence of antibodies to BLV by the immunofluorescent antibody test. At 6-18 mo of age, only one of the 30 calves from leukemia-free herds had a detectable antibody response to BLV. By 43-48 mo of age the number of antibody-positive animals had risen to 17. The foster dam's herd of origin did not significantly affect the rate of BLV infection. These results indicate that BLV can be horizontally transmitted from infected to noninfected animals.

3209   VERTICAL TRANSMISSION OF C-TYPE VIRUSES: THEIR PRESENCE IN BABOON FOLLICULAR OOCYTES AND TUBAL OVA. (Eng.) Kalter, S. S. (Microbiology and Infectious Diseases, Southwest Foundation Res. and Education, San Antonio, Tex. 78284); Heberling, R. L.; Smith, G. C.; Panigel, M.; Kraemer, D. C.; Helmke, R. J.; Hellman, A. *J. Natl. Cancer Inst.* 54(5):1173-1176; 1975.

The follicular oocytes and tubal ova of baboons were studied by electron microscopy to detect C-type virus particles; the presence of such particles would support the hypothesis that either a provirus or virogene-oncogene, or viral information rather than an intact virus acts as a mechanism for inheritance of the oncogenic potential vertically. C-type particles were found in follicular oocytes from 5 of 7 baboons and in tubal ova from 2 of 5 animals. Typically, these particles were adjacent to the plasma membrane in the perivitelline space or along the inner margin of the zona pellucida. The particles were present in both atretic and nonatretic oocytes. No intracytoplasmic or intracisternal A-type particles were observed in follicular cells surrounding the follicular oocytes or in any other ovarian structures. Thus, these C-type viruses may be of germ-cell origin. These findings support the concept of vertical transmission of C-type particles.

3210   LOCALIZATION AND QUANTITATION OF EBV-
        ASSOCIATED NUCLEAR ANTIGEN (EBNA) IN
RAJI CELLS. (Eng.) Bahr, G. F. (Armed Forces
Inst. Pathology, Washington, D. C., 20306); Mikel,
U.; Klein, G. *Beitr. Pathol.* 155(1):72-78; 1975.

The direct localization of Epstein-Barr virus (EBV)-associated nuclear antigen (EBNA) in Raji cells was studied *via* electron microscopy. In experiments employing ferritin-labeled goat antihuman antibody, β_{1C}, no meaningful results were obtained; both positive and negative serum, and unreacted chromatin bound diffusely and in patches. An immunocytochemical method involving a horseradish-peroxidase labeling technique showed a strong reaction on and along every chromatin fibril. This process resulted in encrustation and caking of chromatin strands when positive anti-EBNA serum was used. The sensitivity of the method was tested by a quantitative analysis of the electron microscopic material, based on the relation of contrast and object dry mass. The dry mass of Raji chromatin increased after exposure to both EBV-negative and -positive serum, most significantly with the latter; osmium tetroxide produced a 15% decrease in mass. It is concluded that the complement-fixing EBV-associated antigen is widely distributed throughout the nucleus. It occurs on chromatin fibers at sites so close to each other that available immunocytochemical techniques cannot distinguish them.

3211   STUDIES ON THE PRESENCE OF ANTIBODIES TO
        EB VIRUS AND OTHER HERPESVIRUSES IN NORMAL
CHILDREN AND IN INFECTIOUS MONONUCLEOSIS. (Eng.)
Gergely, L. (Univ. Med. Sch., Debrecen, Hungary);
Czeglédy, J.; Váczi, L.; Szabó, B.; Binder, L.;
Szalka, A. *Acta Microbiol. Acad. Sci. Hung.* 22(2):
75-82; 1975.

The frequency of antibodies to Epstein-Barr virus and of antibodies to herpes simplex virus and cytomegalovirus were studied in children and in patients with infectious mononucleosis. The sera of 100 Hungarian children free from infectious disease were examined by indirect immunofluorescence for the presence of antibodies to Epstein-Barr virus. The sera of 46% of the children in their first year of life was positive. The corresponding value for children 2-6 yr old was 66%, and 91% for children aged 7-14 yr. The percentage of subjects with antibodies to cytomegalovirus and herpes simplex virus was about 50%, irrespective of age. Serum samples from 69 patients with infectious mononucleosis were also tested. It was found that in the Paul-Bunnell-positive cases the

sequences, respectively, and each has inverted
terminally redundant regions that correspond to the
internal duplications observed by electron microscopy. The DNA from other strains of HSV-1 and 2
also consists of equal proportions of all four possible permutations of the L and S segments. These
unusual features of HSV DNA molecules have novel
implications with regard to the genetic map and the
mode of replication and evolution of herpes simplex
viruses.

3215   DNA SYNTHESIS AND DNA POLYMERASE ACTIVITY
        OF HERPES SIMPLEX VIRUS TYPE 1 TEMPERATURE-
SENSITIVE MUTANTS. (Eng.) Aron, G. M. (Southwest
Texas State Univ., Biology Dept., San Marcos, Tex.
78666); Purifoy, D. J. M.; Schaffer*, P. A. J.
Virol. 16(3):498-507; 1975.

Fifteen temperature-sensitive mutants of herpes
simplex virus type 1 were studied with regard to
the relationship between their ability to synthesize
viral DNA and to induce viral DNA polymerase (DP)
activity at permissive (34 C) and nonpermissive
(39 C) temperatures. Human embryonic lung cells
were infected at a multiplicity of 5-10 plaque-
forming units per cell with each mutant and with
KOS strain virus as the wild-type. At 34 C, all
mutants synthesized viral DNA while at 39 C four
mutants demonstrated a DNA$^+$ phenotype, three were
DNA$^\pm$, and eight were DNA$^-$. DNA$^+$ mutants induced
levels of DP activity similar to those of the wild-
type virus at both temperatures, and DNA$^\pm$ mutants
induced reduced levels of DP activity at 39 C but
not at 34 C. Among the DNA$^-$ mutants, three were
DP$^+$, two were DP$^\pm$, and three showed reduced DP
activity at 34 C with no DP activity at 39 C.
DNA$^-$, DP$^-$ mutants induced the synthesis of a
temperature-sensitive DP.

3216   SYNTHESIS OF PROTEINS IN CELLS INFECTED
        WITH HERPESVIRUS. X. PROTEINS EXCRETED
BY CELLS INFECTED WITH HERPES SIMPLEX VIRUS, TYPES
1 AND 2. (Eng.) Kaplan, A. S. (Vanderbilt Univ.
Sch. Med., Nashville, Tenn.); Erickson, J. S.;
Ben-Porat, T. Virology 64(1):132-143, 1975.

The excretion of glycoproteins by cells infected
with herpes simplex virus type 1 (HSV-9) and type
2 (HSV-2) were studied in primary monolayer cultures
of rabbit kidney cells. In all experiments, rabbit
cells were infected with approximately 10-20 plaque-
forming U/cell. Radioactive labels included: [^{3}H]-
L-leucine (46 Ci/mM); [^{14}C]glucosamine (1.8 Ci/mM);
and 6-[^{3}H]glucosamine (12.6 Ci/mM). Antisera were
prepared against excreted proteins and HSV virions.
Polyacrylamide gel electrophoresis was used to
assay profiles of the excreted glycoproteins. Electrophoretic profiles of glycoproteins excreted by
infected and uninfected cells differed from those
of cell-associated glycoproteins. Profiles of
glycoproteins from HSV-1-infected cells differed
from those excreted by HSV-2-infected cells, and
both were unlike the profiles of glycoproteins excreted by uninfected cells. Indirect radioimmuno-
precipitation showed that antisera against glycoproteins excreted by virus-infected cells were

571

highly type-specific. Serum prepared against type 1 (H$_4$) virions reacted significantly with cell-associated glycoproteins synthesized by cells infected with type 2 (333) virus, and precipitated 6.4% of the total glycoproteins. Less than 1% of these glycoproteins was precipitated by serum against H$_4$-excreted proteins. A small amount of glycoproteins excreted from HSV-infected cells reacted with heterologous antisera. Despite the differences in the amount of excreted proteins precipitated by homologous and by heterologous antisera, the proteins in both cases migrated similarly on electrophoresis. While major parts of proteins present in the excreted fraction were type-specific, a small amount of cross-reacting was present in these fractions. The nonspecific precipitates obtained with preimmune sera had featureless profiles. Serum prepared against H$_4$-excreted proteins neutralized 333 virus, but to a lesser extent than serum prepared against H$_4$-virus. The results demonstrate that glycoproteins are excreted from cells infected with both types of HSV and that these proteins are primarily specific for the herpes simplex virus type infecting cells.

3217   EFFECT OF ZINC IONS ON THE SYNTHESIS OF HERPES SIMPLEX VIRUS DNA IN INFECTED BSC-1 CELLS. (Eng.) Shlomai, J. (Hebrew Univ.-Hadassah Medical Sch., Jerusalem, Israel); Asher, Y.; Gordon, Y. J.; Olshevsky, U.; Becker, Y. *Virology* 66(1):330-335; 1975.

The effect of zinc ions on the synthesis of herpes simplex virus (strain HF) DNA in infected BSC-1 cells was tested. At three hours postinfection, the cells were labeled with [^{3}H]thymidine in the absence or presence of ZnSO$_4$ at concentrations of 0.1, 0.2, and 0.3 mM. The cultures were harvested at 18 hr postinfection. ZnSO$_4$ (0.2 mM) inhibited the synthesis of viral DNA. A similar concentration of ZnSO$_4$ had no effect on DNA synthesis in uninfected cells. *In vitro* analyses revealed a marked depression of the DNA-polymerase activity in nuclei obtained from infected cells treated with 0.2 or 0.3 mM ZnSO$_4$, but not in cells treated with 0.1 mM ZnSO$_4$ during a 15-hr incubation. The results indicate that ZnSO$_4$ prevents the synthesis of HSV-DNA by a mechanism involving inhibition of the viral DNA polymerase activity.

3218   ANTIBODY TO HERPES SIMPLEX VIRUS TYPE 2-INDUCED NONSTRUCTURAL PROTEINS IN WOMEN WITH CERVICAL CANCER AND IN CONTROL GROUPS. (Eng.) Anzai, T. (Baylor Coll. Med., Houston, Tex.); Dreesman, G. R.; Courtney, R. J.; Adam, E.; Rawls, W. E.; Benyesh-Melnick, M. *J. Natl. Cancer Inst.* 54(5):1051-1059; 1975.

Sera of patients with cervical cancer and of control groups have been examined for the presence of antibodies of early proteins synthesized in herpes simplex virus type 2-infected cells. Sera from 15 control women, individually matched (age, race, socioeconomic status, sexual- and reproduction-associated factors) with 15 cervical cancer patients, as well as sera from ten patients with breast cancer were ana-

lyzed by an indirect radioimmune precipitation test followed by polyacrylamide gel electrophoretic analysis of immune precipitates. The relative reactivity to a major early nonstructural protein (VP134) was used to compare these selected sera. Approximately 90% of sera from patients with cervical cancer showed higher reactivity to VP134 than the mean of control sera, whereas 30% of sera from patients with breast cancer and less than 60% of sera from control women exceeded the mean value of control sera. The sera of all subjects had previously been tested for the presence of neutralizing antibodies to herpes simplex virus type 2, and it was found that serum reactivity to the protein was independent of the level of neutralizing antibodies. The authors discuss the need for more detailed information to determine if cervical cancer sera preferentially precipitate one or several herpes simplex virus-induced early proteins.

3219   SEROLOGICAL RELATIONSHIP BETWEEN A PATHOGENIC STRAIN OF MAREK'S DISEASE VIRUS, ITS ATTENUATED DERIVATIVE AND HERPES VIRUS OF TURKEYS (Eng.) Ross, L. J. N. (Houghton Poultry Res. Station Houghton, U.K.); Basarab, O.; Walker, D. J.; Whitby, B. *J. Gen. Virol.* 28(1):37-47; 1975.

The serological relationship between a pathogenic strain of Marek's disease virus (MDV), its attenuated derivative and Herpes virus of turkeys (HVI) was studied using precipitating antigens (A, B, and C) as references. Precipitating antigens present in extracts of chick embryo cells infected with the HPRS-16-attenuated strain (HPRS-16/att) of MDV were separated by gel filtration on Sephadex G200, and some of their properties were determined. The two main antigens detected with convalescent MDV serum, referred to as 'B' and 'C' antigens, had mobilities of 0·55 and 0·25, respectively, relative to phenol red on electrophoresis in 7·5% acrylamide gel. The B antigen was relatively stable and of low molecular weight in comparison with the C antigen. B and C antigens were in some instances also detected in culture medium of infected cells, but were distinguishable from the A antigen, a major glycoprotein antigen released into the culture medium of cells infected with HPRS-16. The results of immunodiffusion studies suggested that B antigen is common to MDV and strains of herpes virus of turkeys (HVT), and that at least two antigens (including C) are MDV-specific. The A antigen was also common to MDV and HVT strains. However, that the capacity of HPRS-16/att to synthesize A antigen was considerably reduced in comparison with HPRS-16 and HVT strains, and in some preparations the A antigen could not be detected. Evidence was also obtained for the presence of HVT-specific antigens associated mainly with the cell fraction.

3220   IODINATION OF HERPESVIRUS NUCLEIC ACIDS. (Eng.) Shaw, J. E. (Sch. Medicine, Univ. North Carolina, Chapel Hill, N.C. 27514); Huang, E. S.; Pagano, J. S. *J. Virol.* 16(1):132-140; 1975.

A simple procedure for the iodination of herpes simplex virus (HSV) DNA of high specific activity is presented. Strain 196 of herpes simplex type 2 was

body material and virion were seen in the stratum corneum of the infected birds; this material, carrying infectious virus, is believed to be released into the environment, and to be the main source of environmental contamination.

3222   PATHOGENIC AND SEROLOGIC STUDIES OF JAPAN-
        ESE QUAIL INFECTED WITH JM STRAIN OF MA-
REK'S DISEASE HERPESVIRUS. (Eng.) Mikami, T. (Sapporo Med. Coll., Japan); Onuma, M.; Hayashi, T. T. A.; Narita, M.; Okada, K.; Fujimoto, Y. *J. Natl. Cancer Inst.* 54(3):607-614; 1975.

The pathogenic and serologic effects of the JM strain of Marek's Disease herpesvirus (MDHV) in Japanese quail were investigated. Agar gel-precipitation (AGP) antigen was detected in MDHV-infected birds in both inoculated and contact-exposed birds within 30 days after inoculation. Immunofluorescent antigen for MDHV was found in the cutaneous epidermis and epidermal layers of the follicular epithelium. AGP testing showed that six of 73 quail had antibodies against chicken-positive antigen (CPA). By determing the formation of a continuous line with MD-chicken serum against CPA or quail-positive antigen (QPA), the antigen in quail was found to be MD-specific. Attempts to recover viruses by direct kidney culture failed, but the MD virus was recognized after subpassage of kidney culture cells into embryo fibroblasts. Approximately 87% of the 760 chicken sera had identical reactions to the antigen of infected and/or uninfected fowl; the sera had negative reaction to all four antigens or a positive reaction to CPA and QPA and a negative reaction to CNA and QNA. These results indicate that QPA is specific to MD infection, with most sera reacting specifically with CPA and QPA. The results demonstrate that Japanese quail are susceptible to the JM strain because of serologic and virologic evidence of infection. The specific antigen components could be separated from the nonspecific component of QPA by gel filtration on Sephadex G-150.

3223   POLYPEPTIDES OF MAMMALIAN ONCORNAVIRUSES.
        II. CHARACTERIZATION OF A MURINE LEUKEMIA
VIRUS ·POLYPEPTIDE (p15) BEARING INTERSPECIES REACTIVITY. (Eng.) Schafer, W. (Max-Planck-Institut fur Virusforschung, Tubingen, West Germany); Hunsmann, G.; Moennig, V.; de Noronha, F.; Bolognesi, D. P.; Green, R. W.; Huper, G. *Virology* 63(1):48-59; 1975.

Polypeptide p15 from Friend leukemia virus was isolated by multiple gel filtration steps in guanidine hydrochloride and characterized by serological tests, double-antibody radioimmunoassay, immunoprecipitation, and polyacrylamide gel electrophoresis. Because of its marked tendency to aggregate, renaturation of the protein was performed in the presence of 0.2% sodium deoxycholate. In Ouchterlony tests, antiserum prepared against p15 (in rabbits) was tested for possible cross reactivity with other virion components. No reaction was obtained with p10, p12, p30, or gp71. The antiserum did react with the homologous polypeptide but only in the presence of 0.2-0.5% sodium deoxycholate. The p15 antiserum also reacted with the homologous polypeptide in complement fixation. Radioimmunoassays confirmed the observations obtained with immunodiffusions and complement fixation. Precipitation with the p15 serum occurred only with the homologous polypeptide, and competition of the reaction was observed at limiting antibody dilution by the addition of unlabeled p15 but not by the addition of p10, p12, p30, and gp71. Essentially complete cross reactivity was seen in immunodiffusion of p15 antiserum with Friend, Gross, and Rauscher leukemia viruses. Using this test, it did not appear that p15 of Friend leukemia virus contained determinants distinguishable (type specific) from any in Gross leukemia virus, which belongs to a different serotype. When the reactivity of viruses from several species with p15 antiserum was tested in complement fixation, positive reactivity was obtained with murine, rat, hamster, classical feline, and simian oncornaviruses. A marked variation in the specific complement fixation activity was seen in the viruses studied. The reason for this great variation is not clear, but quantitative and qualitative differences may both play a role.

3224   PROPERTIES OF FELINE LEUKEMIA VIRUS.
        III. ANALYSIS OF THE RNA. (Eng.) Brian,
D. A. (Dépt. Microbiology and Public Health, Michigan
State Univ., East Lansing, Mich. 48824); Thomason,
A. R.; Rottman, F. M.; Velicer*, L. F. *J. Virol.*
16(3):535-545; 1975.

The kinetics of virus labeling was used to study the maturation of viral RNA in the Rickard strain of feline leukemia virus. Viral RNA labeled over differing intervals was characterized by gel electrophoresis and velocity sedimentation in sucrose gradients made up in aqueous buffer and 99% dimethyl sulfoxide. Labeled virus was found within 30 min after adding radioactive uridine to the cells, and production of labeled virus reached a maximum at 4-5 hr after pulse labeling. Native RNA from feline leukemia virus resolved into three size classes when analyzed by electrophoresis on 2.0% polyacrylamide-0.5% agarose gels: a 6.2×10^6 to 7.1×10^6 molecular weight (50-60S) class, an 8.7×10^4 molecular weight (approximately 8S) class, and a 2.5×10^4 molecular weight (4-5S) class. From two experiments during which RNA degradation appeared minimal, these made up 57-76%, 2-5%, and 6-12%, respectively, of the total RNA. The 8S RNA in feline leukemia virus has not previously been reported. The 50-60S RNA from virus harvested after four hours of labeling electrophoretically migrated faster and sedimented more slowly in sucrose gradients than did the same RNA species harvested after 20 hr of labeling. This suggests the existence of an intravirion modification of the high-molecular-weight RNA. The large subunits of denatured viral RNA from both 4- and 20-hr labeled-viral RNA electrophoretically migrated with an estimated molecular weight of 3.2×10^6 but sedimented with 28S ribosomal RNA (2.8×10^6 molecular weight) when analyzed by velocity sedimentation through 99% dimethyl sulfoxide.

3229   EVALUATION OF IRRADIATION-PLUS-URETHAN-
       INDUCED MURINE LEUKEMIA VIRUS "RELEASE"
USING A NEW METHOD FOR QUANTITATION OF ONCORNAVIRUSES
IN TISSUES. (Eng.) Brown, R. C. (Univ. North Caro-
lina Sch. Med., Chapel Hill); Kostyu, J. A.; Kilgore,
A. *Cancer Res.* 35(4):1053-1057; 1975.

A new method for quantitating oncornaviruses in tissues is presented. The method was used to determine
whether virus release associated with leukomogenic
activity produced by X-irradiation-plus-urethan
treatment is related to change in virus quantity.
The new method involves the use of a microchamber
with the sides situated on rotor radii so as to produce a uniform virus-containing sediment of tissue
homogenate-sonicate that is evaluated by electron
microscopic examination of thin sections cut perpen-
dicular to the membrane surface. Samples containing
as little as 10^5 to 10^6 viruses can be relatively
easily counted. Semipurified or purified viruses can
also be counted after mixing with a tissue homogenate-
bovine serum albumin diluent. RFM/Un mice (4-6 wk
old) were given four weekly doses of 170 rads X-irradiation, and C57BL/6J mice received four weekly
doses of 200 rads. Both treatments were followed
three days later with 1 mg/g urethan ip. The virus
quantity was evaluated ten days after the last
irradiation. Treated C57BL/6J mice showed numbers
of virus no greater than controls (8×10^6 virus/ml
in treated mice *versus* 1.1×10^7 virus/ml in controls); similar results were obtained with treated
RFM/Un mice (4.0×10^7 in treated *versus* 3.0×10^7
in controls). These results indicate that the leukemogenic activity, shown to be present in such
thymus-bone marrow homogenates at this time after
irradiation-plus-urethan treatment, is not due to
change in quantity of C-type viruses as has been
previously proposed.

3230   STUDIES ON VIRUS INDUCTION BY 5-BROMODEOXY-
       URIDINE IN NONPRODUCER MURINE SARCOMA VIRUS-
TRANSFORMED 3T3 CELLS. (Eng.) Margalith, M. (St.
Louis Univ. Sch. Med., Mo.); Thornton, H.; Narconis,
R.; Pinkerton, H.; Green, M. *Virology* 65(1):27-39;
1975.

The properties of the nonproducer KA31 cell line
transformed by the Kirsten strain of murine sarcoma
virus were studied. Treatment of cells with 5-bromo-
deoxyuridine (BrdU) induced the synthesis of DNA
polymerase-containing particles which band in sucrose
gradients at a density of 1.16 and utilize poly(I)
$(dC)_{12-18}$, a template-primer specific for RNA-directed
DNA polymerase. The induction process was studied by
analysis of the culture fluid for polymerase activity;
the induced cells were studied for viral proteins by
immunofluoresence. Virus particles were studied with
the electron microscope. Varying the concentration
and the time of exposure to BrdU showed that the
maximal level of polymerase activity was induced when
cells were treated with 20 µg of BrdU for a 24-hr
period, 24 hr after seeding. Under these conditions
about 15% of the cells were induced to synthesize viral
antigens detected with anti-Moloney murine sarcoma
virus rat serum. The level of polymerase activity in
the culture fluid and viral antigen in the cells be-

came maximal at three days after BrdU treatment and decreased rapidly. Electron microscopy showed a large increase in the proportion of cells containing incomplete virus particles and in the number of particles per cell at two and three days after BrdU treatment. At three days after treatment, 82% of the cells contained incomplete virus particles within dilated vesicles of the endoplasmic reticulum, as compared to 18% for untreated cultures. KA31 cells, treated with 20 µg/ml BrdU for 24 hr became flat and fibroblastic, as compared to the rounded, transformed morphology of untreated cells. This effect was even more pronounced when cells were maintained and passaged in the presence of 4 µg/ml BrdU. A high level of virus production could be maintained for at least 15 days by continuous maintenance of cells on 4 µg/ml BrdU. It is not clear whether the mechanism involves continuous reinduction of virus or overcoming the normal inhibition of virus multiplication in these cells.

3231   MURINE SARCOMA VIRUS RELATED NUCLEIC ACID
        SEQUENCES IN A NON-TRANSFORMING VIRUS
DERIVED FROM AN INTERSPECIES PSEUDOTYPE SARCOMA
VIRUS. (Eng.) Okabe, H. (Flow Lab. Inc., Rockville, Md.); Gilden, R. V.; Hatanaka, M. Int. J. Cancer 15(5):849-859; 1975.

The question of whether mouse type-C virus-related sequences detected in the hamster virus GLOH⁻ are derived from Gross passage A virus (GLV) or from Moloney murine sarcoma virus (M-MSV) was investigated. The viruses used included Rauscher leukemia virus (RLV), Harvey-MSV (H-MSV), Akv-1, Graffi-HaLV (G-HaLV), M-MSV(RaLV), and Moloney strain of MuLV, as well as the GLV, M-MSV, and GLOH⁻ viruses. Viral 70S RNA was separated, and single-stranded viral [³H]DNA transcripts complementary to viral RNAs were isolated. [³H]DNA probes were incubated with viral RNA and loaded onto hydroxyapatite (HA) columns. Thermal dissociation (Tm) was also determined. Analysis of transcripts related to M-MSV(RaLV) RNA, revealed a clear relationship to Moloney-related sequences, both in reaction extent and Tm. The GLV transcript showed the same approximate Tm with M-MSV(RaLV) RNA (72.0 C) as with M-MSV RNA (72.6 C). Because of the thermal stability of inter-strain hybrid molecules, and the stability of Moloney virus sequences in a rat-mouse interspecies pseudotype, it is concluded that mouse virus-specific sequences in GLOH⁻ are derived from Moloney virus and not Gross virus.

3232   RELATIONSHIP IN NUCLEIC ACID SEQUENCES
        BETWEEN MOUSE MAMMARY TUMOR VIRUS VARI-
ANTS. (Eng.) Michalides, R. (Meloy Lab., Inc., Springfield, Va. 22151); Schlom, J. Proc. Natl. Acad. Sci. USA 72(11):4635-4639; 1975.

The relationship of the nucleic acid sequences of horizontally transmitted mouse mammary tumor virus (MMTV) variants from four mouse strains to each other and to a vertically transmitted MMTV variant was determined using competition molecular hybridization experiments. Primary cultures of mouse mammary carcinomas were used as a source of both ³H-labeled and unlabeled 60-70S RNA of MMTV. The

hybridization experiments revealed that, within the limits of the assay, the RNAs of the MMTVs synthesized in culture by the tumors of the mouse strains RIII, GR, A, and C3H, are identical. A comparison of the genomes of the milk-transmitted MMTV(C3H) and the vertically transmitted MMTV (C3Hf) revealed that these two viruses are approximately 75% similar. No nucleic acid sequence homology was observed between MMTV(C3H) 60-70S RNA and the RNAs of murine leukemia virus, Mason-Pfizer virus, or the bromodeoxyuridine induced type-B guinea pig virus. This assay considers the total genome of the RNA tumor virus and avoids inconclusive results due to the use of incomplete or preferential complementary DNA transcripts. The only disadvantage of the assay is that the hybrid formation excludes some of the added radioactively labeled RNA.

3233   NON-REPETITIVE DNA TRANSCRIPTS IN NUCLEI
        AND POLYSOMES OF POLYOMA-TRANSFORMED AND
NON-TRANSFORMED MOUSE CELLS. (Eng.) Grady, L. J. (Div. Lab. Res., New York State Dept. Health, Albany) Campbell, W. P. Nature 254(5498)356-358; 1975.

RNA-DNA saturation-hybridization experiments were used to detect the percent of common sequences between non-repetitive DNA and nuclear or polysomal RNA in polyoma-transformed (PY AL/N) or nontransformed (AL/N) cells. Nuclear and cytoplasmic fractions were prepared using 0.05% Triton X-100. Nuclear RNA was obtained and polysomes were purified from the supernatant by discontinuous or linear sucrose density gradient centrifugation. Polysomal RNA was isolated by phenol-chloroform extraction. After further purification, hybridization reactions using the appropriate RNA (12 mg/ml) and labeled, non-repetitive DNA (12 µg/ml) were performed. Time courses of hybridization reactions were plotted and the percent of RNA hybridized to DNA was calculated. In the PY AL/N cells nuclear RNA contained about 30% of the non-repetitive DNA sequences, whereas polysomal RNA contained about 6.4%. In the AL/N cells, the nuclear RNA contained about 20% of the non-repetitive DNA sequences in both subconfluent and confluent growth states. However the polysomes of subconfluent AL/N cells showed a greater percentage of DNA transcripts (about 8.6%) than the polysomes in confluent AL/N cells (5.7%). Both qualitative and quantitative differences appear to exist in the polysomal transcripts in these two growth states.

3234   ELECTROPHORETIC ANALYSIS OF THE STRUCTURAL
        POLYPEPTIDES OF POLYOMA VIRUS MUTANTS.
(Eng.) Frost, E. (Départment de Microbiologie, Centre Hospitalier Universitaire, Université de Sherbrooke, Sherbrooke, Québec, Canada); Bourgaux, P. Virology 65(1):286-288; 1975.

The structural polypeptides of wild-type polyoma virus were compared with those of some of its temperature-sensitive (ts) mutants using polyacrylamide gel electrophoresis. The study was limited to the early mutant, ts-a, two complementing late mutants, ts-10 and ts-1260, the noncomplementing ts-3, and

576

ted complex formation. Many sedimentation peaks
were inhibited by pretreatment of the DNA with DN-
ase, or by treating the capsids with pronase. The
demonstration that empty polyoma capsids form complexes with DNA, while whole virions do not, suggests that the empty capsids contain some site(s)
that have an affinity for DNA. The author suggests that the site(s) may be attached to or part
of the inner surface of the empty capsids or, al-
ternatively, at a possible point of incompleteness
of some of the empty capsids.

3236   A COMPARATIVE STUDY OF BK AND POLYOMA
        VIRUSES. (Eng.) Seehafer, J. (Dept.
Biochemistry, Univ. Alberta, Edmonton, Alberta,
Canada); Salmi, A.; Scraba, D. G.; Colter, J. S.
Virology 66(1):192-205; 1975.

The molecular anatomies of BK and polyoma (Py)
viruses were compared by hemagglutination and hemagglutination inhibition tests, sodium dodecyl
sulfate-polyacrylamide gel electrophoresis, centrifugal analysis of viral DNA in alkaline sucrose
density gradients, and electron microscopy. The
viruses were purified by a procedure involving
precipitation of the virus from crude lysates with
methanol, sequential treatment of the resuspended
methanol precipitate with DNase and RNase, clari-
fication of the suspension by chloroform extraction,
sedimentation of the virus through 20% sucrose
solution, and equilibrium centrifugation in CsCl
density gradient. The capsids of both were found
to be composed of 72 capsomeres arranged in a T = 7d
icosahedral lattice. Analysis of the DNAs of the
two viruses suggested that the full-length DNA of
BK virus may be slightly smaller than that of Py
virus. The two virions contained the same number
of structural polypeptides, but the molecular mass
of BK-VP1 was about 4,000 daltons less, and those of
BK-VP2 and 3 about 4,000 daltons more than those
of the corresponding Py polypeptides. The observation that the major capsid polypeptide, BK-VP1, was
smaller than Py-VP1, was compatible with the observed
difference in the diameters of the two virions
(BK = 405 ± 10 Å; Py = 430 ± 15 Å). The two viruses
differed sharply in the relative efficiencies with
which they hemagglutinated guinea pig and human
RBC. BK virus, unlike Py virus, replicated only
in certain human or monkey cells; human fetal kidney
cells were found to be the most satisfactory for the
propagation of this agent. On the whole, the results
demonstrate a striking similarity in the molecular
anatomies of BK and Py viruses.

3237   RELATIONSHIP BETWEEN A-TYPE AND C-TYPE
        PARTICLES IN CELLS INFECTED BY ROUS SAR-
COMA VIRUS. (Eng.) de Giuli, C. (Rockefeller
Univ., New York, N.Y. 10021); Hanafusa*, H.; Ka-
wai, S.; Dales, S.; Chen, J. H.; Hsu, K. C. *Proc.
Natl. Acad. Sci. USA* 72(9):3706-3710; 1975.

The antigenic relationship between Rous sarcoma
virus (RSV) and cytoplasmic A-type particles occasionally found in RSV-infected chicken embryo
cells (type C/E) and the circumstances surrounding

the formation of such particles were investigated using various RSV strains (Schmidt-Ruppin subgroup A and D, and a temperature sensitive mutant ts68) and several Rous-associated strains (RAV). To detect A particles, cells were washed repeatedly, made permeable by a two-minute exposure to .005% Nonident P40 (NP40), centrifuged (800 x g for ten minutes), and resuspended in Eagle's MEM medium. For 30 min cells were incubated at 37 C with mono-specific antiserum against RAV-2 proteins p27, p19, and p15. Cells were then centrifuged and resuspended twice. Sections were prepared and stained. Because so few sections revealed A particles, the statistical frequency of A particle occurrence was determined by treating cells at 0 C with 0.25 or 0.5% NP40 to partially solubilize organelles, and centrifugation at 1200 x g for 30 min; this produced particle-enrichment. Results demonstrated that A particles contain components immunologically related to the internal proteins of C-type virus particles. RSV-infected cultures contained A particles more consistently than RAV-infected cultures. Normal cells contained no A particles. Particles were present in 5-30% of infected cell profiles, suggesting that they are not an obligatory precursor to C particles, because then a higher percentage would be expected. Cell cultures infected with serial dilutions of RAV-2[4] and passage every three days were examined for gross cytopathological changes and the frequency of A particle appearance; the higher the initial concentration of virus, the sooner both cytopathological change and A particles appeared. It is suggested that A particles are formed as a product of C-type virus infection and develop in cells undergoing cytopathological degeneration.

3238   EFFECT OF ONCOGENIC VIRUS ON MUSCLE DIFFER-
        ENTIATION. (Eng.) Holtzer, H. (Sch. Medicine, Univ. Pennsylvania, Philadelphia, Pa. 19174); Biehl, J.; Yeoh, G.; Meganathan, R.; Kaji, A. *Proc. Natl. Acad. Sci. USA* 72(10):4051-4055; 1975.

The question of whether some events that control differentiation can modulate the course of viral-induced transformation was investigated in breast muscle cells from leukosis-free ten-day chick embryos. Chick muscle cultures infected with the Prague strain of Rous sarcoma virus formed myotubes, but these myotubes vacuolated; by day six, most had degenerated, leaving only large numbers of transformed mononucleated, replicating cells. Muscle cultures infected with a temperature-sensitive mutant (ts) at permissive temperatures (36 C) behaved as cells infected with wild-type Rous sarcoma virus. Temperature-sensitive infected cells reared for eight days at nonpermissive temperature (41 C) formed contracting myotubes, plus large numbers of fibroblastic cells. If these cultures were incubated at the permissive temperature, the myotubes vacuolated and degenerated within 72 hr, whereas the mononucleated cells became transformed. If replicating ts-transformed cells after eight days at permissive temperature were shifted to nonpermissive temperature, many cells fused and formed contracting, post-mitotic myotubes within 72 hr. Creatine kinase (ATP:creatine N-phosphotransferase) levels

paralleled the formation and degeneration of myotubes during these temperature shifts. If the viral transforming gene was expressed in the post-mitotic myotubes, it was lethal; it was not lethal if expressed in replicating precursor myogenic cells. The data contribute to the following points regarding basic controls in the neoplastic cell. The viral gene expression at permissive temperature blocks further myogenesis depending on the position of the cells in the myogenic program. The virus does not cancel the replicating, transformed myogenic cells' commitment t or position in, the myogenic lineage. When the trans forming action of the virus is suppressed, the normal myogenic program resumes.

3239   INFECTIOUS DNA CODING FOR A TEMPERATURE-
        SENSITIVE DNA POLYMERASE OF THE COORDINATE *LA*335 MUTANT OF ROUS SARCOMA VIRUS (RSV). (Eng.) Hillova, J. (Inst. Cancerology and Immunogenetics, Villejuif, France); Mariage, R.; Hill, M. *Virology* 67(1):292-296; 1975.

The DNA from chick embryo fibroblasts (CEF) transformed by Prague strain Rous sarcoma virus, subgroup C (PR-C) or a Rous sarcoma virus (RSV) temperature-sensitive (ts) mutant (*LA*335) was used in transfection experiments to determine if the DNA polymerase of the recovered virus is specified by the infectious DNA of the recipient cell in this assay. Most transfection assays were carried out using a DEAE-dextran pretreatment, but in one experiment the technique using coprecipitation of the DNA with calcium phosphate was employed. The DNA extracted from *LA*335-transformed CEF gave rise in the transfection experiments to transforming viruses. These viruses exhibited a ts lesion in early functions similar to that characteristic of the *LA*335 mutant. One of the recovered viruses was further examined for a DNA polymerase activity at the permissive and nonpermissive temperatures and was found to carry an enzyme with a thermosensitivity similar to that of the DNA polymerase of *LA*335. It is concluded that the infectious viral DNA extracted from RSV-transformed cells carries the genetic information for the DNA polymerase specific for the parent virus. This information is transferred to the progeny virus upon transfection to permissive cells.

3240   RNA OF REPLICATION-DEFECTIVE STRAINS OF
        ROUS SARCOMA VIRUS. (Eng.) Duesberg, P. H. (Dep. Mol. Biol. Virus Lab., Univ. California, Berkeley); Kawai, S.; Wang, L.-H.; Vogt, P. K. Murphy, H. M.; Hanafusa, H. *Proc. Natl. Acad. Sci. USA* 72(4):1569-1573; 1975.

The RNA of a replication-defective (*rd*) mutant [isolated from nondefective (*nd*) Schmidt-Ruppin Rous sarcoma virus of subgroup A (SR-A) and designated · SR-N8] was compared to the RNAs of SR-A, transformation-defective (*td*) SR-A, and to *rd* Bryan Rous sarcoma virus, RSV(-). SR-N8-transformed chick embryo fibroblasts were infected with *td* SR-A to form SR-N8 pseudotypes. The RNAs of all species were subjected to electrophoresis in aqueous and formamide polyacrylamide gel. The 30-40S RNA of

DNA sequences within the purified viral-host DNA
may reflect a higher degree of organization of the
DNA of higher organisms.

3242   INTERACTION OF PARTIALLY PURIFIED SIMIAN
        VIRUS 40 T ANTIGEN WITH CIRCULAR VIRAL DNA
MOLECULES. (Eng.) Jessel, D. (Sidney Farber Cancer
Cent., Boston, Mass.); Hudson, J.; Landau, T.; Tenen,
D.; Livingston, D. M. *Proc. Natl. Acad. Sci. USA*
72(5):1960-1964, 1975.

A DNA binding assay for identifying a DNA binding
activity in T-antigen-containing fractions is presented. T-antigen was extracted and purified from
simian virus 80 (SV80), a line of SV40-transformed
human cells. The small-plaque strain of SV40 (SVS)
was propagated in primary African green monkey kidney cells. SV40 DNA I was isolated from infected
Vero cells, and other viral DNA molecules were isolated and purified. During the DNA binding assay,
nitrocellulose filters were used to filter the basic
reaction mixture of 0.01 M Tris x HCl (pH 7.4), 0.1
mM manganese acetate, 0.1 mM dithiothreitol, 3% dimethylsulfoxide, 10 μg/ml bovine serumalbumin, SV40
[^{3}H]DNA I, and T protein. Two 0.10 ml aliquots were
individually filtered. A complement fixation test for
T-antigen was carried out. T-antigen cotracked with
DNA binding activity on three types of columns:
DEAE-cellulose, agarose A-1.5m, and phosphocellulose. Heat inactivation kinetics of DNA binding
activity and T-antigen immunoreactivity paralleled
each other. The T-antigen fraction did not recognize DNA sequences unique to SV40 DNA. T-antigen
may be the DNA binding protein detected, but it is
possible that it is a non-T protein that cochromato-
graphs and has the same heat lability. It is likely
that T-antigen is the DNA binding protein observed;
a filter binding assay of [^{3}H]DNA may more efficiently measure quantitative aspects of this protein.

3243   *IN VITRO* TRANSFORMATION OF RAT AND MOUSE
        CELLS BY DNA FROM SIMIAN VIRUS 40. *(Eng.)*
Abrahams, P. J. (Lab. Physiological Chemistry, Univ.
Leiden, Wassenaarseweg 72, Leiden, Netherlands);
van der Eb, A. J. *J. Virol.* 16(1):206-209; 1975.

The transformation of rat kidney cells and mouse
3T3 cells by simian virus 40 (SV40) DNA was investigated employing the calcium technique. SV40
clone 307 L was grown in BSC-1 cells; viral DNA
was extracted and banded in CsCl. Only the closed-
circular DNA present in the dense band was used in
infection of kidney cells. The cells were infected
by adding 0.02-8.00 μg viral DNA, 5.0 μg salmon
sperm DNA, and 125 mM CaCl$_2$ at pH 7.05 to each dish.
Several colonies were isolated and established as
cell lines; all lines tested were found to contain
SV40-specific T antigen in 95-100% of the cells.
The dose response curve for transformation of primary rat kidney cells with SV40 DNA was not linear,
and indicated a highest transformation efficiency
at low viral DNA concentrations. A clone of 3T3
cells was also transformed by SV40 DNA. Although
all cell lines tested contained SV40-specific T
antigen in 95-100% of the cells, the 3T3 cells were

found more difficult to handle and maintain. It was thus demonstrated that a reproducible transformation of cells *in vitro* with SV40 DNA could be obtained *via* the calcium technique. In addition, preliminary experiments indicate that transformation of rat kidney cells with linear SV40 molecules and specific DNA fragments is also possible.

3244 REGULATION MECHANISM OF SIMIAN VIRUS 40 LATE GENE EXPRESSION IN PRIMARY MOUSE KIDNEY CELLS AND SIMIAN VIRUS 40 TRANSFORMED 3T3 CELLS. (Eng.) Graessmann, M. (Institut fur Molekularbiologie und Biochemie der Freien Universitaet Berlin, Berlin 33, Arnimallee 22, Germany); Graessmann, A. *Virology* 65(2):591-594; 1975.

Primary and continuous lines of mouse cells are nonpermissive for simian virus 40 (SV40); these cells, infected by the conventional virus absorption method, promote expression of the early but not of the late viral genes. To determine whether a specific factor prevents the expression of the late virus genes, SV40 V-antigen synthesis was studied in primary mouse kidney cells and SV40-transformed BALB/c 3T3 cells (SV-T2) infected with SV40-DNA component I by the microinjection technique. Induction of V antigen was observed in a large number of mouse kidney and SV-T2 cells, and the proportion of cells synthesizing antigen was correlated with the quantity of DNA molecules injected. The results support the hypothesis that restriction of the late SV40 gene expression is directed by a cellular factor(s) in these cells. They do not support the hypothesis that failure of late SV40 gene functions is due to a lack of a factor required for late gene expression.

3245 ACTION OF S_1 NUCLEASE ON NICKED CIRCULAR SIMIAN VIRUS 40 DNA. (Eng.) Chowdhury, K. (Institut für Virusforschung, Deutsches Krebsforschungzentrum, 69 Heidelberg); Gruss, P.; Waldeck, W.; Sauer, G. *Biochem. Biophys. Res. Commun.* 64(2): 709-716; 1975.

Simian virus 40 (SV40) DNA form II, generated either by radiation damage or by DNase and containing on · the average more than one single strand nick/molecule, was treated with S_1 nuclease to test the hypothesis that S_1 might be a reagent to extend single strand nicks into double strand breaks. The reactions were carried out at 37 C for 30 min. Under these conditions 2 µl of S_1 nuclease hydrolyzed denatured ^{3}H-labeled SV40 DNA form II (2 µg) to the extent that 95% of the radioactivity was no longer in acid precipitable form; native SV40 DNA form II was not rendered acid soluble. The reaction products were analyzed by both alkaline sucrose velocity sedimentation and by agarose gel electrophoresis. Linear duplex molecules of unit length (form III) were generated at the enzyme concentration sufficient to achieve 95% hydrolysis of at least 100 times the amount of single-stranded DNA. This reaction was dependent on the concentration of the S_1 nuclease. Over a wide range, comprising a tenfold increase of enzyme concentration, form III was generated exclusively, and no smaller DNA duplex was

tion of the formaldehyde-treated complex, the remaining fragment was treated with Pronase (0.5 mg/ml) to remove the protein; the DNA was sedimented on a neutral sucrose gradient along with a marker of calf thymus DNA of 105 nucleotide pairs. The marker calf thymus DNA sedimented 5.0S, while the nuclease-resistant piece of DNA sedimented at 3.1S. The molecular weight of the protected piece of DNA was estimated at 15,000, which would represent 25 nucleotide base pairs. This figure is within the range of the one derived from the nuclease digestion of the complex, which was 1% of the total SV40 DNA protected or 60 nucleotide base pairs. These results are consistent with the previously published statements that the DNA-protein complex consists of a protein (possibly that specified by the D cistron of SV40) associated with the genome at a unique site.

3249   ENHANCEMENT OF SV40 TRANSFORMATION BY
        TREATMENT OF C3H2K CELLS WITH UV LIGHT
AND CAFFEINE. I. COMBINED EFFECT OF UV LIGHT AND
CAFFEINE. (Eng.) Ide, T. (Inst. Medical Science,
Univ. Tokyo, Shirokanedai 4-6-1, Minatoku, Tokyo,
Japan); Anzai, K.; Andoh, T. $Virology$ 66(2):568-578; 1975.

The effect of caffeine (0-4 mM) in conjunction with UV (0-150 ergs/mm^2/sec) or x-ray (0-500 R/min) irradiation on the transformation frequency produced by simian virus 40 (SV40) was studied in the C3H2K mouse kidney cell line. In some experiments, the cultures were also treated with dibutyryl cyclic-AMP (0.1 mM) or theophylline (0.5 mM). The cell cultures were irradiated, infected with virus suspension, and incubated for four days with caffeine. Plating efficiency and transformation frequency were then determined. In the absence of virus, no foci of transformed cells were seen in cultures treated with UV, x-rays, and/or caffeine. Plating efficiencies of control and SV40-infected cells were essentially the same. UV, x-rays, and caffeine decreased the plating efficiency; increasing dosages of caffeine resulted in increasing potentiation of the effect of irradiation. UV, x-rays, and caffeine enhanced the frequency of SV40-induced transformation. Irradiation of cells with increasing doses of UV just before infection resulted in a 2-fold enhancement of the transformation frequency up to a dose of 90 ergs/mm^2 and a 3.3-fold increase at 150 ergs/mm^2. The addition of 1 mM caffeine to nonirradiated cells also enhanced the transformation frequency by 2-fold. When the cultures were irradiated and treated with 1 mM caffeine, the enhancement was about 4-fold up to a UV dose of 90 ergs/mm^2 and 5.9-fold at 150 ergs/mm^2. When 0.1-4 mM caffeine was added, the absolute number of transformations increased, resulting in an enhancement ratio of 1.3-6.8. After the addition of similar doses of caffeine to UV-irradiated cultures (75 ergs/mm^2), the enhancement of transformation frequency ranged from 2.0-13.3. The transformation frequency was not appreciably affected by the addition of dibutyryl cyclic-AMP or theophylline. The results suggest that irradiation and caffeine had a greater than additive effect on transformation, and that they modified and stimulated different sites in the over-all transformation process. The intimate involvement of DNA repair processes in oncogenic transformation is strongly suggested.

3250   PROPERTIES OF PROSTATIC CULTURES TRANS-
         FORMED BY SV40. (Eng.) Paulson, D. F.
(Duke Univ. Medical Center, Durham, N. C. 27710);
Bonar, R. A.; Sharief, Y.; Vergara, J. R.; Reich,
C.; Shah, K. V. *Cancer Chemother. Rep. (Part 1)*
59(1):51-55; 1975.

The characteristics of Syrian golden hamster prostatic tissue established in explant culture and infected with a 10^6-cell tissue culture infectious dose
(50% effective) of simian virus 40 (SV40) were
studied. After *in vitro* transformation, the cells
were produced in quantity, and 60×10^6 cells were
injected sc into two adult male Syrian golden hamsters 24 hr after 400 rads of whole-body radiation.
After 60-90 days, one small palpable tumor developed.
This tumor could be serially transplanted in adult
male animals without immunosuppression. The tumor
cells were established in tissue culture, and the
cells were returned to adult animals without immunosuppression where they rapidly produced fast-growing
tumors. The solid tumors were composed of sheets
of pleomorphic polygonal cells with large nuclei
and many nucleoli; they resembled undifferentiated
human prostatic carcinoma. *In vitro*, the cultures
contained small, rapidly growing cells with a population doubling time of about 1.3 days. The cells
carried the SV40-specific antigen. The modal chromosome number was 66-68 with a distribution of 47-
120. Cells exposed to 2-bromo-5'-deoxyuridine
(1×10^4 M for two days) in culture did not release
particles with RNA-dependent DNA polymerase activity.
Endocrine sensitivity *in vivo* and *in vitro* is
undetermined to date.

3251   TEMPERATURE-SENSITIVE MUTANTS OF SIMIAN
         VIRUS 40 SELECTED BY TRANSFORMING ABILITY.
(Eng.) Yamaguchi, N. (Inst. Med. Sci., Univ. Tokyo,
Japan);.Kuchino, T. *J. Virol.* 15(6):1297-1301;
1975.

Eight temperature-sensitive (TS) mutants of simian
virus 40 (SV40) were isolated by UV irradiation or
hydroxylamine treatment of wild-type (WT) parental
SV40 virus propagated in cultures of African green
monkey (AGMK) cells or GC7 cells, a clonal AGMK
cell line. The eight mutants were able to transform nonpermissive rat cells at 32.5 C but not at
38.5 C, and were also temperature sensitive for
replication in permissive AGMK or GC7 cells. In
rat cells, WT virus and a control ts mutant with
unimpaired transforming ability formed 1.2- to
2.2-fold more foci at 38.5 C than at 32.5 C, whereas
the eight transformation-defective mutants formed
9- to 125-fold fewer foci at 38.5 than at 32.5 C.
Although these results indicated that a viral function required for formation of dense foci in rat
cells is also necessary for SV40 replication in
monkey cells, the temperature sensitivity of the
ts gene function was expressed differently in transformation and viral replication. For example, the
efficiency of plaque formation of two ts mutants

terone, 1 µg/ml aldosterone, 1 µg/ml dexamethasone
and estrone (0.01, 0.1 and 1 µg/ml). Examination of
changes in enzyme activities of membrane-associated
enzymes and protein patterns in infected susceptible
and resistant cells made susceptible with calcium
showed that membrane protein composition changed during foci formation. In infected resistant cells made
susceptible, alkaline phosphatase activity was virtually undetectable. Rearrangement and/or formation
of additional membrane-associated isoenzymes was also
demonstrated in infected cultures. Resistance to
foci formation occurred during passages 45-50 and was
reversed by the addition of hormones. It is suggested
that this enhancement by hormones may be associated
with alterations in cell membranes.

3256   PURIFICATION OF SV-40 MESSENGER RNA BY
        HYBRIDIZATION TO SV-40 DNA COVALENTLY
BOUND TO SEPHAROSE. (Eng.) Gilboa, E. (Weizmann
Inst. Science, Rehovot, Israel); Prives, C. L.;
Aviv, H. *Biochemistry* 14(19):4215-4220; 1975.

The purification and properties of Simian virus 40
(SV40) DNA covalently bound to Sepharose and its
use in the purification of SV40-specific (mRNA) are
reported. SV40 DNA sheared form was coupled in a
stable covalent bond to cyanogen bromide-activated
Sepharose. The DNA-Sepharose was then hybridized
to RNA complementary to SV40 DNA and to RNA from
SV40-infected BS-C-1 cells. In other experiments,
the hybridization procedures were carried out using
SV40 DNA immobilized on nitrocellulose filters. The
virus-specific RNA selected in the hybridization
experiments was separated by polyacrylamide gel
electrophoresis, and mRNA preparations were translated using cell-free extracts from wheat germ.
Under the conditions used, at least 80% of the SV40
DNA was bound to Sepharose. The $t\frac{1}{2}$ of the hybridization of SV40 complementary RNA to SV40 DNA-Sepharose was one hour, the rate being sufficiently
rapid to purify SV40 sequences from solutions containing as little as 0.05-0.1 µg/ml. Nonspecific
hybridization of RNA was in the range of 0.1-0.2%
of the total input RNA. The DNA-Sepharose was fairly stable and could be reused several times to purify RNA. Although the virus-specific RNA obtained
using SV40 DNA immobilized on nitrocellulose filters
comprised the same percent of input RNA as was observed using SV40 DNA-Sepharose, a greater loss of
virus-specific RNA was observed with the former.
The virus-specific RNA when added to the cell-free
extracts from wheat germ directed the synthesis of
the major viral structural protein vp-1. Thus, the
hybridization of RNA to DNA immobilized on a solid
matrix proceeds with sufficient rapidity to be useful
both for analytical studies of RNA and for the preparation of biologically active mRNA.

3257   ISOLATION AND PARTIAL CHARACTERIZATION
        OF SINGLE-STRANDED ADENOVIRAL DNA PRODUCED
DURING SYNTHESIS OF ADENOVIRUS TYPE 2 DNA. (Eng.)
Lavelle, G. (Oak Ridge Natl. Lab., Oak Ridge, Tenn.
37830); Patch, C.; Khoury, G.; Rose, J. *J. Virol.*
16(4):775-782; 1975.

3258   IMMUNE FLUORESCENT STUDY OF THE MIGRATION
        OF ADENOVIRUS 5 IN HeLa CELL CYTOPLASM
DURING THE PHASE OF PENETRATION. (Fre.) Lyon, M.
(Equipe de Recherche n° 124 du CNRS, Unité de
Virologie, INSERM, 1, place Pr Joseph-Renaut,
69008 Lyon, France); Chardonnet, Y. *C. R. Acad.
Sci. [D] (Paris)* 280(16):1919-1922; 1975.

3259   STUDIES ON SIMIAN VIRUSES AS POSSIBLE
        CONTAMINANTS OF INACTIVATED VIRUS VACCINES.
I. DIRECT AND SEROLOGIC DETECTION OF SIMIAN ADENO-
VIRUS SV_{20}. (Ger.) von Mettenheim, A. E. (Paul-
Ehrlich-Institut, D-6 Frankfurt a.M., Paul-Ehrlich-
Str. 42-44, West Germany). *Zentralbl. Bakteriol.
[Orig. A]* 232(2-3):131-140; 1975.

3260   SEQUENCE RELATEDNESS BETWEEN THE SUBUNITS
        OF AVIAN MYELOBLASTOSIS VIRUS REVERSE
TRANSCRIPTASE. (Eng.) Rho, H. M. (St. Louis Univ.
Sch. Medicine, St. Louis, Mo. 63110); Grandgenett,
D. P.; Green, M. *J. Biol. Chem.* 250(13):5278-
5280; 1975.

3261   INFECTIOUS DNA RECOVERED FROM AVIAN TUMOR-
        VIRUS-PRODUCING CELLS. (Eng.) Vigier, P.
(Institut du Radium, Faculté des Sciences, Orsay,
France); Montagnier, L. *Int. J. Cancer* 15(1):67-
77; 1975.

3262   C TYPE MURINE LEUKAEMIA VIRUS PARTICLES
        IN TETRAPARENTAL AKR ↔ CBA CHIMAERAS.
(Eng.) Wills, E. J. (Clinical Res. Centre, Watford
Road, Harrow HA1 3UJ, Middlesex, England); Tuffrey,
M.; Barnes, R. D. *Clin. Exp. Immunol.* 20(3):563-
569; 1975.

3263   BIOCHEMICAL AND PHYSICO-CHEMICAL PROPER-
        TIES OF C TYPE VIRUS RELEASED BY CELLS OF
SPONTANEOUS LYMPHOSARCOMA OF MICE OF CC57BR STRAIN.
(Rus.) Lovenetskii, A. N. (Inst. Exper. Pathol.
Ther., Acad. Med. Sci. U.S.S.R., Moscow, USSR),
Kiselev, F. L.; Zaretskii, I. Z,; Irlin, I. S.;
Bykovskii, A. F. *Vopr. Virusol.* (1):14-20; 1975.

3264   MURINE TUMOR INDUCTION BY CYTOMEGALOVIRUS.
        (Eng.) Cunningham, B. D. (Dept. Biology,
Texas Woman's Unvi., Box 23971, TWU Station, Denton,
Tex. 76204); Sims, R. A.; Zimmermann, E. R.; Byrd,
D. L. *Oral Surg.* 40(1):130-134; 1975.

3265   LONG-TERM PERSISTENCE OF CYTOMEGALOVIRUS
        GENOME IN CULTURED HUMAN CELLS OF PROSTATIC
ORIGIN. (Eng.) Rapp, F. (Milton S. Hershey Medi-
cal Center of Pennsylvania State Univ., Coll. Medi-
cine, Hershey, Pa. 17033); Geder, L.; Murasko, D.;
Lausch, R.; Ladda, R.; Huang, E. S.; Webber, M. M.
J. Virol. 16(4):982-990; 1975.

pio 10, Finland); Mantyjarvi, R. A. *Acta Pathol.*
Microbiol. Scand. [B] 83B(4):347-352; 1975.

3284   PHYSICAL MAP OF THE BK VIRUS GENOME.
        (Eng.) Howley, P. M. (Natl. Inst. Allergy
and Infectious Diseases, Bethesda, Md. 20014);
Khoury, G.; Byrne, J. C.; Takemoto, K. K.; Martin,
M. A. *J. Virol.* 16(4):959-973; 1975.

3285   THE SPECIFIC IMMUNE-PRECIPITATION OF INTRA-
        CELLULAR PRECURSOR-LIKE POLYPEPTIDES OF
RAUSCHER LEUKEMIA VIRUS [abstract]. (Eng.) Naso,
R. B. (Univ. Texas Syst. Cancer Cent., M. D. Ander-
son Hosp. Tumor Inst., Houston); Arcement, L. J.;
Arlinghaus, R. B. *Fed. Proc.* 34(3):527; 1975.

3286   ISOLATION AND CHARACTERIZATION OF VIRUSES
        FROM NATURAL OUTBREAKS OF RETICULOENDO-
THELIOSIS IN TURKEYS. (Eng.) Sarma, P. S. (Natl.
Cancer Inst., Bethesda, Md. 20014); Jain, D. K.;
Mishra, N. K.; Vernon, M. L.; Paul, P. S.; Pomeroy,
B. S. *J. Natl. Cancer Inst.* 54(6):1355-1359; 1975.

3287   COMPARATIVE ULTRASTRUCTURAL STUDY OF FOUR
        RETICULOENDOTHELIOSIS VIRUSES. (Eng.)
Kang, C. Y. (Univ. of Texas Southwestern Medical
Sch., Dallas, Tex. 75235); Wong, T. C.; Holmes, K.
V. *J. Virol.* 16(4):1027-1038; 1975.

3288   DETECTION OF REVERSE TRANSCRIPTASE IN HUMAN
        BREAST TUMOURS WITH POLY(Cm)·OLIGO(dG).
(Eng.) Gerard, G. F. (St. Louis Univ. Sch. Medicine,
St. Louis, Mo. 63110); Loewenstein, P. M.; Green,
M.; Rottman, F. *Nature* 256(5513):140-143; 1975.

3289   CELL SURFACE CHANGES ACCOMPANYING VIRAL
        TRANSFORMATION: *N*-ACETYLNEURAMINIC ACID
ECTOTRANSFERASE SYSTEM ACTIVITY. (Eng.) Spataro,
A. C. (Univ. Rochester Sch. Medicine and Dentistry,
Rochester, N. Y. 14642); Morgan, H. R.; Bosmann, H.
B. *Proc. Soc. Exp. Biol. Med.* 149(2):486-490; 1975.

3290   AN IMPROVED METHOD OF FIXATION FOR IMMUNO-
        FLUORESCENT DETECTION OF SV40 T-ANTIGEN IN
INFECTED HUMAN FIBROBLASTS. (Eng.) Kaplan, M. M.
(Meloy Lab., Inc., 6715 Electronic Drive, Spring-
field, Va. 22151); Giard, D. J.; Blattner, W. A.;
Lubiniecki, A. S.; Fraumeni, J. F., Jr. *Proc. Soc.*
Exp. Biol. Med. 148(3):660-664; 1975.

3291   HIGH AFFINITY BINDING OF SIMIAN VIRUS 40
        T ANTIGEN TO SV40 DNA [abstract]. (Eng.)
Spillman, T. (Univ. Illinois, Urbana); Spomer, B.;
Hager, L. *Fed. Proc.* 34(3):527; 1975.

3292   POSSIBLE NUCLEASE ACTIVITY OF THE T-
        ANTIGEN OF VIRUS SV-40. (Eng.) Shlian-
kevich, M. A. (Inst. Experimental Clinical Oncology,
Acad. Medical Sciences U.S.S.R., Moscow, U.S.S.R.);

Pirtskhalaishvili, D. S.; Prigozhina, T. B.; Drize,
O. B.; Shapot, V. S. *Bull. Exp. Biol. Med.* 79(1):
46-49; 1975.

3293   DETECTION OF ENDONUCLEASE ACTIVITY IN SV-
        40 VIRUS T-ANTIGEN PREPARATIONS. (Rus.)
Shliankevich, M. A. (Inst. Exper. Clin. Oncol.,
Acad. Med. Sci., Moscow, U.S.S.R.); Drize, O. B.;
Shapot, V. S. *Dokl. Akad. Nauk S.S.S.R.* 221(4):
987-989; 1975.

3294   METHYLATED SV40 mRNAs. (Eng.) Aloni, Y.
        (Dept. Genetics, The Weizmann Inst. Science,
Rehovot, Israel). *FEBS Lett.* 54(3):363-367; 1975.

3295   RELEASE OF A GROWTH FACTOR FOR SV40 VIRUS-
        TRANSFORMED CELLS BY RAT LIVER AND HEMI-
CORPUS PERFUSIONS. (Eng.) Lipton, A. (Milton S.
Hershey Medical Center, Pennsylvania State Univ.,
Hershey, Pa. 17033); Roehm, C. J.; Robertson, J. W.;
Dietz, J. M.; Jefferson, L. S. *Exp. Cell Res.*
93(1):230-234; 1975.

3296   AMINO ACID TRANSPORT SYSTEMS IN 3T3 AND
        SV3T3 MOUSE CELLS [abstract]. (Eng.)
Cecchini, G. (Dep. Biol. Chem., Univ. Michigan,
Ann Arbor); Lee, M.; Oxender, D. L. *Fed. Proc.*
34(3):556; 1975.

3297   INDEPENDENT REGULATION OF CELLULAR PRO-
        PERTIES IN THERMOSENSITIVE TRANSFORMATION
MUTANTS OF MOUSE FIBROBLASTS. (Eng.) Rudland, P.
S. (Imperial Cancer Res. Fund, PO Box 123, Lincoln
Inn Fields, London WC2A 3PX, UK); Pearlstein, E.;
Kamely, D.; Nutt, M.; Eckhart, W. *Nature* 256(5512
43-46; 1975.

See also:

*   (Rev):   3010,   3011,   3012,   3013,   3014,   3015
             3016,   3017,   3018,   3052,   3053,   3054
             3069
*   (Chem):   3123,   3155,   3166
*   (Immun): 3310,   3312,   3326,   3328,   3345,   33
             3352,   3354,   3355,   3356,   3358,   33
             3367,   3369,   3371,   3381,   3390,   33
*   (Path):   3419,   3435,   3464,   3488

B16 was a weaker immunogen than EL4 in H(zl) mice;
Bl/6J mice had no detectable resistance to any of
the three methylcholanthrene-induced fibrosarcomas.
These results indicate less stringent requirements
for host-tumor susceptibility than for skin graft
compatibility between H(zl) mice and C57Bl/6J mice.
The *in vitro* secondary cell-mediated cytolytic
response of H(zl) mice to EL4 was specific and was
of similar magnitude to the Balb/c secondary *in
vitro* cytolytic response to the same tumor. Thus,
cellular immunity, as assayed *in vitro*, need not
correlate with humoral antibody production. H(zl)
IPEC against EL4 tumor were capable of inhibiting
EL4 growth *in vivo* in F_1(H(zl) x C57Bl) mice, which
normally have no resistance to EL4. Further, IPEC
against EL4 were capable of *in vitro* cytolysis of
normal Bl/6J cells, indicating the presence of
common antigens on Bl/6J and EL4 to which H(zl)
responds. Bl/6J lymph node cells, but not H(zl)
cells, were able to competitively inhibit EL4
lysis in the *in vitro* cytolysis of EL4 by Balb/c
IPEC against EL4. This provides an additional means
of distinction between H(zl) and C57Bl/6J normal
cells.

3300   SOME IMMUNOLOGIC CHARACTERISTICS OF CAR-
        CINOMA OF THE COLON AND RECTUM. (Eng.)
Elias, E. G. (Dept. General Surgery, Roswell Park
Memorial Inst., Buffalo, N.Y.); Elias, L. L. *Surg.
Gynecol. Obstet.* 141(5):715-718; 1975.

The WBC migration-inhibition assay was used to
assess *in vitro* cell mediated immunity in patients
with surgically staged carcinoma of the colon and
rectum. Tumor cells and extracts were used in both
the autologous and the homologous systems. The
study was doubly controlled by the use of WBC from
normal healthy volunteers and by autologous intestinal mucosa cells and extracts. The WBC of
patients were significantly sensitized to autologous tumor cells but not to homologous tumor
cells or extracts when compared with the WBC of
normal healthy volunteers. Leukocytes of patients
tested with autologous tumors or mucosa showed that
only those of patients with Dukes C classification
were significantly sensitized to their tumors.
These results suggest that carcinomas of the colon
and rectum are heterogenous tumors.

3301   A LONG-TERM ^{51}CHROMIUM ASSAY FOR *IN VITRO*
        CELL-MEDIATED TUMOR IMMUNITY. CORRELATION
WITH SIMULTANEOUSLY PERFORMED MICROPLATE ASSAYS.
(Eng.) Steele, G., Jr. (Wallenberg Lab., Univ. Lund,
Lund, Sweden); Sjogren, H. O.; Lannerstad, O.; Sta-
denberg, I. *Int. J. Cancer* 16(4):682-693; 1975.

In vitro assays of cell-mediated tumor immunity utilizing ^{51}Chromium (^{51}Cr) labeling of cultured adherent solid tumor cells were designed which allowed
an effector cell/target cell incubation time of 48
hr without overriding spontaneous ^{51}Cr release. In
a series of 16 consecutive experiments, blood lymphocytes from healthy human donors, from patients
with tumors unrelated to the cultured tumor target
cells, and from colon carcinoma and melanoma patients

were tested for their cytotoxic effects on various
target cell pairs, human colon carcinoma, melanoma,
or skin fibroblasts. The same reagents were used in
simultaneously performed microplate and ^{51}Cr assays.
Results obtained by visual counting of microplate
tests and by 24 hr assays of ^{51}Cr release or ^{51}Cr
retention correlated in 20/25 effector-cell/target-
cell combinations. In a series of six consecutive
experiments, lymph-node cells from untreated Wistar/
Furth rats, and rats bearing either chemically-in-
duced colon carcinoma NG-W1 or polyoma virus-induced
sarcoma P-W13 were tested for cytotoxicity on syngeneic rat colon carcinoma and sarcoma target cells.
Criss-cross type experiments were performed by microplate and ^{51}Cr techniques done in parallel. Results
obtained by visual counting of microplate tests and
by 48 hr assays of ^{51}Cr release or ^{51}Cr retention
correlated in 15/18 effector-cell/target-cell combinations. The ^{51}Cr assay is less tedious than microplate tests with visual counting and, at least
with human effector cells on human target cells, requires a shorter incubation period to obtain similar
cytotoxic effects.

3302   SYNERGY BETWEEN SUBPOPULATIONS OF NORMAL
        MOUSE SPLEEN CELLS IN THE *IN VITRO* GENERA-
TION OF CELL-MEDIATED CYTOTOXICITY SPECIFIC FOR "MOD-
IFIED SELF" ANTIGENS. (Eng.) Hodes, R. J. (Immunology Branch, Natl. Cancer Inst., Bethesda, Md.
20014); Hathcock, K. S.; Shearer, G. M. *J. Immunol.*
115(4):1122-1125; 1975.

Responding lymphoid cells were cultured *in vitro* with
irradiated trinotrophenyl (TNP)-modified syngeneic
spleen cells. Direct cell-mediated cytotoxicity
developed. This occurred specifically for target
cells bearing both the TNP moiety and histocompatibility determinants of the modified sensitizing cell.
Two subpopulations of normal mouse spleen cells were
shown to synergize in the *in vitro* generation of specific cell-mediated cytotoxicity to these "modified
self" antigens. The synergizing populations were
nylon wool column-adherent and column-nonadherent
fractions of normal mouse spleen. When mixtures of
these two cell populations were cultured *in vitro*
with irradiated TNP-modified syngeneic spleen cells,
greater cytotoxicity was generated than would be expected from the sum of cytotoxic activities generated
in the two populations sensitized separately. The
synergizing cell in the column-adherent population
was resistant to lysis by rabbit anti-mouse brain
serum, was distinct from the cytotoxic effector T
lymphocyte, and was unresponsive to phytohemagglutinin; its synergizing function could not be replaced
by peritoneal cells. The results suggest that it is
a non-T cell which may be distinct from the macrophage.

3303   CELL-MEDIATED CYTOTOXICITY TO CHEMICALLY-
        INDUCED RAT TUMORS. (Eng.) Zoller, M.
(Deutsches Krebsforschungzentrum Heidelberg, Institut fur Nuklearmedizin, 69 Heidelberg, West Germany);
Price, M. R.; Baldwin, R. W. *Int. J. Cancer* 16(4):
593-606; 1975.

Cell-mediated immune reactions against carcinogen-

were cultured *in vitro*. In agreement with previous
observations, the supernatants suppressed specific
lymph-node cell-mediated cytotoxicity directed a-
gainst the respective neoplasm (blocking factors)
and they induced specific antibody-dependent cellular
cytotoxicity (ADC) mediated by lymph-node cells from
non-immune mice. Spleen cells passed through
columns of Sephadex G-10, which removed plasma cells
and macrophages, no longer synthesized factors with
ADC activity though synthesis of blocking factors
continued. However, lysis of Thy-1-positive cells
with antiserum and complement did abolish the synthesis of blocking factors. Spleen cells from tumor-
bearing mice were then enriched for Thy-1-positive
cells by selective removal of macrophages and plasma
cells on G-10 columns and of immunoglobulin (Ig)-
bearing cells on anti-mouse Ig affinity columns. This
T-cell-enriched population synthesized blocking factors and restored blocking factors synthesis in cultures depleted of Thy-1-positive cells. Similarly
enriched spleen cells from normal control mice did
not synthesize tumor-specific blocking factors but
partially restored synthesis of blocking factor in
tumor-bearer spleen cultures depleted of Thy-1-posi-
tive cells. The findings suggest that thymectomy
might depress the formation of blocking factors.
Cells retained by G-10 columns are involved in *in
vitro* production of ADC. Lymphocytes carrying the
Thy-1 marker are implicated in the synthesis of
blocking factors.

3307   SPONTANEOUS AND INDUCED TUMOR INCIDENCE IN
        GERMFREE "NUDE" MICE. (Eng.) Outzen, H.
C. (Fox Chase Cent. Cancer Med. Sci., Philadelphia,
Pa.); Custer, R. P.; Eaton, G. J.; Prehn, R. T. *J.
Reticuloendothel. Soc.* 17(1):1-9; 1975.

A colony of conventional and germfree "nude" (nu/nu)
mice and their heterozygous (nu/+) littermates were
investigated with respect to spontaneous and 3-meth-
ylcholanthrene (MCA)-induced tumor incidence. A
total of 308 male and female nu/+ and 261 male and
female "nude" mice were maintained germfree for over
20 mo with routine sterility monitoring and observation. Many of the "nude" colony received mouse
skin allografts to demonstrate their general T-cell
incompetence. The grafts themselves did not affect
tumor. incidence or latency. MCA was implanted (5%
pellets) in six "nude" and six nu/+ mice, which were
then observed for six months. The germfree "nude"
mouse colony developed 22 "spontaneous" lymphoreticular tumors with a mean latency $\pm$ standard deviation of 57 wk $\pm$ 1.56 wk. Three mammary tumors were
also observed despite the absence of the murine mammary tumor virus. The spontaneous tumors were not
sex-related. The germfree nu/+ females showed no
tumors at all by the cut-off date. The comparatively short-lived conventional "nude" colony showed
only two lymphoreticular tumors, which was attributed to the long latency of spontaneous tumor incidence. The conventionally maintained nu/+ colony
developed several spontaneous neoplasms, the most
prominent of which were 12 mammary tumors, with a
latency $\pm$ SD of 60 wk $\pm$ 1.26 wk. The absence of
these tumors in the germfree nu/+ females was
thought to be due to the elimination of the mammary

tumor virus. Four of the six mice in both the
"nude" and nu/+ germfree MCA-treated groups developed fibrosarcomas with no significant difference
in the latency periods. The authors suggest that
the germfree "nude" mouse, deficient in all thymus-
derived, cell-mediated (T cell) immunologic responses, is a good model for testing immunosurveillance and/or immunostimulation as active modu-
lators of tumor regression of development. The
large number of spontaneous neoplasms in the germfree "nudes" was not felt to be necessarily related
to, or to preclude, either the immunosurveillance
or immunostimulation hypotheses. Although the results of the MCA-induction experiment implied that
immunosurveillance was not effective, the authors
warn that the dosage of MCA was greater than the
threshold dose, and that further investigation is
needed to clarify the matter.

3308   INCREASED INCIDENCE OF MALIGNANCY DURING
           CHRONIC RENAL FAILURE. (Eng.) Matas,
A. J. (Dept. Surg., Univ. Minnesota, Minneapolis);
Simmons, R. L.; Kjellstrand, C. M.; Buselmeier, T.
J.; Najarian, J. S. *Lancet* 1(7912):883-885; 1975.

The incidence of cancer in 646 dialysis/transplant
patients before uremia developed, during the period
of progressive uremia, and post-transplantation was
compared. Ten tumors (three breast, two kidney, one
leukemia, one lung, one insulinoma, one thyroid, one
cervix *in situ*) developed in nine patients during the
period of progressive uremia, a significant increase
over the expected number in the age-matched general
population (148/100,000/yr). Six of these patients
have received transplants and have no evidence of
recurrent disease six months to four years post-
transplantation. Eleven *de-novo* tumors have developed in 530 transplant recipients (four cervix *in
situ*, two skin, two reticulum-cell sarcomas, one
lip, one dysgerminoma, one colon)--a significant increase over the age-matched general population
(102/100,000/yr). The cancers in the uremic patients are relatively common types of mesenchymal
tumors while the cancers in the transplant recipients are epithelial and lymphoproliferative. This
difference may reflect the presence of the graft in
the transplant patient or may be due to different
patterns of immunosuppression in these two populations.

3309   LOSS OF MARROW ALLOGRAFT RESISTANCE IN
           MICE WITH TRANSPLANTED METHYLCHOLANTHRENE-
INDUCED SARCOMAS. (Eng.) Kumar, V. (Boston Univ.
Sch. Medicine, Boston, Mass. 02118); Bennett, M.
J. Natl. Cancer Inst. 55(2):489-492; 1975.

To determine if the effector cells responsible for
rejection of marrow stem cell allografts are suppressed in mice bearing tumors, C57BL/6 (B6) mice
were subjected to s.c. transplants of 10^5 syngeneic
3-methylcholanthrene (MCA)-induced sarcoma cells;
and when the resulting tumors reached 2.0-2.5 cm in
diameter, these mice and control B6 and (BALB/c x
A)F_1 (CAF$_1$) uninoculated animals were irradiated
and infused i.v. with 2.5 x 10^6 BALB/c marrow cells.

3311 GYNECOLOGIC MALIGNANCIES IN IMMUNOSUP-
PRESSED ORGAN HOMOGRAFT RECIPIENTS.
(Eng.) Porreco, R. (Univ. Colorado Sch. Med., Denver); Penn, I.; Droegemueller, W.; Greer, B.; Makow-
ski, E. Obstet. Gynecol. 45(4):359-364; 1975.

Data collected by the Denver Transplant Tumor Registry show that immunosuppressed organ homograft recipients have a 5 to 6% incidence of de novo malignancies at some time after transplantation. Gynecologic
cancers were encountered in 21 of 224 patients (9%)
with these tumors. The predominant lesion was carcinoma of the cervix (18 cases), of which 16 were
intraepithelial and two were invasive. All 18 patients had been recipients of kidney homografts; the
mean time interval from renal transplantation to
diagnosis of carcinoma was 38 mo (range, 6-97 mo).
Of the major immunosuppressive agents, azathioprine
and prednisone had been administered to all patients.
The lesions were treated by hysterectomy in 12 instances, side conization in six, and cryosurgery in
one, with one patient having wide conization followed
later by hysterectomy for recurrence of the lesion.
Four patients died, two from homograft rejection,
one from metastases, and one from an unrelated infection. Gynecologic malignancies have also been
encountered in nontransplant patients who were
treated with immunosuppressive agents or cancer chemotherapy. All such individuals require gynecologic
examination before commencement of treatment and at
regular intervals thereafter so that malignancies
may be diagnosed at an early stage and treated effectively. Most neoplasms respond well to conventional cancer therapy, but high-grade malignancies
may necessitate reduction or cessation of immunosuppressive therapy as well.

3312 SPECIFIC SUSCEPTIBILITY OF SENSITIZED
(MEMORY) B CELLS TO SUPPRESSION AND ANTI-
GENIC ALTERATION BY MURINE LEUKEMIA VIRUS. (Eng.)
Cerny, J. (Harvard Sch. Public Health, Boston, Mass);
Waner, E. B. J. Immunol. 114(2):571-580; 1975.

To obtain evidence for the hypothesis that Friend
virus (FV) infects and inhibits sensitized (memory)
cells but not virgin immunocompetent cells, experiments were carried out involving BALB/c mice infected
with different leukemia viruses and immunized with
different antigens. The viruses used were Friend
virus and Moloney leukemia virus (MuLV-M); the antigens were Vibrio cholerae vaccine and SRBC. Isologous antiserum against FV-induced cell membrane
antigen (FVMA) was prepared by repeated immunization
of BALB/c mice with virus vaccine and with isologous
FV-infected spleen cells. The kinetics and magnitude
of the specific vibriolytic plaque-forming cells
(PFC) response in the spleen following a single primary immunization with the cholera vaccine showed no
difference in the normal and FV-infected leukemic
mice. However, the secondary response of sensitized
FV-infected mice was greatly inhibited as compared
to that of primed mice that were not infected. As
with FV infection, MuLV-M infection selectively inhibited the secondary PFC response to cholera vaccine, although the primary response was also inhibited in the MuLV-M-infected mice. The degree of

immunosuppression due to FV priming was comparable
at various intervals, indicating that the immunologic memory to cholera antigen and the susceptibility of the memory to viral suppression is a long
term phenomenon. Hemolytic PFC response to a
single immunization with SRBC (an antigen to which
laboratory mice have a natural immune background)
was suppressed by infection with both viruses in a
manner similar to the suppression of the secondary
vibriolytic response. The suppressive effect of FV
on the different immune responses was correlated
with the appearance of FVMA on respective individual
PFC from the spleen. The FVMA was detected by inhi-
ition of PFC in the presence of specific anti-FVMA
antiserum and complement. The findings thus indicated an absolute correlation between viral suppression of a given clone of immunocompetent cells and
antigenic alteration of individual antibody-forming
cells due to virus infection. Furthermore, the absence of both functional inhibition and virus-
induced antigen in the primary immunocompetent cell
and the presence of both of these virus-related
functions in the memory cells strongly suggests that
the latter are selectively susceptible to oncornavirus.

3313 ALTERATION OF IMMUNOGLOBULIN PHENOTYPE I
CELL CULTURE-ADAPTED LINES OF TWO MOUSE
PLASMACYTOMAS. (Eng.) Hausman, S. J. (Gerontology
Res. Center, Baltimore City Hospitals, Baltimore,
Md. 21224); Bosma, M. J. J. Exp. Med. 142(4):998-
1010; 1975.

Murine plasmacytomas were adapted to continuous in
vitro culture by alternate passage between culture
and animal. BALB/c mice congenic for different imm
noglobulin (Ig) allotypes were used in these studie
For animal passage, each plasmacytoma was grown in
congenic BALB/c mice having a different allotype
from that of the tumor Ig. To facilitate the handl
of tumor cells, plasmacytoma were injected into
each new host mouse to induce ascites. Tumor cells
were considered to be fully adapted to culture afte
five successive in vitro passages in spinner cultur
(NH₄)₂SO₄ precipitation and DEAE-cellulose chromato
graphy were used to obtain 7S Ig from pooled sera o
the plasmacytoma-bearing mice. Salt precipitation
and fractionation on Bio-Gel A-1.5m Agarose were us
to obtain the IgM. Appropriate antisera were used
identify the Ig allotypes produced by the tumors be
fore and after adaptation. Studies of the kinetics
of adaptation revealed a selection for growth of
variant plasmacytoma cells. For example, plasmacyt
MOPC 173, when placed in spinner culture for the
third time, showed a decline in cell count from the
level of 6.0×10^8 for the original seeding to les
than 3×10^4 living cells after 17 days of culture
However, over the next 20 days, the number increase
exponentially to greater than 10^7 cells, and extrapolation of the growth back to zero time indicated
that less than 50 cells gave rise to the altered
line. This meant that only about one in 10^7 cells
of the starting population could grow in culture, a
that such cells must have been variant cells. The
inclusion of an altered Ig phenotype among such cel
explained the variant Ig-producing cell lines that

were observed in 2 of 6 different Ig-producing plasma-cytomas that were adapted to culture. The first variant, an IgM-producing cell line (104-76), was adapted from a transplanted line of MOPC 104C that had stopped producing IgM with binding specificity for α1-3 Dextran. Unlike MOPC 104E, the IgM of 104-76 contained κ- instead of λ-light chains and probably contained an altered or different μ-heavy chain. A second variant (352-57) was found in an IgG_{2b}-producing tumor (MOPC 352) which was induced in a BALB/c mouse strain (CB-6) that carried Ig genes of the C57BL/Ka allotype. This cell line apparently switched from producing IgG_{2b} molecules of the C57BL allotype (H[9]) and of a known idiotype to IgG_1 molecules of the BALB/c allotype (F[19]) without the idio-type marker. The propagation of a biclonal plasma-cytoma from the time of original tumor induction is not thought to be a likely explanation for these results. Rather, it appears that plasmacytoma variants were being dealt with, or that secondary tumors of host origin were being induced.

3314   TUMOUR REJECTION PROPERTIES OF SOLUBILIZED TSTA FROM AN SV40-INDUCED NEOPLASM. (Eng.) Law, L. W. (Natl. Cancer Inst., Bethesda, Md. 20014) Henriksen, O.; Appella, E. *Nature* 257(5523):234-235; 1975.

Ten consecutive, solubilized antigen preparations of a simian virus 40 (SV40)-induced sarcoma (mKSA) were examined for their ability to immunize syngeneic BALB/c mice against this neoplasm. The ten prepara-tions constituted crude membrane, crude soluble, or Sephadex G-150-chromatographed F-2 and F-3 fractions from membranes of either the ascitic form (ASC) of mKSA or a tissue culture subline (TC) with a reduced tumor-inducing capacity. BALB/c mice, 8-12 wk old, were given two sc injections of antigen at 2-wk in-tervals. All ten antigen preparations effectively protected the mice against challenge with mKSA cells. mKSA(TC) and mKSA(ASC) were equally effective, indi-cating that they share the same SV40-specific tumor reaction antigens in spite of long-continued passage *in vitro* and *in vivo*. Protection was afforded at cell-challenge concentrations well above the median effective tumor dose of 10^2 cells. The degree of protection was in the range of that provided by a single sc injection of 10^5 or 10^6 mKSA-TC cells. *In vitro* specific lymphocyte stimulation was observed with low concentrations of two ASC antigen prepara-tions. In the Winn assay, sensitized spleen cells from mice immunized with crude soluble mKSA could neutralize mKSA tumor cells. Immunization did not protect BALB/c mice against the syngenic plasma cell tumor Adj-PC5 or against the chemical carcinogen-induced tumor Meth A.

3315   IgM-INDUCED TUMOR CELL CYTOTOXICITY MEDI-ATED BY NORMAL THYMOCYTES. (Eng.) Lamon, E. W. (Sch. Medicine, Univ. Alabama, Birmingham, Ala. 35294); Whitten, H. D.; Lidin, B.; Fudenberg, H. H. *J. Exp. Med.* 142(2):542-547; 1975.

Because immunoglobulin M (IgM) from sera of mice

myeloma cells never formed rosettes in either the
direct or indirect assay. When the lighter, non-
rosetting MOPC-315 cells were removed from the heavier rosetting cells by centrifugation, they were
found on inoculation into unimmunized BALB/c mice
to give rise to variant tumors that produced only
(L^{315}). Similar manipulation gave rise to a MOPC-
460 variant. The rosette assay thus demonstrated
the existence of rare variant cells in tumors that
had been consistently passaged through unimmunized
hosts. Further studies showed that 7 of 8 BALB/c
myeloma proteins could successfully induce Ab responses in the same strain. One of the explanations
that was offered was that most myeloma idiotypes are
too rarely expressed to establish tolerance. With a
radioimmunoassay employing goat Abs that were mono-
specific for different classes of mouse immunoglobulins, it was also found that most of the anti-
idiotypes were immunoglobulin G1 (IgG1) and some
were immunoglobulin G2a (IgG2a) molecules. It was
proposed that determination of whether immunization
with an anti-idiotype stimulates the production of
its target idiotype or a completely different set
of anti-anti-idiotype molecules would now be feasible.

3318   THE ISOLATION OF CARCINOEMBRYONIC ANTIGEN
        FROM TUMOR TISSUE AT NEUTRAL pH. (Eng.)
Carrico, R. J. (Ames Res. Lab., Elkhart, Indiana
46514); Usategui-Gomez, M. *Cancer Res.* 35(11/Part
1):2928-2934; 1975.

Carcinoembryonic antigen (CEA) was purified from
tumor tissue (liver metastases of human colonic
carcinoma) under mild conditions at neutral pH by
a procedure that utilized affinity chromatography
on concanavalin A. Further purification by gel filtration provided CEA in 10-20% yield and 10% purity.
Antibody to this preparation was rendered specific
for CEA by adsorption on a column of normal liver
proteins bound to Sepharose. On reaction by immunodiffusion against a crude tumor extract, the adsorbed
antibody produced two precipitin lines, of which one
was relatively weak. These two precipitin lines
fused completely with the two respective lines produced by antibody to perchloric acid-treated CEA.
The major antigen found in crude tumor extracts and
in CEA preparations purified at neutral pH was nearly
undetectable in perchloric acid extracts of tumor
homogenates. Further investigations showed that 60-
70% of the CEA in crude tumor extracts and in preparations isolated at neutral pH was destroyed and/
or became insoluble under acidic conditions.

3319   CHEMOTHERAPEUTIC DRUGS INCREASE KILLING
        OF TUMOR CELLS BY ANTIBODY AND COMPLEMENT.
(Eng.) Segerling, M. (Natl. Cancer Inst., Bethesda,
Md.); Ohanian, S. H.; Borsos, T. *Science* 188(4):
55-57; 1975.

The *in vitro* effects of chemotherapeutic agents on
the ability of xenogeneic antibody (antibody to Forssman antigen, anti-F) and guinea pig complement to
kill tumor cells were investigated. Ascitic forms
of diethylnitrosamine-induced guinea pig hepatoma
cells (line-1 and line-10) were added to 5.0 ml of

tissue culture medium containing 50 µg/ml adriamycin, 20 µg/ml vincristine sulfate, 20 µg/ml azacytidine, 500 µg/ml 5-fluorouracil (5-FU), 500 µg/ml
methotrexate (MTX), or 500 µg/ml 6-mercaptopurine
(6-MP). Ascitic fluids were collected 8-10 days
after ip injection of 3 x 10^6 tumor cells. Ascites
tumor cells (60 x 10^6) were freed of RBC, washed
twice with barbital-buffered saline and resuspended
in culture medium. After incubation of 2.5 x 10^6
line-1 or line-10 cells for 17 hr at 37 C with one
of the chemotherapeutic agents, 2.0 ml portions were
washed and resuspended to a density of 10^6 cells/ml
culture medium. Cell death was determined by
staining cells with trypan blue after incubation
with anti-F plus C. During this incubation, line-
10 cells were more susceptible when they had been
incubated with adriamycin (46% cell death), vincristine sulfate (50%), and azacytidine (69%) in comparison with controls (16%). Line-1 cells were more
susceptible upon prior incubation with adriamycin
(52% cell death), azacytidine (61%), 5-FU (69%),
MTX (54%), and 6-MP (53%) when compared to controls
(26%). The authors speculate that the beneficial
effects of chemotherapeutic agents may be partially
due to an increased susceptibility of tumor cells to
antibody plus complement.

3320   CARCINOEMBRYONIC ANTIGEN: EVIDENCE FOR
        MULTIPLE ANTIGENIC DETERMINANTS AND ISO-
ANTIGENS. (Eng.) Vrba, R. (Harvard Medical Sch.,
Boston, Mass. 02114); Alpert, E.; Isselbacher, K.
J. *Proc. Natl. Acad. Sci. USA* 72(11):4602-4606;
1975.

Carcinoembryonic antigen (CEA) preparations were
compared by radioimmunoassay. The preparations
studied included four CEA standards (CEA-Roche, CEA-
Montreal, CEA-City of Hope, and CEA-British) and CEA
from serum and liver metastases of a patient with
cancer of the colon who had an extremely high concentration of serum CEA (more than 26,000 ng/ml).
The data indicate that the CEA-Roche standard differs significantly from the other three CEA standards tested, and that the serum CEA from the patient
was antigenically different from currently available
CEA standards as well as from the CEA obtained from
the patient's own liver metastases. These antigenic
differences were reflected in radioimmunoassay inhibition curves that were different and that were
not affected by perchloric acid extraction of CEA.
Because of the antigenic variation in the serum CEA,
markedly different CEA concentrations (varying by
three orders of magnitude) were measurable by two
different antisera (Roche and Montreal). All the
various CEA standards and samples cochromatographed
on columns of Sepharose-6B, despite the large antigenic variation. It is claimed that CEA consists
of a family of "isoantigens" with multiple antigenic
determinants. A serum CEA isoantigen was identified
that was different from currently available standards. Results of radioimmunoassays currently used
for CEA measurement may not represent absolute concentrations of serum "CEA", but may reflect the
binding affinity of different isoantigens to a particular polyvalent CEA antiserum.

of the tumor. The immunofluorescence staining of colon carcinoma by anti-NCA serum showed marked infiltration of the tumor stroma by polymorphonuclear and macrophagic cells.

3325   ISOLATION OF WILM'S TUMOR ANTIGENS BY CHELATION. (Eng.) Beierle, J. W. (Univ. Southern California Sch. Dentistry, Los Angeles, Calif. 90007); Wise, K. S.; Trump, G. N.; Allerton, S. E. *Clin. Chim. Acta* 61(3):411-414; 1975.

A procedure for EDTA extraction of minced Wilm's tumor was assessed as a method for isolating Wilm's tumor antigens. Extraction with 0.02% sodium-EDTA in phosphate-buffered saline was performed at room temperature for two hours with continuous stirring. The mixture was then centrifuged for 15 min at 20,000 x g, and the turbid supernatant (the "EDTA extract") was saved. A single precipitin line formed when rabbit antiserum, adsorbed with pooled human plasma and normal adult kidney EDTA homogenates, was allowed to react with the EDTA extract in Ouchterlony double diffusion tests. A line of complete indentity was observed between the EDTA tumor extract and EDTA extracts of *in vitro* cultures of Wilm's cells. Thus, the use of EDTA resulted in release of immunologically identical antigens from Wilm's tumor and a cultured Wilm's cell line. Chelation should therefore prove useful in isolating such membrane-bound antigens.

3326   INDUCTION OF EPSTEIN-BARR VIRUS-ASSOCIATED NUCLEAR ANTIGEN DURING *IN VITRO* TRANSFORMATION OF HUMAN LYMPHOID CELLS. (Eng.) Leibold, W. (Inst. Pathol., Hannover, East Germany); Flanagan, T. D.; Menezes, J.; Klein, G. *J. Natl. Cancer Inst.* 54 (1):65-68, 1975.

The induction of Epstein-Barr virus-associated nuclear antigen (EBNA) was studied in human lymphoid cells isolated from the peripheral blood of human adults (from cord blood), and from human fetal liver, spleen, bone marrow, and thymus. These were cultivated with or without a cell-free preparation of Epstein-Barr virus line B95-8 (EBV) that demonstrated transforming activity. In some cultures, phytohemagglutinin (PHA), concanavalin A (Con A), or pokeweed mitogen (PWM) were added in doses suboptimal with regard to DNA synthesis stimulation. Incubation was in RPMI-1640 medium, 36 C, 5% CO_2 in air, 80-95% relative humidity. All sera were checked to exclude the presence of antibody to nuclear antigen. A positive control consisted of a similarly prepared known EBNA-positive lymphoblastoid cell line (Raji). Uninfected test cells served as negative controls. The uninfected controls remained EBNA-negative. EBNA-positive lymphoblasts were first observed from days 1-6, mostly on day 3. This was followed by the appearance of some EBNA-positive small lymphoid cells. Transformation into permanent lymphoblastoid cell lines (LCL) was seen from days 12-19, at which point more than 80% of the cells were EBNA-positive. The results were similar in the mitogen-treated EBV-infected cultures, but mitogen alone did not produce EBNA-positive cells.

PHA produced a slightly lower percentage of EBNA-positive cells. Con A showed a longer latency period before EBNA-positive cells appeared, although transformation was observed after a normal latency. PWM caused an early appearance of EBNA-positive cells, the percentage of which was highest on day 2. With respect to the different groups of fetal cells, only the thymus specimens failed to show EBNA induction or establishment of LCL. The authors attribute the results of their mitogen study to the possibility that B-type lymphocytes have EBV receptors but T-type lymphocytes do not. PWM is known to affect both types, whereas PHA is a preferential mitogen for only the T-type. This was supported by the failure to induce EBNA or LCL in the fetal thymus cells. It is suggested that EBNA may be a necessary, although not in itself sufficient, condition for the growth of EBV-transformed lymphoid cells as permanent lines.

3327   A NEW MICROMETHOD FOR THE DETECTION OF HL-A
        ANTIGENS ON CULTURED HUMAN TUMOR CELLS.
(Eng.) Fritze, D. (Harbor General Hosp., 1000 West Carson St., Torrance, Calif. 90509); Kern, D. H.; Drogemuller, C. R.; Pilch, Y. H.* *Transplantation* 20(3):211-218; 1975.

A rapid microcytotoxicity assay for the detection of HL-A antigens on tissue culture cells derived from human solid tumors is described. Tumor cells were prelabeled with ^{125}iododeoxyuridine. Isotopically labeled tumor cells were reacted with up to 37 highly selected HL-A antisera and diluted rabbit complement. Results of the HL-A typing of nine human tumor cell lines are reported. Three melanoma cell lines showed individually distinct HL-A profiles at the first HL-A locus which agreed with the antigenic pattern of the tumor donor's autologous lymphocytes. Less reactivity was noted with HL-A antisera defining second locus specificities on the three melanoma cell lines. Some other cell lines showed more HL-A reactions than were required to present a "full house." This method has obviated the necessity for visually enumerating residual tumor target cells.

3328 . PURIFICATION OF POLYOMA T ANTIGEN FROM
        TRANSFORMED CELLS. (Eng.) Paulin, D.
(Department de Biologie Moleculaire, Institut Pasteur, 75015 Paris, France); Gaudray, P.; Cuzin, F. *Biochem. Biophys. Res. Commun.* 65(4):1418-1426; 1975.

As a preliminary step toward analysis of the function of the polyoma virus gene, ts-a, at the biochemical level, a purification procedure was devised for T antigen, using complement fixation as an assay. Polyoma virus-transformed PY6' cells were radiolabeled, homogenized, and centrifuged 30 min at 30,000 g. The supernatant was treated with $(NH_4)_2SO_4$ to 20% saturation, the resulting precipitate was discarded, the new supernatant was 40% saturated with $(NH_4)_2SO_4$, and the precipitate obtained this time was redissolved in phosphate-buffered 10% glycerol and dialyzed against the same buffer. The preparation thus obtained was separated into two fractions, I and II, by chromatography on DEAE cellulose. The

active fractions were further purified by selective removal of cell protein contaminants on immunoabsorbant columns made of Sepharose 4 B to which purified sheep antibodies to mouse cell protein were coupled. The fraction I now showed a purification factor of 356; the fraction II, a purification factor of 250. By electrophoretic analysis on SDS-polyacrylamide gels, using suitable reference standards, it was found that fraction II contained five main peptides with low background, while fraction I contained four main peptides and was more heterogeneous. The molecular weights (MW) of the main fraction II components were 52,000, 62,000, 68,000, 72,000 and 86,000; those of the fraction I components were 43,000, 60,000, 74,000 and 85,000. When the preparation obtained at the second $(NH_4)_2SO_4$ precipitation step was further purified by adsorptio on insolubilized anti-T antibodies and eluted with pH 2.2 glycine buffer, two components resolvable by polyacrylamide gel electrophoresis and showing MW of 70-75,000 and 80-90,000 were obtained. Two hypotheses were offered to explain the results. One was that the 70,000 MW peptide constituted the only viral gene product, while the 90,000 MW peptide was a cellular component associated with the viral protein. The other was that a mRNA corresponding to 70 80% of the early viral region was translated into the 90,000 MW peptide, which was, in turn, partially converted into a smaller antigenic component.

3329   BLOOD-GROUP PRECURSORS AND CANCER-RELATED
        ANTIGENS. (Eng.) Feizi, T. (Clinical ·
Res. Center, Watford Road, Harrow, Middlesex HA1 3U7, England); Turberville, C.; Westwood, J. H. *an-Lancet* 2(7931):391-393; 1975.

Glycoprotein extracts from four primary colorectal carcinomas and from three liver metastases from colonic carcinomas were tested for blood-group precursor-like activities using human anti-I and anti-i cold agglutinins as reagents. Substantial activity was detected by quantitative precipitin assays in the high-molecular weight fractions of two metastatic tumors. A carcinoembryonic antigen (CEA) fraction from one of these tumors was also active. The CEA fraction with the highest precursor-like activity was incubated with the anti-I and anti-i sera; after centrifugation the supernatants and the washed precipitates were assayed for CEA activity. All the CEA activity was recovered in the supernatants, showing that the CEA and precursor-like molecules are located on separate molecules. These data indicate that levels of precursor-like antigens in endodermal tumors and their possible diagnostic value warrant detailed investigation.

3330   MECHANISMS OF ACTION OF Ir GENES AND THEIR
        POSSIBLE APPLICATION TO TUMORS. (Eng.)
Lonai, P. (Dept. Medicine, Univ. Stanford, Stanford, Calif.); Grumet, F. C. *Transplant. Proc.* 7(2):149-153; 1975.

To obtain evidence for the expression of *Ir* genes in T cells, an *in vitro* system was established to detect T-cell response to antigens which elicit im-

was cultured in a 14 liter fermentor for a 26-day period. During this time TL and Thy-1 expression did not vary significantly, demonstrating that lymphoblastoid cell lines can be cultured on a continuous basis and will continue to express their surface alloantigens.

3332 GENERATION OF CYTOTOXIC T LYMPHOCYTES IN VITRO: V. RESPONSE OF NORMAL AND IMMUNE SPLEEN CELLS TO SUBCELLULAR ALLOANTIGENS. (Eng.) Engers, H. D. (Swiss Inst. Experimental Cancer Res., 1011 Lausanne, Switzerland); Thomas, K.; Cerottini, J.-C.; Brunner, K. T. *J. Immunol.* 115(2):356-360; 1975.

Subcellular particulate membrane fragments prepared from murine lymphoid or tumor cells were used as a cell-free antigen source in order to stimulate the generation of cytolytic thymus-derived effector cells (CTL) *in vitro*. Tumor cell lines syngeneic to C57BL/6 (EL4 leukemia), DBA/2 (P-815 mastocytoma, L-1210 lymphoma) and to BALB/c strains (MOPC-315 plasmocytoma) were maintained both in the ascitic form and *in vitro*. Cell-mediated cytotoxicity was determined by a ^{51}Cr release method. When cultivated with normal spleen cells as a source of responding lymphocytes, particulate antigen preparations induced only low CTL activities. However, normal responses were observed with spleens from C57BL/6 mice that had been made immune by inoculation with P-815 cells (30 x 10^6 cells, ip) 2-4 mo previously were used as a source of responding cells; cultures containing immune spleen cells showed a 3- to 5-fold higher lytic activity than primary cultures containing normal spleen cells when they were stimulated with irradiated DBA/2 spleen cells. In both cases, results were compared to those responses obtained with intact irradiated (1,000 rads) normal spleen cells as stimulating antigen. The effector cells generated with particulate antigen preparations (obtained either by hypotonic shock or by sonication) were shown to be T cells and were characterized with regard to the kinetics of their response and their dose-activity relationship. It was also shown that the responses observed were specific, both at the level of initiation and at the effector level. The results suggest that a fundamental difference exists between normal and alloimmune spleen cells in their ability to respond to stimulation by subcellular antigen preparations *in vitro*.

3333 LYMPHOCYTES IN PATIENTS WITH OVARIAN CANCER. (Eng.) Wolff, J. P. (Dept. Gynaecol., Gustave-Roussy Institute, Villejuif, France); De Oliveira, C. F. *Obstet. Gynecol.* 45(6):656-658; 1975.

The total number of circulating lymphocytes and their morphologic transformation in the presence of several antigens was studied. The subjects were 156 women, 81 of whom currently had an ovarian cancer or had had one in the past. Two control groups were used, including 54 normal women and 21 patients with malignancies other than ovarian cancer. Blood samples were taken and the total number of lymphocytes was determined. WBC suspensions were then incubated in a medium containing phytohemagglutinin (PHA) or poke-weed mitogen (PWM). The number of lymphocytes trans-formed into lymphoblasts in the presence of the two antigens was then counted. There was a significant decrease in total lymphocytes in the pathologic con-trol group and in the patients with ovarian cancer from the figures obtained in healthy women. In the presence of both antigens, lymphocyte transforma-tion was significantly decreased from normal values in the pathologic control group and in patients who currently had an ovarian cancer. For the latter group, the survival at six months and the number of lymphocytes or the rate of transformation with phyto-hemagglutinin were correlated. The authors conclude that total lymphocytes in circulation and the rate of lymphoblastic transformation is a very important point in the prognosis for patients with ovarian cancer.

3334   B AND T CELL LYMPHOMAS: ANALYSIS OF BLOOD
        AND LYMPH NODES IN 87 PATIENTS. (Eng.)
Gajl-Peczalska, K. J. (Box 609, University of Minne-sota Hosp., Minneapolis, Minn. 55455); Bloomfield, C. D.; Coccia, P. F.; Sosin, H.; Brunning, R. D.; Kersey, J. H. *Am. J. Med.* 59(5):674-685; 1975.

B and T cell populations were studied in blood and neoplastic tissues from 64 untreated and 23 treated patients with non-Hodgkin's lymphoma. This study was undertaken to evaluate the relation of B and T cell markers in various lymphomas to the currently ac-cepted morphologic classifications and to determine the utility of various tissues in defining the cell of origin of a lymphoma. When histologically in-volved blood, bone marrow, lymph nodes or body fluids were studied, a B or T cell origin of the lymphoma was identified in 68% of the patients. A B cell ori-gin was found in 18% of adults classified as having nodular (N) or diffuse (D) poorly differentiated lym-phocytic lymphoma (PDLL). One lymphoma of T cell origin was observed in an adult with poorly differen-tiated lymphocytic lymphoma-diffuse (PDLL-D). In contrast, all cases of PDLL-D in children were T cell in origin. The origin of American Burkitt's (stem cell) lymphoma in two children was the B cell. When histologically involved blood was implicated, a B or T cell origin was demonstrated in 10 of 21 (48%) adults. Evidence of a monoclonal proliferation of B lympho-cytes in the blood was found in two adults with more than 7 per cent lymphoma cells in Wright-Giemsa stained blood smears. When neoplastic lymph nodes were stu-died, the diagnosis of a B cell lymphoma was made in 8 of 12 (67%) adults. Study of surface markers on malignant cells in cerebrospinal or serosal fluids frequently revealed a B or T cell origin of the lym-phoma. B and T lymphocyte numbers in the blood did not correlate with immunoglobulin or skin test ab-normalities. Abnormalities in circulating B or T cell percentages at diagnosis were a poor prognostic sign in patients with PDLL-D.

3335   SEPARATION OF T LYMPHOCYTES FROM NORMAL
        INDIVIDUALS AND PATIENTS WITH B LYMPHOCYTE
CHRONIC LYMPHOCYTIC LEUKAEMIA. (Eng.) Fernandez, L. A. (Camp Hill Hosp., Halifax, Canada); MacSween, J. M.; Langley, G. R. *Immunology* 28(2):231-241; 1975.

A technique is reported for obtaining a purified population of viable T lymphocytes from the periph-eral blood of normal and leukemic subjects. Twenty-seven normal individuals and ten patients with chron-ic lymphocytic leukemia were studied. Identifica-tion of B cells was based on detection of membrane-bound immunoglobulins with the aid of fluorescein-conjugated antisera to human immunoglobulin deter-minants. T cells were identified by rosette-forma-tion with SRBC. When lymphocytes collected from nor-mal individuals and containing an average of 16.1% B cells were applied to IgG-anti-IgG-coated Degalan bead columns, held for two hours at 4 C, and eluted with medium 199, the eluted cells contained less than 2% B cells. When lymphocytes collected from leukemic patients and containing an average of 68.6% B cells were similarly fractionated, the eluted cells contained 36.4% B cells. To improve the puri-fication of T lymphocytes from leukemic patients, columns of uncoated Degalan bead columns were first used to remove nonspecifically adherent cells. The lymphocytes eluted from this column were then ap-plied to the IgG-anti-IgG-coated Degalan bead col-umns, and this resulted in a final preparation of T cells containing less than 2% B cells. For three cases of chronic lymphocytic leukemia with low den-sity surface immunoglobulins, however, separation by this method was unsuccessful. Using a column pre-pared with Degalan beads coated with purified IgG which was previously heat-aggregated at 65 C for 45 min, then added to the beads and held at 37 C for 30 min, lymphocytes from a patient which had lost detectable surface immunoglobulins showed an in-creased proportion of rosette-forming T cells (from 4% before separation to 40% after separation).

3336   CYTOTOXICITY OF LYMPHOCYTES FROM HEALTHY
        SUBJECTS AND FROM MELANOMA PATIENTS
AGAINST CULTURED MELANOMA CELLS. (Eng.) Pavie-Fis-cher, J. (Laboratoire d'Immunologie des Tumeurs, Institut de Recherches sur les Maladies du Sang, Centr Hayem, Hopital Saint Louis, 2, Place du Dr. Fournier, 75-010, Paris, France); Kourilsky, F. M.; Picard, F.; Banzet, P.; Puissant, A. *Clin. Exp. Immunol.* 21(3):430-441; 1975.

The *in vitro* cytotoxicity of lymphocytes from 47 melanoma patients (23 with primitive localized tumors and no detectable lymph node involvement; 24 with metastatic and recurrent tumors) and 13 healthy con-trols for cultured melanoma cells was studied in the ^{51}Cr release assay. Two different melanoma cell lines were used as target cells: SK Mell cells cul-tured in suspension and NK 1_1, a tissue culture line growing as a monolayer. Lymphocytes from most healthy subjects were cytotoxic to a variable extent for both cell lines; because of this variability, no correct control figure of a normal 'cytotoxic effect' could be defined. Approximately the same level of cytotoxicity was observed when lymphocytes from the same healthy subjects were retested after a 2-mo interval. A retrospective search for isoimmunization revealed previous transfusions and multiparity in three donors, but failed to explain the toxic effect of lymphocytes from the other subjects. Lymphocytes from the melanoma patients were also cytotoxic to the two melanoma cell lines, but the highest degree

the differentiation of these cells might be blocked at the step where IgD and IgM overlap, with IgD disappearing as soon as the cells start to secrete IgM. The lymphocytes of the CLL patients were therefore investigated for crystalline immunoglobulin inclusions. In none of the 18 cases could the inclusions be found. Further investigation of the two immunoglobulin receptors in malignant lymphocyte proliferations might give a better understanding of the underlying mechanisms.

3339 . EFFECTS OF MITOGENS FOR MOUSE B LYMPHO-
CYTES ON CHICKEN LYMPHOID CELLS. (Eng.)
Tufveson, G. (Biomedical Center, Box 571, S-751 23
Uppsala, Sweden); Alm, G. V. *Immunology* 29(4):
697-707; 1975.

Because *Escherichia coli* lipopolysaccharide (LPS), tuberculin-purified protein derivative (PPD), and dextran sulfate (DxS) were known to be potent mitogens for mouse B lymphocytes and to enhance polyclonal antibody synthesis in mouse spleen cells, a study of the effects of these mitogens and also of the T-cell mitogen, concanavalin A (Con A), on chicken lymphoid cells was carried out for comparison. The effect of bursectomy on the mitogen-induced proliferative response of chicken spleen cells was also studied. The chickens were White Leghorn Babcock B-300 strain; control mice were CBA/J strain. Chemical bursectomy in the chickens was performed by ip injection of cyclophosphamide. Cultures of chicken spleen, thymus, and bursa cells and of mouse spleen cells were prepared in RPMI 1640 medium. [^{3}H]-thymidine uptake was used as an indicator of mitogen stimulation of cell proliferation. Tests for polyclonal antibody synthesis were based on the occurrence of plaque-forming cells (PFC) in unimmunized cultures, using a method involving SRBC and TNP-coupled SRBC as target cells. Tests for primary antibody response were carried out by immunizing cultures with SRBC *in vitro* and subsequently assaying the cultures for direct PFC using SRBC as target cells. LPS and PPD increased the proliferation of chicken spleen cells in culture, but DxS was not stimulatory. Thymus and bursa lymphocytes were not stimulated by any of these compounds. Spleen cells from bursectomized chickens showed decreased responses to LPS and PPD, but responded normally to the T-cell mitogen Con A. None of the tested mitogens induced polyclonal antibody synthesis or directly enhanced the primary antibody response to SRBC in spleen cell cultures. LPS-coated SRBC, however, enhanced the *in vitro* antibody response to SRBC. The results suggested a moderate proliferative response of chicken lymphocytes to LPS and PPD, possibly involving B cells, but no further effects comparable to those observed with mouse B lymphocytes.

3340 BLOCKING OF SPLEEN CELL ACTIVITY AGAINST
TARGET MAMMARY TUMOR CELLS BY VIRAL ANTI-
GENS. (Eng.) Blair, P. B. (Dept. of Bacteriology, Univ. of California, Berkeley, Calif. 94720); Lane, M. A.; Yagi, M. J. *J. Immunol.* 115(1):190-194; 1975.

A series of microcytotoxicity tests were used to

answer the questions of whether antigen alone is sufficient to inhibit spleen cell activity, and whether antigens of the mammary tumor virus (MTV) are effective in blocking spleen cell activity against mammary tumor cells. Virus preparations derived from the milk of six strains of mice and from culture fluids of the FUKU cell line were the source of antigen. Spleen cells from three types of donors were sources of immune cells. Cultures of MTV-induced BALB/cfC3H mammary tumors and a non-MTV-producing FUKU cell line served as sources of target cells. The donors consisted of two adult virgin BALB/cfC3H females, which made a T cell response to the tumor antigen; four multiparous but tumor-free BALB/cfC3H females, which made an even stronger T cell response, and a non-T cell response; and one adult virgin BALB/c female, whose response was entirely non-T cell. Three procedures were used for testing for blocking activity: (1) the target cells were treated with the antigen, (2) virus and spleen cells were added to target cells at the same time, and (3) the spleen cells were incubated with the antigen for 30 min before addition to the target cells. No blocking was observed by the first procedure; it was strongest by the third procedure. The T cell responses of the virgin and multiparous BALB/cfC3H females were effectively blocked, while non-T responses of multiparous BALB/cfC3H tumor-free females or of virgin BALB/c females were blocked by some, but not all, of the MTV antigen preparations. Murine leukemia virus (MuLV) was ineffective in blocking spleen cell activity against the MTV-induced mammary tumor cells. In comparable experiments involving the use of MuLV as the antigen, MuLV-producing FUKU cells as the target cells, and FUKU-sensitized lines as sources of spleen cells, the MuLV antigen was effective as a blocking agent, while MTV was not. The specificity of blocking demonstrates that it is immunologic in nature.

3341   RESISTANCE TO HERPES SIMPLEX TYPE 2 VIRUS INDUCED BY AN IMMUNOPOTENTIATOR (PYRAN) IN IMMUNOSUPPRESSED MICE. (Eng.) Moraban, P. S. (Medical Coll. Virginia, Virginia Commonwealth Univ., Richmond, Va. 23298); McCord, R. S. *J. Immunol.* 115(1):311-313; 1975.

The protective effects of pyran, an interferon inducer and immunopotentiator, against herpes simplex virus type 2 (HSV-2) were investigated in normal female BALB/c mice and in mice depleted of thymus-derived (T) lymphocytes by thymectomy, lethal irradiation (900 R), and bone marrow reconstitution (TXB mice). Adult (16-wk-old) TXB mice were approximately ten times more susceptible than agematched normal mice to iv or intravaginal infection with HSV-2. The LD_{50} for iv HSV-2 in normal adult mice was $10^{3.5}$ plaque-forming units (PFU) compared with $10^{2.4}$ PFU in adult TBX mice. However, 5-wk-old mice were even more susceptible to HSV-2 than were adult TBX mice ($LD_{50}=10^{1.4}$), indicating that factors other than T cells are involved in resistance to HSV-2. Pyran treatment (25 mg/kg, iv) significantly prolonged the mean survival time and reduced the mortality of both normal and TXCB mice infected iv with challenge doses of viruses ranging over 1,000-fold in concentrations. The LD_{50} increased to $10^{3.6}$

PFU in TBX mice and to $10^{4.1}$ PFU in normal mice. Treatment of TXB mice with pyran before intravaginal infection with HSV-2 (10^5 PFU) was also effective in significantly prolonging the mean survival time, although the drug was not as effective against intravaginal HSV-2 as against iv HSV-2. These results suggest a possible role for immunopotentiator-activated macrophages in the increased resistance afforded to mice by pyran treatment. As previously shown, the pyran-induced activation of macrophages is apparently independent of T cells because the drug induced the production of macrophages cytotoxic for tumor cells in TXB mice.

3342   *IN VITRO* LYMPHOCYTE REACTIVITY AND T-CELL LEVELS IN CHRONIC CIGARETTE SMOKERS. (Eng.) Silverman, N. A. (Natl. Cancer Inst., Bethesda, Md. 20014); Potvin, C.; Alexander, J. C., Jr.; Chretien*, P. B. *Clin. Exp. Immunol.* 22(2):285-292; 1975.

Peripheral blood total WBC, lymphocyte and thymus-dependent lymphocyte (T cell) levels and *in vitro* lymphocyte reactivity (LR) to phytohemagglutinin (PHA) were determined in 153 chronic cigarette smokers (58 men, 95 women, 20-78 yr old, mean 35 yr) and 115 nonsmokers (51 men, 64 women who had smoked; 20-78 yr old, mean 37.6 yr). In nonsmokers the WBC, lymphocytes, and T cells/mm³ were 6023, 1981, and 1382, respectively. In smokers the corresponding levels were 6803, 2325, and 1594, respectively. The WBC levels for smokers with 3-10 pack-yr (packs/day x years of smoking) was 6417, and for more than 50 pack/yr, 7662/mm³. The WBC level in nonsmokers over 50 yr old was 6411/mm³. There was no correlation of lymphocyte or T cell levels with age. Lymphocyte reactivity to PHA was significantly higher in smokers less than 40 yr of age or in those with a 20 pack-yr usage level. Older smokers or those with a history of heavier cigarette consumption did not differ from normals in LR to PHA, contrary to previous reports. The elevated WBC levels in smokers may be secondary to chronic subclinical pulmonary inflammation. The selection of older smokers may have been biased by the exclusion of those not in good health.

3343   THE Fc RECEPTOR ON THYMUS-DERIVED LYMPHOCYTES: II. MITOGEN RESPONSIVENESS OF T LYMPHOCYTES BEARING THE Fc RECEPTOR. (Eng.) Stout, R. D. (Stanford Univ. Sch. Medicine, Stanford, Calif. 94305); Herzenberg, L. A. *J. Exp. Med.* 142(5):1041-1051; 1975.

The responsiveness of purified Fc^- and Fc^+ thymus-derived (T) lymphocytes, isolated from normal BALB/cN mouse spleen cell populations by cell sorting on the fluorescence activated cell sorter, was examined. Although both Fc^- and Fc^+ T cells responded to phytohemagglutinin (2.5 µg/ml,) 48 hr, the response to concanavalin A (Con A 1.5 µg/ml, 48 hr) was found to be a characteristic of the Fc^+ T lymphocyte. The poor responsiveness of the Fc^- T cells to Con A was shown not to be due to a requirement for either different concentrations of Con A (0.15-5.0 µg/ml Con A, 0.5-5.0 µg/ml phytohemagglutinin) or for adherent cells. The addition of Fc^+ T cells to the Fc^- T cells in a ratio of 1:3 resulted in a mitotic re-

munization with 0.5 ml of 0.5-5.0% sheep RBC. Immunocompetence was assessed by measurement of spleen cell
antibody formation in a plaque assay, as determined
by localized hemolysis in a suspension of spleen
cells and sheep RBC in agar gel. It was found that
transfer of macrophages from normal mice to leukemia
virus-infected recipients had no effect on the virus-
induced impairment of the immune response. Similar
transfer of macrophages to normal uninfected mice
generally resulted in a moderate depression of the
expected immune response. In no case, however, did
the macrophages enhance the immune responses in normal or virus-infected mice. It appeared likely that
the defect in immunocompetence of the leukemic mice
lay mainly with impairment of antibody forming cells
and their precursors and not with antigen processing
cells per se.

3346   METHODOLOGICAL APPROACHES TO THE STUDY OF
        CARBOHYDRATE SURFACE RECEPTORS ON MACRO-
PHAGES AND TUMOR CELLS. (Eng.) Wollweber, L. (Inst.
Microbiology and Experimental Therapy, 69 Jena, German Democratic Republic); Fritsch, S. Neoplasma
22(2):157-162; 1975.

Specific carbohydrate receptors on cells were shown
to be made visible by ultracytochemical methods using concanavalin A (Con A), followed by peroxidase
(PO); Con A precoupled with PO; Ricinum communis
agglutinin precoupled with PO; or wheat germ agglutinin, followed by ovomucoid precoupled to PO. Pre-
coupling was accomplished with the aid of glutaraldehyde (GA). The catalytic activity of the PO was
revealed in each case by final treatment of the cells
with diaminobenzene and osmium tetroxide. The ra-
tionale behind the use of the different combinations
was based on the specificity of the carbohydrate receptors involved. The cells used in the tests were
macrophages originating from peritoneal cells of AB/
Jena mice and tumor cells originating from L1210 ascites tumors of $NMDF_1$ mice. The reactions were carried out on native cells, suspended in phosphate-
buffered saline, and also on GA-fixed cells in 2%
GS-cacodylate buffer. Lectin-treatment of native
cells resulted in agglutination, while lectin-treatment of GA-fixed cells did not. A relatively continuous dark surface layer of variable thickness
could be demonstrated on GA-fixed cells, while native cells showed a discontinuous surface coat characterized by strongly marked and almost contrastless
membrane segments. Quantitative estimations of the
Con A reaction products were based on two parameters:
area per length of membrane (A/L), as a parameter for
the thickness of the reaction product; and frequency
of features (N/L), as a parameter for the disconti-
nuity of the surface reaction. A/L and N/L were inversely proportional, indicating that the thicker the
surface reaction, the rarer its interruptions. Native macrophages showed greater variations of thickness of stained cell coat from one segment to another
than did GA-fixed macrophages. Different individual
peritoneal cells showed uniform mean values for A/L
over the total circumference, indicating a lack of
loss of glycocalyx of native normal peritoneal cells
during the preparation process, as had previously
been found for certain tumor cells.

3347   INHIBITION OF THE CELLULAR IMMUNE RESPONSE
TO SIMIAN VIRUS 40 TUMOR CELLS IN TUMOR-
BEARING TUMOR-IMMUNE MICE BY CONCANAVALIN A. (Eng.)
Davis, S. (Public Health Serv., U.S. Dep. Health,
Educ. Welfare, Bethesda, Md.); Boone, C. W. *J. Natl
Natl. Cancer Inst.* 54(2):435-438; 1975.

The effects of *in vivo* administration of concanavalin A (Con A) on the primary and secondary immune
responses of male BALB/cAnN mice to syngeneic simian virus 40 (SV40)-induced tumor cells (E4) were
investigated. To demonstrate the effects of Con A
on established tumor immunity, 10^6 E4 tumor cells
were injected into the foot pads of immunized mice
at different times (day 0, 1-5, 9, 11) after *in
vivo* administration of the Con A. The mice were
immunized by excision of growing tumors ten days
after inoculation with 10^6 E4 cells. One group of
E4-immune mice received a single ip injection of
400 μg Con A on day 0; another group received 50 μg
Con A daily from day 0. The single dose of Con A
immediately and markedly suppressed the immune response, which slowly returned to normal levels in
six days. The small, continuous doses given daily
caused a gradual decrease, which tended to reach a
plateau by the fifth day. To determine the effects
of Con A on the cell-mediated immune response in
mice with progressively growing tumors, the radioisotopic foot-pad assay was used on mice at various
intervals (days 0, 2, 6, 8, 12, and 16) after implantation of E4 cells in the foot pad. When the
single dose of Con A was used, the response was significantly depressed, and the depression was greatest
on the eighth day; when the daily dose regimen was
used, the depression on the eighth day was still
greater. When the tumors were excised on day 25,
those from the Con A-treated mice were significantly
greater in size than those of the untreated control
mice. These results showed that Con A suppressed
cellular response to progressively growing tumor and
enhanced tumor growth. To determine the effect of
spleen cells from Con A-treated E4-immune mice on
the local adoptive transfer of E4 immunity, E4-
immune mice were given ip injections of 400 μg
Con A; 16 hr later, the spleen cells were removed
and single cell suspensions were prepared. Then
20 x 10^6 spleen lymphocytes were mixed with 10^6
E4 tumor cells and injected into the foot pads of
treated immune control mice. The reversibility of
this immunosuppression was shown by near total restoration of immune reactivity after incubation of
the lymphocytes with α-methyl-D-pyranosyl sugars.
This suggests that the mechanism by which Con A
suppresses cell immune reponse may be due, in part,
to its ability to bind to receptors on T cell membranes.

3348   LYMPHOCYTE-MEDIATED CYTOTOXICITY AGAINST
TUMOR CELLS. I. CON A ACTIVATED CYTO-
TOXIC EFFECTOR CELLS EXHIBIT IMMUNOLOGICAL SPECI-
FICITY. (Eng.) Waterfield, J. D. (Division of
Immunology, Karolinska Institutet, Wallenberger-
laboratory, Lilla Freskati, 104 05 Stockholm 50,
Sweden); Waterfield, E. M.; Moller, G. *Cell.
Immunol.* 17(2):392-404; 1975.

The phenonenon of concanavalin A (Con A)-induced
cytotoxicity in spleen cells of mice was analyzed
with regard to similarities between *in vivo* and *in
vitro* cytotoxic activation. The cells were observed
for the expression of Con A-induced cytotoxicity when
Con A was present or absent in the assay system.
A/Sn (H-2^a), A/Sn x B10 5M (H-$2^{a/b}$), B10 5M (H-2^b),
CBA (H-2^k), C57 Leaden (H-2^b), and DBA/2J (H-2^d)
mice, both age and sex matched, were used. For
maximum *in vivo* activation, 240 μg Con A in 0.1
ml injected iv was used routinely. To establish
dose-response curves, mice were similarly injected
with 6-480 μg Con A. The tumors used as sources
of target cells were the YAC Molony virus-induced
ascites leukemia (H-2^a) and the DBA/2J P815 (H-2^d).
The target cells were labeled with 5-iodo-2-deoxy-
uridine-I^{125} (^{125}IUdR). Spleen cells from Con A-
injected mice were cytotoxic to allogeneic tumor
target cells *in vitro*, the degree of cytotoxicity
being dependent upon the dose of Con A. Cytotoxicity
did not occur when the lymphocytes were syngeneic
to the target cells. When extraneous Con A was
added to the mixture of syngeneic lymphocytes and
target cells, cytotoxicity occurred to the same
extent as that found in allogeneic combinations
without the addition of Con A. Analogous findings
were observed when spleen cells were activated with
Con A *in vitro*. Only allogeneic lymphocytes caused
cytotoxicity after treatment with the competitive
ligand, methyl-α, D-mannopyranoside to remove surface bound Con A. However, extraneously added Con A
also caused expression of cytotoxicity with syngeneic Con A-activated lymphocytes. It is concluded
that Con A transforms lymphocytes into cytotoxic
effector cells, but that these cells cannot express
cytotoxcity unless they bind to the target cells.
The author suggests that allogeneic Con A-activated
lymphocytes recognize the target cells by clonally
distributed recognition receptors and are therefore
capable of binding to the targets by themselves.
Syngeneic cells, on the other hand, do not possess
such receptors, and consequently can only express
cytotoxicity when agglutinated to the targets by extraneous agents such as Con A.

3349   A COMPARATIVE STUDY OF FOUR CYTOCHEMICAL
DETECTION METHODS OF CONCANAVALIN A BIND-
ING SITES ON THE CELL MEMBRANE. (Eng.) Temmink, J.
H. M. (Netherlands Cancer Inst., Amsterdam, Nether-
lands); Collard, J. G.; Spits, H.; Roos, E. *Exp.
Cell Res.* 92(2):307-322; 1975.

Four different electron cytochemical methods to detect concanavalin A (ConA) binding sites on the
plasma membrane of mouse fibroblasts were compared.
The ConA binding sites were made visible either by
adding ConA, followed by horseradish peroxidase
(HRP) or hemocyanin (HC), or by marking the sites
with complexes of ConA with ferritin (Fer) or with
microperoxidase (MP). HC and Fer were directly visible in the electron microscope; HRP and MP were
detected by their electron-dense reaction production
with diaminobenzidin and H_2O_2. Differences in sensitivity of the ConA binding sites for the different markers were found and resulted in a tentative
interpretation of the labeling reactions. All experiments suggested that normal and transformed murine fibroblasts both have plasma membranes in which

Inst., Moscow, U.S.S.R.); Hollinshead, A. *Science* 190(4212):391-392; 1975.

3358　A STUDY OF IMMUNE-SUPPRESSIVE ACTIVITY OF HERPES SIMPLEX VIRUS TYPE 2 AND THE TUMOR ENHANCING POTENTIAL OF THE VIRUS ON YOSHIDA SARCOMA. (Eng.) Reiss-Gutfreund, R. J. (Institut fur Krebsforschung, Universitat Wien, Austria); Dostal, V. *Osterr. Z. Onkol.* 2(2/3):63-68; 1975.

3359　IMMUNOSUPPRESSIVE EFFECT OF A HUMAN HEPATIC GLYCOFERROPROTEIN, α2H GLOBULIN: A STUDY ON THE TRANSFORMATION OF NORMAL HUMAN LYMPHOCYTES. (Eng.) Buffe, D. (Inst. de Recherches Scientifiques sur le Cancer, 94 Villejuif, B. P. No. 8, France); Rimbaut, C. *Immunology* 29(1):175-184; 1975.

3360　INCREASED ACTIVITY OF SPLENIC SUPPRESSOR CELLS IN MICE UNDER THE CONDITIONS OF TUMOR GROWTH. (Rus.) Gambarov, S. S. (Inst. Biophysics of the Ministry of Public Health of the USSR, Moscow, USSR); Petrov, R. V.; Khaitov, R. M.; Norimov, A. Sh. *Dokl. Akad. Nauk SSSR* 224(5):1195-1197; 1975.

3361　CANCER OF THE OVARY. THE LEVEL OF IMMUNOGLOBULINS IN PATIENTS WHO HAVE PRIMARY TUMORS OF THE OVARY. (Fre.) Wolff, J. P. (Institut Gustave-Roussy, 16 bis, avenue P. Vaillant-Couturier F 94800, Villejuif); de-Oliveira, C. F. *Gynecol. Obstet. Biol. Reprod. (Paris)* (1):75-92; 1975.

3362　THE EFFECT OF PLASMACYTOMAS ON SERUM IMMUNOGLOBULIN LEVELS OF BALB/c MICE. (Eng.) Fenton, M. R. (Temple Univ. Sch. Med., Philadelphia, Pa.); Havas, H. F. *J. Immunol.* 114(2/Part 2):793-801; 1975.

3363　POLYMERIZED MONOCLONAL IgA IN TWO PATIENTS WITH MYELOMATOSIS AND HYPERVISCOSITY SYNDROME. (Eng.) Virella, G. (Gulbenkian Inst. Science, Biological Res. Center, Oeiras, Portugal); Preto, R. V.; Graca, F. *Br. J. Haematol.* 30(4):479-487; 1975.

3364　ANTIBODY RESPONSE TO POLIOVIRUS IMMUNIZATION IN CHILDHOOD LEUKEMIA. (Eng.) Bosu, S. K. (Roswell Park Memorial Inst., Dept. Pediatrics, 666 Elm St., Buffalo, N.Y. 14203); Ciudad, H.; Sinks*, L. F.; Ogra, P. I. *Med. Pediatr. Oncol.* 1(3):217-225; 1975.

3365　SURVEY FOR ANTIBODIES TO LEUKEMIA (C-TYPE) VIRUS IN CATTLE. (Eng.) Baumgartener, L. E. (Dep. Vet. Sci., Univ. Wisconsin, Madison); Olson, C.; Miller, J. M.; Van Der Maaten, M. J. *J. Am. Vet. Med. Assoc.* 166(3):249-251; 1975.

3366   AUTOANTIBODIES WITH ANTILIPOPROTEIN SPECI-
       FICITY AND HYPOLIPOPROTEINEMIA IN PATIENTS
WITH CANCER.   (Eng.) Riesen, W. (Inst. Clin. Exp.
Cancer Res. Univ. Berne, Tiefenau-Hosp., Switzerland); Noseda, G.; Morell, A.; Bütler, R.; Baran-
dun, S.; Nydegger, U. E. *Cancer Res.* 35(3):535-
541; 1975.

3367   IMMUNE COMPLEXES ASSOCIATED WITH NEOPLASIA:
       PRESENCE OF EPSTEIN-BARR VIRUS ANTIGEN-
ANTIBODY COMPLEXES IN BURKITT'S LYMPHOMA.   (Eng.)
Oldstone, M. B. A. (Scripps Clinic and Res. Found-
ation, La Jolla, Calif.); Theofilopoulos, A. N.;
Gunven, P.; Klein, G. *Intervirology* 4(5):292-
302; 1974.

3368   STABILITY AND DISSOCIATION OF P3HR-1 BUR-
       KITT'S LYMPHOMA CELL SOLUBLE COMPLEMENT-
FIXING ANTIGEN IDENTIFIED WITH HUMAN SERUM.   (Eng.)
Weliky, N. (Chem. Sci. Dep., Syst. Group of TRW,
Inc., Redondo Beach, Calif. 90278); Leaman, D. H.,
Jr.; Kallman, B. J. *Cancer Res.* 35(6):1580-1585;
1975.

3369   ANTIGENIC CHARACTERIZATION OF TYPE C RNA
       VIRUS ISOLATES OF GIBBON APES.   (Eng.)
Tronick, S. R. (Natl. Cancer Inst., Bethesda, Md.),
Stephenson, J. R.; Aaronson, S. A.; Kawakami, T.
G. *J. Virol.* 15(1):115-120; 1975.

3370   EVIDENCE OF A SURFACE ANTIGEN ON LYMPHO-
       CYTES IN HODGKIN'S DISEASE.   (Fre.) Car-
cassonne, Y. (Unité 119 de l'INSERM, Laboratoire
d'Hématologie Oncologique, 27, boulevard Lei Roure,
13009 Marseille, France); Favre, R.; Meyer, G. *C.
R. Acad. Sci. [D] (Paris)* 280(12):1505-1507; 1975.

3371   CELL SURFACE ANTIGENS IN VIRUS-TRANSFORMED
       CELLS.   (Eng.) Pasternak, G. (Zentral-
institut fur Krebsforschung der Akademie der Wis-
senschaften der DDR, Abteilung Immunbiologie, Berlin-Buch, East Germany). *Proc. Int. Cancer Congr.
11th.* Vol. 2 *(Chemical and Viral Oncogenesis).*
Florence, Italy, October 20-26, 1974. Edited by
Bucalossi, P.; Veronesi, U.; Cascinelli, N. New
York, American Elsevier, 1975, pp. 237-242.

3372   ANTIGENS ASSOCIATED WITH CHEMICALLY TRANS-
       FORMED CELLS.   (Eng.) Embleton, M. J.
(Cancer Res. Campaign Labs., Univ. Nottingham,
Nottingham, NG7 2RD, U.K.). *Proc. Int. Cancer
Congr. 11th.* Vol. 2 *(Chemical and Viral Oncogene-
sis).* Florence, Italy, October 20-26, 1974. Edit-
ed by Bucalossi, P.; Veronesi, U.; Cascinelli, N.
New York, American Elsevier, 1975, pp. 102-106.

3373   AUSTRALIA ANTIGEN AND MALIGNANT HEPATOMA.
       (Eng.) Theodoropoulos, G. (Med. Sch.,
Univ. Athens, Greece); Archimandritis, A.; Angelo-
poulos, B. *Ann. Intern. Med.* 82(6):809-810; 1975.

3374   SEROLOGICAL CHARACTERIZATION OF MEMBRANE
       ASSOCIATED ANTIGENS ON HUMAN MALIGNANT
MELANOMA [abstract]. (Eng.) Seibert, E. (Abteilung fur Experimentelle Dermatologie, Universitats-
Hautklinik, Munster, West Germany); Sorg, C.; Mul-
ler-Neumann, Ch.; Macher, E. *Z. Immunitatsforsch.*
150(3):240-241; 1975.

3375   BLOOD GROUP SPECIFIC ANTIGENS IN CANCER
       AND THEIR SIGNIFICANCE FOR THE HISTOLOGIC-
AL TUMOUR DIAGNOSIS. (Ger.) Scholz, P. (Pathologisches Institut der Humboldt-Universität zu Berlin,
104 Berlin, Schumannstrasse 20/21, East Germany).
Zentralbl. Allg. Pathol. 119(1/2):22-26; 1975.

3376   PRESENCE OF BLOOD GROUP H ANTIGEN ON A
       CARCINOEMBRYONIC ANTIGEN. ENZYMATIC
MODIFICATION OF THE SPECIFICITY FOR H INTO SPECIFI-
CITY A AND B. (Fre.) Magous, R. (Equipe de Recherche du CNRS n° 62, Ecole Nationale Superieure
de Chimie, 8, rue de l'Ecole Normale, 34075 Mont-
pellier Cedex, France); Lecou, C.; Bali, J. P.;
Mousseron-Canet, M. *C. R. Acad. Sci. [D] (Paris)*
280(21):2513-2516; 1975.

3377   SEARCHING FOR THE CARCINOEMBRYONIC ANTIGEN
       (CEA) AND ASSOCIATED ANTIGENS IN BLOOD
AND HEMATOPOIETIC ORGAN CELLS [abstract]. (Fre.)
Bordes, M. (No affiliation given); Knobel, S.;
Martin, F. *Ann. Immunol. (Paris)* 126C(1):100; 1975.

3378   COMPARATIVE RADIOIMMUNOLOGIC DETERMINATION
       OF THE CARCINOEMBRYONIC ANTIGEN (CEA) AND
OF AN IMMUNOLOGICALLY RELATED ANTIGEN (NCA) IN
PATIENT SERA [abstract]. (Fre.) von Kleist, S.
(No affiliation given); Troupel, S.; Burtin, P.
Ann. Immunol. (Paris) 126C(1):100; 1975.

3379   DETECTION OF A CARCINOEMBRYONIC ANTIGEN ON
       THE MEMBRANE OF RAT INTESTINAL CANCER
CELLS [abstract]. (Fre.) Martin, F. (No affilia-
tion given); Martin, M.; Knobel, S.; Bordes, M.
Ann. Immunol. (Paris) 126C(1):100; 1975.

3380   LYMPHOCYTE ACTIVATION *IN VITRO* TO MURINE
       ONCO-FOETAL ANTIGENS. (Eng.) Chism, S.
(Royal Melbourne Hosp., Parkville, Victoria 3050,
Australia); Burton, R.; Warner, N. L. *Nature*
257(5527):594-596; 1975.

3381   B CELL ANTIGEN MARKERS ON AVIAN LYMPHOID
       LEUKOSIS TUMOUR CELLS. (Eng.) Payne, L.
N. (Houghton Poultry Res. Station, Houghton, Hunt-
ingdon, England); Rennie, M. *Vet. Rec.* 96(20):
454-455; 1975.

3382   ROLE OF T CELL IMMUNOGLOBULIN IN CELL CO-
       OPERATION, T-CELL SUPPRESSION, AND ANTI-
GENIC COMPETITION. (Eng.) Feldmann, M. (Dep.
Zool., Univ. Coll., London, England); Kontiainen,

3390 "NATURAL" KILLER CELLS IN THE MOUSE. I.
CYTOTOXIC CELLS WITH SPECIFICITY FOR
MOUSE MOLONEY LEUKEMIA CELLS. SPECIFICITY AND
DISTRIBUTION ACCORDING TO GENOTYPE. (Eng.) Kies-
sling, R. (Karolinska Inst., Stockholm, Sweden);
Klein, E.; Wigzell, H. *Eur. J. Immunol.* 5(2):112-
117; 1975.

3391 "NATURAL" KILLER CELLS IN THE MOUSE. II.
CYTOTOXIC CELLS WITH SPECIFICITY FOR
MOUSE MOLONEY LEUKEMIA CELLS. CHARACTERISTICS OF
THE KILLER CELL. (Eng.) Kiessling, R. (Karolinska
Inst., Stockholm, Sweden); Klein, E.; Pross, H.;
Wigzell, H. *Eur. J. Immunol.* 5(2):117-121; 1975.

3392 IDENTIFICATION OF MACROPHAGE-LIKE CHAR-
ACTERISTICS IN A CULTURED MURINE TUMOR
LINE. (Eng.) Koren, H. S. (Natl. Cancer Inst.,
Bethesda, Md.); Handwerger, B. S.; Wunderlich, J.
R. *J. Immunol.* 114(2/Part 2):894-897; 1975.

3393 INTERACTION OF *LENS CULINARIS* LECTIN,
CONCANAVALIN A, *RICINUS COMMUNIS* AGGLUTININ
AND WHEAT GERM AGGLUTININ WITH THE CELL SURFACE OF
NORMAL AND TRANSFORMED RAT LIVER CELLS. (Eng.)
Roth, J. (Inst. Pathology, Friedrich-Schiller-Univ.,
DDR--69 Jena, Ziegelmuhlenweg 1, East Germany);
Neupert, G.; Thoss, K. *Exp. Pathol. (Jena)* 10(5/6):
309-317; 1975.

3394 MODULATION OF LYMPHOCYTE RECEPTOR MOBILITY
BY CONCANAVALIN A AND COLCHICINE. (Eng.)
Yahara, I. (Rockefeller Univ., New York, N.Y. 10021);
Edelman, G. M. *Ann. N.Y. Acad. Sci.* 253:455-469;
1975.

3395 A LECTIN MODEL FOR UNRESPONSIVENESS. I.
REVERSIBLE EFFECTS OF HIGH DOSES OF CON-
CANAVALIN A ON SPONTANEOUS AND INDUCED DNA SYNTHE-
SIS. (Eng.) Chauvenet, A. R. (Duke Univ. Med.
Cent., Durham, N.C.); Scott, D. W. *J. Immunol.*
114(1/Part 2):470-475; 1975.

See also:

* (Rev):   3004, 3055, 3056, 3057, 3058, 3059,
           3060, 3061
* (Chem): 3105, 3123, 3140
* (Phys): 3181
* (Viral): 3208, 3211, 3226, 3255, 3266, 3271,
           3272, 3290, 3292
* (Path): 3406, 3407, 3418, 3422, 3425, 3447,
           3450, 3481

3396   HYPERPLASIA OF RAT MAMMARY GLAND *IN VITRO*.
         (Eng.) Palekar, L. (New York Univ. Med.
Cent., N.Y.): Waldo, E. D.; Seidman, I. *J. Natl.*
Cancer Inst. 54(1):235-240; 1975.

Tests were carried out to determine if the intraductal hyperplasia of rat mammary explants in the
presence of insulin is caused by proliferation of
epithelial or of myoepithelial cells, or if sex
hormones affect the process. Mammary explants
were cultured at 37 C, 90% O_2, and 10% CO_2, with
5 µg/ml Sigma bovine insulin and routine additions
of penicillin (100 U/ml) and streptomycin (100 µg/
ml). The cultures were terminated at 1, 2, 3, and
4 wk, with maximum hyperplasia occurring at about
two weeks. The hyperplasia exhibited trabecular,
papillary, and, rarely, concentric morphologic
patterns. Electron microscopic examination of 20
explants for each time period showed the proliferating cell to be epithelial, not myoepithelial. Estrogen at 0.01 µg/ml showed no inhibitory effect
on insulin-induced hyperplasia, thus contradicting
previous reports; testosterone at 6 µg/ml inhibited
the hyperplasia by about half, but only when it was
added at the beginning of the culture period.

3397   STEWART-TREVES SYNDROME. (Ger.) Zschoch,
         H. (Pathologisches Inst., DDR-18 Branden-
burg 12, Anton-Saefkow-Allee 2, East Germany). *Zen-*
tralbl. Chir. 100(14):868-872; 1975.

Stewart-Treves syndrome was observed in a 59-yr-old
woman who had been mastectomized for carcinoma
cribrosum of the right breast 11 yr earlier. She
developed a therapy-resistant lymphedema of the
right arm six months after mastectomy, and lymphangiosarcoma nearly 11 yr after mastectomy. A
statistical analysis of 114 known cases showed that
chronic lymphedema is the only pathogenetic factor
involved in Stewart-Treves syndrome, while neither
sex, nor the course and therapy of the original
carcinoma, nor the age at the manifestation of the
original carcinoma or of the lymphedema play any
pathogenetic role. Most patients affected by the
Stewart-Treves syndrome are cured breast cancer
patients.

3398   LYMPHOSARCOMA OF THE BREAST. (Eng.)
         Ross, C. F. (St. Peter's Hosp., Chertsey,
Surrey, England); Eley*, A. *Br. J. Surg.* 62(8):
651-652; 1975.

A 68-yr-old woman presenting with a lump in the
right breast underwent a simple mastectomy for a
tumor thought to be a medullary carcinoma but which
microscopically proved to be a well-differentiated
lymphosarcoma. The lesion was considered primary.
However, the patient subsequently developed a subcutaneous lump over the left hip followed by an
enlarged lymph node in the left groin. The hip
tumor closely resembled the breast tumor, while
the lymph node tumor resembled reticulum cell sarcoma. The patient responded to chlorambucil but
eventually died of bronchopneumonia after developing
gross hepatomegaly. Autopsy revealed an undifferentiated malignant lymphoma in the liver and spleen
but no deposits elsewhere. Histologically, the

production. These results suggest that increased
intracellular cyclic AMP concentrations may be
related to the observed change of flattened adrenal
cells to a spherical shape in monolayer culture.

3403   MIXED MESENCHYMAL DIFFERENTIATION IN
        MENINGIOMAS. (Eng.) Ball, J. (Royal
Infirmary, Sheffield S6 3DA, England); Cook, T. A.;
Lynch, P. G.; Timperley*, W. R. J. Pathol. 116(3):
253-258; 1975.

Case histories of five patients (two men and three
women, aged 16-55 yr) with intracranial meningeal
tumors showing mixed mesenchymal differentiation are
presented. Three patients survived surgical removal
of the tumor and showed no signs of recurrence
after at least one year; the other two patients
died postoperatively. Three of the intracranial
tumors contained cartilage; three contained bone;
four contained hyaline fibrous strands or nodules,
which were calcified in two instances; three contained angiomatous areas; one contained pericytoma-
like areas; four contained pleomorphic and sometimes multinucleate giant cells; and one contained
a mucoid matrix including spheroidal cells superficially resembling chordoma. These tumors all
appeared to have arisen from the meninges and had
the properties attributed to mesenchymal cells.
These tumors can be misdiagnosed as metastatic
deposits, especially in frozen section.

3404   HUMAN NERVOUS SYSTEM TUMORS: OBSERVA-
        TIONS BY HIGH VOLTAGE ELECTRON MICRO-
SCOPY. (Eng.) Lyser, K. M. (Hunter Coll., Box
1030, 695 Park Ave., New York, N.Y. 10021).
Acta Neuropathol. (Berl.) 32(4):313-324; 1975.

Thick sections (0.5-2μ) of biopsies from human
nervous system tumors (Schwannoma, ependymoma,
medulloblastoma), fixed in aldehydes followed by
osmium, and stained with uranyl acetate and lead,
were studied at 2.5 million V. They were then
compared to thin sections of the same material
observed by ordinary low voltage electron microscopy. High voltage electron microscopy permits
direct observation of cell fine structure in three
dimensions, including the spatial relations of
organelles. Details of contours of nuclear envelopes, shapes of mitochondria, fine aspects of the
structure of cell surfaces and processes, such as
the flat sheet-like and irregular cylindrical processes of Schwannoma cells, the small projections
and ridges on their surfaces, and microvilli and
cilia of ependymoma cells, and other features have
been observed. These initial observations demonstrate the applicability of high voltage electron
microscopy to further study of neural neoplasms.

3405   MALIGNANT MESENCHYMOMA: CASE REPORT WITH
        ELECTRON MICROSCOPIC STUDY. (Eng.)
Klima, M. (Veterans Administration Hosp., 2002
Holcombe Blvd., Houston, Tex. 77031); Smith, M.;
Spjut, H. J.; Root, E. N. Cancer 36(3):1086-1094;
1975.

A malignant mesenchymoma of the thigh was studied

by light and electron microscopy. A 65-yr-old
Caucasian man had experienced thrombophlebitis and
chronic venous stasis, and presented with a mass
in the posterior part of the left thigh. Otherwise,
physical and laboratory examinations were within
normal limits. Surgical excision exposed a well-
circumscribed mass located within the upper parts
of the hamstring muscles; the upper tumor pole was
near the ischial tuberosity, while the pseudocapsule
was in contact with the sciatic nerve. The tumor
was excised, and frozen section examination yielded
a diagnosis of sarcoma. Despite postoperative
radiotherapy and chemotherapy, multiple bilateral
pulmonary metastases, hemorrhagic pleural effusion,
and death occurred. Grossly, the neoplasm appeared
pseudoencapsulated and generally mucoid. Histologically, there were a sparse stroma, focal necrosis,
and many transitional areas from cellular leiomyo-
sarcomatous to bizarre myxomatous. The gelatinous
areas were varied, generally displaying a large,
irregular cell with a large, dark, bizarre nucleus.
Electron microscopy revealed a prevalent cell un-
characteristic of a single cell type and no basic
ultrastructural differences between the leiomyo-
sarcomatous and the gelatinous areas. The three
most differentiated cell types were large ovoid
cells, slender cells with occasional fat droplets,
and multinucleated giant cells. However, most
cells showed transitional patterns, irregular nuclei,
and signs of degeneration. While light microscopy
favored a diagnosis of leiomyosarcoma or liposarcoma,
electron microscopy revealed most cells similar to
osteoblasts and chondroblasts. Prognosis was considered poor, and the name mesenchymoma considered
proper.

3406   MULTIPLE KERATOACANTHOMAS: ASSOCIATION
        WITH DEFICIENT CELL MEDIATED IMMUNITY.
(Eng.) Claudy, A. (Hopital Edouard Herriot, Lyon,
France); Thivolet*, J. *Br. J. Dermatol.* 93(5):
593-595; 1975.

A case of multiple keratoacanthomas associated with
deficient cell mediated immunity in a 44-yr-old
male Caucasian is presented. Many attempts to
restore the lymphocyte function (transfer factor,
levamisole) failed but treatments may have had a
preventive effect on the development of new lesions.
Very few studies of immune defense mechanisms
against keratoacanthoma have been made. Keratoacanthoma should remain isolated and heal spontaneously
if the cellular immune responses are efficient.
This case shows an alteration of *in vivo* and *in
vitro* cellular immune responses without any clinical
evidence of infection or of other tumors. It appears that cell mediated immunity may play a major
role in the defense against keratoacanthoma. Its
deficiency may explain the severity of the evolution of the lesions and their recurrences.

3407   FAMILY STUDIES OF MALIGNANT MELANOMA
        AND ASSOCIATED CANCER. (Eng.) Lynch,
H. T. (Creighton Univ. Sch. Medicine, Omaha, Nebr.);
Frichot, B. C.; Lynch, P.; Lynch, J.; Guirgis, H.
A. *Surg. Gynecol. Obstet.* 141(4):517-522; 1975.

Five cancer-prone families are presented in whom

nizable virus-like particles occurs rarely in these
tumors, due to the high shedding rate.

3411   MULTIPLE PRIMARY CARCINOMAS OF RESPIRATORY
        TRACT: PRIMARY CARCINOMA OF LARYNX FOL-
LOWED BY PRIMARY CARCINOMAS INVOLVING TWO LUNGS
CONSECUTIVELY. (Eng.) Sakula, A. (Redhill General
Hosp., Surrey, England). *Br. J. Dis. Chest* 68:128-
136; 1974.

A case report is presented of a heavy cigarette
smoker who developed multiple primary carcinoma of
the respiratory tract. A 53-yr old man developed
a primary, well-differentiated, squamous-cell carcinoma of the larynx which was successfully treated
by total laryngectomy and block dissection of the
involved cervical lymph nodes. He developed a primary oat-cell carcinoma with areas of squamoid
appearance in the left lower lobe bronchus 2 1/2 yr
later. This condition was successfully treated by
lobectomy and radiotherapy. The patient remained
well for the next ten years before developing
another primary carcinoma of oat-cell type in the
right main bronchus and right upper lobe that
infiltrated the outer wall of the right pulmonary
vein. At autopsy there was no evidence of malignant
disease in any other organ. This is the first
reported example of a primary carcinoma of the
larynx followed by independent development of
primary carcinomas first in one lung and then in
the other. The time intervals between the three
primary carcinomas and their different histological
patterns and the autopsy findings did not suggest
any connection between them. The family history
revealed several cases of primary lung cancer, suggesting that the patient might be prone to cancer.
It is suggested, however, that the more likely
explanation is the exposure of the larynx and bronchi to a common carcinogenic agent, such as cigarette
smoke. This explanation is strengthened by the observation that all reported cases of multiple cancer
of the respiratory tract have occurred in 50-70-yr-
old men who were heavy cigarette smokers.

3412   ON PLEURAL MESOTHELIOMAS. (Rus.) Bogush,
L. K. (Moscow City Hosp. Tuberculosis, No. 7, Moscow,
USSR); Zharakhovich, I. A.; Kubrik, N. E.; Rasshivalov,
S. K.; Ermolaev, K. K.; Shmelev, M. M.; Lavrent'eva,
N. B. *Vopr. Onkol.* 47(6):3-8; 1975.

Pleural mesotheliomas were studied in 19 patients
(ten men and nine women). Their ages were 21-60 yr,
with five older than 40 yr. In 13 patients, the
diagnosis was established by means of the data gained
with biopsy and surgical procedures; in the remaining six, the pathological condition was found at
autopsy. Based on the clinical-radiological picture,
the patients were divided into three groups: (1)
four patients without exudative accumulation during
the development of the mesothelioma; (2) seven patients with exudative accumulation in the pleural
cavity; and (3) eight subjects with the tumor developing at the site of a pneumothorax which had been
induced years before. In patients in group 1, the
pathological process was revealed by fluorography.
The subjects of group 2 had been originally treated

unsuccessfully for tuberculosis or cardiac decompen-
sation; the first symptom of the diease was the
occurrence of serous exudate determined by needle
biopsy, with a sensation of pressure in the affected
area. In group 3, pneumothorax was applied to treat
infiltrative and focal tuberculosis. Subsequently,
the mesothelioma developed over a period of 13-24 yr;
this was based on gradual exacerbation of postpneumo-
thoracic pleural empyema, for which the patients had
been treated. Aspiration of the empyema showed the
presence of serous-purulent or hemorrhagic exudate.
X-rays revealed encapsulation of the empyemic cavity.
The prognosis of mesothelioma is poor, with the diffuse variety of the disease ending in death only
months after surgery. In some cases, postsurgical
life expectancy of the nodular type of the disease
reaches 5-9 yr.

3413   HISTOLOGIC TYPES AND POSSIBLE INITIAL
          STAGES IN EARLY GASTRIC CARCINOMA. (Eng.)
Grundmann, E. (Pathologisches Institut der Universitat, D-44 Munster i. W., Westring 17 West Germany). *Beitr. Pathol.* 154(3):256-280; 1975.

A study was undertaken to determine which zones of
the human stomach may be liable to carcinomatous
degeneration. Among 2003 gastric specimens from
301 patients, a diagnosis of carcinoma was made in
45 cases. Examination of resection preparations
revealed 36 cases of deep invasive stomach carcinoma and nine cases of early gastric cancer confined to mucosa and/or submucosa. Carcinomatous
proliferations limited to mucosa or submucosa were
classified in three histologic types: intestinal
(adeno), mucocellular (signet ring cell), and anaplastic (solid) type of early gastric cancer.
Mixed types were found, which combined the first
and the second, or the second and third types.
One case presented a mixture of all three types.
Possible precursor or initial stages of all types
were found in 31 biopsies. Some of these were
glandular lesions in superfical parts of the mucosa. "Signet ring cell drippings" from lower
parts of tubule necks were recorded as an initial
form of the signet ring cell type. The process is
interpreted as a detaching of isolated signet ring
cells from a gland neck zone in progressing atypical transformation. An early neoplastic stage of
the anaplastic (solid) type of early gastric cancer
was identified in the "gland neck dysplasis" located exclusively in the antrum between surface
mucosa and antrum glands. This lesion was rich in
cells and stretched like a broad ribbon. These
results confirm the suggestion that the region of
gland necks must be a critical zone of transformation and as such, they must be the decisive point
of origin of early gastric cancers.

3414   PREMALIGNANCY OF THE MUCOSAL POLYP IN
          THE LARGE INTESTINE: II. ESTIMATION
OF THE PERIODS REQUIRED FOR MALIGNANT TRANSFORMA-
TION OF MUCOSAL POLYPS. (Eng.) Kozuka, S. (Nagoya
Univ. Sch. Medicine, 65 Tsurumaicho, Showaku,
Nagoya, 466, Japan); Nogaki, M.; Ozeki, T.; Masu-
mori, S. *Dis. Colon Rectum* 18(6):494-499; 1975.

In reply to a letter questioning his identification
of Reed-Sternberg (RS) cells by scanning electron
microscopy, the writer reviews the reasons for apparent differences in the surface of the RS cells described and those tentatively identified as Hodgkin's
cells. The differences are ascribable to differences
in temperature, pH or culture substrate; methods of
fixation and preparation of specimens for scanning
electron microscopy; and stages of activation of the
cell. The macrophage origin of the RS cells was suggested by the appearance of lamellae and particles
on the surface of the cells within 10-20 min after
antigen stimulation and by changes in surface ultrastructure during T cell-mediated cytolysis. Further
support for a macrophage origin came from the observations of (a) lymphocytes within phagocytic vacuoles
of RS cells with phase microscopy, and (b) lymphocytes
partially engulfed by RS cells with scanning electron
microscopy. This finding might explain the impaired
cellular immunity and lymphopenia seen in patients
with advanced disease.

3419   ONCORNAVIRUS-LIKE PARTICLES FROM CULTURED
        BONE MARROW CELLS PRECEDING LEUKEMIA AND
MALIGNANT HISTIOCYTOSIS. (Eng.) Vosika, G. J. (Univ.
Minnesota Health Sciences Center, Minneapolis, Minn.
55455); Krivit, W.; Gerrard, J. M.; Coccia, P. F.;
Nesbit, M. E.; Coalson, J. J.; Kennedy, B. J. *Proc.
Natl. Acad. Sci. USA* 72(7):2804-2808; 1975.

Two patients were studied in whom preleukemia was
among the diagnostic considerations. Bone marrow
from these patients was cultivated for one wk in
conditioned media with dexamethasone, then a high-
speed pellet of the supernatant fluid and disrupted
cells was prepared and analyzed on a sucrose gradient for enzymatic activity characteristic of RNA-
related DNA polymerase (reverse transcriptase).
Peaks of endogenous DNA polymerase activity showed
ribonuclease sensitivity and/or stimulation with
the synthetic template poly(rC)-(dG)$_{12-18}$. This
was demonstrated in both patients at densities of
1.15 to 1.19 and 1.21 to 1.24 g/ml. Diagnosis two
and four months after initial evaluation revealed
acute myelogenous leukemia and malignant histiocytosis, respectively. Prior studies had suggested
a possible etiological significance of such particles
in human leukemia. The demonstration of similar
particles preceding clinically overt disease in
these patients supports this hypothesis and offers
the possibility of early diagnosis and treatment.

3420   ABNORMAL HEMOGLOBIN SYNTHESIS IN SOME
        LEUKEMIC PATIENTS. (Eng.) Pagnier, J.
(Institut de Pathologie Moleculaire, Groupe U 15,
de l'Institut National de la Sante et de la Recherche
Medicale, Laboratoire associe au Centre National de
la Recherche Scientifique, 24, rue du Faubourg Saint-
Jacques, 75014 Paris, France); Labie, D. *Biochimie*
57(1):71-76; 1975.

The hemoglobin chain synthesis during leukemic processes was studied in patients (20-50 ml blood from
each patient) with fetal hemoglobin ranging from
5-20%. [^{14}C]leucine (40-50 mCi/mM) or [^{3}H]leucine or
[^{3}H]lysine (15-40 Ci/mM) served as markers. Regular
blood donors with discrete reticulocytosis who had

not been bled in three months served as controls.
Radioactive precursor incorporation varied with the
reticulocyte count of the sample. After 30-min incubations, the $\beta^A/\alpha A$ specific activity ratios ranged
from 1.25-1.55 for the patients compared to 1.08 for
controls. The decrease in the ratio observed for
the patients after two hours of incubation was caused
by a decrease in chain synthesis and stable α synthesis rate. Specific activity ratios for γ^F/α^F showed
a balanced synthesis. The α chain specific activity
was identical in fetal and adult hemoglobin. In all
cases but one, the total radioactivity of non-α
chains was greater than that of α chains. The precursors were identified by Sephadex-G100 chromatography; the radioactivity incorporated into the di-
meric peak ranged from 2.7-8.7% for the patients, in
contrast with 1-2% for the controls. In this peak,
the labeling was mainly located in the α chain, and,
in four cases, it was all in the α chain. In the
controls, radioactivity was found in both α and β
chains with a ratio of 0.35. These results suggest
a β-thalassemic pattern, in which free α chains are
associated with defective chain synthesis. The
clonal disorder hypothesis cannot be ruled out.

3421   ULTRASTRUCTURE OF ACUTE MYELOID LEUKEMIA
        ARISING IN MULTIPLE MYELOMA. (Eng.) Skin-
nider, L. F. (Dept. Pathol., Univ. Saskatchewan, Saskatoon, Canada); Ghadially, F. N. *Hum. Pathol.* 6(3):
379-384; 1975.

Electron microscopy of bone marrow from a case of
multiple myeloma in which acute myeloid leukemia
supervened four years after diagnosis showed two distinct populations of neoplastic cells. One population was plasmacytoid, the other myeloid. No intermediate cell types were found. The patient (a 73-yr-
old man) was treated first with melphalan 2 mg/day,
which was discontinued if the WBC count fell below
4000/mm^3. Therapy with melphalan discontinued after
diagnosis of leukemia. It appears that the acute
leukemia in this patient was a distinctly new neoplasm arising from the myeloid series of cells. Since
in nearly all previously reported cases, as well as
this patient, alkylating agents had been administered,
this may have been a factor in the development of
acute myeloid leukemia.

3422   CYTOGENIC STUDIES IN MYELOMATOSIS. (Eng.)
        Jensen, M. K. (Division of Haematology,
Department of Medicine B, Aalborg Sygehus Syd, Aalborg,
Denmark); Eriksen, J.; Djernes, B. W. *Scand. J. Hae-
matol.* 14(3):201-209; 1975.

The frequency and types of chromosome abnormalities
were studied in bone marrow aspirates from 25 un-
selected patients with myelomatosis. Their ages
ranged from 29-77 yr; 17 were men. Sixteen patients
were studied before treatment; the remainder had
recieved therapy with either melphalan or cyclophosphamide. Mitoses were classified as plasmocytic only
if the mitotic cells corresponded in size and tinc-
torial characteristics to plasma or myeloma cells.
Three types of abnormalities of the chromosome complement were seen: abnormal stem lines, structural
aberations, abnormal appearance of the chromatin.
Abnormal stem lines were present in seven patients;

specimens were analyzed. Etiologically, numerous
theories have been advanced. Leukoplakia may be
a metaplastic reaction of normal epithelium to
noxious stimuli. Chronic infection probably
serves as the major etiologic factor. Avivi-
minosis A, tuberculosis, syphilis, hormonal changes,
and schistosomiasis have also been associated.
Grossly, leukoplakia is grayish-white and lesions
are described as having a course textural appearance. It may be diffuse or localized and may have
sharp or indistinct borders. The plaques are
free of vessels yet the underlying mucosa is
fryable. Microscopically, lesions have a keratin-
ized squamous epithelial appearance. In the current
study 18 patients are women and 26 are men and
generally a 4:1 ratio is found for lower tract
involvement and an equal ratio for upper tract
involvement. The right kidney pelvis is more frequently involved. Age of susceptibility ranges
from the third to the eighth decade. Radiologi-
cally, the.pyelogram can be normal or show signs
of recurrent infection. Renal washings may show
desquamated cells and patients with upper tract
involvement may note the passage of tissue upon
voiding. Arteriography is nonspecific. Differential diagnosis includes encrusted alkaline cystitis, malacoplakia, or yeast infection. Ten to
20% of cases are associated with carcinoma at the
time of diagnosis. In the current series, five
squamous cell carcinomas are associated with 36
leukoplakias of the bladder, 2 of 7 in the renal
pelvis and 1 of 3 in the urethra. Transitional
cell carcinoma, adenocarcinoma and fibrosarcoma
may also be found. Treatments are many and varied
but no one treatment seems to be effective.
Radical surgery is indicated if symptoms are
severe and may also remove the malignant potential.
Current therapeutic recommendations are to follow
the patient as if he had a low grade malignancy.

3427   SERUM LACTATE DEHYDROGENASE ACTIVITY
         IN PATIENTS WITH TROPHOBLASTIC NEO-
PLASMS. (Eng.) Sekiya, S. (Chiba Univ. Sch.
Med., Japan); Kikuchi, Y.; Kudoh, J.; Takamizawa,
H. *Am. J. Obstet. Gynecol.* 121(8):1119-1120;
1975.

The activity of serum lactate dehydrogenase (LDH)
in twenty-two patients diagnosed as having
histological gestational trophoblastic neoplasms
was investigated to determine its possible use as
an aid in clinical diagnosis. Samples of venous
blood were examined before and after treatment.
LDH values were determined by the method of sequential multiple analysis. In all patients, serum
transaminase and human chronic gonadotropin (HCG)
levels were normal. Total serum LDH activity in
patients with choriocarcinoma was higher (488 $\pm$
83 IU)/1 than levels found in hydatidiform mole
(241 $\pm$ 25 IU) or in chorioadenoma destruens
(161 $\pm$ 7 IU). Serum LDH values in patients with
hydatidiform mole or chorioadenoma destruens did
not show a good correlation with that of urinary
HCG. Facts presented in this investigation suggest the use of LDH values as an aid in the diagnosis of choriocarcinoma and chorioadenoma destruens before treatments are started.

3428   SIGNIFICANCE OF PROLIFERATIVE EPITHELIAL
        LESIONS OF THE UTERINE TUBE. (Eng.)
Moore, S. W. (Hosp. Univ. Pennsylvania, Philadelphia); Enterline, H. T. *Obstet. Gynecol.* 45(4):
385-390; 1975.

In order to assess the incidence and significance
of proliferative epithelial lesions of the oviduct,
entire oviducts from 124 hysterectomies were examined. Twenty-three cases (18.5%) exhibited focal tubal epithelial proliferation in at least one
of an average of 10.5 examined sections. The lesions showed the following characteristics: (1) an
increase in the number of nuclei, which in some
cases was the only abnormality, (2) stratification
of nuclei in a usually thickened mucosa, and (3)
less polarity accompanied by nuclear atypia.
Changes seen most often included nuclear enlargement, hyperchromasia, irregularities in shape, in-
dentations in the nuclear membranes, and uneven
chromatin distribution. A very few mitoses were
seen occurring in sections showing proliferative
lesions; two mitoses occurred in a case of severe
acute and chronic salpingitis. Papillary forma-
tions of various descriptions were found in nine
cases. Salpingitis was present in 11 cases;
where it was mild to moderate, the epithelial alterations were not qualitatively or quantitatively
different from cases in which it was not found.
An additional epithelial abnormality consisted of
an increase in cytoplasmic volume, with a granular
bright acidophilic staining. The frequency of
proliferative epithelial lesions when actively
sought and the rarity of invasive carcinoma of the
oviduct make it unlikely that the lesions represent noninvasive carcinoma. There is no justifi-
cation for diagnosing and treating these lesions
as carcinoma *in situ*.

3429   VAGINAL ADENOSIS: ANALYSIS OF 325
        BIOPSY SPECIMENS FROM 100 PATIENTS.
(Eng.) Antonioli, D. A. (Beth Israel Hosp.,
330 Brookline Ave., Boston, Mass. 02215); Burke,
L. *Am. J. Clin. Pathol.* 64(5):625-638; 1975.

The histologic features of vaginal adenosis and
its malignant potential were assessed based on
a study of 339 vaginal and 30 exocervical biopsy
specimens from 100 women aged 12-28 yr. In 77%
of the cases, the mothers of these patients had
taken diethylstilbestrol (DES) during the pregnancy; 7% of mothers had no such history, and in
16% the history was unknown. For each biopsy
specimen showing adenosis, five histologic
features were evaluated: gland types; mucosal surface involvement by adenosis; squamous metaplasia
of adenosis; inflammation; and epithelial atypia.
Eighty-eight percent of the patients were positive
for adenosis, and approximately 87% of the patients
with positive DES-exposure histories had adenosis.
Of the 183 vaginal specimens in which the glands
could be evaluated, 66% showed a single type of
gland and the rest showed various combinations
of two or all three gland types. Mucosal involvement by adenosis as either columnar or metaplastic
epithelium was found in 60% of the biopsies, and
squamous metaplasia was present in almost 90% of

3433   INTRACISTERNAL A-PARTICLES IN TRANSPLAN-
        TABLE MURINE TERATOMAS. (Eng.) Spence,
A. M. (Stanford Univ. Sch. Medicine, Stanford, Calif.
94305); Vandenberg, S. R.; Herman, M. M. *Beitr.*
Pathol. 155(4):428-434; 1975.

Intracisternal A-particles observed by electron microscopy in transplantable murine teratomas, are reported. Teratoma OTT 2466 consisted of uniform
sheets of large, undifferentiated cells with occasional rosette and tubule formation. Many intracisternal A-particles, 80-100 nm in diameter were
seen and were composed of two concentric electron-
dense shells surrounding a 40 nm electron lucent
core. OTT 6050 demonstrated A-particles with the
same features as above in a few of the stem cells,
in the ecto- and endodermal layers of the embryoid
bodies, in the cells of primitive neural and respiratory epithelial cells, and in the cells of primitive neuroepithelial rosettes. In the tissues of
A/He and 129 strain female mice not receiving tumor
transplants, rare intracisternal A-particles were
found only in the thymus of the A/He mouse. In the
offspring of the cross between these strains and
in 129/Sv strain males, the particles were found
in the thymus and spleen only. No intracytoplasmic
A-, B-, or C-particles were seen in any of the above,
and immunofluorescence failed to demonstrate detectable levels of radiation leukemia virus antigen in
suspensions of OTT 2466 of OTT 6050. Immunodiffusion failed to reveal levels of group-specific antigens to the Moloney sarcoma virus-murine lukemia virus complex in either teratoma. The negative viral
immunoassays indicate the absence of C-particles in
each teratoma and the absence of cross reactivity
between A- and C-particles. These findings add the
OTT 2466 teratoma and the A/He mouse to the list of
experimental mouse strains and tumors that contain
intracisternal A-particles; they also confirm the
presence of these particles in the 129/Sv mouse
strain and in the OTT 6050 teratoma. The intracisternal A-particles are not confined to stem cells
or embryoid bodies of OTT 6050 and occur in other
cell types.

3434   MORPHOLOGICAL OBSERVATIONS OF DYSPLASTIC
        GLIOMAS HETEROTRANSPLANTED TO EXPERIMENTAL
ANIMALS. (Eng.) Gluszez, A. (Inst. Pathol., Med.
Acad. Lodz, Poland); Alwasiak, J.; Papierz, W.;
Lach, B. *Acta Neuropathol. (Berl.)* 31(1):21-28;
1975.

Morphological analysis of malignant dysplastic gliomas during consecutive transplantation in guinea
pigs was carried out. Of 15 transplanted gliomas,
successful tumor growth in the anterior chamber of
the eye was obtained in five cases. These tumors
were examined histologically and explanted *in vitro*.
Each tumor was transplanted in a group of 3-5 animals. When the tumor had grown to fill the entire
anterior chamber, it was retransplanted into other
groups. In three of the successful transplants,
only one passage was obtained. In the other cases,
2 and 4 subsequent passages were obtained. Filling
of the whole anterior chamber by the tumor took an

average of ten months. The tumors contained cells
with abundant cytoplasm and/or monstrous, atypical
cells which outnumbered smaller anaplastic cells.
The tumors showed regressive changes (e.g. necrosis)
and stromal proliferation. Where four successive
passages were obtained, there was an intensive proliferation of the small-cell components and growth
time was unusually short; filling of the chamber
took only four months. Small cells similarly showed
expansive growth while cells rich in cytoplasm showed
slow growth and degeneration in tissue culture. The
small vital potential of the cytoplasm-abundant
cells, regressive changes, and stromal proliferation
thus appear to inhibit survival of dysplastic gliomas. Further investigation of dysplastic changes
and whether or not they are of form of degeneration
which disappears during multiple passages is indicated.

3435   PLASMA MEMBRANE ALTERATION ASSOCIATED WITH
        MALIGNANT TRANSFORMATION IN CULTURE.
(Eng.) Gilula, N. B. (Rockefeller Univ. New York,
N.Y. 10021); Eger, R. R.; Rifkin, D. B. *Proc. Natl.
Acad. Sci. USA* 72(9):3594-3598; 1975.

The intramembrane organization of the plasma membranes
of nonmalignant cells in culture was compared by
freeze-fracturing with that of virally-transformed
malignant cells. No dramatic differences were in the
distribution of intramembrane particles in the plasma
membranes of these cells when the cells were examined
without fixation or with mild fixation (glutaraldehyde
treatment) prior to freezing. However, a redistribution of intramembrane particles into aggregates occurred in the membranes of nontransformed cells after
treatment with glycerol. The aggregation of particles
was extensive in normal chick embryo fibroblasts, and
less extensive in mouse 3T3 cells. The glycerol-
induced particle redistribution was not inhibited at
4C, but was inhibited by pretreatment with 2.5%
glutaraldehyde. A significant number of the cells
remained viable after the glycerol treatment, and
the process was reversible. Particle aggregation
did not appear to be related to either growth rate
or cell density. Transformed Rous sarcoma virus/
chick embryo fibroblasts and simian virus 40/3T3
cells had few particle aggregates after glycerol
treatment. The plasma membranes of chick embryo
fibroblasts transformed with a mutant of Rous sarcoma virus (TS-68) that is temperature sensitive for
transformation, had few particle aggregates when
grown at the permissive temperature (37 C). Extremely
prominent particle aggregates were present in the
plasma membranes of cells grown at the nonpermissive
temperature (41 C). These observations indicate that
there is an alteration in the plasma membrane associated with viral transformation which is related to a
glycerol-sensitive mechanism that controls the distribution of intramembrane particles.

3436   CONTRIBUTION TO THE KNOWLEDGE OF MYOSITIS
        OSSIFICANS LOCALISATA. (Cro.) Silobrcic,
F. (General Hosp., 58000 Split, SFR, Yugoslavia);
Posinkovic, B. *Libri Oncol.* 3(4):295-300; 1974.

3454   FAMILIAL FORM OF HIPPEL-LINDAU'S DISEASE.
(Fre.) Tridon, P. (8, avenue Boffrand,
5400 Nancy, France); Picart, L.; Vidailhet, M. *J.
Genet. Hum.* 23(Suppl.):238-239; 1975.

3455   DERMATOGLYPHICS AND NEOPLASTIC DISEASE
[abstract]. (Eng.) Bierman, H. R. (Inst.
Cancer and Blood Res., Beverly Hills, Calif.),
Holmes, B.; Winton, W. *Proc. Am. Assoc. Cancer Res.*
16:253; 1975.

3456   A COMPARISON OF CHROMOSOMAL RADIOSENSITIVI-
TIES OF SOMATIC CELLS OF MOUSE AND MAN.
(Eng.) Sasaki, M. S. (Tokyo Medical and Dental Univ.,
Yushina, Tokyo 113, Japan). *Mutat. Res.* 29(3):433-
448; 1975.

3457   CANCER OF THE GALLBLADDER. (Rus.) Ar-
tem'eva, N. N. (Faculty of Clinical Surgery, Leningrad Inst. Sanitation, Hygiene and Medicine, Leningrad, U.S.S.R.); Burygina, L. V. *Vestn.
Khir.* 115(7):35-40; 1975.

3458   CARCINOMA OF THE EXOCRINE PANCREAS: PATHO-
LOGIC ASPECTS. (Eng.) Kissane, J. M.
(Washington Univ. Sch. Medicine, St. Louis, Mo.).
J. Surg. Oncol. 7(2):167-174; 1975.

3459   PRIMARY LIVER CARCINOMA AS "ACUTE STOMACH"
[abstract]. (Dut.) Zuidema, P. J. (Amsterdam, Netherlands). *Ned. Tijdschr. Geneeskd.* 119
(37):1423-1424; 1975.

3460   STIMULATION OF INTRAHEPATIC MORRIS HEPA-
TOMA GROWTH BY DIVERSION OF PORTAL VENOUS
BLOOD FROM LIVER [abstract]. (Eng.) Reichle, F. A.
(Temple Univ. Health Sci. Cent., Philadelphia, Pa.);
Morris, H. P.; Reichle, R. M.; Rao, N. S. *Am. Proc.
Assoc. Cancer Res.* 16:192; 1975.

3461   HEPATIC ANGIOSARCOMA [abstract]. (Eng.)
McAllister, H. A., Jr. (Armed Forces Inst.
Pathology, Washington, D.C. 20306); Ishak, K. G.
Am. J. Clin. Pathol. 63(4):601; 1975.

3462   CARDIAC SARCOMA (OF ANITSCHKOW CELL TYPE?)
PROPAGATING ALONG THE PULMONARY ARTERY:
MORPHOLOGICAL AND ULTRASTRUCTURAL STUDY. (Fre.)
Chomette, G. (Service d'Anatomie Pathologique C.H.U.
Pitie-Salpetriere, 83, bd de l'Hopital, 75013 Paris,
France); Auriol, M.; Laborde, J.-P.; Guiraudon, G.;
Vaugier, G. *Arch. Anat. Pathol. (Paris)* 23(3):237-
240; 1975.

3463   CARDIAC GLOMUS TUMOR: LIGHT AND ELECTRON
MICROSCOPIC INVESTIGATIONS. (Ger.) Ries-
ner, K. (Pathologisches Institut der Universitaet
Hamburg, D-2000 Hamburg 20, Martinistrasse 52, West
Germany); Bocker, W. *Z. Krebsforsch.* 84(1):59-66;
1975.

3464   THE ULTRASTRUCTURE OF EXPERIMENTAL VIRUS-
        INDUCED HEMANGIOPERICYTOMA. (Rus.) Kolo-
miets, O. L. (L. A. Tarasevich State Scientific Res.
Inst. for Standardization and Control of Medical
Biological Preparations, Moscow, USSR). *Arkh. Patol.*
37(8):53-60; 1975.

3465   CYSTIC LYMPHANGIOMA: A RARE RETROPERITO-
        NEAL TUMOR. (Spa.) Zuluaga, A. (Servicio
de Patologia Quirurgica I. Facultad de Medicina de
Granada, Spain); Garcia Gil, J. M.; Cabrera, V.;
Ramirez, M. A.; Fernandez, J. A. *Arch. Esp. Urol.*
28(4):435-446; 1975.

3466   DETECTION OF A TISSUE-BOUND BLOOD GROUP
        SUBSTANCE IN ANGIOBLASTIC MENINGIOMAS AND
HEMANGIOBLASTOMAS OF THE CENTRAL NERVOUS SYSTEM [ab-
stract]. (Ger.) Jellinger, K. (Wien, Austria);
Denk, H. *Zentralbl. Allg. Pathol.* 119(3):228; 1975.

3467   MULTICENTRIC AND ISOLATED MULTIFOCAL GLIO-
        BLASTOMA MULTIFORME SIMULATING METASTATIC
DISEASE. (Eng.) Prather, J. L. (Tripler Army Med.
Cent., Honolulu, Hawaii); Long, J. M.; van Heertum,
R.; Hardman, J. *Br. J. Radiol.* 48(565):10-15; 1975.

3468   CHEMODECTOMA OF THE VAGUS NERVE: REPORT
        OF A CASE WITH ULTRASTRUCTURAL STUDY.
(Eng.) Fernandez, B. B. (Lutheran Gen. Hosp., Park
Ridge, Ill.); Hernandez, F. J.; Staley, C. J. *Can-
cer* 35(1):263-269; 1975.

3469   POSTERIOR FOSSA SYNDROME AS TERMINAL COM-
        PLICATION OF HISTIOCYTOSIS X. STUDY OF
C.N.S. LESIONS: ASSOCIATION OF P.V.P. THESAURIS-
MOSIS. (Fre.) Faivre, J. (Hopital Pontchaillou,
35011 Rennes, France); Pecker, J.; Ferrand, B. *Sem.
Hôp. Paris* 51(37/38):2229-2237; 1975.

3470   THE TRACHEO-BRONCHIAL CYLINDROMA (ADENOID
        CYSTIC CARCINOMA): AN ELECTRON MICROSCOPIC
STUDY. (Fre.) Verley, J.-M. (Laboratoire de Mi-
croscopie electronique, Hopital Broussais, 96, rue
Didot, 75014 Paris, France); Hollmann, K.-H. *Arch.
Anat. Pathol. (Paris)* 23(3):177-183; 1975.

3471   CYTOLOGY OF THE SOLITARY PAPILLOMA OF THE
        BRONCHUS. (Eng.) Roglic, M. (Hospital
for Chest Diseases, Zagreb, Yugoslavia); Jukic, S.;
Damjanov, I. *Acta Cytol. (Baltimore)* 19(1):11-13;
1975.

3472   THE RELATIONSHIP OF TUMOR VESSELS, TUMOR
        CELL CLUMPS AND PULMONARY METASTASES [ab-
stract]. (Eng.) Liotta, L. A. (St. Luke's Hosp.,
Cleveland, Ohio); Kleinerman, J.; Saidel, G. *Proc.
Am. Assoc. Cancer Res.* 16:173; 1975.

1, place du Parvis-Notre-Dame, 75181 Paris Cedex 04, France). *Arch. Anat. Pathol. (Paris)* 23(2):97-106; 1975.

3491   BURKITT'S LYMPHOMA. (Eng.) Sidhu, S. S. All India Inst. Medical Sciences, New Delhi); Sukhija, D. S.; Parkash, H. *Oral Surg.* 39(3): 463-468; 1975.

3492   STUDY OF THE ULTRASTRUCTURAL ASPECTS OF A NEPHROBLASTOMA. (Spa.) Virseda, J. M. P. (Hospital de Navar. Pamplona. Servicios de Anatomía Patológica y Urología); Aznar, A. I. *Arch. Esp. Urol.* 28(1):73-88; 1975.

3493   BENIGN POLYPS WITH PROSTATIC-TYPE EPITHE-LIUM OF THE URETHRA. (Eng.) Craig, J. R. (Univ. Michigan Medical Sch., 1335 E. Catherine St., Ann Arbor, Mich. 48104): Hart, W. R. *Am. J. Clin. Pathol.* 63(3):343-347; 1975.

3494   CENTRAL ADENOCARCINOMA OF THE MANDIBLE. (Eng.) Toth, B. B. (Univ. Texas Dental Branch, Houston); Byrne, R. P.; Hinds, E. C. *Oral Surg.* 39(3):436-446; 1975.

3495   CHANGING PATTERNS IN ENDOMETRIAL ADENO-CARCINOMA: A STUDY OF 291 CONSECUTIVE CASES AT A LARGE PRIVATE HOSPITAL, 1960-1973. (Eng.) Quint, B. C. (801 Broadway, Seattle, Wash. 98122). *Am. J. Obstet. Gynecol.* 122(4):498-501; 1975.

3496   MIXED MULLERIAN TUMORS OF THE UTERUS: UL-TRASTRUCTURAL STUDIES ON THE DIFFERENTIA-TION OF RHABDOMYOBLASTS. (Eng.) Bocker, W. (Pathologisches Institut der Universitat, D-2000 Hamburg 20, Martinistrasse 52, West Germany); Stegner, H.-E. *Virchows Arch. [Pathol. Anat.]* 365(4):337-349; 1975.

3497   INCIDENCE OF ADENOCARCINOMA OF THE CERVIX UTERI IN CYTOLOGICAL AND COLPOSCOPIC SCREENING. (Ita.) Pieroni, G. (Provincia di Firenze, Centro Malattie Sociali e Medicina Preventiva, Clinica Obstetrica e Ginecologica Univ.-Firenze); Cariaggi, P.; Ottaviano, M.; Marchionni, M.; Mincione, G.; Bartoloni St. Omer, F. *Minerva Ginecol.* 27(5):418-421; 1975.

3498   THE EXTENT OF THE ATYPICAL SQUAMOUS EPI-THELIUM OF THE CERVIX UTERI. (Ger.) Holzer, E. (Universitats-Frauenklinik, A-8036 Graz, Austria); Pickel, H. *Arch. Geschwulstforsch.* 45(1): 79-91; 1975.

3499   THE HISTOGENESIS OF MIXED CERVICAL CARCI-NOMAS. THE CONCEPT OF ENDOCERVICAL COLUM-NAR-CELL DYSPLASIA. (Eng.) Alva, J. (Health Sciences Centre, 700 William Avenue, Winnipeg, Manitoba, Canada R3E 0Z3); Lauchlan, S. C. *Am. J. Clin. Pathol.* 64(1):20-25; 1975.

3500   CLEAR-CELL ADENOCARCINOMA OF CERVIX AND
        VAGINA. A CLINICAL AND PATHOLOGIC REVIEW
OF 20 CASES [abstract]. (Eng.) Fidler, W. J. (Univ.
Michigan, Ann Arbor, Mich. 48104); Woodruff, J. M.;
Nordqvist, S. R. B. *Am. J. Clin. Pathol.* 63(4)603;
1975.

3501   MELANOMA OF THE VULVA. (Eng.) Karlen, J.
        R. (Roswell Park Mem. Inst., Buffalo, N.Y.);
Piver, M. S.; Barlow, J. J. *Obstet. Gynecol.* 45(2):
181-185; 1975.

3502   ULTRASTRUCTURE OF GONADOBLASTOMA. (Eng.
        Damjanov, I. (Univ. Conn. Health Cent.,
Farmington, Conn. 06032); Drobnjak, P.; Grizelj, V
Arch. Pathol. 99(1):25-31; 1975.

See also:

* (Rev):   3010, 3020, 3021, 3022, 3023, 3062
            3063, 3064, 3065, 3066, 3067, 3068
            3070
* (Chem):   3078, 3080, 3087, 3106, 3107, 311
* (Viral): 3222, 3226, 3250
* (Immun): 3308, 3334, 3360, 3363
* (Epid-Biom): 3507, 3513, 3517, 3525

the prevalence in Caucasian and black South Africans. These diseases all not only have their maximum prevalence in the most economically developed countries, but also their prevalence rises first in the upper socio-economic groups of the developing countries. All available information suggests that the prevalence of these diseases in the Western world has increased markedly during the present century, and that they were significantly less prevalent 40 yr ago in black than in Caucasian Americans. It is suggested that the over-refining of carbohydrate foods, which results in fiber-depleted diets, is an etiological factor that is consistent with available epidemiologic evidence on the diseases.

3505   TRENDS FOR TOTAL CANCER DEATHS IN MISSOURI
        AND THE UNITED STATES: 1950-1967. (Eng.)
Edmonds, L. E. (Environmental Health Surveillance Center, Univ. Missouri, Columbia); Marienfeld, C. J. Mo. Med. 72(8):499-501; 1975.

Cancer mortality trends in Missouri and in the U. S. from 1950-1967 were compared. White male death rates for all malignant neoplasms have increased both in Missouri and in the U. S. for nearly every 10-yr age group between the ages of 25-84. The exception is the 25- to 34-yr age group, in which the very low death rates have remained unchanged. In contrast, white female death rates since 1950 have decreased in every age group, both in Missouri and in the U. S. as a whole. The U. S. male death rates have increased by 1-24% in every age group, except those beyond age 34. The Missouri cancer death rates in these groups have increased from 18-33%. By 1967, the Missouri male death rates had become higher than the U. S. rate in every age group except the 65- to 74-yr group; the increase was due predominantly to deaths from respiratory cancer. White female death rates for both the U. S. and Missouri have declined by 6-21% in the various age groups, the smallest decline occurring in the 45- to 64-yr age group. Total cancer death rates are higher for women until about age 55, after which the male rates become higher and remain higher. This pattern change is attributed to the higher female reproductive cancer death rate in women under 55, and to the greatly increased respiratory cancer death rate among the older male population.

3506   TUMOR FREQUENCY IN AUTOPSY CASES. I:
        FREQUENCY AGE AND SEX DISTRIBUTION, PRI-
MARY LOCALIZATION. (Ger.) Rudiger, K. D. (Pathologisches Institut der Medizinischen Akademie Erfurt, 50 Erfurt, Nordhauser Strasse 74, East Germany); Guthert, H.; Wockel, W.; Hochheim, A. Zentrabl. Allg. Pathol. 119(3):194-207; 1975.

The frequency, age and sex distribution, and localization of primary malignant epithelial and nonepithelial tumors were studied with respect to 22,155 autopsies (12,212 men, 9,943 women) during the period 1950-1966. The overall number of malignant tumors was 7,533 (34.2%), of which 99 cases were eliminated because of the unknown location of the primary tumor. The remaining 7,442 cases (33.6%) were distributed among 3,987 men (32.6%) and 3,455 women (34.7%). The frequency difference was statistically signifi-

cant. Malignant epithelial tumors were found in
5,559 cases (25.3% of autopsies; 75.2% of malignant
tumors), including 2,945 cases in men, and 2,654
cases in women. Malignant nonepithelial tumors were
found in 1,843 cases (8.3% of autopsies, or 24.8% of
all malignant tumors), including 1,042 in men, and
801 in women. The frequency of epithelial tumors
in both sexes and in men separately was highest in
the age range 65-69 yr, and in the range of 60-64 yr
in women. The frequency of nonepithelial tumors was
highest between 60 and 64 yr in both sexes together
and in men separately, and between 55 and 59 yr in
women. In men, the organ-specific frequency decreased in the order: lung, stomach, hemopoietic-
lymphatic system, colon and CNS. The corresponding
order established for women was: reproductive organs (excluding mammae), stomach, colon, hemopoietic-
lymphatic system, and CNS.

3507   PHEOCHROMOCYTOMA IN BLACKS. (Eng.) Chung,
        E. B. (Howard Univ. Coll. Med., Washington,
D.C.); Pressoir, R. *J. Natl. Med. Assoc.* 67(1):22-
26; 1975.

During the period 1958 to 1973 (16 yr) only five
patients with pheochromocytomas were discovered at
the Howard University-Freedmen's Hospital. All of
the patients were black (two men and three women).
They ranged in age from 31 to 66 yr. The tumor was
located in the right adrenal in four cases and in
the left retroperitoneum in one. In two cases, the
tumor was successfully removed surgically and one of
these was diagnosed preoperatively. The remaining
three cases were found at autopsy. The incidence of
pheochromocytoma was 0.08% of autopsies and 0.003%
of routine surgicals which is comparable to that of
whites. In all but one patient, the tumor originated
in the right adrenal gland, the other was extra-ad-
renal. A history of hypertension was present in
four of five patients. Pheochromocytoma should be
considered in the diagnosis of every case of hypertension and appropriate screening tests made since
this form of hypertension is surgically curable. A
brief review of the literature is presented.

3508   MASS SCREENINGS FOR CERVICAL CANCER IN FIN-
        LAND 1963-71; ORGANIZATION, EXTENT, AND EPI-
DEMIOLOGICAL IMPLICATIONS. (Eng.) Hakama, M. (Finnish Cancer Regist., Helsinki); Joutsenlahti, U.;
Virtanen, A.; Rasanen-Virtanen, U. *Am. Clin. Res.*
7(2):101-111; 1975.

A mass screening program to help eradicate cervical
cancer was undertaken by the Finnish government with
the cooperation of individual municipalities. Up to
100,000 smears are taken annually from more than two
million women. About 400,000 smears were taken during the period 1963-71; the detection rates per 10^5
were found to be 69 for invasive cancer, six for
microinvasive carcinoma, 168 for carcinoma *in situ*,
127 for dysplasia of high degree. Those previously
married, and those experiencing bleeding on sexual
intercourse, formed high risk groups (relative risks
2 and 6) whereas those previously electro-coagulated
had a relative risk of 0.13 only. Probably no changes
will occur in the time trends of mortality rates as
a result of the screening. The incidence of invasive

had undergone hysterectomy compared to 4% of the cancer patients. Significantly greater numbers of cancer patients gave histories of earlier breast trauma. However, these patients may be unusually likely to recall or exaggerate breast traumas.

3513   ESOPHAGEAL CANCER IN ISRAEL: SELECTED
        CLINICAL AND EPIDEMIOLOGICAL ASPECTS.
(Eng.) Shani, M. (Chaim Sheba Medical Center, Tel Hashomer, Israel); Modan*, B. *Am. J. Dig. Dis.* 20(10):951-954; 1975.

Clinical and epidemiological aspects of esophageal cancer, based on a nationwide study in Israel, are reported. Records of patients diagnosed as having carcinoma of the esophagus between 1960 and 1966 were reviewed. Three hundred and forty-six newly diagnosed cases yielded a mean incidence of 2.3/100,000. The middle third of the esophagus was the most common site (46%). Seventy-five percent of the tumors were squamous cell carcinoma, 17% were adeno-carcinoma, and 4.5% were anaplastic carcinoma. Mean age was 62.3 yr at diagnosis, and the male/female ratio was 1.6:1. Incidence was highest among the Asian-born group (Iran, Turkey, and Yemen) under the age of 60 yr. The most common symptom was dysphagia (83%). Endoscopic examination coupled with X-ray gave a positive diagnosis in 91.5% of the cases. The disease was localized in 37% of the patients, mediastinal in 40%, and there was generalized spread in 23%. Combined surgery and postoperative therapy yielded the best results, but only in a select group of patients. A high operative mortality of 42% was observed, with overall 5-yr survival at 5.8%. The fact that esophageal cancer incidence in persons under 60 yr is highest in Asian-born patients is emphasized. The lack of excess in the older patients in this ethnic group probably represents a lower rate of detection and referral than a recent introduction of a special carcinogen.

3514   THE EPIDEMIOLOGY OF ESOPHAGEAL CANCER IN
        NORTH CHINA AND PRELIMINARY RESULTS IN THE
INVESTIGATION OF ITS ETIOLOGICAL FACTORS. (Eng.)
The Co-ordinating Group for Res. on the Etiology of Esophageal Cancer of North China. *Sci. Sinica* 18(1): 131-148; 1975.

An epidemiological survey of esophageal carcinoma covering an area of 14 counties and cities was undertaken in North China. The subjects swallowed a balloon covered with netting to collect exfoliated cells for esophageal mucosa cytosmears. Statistical work on mortality rates was based on (1) roentgen examination, (2) cytological smear, (3) medical history, and (4) inquiry from relatives of deceased. In Linhsien County, a high prevalence of malignant tumors in gullets of poultry were found. Deaths due to cancer of the esophagus accounted for 20% of total deaths in Linhsien County. Analyses of nitrosamines, diethylnitrosamine, dimethylnitrosamine, methylphenylnitrosamine, secondary amines, nitrites and nitrates in food items from Linhsien County were made. *Geotrichum candidum* Link, a fungus found in local pickled vegetables, was studied to determine any relationship to the occurrence of esophageal cancer. The fungus provoked epithelial hyperplasia and papillary growths

in the forestomachs of mice and rats. Experiments
with Vitamin C and A were also carried out to determine any relationship between them and esophageal cancer. Vitamin C and A caused a blocking effect when
given to rats intubated with methylbenzylnitrosamine
or methylbenzylamine and sodium nitrite. Viruses and
genetics were not discarded as factors in development
of esophageal cancer by the authors. Experiments on
animals and clinical trials are still being carried
out to find effective methods of neutralizing or
inhibiting the carcinogenic properties discussed in
association with esophageal cancer.

3515   AN ADDITIONAL YOUNGER-AGE PEAK FOR CANCER
        OF THE NASOPHARYNX. (Eng.) Balakrishnan,
V. (Cancer Res. Inst., Tata Mem. Cent., Bombay, India). *Int. J. Cancer* 15(4):651-657; 1975.

An additional younger-age peak for cancer of the
nasopharynx in India is described. Analysis of the
case records of 666 patients with nasopharyngeal
carcinoma was performed. Of the patients, 535 were
male and 131 were female. Age distributions of each
sex showed a younger-age peak for males in the 13-22
yr age group. Age distributions according to religious and linguistic groups showed a younger-age
peak for Maharashtrian Hindus in the 18-22 yr range;
for Gujarati Hindus in the 13-17 yr range; for other
Hindus in the 13-2? yr range; and for non-Hindus in
the 13-17 yr range. When the group was divided according to the year of diagnosis, the peak for those
diagnosed from 1941-1965, and for those diagnosed
from 1965-1972, was 13-22 yr for both groups. This
study confirms earlier reports of the existance of
a younger-age peak for this disease. Because of the
occurrence of this peak at the time of transition
from adolescence to adulthood, it is suggested that
the endocrine changes of maturation may affect the
incidence of this type of cancer.

3516   LUNG CANCER AND AIR POLLUTION IN SOUTHCEN-
        TRAL LOS ANGELES COUNTY. (Eng.) Hender-
son, B. E. (Univ. Southern California, Sch. Medicine,
2025 Zonal Ave., Los Angeles, Calif. 90033); Gordon,
R. J.; Menck, H.; Soohoo, J.; Martin, S. P.; Pike, M.
C. *Am. J. Epidemiol.* 101(6):477-488; 1975.

The possibility that an excess in lung cancer was due
to local air pollution prompted an analysis of mortality figures from 1970. An increased rate of lung
cancer has been consistently observed from 1968-1972
among males in southcentral Los Angeles. This excess
risk occurred across several social classes and occupational categories. No differential excess of oral
cavity, pancreatic, laryngeal and bladder cancer was
observed in the same area, lessening the possibility
that regional variations in smoking habits accounted
for the excess lung cancer. For the entire county,
occupation was an important determinant of lung cancer risk, but within the limits of proportional incidence statistics, it did not explain the excess
male risk in the southcentral area. Air sampling indicated an excess of certain polynuclear aromatic
hydrocarbons (PAH) in southcentral Los Angeles. There
was a correlation between the geographic distribution
of lung cancer cases and the general location of

allow the calculation of reliable morbidity rates by site, sex and age; and a long period of collection and a good follow-up system. These features should be characteristic of a good cancer registry.

3521   EPIDEMIOLOGY OF CANCER IN THE INDIAN SUB CONTINENT. (Eng.) Jussawalla, D. J. (Tata Memorial Centre, Bombay, India); Gangadharan, P. *Indian J. Cancer* 12(1/Suppl.):1-7; 1975.

3522   THE GEOGRAPHICAL DISTRIBUTION OF LEUKEMIA. (Ger.) Mairose, U. B. (I. Medizinische Abteilung des Allgemeinen Krankenhauses Hamburg-Barmbek und I. Medizinische Universitätsklinik Hamburg-Eppendorf, West Germany)'; Dörken, H. *Med. Klin.* 70(2):53-58; 1975.

3523   DIFFERENCES IN DISTRIBUTION OF LEUKEMIAS IN THE CRACOW REGION IN THE YEARS 1961-1968.· (Pol.) Aleksandrowicz, J. (Kliniki Hematologicznej AM w Krakowie, Poland); Janicki, K.; Sliwczynska, B. *Folia Med. Cracov.* 17(1):3-10; . 1975. '

3524   BURKITT-LIKE LYMPHOMA IN CHILDREN [abstract]. . (Eng.) Kaschula, R. O. C. (Children's Tumour Clinic, Groote Schuur, Cape Town, South Africa); Bennett, ·M. B.; Karabus, C. D. *S Afr Med J* 49(26):1062-1063; 1975.

3525   PRIMARY CANCER OF THE LUNG--A COUNTRY-REPRESENTATIVE SERIES. (Swe.) Kallqvist, I. (Eskilstuna centrallasarett, Eskilstuna, Sweden). *Lakartidningen.*72(44):4291-4295; 1975.

·3526   TUMOR FREQUENCY IN AUTOPSY CASES. II. THE MALIGNANT TUMORS OF THE STOMACH. (Ger.) Rudiger, K. -D. (Pathologisches Insitut, Medizinschen Akademie, Erfurt, East Germany); Guthert, H.; Wockel, W.; Saalfeld, J. *Zentralbl. Allg. Pathol.* 119(4):303-315; 1975.

3527   COMPARATIVE CLINICAL-PATHOLOGICAL STUDIES ON PLASMOCYTOMA. (Ger.) Dominok, G. W. (Patholog. Institut Med. Akademie, 8019 Dresden, Fetscherstrasse 74, East Germany); Henker, J.; Heidel, G.; Herold, H. -J. *J. Gesamte Inn. Med.* 30(15):498-502; 1975.

3528   COHORT STUDIES: IN POPULATIONS COVERED BY A COMPREHENSIVE MEDICAL CARE SYSTEM.' (Eng.) Shapiro, S. (Johns Hopkins Univ., Baltimore, Md.). *Proc. Int. Cancer Congr. 11th.* Vol. 3 *(Cancer Epidemiology, Environmental Factors).* Florence, Italy, October 20-26, 1974. Edited by Bucalossi, P.; Veronesi, U.; Cascinelli, N. New York, American Elsevier, 1975, pp. 1-10.

3529   EPIDEMIOLOGICAL STUDIES INTO THE INCIDENCE
        AND LATENT PERIOD OF ASBESTOS-INDUCED
HYALINOSIS, FIBRINOSIS, AND MESOTHELIOMA. (Ger.)
Bittersohl, G. (Betriebspoliklinik Leuna, 422
Leuna, Breitscheidstrasse, East Germany). *Z.
Gesamte Hyg.* 21(5):369-371; 1975.

3530   SURVEILLANCE OF WORKERS WITH HISTORY OF
        EXPOSURE TO VINYL CHLORIDE. (Eng.) Per-
kel, G. (Vinyl Chloride Committee, Industrial
Union Dept., AFL-CIO, Washington, D.C.); Mazzocchi,
A.; Beliczky, L. *Ann. NY Acad. Sci.* 246:311-312;
1975.

3531   RECORD LINKAGE SYSTEM (WITH SPECIAL REFER-
        ENCE TO EPIDEMIOLOGY OF CANCER). (Eng.)
        Adelstein, A. M. (Office Population Cen-
suses and Surveys, Medical Div., London, England).
Proc. Int. Cancer Congr. 11th. Vol. 3 *(Cancer Epi-
demiology, Environmental Factors)*. Florence, Italy,
October 20-26, 1974. Edited by Bucalossi, P.;
Veronesi, U.; Cascinelli, N. New York, American
Elsevier, 1975, pp. 40-43.

3532   LINKAGE OF CENSUS AND DEATH RECORDS TO
        OBTAIN MORTALITY REGISTERS FOR EPIDEMIO-
LOGICAL STUDIES IN SWEDEN. (Eng.) Bolander, A.-M.
(Natl. Central Bureau Statistics, Stockholm, Swe-
den). *Proc. Int. Cancer Congr. 11th.* Vol. 3 *(Can-
cer Epidemiology, Environmental Factors)*. Florence,
Italy, October 20-26, 1974. Edited by Bucalossi,
P.; Veronesi, U.; Cascinelli, N. New York, Ameri-
can Elsevier, 1975, pp. 36-39.

A simple method for incubation of tissue specimens
(1 mm^3 or smaller) with thymidine methyl-^{3}H (^{3}HTdR)
under 3 atm oxygen tension is described. In two
spindle cell carcinomas of mouse salivary gland
derivation, a plasmacytoma in BALB/c mice, and a
mammary carcinoma in C3H mice, this *in vitro* method
gave a *in vitro* labeling index (LI) that corresponded closely with the LI determined by injection
of ^{3}HTdR *in vivo*. The LI of murine tumors was not
affected by storage at 22 C for up to 135 min. The
LI of the human carcinomas was constant for at least
75 min at 22 C. The mean LI for 39 invasive human
mammary carcinomas was 3.8, but the distribution was
skewed to the right, and the most frequently observed LI was between 2 and 3. Larger carcinomas
had significantly larger LI than smaller tumors,
which suggests that the larger tumors were faster-
growing at the time of observation than the smaller
tumors. The data suggest that a high LI is associated with more frequent metastasis to axillary
lymph nodes.

3539   THE PANCREATIC EPITHELIAL CELL *IN VITRO:*
        A POSSIBLE MODEL SYSTEM FOR STUDIES IN CAR-
CINOGENESIS. (Eng.) Hay, R. J. (Dept. of Biol. Sci.,
Wright State Univ., Dayton, Ohio 45431). *Cancer Res.*
35(8):2289-2291; 1975.

A technique is described for the isolation and maintenance in culture of pancreatic epithelial cells from
human embryos or neonatal guinea pigs. Minced tissue was dissociated by agitation (120 rpm) for 15 min
at 37 C in 0.125% crude trypsin and 0.5% collagenase.
Centrifugal fractionation showed that the preparations
from guinea pig pancreas were up to 95% acinar cells.
Dissociated cells from human material were uncharac-
terized. Suspensions of dissociated cells were incubated with gyration for 16 to 18 hr, and cell aggregates that formed were placed in stationary culture.
Two-dimensional colonial aggregates developed on the
plastic substratum. Cells comprising such colonies
formed junctional complexes and could be maintained
for 20 to 40 days *in vitro*. This system may provide
a useful model for studies on pancreatic exocrine
cell physiology and carcinogenesis.

3540   HIGH TERMINAL DEOXYNUCLEOTIDYL TRANSFERASE
        ACTIVITY IN A NEW T-CELL LINE (RPMI 8402)
OF ACUTE LYMPHOBLASTIC LEUKEMIA ORIGIN. (Eng.) Sri-
vastava, B. I. S. (Dep. Exp. Ther., Roswell Park Mem.
Inst., Buffalo, N.Y.); Minowada, J.; Moore, G. E.
J. Natl. Cancer Inst. 55(1):11-14; 1975.

The high activity of terminal transferase in a new
T-cell-dependent lymphocyte line (RPMI 8402) of
acute lymphoblastic leukemic origin is reported.
This enzyme resembled the terminal transferase
from other human cells in all its properties including K_m (0.7×10^{-6} M for dGTP). The high activity of this enzyme in RPMI 8402 and fresh acute leukemia lymphoblasts, in contrast to the low activity
of this enzyme reported for "thymus-independent"
cells, suggests that this cell line may have originated from leukemia cells. Moreover, the high activity of terminal transferase in RPMI 8402 cells
should make feasible large-scale purification of
this enzyme for detailed studies.

3541 . DIFFERENTIATION OF ERYTHROLEUKEMIC CELLS
IN VITRO: IRREVERSIBLE INDUCTION BY DI-
METHYL SULFOXIDE (DMSO). (Eng.) Preisler, H. D.
(Dept. of Medicine A, Roswell Park Memorial Insti-
tute, Buffalo, N. Y. 14203); Giladi, M. *J. Cell.
Physiol.* 85(3):537-546, 1975.

This study was made to determine the shortest exposure time to dimethyl sulfoxide (DMSO) needed for
the induction of differentiation of Friend erythroleukemia cells and to discover what happens to the
differentiating cells when DMSO is removed from the
medium. Exposure to DMSO for less than 24 hours
failed to produce any detectable evidence of erythroid differentiation. On the other hand, culture
in the presence of DMSO for 24 hours followed by
culture in DMSO-free medium for four additional days
produced a small but detectable increment in the
proportion of benzidine positive cells in the culture. Once the differentiation of an individual cell
was initiated, the process continued after removal of
DMSO from the medium. The cell became progressively
more differentiated as evidenced by increases in the
intensity of benzidine staining as well as in the
rate of heme synthesis and heme content. However,
when cells which had been induced to differentiate
by DMSO were cultured in DMSO-free medium for more
than 3-4 days, they became vacuolated and apparently
died. This latter phenomenon, as well as the more
rapid proliferation of the undifferentiated cells
in the culture, accounts for the observation that
when new cultures are established from cultures
which have been grown in the presence of DMSO for
several days, the culture which results ultimately
contains only differentiated cells.

3542 REGULATION OF GROWTH OF MOUSE MASTOCYTOMA
CELLS. (Eng.) Davis, J. (Dept. Cell Biol.,
Univ. Auckland, New Zealand); Ralph, R. K. *Cancer
Res.* 35(6):1495-1504; 1975.

The effects of dibutyryl cyclic AMP and theophylline on the growth of the mouse mastocytoma cell line
PY815 both *in vivo* and *in vitro*, were investigated.
The effect of 4'-(9-acridinylamino)methanesulfon-*m*-
anisidine (m-AMSA) on the cyclic AMP and ATP content
in this cell line was also studied. PY815 mastocytoma cells were passaged in four DBA x C3H F_1 hybrid
mice by serial ip injection of 10^6 cells. Two days
later, two mice were given [5-*methyl*-^{3}H]thymidine
(26 Ci/mM ip) and the other two mice were given
[2-^{14}C]thymidine (62 mCi/mM). After one hour, the
latter mice were injected with 7.5 x 10^{-2} M theophylline and 7.5 x 10^{-3} M dibutyryl cyclic AMP (0.2 ml of
each substance). Aliquots were subjected to velocity
gradient fractionation, and radioactivity was determined. Dibutyryl cyclic AMP (10^{-4} M) and theophylline (10^{-3}M) were added to cultured cells. N^6, O^2-
Dibutyryl cyclic AMP plus theophylline inhibited the
growth of the mast cell tumor line both *in vivo* and
in vitro. The inhibitory effect on growth *in vitro*
was rapidly reversed following removal of the drugs.
Growth inhibition was accompanied by reduced cell
surface activity and increased cell-cell adhesion.
The drug-treated cells accumulated distinct membrane-
bound granules, which are characteristic of more mature mast cells. Treated cells also developed in-

(5 µg/ml, 30 min, 22 C) increased cell-to-cell binding
only if both cells were treated. Although cell-to-
cell binding induced by the lectins from soybean and
wheat germ could be partially reversed by the appropriate competitive saccharide inhibitor, binding induced by Con A could not be reversed. The results
indicate that cell-to-cell binding induced by a lectin can be prevented by an insufficient density of
receptors for the lectin, insufficient receptor mobility, or induced clustering of receptors. This suggests that cell-to-cell binding induced by different
lectins with a variety of cell types is initiated by
a mechanism involving the alignment of complementary
receptors on the colliding cells for the formation of
multiple cell-to-lectin-to-cell bridges.

3547   GROWTH-DEPENDENT ALTERATIONS IN OLIGOMAN-
        NOSYL CORES OF GLYCOPEPTIDES. (Eng.) Cec-
carini, C. (Dept. Biological Sciences, Hunter Coll.
CUNY, New York, N.Y.); Muramatsu, T.; Tsang, J.;
Atkinson, P. H. *Proc. Natl. Acad. Sci. USA* 72(8):
3139-3143: 1975.

In order to study growth-dependent pattern changes
of glycopeptides. radioactive mannose was used as
a precursor to label the surface glycopeptides of
growing and nongrowing human diploid cells (KL-2).
The glycopeptides were separated into neutral glycopeptides and acidic glycopeptides by paper electrophoresis. Growth-dependent alterations occurred
in oligomannosyl cores in the neutral glycopeptides;
namely, the neutral glycopeptides from the surface
of growing cells were more resistant to endo-β-N-
acetylglucosaminidase D but were more susceptible
to α-mannosidase (EC 3.2.1.24; α-D-mannose manno-
hydrolase) than those derived from the surface of
nongrowing cells. Another growth-dependent change
was found when the endoglycosidase-resistant material from acidic glycopeptides was compared by
paper electrophoresis at pH 1.9. The material from
the surface of nongrowing cells contained a component that was absent or greatly reduced in growing cells. The surface material from growing cells
may be enriched in glycopeptides with complex and
exposed oligomannosyl cores such as the majority
of those found in ovalbumin glycopeptides.

3548   NUCLEAR PROTEIN CHANGES IN RAT HEPATOMAS
        CORRELATING WITH GROWTH RATE. (Eng.)
Lea, M. A. (New Jersey Medical Sch., Newark, N. J.
07103); Koch, M. R.; Morris, H. P. *Cancer Res.*
35(7):1693-1697; 1975.

The nature of nuclear proteins that are soluble in
8 M urea-50 mM phosphate, pH 7.6, was compared in
rat liver and Morris hepatomas to examine any progressive changes that correlate with the growth rate
of the tumors. Hepatomas were transplanted bilaterally sc in male Buffalo rats. Partial hepatectomies involved the removal of approximately 66% of
the liver. For isoelectric focusing in polyacrylamide gels, the gels contained 200 µg protein in a
total volume of 2 ml. The pH gradient was measured
by pH determinations of 2-ml aqueous extracts of
4-mm gel slices. Polyacrylamide gel electrophoresis
in 6.25 M urea-0.9 M acetic acid was used for separation of histones. Isoelectric focusing, using

carrier ampholytes for a pH gradient of 3.5-10, indicated that with increasing growth rate of the hepatomas there was a progressive tendency for a decrease in nonhistone nuclear proteins with isoelectric points in the range 7.5-8.9 and an increase in
the range 5.1-6.7. Studies on the influence of time
on the pH gradient revealed that a nonuniform drift
provided a better resolution of the pH range 7.5-8.9
at seven hours than at 24 hr; the latter time for
electrofocusing gave an improved resolution of the
pH range 5.1-6.7. Polyacrylamide gel electrophoresis showed that 8 M urea-50 mM phosphate, pH 7.6,
extracted a small part of the histones from nuclei
of both liver and hepatomas. There was less extraction of histones from the hepatoma nuclei, especially in two rapidly growing hepatomas, with the most
notable difference being seen in the lysine-rich H1
histone. The results suggest that in addition to
qualitative or quantitative changes in nonhistone
nuclear proteins in liver cancer there are alterations in the binding of histones to chromatin.

3549   CHARACTERIZATION OF R-TYPE VITAMIN B_{12}-
        BINDING PROTEINS BY ISOELECTRIC FOCUSING.
III. COBALOPHILIN (R PROTEIN) IN MYELOPROLIFERA-
TIVE STATES AND LEUKOCYTOSIS. (Eng.) Stenman, U.-
H. (The Minerva Foundation Inst. Medical Res., P.
O. Box 819, SF-00101 Helsinki 10, Finland). *Scand.
J. Clin. Lab. Invest.* 35(2):157-161; 1975.

The R-type vitamin B_{12}-binding protein (cobalophilin) in plasma and serum of patients with myelopro-
liferative diseases and leukocytosis was studied
by gel filtration and isoelectric focusing. The
isoproteins composing the cobalophilin were mainly
the same in these disorders as in plasma of five
healthy subjects. In 6 of 12 patients with chronic
myelogenous leukemia and 2 of 14 patients with leukocytosis some of the isoproteins were more acid
than those found in healthy subjects. In chronic
myelogenous leukemia, the high unsaturated vitamin
B_{12}-binding capacity of plasma was due to an increase in the fairly acid isoproteins, whereas in
polycythemia vera and in most cases of leukocytosis
there was an increase in the least acid isoproteins.

3550   COMPARISON OF GLUCOCORTICOID-BINDING PRO-
        TEINS IN NORMAL AND NEOPLASTIC MAMMARY TIS-
SUES OF THE RAT. (Eng.) Goral, J. E. (Univ. of Rochester Sch. of Medicine and Dentistry, Rochester,
N.Y.); Wittliff, J. L. *Biochemistry* 14(13):2944-
2952; 1975.

Kinetic and molecular properties of components binding (^{3}H)triamcinolone acetonide were studied using
105,000 x g supernatants of lactating mammary gland,
R3230AC, and dimethylbenz(a)anthracene (DMBA) induced
mammary tumors of the rat. The relationship between
specific glucocorticoid binding and protein concentration was linear in the range of 0.5-4.0 mg pro-
tein/reaction. These cytoplasmic macromolecules
bound (^{3}H)triamcinolone acetonide with limited capacity (50-400 fmole/mg of cytosol protein) and high
affinity, K_d about 10^{-8}-10^{-9} M. Association of ^{3}H-
triamcinolone acetonide and its binding sites was
maximal by 6-8 hr at 3 C and remained unchanged up to

the ^{32}P uptake per mg protein was about twice as high
in Novikoff hepatoma nucleolar proteins as in normal
rat liver nucleolar proteins, but, generally, the same
proteins were labeled in both tissues. The results
indicate that certain nucleolar proteins from both
tissues differ *in vitro* and *in vivo* with respect to
^{32}P incorporation. It is suggested that highly labeled Spot A1-4 of normal liver may be the AL(2A,
f2a2) histone or a nonhistone protein of low molecular weight.

3554   INITIATION FACTORS IN PROTEIN SYNTHESIS
        BY FREE AND MEMBRANE-BOUND POLYRIBOSOMES
OF LIVER AND HEPATOMA. (Eng.) Murty, C. N. (Univ.
South Florida, Coll. Medicine, Tampa, Fla. 33620);
Verney, E.; Sidransky, H. *Biochem. J.* 152(1):143-
151; 1975.

The activity of initiation factors obtained from
free and membrane-bound polyribosomes of liver and
of transplantable H5123 Morris hepatoma of rats was
investigated by poly(U)-directed polyphenylalanine
synthesis. Initiation factors of membrane-bound
polyribosomes prepared with deoxycholate exhibited
50-75% less activity in incorporating [^{14}C]phenyl-
alanyltRNA into polypeptides than did initiation
factors of free polyribosomes. However, when membrane-bound polyribosomes were prepared after using
Triton X-100, no significant differences in activities were observed between the initiation factors
of free and membrane-bound polyribosomes. These
results suggest that Triton X-100 is preferable to
deoxycholate in the isolation of initiation factors
from polyribosomes. Initiation factors, prepared
by using Triton X-100, of free polyribosomes of
hepatoma exhibited greater activity in the stimulation of polyphenylalanine synthesis than did the
initiation factors of free or membrane-bound polyribosomes of host livers or of membrane-bound polyribosomes of hepatomas.

3555   ISOLATION OF RABBIT·RETICULOCYTE INITIA-
        TION FACTORS BY MEANS OF HEPARIN BOUND TO
SEPHAROSE. (Eng.) Waldman, A. A. (New York Blood
Center, New York, N.Y. 10021): Marx, G.; Goldstein*,
J. *Proc. Nat. Acad. Sci. USA* 72(6):2352-2356; 1975.

Cell-free extracts of New Zealand white rabbit reticulocytes were passed through a heparin-Sepharose
column to confirm the interaction of heparin and
initiation factors. Fraction C, 0.8% of the applied
protein, was recovered from the columns by increasing the KCl concentration of the eluting buffer from
0.12 M to 0.5 M. The initiation factor activity of
the bound protein equaled that of a crude initiation
fraction preparation obtained from rabbit reticulocyte ribosomes. In the presence and absence of added rabbit globin messenger RNA (mRNA), fraction C
was assayed for its ability to stimulate protein
synthesis in a preincubated krebs ascites cell-free
system. Fraction C did not stimulate protein synthesis in the absence of added mRNA; however, in the
presence of added globin mRNA, addition of fraction
C markedly stimulated L-[^{3}H]leucine incorporation
into protein. The initiation factor activity of
fraction C was confirmed in a fractionated cell-free
system. The protein pattern of the fraction elec-

trophoresed on polyacrylamide gels differed sharply
from the parent reticulocyte S-30, but was similar
to the pattern of crude initiation factor. Addition
of either heparin or aurin tricarboxylic acid (ATA)
to assay mixtures containing untreated reticulocyte
S-30 or reticulocyte S-30 passed through a control
column decreased protein synthesis. Protein synthesis of assay mixtures containing reticulocyte
S-30 that passed through a heparin-Sepharose column
was unaffected by the presence of heparin or ATA,
indicating that the L-[^{3}H]leucine incorporation observed with this extract represented only elongation
and termination of protein chains. Thus, by inhibition of initiation, the level of protein synthesis
in both the untreated and control column pass-through
S-30 could be reduced to that of the heparin column
pass-through S-30. Addition of fraction C resulted
in a restoration of protein synthesis in the heparin
column pass-through S-30 to that of the untreated
S-30. The authors suggest that heparin-Sepharose
chromatography simplifies the preparation of initiation factors.

3556   SOME PROPERTIES OF A DNA-UNWINDING PRO-
        TEIN UNIQUE TO LYMPHOCYTES FROM CHRONIC
LYMPHOCYTIC LEUKEMIA. (Eng.) Huang, A. T. F. (Duke
Univ. Med. Cent., Durham, N.C.); Riddle, M. M.;
Koons, L. S. *Cancer Res.* 35(4):981-986; 1975.

Lymphocytes from a common human leukemia, chronic
lymphocytic leukemia, have a greatly enhanced capability of DNA repair and a concomitantly prolonged survival *in vitro* after damage to DNA. A DNA-binding
protein with a molecular wt of 24,000 was isolated
and purified from lymphocytes of four patients with
chronic lymphocytic leukemia. The purification process involved passing crude cell homogenates twice
through a double-stranded DNA column and rechromato-
graphing the flow-through in a UV-irradiated DNA
cellulose column. The protein binds tightly to both
UV-irradiated and single-stranded DNA. At 35° it
enhances the helix-coil transition of poly[d(A-T)]
and the UV-irradiated calf thymus DNA but is ineffi-
cient in ordinary native DNA. This protein also
facilitates the rate of UV-endonuclease incision of
UV DNA but does not induce any nicks by itself. This
finding suggests that the protein may be involved in
DNA repair by enhancing such activity, and also offers an explanation for the increased DNA repair observed in chronic lymphocytic leukemia cells. When
human metaphase chromosomes are exposed to the protein, it induces marked lengthening of chromatids
suggesting that this protein may also act on complex
chromosomes. By quantitative immunochemical determinations, such protein could not be found in lymphocyte extracts of three normal individuals.

3557   PUTRESCINE, SPERMIDINE, N-ACETYLSPERMIDINE
        AND SPERMINE IN THE URINE OF PATIENTS WITH
LEUKEMIAS AND TUMORS. (Eng.) Tsuji, M. (Osaka Univ.
Med. Sch., Japan); Nakajima, T.; and Sano, I. *Clin.
Chim. Acta* 59(2):161-167; 1975.

N-acetylspermidine A and *N*-acetylspermidine B were
isolated from human urine containing 120 g of crea-
tinine by ion exchange chromatography. Structures
of the isolated compounds were assigned to monoacetyl-

1.690 g/cm components contain all the repetitive
sequences which comprise 15% of the total nucleolar
DNA. The ribosomal cistrons are found in components
having buoyant density between 1.707 g/cm^3 and
1.725 g/cm^3. Sodium-p-aminosalicylate-DNA interactions showed that only the 1.700 g/cm^3 fraction has
a destabilized secondary structure. The results
suggest that the sequences with buoyant densities
of 1.707 and 1.690 g/cm^3 are localized in the high
molecular weight DNA of intranucleolar chromatin.
The 1.700 g/cm^3 fraction contains perinucleolar
chromatin.

3561   DAMAGED DNA IN LYMPHOCYTES OF APLASTIC
        ANEMIA. (Eng.) Hashimoto, Y. (Third
Dept. Internal Medicine, Faculty Medicine, Univ.
Tokyo, Tokyo, Japan 113); Takaku, F.; Kosaka, K.
Blood 46(5):735-742; 1975.

The size of single-stranded DNA in lymphocytes in
G_0 stage from 22 patients with acquired primary and
secondary aplastic anemia was estimated by alkaline
sucrose gradient centrifugation. The average size
was 9.3 x 10^7 daltons *vs* 1.23 x 10^8 daltons for
normal persons. The lymphocytes of patients contained significantly more single-strand breaks in
DNA, compared with those of normal persons. The
difference in size of single-stranded DNA which had
been present in nontransformed lymphocytes could
also be observed in transformed lymphocytes. Some
characteristic differences could be observed in the
sedimentation patterns of single-stranded DNA in
the lymphocytes of patients with aplastic anemia and
those of normal persons. For normal persons, a
single main peak was seen in fraction 11 or 12 in
each sedimentation profile; neither a shoulder nor
an accessory peak was observed. In patients with
aplastic anemia, the main peak was slightly shifted
to the right; and a definite shoulder was seen in
two cases. The single-strand breaks in DNA suggested that the repair processes were disturbed in
the DNA molecules of circulating lymphocytes from
patients with acquired primary and secondary aplastic anemia.

3562   DNA CONTENT OF MOUSE HEPATOMA NUCLEI
        DETERMINED BY FEULGEN CYTOPHOTOMETRY.
(Eng.) Enesco, H. E. (Dept. Biological Sciences
Sir George Williams Univ., Montreal, Quebec HSG
1M8, Canada); Aczel, J. *Neoplasma* 22(4):391-394;
1975.

The nuclear DNA content of hepatomas (Jax Code
BW7756) transplanted sc into C57L/J mice was determined by Feulgen cytophotometry. The average
amount of DNA/nucleus expressed in arbitrary units
(AU) was 200.30, 407.70, and 810.00 AU for normal
2N, 4N and 8N hepatic cells, respectively. This
series of DNA values gave points of comparison for
the DNA values in hepatoma cell nuclei. The DNA/
hepatoma cell nucleus was 302-1830 AU with an average of 825.93 $\pm$ 117.20 AU. About 30% of the
hepatoma cells were in the hypertetraploid category, another 30% were in the octaploid and hyper-
octaploid range, and the rest (except a few which
could be considered tetraploid) exhibited a greater
degree of polyploidy and aneuploidy, some in the

16N range and higher. Feulgen cytophotometric results showed that the average amount of DNA in the
hepatoma cell nuclei was at the octaploid level.
The hepatoma cell nuclei showed a wide range of variation
from cell to cell, indicating that the hepatoma cell
population is both polyploid and aneuploid.

3563   THE RELATIVE UTILIZATION OF THE ACYL DIHY-
        DROXYACETONE PHOSPHATE AND GLYCEROL PHOS-
        PHATE PATHWAYS FOR SYNTHESIS OF GLYCEROLIPIDS
IN VARIOUS TUMORS AND NORMAL TISSUES. (Eng.) Pol-
lock, R. J. (Ment. Health Res. Inst., Univ. Michigan,
Ann Arbor); Hajra, A. K.; Agranoff, B. W. *Biochim.
Biophys. Acta* 380(3):421-435; 1975.

Rates of phosphatidate synthesis from dihydroxyacetone phosphate *via* acyl dihydroxyacetone phosphate
or glycerol phosphate were compared in homogenates
of 13 tissues, most of which were deficient in glycerol phosphate dehydrogenase. Sources of the tissue
homogenates included nitrosourea-transformed rat
astroglia, simian virus 40-transformed and normal
astroglia from Syrian hamsters, neuroblastoma (C1300)
cells from A/J mice, splenic leukocytes from leukemic
and normal C58/wm mice, lymphocytes from human peripheral blood. Sprague-Dawley rat thymocytes, male
Swiss mice brain and liver, and fetal brain from a
Sprague-Dawley rat. In all tissues examined, diby-
droxyacetone phosphate entered phosphatidate more
rapidly *via* acyl dihydroxyacetone phosphate than *via*
glycerol phosphate. Tissues with a relatively low
rate of phosphatidate synthesis *via* glycerol phosphate showed no compensating increase in the rate of
synthesis via acyl dihydroxyacetone phosphate. The
rates at which tissue homogenates synthesize phosphatidate from dihydroxyacetone phosphate via glycerol
phosphate increase as glycerol phosphate dehydrogenase increases. Both glycerol phosphate dehydrogenase
and glycerol phosphate:acyl coenzyme A (CoA) acyltransferase are more active than dihydroxyacetone
phosphate:acyl CoA acyltransferase. Thus, all the
tissue homogenates possessed an apparently greater
capability to synthesize phosphatidate via glycerol
phosphate than via acyl dihydroxyacetone phosphate,
but did not express this potential. This result is
discussed in relation to *in vivo* substrate limita-
tions.

3564   COMPARATIVE ADENYLATE CYCLASE ACTIVITIES
        IN HOMOGENATE AND PLASMA MEMBRANE FRAC-
TIONS OF MORRIS HEPATOMA 5123tc (h). (Eng.) Hickie,
R. A. (Univ. Saskatchewan, Saskatoon, Canada); Jan,
S. H.; Datta, A. *Cancer Res.* 35(3):596-600; 1975.

Homogenate and plasma membrane fractions of Morris
hepatoma 5123tc(h) and Buffalo rat liver were studied
with regard to their relative basal activities of
adenylate cyclase and to the comparative responsiveness of this enzyme to glucagon, sodium fluoride,
epinephrine, prostaglandin E_1, and insulin. The basal
adenylate cyclase activities of the hepatoma fractions were similar to those of liver at an ATP concentration of 3.2 mM; if the substrate affinity (K_m-
ATP) of the tumor enzyme is comparable to that of
liver, these findings suggest that the reduced basal
cyclic AMP levels found to occur in hepatoma

5123tc(h) probably are not due to a decreased basal
rate of formation of this cyclic nucleotide. Glucagon (5.6 µM) significantly stimulated adenylate cyclase in both fractions of hepatoma and liver; however
the responsiveness of the tumor enzyme to this hormone was substantially lower than the responsiveness
of liver for both homogenate and plasma membrane
preparations; i.e., activities were enhanced 18-fold
(relative to the basal activity) for liver homogenate
compared with only a six-fold increase for tumor.
With the plasma membrane preparations, glucagon increased the activities 5- and 3.5-fold in liver and
hepatoma, respectively. Sodium fluoride (10 mM), in
contrast to glucagon, increased the adenylate cyclase
activity to approximately the same extent (about 10-
fold) in the liver and hepatoma preparations. Epinephrine (100 µM) enhanced the liver and hepatoma homogenate activities 3- to 4-fold and the hepatoma plasma membrane activities two-fold: however, the liver
plasma membrane activities were not increased. Prostaglandin E_1 (56.6 µM) significantly increased adenylate cyclase activities of liver and hepatoma homogenates (i.e., 1.5- and 3-fold, respectively) but not
of the plasma membrane preparations. Insulin (0.7 µM)
did not significantly alter adenylate cyclase activities in any of the preparations.

3565   INCREASED ACTIVITY OF LOW-K_m CYCLIC
        ADENOSINE 3':5'-MONOPHOSPHATE PHOSPHO-
DIESTERASE IN PLASMA MEMBRANES OF MORRIS HEPATOMA
5123tc (h). (Eng.) Hickie, R. A. (Univ. Saskatchewan, Saskatoon, Canada); Walker, C. M.; Datta, A.
Cancer Res. 35(3):601-605; 1975.

Homogenates and plasma membranes of Morris Hepatoma 5123tc (h) tumors and rat liver were utilized
to demonstrate the increased activity of Low-K_m
cyclic 3'-AMP phosphodiesterase. Tissue homogenates were prepared in sucrose, potassium phosphate and EDTA with one half being used directly
and the other half centrifuged at 10,000 x *g*.
The release of P_i from cAMP was used as an indication of total cAMP phosphodiesterase activity.
The "2-stage" assay method was incorporated to
differentiate between low K_m and high K_m phosphodiesterase activity. Results indicated that cAMP
of hepatoma homogenates and 100,000 x g supernatant fractions were lower than those of liver.
There was a significant increase in low k_m cAMP
phosphodiesterase activity in hepatoma plasma
membranes; theophylline was found to inhibit activity. This increase is thought to be related to an
increase in the metabolic rate of cAMP by low
k_m phosphodiesterase in plasma membranes. Further
investigation into the mechanisms by which this
increase occurs in indicated.

3566   COMPARISON OF TWO ISOZYMES OF CARBONIC
        ANHYDRASE IN THE RAT ANTERIOR PITUITARY
GLAND AND PITUITARY TUMORS. (Eng.) Kimura, H.
(Univ. Virginia Sch. Med., Charlottesville); MacLeod,
R. M. *Cancer Res.* 35(3):796-800; 1975.

Two isozymes of carbonic anhydrase were compared in
the anterior pituitary gland of nontumor-bearing rats
and in the hormone-secreting pituitary tumors. In

klinik der Universitat Munchen, 8 Munchen 15, BRD);
Antonowicz, I.; Lazarus, H.; Shwachman, H. *Exp. Cell
Res.* 91(1):152-158; 1975.

The activities of seven lysosomal and three mitochondrial enzymes from isolated lysosomes and mitochondria of cultivated lymphoid cell lines, obtained
from three patients with acute lymphoblastic leukemia
and from six normal individuals, were studied. The
lysosomal enzymes included α-glucosidase, β-gluco-
sidase, β-glucuronidase, β-galactosidase, aryl sulfa-
tase, *N*-acetyl-β-glucosaminidase, and acid phosphatase. These enzymes are involved in the degradation
of glycoprotein, glycolipids, mucopolysaccharide-
protein complexes, polysaccharides, organic sulfates,
and phosphoric esters. The mitochondrial enzymes
studied were succinate dehydrogenase, glutamate dehydrogenase, and malate dehydrogenase. The specific
activities of the lysosomal enzymes were markedly
higher in the leukemic cell lines than in the control
cell lines, with the possible exception of aryl sulfa-
tase. No marked increase of mitochondrial enzyme
activities was found in mitochondrial fractions from
the leukemic cells. This study demonstrates that
lymphoid cells from patients with acute lymphoblastic
leukemia differ markedly from normal cells in terms
of lysosomal function.

3569   HISTOCHEMICAL STUDIES OF GLYCOGEN META-
        BOLIZING ENZYMES IN NORMAL AND ABNORMAL
HUMAN CERVICAL EPITHELIUM. (Eng.) Skerlavay, M.
A. (State University Hosp., Box 37, 450 Clarkson
Ave., Brooklyn, N. Y. 11203); Bigelow, B.; Demo-
poulus, R. I. *Acta Histochem. (Jena)* 53(2):233-
237; 1975.

The stratified squamous epithelium in eleven cervical cone biopsies was studied for localization
of glycogen, glycogen synthetase, branching enzyme
and amylophosphorylase in areas showing normal,
dysplastic epithelium, and carcinoma *in situ.* In
the normal cervix glycogen was present in the intermediate and superficial cells; amylophosphorylase activity was present in the basal and parabasal
layers; and the branching enzyme was found largely
in the lower and middle zone of the intermediate
layer. As the degree of abnormality increased, a
progressive decrease of glycogen and branching
enzyme and increasing amylophosphorylase was found,
suggesting that amylophosphorylase is a constitutive enzyme, persisting when no glycogen is found,
in normal, dysplastic, and neoplastic cells. Branching enzyme was not shown in any example of severe
dysplasia or carcinoma *in situ;* amylophosphorylase
was present through the full thickness of the epithelium. The method for localization of glycogen
synthetase activity was not reproducible.

3570   AN EVALUATION OF LYSOSOMAL ENZYME LEAKAGE
        AS A FACTOR INFLUENCING THE BEHAVIOUR OF
TRANSFORMED CELLS. (Eng.) Elligsen, J. D. (Dept.
Biology and Physics Univ. Waterloo, Waterloo, Ont.,
Canada N2L3Gl); Thompson, J. E.; Frey, H. E. *Exp.
Cell Res.* 92(1):87-94; 1975.

The extent of lysosomal enzyme leakage from normal
and transformed baby hamster kidney (BHK) cells was

compared in light of knowledge of normal lysosomal
contaminants of fetal calf serum. Commerically,
available fetal calf serum contained substantial
amounts of protease, acid phosphatase, and glycosi-
dase activity that could be inactivated by heating
serum to 70 C for 15 min. Heat inactivation had no
influence on the ability of serum to support growth
of polyoma-transformed BHK (pyBHK) cells and only
marginally affected the growth properties of BHK
cells. Glycosidases and acid phosphatase were also detectable in medium containing heat-inactivated
serum and conditioned by growth of either BHK or
pyBHK cells. This latter activity could be attributed solely to enzymes derived from the growing
cells and, for most of the enzymes monitored, was
present to a greater extent in medium conditioned
by transformed cells as compared to normal cells.
In all cases, however, enzyme activity derived solely
from cells was only 1-10% of that indigenous to commercial tissue culture medium by virtue of its fetal
calf serum component. The observation that these
elevated enzyme levels accompany DNA or RNA tumor
virus-mediated cell transformation indirectly supports the proposal that lysosomal enzymes may cause
uncontrolled cell growth by creating architectural
modifications of the cell surface.

3571   4-AMINOBUTYRATE IN MAMMALIAN PUTRESCINE
        CATABOLISM. (Eng.) Seiler, N. (Max-
Planck-Institut fur Hirnforschung, Arbeitsgruppe
Neurochemie, Frankfurt/Main, West Germany); Eichen-
topf, B. Biochem. J. 152(2):201-210; 1975.

The effects of inhibitors of diamine oxidase, monoamine oxidase and 4-aminobutyrate aminotransferase on
the catabolism of putrescine in mice were studied.
catabolism of putrescine in mice were studied.
Diamine oxidase inhibitors (hydrazine sulfate,
60 mg/kg,ip; aminoguanidine sulfate, 60 mg/kg,ip)
and 50 mg/kg carboxymethoxylamine (aminooxyacetate)
markedly (100 mg/kg, 5.5 hr after administration or
50 mg/kg, 2,3, and 5 hr) the metabolism of $[^{14}C]$
putrescine to $^{14}CO_2$ to about 2%, but they affected
different enzymes. Aminoguanidine specifically
inhibited the mitochondrial and non-mitochondrial
diamine oxidases, whereas carboxymethoxylamine
specifically inhibited 4-aminobutyrate transamina-
tion by the mitochondrial pathway. Hydrazine
inhibited at both sites, and resulted in increased
concentrations of 4-aminobutyrate in brain and liver.
Pretreatment of mice with carboxymethoxylamine
(100 mg/kg, 5.5 hr after administration or 50 mg/kg,
2,3, and 5 hr) and $[^{14}C]$putrescine led to the urinary
excretion of amino$[^{14}C]$butyrate. Carboxymethoxylamine did not affect the non-mitochondrial pathway
of putrescine catabolism, as the product of oxidative
deamination of putrescine in the extramitochondrial
compartment was not further oxidized but was excreted in the urine as derivatives of 4-aminobutyr-
aldehyde. The monoamine oxidase inhibitor, pargy-
line (60 mg/kg) decreased the metabolism of $[^{14}C]$-
putrescine to $^{14}CO_2$ in vivo. Catabolism of putrescine to CO_2 in vivo occurs along different pathways,
both of which have 4-aminobutyrate as a common
intermediate, in contrast with the non-mitochondrial
catabolism of putrescine, which terminates in the

junction with a low-fiber diet. Stool weights and stool fat excretion increased in subjects on both dietary fibers. Bagasse increased the daily loss of acid steroids, but bran failed to affect bile acid excretion. Decreased transit time, as measured by radio-opaque pellets, occurred with bagasse without alteration in fecal flora. The raised excretion of bile acids and fatty acids failed to lower the plasma cholesterol and triglycerides after 12 wk. Thus, different fiber sources with variable components have dissimilar metabolic effects.

3576   STUDIES ON THE DIFFERENTIATION OF HUMAN
        AND EXPERIMENTAL GLIOMAS IN ORGAN CULTURE
SYSTEMS. (Eng.) Rubinstein, L. J. (No affiliation given); Herman, M. M. *Recent Results Cancer Res.* 51:35-51; 1975.

3577   STUDY OF THE PROCESSES OF DEVELOPMENT OF
        MALIGNANT TUMOURS BY HYBRIDIZATION OF
SOMATIC CELLS. (Rus.) Popov, A. S. (Inst. Biology of Evolution USSR Acad. Sci., Moscow, USSR). *Arkh. Patol.* 37(6):84-87; 1975.

3578   INVESTIGATION OF CHORIOCARCINOMA CLONAL
        CELL LINES *IN VITRO* AND CHORIOCARCINOMA
TRANSPLANTS IN THE HAMSTER FOR THE SECRETION OF A THYROID-STIMULATING FACTOR. (Eng.) Ketelslegers, J.-M. (Natl. Inst. Health, Bethesda, Md.); Nisula, B. C.; Kohler, P. O. *Endocrinology* 96(3):808-810; 1975.

3579   MODIFICATION OF TUMORIGENICITY OF L1210
        LEUKEMIA BY FERRIC CHLORIDE. (Eng.)
Khan, A. (Wadley Inst. Molecular Medicine, 9000 Harry Hines, Dallas, Tex. 75235); Thometz, D. *Wadley Med. Bull.* 5(2):161-164; 1975.

3580   COMPARISON OF THE "CHROMATIN FRACTION II"
        NONHISTONE PROTEINS OF NORMAL, PHYTO-
HEMAGGLUTININ STIMULATED AND LEUKEMIC HUMAN LYMPHO-CYTES AND MYELOCYTES [abstract]. (Eng.) Yeoman, L. C. (Baylor Coll. Med., Houston, Tex.); Seeber, S.; Taylor, C. W.; Busch, H. *Proc. Am. Assoc. Cancer Res.* 16:43; 1975.

3581   INCREASED DEGRADATION RATES OF PROTEIN
        SYNTHESIZED IN HEPATOMA CELLS IN THE
PRESENCE OF AMINO ACID ANALOGUES. (Eng.) Knowles, S. E. (CSIRO Division of Nutritional Biochemistry, Adelaide, S.A. 5000, Australia); Gunn, J. M.; Hanson, R. W.; Ballard, F. J. *Biochem. J.* 146(3):595-600; 1975.

3582   RELATIONSHIP BETWEEN IODINATION AND THE
        POLYPEPTIDE CHAIN COMPOSITION OF THYRO-
GLOBULIN. (Eng.) Haeberli, A. (Experimentelle Endokrinologie, Inselspital Bern, Switzerland); Salvatore, G.; Edelhoch, H.; Rall, J. E. *J. Biol. Chem.* 250(19):7836-7841; 1975.

3583   ACTIN ASSOCIATED WITH MEMBRANES FROM 3T3
        MOUSE FIBROBLAST AND HeLa CELLS. (Eng.)
Gruenstein, E. (Dept. Biol., Massachusetts Inst.
Technol., Cambridge); Rich, A.; Weihing, R. R. *J.*
Cell Biol. 64(1):223-234; 1975.

3584   POLYADP-RIBOSYLATION OF HeLa CELL NUCLEAR
        PROTEINS: RELEASE OF TEMPLATE RESTRIC-
TION FOR DNA POLYMERASE. (Eng.) Smulson, M.
(Georgetown Univ. Sch. Med. Dept., Washington, D.
C.); Roberts, J.; Stark, P. *J. Biochem. (Tokyo)*
77(1):7p; 1975.

3585   CYTOCHEMICAL STUDY OF THE DYNAMICS OF
        THE SYNTHESIS OF RNA DETECTED IN THE
METAPHASIC CHROMOSOMES OF HeLa CELLS. (Rus.)
Kucheriavenko, A. A. (Inst. Human Morphology of the
U.S.S.R. Acad. Medical Sciences, Moscow, U.S.S.R.).
Biull. Eksp. Biol. Med. 79(3):94-97; 1975.

3586   STIMULATION OF DNA SYNTHESIS *IN VITRO* BY
        A TUMOR PROTEASE, CANCER PROCOAGULANT A
(CPA) [abstract]. (Eng.) Wood, L. G. (Univ. Colorado Med. Cent., Denver); Gordon, S. G. *Clin. Res.*
23(2):137A; 1975.

3587   MICROFLUOROMETRIC COMPARISONS OF HEAT-
        INDUCED NUCLEAR ACRIDINE ORANGE METACHRO-
MASIA BETWEEN NORMAL CELLS AND NEOPLASTIC CELLS
FROM PRIMARY TUMORS OF DIVERSE ORIGIN. (Eng.) Al-
varez, M. R. (Dept. Biol., Univ. South Florida,
Tampa). *Cancer Res.* 35(1):93-98; 1975.

3588   FREE FATTY ACID METABOLISM OF LEUKEMIC
        AND NORMAL LEUKOCYTES [abstract]. (Eng.)
Burns, C. P. (Univ. Iowa Coll. Med., Iowa City);
Welshman, I. R.; Spector, A. A. *Proc. Am. Assoc.*
Cancer Res. 16:44; 1975.

3589   QUALITATIVE AND QUANTITATIVE HISTOENZYMO-
        CHEMISTRY OF GASTRIC CANCER. (Rus.)
Sklianskaia, O. A. (I. M. Sechenov First Moscow
Medical Inst., Moscow, USSR). *Arkh. Patol.* 37(1):
38-45; 1975.

3590   ALTERATIONS IN THE ISOZYME SPECTRUM AND
        THE ACTIVITY OF LACTATE DEHYDROGENASE IN
DEVELOPMENT OF EHRLICH ASCITES TUMOR. (Rus.) Aga-
tova, A. I. (Inst. Chemical Physics, Acad. Sciences
of U.S.S.R., Moscow, U.S.S.R.). *Vopr. Med. Khim.*
21(3):276-281; 1975.

SCHAERER, R.
    3487*
SCHAFER, W.
    3223
SCHAEFFER, P.A.
    3215
SCHAPIRA, F.
    3591*
SCHERER, E.
    3432
SCHLENKER, R.A.
    3179
SCHLOM, J.
    3232, 3269*
SCHMID, K.O.
    3408
SCHMITT, M.
    3276*
SCHMITT-VERHULST, A.M.
    3383*
SCHNEIDERMAN, M.H.
    3189*
SCHOLES, M.C.
    3489*
SCHOLZ, D.A.
    3479*
SCHOLZ, P.
    3375*
SCHUG, W.G.
    3187*
SCHWEINSBERG, E.
    3165*
SCHWEINSBERG, F.
    3165*
SCOLNICK, E.M.
    3551
SCOTT, D.W.
    3395*
SCRABA, D.G.
    3236
SCRIBNER, J.D.
    3043*
SEBRING, C.
    3144*
SEEBER, S.
    3580*
SEEHAFER, J.
    3236
SEGERLING, M.
    3319
SEIBERT, E.
    3374*
SEIDMAN, H.
    3518
SEIDMAN, I.
    3131*, 3396
SEILER, N.
    3571
SEKIYA, S.
    3427
SELECKY, P.
    3022
SELIKOFF, I.J.
    3518
SELKIRK, J.K.
    3085
SEMAN, G.
    3226

SEN, L.
    3388*
SEVERINI, M.
    3488*
SEYMOUR, R.B.
    3038*
SHAFER, W.G.
    3478*
SHAH, K.V.
    3250
SHAMBERGER, R.J.
    3164*
SHANI, M.
    3513
SHAPIRO, S.
    3528*
SHAPOT, V.S.
    3292*, 3293*
SHARIEF, Y.
    3250
SHARP, P.A.
    3196
SHATTON, J.B.
    3592*
SHAW, J.E.
    3220
SHEARER, G.M.
    3302, 3383*
SHEARER, R.W.
    3155*
SHEARER, W.T.
    3351*
SHEN, L.
    3304
SHERMAN, A.I.
    3430
SHERR, C.J.
    3205
SHETTIGARA, P.T.
    3137*
SHIH, T.-W.
    3084
SHIRAI, T.
    3087
SHLIANKEVICH, M.A.
    3292*, 3293*
SHLOMAI, J.
    3217
SHMELEV, M.M.
    3412
SHOJI, M.
    3595*
SHOOTER, K.V.
    3111
SHUBIK, P.
    3159*
SHUDO, K.
    3167*
SHULLENBERGER, C.C.
    3027
SHUSTER, J.
    3028
SHWACHMAN, H.
    3568
SIDHU, S.S.
    3491*
SIDRANSKY, H.
    3554

SILOBR
    343
SILVA,
    327
SILVER
    340
SILVER
    334
SIMMON
    312
SIMMON
    330
SIMON,
    344
SIMS,
    320
SINKOV
    302
SINKS,
    334
SIPES,
    30
SIPPE
    350
SIROV
    31
SJOGR
    33
SKERL
    35
SKINN
    34
SKLIA
    35
SKOTN
    31
SLIWC
    35
SLOAN
    31
SMITH
    30
SMITH
    32
SMITH
    34
SMITH
    34
SMITH
    35
SMITH
    35
SMULS
    35
SMYK,
    3
SNYDE
    3
SOLOV
    34
SOLOV
    3
SOLTE
    3
SOMER
    30
SOMOG
    30

SONTAG, J.M.
3078
SOOHOO, J.
3516
SORG, C.
3374*
SOSIN, H.
3334
SOTNIKOV, V.M.
3473*
SOTTO, J.-J.
3487*
SOUTHGATE, D.A.T.
3575
SPATARO, A.C.
3289*
SPECTOR, A.A.
3588*
SPEIER, J.L.
3083
SPENCE, A.M.
3433
SPICER, S.S.
3417
SPILLMAN, T.
3291*
SPIRTAS, R.
3519
SPITS, H.
3349
SPJUT, H.J.
3405
SPOMER, B.
3291*
SPRECHER-GOLDBERGER, S.
3354*
SPRINGER, G.F.
3322
SRIVASTAVA, B.I.S.
3540
STADENBERG, I.
3301
STALEY, C.J.
3468*
STAMMINGER, G.M.
3254
STARK, P.
3584*
STEARNS, M.
3449*
STEELE, G., JR.
3301
STEGER, R.W.
3430
STEGNER, H.-E.
3496*
STEINGLASS, M.
3095, 3431
STENMAN, U.-H.
3549
STEPHENSON, J.R.
3204, 3206, 3369*
STERNBERG, S.S.
3449*
STEVENS, W.
3178
STIENON, J.
3354*

STONER, G.D.
3123
STOUT, R.D.
3343
STOVER, B.J.
3178
STRAFFON, R.A.
3425
STRAUSS, R.A.
3447*
STREITWESER, D.
3125
STRIMLAN, C.V.
3425
STRIPP, B.
3091
SUGIHARA, S.
3087
SUGIMURA, T.
3593*
SUKHIJA, D.S.
3491*
SWANN, P.F.
3045*
SWENBERG, J.A.
3042*
SZABO, B.
3211
SZALKA, A.
3211
SZCZUREK, Z.
3594*
TABER, B.Z.
3003
TACK, L.
3354*
TAGUCHI, T.
3440*
TAKAKU, F.
3561
TAKAMIZAWA, H.
3427
TAKEMOTO, K.K.
3284*
TAKEWAKI, K.
3163*
TALAS, M.
3203
TAMM, I.
3010
TANIGUCHI, H.
3440*
TANNENBAUM, S.R.
3154*
TANTU, P.
3533*
TANZARELLA, C.
3190*
TAYLOR, C.W.
3580*
TAYLOR, H.W.
3117
TEEARS, R.J.
3401
TEI, N.
3440*
TEMMINK, J.H.M.
3349

TENEN, D.
3242
TERASHI, S.I.
3113
THAYER, R.E.
3235, 3282*
THEMANN, H.
3484*
THEODOROPOULOS, G.
3373*
THEOFILOPOULOS, A.N.
3367*
THIESS, A.M.
3007
THIRY, L.
3354*
THIVOLET, J.
3406
THOMAS, K.
3332
THOMASON, A.R.
3224
THOMETZ, D.
3579*
THOMPSON, D.J.
3098
THOMPSON, J.E.
3570
THOMPSON, K.H.
3185*
THORNE, R.F.W.
3543
THORNLEY, A.L.
3029
THORNTON, H.
3230
THOSS, K.
3393*
TIMANOVSKAIA, V.V.
3274*
TIMPERLY, W.R.
3403
TODARO, G.J.
3205
TOGNELLA, S.
3058*
TOLMACH, L.J.
3182
TOMATIS, L.
3145*
TOMINAGA, T.
3440*
TOMKINS, G.M.
3574
TOMKINS, W.A.F.
3323
TONDA, G.P.
3387*
TOTANIA, R.
3099
TOTH, B.B.
3494*
TRIDON, P.
3454*
TRKULA, D.
3213
TRONICK, S.R.
3369*

VERA, N.
3228
VERGARA, J.R.
3250
VERLEY, J.-M.
3470*
VERNEY, E.
3554
VERNON, M.L.
3286*
VESSELINOVITCH, S.D.
3089
VESSEY, M.P.
3510
VIALE, G.
3387*
VIDAILHET, M.
3454*
VIGIER, P.
3261*
VINOGRADOVA, O.M.
3451*
VINORES, S.
3170*
VIRELLA, G.
3363*
VIRSEDA, J.M.P.
3492*
VIRTANEN, A.
3508
VITETTA, E.S.
3225
VITORGAN, YU.E.
3201
VOGT, P.K.
3240
VON KLEIST, S.
3378*
VON METTENHEIM, A.E.
3259*
VOSIKA, G.J.
3419
VRBA, R.
3320
VYAS, G.N.
3271*
VYAS, R.
3097
WADA, A.
3440*
WADSWORTH, L.D.
3489*
WADSWORTH, S.
3013
WADSWORTH, S.C.
3214
WAGNER, G.
3533*
WAGNER, R.M.
3123
WAGNER, V.
3181
WALDECK, W.
3241, 3245
WALDMAN, A.A.
3555
WALDO, E.D.
3396

WALKER, C.M.
3565
WALKER, D.J.
3219
WALKER, L.
3304
WALKER, R.
3046*
WALTERS, R.L.
3575
WANEBO, H.J.
3449*
WANER, E.B.
3312
WANG, C.-H.
3027
WANG, H.-H.
3113
WANG, L.-H.
3240
WARD, H.A.
3056*
WARD, J.M.
3078
WARNER, N.L.
3380*
WAPNER, V.
3130*
WATERFIELD, E.M.
3348
WATERFIELD, J.D.
3348
WATERHOUSE, J.A.
3520
WATTENBERG, L.W.
3083
WEBBER, M.M.
3265*
WEBER, A.
3591*
WEBER, G.H.
3254
WEEKES, U.
3158*
WEEKS, C.E.
3120
WEIHING, R.R.
3583*
WEINHOUSE, S.
3592*
WEINSTEIN, I.B.
3001
WEISBURGER, E.K.
3032, 3078
WEISBURGER, J.H.
3034*, 3161*
WELIKY, N.
3368*
WELLINGS, S.R.
3400
WELSHMAN, I.R.
3588*
WERDER, E.A.
3477*
WERNER, B.
3272*
WERSHALL, J.
3410

.A.
.J.
D.H.

.M.

WESTWOOD, J.H.
  3329
WETCHLER, B.B.
  3442*, 3443*
WEYMOUTH, L.
  3166*
WHITAKER, R.H.
  3174*
WHITBY, B.
  3219
WHITTEN, H.O.
  3315
WIEBEL, F.J.
  3085
WIGZELL, H.
  3390*, 3391*
WILKOFF, L.J.
  3150*
WILLIAMS, J.
  3196
WILLIAMS, M.A.
  3091
WILLIAMS, M.L.
  3168*
WILLIS, C.E.
  3164*
WILLS, E.J.
  3262*
WILSON-DAVIS, K.
  3512
WILSON, J.S.
  3180
WILTSHAW, E.
  3489*
WIMHOFER, G.
  3276*
WINTER, R.
  3594*
WINTERS, W.D.
  3270*
WINTON, W.
  3455*
WISE, K.S.
  3325
WISHNOW, R.M.
  3402
WITHYCOMBE, J.F.R.
  3174*
WITTLIFF, J.L.
  3093, 3550
WOCKEL, W.
  3506, 3526*
WOGAN, G.N.
  .3154*
WOLF, A.F.
  3005
WOLFF, J.P.
  3333, 3361*
WOLLWEBER, L.
  3346
WONG, T.C.
  3287*
WOOD, L.G.
  3586*
WOODRUFF, J.M.
  3500*
WRIGHT, C.M.
  3077

WUNDERLICH, J.R.
  3392*
YACHIN, S.
  3344
YAGER, J.D., JR.
  3129*
YAGI, M.J.
  3340
YAHARA, I.
  3394*
YAMADA, M.
  3593*
YAMAGUCHI, N.
  3001, 3251
YAO, R.C.
  3132*
YEOH, G.
  3238
YEOMAN, L.C.
  3580*
YEUNG, K.-Y.
  3416
YOHN, D.S.
  3255
YOSHIDA, M.
  3099
YOUNG, H.A.
  3551
ZACHARY, A.A.
  3425
ZALTA, J.P.
  3560
ZAMCHECK, N.
  3019
ZARETSKII, I.Z.
  3263*
ZHARAKHOVICH, I.A.
  3412
ZIELINSKI, W.L., JR.
  3157*
ZIMMERMANN, E.R.
  3264*
ZOLLER, M.
  3303
ZSCHOCH, H.
  3397
ZUIDEMA, P.J.
  3459*
ZULUAGA, A.
  3465*
ZWERNER, R.K.
  3331
99999999999999999999999999999999
  9999999999999999999999999999-
    99999
      9999999999999999999999-
        9999999999999999999

2-ACETAMIDOFLUORENE
    SEE ACETAMIDE, N-FLUOREN-2-YL-

2-ACETAMINOFLUORENE
    SEE ACETAMIDE, N-FLUOREN-2-YL-

ACETIC ACID, CHLORO-
    ETHYLENE, CHLORO-
        CARCINOGENIC METABOLITE, RAT, 3124

ACETIC ACID, (ETHYLENEBIS
  (OXYETHYLENENITRILO))TETRA-
    CARCINOMA, EHRLICH TUMOR
        CALCIUM, 3543

ACETIC ACID, THIODI-
    ETHYLENE, CHLORO-
        CARCINOGENIC METABOLITE, RAT, 3124

ACETOHYDROXAMIC ACID, N-FLUOREN-2-YL-
    ACYLTRANSFERASES
        ENZYMATIC ACTIVITY, 3082
    CELL MEMBRANE
        ULTRASTRUCTURAL STUDY, RAT, 3080
    LIVER NEOPLASMS
        CELL NUCLEUS, 3080
        ULTRASTRUCTURAL STUDY, RAT, 3080
    LYMPHOCYTES
        DNA REPLICATION, 3097

N-ACETOXY-N-2-ACETYLAMINOFLUORENE
    SEE ACETAMIDE, N-(ACETYLOXY)-N-9H-
        FLUOREN-2-YL-

2-ACETYLAMINOFLUORENE
    SEE ACETAMIDE, N-FLUOREN-2-YL

ACID PHOSPHATASE
    BREAST NEOPLASMS
        ENZYMATIC ACTIVITY, 3445*
    COLONIC NEOPLASMS
        ENZYMATIC ACTIVITY, 3445*
    LEUKEMIA, LYMPHOBLASTIC
        LYSOSOMES, 3568
    PROSTATIC NEOPLASMS
        HYPERPLASIA, 3537
    VIRUS, POLYOMA
        CELL TRANSFORMATION, NEOPLASTIC,
            3570
    VIRUS, POX
        CELL MEMBRANE, 3255

ACRIDINE, 9-AMINO-
    BACTERIOPHAGES
        MUTAGENIC ACTIVITY, 3130*
    ESCHERICHIA COLI
        MUTAGENIC ACTIVITY, 3130*

ACTIN
    HELA CELLS
        CELL MEMBRANE, 3583*
        CYTOCHALASIN B, 3583*
        ISOLATION AND CHARACTERIZATION,
            3583*

ACTINOMYCIN D
    ANTINEOPLASTIC AGENTS
        CARCINOGENIC POTENTIAL, REVIEW,
            3004

VIRUS, FRIEND MURINE LEUKEMIA
RNA POLYMERASE, 3227
SPLENOMEGALY, 3227

ACYLTRANSFERASES
ACETOHYDROXAMIC ACID, N-FLUOREN-2-YL-
ENZYMATIC ACTIVITY, 3082
CHOLESTEROL, 14-METHYLHEXADECANOATE
REVIEW, 3030

ADENOACANTHOMA
SEE ADENOCARCINOMA

ADENOCARCINOMA
BLADDER NEOPLASMS
LEUKOPLAKIA, 3426
BREAST NEOPLASMS
GENETICS, 3511
KIDNEY FAILURE, CHRONIC, 3308
NEOPLASM METASTASIS, 3401
REVERSE TRANSCRIPTASE, 3288*
VIRUS-LIKE PARTICLES, 3288*
CERVIX NEOPLASMS
CASE REPORT, 3499*, 3500*
DNA, NEOPLASM, 3499*
ENDOMETRIOSIS, 3098
EPIDEMIOLOGY, 3497*
ESTROGENS, 3098
HISTOLOGICAL STUDY, 3497*
PRECANCEROUS CONDITIONS, 3499*
COLONIC NEOPLASMS
CARCINOEMBRYONIC ANTIGEN, 3019,
3028, 3323, 3324
CASE REPORT, 3448*
IMMUNITY, CELLULAR, 3300, 3301
ISOANTIGENS, 3323
PRECANCEROUS CONDITIONS, 3448*
COLONIC NEOPLASTS
HISTOCOMPATIBILITY ANTIGENS, 3327
ESOPHAGEAL NEOPLASMS
EPIDEMIOLOGY, ISRAEL, 3513
FIBROSARCOMA
ASBESTOS, 3138*
INTESTINAL NEOPLASMS
ANTIGENIC DETERMINANTS, 3300
PRECANCEROUS CONDITIONS, 3414
RADIATION, IONIZING, 3195*
KIDNEY NEOPLASMS
GENETICS, 3425
HISTOCOMPATIBILITY ANTIGENS, 3425
LARYNGEAL NEOPLASMS
REVIEW, 3066*
LUNG NEOPLASMS
AIR POLLUTANTS, 3516
CASE REPORT, REVIEW, 3473*
FOLIC ACID, N-NITROSO-, 3154*
KIDNEY FAILURE, CHRONIC, 3308
NEOPLASM METASTASIS, 3401
SCLERODERMA, SYSTEMIC, 3473*
MAMMARY NEOPLASMS, EXPERIMENTAL
BENZ(A)ANTHRACENE, 7,12-DIMETHYL-,
3092
CHROMATIN, 3587*
GLUCOCORTICOID-BINDING PROTEINS,
3550
PROLACTIN, 3092
OVARIAN NEOPLASMS

ASPERGILLUS PARASITICUS
    AVERUFIN, 3132*
    CARCINOGENIC METABOLITE, 3132*
BARBITURIC ACID, 5-ETHYL-5-PHENYL-
    COCARCINOGENIC EFFECT, RAT, 3079
CHOLANTHRENE, 3-METHYL-
    COCARCINOGENIC EFFECT, RAT, 3079
COLONIC NEOPLASMS
    TRANSPLACENTAL CARCINOGENESIS,
        3078
DNA
    BINDING, RAT, KIDNEY, 3077
    BINDING, RAT, LIVER,, 3077
    DENATURAION, HEAT, 3076
    HISTONES, 3076
GLUTAMYL TRANSPEPTIDASE
    CARCINOGENIC ACTIVITY, RAT, 3075
GLUTATHIONE
    PROTECTIVE EFFECT, RAT, 3079
HEPATOMA
    LIPOTROPIC DEFICIENT, HIGH-FAT
        DIET, 3096
KIDNEY NEOPLASMS
    TRANSPLACENTAL CARCINOGENESIS,
        3078
LIVER NEOPLASMS
    CHOLANGIOMA, 3075
MALEIC ACID, DIETHYL ESTER
    COCARCINOGENIC EFFECT, RAT, 3079
MOUTH NEOPLASMS
    TRANSPLACENTAL CARCINOGENESIS,
        3078
PRECANCEROUS CONDITIONS
    CELL TRANSFORMATION, NEOPLASTIC,
        3075
RNA
    BINDING, RAT, KIDNEY, 3077
    BINDING, RAT, LIVER,, 3077
SELENIOUS ACID, DISODIUM SALT
    CELL TRANSFORMATION, NEOPLASTIC,
        3133*

AGGLUTINATION
    LEUKEMIA L1210
        GLUTARALDEHYDE, 3346
    MACROPHAGES
        GLUTARALDEHYDE, 3346

AIR POLLUTANTS
    BLADDER NEOPLASMS
        EPIDEMIOLOGY, SOCIOECONOMIC
            FACTORS, 3516
    ESOPHAGEAL NEOPLASMS
        EPIDEMIOLOGY, SOCIOECONOMIC
            FACTORS, 3516
    LARYNGEAL NEOPLASMS
        EPIDEMIOLOGY, SOCIOECONOMIC
            FACTORS, 3516
    LUNG NEOPLASMS
        ADENOCARCINOMA, 3516
        CARCINOMA, 3516
        CARCINOMA, EPIDERMOID, 3516
        EPIDEMIOLOGY, SOCIOECONOMIC
            FACTORS, 3516
    MOUTH NEOPLASM
        EPIDEMIOLOGY, SOCIOECONOMIC
            FACTORS, 3516
    PANCREATIC NEOPLASMS
        EPIDEMIOLOGY, SOCIOECONOMIC

LI

3

FACTORS, 3516

ALANINE
  VIRUS, SV40
    AMINO ACID TRANSPORT, 3296*

ALANINE AMINOTRANSFERASE
  ACETAMIDE, THIO-
    ETHYL ALCOHOL, 3091
  CARBON TETRACHLORIDE
    ETHYL ALCOHOL, 3091
  DIMETHYLAMINE, N-NITROSO-
    ETHYL ALCOHOL, 3091

ALANINE, 3-((2-CHLOROETHYL)THIO-
  ETHYLENE, CHLORO-
    CARCINOGENIC METABOLITE, RAT, 3124

ALANINE, 3-(3,4-DIHYDROXYPHENYL)-
  NEUROBLASTOMA
    CARCINOGENIC ACTIVITY, 3134*

ALANINE, N-ACETYL-3((2-CHLOROETHYL)THIO)-
  ETHYLENE, CHLORO-
    CARCINOGENIC METABOLITE, RAT, 3124

ALANINE, 3-(P-(BIS(2-CHLOROETHYL)AMINO)
PHENYL)-, L-
  MULTIPLE MYELOMA
    CHROMOSOME ABERRATIONS, 3422
    LEUKEMIA, MYELOBLASTIC, 3421

ALCOHOLISM
  MOUTH NEOPLASMS
    EPIDEMIOLOGY, 3531*
  PHARYNGEAL NEOPLASMS
    EPIDEMIOLOGY, 3531*

ALDOSTERONE
  ACETAMIDE, 2-IODO-
    CARRIER PROTEINS, 3551
  MAMMARY NEOPLASMS, EXPERIMENTAL
    BINDING, COMPETITIVE, 3550
  14-PYRROLE-2,5-DIONE, 1-ETHYL-
    CARRIER PROTEINS, 3551
  VIRUS, POX
    VIRUS REPLICATION, 3255

ALDRICH SYNDROME
  IMMUNOSUPPRESSION
    IMMUNOGLOBULINS, SURFACE, REVIEW,
    3059*
  VIRUS, POLYOMA, BK
    DNA, VIRAL, 3284*
    ISOLATION AND CHARACTERIZATION,
    3284*

ALKALINE PHOSPHATASE
  BREAST NEOPLASMS
    ENZYMATIC ACTIVITY, 3445*
  COLONIC NEOPLASMS
    ENZYMATIC ACTIVITY, 3445*
  VIRUS, POX
    CELL MEMBRANE, 3255

ALKYLATING AGENTS
  DIMETHYLAMINE, N-NITROSO-
    CHROMATIN, 3156*
    DNA, 3156*, 3157*

PROTEINS, 3157*
    RNA, 3157*
  ETHANE, IODO-
    STRUCTURE-ACTIVITY RELATIONSHIP
    3109
  METHANE, IODO-
    STRUCTURE-ACTIVITY RELATIONSHIP
    3109
  METHANESULFONIC ACID, METHYL ESTER
    STRUCTURE-ACTIVITY RELATIONSHIP
    3109
  PYRIDINE, 4-(P-NITROBENZYL)-
    STRUCTURE-ACTIVITY RELATIONSHIP
    3109
  SULFURIC ACID, DIETHYL ESTER
    STRUCTURE-ACTIVITY RELATIONSHIP
    3109
  SULFURIC ACID, DIMETHYL ESTER
    STRUCTURE-ACTIVITY RELATIONSHIP
    3109
  UREA, METHYL NITROSO-
    METHANESULFONIC ACID, METHYL ES
    3112

ALLIUM CEPA
  CHROMOSOME ABNORMALITIES
    SMOKING, 3121

ALPHA-FETOPROTEIN
  SEE FETAL GLOBULINS

ALUMINUM HYDROXIDE
  BENZO(A)PYRENE
    CARCINOGENIC ACTIVITY, 3141*

AMELOBLASTOMA
  BONE NEOPLASMS
    DIFFERENTIATION, 3438*
  SARCOMA, EWING'S
    DIFFERENTIATION, 3438*

AMINO ACIDS
  HEPATOMA
    5-BETA-CHOLAN-24-OIC ACID, 3-AL
    12-ALPHA-DIHYDROXY, 3554
    POLYRIBOSOMES, 3554
    TRITON X-100, 3554
  VIRUS, SV40
    CONCANAVALIN A, 3296*

AMINO ACYL T RNA SYNTHETASES
  CHOLESTEROL, 14-METHYLHEXADECANOATE
    REVIEW, 3030

AMINOGLUTETHIMIDE
  ADRENAL GLAND NEOPLASMS
    STEROIDS, 3402

AMINOPEPTIDASES
  BREAST NEOPLASMS
    ENZYMATIC ACTIVITY, 3445*
  COLONIC NEOPLASMS
    ENZYMATIC ACTIVITY, 3445*

AMNION
  ACETAMIDE N-(ACETYLOXY)-N-9H-FLUORE
  YL-
    ADENOSINE CYCLIC 3',5'
    MONOPHOSPHATE, 3129*

EPIDEMIOLOGY, 3518
OCCUPATIONAL HAZARD, 3038*, 3518
LARYNGEAL NEOPLASMS
HISTOLOGICAL STUDY, REVIEW, 3023
LIVER NEOPLASMS
EPIDEMIOLOGY, REVIEW, 3007
ETHYLENE, CHLORO-, 3007, 3461*
ETHYLENE, CHLORO- POLYMER, 3007
PIPERIDINE, 1-NITROSO-, 3117
3-PIPERIDINOL, 1-NITROSO-, 3117
NERVOUS SYSTEM NEOPLASMS
BLOOD GROUPS, 3465*
ORBITAL NEOPLASMS
ULTRASTRUCTURAL STUDY, REVIEW,
3064*
4-PIPERIDINOL, 1-NITROSO-
SEX FACTORS, 3117
4-PIPERIDONE, 1-NITROSO-
SEX FACTORS, 3117
VIRUS, HERPES
HISTOLOGICAL STUDY, BABOON, 3274*

ANILINE, 4,4'-(IMIDO CARBONYL)BIS(N,N-
DIMETHYL)-
CELL LINE
CARCINOGENESIS SCREENING, MODEL
SYSTEM, 3127*

ANILINE, 4,4'-METHYLENEBIS-2-CHLORO-
OCCUPATIONAL HAZARD
CARCINOGENIC POTENTIAL, REVIEW,
3040*

ANILINE, N-METHYL-N-NITROSO-
ESOPHAGEAL NEOPLASMS
DIETARY FACTORS, CHINA, 3514

ANILINE, N-METHYL-P-(PHENYLAZO)-
MICROSOMES, LIVER
MIXED-FUNCTION AMINE OXIDASE,
3136*

ANILINE, N,N-DIMETHYL-P-NITROSO-
NADH, NADPH OXIDOREDUCTASES
ENZYMATIC ACTIVITY, 3135*

ANILINE, N,N-DIMETHYL-P-PHENYLAZO-
HEPATOMA
ANTIGENS, NEOPLASM, 3303
DNA, 3128*
IMMUNITY, CELLULAR, 3303
NUCLEOTIDES, 3128*

ANTHOPHYLLITE
SEE ASBESTOS

1,8,9-ANTHRACENETRIOL
FIBROBLASTS
DNA REPAIR, 3081
DNA REPLICATION, 3081

ANTHRACOSILICOSIS
PRECANCEROUS CONDITIONS
OCCUPATIONAL HAZARD, 3035

ANTHRAQUINONE, 1,8-DIHYDROXY-
FIBROBLASTS
DNA REPAIR, 3081
DNA REPLICATION, 3081

5

ANTI-ANTIBODIES
    IGE
        HISTAMINE LIBERATION, 3060*

ANTIBODIES
    BREAST NEOPLASMS
        ADENOFIBROMA, 3322
        CARCINOMA, 3322
        CYSTS, 3322
        SCLEROSIS, 3322
    COMPLEMENT
        CELL SURVIVAL, 3319
    FORSSMAN ANTIGEN
        CELL SURVIVAL, 3319
    INFECTIOUS MONONUCLEOSIS
        VIRUS, CYTOMEGALO, 3211
        VIRUS, EPSTEIN-BARR, 3211
        VIRUS, HERPES SIMPLEX, 3211
    LIVER NEOPLASMS
        AUSTRALIA ANTIGEN, 3273*
    VIRUS, C-TYPE BOVINE LEUKEMIA
        ASSAY, BEEF AND DAIRY CATTLE,
        3365*

ANTIBODIES, NEOPLASM
    FIBROMA
        VIRUS, BOVINE PAPILLOMA, 3281*
    FIBROSARCOMA
        VIRUS, BOVINE PAPILLOMA, 3281*

    MELANOMA
        IGA, 3321
        IGG, 3321
        IGM, 3321
        IMMUNOGLOBULINS, 3321

ANTIBODIES, VIRAL
    FIBROMA
        VIRUS, BOVINE PAPILLOMA, 3281*
    FIBROSARCOMA
        VIRUS, BOVINE PAPILLOMA, 3281*
        VIRUS, POLYOMA, BK, 3283*
    LEUKEMIA, LYMPHOBLASTIC
        POLIOVIRUS VACCINE, 3364*
    SARCOMA, KAPOSI'S
        VIRUS, CYTOMEGALO, 3212
        VIRUS, EPSTEIN-BARR, 3212
        VIRUS, HERPES SIMPLEX 1, 3212
        VIRUS, HERPES SIMPLEX 2, 3212
    VIRUS, C-TYPE LEUKEMIA
        TRANSMISSION, HORIZONTAL, 3208
    VIRUS, SV40
        VACCINES, 3259*

ANTIGEN-ANTIBODY COMPLEX
    NEOPLASMS
        IMMUNITY, CELLULAR, 3061*
        IMMUNOCHEMISTRY, 3061*

ANTIGEN-ANTIBODY REACTIONS
    BLADDER NEOPLASMS
        ANTIGENS, NEOPLASM, 3226
    BREAST NEOPLASMS
        VIRUS, HERPES SIMPLEX 2, 3218
    CERVIX NEOPLASMS
        VIRUS, HERPES SIMPLEX 2, 3218
    NEOPLASMS
        IMMUNOFLUORESCENT STUDY, REVIEW,

        3056*
    PROSTATIC NEOPLASMS
        ANTIGENS, NEOPLASM, 3226

ANTIGENIC DETERMINANTS
    COLONIC NEOPLASMS
        CARCINOEMBRYONIC ANTIGEN,
    FIBROSARCOMA
        VIRUS, C-TYPE RNA TUMOR, 3
    INTESTINAL NEOPLASMS
        ADENOCARCINOMA, 3300
    LYMPHOSARCOMA
        VIRUS, C-TYPE RNA TUMOR, 3
    PLASMACYTOMA
        MYELOMA PROTEINS, 3317
    PLASMACYTOMA
        IMMUNOGLOBULIN FRAGMENTS,
    VIRUS, MAREK'S DISEASE HERPES
        VIRUS, TURKEY HERPES, 3219
    VIRUS, ROUS SARCOMA
        INCLUSION BODIES, VIRAL, 3

ANTIGENS, NEOPLASM
    ADENOCARCINOMA
        RADIATION, IONIZING, 3195*
    ANTIGENS, VIRAL
        LYMPHOCYTES, 3380*
    BENZ(A)ANTHRACENE, 7,12-DIMETH
        CELL TRANSFORMATION, NEOPL
        3372*
    BENZO(A)PYRENE
        CELL TRANSFORMATION, NEOPL
        3140*
    BLADDER NEOPLASMS
        ANTIGEN-ANTIBODY REACTIONS
    BREAST NEOPLASMS
        IMMUNE RESPONSE, 3298
    BURKITT'S LYMPHOMA
        STABILITY AND DISSOCIATION
    CHOLANTHRENE, 3-METHYL-
        CELL TRANSFORMATION, NEOPL
        3140*, 3372*
    COLONIC NEOPLASMS
        BLOOD-GROUP PRECURSOR ACTI
        3329
        CARCINOEMBRYONIC ANTIGEN,
        CELL MEMBRANE, 3377*
        ISOLATION AND CHARACTERIZA
        3318
    FIBROBLASTS
        CELL TRANSFORMATION, NEOPL
        3372*
    GUANIDINE, 1-METHYL-3-NITRO-1-
        CELL TRANSFORMATION, NEOPL
        3140*, 3372*
    HEPATOMA
        ANILINE, N,N-DIMETHYL-P-PH
        3303
    HODGKIN'S DISEASE
        LYMPHOCYTES, 3370*
    INTESTINAL NEOPLASMS
        RADIATION, IONIZING, 3195*
    LEUKEMIA
        IMMUNITY, CELLULAR, 3055*
        IMMUNOLOGIC TECHNICS, REVI
        3055*

CELLS, CULTURED, 3265*
VIRUS, EPSTEIN-BARR
CELL TRANSFORMATION, NEOPLASTIC,
3326
CHROMATIN, 3210
CONCANAVALIN A, 3326
ISOLATION AND CHARACTERIZATION,
3210
B-LYMPHOCYTES, 3326
T-LYMPHOCYTES, 3326
PLANT AGGLUTININS, 3326
ULTRASTRUCTURAL STUDY, 3210
VIRUS, FELINE LEUKEMIA
EPIDEMIOLOGY, REVIEW, 3015
VIRUS, FRIEND MURINE LEUKEMIA
IMMUNOGLOBULINS, SURFACE, 3310
ISOLATION AND CHARACTERIZATION,
3223
B-LYMPHOCYTES, 3310
T-LYMPHOCYTES, 3310
TRANS-SPECIES REACTIVITY, 3223
VIRUS, GRAFFI MURINE LEUKEMIA
CELL SURFACE ANTIGEN, 3371*
VIRUS, GROSS MURINE LEUKEMIA
GLYCOPROTEINS, 3225
T-LYMPHOCYTES, 3225
VIRUS, HEPATITIS
DETECTION METHOD, 3271*
VIRUS, KIRSTEN MURINE SARCOMA
URIDINE, 5-BROMO-2'-DEOXY-, 3230
VIRUS, MAREK'S DISEASE HERPES
IMMUNE SERUMS, 3219
INCLUSION BODIES, VIRAL, 3221
PATHOGENESIS, 3222
VIRUS, MOLONEY MURINE SARCOMA
T-LYMPHOCYTES, 3356*
VIRUS, MURINE MAMMARY TUMOR
IMMUNOSUPPRESSION, 3340
T-LYMPHOCYTES, 3340
VIRUS, SV40
DEOXYRIBONUCLEASE, 3292*
DNA, VIRAL, 3242, 3244, 3291*
ENDONUCLEASES, 3293*
FLUORESCENT ANTIBODY TECHNIC,
3290*
VACCINES, 3259*

ANTILYMPHOCYTE SERUM
5   VIRUS, FRIEND MURINE LEUKEMIA
IMMUNE RESPONSE, 3310

ANTINEOPLASTIC AGENTS
ACTINOMYCIN D
5   CARCINOGENIC POTENTIAL, REVIEW,
3004
DIETHYLAMINE, 2,2'-DICHLORO-N-METHYL-
CARCINOGENIC POTENTIAL, REVIEW,
3004
METHOTREXATE
CARCINOGENIC POTENTIAL, REVIEW,
3004
NEOPLASMS, EXPERIMENTAL
CARCINOGENIC POTENTIAL, REVIEW,
3032

ANTISERUM

SEE IMMUNE SERUMS

ARGININE
    HEPATOMA
        GUANOSINE TRIPHOSPHATE, 3596*
        PROTEIN DENATURATION, 3581*

ARSENIC
    LUNG NEOPLASMS
        CARCINOGENIC POTENTIAL, REIVEW,
        3026
    SKIN NEOPLASMS
        CARCINOGENIC POTENTIAL, REVIEW,
        3026

ARYL HYDROCARBON HYDROXYLASES
    BENZO(A)PYRENE
        CHOLANTHRENE, 3-METHYL-, 3144*
    CARCINOGEN, ENVIRONMENTAL
        CARCINOGENIC ACTIVITY, MOUSE,
        REVIEW, 3006
    CHOLANTHRENE, 3-METHYL-
        LYMPHOCYTES, 3151*
    DIBENZ(A,H)ANTHRACENE
        LYMPHOCYTES, 3151*
    MICROSOMES
        BENZO(A)PYRENE, 3084
    MICROSOMES, LIVER
        BENZO(A)PYRENE, 3083
        PHENOL, (1,1-DIMETHYL)-4-METHOXY-,
        3083
    POLYCYCLIC HYDROCARBONS
        MECHANISM OF ACTION, 3175*

ASBESTOS
    LARYNGEAL NEOPLASMS
        SMOKING, 3137*
    LUNG NEOPLASMS
        MESOTHELIOMA, 3048*
    MESOTHELIOMA
        EPIDEMIOLOGY, 3529*
        OCCUPATIONAL HAZARD, 3529*
        PRECANCEROUS CONDITIONS, 3529*
    PRECANCEROUS CONDITIONS
        EPIDEMIOLOGY, 3529*
    THORACIC NEOPLASMS
        FIBROSARCOMA, 3138*
        LYMPHOMA, 3138*
        SARCOMA, RETICULUM CELL, 3138*

ASCORBIC ACID
    DIMETHYLAMINE, N-NITROSO-
        PROTECTIVE EFFECT, 3159*
    LIVER NEOPLASMS
        DIMETHYLAMINE, N-NITROSO-, 3159*
        UREA, ETHYL NITROSO-, 3159*

ASPERGILLUS FLAVUS
    AFLATOXIN B1
        REVIEW, 3033*
    SELENIOUS ACID, DISODIUM SALT
        CELL TRANSFORMATION, NEOPLASTIC,
        3133*

ASPERGILLUS FUMIGATUS
    SELENIOUS ACID, DISODIUM SALT
        CELL TRANSFORMATION, NEOPLASTIC,
        3133*

CAPSELLA BURSA-PASTORIS, 3086

BENZENE, 4-ALLYL-1,2-(METHYLENEDIOXY)-
    MICROSOMES, LIVER
        ABSORPTION SPECTRA, RAT, 3088
        CYTOCHROME P-450, 3088

BENZENE, BROMO-
    CARCINOGEN, ENVIRONMENTAL
        CARCINOGENIC ACTIVITY, MOUSE,
            REVIEW, 3006

BENZENE, 1-FLUORO-2,4-DINITRO-
    KERATOACANTHOMA
        IMMUNE RESPONSE, 3406

BENZENEACETAMIDE, ALPHA-ETHYNYL-N,N-
    DIMETHYL-ALPHA-PHENYL-
    HEPATOMA
        LIPOTROPIC DEFICIENT, HIGH-FAT
            DIET, 3096
    STOMACH NEOPLASMS
        LIPOTROPIC DEFICIENT, HIGH-FAT
            DIET, 3096

BENZIDINE
    BLADDER NEOPLASMS
        CARCINOGENIC POTENTIAL, REIVEW,
            3026
    LUNG NEOPLASMS
        ADENOMA, 3089
    LYMPHATIC NEOPLASMS
        RETICULOENDOTHELIAL SYSTEM, 3089
    NEOPLASMS, EXPERIMENTAL
        LACRIMAL APPARATUS, MOUSE, 3089

BENZIMIDAZOLE, 2-(ALPHA-HYDROXYBENZYL)-
    VIRUS, ECHO
        VIRUS REPLICATION, 3010

BENZO(A)PYREN-3-OL
    BENZO(A)PYRENE
        CYTOCHROME P-450, 3085

BENZO(A)PYREN-6-OL
    BENZO(A)PYRENE
        KINETICS, 3142*

BENZO(A)PYREN-9-OL
    BENZO(A)PYRENE
        CYTOCHROME P-450, 3085

BENZO(A)PYREN-6-YLOXY
    BENZO(A)PYRENE
        NADH, NADPH OXIDOREDUCTASES, 3142*

BENZO(A)PYRENE
    ALUMINUM HYDROXIDE
        CARCINOGENIC ACTIVITY, 3141*
    BENZO(A)PYREN-6-OL
        KINETICS, 3142*
    BENZO(A)PYREN-6-YLOXY
        NADH, NADPH OXIDOREDUCTASES, 3142*
    CARCINOGEN, ENVIRONMENTAL
        CARCINOGENIC ACTIVITY, MOUSE,
            REVIEW, 3006
    CARCINOMA
        CELLS, MUTANT RATS, 3431
        KARYOTYPING, 3431

NUCLEOTIDYLTRANSFERASES, 3431
CELL TRANSFORMATION, NEOPLASTIC
   ANTIGENS, NEOPLASM, 3140*
   DOSE-RESPONSE STUDY, 3102
   NEOPLASM TRANSPLANTATION, 3139*
   TRANSPLACENTAL ACTIVITY, 3140*
CHOLANTHRENE, 3-METHYL-
   ARYL HYDROCARBON HYDROXYLASES,
    3144*
   HYDROXYLASES, 3144*
   MICROSOMES, LIVER, 3144*
CYTOCHROME P-450
   BENZO(A)PYREN-3-OL, 3085
   BENZO(A)PYREN-9-OL, 3085
   POSITION-SPECIFIC OXYGENATION,
    3085
LUNG NEOPLASMS
   AIR POLLUTANTS, 3516
MICROSOMES
   ARYL HYDROCARBON HYDROXYLASES,
    3084
MICROSOMES, LIVER
   ARYL HYDROCARBON HYDROXYLASES,
    3083
MUTAGENS
   AZAGUANINE RESISTANCE, HAMSTER,
    3102
SARCOMA
   CHROMOSOME ABERRATIONS, 3101
   KARYOTYPING, 3101

BENZO(E)PYRENE
CELL TRANSFORMATION, NEOPLASTIC
   NEOPLASM TRANSPLANTATION, 3139*
LUNG NEOPLASMS
   AIR POLLUTANTS, 3516

BENZO(G,H,I)PERYLENE
LUNG NEOPLASMS
   AIR POLLUTANTS, 3516

BENZO(K)FLUORANTHENE
LUNG NEOPLASMS
   AIR POLLUTANTS, 3516

5,6-BENZOFLAVONE
CARCINOGEN, ENVIRONMENTAL
   CARCINOGENIC ACTIVITY, MOUSE,
   REVIEW, 3006

7,8-BENZOFLAVONE
MICROSOMES, LIVER
   ENZYMATIC ACTIVITY, 3143*

BENZOPYRENE HYDROXYLASE
SEE ALSO HYDROXYLASES/ARYL HYDROCARBON
  HYROXYLASES

BENZOXAZOLE, 2-AMINO-5-CHLORO-
CARCINOGEN, ENVIRONMENTAL
   CARCINOGENIC ACTIVITY, MOUSE,
   REVIEW, 3006

3',4'-BENZOXYLIDIDE, ALPHA-ETHYNYL-ALPHA-
PHENYL-
LYMPHOCYTES
   DNA REPLICATION, 3097

BENZYL ALCOHOL, ALPHA-(1-(

(METHYLNITROSOAMINO)ETHYL)-
HEPATOMA
   CARCINOGENIC ACTIVITY, 3154*

BENZYL ALCOHOL, 3,4-DIHYDROXY-ALPHA-(
(METHYLAMINO)-METHYL)-
SEE EPINEPHRINE

BENZYLAMINE, N-METHYL-
GASTRIC JUICE
   CARCINOGENIC METABOLITE, 3165*

BENZYLAMINE, N-METHYL-N-NITROSO-
ESOPHAGEAL NEOPLASMS
   DIETARY FACTORS, CHINA, 3514
   PAPILLOMA, 3514

BILE ACIDS AND SALTS
GASTROINTESTINAL TRANSIT TIME
   DIETARY FIBER, 3575

BLADDER NEOPLASMS
ADENOCARCINOMA
   LEUKOPLAKIA, 3426
AIR POLLUTANTS
   EPIDEMIOLOGY, SOCIOECONOMIC
   FACTORS, 3516
ANTIGENS, NEOPLASM
   ANTIGEN-ANTIBODY REACTIONS, 32
BENZIDINE
   CARCINOGENIC POTENTIAL, REIVEW
   3026
CADMIUM OXIDE
   OCCUPATIONAL HAZARD, 3520
CARCINOMA, EPIDERMOID
   LEUKOPLAKIA, 3426
CARCINOMA, TRANSITIONAL CELL
   LEUKOPLAKIA, 3426
CYCLOPHOSPHAMIDE
   ANTINEOPLASTIC AGENTS, 3152*
   CARCINOGENIC POTENTIAL, REVIEW
   3032
FIBROSARCOMA
   LEUKOPLAKIA, 3426
FORMAMIDE, N-(4-(5-NITRO-2-FURYL)-
THIAZOLYL)-
   LIPOTROPIC DEFICIENT, HIGH-FAT
   DIET, 3096
INCLUSION BODIES, VIRAL
   ULTRASTRUCTURAL STUDY, 3226
TOBACCO
   CARCINOGENIC POTENTIAL, REVIEW
   3026
UREA, METHYL NITROSO-
   MODEL TEST SYSTEM, MOUSE, 3152

BLOOD
CARCINOEMBRYONIC ANTIGEN
   ANEMIA, 3324

BLOOD GROUPS
BREAST NEOPLASMS
   ISOANTIGENS, 3375*
COLONIC NEOPLASMS
   CARCINOEMBRYONIC ANTIGEN, 3376
GASTROINTESTINAL NEOPLASMS
   ISOANTIGENS, 3375*
MOUTH NEOPLASMS
   ISOANTIGENS, 3375*

VOUS SYSTEM NEOPLASMS
  ANGIOSARCOMA, 3466*
RYNGEAL NEOPLASMS
  ISOANTIGENS, 3375*
ROID NEOPLASMS
  ISOANTIGENS, 3375*

LATELETS
  S, ROUS SARCOMA
  TRANSFERASES, 3289*

ISCOSITY
TIPLE MYELOMA
  ANEMIA, SICKLE CELL, 3416

RROW
YMPHOCYTES
  CONCANAVALIN A, 3384*
  DIBUTYRYL CYCLIC AMP, 3384*
  THYMUS EXTRACTS, 3384*

RROW CELLS
CINOEMBRYONIC ANTIGEN
  ANEMIA, 3324
KEMIA
  MYELOPEROXIDASE, 3483*
  PRECANCEROUS CONDITIONS, 3483*
  ULTRASTRUCTURAL STUDY, 3483*
KEMIA, MYELOCYTIC
  DIAGNOSIS AND PROGNOSIS, 3487*
  VIRUS-LIKE PARTICLES, 3419
TIPLE MYELOMA
  CHROMOSOME ABERRATIONS, 3422

OPLASMS
LOBLASTOMA
  DIFFERENTIATION, 3438*
IUM
  RADIATION DOSAGE, REVIEW, 3009
OMA
  CASE REPORT, REVIEW, 3437*
SMACYTOMA
  BENCE-JONES PROTEIN, 3527*
  EPIDEMIOLOGY, GERMANY, 3527*
TONIUM
  RADIATION DOSAGE, REVIEW, 3009
CANCEROUS CONDITIONS
  CASE REPORT, 3436*
COMA
  CASE REPORT, 3436*

EOPLASMS
ROCYTOMA
  4-PIPERIDINOL, 1-NITROSO-, 3117
  4-PIPERIDONE, 1-NITROSO-, 3117
NULOMA
  CASE REPORT, 3469*
PHOSARCOMA
  VIRUS, HERPES SAIMIRI, 3277*
ROPHAGES
  ULTRASTRUCTURAL STUDY, RAT, 3114
  UREA, ETHYL NITROSO-, 3114
INGIOMA
  CASE REPORT, 3403
  HISTOLOGIC STUDY, 3403
ED MESENCHYMAL TUMOR
  CASE REPORT, 3403
  HISTOLOGIC STUDY, 3403
ICULOENDOTHELIOSIS

  CASE REPORT, 3469*
UREA, ETHYL NITROSO-
  TRANSPLACENTAL CARCINOGENESIS, RAT,
  3114

BREAST NEOPLASMS
ACID PHOSPHATASE
  ENZYMATIC ACTIVITY, 3445*
ADENOCARCINOMA
  GENETICS, 3511
  KIDNEY FAILURE, CHRONIC, 3338
  NEOPLASM METASTASIS, 3401
  REVERSE TRANSCRIPTASE, 3288*
  VIRUS-LIKE PARTICLES, 3288*
ADENOFIBROMA
  ANTIBODIES, 3322
  HISTOLOGICAL STUDY, 3400
ADENOSINE TRIPHOSPHATASE
  ENZYMATIC ACTIVITY, 3445*
ALKALINE PHOSPHATASE
  ENZYMATIC ACTIVITY, 3445*
AMINOPEPTIDASES
  ENZYMATIC ACTIVITY, 3445*
ANTIGENS, NEOPLASM
  IMMUNE RESPONSE, 3298
BLOOD GROUPS
  ISOANTIGENS, 3375*
CARCINOEMBRYONIC ANTIGEN
  DIAGNOSIS AND PROGNOSIS, REVIEW,
  3019
  NEOPLASM METASTASIS, 3019
CARCINOMA
  ANTIBODIES, 3322
  CASE REPORT, MALE, 3442*
  EPIDEMIOLOGY, MALE, 3442*
  NEOPLASM METASTASIS, 3298
  REVERSE TRANSCRIPTASE, 3288*
  REVIEW, 3439*
  THYMIDINE INCORPORATION, 3538
  VIRUS-LIKE PARTICLES, 3288*
CARCINOMA, DUCTAL
  CALCINOSIS, 3399
  GENETICS, 3511
  HISTOLOGICAL STUDY, 3400
  MAMMOGRAPHY, DIAGNOSTIC VALUE,
  3399
  SIMULTANEOUS OCCURRENCE, HUSBAND-
  WIFE, 3443*
CARCINOMA, EPIDERMOID
  CYSTOSARCOMA PHYLLODES, 3440*
CARCINOMA IN SITU
  HISTOLOGICAL STUDY, 3400
CARCINOMA, PAPILLARY
  REVERSE TRANSCRIPTASE, 3288*
  VIRUS-LIKE PARTICLES, 3288*
CARCINOMA, SCIRRHOUS
  NEOPLASM METASTASIS, 3401
CONTRACEPTIVES, ORAL
  EPIDEMIOLOGY, ENGLAND, 3510
  REVIEW, 3003
CYSTOSARCOMA PHYLLODES
  CASE REPORT, 3440*
CYSTS
  ANTIBODIES, 3322
DERMATOGLYPHICS
  CARCINOGENESIS SCREENING, 3455*
DIAGNOSIS AND PROGNOSIS
  REVIEW, 3439*
EPIDEMIOLOGY

GERMANY, 3506
REVIEW, 3439*
STATISTICAL ANALYSIS, 3528*
ESTERASES
ENZYMATIC ACTIVITY, 3445*
HISTCOMPATIBILITY ANTIGENS
MICROCYTOTOXICITY ASSAY, 3327
HYDROXYBUTYRATE DEHYDROGENASE
ENZYMATIC ACTIVITY, 3446*
ISOCITRATE DEHYDROGENASE
ENZYMATIC ACTIVITY, 3446*
LYMPHANGIOSARCOMA
CASE REPORT, 3397
LYMPHOCYTES
IMMUNITY, CELLULAR, 3298, 3333
PLANT AGGLUTININS, 3333
LYMPHOSARCOMA
CASE REPORT, 3398
NEOPLASM METASTASIS, 3398
MAMMOGRAPHY
EPIDEMIOLOGY, 3528*
MELANOMA
GENETICS, 3407
MENOPAUSE
EPIDEMIOLOGY, 3512
NADH, NADPH OXIDOREDUCTASES
ENZYMATIC ACTIVITY, 3446*
NEOPLASM METASTASIS
CAPILLARITY, 3444*
CLINICOPATHOLOGIC STUDY,
LOCALIZATION, 3401
THYMIDINE INCORPORATION, 3538
PRECANCEROUS CONDITIONS
HISTOLOGICAL STUDY, 3400
REVIEW, 3026
PREGNANCY
EPIDEMIOLOGY, 3512
PUTRESCINE
URINE, 3557
RESERPINE
CARCINOGENIC POTENTIAL, REVIEW,
3035*
SCLEROSIS
ANTIBODIES, 3322
SPERMIDINE
URINE, 3557
SPERMINE
URINE, 3557
STILBENES
ESTRADIOL, 3169*
SUCCINATE DEHYDROGENASE
ENZYMATIC ACTIVITY, 3446*
UTERINE NEOPLASMS
MENOPAUSE, 3099
PREGNANETRIOL CORTISONE RATIO,
3399
STEROID EXCRETION, URINARY, 3099
VIRUS, HERPES SIMPLEX 2
ANTIGEN-ANTIBODY REACTIONS, 3218
ANTIGENS, VIRAL, 3218
WOUNDS AND INJURIES
EPIDEMIOLOGY, 3512

BROMODEOXYURIDINE
SEE URIDINE, 5-BROMO-2'-DEOXY'

5-BROMODEOXYURIDINE
SEE URIDINE, 5-BROMO-2'-DEOXY'

BRONCHIAL NEOPLASMS
CARCINOMA, OAT CELL
NEOPLASMS, MULTIPLE PRIMARY, 3411
SMOKING, 3411
FETAL GLOBULINS
DIAGNOSIS AND PROGNOSIS, 3019
PAPILLOMA
CASE REPORT, 3471*
ULTRASTRUCTURAL STUDY, 3471*

BRONCHITIS
LUNG NEOPLASMS
EPIDEMIOLOGY, 3517

BURKITT'S LYMPHOMA
ANTIGENS, NEOPLASM
STABILITY AND DISSOCIATION, 3368*
ANTIGENS, VIRAL
IGG, 3367*
IMMUNOGLOBULINS, 3367*
LYMPHATIC NEOPLASMS
CLASSIFICATION, 3490*
B-LYMPHOCYTES
CELL DIFFERENTIATION, REVIEW,
3059*
IMMUNOFLUORESCENCE, 3334
IMMUNOGLOBULINS, SURFACE, REVIEW,
3059*
T-LYMPHOCYTES
ROSETTE FORMATION, 3334
LYMPHOSARCOMA
CASE REPORT, 3491*
EPIDEMIOLOGY, 3524*
VIRUS, EPSTEIN-BARR
ANTIGENS, VIRAL, 3011, 3367*
CELL LINE, 3012

1,3-BUTADIENE, 2-CHLORO-
SALMONELLA TYPHIMURIUM
BARBITURIC ACID, 5-ETHYL-5-PHENYL
3145*
MUTAGENESIS, 3145*

2,3-BUTANEDIOL, 1,4-DIMERCAPTO-
MAMMARY NEOPLASMS, EXPERIMENTAL
GLUCOSEPHOSPHATE DEHYDROGENASE,
3567
PRECANCEROUS CONDITIONS
GLUCOSEPHOSPHATE DEHYDROGENASE,
3567

BUTYRIC ACID, 4-AMINO-
PUTRESCINE
METABOLISM, MOUSE, 3571

CADMIUM OXIDE
BLADDER NEOPLASMS
OCCUPATIONAL HAZARD, 3520
LUNG NEOPLASMS
OCCUPATIONAL HAZARD, 3520
PROSTATIC NEOPLASMS
OCCUPATIONAL HAZARD, 3520
TESTICULAR NEOPLASMS
OCCUPATIONAL HAZARD, 3520

CAFFEINE
SKIN NEOPLASMS
ULTRAVIOLET RAYS, 3031
XERODERMA PIGMENTOSUM, 3031

CARBAMIC ACID, N-METHYL-N-NITROSO- ETHYL
ESTER
    PANCREATIC NEOPLASMS
        DNA REPAIR, 3146*

CARBOHYDRATES
    LEUKEMIA L1210
        CELL MEMBRANE, 3346
        CONCANAVALIN A, 3346
        PLANT AGGLUTININS, 3346
    MACROPHAGES
        CELL MEMBRANE, 3346
        CONCANAVALIN A, 3346
        PLANT AGGLUTININS, 3346

CARBON TETRACHLORIDE
    ETHYL ALCOHOL
        ALANINE AMINOTRANSFERASE, 3091
        HEPATOTOXICITY, 3091
    ISOPROPYL ALCOHOL
        HEPATOTOXICITY, 3091

CARBONIC ANHYDRASE
    PITUITARY NEOPLASMS
        ISOENZYMES, 3566
        PROLACTIN, 3566

CARBOXY-LYASES
    PENTANOIC ACID, 5-AMINO-2-HYDRAZINO-2-
    METHYL-
        ENZYME INHIBITORS, 3120

CARCINOEMBRYONIC ANTIGEN
    ANEMIA
        BLOOD, 3324
        BONE MARROW CELLS, 3324
    BREAST NEOPLASMS
        DIAGNOSIS AND PROGNOSIS, REVIEW,
            3019
        NEOPLASM METASTASIS, 3019
    CELL MEMBRANE
        ANTIBODY-INDUCED REDISTRIBUTION,
            3323
    COLONIC NEOPLASMS
        ADENOCARCINOMA, 3019, 3028, 3323,
            3324
        ANTIGENIC DETERMINANTS, 3320
        ANTIGENS, NEOPLASM, 3377*
        BLOOD-GROUP PRECURSOR ACTIVITY,
            3329
        BLOOD GROUPS, 3376*
        CELL MEMBRANE, 3379*
        DIAGNOSIS AND PROGNOSIS, REVIEW,
            3019
        ISOANTIGENS, 3320
        ISOLATION AND CHARACTERIZATION,
            3318, 3320, 3376*
        NEOPLASM METASTASIS, 3019, 3329
    DIGESTIVE SYSTEM
        ISOLATION AND CHARACTERIZATION,
            3028
    LEUKEMIA
        DIAGNOSIS AND PROGNOSIS, REVIEW,
            3019
    LIVER CIRRHOSIS
        DIAGNOSIS AND PROGNOSIS, REVIEW,
            3019
    LYMPHOMA
        DIAGNOSIS AND PROGNOSIS, REVIEW,

3019
NEOPLASMS
ANTIGENS, NEOPLASM, 3378*
IMMUNOFLUORESCENT STUDY, REVIEW,
3356*
REVIEW, 3028
NEUROBLASTOMA
DIAGNOSIS AND PROGNOSIS, REVIEW,
3019
PANCREATIC NEOPLASMS
CARCINOMA, 3019
PULMONARY NEOPLASMS
DIAGNOSIS AND PROGNOSIS, REVIEW,
3019
RECTAL NEOPLASMS
ADENOCARCINOMA, 3323
BLOOD-GROUP PRECURSOR ACTIVITY,
3329
CARCINOMA, 3450*
UROGENITAL NEOPLASMS
DIAGNOSIS AND PROGNOSIS, REVIEW,
3019

CARCINOGEN, CHEMICAL
CELL LINE
CARCINOGENESIS SCREENING, MODEL
SYSTEM, 3127*
CELL TRANSFORMATION, NEOPLASTIC
CELL STRUCTURE, REVIEW, 3001
MAXIMUM TEST DOSE
DOSE-RESPONSE STUDY, 3177*
NEOPLASMS
BIOLOGICAL ASSAY, REVIEW, 3041*
SALMONELLA TYPHIMURIUM
MUTATION, 3044*

CARCINOGEN, ENVIRONMENTAL
ARYL HYDROCARBON HYDROXYLASES
CARCINOGENIC ACTIVITY, MOUSE,
REVIEW, 3006
5,6-BENZOFLAVONE
CARCINOGENIC ACTIVITY, MOUSE,
REVIEW, 3006
BENZOXAZOLE, 2-AMINO-5-CHLORO-
CARCINOGENIC ACTIVITY, MOUSE,
REVIEW, 3006
CYTOCHROME P-450
CARCINOGENIC ACTIVITY, MOUSE,
REVIEW, 3006
GENETICS
CARCINOGENIC ACTIVITY, MOUSE,
REVIEW, 3006
OXYGENASES
CARCINOGENIC ACTIVITY, MOUSE,
REVIEW, 3006

CARCINOMA
BENZO(A)PYRENE
CELLS, MUTANT RATS, 3431
KARYOTYPING, 3431
NUCLEOTIDYLTRANSFERASES, 3431
BREAST NEOPLASMS
ANTIBODIES, 3322
CASE REPORT, MALE, 3442*
EPIDEMIOLOGY, MALE, 3442*
NEOPLASM METASTASIS, 3298

TER, 3250   NEOPLASMS, MULTIPLE PRIMARY, 3411
                                SMOKING, 3411
                        LUNG NEOPLASMS
                                AIR POLLUTANTS, 3516
L, 3513   PELVIC NEOPLASMS
                                LEUKOPLAKIA, 3426
                        TONGUE NEOPLASMS
GOUS   3-PIPERIDINOL, 1-NITROSO-, 3117
3308   TRANSPLANTATION, HOMOLOGOUS
                                IMMUNOSUPPRESSION, 3308
                                KIDNEY, 3308
                      URETHRAL NEOPLASMS
                                LEUKOPLAKIA, 3426
                      VAGINAL NEOPLASMS
                                PRECANCEROUS CONDITIONS, 3430
 3400
STIC VALUE,   CARCINOMA IN SITU
                      BREAST NEOPLASMS
ENCE, HUSBAND-   HISTOLOGICAL STUDY, 3400
                      CERVIX NEOPLASMS
                                EPIDEMIOLOGY, FINLAND, 3508
IS   GLYCOGEN, 3569
TETRA-   GLYCOGEN PHOSPHORYLASE, 3569
                                GLYCOGEN SYNTHETASE, 3569
                                KIDNEY FAILURE, CHRONIC, 3308
                                PRECANCEROUS CONDITIONS, 3498*
 3543   TRANSPLANTATION, HOMOLOGOUS, 3308
EXADECANOATE   VIRUS, HERPES SIMPLEX 2, 3014
030   FALLOPIAN TUBE NEOPLASMS
-(N-2-   ULTRASTRUCTURAL STUDY, 3428
N-N-YL)-   MOUTH NEOPLASMS
                                HISTOLOGICAL STUDY, 3478*

DY, 3160*
, SEMICARBAZONE   CARCINOMA, MUCINOUS
                      STOMACH NEOPLASMS
                                ENZYMATIC ACTIVITY, 3589*
                                HISTOLOGICAL STUDY, 3413
MOUSE, 3590*
E
 3594*   CARCINOMA, OAT CELL
                      BRONCHIAL NEOPLASMS
                                NEOPLASMS, MULTIPLE PRIMARY, 3411
YL)CHLORO-   SMOKING, 3411

                    CARCINOMA, PAPILLARY
                      BREAST NEOPLASMS
                                REVERSE TRANSCRIPTASE, 3288*
                                VIRUS-LIKE PARTICLES, 3288*
                      THYROID NEOPLASMS
3543   KIDNEY FAILURE, CHRONIC, 3308

                    CARCINOMA, RENAL CELL
                      SEE ADENOCARCINOMA

                    CARCINOMA, SCIRRHOUS
                      BREAST NEOPLASMS
                                NEOPLASM METASTASIS, 3401
                      STOMACH NEOPLASMS
DES, 3440*   ENZYMATIC ACTIVITY, 3589*

                    CARCINOMA, SMALL CELL, LUNG
, REVIEW, 3002   SEE CARCINOMA, OAT CELL

L, 3513   CARCINOMA, SQUAMOUS CELL
                    SEE CARCINOMA, EPIDERMOID
TROSO-, 3117
                    CARCINOMA, TRANSITIONAL CELL

BLADDER NEOPLASMS
    LEUKOPLAKIA, 3426
PELVIC NEOPLASMS
    LEUKOPLAKIA, 3426

CARCINOMA 256, WALKER
    CHOLESTEROL, 14-METHYLHEXADECANOATE
        ISOLATION, RAT, 3030

CARCINOSARCOMA
    FALLOPIAN TUBE NEOPLASMS
        ULTRASTRUCTURAL STUDY, 3428

CARDIOSPASM
    ESOPHAGEAL NEOPLASMS
        CASE REPORT, 3409
        DIAGNOSIS AND PROGNOSIS, 3409

CARRIER PROTEINS
    ACETAMIDE, 2-IODO-
        ALDOSTERONE, 3551
        CORTISOL, 3551
        DEXAMETHASONE, 3551
        PROGESTERONE, 3551
    LEUKEMIA, MYELOCYTIC
        VITAMIN B12, 3549
    LEUKOCYTOSIS
        VITAMIN B12, 3549
    POLYCYTHEMIA VERA
        VITAMIN B12, 3549
    1H-PYRROLE-2,5-DIONE, 1-ETHYL-
        ALDOSTERONE, 3551
        CORTISOL, 3551
        DEXAMETHASONE, 3551
        PROGESTERONE, 3551

CELL DIFFERENTIATION
    SEE ALSO GROWTH
    ERYTHROLEUKEMIA
        METHANE, SULFINYLBIS-, 3541
    GLIOMA
        TISSUE CULTURE, 3576*
    INTESTINE, SMALL
        RADIATION, IONIZING, 3184*
    LEUKEMIA, MYELOBLASTIC
        CLINICAL STAGE, 3063*
        GLYCOPROTEINS, 3063*
        LEUKOCYTES, 3063*
        MACROPHAGES, 3063*
    LEUKEMIA, MYELOCYTIC
        LEUKOCYTES, 3063*
        MACROPHAGES, 3063*
    UREA, ETHYL NITROSO-
        NERVE GROWTH FACTOR, 3170*
    VIRUS, FRIEND MURINE LEUKEMIA
        METHANE, SULFINYLBIS-, 3268*
    VIRUS, ROUS SARCOMA
        CELL TRANSFORMATION, NEOPLASTIC,
        3238
        MUSCLE CULTURES, 3238
        TEMPERATURE SENSITIVE MUTANTS,
        3238

CELL DIVISION
    SEE ALSO GROWTH

CELL TRANSFORMATION, NEOPLASTIC
    HYBRIDIZATION, 3577*
CYCLOHEXIMIDE
    RADIATION, 3189*
INTESTINE, SMALL
    RADIATION, IONIZING, 3184*
LYMPHOCYTES
    NUCLEOTIDYLTRANSFERASES, 3572
VIRUS, ADENO 12
    CELL TRANSFORMATION, NEOPLASTI
    3202
VIRUS, SV40
    CELL TRANSFORMATION, NEOPLASTI
    3202

CELL LINE
    ACETAMIDE, N-FLUOREN-2-YL-
        CARCINOGENESIS SCREENING, MODE
        SYSTEM, 3127*
    ANILINE, 4,4'-(IMIDO CARBONYL)BIS(
        DIMETHYL)-
        CARCINOGENESIS SCREENING, MODE
        SYSTEM, 3127*
    CARBAMIC ACID, ETHYL ESTER
        CARCINOGENESIS SCREENING, MODE
        SYSTEM, 3127*
    CARCINOGEN, CHEMICAL
        CARCINOGENESIS SCREENING, MODE
        SYSTEM, 3127*
    CELL TRANSFORMATION, NEOPLASTIC
        CHROMIC ACID, CALCIUM SALT, 30
    DIETHYLAMINE, N-NITROSO-
        CARCINOGENESIS SCREENING, MODE
        SYSTEM, 3127*
    LYMPHOCYTE TRANSFORMATION
        ULTRASTRUCTURAL STUDY, REVIEW,
        3027
    MACROPHAGES
        ISOLATION AND CHARACTERIZATION
        3392*
    PHAGOCYTES
        ISOLATION AND CHARACTERIZATION
        3392*
    VIRUS, EPSTEIN-BARR
        BURKITT'S LYMPHOMA, 3012
        LYMPHOCYTES, 3012

CELL MEMBRANE
    ACETOHYDROXAMIC ACID, N-FLUOREN-2-
        ULTRASTRUCTURAL STUDY, RAT, 30
    CARCINOEMBRYONIC ANTIGEN
        ANTIBODY-INDUCED REDISTRIBUTIO
        3323
    COLONIC NEOPLASMS
        ANTIGENS, NEOPLASM, 3377*
        CARCINOEMBRYONIC ANTIGEN, 3379
    COMPLEMENT
        DNA, 3351*
    FIBROBLASTS
        CONCANAVALIN A, 3349
    FORMAMIDE, N,N-DIMETHYL-
        GLYCOPROTEINS, 3095
    GLYCOPROTEINS
        GROWTH-DEPENDENT ALTERATIONS,
        ISOLATION AND CHARACTERIZATION
        3547
    HELA CELLS
        ACTIN, 3583*

CELL LINE
    CHROMIC ACID, CALCIUM SALT, 3090
CHOLANTHRENE, 3-METHYL-
    ANTIGENS, NEOPLASM, 3143*, 3372*
    CELL STRUCTURE, REVIEW, 3001
    DOSE-RESPONSE STUDY, 3135
    NEOPLASM TRANSPLANTATION, 3139*
    TRANSPLACENTAL ACTIVITY, 3140*
DIETHYLAMINE, N-NITROSO-
    CONCANAVALIN A, 3393*
    PLANT AGGLUTININS, 3393*
DNA
    CARCINOGENESIS SCREENING, 3042*
FIBROBLASTS
    ANTIGENS, NEOPLASM, 3372*
    TEMPERATURE SENSITIVE MUTANTS,
        3297*
    VIRUS, MOLONEY MURINE SARCOMA,
        3356*
GUANIDINE, 1-METHYL-3-NITRO-1-NITROSO-
    ANTIGENS, NEOPLASM, 3143*, 3372*
    TRANSPLACENTAL ACTIVITY, 3140*
LUNG NEOPLASMS
    CHOLANTHRENE, 3-METHYL-, 3474*
ONCOGENIC VIRUSES
    CELL STRUCTURE, REVIEW, 3001
    REVIEW, 3010, 3053*
PITUITARY NEOPLASMS
    ULTRASTRUCTURAL STUDY, REVIEW,
        3020
QUINOLINE, 4-NITRO-, 1-OXIDE
    CELL STRUCTURE, REVIEW, 3001
SARCOMA
    VIRUS, MOLONEY MURINE LEUKEMIA,
        3228
SELENIOUS ACID, DISODIUM SALT
    AFLATOXIN B1, 3133*
    ASPERGILLUS FLAVUS, 3133*
    ASPERGILLUS FUMIGATUS, 3133*
    PENICILLIUM, 3133*
TERATOID TUMOR
    HISTOLOGICAL STUDY, 3536
VIRUS, ADENO 12
    CELL DIVISION, 3202
    CHROMOSOMES, 3202
    CYTOCHALASIN B, 3202
    LACTATE DEHYDROGENASE, 3201
    MALATE DEHYDROGENASE, 3201
    METHANE, SULFINYLBIS-, 3202
    VACCINE CELLS, HAMSTERS, 3354*
VIRUS, C-TYPE RNA TUMOR
    REVIEW, 3053*
VIRUS, CYTOMEGALO
    CELLS, CULTURED, 3265*
VIRUS, EPSTEIN-BARR
    ANTIGENS, VIRAL, 3326
    CARCINOGENIC POTENTIAL, PRIMATES,
        REVIEW, 3052*
VIRUS, FELINE FIBROSARCOMA
    CARCINOGENIC POTENTIAL, PRIMATES,
        REVIEW, 3052*
VIRUS, FRIEND MURINE LEUKEMIA
    VIRUS, MOLONEY MURINE LEUKEMIA,
        3228
VIRUS, HERPES ATELES
    CARCINOGENIC POTENTIAL, PRIMATES,
        REVIEW, 3052*
VIRUS, HERPES SAIMIRI
    CARCINOGENIC POTENTIAL, PRIMATES,

REVIEW, 3052*
VIRUS, HERPES SIMPLEX
    CARCINOGENIC POTENTIAL, PRIMATES,
        REVIEW, 3052*
VIRUS, HERPES SIMPLEX 1
    LEVAMISOLE, 3275*
    NEOPLASM METASTASIS, 3275*
    VACCINE CELLS, HAMSTERS, 3354*
VIRUS, HERPES SIMPLEX 2
    VACCINE CELLS, HAMSTERS, 3354*
VIRUS, KIRSTEN MURINE LEUKEMIA
    VIRUS, MOLONEY MURINE LEUKEMIA,
        3228
VIRUS, MASON-PFIZER MONKEY
    CARCINOGENIC POTENTIAL, PRIMATES,
        REVIEW, 3052*
VIRUS, MOLONEY MURINE SARCOMA-LEUKEMIA
    VIRUS, MOLONEY MURINE LEUKEMIA,
        3228
VIRUS, PAPILLOMA
    CARCINOGENIC POTENTIAL, PRIMATES,
        REVIEW, 3052*
VIRUS, PAPOVA
    GENE PRODUCTS, 3017
VIRUS, POLYOMA
    ACID PHOSPHATASE, 3570
    GALACTOSAMINIDASE, 3570
    GALACTOSIDASES, 3570
    GLUCOSAMINIDASE, 3570
    GLUCOSIDASES, 3570
    LYSOSOMES, 3570
VIRUS, RAUSCHER MURINE LEUKEMIA
    VIRUS, MOLONEY MURINE LEUKEMIA,
        3228
VIRUS, ROUS AVIAN
    DNA, VIRAL, 3261*
VIRUS, ROUS SARCOMA
    CARCINOGENIC POTENTIAL, PRIMATES,
        REVIEW, 3052*
    CELL DIFFERENTIATION, 3238
    CELL MEMBRANE, 3289*, 3435
    DNA, VIRAL, 3261*
    GLYCEROL, 3435
    TEMPERATURE SENSITIVE MUTANTS,
        3435
VIRUS, SIMIAN ADENO
    CARCINOGENIC POTENTIAL, PRIMATES,
        REVIEW, 3052*
VIRUS, SV40
    CAFFEINE, 3249
    CARCINOGENIC POTENTIAL, PRIMATES,
        REVIEW, 3052*
    CELL DIVISION, 3202
    CELL MEMBRANE, 3435
    CHROMOSOMES, 3202
    CYTOCHALASIN B, 3202
    DIBUTYRYL CYCLIC AMP, 3249
    DNA, VIRAL, 3243
    GLYCEROL, 3435
    METHANE, SULFINYLBIS-, 3202
    RADIATION, 3249
    TEMPERATURE SENSITIVE MUTANTS,
        3251
    THEOPHYLLINE, 3249
    ULTRAVIOLET RAYS, 3249

ELLS, CULTURED
PANCREAS
    MODEL SYSTEM, 3539

CELLULAR INCLUSIONS
    LEUKEMIA, LYMPHOCYTIC
        IMMUNOGLOBULINS, 3338

CEREBELLAR NEOPLASMS
    ANGIOMATOSIS
        GENETICS, 3454*

CERVIX NEOPLASMS
    ADENOCARCINOMA
        CASE REPORT, 3499*, 3500*
        DNA, NEOPLASM, 3499*
        ENDOMETRIOSIS, 3098
        EPIDEMIOLOGY, 3497*
        ESTROGENS, 3098
        HISTOLOGICAL STUDY, 3497*
        PRECANCEROUS CONDITIONS, 3499*
    CARCINOMA
        IMMUNOSUPPRESSION, 3311
        VIRUS, HERPES SIMPLEX 2, 3276*
    CARCINOMA IN SITU
        EPIDEMIOLOGY, FINLAND, 3508
        GLYCOGEN, 3569
        GLYCOGEN PHOSPHORYLASE, 3569
        GLYCOGEN SYNTHETASE, 3569
        PRECANCEROUS CONDITIONS, 3498*
        TRANSPLANTATION, HOMOLOGOUS, 3
        VIRUS, HERPES SIMPLEX 2, 3014
    CARCINOMA IN STIU
        KIDNEY FAILURE, CHRONIC, 3308
    EPIDEMIOLOGY
        STATISTICAL ANALYSIS, REVIEW,
    ESTROGENS
        HORMONE DEPENDENT, 3173*
    PRECANCEROUS CONDITIONS
        EPIDEMIOLOGY, FINLAND, 3508
        GLYCOGEN, 3569
        GLYCOGEN PHOSPHORYLASE, 3569
        GLYCOGEN SYNTHETASE, 3569
        REVIEW, 3026
    VIRUS, HERPES SIMPLEX 1
        EPIDEMIOLOGY, REVIEW, 3014
    VIRUS, HERPES SIMPLEX 2
        ANTIGEN-ANTIBODY REACTIONS, 32
        ANTIGENS, VIRAL, 3218
        CYTOPATHOLOGICAL STUDY, REVIEW
            3014
        DNA, VIRAL, 3276*
        EPIDEMIOLOGY, REVIEW, 3014
        IMMUNOLOGICAL STUDY, REVIEW, 3

CESIUM
    ACCUMULATION
        CROP EFFECT, 3148*

CHALONES
    ADRENALINE
        CELL CYCLE KINETICS, 3029
    ENZYMES
        CELL CYCLE KINETICS, 3029
    B-LYMPHOCYTES
        SIALIC ACIDS, 3029
    T-LYMPHOCYTES
        SIALIC ACIDS, 3029
    MELANOMA
        CORTISOL, 3029
        DNA REPLICATION, 3029
        RNA, 3029

LYMPHATIC NEOPLASMS
    MACROPHAGES, 3392*
    PHAGOCYTES, 3392*
METHANE, SULFINYLBIS-
    CELLS, CULTURED, 3095
MUTAGENS
    AZAGUANINE RESISTANCE, HAMSTER,
      3132
PROSTATIC NEOPLASMS
    PRECANCEROUS CONDITIONS, 3136
    RETINOIC ACID, 3150*
PUROMYCIN
    CELLS, CULTURED, 3095
SARCOMA
    BLOCKING FACTOR SYNTHESIS, 3306
    CHROMOSOME ABERRATIONS, 3101
    IMMUNITY, CELLULAR, 3305, 3306
    IMMUNOSUPPRESSION, 3309
    KARYOTYPING, 3101
    TRANSPLANTATION IMMUNOLOGY, 3309
    VIRUS-LIKE PARTICLES, 3270*
SKIN NEOPLASMS
    HISTOLOGICAL STUDY, MOUSE, 3107
THEOPHYLLINE
    CELLS, CULTURED, 3095
URIDINE, 5-BROMO-2'-DEOXY-
    CELLS, CULTURED, 3095
VIRUS, MOLONEY MURINE SARCOMA
    DNA-RNA HYBRIDIZATION, 3207
VIRUS, POLYOMA
    DNA-RNA HYBRIDIZATION, 3207
VIRUS, RAUSCHER LEUKEMIA
    DNA-RNA HYBRIDIZATION, 3207

CHOLESTEROL
    SERUM LEVELS
      DIETARY FIBER, 3575

CHOLESTEROL, 14-METHYLHEXADECANOATE
    ACYLTRANSFERASES
      REVIEW, 3030
    AMINO ACYL T RNA SYNTHETASES
      REVIEW, 3030
    CARCINOMA, EHRLICH TUMOR
      ISOLATION, MOUSE, 3030
    CARCINOMA 256, WALKER
      ISOLATION, RAT, 3030
    GENETIC CODE
      REVIEW, 3030
    NEOPLASMS
      CHOLESTEROL METABOLISM, REVIEW,
        3030
    OVARIAN NEOPLASMS
      ISOLATION, RAT, 3030
    PROTEINS
      REVIEW, 3030
    REVERSE TRANSCRIPTASE
      REVIEW, 3030
    RNA MESSENGER
      REVIEW, 3030
    SARCOMA
      ISOLATION, MOUSE, 3030
    SARCOMA 180, CROCKER
      ISOLATION, MOUSE, 3030

CHONDROID SYRINGOMA
    SEE NEOPLASMS, EMBRYONAL AND MIXED

CHONDROMA

LARYNGEAL NEOPLASMS
  HISTOLOGICAL STUDY, REVIEW, 3023

CHONDROMYXOID FIBROMA
  SEE CHONDROMA

CHONDROSARCOMA
  LARYNGEAL NEOPLASMS
    HISTOLOGICAL STUDY, REVIEW, 3023
  ORBITAL NEOPLASMS
    ULTRASTRUCTURAL STUDY, REVIEW,
      3064*

CHORIOADENOMA
  UROGENITAL NEOPLASMS
    GONADOTROPINS, CHORIONIC, 3427
    LACTATE DEHYDROGENASE, 3427

CHORIOCARCINOMA
  HYDATIDFORM MOLE
    EPIDEMIOLOGY, 3071*
  THYROTROPIN
    NEOPLASM TRANSPLANTATION, 3578*
  UROGENITAL NEOPLASMS
    GONADOTROPINS, CHORIONIC, 3427
    LACTATE DEHYDROGENASE, 3427

CHROMATIN
  DIMETHYLAMINE, N-NITROSO-
    ALKYLATING AGENTS, 3156*
  HEPATOMA
    DNA, 3560
  LEUKEMIA
    CELL NUCLEUS, 3587*
  LEUKEMIA, LYMPHOBLASTIC
    PROTEINS, 3580*
  LEUKEMIA, LYMPHOCYTIC
    PROTEINS, 3580*
  LEUKEMIA, MYELOCYTIC
    PROTEINS, 3580*
  MAMMARY NEOPLASMS, EXPERIMENTAL
    ADENOCARCINOMA, 3587*
    UREA, METHYLNITROSO-, 3587*
  VAGINAL NEOPLASMS
    CELL NUCLEUS, 3587*
  VIRUS, EPSTEIN-BARR
    ANTIGENS, VIRAL, 3210

CHROMIC ACID, CALCIUM SALT
  CELL LINE
    CELL TRANSFORMATION, NEOPLASTIC,
      3090

CHROMIUM
  LUNG NEOPLASMS
    CARCINOGENIC POTENTIAL, REIVEW,
      3026

CHROMOSOME ABERRATIONS
  ADRENALECTOMY
    MUTAGENIC ACTIVITY, REVIEW, 3049*
  CORTISOL
    MUTAGENIC ACTIVITY, REVIEW, 3049*
  DROSOPHILA MELANOGASTER
    RADIATION, IONIZING, 3190*, 3193*
  ESTRADIOL
    MUTAGENIC ACTIVITY, REVIEW, 3049*
  ETHANOL, 2-MERCAPTO-
    LYMPHOCYTES, 3424

PLANT AGGLUTININS, 3424
LEUKEMIA, MYELOCYTIC
  ULTRASTRUCTURAL STUDY, 3489*
LYMPHOCYTES
  RADIATION, IONIZING, 3194*
LYMPHOCYTES
  MITOMYCIN C, 3103
MITOMYCIN C
  ANALYTICAL METHOD, 3103
MULTIPLE MYELOMA
  ALANINE, 3-(P-(BIS(2-CHLOROETH
    AMINO)PHENYL)-, L-, 3422
  BONE MARROW CELLS, 3422
  CYCLOPHOSPHAMIDE, 3422
PROGESTERONE
  MUTAGENIC ACTIVITY, REVIEW, 30
RADIATION, IONIZING
  CELL CYCLE KINETICS, 3183, 345
  GENETICS, 3456*
  PLANT AGGLUTININS, 3183
  RADIATION DOSAGE, 3191*
SARCOMA
  BENZO(A)PYRENE, 3101
  CHOLANTHRENE, 3-METHYL-, 3101
TESTOSTERONE
  MUTAGENIC ACTIVITY, REVIEW, 30
TRANSPLANTATION, HOMOLOGOUS
  MUTAGENIC ACTIVITY, REVIEW, 30
VIRUS, HERPES SIMPLEX
  MUTAGENIC ACTIVITY, REVIEW, 30

CHROMOSOME ABNORMALITIES
  ALLIUM CEPA
    SMOKING, 3121
  CHOLANTHRENE, 3-METHYL-
    DOSE-RESPONSE STUDY, 3105
  LEUKEMIA, MYELOBLASTIC
    CASE REPORT, 3423
    KARYOTYPING, 3423
  PLANTS
    SMOKING, 3121

CHROMOSOME MAPPING
  VIRUS, SV40
    TEMPERATURE SENSITIVE MUTANTS,
      3018

CHROMOSOMES
  HELA CELLS
    DNA REPLICATION, 3585*
    RNA, 3585*
  LEUKEMIA, LYMPHOCYTIC
    DNA, 3556
  VIRUS, ADENO 12
    CELL TRANSFORMATION, NEOPLASTI
      3202
  VIRUS, CYTOMEGALO
    CELLS, CULTURED, 3265*
  VIRUS, SV40
    CELL TRANSFORMATION, NEOPLASTI
      3202

CHROMSOME ABERRATIONS
  OVUM
    RADIATION, 3191*
    RADIATION, IONIZING, 3191*

CHRONIC DISEASE
  LEUKEMIA

HISTCOMPATIBILITY ANTIGENS
    MICROCYTOTOXCITY ASSAY, 3327
HYDRAZINE, 1,2-DIMETHYL-
    AUTORADIOGRAPHY, 3162*
    DNA REPLICATION, 3162*
HYDROXYBUTYRATE DEHYDROGENASE
    ENZYMATIC ACTIVITY, 3446*
IMMUNITY, CELLULAR
    CHROMIUM RELEASE ASSAY, 3301
INTESTINAL POLYPS
    CASE REPORT, 3447*
    NEOPLASM METASTASIS, 3447*
ISOCITRATE DEHYDROGENASE
    ENZYMATIC ACTIVITY, 3446*
LYMPH NODES
    PROGNOSIS, 3449*
NADH, NADPH OXIDOREDUCTASES
    ENZYMATIC ACTIVITY, 3446*
POLYPS
    GENETICS, 3367*
SUCCINATE DEHYDROGENASE
    ENZYMATIC ACTIVITY, 3446*

COLORING AGENTS
    CARCINOGEN, CHEMICAL
        REVIEW, 3176*

COMPLEMENT
    ANTIBODIES
        CELL SURVIVAL, 3319
    DNA
        CELL MEMBRANE, 3351*
    LYMPHOCYTES
        IMMUNE RESPONSE, K CELL, 3304

CONCANAVALIN A
    BONE MARROW
        T-LYMPHOCYTES, 3384*
    CELL TRANSFORMATION, NEOPLASTIC
        DIETHYLAMINE, N-NITROSO-, 3393*
    DNA REPLICATION
        LYMPHOCYTE TRANSFORMATION, 3395*
    FIBROBLASTS
        CELL MEMBRANE, 3349
    FORMAMIDE, N,N-DIMETHYL-
        CELLS, CULTURED, 3095
    L CELLS
        BINDING, 3546
    LEUKEMIA L1210
        BINDING, 3546
        CARBOHYDRATES, 3346
    LEUKEMIA, MYELOCYTIC
        BINDING, 3546
    LYMPHOCYTE TRANSFORMATION
        PUTRESCINE, 3120
    LYMPHOCYTES
        IMMUNITY, CELLULAR, 3348
        RECEPTOR MOBILITY, 3394*
    T-LYMPHOCYTES
        IMMUNE RESPONSE, CHICKEN, 3339
        IMMUNOGLOBULINS, FC, 3343
    MACROPHAGES
        CARBOHYDRATES, 3346
    METHANE, SULFINYLBIS-
        CELLS, CULTURED, 3095
    URIDINE, 5-BROMO-2'DEOXY-

CELLS, CULTURED, 3095
VIRUS, C-TYPE MURINE
VIRUS, KIRSTEN MURINE SARCOMA,
3206
VIRUS, EPSTEIN-BARR
ANTIGENS, VIRAL, 3326
VIRUS, SV40
AMINO ACIDS, 3296*
CELL MEMBRANE, 3349
IMMUNITY, CELLULAR, 3347
IMMUNOSUPPRESSION, 3347
TRANSPLANTATION IMMUNOLOGY, 3347

CONTRACEPTIVES, ORAL
BREAST NEOPLASMS
EPIDEMIOLOGY, ENGLAND, 3510
REVIEW, 3003

CORONARY DISEASE
EPIDEMIOLOGY
DIETARY FACTORS, REVIEW, 3504

CORONENE
LUNG NEOPLASMS
AIR POLLUTANTS, 3516

CORTICOSTERONE
MAMMARY NEOPLASMS, EXPERIMENTAL
BINDING, COMPETITIVE, 3550

CORTICOTROPIN
ADRENAL GLAND NEOPLASMS
STEROIDS, 3402
BENZ(A)ANTHRACENE
PENTANAMIDE, 2-ISOPROPYL-, 3094
4-PENTENAMIDE, 2-ISOPROPYL-, 3094

CORTISOL
ACETAMIDE, 2-IODO-
CARRIER PROTEINS, 3551
CHROMOSOME ABERRATIONS
MUTAGENIC ACTIVITY, REVIEW, 3049*
MAMMARY NEOPLASMS, EXPERIMENTAL
BINDING, COMPETITIVE, 3550
MELANOMA
CHALONES, 3029
1H-PYRROLE-2,5-DIONE, 1-ETHYL-
CARRIER PROTEINS, 3551
VIRUS, MURINE SARCOMA
VIRUS REPLICATION, 3280*
VIRUS, POX
VIRUS REPLICATION, 3255

CRANIOPHARYNGIOMA
PUTRESCINE
URINE, 3557
SPERMIDINE
URINE, 3557
SPERMINE
URINE, 3557

CRESOL, 2-(2-BENZIMIDAZOLYL)-
VIRUS, POLIO
CYTOTOXICITY, REVIEW, 3010

CROCIDOLITE
SEE ASBESTOS

CROTON OIL

AMNION
ADENOSINE CYCLIC 3',5'
MONOPHOSPHATE, 3129*
DNA REPAIR, 3129*
FIBROBLASTS
DNA REPAIR, 3081
DNA REPLICATION, 3081
NEOPLASMS
CARCINOGENIC ACTIVITY, 3164*
NEOPLASMS, EXPERIMENTAL
DNA REPLICATION, 3031

CYCLIC AMP
SEE ADENOSINE CYCLIC 3',5'
MONOPHOSPHATE

CYCLIC AMP PHOSPHODIESTERASE
SEE PHOSPHODIESTERASES

CYCLOHEXAMIDE
CHOLANTHRENE, 3-METHYL-
CELLS, CULTURED, 3095

CYCLOHEXANE, 1,2,3,4,5,6-HEXACHLORO-
LIVER NEOPLASMS
HEPATOMA, 3087

CYCLOHEXANE, 1,2,3,4,5,6-HEXACHLORO-
GAMMA ISOMER
CARCINOGEN, ENVIRONMENTAL
CARCINOGENIC ACTIVITY, MOUSE
REVIEW, 3006

CYCLOHEXIMIDE
ADRENAL GLAND NEOPLASMS
STEROIDS, 3402
RADIATION
CELL DIVISION, 3189*

CYCLOPHOSPHAMIDE
ACETALDEHYDE, CHLORO-
CARCINOGENIC METABOLITE, 312
BLADDER NEOPLASMS
ANTINEOPLASTIC AGENTS, 3152*
CARCINOGENIC POTENTIAL, REVI
3032
LUNG NEOPLASMS
CARCINOGENIC POTENTIAL, REVI
3032
MAMMARY NEOPLASMS, EXPERIMENTAL
CARCINOGENIC POTENTIAL, REVI
3032
MULTIPLE MYELOMA
CHROMOSOME ABERRATIONS, 3422

CYLINDROMA
LARYNGEAL NEOPLASMS
REVIEW, 3066*
TRACHEAL NEOPLASMS
CASE REPORT, 3470*
ULTRASTRUCTURAL STUDY, 3470*

CYSTADENOCARCINOMA
OVARIAN NEOPLASMS
IMMUNOGLOBULINS, 3361*

CYSTADENOMA, MUCINOUS
OVARIAN NEOPLASMS
IMMUNOGLOBULINS, 3361*

DIAGNOSIS AND PROGNOSIS, 3451*

DEXAMETHASONE
    ACETAMIDE, 2-IODO-
        CARRIER PROTEINS, 3551
    MAMMARY NEOPLASMS, EXPERIMENTAL
        BINDING, COMPETITIVE, 3550
    1H-PYRROLE-2,5-DIONE, 1-ETHYL-
        CARRIER PROTEINS, 3551
    VIRUS, MURINE SARCOMA
        VIRUS REPLICATION, 3280*
    VIRUS, POX
        VIRUS REPLICATION, 3255

DEXTRAN
    B-LYMPHOCYTES
        IMMUNE RESPONSE, CHICKEN, 3339

DIABETES INSIPIDUS
    PITUITARY NEOPLASMS
        NEOPLASM METASTASIS, 3401

IC,

DIAPHRAGMATIC HERNIA
    EPIDEMIOLOGY
IC,   DIETARY FACTORS, REVIEW, 3504

DIBENZ(A,H)ANTHRACENE
    ARYL HYDROCARBON HYDROXYLASES
        LYMPHOCYTES, 3151*

DIBENZO(DEF,MNO)CHRYSENE
    LUNG NEOPLASMS
        AIR POLLUTANTS, 3516

DIBUTYRYL CYCLIC AMP
    BONE MARROW
        T-LYMPHOCYTES, 3384*
    CHOLANTHRENE, 3-METHYL-
        CELLS, CULTURED, 3095
    SARCOMA, MAST CELL
        GROWTH, 3542
    VIRUS, SV40
        CELL TRANSFORMATION, NEOPLASTIC,
        3249

DIET
    SALMONELLA TYPHIMURIUM
        MUTAGENIC EFFECT, IRRADIATION,
        3188*
    STOMACH NEOPLASMS
        PRECANCEROUS CONDITIONS, 3026

DIETARY FATS
    COLONIC NEOPLASMS
        REVIEW, 3034*

DIETHYLAMINE, 2,2'-DICHLORO-N-METHYL-
    ANTINEOPLASTIC AGENTS
        CARCINOGENIC POTENTIAL, REVIEW,
        3004
R   BACTERIOPHAGES
        RNA, 3111
    METHANESULFONIC, ACID, METHYL ESTER
        URINARY EXCRETION, 3108

DIETHYLAMINE, N-NITROSO-
    CARCINOMA
        CELLS, MUTANT RATS, 3431
        KARYOTYPING, 3431

NUCLEOTIDYLTRANSFERASES, 3431
CELL LINE
  CARCINOGENESIS SCREENING, MODEL
    SYSTEM, 3127*
CELL TRANSFORMATION, NEOPLASTIC
  CONCANAVALIN A, 3393*
  PLANT AGGLUTININS, 3393*
ESOPHAGEAL NEOPLASMS
  DIETARY FACTORS, CHINA, 3514
HEPATOMA
  ADENOSINE TRIPHOSPHATASE, 3432
  CARCINOGENIC ACTIVITY, 3154*
  GLUCOSEPHOSPHATASE, 3432
  GLYCOGEN, 3432
  GROWTH, 3432
  IMMUNE RESPONSE, 3319
LIVER NEOPLASMS
  ADENOSINE TRIPHOSPHATASE, 3432
  GLUCOSEPHOSPHATASE, 3432
  GLYCOGEN, 3432
  GROWTH, 3432
  PRECANCEROUS CONDITIONS, 3432
SALMONELLA TYPHIMURIUM
  MICROSOMES, LIVER, 3158*
  MUTAGENESIS, 3158*

DIETHYLNITROSAMINE
  SEE DIETHYLAMINE, N-NITROSO-

DIETHYLSTILBESTROL
  SEE 4,4'-STILBENEDIOL, ALPHA, ALPHA'-
    DIETHYL-

DIGESTIVE SYSTEM
  CARCINOEMBRYONIC ANTIGEN
    ISOLATION AND CHARACTERIZATION,
      3028

DIGESTIVE SYSTEM NEOPLASMS
  PETROLEUM
    OCCUPATIONAL HAZARD, 3520

DIHYDROAFLATOXIN B1
  SEE AFLATOXIN B2

DIMETHYL BENZANTHRACENE
  SEE BENZ(A)ANTHRACENE, 7,12-DIMETHYL-

DIMETHYL SULFOXIDE
  SEE METHANE, SULFINYLBIS-

DIMETHYLAMINE, N-NITROSO-
  ASCORBIC ACID
    PROTECTIVE EFFECT, 3159*
  CHROMATIN
    ALKYLATING AGENTS, 3156*
  DNA
    ALKYLATING AGENTS, 3156*, 3157*
  ESOPHAGEAL NEOPLASMS
    DIETARY FACTORS, CHINA, 3514
  ETHYL ALCOHOL
    ALANINE AMINOTRANSFERASE, 3091
  KIDNEY NEOPLASMS
    RNA, 3155*
  LIVER NEOPLASMS
    ASCORBIC ACID, 3159*
    RNA, 3115
  PROTEINS
    ALKYLATING AGENTS, 3157*

RNA
  ALKYLATING AGENTS, 3157*
SALMONELLA TYPHIMURIUM
  MICROSOMES, LIVER, 3158*
  MUTAGENESIS, 3158*

DIMETHYLAMINOAZOBENZENE
  SEE ANILINE, N,N-DIMETHYL-P-PHENY

1,2-DIMETHYLHYDRAZINE
  SEE HYDRAZINE, 1,2-DIMETHYL-

DIMETHYLNITROSAMINE
  SEE DIMETHYLAMINE, N-NITROSO-

DISGERMINOMA
  ORBITAL NEOPLASMS
    ULTRASTRUCTURAL STUDY, REVIEW,
      3364*
  OVARIAN NEOPLASMS, 3311
    IMMUNOGLOBULINS, 3361*
    IMMUNOSUPPRESSION, 3311
    TRANSPLANTATION, HOMOLOGOUS,
    ULTRASTRUCTURAL STUDY, 3502*
  TESTICULAR NEOPLASMS
    CASE REPORT, 3475*
    KIDNEY FAILURE, CHRONIC, 3308

DIVERTICULITIS, COLONIC
  EPIDEMIOLOGY
    DIETARY FACTORS, REVIEW, 3504

DIVERTICULOSIS, COLONIC
  EPIDEMIOLOGY
    DIETARY FACTORS, REVIEW, 3504

DNA
  ACETAMIDE, 2-IODO-
    BINDING SITES, 3551
  AFLATOXIN B1
    BINDING, RAT, KIDNEY, 3077
    BINDING, RAT, LIVER,, 3077
    DENATURATION, HEAT, 3076
    HISTONES, 3076
  CARCINOMA, EHRLICH TUMOR
    2-FURALDEHYDE, 5-NITRO-,
      SEMICARBAZONE, 3118
  CARCINOMA, EPIDERMOID
    CELL CYCLE KINETICS, REVIEW, 3
  CELL TRANSFORMATION, NEOPLASTIC
    CARCINOGENESIS SCREENING, 3042
  COMPLEMENT
    CELL MEMBRANE, 3351*
  DIMETHYLAMINE, N-NITROSO-
    ALKYLATING AGENTS, 3156*, 3157
  DUODENAL ULCER
    TOBACCO, 3172*
  GASTRIC JUICE
    TOBACCO, 3172*
  GASTRITIS
    TOBACCO, 3172*
  HELA CELLS
    DNA POLYMERASE, 3584*
    NAD NUCLEOSIDASE, 3584*
  HEPATOMA
    ACETAMIDE, N-FLUOREN-2-YLOI-,
      3128*
    ANILINE, N,N-DIMETHYL-P-PHENYL
      3128*

AMNION
    ACETAMIDE N-(ACETYLOXY)-N-9H-
        FLUOREN-2-YL-, 3129*
    CROTON OIL, 3129*
    12-O-TETRADECANOYLPHORBOL-13-
        ACETATE, 3129*
    ULTRAVIOLET RAYS, 3129*
FIBROBLASTS
    ACETAMIDE, N-(ACETYLOXY)-N-9H-
        FLUOREN-2-YL-, 3081
    1,8,9-ANTHRACENETRIOL, 3081
    ANTHRAQUINONE, 1,8-DIHYDROXY-,
        3081
    CROTON OIL, 3081
    PHENOL, 3081
    PHORBOL, 3081
    12-O-TETRADECANOYLPHORBOL-13-
        ACETATE, 3081
HELA CELLS
    VIRUS, SENDAI, 3031
LEUKEMIA
    DNA, SINGLE STRANDED, 3561
LEUKEMIA, LYMPHOCYTIC
    DNA, 3556
NEOPLASMS
    REVIEW, 3031
NEOPLASMS, EXPERIMENTAL
    12-O-TETRADECANOYLPHORBOL-13-
        ACETATE, 3031
PANCREATIC NEOPLASMS
    CARBAMIC ACID, N-METHYL-N-NITROSO-
        ETHYL ESTER, 3146*
STRUCTURE-ACTIVITY RELATIONSHIP
    CARCINOGENESIS SCREENING, 3039*
    MUTAGENESIS SCREENING, 3039*
VIRUS, CYTOMEGALO
    DNA, VIRAL, 3220

DNA REPLICATION
    COLONIC NEOPLASMS
        HYDRAZINE, 1,2-DIMETHYL-, 3162*
    CONCANAVALIN A
        LYMPHOCYTE TRANSFORMATION, 3395*
    FIBROBLASTS
        ACETAMIDE, N-(ACETYLOXY)-N-9H-
           FLUOREN-2-YL-, 3081
        1,8,9-ANTHRACENETRIOL, 3081
        ANTHRAQUINONE, 1,8-DIHYDROXY-,
           3081
        CROTON OIL, 3081
        PEPTIDE HYDROLASES, 3586*
        PHENOL, 3081
        PHENOL, P-NITRO-, 3081
        PHORBOL, 3081
        12-O-TETRADECANOYLPHORBOL-13-
           ACETATE, 3081
    HEAD AND NECK NEOPLASMS
        GLIOMA, 3534*
    HELA CELLS
        CELL CYCLE KINETICS, 3585*
        CHROMOSOMES, 3585*
        PULSE CHASE STUDY, CELL NUCLEI,
           3559
        RADIATION, IONIZING, 3182
    LYMPHOCYTES
        ACETAMIDE, N-FLUOREN-2-YL-, 3097
        ACETOHYDROXAMIC ACID, N-FLUOREN-2-
           YL-, 3097
        3',4'-BENZOXYLIDIDE, ALPHA-ETHYNYL-

ALPHA-PHENYL-, 3097
HYDANTOIN, 5,5-DIPHENYL-, 3097
HYDANTOIN, 3-HYDROXY-, 3097
HYDANTOIN, 3-HYDROXY-1-ACETYL-5,5-
  DIPHENYL-, 3097
HYDANTOIN, 3-HYDROXY-5,5-DIPHENYL-,
  3097
HYDANTOIN, 3-HYDROXY-5-ISOPROPYL-,
  3097
HYDANTOIN, 3-HYDROXY-5-METHYL-,
  3097
HYDANTOIN, 5-(4-HYDROXYPHENYL)-5-
  PHENYL-, 3097
MELANOMA
  CHALONES, 3029
NEOPLASMS, EXPERIMENTAL
  CROTON OIL, 3031
RADIATION, IONIZING
  CELL CYCLE KINETICS, 3456*
  GENETICS, 3456*
VIRUS, HERPES SIMPLEX 1
  ZINC SULFATE, 3217
VIRUS, MENGO
  STREPTOVARICINS, 3010
VIRUS, SV40
  HYBRIDIZATION, 3241

DNA, SINGLE STRANDED
  ANEMIA, APLASTIC
    LYMPHOCYTE TRANSFORMATION, 3561
    PRECANCEROUS CONDITIONS, 3561
  LEUKEMIA
    DNA REPAIR, 3561
  VIRUS, ADENO 2
    ISOLATION AND CHARACTERIZATION,
    3257*

DNA, VIRAL
  ALDRICH SYNDROME
    VIRUS, POLYOMA, BK, 3284*
  BACTERIOPHAGES
    BINDING ASSAY, 3242
  CERVIX NEOPLASMS
    VIRUS, HERPES SIMPLEX 2, 3276*
  LEUKEMIA
    REVIEW, 3054*
  NEOPLASMS
    VIRUS, SIMIAN ADENO, 3203
  2-OXETANONE
    MUTAGENIC ACTIVITY, 3166*
  VIRUS, ADENO 2
    ENDONUCLEASES, 3198
    RNA, MESSENGER, 3198
  VIRUS, ADENO 5
    VIRAL PROTEINS, 3196
  VIRUS, CYTOMEGALO
    DNA REPAIR, 3220
    IODINATION, 3220
  VIRUS, HERPES SIMPLEX 1
    ENDONUCLEASES, 3214
    STRUCTURAL ANALYSIS, 3214
    TEMPERATURE SENSITIVE MUTANTS,
    3215
  VIRUS, HERPES SIMPLEX 2
    ENDONUCLEASES, 3220
    IODINATION, 3220
  VIRUS, PAPOVA, BK
    ISOLATION AND CHARACTERIZATION,
    3236

VIRUS, POLYOMA
  BINDING ASSAY, 3242
  COMPLEX FORMATION, 3235
  DNA-RNA HYBRIDIZATION, 3233
  VIRUS REPLICATION, 3236
VIRUS, POLYOMA 1
  VIRAL PROTEINS, 3282*
VIRUS, POLYOMA 2
  VIRAL PROTEINS, 3282*
VIRUS, ROUS AVIAN
  CELL TRANSFORMATION, NEOPLAST
  3261*
VIRUS, ROUS SARCOMA
  CELL TRANSFORMATION, NEOPLAST
  3261*
  DNA POLYMERASE, 3239
  TEMPERATURE SENSITIVE MUTANTS
  3239
VIRUS, SIMIAN ADENO
  TRANS-SPECIES INFECTION, HAMS
  3203
  VIRUS REPLICATION, 3203
VIRUS, SV40
  ANTIGENS, VIRAL, 3242, 3244,
  BINDING ASSAY, 3242
  CELL TRANSFORMATION, NEOPLAST
  3243
  DEOXYRIBONUCLEASE, 3245
  DNA-RNA HYBRIDIZATION, 3246
  ENDONUCLEASES, 3246, 3248
  GENETICS, 3244
  ISOLATION AND CHARACTERIZATIO
  3248
  RADIATION, IONIZING, 3245
  RNA POLYMERASE, 3246
  RNA, VIRAL, 3246

DROSOPHILA MELANOGASTER
  RADIATION, IONIZING
    CHROMOSOME ABERRATIONS, 3190*
    3193*

DRUG THERAPY
  IMMUNOSUPPRESSIVE AGENTS
    COCARCINOGENIC EFFECT, REVIEW
    3004
  RADIATION
    COCARCINOGENIC EFFECT, REVIEW
    3004

DUODENAL ULCER
  DNA
    TOBACCO, 3172*

EHRLICH ASCITES CARCINOMA
  SEE CARCINOMA, EHRLICH TUMOR

EHRLICH ASCITES TUMOR (CELL)
  SEE CARCINOMA, EHRLICH TUMOR

ENDOMETRIOSIS
  CERVIX NEOPLASMS
    ADENOCARCINOMA, 3098
  FALLOPIAN TUBE NEOPLASMS
    ULTRASTRUCTURAL STUDY, 3428
  HORMONES, 3599*

ENDONUCLEASES
  VIRUS, ADENO 2

ESCHERICHIA COLI
    ACRIDINE, 9-AMINO-
        MUTAGENIC ACTIVITY, 3130*

ESOPHAGEAL NEOPLASMS
    ADENOCARCINOMA
        EPIDEMIOLOGY, ISRAEL, 3513
    AIR POLLUTANTS
        EPIDEMIOLOGY, SOCIOECONOMIC
        FACTORS, 3516
    ANILINE, N-METHYL-N-NITROSO-
        DIETARY FACTORS, CHINA, 3514
    BENZYLAMINE, N-METHYL-N-NITROSO-
        DIETARY FACTORS, CHINA, 3514
    CARCINOMA
        PIPERIDINE, 1-NITROSO-, 3117
    CARCINOMA, ANAPLASTIC
        EPIDEMIOLOGY, ISRAEL, 3513
    CARCINOMA, EPIDERMOID
        EPIDEMIOLOGY, ISRAEL, 3513
    CARDIOSPASM
        CASE REPORT, 3409
        DIAGNOSIS AND PROGNOSIS, 3409
    DEGLUTITION DISORDERS
        CASE REPORT, 3409
    DIETHYLAMINE, N-NITROSO-
        DIETARY FACTORS, CHINA, 3514
    DIMETHYLAMINE, N-NITROSO-
        DIETARY FACTORS, CHINA, 3514
    EPIDEMIOLOGY
        DIETARY FACTORS, CHINA, 3514
    PAPILLOMA
        BENZYLAMINE, N-METHYL-N-NITROSO-,
        3514
    PRECANCEROUS CONDITIONS
        GENETICS, 3067*
    TOBACCO
        CARCINOGENIC POTENTIAL, REIVEW,
        3026

ESOPHGEAL NEOPLASMS
    CARCINOMA, EPIDERMOID
        3-PIPERIDINOL, 1-NITROSO-, 3117
    PAPILLOMA
        3-PIPERIDINOL, 1-NITROSO-, 3117

ESTERASES
    BREAST NEOPLASMS
        ENZYMATIC ACTIVITY, 3445*

ESTRADIOL
    BREAST NEOPLASMS
        STILBENES, 3169*
    CHROMOSOME ABERRATIONS
        MUTAGENIC ACTIVITY, REVIEW, 3049*
    MAMMARY NEOPLASMS, EXPERIMENTAL
        BENZ(A)ANTHRACENE, 7,12-DIMETHYL-,
        3393
        BINDING, COMPETITIVE, 3550
    PROSTATIC NEOPLASMS, 3174*
    UTERINE NEOPLASMS
        PROGESTERONE, 3163*
        TESTOSTERONE, 3163*
    VIRUS, MURINE SARCOMA
        VIRUS REPLICATION, 3280*
    VIRUS, POX
        VIRUS REPLICATION, 3255

17-BETA ESTRADIOL

        SEE ESTRADIOL

ESTROGENS
    CERVIX NEOPLASMS
        ADENOCARCINOMA, 3098
    MAMMARY NEOPLASMS, EXPERIMENTAL
        HYPERPLASIA, 3396
    NEOPLASM TRANSPLANTATION, 3173*

ESTRONE
    MAMMARY NEOPLASMS, EXPERIMENTAL
        BINDING, COMPETITIVE, 3550
    VIRUS, POX
        VIRUS REPLICATION, 3255

ETHANE, 1,2,-DICHLORO-
    ACETALDEHYDE, CHLORO-
        CARCINOGENIC METABOLITE, 3125

ETHANE, IODO-
    ALKYLATING AGENTS
        STRUCTURE-ACTIVITY RELATIONSHIP,
            3109

ETHANESULPHONIC ACID, 2-(N-2-HYDROXYETHYL-
    PIPERAZIN-N-YL)-
        CARCINOMA, EHRLICH TUMOR
            CALCIUM, 3543

ETHANOL, 2-CHLORO-
    ACETALDEHYDE, CHLORO-
        CARCINOGENIC METABOLITE, 3125

ETHANOL, 2-MERCAPTO-
    LYMPHOCYTES
        CHROMOSOME ABERRATIONS, 3424
    PLANT AGGLUTININS
        CHROMOSOME ABERRATIONS, 3424

ETHER, BIS(CHLOROMETHYL)-
    BACTERIOPHAGES
        RNA, 3111

ETHIDIUM BROMIDE
    CARCINOMA, EHRLICH TUMOR
        ULTRASTRUCTURAL STUDY, 3160*
    FIBROBLASTS
        ULTRASTRUCTURAL STUDY, 3160*
    MITOCHONDRIA
        ULTRASTRUCTURAL STUDY, 3160*

ETHIDIUM CHLORIDE
    CHOLANTHRENE, 3-METHYL-
        CELLS, CULTURED, 3095

ETHIONINE
    SEE BUTYRIC ACID, 2-AMINO-4-(ETHYLTHIO)-

ETHYL ALCOHOL
    ACETAMIDE, THIO-
        ALANINE AMINOTRANSFERASE, 3091
    CARBON TETRACHLORIDE
        ALANINE AMINOTRANSFERASE, 3091
        HEPATOTOXICITY, 3091
    DIMETHYLAMINE, N-NITROSO-
        ALANINE AMINOTRANSFERASE, 3091

ETHYL CARBAMATE

```
                GROWTH, 3544
        CELL MEMBRANE
                CONCANAVALIN A, 3349
        CELL TRANSFORMATION, NEOPLASTIC
                ANTIGENS, NEOPLASM, 3372*
                TEMPERATURE SENSITIVE MUTANTS,
                3297*
        CROTON OIL
                DNA REPAIR, 3081
                DNA REPLICATION, 3081
        DNA REPLICATION
                PEPTIDE HYDROLASES, 3586*
        ETHIDIUM BROMIDE
                ULTRASTRUCTURAL STUDY, 3160*
        GROWTH SUBSTANCES
                PERFUSED RAT LIVER, 3295*
        IMMUNITY, CELLULAR
                CHROMIUM RELEASE ASSAY, 3301
        LYMPHOCYTES
                MITOMYCIN C, 3103
        MAGNESIUM CHLORIDE
                GROWTH, 3544
        PHENOL
                DNA REPAIR, 3081
                DNA REPLICATION, 3081
        PHENOL, P-NITRO-
                DNA REPLICATION, 3081
        PHORBOL
                DNA REPAIR, 3081
                DNA REPLICATION, 3081
        PHOSPHATES
                GROWTH, 3544
        POTASSIUM CHLORIDE
                GROWTH, 3544
        12-O-TETRADECANOYLPHORBOL-13-ACETATE
                DNA REPAIR, 3081
                DNA REPLICATION, 3081
        VIRUS, MOLONEY MURINE SARCOMA
                CELL TRANSFORMATION, NEOPLASTIC,
                3356*
        VIRUS, MURINE SARCOMA
                VIRUS REPLICATION, 3280*
        VIRUS, ROUS SARCOMA
                TRANSFERASES, 3289*
        VIRUS, SV40
                FLUORESCENT ANTIBODY TECHNIC,
                .3290*
        VITAMINS
                GROWTH, 3544

FIBROMA
        LARYNGEAL NEOPLASMS
                ULTRASTRUCTURAL STUDY, 3410
        STOMACH NEOPLASMS
                ENZYMATIC ACTIVITY, 3589*
        TONSILLAR NEOPLASMS
                CASE REPORT, CONGENITAL, 3408
        VIRUS, BOVINE PAPILLOMA
                ANTIBODIES, NEOPLAS, 3281*
                ANTIBODIES, VIRAL, 3281*
                TRANSPLANTATION IMMUNOLOGY, 3281*

FIBROSARCOMA
        BLADDER NEOPLASMS
                LEUKOPLAKIA, 3426
        CHOLANTHRENE, 3-METHYL-
                GRAFT REJECTION, 3299
                IMMUNITY, CELLULAR, 3299, 3303,
                3307
```

LARYNGEAL NEOPLASMS
    HISTOLOGICAL STUDY, REVIEW, 3023
ORBITAL NEOPLASMS
    ULTRASTRUCTURAL STUDY, REVIEW,
    3064*
THORACIC NEOPLASMS
    ASBESTOS, 3138*
VIRUS, BOVINE PAPILLOMA
    ANTIBODIES, NEOPLAS, 3281*
    ANTIBODIES, VIRAL, 3281*
    TRANSPLANTATION IMMUNOLOGY, 3281*
VIRUS, C-TYPE RNA TUMOR
    ANTIGENIC DETERMINANTS, 3369*
    PEPTIDES, 3369*
VIRUS, POLYOMA, BK
    ANTIBODIES, VIRAL, 3283*
    ANTIGENS, VIRAL, 3283*
    NEOPLASM TRANSPLANTATION, 3283*

FICUSIN
    NEOPLASMS, EXPERIMENTAL
        PHOTOADDITION, REVIEW, 3074*

FLUORANTHENE
    LUNG NEOPLASMS
        AIR POLLUTANTS, 3516

FLUORENYL ACETAMIDE
    SEE ACETAMIDE, N-FLUOREN-2-YL-

FOLIC ACID, N-NITROSO-
    LUNG NEOPLASMS
        ADENOCARCINOMA, 3154*

FOLLICLE-STIMULATING HORMONE
    SEE FSH

FOOD ANALYSIS
    NITRIC ACID, SODIUM SALT
        VEGETABLES, 3046*
    NITROUS ACID, SODIUM SALT
        VEGETABLES, 3046*

FOOD CONTAMINATION
    NITRIC ACID, SODIUM SALT
        CARCINOGENIC POTENTIAL, REVIEW,
        3045*·
        METHEMOGLOBINEMIA, 3045*
    NITROUS ACID, SODIUM SALT
        CARCINOGENIC POTENTIAL, REVIEW,
        3045*
        METHEMOGLOBINEMIA, 3045*
    PHENOL
        NITROSATION, 3047*

FOOD INSPECTION
    NITROSO COMPOUNDS
        ANALYSIS, 3116

FORMALDEHYDE
    ETHYLENE, CHLORO-
        CARCINOGENIC METABOLITE, RAT, 3124

FORMAMIDE, N,N-DIMETHYL-
    CELL MEMBRANE
        GLYCOPROTEINS, 3095
    CHOLANTHRENE, 3-METHYL-
        CELLS, CULTURED, 3095
    CONCANAVALIN A

        CELL DIFFERENTIATION
            TISSUE CULTURE, 3576*
        EYE NEOPLASMS
            TRANSPLANTATION, HETEROLOGOUS,
            3434
        HEAD AND NECK NEOPLASMS
            CELL CYCLE KINETICS, 3534*
            DNA REPLICATION, 3534*
        TRANSPLANTATION, HETEROLOGOUS
            HISTOLOGICAL STUDY, GUINEA PIG,
            3434

GLOMANGIOMA
    HEART NEOPLASMS
        CASE REPORT, 3463*
        ULTRASTRUCTURAL STUDY, 3463*

GLUCAGON
    HEPATOMA
        ADENYL CYCLASE, 3564

GLUCOCORTICOIDS
    HEPATOMA
        RECEPTOR-STEROID COMPLEX, 3574

GLUCOSAMINIDASE
    LEUKEMIA, LYMPHOBLASTIC
        LYSOSOMES, 3568
    VIRUS, POLYOMA
        CELL TRANSFORMATION, NEOPLASTIC,
        3570

GLUCOSEPHOSPHATASE
    HEPATOMA
        DIETHYLAMINE, N-NITROSO-, 3432
    LIVER NEOPLASMS
        DIETHYLAMINE, N-NITROSO-, 3432

GLUCOSEPHOSPHATE DEHYDROGENASE
    MAMMARY NEOPLASMS, EXPERIMENTAL
        2,3-BUTANEDIOL, 1,4-DIMERCAPTO-,
        3567
        GLUTATHIONE REDUCTASE, 3567
        NADP, 3567
    PRECANCEROUS CONDITIONS
        2,3-BUTANEDIOL, 1,4-DIMERCAPTO-,
        3567
        GLUTATHIONE REDUCTASE, 3567
        NADP, 3567

GLUCOSIDASES
    LEUKEMIA, LYMPHOBLASTIC
        LYSOSOMES, 3568
    VIRUS, POLYOMA
        CELL TRANSFORMATION, NEOPLASTIC,
        3570

GLUCURONIDASE
    LEUKEMIA, LYMPHOBLASTIC
        LYSOSOMES, 3568
    VIRUS, POX
        CELL MEMBRANE, 3255

GLUTAMATE DEHYDROGENASE
    LEUKEMIA, LYMPHOBLASTIC
        MITOCHONDRIA, 3568

GLUTAMIC ACID
    ETHYLENE, CHLORO-

CARCINOGENIC METABOLITE, RAT, 3124

GLUTAMYL TRANSPEPTIDASE
    AFLATOXIN B1
        CARCINOGENIC ACTIVITY, RAT, 3075

GLUTARALDEHYDE
    LEJKEMIA L1210
        AGGLUTINATION, 3346
    MACROPHAGES
        AGGLUTINATION, 3346

GLUTATHIONE
    AFLATOXIN B1
        PROTECTIVE EFFECT, RAT, 3079

GLUTATHIONE REDUCTASE
    MAMMARY NEOPLASMS, EXPERIMENTAL
        GLUCOSEPHOSPHATE DEHYDROGENASE,
            3567
    PRECANCEROUS CONDITIONS
        GLUCOSEPHOSPHATE DEHYDROGENASE,
            3567

GLYCEROL
    VIRUS, ROUS SARCOMA
        CELL TRANSFORMATION, NEOPLASTIC,
            3435
    VIRUS, SV40
        CELL TRANSFORMATION, NEOPLASTIC,
            3435

GLYCIDALDEHYDE
    NEOPLASMS
        CARCINOGENIC ACTIVITY, 3164*

GLYCINE
    VIRUS, SV40
        AMINO ACID TRANSPORT, 3296*

GLYCOGEN
    CERVIX NEOPLASMS
        CARCINOMA IN SITU, 3569
        PRECANCEROUS CONDITIONS, 3569
    HEPATOMA
        DIETHYLAMINE, N-NITROSO-, 3432
    LIVER NEOPLASMS
        DIETHYLAMINE, N-NITROSO-, 3432

GLYCOGEN PHOSPHORYLASE
    CERVIX NEOPLASMS
        CARCINOMA IN SITU, 3569
        PRECANCEROUS CONDITIONS, 3569

GLYCOGEN SYNTHETASE
    CERVIX NEOPLASMS
        CARCINOMA IN SITU, 3569
        PRECANCEROUS CONDITIONS, 3569

GLYCOLIPIDS
    ASTROCYTOMA
        UREA, NITROSO-, 3563
        VIRUS, SV40, 3563
    LEJKEMIA
        TRIOSES, 3563
    LEJKEMIA, LYMPHOBLASTIC
        LYSOSOMES, 3568
    NEUROBLASTOMA
        TRIOSES, 3563

                    ULTRASTRUCTURAL STUDY, 3463*
            MESENCHYMOMA
                CASE REPORT, 3462*
                ULTRASTRUCTURAL STUDY, 3462*
            MESOTHELIOMA
                HISTOLOGICAL STUDY, RAT, 3415
                ULTRASTRUCTURAL STUDY, RAT, 3415

        HELA CELLS
            ACTIN
'L)   CELL MEMBRANE, 3583*
                CYTOCHALASIN B, 3583*
                ISOLATION AND CHARACTERIZATION,
                    3583*
            DNA
                DNA POLYMERASE, 3584*
                NAD NUCLEOSIDASE, 3584*
            DNA REPLICATION
                CELL CYCLE KINETICS, 3585*
                CHROMOSOMES, 3585*
                PULSE CHASE STUDY, CELL NUCLEI,
                    3559
            POLYRIBOSOMES
                IGG, 3316
                IMMUNOGLOBULINS, 3316
            RADIATION, IONIZING
                CELL CYCLE KINETICS, 3182
                DNA REPLICATION, 3182
            RNA
                CELL CYCLE KINETICS, 3585*
                CHROMOSOMES, 3585*
            VIRUS, ADENO 5
                IMMUNE SERUMS, 3258*
            VIRUS, POLIO
                PHOSPHOLIPIDS, 3010
                RNA, VIRAL, 3253
            VIRUS, SENDAI
                DNA REPAIR, 3031
            VIRUS, VESICULAR STOMATITIS
                RNA, VIRAL, 3253

R,   HEMAGGLUTINATION
            RESPIRATORY TRACT NEOPLASMS
                LIPOPROTEINS, 3366*
            UROGENITAL NEOPLASMS
                LIPOPROTEINS, 3366*

        HEMANGIOBLASTOMA
            SEE ANGIOSARCOMA

        HEMANGIOMA
            ORBITAL NEOPLASMS
                ULTRASTRUCTURAL STUDY, REVIEW,
                    3064*

        HEMANGIOPERICYTOMA
            LEIOMYOSARCOMA
                ULTRASTRUCTURAL STUDY, 3464*
            ORBITAL NEOPLASMS
                ULTRASTRUCTURAL STUDY, REVIEW,
                    3064*
            VIRUS, ADENO 3
                ULTRASTRUCTURAL STUDY, 3464*

        HEMATOSARCOMA
            SEE ANGIOSARCOMA

        HEMOGLOBULINS
            LEUKEMIA

FETAL GLOBULINS, 3420

HEPARIN
    PROTEIN
        RETICULOCYTES, 3555

HEPATOCARCINOMA
    SEE HEPATOMA

HEPATOCELLULAR CARCINOMA
    SEE HEPATOMA

HEPATOMA
    ABDOMEN, ACUTE
        DIAGNOSIS AND PROGNOSIS, 3459*
    ACETAMIDE, N-FLUOREN-2-YL-
        LIPOTROPIC DEFICIENT, HIGH-FAT
            DIET, 3096
    ACETAMIDE, N-FLUOREN-2-YLDI-
        DNA, 3128*
        NUCLEOTIDES, 3128*
    ADENOSINE CYCLIC 3',5' MONOPHOPHATE
        PROTEIN KINASE, 3595*
    ADRIAMYCIN
        IMMUNE RESPONSE, 3319
    AFLATOXIN B1
        LIPOTROPIC DEFICIENT, HIGH-FAT
            DIET, 3096
    ANILINE, N,N-DIMETHYL-P-PHENYLAZO-
        ANTIGENS, NEOPLASM, 3303
        DNA, 3128*
        IMMUNITY, CELLULAR, 3303
        NUCLEOTIDES, 3128*
    AUSTRALIA ANTIGEN
        EPIDEMIOLOGY, REVIEW, 3373*
    BENZENAMINE, N,N-DIMETHYL-4-((3-
        METHYLPHENYL)AZO)-
        CAPSELLA BURSA-PASTORIS, 3086
    BENZENEACETAMIDE, ALPHA-ETHYNYL-N,N-
        DIMETHYL-ALPHA-PHENYL-
        LIPOTROPIC DEFICIENT, HIGH-FAT
            DIET, 3096
    BENZYL ALCOHOL, ALPHA-(1-(
        (METHYLNITROSOAMINO)ETHYL)-
        CARCINOGENIC ACTIVITY, 3154*
    CELL MEMBRANE
        PHOSPHODIESTERASES, 3565
    5-BETA-CHOLAN-24-OIC ACID, 3-ALPHA,12-
        ALPHA-DIHYDROXY
        AMINO ACIDS, 3554
        POLYRIBOSOMES, 3554
    DIETHYLAMINE, N-NITROSO-
        ADENOSINE TRIPHOSPHATASE, 3432
        CARCINOGENIC ACTIVITY, 3154*
        GLUCOSEPHOSPHATASE, 3432
        GLYCOGEN, 3432
        GROWTH, 3432
        IMMUNE RESPONSE, 3319
    DNA
        CHROMATIN, 3560
    DNA, NEOPLASM
        CYTOPHOTOMETRIC STUDY, 3562
    DNA NUCLEOTIDYLTRANSFERASES
        NUCLEOSIDE DIPHOSPHATE SUGARS,
            3593*
    EPINEPHRINE
        ADENYL CYCLASE, 3564
    FRUCTOSEDIPHOSPHATE ALDOLASE
        ISOLATION AND CHARACTERIZATION,

```
                    ULTRASTRUCTURAL STUDY, 3417
            EPIDEMIOLOGY
                    HISTOLOGICAL STUDY, REVIEW, 3025
                    STATISTICAL ANALYSIS, REVIEW, 3024
            LEUKEMIA, MYELOBLASTIC
                    RADIATION, IONIZING, 3192*
                    RADIOTHERAPY, CASE REPORT, 3192*
            LYMPHOCYTES
                    ANTIGENS, NEOPLASM, 3370*
            B-LYMPHOCYTES
                    IGG, 3058*
                    IMMUNOGLOBULINS, SURFACE, 3058*
                    ROSETTE FORMATION, 3386*
            T-LYMPHOCYTES
                    IGG, 3058*
                    IMMUNITY, CELLULAR, 3418
                    IMMUNOGLOBULINS, SURFACE, 3058*
                    ROSETTE FORMATION, 3386*
                    ULTRASTRUCTURAL STUDY, 3418
            MACROPHAGES
                    ULTRASTRUCTURAL STUDY, 3418


HORMONES
        ENDOMETRIOSIS, 3599*
        PROSTATIC NEOPLASMS
                GROWTH, 3174*


HYDANTOIN, 5,5-DIPHENYL-
        LYMPHOCYTES
                DNA REPLICATION, 3097


HYDANTOIN, 3-HYDROXY-
        LYMPHOCYTES
                DNA REPLICATION, 3097


HYDANTOIN, 3-HYDROXY-1-ACETYL-5,5-DIPHENYL-

        LYMPHOCYTES
                DNA REPLICATION, 3097


HYDANTOIN, 3-HYDROXY-5,5-DIPHENYL-
        LYMPHOCYTES
                DNA REPLICATION, 3097


HYDANTOIN, 3-HYDROXY-5-ISOPROPYL-
        LYMPHOCYTES
                DNA REPLICATION, 3097


HYDANTOIN, 3-HYDROXY-5-METHYL-
        LYMPHOCYTES
                DNA REPLICATION, 3097


HYDANTOIN, 5-(4-HYDROXYPHENYL)-5-PHENYL-
        LYMPHOCYTES
                DNA REPLICATION, 3097


HYDATIDFORM MOLE
        CHORIOCARCINOMA
                EPIDEMIOLOGY, 3071*


HYDATIFORM MOLE
        UROGENITAL NEOPLASMS
                GONADOTROPINS, CHORIONIC, 3427
                LACTATE DEHYDROGENASE, 3427


HYDRAZINE, 1,2-DIMETHYL-
        COLONIC NEOPLASMS
                AUTORADIOGRAPHY, 3162*
                DNA REPLICATION, 3162*
```

HYDRAZINE, METHYL-
    ACTIVE INTERMEDIATE, 3161*
METHANE, AZOXY-
    ACTIVE INTERMEDIATE, 3161*

HYDRAZINE, METHYL-
    HYDRAZINE, 1,2-DIMETHYL-
        ACTIVE INTERMEDIATE, 3161*

HYDROCORTISONE
    SEE CORTISOL

HYDROLASES
    CHOLANTHRENE, 11,12-EPOXY-3-METHYL-
        ACTIVITY ASSAY, 3104
        CHOLANTHRENE-11,12-DIOL, 11,12-
        DIHYDRO-3-METHYL-, 3104

N-HYDROXY-2-ACETYLAMINOFLUORENE
    SEE ACETOHYDROXAMIC ACID, N-FLUOREN-2-
    YL-

HYDROXYBUTYRATE DEHYDROGENASE
    BREAST NEOPLASMS
        ENZYMATIC ACTIVITY, 3446*
    COLONIC NEOPLASMS
        ENZYMATIC ACTIVITY, 3446*

HYDROXYLAMINE, N-(P-(PHENYLAZO)PHENYL)-
    MICROSOMES, LIVER
        MIXED-FUNCTION AMINE OXIDASE,
        3136*

HYDROXYLASES
    BENZO(A)PYRENE
        CHOLANTHRENE, 3-METHYL-, 3144*

HYPERCALCEMIA
    HEPATOMA
        CASE REPORT, 3479*

HYPERNEPHROID CARCINOMA
    SEE ADENOCARCINOMA

HYPERNEPHROMA
    SEE ADENOCARCINOMA

HYPERPLASIA
    FALLOPIAN TUBE NEOPLASMS
        ULTRASTRUCTURAL STUDY, 3428
    MAMMARY NEOPLASMS, EXPERIMENTAL
        ESTROGENS, 3396
        INSULIN, 3396
        TESTOSTERONE, 3396
        ULTRASTRUCTURAL STUDY, 3396
    PROSTATIC NEOPLASMS
        ACID PHOSPHATASE, 3537
    RNA, BACTERIAL
        CARCINOGENIC EFFECT, 3171*

HYPERSENSITIVITY, IMMEDIATE
    T-LYMPHOCYTES
        IMMUNITY, CELLULAR, 3389*

HYPERTENSION
    ADRENAL GLAND NEOPLASMS
        PHEOCHROMOCYTOMA, 3507

IGA

LEUKEMIA, LYMPHOBLASTIC
    B-LYMPHOCYTES, 3388*
    POLIOVIRUS VACCINE, 3364*
MELANOMA
    ANTIBODIES, NEOPLASM, 3321
MULTIPLE MYELOMA
    HYPERVISCOSITY SYNDROME, 3
PLASMACYTOMA
    IMMUNOGLOBULINS, 3362*
RADON
    RADIATION, IONIZING, 3181
RECTAL NEOPLASMS
    CARCINOMA, 3450*
SARCOMA
    IMMUNOGLOBULINS, 3362*
URANIUM
    RADIATION, IONIZING, 3181

IGD
    LEUKEMIA, LYMPHOCYTIC
        B-LYMPHOCYTES, 3338

IGE
    ANTI-ANTIBODIES
        HISTAMINE LIBERATION, 3060

IGG
    BURKITT'S LYMPHOMA
        ANTIGENS, VIRAL, 3367*
    HELA CELLS
        POLYRIBOSOMES, 3316
    HODGKIN'S DISEASE
        B-LYMPHOCYTES, 3058*
        T-LYMPHOCYTES, 3058*
    LEUKEMIA, LYMPHOBLASTIC
        B-LYMPHOCYTES, 3387*, 3388
    LEUKEMIA, LYMPHOCYTIC
        B-LYMPHOCYTES, 3335, 3387*
    MELANOMA
        ANTIBODIES, NEOPLASM, 3321
    MULTIPLE MYELOMA
        POLYRIBOSOMES, 3316
        RNA, MESSENGER, 3316
    PLASMACYTOMA
        IMMUNOGLOBULINS, 3362*
    RADON
        RADIATION, IONIZING, 3181
    SARCOMA
        IMMUNOGLOBULINS, 3362*
    URANIUM
        RADIATION, IONIZING, 3181

IGM
    LEUKEMIA, LYMPHOBLASTIC
        B-LYMPHOCYTES, 3388*
        POLIOVIRUS VACCINE, 3364*
    LEUKEMIA, LYMPHOCYTIC
        B-LYMPHOCYTES, 3335, 3338
    MELANOMA
        ANTIBODIES, NEOPLASM, 3321
    PLASMACYTOMA
        IMMUNOGLOBULINS, 3362*
    RADON
        RADIATION, IONIZING, 3181
    SARCOMA
        IMMUNOGLOBULINS, 3362*
    URANIUM
        RADIATION, IONIZING, 3181
    VIRUS, MOLONEY MURINE SARCOMA

                CELL, 3304
        B-LYMPHOCYTES
            TRANSPLANTATION IMMUNOLOGY, 3360*
        T-LYMPHOCYTES
            HYPERSENSITIVITY, IMMEDIATE, 3389*
            B-LYMPHOCYTES, 3382*
            PICRIC ACID, 3302
            TRANSPLANTATION IMMUNOLOGY, 3360*
            TRINITROPHENYL-MODIFIED CELLS,
                3383*
        LYMPHOMA
            T-LYMPHOCYTES, 3332
            VIRUS, MURINE LEUKEMIA, 3390*
        MELANOMA
            ANTIGENS, NEOPLASM, 3357*
            CHROMIUM RELEASE ASSAY, 3301
            GRAFT REJECTION, 3299
            LYMPHOCYTES, 3266*
            T-LYMPHOCYTES, 3336
            NEOPLASM METASTASIS, 3336
            NEOPLASM REGRESSION, SPONTANEOUS,
                3481*
            ULTRASTRUCTURAL STUDY, 3481*
        NEOPLASMS
            ANTIGEN-ANTIBODY COMPLEX, 3061*
            ANTIGENS, NEOPLASM, 3061*
        OVARIAN NEOPLASMS
            LYMPHOCYTES, 3333
        PLASMACYTOMA
            T-LYMPHOCYTES, 3332
        RECTAL NEOPLASMS
            ADENOCARCINOMA, 3300
        SARCOMA
            CHOLANTHRENE, 3-METHYL-, 3305,
                3306
        SARCOMA 180, CROCKER
            REGIONAL NODE IMMUNITY, MOUSE,
                3353*
        SARCOMA, MAST CELL
            T-LYMPHOCYTES, 3332
        SKIN NEOPLASMS
            KERATOACANTHOMA, 3406
        UROGENITAL NEOPLASMS
            LYMPHOCYTES, 3333
        VIRUS, ADENO
            VIRUS REPLICATION, 3355*
        VIRUS, EPSTEIN-BARR
            LYMPHOCYTES, 3266*
        VIRUS, MOLONEY MURINE LEUKEMIA
            KILLER CELLS, 3391*
        VIRUS, MOLONEY MURINE SARCOMA
            IGM, 3315
        VIRUS, SV40
            CONCANAVALIN A, 3347

IMMUNIZATION
    SARCOMA
        VIRUS, SV40, 3314

IMMUNOGLOBULIN FRAGMENTS
    LEUKEMIA, LYMPHOCYTIC
        CASE REPORT, 3485*
    PLASMACYTOMA
        MYELOMA PROTEINS, 3317
    PLASMACYTOMA
        ANTIGENIC DETERMINANTS, 3313

IMMUNOGLOBULINS
    SEE ALSO IGA/IGD/IGE/IGG/IGM

BURKITT'S LYMPHOMA
    ANTIGENS, VIRAL, 3367*
HELA CELLS
    POLYRIBOSOMES, 3316
LEUKEMIA, LYMPHOBLASTIC
    B-LYMPHOCYTES, 3388*
    POLIOVIRUS VACCINE, 3364*
LEUKEMIA, LYMPHOCYTIC
    CELLULAR INCLUSIONS, 3338
    LYMPHOCYTES, 3385*
    B-LYMPHOCYTES, 3335
LYMPHOCYTES
    IMMUNE RESPONSE, K CELL, 3304
T-LYMPHOCYTES
    IMMUNOSUPPRESSION, 3382*
MELANOMA
    ANTIBODIES, NEOPLASM, 3321
MULTIPLE MYELOMA
    POLYRIBOSOMES, 3316
OVARIAN NEOPLASMS
    ADENOCARCINOMA, 3361*
    CYSTADENOCARCINOMA, 3361*
    CYSTADENOMA, MUCINOUS, 3361*
    DISGERMINOMA, 3361*
PLASMACYTOMA
    IGA, 3362*
    IGG, 3362*
    IGM, 3362*
RADON
    RADIATION, IONIZING, 3181
RECTAL NEOPLASMS
    CARCINOMA, 3450*
SARCOMA
    IGA, 3362*
    IGG, 3362*
    IGM, 3362*
URANIUM
    RADIATION, IONIZING, 3181
VIRUS, AVIAN LEUKOSIS
    B-LYMPHOCYTES, 3381*
    T-LYMPHOCYTES, 3381*

IMMUNOGLOBULINS, FC
    LEUKEMIA, LYMPHOBLASTIC
        B-LYMPHOCYTES, 3387*
    LEUKEMIA, LYMPHOCYTIC
        B-LYMPHOCYTES, 3387*
    LYMPHOCYTES
        IMMUNE RESPONSE, K CELL, 3304
    T-LYMPHOCYTES
        CONCANAVALIN A, 3343
        MITOSIS, 3343
        PLANT AGGLUTININS, 3343

IMMUNOGLOBULINS, SURFACE
    HODGKIN'S DISEASE
        B-LYMPHOCYTES, 3058*
        T-LYMPHOCYTES, 3058*
    LEUKEMIA, LYMPHOCYTIC
        B-LYMPHOCYTES, 3338
    VIRUS, FRIEND MURINE LEUKEMIA
        ANTIGENS, VIRAL, 3310
        B-LYMPHOCYTES, 3310
        T-LYMPHOCYTES, 3310

IMMUNOSUPPRESSION
    ALDRICH SYNDROME
        IMMUNOGLOBULINS, SURFACE, REVIEW,
        3059*

ATAXIA TELANGIECTASIA
    IMMUNOGLOBULINS, SURFACE, REVI
    3059*
CARCINOMA
    TRANSPLANTATION, HOMOLOGOUS, 3
CARCINOMA, BASAL CELL
    TRANSPLANTATION, HOMOLOGOUS, 3
CARCINOMA, EPIDERMOID
    TRANSPLANTATION, HOMOLOGOUS, 3
CERVIX NEOPLASMS
    CARCINOMA, 3311
LYMPHATIC NEOPLASMS
    INCIDENCE, ATHYMIC MOUSE, 3307
LYMPHOCYTE TRANSFORMATION
    GLYCOPROTEINS, 3359*
T-LYMPHOCYTES
    IMMUNOGLOBULINS, 3382*
OVARIAN NEOPLASMS
    DISGERMINOMA, 3311
SARCOMA
    CHOLANTHRENE, 3-METHYL-, 3309
    VIRUS, HERPES SIMPLEX 2, 3358*
SARCOMA, RETICULUM CELL
    TRANSPLANTATION, HOMOLOGOUS, 3
UTERINE NEOPLASMS
    ADENOCARCINOMA, 3311
VIRUS, FRIEND MURINE LEUKEMIA
    VIBRIO INFECTIONS, 3312
VIRUS, HERPES SIMPLEX 2
    PYRANS, 3341
VIRUS, MOLONEY MURINE LEUKEMIA
    VIBRIO INFECTIONS, 3312
VIRUS, MURINE MAMMARY TUMOR
    ANTIGENS, VIRAL, 3340
    T-LYMPHOCYTES, 3340
    VIRUS, MURINE LEUKEMIA, 3340
VIRUS, SV40
    CONCANAVALIN A, 3347

IMMUNOSUPPRESSIVE AGENTS
    DRUG THERAPY
        COCARCINOGENIC EFFECT, REVIEW,
        3304

IMMUNOTHERAPY
    LEUKEMIA L1210
        IRON CHLORIDE, 3579*

INCLUSION BODIES, VIRAL
    BLADDER NEOPLASMS
        ULTRASTRUCTURE STUDY, 3226
    PROSTATIC NEOPLASMS
        ULTRASTRUCTURAL STUDY, 3226
    TERATOID TUMOR
        ULTRASTRUCTURAL STUDY, MOUSE,
    VIRUS, C-TYPE MURINE LEUKEMIA
        GENETICS, 3262*
    VIRUS, C-TYPE RNA TUMOR
        CARBAMIC ACID, ETHYL ESTER, 32
        RADIATION, IONIZING, 3229
    VIRUS, EQUINE ENCEPHALITIS
        ULTRASTRUCTURAL STUDY, 3252
    VIRUS, MAREK'S DISEASE HERPES
        ANTIGENS, VIRAL, 3221
        VIRUS REPLICATION, 3221
    VIRUS, RAUSCHER MURINE LEUKEMIA
        CARBAMIC ACID, ETHYL ESTER, 32
        RADIATION, IONIZING, 3229
    VIRUS, ROUS SARCOMA

```
                    BLOOD GROUPS, 3375*
            COLONIC NEOPLASMS
                ADENOCARCINOMA, 3323
                CARCINOEMBRYONIC ANTIGEN, 3320
            GASTROINTESTINAL NEOPLASMS
                BLOOD GROUPS, 3375*
            LEUKEMIA
                T-LYMPHOCYTES, 3331
            LEUKEMIA L1210
                T-LYMPHOCYTES, 3332
            LEUKEMIA, RADIATION-INDUCED
                T-LYMPHOCYTES, 3331
            LYMPHOMA
                T-LYMPHOCYTES, 3332
            MOUTH NEOPLASMS
                BLOOD GROUPS, 3375*
            PHARYNGEAL NEOPLASMS
                BLOOD GROUPS, 3375*
            PLASMACYTOMA
                T-LYMPHOCYTES, 3332
            RECTAL NEOPLASMS
                ADENOCARCINOMA, 3323
            SARCOMA, MAST CELL
                T-LYMPHOCYTES, 3332
            THYROID NEOPLASMS
                BLOOD GROUPS, 3375*
            VIRUS, GROSS MURINE LEUKEMIA
                T-LYMPHOCYTES, 3331

        ISOCITRATE DEHYDROGENASE
            BREAST NEOPLASMS
                ENZYMATIC ACTIVITY, 3446*
            COLONIC NEOPLASMS
                ENZYMATIC ACTIVITY, 3446*

        ISOENZYMES
            CARCINOMA, EHRLICH TUMOR
                LACTATE DEHYDROGENASE, 3590*
            PITUITARY NEOPLASMS
                CARBONIC ANHYDRASE, 3566

        ISOGENIC TRANSPLANTATION
            SEE TRANSPLANTATION, HOMOLOGOUS

        ISOPROPYL ALCOHOL
            CARBON TETRACHLORIDE
                HEPATOTOXICITY, 3091

        JAW NEOPLASMS
            ODONTOGENIC TUMOR
                ULTRASTRUCTURAL STUDY, RAT, 3113
                UREA, 1-BUTYL-1-NITROSO-, 3113

        KARYOTYPING
            CARCINOMA
                BENZO(A)PYRENE, 3431
                DIETHYLAMINE, N-NITROSO-, 3431
                METHANOL, (METHYL-ONN-AZOXY)-,
                  3431
                VIRUS, SV40, 3250
            LEUKEMIA, MYELOBLASTIC
                CHROMOSOME ABNORMALITIES, 3423
            SARCOMA
                BENZO(A)PYRENE, 3101
                CHOLANTHRENE, 3-METHYL-, 3101

        KERATOACANTHOMA
            BENZENE, 1-FLUORO-2,4-DINITRO-
                IMMUNE RESPONSE, 3406
```

LEVAMISOLE
  IMMUNE RESPONSE, 3406
PLANT AGGLUTININS
  IMMUNE RESPONSE, 3406
SKIN NEOPLASMS
  CASE REPORT, 3406
  IMMUNITY, CELLULAR, 3406
TRANSFER FACTOR
  IMMUNE RESPONSE, 3406
TUBERCULIN
  IMMUNE RESPONSE, 3406

KERATOSIS
SKIN NEOPLASMS
  PRECANCEROUS CONDITIONS, 3452*

KIDNEY FAILURE, CHRONIC
BREAST NEOPLASMS
  ADENOCARCINOMA, 3308
CERVIX NEOPLASMS
  CARCINOMA IN STIU, 3308
KIDNEY NEOPLASMS
  ADENOMA, 3308
LUNG NEOPLASMS
  ADENOCARCINOMA, 3308
PANCREATIC NEOPLASMS
  ISLET CELL TUMOR, 3308
TESTICULAR NEOPLASMS
  DISGERMINOMA, 3308
THYROID NEOPLASMS
  CARCINOMA, PAPILLARY, 3308

KIDNEY NEOPLASMS
ADENOCARCINOMA
  GENETICS, 3425
  HISTOCOMPATIBILITY ANTIGENS, 3425
ADENOMA
  KIDNEY FAILURE, CHRONIC, 3308
AFLATOXIN B1
  TRANSPLACENTAL CARCINOGENESIS,
  3078
DIMETHYLAMINE, N-NITROSO-
  RNA, 3155*
LYMPHOSARCOMA
  VIRUS, HERPES SAIMIRI, 3277*
TERATOID TUMOR
  TRANSPLANTATION, HOMOLOGOUS, 3536
THORIUM DIOXIDE
  CARCINOGENIC POTENTIAL, REIVEW,
  3026
VIRUS, POLYOMA
  RNA, 3155*

L CELLS
CONCANAVALIN A
  BINDING, 3546
NEURAMINIDASE
  BINDING, 3546
PLANT AGGLUTININS
  BINDING, 3546
TRYPSIN
  BINDING, 3546

LACTATE DEHYDROGENASE
CARCINOMA, EHRLICH TUMOR
  ISOENZYMES, 3590*
  ORGAN DISTRUBTION, MOUSE, 3590*
UROGENITAL NEOPLASMS
  CHORIOADENOMA, 3427

CHORIOCARCINOMA, 3427
HYDATIFORM MOLE, 3427
VIRUS, ADENO 12
  CELL TRANSFORMATION, NEOPLASTIC,
  3201
  ISOLATION AND CHARACTERIZATION,
  3201

LACTOYLGLUTATHIONE LYASE
ADENO FIBROMA
  ENZYMATIC ACTIVITY, 3594*
CARCINOMA, EHRLICH TUMOR
  ENZYMATIC ACTIVITY, 3594*
HEPATOMA
  ENZYMATIC ACTIVITY, 3594*
MELANOMA
  ENZYMATIC ACTIVITY, 3594*
SARCOMA
  ENZYMATIC ACTIVITY, 3594*

LANTHANUM
CARCINOMA, EHRLICH TUMOR
  CALCIUM, 3543

LARYNGEAL NEOPLASMS
ADENOCARCINOMA
  REVIEW, 3066*
AIR POLLUTANTS
  EPIDEMIOLOGY, SOCIOECONOMIC
  FACTORS, 3516
ANGIOSARCOMA
  HISTOLOGICAL STUDY, REVIEW, 3023
ASBESTOS
  SMOKING, 3137*
CARCINOMA, EPIDERMOID
  NEOPLASMS, MULTIPLE PRIMARY, 341
  SMOKING, 3411
CHONDROMA
  HISTOLOGICAL STUDY, REVIEW, 3023
CHONDROSARCOMA
  HISTOLOGICAL STUDY, REVIEW, 3023
CYLINDROMA
  REVIEW, 3066*
FIBROMA
  ULTRASTRUCTURAL STUDY, 3410
FIBROSARCOMA
  HISTOLOGICAL STUDY, REVIEW, 3023
LEIOMYOSARCOMA
  HISTOLOGICAL STUDY, REVIEW, 3023
RHABDOMYOSARCOMA
  HISTOLOGICAL STUDY, REVIEW, 3023
SARCOMA, KAPOSI'S
  HISTOLOGICAL STUDY, REVIEW, 3023
SARCOMA, OSTEOGENIC
  HISTOLOGICAL STUDY, REVIEW, 3023
VIRUS-LIKE PARTICLES
  VIRUS, PAPILLOMA, 3410

LEIOMYOSARCOMA
HEMANGIOPERICYTOMA
  ULTRASTRUCTURAL STUDY, 3464*
LARYNGEAL NEOPLASMS
  HISTOLOGICAL STUDY, REVIEW, 3023
ORBITAL NEOPLASMS
  ULTRASTRUCTURAL STUDY, REVIEW,
  3064*
VIRUS, ADENO 3
  ULTRASTRUCTURAL STUDY, 3464*

SEE LEUKEMIA, MYELOCYTIC

LEUKEMIA L1210
    CARBOHYDRATES
        CONCANAVALIN A, 3346
        PLANT AGGLUTININS, 3346
    CELL MEMBRANE
        CARBOHYDRATES, 3346
    CONCANAVALIN A
        BINDING, 3546
    GLUTARALDEHYDE
        AGGLUTINATION, 3346
    IRON CHLORIDE
        IMMUNOTHERAPY, 3579*
        TUMOR CELL MODIFICATION, 3579*
    T-LYMPHOCYTES
        IMMUNITY, CELLULAR, 3332
        ISOANTIGENS, 3332
        RADIATION, 3332
    NEURAMINIDASE
        BINDING, 3546
    PLANT AGGLUTININS
        BINDING, 3546
    TRYPSIN
        BINDING, 3546

LEUKEMIA, LYMPHOBLASTIC
    ACID PHOSPHATASE
        LYSOSOMES, 3568
    CELL CYCLE KINETICS
        DRUG THERAPY, 3072*
        REVIEW, 3072*
    CHROMATIN
        PROTEINS, 3580*
    GALACTOSIDASES
        LYSOSOMES, 3568
    GLUCOSAMINIDASE
        LYSOSOMES, 3568
    GLUCOSIDASES
        LYSOSOMES, 3568
    GLUCURONIDASE
        LYSOSOMES, 3568
    GLUTAMATE DEHYDROGENASE
        MITOCHONDRIA, 3568
    GLYCOLIPIDS
        LYSOSOMES, 3568
    GLYCOPROTEINS
        LYSOSOMES, 3568
    IGA
        B-LYMPHOCYTES, 3388*
    IGG
        B-LYMPHOCYTES, 3388*
    IGM
        B-LYMPHOCYTES, 3388*
    IMMUNITY, CELLULAR
        T-LYMPHOCYTES, 3388*
    IMMUNOGLOBULINS
        B-LYMPHOCYTES, 3388*
    LYMPHATIC NEOPLASMS
        CLASSIFICATION, 3490*
    B-LYMPHOCYTES
        CELL DIFFERENTIATION, REVIEW,
            3059*
        IGG, 3387*
        IMMUNOGLOBULINS, FC, 3387*
        IMMUNOGLOBULINS, SURFACE, REVIEW,
            3059*
        ROSETTE FORMATION, 3386*
    T-LYMPHOCYTES

CELL DIFFERENTIATION, REVIEW,
3059*
IMMUNOGLOBULINS, SURFACE, REVIEW,
3059*
ROSETTE FORMATION, 3386*, 3387*
MALATE DEHYDROGENASE
MITOCHONDRIA, 3568
NUCLEOTIDYLTRANSFERASES
T-LYMPHOCYTES, 3540
POLIOVIRUS VACCINE
ANTIBODIES, VIRAL, 3364*
IGA, 3364*
IGM, 3364*
IMMUNE RESPONSE, 3364*
IMMUNITY, ACTIVE, 3364*
IMMUNOGLOBULINS, 3364*
SUCCINATE DEHYDROGENASE
MITOCHONDRIA, 3568
SULFATASES
LYSOSOMES, 3568
VIRUS, BOVINE LEUKEMIA
ULTRASTRUCTURAL STUDY, 3488*
VIRUS, C-TYPE MURINE LEUKEMIA
TRANSPLANTATION, HOMOLOGOUS, 3204

LEUKEMIA, LYMPHOCYTIC
CELL CYCLE KINETICS
REVIEW, 3072*
CELLULAR INCLUSIONS
IMMUNOGLOBULINS, 3338
CHROMATIN
PROTEINS, 3580*
DNA
CHROMOSOMES, 3556
DNA REPAIR, 3556
EPIDEMIOLOGY
POLAND, 3523*
ERYTHROCYTES
PORPHYRINS, 3486*
IMMUNOGLOBULIN FRAGMENTS
CASE REPORT, 3485*
LYMPHATIC NEOPLASMS
CLASSIFICATION, 3490*
LYMPHOCYTES
IMMUNOGLOBULINS, 3385*
REVIEW, 3057*
B-LYMPHOCYTES
CELL DIFFERENTIATION, REVIEW,
3059*
IGD, 3338
IGG, 3335, 3387*
IGM, 3335, 3338
IMMUNOGLOBULINS, 3335
IMMUNOGLOBULINS, FC, 3387*
IMMUNOGLOBULINS, SURFACE, 3338
IMMUNOGLOBULINS, SURFACE, REVIEW,
3059*
ROSETTE FORMATION, 3386*
T-LYMPHOCYTES
CELL DIFFERENTIATION, REVIEW,
3059*
ROSETTE FORMATION, 3335, 3386*,
3387*
POLYRIBONUCLEOTIDES
DNA, 3556
ULTRAVIOLET RAYS
DNA, 3556

LEUKEMIA, MONOCYTIC
REVERSE TRANSCRIPTASE
PRECANCEROUS CONDITIONS, 3419
VIRUS, FELINE LEUKEMIA
ANTIGENS, VIRAL, 3205
VIRUS, GIBBON APE LEUKEMIA
ANTIGENS, VIRAL, 3205
VIRUS-LIKE PARTICLES
PRECANCEROUS CONDITIONS, 3419
VIRUS, M7 BABOON
ANTIGENS, VIRAL, 3205
VIRUS, RAUSCHER MURINE LEUKEMIA
ANTIGENS, VIRAL, 3205
VIRUS, WOOLLY MONKEY SARCOMA
ANTIGENS, VIRAL, 3205

LEUKEMIA, MYELOBLASTIC
CELL CYCLE KINETICS
DRUG THERAPY, 3072*
REVIEW, 3072*
CELL DIFFERENTATION
CLINICAL STAGE, 3063*
CHROMOSOME ABNORMALITIES
CASE REPORT, 3423
KARYOTYPING, 3423
ERYTHROCYTES
PORPHYRINS, 3486*
GLYCOPROTEINS
CELL DIFFERENTIATION, 3063*
HODGKIN'S DISEASE
RADIATION, IONIZING, 3192*
RADIOTHERAPY, CASE REPORT, 3192
LEUKOCYTES
CELL DIFFERENTIATION, 3063*
B-LYMPHOCYTES
ROSETTE FORMATION, 3386*
T-LYMPHOCYTES
ROSETTE FORMATION, 3386*
MACROPHAGES
CELL DIFFERENTIATION, 3063*
MULTIPLE MYELOMA
ALANINE, 3-(P-(BIS(2-CHLOROETHY
AMINO)PHENYL)-, L-, 3421
CASE REPORT, 3421
REVERSE TRANSCRIPTASE
PRECANCEROUS CONDITIONS, 3419
VIRUS, C-TYPE
REVERSE TRANSCRIPTASE, 3054*
VIRUS, FELINE LEUKEMIA
ANTIGENS, VIRAL, 3205
VIRUS, GIBBON APE LEUKEMIA
ANTIGENS, VIRAL, 3205
VIRUS-LIKE PARTICLES
PRECANCEROUS CONDITIONS, 3419
VIRUS, M7 BABOON
ANTIGENS, VIRAL, 3205
VIRUS, RAUSCHER MURINE LEUKEMIA
ANTIGENS, VIRAL, 3205
VIRUS, WOOLLY MONKEY SARCOMA
ANTIGENS, VIRAL, 3205

LEUKEMIA, MYELOCYTIC
BONE MARROW CELLS
DIAGNOSIS AND PROGNOSIS, 3487*
VIRUS-LIKE PARTICLES, 3419
CELL CYCLE KINETICS
REVIEW, 3072*

                    CARCINOMA, EPIDERMOID, 3426
                    CARCINOMA, TRANSITIONAL CELL, 3426
           URETHRAL NEOPLASMS
                    CARCINOMA, EPIDERMOID, 3426
           URINARY TRACT
                    DIAGNOSIS AND PROGNOSIS, 3426

        LEUKOPLAKIA, ORAL
           MOUTH NEOPLASMS
                    CARCINOMA, 3065*
                    ETIOLOGICAL ASPECTS, 3065*
                    PRECANCEROUS CONDITIONS, 3065*

        LEUROCRISTINE
           HEPATOMA
                    IMMUNE RESPONSE, 3319

        LEVAMISOLE
           KERATOACANTHOMA
                    IMMUNE RESPONSE, 3406
           VIRUS, HERPES SIMPLEX 1
                    CELL TRANSFORMATION, NEOPLASTIC,
                       3275*

        LIP NEOPLASMS
           ULTRAVIOLET RAYS
                    CARCINOGENIC POTENTIAL, REVIEW,
                       3026
3204
        LIPOPOLYSACCHARIDES
           B-LYMPHOCYTES
                    IMMUNE RESPONSE, CHICKEN, 3339
           VIRUS, C-TYPE MURINE
                    VIRUS, KIRSTEN MURINE SARCOMA,
                       3206

        LIPOPROTEINS
           RESPIRATORY TRACT NEOPLASMS
                    AUTOANTIBODIES, 3366*
                    HEMAGGLUTINATION, 3366*
           UROGENITAL NEOPLASMS
                    AUTOANTIBODIES, 3366*
                    HEMAGGLUTINATION, 3366*
        LIPOSARCOMA
           ESTROGENS
                    HORMONE DEPENDENT, 3173*
           ORBITAL NEOPLASMS
                    ULTRASTRUCTURAL STUDY, REVIEW,3064*

        LIVER CELL CARCINOMA
00
                    SEE HEPATOMA

        LIVER CIRRHOSIS
           CARCINOEMBRYONIC ANTIGEN
                    DIAGNOSIS AND PROGNOSIS, REVIEW,
                       3019

        LIVER NEOPLASMS
           ACETAMIDE, N-FLUOREN-2-YL-
                    DNA, 3126*
                    PRECANCEROUS CONDITIONS, 3126*
           ACETOHYDROXAMIC ACID, N-FLUOREN-2-YL-
                    CELL NUCLEUS, 3080
                    ULTRASTRUCTURAL STUDY, RAT, 3080
           ANGIOSARCOMA
                    EPIDEMIOLOGY, REVIEW, 3007
3426
                    ETHYLENE, CHLORO-, 3007, 3461*
                    ETHYLENE, CHLORO- POLYMER, 3007
                    PIPERIDINE, 1-NITROSO-, 3117

3-PIPERIDINOL, 1-NITROSO-, 3117
ASCORBIC ACID
    DIMETHYLAMINE, N-NITROSO-, 3159*
    UREA, ETHYL NITROSO-, 3159*
AUSTRALIA ANTIGEN
    ANTIBODIES, 3273*
CARCINOMA
    AUSTRALIA ANTIGEN, 3272*
    FERRITIN, 3019
    FETAL GLOBULINS, 3019
CHOLANGIOMA
    AFLATOXIN B1, 3075
DIETHYLAMINE, N-NITROSO-
    ADENOSINE TRIPHOSPHATASE, 3432
    GLUCOSEPHOSPHATASE, 3432
    GLYCOGEN, 3432
    GROWTH, 3432
    PRECANCEROUS CONDITIONS, 3432
DIMETHYLAMINE, N-NITROSO-
    RNA, 3115
HEPATOMA
    CYCLOHEXANE, 1,2,3,4,5,6-
      HEXACHLORO-, 3087
OCCUPATIONAL HAZARD
    ETHYLENE, CHLORO-, 3007
    ETHYLENE, CHLORO- POLYMER, 3007

LUNG NEOPLASMS
ADENOCARCINOMA
    AIR POLLUTANTS, 3516
    CASE REPORT, REVIEW, 3473*
    FOLIC ACID, N-NITROSO-, 3154*
    KIDNEY FAILURE, CHRONIC, 3308
    NEOPLASM METASTASIS, 3401
    SCLERODERMA, SYSTEMIC, 3473*
ADENOMA
    BENZIDINE, 3089
    CARBAMIC ACID, ETHYL ESTER, 3123
AIR POLLUTANTS
    EPIDEMIOLOGY, SOCIOECONOMIC
      FACTORS, 3516
AIRWAY OBSTRUCTION
    EPIDEMIOLOGY, 3517
ARSENIC
    CARCINOGENIC POTENTIAL, REIVEW,
      3026
BENZ(A)ANTHRACENE
    AIR POLLUTANTS, 3516
BENZO(A)PYRENE
    AIR POLLUTANTS, 3516
BENZO(E)PYRENE
    AIR POLLUTANTS, 3516
BENZO(G,H,I)PERYLENE
    AIR POLLUTANTS, 3516
BENZO(K)FLUORANTHENE
    AIR POLLUTANTS, 3516
BRONCHITIS
    EPIDEMIOLOGY, 3517
CADMIUM OXIDE
    OCCUPATIONAL HAZARD, 3520
CARBAMIC ACID, ETHYL ESTER
    ISOLATION AND CHARACTERIZATION,
      CLONE CELLS, 3123
CARCINOMA
    AIR POLLUTANTS, 3516
CARCINOMA, EPIDERMOID
    AIR POLLUTANTS, 3516
CHOLANTHRENE, 3-METHYL-
    CELL TRANSFORMATION, NEOPLASTIC,

    3474*
    TISSUE CULTURE, 3474*
CHROMIUM
    CARCINOGENIC POTENTIAL, REIVEW
      3026
CHRYSENE
    AIR POLLUTANTS, 3516
CORONENE
    AIR POLLUTANTS, 3516
CYCLOPHOSPHAMIDE
    CARCINOGENIC POTENTIAL, REVIEW
      3032
DIAGNOSIS AND PROGNOSIS
    REVIEW, 3022
DIBENZO(DEF,MNO)CHRYSENE
    AIR POLLUTANTS, 3516
EPIDEMIOLOGY
    GERMANY, 3506
FLUORANTHENE
    AIR POLLUTANTS, 3516
MELANOMA
    GENETICS, 3407
MESOTHELIOMA
    ASBESTOS, 3048*
NEOPLASM METASTASIS
    NEOPLASM METASTASIS, 3401
NICKEL
    CARCINOGENIC POTENTIAL, REIVEW
      3026
PERYLENE
    AIR POLLUTANTS, 3516
PURINE, 6-((METHYL-4-NITROIMADAZOL
    YL)THIO)-
    CARCINOGENIC POTENTIAL, REVIEW
      3032
PYRENE
    AIR POLLUTANTS, 3516
RADIATION
    CARCINOGENIC POTENTIAL, REIVEW
      3026
RETINOL
    PRECANCEROUS CONDITIONS, 3168*
SCLERODERMA, SYSTEMIC
    CASE REPORT, REVIEW, 3473*
SMOKING
    EPIDEMIOLOGY, 3525*
    REVIEW, 3034*
TOBACCO
    CARCINOGENIC POTENTIAL, REVIEW
      3026

LYMPH NODES
COLONIC NEOPLASMS
    PROGNOSIS, 3449*
RECTAL NEOPLASMS
    PROGNOSIS, 3449*

LYMPHANGIOMA
ORBITAL NEOPLASMS
    ULTRASTRUCTURAL STUDY, REVIEW,
      3064*
RETROPERITONEAL NEOPLASMS
    CASE REPORT, 3465*
    DIAGNOSIS AND PROGNOSIS, 3465*

LYMPHANGIOSARCOMA
BREAST NEOPLASMS
    CASE REPORT, 3397
MASTECTOMY

COLCHICINE
    RECEPTOR MOBILITY, 3394*
COMPLEMENT
    IMMUNE RESPONSE, K CELL, 3304
CONCANAVALIN A
    IMMUNITY, CELLULAR, 3348
    RECEPTOR MOBILITY, 3394*
CYTOCHALASIN B
    IMMUNE RESPONSE, K CELL, 3304
DIBENZ(A,H)ANTHRACENE
    ARYL HYDROCARBON HYDROXYLASES, 3151*
DNA REPLICATION
    ACETAMIDE, N-FLUOREN-2-YL-, 3097
    ACETOHYDROXAMIC ACID, N-FLUOREN-2-
        YL-, 3097
    3',4'-BENZOXYLIDIDE, ALPHA-ETHYNYL-
        ALPHA-PHENYL-, 3097
    HYDANTOIN, 5,5-DIPHENYL-, 3097
    HYDANTOIN, 3-HYDROXY-, 3097
    HYDANTOIN, 3-HYDROXY-1-ACETYL-5,5-
        DIPHENYL-, 3097
    HYDANTOIN, 3-HYDROXY-5,5-DIPHENYL-,
        3097
    HYDANTOIN, 3-HYDROXY-5-ISOPROPYL-,
        3097
    HYDANTOIN, 3-HYDROXY-5-METHYL-,
        3097
    HYDANTOIN, 5-(4-HYDROXYPHENYL)-5-
        PHENYL-, 3097
ETHANOL, 2-MERCAPTO-
    CHROMOSOME ABERRATIONS, 3424
HODGKIN'S DISEASE
    ANTIGENS, NEOPLASM, 3370*
IMMUNITY, CELLULAR
    ISOLATION AND CHARACTERIZATION, K
        CELL, 3304
IMMUNOGLOBULINS
    IMMUNE RESPONSE, K CELL, 3304
IMMUNOGLOBULINS, FC
    IMMUNE RESPONSE, K CELL, 3304
LEUKEMIA, LYMPHOCYTIC
    IMMUNOGLOBULINS, 3385*
    REVIEW, 3057*
MELANOMA
    IMMUNITY, CELLULAR, 3266*
MITOMYCIN C
    CHROMOSOME ABERRATIONS, 3103
    FIBROBLASTS, 3103
OVARIAN NEOPLASMS
    IMMUNITY, CELLULAR, 3333
    PLANT AGGLUTININS, 3333
PEPTIDES
    PLANT AGGLUTININS, 3552
PLANT AGGLUTININS
    NUCLEOTIDYLTRANSFERASES, 3572
RADIATION, IONIZING
    CHROMOSOME ABERRATIONS, 3194*
RNA, TRANSFER
    PLANT AGGLUTININS, 3552
SMOKING
    PLANT AGGLUTININS, 3342
UROGENITAL NEOPLASMS
    IMMUNITY, CELLULAR, 3333
    PLANT AGGLUTININS, 3333
VIRUS, EPSTEIN-BARR
    CELL LINE, 3312
    IMMUNITY, CELLULAR, 3266*
VIRUS, FRIEND MURINE LEUKEMIA
    IMMUNE RESPONSE, 3310, 3345

SUBJECT 45

3307

3307

5

-

VIEW,

RAZINO-2-

3333

3572

ASES,

-LYMPHOCYTES
  BURKITT'S LYMPHOMA
    CELL DIFFERENTIATION, REVIEW, 3059*
    IMMUNOFLUORESCENCE, 3334
    IMMUNOGLOBULINS, SURFACE, REVIEW,
    3059*
  CHALONES
    SIALIC ACIDS, 3029
  DEXTRAN
    IMMUNE RESPONSE, CHICKEN, 3339
  HISTOCOMPATIBILITY ANTIGENS
    GENETICS, 3330
    IMMUNE SERUMS, 3330
  HODGKIN'S DISEASE
    IGG, 3058*
    IMMUNOGLOBULINS, SURFACE, 3058*
    ROSETTE FORMATION, 3386*
  LEUKEMIA, LYMPHOBLASTIC
    CELL DIFFERENTIATION, REVIEW,
    3059*
    IGA, 3388*
    IGG, 3387*, 3388*
    IGM, 3388*
    IMMUNOGLOBULINS, 3388*
    IMMUNOGLOBULINS, FC, 3387*
    IMMUNOGLOBULINS, SURFACE, REVIEW,
    3059*
    ROSETTE FORMATION, 3386*
  LEUKEMIA, LYMPHOCYTIC
    CELL DIFFERENTIATION, REVIEW, 3059*
    IGD, 3338
    IGG, 3335, 3387*
    IGM, 3335, 3338
    IMMUNOGLOBULINS, 3335
    IMMUNOGLOBULINS, FC, 3387*
    IMMUNOGLOBULINS, SURFACE, 3338
    IMMUNOGLOBULINS, SURFACE, REVIEW,
    3059*
    ROSETTE FORMATION, 3555

  LIPOPOLYSACCHARIDES
    IMMUNE RESPONSE, CHICKEN, 3339
  LYMPHATIC NEOPLASMS
    CLASSIFICATION, 3490*
  T-LYMPHOCYTES
    IMMUNE RESPONSE, REVIEW, 3382*
    IMMUNITY, CELLULAR, 3382*
  LYMPHOMA
    CELL DIFFERENTIATION, REVIEW,
    3059*
    IMMUNOFLUORESCENCE, 3334
    IMMUNOGLOBULINS, SURFACE, REVIEW,
    3059*
  LYMPHOSARCOMA
    IMMUNOFLUORESCENCE, 3334
    ROSETTE FORMATION, 3386*
  MITOGENS
    IMMUNE RESPONSE, CHICKEN, 3339
  PLANT AGGLUTININS
    LYMPHOCYTE TRANSFORMATION, 3337
  SARCOMA, RETICULUM CELL
    ROSETTE FORMATION, 3386*
  TRANSPLANTATION IMMUNOLOGY
    IMMUNITY, CELLULAR, 3360*
  TUBERCULIN
    IMMUNE RESPONSE, CHICKEN, 3339
  VIRUS, AVIAN LEUKOSIS
    ANTIGENS, NEOPLASM, 3381*
    IMMUNOGLOBULINS, 3381*

  VIRUS, EPSTEIN-BARR
    ANTIGENS, VIRAL, 3326
  VIRUS, FRIEND MURINE LEUKEMIA
    ANTIGENS, VIRAL, 3310
    IMMUNOGLOBULINS, SURFACE, 3310
    VIBRIO INFECTIONS, 3312
  VIRUS, MOLONEY MURINE LEUKEMIA
    VIBRIO INFECTIONS, 3312

T-LYMPHOCYTES
  BONE MARROW
    CONCANAVALIN A, 3384*
    DIBUTYRYL CYCLIC AMP, 3384*
    THYMUS EXTRACTS, 3384*
  BURKITT'S LYMPHOMA
    ROSETTE FORMATION, 3334
  CHALONES
    SIALIC ACIDS, 3029
  CONCANAVALIN A
    IMMUNE RESPONSE, CHICKEN, 3339
  HISTOCOMPATIBILITY ANTIGENS
    GENETICS, 3330
    IMMUNE SERUMS, 3330
  HODGKIN'S DISEASE
    IGG, 3058*
    IMMUNITY, CELLULAR, 3418
    IMMUNOGLOBULINS, SURFACE, 3058*
    ROSETTE FORMATION, 3386*
    ULTRASTRUCTURAL STUDY, 3418
  HYPERSENSITIVITY, IMMEDIATE
    IMMUNITY, CELLULAR, 3389*
  IMMUNITY, CELLULAR
    TRINITROPHENYL-MODIFIED CELLS,
    3383*
  IMMUNOGLOBULINS
    IMMUNOSUPPRESSION, 3382*
  IMMUNOGLOBULINS, FC
    CONCANAVALIN A, 3343
    MITOSIS, 3343
    PLANT AGGLUTININS, 3343
  LEUKEMIA
    CLASSIFICATION, REVIEW, 3321
    GROWTH, 3331
    ISOANTIGENS, 3331
  LEUKEMIA L1210
    IMMUNITY, CELLULAR, 3332
    ISOANTIGENS, 3332
    RADIATION, 3332
  LEUKEMIA, LYMPHOBLASTIC
    CELL DIFFERENTIATION, REVIEW,
    3059*
    IMMUNITY, CELLULAR, 3388*
    IMMUNOGLOBULINS, SURFACE, REVIE
    3059*
    NUCLEOTIDYLTRANSFERASES, 3540
    ROSETTE FORMATION, 3386*, 3387*
  LEUKEMIA, LYMPHOCYTIC
    CELL DIFFERENTIATION, REVIEW,
    3059*
    ROSETTE FORMATION, 3335, 3386*,
    3387*
  LEUKEMIA, MYELOBLASTIC
    ROSETTE FORMATION, 3386*
  LEUKEMIA, RADIATION-INDUCED
    GROWTH, 3331
    ISOANTIGENS, 3331
  LYMPHATIC NEOPLASMS
    CLASSIFICATION, 3490*
  B-LYMPHOCYTES

LYMPHOMA (GENERAL AND UNSPECIFIED)
    SEE ALSO HODGKIN'S DISEASE/BURKITT'S
    LYMPHOMA/CELL/LYMPHOSARCOMA,
    RETICULUM, SARCOMA
    CARCINOEMBRYONIC ANTIGEN
        DIAGNOSIS AND PROGNOSIS, REVIEW,
        3019
    DIAGNOSIS AND PROGNOSIS
        CLASSIFICATION, 3062*
    ESTROGENS
        HORMONE DEPENDENT, 3173*
    ETHYLENE, CHLORO-
        EPIDEMIOLOGY, 3518
    ETHYLENE, CHLORO- POLYMER
        EPIDEMIOLOGY, 3518
    LYMPHATIC NEOPLASMS
        CLASSIFICATION, 3490*
    B-LYMPHOCYTES
        CELL DIFFERENTIATION, REVIEW,
        3059*
        IMMUNOFLUORESCENCE, 3334
        IMMUNOGLOBULINS, SURFACE, REVIEW,
        3059*
    T-LYMPHOCYTES
        CELL DIFFERENTIATION, REVIEW,
        3059*
        CLASSIFICATION, REVIEW, 3021
        IMMUNITY, CELLULAR, 3332
        IMMUNOGLOBULINS, SURFACE, REVIEW,
        3059*
        ISOANTIGENS, 3332
        RADIATION, 3332
        ROSETTE FORMATION, 3334
    PURINE, 6-((METHYL-4-NITROIMADAZOL-5-
    YL)THIO)-
        CARCINOGENIC POTENTIAL, REVIEW,
        3032
    THORACIC NEOPLASMS
        ASBESTOS, 3138*
    VIRUS, C-TYPE MURINE LEUKEMIA
        GENETICS, 3262*
    VIRUS, MURINE LEUKEMIA
        IMMUNITY, CELLULAR, 3390*

LYMPHOSARCOMA
    ADRENAL GLAND NEOPLASMS
        VIRUS, HERPES SAIMIRI, 3277*
    BRAIN NEOPLASMS
        VIRUS, HERPES SAIMIRI, 3277*
    BREAST NEOPLASMS
        CASE REPORT, 3398
        NEOPLASM METASTASIS, 3398
    BURKITT'S LYMPHOMA
        CASE REPORT, 3491*
        EPIDEMIOLOGY, 3524*
    EYE NEOPLASMS
        VIRUS, HERPES SAIMIRI, 3277*
    KIDNEY NEOPLASMS
        VIRUS, HERPES SAIMIRI, 3277*
    LYMPHATIC NEOPLASMS
        CLASSIFICATION, 3490*
    B-LYMPHOCYTES
        IMMUNOFLUORESCENCE, 3334
        ROSETTE FORMATION, 3386*
    T-LYMPHOCYTES
        ROSETTE FORMATION, 3334, 3386*
    METHOTREXATE
        CARCINOGENIC POTENTIAL, REVIEW,

3032
NASOPHARYNGEAL NEOPLASMS
    EPIDEMIOLOGY, AGE FACTORS, 3515
    EPIDEMIOLOGY, INDIA, 3515
    PURINE, 6-((METHYL-4-NITROIMADAZOL-5-
    YL)THIO)-
        CARCINOGENIC POTENTIAL, REVIEW,
        3032
    SKIN NEOPLASMS
        VIRUS, HERPES SAIMIRI, 3277*
    VIRUS, C-TYPE MURINE
        ISOLATION AND CHARACTERIZATION,
        3263*
    VIRUS, C-TYPE RNA TUMOR
        ANTIGENIC DETERMINANTS, 3369*
        PEPTIDES, 3369*
    VIRUS, FELINE LEUKEMIA
        EPIDEMIOLOGY, REVIEW, 3015
    VIRUS, HERPES SAIMIRI
        CLINICOPATHOLOGIC STUDY, RABBIT,
        3277*

LYSOSOMES
    LEUKEMIA, LYMPHOBLASTIC
        ACID PHOSPHATASE, 3568
        GALACTOSIDASES, 3568
        GLUCOSAMINIDASE, 3568
        GLUCOSIDASES, 3568
        GLUCURONIDASE, 3568
        GLYCOLIPIDS, 3568
        GLYCOPROTEINS, 3568
        SULFATASES, 3568
    VIRUS, POLYOMA
        CELL TRANSFORMATION, NEOPLASTIC,
        3570

MACROPHAGES
    BRAIN NEOPLASMS
        ULTRASTRUCTURAL STUDY, RAT, 3114
        UREA, ETHYL NITROSO-, 3114
    CARBOHYDRATES
        CONCANAVALIN A, 3346
        PLANT AGGLUTININS, 3346
    CELL LINE
        ISOLATION AND CHARACTERIZATION,
        3392*
    CELL MEMBRANE
        CARBOHYDRATES, 3346
    GLUTARALDEHYDE
        AGGLUTINATION, 3346
    HODGKIN'S DISEASE
        ULTRASTRUCTURAL STUDY, 3418
    LEUKEMIA, MYELOBLASTIC
        CELL DIFFERENTATION, 3063*
    LEUKEMIA, MYELOCYTIC
        CELL DIFFERENTATION, 3063*
    LYMPHATIC NEOPLASMS
        CHOLANTHRENE, 3-METHYL-, 3392*
    PICRIC ACID
        IMMUNE RESPONSE, 3302
    RADIATION
        IMMUNE RESPONSE, 3302
    VIRUS, FRIEND MURINE LEUKEMIA
        ANTIBODY-PRODUCING CELLS, 3345
        IMMUNE RESPONSE, 3345
    VIRUS, HERPES SIMPLEX 2
        PYRANS, 3341

MAGNESIUM

VIRUS, POX
    VIRUS REPLICATION, 3255

MAGNESIUM CHLORIDE
    FIBROBLASTS
        GROWTH, 3544

MALATE DEHYDROGENASE
    LEUKEMIA, LYMPHOBLASTIC
        MITOCHONDRIA, 3568
    VIRUS, ADENO 12
        CELL TRANSFORMATION, NEOPLASTIC,
        3201
        ISOLATION AND CHARACTERIZATION,
        3201

MALEIC ACID, DIETHYL ESTER
    AFLATOXIN B1
        COCARCINOGENIC EFFECT, RAT, 3079

MAMMARY NEOPLASMS, EXPERIMENTAL
    ADENOCARCINOMA
        BENZ(A)ANTHRACENE, 7,12-DIMETHYL
        3092
        CHROMATIN, 3587*
        GLUCOCORTICOID-BINDING PROTEINS,
        3550
        PROLACTIN, 3092
    ALDOSTERONE
        BINDING, COMPETITIVE, 3550
        5ALPHA-ANDROSTAN-3-ONE, 17BETA-HYDRO

        BINDING, COMPETITIVE, 3550
    BENZ(A)ANTHRACENE, 7,12-DIMETHYL-
        ESTRADIOL, 3093
        GLUCOCORTICOID-BINDING PROTEINS,
        3550
    CARCINOMA
        THYMIDINE INCORPORATION, 3538
    CORTICOSTERONE
        BINDING, COMPETITIVE, 3550
    CORTISOL
        BINDING, COMPETITIVE, 3550
    CYCLOPHOSPHAMIDE
        CARCINOGENIC POTENTIAL, REVIEW,
        3032
    DEXAMETHASONE
        BINDING, COMPETITIVE, 3550
    ESTRADIOL
        BINDING, COMPETITIVE, 3550
    ESTROGENS
        HYPERPLASIA, 3396
    ESTRONE
        BINDING, COMPETITIVE, 3550
    GLUCOSEPHOSPHATE DEHYDROGENASE
        2,3-BUTANEDIOL, 1,4-DIMERCAPTO-,
        3567
        GLUTATHIONE REDUCTASE, 3567
        NADP, 3567
    HYPERPLASIA
        ULTRASTRUCTURAL STUDY, 3396
    INSULIN
        HYPERPLASIA, 3396
    PREGN-4-ENE-3,20-DIONE, 17,21-
        DIHYDROXY-
        BINDING, COMPETITIVE, 3550
    PROGESTERONE
        BINDING, COMPETITIVE, 3550
    TESTOSTERONE

ENZYMATIC ACTIVITY, 3594*
LUNG NEOPLASMS
GENETICS, 3407
LYMPHATIC NEOPLASMS
GENETICS, 3407
LYMPHOCYTES
IMMUNITY, CELLULAR, 3266*
T-LYMPHOCYTES
IMMUNITY, CELLULAR, 3336
NEOPLASM METASTASIS
IMMUNITY, CELLULAR, 3336
NEOPLASMS, MULTIPLE PRIMARY
GENETICS, 3407
SARCOMA
ANTIGENS, NEOPLASM, 3407
VAGINAL NEOPLASMS
EPIDEMIOLOGY, 3501*

MENINGIOMA
BRAIN NEOPLASMS
CASE REPORT, 3403
HISTOLOGIC STUDY, 3403
ORBITAL NEOPLASMS
ULTRASTRUCTURAL STUDY, REVIEW,
3064*

MENOPAUSE
BREAST NEOPLASMS
EPIDEMIOLOGY, 3512
UTERINE NEOPLASMS, 3099

MERCURY, (P-CARBOXYPHENYL)CHLORO-
CARCINOMA, EHRLICH TUMOR
CALCIUM, 3543
PHOSPHATES, 3543

MESENCHYMOMA
HEART NEOPLASMS
CASE REPORT, 3462*
ULTRASTRUCTURAL STUDY, 3462*

MESOTHELIOMA
ASBESTOS
EPIDEMIOLOGY, 3529*
OCCUPATIONAL HAZARD, 3529*
PRECANCEROUS CONDITIONS, 3529*
HEART NEOPLASMS
HISTOLOGICAL STUDY, RAT, 3415
ULTRASTRUCTURAL STUDY, RAT, 3415
LUNG NEOPLASMS
ASBESTOS, 3048*
PLEURAL NEOPLASMS
DIAGNOSIS AND PROGNOSIS, 3412

MESOXALONITRILE
CARCINOMA, EHRLICH TUMOR
CALCIUM, 3543

METHANE, AZOXY-
HYDRAZINE, 1,2-DIMETHYL-
ACTIVE INTERMEDIATE, 3161*

METHANE, IODO-
ALKYLATING AGENTS
STRUCTURE-ACTIVITY RELATIONSHIP,
OUS,   3109

METHANE, SULFINYLBIS-
CELL MEMBRANE

GLYCOPROTEINS, 3095
CHOLANTHRENE, 3-METHYL-
    CELLS, CULTURED, 3095
CONCANAVALIN A
    CELLS, CULTURED, 3095
ERYTHROLEUKEMIA
    CELL DIFFERENTIATION, 3541
    GROWTH, 3541
PHENANTHRENE, 9,10-EPOXY-9,10-DIHYDRO-
    ALKYLATION, 3167*
PLANT AGGLUTININS
    CELLS, CULTURED, 3095
VIRUS, ADENO 12
    CELL TRANSFORMATION, NEOPLASTIC,
    3202
VIRUS, FRIEND MURINE LEUKEMIA
    CELL DIFFERENTIATION, 3268*
    RNA, MESSENGER, 3268*
VIRUS, SV40
    CELL TRANSFORMATION, NEOPLASTIC,
    3202

METHANESULFONAMIDE, N-(4-(ACRIDINYLAMINO)-
3-METHOXYPHENYL)-
    SARCOMA, MAST CELL
        ADENOSINE CYCLIC 3',5'
        MONOPHOSPHATE, 3542
        ADENOSINE TRIPHOSPHATE, 3542
        GROWTH, 3542

METHANESULFONIC ACID, 2-HYDROXYETHYL ESTER
    BACTERIOPHAGES
        RNA, 3111

METHANESULFONIC ACID, METHYL ESTER
    ALKYLATING AGENTS
        STRUCTURE-ACTIVITY RELATIONSHIP,
        3109
    DEOXYRIBONUCLEOSIDES
        URINARY EXCRETION, 3108
    DIETHYLAMINE, 2,2'-DICHLORO-N-METHYL-
        URINARY EXCRETION, 3108
    DNA
        GUANINE, 7-METHYL-, 3112
    SKIN NEOPLASMS
        XERODERMA PIGMENTOSUM, 3031
    UREA, METHYL NITROSO-
        ALKYLATING AGENTS, 3112

METHANOL, (METHYL-ONN-AZOXY)-
    CARCINOMA
        CELLS, MUTANT RATS, 3431
        KARYOTYPING, 3431
        NUCLEOTIDYLTRANSFERASES, 3431

METHOTREXATE
    ANTINEOPLASTIC AGENTS
        CARCINOGENIC POTENTIAL, REVIEW,
        3304
    LYMPHOSARCOMA
        CARCINOGENIC POTENTIAL, REVIEW,
        3032

N-METHYL-N'-NITRO-N-NITROSOGUANIDINE
    SEE GUANIDINE, 1-METHYL-3-NITRO-1-

NITROSO-

N-METHYL-N-NITROSOUREA
    SEE UREA, METHYL NITROSO-

1-METHYL-1-NITROSOUREA
    SEE UREA, METHYL NITROSO-

METHYL SULFOXIDE
    SEE METHANE, SULFINYLBIS-

3-METHYLCHOLANTHRENE
    SEE CHOLANTHRENE, 3-METHYL-

20-METHYLCHOLANTHRENE
    SEE CHOLANTHRENE, 3-METHYL-

METHYLNITROSOUREA
    SEE UREA, METHYL NITROSO-

MICROSOMES
    BENZO(A)PYRENE
        ARYL HYDROCARBON HYDROXYLASES,
        3084

MICROSOMES, LIVER
    ANILINE, N-METHYL-P-(PHENYLAZO)-
        MIXED-FUNCTION AMINE OXIDASE,
        3136*
    BENZENE, 4-ALLYL-1,2-(METHYLENEDIOXY
        ABSORPTION SPECTRA, RAT, 3088
        CYTOCHROME P-450, 3088
    BENZO(A)PYRENE
        ARYL HYDROCARBON HYDROXYLASES,
        3083
        CHOLANTHRENE, 3-METHYL-, 3144*
    7,8-BENZOFLAVONE
        ENZYMATIC ACTIVITY, 3143*
    HYDROXYLAMINE, N-(P-(PHENYLAZO)PHENY

        MIXED-FUNCTION AMINE OXIDASE,
        3136*
    PHENOL, (1,1-DIMETHYL)-4-METHOXY-
        ARYL HYDROCARBON HYDROXYLASES,
        3083
    SALMONELLA TYPHIMURIUM
        DIETHYLAMINE, N-NITROSO-, 3158*
        DIMETHYLAMINE, N-NITROSO-, 3158*

MIGRATION INHIBITORY FACTOR
    VIRUS, ADENO 12
        HISTOCOMPATIBILITY ANTIGENS, 32

MITOCHONDRIA
    ETHIDIUM BROMIDE
        ULTRASTRUCTURAL STUDY, 3160*
    LEUKEMIA, LYMPHOBLASTIC
        GLUTAMATE DEHYDROGENASE, 3568
        MALATE DEHYDROGENASE, 3568
        SUCCINATE DEHYDROGENASE, 3568

MITOGENS
    B-LYMPHOCYTES
        IMMUNE RESPONSE, CHICKEN, 3339

MITOMYCIN C
    CHROMOSOME ABERRATIONS
        ANALYTICAL METHOD, 3103
    LYMPHOCYTES

                    URINE, 3557
                SPERMINE
                    URINE, 3557

            MUTAGENS
                ACETAMIDE, N-(ACETYLOXY)-N-9H-FLUOREN-
                2-YL-
                    AZAGUANINE RESISTANCE, HAMSTER,
                    3102
                BENZ(A)ANTHRACENE, 7,12-DIMETHYL-
                    AZAGUANINE RESISTANCE, HAMSTER,
                    3102
                BENZ(A)ANTHRACENE, 7-METHYL-
                    AZAGUANINE RESISTANCE, HAMSTER,
                    3102
                BENZO(A)PYRENE
                    AZAGUANINE RESISTANCE, HAMSTER,
                    3102
                CHOLANTHRENE, 3-METHYL-
                    AZAGUANINE RESISTANCE, HAMSTER,
                    3102
                GUANIDINE, 1-METHYL-3-NITRO-1-NITROSO-
                    AZAGUANINE RESISTANCE, HAMSTER,
                    3102
                K-REGION EPXIDES
                    AZAGUANINE RESISTANCE, HAMSTER,
                    3102

        MUTATION
            NEOPLASMS
                RADIATION, IONIZING, 3031
                REVIEW, 3031
            ONCOGENIC VIRUSES
                REVIEW, 3069*
            SALMONELLA TYPHIMURIUM
                CARCINOGEN, CHEMICAL, 3044*

        MYCOSIS FUNGOIDES
            T-LYMPHOCYTES
                ROSETTE FORMATION, 3344
                ULTRASTRUCTURAL STUDY, SEZARY
                CELLS, 3344

01   MYELOID METAPLASIA
            VIRUS, C-TYPE MURINE LEUKEMIA
                TRANSPLANTATION, HOMOLOGOUS, 3204

        MYELOMA
            SEE MULTIPLE MYELOMA

        MYELOMA PROTEINS
            PLASMACYTOMA
                ANTIGENIC DETERMINANTS, 3317
                ANTIGENS, NEOPLASM, 3317
                IMMUNOGLOBULIN FRAGMENTS, 3317

        MYELOPEROXIDASE
            LEUKEMIA
                BONE MARROW CELLS, 3483*
                PRECANCEROUS CONDITIONS, 3483*

        MYOBLASTOMA
            ORBITAL NEOPLASMS
                ULTRASTRUCTURAL STUDY, REVIEW,
                3064*

        MYXOMA
            BONE NEOPLASMS
                CASE REPORT, REVIEW, 3437*

NAD NUCLEOSIDASE
  HELA CELLS
    DNA, 3584*

NADH, NADPH OXIDOREDUCTASES
  ANILINE, N,N-DIMETHYL-P-NITROSO-
    ENZYMATIC ACTIVITY, 3135*
  BENZO(A)PYRENE
    BENZO(A)PYREN-6-YLOXY, 3142*
  BREAST NEOPLASMS
    ENZYMATIC ACTIVITY, 3446*
  COLONIC NEOPLASMS
    ENZYMATIC ACTIVITY, 3446*

NADP
  MAMMARY NEOPLASMS, EXPERIMENTAL
    GLUCOSEPHOSPHATE DEHYDROGENASE,
    3567
  PRE INCEROUS CONDITIONS
    GLUCOSEPHOSPHATE DEHYDROGENASE,
    3567

ALPHA-NAPHTHOFLAVONE
  SEE 7,8-BENZOFLAVONE

BETA-NAPHTHOFLAVONE
  SEE 5,6-BENZOFLAVONE

NASOPHARYNGEAL NEOPLASMS
  CARCINOMA
    EPIDEMIOLOGY, AGE FACTORS, 3515
    EPIDEMIOLOGY, INDIA, 3515
    VIRUS, EPSTEIN-BARR, 3012
  LYMPHOSARCOMA
    EPIDEMIOLOGY, AGE FACTORS, 3515
    EPIDEMIOLOGY, INDIA, 3515
  VIRUS, EPSTEIN-BARR
    ANTIGENS, VIRAL, 3011

NEOPLASM METASTASIS
  BREAST NEOPLASMS
    ADENOCARCINOMA, 3401
    CAPILLARITY, 3444*
    CARCINOEMBRYONIC ANTIGEN, 3019
    CARCINOMA, 3298
    CARCINOMA, SCIRRHOUS, 3401
    CLINICOPATHOLOGIC STUDY,
      LOCALIZATION, 3401
    THYMIDINE INCORPORATION, 3538
  COLONIC NEOPLASMS
    CARCINOEMBRYONIC ANTIGEN, 3019,
      3329
    CARCINOMA, 3318, 3320
    INTESTINAL POLYPS, 3447*
  LUNG NEOPLASMS
    ADENOCARCINOMA, 3401
    NEOPLASM METASTASIS, 3401
  MELANOMA
    IMMUNITY, CELLULAR, 3336
  NEOPLASM TRANSPLANTATION
    CAPILLARITY, 3472*
  PANCREATIC NEOPLASMS
    ADENOCARCINOMA, 3401
  PITUITARY NEOPLASMS
    CLINICOPATHOLOGIC STUDY,
      LOCALIZATION, 3401
    DIABETES INSIPIDUS, 3401
  SEED-SOIL FACTORS
    MODEL TEST SYSTEM, 3482*

STOMACH NEOPLASMS
  EPIDEMIOLOGY, GERMANY, 3526*
  VIRUS, HERPES SIMPLEX 1
    CELL TRANSFORMATION, NEOPLASTIC,
      3275*

NEOPLASM REGRESSION, SPONTANEOUS
  MELANOMA
    IMMUNITY, CELLULAR, 3481*

NEOPLASM TRANSPLANTATION
  BENZ(A)ANTHRACENE, 7,12-DIMETHYL-
    CELL TRANSFORMATION, NEOPLASTIC,
      3139*
  BENZ(A)ANTHRACENE, 7-METHYL-
    CELL TRANSFORMATION, NEOPLASTIC,
      3139*
  BENZ(C)ACRIDINE, 7,9-DIMETHYL-
    CELL TRANSFORMATION, NEOPLASTIC,
      3139*
  BENZO(A)PYRENE
    CELL TRANSFORMATION, NEOPLASTIC,
      3139*
  BENZO(E)PYRENE
    CELL TRANSFORMATION, NEOPLASTIC,
      3139*
  CARCINOMA
    VIRUS, SV40, 3250
  CHOLANTHRENE, 3-METHYL-
    CELL TRANSFORMATION, NEOPLASTIC,
      3139*
  CHORIOCARCINOMA
    THYROTROPIN, 3578*
  ESTROGENS, 3173*
  FIBROSARCOMA
    VIRUS, POLYOMA, BK, 3283*
  HEPATOMA
    PORTACAVAL ANASTOMOSIS, 3460*
  NEOPLASM METASTASIS
    CAPILLARITY, 3472*
  SARCOMA
    ADRENAL CORTEX HORMONES, 3600*
    VIRUS, HERPES SIMPLEX 2, 3358*

NEOPLASMS (GENERAL AND UNSPECIFIED)
  SEE ALSO UNDER PARTICULAR SITE
  ANTIGEN-ANTIBODY REACTIONS
    IMMUNOFLUORESCENT STUDY, REVIEW,
      3356*
  AUTOANTIBODIES
    IMMUNOFLUORESCENT STUDY, REVIEW,
      3056*
  BENZ(A)ANTHRACENE, 7,12-DIMETHYL-
    CARCINOGENIC ACTIVITY, 3164*
  CARCINOEMBRYONIC ANTIGEN
    ANTIGENS, NEOPLASM, 3378*
    IMMUNOFLUORESCENT STUDY, REVIEW,
      3056*
    REVIEW, 3028
  CARCINOGEN, CHEMICAL
    BIOLOGICAL ASSAY, REVIEW, 3041*
  CHOLESTEROL, 14-METHYLHEXADECANOATE
    CHOLESTEROL METABOLISM, REVIEW,3
  CROTON OIL
    CARCINOGENIC ACTIVITY, 3164*
  DNA REPAIR
    REVIEW, 3031
  ENVIRONMENTAL HAZARD
    COCARCINOGEN, CHEMICAL, 3043*

GENETICS, 3407

NEPHROBLASTOMA
    ANTIGENS, NEOPLASM
        ISOLATION AND CHARACTERIZATION,
            RABBIT, 3325
    CELL MEMBRANE
        ANTIGENS, NEOPLASM, 3325
    ULTRASTRUCTURAL STUDY
        CASE REPORT, 3492*

NERVE GROWTH FACTOR
    UREA, ETHYL NITROSO-
        CELL DIFFERENTIATION, 3170*

NERVOUS SYSTEM NEOPLASMS
    ANGIOSARCOMA
        BLOOD GROUPS, 3466*
    EPENDYMOMA
        ULTRASTRUCTURAL STUDY, 3404
    EPIDEMIOLOGY
        GERMANY, 3506
    MEDULLOBLASTOMA
        ULTRASTRUCTURAL STUDY, 3404
    NEURILEMMOMA
        ULTRASTRUCTURAL STUDY, 3404
    PARAGANGLIOMA, NONCHROMAFFIN
        CASE REPORT, 3468*
    TERATOID TUMOR
        TRANSPLANTATION, HOMOLOGOUS, 3536

NEURAMINIDASE
    L CELLS
        BINDING, 3546
    LEUKEMIA
        IMMUNITY, CELLULAR, 3350*
        IMMUNOTHERAPY, 3350*
        SIALIC ACIDS, 3350*
        VIRUS, GROSS LEUKEMIA, 3350*
    LEUKEMIA L1210
        BINDING, 3546
    LEUKEMIA, MYELOCYTIC
        BINDING, 3546

NEURILEMMOMA
    NERVOUS SYSTEM NEOPLASMS
        ULTRASTRUCTURAL STUDY, 3404
    ORBITAL NEOPLASMS
        ULTRASTRUCTURAL STUDY, REVIEW,
            3064*

NEUROBLASTOMA
    ALANINE, 3-(3,4-DIHYDROXYPHENYL)-
        CARCINOGENIC ACTIVITY, 3134*
    CARCINOEMBRYONIC ANTIGEN
        DIAGNOSIS AND PROGNOSIS, REVIEW,
            3019
    GLYCOLIPIDS
        TRIOSES, 3563
    PHOSPHOLIPIDS
        TRIOSES, 3563

NEUROFIBROMA
    ORBITAL NEOPLASMS
        ULTRASTRUCTURAL STUDY, REVIEW,
            3064*

NEUROFIBROMATOSIS
    EYE NEOPLASMS

GENETICS, 3070*
REVIEW, 3070*

NICKEL
LUNG NEOPLASMS
CARCINOGENIC POTENTIAL, REIVEW,
3026

NICOTINE
SMOKING
CARCINOGENIC METABOLITE, 3165*
GASTRIC JUICE, 3165*

NICOTINE, 1'-DEMETHYL-
SMOKING
CARCINOGENIC METABOLITE, 3165*
GASTRIC JUICE, 3165*

NITRIC ACID, SODIUM SALT
FOOD ANALYSIS
VEGETABLES, 3046*
FOOD CONTAMINATION
CARCINOGENIC POTENTIAL, REVIEW,
3045*
METHEMOGLOBINEMIA, 3045*

NITROGEN OXIDE
RESPIRATORY TRACT NEOPLASMS
PRECANCEROUS CONDITIONS, 3119

4-NITROQUINOLINE-1-OXIDE
SEE QUINOLINE, 4-NITRO-, 1-OXIDE

NITROSO COMPOUNDS
FOOD INSPECTION
ANALYSIS, 3116

N-NITROSO-N-METHYL UREA
SEE UREA, METHYL NITROSO-

N-NITROSOMETHYLUREA
SEE UREA, METHYL NITROSO-

NITROUS ACID, SODIUM SALT
FOOD ANALYSIS
VEGETABLES, 3046*
FOOD CONTAMINATION
CARCINOGENIC POTENTIAL, REVIEW,
3045*
METHEMOGLOBINEMIA, 3045*

NOSE NEOPLASMS
CARCINOMA
PIPERIDINE, 1-NITROSO-, 3117
3-PIPERIDINOL, 1-NITROSO-, 3117
4-PIPERIDINOL, 1-NITROSO-, 3117
4-PIPERIDONE, 1-NITROSO-, 3117

NUCLEIC ACIDS
SEE ALSO DNA/RNA
NEOPLASMS, EXPERIMENTAL
PHOTOADDITION, REVIEW, 3074*
VIRUS, POLYOMA
ISOLATION AND CHARACTERIZATION,
3233

NUCLEOSIDE DIPHOSPHATE SUGARS
HEPATOMA
DNA NUCLEOTIDYLTRANSFERASES, 3593*

NUCLEOTIDES
HEPATOMA
ACETAMIDE, N-FLUOREN-2-YLDI-,
3128*
ANILINE, N,N-DIMETHYL-P-PHENYLAZ
3128*

NUCLEOTIDYLTRANSFERASES
CARCINOMA
BENZO(A)PYRENE, 3431
DIETHYLAMINE, N-NITROSO-, 3431
METHANOL, (METHYL-ONN-AZOXY)-,
3431.
LEUKEMIA, LYMPHOBLASTIC
T-LYMPHOCYTES, 3540
LYMPHOCYTES
CELL DIVISION, 3572
PLANT AGGLUTININS, 3572

NUTRITION
SEE DIET

OCCUPATIOANL HAZARD
ETHYLENE, CHLORO-
CARCINOGENIC ACTIVITY, REVIEW,
3038*
ETHYLENE, CHLORO- POLYMER
CARCINOGENIC ACTIVITY, REVIEW,
3038*

OCCUPATIONAL HAZARD
ANGIOSARCOMA
ETHYLENE, CHLORO-, 3038*, 3518
ETHYLENE, CHLORO- POLYMER, 3038*
3518
ANILINE, 4,4'-METHYLENEBIS-2-CHLORO-
CARCINOGENIC POTENTIAL, REVIEW,
3040*
BLADDER NEOPLASMS
CADMIUM OXIDE, 3520
DIGESTIVE SYSTEM NEOPLASMS
PETROLEUM, 3520
ETHYLENE, CHLORO-
ANGIOSARCOMA, 3530*
ETHYLENE, CHLORO- POLYMER
PRECANCEROUS CONDITIONS, 3519
GLIOBLASTOMA MULTIFORME
ETHYLENE, CHLORO-, 3518
2-IMIDAZOLIDINETHIONE
CARCINOGENIC POTENTIAL, REVIEW,
3040*
LIVER NEOPLASMS
ETHYLENE, CHLORO-, 3007
ETHYLENE, CHLORO- POLYMER, 3007
LUNG NEOPLASMS
CADMIUM OXIDE, 3520
MESOTHELIOMA
ASBESTOS, 3529*
PRECANCEROUS CONDITIONS
ANTHRACOSILICOSIS, 3005
PNEUMOCONIOSIS, 3005
SILICOSIS, 3005
PROSTATIC NEOPLASMS
CADMIUM OXIDE, 3520
RESPIRATORY TRACT NEOPLASMS
PETROLEUM, 3520
SKIN NEOPLASMS
PETROLEUM, 3520
TESTICULAR NEOPLASMS

ULTRASTRUCTURAL STUDY, REVIEW,
3064*
SARCOMA
ULTRASTRUCTURAL STUDY, REVIEW,
3064*
SARCOMA, OSTEOGENIC
ULTRASTRUCTURAL STUDY, REVIEW,
3064*

OVARIAN NEOPLASMS
ADENOCARCINOMA
IMMUNOGLOBULINS, 3361*
CHOLESTEROL, 14-METHYLHEXADECANOATE
ISOLATION, RAT, 3030
CLASSIFICATION
REVIEW, 3068*
CYSTADENOCARCINOMA
IMMUNOGLOBULINS, 3361*
CYSTADENOMA, MUCINOUS
IMMUNOGLOBULINS, 3361*
DISGERMINOMA, 3311
IMMUNOGLOBULINS, 3361*
IMMUNOSUPPRESSION, 3311
TRANSPLANTATION, HOMOLOGOUS, 3308
ULTRASTRUCTURAL STUDY, 3502*
LYMPHOCYTES
IMMUNITY, CELLULAR, 3333
PLANT AGGLUTININS, 3333

OVUM
RADIATION
CHROMSOME ABERRATIONS, 3191*
RADIATION, IONIZING
CHROMSOME ABERRATIONS, 3191*

2-OXETANONE
DNA, VIRAL
MUTAGENIC ACTIVITY, 3166*
NEOPLASMS
CARCINOGENIC ACTIVITY, 3164*
VIRUS, AVIAN MYELOBLASTOSIS
DNA NUCLEOTIDYLTRANSFERASES, 3166*
VIRUS, RAUSCHER MURINE LEUKEMIA
DNA NUCLEOTIDYLTRANSFERASES, 3166*
VIRUS, ROUS SARCOMA
DNA NUCLEOTIDYLTRANSFERASES, 3166*

OXYGENASES
CARCINOGEN, ENVIRONMENTAL
CARCINOGENIC ACTIVITY, MOUSE,
REVIEW, 3006

PANCREAS
CELLS, CULTURED
MODEL SYSTEM, 3539

PANCREATIC NEOPLASMS, 3173*
ADENOCARCINOMA
NEOPLASM METASTASIS, 3401
REVIEW, 3458*
AIR POLLUTANTS
EPIDEMIOLOGY, SOCIOECONOMIC
FACTORS, 3516
CARBAMIC ACID, N-METHYL-N-NITROSO-
ETHYL ESTER
DNA REPAIR, 3146*
CARCINOMA
CARCINOEMBRYONIC ANTIGEN, 3019
FERRITIN, 3019

REVIEW, 3458*
EPIDEMIOLOGY
  GERMANY, 3506
ESTROGENS
  HORMONE DEPENDENT, 3173*
ISLET CELL TUMOR
  KIDNEY FAILURE, CHRONIC, 3308

PAPILLOMA
  BRONCHIAL NEOPLASMS
    CASE REPORT, 3471*
    ULTRASTRUCTURAL STUDY, 3471*
  ESOPHAGEAL NEOPLASMS
    BENZYLAMINE, N-METHYL-N-NITROSO-,
    3514
  ESOPHGEAL NEOPLASMS
    3-PIPERIDINOL, 1-NITROSO-, 3117
  TONGUE NEOPLASMS
    PIPERIDINE, 1-NITROSO-, 3117
    3-PIPERIDINOL, 1-NITROSO-, 3117

PARAGANGLIOMA, NONCHROMAFFIN
  NERVOUS SYSTEM NEOPLASMS
    CASE REPORT, 3468*
  ORBITAL NEOPLASMS
    ULTRASTRUCTURAL STUDY, REVIEW, 3064*

PARATHYROID NEOPLASMS
  ADENOMA
    CASE REPORT, 3477*
    HYPOGONADISM, 3477*
    SKIN DISEASES, 3477*
  THYROID NEOPLASMS
    CASE REPORT, 3476*

PELVIC NEOPLASMS
  CARCINOMA, EPIDERMOID
    LEUKOPLAKIA, 3426
  CARCINOMA, TRANSITIONAL CELL
    LEUKOPLAKIA, 3426

PENICILLIUM
  SELENIOUS ACID, DISODIUM SALT
    CELL TRANSFORMATION, NEOPLASTIC,
    3133*

PENTANAMIDE, 2-ISOPROPYL-
  BENZ(A)ANTHRACENE
    CORTICOTROPIN, 3094
    PROTECTIVE EFFECT, ADRENAL
    NECROSIS, 3094

PENTANOIC ACID, 5-AMINO-2-HYDRAZINO-2-
  METHYL-
    CARBOXY-LYASES
      ENZYME INHIBITORS, 3120
    LYMPHOCYTE TRANSFORMATION
      PUTRESCINE, 3120

4-PENTENAMIDE, 2-ISOPROPYL-
  BENZ(A)ANTHRACENE
    CORTICOTROPIN, 3094
    PROTECTIVE EFFECT, ADRENAL
    NECROSIS, 3094

PEPTIDE HYDROLASES
  FIBROBLASTS
    DNA REPLICATION, 3586*

PEPTIDES

FIBROSARCOMA
  VIRUS, C-TYPE RNA TUMOR, 3369
LYMPHOCYTES
  PLANT AGGLUTININS, 3552
LYMPHOSARCOMA
  VIRUS, C-TYPE RNA TUMOR, 3369
VIRUS, C-TYPE MURINE LEUKEMIA
  RADIOIMMUNOASSAY, 3204
VIRUS, FRIEND MURINE LEUKEMIA
  ISOLATION AND CHARACTERIZATIO
  3223
VIRUS, PAPOVA, BK
  ISOLATION AND CHARACTERIZATIO
  3236
VIRUS, POLYOMA
  ISOLATION AND CHARACTERIZATIO
  3236, 3328
VIRUS, RAUSCHER MURINE LEUKEMIA
  ISOLATION AND CHARACTERIZATIO
  3252*

PERYLENE
  LUNG NEOPLASMS
    AIR POLLUTANTS, 3516

PETROLEUM
  DIGESTIVE SYSTEM NEOPLASMS
    OCCUPATIONAL HAZARD, 3520
  RESPIRATORY TRACT NEOPLASMS
    OCCUPATIONAL HAZARD, 3520
  SKIN NEOPLASMS
    OCCUPATIONAL HAZARD, 3520
  TESTICULAR NEOPLASMS
    OCCUPATIONAL HAZARD, 3520

PEUTZ-JEGHERS SYNDROME
  INTESTINAL NEOPLASMS
    GENETICS, 3067*

PHAGOCYTES
  CELL LINE
    ISOLATION AND CHARACTERIZATIO
    3392*
  LYMPHATIC NEOPLASMS
    CHOLANTHRENE, 3-METHYL-, 3392

PHARYNGEAL NEOPLASMS
  ALCOHOLISM
    EPIDEMIOLOGY, 3531*
  BLOOD GROUPS
    ISOANTIGENS, 3375*
  PRECANCEROUS CONDITIONS
    GENETICS, 3067*
  TOBACCO
    CARCINOGENIC POTENTIAL, REVIE
    3026

PHENANTHRENE, 9,10-EPOXY-9,10-DIHYDRO
  FORMAMIDE, N,N-DIMETHYL-
    ALKYLATION, 3167*
  METHANE, SULFINYLBIS-
    ALKYLATION, 3167*

PHENOBARBITAL
  SEE BARBITURIC ACID, 5-ETHYL-5-PH

PHENOBARBITONE
  SEE BARBITURIC ACID, 5-ETHYL-5-PH

PHOSPHORIBOSYL PYROPHOSPHATE SYNTHETASE
SEE PHOSPHOTRANSFERASES, ATP

PHOSPHORIC ACID, 2,2-DICHLOROETHENYL
DIMETHYL ESTER
BACTERIOPHAGES
RNA, 3111

PHOSPHORIC ACID, 2,2-DICHLOROVINYL
DIMETHYL ESTER
PHOSPHOROLATING AGENTS
STRUCTURE-ACTIVITY RELATIONSHIP,
3109

PHOSPHORIC ACID, TRIETHYL ESTER
PHOSPHOROLATING AGENTS
STRUCTURE-ACTIVITY RELATIONSHIP,
3109

PHOSPHORIC ACID, TRIMETHYL ESTER
PHOSPHOROLATING AGENTS
STRUCTURE-ACTIVITY RELATIONSHIP,
3109

PHOSPHORYLCHOLINE
SEE CHOLINE

PHOSPHOTRANSFERASES
VIRUS, HERPES AOTUS
ISOLATION AND CHARACTERIZATION,
3213
VIRUS, HERPES MARMOSET
ISOLATION AND CHARACTERIZATION,
3213
VIRUS, HERPES SIMPLEX 1
ISOLATION AND CHARACTERIZATION,
3213
VIRUS, HERPES SIMPLEX 2
ISOLATION AND CHARACTERIZATION,
3213
VIRUS, PSEUDORABIES
ISOLATION AND CHARACTERIZATION,
3213

PHYTOHEMAGGLUTININ
SEE PLANT AGGLUTININS

PICRIC ACID
T-LYMPHOCYTES
IMMUNE RESPONSE, 3302
IMMUNITY, CELLULAR, 3302
MACROPHAGES
IMMUNE RESPONSE, 3302

PIPERIDINE, 1-NITROSO-
ESOPHAGEAL NEOPLASMS
CARCINOMA, 3117
LIVER NEOPLASMS
ANGIOSARCOMA, 3117
NOSE NEOPLASMS
CARCINOMA, 3117
TONGUE NEOPLASMS
PAPILLOMA, 3117

3-PIPERIDINOL, 1-NITROSO-
ESOPHGEAL NEOPLASMS
CARCINOMA, EPIDERMOID, 3117
PAPILLOMA, 3117
LIVER NEOPLASMS

ANGIOSARCOMA, 3117
NOSE NEOPLASMS
　　CARCINOMA, 3117
TONGUE NEOPLASMS
　　CARCINOMA, EPIDERMOID, 3117
　　PAPILLOMA, 3117

4-PIPERIDINOL, 1-NITROSO-
ANGIOSARCOMA
　　SEX FACTORS, 3117
BRAIN NEOPLASMS
　　ASTROCYTOMA, 3117
NOSE NEOPLASMS
　　CARCINOMA, 3117

4-PIPERIDONE, 1-NITROSO-
ANGIOSARCOMA
　　SEX FACTORS, 3117
BRAIN NEOPLASMS
　　ASTROCYTOMA, 3117
NOSE NEOPLASMS
　　CARCINOMA, 3117

PITUITARY GROWTH HORMONE
SEE SOMATOTROPIN

PITUITARY NEOPLASMS
CARBONIC ANHYDRASE
　　ISOENZYMES, 3566
　　PROLACTIN, 3566
CELL TRANSFORMATION, NEOPLASTIC
　　ULTRASTRUCTURAL STUDY, REVIEW,
　　　3320
NEOPLASM METASTASIS
　　CLINICOPATHOLOGIC STUDY,
　　　LOCALIZATION, 3401
　　DIABETES INSIPIDUS, 3401

PLANT AGGLUTININS
BREAST NEOPLASMS
　　LYMPHOCYTES, 3333
CELL TRANSFORMATION, NEOPLASTIC
　　DIETHYLAMINE, N-NITROSO-, 3393*
ETHANOL, 2-MERCAPTO-
　　CHROMOSOME ABERRATIONS, 3424
FORMAMIDE, N,N-DIMETHYL-
　　CELLS, CULTURED, 3095
KERATOACANTHOMA
　　IMMUNE RESPONSE, 3406
L CELLS
　　BINDING, 3546
LEUKEMIA L1210
　　BINDING, 3546
　　CARBOHYDRATES, 3346
LEUKEMIA, MYELOCYTIC
　　BINDING, 3546
LYMPHOCYTES
　　NUCLEOTIDYLTRANSFERASES, 3572
B-LYMPHOCYTES
　　LYMPHOCYTE TRANSFORMATION, 3337
T-LYMPHOCYTES
　　IMMUNOGLOBULINS, FC, 3343
　　LYMPHOCYTE TRANSFORMATION, 3337
MACROPHAGES
　　CARBOHYDRATES, 3346
METHANE, SULFINYLBIS-
　　CELLS, CULTURED, 3095
OVARIAN NEOPLASMS
　　LYMPHOCYTES, 3333

PEPTIDES
LYMPHOCYTES, 3552
RADIATION, IONIZING
　　CHROMOSOME ABERRATIONS, 3183
RNA, TRANSFER
　　LYMPHOCYTES, 3552
SMOKING
　　LYMPHOCYTES, 3342
　　T-LYMPHOCYTES, 3342
URIDINE, 5-BROMO-2'DEOXY-
　　CELLS, CULTURED, 3095
UROGENITAL NEOPLASMS
　　LYMPHOCYTES, 3333
VIRUS, BOVINE LEUKEMIA
　　C-TYPE PARTICLES, 3488*
VIRUS, EPSTEIN-BARR
　　ANTIGENS, VIRAL, 3326

PLANTS
CHROMOSOME ABNORMALITIES
　　SMOKING, 3121
RADIATION, IONIZING
　　ARABIDOPSIS THALINA, 3185*
　　DOSE-RESPONSE STUDY, 3185*
　　GROWTH, 3185*

PLASMACYTOMA
BONE NEOPLASMS
　　BENCE-JONES PROTEIN, 3527*
　　EPIDEMIOLOGY, GERMANY, 3527*
IMMUNOGLOBULIN FRAGMENTS
　　ANTIGENIC DETERMINANTS, 3313
IMMUNOGLOBULINS
　　IGA, 3362*
　　IGG, 3362*
　　IGM, 3362*
T-LYMPHOCYTES
　　IMMUNITY, CELLULAR, 3332
　　ISOANTIGENS, 3332
　　RADIATION, 3332
MYELOMA PROTEINS
　　ANTIGENIC DETERMINANTS, 3317
　　ANTIGENS, NEOPLASM, 3317
　　IMMUNOGLOBULIN FRAGMENTS, 331
THYMIDINE INCORPORATION, 3538

PLEURAL NEOPLASMS
MESOTHELIOMA
　　DIAGNOSIS AND PROGNOSIS, 3412

PLUTONIUM
BONE NEOPLASMS
　　RADIATION DOSAGE, REVIEW, 300
RADIATION DOSAGE
　　REVIEW, 3009
RADIATION, IONIZING
　　RADIATION EFFECT, BONE,DOG, 3
RADIOISOTOPES
　　DISTRIBUTION, 3178

PNEUMOCONIOSIS
PRECANCEROUS CONDITIONS
　　OCCUPATIONAL HAZARD, 3005

POLIOVIRUS VACCINE
LEUKEMIA, LYMPHOBLASTIC
　　ANTIBODIES, VIRAL, 3364*

CELL TRANSFORMATION, NEOPLASTIC, 3075
ANEMIA, APLASTIC
    DNA, SINGLE STRANDED, 3561
ANTHRACOSILICOSIS
    OCCUPATIONAL HAZARD, 3005
ASBESTOS
    EPIDEMIOLOGY, 3529*
BONE NEOPLASMS
    CASE REPORT, 3436*
BREAST NEOPLASMS
    HISTOLOGICAL STUDY, 3400
    REVIEW, 3026
CERVIX NEOPLASMS
    ADENOCARCINOMA, 3499*
    CARCINOMA IN SITU, 3498*
    EPIDEMIOLOGY, FINLAND, 3508
    GLYCOGEN, 3569
    GLYCOGEN PHOSPHORYLASE, 3569
    GLYCOGEN SYNTHETASE, 3569
    REVIEW, 3026
COLONIC NEOPLASMS
    ADENOCARCINOMA, 3448*
ESOPHAGEAL NEOPLASMS
    GENETICS, 3067*
ETHYLENE, CHLORO- POLYMER
    EPIDEMIOLOGY, 3519
    OCCUPATIONAL HAZARD, 3519
GALLBLADDER NEOPLASMS
    DIAGNOSIS AND PROGNOSIS, 3457*
GLUCOSEPHOSPHATE DEHYDROGENASE
    2,3-BUTANEDIOL, 1,4-DIMERCAPTO-, 3567
    GLUTATHIONE REDUCTASE, 3567
    NADP, 3567
HERNIA
PHA,   ULTRASTRUCTURAL STUDY, 3480*
INTESTINAL NEOPLASMS
    ADENOCARCINOMA, 3414
LEUKEMIA
    BONE MARROW CELLS, 3483*
    MYELOPEROXIDASE, 3483*
    RADIATION, IONIZING, 3187*
LEUKEMIA, MONOCYTIC
    REVERSE TRANSCRIPTASE, 3419
    VIRUS-LIKE PARTICLES, 3419
LEUKEMIA, MYELOBLASTIC
    REVERSE TRANSCRIPTASE, 3419
    VIRUS-LIKE PARTICLES, 3419
LEUKEMIA, MYELOCYTIC
    ULTRASTRUCTURAL STUDY, 3484*
LIVER NEOPLASMS
    ACETAMIDE, N-FLUOREN-2-YL-, 3126*
    DIETHYLAMINE, N-NITROSO-, 3432
LUNG NEOPLASMS
    RETINOL, 3168*
MESOTHELIOMA
    ASBESTOS, 3529*
MOUTH NEOPLASMS
    ERYTHROPLASIA, 3065*
    LEUKOPLAKIA, ORAL, 3065*
PHARYNGEAL NEOPLASMS
    GENETICS, 3067*
PNEUMOCONIOSIS
    OCCUPATIONAL HAZARD, 3005
PROSTATIC NEOPLASMS
    CHOLANTHRENE, 3-METHYL-, 3136
    GROWTH, 3174*
    TESTOSTERONE, 3573

ULTRASTRUCTURAL STUDY, 3226
RESPIRATORY TRACT NEOPLASMS
   NITROGEN OXIDE, 3119
RUBBER
   EPIDEMIOLOGY, 3519
SALIVARY GLAND NEOPLASMS
   VIRUS, CYTOMEGALO, 3264*
SILICOSIS
   OCCUPATIONAL HAZARD, 3005
SKIN NEOPLASMS
   DNA, 3452*
   KERATOSIS, 3452*
   RNA, 3452*
STOMACH NEOPLASMS
   ANEMIA, PERNICIOUS, 3026
   DIET, 3026
   HISTOLOGICAL STUDY, 3413
UTERINE NEOPLASMS
   ADENOCARCINOMA, 3428
VAGINAL NEOPLASMS
   ADENOCARCINOMA, 3430
   CARCINOMA, EPIDERMOID, 3430
   4,4'-STILBENEDIOL, ALPHA,ALPHA'-
     DIETHYL-, 3429, 3430

PREDNISONE
   SEE PREGNA-1,4-DIENE-3,11,20-TRIONE,
   17,21-DIHYDROXY-

PREGN-4-ENE-3,20-DIONE, 17,21-DIHYDROXY-
   MAMMARY NEOPLASMS, EXPERIMENTAL
     BINDING, COMPETITIVE, 3550

PREGNANCY
   BREAST NEOPLASMS
     EPIDEMIOLOGY, 3512
   RADIATION, IONIZING
     RADIATION DOSAGE, 3050*

PROGESTERONE
   ACETAMIDE, 2-IODO-
     CARRIER PROTEINS, 3551
   CHROMOSOME ABERRATIONS
     MUTAGENIC ACTIVITY, REVIEW, 3049*
   MAMMARY NEOPLASMS, EXPERIMENTAL
     BINDING, COMPETITIVE, 3550
   1H-PYRROLE-2,5-DIONE, 1-ETHYL-
     CARRIER PROTEINS, 3551
   UTERINE NEOPLASMS
     ESTRADIOL, 3163*
     TESTOSTERONE, 3163*
   VIRUS, MURINE SARCOMA
     VIRUS REPLICATION, 3280*
   VIRUS, POX
     VIRUS REPLICATION, 3255

PROLACTIN
   MAMMARY NEOPLASMS, EXPERIMENTAL
     ADENOCARCINOMA, 3092
   PITUITARY NEOPLASMS
     CARBONIC ANHYDRASE, 3566

PROPANE, 1,2-EPOXY-
   BACTERIOPHAGES
     RNA, 3111

PROPANEDIAL
   NEOPLASMS
     CARCINOGENIC ACTIVITY, 3164*

PROSTAGLANDINS
   HEPATOMA
     ADENYL CYCLASE, 3564

PROSTATIC NEOPLASMS
   ADENOCARCINOMA
     ULTRASTRUCTURAL STUDY, 3226
   ADENOMA
     ANDROGENS, 3100
   ANTIGENS, NEOPLASM
     ANTIGEN-ANTIBODY REACTIONS, 3
   CADMIUM OXIDE
     OCCUPATIONAL HAZARD, 3520
   CARCINOMA
     VIRUS, SV40, 3250
   CHOLANTHRENE, 3-METHYL-
     PRECANCEROUS CONDITIONS, 3106
     RETINOIC ACID, 3150*
   ESTRADIOL, 3174*
   GUANIDINE, 1-METHYL-3-NITRO-1-NIT
     RETINOIC ACID, 3150*
   HORMONES
     GROWTH, 3174*
   HYPERPLASIA
     ACID PHOSPHATASE, 3537
   INCLUSION BODIES, VIRAL
     ULTRASTRUCTURAL STUDY, 3226
   PRECANCEROUS CONDITIONS
     GROWTH, 3174*
     ULTRASTRUCTURAL STUDY, 3226
   STEROIDS
     GROWTH, 3174*
     ISOLATION AND CHARACTERIZATIO
     3598*
   TESTOSTERONE
     PRECANCEROUS CONDITIONS, 3573
   VIRUS, MURINE LEUKEMIA
     DNA-RNA HYBRIDIZATION, 3226

PROTEIN DENATURATION
   HEPATOMA
     ARGININE, 3581*
     CANAVANINE, 3581*

PROTEIN KINASE
   HEPATOMA
     ADENOSINE CYCLIC 3',5'
      MONOPHOPHATE, 3595*
     GUANOSINE CYCLIC 3',5'
      MONOPHOSPHATE, 3595*

PROTEINS
   CHOLESTEROL, 14-METHYLHEXADECANOA
     REVIEW, 3030
   DIMETHYLAMINE, N-NITROSO-
     ALKYLATING AGENTS, 3157*
   HEPARIN
     RETICULOCYTES, 3555
   HEPATOMA
     CELL NUCLEUS, 3548
   LEUKEMIA, LYMPHOBLASTIC
     CHROMATIN, 3580*
   LEUKEMIA, LYMPHOCYTIC
     CHROMATIN, 3580*
   LEUKEMIA, MYELOCYTIC

ALKYLATING AGENTS
    STRUCTURE-ACTIVITY RELATIONSHIP,
       3109

PYRIDOXINE
    HEPATOMA
       SERINE DEHYDRATASE, 3597*

1H-PYRROLE-2,5-DIONE, 1-ETHYL-
    ALDOSTERONE
       CARRIER PROTEINS, 3551
    CORTISOL
       CARRIER PROTEINS, 3551
    DEXAMETHASONE
       CARRIER PROTEINS, 3551
    DNA
       BINDING SITES, 3551
    PROGESTERONE
       CARRIER PROTEINS, 3551

PYRUVATE KINASE
    HEPATOMA
       ISOLATION AND CHARACTERIZATION,
        ISOENZYMES, 3592*

QUINOLINE, 4-NITRO-, 1-OXIDE
    CELL TRANSFORMATION, NEOPLASTIC
       CELL STRUCTURE, REVIEW, 3001
    SKIN NEOPLASMS
       XERODERMA PIGMENTOSUM, 3031

RADIATION
    CYCLOHEXIMIDE
       CELL DIVISION, 3189*
    DRUG THERAPY
       COCARCINOGENIC EFFECT, REVIEW,
        3004
    LEUKEMIA L1210
       T-LYMPHOCYTES, 3332
    LUNG NEOPLASMS
       CARCINOGENIC POTENTIAL, REIVEW,
        3026
    T-LYMPHOCYTES
       IMMUNE RESPONSE, 3302
    LYMPHOMA
       T-LYMPHOCYTES, 3332
    MACROPHAGES
       IMMUNE RESPONSE, 3302
    OVUM
       CHROMSOME ABERRATIONS, 3191*
    PLASMACYTOMA
       T-LYMPHOCYTES, 3332
    SARCOMA, MAST CELL
       T-LYMPHOCYTES, 3332
    SELENIUM
       RETENTION, MAMMALS, 3180
    THYROID NEOPLASMS
       CARCINOMA, 3051*
       RADIATION DOSAGE, 3051*
    VIRUS, SV40
       CELL TRANSFORMATION, NEOPLASTIC,
        3249

RADIATION DOSAGE
    PLUTONIUM
       REVIEW, 3009

RADIATION, IONIZING
    ADENOCARCINOMA

REVIEW,

IL-5-YL)

EVIEW, 3032

.EVIEW, 3032

.EVIEW,

—
20

ANTIGENS, NEOPLASMS, 3195*
CHROMOSOME ABERRATIONS
    CELL CYCLE KINETICS, 3183, 3456*
    GENETICS, 3456*
    RADIATION DOSAGE, 3191*
DNA REPLICATION
    CELL CYCLE KINETICS, 3456*
    GENETICS, 3456*
DROSOPHILA MELANOGASTER
    CHROMOSOME ABERRATIONS, 3190*,
        3193*
HELA CELLS
    CELL CYCLE KINETICS, 3182
    DNA REPLICATION, 3182
HODGKIN'S DISEASE
    LEUKEMIA, MYELOBLASTIC, 3192*
INTESTINAL NEOPLASMS
    ADENOCARCINOMA, 3195*
    ANTIGENS, NEOPLASMS, 3195*
INTESTINE, SMALL
    CELL DIFFERENTIATION, 3184*
    CELL DIVISION, 3184*
LEUKEMIA
    PRECANCEROUS CONDITIONS, 3187*
LYMPHOCYTES
    CHROMOSOME ABERRATIONS, 3194*
NEOPLASMS
    MUTATION, 3031
OVUM
    CHROMSOME ABERRATIONS, 3191*
PLANT AGGLUTININS
    CHROMOSOME ABERRATIONS, 3183
PLANTS
    ARABIDOPSIS THALINA, 3185*
    DOSE-RESPONSE STUDY, 3185*
    GROWTH, 3185*
PLUTONIUM
    RADIATION EFFECT, BONE,DOG, 3179
PREGNANCY
    RADIATION DOSAGE, 3050*
RADIUM
    RADIATION EFFECT, BONE,DOG, 3179
RADON
    IGA, 3181
    IGG, 3181
    IGM, 3181
    IMMUNOGLOBULINS, 3181
SKIN NEOPLASMS
    CARCINOGENIC POTENTIAL, REVIEW,
        3026
THYROID NEOPLASMS
    CARCINOMA, 3186*
URANIUM
    IGA, 3181
    IGG, 3181
    IGM, 3181
    IMMUNOGLOBULINS, 3181
VIRUS, C-TYPE RNA TUMOR
    CARBAMIC ACID, ETHYL ESTER, 3229
    INCLUSION BODIES, VIRAL, 3229
VIRUS, RAUSCHER MURINE LEUKEMIA
    CARBAMIC ACID, ETHYL ESTER, 3229
    INCLUSION BODIES, VIRAL, 3229
VIRUS, SV40
    DNA, CIRCULAR, 3245
    DNA, VIRAL, 3245

RADIOISOTOPES
    PLUTONIUM

    DISTRIBUTION, 3178

RADIUM
    RADIATION, IONIZING
        RADIATION EFFECT, BONE,DOG,

RADON
    RADIATION, IONIZING
        IGA, 3181
        IGG, 3181
        IGM, 3181
        IMMUNOGLOBULINS, 3181

RECEPTORS, HORMONE
    HEPATOMA
        ISOLATION AND CHARACTERIZATI
        3574

RECTAL NEOPLASMS
    ADENOCARCINOMA
        CARCINOEMBRYONIC ANTIGEN, 33
        IMMUNITY, CELLULAR, 3300
        ISOANTIGENS, 3323
    CARCINOEMBRYONIC ANTIGEN
        BLOOD-GROUP PRECURSOR ACTIVI
        3329
    CARCINOMA
        CARCINOEMBRYONIC ANTIGEN, 34
        GLYCOPROTEINS, 3450*
        IGA, 3450*
        IMMUNOGLOBULINS, 3450*
        ULTRASTRUCTURAL STUDY, 3450*
    EPIDEMIOLOGY
        DIETARY FACTORS, REVIEW, 350
    LYMPH NODES
        PROGNOSIS, 3449*

RESERPINE
    BREAST NEOPLASMS
        CARCINOGENIC POTENTIAL, REVI
        3035*

RESPIRATORY TRACT NEOPLASMS
    LIPOPROTEINS
        AUTOANTIBODIES, 3366*
        HEMAGGLUTINATION, 3366*
    NITROGEN OXIDE
        PRECANCEROUS CONDITIONS, 311
    PETROLEUM
        OCCUPATIONAL HAZARD, 3520

RETICULOCYTES
    HEPARIN
        PROTEIN, 3555
    RNA, MESSENGER
        PROTEIN, 3555

RETICULOENDOTHELIAL SYSTEM
    LYMPHATIC NEOPLASMS
        BENZIDINE, 3089

RETICULOENDOTHELIOSIS
    BRAIN NEOPLASMS
        CASE REPORT, 3469*
    VIRUS, AVIAN RETICULOENDOTHELIOSI
        ISOLATION AND CHARACTERIZATI
        3286*

RETICULOSARCOMA

RNA, BACTERIAL
    HYPERPLASIA
        CARCINOGENIC EFFECT, 3171*
    STRAMONIUM
        CARCINOGENIC EFFECT, 3171*

RNA, MESSENGER
    CHOLESTEROL, 14-METHYLHEXADECANOATE
        REVIEW, 3030
    MULTIPLE MYELOMA
        IGG, 3316
        POLYRIBOSOMES, 3316
    PROTEIN
        RETICULOCYTES, 3555
    VIRUS, ADENO 2
        DNA, VIRAL, 3198
        TRANSCRIPTION, 3197
    VIRUS, FRIEND MURINE LEUKEMIA
        METHANE, SULFINYLBIS-, 3268*
        PEPTIDE CHAIN ELONGATION, 3558
    VIRUS, SV40
        DNA-RNA HYBRIDIZATION, 3256

RNA POLYMERASE
    VIRUS, FRIEND MURINE LEUKEMIA
        ACTINOMYCIN D, 3227
        ERYTHROPOIETIN, 3227
    VIRUS, MENGO
        VIRUS REPLICATION, 3010
    VIRUS, SV40
        DNA-RNA HYBRIDIZATION, 3247
        DNA, VIRAL, 3246
        ISOLATION AND CHARACTERIZATION,
            3247

RNA, TRANSFER
    LYMPHOCYTES
        PLANT AGGLUTININS, 3552

RNA, VIRAL
    HELA CELLS
        VIRUS, POLIO, 3253
        VIRUS, VESICULAR STOMATITIS, 3253
    LEUKEMIA
        REVIEW, 3054*
    MAMMARY NEOPLASMS, EXPERIMENTAL
        VIRUS, MURINE MAMMARY TUMOR, 3232
    VIRUS, ADENO 2
        HYBRIDIZATION, 3197
    VIRUS, B-TYPE GUINEA PIG
        URIDINE, 5-BROMO-2'-DEOXY-, 3232
    VIRUS, C-TYPE GUINEA PIG
        VIRUS, MURINE MAMMARY TUMOR, 3269*
        VIRUS, RAUSCHER MURINE LEUKEMIA,
            3269*
    VIRUS, FELINE LEUKEMIA
        ISOLATION AND CHARACTERIZATION,
            3224
    VIRUS, MASON-PFIZER MONKEY
        HYBRIDIZATION, 3232
    VIRUS, MURINE LEUKEMIA
        HYBRIDIZATION, 3232
    VIRUS, POLIO
        CYTOTOXICITY, REVIEW, 3010
        SODIUM CHLORIDE, 3253
    VIRUS, ROUS SARCOMA
        GENETICS, 3240

    ISOLATION AND CHARACTERIZATION,
        3240
  VIRUS, SIMIAN 5
    VIRUS REPLICATION, 3010
  VIRUS, SV40
    DNA-RNA HYBRIDIZATION, 3246
    DNA, VIRAL, 3246
  VIRUS, VESICULAR STOMATITIS
    ISOLATION AND CHARACTERIZATION, DI
        PARTICLES, 3254
    SODIUM CHLORIDE. 3253
    ULTRASTRUCTURAL STUDY, FI
        PARTICLES, 3254

RUBBER
  PRECANCEROUS CONDITIONS
    EPIDEMIOLOGY, 3519

RUTHENIUM RED
  CARCINOMA, EHRLICH TUMOR
    CALCIUM, 3543

SALIVARY GLAND NEOPLASMS, 3173*
  ADENOCARCINOMA
    CASE REPORT, 3494*
  CARCINOMA
    THYMIDINE INCORPORATION, 3538
  ESTROGENS
    HORMONE DEPENDENT, 3173*
  VIRUS, CYTOMEGALO
    PRECANCEROUS CONDITIONS, 3264*

SALMONELLA TYPHIMURIUM
  ACETALDEHYDE, CHLORO-
    MUTAGENIC ACTIVITY, 3125
  1,3-BUTADIENE, 2-CHLORO-
    BARBITURIC ACID, 5-ETHYL-5-PHENYL-,
        3145*
    MUTAGENESIS, 3145*
  ETHYLENE, 1,1-DICHLORO-
    BARBITURIC ACID, 5-ETHYL-5-PHENYL-,
        3145*
    MUTAGENESIS, 3145*
  MUTATION
    CARCINOGEN, CHEMICAL, 3044*
  DIET
    MUTAGENIC EFFECT, IRRADIATION,
        3188*
  DIETHYLAMINE, N-NITROSO-
    MICROSOMES, LIVER, 3158*
    MUTAGENESIS, 3158*
  DIMETHYLAMINE, N-NITROSO-
    MICROSOMES, LIVER, 3158*
    MUTAGENESIS, 3158*

SALPINGITIS
  FALLOPIAN TUBE NEOPLASMS
    ULTRASTRUCTURAL STUDY, 3428

SARCOMA
  ADRENAL CORTEX HORMONES
    NEOPLASM TRANSPLANTATION, 3600*
  ANTIGENS, NEOPLASM
    ISOLATION AND CHARACTERIZATION,
        3314
  BENZO(A)PYRENE
    CHROMOSOME ABERRATIONS, 3101
    KARYOTYPING, 3101
  BONE NEOPLASMS

                CELL TRANSFORMATION, NEOPLASTIC,
                    3133*
        ASPERGILLUS FUMIGATUS
                CELL TRANSFORMATION, NEOPLASTIC,
                    3133*
        PENICILLIUM
                CELL TRANSFORMATION, NEOPLASTIC,
                    3133*

SELENIUM
    RADIATION
        RETENTION, MAMMALS, 3180

SEMINOMA
    SEE DISGERMINOMA

SERINE DEHYDRATASE
    HEPATOMA
        PYRIDOXINE, 3597*

,4*   SIALIC ACIDS
    LEUKEMIA
        NEURAMINIDASE, 3350*
    B-LYMPHOCYTES
        CHALONES, 3029
    T-LYMPHOCYTES
        CHALONES, 3029

SILICOSIS
    PRECANCEROUS CONDITIONS
        OCCUPATIONAL HAZARD, 3005

SKIN NEOPLASMS
    ARSENIC
        CARCINOGENIC POTENTIAL, REVIEW,
            3026
    CHOLANTHRENE, 11,12-EPOXY-3-METHYL-
        HISTOLOGICAL STUDY, MOUSE, 3107
    CHOLANTHRENE, 3-METHYL-
        HISTOLOGICAL STUDY, MOUSE, 3107
    DERMATOMYOSITIS
        AUTOIMMUNE DISEASES, 3451*
        DIAGNOSIS AND PROGNOSIS, 3451*
    DNA
        PRECANCEROUS CONDITIONS, 3452*
    KERATOACANTHOMA
        CASE REPORT, 3406
        IMMUNITY, CELLULAR, 3406
    KERATOSIS
        PRECANCEROUS CONDITIONS, 3452*
    LYMPHOSARCOMA
        VIRUS, HERPES SAIMIRI, 3277*
    PETROLEUM
        OCCUPATIONAL HAZARD, 3520
    RADIATION, IONIZING
        CARCINOGENIC POTENTIAL, REVIEW,
            3026
    RNA
        PRECANCEROUS CONDITIONS, 3452*
    ULTRAVIOLET RAYS
        CAFFEINE, 3031
        CARCINOGENIC EFFECT, REVIEW, 3036*
        CARCINOGENIC POTENTIAL, REVIEW,
            3026
    XERODERMA PIGMENTOSUM
        ACETAMIDE, N-(ACETYLOXY)-N-9H-
            FLUOREN-2-YL-, 3031
        CAFFEINE, 3031
        GUANIDINE, 1-METHYL-3-NITRO-1-

NITROSO-, 3031
METHANESULFONIC ACID, METHYL ESTER,
3031
QUINOLINE, 4-NITRO-, 1-OXIDE, 3031

SMOKING
ALLIUM CEPA
CHROMOSOME ABNORMALITIES, 3121
BRONCHIAL NEOPLASMS
CARCINOMA, OAT CELL, 3411
INHALATION TESTS
MODEL TEST SYSTEM, ANIMAL, 3122
LARYNGEAL NEOPLASMS
ASBESTOS, 3137*
CARCINOMA, EPIDERMOID, 3411
LUNG NEOPLASMS
EPIDEMIOLOGY, 3525*
REVIEW, 3034*
LYMPHOCYTES
PLANT AGGLUTININS, 3342
T-LYMPHOCYTES
PLANT AGGLUTININS, 3342
NICOTINE
CARCINOGENIC METABOLITE, 3165*
GASTRIC JUICE, 3165*
NICOTINE, 1'-DEMETHYL-
CARCINOGENIC METABOLITE, 3165*
GASTRIC JUICE, 3165*
PLANTS
CHROMOSOME ABNORMALITIES, 3121

SODIUM CHLORIDE
VIRUS, POLIO
RNA, VIRAL, 3253
VIRUS, VESICULAR STOMATITIS
RNA, VIRAL, 3253

SODIUM FLUORIDE
HEPATOMA
ADENYL CYCLASE, 3564

SPERMIDINE
BREAST NEOPLASMS
URINE, 3557
CRANIOPHARYNGIOMA
URINE, 3557
LEUKEMIA
CHRONIC DISEASE, 3557
MULTIPLE MYELOMA
URINE, 3557
SARCOMA, OSTEOGENIC
URINE, 3557
SARCOMA, RETICULUM CELL
URINE, 3557
STOMACH NEOPLASMS
URINE, 3557

SPERMINE
BREAST NEOPLASMS
URINE, 3557
CRANIOPHARYNGIOMA
URINE, 3557
LEUKEMIA
CHRONIC DISEASE, 3557
MULTIPLE MYELOMA
URINE, 3557
SARCOMA, OSTEOGENIC
URINE, 3557
SARCOMA, RETICULUM CELL

URINE, 3557
STOMACH NEOPLASMS
URINE, 3557

STEROIDS
ADRENAL GLAND NEOPLASMS
ADENOSINE CYCLIC 3',5'
MONOPHOSPHATE, 3402
AMINOGLUTETHIMIDE, 3402
CORTICOTROPIN, 3402
CYCLOHEXIMIDE, 3402
ENTEROTOXINS, 3402
THEOPHYLLINE, 3402
PROSTATIC NEOPLASMS
GROWTH, 3174*
ISOLATION AND CHARACTERIZATION
3598*

4,4'-STILBENEDIOL, ALPHA,ALPHA'-DIETHY
VAGINAL NEOPLASMS
IN UTERO EXPOSURE, 3037*
PRECANCEROUS CONDITIONS, 3429,
3430
TRANSPLACENTAL CARCINOGENESIS,
3429

STILBENES
BREAST NEOPLASMS
ESTRADIOL, 3169*
UTERINE NEOPLASMS
ADENOCARCINOMA, 3169*

STILBESTROL
SEE 4,4'-STILBENEDIOL, ALPHA, ALPH
DIETHYL-

STOMACH NEOPLASMS
ADENOCARCINOMA
ENZYMATIC ACTIVITY, 3589*
HISTOCOMPATIBILITY ANTIGENS, 3
HISTOLOGICAL STUDY, 3413
ANEMIA, PERNICIOUS
PRECANCEROUS CONDITIONS, 3026
BENZENEACETAMIDE, ALPHA-ETHYNYL-N,
DIMETHYL-ALPHA-PHENYL-
LIPOTROPIC DEFICIENT, HIGH-FAT
DIET, 3096
CARCINOMA
FETAL GLOBULINS, 3019
CARCINOMA, MUCINOUS
ENZYMATIC ACTIVITY, 3589*
HISTOLOGICAL STUDY, 3413
CARCINOMA, SCIRRHOUS
ENZYMATIC ACTIVITY, 3589*
DIET
PRECANCEROUS CONDITIONS, 3026
EPIDEMIOLOGY
GERMANY, 3506
FIBROMA
ENZYMATIC ACTIVITY, 3589*
GASTRITIS
GENETICS, 3067*
GUANIDINE, 1-METHYL-3-NITRO-1-NITR
LIPOTROPIC DEFICIENT, HIGH-FAT
DIET, 3096
NEOPLASM METASTASIS
EPIDEMIOLOGY, GERMANY, 3526*
PRECANCEROUS CONDITIONS
HISTOLOGICAL STUDY, 3413

BINDING, COMPETITIVE, 3550
HYPERPLASIA, 3396
PROSTATIC NEOPLASMS
PRECANCEROUS CONDITIONS, 3573
TRANSPLANTATION, HOMOLOGOUS
CARCINOGENIC EFFECT, PROSTATE,
3573
UTERINE NEOPLASMS
ESTRADIOL, 3163*
PROGESTERONE, 3163*
VIRUS, MURINE SARCOMA
VIRUS REPLICATION, 3280*

12-O-TETRADECANOYLPHORBOL-13-ACETATE
AMNION
ADENOSINE CYCLIC 3',5'
MONOPHOSPHATE, 3129*
DNA REPAIR, 3129*
FIBROBLASTS
DNA REPAIR, 3081
DNA REPLICATION, 3081
NEOPLASMS, EXPERIMENTAL
DNA REPAIR, 3031

THEOPHYLLINE
ADRENAL GLAND NEOPLASMS
STEROIDS, 3402
CHOLANTHRENE, 3-METHYL-
CELLS, CULTURED, 3095
HEPATOMA
PHOSPHODIESTERASES, 3565
SARCOMA, MAST CELL
GROWTH, 3542
VIRUS, SV40
CELL TRANSFORMATION, NEOPLASTIC,
3249

THORACIC NEOPLASMS
FIBROSARCOMA
ASBESTOS, 3138*
LYMPHOMA
ASBESTOS, 3138*
SARCOMA, RETICULUM CELL
ASBESTOS, 3138*

THORIUM DIOXIDE
KIDNEY NEOPLASMS
CARCINOGENIC POTENTIAL, REIVEW,
3026

THOROTRAST
SEE THORIUM DIOXIDE

THYMIDINE KINASE
VIRUS, HERPES AOTUS
ISOLATION AND CHARACTERIZATION,
3213
VIRUS, HERPES MARMOSET
ISOLATION AND CHARACTERIZATION,
3213
VIRUS, HERPES SIMPLEX 1
ISOLATION AND CHARACTERIZATION,
3213
VIRUS, HERPES SIMPLEX 2
ISOLATION AND CHARACTERIZATION,
3213
VIRUS, HERPES TURKEY
ISOLATION AND CHARACTERIZATION,
3213

VIRUS, PSEUDORABIES
    ISOLATION AND CHARACTERIZATION,
    3213

THYMUS EXTRACTS
    BONE MARROW
        T-LYMPHOCYTES, 3384*

THYROID NEOPLASMS
    BLOOD GROUPS
        ISOANTIGENS, 3375*
    CARCINOMA
        IODINE RADIOISOTOPES, 3051*
        RADIATION, 3051*
        RADIATION, IONIZING, 3186*
    CARCINOMA, PAPILLARY
        KIDNEY FAILURE, CHRONIC, 3308
    IODINE RADIOISOTOPES
        RADIATION DOSAGE, 3051*
    PARATHYROID NEOPLASMS
        CASE REPORT, 3476*
    RADIATION
        RADIATION DOSAGE, 3051*

THYROTROPIN
    CHORIOCARCINOMA
        NEOPLASM TRANSPLANTATION, 3578*

TISSUE CULTURE
    LUNG NEOPLASMS
        CHOLANTHRENE, 3-METHYL-, 3474*
    VIRUS, C-TYPE MURINE LEUKEMIA
        VIRUS REPLICATION, 3204

TOBACCO
    BLADDER NEOPLASMS
        CARCINOGENIC POTENTIAL, REVIEW,
        3026
    DUODENAL ULCER
        DNA, 3172*
    ESOPHAGEAL NEOPLASMS
        CARCINOGENIC POTENTIAL, REIVEW,
        3026
    GASTRIC JUICE
        DNA, 3172*
    GASTRITIS
        DNA, 3172*
    LUNG NEOPLASMS
        CARCINOGENIC POTENTIAL, REVIEW,
        3026
    PHARYNGEAL NEOPLASMS
        CARCINOGENIC POTENTIAL, REVIEW,
        3026
    TONGUE NEOPLASMS
        CARCINOGENIC POTENTIAL, REVIEW,
        3026

TOBACCO SMOKE
    SEE SMOKING

TONGUE NEOPLASMS
    CARCINOMA, EPIDERMOID
        3-PIPERIDINOL, 1-NITROSO-, 3117
    PAPILLOMA
        PIPERIDINE, 1-NITROSO-, 3117
        3-PIPERIDINOL, 1-NITROSO-, 3117
    TOBACCO
        CARCINOGENIC POTENTIAL, REVIEW,
        3026

TONSILLAR NEOPLASMS
    FIBROMA
        CASE REPORT, CONGENITAL, 3408
    NEOPLASMS, EMBRYONAL AND MIXED
        CASE REPORT, 3408

TRACHEAL NEOPLASMS
    CYLINDROMA
        CASE REPORT, 3470*
        ULTRASTRUCTURAL STUDY, 3470*

TRANSFERASES
    VIRUS, ROUS SARCOMA
        BLOOD PLATELETS, 3289*
        CELL MEMBRANE, 3289*
        FIBROBLASTS, 3289*

TRANSPLANTATION ANTIGENS
    SEE HISTOCOMPATIBILITY ANTIGENS

TRANSPLANTATION, HETEROLOGOUS
    EYE NEOPLASMS
        GLIOMA, 3434
    GLIOMA
        HISTOLOGICAL STUDY, GUINEA PIG
        3434

TRANSPLANTATION, HOMOLOGOUS
    CARCINOMA
        IMMUNOSUPPRESSION, 3308
    CARCINOMA, BASAL CELL
        IMMUNOSUPPRESSION, 3308
        KIDNEY, 3308
    CARCINOMA, EPIDERMOID
        IMMUNOSUPPRESSION, 3308
        KIDNEY, 3308
    CERVIX NEOPLASMS
        CARCINOMA IN SITU, 3308
    CHROMOSOME ABERRATIONS
        MUTAGENIC ACTIVITY, REVIEW, 30
    KIDNEY NEOPLASMS
        TERATOID TUMOR, 3536
    LEUKEMIA, LYMPHOBLASTIC
        VIRUS, C-TYPE MURINE LEUKEMIA,
        3204
    LEUKEMIA, MYELOCYTIC
        VIRUS, C-TYPE MURINE LEUKEMIA,
        3204
    MYELOID METAPLASIA
        VIRUS, C-TYPE MURINE LEUKEMIA,
        3204
    NERVOUS SYSTEM NEOPLASMS
        TERATOID TUMOR, 3536
    OVARIAN NEOPLASMS
        DISGERMINOMA, 3308
    SARCOMA, RETICULUM CELL
        IMMUNOSUPPRESSION, 3308
        KIDNEY, 3308
    TESTOSTERONE
        CARCINOGENIC EFFECT, PROSTATE,
        3573

TRANSPLANTATION IMMUNOLOGY
    FIBROMA
        VIRUS, BOVINE PAPILLOMA, 3281*
    FIBROSARCOMA
        VIRUS, BOVINE PAPILLOMA, 3281*
    B-LYMPHOCYTES
        IMMUNITY, CELLULAR, 3360*

URACIL, 5-FLUORO-
    HEPATOMA
        IMMUNE RESPONSE, 3319

URANIUM
    RADIATION, IONIZING
        IGA, 3181
        IGG, 3181
        IGM, 3181
        IMMUNOGLOBULINS, 3181

UREA
    ETHYLENE, CHLORO-
        CARCINOGENIC METABOLITE, RAT, 3124

UREA, 1-BUTYL-1-NITROSO-
    JAW NEOPLASMS
        ODONTOGENIC TUMOR, 3113

UREA, ETHYL NITROSO-
    BRAIN NEOPLASMS
        MACROPHAGES, 3114
        TRANSPLACENTAL CARCINOGENESIS, RAT,
        3114
    LIVER NEOPLASMS
        ASCORBIC ACID, 3159*
    NERVE GROWTH FACTOR
        CELL DIFFERENTIATION, 3170*

UREA, METHYL NITROSO-
    BLADDER NEOPLASMS
        MODEL TEST SYSTEM, MOUSE, 3152*
    DNA
        GUANINE, 7-METHYL-, 3112
        METHYLATION, 3110
    MAMMARY NEOPLASMS, EXPERIMENTAL
        CHROMATIN, 3587*
    METHANESULFONIC ACID, METHYL ESTER
        ALKYLATING AGENTS, 3112

UREA, NITROSO-
    ASTROCYTOMA
        GLYCOLIPIDS, 3563

UREA, 1,1,3,3,-TETRAMETHYL-2-THIO-
    OCCUPATIONAL HAZARD
        CARCINOGENIC POTENTIAL, REVIEW,
        3040*

URETHANE
    SEE CARBAMIC ACID, ETHYL ESTER

URETHRAL NEOPLASMS
    CARCINOMA, EPIDERMOID
        LEUKOPLAKIA, 3426
    POLYPS
        ULTRASTRUCTURAL STUDY, 3493*

URIDINE, 5-BROMO-2'-DEOXY-
    CELL MEMBRANE
        GLYCOPROTEINS, 3095
    CHOLANTHRENE, 3-METHYL-
        CELLS, CULTURED, 3095

CONCANAVALIN A
    CELLS, CULTURED, 3095
PLANT AGGLUTININS
    CELLS, CULTURED, 3095
VIRUS, B-TYPE, GUINEA PIG
    RNA, VIRAL, 3232
VIRUS, KIRSTEN MURINE SARCOMA
    ANTIGENS, VIRAL, 3230
    DNA NUCLEOTIDYLTRANSFERASES, 3230
    ULTRASTRUCTURAL STUDY, 3230
    VIRAL PROTEINS, 3230
VIRUS, MOLONEY MURINE SARCOMA
    DNA-RNA HYBRIDIZATION, 3207
VIRUS, POLYOMA
    DNA-RNA HYBRIDIZATION, 3207
VIRUS, RAUSCHER LEUKEMIA
    DNA-RNA HYBRIDIZATION, 3207
VIRUS, SV40
    REVERSE TRANSCRIPTASE, 3250

URIDINE, 2'-DEOXY-5-IODO-
  DNA
    AUTORADIOGRAPHY, 3535*

URIDINE KINASE
  SEE PHOSPHOTRANSFERASES, ATP

URINARY TRACT
  LEUKOPLAKIA
    DIAGNOSIS AND PROGNOSIS, 3426

UROGENITAL NEOPLASMS
  CARCINOEMBRYONIC ANTIGEN
    DIAGNOSIS AND PROGNOSIS, REVIEW,
    3019
  CHORIOADENOMA
    GONADOTROPINS, CHORIONIC, 3427
    LACTATE DEHYDROGENASE, 3427
  CHORIOCARCINOMA
    GONADOTROPINS, CHORIONIC, 3427
    LACTATE DEHYDROGENASE, 3427
  EPIDEMIOLOGY
    GERMANY, 3506
  FETAL GLOBULINS
    DIAGNOSIS AND PROGNOSIS, 3019
  HYDATIFORM MOLE
    GONADOTROPINS, CHORIONIC, 3427
    LACTATE DEHYDROGENASE, 3427
  LIPOPROTEINS
    AUTOANTIBODIES, 3366*
    HEMAGGLUTINATION, 3366*
  LYMPHOCYTES
    IMMUNITY, CELLULAR, 3333
    PLANT AGGLUTININS, 3333

UTERINE NEOPLASMS
  ADENOCARCINOMA, 3311
    EPIDEMIOLOGY, 3495*
    IMMUNOSUPPRESSION, 3311
    PRECANCEROUS CONDITIONS, 3428
    STILBENES, 3169*
  BREAST NEOPLASMS
    MENOPAUSE, 3099
    PREGNANETRIOL CORTISONE RATIO,
    3099
    STEROID EXCRETION, URINARY, 3099
  CARCINOMA

    ULTRASTRUCTURAL STUDY,
    RHABDOMYOBLASTS, 3496*
ESTRADIOL
    PROGESTERONE, 3163*
    TESTOSTERONE, 3163*
PROGESTERONE
    TESTOSTERONE, 3163*

VACCINES
  VIRUS, SV40
    ANTIBODIES, VIRAL, 3259*
    ANTIGENS, VIRAL, 3259*

VAGINAL NEOPLASMS
  ADENOCARCINOMA
    CASE REPORT, 3500*
    PRECANCEROUS CONDITIONS, 343
  CARCINOMA, EPIDERMOID
    PRECANCEROUS CONDITIONS, 343
  CHROMATIN
    CELL NUCLEUS, 3587*
  MELANOMA
    EPIDEMIOLOGY, 3501*
  4,4'-STILBENEDIOL, ALPHA,ALPHA'-
    DIETHYL-
    IN UTERO EXPOSURE, 3037*
    PRECANCEROUS CONDITIONS, 342
    3430
    TRANSPLACENTAL CARCINOGENESI
    3429

VERTICAL TRANSMISSION
  VIRUS, C-TYPE RNA TUMOR
    VIRUS-LIKE PARTICLES, 3209

VIBRIO INFECTIONS
  VIRUS, FRIEND MURINE LEUKEMIA
    IMMUNOSUPPRESSION, 3312
    B-LYMPHOCYTES, 3312
  VIRUS, MOLONEY MURINE LEUKEMIA
    IMMUNOSUPPRESSION, 3312
    B-LYMPHOCYTES, 3312

VINCRISTINE
  SEE LEUROCRISTINE

VINYL CHLORIDE
  SEE ETHYLENE, CHLORO-

VIRAL PROTEINS
  VIRUS, ADENO 5
    CHROMOSOME MAPPING, 3196
    DNA, VIRAL, 3196
    TEMPERATURE SENSITIVE MUTANT
    3196
  VIRUS, KIRSTEN MURINE SARCOMA
    URIDINE, 5-BROMO-2'-DEOXY-,
  VIRUS, POLYOMA
    COMPLEX FORMATION, 3235
    ISOLATION AND CHARACTERIZATI
    3234
    TEMPERATURE SENSITIVE MUTANT
    3234
  VIRUS, POLYOMA 1
    DNA BINDING, 3282*
    DNA, VIRAL, 3282*
  VIRUS, POLYOMA 2
    DNA BINDING, 3282*
    DNA, VIRAL, 3282*

VIRUS, AVIAN MYELOBLASTOSIS
DNA NUCLEOTIDYLTRANSFERASES
2-OXETANONE, 3166*
REVERSE TRANSCRIPTASE
SUBUNIT SEQUENCE, 3260*

VIRUS, AVIAN RETICULOENDOTHELIOSIS
DNA NUCLEOTIDYTRANSFERASES
ISOLATION AND CHARACTERIZATION,
3286*
RETICULOENDOTHELIOSIS
ISOLATION AND CHARACTERIZATION,
3286*

VIRUS, AVIAN SPLEEN NECROSIS
VIRUS REPLICATION
ULTRASTRUCTURAL STUDY, 3287*

VIRUS, B-TYPE GUINEA PIG
URIDINE, 5-BROMO-2'-DEOXY-
RNA, VIRAL, 3232

VIRUS, BOVINE LEUKEMIA
LEUKEMIA, LYMPHOBLASTIC
ULTRASTRUCTURAL STUDY, 3488*
PLANT AGGLUTININS
C-TYPE PARTICLES, 3488*

VIRUS, BOVINE PAPILLOMA
FIBROMA
ANTIBODIES, NEOPLAS, 3281*
ANTIBODIES, VIRAL, 3281*
TRANSPLANTATION IMMUNOLOGY, 3281*
FIBROSARCOMA
ANTIBODIES, NEOPLAS, 3281*
ANTIBODIES, VIRAL, 3281*
TRANSPLANTATION IMMUNOLOGY, 3281*

VIRUS, C-TYPE
ANTIGENS, VIRAL
REVERSE TRANSCRIPTASE, 3207
LEUKEMIA, MYELOBLASTIC
REVERSE TRANSCRIPTASE, 3054*

VIRUS, C-TYPE BOVINE LEUKEMIA
ANTIBODIES
ASSAY, BEEF AND DAIRY CATTLE,
3365*

VIRUS, C-TYPE GUINEA PIG
VIRUS, MURINE MAMMARY TUMOR
DNA-RNA HYBRIDIZATION, 3269*
RNA, VIRAL, 3269*
VIRUS, RAUSCHER MURINE LEUKEMIA
DNA-RNA HYBRIDIZATION, 3269*
RNA, VIRAL, 3269*

VIRUS, C-TYPE LEUKEMIA
ANTIBODIES, VIRAL
TRANSMISSION, HORIZONTAL, 3208

VIRUS, C-TYPE MURINE
LYMPHOSARCOMA
ISOLATION AND CHARACTERIZATION,
3263*
VIRUS, KIRSTEN MURINE SARCOMA
CONCANAVALIN A, 3206

LIPOPOLYSACCHARIDES, 3206

VIRUS, C-TYPE MURINE LEUKEMIA
  INCLUSION BODIES, VIRAL
    GENETICS, 3262*
  LEUKEMIA, LYMPHOBLASTIC
    TRANSPLANTATION, HOMOLOGOUS, 3204
  LEUKEMIA, MYELOCYTIC
    TRANSPLANTATION, HOMOLOGOUS, 3204
  LYMPHOMA
    GENETICS, 3262*
  MYELOID METAPLASIA
    TRANSPLANTATION, HOMOLOGOUS, 3204
  PEPTIDES
    RADIOIMMUNOASSAY, 3204
  VIRUS, KIRSTEN MURINE SARCOMA
    VIRUS RESCUE, 3204
  VIRUS REPLICATION
    TISSUE CULTURE, 3204

VIRUS, C-TYPE RNA TUMOR
  CARBAMIC ACID, ETHYL ESTER
    INCLUSION BODIES, VIRAL, 3229
  CELL TRANSFORMATION, NEOPLASTIC
    REVIEW, 3053*
  FIBROSARCOMA
    ANTIGENIC DETERMINANTS, 3369*
    PEPTIDES, 3369*
  LYMPHOSARCOMA
    ANTIGENIC DETERMINANTS, 3369*
    PEPTIDES, 3369*
  RADIATION, IONIZING
    CARBAMIC ACID, ETHYL ESTER, 3229
    INCLUSION BODIES, VIRAL, 3229
  VIRUS-LIKE PARTICLES
    ULTRASTRUCTURAL STUDY, 3209
    VERTICAL TRANSMISSION, 3209

VIRUS, CYTOMEGALO
  ANTIGENS, VIRAL
    CELLS, CULTURED, 3265*
  CELL TRANSFORMATION, NEOPLASTIC
    CELLS, CULTURED, 3265*
  CHROMOSOMES
    CELLS, CULTURED, 3265*
  DNA, VIRAL
    DNA REPAIR, 3220
    IODINATION, 3220
  INFECTIOUS MONONUCLEOSIS
    ANTIBODIES, 3211
  SALIVARY GLAND NEOPLASMS
    PRECANCEROUS CONDITIONS, 3264*
  SARCOMA, KAPOSI'S
    ANTIBODIES, VIRAL, 3212
  VIRUS, REPLICATION
    CELLS, CULTURED, 3265*

VIRUS, ECHO
  BENZIMIDAZOLE, 2-(ALPHA-HYDROXYBENZYL)-

    VIRUS REPLICATION, 3010

VIRUS, EPSTEIN-BARR
  ANTIGENS, VIRAL
    CHROMATIN, 3210
    ISOLATION AND CHARACTERIZATION,
      3210
    ULTRASTRUCTURAL STUDY, 3210
  BURKITT'S LYMPHOMA

ANTIGENS, VIRAL, 3011, 3367*
  CELL LINE, 3012
CELL TRANSFORMATION, NEOPLASTIC
  ANTIGENS, VIRAL, 3326
  CARCINOGENIC POTENTIAL, PRIMATE
    REVIEW, 3352*
CONCANAVALIN A
  ANTIGENS, VIRAL, 3326
INFECTIOUS MONONUCLEOSIS
  ANTIBODIES, 3211
  ANTIGENS, VIRAL, 3011
LYMPHOCYTES
  CELL LINE, 3012
  IMMUNITY, CELLULAR, 3266*
B-LYMPHOCYTES
  ANTIGENS, VIRAL, 3326
T-LYMPHOCYTES
  ANTIGENS, VIRAL, 3326
NASOPHARYNGEAL NEOPLASMS
  ANTIGENS, VIRAL, 3011
  CARCINOMA, 3012
PLANT AGGLUTININS
  ANTIGENS, VIRAL, 3326
SARCOMA, KAPOSI'S
  ANTIBODIES, VIRAL, 3212

VIRUS, EQUINE ENCEPHALITIS
  INCLUSION BODIES, VIRAL
    ULTRASTRUCTURAL STUDY, 3252
  VIRUS REPLICATION
    ULTRASTRUCTURAL STUDY, 3252

VIRUS, FELINE FIBROSARCOMA
  CELL TRANSFORMATION, NEOPLASTIC
    CARCINOGENIC POTENTIAL, PRIMATE
      REVIEW, 3352*

VIRUS, FELINE LEUKEMIA
  ANTIGENS, VIRAL
    EPIDEMIOLOGY, REVIEW, 3015
  ERYTHREMIC MYELOSIS
    EPIDEMIOLOGY, REVIEW, 3015
  ERYTHROLEUKEMIA
    EPIDEMIOLOGY, REVIEW, 3015
  LEUKEMIA
    EPIDEMIOLOGY, REVIEW, 3015
  LEUKEMIA, MONOCYTIC
    ANTIGENS, VIRAL, 3205
  LEUKEMIA, MYELOBLASTIC
    ANTIGENS, VIRAL, 3205
  LEUKEMIA, MYELOCYTIC
    EPIDEMIOLOGY, REVIEW, 3015
  LYMPHATIC NEOPLASMS
    EPIDEMIOLOGY, REVIEW, 3015
  LYMPHOSARCOMA
    EPIDEMIOLOGY, REVIEW, 3015
  RNA, VIRAL
    ISOLATION AND CHARACTERIZATION,
      3224

VIRUS, FRIEND MURINE LEUKEMIA
  ACTINOMYCIN D
    RNA POLYMERASE, 3227
    SPLENOMEGALY, 3227
  ANTIGENS, VIRAL
    IMMUNOGLOBULINS, SURFACE, 3313
    ISOLATION AND CHARACTERIZATION,
      3223
    TRANS-SPECIES REACTIVITY, 3223

ANTILYMPHOCYTE SERUM
    IMMUNE RESPONSE, 3310
ERYTHROPOIETIN
    RNA POLYMERASE, 3227
LEUKEMIA
    INTERFERON, 3267*
LYMPHOCYTES
    IMMUNE RESPONSE, 3310, 3345
B-LYMPHOCYTES
    ANTIGENS, VIRAL, 3310
    IMMUNOGLOBULINS, SURFACE, 3310
T-LYMPHOCYTES
    ANTIGENS, VIRAL, 3310
    IMMUNOGLOBULINS, SURFACE, 3310
MACROPHAGES
    ANTIBODY-PRODUCING CELLS, 3345
    IMMUNE RESPONSE, 3345
METHANE, SULFINYLBIS-
    CELL DIFFERENTIATION, 3268*
    RNA, MESSENGER, 3268*
PEPTIDES
    ISOLATION AND CHARACTERIZATION,
    3223
RNA, MESSENGER
    PEPTIDE CHAIN ELONGATION, 3558
VIBRIO INFECTIONS
    IMMUNOSUPPRESSION, 3312
    B-LYMPHOCYTES, 3312
VIRUS, MOLONEY MURINE LEUKEMIA
    CELL TRANSFORMATION, NEOPLASTIC,
    3228

S, GIBBON APE LEUKEMIA
LEUKEMIA, MONOCYTIC
    ANTIGENS, VIRAL, 3205
LEUKEMIA, MYELOBLASTIC
    ANTIGENS, VIRAL, 3205

S, GRAFFI HAMSTER LEUKEMIA
VIRUS, HAMSTER LEUKEMIA
    HYBRIDIZATION, 3231

S, GRAFFI MURINE LEUKEMIA
ANTIGENS, VIRAL
    CELL SURFACE ANTIGEN, 3371*

S, GROSS LEUKEMIA
LEUKEMIA
    IMMUNOTHERAPY, 3350*
    NEURAMINIDASE, 3350*

S, GROSS MURINE LEUKEMIA
GLYCOPROTEINS
    ANTIGENS, VIRAL, 3225
T-LYMPHOCYTES
    ANTIGENS, VIRAL, 3225
    GLYCOPROTEINS, 3225
    GROWTH, 3331
    ISOANTIGENS, 3331

S, GROSS PASSAGE A
VIRUS, HAMSTER LEUKEMIA
    HYBRIDIZATION, 3231

S, HAMSTER LEUKEMIA
VIRUS, GRAFFI HAMSTER LEUKEMIA
    HYBRIDIZATION, 3231
VIRUS, GROSS PASSAGE A
    HYBRIDIZATION, 3231

VIRUS, HARVEY MURINE SARCOMA
    HYBRIDIZATION, 3231
VIRUS, MOLONEY MURINE LEUKEMIA
    HYBRIDIZATION, 3231
VIRUS, MOLONEY MURINE SARCOMA
    HYBRIDIZATION, 3231
VIRUS, RAUSCHER MURINE LEUKEMIA
    HYBRIDIZATION, 3231

VIRUS, HARVEY MURINE SARCOMA
VIRUS, HAMSTER LEUKEMIA
    HYBRIDIZATION, 3231

VIRUS, HEPATITIS
ANTIGENS, VIRAL
    DETECTION METHOD, 3271*
AUSTRALIA ANTIGEN
    DETECTION METHOD, 3271*

VIRUS, HERPES
ANGIOSARCOMA
    HISTOLOGICAL STUDY, BABOON, 3274*

VIRUS, HERPES AOTUS
PHOSPHOTRANSFERASES
    ISOLATION AND CHARACTERIZATION,
    3213
THYMIDINE KINASE
    ISOLATION AND CHARACTERIZATION,
    3213

VIRUS, HERPES ATELES
CELL TRANSFORMATION, NEOPLASTIC
    CARCINOGENIC POTENTIAL, PRIMATES,
    REVIEW, 3052*

VIRUS, HERPES MARMOSET
PHOSPHOTRANSFERASES
    ISOLATION AND CHARACTERIZATION,
    3213
THYMIDINE KINASE
    ISOLATION AND CHARACTERIZATION,
    3213

VIRUS, HERPES SAIMIRI
ADRENAL GLAND NEOPLASMS
    LYMPHOSARCOMA, 3277*
BRAIN NEOPLASMS
    LYMPHOSARCOMA, 3277*
CELL TRANSFORMATION, NEOPLASTIC
    CARCINOGENIC POTENTIAL, PRIMATES,
    REVIEW, 3052*
EYE NEOPLASMS
    LYMPHOSARCOMA, 3277*
KIDNEY NEOPLASMS
    LYMPHOSARCOMA, 3277*
LYMPHOSARCOMA
    CLINICOPATHOLOGIC STUDY, RABBIT,
    3277*
SKIN NEOPLASMS
    LYMPHOSARCOMA, 3277*

VIRUS, HERPES SIMPLEX
CELL TRANSFORMATION, NEOPLASTIC
    CARCINOGENIC POTENTIAL, PRIMATES,
    REVIEW, 3052*
CHROMOSOME ABERRATIONS
    MUTAGENIC ACTIVITY, REVIEW, 3049*
INFECTIOUS MONONUCLEOSIS

ANTIBODIES, 3211

VIRUS, HERPES SIMPLEX 1
  CELL TRANSFORMATION, NEOPLASTIC
    LEVAMISOLE, 3275*
    NEOPLASM METASTASIS, 3275*
    VACCINE CELLS, HAMSTERS, 3354*
  CERVIX NEOPLASMS
    EPIDEMIOLOGY, REVIEW, 3014
  DNA
    ISOLATION AND CHARACTERIZATION,
    3013
  DNA POLYMERASE
    TEMPERATURE SENSITIVE MUTANTS,
    3215
  DNA, VIRAL
    ENDONUCLEASES, 3214
    STRUCTURAL ANALYSIS, 3214
    TEMPERATURE SENSITIVE MUTANTS,
    3215
  GLYCOPROTEINS
    ISOLATION AND CHARACTERIZATION,
    3216
  PHOSPHOTRANSFERASES
    ISOLATION AND CHARACTERIZATION,
    3213
  SARCOMA, KAPOSI'S
    ANTIBODIES, VIRAL, 3212
  THYMIDINE KINASE
    ISOLATION AND CHARACTERIZATION,
    3213
  VIRUS-LIKE PARTICLES
    DNA, 3013
  ZINC SULFATE
    DNA NUCLEOTIDYLTRANSFERASES, 3217
    DNA REPLICATION, 3217

VIRUS, HERPES SIMPLEX 2
  BREAST NEOPLASMS
    ANTIGEN-ANTIBODY REACTIONS, 3218
    ANTIGENS, VIRAL, 3218
  CELL TRANSFORMATION, NEOPLASTIC
    VACCINE CELLS, HAMSTERS, 3354*
  CERVIX NEOPLASMS
    ANTIGEN-ANTIBODY REACTIONS, 3218
    ANTIGENS, VIRAL, 3218
    CARCINOMA, 3276*
    CARCINOMA IN SITU, 3014
    CYTOPATHOLOGICAL STUDY, REVIEW,
    3014
    DNA, VIRAL, 3276*
    EPIDEMIOLOGY, REVIEW, 3014
    IMMUNOLOGICAL STUDY, REVIEW, 3014
  DNA, VIRAL
    ENDONUCLEASES, 3220
    IODINATION, 3220
  GLYCOPROTEINS
    ISOLATION AND CHARACTERIZATION,
    3216
  PHOSPHOTRANSFERASES
    ISOLATION AND CHARACTERIZATION,
    3213
  PYRANS
    IMMUNOSUPPRESSION, 3341
    T-LYMPHOCYTES, 3341
    MACROPHAGES, 3341
  SARCOMA
    ANTIGENS, VIRAL, 3014
    IMMUNOSUPPRESSION, 3358*

  NEOPLASM TRANSPLANTATION, 3358*
  SARCOMA, KAPOSI'S
    ANTIBODIES, VIRAL, 3212
  THYMIDINE KINASE
    ISOLATION AND CHARACTERIZATION,
    3213

VIRUS, HERPES TURKEY
  THYMIDINE KINASE
    ISOLATION AND CHARACTERIZATION,
    3213

VIRUS, KIRSTEN MURINE LEUKEMIA
  VIRUS, MOLONEY MURINE LEUKEMIA
    CELL TRANSFORMATION, NEOPLASTIC,
    3228

VIRUS, KIRSTEN MURINE SARCOMA
  URIDINE, 5-BROMO-2'-DEOXY-
    ANTIGENS, VIRAL, 3230
    DNA NUCLEOTIDYLTRANSFERASES, 323
    ULTRASTRUCTURAL STUDY, 3230
    VIRAL PROTEINS, 3230
  VIRUS, C-TYPE MURINE
    CONCANAVALIN A, 3206
    LIPOPOLYSACCHARIDES, 3206
  VIRUS, C-TYPE MURINE LEUKEMIA
    VIRUS RESCUE, 3204

VIRUS-LIKE PARTICLES
  BREAST NEOPLASMS
    ADENOCARCINOMA, 3288*
    CARCINOMA, 3288*
    CARCINOMA, PAPILLARY, 3288*
  LARYNGEAL NEOPLASMS
    VIRUS, PAPILLOMA, 3410
  LEUKEMIA, MONOCYTIC
    PRECANCEROUS CONDITIONS, 3419
  LEUKEMIA, MYELOBLASTIC
    PRECANCEROUS CONDITIONS, 3419
  LEUKEMIA, MYELOCYTIC
    BONE MARROW CELLS, 3419
  SARCOMA
    CHOLANTHRENE, 3-METHYL-, 3270*
  TERATOID TUMOR
    ULTRASTRUCTURAL STUDY, MOUSE, 34
  VIRUS, C-TYPE RNA TUMOR
    ULTRASTRUCTURAL STUDY, 3209
    VERTICAL TRANSMISSION, 3209
  VIRUS, HERPES SIMPLEX 1
    DNA, 3013

VIRUS, M7 BABOON
  LEUKEMIA, MONOCYTIC
    ANTIGENS, VIRAL, 3205
  LEUKEMIA, MYELOBLASTIC
    ANTIGENS, VIRAL, 3205

VIRUS, MAREK'S DISEASE HERPES
  ANTIGENS, VIRAL
    IMMUNE SERUMS, 3219
    PATHOGENESIS, 3222
  AVIAN LEUKOSIS
    REVIEW, 3016
  INCLUSION BODIES, VIRAL
    ANTIGENS, VIRAL, 3221
    VIRUS REPLICATION, 3221
  VIRUS, TURKEY HERPES
    ANTIGENIC DETERMINANTS, 3219

VIRUS, MURINE MAMMARY TUMOR
ANTIGENS, VIRAL
IMMUNOSUPPRESSION, 3340
T-LYMPHOCYTES
ANTIGENS, VIRAL, 3340
IMMUNOSUPPRESSION, 3340
MAMMARY NEOPLASMS, EXPERIMENTAL
IMMUNITY, ACTIVE, 3352*
RNA, VIRAL, 3232
VIRUS, C-TYPE GUINEA PIG
DNA-RNA HYBRIDIZATION, 3269*
RNA, VIRAL, 3269*
VIRUS, MURINE LEUKEMIA
IMMUNOSUPPRESSION, 3340

VIRUS, MURINE SARCOMA
VIRUS REPLICATION
CORTISOL, 3280*
DEXAMETHASONE, 3280*
ESTRADIOL, 3280*
FIBROBLASTS, 3280*
PROGESTERONE, 3280*
TESTOSTERONE, 3280*

VIRUS, N-TROPIC MURINE LEUKEMIA
REVERSE TRANSCRIPTASE
CLONAL VARIANTS, 3278*
VIRUS REPLICATION
CLONAL VARIANTS, 3278*

VIRUS, PAPILLOMA
CELL TRANSFORMATION, NEOPLASTIC
CARCINOGENIC POTENTIAL, PRIMATES,
REVIEW, 3052*
LARYNGEAL NEOPLASMS
VIRUS-LIKE PARTICLES, 3410

VIRUS, PAPOVA
CELL TRANSFORMATION, NEOPLASTIC
GENE PRODUCTS, 3017

VIRUS, PAPOVA, BK
DNA, VIRAL
ISOLATION AND CHARACTERIZATION,
3236
PEPTIDES
ISOLATION AND CHARACTERIZATION,
3236

VIRUS, POLIO
CRESOL, 2-(2-BENZIMIDAZOLYL)-
CYTOTOXICITY, REVIEW, 3010
HELA CELLS
PHOSPHOLIPIDS, 3010
RNA, VIRAL, 3253
POLYRIBOSOMES
PHOSPHOLIPIDS, 3010
VIRUS REPLICATION, 3010
RNA, VIRAL
CYTOTOXICITY, REVIEW, 3010
SODIUM CHLORIDE
RNA, VIRAL, 3253

VIRUS, POLYOMA
ACID PHOSPHATASE
CELL TRANSFORMATION, NEOPLASTIC,
3570
ANTIGENS, NEOPLASM
ISOLATION AND CHARACTERIZATION,

          3328
     CHO.ANTHRENE, 3-METHYL-
          DNA-RNA HYBRIDIZATION, 3207
     DNA, VIRAL
          BINDING ASSAY, 3242
          COMPLEX FORMATION, 3235
          DNA-RNA HYBRIDIZATION, 3233
          VIRUS REPLICATION, 3236
     GALACTOSAMINIDASE
          CELL TRANSFORMATION, NEOPLASTIC,
          3570
     GALACTOSIDASES
          CELL TRANSFORMATION, NEOPLASTIC,
          3570
     GLUCOSAMINIDASE
          CELL TRANSFORMATION, NEOPLASTIC,
          3570
     GLUCOSIDASES
          CELL TRANSFORMATION, NEOPLASTIC,
          3570
     KIDNEY NEOPLASMS
          RNA, 3155*
     LYSOSOMES
          CELL TRANSFORMATION, NEOPLASTIC,
          3570
     NUCLEIC ACIDS
          ISOLATION AND CHARACTERIZATION,
          3233
     PEPTIDES
          ISOLATION AND CHARACTERIZATION,
          3236, 3328
     URIDINE, 5-BROMO-2'-DEOXY-
          DNA-RNA HYBRIDIZATION, 3207
     VIRAL PROTEINS
          COMPLEX FORMATION, 3235
          ISOLATION AND CHARACTERIZATION,
          3234
          TEMPERATURE SENSITIVE MUTANTS,
          3234

VIRUS, POLYOMA 1
     VIRAL PROTEINS
          DNA BINDING, 3282*
          DNA, VIRAL, 3282*

VIRUS, POLYOMA 2
     VIRAL PROTEINS
          DNA BINDING, 3282*
          DNA, VIRAL, 3282*

VIRUS, POLYOMA, BK
     ALDRICH SYNDROME
          DNA, VIRAL, 3284*
          ISOLATION AND CHARACTERIZATION,
          3284*
     FIBROSARCOMA
          ANTIBODIES, VIRAL, 3283*
          ANTIGENS, VIRAL, 3283*
          NEOPLASM TRANSPLANTATION, 3283*
     VIRUS, SV40
          HYBRIDIZATION, 3284*

VIRUS, POX
     ACID PHOSPHATASE
          CELL MEMBRANE, 3255
     ADENOSINE TRIPHOSPHATASE
          CELL MEMBRANE, 3255
     ALDOSTERONE
          VIRUS REPLICATION, 3255

CARCINOGENIC POTENTIAL, PRIMATES,
REVIEW, 3052*
CELL MEMBRANE, 3289*
DNA, VIRAL, 3261*
TEMPERATURE SENSITIVE MUTANTS,
3435
DNA NUCLEOTIDYLTRANSFERASES
2-OXETANONE, 3166*
DNA POLYMERASE
TEMPERATURE SENSITIVE MUTANTS,
3239
DNA, VIRAL
DNA POLYMERASE, 3239
TEMPERATURE SENSITIVE MUTANTS,
3239
GLYCEROL
CELL TRANSFORMATION, NEOPLASTIC,
3435
INCLUSION BODIES, VIRAL
ANTIGENIC DETERMINANTS, 3237
ISOLATION AND CHARACTERIZATION, A-
TYPE, 3237
ULTRASTRUCTURAL STUDY, 3237
RNA, VIRAL
GENETICS, 3240
ISOLATION AND CHARACTERIZATION,
3240
TRANSFERASES
BLOOD PLATELETS, 3289*
CELL MEMBRANE, 3289*
FIBROBLASTS, 3289*

VIRUS, SENDAI
HELA CELLS
DNA REPAIR, 3031

VIRUS, SIMIAN 5
RNA, VIRAL
VIRUS REPLICATION, 3010

VIRUS, SIMIAN ADENO
CELL TRANSFORMATION, NEOPLASTIC
CARCINOGENIC POTENTIAL, PRIMATES,
REVIEW, 3052*
DNA, VIRAL
TRANS-SPECIES INFECTION, HAMSTER,
3203
VIRUS REPLICATION, 3203
NEOPLASMS
DNA, VIRAL, 3203

VIRUS, SIMIAN SARCOMA
SARCOMA
ULTRASTRUCTURAL STUDY, 3279*

VIRUS, SV40
ALANINE
AMINO ACID TRANSPORT, 3296*
ANTIGENS, VIRAL
DEOXYRIBONUCLEASE, 3292*
DNA, VIRAL, 3242, 3291*
FLUORESCENT ANTIBODY TECHNIC,
3290*
ASTROCYTOMA
GLYCOLIPIDS, 3563
CAFFEINE
CELL TRANSFORMATION, NEOPLASTIC,
3249
CARCINOMA

CELL TRANSFORMATION, NEOPLASTIC,
  3250
  KARYOTYPING, 3250
  NEOPLASM TRANSPLANTATION, 3250
  TRANS-SPECIES, HAMSTER, 3250
CELL MEMBRANE
  CELL TRANSFORMATION, NEOPLASTIC,
  3435
  CONCANAVALIN A, 3349
CELL TRANSFORMATION, NEOPLASTIC
  CARCINOGENIC POTENTIAL, PRIMATES,
  REVIEW, 3052*
  CELL DIVISION, 3202
  CHROMOSOMES, 3202
  CYTOCHALASIN B, 3202
  METHANE, SULFINYLBIS-, 3202
  TEMPERATURE SENSITIVE MUTANTS,
  3251
CHROMOSOME MAPPING
  TEMPERATURE SENSITIVE MUTANTS,
  3018
CONCANAVALIN A
  AMINO ACIDS, 3296*
  IMMUNITY, CELLULAR, 3347
  IMMUNOSUPPRESSION, 3347
  TRANSPLANTATION IMMUNOLOGY, 3347
DEOXYRIBONUCLEASE
  DNA, CIRCULAR, 3245
  DNA, VIRAL, 3245
DIBUTYRYL CYCLIC AMP
  CELL TRANSFORMATION, NEOPLASTIC,
  3249
DNA REPLICATION
  HYBRIDIZATION, 3241
DNA, VIRAL
  ANTIGENS, VIRAL, 3244
  BINDING ASSAY, 3242
  CELL TRANSFORMATION, NEOPLASTIC,
  3243
  DNA-RNA HYBRIDIZATION, 3246
  GENETICS, 3244
  ISOLATION AND CHARACTERIZATION,
  3248
ENDONUCLEASES
  ANTIGENS, VIRAL, 3293*
  DNA, VIRAL, 3246, 3248
FIBROBLASTS
  FLUORESCENT ANTIBODY TECHNIC,
  3290*
GLYCEROL
  CELL TRANSFORMATION, NEOPLASTIC,
  3435
GLYCINE
  AMINO ACID TRANSPORT, 3296*
GROWTH SUBSTANCES
  PERFUSED RAT LIVER, 3295*
LEUCINE
  AMINO ACID TRANSPORT, 3296*
PHENYLALANINE
  AMINO ACID TRANSPORT, 3296*
PROSTATIC NEOPLASMS
  CARCINOMA, 3250
RADIATION
  CELL TRANSFORMATION, NEOPLASTIC,
  3249
RADIATION, IONIZING
  DNA, CIRCULAR, 3245
  DNA, VIRAL, 3245
RNA

METHYLATION, 3294*
RNA, MESSENGER
  DNA-RNA HYBRIDIZATION, 3256
RNA POLYMERASE
  DNA-RNA HYBRIDIZATION, 3247
  DNA, VIRAL, 3246
  ISOLATION AND CHARACTERIZATION,
  3247
RNA, VIRAL
  DNA-RNA HYBRIDIZATION, 3246
  DNA, VIRAL, 3246
SARCOMA
  ANTIGENS, NEOPLASM, 3314
  HISTOCOMPATIBILITY ANTIGENS, 331
  IMMUNIZATION, 3314
THEOPHYLLINE
  CELL TRANSFORMATION, NEOPLASTIC,
  3249
ULTRAVIOLET RAYS
  CELL TRANSFORMATION, NEOPLASTIC,
  3249
URIDINE, 5-BROMO-2'-DEOXY-
  REVERSE TRANSCRIPTASE, 3250
VACCINES
  ANTIBODIES, VIRAL, 3259*
  ANTIGENS, VIRAL, 3259*
VIRUS, POLYOMA, BK
  HYBRIDIZATION, 3284*
VIRUS REPLICATION
  TEMPERATURE SENSITIVE MUTANTS,
  3251

VIRUS, TURKEY HERPES
VIRUS, MAREK'S DISEASE HERPES
  ANTIGENIC DETERMINANTS, 3219

VIRUS, VESICULAR STOMATITIS
  HELA CELLS
    RNA, VIRAL, 3253
  RNA, VIRAL
    ISOLATION AND CHARACTERIZATION,
    PARTICLES, 3254
    ULTRASTRUCTURAL STUDY, FI
    PARTICLES, 3254
  SODIUM CHLORIDE
    RNA, VIRAL, 3253

VIRUS, WOOLLY MONKEY SARCOMA
  LEUKEMIA, MONOCYTIC
    ANTIGENS, VIRAL, 3205
  LEUKEMIA, MYELOBLASTIC
    ANTIGENS, VIRAL, 3205

VITAMIN A
  SEE RETINOL

VITAMIN B12
  LEUKEMIA, MYELOCYTIC
    CARRIER PROTEINS, 3549
  LEUKOCYTOSIS
    CARRIER PROTEINS, 3549
  POLYCYTHEMIA VERA
    CARRIER PROTEINS, 3549

VITAMIN C
  SEE ASCORBIC ACID

VITAMINS
  FIBROBLASTS

TATION WITHOUT ACCOMPANYING ABSTRACT

CAS REGISTRY NUMBER	ABSTRACT NUMBER
55801............................	3086
56122............................	3571
56235............................	3091
56495............................	3001, 3006, 3079, 3095, 3101, 3102, 3105, 3106, 3107, 3207, 3299, 3303, 3305, 3306, 3307, 3309
56531............................	3429, 3430
56553............................	3094, 3516
56575............................	3001, 3031
56815............................	3435
57136............................	3124
57227............................	3319
57410............................	3097
57830............................	3255, 3550, 3551
57885............................	3575
57976............................	3001, 3006, 3092, 3093, 3102, 3550
58082............................	3031, 3249
58220............................	3396, 3550, 3573
58559............................	3095, 3249, 3402, 3542, 3565
58899............................	3006
59052............................	3004, 3032
59143............................	3095, 3207, 3230, 3232, 3250

CHEMICAL ABSTRACT SERVICES REGISTRY NUMBER INDEX

CAS REGISTRY NUMBER	ABSTRACT NUMBER	CAS REGISTRY NUMBER	ABSTRACT NUMBER
59859.........................	3543	75014.........................	3006, 3124, 3518
59870.........................	3118	75569.........................	3111
60117.........................	3303	76255.........................	3550
60242.........................	3424	77781.........................	3109
61803.........................	3006	79118.........................	3124
62555.........................	3091	83443.........................	3554
62759.........................	3091, 3115, 3514	88891.........................	3302
64175.........................	3091	92851.........................	3021, 3089
64675.........................	3190	94597.........................	3088
66273.........................	3031, 3108, 3109, 3112	100027.........................	3081
66819.........................	3095, 3402	100754.........................	3117
67425.........................	3543	107062.........................	3125
67630.........................	3091	107073.........................	3125
67685.........................	3095, 3202, 3431, 3541	107200.........................	3125
68122.........................	3095	108592.........................	3081
68199.........................	3549	108861.........................	3006
70188.........................	3079	110601.........................	3120, 3557,
70257.........................	3031, 3096, 3102	111308.........................	3346
70348.........................	3406	117102.........................	3081
71443.........................	3557	124209.........................	3557
74884.........................	3109	128530.........................	3551
		129000.........................	3516

ACT NUMBER	CAS REGISTRY NUMBER	ABSTRACT NUMBER
	684935..........................	3110, 3112
	759739..........................	3114
3422	773302..........................	3217
	869012..........................	3113
	937406..........................	3514
	1162658..........................	3075, 3076, 3077, 3078, 3079, 3096
	1239330..........................	3124
	1306190..........................	3520
	1314201..........................	3026
	1404746..........................	3010
	2001958..........................	3545
	2541697..........................	3102
	3483123..........................	3567
	4431009..........................	3555
	6051872..........................	3006
	6098448..........................	3001, 3031, 3081, 3102
	7439910..........................	3542
	7439954..........................	3255
	7439965..........................	3255
	7440020..........................	3026
	7440075..........................	3009, 3178, 3179

CHEMICAL ABSTRACT SERVICES REGISTRY NUMBER INDEX

CAS REGISTRY NUMBER	ABSTRACT NUMBER	CAS REGISTRY NUMBER	ABSTRACT NU
7440382	3026	10043922	3181
7440473	3026	13345216	3085
7440702	3255, 3543	1651298	3031, 3081
7441044	3179	17573216	3085
7447407	3544	23214928	3319
7647145	3253	24554265	3096
7681494	3564		
7786303	3544		
7782492	3180		
8001283	3031, 3081		
8002059	3520		
8063943	3007, 3518, 3519		
9002077	3546		
9002624	3092, 3566		
9002862	3007, 3518, 3519		
9004108	3396, 3564		
9005496	3555		
9005792	3432, 3569		
9006046	3519		
9007732	3019		
10024972	3119		
10043524	3544		

UMBER   WLN   ABSTRACT NUMBER

               L B656 HHJ EMV1 ....
                                                3096, 3097

               L B656 HHJ ENQV1 ....
                                                3080, 3082, 3097

               L C6566 1A PJ ....
                                                3516

               L C666 BV IVJ DQ NQ ....
                                                3081

               L D6 B666J ....
                                                3094, 3516

               L D6 B666J C J ....
                                                3001, 3006, 3092,
                                                3093, 3102, 3550

               L D6 B666J J ....
                                                3102

               L D6 B6666 2AB TJ ....
                                                3006, 3083, 3084,
                                                3085, 3101, 3102,
                                                3431, 3516

               L E5 B666 FVTTT&J E OQ ....
                                                3255, 3550

               L E5 B666 LUTJ A E FY&3Y OQ -B&AEFO ...
                                                3575

               L E5 B666 OV MUTJ A CQ E FV1Q FQ -B&ACEF
                                                3029, 3255, 3550,
                                                3551

               L E5 B666 OV MUTJ A E FQ -B&AEF ....
                                                3396, 3550, 3573

               L E5 B666 OV MUTJ A E FV1 -B&AEF ......
                                                3255, 3550, 3551

               L E5 B666TJ A DQ E FY2VR OQ ....
                                                3554

               L E5 B666TJ A1Q CQ E IQ MQ QQ F- DT5OV
                 EHJ& OO- BT6OTJ CQ DQ EQ F 3545

               L E5 B666TTT&J E FQ OQ ....
                                                3093, 3255, 3550
3124, 3125,
               L E6 B6656 1A T&&&T&J R ....
                                                3001, 3006, 3079,
                                                3095, 3101, 3102,
                                                3105, 3106, 3107,
                                                3207, 3299, 3303,
                                                3305, 3306, 3307,
3108, 3111   3309

               L E6 B666J ....
                                                3516

        WLN 1

√LN|	ABSTRACT NUMBER	√LN		ABSTRACT NUMBE

L6TJ AG BG CG DG EG FG *GAMMA 3006

L6T5 AG BG CG DG EG FG 3087

L666 B6 2AB PJ 3516

ONN1&R 3514

ONN1&1 .... 3091, 3115, 3514

ONN1&1R 3514

ONN2&2 .... 3319, 3431, 343? 3514

OS1&1 .... 3095, 3202, 343? 3541

QR 3081

QR BQ DYQ1M1 -LQR BQ DYQ1M1 3564

QR DY2& 2U 3429, 3430

QVR BQ E·· 3Y* &Q 3555

QVYZ1R DN2G2G -L 3421, 3422

QV1G 3124

QV1N1VQ2O 22 3543

QV1S1VQ 3124

QY 3091

Q1NUNO&1 3431

Q1YQ1Q............................... 3435

Q2 3091

Q2G 3125

5H2Q! 3424

T B666 HKJ EZ H2 IR& LZ &G &9/26 3095

T C6 B5665 2AB S BX IN ON NU JH&&TTTJ FO1
   IVH KVO1 KQ LOV1 M2 E- NT F6 E596 A BA
   LM&&TTJ NVO1RQ R2................. 3319

T C666 BO EV INJ D FZ N G- K-/VM- OT5-16
   AN FVN IVN LVO PVM SVTJ G J KY N RY 2
                                   3004, 3227

T D3 B556 BN EM JV MVTTT&J GO1 H1OVZ KZ L
                                   3103

T F5 C6 B655 DOV GV OO QO RUT&&TTJ LO1
                                   3075, 3076,
                                   3078, 3079,

T3OTJ B 3111

T5M CNJ 3565

T5MVMV EHJ ER& ER 3097

T5OJ BNW E- ET5N CSJ BMVH 3096

T5OJ BNW E1UNMVZ 3118

T5VNVJ 3551

T56 BM DN FN HNJ ISH 3319

T56 BM DN FNVNVJ F H 3095, 3249, 3542, 3565

T56 BN DN FN HNJ IN1&1 D- BT5OTJ CQ
   DMVYZ1R DO1& E1Q................... 3095

T56 BN DN FNVNVJ B F H 3031, 3249

T56 BN DOJ CZ HG 3006

T56 BO DO CHJ G2U1 3088

T6MPOTJ BO BN2G2G 3032, 3125,

T6MVMTJ E1YQ- BL6VTJ D F 3095, 3402

LINE NOTATION INDEX

WLN   ABSTRACT NUMBER

ZVZ ....
3124

ZV1I ....
3551

ZYUS ....
3091
3230,

Z3M4M3Z ....
3557

Z3VQ ....
3571

1N1&R DNUNR ....
3303

1N1&R DNUNR C ....
3086

1OPO&O1&O1 ....
3109

1OSWO1 ....
3109

2OSWO2..............................
3109

2OV1U1VO2 ....
3079

3102

3109,

WLN 3

CARCINOGENESIS ABSTRACTS

A monthly publication of the

National Cancer Institute

Editor

George P. Studzinski, M.D.

College of Medicine and Dentistry

of New Jersey, Newark

Associate Editor

Jussi J. Saukkonen, M.D.

Jefferson Medical College, Philadelphia

NCI Staff Consultants

Elizabeth Weisburger, Ph.D.

Joan W. Chase, M.S.

DEPOSITORY

OCT 28 1976

UNIV. OF ILL. LIBRARY
'AT URBANA-CHAMPAIGN

Literature Selected, Abstracted, and Indexed

by

The Franklin Institute Research Laboratories
Science Information Services
Biomedical Section

Bruce H. Kleinstein, Ph.D., J.D., Group Manager,

Biomedical Projects

Ruthann E. Auchinleck
Production Editor

Contract Number NO1-CP-43293

Public Health Service, USDHEW

DHEW Publication No. (NIH) 76-301

PREFACE

Carcinogenesis Abstracts is a publication of the
National Cancer Institute. The journal serves as a
vehicle through which current documentation of carcinogenesis research highlights are compiled, condensed, and disseminated on a regular basis. It
represents an integral part of the Institute's program of fostering and supporting coordinated research
into cancer etiology. Issues of *Carcinogenesis Abstracts* normally contain three-hundred abstracts and
three-hundred citations (unaccompanied by corresponding abstracts). Abstracts and citations refer to
the current scientific literature that describes the
most significant carcinogenesis research carried on
at the National Cancer Institute, other governmental
agencies, and private institutions. *Carcinogenesis
Abstracts* is intended to be a highly useful current
awareness tool for scientists engaged in carcinogenesis research or related areas. The great number and
diversity of publications relevant to carcinogenesis
make imperative the availability of this service to
investigators whose work requires that they keep
abreast with current developments in the field.

Carcinogenesis Abstracts is normally published monthly. Volume XIII covers the scientific literature published from Jan 1975 through Dec 1975. To increase.
the usefulness of *Carcinogenesis Abstracts,* Volume
XIII, a Wiswesser Line Notation index and a Chemical
Abstracts Service Registry Number index have been
provided. These indexes reference compounds described in abstracted articles. A cumulative subject,
author, CAS Registry Number, and Wiswesser Line Notation index for Volume XIII will be published shortly
after the final regular issue.

Carcinogenesis Abstracts is available free of charge
to libraries and to individuals who have a professional interest in carcinogenesis. Requests for
Carcinogenesis Abstracts from qualified individuals
should include statements of their relationship to
carcinogenesis research. All correspondence should
be addressed as follows.

> *Carcinogenesis Abstracts*
> Room C-325
> Landow Building
> National Cancer Institute
> National Institutes of Health
> Bethesda, Maryland 20014

The Secretary of Health, Education and Welfare has
determined that the publication of this periodical
is necessary in the transaction of the public business required by law of this Department. Use of
funds for printing this periodical has been approved
by the Director of the Office of Management and Budget through November 30, 1977.

NOTE

ournal names are abbreviated according to the list of abbreviations used by *Index Medicus*. For
t covered by *Index Medicus*, the abbreviations found in *Chemical Abstracts Service Source Index*,
umulative, are used. New journals are verified in *New Serial Titles* and abbreviated according
ional Standard ISO 833. An asterisk indicates the author to address (other than the primary) in
reprints.

LANGUAGE ABBREVIATIONS

Afr.	Afrikaans	Ind.	Indonesian
Ara.	Arabic	Ita.	Italian
Bul.	Bulgarian	Jpn.	Japanese
Chi.	Chinese	Kor.	Korean
Cro.	Croation	Lav.	Latvian
Cze.	Czech	Lit.	Lithuanian
Dan.	Danish	Nor.	Norwegian
Dut.	Dutch	Pol.	Polish
Eng.	English	Por.	Portuguese
Est.	Estonian	Rum.	Rumanian
Fin.	Finnish	Rus.	Russian
Fle.	Flemish	Ser.	Serbo-Croatian
Fre.	French	Slo.	Slovak
Geo.	Georgian	Spa.	Spanish
Ger.	German	Swe.	Swedish
Gre.	Greek	Tha.	Thai
Heb.	Hebrew	Tur.	Turkish
Hun.	Hungarian	Ukr.	Ukrainian
Ice.	Icelandic	Vie.	Vietnamese

ABBREVIATIONS USED IN ABSTRACTS

angstrom(s)		M	molar
adrenocorticotropic hormone		mM	millimolar
adenosine diphosphate		μM	micromolar
adenosine monophosphate		mOsm	milliosmolar
adenosine triphosphate		mEq	milliequivalents
Bacillus Calmette Guerin		min	minute(s)
twice daily		mo	month(s)
degrees centigrade		MTD	maximum tolerated dose
calorie(s)		N	normal concentration
kilocalorie(s)		NAD	nicotinamide adenine dinucleotide
cubic centimeter(s)		NADH	reduced nicotinamide adenine dinucleotide
curie(s)		NADP	nicotinamide adenine dinucleotidephosphate
millicurie(s)		NADPH	reduced nicotinamide adenine dinucleotide-
microcurie(s)			phosphate
centimeter(s)		ng	nanogram(s) (10^{-9})
central nervous system		od	once daily
counts per minute		Pa	ambient pressure
deciliter(s)		PAS	periodic acid-Schiff
milliliter(s)		pg	picogram(s) (10^{-12})
microliter(s)		pgEq	picogram equivalent
deoxyribonucleic acid		po	orally
median effective dose		ppb	parts per billion
ethylenediamine tetraacetic acid		ppm	parts per million
erythrocyte sedimentation rate		qid	four times daily
gram(s)		qod	every other day
kilogram(s)		QO_2	oxygen quotient
milligram(s)		R	roentgen(s)
microgram(s)		RBC	red blood cells (erythrocytes)
hemoglobin		RNA	ribonucleic acid
hour(s)		sc	subcutaneous
intra-arterial		sec	second(s)
intracerebral		SGOT	serum glutamic-oxalacetic transaminase
intracavitary		SGPT	serum glutamic-pyruvic transaminase
intradermal		SRBS	sheep red blood cells
increased life span		TCD	tissue culture dose
intramuscular		TCD_{50}	median tissue culture dose
intraperitoneal		tid	three times daily
intrapleural		U	unit(s)
intratumorous		mU	milliunit(s)
International Unit		UV	ultraviolet
intravenous		vol	volume
Michaelis constant		WBC	white blood cells (leukocytes)
lethal dose		wk	week(s)
median lethal dose		wt	weight
meter(s)		x	times
millimeter(s)		yr	year(s)

CONTENTS

	Cross Reference Abbreviations	Abstracts, Citations	Page
REVIEW.............................(Rev)...........3601-3667			641
CHEMICAL CARCINOGENESIS.............(Chem).........3668-3770			654
PHYSICAL CARCINOGENESIS.............(Phys).........3771-3787			671
VIRAL CARCINOGENESIS................(Viral)........3788-3884			674
IMMUNOLOGY..........................(Immun)........3885-3988			692
PATHOGENESIS........................(Path).........3989-4112			709
EPIDEMIOLOGY AND BIOMETRY...........(Epid-Biom).....4113-4147			729
MISCELLANEOUS.......................(Misc).........4148-4200			737
AUTHOR INDEX..			
SUBJECT INDEX...			
CHEMICAL ABSTRACT SERVICES REGISTRY NUMBER INDEX.......................			
WISWESSER LINE NOTATION INDEX..			

caused by aflatoxins have demonstrated abnormalities
of chromosomal fragments, stickiness, bridges, chromatid breakage, and rearrangements. The most sensitive region is the heterochromatic area of the X
chromosomes. The cytogenetic effects of aflatoxin
in human WBC and in cells from rat kidney appear to
be delay type, suggesting that aflatoxin is an alkylating agent. Studies on the induction of gene mutations have shown comparable relative frequencies of
genotypes, allelic complementation, and nonpolarized
complementation patterns induced by aflatoxin B_1 and
G_1. Especially noteworthy is a quantitative relation
of mutagenicity to interaction with DNA. While some
studies on tumor induction have suggested that aflatoxin is active *per se*, others have indicated that
metabolic activation is necessary. Tumor induction
in only metabolically active cells, species-related
teratogenicity, and definite organ specificity are
cited. Other genetically related activities observed
include induced phage production in lysogenic bacteria, inhibited DNA synthesis and inhibition of
mitosis. The mutagenic hazard of aflatoxin is discussed in relation to aflatoxin distribution and
contamination. A high cancer incidence has been
found in certain geographic areas, and several studies have indicated that aflatoxin is responsible.
Although direct evidence is still lacking, the induction of chromosomal aberrations in cultured human
cells and the conversion of aflatoxin to mutagenic
metabolites by human liver microsomes suggest a
potential mutagenic hazard to man. (108 references)

3603   MORE ON THE AFLATOXIN-HEPATOMA STORY.
        (Eng.) Anonymous *Br. Med. J.* 2(5972):
647-648; 1975.

The role of aflatoxins and hepatitis B antigen in
the etiology of primary hepatocellular carcinoma
is reviewed. The geographical and cultural distribution of primary hepatoma in Africa suggests that
a toxic environmental factor may be implicated in
its pathogenesis. Surveys in Africa and Thailand,
where primary hepatoma is common, have shown that
up to 65% of groundnut (the staple diet of rural
Africans) and 30% of the rice sampled may be contaminated with aflatoxin. Three separate surveys
indicated that the greater the quantity of aflatoxin consumed, the greater the observed incidence
of primary hepatoma. When data from these three
studies were pooled, there was a definite correlation between the risk of developing primary hepatocellular carcinoma and the degree to which food-
stuffs were contaminated with aflatoxin. Other
studies have revealed a high frequency of hepatitis
B antigen both in the general population and as a
cause of cirrhosis in areas such as Africa, where
hepatoma is common. It is possible that it is infection with hepatitis B antigen *in utero* or in early
infancy that leads to hepatoma. In three Japanese
families an association was uncovered between chronic
presence of hepatitis B antigen acquired by transmission from mother to infant and primary hepatoma,
usually associated with cirrhosis. A high proportion of hepatitis B antigen-positive patients with
hepatoma have been found to have cirrhosis, leading
to the suggestion that the neoplasm arises as a
late manifestation of cirrhosis after hepatitis

due to hepatitis B antigen. It has also been suggested that hepatitis B antigen may act synergistically with aflatoxins to produce liver cancer. (28 references)

3604   MYCOTOXINS (Eng.) Austwick, P. K. C.
(Nuffield Inst. Comparative Medicine,
The Zoological Society of London, London, England).
Br. Med. Bull. 31(3):222-229; 1975.

An assessment of mycotoxins as environmental chemicals is presented. The chemical nature and representative structure of the myctotoxins are noted; many are found to be well-known fungal metabolites. A discussion of the production of mycotoxins in the environment notes their origin from mostly pathogen or plant-decomposing fungi species, which in turn are dependent upon relative humidity, temperatures, general climate, the conditions of crop storage after harvest, and the geographical distribution of toxigenic strains of the fungi. The primary route of mycotoxin absorption is the consumption of an affected crop, while a secondary route is via consumption of the flesh or other products of an affected animal. Although the diagnosis of mycotoxicosis is dependent upon the elimination of other disease possibilities and the demonstration of the presence of an effective level of mycotoxin, the effects of mycotoxins in animals are well documented. Non-specific symptoms of fungal hepatotoxicity are noted, and histopathological changes show distinct differences between the individual mycotoxins. Mycotoxin predilection for the kidney is also observed, rare direct neurotoxicity is noted, and few resultant outbreaks of skin lesions are found. Alimentary mycotoxicoses are characterized by ulceration and hemorrhage of the mucosa, the existence of pulmonary mycotoxicosis is suggested, and genitotoxins are demonstrated in pigs. Evidence of teratogenic change by fungal metabolites is accumulating, and especially cites cytochalasin B and aflatoxin B₁; there is also a positive correlation between aflatoxin consumption and high primary liver cancer rates. Furthermore, fungi are linked indirectly with toxicoses through their ability to induce plants to form systemic or local substances protective against invasion of the tissues. Rapid and accurate quantitative estimates of mycotoxin presence are available and their levels of control are suggested: (1) prevention of mycotoxin formation (2) prevention of mycotoxin consumption, and (3) prevention of action on the largest tissue via "protective feeding". (91 references)

3605   POLYUNSATURATED FATTY ACIDS AND COLON CANCER. (Eng.) Heyden, S. (Duke Medical Center, Duke Univ., Durham, N.C. 27710). Nutr. Metab. 17(6):321-328; 1974.

The possible co-carcinogenic effects of a highly polyunsaturated fatty acid diet are reviewed and disputed. An eight-year double-blind trial shows significantly more cancer deaths in the experimental group (31) than in the control group (17 cancer deaths), and implies a co-carcinogenic effect of the diet high in polyunsaturated fat. However, the revelation that only 19 of the 31 cancer victims in the experimental group had actually adhered to the prescribed diet negates that statistical difference, and is consistent with the hypothesis that the cholesterol-lowering diets do not influence cancer risk. Other studies also note no differences in the occurrence of colon-rectum cancer between controls and experimental persons, either in the dietary phase or post-diet phase. A retrospective study of 90 cases of colon cancer notes lower-than-expected initial levels of blood-cholesterol in the colon cancer patient; however, rather than being especially significant, the lowered cholesterol levels are viewed as obvious manifestations of a profound systemic derangement of the host metabolism. A literature review reveals no difference in dietary fat intake between colon cancer patients and controls, while an additional study finds no differences in the consumption of vegetables vs animal fat. The evidence presented thus exculpates polyunsaturated fatty acid diets from causing cancer, but still fails to interpret the cholesterol-colon cancer data. While there is abundant support of the hypothesis that the decreased intake of fiber and consequently decreased bowel-transit time is markedly associated with the high incidence of colonic cancer in the Western countries, the implication of polyunsaturated fatty acid diets is not supported by clinical experience, epidemiological research, or experimental studies. (9 references)

3606   DIETARY FAT IN RELATION TO TUMORIGENESIS.
(Eng.) Carroll, K. K. (Dept. Biochemistry, Univ. Western Ontario, London, Ontario, N6A 3K7, Canada); Khor, H. T. Prog. Biochem. Pharmacol. 10:308-353; 1975.

The effects of dietary fat on tumorigenesis in animals, and epidemiologic data on dietary fat intake in relation to cancer mortality in humans are reviewed. Studies with experimental animals show an enhanced incidence of induced skin tumors in mice fed high fat diets; various sources of fats are utilized, and attempts to determine the effective fat components are noted. It is found that a high fat diet is generally more effective when fed after exposure to the carcinogen. Likewise, dietary fat also increases the incidence of spontaneous mammary tumors in DBA and C3H mice and in rats. While such enhancement is related to the amount of dietary fat, unsaturated fats have a greater effect than saturated fats. Similar results were found in studies on spontaneous mice hepatomas, aminoazo-dye-induced hepatomas, ¹³¹I-induced hypophyseal tumors, and spontaneous intracranial tumors. Mouse tumors not enhanced by dietary fat include skin sarcomas and spontaneous and induced leukemia. A consideration of the possible mode of action of dietary fat suggests enhanced tumor incidence via facilitated carcinogen absorption, and/or augmented tumor induction via metabolic stimulation of potential tumor cells. Despite numerous complicating and contradictory factors, epidemiological data on humans indicate a selectively strong positive correlation between dietary fat intake and mortality from certain kinds of cancer, such as breast cancer, prostatic cancer, and cancer of the colon. However, other existing

section, exenteration, local excision, and/or cancericidal radiation doses. The risk of carcinoma in the exposed population is evaluated, and screening of asymptomatic exposed girls by the age of 14 yrs is suggested. Screening examinations thus far reveal nonneoplastic vaginal abnormalities in 66%, and cervical abnormalities in 95% of DES-exposed subjects. Detection and management of vaginal adenosis and cervical erosion are discussed. It is established that prenatal exposure to DES and similar nonsteroidal estrogens results in frequent extensive cervical erosion and vaginal adenosis, but only rare development of clear-cell adenocarcinoma of the vagina and cervix. (41 references)

3609   RAUWOLFIA DERIVATIVES AND BREAST CANCER.
        (Eng.) O'Fallon, W. M. (Mayo Clinic, Rochester, Minn. 55901); Larbarthe, D. R. *Lancet* 2(7938):773; 1975.

An amendment to a previous report on the use of rauwolfia derivatives and the occurrence of breast cancer is presented. The rate of use of rauwolfia derivatives did not greatly change during the late 1960's and early 1970's. Of the whole cohort of hypertensive cases with breast cancer, 16.8% and 16.7% had been exposed to rauwolfia derivatives during the respective time periods. However, it is belatedly noted that while 29.4% of those diagnosed before 1971 had used other hypertensive agents, 35.3% of the whole cohort, i.e. until 1973, had such use. In contrasting hypertensive cases and controls diagnosed in the same time period, it was discovered that the breast cancer patients had not used thiazides, triamterene, or methyldopa more frequently than the controls. The general conclusion of the prior report, i.e., the lack of association between the use of reserpine and related drugs and breast cancer, is unaffected. (No references)

3610   AIR POLLUTION, SMOKING AND LUNG CANCER.
        (Eng.) Butler, J. D. (Univ. Aston, Birmingham B4 7ET, England). *Chem. Br.* 11(10):358-363; 1975.

The possible synergistic influences of cigarette smoking, air pollution, heavy metals, and carcinogenic polycyclic hydrocarbons on the development of lung cancer are discussed. In demonstrating the carcinogenic properties of many polynuclear hydrocarbons, a carcinogenic mechanism implicating hydroxylation of a methyl group is suggested. While atmospheric concentrations of benzo(a)pyrene are regarded as a measure of the total polynuclear hydrocarbons present, coronene concentration is considered a useful index of air pollution caused by motor traffic. The majority of such polynuclear hydrocarbons are found derived from the partial combustion of organic matter, *via* pyrolysis and subsequent rearrangement of the free radicals thus formed. Studies of the occurrence of polycyclic hydrocarbons note their proliferation in urban areas and their origination from heating oil and coal burning, refuse incineration, and combustion of propulsion fuel. British government records of atmospheric smoke and sulfur dioxide levels indicate that, provided annual mean smoke concentrations are kept low, an average annual limit of 75-100 µg/M^3 of sulfur dioxide is acceptable. While only limited information on the distribution of polycyclic compounds near known sources is noted, seasonal fluctuations are found. A discussion of the isolation of such polycyclic hydrocarbons notes the required separation of particulate from adsorbate, followed by identification of individual compounds via alumina chromatographic columns, paper chromatography with fluorescence emission, spectrophotometry, or thin layer and gas chromatography. A survey of mortality and smoking habits reveals a linear relationship between average number of cigarettes smoked per day and the annual death rate, but notes a rapid reduction of the death rate following curtailment of cigarette smoking. While incompatible with the supposed 17 yr cancer development process, it suggests a co-carcinogenic effect of polycyclic hydrocarbons and other environmental factors. A multiple regression analysis correlating lung cancer mortality, cigarette smoking, solid fuel consumption, and benzo[a]pyrene concentrations is described. While acknowledging the possible synergistic influence of heavy metals, it is noted that the air pollution factor may be less significant, except for the premise that cigarette smoking is largely responsible for lung cancer. (24 references)

3611   THYROID IRRADIATION AND CARCINOGENESIS:
        REVIEW WITH ASSESSMENT OF CLINICAL IMPLI-
CATIONS. (Eng.) Foster, R. S., Jr. (Univ. Vermont, Coll. Medicine, Burlington, Vt. 05401). *Am. J. Surg.* 130(5):608-611; 1975.

Data relative to carcinogenesis from thyroid irradiation are reviewed. Investigations of the clinical significance of the development of thyroid carcinoma have revealed a 75-80% incidence as a result of previous irradiation. Development of thyroid carcinoma after upper thoracic irradiation in infants is reported most commonly. Thyroid carcinoma is also related to irradiation for various benign conditions, and has a mean latent period of 20 yr. Both external irradiation and radioactive iodine increase the frequency of thyroid adenomas and thyroid carcinomas in rats. Goitrogen administration and partial thyroidectomy also increase the incidence of adenoma and carcinoma after irradiation. The dose-response relationship of irradiation induction of thyroid carcinoma is roughly bell-shaped; lower doses are less likely to induce malignancy. The risk of thyroid neoplasia increases linearly over the dose range of 20-1,125 rads to the thyroid. Accidental exposure to fallout containing radioactive iodine (150 rads) also results in a significant increase of benign and malignant thyroid neoplasms. Furthermore, a low incidence of neoplasia is anticipated due to the higher therapeutic dose (1,000-20,000 rads) of ^{131}I used for treatment of diffuse toxic goiter. However, surgical resection for nontoxic goiter has a much greater immediate risk than ^{131}I thyroid ablation in children, adolescents, and adults with Grave's disease. Whereas ^{131}I thyroid scans are noted capable of delivering a radiation dose of

potential risk (50 µCi), use of technetium 99m is advocated. (36 references)

3612   THE CONCEPTS OF CRITICAL ORGAN AND
        RADIATION DOSE AS APPLIED TO PLUTONIUM.
(Eng.) Stannard, J. N. (Univ. Rochester Sch. Medicine and Dentistry, Rochester, N. Y. 14642). *Health Phys.* 29(4):539-550; 1975.

The concepts of critical organ or tissue and calculation of radiation dose are discussed, and radiation and chemical hazard evaluation procedures are compared. The derivation of classical exposure standards for plutonium, with bone as the critical organ, are not based on calculation of a maximum permissible dose rate to that organ; rather, the plutonium body burden figure is related to the radium standard in man through comparison of the relative toxicity of plutonium and radium in animals. A comparison of maximum permissible body or organ burdens for plutonium-239 and polonium-210, based on several different criteria, shows that choice of criterion for a critical organ makes a large and significant difference in the resulting plutonium standards. Changes in models and modifications of metabolic parameters result in only slightly modified permissible plutonium levels. Consideration of the organ weight or volume and the possible significance of "hot particles" in such calculations is also discussed. On the basis of highest concentration and highest calculated doses, the respiratory lymph nodes are considered the critical organ for inhalation exposure to plutonium; however, it is further suggested that the lung and respiratory lymph nodes be taken together as a critical organ, with a combined weight of 1000 g. In the transfer from occupational to general population exposure standards, recommended maximum permissible annual dose rates attempt to employ a scaling factor and do not distinguish among organs. The lifetime and annual dose commitment concept is discussed, and the relationships between the procedures of risk estimation and the critical organ convention are examined briefly. It is noted that standards derived from plutonium risk estimates are in a different setting than those derived from the classical critical organ convention and from external radiation exposures. (43 references)

3613   LOW-LEVEL IONIZING RADIATION AND MAN.
        (Eng.) Diesendorf, M. (Dept. Applied Mathematics, SGS, Australian Natl. Univ., Canberra, Australian Capital Territory, Australia). *Search* 6(8):328-334; 1975.

Evidence that low doses of ionizing radiation may be harmful to man is reviewed. Medical and dental x-rays are the main source of man-made radiation exposure in Western countries. Selective exposure of individuals is a valuable tool of modern medicine, but in the case of unselective mass irradiations the risks may be greater than the benefits for the majority of people exposed. For most people exposed to compulsory mass chest x-rays in Australia, the estimated average risk of death from radiation induced cancer is greater than that from undetected

with appropriate antiserum, unknown oncogenic RNA
viruses may be detected in certain human tumors,
although some mechanism may suppress the budding
and/or maturation of the virus. Type C virus-
induced CSA, other than those associated with MuLV,
are the sarcoma virus-induced CSA. Although murine
sarcoma virus (MSV) obtains its viral envelope from
MuLV and/or endogenous type C viruses, immunoelectron microscopy demonstrated that *in vitro* MSV-
transformed and subcloned mouse- and rat-nonproducer
cells induce a small amount of common specific CSA
different from the MuLV and its induced antigens.
MSV requires type C viral envelope from helper
viruses to replicate *in vitro* and *in vivo*. Natural
antibodies, IgG and IgM classes, to CSA and VEA were
of type C virus-induced tumors and occur naturally
in a variety of inbred strains and F_1 hybrids of
mice. Natural antibodies to these antigens were
found in the kidney antigen-antibody complexes in
mouse strains with high incidences of spontaneous
leukemia. Consequently, the presence of natural
antibodies to VEA in free form suggests that these
antibodies play a role in the suppressive mechanisms
of type C virus maturation and oncogenesis in resistant mice strains. Since, these antibodies have
been detected in several species other than the
mouse, it seems possible that an antibody occurring
in humans suppresses the maturation of possible
human type C viruses. By passive immunization with
antiserum to VEA, both prophylaxis and treatment of
virus-induced tumors may be achieved. (32 references)

3616   VACCINATION AGAINST HERPES GROUP VIRUSES.
        (Eng.) Plotkin, S. A. (Children's Hosp.
Philadelphia, 34 Civic Center Blvd., Philadelphia,
Pa. 19104). *Pediatrics* 56(4):494-496; 1975.

Problems involved in the development of herpes
virus group vaccines are summarized. A major factor
is the possible oncogenicity of human herpes viruses.
Several herpes viruses of animals have been shown
to be oncogenic in homologous or heterologous species. Although irradiated human cytomegalovirus
(CMV) was reported to transform hamster embryo
fibroblasts, no epidemiologic evidence yet impli-
cates either human CMV or varicella-zoster (V-Z)
in the causation of malignancy. Whereas almost
all herpes viruses have the capacity to maintain
the viral genome in a noninfectious state and subsequently reactivate, the persistence of herpes
viruses in latent form is a second major concern.
However, if their replication is localized, attenuated vaccines may not result in enough virus to
cause widespread infection. A balancing of the
possible ill effects of attenuated CMV or V-Z
viruses against the actual effects of the wild
viruses is suggested. The importance of cellular
immunity in protection against herpes viruses is
also discussed. Live vaccines are expected to
induce cellular and humoral immunity equal to that
derived from infection. Effective veterinary vaccines thus developed include attenuated Marek's
disease vaccine, killed *Herpes virus saimiri* vaccine,
and partly attenuated mouse CMV vaccine. Arguments
for cautious optimism in the development of herpes
virus vaccines note the resultant brain damage of
CMV infected infants, the mortality associated

with complications of varicella, and the enormous morbidity of zoster. (29 references)

3617   IDIOTYPE-POSITIVE T LYMPHOCYTES. (Eng.)
        Binz, H. (Uppsala Univ. Medical Sch.,
Box 562, 751 22 Uppsala, Sweden); Kimura, A.;
Wigzell, H. *Scand. J. Immunol.* 4(5/6):413-420;
1975.

Studies suggesting that B and T lymphocytes may use identical genetic material for the buildup of their respective antigen-binding sites on the receptor molecules are reviewed. Several sets of observations have indicated that B and T lymphocytes reactive against the same antigenic determinants express shared idiotypic determinants on their respective antigen-binding receptors. Two systems especially discussed are in the rat and the mouse, using the major histocompatibility Ag-B and the group A streptococcal carbohydrate antigenic determinants as the immunogen, respectively. Data indicate that the antigen-binding receptors used by T and B lymphocytes use similar if not identical genetic material in the construction of their respective antigen-binding areas. The favored interpretation is that of shared idiotypes. Anti-idiotypic antibodies have been used to directly visualize and enumerate idiotype-positive lymphocytes, *via* fluorescent antibody techniques, autoradiography, or cytofluorimetry. The idiotypic-positive T lymphocytes have thus been shown to constitute a clearly distinct group of cells. In addition, separation data from affinity chromatography have demonstrated the existence of physically separable normal T lymphocytes with largely non-overlapping immune reactivity against Ag-B antigens. Preliminary findings suggest that while the B-cell product is a two-strand chain structure, the released T-cell receptor behaves like a single chain under reducing-alkylating conditions. The released receptor molecule has a size of around 35,000 daltons, and probably involves additional genetic material to that of the heavy-chain V genes. (41 references)

3618   THE RELATIONSHIP BETWEEN CUTANEOUS AND
        VISCERAL CARCINOMAS [editorial]. (Eng.)
Maize, J. C. (State Univ. New York at Buffalo,
Sch. Medicine, Buffalo, N. Y.). *J.A.M.A.* 233(9):
986; 1975.

The relationship between cutaneous carcinoma and visceral malignant neoplasms has long been a subject of controversy. Researchers at the Finsen Institute in Denmark have recently found no appreciable deviation from the expected incidence of visceral neoplasms in patients with either Bowen disease or multiple basal cell carcinomas. Their conclusion in the case of Bowen disease differs considerably from that of U.S. authors. In more than one quarter of 155 patients with Bowen disease, a primary extracutaneous malignant neoplasm developed an average of five to six years after onset of the cutaneous tumor. Treatment of the Bowen disease had no effect on the subsequent development of systemic malignant neoplasm. An incidence of internal carcinoma of 15% to 22.6% in patients

with Bowen disease has been reported by others. The yield of internal neoplasms generated by a thorough evaluation of patients with basal cell or squamous cell carcinomas of the skin is not appreciably higher than in the uninvolved population of similar age. It would seem advisable to assess and observe patients with Bowen disease carefully, for the development of internal neoplasms.

3619   IMMUNITY IN CANCER. (Eng.) Seibert,
        F. B. (Res. Lab., Veterans Administration
Center, Bay Pines, Fla. 33504). *Science* 190(4216):
809; 1975.

The reduction in the occurrence of mammary tumors in C3H/He mice given injections of poly(A)·poly(U), living LDH-virus, or heat-killed C3H ♀ 27 Brtu bacterium vaccine is briefly reviewed. In all three sets of experiments, a 20% increment of survival was effected. A subsequent 90-100% tumor incidence in treated mice, after 380 days, suggested a loss of cellular immunity and/or delay of spontaneous tumor development. The advantages of a sterile autogenous or homologous vaccine are noted, and the problem of live or dead bacterial contamination is acknowledged. (5 references)

3620   ULTRASTRUCTURE OF MAMMALIAN CHROMOSOME
        ABERRATIONS. (Eng.) Brinkley, B. R.
(Univ. Texas Medical Branch, Galveston, Tex.);
Hittelman, W. N. *Int. Rev. Cytol.* 42:49-101; 1975.

The spectrum of chromosome damage produced by various natural and synthetic clastogens as evaluated by both light and electron microscopy is described. Various aspects of detectable chromosome damage are compared. These include: chromosome-type aberrations, chromatid-type aberrations, gaps, exchanges, subchromatid aberrations, chromosome stickiness, and damage to specialized chromosome regions. Ultrastructural studies generally confirm light microscope analyses. There is some controversy between the two techniques, over aberrations scored as breaks *vs* gaps. In chromatid breaks viewed by the electron microscope, the broken region is characterized by an absence of interconnecting fibrils between the two broken parts. Ultrastructural studies suggest that such breaks may be either aligned or unaligned. Chromatid gaps appear composed of several size classes of chromatin fibrils, as indicated by ultrastructural analysis on isolated chromosomes and ultrathin sections. Exchanges effected by chromatid breaks, and subchromatid aberrations such as side-arm bridges, are also more clearly defined by electron optics. In both side-arm bridges and chromosome stickiness, thin "subchromatid" fibrils are observed between daughter chromatids at metaphase, anaphase, and telophase. While somewhat ineffective as a tool for evaluating metaphase chromosome strandedness in eukaryotic cells, transmission electron microscopy reveals damage to specialized regions such as the kinetochore, and nucleolar organizer. Analysis of the chromosome target of ionizing radiations,

a more advanced age is consequently due to an epithelial cell transformation by the strongly accumulated carcinogens at an age when the dilution effect resulting from cell division can no longer stem the cell transformation and dedifferentiation.

3623   PREDISPOSING FACTORS IN CARCINOMA OF THE COLON. (Eng.) Sherlock, P. (Mem. Sloan-Kettering Cancer Cent., New York, N.Y.); Lipkin, M.; Winawer, S. J. *Adv. Int. Med.* 20:121-150; 1975.

The predisposing factors in carcinoma of the colon including environmental, genetic, immunologic, and cellular aspects are reviewed and the management of premalignant clinical conditions is discussed. Colorectal carcinoma is a disease of "advancing civilization," accounting for 15% of all malignant neoplasms and the second cause of death from cancer in the U. S. Geographically, the incidence of colonic cancer is high in north and west Europe, New Zealand, and North America and low in South America, Africa, and Asia. Epidemiologic investigations suggest that diet and bacteria produce metabolic products that can act as carcinogens or co-carcinogens within the lumen of the colon; i.e., feces from populations with a high colonic cancer incidence contain significantly more *Bacteroides* and *Clostridium* species and fewer streptococci and lactobacilli. High incidence of colonic cancer is found in individuals consuming diets rich in fat and animal protein and in whom fecal matter contains a high concentration of neutral and acid sterols and bile acid derivatives. Experiments with animal models suggest that specific changes occur during neoplastic transformation of colonic cells of man. These changes develop in areas of cell metabolism that increase DNA synthesis within the cells and enhance the cells' ability to proliferate. In addition, immunologic investigations have led to the development of procedures involving the detection of tumor antigens. Although cancer of the colon and rectum are largely influenced by environmental factors, the genetic susceptibility is considered and may be helpful in assessing mucosa at risk. Several genetic syndromes with a predisposition to malignant change are transmitted by a single gene; however, there are some familial aggregations of colonic cancer that do not fit the predicted mendelian ratio. Thus far, carcinoma of the colon appears to have a genetic basis that resists analytic efforts to prove a single-gene difference, except for the familial polyposis syndrome. Continued exposure to carcinogenic influences (as in chronic inflammations) may be responsible for the development of malignancy in ulcerative colitis; definite criteria have been established for assessing the potential malignancy of the disease. A controversy continues to surround the premalignant character of adenomatous polyps; arguments for and against adenomatous polyps as precursors of cancer of the colon and rectum are presented. Although sigmoidoscopy, biopsy, and barium enema with air contrast remain important diagnostic techniques, fiberoptic colonoscopy better defines suspicious areas and permits the removal of many of the lesions without major surgery. The authors conclude that

a technique employing biochemical or immunologic
measurements to indicate malignancy potential of
colonic lesions would be most useful for the prevention and management of cases of colonic cancer.
(96 references)

**3624   NON-MYOGENIC TUMORS INVOLVING SKELETAL
MUSCLE: A SURVEY WITH SPECIAL REFERENCE
TO ALVEOLAR SOFT PART SARCOMA.** (Eng.) Delaney,
W. E. (St. Vincent's Hosp. and Medical Center of
New York, New York, N. Y. 10011). *Ann. Clin. Lab.
Sci.* 5(4):236-241; 1975.

The distinctive non-myogenic tumor which occurs
exclusively in skeletal muscle, alveolar soft part
sarcoma, is the subject of this review. This tumor
has ultrastructural features of a secretory lesion
with distinctive membrane-bound crystalloids which
may be lipid in nature. The ultrastructural features
of alveolar soft part sarcoma do not support a myo-
genous derivation. This clinically indolent, malignant tumor, occurring exclusively in skeletal muscle,
probably originates in the paraganglia. (14 references)

**3625   MORPHOLOGY AND CLASSIFICATION OF MALIGNANT
LYMPHOMAS AND SO-CALLED RETICULOSES.**
(Eng.) Lennert, K. (Dept. Pathology, Univ. Kiel,
Postfach 4324, D-2300 Kiel, Germany). *Acta Neuro-
pathol. [Suppl.] (Berl.)* 6:1-16; 1975.

An introductory survey of the classification and
morphology of Hodgkin and nonHodgkin malignant
lymphomas and reticuloses is presented. Such a
reclassification involves the application of immunochemical, immunomorphological, histological, histochemical, cytological, and ultrastructural techniques.
Hodgkin's disease is classified into four categories:
(1) lymphocytic predominance, (2) nodular sclerosing,
(3) mixed, and (4) lymphocytic depletion. Three
types of reticuloses originally distinguished as
small, medium-sized, and large cell variants are
now interpreted as hairy cell leukemia, myeloid
leukemia, and variant of immunoblastic sarcoma,
respectively. A simple scheme demonstrating the
basic cytological correlations of thymus-derived
(T-) and bone marrow-derived (B-) lymphocytes is
discussed in detail. Several classifications of
malignant lymphomas are compared, and the essential
points of the Kiel classification are noted. This
classification distinguishes two main groups of
malignant lymphomas, omits the use of the words
"lymphosarcoma" and "leukemia", considers the so-
called reticulosarcomas as malignant lymphoma, immunoblastic, and substitutes the terms centroblast
and centrocyte for germinoblast and germinocyte,
respectively. The low-grade malignant lymphomas,
suffixed "-cytic", include lymphocytic lymphomas of
the subtypes chronic lymphoid leukemia, hairy cell
leukemia, and mycosis fungoides, lymphoplasmacytoid
or immunocytic lymphoma, centrocytic lymphoma, and
centroblastic/centrocytic lymphoma. The high-grade
malignancies, suffixed "-blastic", include centro-
blastic malignant lymphoma, lymphoblastic lymphoma,
including Burkitt and convoluted cell types, and
immunoblastic lymphoma. In addition, true reticulo-

sarcoma is noted to exist. A correlation of the
different non-Hodgkin lymphomas with the cytological
scheme reveals that the T-cell series includes
lymphoblastic lymphoma of the convoluted type and
Sezary syndrome, while most non-Hodgkin lymphomas
are of B-cell origin. (49 references)

**3626   THE ROLE OF THE SPLEEN IN TUMORS AND
LEUKEMIAS (LITERATURE REVIEW).** (Rus.)
Zverkova, A. S. (Kiev Scientific Res. Inst. Hemato-
logy and Blood Transfusion, Kiev, USSR). *Vrach.
Delo* (7):80-83; 1975.

Observations on the role of the spleen in malignant
processes are reviewed. A very low incidence of primary tumors and metastases is characteristic for the
spleen as compared with other inner organs. Splenectomy was found to lead to the loss of immunity to
tumors in certain tumor-resistant animal species, and
splenectomy in humans and animals was found to
accelerate and aggravate the development and course
of malignant tumors and leukemia, and to provoke the
transformation of one tumor type in another, more
malignant one, e.g., of chronic reticulosis into
reticulosarcomatosis. The transplantability of
certain leukemia cell lines was found to increase
considerably following splenectomy in animals. The
implantation iv or ip injection of normal or immunized spleen cells or cell extracts has been demonstrated to inhibit the tumor growth and to prolong
the survival in humans and animals. The literature
data indicate the important immunological and antineoplastic role of the spleen, and the impairment
of certain immunological and antitumor functions of
the spleen in leukemia (reduced antibody formation
and phagocytic activity of the spleen). The
mechanism of the antineoplastic action of the normal
spleen may be due to the living cell or to a chemical
agent. (52 references)

3627   RECENT PERSPECTIVES IN OCCUPATIONAL CANCER.
(Eng.) Selikoff, I. J. (Mount Sinai Sch.
Medicine, Fifth Ave. and 100 St., New York, N.Y.
10029). *Ambio* 4(1):14-17; 1975.

Recent developments in the study of occupational
cancer are reviewed, including the identification
of previously unsuspected carcinogenic agents and
the evaluation of materials used in industry and/or
agriculture. A considerable amount of time (usually
20-35 yr) must elapse between exposure to an occupational carcinogen and the onset of disease. Therefore, epidemiologists must depend on other disci-
plines to identify carcinogenic substances that have
been introduced within the last 50 yr. Other investigations indicate the possibility of the interaction
of two or more agents. Asbestos workers with a
history of cigarette smoking had 92 times the risk
of death from lung cancer compared with individuals
of the same age who neither worked with asbestos
or smoked. Substances that may have no carcinogenic effect by themselves may, in concert with
others, have an additive or multiplicative effect
on malignancy and other substances, deemed innocuous
in one environment, may be carcinogenic in a new
environment. Two neoplasms, hemangiosarcoma of the

genic sarcoma during adolescence are correlated with skeletal growth patterns, pubertal growth patterns, and skeletal stress. Whereas the role of genetic susceptibility and familial susceptibility mechanisms is suggested, regular follow up of persons with these and other predisposing conditions is advised. (49 references)

3629   CUTTING OUT CANCERS. (Eng.) Swaffield,
        L. (No affiliation given). *Nature*
258(5531):94-95; 1975.

The international standards of protection of workers against carcinogens, as constructed at the International Labour Conference (ILC), are reviewed and evaluated. Major inadequacies arise from the preoccupation with finding a dramatic breakthrough cure, the great latitude allowed individual nations, and the economic realities. Two case histories illustrate the extremes of negligent and diligent industrial concern: production of an anti-oxidant containing 2-naphthylamine (Nonox S), and vinyl chloride monomer production, respectively. Whereas attempts to trace and control all carcinogens appear economically unfeasible, the individual nations are to decide the acceptable social and economic risk. The ILO convention requires replacement or control of carcinogens and a system of records kept. Further recommendations include the promotion of epidemiological studies and production of educational guides. However, international cooperation on occupational health research is acknowledged to be very poor. (no references)

3630   INCIDENCE OF NEOPLASIAS OF THE BLADDER
        IN THE TWO SEXES. (A SURVEY CONDUCTED
AMONG UROLOGISTS IN SPAIN). (Spa.) Cifuentes
Delatte, L. (No affiliation given). *An. R. Acad.
Nacl. Med. (Madr.)* 92(3):483-499; 1975.

A review of 17,416 cases of bladder neoplasms in Spaniards between 1950-1974, revealed a male:female incidence ratio of 7.24:1. Ranging from a high of 14.1:1 in the Canary Islands to a low of 5.25:1 in the Basque Provinces. This ratio is higher than other comparative studies published in England, Japan, Sweden and the U.S.A. These results ranged between (2.51:1 to 4.74:1). The high incidence of bladder neoplasms in Spanish males is attributed to: 1) increased use of tobacco among men, and 2) greater exposure to industrial toxins. i.e. aniline and alpha- and beta-naphthylamine. Other compounds associated with bladder neoplasms were: food dyes, coffee contaminated with 3,4-benzpyrene, phenacetin, cyclamates, and schistosomiasis. A panel discussion of Spanish urologists urged: 1) greater study of industrial compounds (e.g. plastics) as possible carcinogens. 2) a campaign against the use of tobacco. (no references)

3631   *IN SITU* CYTOLOGIC NUCLEIC ACID HYBRIDI-
        ZATION AND ITS RELATION TO CANCER. (Eng.)
McDougall, J. K. (Medical Sch., Univ. Birmingham,
Birmingham B15 2TJ, England). *Proc. Int. Cancer
Congr. 11th.* Vol. 1 *(Cell Biology and Tumor Immun-*

ology). Florence, Italy, October 20-26, 1974. Edited by Bucalossi, P.; Veronesi, U.; Cascinelli, N. New York, American Elsevier, 1975, pp. 100-102.

The use of molecular hybridization methods and autoradiographic detection in the study of satellite and viral DNA sequences is described. The techniques employ purified nucleic acids from cells or viruses in the examination of sequence complexity by renaturation in solution or detection of complementary sequences immobilized on membranes. Satellite DNA sequences are localized on the chromosomes of many species, ribosomal cistrons are mapped in man, and globin messenger RNA is also localized by *in situ* hybridization with *in vitro* transcribed complementary DNA. Despite the confusion of polymorphisms, the persistence of unique sequences in tumor cells representing viral genes has been demonstrated. *In situ* hybridization studies have detected virus DNA in virus-transformed and tumor cells, and in situations where the virus may exist in a latent infection state. Membrane hybridization has revealed Epstein-Barr virus DNA in nasopharyngeal carcinoma, while *in situ* hybridization has shown that the viral genomes are present in the epithelial cells of the tumor. An association of Epstein-Barr virus genomes with host cell chromosomes has also been indicated. The *in situ* method showed the presence of herpesvirus DNA in four specimens from 11 biopsies of cervical carcinoma specimens. The hybridization of adenovirus cRNA to virus-infected or transformed rodent interphase and metaphase cells also indicates a random association of viral DNA with the chromosomes. Thus it is noted that *in situ* hybridization studies can demonstrate the localization of viral genes in tumor cells. (30 references)

3632   .LIPOPROTEINS IN RELATION TO CANCER.   (Eng.)
          Barclay, M. (Memorial Sloan-Kettering Cancer Center, New York, N.Y. 10021); Skipski, V. P. *Prog. Biochem. Pharmacol.* 10:76-111; 1975.

The measurement, quantitation, and patterns of serum lipoprotein levels in normal humans and cancer patients is reviewed. While there is presently no evidence that the lipoprotein fractions from various groups of normal fasting subjects have remarkably different amounts or kinds of lipids or proteins, lipoprotein levels are influenced by a variety of factors in health and disease. Therefore, the careful selection and description of "normal" subjects and cancer patients is discussed. Normal (fasting) women have values for total very low-density lipoproteins (VLDL) of 7-12 mg/100 ml serum, with very low standard deviations. However, normal men with a positive family history of cancer have elevated values for total VLDL (73-296 mg/100 ml serum) and more variability ($\pm$ 91); likewise, all cancer patients report significantly higher levels of total VLDL, resulting primarily from the components with S_f20-100 and S_f0-20. Of the three classes of low-density lipoproteins (LDL) recognized, ultracentrifugal techniques show low and similar values (0.84 mg/100 ml serum) for the S_f10-20 class in normal subjects, very high and variable S_f10-20 values (0-433 mg/100 ml serum) in women with breast cancer, and high S_f0-3 values (15-27 mg/100 ml serum) in

men with various cancers. Such LDLs are seen less sensitive to the diet than VLDLs, and rarely elevated in children with various kinds of cancer. Use of ultracentrifugation, electrophoretic, and precipitation techniques show lower average levels of high-density lipoproteins (HDL_2 and HDL_3) in both normal women and men with pronounced family histories of cancer (26-86 mg/100 ml serum and 16-46 mg/100 ml serum, resp.). While normal HDL_2 values range from 93-186 and 70-122 mg/100 ml serum resp., unusually low HDL_2 values of 15-54 mg/100 ml serum are found in men with various cancers. Lowered α-lipoprotein (HDL) values are also reported with the growth of pregnancy, but are assumed to be related to cancerous growth. Patients with gynecological malignancies also have higher average serum concentrations of free fatty acids than normal subjects, i.e. 20.4 *vs* 13.6 mg/100 ml serum, with especially notable decreases in linoleic acid and increases in stearic and palmitic acid levels. Studies on animals with experimental tumors report similar observations, but qualitative and quantitative differences in VLDL, LDL, HDL_2 and HDL_3 levels than observed in human subjects. It is noted that VLDL elevates concurrently with significant decreases in HDL_2 and that unusually high HDL_1 values concur with significantly lowered values for other HDLs. The patterns occur especially and strikingly in male cancer patients, but are also observed in women with cancer. (66 references)

3633   ISOENZYMES IN CANCER. (Rus.) Komov,
          V. P. (Inst. Chem. Pharmaceut., Leningrad, USSR). *Vopr. Onkol.* 21(4):80-87; 1975.

Studies on changes of different isoenzyme spectra in connection with benign and malignant tumors in the human or animal organism are reviewed. Malignant transformation was found to be accompanied by an increase in the level of muscular and hepatic forms of lactate dehydrogenase (LDH) isoenzymes. Reduction of the H-form of LDH in leukocytes was observed in human leukemias. Considerable changes were found in the isoenzyme spectra and catalytic functions of aldolase, hexokinase, catalase, alkaline phosphatase, DNA-polymerase, uridine cytidyl kinase, serine transoxymethylase, desoxymethydyl kinase, pyruvate kinase, and other enzymes in the presence of malignant process. The presence of hexokinase in the serum of cancer patients is believed to be due to the increased permeability of the cell membrane. The decreased catalase activity in the liver of tumor-bearing animals was found to return to normal after tumor extirpation. One of the four catalase isoenzymes was fully inhibited in patients with lung cancer, but qualitative normalization of the isoenzyme spectrum was observed 22 days after radical surgery. Two catalase isoenzymes were inhibited in patients with gastric carcinoma. The studies indicate the inhibition of one or more isoenzymes or the appearance of additional isoenzymes without any qualitative change in the isoenzyme spectra in the presence of malignant process. The different changes in the isoenzyme spectra are very likely to be dependent on the nature of the enzyme, the mode of the synthesis of the isoenzymes, of the tissue specificity, and sometimes even of the localization of the tumor. Changes in the quantitative

3635   EXPERIMENTAL GASTRIC CANCER IN THE DOG AND
        OTHER EXPERIMENTAL ANIMALS. A SURVEY OF
THE PRESENT STATUS. (Eng.) Nagayo, T. (Aichi Cancer
Center, Res. Inst., Nagoya, Japan). *Proc. Int. Cancer
Congr. 11th.* Vol. 6 *(Tumors of Specific Sites)*.
Florence, Italy, October 20-26, 1974. Edited by
Bucalossi, P.; Veronesi, U.; Cascinelli, N. New
York, American Elsevier, 1975, pp. 215-219. (16
references)

3636   HEPATOMA ASSOCIATED WITH ANDROGENIC
        STEROIDS. (Eng.) Anthony, P. P. (Mid-
dlesex Hosp., London W1N 8AA, England). *Lancet*
1(7908):685-686; 1975. (12 references)

3637   STEROIDAL CONTRACEPTIVE THERAPY: HEPATOMAS
        AND VITAMIN A. (Eng.) Gal, I. (Dept.
Biological Sciences, Hatfield Polytechnic, Hatfield,
Herts, England). *Lancet* 1(7908):684; 1975. (14
references)

3638   PROLACTIN AND BREAST CANCER. (Fre.)
        Cappelaere, P. (Institut de Recherches
sur le Cancer de Lille, Boite Postale 3569, 59020
Lille Cedex, France). *Pathol. Biol. (Paris)* 23(2):
161-170; 1975. (79 references)

3639   CHEMICAL CARCINOGENESIS AND THE PANCREAS.
        (Eng.) Bates, R. R. (Natl. Cancer Inst.,
Fort Detrick, Frederick, Md.). *J. Surg. Oncol.*
7(2):143-149; 1975. (45 references)

3640   CHEMICALS IN THE INDUCTION OF RESPIRATORY
        TRACT TUMORS. (Eng.) Fraumeni, J. F., Jr.
(Natl. Cancer Inst., Bethesda, Md.). *Proc. Int.
Cancer Congr. 11th.* Vol. 3 *(Cancer Epidemiology,
Environmental Factors)*. Florence, Italy, October
20-26, 1974. Edited by Bucalossi, P.; Veronesi, U.;
Cascinelli, N. New York, American Elsevier, 1975,
pp. 327-335. (74 references)

3641   RADIOLOGICAL INVESTIGATIONS AND THERA-
        PEUTIC ASPECTS OF CHEMICALLY INDUCED
TUMORS. (Ger.) Burkle, G. (Medizinisches
Strahleninstitut der Universitat Tubingen, 7400
Tubingen, Rontgenweg 11, West Germany). *Fortschr.
Geb. Roentgenstr. Nuklearmed.* 122(4):352-364; 1975.
(82 references)

3642   REVIEW OF THE ENVIRONMENTAL FATE OF
        SELECTED CHEMICALS. (Eng.) Radding, S. B.
(Stanford Res. Inst., Menlo Park, Calif.); Holt,
B. R.; Jones, J. L.; Liu, D. H.; Mill, T. 44 pp.,
1975. [available through National Technical Information Services, Washington D.C. Document No. PB-
238 908/8WP]

3643   ASBESTOS CARCINOGENESIS [abstract]. (Eng.)
         Wagner, J. C. (Medical Res. Council,
Penarth, England). *Br. J. Cancer* 32(2):258-259;
1975. (No references)

3644   THE TOXICITY OF PLUTONIUM. (Eng.) Anony-
         mous. *Medical Res. Council Report.* Lon-
don, HMSO, 1975, 45 pp.

3645   PLUTONIUM PARTICLES: SOME LIKE THEM HOT.
         (Eng.) Lovins, A. B. (Friends of the
Earth); Patterson, W. C. *Nature* 254(5498): 278-280;
1975. (No references)

3646   TUMOR VIRUSES (A BIBLIOGRAPHY WITH AB-
         STRACTS). (Eng.) Crockett, P. W. (Natl.
Technical Information Service, Springfield, Va.).
75 pp., 1975. [available through National Technical
Information Services, Washington D. C. Document No.
NTIS/PS-75/263/4WJ].

3647   THE FELINE LEUKEMOGENIC AND SARCOMATOGENIC
         VIRUSES. (Fre.) Parodi, A. L. (Labora-
toire d'Anatomie Pathologique, Ecole Nationale
Veterinaire, 94701 Alfort, France). *C. R. Soc. Biol.*
(Paris) 169(3/Suppl.):794-806; 1975. (89 references)

3648   VIROLOGY OF PROSTATIC CANCER. (Eng.)
         Sanford, E. J. (Pennsylvania State Univ.
Coll. Medicine, Hershey, Pa. 17033); Rohner, T. J.;
Rapp, F. *Cancer Chemother. Rep. (Part 1)* 59(1):33-
38; 1975.

3649   THE VIRAL ETIOLOGY OF LEUKEMIA. (Eng.)
         Levine, P. H. (Natl. Cancer Inst., Landow
Building, Room C306, Bethesda, Md. 20014); Gravell,
M. *Mod. Probl. Paediatr.* 16:137-166; 1975. (172
references)

3650   THE ETIOLOGY OF LYMPHORETICULOSARCOMA.
         (Fre.) Hoerni, B. (Fondation Bergonie,
180, rue Saint-Genes, F 33076 Bordeaux, France);
Chauvergne, J.; Durand, M. *Bordeaux Med.* 8(4):335-
340; 1975. (43 references)

3651   CANCER IMMUNOLOGY. (Eng.) Hersh, E. M.
         (M. D. Anderson Hosp. Tumor Inst., Houston,
Tex. 77025). *Trans. Stud. Coll. Physicians Phila.*
42(3):234-236; 1975. (No references)

3652   ANOTHER LOOK AT IMMUNOLOGIC SURVEILLANCE.
         (Eng.) Schwartz, R. S. (Tufts Univ. Sch.
of Medicine, Boston, Mass. 02111). *N. Engl. J. Med.*
293(4):181-184; 1975. (36 references)

3666   METABOLIC AND BIOCHEMICAL CHANGES IN
        LEUKEMIA. (Eng.) Jaffe, N. (Harvard
Medical School, Boston, Mass. 02115). *Mod. Probl.*
Paediatr. 16:113-136; 1975. (114 references)

3667   MAMMALIAN PLASMA MEMBRANE. (Eng.) Bret-
        scher, M. S. (Medical Res. Council Lab.
Molecular Biology, Hills Road, Cambridge CB2 2QH,
U.K.); Raff, M. C. *Nature* 258(5530):43-49; 1975.
(84 references)

3668   TRANSLATIONAL STEP INHIBITED *IN VIVO* BY
        AFLATOXIN B₁ IN RAT-LIVER POLYSOMES.
(Eng.) Sarasin, A. (Institut de Recherches Scientifiques sur le Cancer, Bôîte postale 8, F-94800
Villejuif, France); Moulé, Y. *Eur. J. Biochem.* 54(2):
329-340; 1975.

An attempt was made to localize the translational
step inhibited *in vivo* by aflatoxin B₁ in rat liver
cells. A simulation study was used to determine
precisely the site inhibited *in vivo* after drug in-
toxication. It was based on two parameters: the
kinetics of polysome labeling to follow the nascent
peptide synthesis, and the kinetics of supernatant
labeling to follow the completed protein synthesis.
Wistar rats were given aflatoxin B₁ (1 mg/kg ip),
followed by 12.5 μCi DL-(*carboxy*-¹⁴C) leucine ip
ten minutes before killing. Up to five hours after
dosing, aflatoxin specifically inhibited the elongation and/or termination steps during protein synthesis; after longer periods of time, inhibition occurred essentially at the initiation step. When the
intracellular concentration of aflatoxin was too
high, particularly two hours after dosing, each step
of protein synthesis was blocked. Polypeptide synthesis by the postmitochondrial supernatants isolated from aflatoxin-treated animals was impaired in
the same proportion as protein synthesis *in vivo*.
The damage caused by aflatoxin was mostly observed
on microsomes. However, purified polysomes isolated
from aflatoxin-treated rats synthesized proteins
in vitro to the same extent as those from controls.
These results suggest that aflatoxin metabolite(s)
are bound to polysomes with noncovalent bonds. These
active metabolites are probably lost during polysome
isolation procedures. Early protein synthesis inhibition may be related to the carcinogenic properties of aflatoxin B₁.

3669   AFLATOXIN B₁ HYDROXYLATION: SENSITIVE
        MONOOXYGENASE MICROASSAY SUITABLE FOR
LIVER BIOPSY SPECIMENS. (Eng.) Krieger, R. I.
(Dept. Environmental Toxicology, Univ. California,
Davis, Calif. 95616); Daine, K.; Hsieh, D. P. H.;
Miller, J. L.; Thongsinthusak, T. *Res. Commun.
Chem. Pathol. Pharmacol.* 12(2):287-296; 1975.

A method for measuring the rate of monooxygenase
catalyzed hydroxylation of aflatoxin B₁ to aflatoxin Q₁ (AFQ) has been developed. It is suitable
for use with small amounts of tissue obtained by
needle or surgical biopsy of liver. Liver samples
are homogenized in buffered saline containing a
glucose-6-phosphate:NADPH generating system.
¹⁴C-AFB is added to initiate a 10-min oxidation
period. Quick-freezing stops oxidations, and
AFB₁ and AFQ₁ are extracted using CHCl₃. After
solvent removal, the residue is resuspended in ben-
zene-acetonitrile (98:2) and applied to silica gel
thin-layer chromatography plates, which are developed
twice in chloroform-acetone-*n*-hexane (85:15:20).
AFB₁ and AFQ₁ were located by fluorescence and
scraped into vials for scintillation counting. The
procedure resulted in a reaction which was linear
for ten minutes using 5-10 mg liver tissue/0.2 ml
incubation. Aflatoxin M₁ and three other metabolites

(Eng.) Maher, V. M. (Michigan Cancer Foundation, 110 E. Warren Ave., Detroit, Mich. 48201); Birch, N.; Otto, J. R.; McCormick, J. J. *J. Natl. Cancer Inst.* 54(6):1287-1294; 1975.

The effect of exposure to UV irradiation or to the *N*-acetoxy-ester derivatives of four carcinogenic aromatic amides (4-acetylaminobiphenyl (AABP), 2-acetylaminofluorene (AAF), 2-acetylaminophenanthrene, and 4-acetylaminostilbene) on cell survival was compared in strains of cultured human fibroblasts possessing normal rates of excision repair of DNA and in three strains of xeroderma pigmentosum (XP) cells, each differing in its rate of excision repair. These fibroblasts (XP2BE, XP4BE, XP12BE, and XP13BE) were derived from skin of XP patients. The slope of the survival curve for the XP strain with the poorest capacity for excision repair after UV exposure (XP-12BE complementation group A) was 5.8-fold steeper than the exponential portion of the curve for the normally repairing strains; that of XP2BE (complementation group C) was 1.95-fold; and that of XP4BE (a variant capable of a normal rate of dimer excision) was only 1.3-fold steeper. The slope of the survival curves after exposure to each *N*-acetoxy ester derivative for these same XP strains averaged 6.4, 2.0, and 1.4 times steeper, respectively, than that of the normal strains tested. The excision repair capacity of these lines after exposure to *N*-acetoxy-AAF (50 µM/ml) was tested with alkaline cesium chloride density gradient centrifugation to detect incorporation of tritiated thymidine into nonreplicated DNA. The normal strains and XP4BE exhibited DNA excision repair, whereas XP patients 2 and 12 did not. The cytotoxic effect of the four parent aromatic amide carcinogens and some of their derivatives, in the XP2BE strain, was compared with their effect on the normal fibroblasts. The parent amides proved to be noncytotoxic, while the *N*-hydroxy derivatives of each aromatic amide were highly cytotoxic, as were the ester compounds. For each active derivative, the slope of the survival curve for XP2BE was 2-2.5 times steeper than that of the normally repairing strain.

3675   ACTIVATION OF CARCINOGENIC ARYLHYDROXAMIC ACIDS BY HUMAN TISSUES. (Eng.) King, C. M. (Michael Reese Hosp. and Medical Center, 29th St. and Ellis Ave., Chicago, Ill. 60616); Olive, C. W.; Cardona, R. A. *J. Natl. Cancer Inst.* 55(2):285-287; 1975.

The abilities of cytosols of human tissues obtained at surgery or autopsy to promote the introduction of the arylamine moiety of *N*-hydroxy-*N*-2-fluorenylacetamide or *N*-hydroxy-*N*-4-biphenylacetamide into tRNA were determined. Incubation of the arylhydroxamic acids (0.042 µM) and tRNA with 105,000 x g supernatants of homogenates of small intestine, liver, or colon led to formation of arylamine-substituted nucleic acid adducts. These results indicate that enzymes of human tissues can activate arylhydroxamic acids by $N \to O$ acyl transfer. The unstable *N*-acetoxy-arylamines formed by these enzymes reacted spontaneously with the tRNA to give covalently linked adducts with the nucleic acid. Biphenylamine incorporation was only 17% of that of fluorenylamine.under conditions most favorable for adduct formation with the fluorene derivative. The activation of these arylhydroxamic acids demonstrates that an enzyme of human tissue can transform metabolites of carcinogenic arylamines into derivatives capable of covalent reaction with nucleic acids.

3676   GENETIC DIFFERENCES IN THE INDUCTION OF ARYL HYDROCARBON HYDROXYLASE AND BENZO[a]-PYRENE CARCINOGENESIS IN C3H/He AND DBA/2 STRAINS OF MICE. (Eng.) Watanabe, M. (Res. Inst. for Tuberculosis, Leprosy and Cancer, Tohoku Univ., Hirosemachi 4-12, Sendai 980, Japan); Watanabe, K.; Konno, K.; Sato, H. *Gann* 66(3):217-226; 1975.

The effects of 3-methylcholanthrene and 5,6- and 7,8-benzoflavones on aryl hydrocarbon (benzo[a]-pyrene) hydroxylase activity in the liver, small intestine, lung, and skin tissues from C3H/He and DBA/2 strains of mice under controlled lighting conditions were examined. The animals were kept in an environment in which the lights were on from 10 p.m. to 10 a.m. in a 24-hr cycle. 3-Methylcholanthrene and 5,6- or 7,8-benzoflavone were administered intragastrically at a dose of 1 mg in 0.1 ml of corn oil per 10 g body weight at 9 a.m. Apparent differences were observed in the inducibility of the hepatic enzyme by the inducers between the two strains of mice, showing that the enzyme in the C3H/He mice is inducible but not in the DBA/2 mice. Comparison of the results on the enzyme induction by 3-methylcholanthrene in the liver among the parent, intercrossed, and backcrossed mice suggested that the inducibility may be inherited as a single autosomal dominant trait. However, different genetic responses to 3-methylcholanthrene and benzoflavones in the enzyme of small intestine might be considered, because a discrepancy in the inducibility of the enzyme between the liver and the small intestine from the identical mice was demonstrated when the progeny of F_2 and (DBA/2 x F_1) backcross were used. On the other hand, no apparent difference was found in the inducibility of the enzyme in the lung and skin between the C3H/He and the DBA/2 mice. It is assumed that the genetic regulation of the induction of aryl hydrocarbon hydroxylase was separately controlled in the respective tissues from the identical mice. By the repeated topical applications of benzo[a]pyrene (100 nM/0.2 ml acetone), the incidence of skin cancer was higher in the DBA/2 mice (4.6%) than in the C3H/He mice (1.5%). The relationship between the induction of aryl hydrocarbon hydroxylase and carcinogenicity in the skin is discussed.

3677   THE AUTOXIDATION OF 6-HYDROXYBENZO[a]-PYRENE AND 6-OXOBENZO[a]PYRENE RADICAL, REACTIVE METABOLITES OF BENZO[a]PYRENE. (Eng.) Lorentzen, R. J. (Johns Hopkins Univ., Baltimore, Md., 21205); Caspary, W. J.; Lesko, S. A.; Ts'o, P. O. P. *Biochemistry* 14(18):3970-3977; 1975.

A labile metabolite of benzo(a)pyrene, 6-hydroxybenzo(a)pyrene, was autoxidized in aqueous buffer-ethanol solutions to produce three stable diones

whereas 17-β-estradiol was a strong inhibitor at
20 µg/ml. Progesterone and testosterone gave about
half the effect of 17-β-estradiol, whereas 17-α-
esteradiol had a smaller but significant effect.
It is concluded that hormonal responsiveness *in
vitro* may not mirror actual hormonal responsiveness
in vivo of DMBA-induced mammary tumor in the rat.

3681   CARCINOGENICITY OF 6-AMINOCHRYSENE IN
        MICE. (Eng.) Lambelin, G. (Cont. Pharma
Res. Lab., Machelen, Belgium); Roba, J.; Roncucci,
R.; Parmentier, R. *Eur. J. Cancer* 11(5):327-334;
1975.

The carcinogenicity of 6-aminochrysene was studied
in mice. Bioavailability studies were conducted
in 60 male CFLP mice. 6-Aminochrysene (5 mg) was
administered either p.o. or topically. Four groups
of 10 animals each were used for the determination
of the 0/48 hr ^{3}H-urinary excretion after single
skin application of the ointment to either the dorsal or ventral skin, with or without a collar which
prevented the animals from licking up the ointment.
Two other groups of ten animals each received the
compound by stomach tube; one group received it as
1% tragacanth gum mucilage (0.3 ml/animal), and
the other received it in the ointment form. The
assay of unchanged compound was performed by gas
liquid chromatography. Urinary excretion was about
twice as high after skin application (4.19-7.58%
of the dose administered) than after p.o. administration (1.85-3.32% of the dose administered).
Three hundred SPF CFLP strain mice were then used
in carcinogenicity studies. About 100 mg of an
ointment containing 5% 6-aminochrysene was rubbed
daily on either the dorsal or ventral skin. Mice
were examined regularly for skin lesions. Half of
the mice were treated for six months; the others
were treated for an additional three months. Benign
skin tumors developed after three months and skin
malignancies developed after seven months. Primitive cutaneous tumors arose at the site of application and other lesions arose elsewhere on the skin
and in the viscera. Gross and histopathological
examination revealed that 92% of mice in the 6-
aminochrysene-treated groups had tumors, compared
to 6% and 5%, resp. in vehicle-treated and untreated
controls. In the 6-aminochrysene treated mice,
lung tumors were observed in 71% of the animals
and liver lesions in 38%. Female mice responded
earlier than males. Induction of skin tumors was
more rapid when 6-aminochrysene was applied ventrally
rather than dorsally. The bioavailability studies
indicate that the bioavailability of 6-aminochrysene
is low and that skin absorption of the compound is
more effective than oral absorption. Like the
other polycyclic hydrocarbons, 6-aminochrysene
exhibits a carcinogenic activity in mice.

3682   SPECIFIC CRYSTAL STRUCTURES OF CARCINO- ·
        GENIC HYDROCARBONS. (Ger.) Contag, B.
(Fachbereich Chemie, Technische Fachhochschule
Berlin, West Germany). *Naturwissenschaften* 62(9):
434-435; 1975.

All of 14 carcinogenic polycyclic hydrocarbons studied showed a particular two-dimensional structure of specific crystal faces. Examples are: benz[a]anthracene, benzo[a]pyrene, 3-methylcholanthrene, dibenz[a,h]acridin, and benzo[g,h,i]perylene. None of 15 similar but noncarcinogenic compounds demonstrated this specific structure. The significance of this structural pattern is envisaged as a matrix for the specific two-dimensional adsorption of biogenic macromolecules. It is assumed that this specific adsorption may be the primary reaction of these carcinogenic compounds.

3683   CHROMATOGRAPHIC ANALYSES OF 3-METHYLCHOL-ANTHRENE METABOLISM IN ADULT AND FETAL MICE AND THE OCCURRENCE OF CONJUGATING ENZYMES IN THE FETUS.   (Eng.)   Takahashi, G. (Chest Dis. Res. Inst., Kyoto Univ., Japan); Yasuhira, K. *Cancer Res.* 35(3): 613-620; 1975.

Metabolites of 3-methylcholanthrene (3MC) were studied in adult and fetal DDD strain mice after i.v. injection of ^{3}H-3MC or ^{14}C-3MC. Sephadex LH-20 chromatography yielded three groups of metabolites: one unidentified, one containing the hydrocarbon and hydroxylated derivates, and one consisting of conjugated metabolites. The latter was further separated into a fraction susceptible to the action of β-glucuronidase (and to a lesser extent arylsulfatase) and a fraction resistant to these enzymes but completely acid-hydrolyzable. The enzyme-resistant conjugates predominated in brain, muscle, and lung, while those sensitive to the enzymes predominated in kidney, liver, and bile in adult mice. In fetal mice, the enzyme-resistant conjugates were more characteristic of the early stages of development, while the enzyme-labile conjugates in kidney, liver, and intestinal tract increased as fetuses approached term. Since conjugated metabolites of 3MC did not cross the placental barrier (although 3MC and non-conjugated metabolites did) it is concluded that hydrolases and conjugases are active in fetal mouse tissues.

3684   CARCINOGENICITY OF THREE DOSE LEVELS OF 1,4-BIS(4-FLUOROPHENYL)-2-PROPYNYL-*N*-CYCLOOCTYL CARBAMATE IN MALE SPRAGUE-DAWLEY AND F344 RATS.   (Eng.)   Weisburger, E. K. (Natl. Cancer Inst., Bethesda, Md.); Ulland, B. M.; Schueler, R. L.; Weisburger, J. H.; Harris, P. N. *J. Natl. Cancer Inst.* 54(4):975-979; 1975.

1,4-Bis(4-fluorophenyl)-2-propynyl-*N*-cyclooctyl carbamate (FPOC) was incorporated into the diet of male Sprague-Dawley and F344 rats at 125, 250 and 500 ppm in order to determine if lower doses permitted longer survival and greater incidence of intestinal carcinomas. The animals were fed the diet continuously for the first four wk; the protocol was then changed to three wk of compound administration followed by one wk of basal diet and this cycle was repeated every fourth wk. The experiment was terminated at 30 wk in F344 rats and at 42 wk in Sprague-Dawley rats. The entire gastrointestinal tract was examined at complete necropsy. F344 male rats developed scirrhous, papillary and cystic mucinous carcinoma in the small

intestine. Mortality in high dose groups was attributed to dose-related chronic glomerolonephritis. Intestinal tumors in the Sprague-Dawley rats were similar to those of F344 except for two colon carcinomas. Mammary tumors, cystic and papillary adenocarcinomas were observed. Hematopoietic neoplasia developed in Sprague-Dawley rats and hepatocellular carcinomas were observed in one control and two experimental Sprague-Dawley rats. After FPOC feedings, a 68% survival rate occurred in Sprague-Dawley rats after 18 wk and no F344 rats lived after 12 wk. A high incidence of adenocarcinoma in mammary glands of Sprague-Dawley males is induced by FPOC. Using a lower dose of FPOC does not permit longer survival or higher incidence of tumors.

3685   THE METABOLISM OF ETHIONINE IN RATS.   (Eng.) Brada, Z. (Papanicolaou Cancer Res. Inst., Miami, Fla. 33136); Bulba, S.; Cohen, J. *Cancer Res.* 35(10):2674-2683; 1975.

The L-[ethyl-1-^{14}C]ethionine metabolites soluble in trichloroacetic acid were studied in female CFN rats by the use of column chromatography. After po application of 12.5 mg/100 g ethionine, its absorption from intestinal lumen was rapid, and about 60% was absorbed within 90 min. Any unabsorbed ethionine was later excreted in the feces. During the passage through the gastrointestinal tract, a portion of ethionine was metabolized. The fate of absorbed ethionine was investigated in the small intestine, liver, blood, kidney, and urine as a function of time after application. A great part of ethionine was quickly oxidized to ethionine sulfoxide. In liver and kidney, the concentration of ethionine sulfoxide was higher than that of free ethionine. In all organs, the presence of *N*-acetylethionine sulfoxide was also demonstrated. Ethionine sulfoxide can be reduced, and *N*-acetylethionine can be deacetylated *in vivo* as demonstrated by the formation of *S*-adenosylethionine from ethionine sulfoxide and *N*-acetylethionine. In urine, four main components were observed: *N*-acetylethionine sulfoxide, *S*-adenosylethionine, ethionine sulfoxide, and free ethionine. Some minor unidentified components were also present in the urine and in different organs. The probable site of origin of urinary *S*-adenosylethionine is the kidney.

3686   MUTAGENICITY *IN VITRO* AND POTENTIAL CARCINOGENICITY OF CHLORINATED ETHYLENES AS A FUNCTION OF METABOLIC OXIRANE FORMATION.   (Eng.) Greim, H. (Dept. Toxicology of Gesellschaft fur Strahlenund Umweltforschung, 8042 Munchen-Neuherberg, West Germany); Bonse, G.; Radwan, Z.; Reichert, D.; Henschler, D. *Biochem. Pharmacol.* 24(21): 2013-2017; 1975.

The mutagenicity of six chlorinated ethylenes formed during microsomal activation was tested in a metabolizing *in vitro* system with *Escherichia coli* K₁₂. Mutation rates were determined for three back mutation systems gal+, arg+, and nad+, and for the MTR system where forward mutation leads to resistance to 5-methyl-DL-tryptophan. Cells were incubated

dence strains, had maximum PRL levels of about
200 ng/ml; C57BL/ST and BALB/cST, the two low-
incidence strains, had about 600 ng/ml; and DBA/2St,
the medium-incidence strain, had an intermediate
level. The rate of metabolism of PRL in strains
with high incidence of mammary tumors may be faster
than in those with low incidences. Perphenazine
had no influence on GH secretion in most mice.
The strain-specific differences in PRL and GH concentrations were usually present even during cyc-
lical and diurnal fluctuations. Serum PRL and GH
levels were generally higher during the follicular
phase and lower during the luteal phase of the
estrous cycle in both C3H/St and C57BL/St strains.
It is possible that these strain-specific differences
may be an important factor in the development of
mammary tumors.

3689   *IN VITRO* STIMULATION OF HUMAN BREAST TIS-
        SUE BY HUMAN PROLACTIN. (Eng.) Dilley,
W. G. (Coll. Physicians Surg., Columbia Univ., New
York, N.Y.); Kister, S. J. *J. Natl. Cancer Inst.*
55(1):35-36; 1975.

The growth effects of insulin, ovine prolactin, and
human prolactin on four-day cultures of normal breast
tissue from 30 women were studied. All hormones were
used at 5 µg/ml. Eight hours before termination, colchicine (0.1 µg/ml) or ^{3}H-thymidine (1 µCi) was added
to each culture, and mitotic figures or labeled cells
were counted in histologic sections. Although the
epithelium survived four days in culture, there was
little mitosis or DNA synthesis in the absence of
hormones *in vitro*. Insulin increased the mitotic
activity of cultures by 390% and the further addition
of human prolactin resulted in a 36% increase in activity over that obtained with insulin alone. The
addition of ovine prolactin caused only a minimal increase in the insulin-induced mitotic activity. These
results suggest that maximal effects on normal human
breast tissue can be obtained with human hormones;
further studies to specify the carcinogenic function
of prolactin *in vitro* on human tissue should use human prolactin.

3690   INCREASED RISK OF ENDOMETRIAL CARCINOMA
        AMONG USERS OF CONJUGATED ESTROGENS. (Eng.)
Ziel, H. K. (4900 Sunset Blvd., Los Angeles, Calif.
90027); Finkle, W. D. *N. Engl. J. Med.* 293(23):
1167-1170; 1975.

The possibility that the use of conjugated estrogens increases the risk of endometrial carcinoma
was investigated in patients and a two-fold age-
matched control series from the same population.
Conjugated estrogens (principally sodium estrone
sulfate) use was recorded for 57% of 94 patients
with endometrial carcinoma, and for 15% of controls.
The corresponding point estimate of the (instan-
taneous) risk ratio was 7.6 with one-sided 95% lower
confidence limit of 4.7. The risk-ratio estimate
increased with duration of exposure: from 5.6 for
1 to 4.9 yr exposure to 13.9 for seven or more
years. The estimated proportion of cases related
to conjugated estrogens, the etiologic fraction,
was 50% with a one-sided 95% lower confidence limit

of 41%. These data suggest that conjugated estrogens have an etiologic role in endometrial carcinoma.

3691   CARCINOGENICITY OF THE ANTINEOPLASTIC
        AGENT, 5-(3,3-DIMETHYL-1-TRIAZENO)-IMID-
AZOLE-4-CARBOXAMIDE, AND ITS METABOLITES IN RATS.
(Eng.) Beal, D. D. (St. Vincent Hosp., Worcester,
Mass.); Skibba, J. L.; Croft, W. A.; Cohen, S. M.;
Bryan, G. T. J. Natl. Cancer Inst. 54(4):951-957;
1975.

The carcinogenicities of the antineoplastic agent,
5-(3,3-dimethyl-1-triazeno)imidazole-4-carboxamide
(DTIC), and its metabolites 5-(3-methyl-1-triazeno)-
imidazole-4-carboxamide (MTIC), 5-aminoimidazole-
4-carboxamide (AIC), 2-azahypoxanthine (2-AH), and
5-diazoimidazole-4-carboxamide (Diazo-ICA) were
studied. Test animals including weanling Sprague-
Dawley rats and female Buffalo rats were fed diets
supplemented with DTIC (346-974 mg/rat) and DTIC
metabolites. Cumulative doses were: 9.5-435 mg/
rat Diazo-ICA; 601 mg/rat 2-AH; and 1-890 mg/rat
MTIC. DTIC-induced thymic and mammary tumors were
injected sc and ip, respectively, into female
Sprague-Dawley rats. Male Sprague-Dawley rats
were given ^{14}C-DTIC-2 (6 μCi) or ^{14}C-DTIC-methyl
(10 μCi) ip prior to analysis of feces, urine and
expired CO_2 to determine ^{14}C retention over eight
hours. Rats fed DTIC developed tumors of the mammary gland, thymus and spleen. Animals fed a cumulative dose of 740 mg/rat/14 wk DTIC developed a
100% incidence of mammary adenocarcinomas and thymic lymphosarcomas within 18 wk. Both thymic and
mammary tumors were transplantable. Metabolites
of DTIC-Diazo-ICA pathway were not active. MTIC
(890 mg/rat/14 wk) caused a high incidence (13/16)
of mammary adenofibromas and a low incidence of
leiomyosarcomas of the uterus (3/16); this was not
altered by single or multiple ip injections. Twenty animals fed 2-AH (601 mg/rat/21 wk) developed
three mammary adenocarcinomas and one adrenal cortical adenoma. Twenty rats fed AIC (1,278 mg/rat/
21 wk) developed various malignant tumors. DTIC
induced thymic lymphosarcomas in greater than 90%
incidence after feeding, and 44% incidence after a
single ip injection (400 mg/kg). It also induced
a 50% incidence of mammary adenocarcinomas in male
Sprague-Dawley rats within 18 wk. No correlation
was found between DTIC uptake by a given organ and
its susceptibility to tumors. These studies show
that none of DTIC's metabolites are as carcinogenic
as DTIC alone. Because the data show carcinogenic
properties of DTIC resembling those of carcinogenic
N-nitroso compounds, these may have similar mecha-
nism(s) of action.

3692   MEGALOCYTOSIS AND OTHER ABNORMALITIES
        EXPRESSED DURING PROLIFERATION IN REGEN-
ERATING LIVER OF RATS TREATED WITH METHYLAZOXYMETH-
ANOL ACETATE PRIOR TO PARTIAL HEPATECTOMY. (Eng.)
Zedeck, M. S. (Memorial Sloan-Kettering Cancer Cen-
ter, New York, N. Y. 10021); Sternberg, S. S.
Cancer Res. 35(8):2117-2122; 1975.

The effect of a single dose of methylazoxymethanol
acetate on regenerating liver cells was studied in
male SD rats. Methylazoxymethanol acetate (35 mg/kg
in normal saline was injected iv into 4-wk-old animals, which were partially hepatectomized seven days
later. At 20 and 28 hr after surgery, the hepatocytes (and their nuclei) of treated rats were larger
than those of untreated rats. By 48 hr, nearly all
hepatocytes of treated rats were megalocytic. At
72 and 96 hr, changes in liver architecture were
found. No changes were found in the controls. At
3-4 mo, most changes were reversed. Mitotic indices
of hepatocytes were less in treated vs untreated
rats at 28 hr, but were higher by day 2-14. At 28
hr, eight of ten rats showed abnormal mitotic figures. The percentage of abnormal mitoses at 28,
48, and 96 hr were 16.2, 11.1, and 19.5%, respectively, in treated rats and 0.15, 0.7, and 0.43%
in control rats, respectively. Rats that were partially hepatectomized eight weeks after injection
showed 6.5% abnormal mitoses at 48 hr. Seven 3-mo-
old rats were given the same treatment as the original group. The livers of each showed 2/3 of the
hepatocytes to be enlarged, and 50% abnormal mitoses. Another five rats were treated similarly, and
autopsy at eight months disclosed neither hyperplasia nor carcinoma. At 20 hr after surgery [methyl-
^{3}H]TdR incorporation was significantly less in the
treated rats. At 48 hr, when control levels were
decreasing, those of treated rats were maximum and
were significantly higher at four and seven days.
Total liver DNA, RNA, and protein synthesis were
also significantly reduced in treated vs control
rats. The results suggest that methylazoxymethanol
acetate causes a reduction in the number of regenerating cells after partial hepatectomy. The presence of megalocytes and abnormal mitoses do not seem
to result in increased incidence of tumors. The
latent effects of the carcinogen appear to be long
lasting and are evident in regenerative proliferation.

3693   ENZYMATIC FORMATION OF CHEMICALLY REACTIVE
        METABOLITES OF N-NITROSO-DESMETHYL TRIPEL-
ENNAMINE BY A MECHANSIM OTHER THAN N-DEALKYLATION.
(Eng.) Rao, G. S. (Building 10, Room 8N-107, Natl.
Heart Lung Inst., Bethesda, Md. 20014); Krishna, G.;
Gillette, J. R. Biochem. Pharmacol. 24(18):1707-
1711; 1975.

Benzyl-^{14}C and N-methyl-^{3}H labeled N-nitroso derivatives (NDT) of tripelennamine were synthesized
by reacting the corresponding labeled parent drug
with sodium nitrite at pH 1-2, and their covalent
binding to rat liver microsomes was compared with
that of radiolabeled tripelennamine. Liver microsomes were prepared from 10,000 x g supernatants
of liver homogenates from male Sprague-Dawley rats
by a calcium precipitation method. The covalent
binding of these substances is mediated by liver
microsomal cytochrome P-450 enzymes; it required
an NADPH-generating system and oxygen, and was inhibited by CO:O_2 (8:2). Reduced glutathione (1 mM)
also inhibited the covalent binding. The covalent
binding of NDT to liver microsomal proteins from
phenobarbital-pretreated rats (80 mg/kg, ip daily
for three days) was ten times greater than that of
tripelennamine. A K_m of 60 μM and a V_{max} of 1 nM/
mg of protein/min were obtained for the covalent

The results suggest that nuclear DNase II may be more specifically implicated in regulating DNA synthesis.

3696   TRANSPLACENTAL EFFECTS OF DIETHYLNITRO-
        SAMINE IN SYRIAN HAMSTERS AS RELATED TO
DIFFERENT DAYS OF ADMINISTRATION DURING PREGNANCY.
(Eng.) Mohr, U. (Abteilung fur Experimentelle Path-
ologie, Medizinische Hochschule Hannover, 3000 Han-
nover-Kleefeld, Karl-Wiechert-Allee 9, West Germany);
Reznik-Schuller, H.; Reznik, G.; Hilfrich, J. J.
Natl. Cancer Inst. 55(3):681-683; 1975.

The precise day or days of pregnancy on which a single dose of diethylnitrosamine (DEN) induces the
highest tumor incidence in the offspring of Syrian
hamsters was investigated. Female hamsters were
given a single sc dose of 45 mg DEN/kg on one of the
15 days of pregnancy. In the offspring of females
treated during the first 11 days of pregnancy, no
respiratory tract tumors were found. The offspring
of mothers given DEN during the last 4 days (12-15)
of pregnancy developed respiratory tract neoplasms
at a rate of up to 95%. A lower incidence of tumors
in other organs seemed independent of the day of DEN
treatment. It is concluded that days 12-15 of gestation are the decisive period during which the fetus
is most susceptible to postnatal development of respiratory tract tumors when the mothers are exposed
to DEN.

3697   IN VITRO METABOLISM AND MICROSOME-MEDIATED
        MUTAGENICITY OF DIALKYLNITROSAMINES IN RAT,
HAMSTER, AND MOUSE TISSUES. (Eng.) Bartsch, H. (Int.
Agency Res. Cancer, Lyon, France); Malaveille, C.;
Montesano, R. Cancer Res. 35(3):644-651; 1975.

In an effort to understand the carcinogenic organotropism of dimethylnitrosamine (DMN) and diethylnitrosamine (DEN) in rats and hamster, the relationship between the site of metabolic conversion and
the formation of mutagenic reactants was examined.
Rates of conversion of $^{14}CO_2$ and/or into mutagenic
reactants were measured using Salmonella typhimurium
G-46 or TA 1530 and fortified tissue fractions in
vitro. In general, there was a correlation between
the rate of CO_2 production from DMN and DEN in tissue slices and the production of mutagenic reactants
with the organ distribution of induced tumors. One
exception was the hamster lung, which, while a major
target organ in DEN carcinogenesis, did not convert
DEN into metabolites mutagenic for S. typhimurium TA
1530, although the rate of CO_2 production from DEN
was even higher in hamster lung slices than in liver
slices. Pretreatment with phenobarbitone (1 mg/ml
in drinking water for seven days) doubled the mutation frequency with DMN in liver fractions from
rats, hamsters, and OF-1 mice; with DEN, the mutation frequency increased 40-fold in rats, 10-fold in
mice, and 4-fold in hamster compared to untreated
animals. The results support the notion that tumor
localization following exposure to DMN and DEN is
largely determined by the ability of the target tissue to metabolize the carcinogens.

3698   SYNCHRONIZED LIVER CELLS: A NEW TOOL FOR
        ANALYSIS OF CELL-CYCLE DEPENDENT CARCINO-
GEN-BINDING TO DNA *IN VIVO*. (Eng.) Rabes, H. M.
(Pathologisches Institut der Universitat, Thalkirch-
ner Strasse 36, D-8 Munchen 2, German Federal Repub-
lic, BRD); Iseler, G.; Tuczek, H. V.; Kerler, R.
Experientia 31(6):686-687; 1975.

The degree of liver DNA alkylation prior to, during,
and after DNA replication *in vivo* was measured in
male Wistar rats subjected to two-thirds hepatectomy
and injected ip with ^{14}C-labeled N,N-dimethylnitros-
amine (DMNA, 0.1 μCi/g) and ^{3}H-thymidine (^{3}H-TdR, 0.8
μCi/g). In some rats, initiation of DNA synthesis
in regenerating liver was blocked by a continuous
infusion of hydroxyurea (HU, 1.25 x 10^{-3} M/kg/hr).
These rats received ^{14}C-DMNA and ^{3}H-TdR either during the continuous HU infusion (beginning 14 hr after partial hepatectomy), four hours after the infusion (14-39 hr after partial hepatectomy), or
eight hours after the infusion (47 hr after partial
hepatectomy). Specific activity of DNA from ^{3}H-
TdR was low in the presynthetic G1 phase of the
cell cycle and during the HU-induced inhibition of
DNA synthesis at 28 hr after partial hepatectomy.
DNA synthesis increased markedly four hours after
release from the HU block, but declined again after
another four hours. Concomitant with the upsurge
in DNA synthesis at four hours after release from
the HU block, incorporation of radioactivity from
^{14}C-DMNA increased significantly. The specific activity of DNA from ^{14}C-DMNA did not decline after
the decline of DNA synthesis at eight hours after
the HU block. These results indicate that the uptake of ^{14}C radioactivity from the labeled methyl
groups in DMNA is low in G1 phase, enhanced in G2
phase, and highest in the S phase. The site, degree,
and persistence of possible alkylation in specific
carcinogen-binding to DNA remains to be evaluated.

3699   MUTAGENICITY OF DIMETHYLNITROSAMINE TO
        MAMMALIAN CELLS AS DETERMINED BY THE USE
OF MOUSE LIVER MICROSOMES. (Eng.) Umeda, M.
(Yokohama City Univ. Sch. Medicine, Urafune-cho,
Minami-ku, Yokohama 232, Japan); Saito, M. *Mutat.
Res.* 30(2):249-254; 1975.

The mutagenicity of dimethylnitrosamine (DMN) to
mammalian cells was investigated using a metabolic
activation system. 8-Azaguanine (8-AG)-sensitive
FM3A cells derived from a C3H mouse mammary carcinoma
were incubated with DMN (225μM), NADPH (2.5-18μM)
and a microsomal suspension prepared from the livers
of 8-12-wk-old male DDD mice. The treated cells
were overlayed on agarose plates with or without
20 μg/ml of 8 AG. After 12-14 days, the colonies
were counted. DMN alone caused little change in the
mutation frequency of the FM3A cells over control
values. However, when the cells were incubated with
DMN plus NADPH and the microsomal solution, the
mutation frequency was increased 7.4-4.9 times compared
with controls. The mutation frequency in the complete
reaction mixture increased proportionally to the
incubation time up to 30 min and then leveled off.
The mutation frequency was highest after incubation
in complete medium for 30 min followed by a 2-day

expression period in growth medium. Concentrations
of NADPH greater than 0.3 mM were sufficient to
support the high mutation frequency. The results
show that mutagenic frequencies induced by liver
microsomes from DDD and C3H mice are similar, but
AKR microsomes induce a lower mutagenic frequency.

3700   DEGRADATION OF *N*-NITROSAMINES BY INTESTI-
        NAL BACTERIA. (Eng.) Rowland, I. R. (Br.
Ind. Biol. Res. Assoc., Carshalton, England); Grasso
P. *Appl. Microbiol.* 29(1):7-12; 1975.

Bacteria, commonly found in the alimentary tracts of
man and animals, and capable of degrading diphenyl-
nitrosamine (DPN) and dimethylnitrosamine (DMN) were
examined for their importance in reducing nitrosamine levels in man. Five main groups of bacteria
were isolated from the gastrointestinal tract of
male Wistar rats: Escherichia, bacteriodes, bifido-
bacterium, *Lactobacillus*, and Streptococcus *faecalis*
The experiments were performed at ph 7.0 in the absence of glucose. All were under anaerobic conditions, except for *E. coli* strains and *Proteus* species. The ability to degrade DPN and DMN appeared
to be a common property of bacteria in most of the
bacterial types tested, although the rate of nitrosamine breakdown varied considerably both within and
between groups. Strains of *Lactobacilli* and *E. coli*
were the most active in degrading nitrosamines.
Several strains of bifidobacteria and bacteroides
which constitute more than 99% of the total number
of bacteria in the alimentary tracts of man and animals were able to degrade DPN and DMN, but their
rates of breakdown were lower than other bacteria
tested. The observation that several species could
degrade DPN but not DMN suggests the two processes
may be mediated by different mechanisms. In situations where small quantities of nitrosamines in in-
gested food come into contact with intestinal bacterial synthesis of nitrosamines is possible, it is
feasible that net synthesis of nitrosamines may be
reduced by the degradative action of bacterial enzymes.

3701   CARCINOGENICITY OF 4-HYDROXYBUTYL-BUTYL-
        NITROSAMINE IN SYRIAN HAMSTERS. (Eng.)
Althoff, J. (Eppley Inst. Res. Cancer, 42nd and
Dewey Ave., Omaha, Nebr. 68105); Kruger, F. W.
Cancer Lett. 1(1): 15-19; 1975.

The biological effects of chronic treatment with
the dibutylnitrosamine (DBN), metabolite, 4-hydroxy-
butyl-butylnitrosamine (HBBN), were studied in
Syrian hamsters to determine whether their response
to DBN is similar to that previously reported for
rats. The LD$_{50}$ of HBBN was estimated to be 3.0
g/kg after eight days of observation of ten hamsters
given sc doses of HBBN (1, 2, 4, and 8 g/kg). A
positive dose-effect relationship was demonstrated
after 75 wk of HBBN administration (0.6, 0.3, and
0.15 g/kg, sc, once weekly) to 90 animals. This
dose-effect relationship was mainly due to the
occurrence of neoplasms in the trachea, bile ducts,
and urinary bladder. The occurrence of neoplasms
in organs other than the urinary bladder contrasts

and nitrosomorpholine and analyzed by sucrose gradient centrifugation. Liver sections from rats killed at 0.5, 1, 2, 4, 6, 24, and 48 hr after treatment were examined by light and electron microscopy. Hepatotoxic doses of the nitrosamines caused inhibition of incorporation of $[^{14}C]$leucine into hepatic proteins; this was accompanied by progressive disaggregation of polysomes, which paralleled the known time course of metabolism of each compound. Dimethylnitrosamine and N-nitrosomorpholine inhibited incorporation of $[^{14}C]$orotate into liver RNA, but diethylnitrosamine caused a slight stimulation. Electron microscopy revealed similar hepatic cytoplasmic changes induced by each nitrosamine, including dilation and degranulation of the rough endoplasmic reticulum. Nuclear changes differed with each compound, N-nitrosomorpholine having more marked effects than either of the other two compounds. It is suggested that the differences between dimethylnitrosamine, diethylnitrosamine, and nitrosomorpholine in their effect on nuclear morphology is a consequence of differences in the initial reaction of these carcinogens with DNA. It is also suggested that the similar pattern of cytoplasmic changes induced by the three nitrosamines is a consequence of their metabolism by the liver. Their acute toxicity is probably dependent on this metabolism, and it is possible that intermediates produced from either the dialkyl or cyclic nitrosamines may be involved in the carcinogenic process.

3704   EFFECTS OF NITROSOCARBARYL ON BALB/3T3
        CELLS. (Eng.) Quarles, J. M. (Univ.
Tennessee--Oak Ridge Graduate Sch. Biomedical
Sciences, Oak Ridge, Tenn. 37830); Tennant, R. W.
Cancer Res. 35(10):2637-2645; 1975.

Transformation of mammalian cells in culture and activation of endogenous leukemia viruses was attempted with *N*-nitrosocarbaryl. Carbaryl (*N*-methyl-1-naphthylcarbamate) and *N*-nitrosocarbaryl were tested on BALB/3T3 (clone A31) cells in culture. Nitrosocarbaryl (10 or 20 µg/ml), but not carbaryl (10, 20, or 40 µg/ml), caused transformation of the BALB/3T3 fibroblasts, but neither induced the complete expression of endogenous murine leukemia virus. Transformed cells differed from the parental control cells by loss of contact inhibition, change in morphology, growth in soft agar, growth to higher saturation densities, and tumorigenicity in normal newborn and irradiated weanling BALB/c mice and athymic (nude) mice. Transformed clones were negative for expression of RNA tumor virus antigens, viral reverse transcriptase, and infectious virus. Thus, it appears that nitrosocarbaryl can transform BALB/3T3 cells to tumorigenic cells with altered biological properties but without complete activation of RNA tumor viruses in the transformed cells. Expression of viral antigen in the transformed cells was inducible by iododeoxyuridine, indicating that the endogenous viral genome was retained in an unexpressed state. Nitrosocarbaryl may be a weak carcinogen.

3705   TRANSFORMATION OF HUMAN CELLS IN CULTURE
        BY N-METHYL-N'-NITRO-N-NITROSOGUANIDINE.
(Eng.) Rhim, J. S. (Microbiological Assoc., Beth-

esda, Md. 20014); Park, D. K.; Arnstein, P.; Huebner, R. J.; Weisburger, E. K.; Nelson-Rees, W. A. *Nature* 256(5520):751-753; 1975.

The possibility of using continuous lines of human sarcoma cells for chemical transformation was investigated. The human osteosarcoma clonal line was maintained in Eagle's minimal essential medium with 10% fetal calf serum until the 31st subculture, when the medium was replaced with media containing 5.0-0.01 g/ml N-methyl-N'-nitro-N-nitrosoguanidine (MNNG) in 0.5% dimethylsulfoxide. Doses of 1.0 g/ml MNNG or greater were fatal. In cultures treated with 1.0 μg/ml MNNG, morphological alterations of cells and abnormal growth patterns were noted in the sixth subculture, 55-59 days after treatment. Alterations induced by 0.01 μg/ml of the carcinogen became more pronounced after further subcultivation. The morphological changes in these cultures were similar to those previously observed in human osteosarcoma cells transformed by Kirsten sarcoma virus. Foci of transformed cells consisted mainly of randomly oriented, spindle-shaped cells with nuclear and cytoplasmic overlapping that stained heavily with Giemsa. The growth rate of the cells was double that of control cells treated with dimethyl sulfoxide alone. Chromosome analysis showed that the MNNG-transformed cells had the same markers as reported for untreated parental cells. All five NIH nude athymic mice injected sc with 5 x 10⁶ cells from cultures treated with 0.01 g/ml MNNG developed poorly differentiated sarcoma within two wk of inoculation. Three of five mice given cells from cultures treated with 0.1 g/ml MNNG developed small, persisting nodules that resembled osteosarcomas. The sarcomas were progressive and transplantable. The system used in this study represents an initial advance in the direction of the *in vitro* transformation of human cells with chemical carcinogens.

3706   IMMUNOFLUORESCENT STAINING OF GASTRIC MU-
        COSAL GLYCOPROTEIN IN GASTRIC CARCINOMA
OF DOGS AND RATS INDUCED BY N-METHYL-N'-NITRO-N-
NITROSOGUANIDINE. (Eng.) Kawasaki, H. (Kurume Univ. Sch. Med., 67 Asahi-machi, Kurume 830, Japan); Nakayama, K.; Kimoto, E. *Gann* 66(4):427-431; 1975.

Immunohistological studies, using the fluorescein isothiocyanate-labeled antibody against gastric mucosal glycoprotein, were made during the development of gastric cancer, induced in pointer dogs and Wistar rats by po administration of N-methyl-N'-nitro-N-nitrosoguanidine (MNNG). The dogs and rats were given 167 g/ml MNNG in their drinking water. In an early stage, the regenerative glands were lined by fluorescent mucus cells. Carcinoma cells of orderly glandular structure, produced in the dogs after 485 days, were devoid of fluorescence. Carcinoma cells of less differentiation, produced in the rats during a further advanced stage (540 days), were well fluorescent. The immunofluorescent profiles of such experimentally induced gastric carcinoma were found to be the same as those of human gastric adenocarcinoma.

3707   CARCINOGENICITY OF N-NITROSO-3,4-DICHLORO-
        AND N-NITROSO-3,4-DIBROMOPIPERIDINE IN
RATS. (Eng.) Lijinsky, W. (Biology Div., Oak

Ridge Natl. Lab., Oak Ridge, Tenn. 37830); Taylor, H. W. *Cancer Res.* 35(11/Part 1):3209-3211; 1975.

Comparative carcinogenic activity of nitrosopiperidine, and two of its halogenated derivatives, 3,4-dichloro- and 3,4-dibromonitrosopiperidine, was studied in male Sprague-Dawley rats. Using distilled water and alcohol as solvents, equimolar concentrations (0.35 mM) or 3,4-dichloro- and 3,4-dibromonitrosopiperidine and a higher concentration (0.88 mM) of nitrosopiperidine were administered po to groups of 15 rats. Each group received 60 ml of a solution containing one of the compounds daily five days a week; tap water was given on the other two days. When rats in a group began to die, dosing was discontinued. Animals were kept until they died naturally, or were killed when moribund. Necropsies were then performed, and organs with tumors or lesions were fixed for histological examination. Administration of the dichloro compound (total dose 0.5 mM) for 15 wk caused death of all 15 rats at 24 wk; 27 wk of the dibromo compound (total dose 1.0 mM) caused the death of 15 rats at 41 wk. Animals died of tumors of the tongue, esophagus, stomach, and respiratory system. After the three-fold higher daily dose of nitrosopiperidine (total dose 3.9 mM), 8 of 15 rats were still alive at 40 wk. Not until 55 wk were all in this group dead of tumors similar to those induced by the halogenated piperidines except for absence of respiratory system tumors. Whereas substitution of bromine in the position β to the nitroso function of nitrosopiperidine increases carcinogenicity to a lesser extent than substitution of chlorine, both derivatives are far more active than the parent compound.

3708   SPECIFIC UPTAKE OF LABELED N-NITROSO-
        METHYLUREA IN THE PANCREATIC ISLETS OF
CHINESE HAMSTERS. (Eng.) Tjalve, H. (Dept. Toxicology and Dept. Pathology, Univ. Uppsala, Box 573, S-751 23 Uppsala, Sweden); Wilander, E. *Experientia* 31(9):1061-1062; 1975.

The distribution of labeled N-nitrosomethylurea (NMU) was studied by whole-body autoradiography in mice and Chinese hamsters. In hamsters injected (40 mg/kg, iv) with labeled NMU, a selective uptake of radioactivity occurred in the pancreatic islets after 30 min. At the 1 hr survival interval, the radioactivity in the pancreatic islets and in the liver was the highest in the body. Later the radioactivity in the pancreatic islets decreased and did not exceed that in the exocrine pancreas or in the blood. The radioactivity in the pancreatic islets of the mice (C57-Bl) was low and did not exceed the level of the blood. The accumulated isotope may be a metabolite of NMU in the Chinese hamsters. Previous work showed that pancreatic islet β-cell destruction in mice has been obtained only with high doses of NMU (230 mg/kg). In Chinese hamsters, doses of 50 mg/kg of NMU resulted in diabetes. It appears that the diabetogenic property of NMU is related to the selective accumulation of NMU or its metabolites in the pancreatic islets.

bioassays. The specific compounds in the condensate fractions which are responsible for their activity have not been identified.

3711   INHALATION EXPERIMENTS WITH ^{14}C-LABELED CIGARETTE SMOKE. II. THE DISTRIBUTION OF CIGARETTE SMOKE PARTICLES IN THE HAMSTER RESPIRATORY TRACT AFTER EXPOSURE IN TWO DIFFERENT SMOKING SYSTEMS. (Eng.) Kmoch, N. (Medizinische Hochschule Hannover, 3000 Hannover-Kleefeld, Karl-Wiechert-Allee 9, W. Germany); Reznik, G.; Mohr, U. *Toxicology* 4(3):373-383; 1975.

Thirty-two male and 37 female Syrian golden hamsters were exposed to the smoke of [^{14}C]dotriacontane-16,17 ([^{14}C]DOT)-labeled cigarettes in two different exposure systems differing in terms of the smoke concentration drawn into the exposure chamber. To compare the effectiveness of the two systems, the inhaled dose of ^{14}C-labeled cigarette smoke was determined in the different parts of the hamster respiratory tract. About six times more smoke particles were deposited in the respiratory tract after exposure to high smoke concentration with intermittent puffs of fresh air (closed system) than after exposure to smoke diluted with air (open system). About 80% of the diluted smoke reached the bronchi and lung compared to approximately 60% of the concentrated smoke. The remainder of the dose was trapped in the upper respiratory tract, primarily in the nose and larynx. In the open system the total dose of inhaled smoke was dependent upon the position in the exposure chamber. The higher activity recovered from the hamsters in the closed system is possibly due to the higher concentration of smoke in the exposure chamber of the closed system. The results show that the open system can increase the inhaled dose to about six times that of the closed system.

3712   SUPPRESSION OF LYMPHOCYTE FUNCTION BY PRODUCTS DERIVED FROM CIGARETTE SMOKE. (Eng.) Roszman, T. L. (Coll. Med., Univ. Kentucky, Lexington); Elliott, L. H.; Rogers, A. S. *Am. Rev. Respir. Dis.* 111(4):453-457; 1975.

The effects of various concentrations of nicotine and a water soluble fraction (WSF) from whole cigarette smoke on the transformation of rabbit peripheral lymphocytes stimulated by concanavalin A (Con A) or goat anti-rabbit Fab were determined. New Zealand adult female rabbits were used. After exposure to [H^3]-thymidine for 16 hr, lymphocytes stimulated with optimal doses of the mitogens showed an inhibition of [H^3]-thymidine incorporation into the DNA. The concentration of nicotine required to reduce the Con A and anti-Fab responses to 50% of control levels was 120 μg/ml and 90 μg/ml, respectively. About 48 μg/ml WSF was required to suppress DNA synthesis of cultures stimulated with Con A to 50% of control values, and 27 μg/ml for cultures stimulated with anti-Fab. Because cell viabilities of cultures exposed to various concentrations of nicotine or WSF were essentially the same as the viability of unexposed cultures, mitogen suppression could not be attributed to cytotoxicity. Neither nicotine nor WSF were mitogenic by themselves. Because previous studies have indicated that Con A specifically stimulates rabbit T-cells, whereas anti-Fab is specific for B-cells, this study indicates that both cell functions are susceptible to the action of nicotine and WSF. The authors conclude that *in vitro* systems can be used in investigating mechanisms involved in the effects of cigarette smoke on the immune response.

3713   ACTION OF SULPIRIDE ON SPONTANEOUS DEVELOP-
        MENT OF BREAST CANCER IN MICE. (Fre.)
Rudali, G. (Lab. Genet., Curie Foundation, Inst. Radium, Paris, France); Aussepe, L. *C. R. Acad. Sci. Paris Ser.* D 280:1209-1211; 1975.

The effect of sulpiride on the incidence of spontaneous mammary tumors in male and female (C3H x RIII)F_1 mice subjected to castration and/or adrenalectomy was observed. Sulpiride stimulates lactation, probably by inducing the release of prolactin. Beginning at 30 days of age, sulpiride was administered (120 µg, po) until the mice died or were killed *in extremis*. Mice were examined weekly for the appearance of tumors. Castrated, adrenalectomized females had a much lower incidence of tumors (14%) compared to intact females (91.5%). However, sulpiride did not increase the incidence. Compared to untreated mice, sulpiride did not alter the incidence of mammary tumors in castrated males or females. Histological examination of female mice on sulpiride revealed mammary development and milk in the acini and ducts. An increase in the number of carmine-stained cells in the pituitary was also observed. It is concluded that a chemical compound that stimulates secretion of prolactin may not necessarily be carcinogenic for the mammary gland.

3714   NEOPLASTIC AND LIFE-SPAN EFFECTS OF
        CHRONIC EXPOSURE TO TRITIUM. I. EF-
FECTS ON ADULT RATS EXPOSED DURING PREGNANCY.
(Eng.) Cahill, D. F. (Natl. Environmental Res. Center, Research Triangle Park, N.C. 27711); Wright, J. F.; Godbold, J. H.; Ward, J. M.; Laskey, J. W.; Tompkins, E. A. *J. Natl. Cancer Inst.* 55(2):371-374; 1975.

Observations were made on the late effects of irradiation during pregnancy in order to determine how changes in hormone patterns may influence the neoplastic potential or irradiation on mammary tissue. One hundred and nine pregnant female Sprague-Dawley rats were inoculated ip with tritiated water (HTO), and equilibrium levels were maintained throughout pregnancy by HTO in drinking water. The tritium activities were 1, 10, 50, and 100 µCi HTO/ml body water, which provided cumulative, whole-body radiation doses of approximately 6.6, 66, 330, and 660 rads. Administration of the radioisotope was terminated at parturition. Throughout their life-spans and at autopsy, the dams showed an increased incidence of mammary fibroadenomas at exposure to 330 and 660 rads. Although the data for the incidence of malignant mammary neoplasms were consistent with a linear dose response, the small numbers of tumors preclude specific definition of the dose-response curve. Postexposure

life-spans for dams chronically exposed to 66, 330, and 660 rads during pregnancy were reduced by 14, 24, and 22%, respectively. Accelerated aging was also demonstrated in these rats: The mean age for mammary fibroadenoma onset decreased with an increasing dose of radiation. A more comprehensive investigation of the radiosensitivity of pregnant animals in advisable.

3715   EFFECTS OF THE CARCINOGEN URETHANE ON
        NUCLEAR RNA POLYMERASE ACTIVITIES.
(Eng.) Eker, P. (The Norwegian Radium Hosp., Montebello, Oslo 3, Norway). *Eur. J. Cancer* 11(7):493-497; 1975.

The effect of urethane on RNA polymerase activity of nuclei isolated from liver and lung of male Wistar rats was investigated. The effect of the carcinogen on *in vitro* RNA synthesis with *Escherichia coli* RNA polymerase was also studied. Intraperitoneal injection of 1 mg/g urethane inhibited the Mg^{2+}-dependent RNA polymerase activity of isolated lung nuclei by about 50%. The inhibition was maximal one hour after the injection. Twenty hours after administration, the Mg^{2+}-stimulated activity was again normal. The template activity of chromatin isolated from lung nuclei was measured using *E. coli* RNA; this showed that the normal chromatin and the chromatin from urethane-treated rats was identical. The activity of Mn^{2+} plus $(NH_4)_2SO_4$-dependent RNA polymerase activity of isolated lung nuclei was not affected by urethane injection. The two RNA polymerase activities of liver nuclei were not inhibited by the carcinogen. Mg^{2+}-stimulated nuclear RNA synthesis *in vitro* was inhibited markedly by urethane at concentrations above 50 mM, whereas the Mn^{2+} plus $(NH_4)_2SO_4$-dependent activity was only slightly affected. *In vitro* RNA synthesis catalyzed by *E. coli* RNA polymerase was also inhibited by high concentrations of the carcinogen.

3716   A GAS-CHROMATOGRAPHIC METHOD FOR THE
        PREPARATION OF ^{14}C-LABELED VINYL CHLOR-
IDE. (Eng.) Wagner, E. R. (Dow Chemical Co., Midland, Mich. 48640); Muelder, W. M.; Watanabe, P. G.; Hefner, R. E., Jr.; Braun, W. H.; Gehring, P. J. *J. Labelled Compd.* 11(4):535-542; 1975.

A method for producing vinyl-$^{14}C_2$ chloride (^{14}C-VC) by the dehydrohalogenation of 1,2-dichlorethane-$^{14}C_2$ (^{14}C-DCE) is described. ^{14}C-VC is produced as a gas mixed with helium by passing ^{14}C-DCE over a carbon catalyst at high temperatures. A modified Hewlett-Packard model 5750 gas chromatographic column is used. The large separation coil (2.4 m x 3 mm) is packed with 80-90 mesh Dow Corning 410 Gum on Chromosorb W (acid washed). The smaller pyrolysis tube (90 cm x 3 mm) is packed with a charcoal mixture of 20 g Chromosorb W and 10 g Darco G-60 charcoal powder. An external thermocouple allows independent heating of this portion of the column. The tube is connected to a thermal conductivity detector with a collecting syringe at the terminal end. The catalyst is heated to 470-480 C, and up to 10 µl of liquid

3722   STUDIES ON HIGH SUSCEPTIBILITY OF NURSING
        MOTHER MICE IN 2,7-FAA HEPATOCARCINOGENESIS
[abstract]. (Jpn.) Kimura, I. (Aichi Cancer Center
Res. Inst.. Nagoya, Japan); Miyake, T., Furukawa, E.;
Nishio, O. *Gann, Proc. Jpn. Cancer Assoc.*, *34th
Annual Meeting, October 1975.* p. 18.

3723   THE RELATIONSHIP OF EPIDERMAL HISTADASE
        LEVELS TO TUMOR PROMOTION, HYPERPLASIA AND
NEOPLASIA [abstract]. (Eng.) Colburn, N. H. (Dept.
Dermatology, Univ. Michigan, Ann Arbor, Mich. 48104);
Head, R. A.; Lau, S. *Proc. Am. Assoc. Cancer Res.*
16:46; 1975.

3724   RAPID FORMATION OF IMINODIACETATE FROM
        PHOTOCHEMICAL DEGRADATION OF Fe(III)NI-
TRILOTRIACETATE SOLUTIONS. (Eng.) Stolzberg, R. J.
(The New England Aquarium, Boston, Mass.); Hume, D.
N. *Environ. Sci. Technol.* 9(7):654-656; 1975.

3725   PHORBOLOL MYRISTATE ACETATE: A NEW META-
        BOLITE OF PHORBOL MYRISTATE ACETATE IN
MOUSE SKIN [abstract]. (Eng.) Segal, A. (New York
Univ. Medical Center, New York, N.Y. 10016); Van
Duuren, B. L.; Mate, U. *Proc. Am. Assoc. Cancer Res.*
16:21; 1975.

3726   THE INTERACTION OF PHORBOL MYRISTATE ACE-
        TATE, A TUMOR PROMOTER, WITH RAT LIVER
PLASMA MEMBRANES: A FLUORESCENCE STUDY [abstract].
(Eng.) Witz, G. (New York Univ. Medical Center,
New York, N.Y. 10016); Van Duuren, B. L.; Banerjee,
S. *Proc. Am. Assoc. Cancer Res.* 16:30; 1975.

3727   REDUCTION OF LEAD ACETATE TERATOGENICITY
        IN THE CHICK BY ASCORBIC ACID [abstract].
(Eng.) King, D. W. (Dept. Zool., Natl. Taiwan Univ.,
Taipei); Liu, J. *Fed. Proc.* 34(3):1049; 1975.

3728   ANALYTICAL STUDY ON THE ROLE OF POLY-
        CHLORINATED BIPHENYLS (PCBs) IN THE
MECHANISM OF AZO DYE HEPATOCARCINOGENESIS IN RATS
[abstract]. (Jpn.) Kimura, N. (Cancer Res. Inst.,
Kyushu Univ., Fukuoka, Japan); Kanematsu, T.; Baba,
T. *Proc. Jpn. Cancer Assoc. 34th Annual Meeting,
October 1975.* p. 17.

3729   THE EFFECT OF THE POTENT MUTAGEN AZIDE ON
        DEOXYRIBONUCLEIC ACID [abstract]. (Eng.)
Sideris, E. G. (Bio. Div., Nucl. Res. Cent. Democritos,
Athens, Greece); Argyrakis, M. *Mutat. Res.* 29(2):
239; 1975.

3730   BIOCHEMICAL AND HISTOCHEMICAL STUDY OF
        THE λ-GLUTAMYL TRANSPEPTIDASE ACTIVITY IN
RAT LIVER DURING 3'-Me-DAB CARCINOGENESIS [abstract].
(Jpn.) Chisaka, N. (Sapporo Medical Coll., Sapporo,
Japan); Kaneko, A.; Yoshida, Y.; Yokoyama, S.; Dempp,
K.; Onoe, T. *Gann, Proc. Jpn. Cancer Assoc., 34th
Annual Meeting, October 1975.* p. 17.

3731　INDUCTION OF HEPATOMA AND HYPERPLASIA OF
THE URINARY BLADDER IN FEMALE MICE BY
FEEDING BENZIDINE DYES [abstract]. (Jpn.) Miya-
kawa, M. (Sch. Medicine, Kyoto Univ. Kyoto, Japan);
Yoshida, O. *Gann, Proc. Jpn. Cancer Assoc., 34th
Annual Meeting, October 1975.* p. 17.

3732　PROTECTIVE EFFECT OF REDUCDYN AGAINST
THE CHROMOSOME DAMAGING ACTIVITY OF 2,3,5-
TRIETHYLENEIMINO-BENZOQUINONE-1,4 ON HUMAN LYMPHO-
CYTES *IN VITRO*. (Eng.) Stosiek, M. (Institut fur
Humangenetik und Anthropologie der Universitat
Erlangen-Nurnberg, Erlangen, West Germany); Gebhart,
E. *Mutat. Res.* 29(2):284; 1975.

3733　INFLUENCE OF CROTON OIL IN HAMSTER LUNG
CARCINOGENESIS [abstract]. (Eng.) Sella-
kumar, A. R. (New York Univ. Medical Center, New
York, N.Y. 10016); Kuschner, M.; Laskin, S. *Proc.
Am. Assoc. Cancer Res.* 16:57; 1975.

3734　SEX DIFFERENCES IN THE EFFECTS OF ETHIONINE
ON RNA AND PROTEIN SYNTHESIS IN SWISS MICE
[abstract]. (Eng.) Berry, D. E. (Med. Coll. Virginia, Richmond); Friedman, M. A. *Fed. Proc.* 34(3):
872; 1975.

3735　MICROSOMAL MEMBRANE ALTERATIONS AFTER
ACUTE AND CHRONIC ETHIONINE ADMINISTRA-
TION [abstract]. (Eng.) Kisilevsky, R. (Queen's
Univ., Kingston, Canada); Weiler, L. *Fed. Proc.*
34(3):872; 1975.

3736　SCANNING ELECTRON MICROSCOPY (SEM) OBSER-
VATIONS OF URINARY CYTOLOGY FROM RATS
FED N-[4-(5-NITRO-2-FURYL)-2-THIAZOLYL] FORMAMIDE
(FANFT) [abstract]. (Eng.) Jacobs, J. (St. Vincent
Hosp., Worcester, Mass. 01610); Cohen, S. M.; Arai,
M.; Friedell, G. H. *Proc. Am. Assoc. Cancer Res.*
16:49; 1975.

3737　CARCINOGENICITY OF THE ANTINEOPLASTIC
AGENT 4(5)-(3,3-DI-METHYL-1-TRIAZENO)
IMIDAZOLE-5(4)-CARBOXAMIDE (NSC-45388, DTIC) IN
GERMFREE RATS [abstract]. (Eng.) Croft, W. A.,
Jr. (Univ. Wisconsin Medical Sch., Madison);
Skibba, J. L.; Bryan, G. T. *Fed. Proc.* 34(3):229;
1975.

3738　MECHANISM OF ACTION OF STEROIDAL ANTI-
INFLAMMATORY AGENTS THAT INHIBIT SKIN
CARCINOGENESIS [abstract]. (Eng.) Slaga, T. J.
(Fred Hutchinson Cancer Res. Center, Seattle, Wash.
98104); Thompson, S.; Schwarz, J. A. *Proc. Am. Assoc.
Cancer Res.* 16:37; 1975.

3739　ANTIVIRAL ANTIESTROGEN STUDIES OF DES-
INDUCED HAMSTER RENAL TUMORS [abstract].
(Eng.) Dodge, A. H. (Stanford Sch. Medicine, Stan-
ford, Calif. 94305); Kirkman, H. *Proc. Am. Assoc.
Cancer Res.* 16:20; 1975.

3740　INFLUENCE OF ANTI-OESTROGENS ON THE SPECI-
FIC BINDING *IN VITRO* OF [^{3}H]OESTRADIOL BY
CYTOSOL OF RAT MAMMARY TUMOURS AND HUMAN BREAST
CARCINOMATA. (Eng.) Powell-Jones, W. (Welsh Natl.
Sch. of Medicine, Health Park, Cardiff CF4 4XX, U.K.)
Jenner, D. A.; Blamey, R. W.; Davies, P.; Griffiths,
K. *Biochem. J.* 150(1):71-75; 1975.

3741　PROLACTIN AND ESTRADIOL CONTROL GROWTH OF
RAT MAMMARY ADENOCARCINOMA CELLS *IN VITRO*
[abstract]. (Eng.) Chan, P.-C. (American Health
Foundation, Valhalla, N.Y. 10595); Cohen, L. A.;
Wynder, E. L. *Proc. Am. Assoc. Cancer Res.* 16:40;
1975.

3742　COMPARISON OF CYTOCHROME P450 AND ARYL
HYDROCARBON HYDROXYLASE LEVELS WITH BINDING
OF AROMATIC HYDROCARBONS TO DNA IN UNINDUCED AND
INDUCED RAT LIVER NUCLEI [abstract]. (Eng.) Rogan,
E. G. (Eppley Inst. for Res. in Cancer, Univ.
Nebraska, Omaha, Nebr. 68105); Cavalieri, E. *Proc.
Am. Assoc. Cancer Res.* 16:60; 1975.

3743　INHIBITION OF ENZYME-CATALYZED, IRREVER-
SIBLE BINDING OF BENZO(a)-PYRENE BY MOUSE
LUNG MICROSOMES [abstract]. (Eng.) Shih, T.-W.
(Southern Res. Inst., Birmingham, Ala. 35205).
Proc. Am. Assoc. Cancer Res. 16:25; 1975.

3744　BINDING OF LIGHT-EXPOSED BENZ(a)ANTHRACENE
(BA) DERIVATIVES TO GLYCERALDEHYDE-3-PHOS-
PHATE DEHYDROGENASE (GPDH) AND MAMMARY TISSUE [ab-
stract]. (Eng.) Grubbs, C. J. (Univ. Tennessee
Center Health Sciences, Memphis, Tenn. 38163);
Wood, J. L. *Proc. Am. Assoc. Cancer Res.* 16:16;
1975.

3745　HISTOCHEMICAL-DEMONSTRATION OF POLYNUCLEO-
TIDE PHOSPHORYLASE SYNTHETIC ACTIVITY IN
RAT AND HUMAN MAMMARY CARCINOMA [abstract]. (Eng.)
Tsou, K. C. (Univ. Pennsylvania Sch. Medicine, Phila-
delphia, Pa. 19174); Yip, K. F.; Lo, K. W.; Levin,
J. M.; Rhoads, J. E.; Rosato, F. E.; Ruberg, R.;
Randall, P. *Proc. Am. Assoc. Cancer Res.* 16:26;
1975.

3746　THE EFFECT OF ISOPROPYLVALERAMIDE (IVA)
AND ALLYLISOPROPYLACETAMIDE (AIA) ON
CARCINOGEN-INDUCED DEPRESSION OF DNA SYNTHESIS *IN
VIVO* AND CYTOTOXICITY AND MALIGNANT TRANSFORMATION
IN VITRO [abstract]. (Eng.) Langenbach. R. (Univ.
Nebraska Medical Center, Omaha, Nebr. 68105);
Somogyi, A. *Proc. Am. Assoc. Cancer Res.* 16:41;
1975.

3747　BINDING OF DIMETHYLBENZANTHRACINE (DMBA)
TO ESTROGEN RECEPTOR PROTEIN IN RAT UTERI
AND MAMMARY GLANDS [abstract]. (Eng.) Lucid, S. W.
(Univ. Oklahoma Health Sciences Center, Oklahoma
City, Okla. 73104); Shaw, M. T. *Proc. Am. Assoc.
Cancer Res.* 16:50; 1975.

3757   CHROMOSOME ANALYSIS OF BONE MARROW AND
        NUCLEUS ANOMALY TEST IN MAMMALS AFTER
TREATMENT WITH INH [abstract]. (Eng.) Muller, D.
(No affiliation); Grafe, A.; Miltenburger, H. G.;
Rohrborn, G.; Schulze-Schenking, M. *Mutat. Res.*
29(2):256-257; 1975.

3758   EFFECT OF THYMECTOMY ON CARCINOGENICITY
        OF 4(5)-(3,3-DIMETHYL-1-TRIAZENO)IMIDA-
ZOLE-5(4)-CARBOXAMIDE (DTIC, NSC-45388) IN GNOTO-
BIOTIC RATS [abstract]. (Eng.) Croft, W. A., Jr.
(Univ. Wisconsin Medical Sch., Madison, Wis. 53706);
Skibba, J. L.; Bryan, G. T. *Proc. Am. Assoc. Cancer
Res.* 16:40; 1975.

3759   THE EFFECT OF CYTOSINE ARABINOSIDE ON
        METHYL METHANESULFONATE INDUCED DNA "REPAIR"
[abstract]. (Eng.) Bottomley, R. (Oklahoma Medical
Res. Foundation, Oklahoma City, Okla. 73104);
Garner, W.; Hawrylko, J. *Proc. Am. Assoc. Cancer
Res.* 16:53; 1975.

3760   COLLABORATIVE STUDY OF MMS MUTAGENICITY
        WITH THE DOMINANT LETHAL ASSAY IN MICE
[abstract]. (Eng.) Ehling, U. H. (Gesellschaft
fur Strahlen- und Umweltforschung, Neuherberg);
Frohberg, H.; Schulze-Schencking, M.; Merck, E.;
Lang, R.; Lorke, D.; Machemer, L.; Matter, B. E.;
Muller, D.; Rohrborn, G.; Buselmaier, W. *Mutat. Res.*
29(2):261; 1975.

3761   EXPERIMENTAL ISLET-CELL ADENOMA OF THE
        PANCREAS IN RATS BY MONOCROTALINE [abstract].
(Jpn.) Hayashi, Y. (Shionogi Res. Lab., Osaka,
Japan); Hasegawa, T.; Imai, K.; Katayama, H. *Gann,
Proc. Jpn. Cancer Assoc., 34th Annual Meeting,*   er
October 1975. p. 19.

3762   A POLAROGRAPHIC AND SPECTRAL STUDY OF SOME
        C- AND N-NITROSO COMPOUNDS. (Eng.) Smyth,
W. F. (Chem. Dept., Chelsea Coll., London, England);
Watkiss, P.; Burmicz, J. S.; Hanley, H. O. *Anal.
Chim. Acta* 78(1):81-92; 1975.

3763   HISTOGENESIS OF LIVER CANCER IN MEDAKAS
        (*ORYZIAS LATIPES*) BY DIETHYLNITROSAMINE
(DENA) [abstract]. (Jpn.) Ishikawa, T. (Cancer
Inst., Tokyo, Japan); Takayama, S. *Gann, Proc.
Jpn. Cancer Assoc., 34th Annual Meeting, October
1975.* p. 16.

3764   DIAGNOSIS OF EXPERIMENTALLY INDUCED BRON-
        CHOGENIC TUMOURS IN THE EUROPEAN HAMSTER
WITH BRONCHOGRAPHS. (Eng.) Eckel, H. (Med. Hoch-
schule Hannover Abt. f. experimentelle Pathologie
D-3000 Hannover-Kleefeld Karl-Weichert-Allee 9
West Germany); Reznik-Schuller, H.; Reznik, G.;
Ohse, G.; Mohr, U. *Z. Krebsforsch.* 83(3):207-212;
1975.

3765   CARBAMOYLATION OF AMINO ACIDS, PEPTIDES,
        AND PROTEINS BY NITROSOUREAS [abstract].
(Eng.) Bowdon, B. J. (Southern Res. Inst., Birming-
ham, Ala. 35205); Wheeler, G. P. *Proc. Am. Assoc.
Cancer Res.* 16:38; 1975.

3766   ISOLATION OF CENTROLOBULAR AND PERILOBULAR
        HEPATOCYTES AFTER PHENOBARBITAL TREATMENT.
(Eng.) Wanson, J. C. (Laboratoire de Cytologie et
de Cancerologie Experimentale, Universite Libre de
Bruxelles, 1000 Bruxelles, Belgium); Drochmans, P.;
May, C.; Penasse, W.; Popowski, A. *J. Cell Biol.*
66(1):23-41; 1975.

3767   EXPERIMENTAL STUDIES ON CARCINOGENICITY OF
        QUINOLINE AND 2-CHLOROQUINOLINE [abstract].
(Jpn.) Hirao, K. (1st Dept. Pathology, Nagoya City
Univ., Nagoya, Japan); Shinohara, Y.; Hananouchi, M.;
Ogiso, T.; Kawabata, H.; Takahashi, M.; Ito, N.
*Gann, Proc. Jpn. Cancer Assoc., 34th Annual Meeting,
October 1975.* p. 17.

3768   CARCINOEMBRYONIC ANTIGEN AND T-LYMPHOCYTE
        LEVELS IN THE CHRONIC CIGARETTE SMOKER
[abstract]. (Eng.) Vandevoorde, J. P. (Hoffmann-
La Roche Inc., Nutley, N. J. 07110); Hainsselin,
L. M.; Fresolone, J.; Snyder, J. J.; Hansen, H. J.
Proc. Am. Assoc. Cancer Res. 16:55; 1975.

3769   INHIBITION OF DNA SYNTHESIS AND DNA REPAIR
        BY TOBACCO SMOKE CONDENSATE FRACTIONS AND
OTHER TUMOR PROMOTERS [abstract]. (Eng.) Rasmussen,
R. E. (Univ. California, San Francisco, Calif. 94143
Proc. Am. Assoc. Cancer Res. 16:60; 1975.

3770   ISOLATION OF CARCINOGENIC TANNIN FROM
        BRACKEN FERN (BF) [abstract]. (Eng.)
Wang, C. Y. (Univ. Wisconsin Medical Sch., Madison,
Wis. 53706); Chiu, C. W.; Pamukcu, A. M.; Bryan,
G. T. *Proc. Am. Assoc. Cancer Res.* 16:54; 1975.

See also:

* (Rev):   3602, 3603, 3604, 3605, 3606, 3607,
           3608, 3609, 3610, 3618, 3623, 3627,
           3629, 3630, 3635, 3636, 3637, 3638,
           3639, 3640, 3641, 3642, 3643, 3660,
           3661, 3662, 3663, 3664, 3665
* (Phys): 3778
* (Viral): 3870
* (Immun): 3896, 3939, 3945, 3956, 3959
* (Path): 4024, 4036, 4077, 4096, 4097, 4098,
           4099
* (Epid-Biom): 4116, 4121, 4124, 4125, 4126,
               4127, 4128, 4135, 4136, 4137,
               4138, 4141

an abnormal thyroid scan or thyroid examination,
or both should have a total thyroidectomy.

3773   CANCER MORTALITY FOLLOWING IRRADIATION
         IN INFANCY FOR HEMANGIOMA. (Eng.) Li,
F. P. (Natl. Cancer Inst., Field Station, 35 Binney
St., Boston, Mass. 02115); Cassady, J. R.; Barnett,
E. *Radiology* 113(1):177-178; 1974.

Three cancer deaths were found (*versus* 2.4 expected)
in a record-linkage study of 4,746 patients irradiated in infancy for hemangioma of the skin. The
roster came from the Children's Hospital Medical
Center, Boston, 1946-1968 and was matched against
the names of children who died from cancer, 1960-
1968. The total dose in nearly all cases was 300-
600 R from a 50 kVp orthovoltage unit. Only one
tumor, a testicular teratoma, arose in proximity
to the treatment field and may have been induced by
radiation. The findings show no significant excess
cancer deaths following low-level irradiation, but
occurrence of non-lethal cancers of the skin or
other organs was not evaluated.

3774   THE SUBCELLULAR DISTRIBUTION OF ^{239}Pu IN
         RAT LIVER PARENCHYMAL CELLS AFTER EXPOSURE
IN VIVO AND *IN VITRO*. (Eng.) Gurney, J. R. (Inst.
Cancer Res., Sutton, Surrey SM2 5PT, England);
Taylor, D. M. *Health Phys.* 29(5):655-661; 1975.

The uptake and subcellular distribution of pluto-
nium-239 in L5178Y murine leukemia cells and in
suspensions of rat liver parenchymal cells exposed
to the nuclide *in vitro* and in similar suspensions
prepared from the livers of rats previously injected with plutonium-239 were studied. The stock
solution of plutonium-239 had a concentration of
6.5 mg (400 μCi)/ml 3N HNO_3. To 5-μCi aliquots of
this solution were added 5 ml of 1.3 mM sodium
citrate, and the solution was allowed to stand for
24 hr at 4 C. The solution was either injected
iv into rats (4 μCi/animal) four days before sacrifice for the *in vivo* experiments, or added at a
concentration of 0.2 μCi/100 ml to the culture
medium one hour after the start of incubation. The
L5178Y cells were homogenized in 1 mM sodium bi-
carbonate. The liver cells were suspended in ice-
cold 0.25 M sucrose; 0.1 ml of the 105,000 x g
supernatant from a 20% liver homogenate was added
to reduce mitochondrial fragility; and the cells
were homogenized in a sonicator at an amplitude of
9 μm for four seconds. The homogenates were subjected to fractionation by differential centrifugation at 2-4 C, and radioactivity was assayed by
liquid scintillation counting. The uptake of plu-
tonium-239 by isolated liver cells or by L5178Y
cells after being exposed to the nuclide for 24
hr in culture ranged from 0.6-1.5% of the total
plutonium-239 added to the medium/10^6 cells. There
were considerable differences in the subcellular
distribution between liver cells exposed to pluto-
nium-239 *in vitro* and *in vivo*. Following *in vivo*
exposure the highest specific activity was found in
the lysosomal fraction; after exposure *in vitro*,
however, the highest specific activity was found in
the nuclear fraction. A similar association of

671

plutonium-239 with the nuclear and cell debris fractions was observed in the L5178Y cells following exposure to the nuclide *in vitro*. It is concluded that it is not possible to extrapolate from results obtained from studies on the interaction between plutonium and tissues *in vitro* to mechanisms by which cells and tissues handle plutonium *in vivo*.

3775   INDUCTION OF SKIN TUMORS IN HAIRLESS MICE
        BY A SINGLE EXPOSURE TO UV RADIATION.
(Eng.) Hsu, J. (Coll. Phys. Surg., Columbia Univ., New York, N.Y.), Forbes, P. D.; Harber, L. C.; Lakow, E. *Photochem. Photobiol.* 21(3):185-188; 1975.

Circumstances under which UV treatment alone might produce skin tumors in hairless mice were investigated. In total, 127 hairless mutant mice were irradiated and 59 others served as controls. The UV sources were fluorescent sun lamps emitting UV radiation principally in the 280-320 nm range. Single (skin surface) doses of 3×10^4 J/m^2 to 24×10^4 J/m^2 were delivered in three hr or less. Following exposure, mice were examined daily for 14 days, and once wkly thereafter for periods ranging from 70 to 280 days. Mice exposed to low doses of UV radiation developed erythema and edema within 24 hr. Ulcerations were noted after 48 hr on the dorsal surface and head of those mice receiving more than 6×10^4 J/m^2. The higher doses resulted in more severe acute damage as well as in greater tumor yield. Most of the tumors were benign hyperplastic epithelial papillomas; 4/96 tumors examined histologically proved to be squamous cell carcinomas. It is concluded that tumors can be induced by a single dose of UV, but that the dose must be great enough to produce significant tissue damage. Qualitatively, this study indicates that UV can act as a complete carcinogen without an absolute requirement for exogenous promotion; quantitatively, the data show that induction of malignant growths by a single UV exposure is an event of low probability.

3776   THE EFFECT OF A MAMMALIAN REPAIR ENDO-
        NUCLEASE ON X-IRRADIATED DNA. (Eng.)
Tomura, T. (ULCA Center for Health Sciences, Los Angeles, Calif. 90024); Van Lancker, J. L. *Biochim. Biophys. Acta* 402(3):343-350; 1975.

The effects of mammalian repair endonuclease on DNA extracted from the livers of X-irradiated (1,500 or 3,000 rads total body irradiation) and on purified, X-irradiated (20,000 rads) DNA *in vitro* were studied. The effect of endonuclease on irradiated and nonirradiated DNA was tested by: sedimentation of the DNA in neutral and alkaline sucrose gradients; ^{32}P release after the sequential action of endonuclease and alkaline phosphatase, release of the acid-soluble material after sequential attack of the DNA with endonuclease, alkaline phosphatase, and DNA polymerase in the absence of triphosphates; and incorporation of triphosphates after the sequential treatment of DNA with endonuclease, alkaline phosphatase, and DNA polymerase. Treatment of non-

irradiated DNA with endonuclease did not alter the sedimentation pattern on either alkaline or neutral sucrose gradients. The altered patterns seen after endonuclease treatment of irradiated DNA suggested the formation of single strand breaks and an increase in double strand breaks. A dose-dependent release of acid-soluble material was observed following treatment of irradiated DNA with endonuclease, alkaline phosphatase, and DNA polymerase. ^{32}P-labeled DNA irradiated *in vitro* and *in vivo* also acted as a substrate for the repair endonuclease, and the ability of irradiated DNA to serve as a primer for DNA polymerase was increased by repair endonuclease, alkaline phosphatase, and DNA polymerase treatment. The data suggest that in addition to strand breaks, X-irradiation causes base damage, which may be the target of the endonuclease.

3777   COMPARISON OF FREQUENCIES OF RADIATION-
        INDUCED CHROMOSOME ABERRATIONS IN SOMATIC
AND GERM CELLS OF THE RHESUS MONKEY [abstract].
(Eng.) van Buul, P. P. W. (Dept. Radiation Genetics and Chemical Mutagenesis, State Univ. Leiden, 62 Wassenaarseweg, Leiden, The Netherlands). *Int. J. Radiat. Biol.* 27(6):589-590; 1975.

3778   REPAIR REPLICATION AND ITS DISTRIBUTION
        IN "NORMAL" AND CANCEROUS HUMAN LYMPHO-
BLASTOID CELL LINES AFTER PHYSICAL AND CHEMICAL
DAMAGE [abstract]. (Eng.) Meltz, M. L. (Southwest Found. Res. Educ., San Antonio, Tex.); Thornburg, W. H. *Radiat. Res.* 62(3):533; 1975.

3779   EFFECT OF INHIBITION OF RNA AND OF PROTEIN
        SYNTHESIS ON RECOVERY FROM RADIATION-
INDUCED DIVISION DELAY [abstract]. (Eng.) Leeper, D. B. (Thomas Jefferson Univ., Philadelphia, Pa.) *Radiat. Res.* 62(3):538; 1975.

3780   AGE-SPECIFIC MORTALITY RATES AND PREVALENCE
        FREQUENCIES FOR TWO LYMPHOMA TYPES, AND
FOR LETHAL AND NONLETHAL PULMONARY TUMORS IN THE
BCF$_1$ MOUSE [abstract]. (Eng.) Sacher, G. A. (Div. Biol. Med. Res., Argonne Natl. Lab., Ill.); Fry, R. J. M.; Tyler, S. A.; Ainsworth, E. J.; Lombard, L.; Staffeldt, E. F.; Allen, K. H. *Radiat. Res.* 62(3):595; 1975.

3781   THE TOXICITY OF INHALED ^{239}PuO$_2$ IN IMMATURE,
        YOUNG ADULT, AND AGED SYRIAN HAMSTERS [ab-
stract]. (Eng.) Hobbs, C. H. (Inhalation Toxicology Res. Inst., Lovelace Foundation, Albuquerque, N.M.); Mewhinney, J. A.; Slauson, D. O.; McClellan, R. O.; Miglio, J. J. *Toxicol. Appl. Pharmacol.* 33(1):194-195; 1975.

3782   LEUKEMOGENIC EFFECT OF HEAVY IONS IN RF
        MICE (CARBON-ION BEAM 250 MeV/amu) [ab-
stract]. (Eng.) Kelly, L. S. (Lawrence Berkeley Lab., Univ. Calif.); Daniels, S. J.; LaPlant, P. R.; Thomas, R. H. *Radiat. Res.* 62(3):595; 1975.

3786   IRRADIATION OF THE NECK DURING CHILDHOOD:
        AN ASSESSMENT OF FACTORS WHICH INFLUENCE
THE DEVELOPMENT OF THYROID TUMORS IN PERSONS AT RISK
[abstract]. (Eng.) Colman, M. (Michael Reese Med.
Cent., Chicago, Ill.); Simpson, L. R.; Patterson,
L. K.; Ovadia, J. *Radiat. Res.* 62(3):599; 1975.

3787   HEPATIC RADIATION INJURY: EARLY AND LATE
        [abstract]. (Eng.) Kinzie, J. J. (Dept.
Radiol., Univ. Chicago, Ill.); Swartz, H.; Hensley,
G.; Griem, M. L. *Radiat. Res.* 62(3):543; 1975.

See also:

*   (Rev):   3611,   3612,   3613,   3628,   3643,   3644,
             3645,   3650,   3653
*   (Immun): 3911,   3935,   3947,   3982
*   (Path):   3991,   4003,   4026,   4037
*   (Epid-Biom):   4113

3788   STRUCTURAL PROTEINS OF ADENOVIRUSES: XII. LOCATION AND NEIGHBOR RELATIONSHIP AMONG PROTEINS OF ADENOVIRION TYPE 2 AS REVEALED BY ENZYMATIC IODINATION, IMMUNOPRECIPITATION AND CHEMICAL CROSS-LINKING. (Eng.) Everitt, E. (Wallenberg Lab., Univ. Uppsala, Uppsala, Sweden); Lutter, L.; Philipson, L. *Virology* 67(1):197-208; 1975.

A typographical model of adenovirus type 2 was constructed based on results with enzymatic iodination of intact and disrupted virions, immunoprecipitation of intact virions with specific antisera, and crosslinking with reversible chemical cross-linkers to reveal adjacent proteins. Material from sonically treated isolated nuclei of productively infected HeLa cells was iodinated and the labeled virus purified by a procedure causing a minimum of damage to the virions. The sodium dodecyl sulfate-polyacrylamide-gel electrophoresis pattern of labeled polypeptides of intact virions was compared to the pattern obtained after labeling pyridine-disrupted virions. In the immunoprecipitation experiments, the virus material was labeled with [^{3}H]thymidine and immunoprecipitation was carried out with a direct and a double antibody technique. Crosslinking experiments were performed using the cleavable protein crosslinker, tartaryl diazide, which is capable of crosslinking protein amino groups which are at most 0.6 nm from each other. In iodinated intact virions most of the label was confined to hexons, fibers and to some extent to penton bases. After the virions had been frozen and thawed once, the iodine label was also present in polypeptides V and VII and to a lesser extent in polypeptide VI, suggesting that adenovirions are not rigid entities. Virions disrupted with pyridine were labeled in all polypeptides except in polypeptide IX. Antisera against proteins IIIa and IX precipitated virions with the double antibody technique but these proteins failed to become labeled in intact virions. The efficient precipitation of virions with anti-IIIa suggests an external location. Antisera against proteins VI-VIII did not precipitate virions and the corresponding polypeptides were not iodinated in intact virions. Two new electrophoretic components appeared following crosslinking: one composed of polypeptides V and VII and the other a dimer of polypeptide VII. A complex of IIIa and VII, a dimer of polypeptide VI, and a complex consisting of polypeptides V and VI were also detected. These results suggest that proteins VI, VII, and VIII are located within the virion, that the core protein V is also an internal protein probably located at the vertices in close connection with the penton bases, hexons, and protein IIIa molecules, and that proteins IIIa and IX lack tyrosine residues exposed on the outside of the virion but have external antigenic determinants.

3789   HUMAN CELLS REPAIR DNA DAMAGED BY NITROUS ACID. (Eng.) Day, R. S., III, (Natl. Cancer Inst., Bethesda, Md.). *Mutat. Res.* 27(3):407-409; 1975.

The ability of human cells to repair DNA damaged by nitrous acid was tested. Adenovirus 2 stock (0.1

ml) containing 3 x 10^7 plaque-forming U/ml was treated at 37 C for periods up to 16-20 min with 1M NaNO$_2$ (0.6 ml) at pH 5.0, in 0.25 M NaC$_2$H$_3$O$_2$. Samples were titered on cultured fibroblasts obtained from normal subjects (strains KD, ND, and HG800) and from subjects with xeroderma pigmentosum (strains XP, LO, XP$_{12}$BE, and XP$_4$BE). With three of the four xeroderma pigmentosum strains at hosts, NaNO$_2$-damaged virus showed only half the plaque-forming ability of the normal strains. This indicates that DNA damaged by nitrous acid is at least partly repaired by normal human cells.

3790   STRUCTURE AND COMPOSITION OF THE ADENOVIRUS TYPE 2 CORE. (Eng.) Brown, D. T. (Inst. Genetics, Univ. Cologne, Cologne, Germany); Westphal, M.; Burlingham, B. T.; Winterhoff, U.; Doerfler, W. *J. Virol.* 16(2):366-387; 1975.

The structure and composition of the core of adenovirus type 2 were analyzed by electron microscopy and biochemical techniques after differential degradation of the virion by heat, by pyridine, or by sarcosyl treatment. In negatively stained preparations purified sarcosyl cores revealed spherical subunits 21.6 nm in diameter in the electron microscope. It is suggested that these subunits are organized as an icosahedron that has its axes of symmetry coincident with those of the viral capsid. The subunits were connected by the viral DNA molecule. The sarcosyl cores contained the viral DNA and predominantly the arginine/alanine-rich core polypeptide VII. When sarcosyl cores were spread on a protein film, tightly coiled particles were observed; these gradually unfolded giving rise to a rosette-like pattern due to the uncoiling DNA molecule. Completely unfolded DNA molecules were circular. Pyridine cores consisted of the viral DNA and polypeptides V and VII. In negatively stained preparations of pyridine cores, the subunit arrangement apparent in the sarcosyl cores was masked by an additional shell which was probably formed by polypeptide V. In freeze-cleaved preparations of the adenovirion, two fracture planes were recognized. One fracture plane probably passed between the outer capsid of the virion and polypeptide V exposing a subviral particle that corresponded to the pyridine core. The second fracture plan observed could be located between polypeptide V and the polypeptide VII-DNA complex, thus uncovering a subviral structure that corresponded to the sarcosyl core. In the sarcosyl core, polypeptide VII was tightly bound to the viral DNA, which was susceptible to digestion with DNase (1 μg/ml). The restriction endonuclease EcoRI cleaved the viral DNA in the sarcosyl cores into the six specific fragments. These fragments could be resolved on polyacrylamide-agarose gels provided the sarcosyl cores were treated with pronase (5 mg/ml) after incubation with the restriction endonuclease. When pronase digestion was omitted, a complex of the terminal EcoRI fragments of adenovirus DNA and protein could be isolated. From this complex, the terminal DNA fragments could be liberated after pronase treatment. The complex described is presumably responsible for the circularization of the viral DNA inside the virion. The nature of the

fractions assayed for [^{3}H] radioactivity by liquid scintillation counting. Virus particles were identified and categorized by electron microscopy. No complete adenoviruses were seen in the electron microscope in harvests prepared under arginine-free conditions. Although fewer physical satellite particles (5 x 10^{10} satellite virus particles/ml compared to 1.3 x 10^{12} in arginine-rich medium) were produced under arginine-free conditions and their morphology was less regular, their particle/-infectivity ratio (10$^{6.8}$ hemagglutinin-producing U/ml) was not significantly different from that of satellite progeny (10$^{8.8}$ hemagglutinin-producing U/ml) isolated from arginine-rich conditions in which adenoviruses were also demonstrated. The satellite particles isolated from arginine-free medium, although originally infectious, lost infectivity rapidly and were less stable morphologically than satellite viruses produced under standard nutritional conditions. Satellite particles were visualized in both the nucleus and cytoplasm indicating that transport of viral components is not inhibited. Type 4 satellite virus particles harvested from arginine-free growth conditions lacked the hemagglutinin normally associated with the mature type 4 virion, which may indicate that certain structural proteins are missing.

3793   REGIONAL LOCALIZATION OF HUMAN GENES IN
        VIRUS-INDUCED UNCOILER REGIONS. (Eng.)
McDougall, J. K. (Dept. Cancer Studies, Univ. Birmingham, Birmingham, Conn.); Elsevier, S. M.; Kucherlapati, R. S.; Ruddle, F. H. *Cytogenet. Cell Genet.* 14(3-6):202-204; 1975.

Previous studies on adenovirus 12-infected mouse-man hybrid cells containing only the long arm and the centromere region of human chromosome 17 enabled localization of the thymidine kinase (TK) gene to the virus-induced uncoiler region. In the present study, the localization of galactokinase (GaK) of an adenovirus 12-infected hybrid cell line derived by Sendai virus-mediated fusion of normal human fibroblasts and mouse L cells deficient in TK was investigated. Subclones previously shown to have lost TK activity and the portion of the chromosome 17 distal to 17q21 were negative for human GaK. Those subclones retaining a larger portion of the long arms, but which are deleted distal to the break in 17q22, retained both human TK and GaK activity. Thus, the genes for both these enzymes must be located in the uncoiler regions between the two break points and are therefore assigned to human chromosome 17, band q21-22. Studies are in progress to determine whether a virus-coded DNA-binding protein may be the mediator of the uncoiling effect.

3794   CELL-TO-CELL ATTACHMENTS AND ASSOCIA-
        TIONS IN TUMORS INDUCED BY CELO VIRUS
OR VIRUS-TRANSFORMED CELLS IN HAMSTERS. (Eng.)
Jasty, V. (Dept. Toxicology, Sterling-Winthrop Res. Inst., Rensselaer, N.Y. 12144); Pendola, R.; Yates, V. J. *Am. J. Vet. Res.* 36(11):1643-1648; 1975.

Cell-to-cell attachments or associations in experimentally induced sarcomas of chicken embryo lethal orphan virus (avian adenovirus) origin were studied in hamsters by electron microscopy. The tumors were induced by sc inoculation of the small-plaque variant of the virus into newborn hamsters or by sc inoculation of hamster embryo cells transformed *in vitro* by palque variants of the virus into weanling hamsters. In virus-induced tumors, the cell-to-cell attachments and associations depended mainly on the cell density present in a given area of the tumor. Where tumor cells were closely packed, the adjacent cells were related to each other by noninterdigitating plasma membranes of uniform thickness, closely apposed to each other but separated by a distance of 20-25 nm. Specialized focal cell-to-cell attachments were frequently observed along the parallel-running plasma membranes; in many instances, the attachment sites resembled desmosome-like structures. Less frequently, the cells were attached by button-like projections between the cell surface. In transformed cell-induced tumors, the most common intercellular relationship was the interlocking of the apposing plasma membranes. Cell-to-cell attachments resembling desmosome-like structures and button-like projections were also observed in these tumors. In rare instances, adjacent cells were held together by interdigitating filopodia present on their free surfaces. The wide spectrum of cell-to-cell associations observed in this study is indicative of a poor cell-to-cell communication system that distinguishes transformed cells from their normal counterparts.

3795 PRODUCTION BY EBV INFECTION OF AN EBNA-POSITIVE SUBLINE FROM AN EBNA-NEGATIVE HUMAN LYMPHOMA CELL LINE WITHOUT DETECTABLE EBV DNA. (Eng.) Clements, G. B. (Karolinska Inst., S-104 01 Stockholm 60, Sweden); Klein, G.; Povey, S. *Int. J. Cancer* 16(1):125-133; 1975.

The establishment and characteristics of a continuous lymphoma cell line, BJAB, derived from the tumor of an exceptional African case of Burkitt's lymphoma (BL), are described. Unlike 97% of African BL cases previously studied, neither the original tumor cells nor the cell line contained detectable amounts of Epstein-Barr virus (EBV) DNA, nor did they express the EBV-determined nuclear antigen (EBNA). The cells of the established line had the characteristics of B-type lymphocytes and they carried receptors for EBV. EBNA was induced in the majority of BJAB cells after EBV infection. Usually the cells died within ten days of infection, but it was possible to establish a permanent EBNA-positive variant (GC-BJAB) of BJAB. The patient from whose tumor the original BJAB line was established was seropositive for EBV antigens, indicating previous exposure to and continuing presence of the virus; yet the tumor had not become infected by EBV. This evidence shows that EBV is not readily "picked up" by the lymphoma.

3796 TARGET CELL FOR ONCOGENIC ACTION OF POLY-CYTHAEMIA-INDUCING FRIEND VIRUS. (Eng.) Tambourin, P. E. (Unite de Physiologie Cellulaire de

The presence of poly(A) polymerases was studied in a clone of the established hamster embryo fibroblast line NIL, and in a subclone of the line transformed by virus hamster sarcoma (HS) virus. DEAE-cellulose chromatography revealed three peaks of activity of the DNA-cellular wash fractions from stationary NIL and NIL-HS virus cell; differences in their catalytic activities suggested the existence of three distinct poly(A) polymerases, designated I, IIA, and IIB. Phosphocellulose chromatography of polymerases I and II from growing NIL-HS virus cells revealed polymerase I binding to the column; polymerase IIA did not bind, while polymerase IIB was bound to the phosphocellulose and was eluted. Experiments performed with the enzyme from growing NIL-HS virus illustrated the 6-fold higher incorporation of $[^3H]ATP$ by nucleotidyltransferase with MN^{2+} than with Mg^{2+}. In contrast, the incorporation of $[^3H]$ cytosine triphosphate (CTP) in the presence of Mn^{2+} was only 80% of that with Mg^{2+}. The average chain length of the products of the various poly(a) polymerases was calculated as 4, 50, and 20 nucleotide residues for polymerase I, IIA, and IIB, respectively. The product synthesized by nucleotidyltransferase, in the presence of Mg^{2+} or Mn^{2+}, was almost completely degraded by RNase. The products of polymerase I, IIA, and IIB were highly RNase-resistant. The three polymerases illustrated different requirements for divalent cations; all of the poly(A) polymerases were highly specific for ATP. The finding that poly(A) was an effective primer for the three polymerases indicate the importance of the length of the poly(A) sequence for printing activity. It is suggested that poly(A) polymerases may play a role in the post-transcriptional addition of poly(A) sequences to the 3' terminus of messenger RNAs.

3800   BASE SEQUENCE DIFFERENCES BETWEEN THE
        RNA COMPONENTS OF HARVEY SARCOMA VIRUS.
(Eng.) Maisel, J. (Dept. Molecular Biology and Virus Lab., Univ. California, Berkeley, Calif. 94720); Scolnick, E. M.; Duesberg*, P. *J. Virol.* 16(3):749-753; 1975.

The large (component 1) and the small (component 2) 30-40S RNA species were isolated from the Harvey sarcoma-Moloney leukemia virus complex by preparative gel electrophoresis. Harvey RNA component 1 was completely complementary to DNA transcribed from Moloney leukemia virus (MLV) RNA and showed no homology to DNA transcribed from a C-type rat virus (NRK) when annealed under conditions of DNA excess. Harvey RNA component 2 was about 65% complementary to MLV DNA and about 33% complementary to NRK virus DNA. Approximately 60-80% of the MLV-specific sequences in RNA component 2 was either a distinct molecular species or was part of a hybrid molecule including NRK virus- and MLV-specific sequences. The rest of the MLV sequences in component 2 could be accounted for by degraded component 1 copurifying with component 2. The possible role of these sequences in the ability of the virus to transform cells is discussed.

3801   POLYPEPTIDES SYNTHESIZED IN HERPES SIMPLEX
        VIRUS TYPE 2-INFECTED HEp-2 CELLS. (Eng.)
Powell, K. L. (Baylor Coll. Medicine, Houston, Tex.

77025); Courtney, R. J. *Virology* 66(1):217-228; 1975.

The proteins induced by herpes simplex virus type 2 in infected HEp-2 cells were studied by high-resolution polyacrylamide slab-gel electrophoresis. A total of 51 infected cell-specific polypeptides (ICSP) were detected, accounting for three-fourths of the coding capacity of the virus. The induced polypeptides were allocated to three groups based on their time of synthesis in the virus growth cycle. Cycloheximide treatment (50 μg/ml) of infected cells during the early stage of virus growth was found to cause irreversible inhibition of protein synthesis. Herpes simplex virus 2 polypeptides that underwent considerable posttranslational modification were detected.

3802   *IN VITRO* SYNTHESIS OF DNA IN NUCLEI ISOLATED FROM HUMAN LUNG CELLS INFECTED WITH HERPES SIMPLEX TYPE II VIRUS. (Eng.) Kolber, A. R. (German Cancer Res. Cent., Heidelberg, West Germany). *J Virol.* 15(2):322-331; 1975.

The characteristics of an isolated nuclear system prepared from herpes type II-and mock-infected human embryonic lung cells is described. Human embryonic lung cells were infected with plaque forming unit (PFU) of herpes Type II virus/cell. The cells were incubated and nuclei were prepared and suspended in isoionic HEPES buffer. The *in vitro* reaction mixture consisted of 7 μCi of [³H]dTTP. Incorporation of [³H]TTP in the *in vitro* reaction mixture required Mg²⁺ and ATP. Overall *in vitro* DNA synthesis in nuclei isolated from herpes-infected cells was semiconservative, as demonstrated by bromodeoxyuridine-substituted DNA density transfer experiments, but exhibited a significant fraction of repair type replication. Relative rates of total DNA synthesis *in vitro* and *in vivo* were the same any time after infection with 2 PFU herpes simplex type II virus/cell. Isolated nuclei synthesized cell and viral DNA for a length of time and rate dependent upon the incubation temperature, but there were differences in the length of time of linear *in vitro* DNA synthesis between herpes and mock-infected cells. The temperature optima for *in vitro* DNA synthesis differed significantly for herpes-and mock-infected cells, and were the same for cells abortively infected with herpes type II as for mock-infected cells. An explanation for the stimulation of overall synthesis observed in the herpes isolated nuclei system and *in vivo* may be the induction of a new DNA polymerase coded for by the herpes genome and localized in the nucleus.

3803   FAILURE OF NEUTRAL-RED PHOTODYNAMIC INACTIVATION IN RECURRENT HERPES SIMPLEX VIRUS INFECTIONS. (Eng.) Myers, M. G. (Univ. Iowa Hosp., Iowa City, Iowa 52242); Oxman, M. N.; Clark, J. E.; Arndt, K. A. *N. Engl. J. Med.* 293(19):945-949; 1975.

Because photodynamic inactivation of herpes simplex virus infections may not be free of hazard, the efficacy of photodynamic inactivation with neutral red and light was evaluated in a placebo (phenolsul-fonphthalein)-controlled study of 170 episodes of recurrent herpes simplex virus infection in 96 patients. Herpes simplex vesicles were painted with aqueous 0.1% solutions of neutral red or phenolsulfonphthalein and exposed to a 100-watt white incandescent lamp for 15 min at a distance of 26.5 cm. The patients were instructed to repeat the light exposure after 4-6 hr and then again 24 hr later. If new vesicles appeared within 48 hr, they were similarly treated. The neutral red was shown to photoinactivate herpes simplex virus *in vitro*, but had no significant effect on the rate of resolution of herpetic lesions (P > 0.10) or on the frequency of subsequent recurrences (P > 0.10), except for orolabial lesions, in which an adverse effect on the rate of subsequent recurrences was observed (P < 0.05). In the absence of demonstrated efficacy, the routine use of neutral red and light in patients with recurrent herpes simplex virus infections should be discontinued. Furthermore, other photoactive dyes should not be used until their efficacy has been demonstrated by suitably controlled clinical trials.

3804   EFFECTS OF DISCRETE NUCLEAR U.V.-MICROBEAM IRRADIATION ON HERPES VIRUS AND SV40 INFECTION. (Eng.) Deak, I. I. (Sir William Dunn Sch. Pathol., Univ. Oxford, England); Defendi, V. *J. Cell Sci.* 17(3):531-538; 1975.

The requirement for a nucleolus in the expression of structural genes of nuclear viruses was examined in BSC-I cells. The nucleoli and other nuclear components were irradiated by a 4 μm UV beam for seven seconds; whole nuclei were irradiated by a 25 μm beam for the same time. The cells were then exposed to herpes virus type I (20 plaque-forming U/cell) or to 10 plaque-forming U/cell of simian virus 40 (SV40) for one hour at 37 C. Embryonic chick RBC nuclei were fused to BSC-I cells; SV40 viral markers were studied after various nuclear components of the heterokaryons were irradiated. Indirect immunofluorescence measurement of T and V antigens, as well as the measurement of tritiated thymidine uptake in the SV40-infected cells, revealed complete viral-marker inhibition of cells in which the whole nucleus was irradiated. Irradiation of limited regions of the nucleus, including irradiation of a solitary nucleolus within a nucleus, had no effect on the expression of SV40 viral markers. Similarly, whole nucleus irradiation of herpes virus-infected cells prevented the appearance of viral antigens and also of cell surface changes demonstrated by concanavalin A-mediated hemadsorption. Irradiation of any site within the nucleus caused some reduction in the proportion of cells containing viral antigens, but not in the proportion of those expressing other viral antigen markers. There was no significant difference between cells in which a single nucleous was irradiated, and those in which one nucleolus of two within the same nucleus was irradiated or those in which a nucleoplasmic area was irradiated. The proportion of cells synthesizing V antigen did not fall markedly in the irradiated nuclei of heterokaryons, and V antigen appeared in both the BSC-I cells and the RBC nuclei. This suggests that infecting SV40 DNA can enter the RBC nuclei and direct V antigen synthesis in the cytoplasm of the heterokaryons. The results show that nuclear virus gene expression, unlike cellular gene expression, is not dependent on the nucleolus. SV40 DNA may also enter reactivated chick RBC nuclei in heterokaryons and direct the synthesis of both viral antigen and DNA; these nuclei may thus be rendered permissive to SV40 infection by the BSC-I cytoplasm.

3805   ANATOMY OF HERPES SIMPLEX VIRUS DNA:
       STRAIN DIFFERENCES AND HETEROGENEITY IN
THE LOCATIONS OF RESTRICTION ENDONUCLEASE CLEAVAGE
SITES. (Eng.) Hayward, G. S. (Dep. Microbiol.,
Univ. Chicago, Ill.); Frenkel, N.; Roizman, B.
Proc. Natl. Acad. Sci. USA 72(5):1768-1772; 1975.

Herpes simplex virus (HSV) was digested by the restriction endonucleases from *Hemophilus influenzae* strain RY13 (*Eco*RI). Electrophoretic study showed that the digests contained fragments with molecular weights from 1×10^6 to 28×10^6. The electrophoretic profiles obtained in 0.3% agarose gels with DNA fragments from nine different strains of HSV-1 could be readily differentiated from the patterns exhibited by the corresponding fragments from four separate strains of type 2 virus; however, within each serotype, the laboratory strains differed significantly among themselves and also from isolates passaged a minimum number of times outside the human host. Digestion of all DNAs of HSV with either enzyme reproducibly generated two classes of fragments (major and minor) which differed in M concentration. Although the molecular wt of an intact HSV-1(F1) DNA molecule is approximately 98×10^6, the summed molecular weights of all major and minor *Hin*III fragments totalled 160×10^6, and the seven major fragments alone accounted for only 60×10^6. These unusual features indicate the existence of limited heterogeneity in the positions of cleavage sites along individual molecules. Minor fragment patterns were reproducible and remained after plaque purification, whereas defective DNA molecules of high buoyant density appeared only after serial undiluted passaging and were easily eliminated. Thus the minor fragments cannot be attributed to the presence of these defective DNA molecules. This type of DNA was resistant to cleavage by *Hin*III and gave large amounts of only two species of *Eco*RI fragments, suggesting that the defective molecules consist of many tandem repeats of a small segment of viral DNA. The heterogeneity in the viral DNA of normal density appears to be related to the structural organization of the molecules and does not necessarily imply differences in genetic content.

3806   UPTAKE OF [³H] THYMIDINE AND CELL DNA
       SYNTHESIS DURING THE EARLY MULTIPLICATION
PHASE OF HERPESVIRUS HOMINIS IN BHK CELLS. (Eng.)
Bittlingmaier, K. (Inst. Medical Microbiology, Univ.
Mainz, Mainz, West Germany); Schneider, D.; Falke,
D. *Biochim. Biophys. Acta* 407(4):384-391; 1975.

The uptake of ³H-thymidine and its incorporation in cellular DNA of BHK cells during the first six hours after infection with herpesvirus hominis was studied in an attempt to determine if changes in

the uptake of precursors might indicate a function alteration of the cell membrane and consequently might influence cellular metabolism. For most experiments, cells were used 32 hr after seeding in Eagle's medium containing 5% calf serum, 0.5% tryptose phosphate, and buffers. The cells were infected with herpesvirus strains Lennette and IES for 30 min or one hour. The multiplicity of infection was 5-10/cell. After varying periods of time, cells were labeled for 30 or 60 min with ³H-thymidine. The supernatant was discarded, and the cell pellet was resuspended in 0.5 ml 5% trichloroacetic acid solution. After ultrasonication for ten seconds, the material was centrifuged for 20 min at 4,000 revolutions/min. Radioactivity wa determined in a liquid scintillation spectrometer. During the absorption and penetration period, the inhibition of uptake and of incorporation of ³H-thymidine was sensitive to heat, but not to UV irradiation or cycloheximide treatment. The viral eclipse period was characterized by a strongly increased uptake (8-fold) of ³H-thymidine and by inhibition of cellular DNA synthesis. Both were sensitive to UV irradiation of the particles and cycloheximide treatment of the cells. It is conclud that events during adsorption and penetration are dependent on the particles themselves; events during the eclipse phase, however, depend on the activity of the viral genome. Increased uptake of ³H-thymidine during the eclipse could be the consequence of a general change in pyrimidine metabolism, or could be caused by some virus genome-induced alterations of the cellular membrane.

3807   CHRONIC ANTIGENIC STIMULATION, HERPESVIR
       INFECTION, AND CANCER IN TRANSPLANT RECI-
PIENTS. (Eng.) Matas, A. J. (Univ. Minnesota Hos
Minneapolis, Minn. 55455); Simmons*, R. L.; Najari
J. S. *Lancet* 1(7919):1277-1279; 1975.

The possible relationships of immunosuppression, chronic antigenic stimulation, herpesvirus infection, and increased incidences of lymphoproliferative disorders and carcinomas in transplant recipients are discussed. Unlike cancers in the genera population, 75% of the tumors reported in transplar recipients are lymphoproliferative tumors or skin, lip, or cervical carcinomas. The preponderance of the tumors, and the unusual location, behavior, an histology of the lymphoproliferative disorders, sug gest that the pathogenesis is not simply due to dr induced loss of systemic immunosurveillance nor to the immunological privilege of a special site. After noting the experimental demonstration of laten oncogenic RNA viruses, clinical reactivation of la tent DNA viruses, especially the herpesvirus, is reported in 70-90% of renal transplant patients. Evidence linking the Epstein Barr virus with Burkitt's lymphoma is cited, and the association of Epstein Barr virus with 85% of the squamous cell nasopharyngeal cancers is also noted. It is also shown that antibody to herpes simplex virus type I is found in 90% of patients with cervical carcinom while both previous herpes zoster and herpes simpl virus type I infection are related to the subseque development of skin cancer by anecdotal evidence. Thus the lymphoproliferative tumors, carcinoma of the skin, and carcinoma of the cervix are all pre-viously associated with herpesvirus infections whose persistence in the transplant patient is indicated by frequent clinical exacerbations or constant shed-ding. It is suggested that (oncogenic) herpesviruses rest in lymphocytes and/or in neural tissue during latency, and are reactivated by the chronic anti-genic stimulation of the graft in the immunosuppres-sed patient. While the development of vaccines to prevent herpesvirus infections in such patients is suggested, the development of substances to prevent reactivation of the viruses is considered more im-portant.

3808   ENHANCEMENT OF MOLONEY SARCOMA VIRUS-
       INDUCED TUMOR GROWTH BY CONCANAVALIN A.
(Eng.) Davis, S. (Natl. Cancer Inst., Bethesda, Md.);
Redmon, L. W.; Pearson, G. R. *Immunol. Commun.* 4(1):
29-37; 1975.

In an attempt to further delineate the role of the T cell in the immune response to Moloney sarcoma virus (MSV) *in vivo*, the immunosuppressive effects of concanavalin A (Con A) on tumor growth was studied. In mice inoculated with 10^6 focus forming units/ml MSV. Adult male Balb/c mice were injected ip with Con A (100 µg) three times weekly, beginning either on day zero relative to virus inoculation or when the tumors reached 6 mm in size. In mice treated with Con A commensurate with the appearance of 6 mm tumors, a significantly higher percentage of pro-gressively growing tumors (79%) than in the mice treated only with virus (39%) was observed. Treat-ment with Con A on day zero relative to MSV inocula-tion had no significant effect. Antibody production as measured by indirect membrane fluorescence was not affected by Con A. Because *in vivo*-administered Con A is known to have a specific suppressive effect on T lymphocytes, it is suggested that the primary immunologic response to MSV antigens is a function of T cells.

3809   STUDIES ON THE BUDDING PROCESS OF A TEM-
       PERATURE-SENSITIVE MUTANT OF MURINE LEUK-
EMIA VIRUS WITH A SCANNING ELECTRON MICROSCOPE.
(Eng.) Wong, P. K. Y. (Sch. Basic Medical Science,
Univ. Illinois, Urbana, Ill. 61801); MacLeod, R.
J. Virol. 16(2):434-442; 1975.

The scanning electron microscope was used to study the budding process of the wild-type Moloney murine leukemia virus and one of its temperature-sensitive mutants, designated ts 3. A considerably larger number of budding particles was observed on TB cells infected with ts 3 at the nonpermissive temperature (39 C) than at the permissive temperature (34 C). No apparent difference was noted between the number of particles on ts 3-infected cells at 34 C and wild-type-infected cells at 34 or 39 C. Virions were detected at the cell membrane of ts 3-infected cells at 39 C as early as eight hours postinfection. Virion density increased progressively up to 48 hr, after which no increase was observed. An average of 1,600 virus particles was observed at the cell surface at the peak of virus production. The dis-tribution of these on the cell membrane appeared

to be random. The maximum proportion of the cell surface occupied by the viral particles did not exceed 10%. After temperature shift from 39 to 34 C, approximately 90% of the particles had dissociated from the cell membrane within one hour.

3810   SARCOMA IN A HAMSTER INOCULATED WITH BK
       VIRUS, A HUMAN PAPOVAVIRUS. (Eng.) Shah,
K. V. (Johns Hopkins Univ., Sch. Hyg. Public Health,
Baltimore, Md.); Daniel, R. W.; Strandberg, J. D.
J. Natl. Cancer Inst. 54(4):945-950; 1975.

A sarcoma occurring in one hamster inoculated with a human papovavirus, BK virus (BKV), is described. Outbred Syrian hamsters LAK:LGV (SYR) were used, with half being neonatally thymectomized. They were inoculated either within 24 hr of birth with 0.1 ml (6.5 $\log_{10}$ units) virus sc, or at five weeks of age with 0.5 ml (8.8 $\log_{10}$ units) concentrated vi-rus iv; the latter mice were examined every two weeks for lymph gland tumors. Tumors appeared, de-pending on the route of inoculum, but only one was detected after one year. The tumor was maintained by transfer *in vivo* or by passage *in vitro*. The cells of the third passage (CC-3) from the original tumor were reacted with 52 sera from hamsters with large, second- to fifth-passage tumors. Only four of these sera stained the tumor in indirect immuno-fluorescence tests. The sera reactive to the CC-3 tumor cells were also reactive to BKV-infected WI-38 cells, with 63 sera being reactive. There was a high correlation between readings of the sera to BKV and simian virus 40 (SV40)-infected cells. Re-activity of the tumor cells was increased by re-ducing fixation time to five minutes and storage at -70 C. There was considerable cross-reactivity between the T antigens of BKV and SV40 viruses. In repeated immunofluorescence tests with BKV-immune rabbit sera, tumor cells were negative for BKV vi-rion antigens. BKV was not rescued by fusion of CC and CS tumor cell lines with VERO and WI-38 cells. BKV immunization also failed to protect against challenge with sarcoma virus or SV40 tumor cells. These data indicate that at least part of the BKV genome is present in these tumor cells, and that the integrated BKV genome is defective.

3811   SEROLOGICAL ANALYSIS OF AN ONCORNAVIRUS
       (PMF VIRUS) DETECTED IN MALIGNANT PERMA-
NENT HUMAN CELL LINES. (Eng.) Micheel, B. (Cen-
tral Inst. for Cancer Res. of the Acad. Sciences of
the GDR, Dept. Virology and Experimental and Clini-
cal Immunology, 1115 Berlin-Buch, East Germany);
Papsdorf, G.; Wunderlich, V.; Niezabitowski, A.;
Widmaier, R.; Bender, E.; Graffi, A. *Acta Biol.
Med. Ger.* 34(4):K39-K45; 1975.

A serological analysis of the PMF virus detected in a permanent malignant human cell line was con-ducted by means of an immunodiffusion test using antisera against different known oncornaviruses. Most experiments were performed with antigen pre-parations from permanently cultivated ovarian car-cinoma cells (line Tu 197) infected with PMF virus. The virus showed a positive reaction with antisera against Mason Pfizer monkey virus (MPMV).

lated from the disrupted virus and was purified by
discontinuous sucrose gradient centrifugation. The
in vitro protein was measured by incorporation of
^{3}H-leucine in the presence of ATP, CTP, GTP, phos-
phocreatine, creatine phosphokinase, 19 unlabeled
amino acids, and known units of viral core. The
reaction was incubated and spotted on Whatman filters and dropped into cold trichloroacetic acid.
After boiling and processing the samples were counted in a liquid scintillation counter. The *in vitro*
products were incubated with ribonuclease. The
solubilized proteins were immunoprecipitated by a
double antibody technique, and were analyzed by
polyacrylamide gel electrophoresis and by guanidine hydrochloride agarose column chromatography.
The uptake of ^{3}H-leucine by the AMV core component
was linear for 120 min. The incorporation of ^{3}H-
leucine was proportional to the concentration of
viral cores. Puromycin and aurintricarboxylic acid
inhibited the *in vitro* reaction. The *in vitro*-
synthesized proteins coelectrophoresed and cochromatographed with known proteins, and were immunoprecipitated by total and monospecific antibodies to
known AMV proteins. The authors conclude that the
protein synthesizing system may be an integral part
of the structure of the AMV core and may serve to
translate *in vivo* the viral genome.

3814   INHIBITION OF ROUS SARCOMA VIRUS BY α-
        AMANITIN: POSSIBLE ROLE OF CELL DNA-
DEPENDENT RNA POLYMERASE FORM II. (Eng.) Dinowitz,
M. (Arizona Medical Center, Univ. Arizona, Tucson,
Ariz. 85724). *Virology* 66(1):1-9; 1975.

The effect of α-amanitin, an inhibitor of the cell
nuclear form II DNA-dependent RNA polymerase, on
Rous sarcoma virus (RSV) synthesis in RSV-transformed chick embryo cells was examined. Utilizing
incorporation of [^{3}H]uridine into virions as a
sensitive measure of RSV synthesis and of inhibition
of synthesis by α-amanitin, inhibition of virus
production could be detected within two hours after
addition of the toxin to the cells. The dose response of transformed cells showed about a 15% reduction in virus production at levels of 1-3 µg/ml
α-amanitin within three hours. Analysis of the
appearance of newly labeled [^{3}H]RNA onto polyribosomes showed early partial inhibition by the toxin
at a rate similar to the rate of inhibition of RSV
synthesis. Little if any inhibition of ribosomal
RNA (rRNA), the product of the nucleolar form I
DNA-dependent RNA polymerase, was detectable until
several hours after the initial inhibition of RSV
was observed. These data indicate that RSV requires
the form II polymerase, or an enzyme with a similar
sensitivity to α-amanitin, for replication.

3815   FURTHER EVIDENCE FOR THE EXISTENCE OF A
        VIRAL ENVELOPE PROTEIN DEFECT IN THE BRYAN
HIGH-TITER STRAIN OF ROUS SARCOMA VIRUS. (Eng.)
Ogura, H. (Institut fur Virologie Fachbereich Human-
medizin, Justus-Liebig-Universitat, 6300 Giessen,
West Germany); Friis, R. *J. Virol.* 16(2):443-446;
1975.

The hypothesis that Bryan high-titer Rous sarcoma

virus (BH RSV) is unable to make functional viral
envelope proteins was further investigated. BH
RSV was observed with the electron microscope following freeze-drying. Progeny made in the absence
of a helper virus lacked visible surface projections
or spikes. Phenotypic mixing experiments employing
BH RSV and a thermolabile mutant of vesicular stomatitis virus, tl 17, yielded no evidence of pseudotype formation. Since tl 17 is known to be defective for an envelope glycoprotein, the lack of
successful phenotypic mixing with BH RSV is consistent with the observed absence of viral spikes.

3816   *IN VITRO* TRANSCRIPTION OF DNA FROM THE
        70S RNA OF ROUS SARCOMA VIRUS: IDENTI-
FICATION AND CHARACTERIZATION OF VARIOUS SIZE
CLASSES OF DNA TRANSCRIPTS. (Eng.) Collett, M. S.
(Univ. Minnesota Medical Sch., Minneapolis, Minn.
55455); Faras, A. J. *J. Virol.* 16(5):1220-1228;
1975.

Data indicating that the RNA-directed DNA polymerase of Rous sarcoma virus (RSV) is capable of
transcribing large regions of RSV 70S RNA into DNA
is presented. The synthesis of DNA by detergent-
disrupted virions of the Schmidt-Ruppin and B77
strains of RSV was performed using standard reaction mixtures containing rate-limiting concentrations of one of the deoxynucleoside triphosphates
or equimolar concentrations of deoxynucleoside triphosphate precursors. The DNA products were analyzed by polyacrylamide gel electrophoresis, alkaline sucrose sedimentation, and nucleic acid hybridization. Although the bulk (75%) of the DNA
synthesized under rate-limiting concentrations of
precursor migrated in the range of 4-7S RNA markers in the polyacrylamide gels, a portion of the
DNA migrated between the 7S and 18S markers. However, less than 40% of the DNA transcripts synthesized in the presence of equimolar concentrations
of deoxynucleoside triphosphates were found in the
regions containing the 4S and 7S markers. Larger
DNA chains, ranging between 1,500 and 4,500 nucleotides in length, represented less than 5% of the
total DNA product synthesized in the presence of
rate-limiting concentrations of TTP. In the presence of equimolar concentrations of precursors,
however, about 20% of the total DNA product exhibited a sedimentation coefficient of 10S in alkaline sucrose gradients. Nucleic acid hybridization studies confirmed the length and genetic complexity of the larger DNA transcripts. The identification of large DNA chains in the DNA product
synthesized *in vitro* by the RSV RNA-directed DNA
polymerase provides an explanation for the paradox
existing between the limited number of primer sites
per 70S RNA genome, the small size of the bulk of
the DNA product, and the extent of the RSV genome
represented by the DNA product.

3817   THE ROLE OF THE *tvb* LOCUS IN GENETIC
        RESISTANCE TO RSV(RAV-O). (Eng.) Crit-
tenden, L. B. (ARS Animal Physiology and Genetics
Inst., Beltsville, Md. 20705). Motta, J. V. *Virology* 67(2):327-334; 1975.

To identify the loci controlling recessive resis-

part of the late region of the viral genome plus
10-20% of the early SV40 19S messenger RNA transcribed from the early region of the viral genome.
The results also suggest that the SV40 16S messenger
RNA preparation contains a late SV40 16S RNA species
transcribed from a region including the *Hind*-frag-
ments K,F,J, and G and that this region is located
between 0.945 and 0.175 map units.

3820   RNAs OF SIMIAN VIRUS 40 IN PRODUCTIVELY
        INFECTED MONKEY CELLS: KINETICS OF FOR-
MATION AND DECAY IN ENUCLEATE CELLS. (Eng.) Aloni,
Y. (Weizmann Inst. of Science, Rehovot, Israel);
Shani, M.; Reuveni, Y. *Proc. Natl. Acad. Sci. USA*
72(7):2587-2591; 1975.

The usefulness of cytochalasin B enucleate cells
for the study of metabolism of cytoplasmic messenger RNA (mRNA) and for determining its half-life
in animal cells was investigated. BSC-1 monkey
cell cultures were infected with plaque-purified
simian virus 40 (SV40, strain 777); the RNA was
labeled with [^{3}H]uridine 3-48 hr after infection.
After infection, the cells were enucleated by exposure to cytochalasin B (10 μg/ml) in Eagle's
medium supplemented with amino acids, vitamins,
and 2% calf serum. The ^{3}H-labeled RNA was extracted from the cells and digested with 50 μg/ml
of DNase I, after which the digest was extracted
with sodium dodecyl sulfate/phenol and fractionated
on sucrose gradients. RNA-DNA hybridization was
used to distinguish the labeled SV40 RNA from the
background of cellular RNA. About 80% of the virus-
specific RNA in the cells was distributed with several reproducible peaks between the 28S and 18S
ribosomal RNA markers on the sucrose gradients.
The molecules present in the first recognizable
viral RNA peak (26S) appeared to be the primary
viral RNA transcripts, while the RNA molecules present in the third (19S) peak were the final products of the processing of the viral RNA in the
nucleus. The major viral component in the cytoplasm was 16S RNA, but it was not detected in the
nucleus. The 19S species was also found in the
cytoplasm, and the change in the proportion of the
19S and 16S RNAs took place after they entered the
cytoplasm. The decay rate of the 19S component
approximated first-order kinetics, with a mean
half-life of about three hours. The decay curve
for the 16S component, on the other hand, could
be divided into two periods: up to 5-6 hr after
infection, there was a slight increase in the
radioactivity incorporated into the 16S RNA; after that, decay proceeded exponentially with a half-
life of about six hours. The results indicate a
precursor-product relationship between the 19S and
16S species.

3821   TEMPERATURE DEPENDENCE OF STRAND SEPARA-
        TION OF THE DNA MOLECULES CONTAINING IN-
TEGRATED SV40 DNA IN TRANSFORMED CELLS. (Eng.)
Fried, A. H. (Institut fur Virusforschung, Deut-
sches Krebsforschungszentrum, 69 Heidelberg, West
Germany). *Nucleic Acids Res.* 2(9):1591-1608; 1975.

A procedure was established for the analysis of

temperature dependence in DNA molecules which contain the integrated simian virus 40 (SV40) DNA in transformed mouse cells. The size of the DNA molecules isolated from SVT2 cells was measured after their exposure to different temperatures by employing sedimentation analysis at alkaline pH. The T_m of linear SV40 DNA was then measured using SV40 DNA III molecules (generated using the R_1 restriction endonuclease) and SV40 DNA II molecules (generated using pancreatic DNase). The melting curves of these populations were determined using an RNA-DNA hybridization assay specific for single-stranded DNA, a nitrocellulose filter binding assay (which is also specific for single-stranded DNA), and by assaying for the amount of single-stranded DNA with the S_1 endonuclease after heating and cooling the DNA. Finally, the melting curves of the SV40-cellular DNA molecules from three cloned transformed mouse cell lines (SVT2, 11A8, and SV3T3-47) were measured. The size of the cellular DNA molecules was about 17 times that of a unit length of the SV40 DNA molecule even when the DNA was heated to 65 C for ten minutes. The RNA-DNA hybridization assay gave the same T_m for a unit length of SV40 DNA molecule as the S_1 nuclease assay and the nitrocellulose filter binding assay. Also, the width of the transition in the melting curve was of comparable sharpness, not more than two degrees, using all three assays. The T_m of almost all of the SV40-cellular DNA molecules was greater than that of unit length SV40 DNA. Combining the results from all three transformed cell lines, SV40-cellular DNA molecules that underwent complete strand separation at 55-68 C were found. Thus, SV40-cellular DNA molecules that undergo complete strand separation with a range of 1-2 degrees would represent a homogenous group of molecules with respect to those cellular base sequences which determine the T_m of the molecules. In the 11A8 cell line, about 40% of the integrated SV40 DNA could be accounted for by such a homogenous group. The results indicate that RNA-DNA hybridization is a reliable method for deciding whether or not a molecule containing SV40 base sequences has undergone complete strand separation.

3822   INITIATION AND MAINTENANCE OF CELL TRANS-
        FORMATION BY SIMIAN VIRUS 40: A VIRAL
GENETIC PROPERTY. (Eng.) Kimura, G. (Tottori Univ. Sch. Med., Yonago 683, Japan); Itagaki, A. *Proc. Natl. Acad. Sci. USA* 72(2):673-677; 1975.

The role of simian virus 40 (SV40) genes in biological phenomena induced in cells after infection and transformation by this virus was studied. The transforming ability in 10% serum medium of the temperature-sensitive mutants of SV40 in the complementation group III (ts6-40 type mutants) was greatly reduced when infected rat 3Y1 cells were incubated at the restrictive temperature of 40 C or incubated first at 40 C for three days and then shifted to the permissive temperature of 33 C. Transformation did occur efficiently after incubation at 33 C or after an initial incubation at 33 C for five days followed by a shift to 40 C. When growth properties of 3Y1 cells transformed at 33 C by the group III mutants were examined at 40 C, several aspects of the transformed state were rendered temperature-sensitive.

These aspects were the ability of cells to grow in low serum (1.5%) medium and to make colonies, in 10% serum medium, on monolayers of untransformed 3Y1 cells and in soft agar. It is concluded that a simian virus 40 gene (cistron III) controls the initiation, as well as at least some aspects of the maintenance, of transformation and that the initiation reaction is a more heat-labile event than the maintenance reaction(s) under the experimental conditions.

3823   CIRCULAR DNA FROM HETEROKARYONS OF SV40-
        TRANSFORMED MOUSE AND AFRICAN GREEN MONKEY
CELLS. *(Eng.)* Yoshiike, K. (Natl. Inst. Health, 2-10-35 Kamiosaki, Shinagawa-ku, Tokyo 141, Japan); Watanabe, S.; Suzuki, K.; Uchida, S. *Jpn. J. Microbiol.* 19(3):237-240; 1975.

In an attempt to explore a method for the study of simian virus 40 (SV40) genome excision, closed circular DNA was extracted from heterokaryon cultures of SV40-transformed mouse and susceptible monkey cells and its length determined by electron microscopy. Dish cultures and coverslip cultures were prepared from incubation mixtures of African green monkey kidney (GMK) cells, transformed mouse 3T3-L or 3T3-M cells, and UV-inactivated Sendai virus. After 72 hr of cultivation, the coverslips were stained with fluorescent antibodies and circular DNA was extracted from the dish cultures. The circular DNA molecules from the heterokaryon cultures appeared to be heterogeneous and ranged in size from 1.0-3.0 μm. The molecules constituting the largest proportion of the population had a length similar to that of the viral genome. The viral DNA from transformed 3T3-M cells, which is believed to have been transformed by defective SV40, was approximately 12% shorter than that from 3T3-L cells containing the complete genome. V- and T-antigen positive cells were found in the heterokaryon cultures of GMK cells and transformed 3T3-L cells, but cultures of GMK and transformed 3T3-M cells contained only T-antigen positive cells. The method used in this study, if applied to various transformed cell lines including those transformed by defective deletion mutants, may provide information about the excision of the SV40 genome.

3824   DNA REPLICATION IN SYNCHRONIZED CULTURED
        MAMMALIAN CELLS. V. THE TEMPORAL ORDER
OF SYNTHESIS OF COMPONENT α DNA DURING MONKEY DNA SYNTHESIS INDUCED BY SV40 VIRUS. (Eng.) Parker, R. J. (Albert Einstein Coll. Medicine, 1300 Morris Park Ave., Bronx, N. Y. 10461); Tobia, A. M.; Baum, S. G.; Schildkraut*, C. L. *Virology* 66(1):82-93; 1975.

Purified simian virus 40 (SV40) was used to induce host DNA replication in contact-inhibited monolayer cultures of African green monkey kidney cells. Approximately 20% of the nuclear DNA of these cells is the simple sequence, component α DNA. At the beginning of DNA synthesis induced by SV40 viral infection, the average ratio of the specific radioactivity of component α DNA to that of bulk DNA was 0.18. Similar low specific activity ratios

RUS. (Eng.) May, J. T. (John Curtin Sch. Med. Res.,
Australian Natl. Univ., Canberra); Robinson, A. J.
J. Gen. Virol. 26(2):209-213; 1975.

The effect of exposure to a pulse of uridine on self-
annealing and self-complementary RNA was investigated.
RNA was extracted from primary chicken embryo kidney
cells infected with chicken embryo lethal orphan virus and exposed to a pulse of $[5-^3H]$-uridine late in
infection (i.e., after 19 hr). The cells were lysed,
and phenol was extracted without using chloroform.
Pulse-labeled RNA from the virus infected cells was
2.5% resistant to pancreatic ribonuclease digestion,
as compared to a 1.4% resistance of the uninfected
cells. Upon self-annealing, the amount of ribonuclease-resistant RNA increased in both the infected and
uninfected cell extracts (to levels of 4.5% and 2.0%,
respectively). Upon chromatography on a Sephadex
G-100 column, labeled RNA eluted with the void volume
in both the infected and uninfected cell RNA; the
virus-infected cell RNA contained a 3-fold higher proportion of the loaded sample in the void volume. Experiments done to establish the double-stranded nature
of the RNA eluting in the void volume illustrated both
the melting profile and cesium sulfate density characteristics of double-stranded RNA; once heated, the
RNA behaved like single-stranded RNA. Furthermore,
the ribonuclease-resistant RNA extracted from both
uninfected and infected RNA had identical sedimentation values of 8S in sucrose. About 50% of the virus-
infected RNA was found to anneal to chick embryo lethal orphan DNA, while only 1% of uninfected RNA was
bound; similar values were obtained in four separate
preparations. Of the RNA that hybridized to the
chick embryo lethal orphan DNA, about equal amounts
bound to the heavy and light strands. The experiments
indicate that self-complementary virus-specific RNA,
forming double-stranded RNA under self-annealing conditions, is produced late in the chick embryo lethal
orphan virus infection of chick embryo kidney cells;
this represents 0.45% of the RNA synthesized during
a one-hour period late in infection.

3828   ELECTROPHORETIC MOBILITIES OF RNA TUMOR
        VIRUSES. STUDIES BY DOPPLER-SHIFTED
LIGHT SCATTERING SPECTROSCOPY. (Eng.) Rimai, L.
(Scientific Res. Staff, Ford Motor Co., Dearborn,
Mich. 48121); Salmeen, I.; Hart, D.; Liebes, L.;
Rich, M. A.; McCormick, J. J. *Biochemistry* 14(21)
4621-4627; 1975.

The electrophoretic mobilities, at several pH
values, of purified avian myeloblastosis (AMV),
murine leukemia (MuLV), murine mammary tumor (Mu-
MTV), and feline leukemia (FeLV) viruses were
measured by laser beat frequency light scattering
spectroscopy. The mobilities of these viruses are
similar at pH $\geq$ 7 (-2.7 to -3.2 x 10^{-4} cm/sec V/cm).
The isoelectric points of MuLV and AMV are apparently less than pH 3, whereas for FeLV the data
suggest an isoelectric point between 3 and 5. By
using a Debye-Huckel model to describe the interactions between electrolytes and virus, it was
shown that the values for the mobility of MuMTV
obtained in low ionic strength buffer (0.005) are
consistent with the values previously obtained in
high ionic strength buffer (0.10). This model

was also used to calculate the surface charge
densities of the virions. A comparison between
virus and RBC data at pH > 7 indicates that in
terms of the surface charge density, the envelope
of the RNA tumor viruses is not very different from
the RBC membrane. It is suggested that the mobility
measured at pH $\geq$ 7 may be a general characteristic
of RNA tumor viruses and, in combination with the
diffusion constant and the bouyant density, may be
useful in assaying the presence of such viruses.

3829   INTRACELLULAR SYNTHESIS OF MOUSE MAMMARY
        TUMOR VIRUS POLYPEPTIDES: INDICATION
OF A PRECURSOR GLYCOPROTEIN. (Eng.) Dickson, C.
(Cancer Res. Lab., Univ. California, Berkeley,
Calif. 94720); Puma, J. P.; Nandi, S. J. Virol.
16(2):250-258; 1975. .

Mouse mammary tumor virus polypeptides were detected
in the cytoplasm of mouse mammary tumor cell cultures (BALB/cfC3HCrgl spontaneous mammary adenocarcinomas) using immunological precipitation techniques. The anti-mouse mammary tumor virus serum
precipitated the major virion glycoproteins gp49
and gp37.5/33.5 and a viral-related nonvirion glycoprotein of 76,000 daltons. Subcellular fractionation
studies revealed that the cell-associated virion
glycoproteins were present in the membrane fraction.
Pulse-chase experiments indicated that a viral-
related nonvirion glycoprotein of 76,000 daltons
may be a precursor to one or more of the virion
glycoproteins.

3830   C-TYPE VIRUS ANTIGENS DETECTED BY IMMUNO-
        FLUORESCENCE IN HUMAN BONE TUMOUR CUL-
TURES. (Eng.) Zurcher, C. (Inst. for Experimental
Gerontology TNO, Rijswijk, Z.H., The Netherlands);
Brinkhof, J.; Bentvelzen, P.; de Man, J. C. H.
Nature 254(5499):457-459; 1975.

Indirect immunofluorescence was used to detect
antigens which have determinants in common with
the Rauscher murine and simian leukemia viruses
(RLV and SLV) in human bone tumor cultures. Rat
anti-RLV and guinea pig anti-SLV did not react
with unfected animal cell lines, but gave a
very strong reaction with cultures infected with
the viruses to which the antisera were directed;
there was a significant cross reaction between
RLV and SLV. The bovine leukosis virus (BoLV)
did not show a cross reaction with either RLV
or SLV. One of six rat antisera to RLV reacted
with several human bone tumor cultures and a
woolly monkey virus antiserum also gave a positive reaction with several human bone tumor cultures. Rat anti-RLV serum reacted with a SLV-
producing culture but not with any of the human
bone tumor cultures, and tumor lines which reacted with anti-RLV or anti-SLV did not react
with a goat serum which specifically reacts with
the BoLV. The fluorescence with anti-RLV and
anti-SLV varied per passage, especially early
passages being highly positive. Absorption of
the antisera with RLV isolated from the plasma
of leukemic mice or with murine mammary tumor
virus markedly reduced the reaction in the im-

munofluorescence assay on RLV-infected cells and
completely blocked the reaction with human tumors;
similar results were obtained after in vivo absorption and after absorption of the antiserum
to SLV with SLV. Control absorptions only slightly
diminished the reaction in both systems. The results suggest that cultures from human bone tumors
harbor a viral entity that occasionally produces
antigens which have determinants in common with
known animal C-type oncornaviruses.

3831   C-TYPE VIRUS PARTICLES IN PLACENTA OF
        NORMAL HEALTHY SPRAGUE-DAWLEY RATS.
(Eng.) Gross, L. (Veterans Admin. Hosp., Bronx,
N.Y. 10468); Schidlovsky, G.; Feldman, D.; Dreyfuss,
Y.; Moore, L. A. Proc. Natl. Acad. Sci. USA 72(8):
3240-3244; 1975.

Placentas from four young, healthy Sprague-Dawley
rats were examined with the electron microscope
for the presence of C-type virus particles. These
particles were found in two specimens. One specimen contained virus particles budding from the
plasma membrane of cells in the junctional zone of
the placenta, i.e., the region where the fetal and
maternal cell layers meet. In the other placenta,
immature and mature C-type virus particles were
observed among cell debris in the junctional region. These findings place the rat placenta in
the same group as those of several other species,
such as rhesus, baboon, and marmoset monkeys, as
well as humans, in which C-type virus particles
have been detected by electron microscopy. The
presence of the C-type particles in the placenta
of Sprague-Dawley rats is particularly significant
because a considerable number of these animals
spontaneously develop a variety of malignant tumors in addition to occasional spontaneous leukemias and malignant lymphomas. However, none of
these spontaneous tumors reveals the presence of
virus particles on electron microscopic examination. The nature of virus particles detected in
the rat placenta remains to be determined. It is
hypothesized that they may represent the passage
of latent, presumably oncogenic, viruses that are
transmitted vertically from parents to offspring.
In the course of this passage, some particles may
be formed, temporarily emerging from their latency
before losing their identity, and again being incorporated into the cell genetic components.

3832   AN ANTIGENICALLY RELATED TRYPTIC POLY-
        PEPTIDE FROM SEVERAL MAMMALIAN TYPE C
RNA VIRUS p30s. (Eng.) Davis, J. (Flow Lab., Inc.,
P.O. Box 2226, 1710 Chapman Ave., Rockville, Md.
20852); Gilden, R. V.; Oroszlan, S. Intervirology
5(1/2):21-30; 1975.

Antigenic polypeptides were studied to identify
the immunogenic regions of the p30 sequence. Type
C p30 homologs from rat (RaLV), hamster (HaLV),
mouse (MuLV), and gibbon (GaLV) type C viruses
were purified by isoelectric focusing. The MuLV
isoelectric focused polypeptide (IEFP) was also
purified from trypsinized MuLV p30 by isoelectric
focusing. The antisera used were from goats im-

from a patient with acute myelogenous leukemia were characterized and compared with the C-type simian sarcoma-associated virus (SSAV). In addition, the infectious nature of these particles was examined. The establishment of chronically infected virus producing cells required three months and was then only established in four cell lines (human rhabdomyosarcoma cells, fibroblasts from a human embryo, canine thymus cells, and normal rat kidney cells) after inoculation into a wide variety of animal and human cell lines. A more rapid transmission of the virus to two cell lines (a secondary virus) was obtained within three weeks of exposure of these cells to β-propiolactone-inactivated sendai virus. Secondary HL23V-1 infected a wider variety of cells than the primary virus. XC syncytial plaque assay demonstrated the greatest susceptibility to this virus infection in man and rhesus monkey cells; however, a linear dose-response curve was obtained for all cell lines tested. This assay plus radio-immunoassay demonstrated the frequent occurrence of antibodies to HL23V-1 in the human population. In a virus interference test, fetal human lung fibroblasts were infected with HL23V-1 or gibbon ape leukemia virus (GALV) and challenged with murine sarcoma virus pseudotypes, MSV-(HL23V-1) and MSV-(SSAV). Both leukemic viruses reduced the plating efficiency of the two pseudotypes to less than 1% that of uninfected lung fibroblasts indicating the similarity of receptor sites for HL23V-1, SSAV, and GALV. HL23V-1 and HL23V-5 showed early identical patterns in the assays for inhibition of reverse transcriptase activity; they were most strongly inhibited by antiserum prepared against SSV enzyme and less strongly inhibited by anti-GALV enzume. Nucleic acid hybridization experiments demonstrated a relation between viral RNAs and cellular DNA of SSV-infected Kirsten-transformed normal rat kidney cells; t_m values indicated that the hybrids were well-matched in base pairing. The results demonstrate that although HL23V-1 and GALV, and particularly, SSV, are members of the same family, they are not identical.

3835   PARTIAL CHARACTERIZATION OF C-TYPE PAR-
        TICLES IN A CELL LINE (WR-9) DERIVED FROM
A RAT EPIDERMOID CARCINOMA OF SPONTANEOUS ORIGIN.
(Eng.) Sottong, P. (Electro-Nucleonics Lab., Inc.
Bethesda, Md. 20014); Woo, J.; Sarma, P. S.; Vai-
tuzis, Z. *Cancer Res.* 35(10):2864-2871; 1975.

A C-type virus that is continuously released from the WR-9 cell line, and was derived from a sponta-neous epidermoid carcinoma was purified by means of large-scale tissue culture techniques and high-volume zonal centrifuges. Using relatively pure virus concentrates, partial characterization of the virus was accomplished. Up to 60 l of spent culture medium from relatively low virus-yielding cultures were processed at a time through the Model K ultra-centrifuge in order to obtain quantities of virus sufficient for convenient Tween-ether extraction of major polypeptide (30,000 daltons). This structural protein, having group-specific (gs) reactivity, was purified and isolated by isoelectric-focusing tech-niques. A UV absorption peak (A_{280}) was found to be coincident with a major peak of radioactivity at pH 8.6, the isoelectric point (pI) for rat virus gs

antigen previously reported by other indicators.
Because species-specific (*gs*-1) and cross-reactive
(*gs*-3) determinants coexist on this protein, fractions containing the group-specific antigen were
identified on the basis of the mammalian interspecies
determinant (*gs*-3), using antiserum prepared against
Tween-ether-disrupted feline leukemia virus. At the
same time, reactivity to the *gs*-1 determinants in
identical fractions was observed in complement fixation and gel diffusion assays, using guinea pig antiserum known to contain principally antibodies to rat
gs-1 determinants. The WR-9 rat virus line may be of
use in providing an additional source of C-type particles that are capable of yielding good *gs* reagents.

3836   COMPARISON OF RNA-DIRECTED DNA POLYMERASES
        FROM XENOTROPIC AND ECOTROPIC VIRUSES.
(Eng.) Strickland, J. E. (National Cancer Inst.,
Frederick Cancer Res. Center, Frederick, Md. 21701);
Fowler, A. K.; Hellman, A. *Biochem. Biophys. Res.
Commun.* 65(3):1123-1129; 1975.

RNA-directed DNA polymerases from several xenotropic
and ecotropic type-C murine viruses were compared.
On glycerol sedimentation velocity gradients polymerases from Rauscher murine leukemia virus, AT-124
virus, New Zealand Black virus, and BALB: virus 2
appeared at identical positions as single peaks,
indicating similar molecular weights. Immunological
comparisons using antisera against DNA polymerases
from Rauscher virus and New Zealand Black virus
failed to demonstrate significant differences among
the viral enzymes tested.

3837   SMOOTH MUSCLE ANTIBODY IN PATIENTS WITH
        WARTS. (Eng.) McMillan, S. A. (The
Laboratories, Belfast City Hosp., Belfast BT9 7AD,
Northern Ireland); Haire, M. *Clin. Exp. Immunol.*
21(2):339-344; 1975.

Sera from 54 patients with active common and plantar
warts were tested for the presence of commonly
found autoantibodies (smooth muscle antibody, antinuclear antibody, antireticulin antibody, and gastric parietal cell antibody). Smooth muscle antibody (SMA) was detected at a significantly higher
level (25.9%) than in a control group of an equal
number of healthy blood donors matched for age and
sex (7.4%), and was in the IgM class. The SMA-
positive sera gave a microfilamentous staining
pattern, also in the IgM class, on fixed HEp$_2$ tissue
culture cells. Absorption of positive sera with
smooth muscle removed both SMA staining and the
anti-cellular staining. The finding of SMA and
anticellular antibodies of IgM class implies that
they are a response to a persistent antigenic
stimulus.

3838   METHOD FOR DETERMINATION OF NUCLEOTIDE
        SEQUENCE HOMOLOGY BETWEEN VIRAL GENOMES
BY DNA REASSOCIATION KINETICS. (Eng.) Fujinaga,
K. (Sapporo Med. Coll., Japan); Sekikawa, K.; Yama-
zaki, H. *J. Virol.* 15(3):466-470; 1975.

D.; Ghysdael, J.; Portetelle, D.; Burny*, A. *Acta Haematol. (Basel)* 54(4):201-209; 1975.

3858   DETECTION OF VIRAL ANTIGENS IN FELINE LEU-
        KEMIAS BY IMMUNE FLUORESCENCE TEST. (Por.)
Terrinha, A. M. (Cadeira de Bacteriologia e Virologia,
Instituto de Higiene e Medicina Tropical, Lisbon,
Portugal); Vigario, J. D.; Moura Nunes, J. F.; Noronha,
F. *Arq. Patol.* 46(2/3):299-307; 1974.

3859   GENETIC TRANSMISSION OF ENDOGENOUS N- AND
        B-TROPIC MURINE LEUKEMIA VIRUSES IN LOW-
LEUKEMIC STRAIN C57BL/6. (Eng.) Odaka, T. (Inst.
Medical Science, P. O. Takanawa, Tokyo 108, Japan).
J. Virol. 15(2):332-337; 1975.

3860   THE EFFECT OF INTERFERON ON EXOGENOUS AND
        ENDOGENOUS MLV INFECTION [abstract]. (Eng.)
Pitha, P. M. (Johns Hopkins Oncol. Cent., Baltimore,
Md.); Rowe, W. P. *Proc. Am. Assoc. Cancer Res.*
16:159; 1975.

3861   VERTICAL TRANSMISSION OF MULV (SCRIPPS) IN
        MICE [abstract]. (Eng.) Jenson, A. B.
(Scripps Clin. Res. Found., La Jolla, Calif.); Groff,
D. E.; Byers, M. J.; McConahey, P. J. *Fed. Proc.*
34(3):974; 1975.

3862   BIOCHEMICAL AND IMMUNOLOGICAL COMPARISON
        OF THE ALKALINE PHOSPHATASE OF MURINE
LYMPHOMA AND PLACENTA [abstract]. (Eng.) Flanigan,
G., III. (Biol. Dept. Atlanta Univ., Ga.), Floyd,
R. A.; Tisdale, V. G.; Lumb, J. R. *Proc. Am. Assoc.
Cancer Res.* 16:117; 1975.

3863   CHARACTERIZATION OF A- AND B-TYPE PARTICLES
        PRODUCED BY MURINE MYELOMA CELLS [abstract]
(Eng.) Robertson, D. (Div. Biol. Biomed. Sci.,
Washington Univ., St. Louis, Mo.); Dobbertin, D.;
Baenziger, N.; Thach, R. E. *Fed. Proc.* 34(3):527;
1975.

3864   CYTOPLASMIC SYNTHESIS OF THE VIRAL DNA EARLY
        DURING INFECTION AND CELL TRANSFORMATION
BY THE MURINE SARCOMA-LEUKEMIA VIRUS. (Eng.) Robin,
M. S. (St. Louis Univ. Sch. of Medicine, St. Louis,
Mo.); Salzberg, S,; Green, M. *Intervirology* 4(5):
268-278; 1974.

3865   INDUCTION OF YOLK-SAC TUMORS IN THE PREG-
        NANT RAT. (Eng.) Vandeputte, M. (Rega
Inst., Cathol. Univ., Leuven, Belgium); Sobis, H.
Eur. J. Obstet. Gynecol. Reprod. Biol. 5(1/2): 155-
159; 1975.

3866   FURTHER STUDIES ON THE RADIORESISTANCE OF
        HIGHLY PURIFIED MOUSE AND MAMMARY TUMOR
VIRUS. (Eng.) Gorka, C. (Laboratoire de Biologie
Cellulaire, Departement de Recherche Fondamentale,
Centre d'Etudes Nucleaires de Grenoble, B.P. 85

Centre de tri-38041 Grenoble Cedex, France); Mouri-quand, J. *Eur. J. Cancer* 11(6):397-402; 1975.

3867　　ISOLATION AND CHARACTERIZATION OF MOUSE
　　　　MAMMARY TUMOR VIRUS CORES [abstract].
(Eng.) Teramoto, Y. A. (Sch. Med., Univ. California, Davis, Calif. 95616); Cardiff, R. D. *Proc. Am. Assoc. Cancer Res.* 16:133; 1975.

3868　　THE ROLE OF PHYSIOLOGICAL STRESS ON BREAST
　　　　TUMOR INCIDENCE IN MICE. [abstract].
(Eng.) Riley, V. (Pacific Northwest Res. Found., Seattle, Wash.); Spackman, D.; Santisteban, G. *Proc. Am. Assoc. Cancer Res.* 16:152; 1975.

3869　　EFFECT OF LONG-TERM INCUBATION ON RNA
　　　　DEPENDENT DNA POLYMERASE (RDP) ACTIVITY
IN MOUSE MAMMARY TUMOR VIRUS (MMTV) [abstract]. (Eng.) Ashley, R. L. (Dept. Med. Pathol., Univ. California, Davis, Calif.); Manning, J. S. *Proc. Am. Assoc. Cancer Res.* 16:144; 1975.

3870　　STUDIES ON THE INTRACISTERNAL A-TYPE
　　　　PARTICLES IN MOUSE PLASMA CELL TUMORS:
INDUCTION OF MATURATION OF THE PARTICLES. (Eng.) Stewart, S. E. (Sch. Med., Georgetown Univ., Washington, D. C.); Kasnic, G., Jr., Urbanski, C.; Myers, M.; Sreevalsan, T. *Ann. N. Y. Acad. Sci.* 243:172-184; 1975.

3871　　ACTIVATION BY 5-IODODEOXYRUIDINE OF SHOPE
　　　　PAPILLOMA VIRAL GENOME IN CULTURED VX2
AND VX7 CARCINOMAS. (Eng.) Inokuchi, T. (Kyushu Dental Coll., Kita-Kyushu, Japan); Ikejiri, S.; Mizuno, F.; Osato, T. *Arch. Virol.* 48(3):275-277; 1975. *Arch. Virol.* 48(2):275-277; 1975.

3872　　EXCRETION OF MORPHOLOGICAL VARIANTS OF
　　　　HUMAN POLYOMA VIRUS. BRIEF REPORT. (Eng.) Lecatsas, G. (Inst. Pathology Univ. Pretoria, P.O. Box 2034, Pretoria, South Africa); Prozesky, O. W. *Arch. Virol.* 47(4):393-397; 1975.

3873　　EVIDENCE OF CHRONIC PERSISTENT INFECTIONS
　　　　WITH POLYOMAVIRUSES (BK TYPE) IN RENAL
TRANSPLANT RECIPIENTS.. (Eng.) Jung, M. (Inst. Med. Microbiol., St. Gallen, Switzerland); Krech, U.; Price, P. C.; Pyndiah, M. N. *Arch. Virol.* 47(1): 39-46; 1975.

3874　　INFLUENCE OF THE HOSTS'S SEX ON THE PATTERN
　　　　OF TUMOR INDUCTION BY INJECTION OF POLYOMA
VIRUS IN "NUDE" THYMUS DEFICIENT MICE [abstract]. (Eng.) Giovanella, B. C. (St. Joseph Hosp., Houston, Tex.); Stehlin, H. S,; Williams, L. J.; Klein, G. *Proc. Am. Assoc. Cancer Res.* 16:141; 1975.

3875　　RELEASE OF POLYOMA VIRUS FROM TRANSFORMED
　　　　PRIMATE CELLS BY CELL FUSION TECHNIQUES
[abstract]. (Eng.) Major, E. O. (Univ. Illinois Med. Cent., Chicago); Wright, P. J.; Bouck, N. P.; di Mayorca, G. *Proc. Am. Assoc. Cancer Res.* 16:134; 1975.

3876　　DIFFUSION AND ELECTROPHORESIS OF RNA TUMOR
　　　　VIRUSES [abstract]. (Eng.) McCormick, J. J. (Michigan Cancer Found., Detroit); Rich, M. A.; Liebes, L.; Rimai, L.; Salmeen, I. *Proc. Am. Assoc. Cancer Res.* 16:152; 1975.

3877　　NUTRITIONAL CONTROL OF GROWTH AND MORPHOLOG
　　　　OF CELLS TRANSFORMED BY AN RNA TUMOR VIRUS
[abstract]. (Eng.) Humphrey, L. P. (Boston Univ. Sch. Med., Mass.); Corwin, L. M. *Fed. Proc.* 34(3): 881; 1975.

3878　　INHIBITION OF GRAFT-VERSUS-HOST-REACTION
　　　　(GVHR) TUMOR INDUCTION BY STREPTOVARCIN
(SV), A REVERSE TRANSCRIPTASE INHIBITOR [abstract]. (Eng.) Cornelius, E. A. (Yale Univ. Sch. Med., New Haven, Conn.); Gray, G. D. *Fed. Proc.* 34(3): 960; 1975.

3879　　VIROGENICITY OF K3b STRAIN OF ROUS SARCOMA
　　　　VIRUS CONVERTED AKR MOUSE CELL [abstract].
(Jpn.) Kowata, J. (Sch. Microbiology, Chiba Univ., Chiba, Japan); Miki, T.; Okazaki, T. *Virus (Tokyo)* 26(3):282; 1975.

3880　　PROTEIN KINASE AND ITS REGULATORY EFFECT
　　　　ON REVERSE TRANSCRIPTASE ACTIVITY OF ROUS
SARCOMA VIRUS [abstract]. (Eng.) Lee, S. G. (Northwestern Univ. Med. Sch., Chicago, Ill.); Jungmann, R. A.; Hung, P. P. *Proc. Am. Assoc. Cancer Res.* 16:116; 1975.

3881　　SPECIES AND TISSUE SPECIFICITY OF AN
　　　　INHIBITORY ANTIBODY PREPARED AGAINST PURI-
FIED PLASMINOGEN ACTIVATOR FROM SV-40 TRANSFORMED HAMSTER CELLS [abstract]. (Eng.) Christman, J. K. (Mt. Sinai Sch. Med., New York, N.Y.); Silverstein, S. C.; Acs, G. *Fed. Proc.* 34(3):532; 1975.

UMOR ANTIGEN: A VIRUS-
PREEXISTING CELL PROTEIN.
Institut fur Virusforschung
szentrum, 69 Heidelberg,
West Germany); Fischer, H.
(4):899-901; 1975.

GLIOMAS IN MARMOSETS BY SIMIAN
, TYPE 1 (SSV-1) [abstract].
. (Rush-Presbyt.-St. Luke's
L.); Wolfe, L. G.; Whisler,
car, B.; Deinhardt, F. *Proc.*
16:119; 1975.

3884   GROWTH OF SV 40, ADENO 7 AND SV 40-ADENO 7
        VIRUSES IN MONKEY AND HUMAN CELLS AT 29
 AND 37 C. (Eng.) Kutinova, L. (Inst. Sera Vaccines,
Prague, Czechoslovakia). *Arch. Virol.* 47(3):257-268;
1975.

See also:

*   (Rev):   3614,   3615,   3616,   3646,   3647,   3648,
              3649,   3653,   3672
*   (Immun): 3894,   3895,   3896,   3897,   3898,   3904,
               3905,   3913,   3916,   3922,   3949,   3975,
               3980
*   (Path): 4012,   4028,   4038,   4079,   4088,   4111
*   (Epid-Biom): 4144

3885   AUTOLOGOUS STIMULATION OF HUMAN LYMPHOCYTE SUBPOPULATIONS. (Eng.) Opelz, G. (Sch. Medicine, Univ. California, Los Angeles, Calif. 90024); Kiuchi, M.; Takasugi, M.; Terasaki, P. I. *J. Exp. Med.* 142(5):1327-1333; 1975.

Background stimulation of autologous lymphocytes was studied *in vitro* in a mixed lymphocyte culture (MLC) test. Peripheral blood cells were obtained from healthy blood donors, 20-30 yr of age. T and B cell lymphocyte suspensions were fractionated into B-rich and T-rich subpopulations by the sheep RBC rosetting technique and the goat-antihuman immunoglobulin column fractionation. Unfractionated lymphocytes responded strongly to autologous B-rich cells (stimulation ratio of 15.35, and 3,401 cpm as measured by liquid scintillation spectrophotometry of [^{3}H]thymidine uptake); the response was half of that to allogeneic cells (30.31 ratio, and 6,718 cpm). The B-rich cell background was high (1,536 cpm) while that of T-rich cells was extremely low (98 cpm). The B-rich fraction stimulated the T-rich subpopulation extremely well (ratio, 29.06); autologons unfractionated cells showed less effect (ratio, 7.64). B-rich cells were also more effective than T-rich cells in stimulating allogeneic unfractionated lymphocytes. Similar findings were observed in 11 separate experiments. In an additional eight experiments, T-cell-rich populations, obtained by goat-antihuman immunoglobulin column fractionation, gave a background cpm of 123 $\pm$ 23; unfractionated cells gave 1,158 $\pm$ 187 cpm. Autologous stimulation of T-rich responders by unfractionated stimulators was observed in each instance (mean stimulation ratio, 12 $\pm$ 5). Thus, the results of autologous MLC tests with cells fractionated by both methods were comparable. The results indicate that the background [^{3}H]thymidine uptake in MLC tests can be altered by changing the proportion of T and B cells in the autologous lymphocyte preparation. Reduction of B cells results in a background lowering and an increase of B cells results in a heightening of the background. The authors conclude that stimulation of T cells; probably by autologous B cells, provides the most probable explanation for the alteration of background stimulation.

3886   ANTILEUKOTACTIC PROPERTIES OF TUMOR CELLS. (Eng.) Brozna, J. P. (Univ. Connecticut Health Center, Farmington, Conn. 06032); Ward, P. A. *J. Clin. Invest.* 56(3):616-623; 1975.

To explain why cellular inflammatory responses are defective in the face of malignant tumors and why malignant tumors fail to excite cellular inflammatory responses, a study was made of leukotactic inactivation due to a tumor cell product. Culture medium of *Escherichia coli* was used as the source of chemotactic factor (CF) against which tumor cell fractions were tested for chemotactic factor inactivator (CFI). The tumor cell fractions were obtained by subjecting extracts of sonically disrupted Walker carcinosarcoma and Novikoff hepatoma cells to fractional centrifugation in a discontinuous sucrose gradient. Inactivation experiments were carried out by mixing CF and tumor cell fraction,

incubating 30 min at 37 C, and assaying the final mixture for residual chemotactic activity. CFI was found in extracts of both tumors. The CFI directly inactivated the bacterial CF as well as the lenkotactic activity associated with C3 and C5 complement fragments and with culture fluids of lectin-stimulated lymphoid cells. The inactivation of the bacterial CF was temperature- and pH-dependent. The CFI was largely associated with the microsomal and cytosol fractions of the tumor cells. CFI activity was also found in neutrophils, alveolar macrophages, and extracts of liver, spleen, and kidney of normal rats. Although CFI was found to be present in ascitic fluids of animals bearing tumors, it was relatively absent in acute inflammatory exudates. The finding of the tumor-associated CFI may explain, in part at least, the tendency of malignant tumor cells to suppress cellular inflammatory reactions.

3887   RESTORATION OF IN VITRO IMMUNE RESPONSIVENESS OF MASTOCYTOMA-SUPPRESSED SPLENOCYTES BY ACTIVATED T CELLS. (Eng.) Kamo, I. (Albert Einstein Medical Center, Philadelphia, Pa. 19141); Patel, C.; Patel, N.; Friedman, H. *J. Immunol.* 115 (2):382-386; 1975.

Because mastocytoma cells are frequently used as an indicator for cell-mediated immune responses, and since the mastocytoma-induced immunosuppression appears to be mediated by a soluble factor(s), the mechanisms involved were investigated in detail. Spleen cells from normal DBA/2 mice pretreated with a soluble factor from mastocytoma cells or from ascitic fluid of mastocytoma-bearing mice were markedly impaired in terms of antibody formation to SRBC *in vitro*. Such immunosuppression by mastocytoma homogenates or ascitic fluid was reversed when syngeneic T cells activated to SRBC were added to the cultures, but not when peritoneal exudate cells or anti-θ-treated normal splenocytes were used. Activated T cells, as well as normal B lymphocytes prepared from spleens of lethally irradiated mice reconstituted with bone marrow cells, were less sensitive to the immunosuppressive factor than nonactivated T cells. The ability of educated T cells to restore immunocompetence of suppressed spleen cells *in vitro* suggests that the target of the immunosuppressive factor from mastocytoma cells may be nonactivated T cells, especially those involved in T cell helper function.

3888   IMMUNOLOGIC INDUCTION OF MALIGNANT LYMPHOMA: IDENTIFICATION OF DONOR AND HOST TUMORS IN THE GRAFT-VERSUS-HOST MODEL. (Eng.) Gleichmann, E. (Abteilung fur Experimentelle Pathologie, Medizinische Hochschule Hannover, 3 Hannover, Karl-Wiechert-Allee 9, West Germany); Gleichmann, H.; Schwartz, R. S.; Weinblatt, A.; Armstrong, M. Y. K. *J. Natl. Cancer Inst.* 54(1):107-116; 1975.

Primary lymphoid tumors were induced by inoculation of parental line spleen cells into H-2-incompatible F_1 hybrid mice. The F_1 recipients received ip injections of approximately 50 x 10^6 parental spleen cells weekly for four weeks. The genotypes of lym-

namely, enhanced recovery of the post-irradiation
immunological deficiency, transfer of virus by the
AKR thymic cells, and influence of thymic cells on
the maturation and/or differentiation of the lymphoid cells.

3890   NONSPECIFIC RESISTANCE TO TUMOR ALLOGRAFT.
            (Eng.) Rejthar, A. (Faculty Medicine, J.
E. Purkyne Univ., 656 91 Brno, Czechoslovakia); Wotke,
J.; Rejtharova, A.; Jaskova, J. *Neoplasma* 22(3):273-
278; 1975.

The effects of an increased volume of reticulohistio-
cyte system and macrophage activation in the recipient
after treatment with BCG and Zymosan on the course of
antitumor concomitant allograft-immunity reaction were
studied. The results were verified on C57B1/6J mice
with immunosuppression due to whole body irradiation
(^{60}Co, 500 rads). A substantially lower host resistance against the tumor allograft was shown by prophylactic administration of BCG and Zymosan. Transplantation of macrophages syngeneic to the host failed
to affect the latter's resistance to subsequent tumor
allograft. The irradiated host's reactivity was regenerated with a suspension of spleen cells or purified small lymphocytes, but transplantation of syngeneic macrophages remained without effect. The difference observed in the host's reactivity followed
the increase in lymphocyte volume and that following
BCG and Zymosan activation of the macrophage system
indicates a relatively small role for the macrophage
system in tumor allograft resistance in contrast to
the essential role played by specific immunity mechanisms mediated in the transplantation reaction mainly
by lymphocytes.

3891   DEFECTS IN CELL-MEDIATED IMMUNITY DURING
            GROWTH OF A SYNGENEIC SIMIAN VIRUS-IN-
DUCED TUMOR. (Eng.) Howell, S. B. (Natl. Cancer
Inst., Bethesda, Md. 20014); Dean, J. H.; Law, L.
W. *Int. J. Cancer* 15(1):152-169; 1975.

Four assays of tumor-specific cell-mediated immunity in BALB/c mice were compared during growth
of a syngeneic simian-virus-40-induced tumor, mKSA.
Major differences were found in immunity detected
by the tumor neutralization test (the Winn test),
direct tumor challenge, the microcytotoxicity assay, and the ^{51}chromium-release lymphocytotoxicity
assay. Progressive growth of the neoplasm followed
by loss of immunity (the eclipse phenomenon) was
documented with the Winn test. It was established
that this eclipse pehnomenon represented a lesion
in the T-cell system of tumor-bearing hosts. This
lesion was found to be specific and unrelated to
anatomic tumor location. The ability to produce
graft-*vs*-host reactions and the ability to respond
to mitogens were generally intact in tumor-bearing
animals. Cells capable of recognizing mKSA tumor
antigens and reconstituting an immune response following surgical removal of tumor or upon transfer
to normal mice were found in the spleens of mice
bearing advanced tumors. No suppressor cells that
might account for the eclipse phenomenon could be
demonstrated. Tumor-bearer serum did not block
neutralizing activity of immune spleen cells in

the Winn test, and immune cells were capable of
neutralizing tumor even in tumor-bearing hosts.
The possibility that the lesion is intrinsic to
T-cell precursors of effector cells is suggested
by the ability of T-cells from immune mice to
generate neutralizing activity in tumor-bearing
hosts.

3892   IMMUNE SURVEILLANCE DIRECTED AGAINST DE-
        REPRESSED CELLULAR AND VIRAL ALLOANTIGENS.
(Eng.) Martin, W. J. (Natl. Cancer Inst., Bethesda,
Md.). *Cell. Immunol.* 15(1):1-10; 1975.

A commentary on genetically-determined tumor susceptibility is presented and attempts are made to explain such susceptibility in terms of impaired immune surveillance due to a failure of normal repression of genetic loci coding for cell surface components. Certain cell surface antigens are found
on tumor cells that are not found on normal cells
of most strains of mice. Some of the cell surface
antigens are believed to be products of endogenous
C-type viruses. Successful repression of these
viral or cellular cell surface antigens by normal
cells, and derepression of the alloantigen on malignant cells could provide anti-tumor immunity. It
has been shown that endogenous C-type virus expression correlates with increased susceptibility to
tumors. An explanation for this observation is
presented which is based on immune surveillance
against derepressed viral alloantigens on a restricted basis.

3893   IMMUNODEPRESSION AND MALIGNANCY. (Eng.)
        Stutman, O. (Memorial Sloan-Kettering
Cancer Center, New York, N.Y. 10021). *Proc. Int.
Cancer Congr. 11th.* Vol. 1 *(Cell Biology and Tumor Immunology).* Florence, Italy, October 20-26,
1974. Edited by Bucalossi, P.; Veronesi, U.;
Coscinelli, N. New York, American Elsevier, 1975,
pp. 275-279.

Evidence supporting the view that malignant development, and not tumor growth or spread, is controlled by an immune mechanism is reviewed. Athymic nude (nu/nu) mice were used to test the effects
of immune deficiencies on tumor development and on
tumor induction by several oncogenic and viral
agents. In a pilot study, the spontaneous tumor
incidence of 35 male and 30 female nude mice in a
CBA/H background was 26 and 13%, respectively.
When the viable yellow gene that favors spontaneous
tumor development was inserted into the nude background, no significant difference in lung adenoma
incidence was observed between nude and control
mice. The incidence of spontaneous malignancies
in nude mice was comparable to that in immunologically normal controls. Experiments with chemical
carcinogens demonstrated no significant differences
in tumor incidence (except for thymic lymphomas)
between immunologically incompetent nude mice exposed to 3-methylcholanthrene or urethan and normal
heterozygotes (n/+). In another series of experiments (ongenic virus-induced tumors), no significant differences were observed between nu/nu and
nu/+ mice (both of C57BL/6 background) when the

animals were tested for susceptibility to leukemia
by Friend virus; both strains remained resistant
to leukemia development. When nude mice in BALB/c
or CBA/H background were infected with the Moloney
strain of murine sarcoma virus, all mice developed
tumors; however, nu/+ mice induced regression of the
growing tumor within 10-12 days after tumor appearance, demonstrating that tumor regression is a thymus-dependent event. In addition, a significant
difference in tumor development occurred in nude
mice infected with polyoma virus and nu/+-CBA/H
controls. Thus, the results with chemical and viral
carcinogens support those obtained with immunodepression procedures; no significant differences in
tumor incidence or latent periods for tumor development are observed between immunologically deficient
mice and their normal nu/+ siblings. The authors
conclude that the role of "immune surveillance"
in tumor development is not fully substantiated by
clinical and experimental data.

3894   DISSEMINATED HERPES SIMPLEX IN UNTREATED
        MULTIPLE MYELOMA. PARADOX OF PRESENT CON-
CEPTS OF IMMUNE DEFECTS. (Eng.) Deresinski, S. C.
(Santa Clara Valley Medical Center, 751 S. Bascom
Ave., San Jose, Calif. 95128); Stevens, D. A. *On-
cology* 30(4):318-323; 1974.

The case of a 64-yr-old man with disseminated herpes
simplex virus infection documented by direct immunofluorescence, and untreated multiple myeloma with
abnormal immunoglobulins is presented. Skin tests
performed on the 13th hospital day with purified
protein derivative of *Mycobacterium tuberculosis*
(5 TU), coccidioidin (1:100), histoplasmin, and
candida antigens were nonreactive. Reports of infections with intracellular pathogens in myeloma
patients are rare; pyogenic infections have been
amply documented. The untreated disease has been
considered a relatively pure defect in humoral immunity. Review of present knowledge suggests that
cell-mediated immunity is of paramount importance
in treating this virus infection. Immune defects
in multiple myeloma, and its infectious complications, may be more complex than suggested by current concepts of this disease.

3895   LEUKEMIA VIRUS-INDUCED IMMUNOSUPPRES-
        SION: SCANNING ELECTRON MICROSCOPY OF
INFECTED SPLEEN CELLS. (Eng.) Farber, P. (Albert
Einstein Medical Center, Philadelphia, Pa.);
Specter, S.; Friedman, H. *Science* 190(4213):469-
471; 1975.

Spleen cells, as well as other lymphoid tissue of
Friend leukemia virus (FLV)-infected mice were
scanned by electron microscopy (EM) to determine
the lymphocyte populations that might be related
to the infectious process and to the development
of immunosuppression. Balb/c mice were infected
by iv injection of FLV. At various times thereafter, representative mice were killed, and their
spleen lymphocytes were processed by Ficoll-Hypaque centrifugation and prepared for EM scanning.
In a parallel series of experiments, representative animals from each group were immunized with

694

response compromised by a strong humoral blocking
component in the C3H/HeJ (MTV-positive) mice and
that the observed diminished response to MBSA in
MTV-infected animals is not due to direct depressing
of the CMI reactivity. MTV infection may facilitate
the production of interfering or blocking humoral
immunity.

3897   IMMUNIZATION AGAINST MAREK'S DISEASE USING
        MAREK'S DISEASE VIRUS-SPECIFIC ANTIGENS
FREE FROM INFECTIOUS VIRUS. (Eng.) Lesnik, F.
(Veterinary Univ., Komenskeho 71 Kosice, Czechoslovakia); Ross, L. J. N. *Int. J. Cancer* 16(1):
153-163; 1975.

Immunity against Marek's disease was conferred by
the use of noninfectious materials extracted with
nonionic detergents from cells infected with the
attenuated strain of Marek's disease virus (MDV).
Antibody-free Rhode Island Red chicks were inoculated im at one wk of age with 0.1 ml cell extracts
emulsified in Freund's complete adjuvant and were
given a second inoculation one wk later without
adjuvant. Protection against natural infection was
obtained in groups inoculated with both soluble
(not sedimented at 100,000 x g/2 hr) and insoluble
antigens present in Nonidet P40 (NP40) extract, but
only with the insoluble fraction of deoxycholate
extract. The results suggest that the immunizing
antigens can be partially solubilized with 0.5%
NP40 and that the growth and spread of MDV are
reduced in immunized chickens.

3898   REGRESSION OF WARTS: AN IMMUNOLOGICAL
        STUDY. (Eng.) Pyrhonen, S. (Dep. Virol.,
Univ. Helsinki, Finland); Johansson, E. *Lancet* 1
(7907):592-596; 1975.

The duration and regression of warts in relation to
the occurrence of wart-virus antibodies were studied
in 173 patients with warts who were observed for 3-
6 mo. In 80% of the patients, wart-virus antibodies
were present and could be measured by immunodiffusion; in 20% these could also be detected by complement-fixation (CF) techniques. The mean duration of
the warts in patients with CF antibodies was 0·6 yr
and in the others 1·9 yr. The occurrence of CF antibodies (IgG) was associated with rapid healing; 75%
of these patients were cured during the first two
months of the observation period. In contrast, of
the patients with antibodies measurable only by the
immunodiffusion technique (IgM and/or low titers of
IgG), only 16% were cured during the first two
months, and they had a fairly constant cure-rate
(approximately 9%/mo) during six months of observation. The results indicate that the cure of warts
is partly connected with immunological phenomena,
especially with the presence of CF antibodies. In
other cases wart regression may be mainly a non-
immune process, perhaps due to a limited lifespan
of wart cells.

3899   PROLIFERATIVE CHARACTERISTICS OF MONO-
        BLASTS GROWN *IN VITRO*. (Eng.) Goud,
T. J. L. M. (Univ. Hosp., Leiden, The Netherlands);
van Furth, R. *J. Exp. Med.* 142(5):1200-1217; 1975.

A previous study in which mononuclear phagocyte colonies were grown in a liquid culture system was continued; the present study concerned the proliferative behavior of the monoblast and promonocyte in colonies. The cell-cycle times of both cell types were determined on the basis of four independent methods. The resulting values all showed excellent agreement: for the monoblast 11.0-11.9 hr and for the promonocyte 11.4-12.8 hr. The DNA synthesis time found for the two cell types amounted to 5.7 hr for the monoblast and 5.5 hr for the promonocyte. The duration of the other phases of the cell cycle of the proliferating mononuclear phagocytes proved to be: G2 phase, 0.6 hr; mitosis phase, 1.8 hr; and G1 phase, 3.5-3.8 hr. The individual colonies showed a biphasic pattern of colony growth, an initial phase of rapid proliferation followed by a stage with markedly decreased growth rate. In the initial stage only monoblasts were present in the colony; when the growth rate decreased, promonocytes and macrophages appeared. These observations support an earlier conclusion that the monoblast is without doubt the precursor of the promonocyte. Colony size varied widely. The main factor underlying this variation proved to be the lag time between the start of the culture and the time point at which the colony-forming cells began to divide. Mathematical analysis showed that the variation in colony size probably did not arise from heterogeneity of the population of colony-forming cells. A mathematical approach was used to determine the proportion of self-replicating and differentiating cells among the dividing monoblasts and promonocytes in the colony. The results indicated that initially *in vitro* the majority of the cells of both types were self-replicating cells, but later an increasing proportion of the dividing cells gave rise to another, more mature type of cell. On the basis of the conclusion that the monoblast initiates the mononuclear phagocyte colony, the number of monoblasts (2.5×10^5) present *in vivo* was estimated to be half the number of the mononocytes. In view of this ratio, the most likely pattern for the proliferation of mononuclear phagocytes in the bone marrow is that a monoblast divides once, giving rise to two promonocytes which in their turn divide once and form two nonproliferating monocytes.

3900   THE ORIGIN AND ROLE OF BLOOD-BORNE MONO-
        CYTES IN RATS WITH A TRANSPLANTED MYELO-
GENOUS LEUKAEMIA. (Eng.) Gauci, C. L. (Chester Beatty Res. Inst., Clifton Ave., Belmont, Sutton, Surrey, SM2 5PX U.K.); Wrathmell, A.; Alexander*, P. *Cancer Lett.* 1(1):33-37; 1975.

The origin and role of blood-borne monocytes associated with the growth of transplantable myelogenous leukemia (SAL) in August rats were investigated. Nonadherent leukemic blast cells of leukocyte enriched buffy coat preparations from heparinized blood of rats with SAL were separated from the macrophage-like blast cells which adhered to the culture vessel. Approximately 2.5% of the cells recovered from 10^5 cell/mm^3 of white cell populations of leukemic rats were macrophages. Leukemia did not result in normal rats from the injection of 2×10^5 of these detached macrophages.

SAL (2×10^5) was injected iv into a group of 18 rats at varying intervals; three rats from each group were then killed and their blood analyzed. In these rats, the number of blood monocytes rose progressively (20 times that of normal rats a day before their death) as the disease progressed. When the buffy coat of August x Marshall F$_1$ hybrids bearing SAL leukemia was cultured and the adherent and nonadherent cells treated with August anti-Marshall serum, only the macrophages bound to the allo-antisera. Binding of the immunoglobulin to the surface of living cells was determined by immunofluorescence. A similar experiment was performed with August rats that had been irradiated with a total body dose of 900 R x-rays and in whom the bone marrow had been reconstituted by the iv injection of 5×10^7 August x Marshall F$_1$ bone marrow cells. Seven days later, 10^5 SAL cells were injected into the rats; ten days later the blood was found to contain more than 10^5 leukemic blast cells/mm^3. The buffy coat gave rise from 5 to 6% macrophages and all the cells were found to be of the F$_1$ phenotype using the anti-Marshall serum assay. These experiments prove that monocytes present in the leukemic blood are of host origin and not derived from the SAL leukemic cells. In addition, the inflammatory reaction and the delayed hypersensitivity response to the ip injection of oyster glycogen of rats bearing SAL leukemia showed no difference from that of normal rats, demonstrating that blood monocytes of rats with SAL function normally in inflammatory situations. The authors conclude that the host response which is responsible for the monocytosis differs from that associated with the entry of monocytes into immunogenic sarcomas.

3901   RELEASE AND SPECIFICITY OF INHIBITORS FOR
        MACROMOLECULAR SYNTHESES FROM VARIOUS
SUSPENSION TUMOR STRAINS. (Eng.) Werner, D. (German Cancer Research Center, Heidelberg, West Germany); Schulte, M. *Eur. J. Cancer* 11(7):521-522; 1975.

Evidence for the release of a peptide that inhibits DNA and protein synthesis, from cells of various tumor strains is presented. Freshly-harvested cells, at the end of a first 90 min incubation, showed no incorporation of ^{3}H-thymidine and ^{14}C-amino acids. Ehrlich ascites tumor cells (in G0, G+, or Karzel) Walker 256 ascites cells, L-929 murine fibrocytes, and Zajdela rat hepatoma cells were studied. The cells immediately recovered their high rates of DNA and protein syntheses when the "conditioned" medium was renewed at the beginning of the second 90 min incubation. During the second 90 min incubation period, a second cessation of the macromolecular syntheses was seen in all the cell strains studied. The cells did not recover the high rates of syntheses when incubated in the medium "conditioned" by another cell strain, indicating that the inhibitors released are not cell-specific. The unspecific nature of the released factors is not consistent with the previously suggested involvement in autoregulation. The factors described here could cooperate with unknown cell specific factors to produce autoregulation *in vivo*.

tween autoantibody titers and the various clinical features (including age, sex, smoking habits, duration and nature of symptoms, presence of hemoptysis, neuropathy, finger clubbing, weight loss, lymph node enlargement, and liver involvement).

3904   ANTIBODY TO HERPES-LIKE VIRUS IN SARCOID-
OSIS. (Eng.) Mitchell, D. N. (Brompton Hosp., Fulham Road, London SW3 6HP, England); McSwiggan, D. A.; Mikhail, J. R.; Heimer, G. V. Am. Rev. Respir. 111(6):880-882; 1975.

Anti-herpes-like virus (HLV) antibody titers in serum samples from 70 consecutive patients with sarcoidosis were measured and compared with titers found in 70 age-, sex-, and race-matched controls. Samples were drawn at the time of diagnosis, and antibody titers were determined by indirect immunofluorescence. Antibody was detected in all 70 sarcoid and control sera. Of the 70 sarcoid patients, 49 (70%) had anti-HLV antibody titers < 1:600; 21 (30%) had titers ≥ 1:600. Of the 70 matched controls, 62 (89%) had titers < 1: 600, and only eight (11%) had titers ≥ 1:600. Although 30 sarcoid patients (43%) had titers higher than the matched control, the difference in the geometric mean titer of the patient and control groups was only 0.5 of a two-fold dilution. The difference in the geometric mean titer between patient and control subject did not vary significantly with sex, race, or type of sarcoidosis. However, differences between sarcoid and control sera were greater among nonwhite patients, who often had a more florid sarcoidosis. Among the 45 sarcoid patients tested for anti-HLV antibody titers after an interval of 18-60 mos, 39 (74%) had unchanged titers; four had a higher titer (by one dilution); and three had a lower titer (by one dilution). There was no evidence of changes in titer associated with progression, regression, or attainment of inactivity of the sarcoidosis. It is concluded that raised anti-HLV antibody titers in patients with sarcoidosis are most likely to be related to the ability of these patients to show enhanced humoral antibody responses.

3905   THE ROLE OF INTERFERON IN THE SPONTANEOUS
REGRESSION OF FRIEND VIRUS INDUCED LEU-
KEMIA. (Eng.) Furmanski, P. (Michigan Cancer Foundation, 110 East Warren Ave., Detroit, Mich. 48201); Juni, S.; Hall, L.; Rich, M. A. Proc. Soc. Exp. Biol. Med. 150(1):11-13; 1975.

The effect of Statolon and polyinosinic-polycytidilic acid (Poly-IC), and the role of interferon in spontaneous regression of Friend virus-induced murine leukemia was investigated. Random-bred male Swiss/ICRHa weanling mice inoculated with conventional Friend murine leukemia virus (MuLV) develop splenic leukemia, which only very rarely regresses spontaneously. A stable variant of Friend MuLV induces leukemia that spontaneously regresses in a large proportion of the infected animals. Conventional or variant Friend MuLV-infected mice were treated with either 4 mg Statolon or with 25 μg Poly-IC inoculated iv three days postinfection. In mice infected with conventional Friend MuLV and treated with Statolon or Poly-IC, the regression rate of leukemia was 10 and 23%, respectively. The rate of spontaneous regression was 38% of the variant strain-induced leukemia, but neither Statolon nor Poly-IC treatment (47 and 34%, respectively) had any significant effect on the spontaneous regression. No significant effect of Statolon or Poly-IC on the incidence of leukemia in conventional or variant Friend MuLV-infected mice was observed. Assays of interferon activity in sera from regressed mice was observed. Assays of interferon activity in sera from regressed mice were made using both the plaque reduction and yield reduction tests. No significant amounts of interferon were detected in the regressed sera above that observed in the control sera. The conventional or variant Friend MuLV did not induce interferon activity above the level in the control sera 6, 30, and 50 hr postvirus inoculation. Sera from variant Friend MuLV-induced leukemic mice at 28-35 days postvirus inoculation, the time just prior to spontaneous regression, did not contain any significant interferon activity. The authors conclude that interferon does not play a significant role in the spontaneous regression of Friend virus leukemia.

3906   SYNTHESIS OF α-FETOPROTEIN BY RAT ASCITES
        HEPATOMA CELLS. (Eng.) Kanai, K. (First
Dept. Medicine, Faculty of Medicine, Univ. Tokyo, Tokyo, Japan); Endo, Y.; Oda, T.; Kaneko, Y. *Ann. NY Acad. Sci.* 259:29-36; 1975.

Two aspects of α-fetoprotein (AFP) synthesis in rat hepatomas were examined. In experiment 1, the intracellular site of AFP synthesis in hepatoma cells was studied using a cell-free protein synthesizing system; in experiment 2, the effect of cyclic nucleotides on AFP production in cultured cells was studied. Rat ascites hepatoma cells were derived from the AH-66 strain transplanted ip to male Donryu rats. In experiment 1, [^{14}C]leucine was incorporated into the AFP fraction when hepatoma ribosomes were used in the system; ribosomes prepared from normal liver incorporated little radioactivity into AFP. The incorporation of [^{14}C]leucine into the AFP fraction by membrane-bound polysomes was 20-90 times higher than by free polysomes. In experiment 2, 0.5mM of dibutyryl-3',5'-adenosine monophosphate (DBcAMP) inhibited the growth of tumor cells; this inhibitory effect was enhanced by theophylline. The incorporation of [^{14}C]leucine into the proteins of tumor cells increased by 60% in the presence of DBcAMP. DBcAMP also increased the production of AFP by hepatoma cells. These experiments confirm the synthesis of AFP by tumor cells. Labeled AFP in the incubation medium indicated that AFP was synthesized mainly by the membrane-bound ribosomes of tumor cells and secreted into the medium, suggesting that AFP may have some function in the circulating blood.

3907   α-FETOPROTEIN IN YOLK SAC TUMOR. (Eng.)
        Tsuchida, Y. (Dept. Pathology, Faculty
of Medicine, Univ. Tokyo, Tokyo, Japan); Endo, Y.; Urano, Y.; Ishida, M. *Ann. N.Y. Acad. Sci.* 259: 221-233; 1975.

The phenomenon of α-fetoprotein (AFP) synthesis in

benign and malignant teratomas was studied. Only those serum AFP levels over 200 ng/ml were considered positive. A survey of AFP occurrence in 88 cases revealed that 33 were malignant; the tumor appeared in the mediastinum in one case, in the retroperitoneum in one, in the sacrococcygeal region in 14, in the testicle in 11, and in the ovary in six. Of the malignant teratomas, 25 were AFP-positive and eight were AFP-negative. Of the 55 benign cases, nine were AFP-positive and 46 were AFP-negative. AFP-positive benign tumors were seen only at the age of one month or younger. There were no AFP-positive cases in patients older than two months of age. Thus, the determination of AFP concentrations appears valuable in the differential diagnosis between benign and malignant teratomas, unless the patient is at the age of one month or younger. Of 19 cases of AFP-positive teratocarcinomas for which histological studies were carried out, 16 showed a typical yolk sac tumor (endodermal sinus tumor). It is proposed that the occurrence of AFP in teratocarcinomas is best explained by the concept of yolk sac tumor, because it is known that large amounts of AFP are synthesized not only by the fetal liver but also by the yolk sac during early embryonic life, and because the diagnosis of yolk sac tumor implies that the tumor is of yolk sac origin morphologically. More direct evidence of AFP synthesis was demonstrated by immunofluorescent techniques in one of the cases. A bright immunofluorescence was seen only in that cell layer that has long been defined as of yolk sac origin.

3908   THE ASSOCIATION OF CARCINOEMBRYONIC ANTI-
        GEN AND PERIPHERAL LYMPHOCYTES. (Eng.)
Papatestas, A. E. (Mount Sinai Hosp., 100th St., and Fifth Ave., New York, N.Y. 10029); Kim, U.; Genkins, G.; Kornfeld, P.; Horowitz, S. H.; Aufses, A. H., Jr. *Surgery* 78(3):343-348; 1975.

The potential value of combined carcinoembryonic antigen (CEA) assays and lymphocyte counts in cancer detection was investigated. A significant association between (CEA) titers and peripheral lymphocyte counts was observed in 148 simultaneous determinations. The association was present in a high cancer risk study group of 97 patients with myasthenia gravis and in a control group of 51 patients with granulomatous disease of the bowel, ulcerative colitis, and colonic neoplasms. A highly significant difference in the percentage of positive CEA assays in relation to lymphocyte counts was noted both in the study and control groups. In the study group CEA was positive in seven of 41 patients (17%) with counts above 2000 mm^3, and in 26 of 56 (43%) of those with lower counts (X^2=9.06, p < 0.005). The corresponding percentages in the control group were 20 and 61% (X^2 = 5.60, p < 0.025). A significant difference between the means of lymphocytes in groups with different CEA titers also was found (F = 6.77, p < 0.05). The finding of lower peripheral lymphocytes and/or higher titers of CEA in groups with increased risk of cancer, i.e., severe myasthenia gravis, patients with thymomas and patients with long history of granulomatous disease of the bowel, suggests an association between the results of the two tests and

698

reactivity of the antiserum with the CEA. The possibility that anti-CEA antibodies contaminating the anti-blood group A serum could have been responsible for the binding with CEA V was excluded by the finding that three different CEA preparations failed to inhibit the binding. When ^{125}I-labeled CEA 101, containing blood group B activity, was eluted from a Sepharose 6B column and the different fractions were tested for radioactivity and for ability to bind anti-CEA and anti-blood group B antisera, it was found that the peak of radioactivity was identical with that observed for nine other CEA preparations fractionated on this column. If the blood group specificity were due to aggregation with contaminating blood glycoproteins, the CEA molecules associated with these aggregated glycoproteins should have been eluted earlier from the column; this was not observed. Fractions of CEA separated by isoelectric focusing showed, despite great charge heterogeneity of the preparations, that the percentage binding by both anti-CEA and anti-blood group antisera remained the same in all fractions. Blood group A antigenic determinants of CEA III preparation were sensitive to periodate oxidation, while CEA-antigenic determinants were almost unaltered by this oxidative procedure. Further experiments revealed no evidence for the presence in test sera of complexes between CEA preparations and corresponding anti-blood group antibody. The findings were all consistent in indicating that different blood group antigenic sites are linked to CEA molecules. At the same time, not all CEA preparations appeared to carry blood group sites. It was not clear whether the antigenic determinants reacting with specific anti-blood group antisera are authentic blood group heterosaccharide side chains or are "blood group-like" antigens.

3911   ONCOFETAL PROTEIN ACCOMPANYING IRRADIATION-
        INDUCED SMALL-BOWEL ADENOCARCINOMA IN THE
RAT. (Eng.) Stevens, R. H. (Radiation Res. Lab.,
14 Medical Lab., Univ. Iowa, Iowa City, Iowa 52242);
England, C. W.; Osborne, J. W.; Cheng, H. F.;
Richerson, H. B. *J. Natl. Cancer Inst.* 55(4):1011-
1013; 1975.

A tumor-associated protein was found in tissue derived from an X-irradiation-induced adenocarcinoma in the small bowel of male Holtzman rats. The protein was associated with the cell membranes of the tumor tissue. It shared common antigenic determinants both with a protein similarly prepared from 17-19 day rat embryos, and with a perchloric acid-soluble protein isolated from the serum of these tumor-bearing rats.

3912   LOCATION BY IMMUNOELECTRON MICROSCOPY OF
        CARCINOEMBRYONIC ANTIGEN ON CULTURED ADENO-
CARCINOMA CELLS. (Eng.) Herberman, R. B. (Natl.
Cancer Inst., Bethesda, Md. 20014); Aoki, T.; Cannon,
G.; Liu, M.; Sturm, M. *J. Natl. Cancer Inst.* 55(4):
797-799; 1975.

The tissue-culture cell line HT-29, derived from a
primary colon adenocarcinoma and previously shown

to secrete carcinoembryonic antigen (CEA), was examined by immunoelectron microscopy for the location
of CEA in relation to the cells. CEA was closely
adjacent to the plasma membrane within 10 nm, a location indistinguishable from that of alloantigens
and other tumor-associated antigens. Previous reports that CEA was rarely demonstrated on the cell
surface may have resulted from interference with
labeling by the heavy coating on the cells used
previously.

3913   A MIXED-IMMUNOGLOBULIN ROSETTE TECHNIQUE
        FOR THE DETECTION OF ANTIBODY TO FELINE
ONCORNAVIRUS-ASSOCIATED CELL MEMBRANE ANTIGEN.
(Eng.) Mackey, L. (Leukemia Res. Unit, Univ. Glasgow,
Scotland); Jarrett, W.; Wilson, L. *Cancer Res.*
35(4):1064-1068; 1975.

A novel technique for detecting immunoglobulin
attached to cell membrane antigens is presented.
Cultured feline leukemia virus (FeLV)-infected lymphoid cells are incubated with the test serum, exposed to a rabbit anti-cat IgG serum, and the antigen-antibody reaction is detected by mixing the
lymphoid cells with SRBC that have been sensitized
with cat anti-SRBC serum. The presence of cat
immunoglobulin on the lymphoid cells is registered
by the formation of SRBC rosettes. The experiments
involved feline lymphoid cells from the FL74 cell
line chronically infected with FeLV of subgroups A,
B, and C, and dog lymphoid cells from the CT45S line
infected with FeLV5. An experiment to detect antibodies to feline oncornavirus-associated cell membrane antigens (FOCMA) by the mixed immunoglobulin
rosette (MIR) method indicated a sigmoidal relationship between serum dilution and rosette formation.
A comparison of the indirect immunofluorescence and
MIR tests showed a close correlation, with the MIR
method exhibiting a higher sensitivity. Sera from
22 uninfected kittens were examined at a 1:4 dilution; when the mean percentage of rosettes was $\pm$
2.7 at this dilution. This indicates that 25% rosette formation lies above the nonspecific range.
When sera from 32 cats with naturally occurring lymphoid cancers were examined, an additional seven
containing positive antibody were detected with the
MIR test. There was also good correlation of the
indirect immunofluorescence titer and the MIR titer
when FeLV-infected dog cells were employed. The
results showed that MIR is approximately ten times
more sensitive than indirect immunofluorescence for
the detection of antibody to FOCMA. Its advantages
include the accurate quantitation of rosettes and
the lack of expensive equipment of labeled sera; the
major disadvantage is the greater length of time
required for the 2-stage process.

3914   HETEROPHILE ANTIGENS AND ANTIBODIES IN
        TRANSPLANTATION AND TUMORS. (Eng.) Mil-
grom, F. (Dept. Microbiology, State Univ. New York,
Buffalo, N.Y.); Kano, K.; Fjelde, A.; Bloom, M. L.
Transplant. Proc. 7(2):201-207; 1975.

Paul-Bunnell antigen, identified by the agglutination of Forssman-negative bovine RBC by infectious-
mononucleosis sera, was examined for its occurrence

complex formed with the TLa-1,3 antiserum and thymus cell antigens contained two types of molecules of molecular weights 50,000 and 12,000, respectively. The large TL antigen polypeptide chain was of a size similar to the H-2 alloantigen-carrying polypeptide chain. In addition, the immune complex induced by TLa-1,3 antiserum contained a small polypeptide chain of a size similar to β_2-microglobulin. Experiments to investigate this relationship involved the isolation and adsorption of TL antigens. The immunosorbent-purified thymus cell surface antigens contained a small amount of aggregated material and antigens of three distinct molecular weights: 50,000, 37,000, and 12,000. The immune complexes were formed with the TLa-1,3 antiserum and contained only the 50,000 and 12,000 dalton components. Subjection of the TLa-1,3 ^{125}I-labeled immune complexes and ^{131}I-labeled β_2-microglobulin to polyacrylamide gel electrophoresis illustrated indistinguishable electrophoretic mobility and apparent homogeneity, as did two-dimensional electrophoresis. The results demonstrate that H-2 and TL-antigens share β_2-microglobulin as a common subunit, with a similar size large polypeptide chain carrying the distinctive antigenic determinants.

3916   TUMOR-SPECIFIC TRANSPLANTATION ANTIGEN(S) OF BOVINE ADENOVIRUSES. (Eng.) Panigrahy, B. (Baylor Coll. Med., Houston, Tex.); McCormick, K. J.; Trentin, J. J. *J. Natl. Cancer Inst.* 54(2):449-451; 1975.

Protection against Bovine Adenovirus type 3 tumor formation (both induced primary and transplantable tumor cells) was studied. Syrian hamsters (LSH/Lak) were immunized with bovine adenovirus types 1, 2, 3 and (BAV-1,-2,-3); human adenovirus type 12 (A-12); simian adenovirus type 7 (SA7); and chicken-embryo-lethal-orphan (CELO) virus. In the primary tumor prevention experiments, newborn hamsters were given 2.3×10^5 plaque-forming units (pfu) of BAV-3sc. They were then injected three times with either BAV-1 (1.3×10^5), BAV-2 (1.5×10^6), BAV-3 (2.3×10^6), A-12 (3.2×10^7), SA7 (3.2×10^7), CELO (3.9×10^8), or with control Madin-Darby bovine kidney cell (MDBK) extracts. (Viral titers were expressed as either pfu or 50% tissue culture infective dose.) No animal inoculated and immunized with BAV-3 developed tumors; immunization with BAV-1 and -2 allowed 44% and 50%, respectively, of the animals to develop tumors as compared with control MDBK extract-immunized animals, whose tumor incidence was 90%. Immunization with A-12, SA7, or CELO did not inhibit primary tumor formation. In the transplantation immunity tests, weanling hamsters were immunized with either BAV-1, -2, -3, or control MDBK cells. When these animals were challenged with 10^3 to 10^7 cells from BAV-3 tumor, tumor formation was observed. The animals immunized with BAV-3 gave a 50% tumor-producing dose of cells 2-3 logs above control, while BAV-1 or -2 immunization gave a 50% tumor-producing dose of cells of about 2 logs above control. These studies indicate that BAV-1, -2, -3 induce cross-reactive transplantation antigens that do not cross-react with those induced by oncogenic adenoviruses of either avian, simian, or human origin.

3917   T CELLS IN CHRONIC LYMPHOCYTIC LEUKAEM• (Eng.) Nowell, P. C. (Univ. Pennsylva• Sch. Med., Philadelphia); Daniele, R.; Winger, L. Rowlands, D. T., Jr.; Gardner, F. *Lancet* (7912):9• 1975.

The occurrence of T cells in a 38-yr-old woman wi• chronic lymphocytic leukemia (CLL) was investigat• Except for an episode of hemolytic anemia in 1969 requiring steroid therapy for six weeks, the dise• was quiescent for 12 years. Thereafter, a series viral and bacterial infections occurred, and ulti• mately resulted in her death one year later. Stu• of peripheral lymphocytes, including phytohemaggl• tinin (PHA) responsiveness, E rosette formation, rosette formation, and surface immunoglobulins we• performed on four occasions during the last two y• of life. Three studies exhibited a predominance • B-cells (i.e., a pattern consistent with typical cell). The other study was performed during an e• sode of acute pseudomonas sinusitis. She was re• ceiving antibiotics and prednisone which had been gradually reduced from 45 mg/day to 30 mg/day dur• the preceding two weeks. The study showed few B cells (EAC rosettes 8%) and predominance of T cel• (E rosettes 79%), with 4-5 times the normal numbe• in the circulation. These cells did not respond PHA. The acute infection may have been responsib• for the increase in the number of T cells and the failure to respond to PHA. It is also postulated that the steroid therapy was responsible for the • responsiveness to PHA. The temporary presence of neoplastic T cells in the blood, capable of formi• E rosettes but unresponsive to PHA, cannot be rul• out. Further investigation on the production of cells in CLL is indicated.

3918   CHRONIC LYMPHOCYTIC LEUKAEMIA OF T-CEL• ORIGIN: IMMUNOLOGICAL AND CLINICAL EVALUATION IN ELEVEN PATIENTS. (Eng.) Brouet, J.-C (Hospital Saint-Louis, 75475 Paris Cedex 10, France); Sasportes, M.; Flandrin, G.; Preud'Homme J.-L.; Seligmann, M. *Lancet* 2(7941):890-893; 197•

Eleven patients with chronic lymphocytic leukemia are reported. The identification of the leukemic cells was performed with seven different membrane markers for either T or B lymphocytes. The react• ity of the leukemic T cells with three different heteroantisera to T cells differed from patient t• patient but was homogeneous in individual cases. This finding suggests that the leukemic lymphocyt• belonged to a single subset of T cells. These lymphocytes responded to allogeneic cells in some these patients. In contrast, stimulation by non-specific mitogens (phytohemagglutinin or pokeweed• was poor in most patients. Two patients were affected with the prolymphocytic type of chronic ly• phocytic leukemia, but a characteristic clinical• hematological pattern was found in nine patients. The blood and marrow infiltration was moderate an• the proliferating T lymphocytes had a high conten• of lysosomal enzymes in all patients and cytoplas• granules in six cases. Other unusual features in-cluded massive splenomegaly (five patients) skin lesions (four patients), and major neutropenia (f•

patients). The results suggest that chronic lymphocytic leukemia of T-cell origin may not be as rare
as it has been previously considered.

3919   A SURFACE MEMBRANE DETERMINANT SHARED BY
         SUBPOPULATIONS OF THYMOCYTES AND B LYMPH-
OCYTES. (Eng.) Stout, R. D. (Stanford Univ. Sch.
Medicine, Stanford, Calif. 94305); Yutoku, M.; Grossberg, A.; Pressman, D.; Herzenberg, L. A. J. Immunol.
115(2):508-512; 1975.

Utilizing a quantitative fluorescence assay with the
fluorescence-activated cell sorter (FACS), it was
demonstrated that a rabbit antiserum obtained by immunization with cells of a mouse IgM-producing plasma
cell tumor (MOPC104E) is reactive with at least two
surface determinants (designated Th-B and ML2) on
subpopulations of normal murine lymphocytes. The
ML12 determinant was restricted to B lymphocytes. The
Th-B determinant was shared by splenic B lymphocytes
and a large subpopulation of thymocytes, the latter
of which expressed a 3-fold higher density of Th-B on
their surface than did the B lymphocytes. Neither
Th-B nor ML2 were found on peripheral T cells or on
brain, liver, or kidney cells. The available evidence
suggesting that Th-B may be a stem cell determinant
that is lost upon maturation is discussed.

3920   POSSIBLE T-CELL ORIGIN OF LYMPHOBLASTS
         IN ACID-PHOSPHATASE-POSITIVE ACUTE LYM-
PHATIC LEUKAEMIA. (Eng.) Ritter, J. (Universi-
tatskinderklinik, Martinistrasse 52, D-2 Hamburg,
West Germany); Gaedicke, G.; Winkler, K.; Beckmann, H.; Landbeck, G. Lancet 2(7924):75; 1975.

Three patients with acid-phosphatase-positive leukemia were tested with immunological cell-surface
markers. Blast cells from all patients formed
spontaneous rosettes with SRBC and with Vibrio
cholerae neuraminidase (VCN) SRBC. These results
plus the response of these patients to therapy
recommended for acute lymphatic leukemia (ALL) indicate that acid-phosphatase-positive leukemia is
a subgroup of ALL probably involving thymus-derived
lymphocytes. This disease is likely to be identical with Sternberg's sarcoma. We suggest the term
acid-phosphatase-positive acute lymphatic leuke-
mia/lymphosarcoma for this type of acute leukemia.

3921   CLASSIFICATION OF MALIGNANT LYMPHOMAS BY
         MEANS OF MEMBRANE MARKERS. (Eng.) Huber,
H. (Medizinische Univ.-Klinik, Anischstrasse 35, A-
6020 Innsbruck, Austria); Michlamayr, G.; Huber, C.
Acta Neuropathol. [Suppl.] (Berlin). 6:31-36; 1975.

The distribution of B and T lymphocytes was studied
in patients with chronic lymphocytic leukemia (CLL),
non-Hodgkin's lymphomas, and Hodgkin's disease. B
lymphocytes were evaluated autoradiographically with
^{125}I-labeled anti-immunoglobulin (Ig) antisera and
^{125}I-labeled heat-aggregated gamma globulin (AGG).
T lymphocytes were assessed by their capacity to form
rosettes with unsensitized neuraminidase-treated SRBC.
In some studies, the proliferating capacity of lymph-

ocytes from patients with Hodgkin's disease or infectious mononucleosis was evaluated by the combined use
of ^{3}H-thymidine labeling and membrane markers. In
most CLL patients, the leukemic lymphocytes behaved
as B cells; thus, 17 of 18 patients had very high
percentages of blood lymphocytes reacting with AGG as
well as with anti-Ig sera, while the percentage of T
lymphocytes was inversely related to the lymphocyte
count. The density of membrane receptors on B lymphocytes was reduced in these patients, and a monoclonal
proliferation of neoplastic cells was observed in sev-
en of nine patients studied. A marked increase of
blood lymphocytes with membrane-bound Ig and receptors
for AGG was noted, however, in only four of 16 patients with non-Hodgkin's lymphoma. In contrast, cell
suspensions of affected lymph nodes showed high numbers of lymphocytes reacting with AGG and anti-Ig
antisera in ten of 12 patients. The density of membrane receptors was not reduced in this group although
lymphoma cells were monoclonal with respect to their
Ig coat. The B/T cell ratio of blood lymphocytes of
patients with Hodgkin's disease was comparable to that
of normal controls. The percentage of B lymphocytes
was slightly higher in affected lymph nodes than in
uninvolved lymph nodes. A preferential proliferation
of T lymphocytes was seen in patients with Hodgkin's
disease and in patients with infectious mononucleosis.
A T lymphocyte response to an as yet unknown stimulus,
perhaps accompanied by a reduced lymphocyte life span,
may play an important role in the pathogenesis of
Hodgkin's disease.

3922   INHIBITION OF THE MITOGENIC RESPONSE OF
         NORMAL PERIPHERAL LYMPHOCYTES BY EXTRACTS
OR SUPERNATANT FLUIDS OF A HERPESVIRUS SAIMIRI LYM-
PHOID TUMOR CELL LINE. (Eng.) Neubauer, R. H.
(Litton Bionetics, Inc., Kensington, Md. 20795);
Wallen, W. C.; Rabin, H. Infect. Immun. 12(5):
1021-1028; 1975.

A Herpesvirus saimiri-infected marmoset lymphoid
cell line (MLC-1) was examined for the presence of
soluble factors which might affect lymphocyte functions and, therefore, relate to the pathogenesis of
lymphoma in vivo. MLC-1 cells, cell extracts, and
culture fluids reduced the spontaneous DNA synthesis
of normal peripheral blood lymphocytes and completely
inhibited their response to phytohemagglutinin
(PHA). Suppression of PHA response was demonstrated
against a variety of human and nonhuman primate
species, with 90-95% inhibition occurring at dilutions of extract as great as 1:5,120. Inhibition of
this type was also demonstrated using extracts of
2 of 5 other lymphoblastoid cell lines. Physical-
chemical characteristics of the active factor(s)
revealed a non-sedimentable, non-dialyzable, trypsin-
resistant molecule that was stable at 56 C for 30 min
but inactivated at 80 C for 30 min. The factor(s)
also exerted an effect on some but not all established
lymphoblastoid cell lines, where DNA, RNA, and protein
synthesis were all inhibited, with DNA synthesis being
the most affected (95% suppression). Cellular respir-
ation was not affected by the presence of the fac-
tor(s), and the inhibition of DNA synthesis was reversible after 24 hr. Purified human interferon did not
reduce the PHA response of normal owl monkey peripheral blood lymphocytes and was less effective

against an established lymphoblastoid cell line than the MLC-1 extract. Antiviral activity was also demonstrated in the preparations and may represent interferon, which these cells are known to produce at low levels.

3923   CHANGES IN LYMPHOCYTE SURFACE IMMUNOGLO-
        BULINS IN MYELOMA AND THE EFFECT OF AN
RNA-CONTAINING PLASMA FACTOR. (Eng.) Chen, Y.
(Veterans Administration West Side Hosp., P.O. Box
8195, Chicago, Ill. 60680); Bhoopalam, N.; Yakulis,
V.; Heller, P. *Ann. Intern. Med.* 83(5):625-631;
1975.

A study was carried out on the mechanism of conversion of the surface immunoglobulins on circulating lymphocytes of patients with multiple myeloma. The patients consisted of 13 men and one woman; ten had IgG, and four had IgA myeloma. Seven of the globulins were of the kappa type; the other seven were lambda. Normal human IgG, IgA, IgM, and kappa and lambda Bence Jones proteins were purified and used for preparation of rabbit antisera, rendered anti-isotypically specific by immunoabsorption procedures. Antisera to myeloma proteins of five patients were prepared and rendered anti-idiotypically specific by similar procedures. The anti-isotypic sera were used for visualization of surface immunoglobulin by direct immunofluorescence; the anti-idiotypic sera, for showing monoclonal myeloma globulins by indirect immunofluorescence. Peripheral blood lymphocytes were isolated by differential centrifugation on Ficoll-Hypaque gradients. The number of B cells with surface immunoglobulins was expressed as percentage of the total number of lymphocytes per cubic millimeter. RNA-rich extracts were prepared from plasma of myeloma patients by a phenol extraction procedure. Experiments on digestion of RNA-rich preparations were performed with pancreatic RNase, DNase, and pronase. *In vitro* tests for conversion of surface immunoglobulins by the RNA preparations were carried out by incubating 10^6 lymphocytes in 0.1 ml barbital-buffered saline (BBS) with 1 mg RNA in 0.1 ml BBS for 30 min at 37 C. After three washings, surface immunoglobulin of the lymphocytes was determined by indirect immunofluorescence. Concentrations of actinomycin D or cycloheximide that are known to suppress synthesis of RNA and of protein, respectively, were added in some tests. In the 14 patients with myeloma, there was a statistically significant reduction in the percentage of lymphocytes with surface immunoglobulins of the various isotypes to about one-third of that in normal controls. When, however, antiidiotypic antisera to the respective myeloma globulins were used for visualization of surface immunoglobulins, a large number of surface immunoglobulin-carrying lymphocytes were detected. The possibility of absorption of these monoclonal surface immunoglobulins from the surrounding plasma was excluded by showing their resynthesis after removal from the cells by trypsinization. The change in the character of surface immunoglobulins was reproduced on normal lymphocytes by the RNA-rich extracts; this effect was inhibited by RNase and cycloheximide. These findings suggest the possiblity that an RNA-

containing plasma factor transmits information for synthesis of surface immunoglobulins between myeloma cells and normal lymphocytes. This mechanism may contribute to the dysfunction of B lymphocytes in patients with myeloma, leading to immunologic deficiency.

3924   LIGAND-INDUCED REDISTRIBUTION OF LYMPHO-
        CYTE MEMBRANE GANGLIOSIDE GM1. (Eng.)
Revesz, T. (ICRF Tumour Immunology Unit, Zoology
Dept., University Coll., London WC1E 6BT, UK);
Greaves, M. *Nature* 257(5522):103-106; 1975.

The redistribution of monosialo-gangliotetraosyl-ceramide (GM1) into aggregates and caps, using cholera toxin (choleragen) and labeled antibodies as ligands, was studied. All lymphocytes, monocytes and granulocytes studied bound choleragen to their surface and uptake was rapidly saturated at 20 C. Binding was inhibited by purified GM1. The biologically inactive choleragenoid bound almost as effectively as choleragen. Human lymphocytes showed pronounced redistribution into discrete caps over a wide range of concentrations. Mouse lymphocytes showed caps only with the lower concentrations used. By incubating mouse spleen lymphocytes with low concentrations (10^{-4}-10^{-2} µg ml^{-1}) of choleragen at 37 C followed by the two antibody layers at 4 C in the presence of azide it was possible to demonstrate capping on approximately 50% of reactive cells. Creation of more GM1 binding sites for choleragen by neuraminidase pretreatment, into GM1, or by direct insertion of GM1 into cells, inhibited capping of GM1 on human tonsil cells. Capping of immunoglobulin molecules on lymphocytes has been reported to be inhibited by a combination of colchicine and cytochalasin. These drugs had precisely the same effect on GM1 redistribution. After either GM1 insertion or neuraminidase treatment acute leukemia cells showed pronounced capping at 10^3 µg GM1/ml but not at higher concentrations.

3925   HUMAN TUMOR-LYMPHOCYTE INTERACTION *IN*
        VITRO: BLASTOGENESIS CORRELATED TO
DETECTABLE IMMUNOGLOBULIN IN THE BIOPSY. (Eng.)
Vanky, F. (Dept. Tumor Biology, Karolinska Inst.,
Stockholm, Sweden); Trempe, G.; Klein, E.;
Stjernsward, J. *Int. J. Cancer* 16(1):113-124;
1975.

Binding of radioiodinated anti-immunoglobulin (Ig) reagent assayed by acid elutable radioactivity was shown in 18 of 44 cell suspensions (41%) prepared from surgical specimens of human tumors. Aliquots of these biopsies were admixed to autologous lymphocytes and in 13 cases they induced stimulation of DNA synthesis. In only one of these 13 cases was the anti-Ig reagent bound, while among the 26 biopsies with low or no binding capacity 12 (46%) were stimulatory, indicating that immunoglobulin-containing biopsies are not stimulatory. Different tiated tumors were among those which had low mounts of Ig or none at all. Malignant tumors such as the myeloma and osteosarcomas contained Ig. Experiments on 32 lymphocyte preparations from different lymphoid organs suggest that the immunoglobulin detected in the tumor cell suspension is not derived from the infiltrating lymphoid cells. The results support previous studies indicating that the presence of Ig in tumor cells is related to the degree of malignancy.

3926　E AND EAC ROSETTING LYMPHOCYTES IN PATIENTS WITH CARCINOMA OF BRONCHUS. I. SOME PARAMETERS OF THE TEST AND OF ITS PROGNOSTIC SIGNIFICANCE. (Eng.) Anthony, H. M. (Sch. Med., Leeds, England); Kirk, J. A.; Madsen, K. E.; Mason, M. K.; Templeman, G. H. *Clin. Exp. Immunol.* 20(1): 29-40; 1975.

T and C3 receptor B lymphocytes in normal donors and in patients with bronchial carcinoma were estimated. Using peripheral blood lymphocytes separated by a Ficoll method and suspended in saline, means of 77.1%, standard deviation (s.d.)= 5.2, E rosettes (T lymphocytes), and 20.1% (s.d. 6.7) SRBC rosettes (B lymphocytes) were obtained with 11 healthy donors (nine men, two women; 22-59 yr). Poorer SRBC-rosette formation resulted from higher centrifugation speeds during the washing of lymphocytes or RBC, insufficient chilling, or rough handling. The presence of 5% albumin in the final mixture stabilized the rosettes and brought a constant subpopulation of B lymphocytes into rosetting. In 31 patients with bronchial carcinoma who, at the time of diagnosis, had SRBC-rosette percentages below 1 s.d. of the mean for normal donors, the length of survival was significantly shorter than in those with normal or high values. The same was true for those in whom null cells were detected. In each case the correlation effect was mainly found in the group of 17 patients with squamous carcinoma, as opposed to those with oat cell carcinoma (eight patients). The complete enumeration of T and B lymphocytes may contribute to the *in vitro* investigation of immune status in cancer patients.

3927　ACTIVATION OF HUMAN PERIPHERAL BLOOD LYMPHOCYTES BY CONCANAVALIN A DEPENDENCE OF MONOCYTES. (Eng.) Hedfors, E. (Serafimer Hosp., P.O. Box 12700, S-112 83 Stockholm, Sweden); Holm, G.; Pettersson, D. *Clin. Exp. Immunol.* 22(2):223-229; 1975.

The potentiation of *in vitro* concanavalin A (Con A)-induced DNA synthesis in human peripheral lymphocytes by blood monocytes was subjected to further investigation. The lymphocytes were prepared from defibrinated venous blood by sedimentation through gelatine, followed by iron treatment to remove phagocytic cells. Acid-treated glass beads were coated with human IgG followed by rabbit anti-human IgG antiserum and were used for absorption and elution of the desired T cells fraction from the lymphocyte preparation. The monocytes were prepared from heparinized venous blood by sedimentation through gelatin, followed by centrifugation on a bovine serum albumin gradient. Monocytes were separated from the top layer fraction by adherence methods using plastic tissue culture flasks. The monocytes were treated with mitomycin C and were used in combination with autologous lympho-

cytes at varying ratios and varying total numbers of cells per culture tube. The culture medium consisted of supplemented RPMI 1640 containing 15% human AB serum. DNA synthesis was determined on the basis of ^{14}C-thymidine incorporation. It was found that the activation of the human lymphocytes or of purified lymphocytes by Con A was highly potentiated by the presence of the autologous monocytes. The optimal lymphocyte:monocyte ratio was 1:1 when the incorporation of thymidine was expressed as total incorporation per culture tube and 1:10 when expressed per lymphocyte. A 5- to 10-fold increase of total DNA synthesis was noted in the presence of 10-90% monocytes. The mechanism responsible for the effect was not known but among those offered were reduced survival of lymphocytes in monocyte-depleted cultures, increased stimulating potency of Con A when bound to monocytes, shortening of cell cycle of Con A-activated lymphocytes by the presence of monocytes, or selective removal of T cells participating in primary response to Con A during fractionation of lymphocytes on the Ig-anti-Ig column. The findings may help to explain the wide variations in Con A responsiveness of human peripheral lymphocytes on the basis of contamination with varying amounts of monocytes.

3928　DISTRIBUTION OF CONCANAVALIN A RECEPTOR SITES ON SPECIFIC POPULATIONS OF EMBRYONIC CELLS. (Eng.) Roberson, M. (Dept. Biology, California State Univ., Northridge, Calif. 91324); Neri, A.; Oppenheimer, S. B. *Science* 189(4203):639-640; 1975.

The distribution of concanavalin A (Con A) receptor sites on the micromere and nonmigratory embryonic cell types (macromeres and mesomeres) of the sea urchin embryo were identified by phase contrast fluorescence microscopy. The distribution of Con A receptor sites was random on all three embryonic cell populations. However, 95% of the cells treated with fluorescein isothiocyanate-conjugated Con A before fixation manifested capped or highly clustered receptor sites on the micromeres. The distribution of receptor sites remained random on the mesomeres and macromeres. The results indicate that Con A receptor sites are more mobile on specific populations of malignant-like migratory embryonic cells.

3929　LEUKAEMIA IN CHILDREN AND THEIR GRANDPARENTS: STUDIES OF IMMUNE FUNCTION IN SIX FAMILIES. (Eng.) Till, M. M. (Inst. Child Health, London, England); Jones, L. H.; Pentycross, C. R.; Hardisty, R. M.; Lawler, S. D.; Harvey, B. A. M.; Soothill, J. F. *Br. J. Haematol.* 29(4):575-586; 1975.

Seven of 500 children with acute leukemia seen over a 15-yr period were known to have a close relative with leukemia or lymphoma. In each case the affected relative was a grandparent of the child, 6 of the 7 being paternal grandparents. The following laboratory tests were performed on the sera or blood cells of most of the parents, sibs, and/or children: total and differential WBC count, serum immunoglobu-

sue, and that human tumor cells may produce characteristic isoferritins which are not present in normal adult liver ferritin. The authors suggest that normal tissue ferritins are hybrid molecules consisting of different portions of at least two dissimilar subunits; the shift in isoferritin profiles in cancer cells may result from a synthesis of a unique subunit or by alterations in the proportions of two normal subunits.

3931   SIGNIFICANCE OF CELLULAR IMMUNOLOGY IN
        NEOPLASTIC DISEASES. (Pol.) Urasinski,
I. .(Klinika Hematologiczna Instytutu Chorob Wewn.
PAM, ul. Unii Lubelskiej 71-344 Szczecin, Poland).
Wiad. Lek. 28(8):653-656; 1975.

3932   THE IMMUNOLOGY OF BREAST CANCER. (Fre.)
        Hoerni, M. B. (Fondation Bergonie, 180, rue
de Saint-Genes, F 33076 Bordeaux-Cedex., France);
Chauvergne, J. *Bordeaux Med.* 8(5):483-491; 1975.

3933   GROWTH OF HUMAN CANCER CELLS AS LUNG
        METASTASES IN IMMUNOLOGICALLY TOLERANT
RATS. (Eng.) Southam, C. M. (Jefferson Medical
Coll., Philadelphia, Pa. 19107); Babcock, V. *Proc.
Soc. Exp. Biol. Med.* 149(1):136-141; 1975.

3934   ANTICANCER ACTIVITY OF REGIONAL LYMPH
        NODES. (Eng.) Orita, K. (Okayama Univ.
Medical Sch., 2-5-1 Shikata, Okayama City, Japan).
Panminerva Med. 17(5/6):170-173; 1975.

3935   THE IMMUNOLOGICAL POTENTIAL OF HUMAN LYMPH
        NODES IN UREMIA AND DURING EXTRACORPOREAL
IRRADIATION OF THE BLOOD. (Eng.) Birkeland, S. A.
(Klovervaenget 26 C, DK-5000 Odense, Denmark);
Moesner, J.; Amtrup, F.; Svendsen, E. V. *Acta
Pathol. Microbiol. Scand.* [C] 83(4):289-297; 1975.

3936   REPRODUCIBILITY OF LYMPHOCYTOTOXICITY AND
    ·   BLOCKING USING A ^{51}Cr-MICROCYTOTOXICITY
ASSAY IN BREAST CANCER PATIENTS AND NORMAL CONTROLS
[abstract]. (Eng.) Levin, A. C. (Rush-Presbyt.-
St. Luke's Med. Cent., Chicago, Ill.); Massey, R. J.;
Kruse, R.; Schewitz, D.; Deinhardt, F. *Fed. Proc.*
34(3):846; 1975.

3937   BIPHASIC TUMOUR-ENHANCING AND TUMOUR-
    ·   INHIBITING EFFECTS OF NON-SPECIFIC IMMUNO-
THERAPY. (Eng.) Siegel, I. (Roosevelt Hosp., New
York, N.Y.); Grieco, M. H.; Rice, J. V. *Lancet*
(7916):1138; 1975.

3938   STUDIES CONCERNING THE IMMUNE RESPONSES OF
        THE CANCER PATIENTS--ESPECIALLY ON THE SKIN
TEST BY PPD, DNCB AND THE VARIANTS IN T-CELL [abstract]. (Jpn.) Chiba, H. (Jikei Univ. Sch. Medicine, Tokyo, Japan); Takahashi, N.; Ohtsuka, A.;
Nagao, F. *Gann, Proc. Jpn. Cancer Assoc., 34th Annual
Meeting, October 1975.* p. 51.

3939   EXPERIMENTAL STUDIES ON THE ALTERATION
        OF IMMUNE RESPONSE IN RATS DURING GASTRIC
CARCINOGENESIS BY PERORAL ADMINISTRATION OF MNNG
[abstract]. (Jpn.) Takahashi, M. (Jikei Univ. Sch.
Medicine, Tokyo, Japan); Takahashi, N.; Chiba, H.;
Ohtsuka, A.; Nagao, F. *Gann, Proc. Jpn. Cancer
Assoc., 34th Annual Meeting, October 1975.* p. 49.

3940   AUTOCHTHONOUS GRAFTS OF SPONTANEOUS MAMMARY
        TUMORS IMPLANTED TOGETHER WITH ALLOGENEIC
TISSUES. (Eng.) Tokuzen, R. (*Natl. Cancer Center
Res. Inst., Tsukiji, Chuo-ku, Tokyo, Japan); Naka-
hara, *W. *Z. Krebsforsch.* 81(3/4):239-242; 1974.

3941   IMMUNOLOGICAL DEVIATION BETWEEN A TRANS-
        PLANTABLE TUMOR AND ITS ARTIFICIALLY CON-
VERTED ASCITIC VARIANT. (Eng.) Tokuzen, R. (*Natl.
Cancer Center Res. Inst., Tsukiji, Chuo-ku, Tokyo,
Japan); Nakahara, *W. *Z. Krebsforsch.* 81(3/4):235-
238; 1974.

3942   ACCELERATED DEATH OF AKR/J MICE BY ASCITIC
        TUMOR OF C3H MICE. (Eng.) Okada, H.
(Natl. Cancer Cent. Res. Inst., Tokyo, Japan). *Jpn.
J. Exp. Med.* 45(1):47-48; 1975.

3943   THE ROLE OF THYMUS ON AUTOSENSITIZATION
        AGAINST SYNGENEIC NORMAL AND MALIGNANT
CELLS. (Eng.) Carnaud, C. (Dept. Tumor Biology,
Karolinska Inst., S-104 01 Stockholm 60, Sweden);
Ilfeld, D.; Petranyi, G.; Klein*, E. *Eur. J. Immunol.*
5(8):575-579; 1975.

3944   CANCER IN IMMUNOSUPPRESSED PATIENTS.
        (Eng.) Penn, I. (Univ. Colorado Medi-
cal Center, Denver, Colo.). *Transplant. Proc.*
7(1/Suppl. 1):553-555; 1975.

3945   EFFECT OF HOST SENSITIZATION ON PATTERNS
        OF METASTASIS. (Eng.) Glaves, D. (Dept.
Experimental Pathology, Roswell Park Memorial Inst.,
Buffalo, N.Y.); Weiss, L. *Transplant. Proc.* 7(2):
253-257; 1975.

3946   PRODUCTION OF DIALYZABLE TRANSFER FACTOR
        OF CELLULARLY MEDIATED IMMUNITY BY CON-
TINOUSLY PROLIFERATING LYMPHOBLASTOID CELLS. (Fre.)
Goust, J.-M. (Laboratoire d'Immunologie, CHU Pitie-
Salpetriere, 91-105, boulevard de l'Hopital, 75634
Paris Cedex 13, France); Viza, D.; Moulias, R.;
Trejdosiewicz, L.; Lesourd, B.; Marescot, M.-R.
C. R. Acad. Sci. [D] *(Paris)* 280(3):371-374; 1975.

3947   IMMUNOCYTE ADHESION AND SUBLETHAL IRRADIA-
        TION. (Fre.) Beaumariage, M. L. (Labora-
toire d'Anatomie pathologique, Universite de Liege,
1, rue des Bonnes Villes, B-4000 Liege, Belgium);
Hiesche, K.; Revesz, L.; Haot, J.; Betz*, E. H.
C. R. Soc. Biol. (Paris) 169(2):459-463; 1975.

3948   YOSHIDA ASCITES SARCOMA GROWN IN MICE.
        (Eng.) Marusic, M. (Dept. Physiology,
Univ. Zagreb, Salata 3, p.o.b. 925, Y-41001 Zagreb,
Yugoslavia); Allegretti, N.; Culo, F. *Int. Arch.
Allergy Appl. Immunol.* 49(4):568-572; 1975.

3949   ACTIVATION OF SUPPRESSOR CELLS AFTER TUMOR
        INDUCTION AND AFTER INJECTION OF *CORYNE-
BACTERIUM PARVUM (CP)* [abstract]. (Eng.) Kirchner,
H. (Natl. Cancer Inst., Bethesda, Md.); Chused, T. M.;
Holden, H. T.; Herberman, R. B. *Fed. Proc.* 34(3):
990; 1975.

3950   IMMUNOSUPPRESSIVE ACTIVITY OF HUMAN ASCITIC
        FLUID [abstract]. (Eng.) Badger, A. M.
(Boston Univ. Med. Sch., Mass.); Cooperband, S. R.;
Glasgow, A. H. *Fed. Proc.* 34(3):1003; 1975.

3951   TUMOR INDUCED IMMUNOSUPPRESSIVE FACTOR(S)
        [abstract]. (Eng.) Cooperband, S. R.
(Boston Univ. Sch. Med., Mass.); Badger, A. M.;
Glasgow, A. H. *Clin. Res.* 23(3):424A; 1975.

3952   SUPPRESSION OF IMMUNE RESPONSE TO SHEEP
        RED BLOOD CELLS IN MICE TREATED WITH
PREPARATIONS OF A TUMOR CELL COMPONENT AND IN TUMOR-
BEARING MICE. (Eng.) Pikovski, M. A. (Heb. Univ.
Hadassah Med. Sch., Jerusalem, Israel); Ziffroni-
Gallon, Y.; Witz, I. P. *Eur. J. Immunol.* 5(7):447-
450; 1975.

3953   CHRONIC MYELOGENOUS LEUKEMIA AS A POSSIBLE
        CONSEQUENCE OF IMMUNOSUPPRESSIVE TREATMENT
OF NEPHROTIC SYNDROME. (Ger.) Cap, J. (Fac. Medicine, Komensky Univ., Bratislava, Czechoslovakia);
Misikova, Z. *Monatsschr. Kinderheilkd.* 123(10):718-
720; 1975.

3954   IMMUNE SUPPRESSING EFFECT OF A HUMAN SERUM
        FERROPROTEIN OF HEPATIC ORIGIN: α_2 H
GLOBULIN. STUDY OF THE BLAST TRANSFORMATION OF NORMAL
LYMPHOCYTES IN THE PRESENCE OF PHYTOHEMAGGLUTININ.
(Fre.) Buffe, D. (Laboratoire d'Immunochimie, Insti-
tut de Recherches sur le Cancer, B. P. n° 8, 94800
Villejuif, France); Rimbaut, C. *C. R. Acad. Sci.*
[D] *(Paris)* 280(3):367-370; 1975.

3955   SERA FROM IRRADIATED RABBITS AS A SOURCE
        OF COMPLEMENT FOR IMMUNE CYTOLYSIS ASSAYS.
(Eng.) Walker, W. S. (St. Jude Children's Res. Hosp.,
Memphis, Tenn.   38101); Mills, B.; Mohanakumar, T.
Transplantation 20(3):260-262; 1975.

3956   SERUM IMMUNOGLOBULIN LEVELS IN MICE BEARING
        A CHEMICALLY INDUCED SARCOMA. (Eng.)
Merino, F. (Departamento de Medicina Experimental,
Instituto Venezolano de Investigaciones Cientificas,
Apartado 1827, Caracas, Venezuela). *Oncology* 30(2):
141-146; 1974.

Hsu, R.; Rochman, H.; Cifonelli, J. A. *Fed. Proc.*
34(3):846; 1975.

3966   CHARACTERIZATION OF CEA-ACTIVE MATERIAL
        ISOLATED FROM HUMAN PANCREATIC JUICE
[abstract]. (Eng.) McCabe, R. P. (Boston City
Hosp., Mass.); Kupchik, H. Z.; Saravis, C. A.;
Gregg, J. A.; Broitman, S. A.; Zamcheck, N. *Fed.
Proc.* 34(3):845; 1975.

3967   CEA-S: A DISTINCTIVE ISOMERIC VARIANT
        OF CARCINOEMBRYONIC ANTIGEN [abstract].
(Eng.) Edgington, T. S. (Scripps Clin. Res. Found.,
La Jolla, Calif.). *Fed. Proc.* 34(3):845; 1975.

3968   CEA-S: A HOMOGENEOUS SPECIES OF CEA [ab-
stract]. (Eng.) Plow, E. F. (Scripps Clin. Res.
Found., La Jolla, Calif.); Edgington, T. S.; Nakamura, R. M. *Fed. Proc.* 34(3):845; 1975.

3969   IMMUNOCHEMICAL DIFFERENCE BETWEEN PLASMA
        AND TUMOR CARCINOEMBRYONIC ANTIGEN (CEA)
[abstract]. (Eng.) Vrba, R. (Harvard Med. Sch.,
Boston, Mass.); Alpert, E.; Isselbacher, K. J. *Fed.
Proc.* 34(3):513; 1975.

3970   CARCINOFETAL ANTIGENS IN SERA OF TUMOR
        PATIENTS. (Ger.) Schultze-Mosgau, H.
(Universitats-Frauenklinik Hamburg-Eppendorf,
BRD-2 Hamburg 20, Martinistr. 52, West Germany);
Fischer K. *Zentralb. Gynaekol.* 97(9):563-567; 1975.

3971   GENETICS OF SOMATIC CELL SURFACE ANTIGENS.
        III. FURTHER ANALYSIS OF THE A_L MARKER.
(Eng.) Jones, C. (Univ. of Colorado Medical Center,
Denver, Colo.); Wuthier, P.; Puck, T. T. *Somatic
Cell Genet.* 1(3):235-246; 1975.

3972   NORMAL TISSUE ANTIGENS IN OVARIAN CARCINOMA
        [abstract]. (Eng.) Hope, N. J. (Univ.
Louisville Sch. Med., Ky.); Burton, R. M.; Espinosa,
E. *Fed. Proc.* 34(3):842; 1975.

3973   TUMOR ASSOCIATED ANTIGENS IN HUMAN MYELOMA
        [abstract]. (Eng.) Krueger, R. G. (Mayo
Clin., Rochester, Minn.); Staneck, L. D.; Kyle, R. A.
Fed. Proc. 34(3):846; 1975.

3974   DEMONSTRATION OF CROSS-REACTING HUMAN TUMOR
        ANTIGENS BY IMMUNOPEROXIDASE STAINING
TECHNIQUE [abstract]. (Eng.) Berkman, J. I. (Long
Isl. Jew.-Hillside Med. Cent., New Hyde Park, N.Y.);
Mesa-Tejada, R.; Klavins, J. V.; Weiss, M. *Fed.
Proc.* 34(3):834; 1975.

3975   TSTA AND S ANTIGEN EXPRESSION IN HAMSTER
        CELLS TRANSFORMED BY PARA-ADENOVIRUS 7.
(Ita.) Santoni, A. (Istituto di Igiene, Universita

di Perugia, Italy); Campanile, F.; Rivosecchi, P.;
Iorio, A. M. *Boll. Ist. Sieroter. Milan* 53(4):578-
582; 1974.

3976   INFLUENCE OF METHYLPREDNISOLONE AND AZA-
         THIOPRINE ON POLYMORPHONUCLEAR NEUTROPHILS
(PMN) AND LYMPHOCYTES IN GERMFREE, MONOCONTAMINATED
AND CONVENTIONAL RATS. (Eng.) Baardsen, A. (Kaptein
Wilhelm Wilhelmsen og Frues Bakteriologiske Institutt,
University Hospital, Rikshospitalet, Oslo, Norway);
Midtvedt, T.; Trippestad, A. *Acta Pathol. Microbiol.
Scand.* [C] 83(3):210-214; 1975.

3977   AGE-RELATED REFRACTORINESS OF PHA-INDUCED
         LYMPHOCYTE TRANSFORMATION. II. ^{125}I-PHA
BINDING TO SPLEEN CELLS FROM YOUNG AND OLD MICE.
(Eng.) Hung, C.-Y. (Oak Ridge Graduate Sch. Biomedi-
cal Sciences, Oak Ridge, Tenn. 37830); Perkins, E.
H.; Yang, W.-K. *Mech. Ageing Dev.* 4(2):103-112;
1975.

3978   LYMPHOCYTE POPULATIONS OF AKR/J MICE. III.
         CHANGES IN THE PRELEUKEMIC STATE. (Eng.)
Zatz, M. M. (Natl. Cancer Inst., Immunology Branch,
Bethesda, Md. 20014). *J. Immunol.* 115(4):1168-1170;
1975.

3979   EFFECT OF TUMOR ASCITIC FLUID ON LYMPHOCYTE
         TRAPPING AND MITOGEN RESPONSIVENESS OF
LYMPH NODE CELLS [abstract] . (Eng.) Mongini, P.
K. (Stanford Univ. Sch. Med., Calif.); Rosenberg,
L. T. *Fed. Proc.* 34(3):1003; 1975.

3980   MODIFICATION BY HERPESVIRUS OF HEREDITARY
         GVHR COMPETENCY. (Eng.) Longenecker, B.
M. (Medical Sciences Building, Univ. Alberta, Edmon-
ton, Alberta T6G 2E1, Canada); Pazderka, F.; Ruth,
R. F. *J. Immunogenet.* 2(1):59-64; 1975.

3981   EFFECT OF HYPOPHYSECTOMY ON THE DEVELOPMENT
         OF A T-CELL LEUKEMIA [abstract]. (Eng.)
Bentley, H. P. (Kiwanis Club Cancer Cent., Univ.
South Alabama, Mobile); Hughes, E. R.; Peterson, R.
D. A. *Clin. Res.* 23(1):72A; 1975.

3982   ALTERATION OF T CELL FUNCTION IN HEALTHY
         PERSONS WITH A HISTORY OF THYMIC X-IRRADI-
ATION. (Eng.) Rieger, C. H. L. (Dept. Pediatrics,
P. O. Box 244, Univ. of Chicago, 950 E. 59th St.,
Chicago, Ill. 60637); Kraft, S. C.; Rothberg, R. M.
J. Allergy Clin. Immunol. 56(4):273-281; 1975.

3983   TUMOR-DERIVED T CELL INHIBITORY PRINCIPLE
         [abstract]. (Eng.) Blom, D. H. (Dept.
Med., Univ. Pennsylvania, Philadelphia); Brody, J. I.
Clin. Res. 23(3):424A; 1975.

3984   CORRELATION OF EXPRESSION OF SURFACE MEM-
         BRANE MARKERS WITH DIFFERENT STAGES OF B
CELL DIFFERENTIATION IN MALIGNANT LYMPHOMAS IN MAN
[abstract]. (Eng.) Augener, W. (Dept. Med., Univ.
Essen, West Germany); Cohnen, G.; Brittinger, G.
Fed. Proc. 34(3):842; 1975.

3985   SPONTANEOUS LYMPHOKINE SYNTHESIS BY HUMAN
         BLOOD MONONUCLEAR CELLS. (Eng.) Arvilommi,
H. (Public Health Lab., SF-40620 Jyvaskyla 62, Finland); Rasanen, L. *Nature* 257(5522):144-146; 1975.

3986   AUGMENTED UPTAKE OF NEURAMINIDASE-TREATED
         SHEEP RED BLOOD CELLS: PARTICIPATION OF
OPSONIC FACTORS. (Eng.) Schmidtke, J. R. (Dept.
Surg., Univ. Minnesota, Box 306 Mayo, Minneapolis,
Minn. 55455); Simmons, R. L. *J. Natl. Cancer Inst.*
54(6):1379-1384; 1975.

3987   SCANNING ELECTRON MICROSCOPY OF MURINE
         MACROPHAGES: SURFACE CHARACTERISTICS
DURING MATURATION, ACTIVATION, AND PHAGOCYTOSIS.
(Eng.) Polliack, A. (Hadassah University Hosp.,
Jerusalem, Israel); Gordon, S. *Lab. Invest.*
33(5):469-477; 1975.

3988   CONCANAVALIN A BINDING OF HUMAN COLON
         AND COLONIC TUMOR. (Eng.) Pittman, J.
(Univ. Alabama Med. Cent., Birmingham, Ala.);
Brattain, M. G.; Weiler, A.; Pretlow, T. G. II.
Fed. Proc. 34(3):846; 1975.

See also:

★ (Rev):   3615, 3616, 3617, 3619, 3623, 3647,
            3650, 3651, 3652, 3653, 3654
★ (Chem): 3712, 3768
★ (Viral): 3807, 3808, 3811, 3837, 3858, 3874,
            3878, 3882
★ (Path): 4033, 4035, 4038, 4043, 4069

ticular interest, as it is considered to be a precancerous lesion.

55;

3991   AUGMENTATION MAMMOPLASTY, IRRADIATION,
        AND BREAST CANCER: A CASE REPORT. (Eng.)
Frantz, P. (Univ. North Carolina Sch. Medicine, Chapel
Hill, N. C. 27514); Herbst*, C. A., Jr. *Cancer* 36(3):
1147-1150; 1975.

A case of a mammary adenocarcinoma associated with
a left Cronin Silastic prosthesis in a 32-yr-old
white woman is reported. Between the ages of 3-15
mo this patient had received six doses (quantity
unknown) of irradiation to the left breast and back
for two congenital hemangiomas. The lack of left-
breast development and residual skin changes indicate that the irradiation exposure was significant.
The prosthesis was implanted only two yr before the
detection of cancer; it may have acted as a promoting factor. Study of this case indicates that:
during mammoplasty, a random biopsy should be taken;
all breast tissue should be placed superficial to the
prosthesis to facilitate frequent examination; and
open-excisional biopsy not violating the prosthetic
pseudocapsule should be used to examine any mass.
A standard radical mastectomy including removal of
the pectoralis major and minor should be performed
without violation of the pseudocapsule if a malignancy is found.

3992   MALE BREAST CANCER. (Eng.) Anonymous
        (No affiliation given). *Acta Pathol.
Microbiol. Scand.* (Suppl. 251):13-35; 1975.

Male breast cancer was evaluated on the basis of
data concerning 265 cases registered in Denmark
over a period of 29 1/2 yr. The data were obtained
by review of hospital records, survey of microscopic
preparations, and examinations of surviving patients. The results are compared to results reported in the literature. Male breast cancer comprises 0.8% of all cases of breast cancer in Denmark. The average age at establishment of diagnosis
was 65.2 yr in 257 cases. The duration of symptoms
in breast cancer is considerably longer in males
than in females, 16% having a duration of symptoms
of two years or more. In only 13% was a palpable
tumor the only symptom on admission. Twenty-seven
percent had ulceration. Ulceration and fixation
to the underlying tissue are not early symptoms in
male breast cancer. According to the TNM classification, 35% of 253 cases were in clinical stage
I, 11% in stage II, 42% in stage III, and 12% in
stage IV. The clinical appearance of the disease
showed a significant improvement from the period
1943-1957 to the period 1958-1972. This improvement is presumed to be related to the significant
change in the clinical stage during the later period. The longer duration of symptoms in the male
may give rise to the somewhat poorer prognosis in
male breast cancer. Of the 265 cases, ten had a
history of gynecomastia; this may support the theory
that gynecomastia may be a premalignant state. Of
30 patients with breast cancer, one had the Klinefelter syndrome. By pooling series of male breast
cancer, in which such screening has been made, it

709

was found that the incidence of the Klinefelter syndrome is higher among men with breast cancer than in the normal male population. Calculated on the basis of personal findings and the results reported by others, the incidence of breast cancer in Klinefelter patients was found to be about 20 times higher than in normal men and about one-fifth of the incidence in females. Men with breast cancer had no specific metabolism of iv administered estradiol-17β or a specific urinary excretion of hormones in this series. However, 25 patients did excrete a significantly smaller quantity of pituitary gonadotrophins than did a control group comprising 60 healthy men of comparable age. The observed and corrected 5-yr survival rates for this disease were 36% and 46%, respectively. The findings in this and other studies give no indication that radical mastectomy is to be preferred to simple mastectomy, since the latter gives rise to fewer postoperative complications and is less strenuous. Therefore, simple mastectomy is recommended for these patients, 40% of whom are 70 yr or more. It is also recommended that the treatment of this rare disease be centralized as much as possible.

3993   INAPPROPRIATE PRODUCTION OF COLLAGEN AND PROLYL HYDROXYLASE BY HUMAN BREAST CANCER CELLS *IN VIVO*. (Eng.) Al-Adnani, M. S. (Univ. Dept. Pathology, Royal Infirmary, Glasgow, Scotland); Kirrane, J. A.; McGee*, J. O. *Br. J. Cancer* 31(6):653-660; 1975.

Thirty-two scirrhous cancers of breast were examined to determine the origin of the collagen stroma in these tumors. The capacity of normal, nonneoplastic, and neoplastic breast epithelium to synthesize collagen was identified by immunoperoxidase procedure and by an immunoglobulin-enzyme-bridge method, using antibodies to collagen and to prolyl hydroxylase. The malignant epithelial cells in 30 of the tumor tissues contained not only collagen, but also prolyl hydroxylase. Neither this collagen-biosynthetic enzyme nor collagen itself was detectable in the spindle cells in the stroma of the tumors. Neither the epithelium in normal breast, that in fibrocystic disease and in fibroadenomata, nor the malignant epithelium in two medullary cancers of breast contained either collagen or prolyl hydroxylase. These results strongly suggest that the malignant epithelium of scirrhous breast cancers produces its own collagen stroma, and that the scirrhous reaction in these tumors is not a host response to tumor invasion. The production of collagen and prolyl hydroxylase by breast cancer cells (of the scirrhous type) therefore represents another example of inappropriate protein production by a human tumor.

3994   FINE STRUCTURE OF A CEREBELLAR "FIBROMA". (Eng.) Hirano, A. (Montefiore Hosp. and Medical Center, 111 East 210 St., Bronx, N.Y. 10461); Llena, J. F.; Chung, H. D. *Acta Neuropathol. (Berl.)* 32(3):175-186; 1975.

The fine structure of an intracerebral "fibroma" was examined. The golf ball-sized tumor was totally

removed from the left cerebellar hemisphere of a 19-yr-old man. It was not attached to the dura mater and was encapsulated. A portion was examined with the electron microscope. The tumor consisted of irregularly-shaped cells connected by well developed junctional complexes. Unusual, fenestrated capillaries with extremely narrow and irregular lumens were frequent. Collagen fibers were uncommon, but the wide extracellular spaces contained large amounts of dense, granular or fibrillar material. The dense material coated the tumor cells and was apparently secreted from small vesicles found within these cells. This tumor is compared with other previously reported tumors with similar characteristics.

3995   NERVOUS SYSTEM NEOPLASMS AND PRIMARY MALIGNANCIES OF OTHER SITES: THE UNIQUE ASSOCIATION BETWEEN MENINGIOMAS AND BREAST CANCER. (Eng.) Schoenberg, B. S. (Section Publications, Mayo Clinic, 200 First St., S.W., Rochester, Minn. 55901); Christine, B. W.; Whisnant, J. P. *Neurology* 25(8)705-712; 1975.

To determine whether nervous system neoplasms are associated with primary malignancies elsewhere, the frequency of multiple primary tumors in patients in whom at least one of the primary tumors was within the nervous system was studied. The patients were Connecticut residents with tumors diagnosed between 1935 and 1964. Of 135 patients, 130 had two primary tumors, four had three primary tumors, and one had four primary tumors. Only with multiple primary tumors involving the brain and breast did the number of observed cases significantly exceed the number of expected cases; eight patients who had a meningioma associated with a breast cancer accounted for this excess. Patients with breast cancer presenting with signs or symptoms of an intracranial neoplasm should be carefully evaluated as the intracranial lesion may be a potentially curable meningioma.

3996   CYTOPHOTOMETRICALLY DETERMINED DNA-PATTERN OF PREMALIGNANT CASES OF DYSPLASIA OF THE CERVIX UTERI: CONTRIBUTION TO THE PROBLEM OF MORPHOLOGIC DEFINITION AND CLINICAL SIGNIFICANCE OF THESE LESIONS. (Ger.) Sachs, H. (Universitäts-Frauenklinik D-6650 Homburg/Saar Bundesrepublik Deutschland); Schittko, G. *Arch. Gynaekol.* 218 (2):95-112; 1975.

Tissue samples from 12 cases of possibly benign epithelial lesions of the uterine neck (metaplasia, regenerating epithelium, atrophic and parakeratotic epithelium, basal cell hyperplasia, and epithelium irritated by inflammation), as well as from eight cases of slight to serious epithelial dysplasia, and from four cases of carcinoma *in situ* were analyzed cytophotometrically. Samples from benign epithelial lesions had a normal DNA pattern. Premalignant dysplasias of all stages showed, however, an aneuploid DNA content in all cases. There was no diploid mode, and atypical proliferation with polydiploidization, DNA stem lines, and a broad scattering of the DNA levels were the characteristic features of premalignant dysplasia and of carcinoma *in situ*. More or less extended loss of the epithelial stratification,

with previously described collagenases, could only
be demonstrated *in vitro* and required protein synthesis. Maximum tumor collagenase occurred at 24
hr *in vitro* and then declined as compared with the
maximum collagenase at 72 hr *in vitro* produced by
oral cavity mucosa. The 14 patients in the series
were ranked in order of the collagenase activity
of their tumors. At 18 mo after the diagnosis, 4
of the 6 patients with the most active tumors were
dead of cancer, and one patient was alive with persistent cancer. High collagenase activity may be
a factor in the clinical aggressiveness of epidermoid carcinomas of the head and neck.

3999   CARCINOMA *IN SITU* OF THE LARYNX. (Eng.)
            Holinger, P. H. (700 North Michigan,
Chicago, Ill. 60611); Schild, J. A. *Laryngoscope*
85(10):1707-1708; 1975.

To provide a clinical evaluation of the management
of keratosis, keratosis with atypia, and carcinoma
in situ with or without microinvasion, 23 patients
"in the all inclusive category of carcinoma *in
situ*" of the larynx were studied. The results indicate that carcinoma *in situ* should be treated with
a full course of radiation or "adequate excisional
surgery". No conclusions on the benign or malignant
potential of keratosis and carcinoma *in situ* can
be drawn from the information available.

4000   BENIGN SMOOTH MUSCLE TUMOURS. (Eng.)
            Farman, A. G. (Dept. Oral Pathology,
Univ. Stellenbosch, Parowvallei, CP South Africa).
S. Afr. Med. J. 49(33):1333-1340; 1975.

A series of 7748 benign smooth muscle tumors, diagnosed in South Africa during a period of 24.5 yr, is
presented. Ninety-five percent of these leiomyomas
occurred in the female genital tract; in another 61
cases there was inadequate histologic material or
uncertainty about the site of origin. The remaining
310 lesions were analyzed for histologic, sex, race,
site, and age distributions. There were 230 skin
lesions, 42 stomach lesions, 18 small intestinal
and duodenal lesions, 7 esophageal lesions, 5 oral
lesions, 5 bladder lesions, and 4 lesions of the
large intestine. The larynx, kidney, and liver were
each involved once. Males were affected in 138 instances and females in 161. The lesions tended to
occur slightly earlier in black than in white females.
Simple leiomyomas comprised 201 (65%) of the lesions,
angiomyomas 91 (29%), and epithelioid leiomyomas 18
(6%). Comparisons were made between the skin, gastric, and oral lesions. Of the five oral cases,
four were angiomyomas and none was a simple leiomyoma. In contrast, simple leiomyomas accounted for
83% of the stomach lesions and 62% of the skin
lesions. Only 7% of the gastric lesions were angiomyomas compared with 34% of the skin lesions. Although there was no significant variation of age
distribution between simple leiomyomas and angiomyomas, the cutaneous simple leiomyomas occurred in
a slightly younger age group than did the angiomyomas. The data show that smooth muscle tumors may
occur at any site where smooth muscle is present.
Angiomyomas, rather than simple leiomyomas, are more

likely to occur in regions where smooth muscle is
found mainly in blood vessels. Simple leiomyomas
predominate in the stomach, which is a hollow viscus
filled with a band of smooth muscle.

4001   SUPPURATIVE MIXED TUMOR OF THE PHARYNX.
        (Rus.) Parkhomovskii, M. A. (Dept. Oto-
rhinolaryngology, MSCh No. 61, Moscow, U.S.S.R.).
Zh. Ushn. Nos. Gorl. Bolezn. (1):105-106; 1975.

Suppurative mixed tumor of the pharynx was removed
from a 58-yr-old man. The tumor was originally
diagnosed as a suppurative cyst. It was composed
of cartilage with a pus-containing central cavity.
Histological examinations revealed chronic tonsillitis and chronic inflammation of the paratonsillar
tissues. The patient had undergone ophthalmectomy
for a sarcoma about 20 yr previously. He is free
from recurrence for over five years despite incomplete excision of the tumor capsule.

4002   ULTRASTRUCTURE OF ANAPLASTIC (SPINDLE AND
        GIANT CELL) CARCINOMA OF THE THYROID.
(Eng.) Jao, W. (Michael Reese Medical Center, Chicago, Ill. 60616); Gould, V. E. Cancer 35(5):1280-
1292; 1975.

Three anaplastic (spindle and giant cell) carcinomas
of the thyroid were studied by light and electron
microscopy; two of the tumors also included foci of
recognizable follicular carcinoma. The patients were
women, 66-71 yr old. The follicular carcinoma cells
displayed prominent mitochondria and rough endoplasmic reticulum, and showed evidence of secretory activity. Desmosomes and complex cellular interdigitations
were evident. Basal laminae were present, with conspicuous reduplication in the well-differentiated
foci. However, some epithelial clusters were surrounded by basal lamina, showing focal discontinuities
through which epithelial cells protruded into the
stroma. The pleomorphic spindle and giant cells
showed cytoplasmic and nuclear characteristics similar to the better differentiated carcinomatous follicular elements, but showed rare desmosomes and no
basal laminae. The basic ultrastructural similarity
between follicular and anaplastic tumor cells confirms
their common epithelial origin. However, while partially retaining their secretory capability, the anaplastic cells progressively lose their capacity to
synthesize basal lamina and develop complex cellular
attachments.

4003   FINE STRUCTURE OF A RADIATION-INDUCED OS-
        TEOGENIC SARCOMA. (Eng.) Lee, W. R.
(Western Infirmary, Glasgow, Scotland); Laurie, J.;
Townsend, A. L. Cancer 36(4):1414-1425; 1975.

An osteogenic sarcoma arose in the right orbit of a
seven-yr-old boy five yr after the right orbit had
been treated by four courses of radiotherapy (total
dose approximately 13,000 rads) for a multicentric
retinoblastoma. Death occurred six months after
the orbital tumor was first detected. Study of the
tumor by electron microscopy showed a varied morphology of cells in which two main types were identi-

fied. In one group, the cells were large with ra-
diolucent cytoplasm, which contained long branching
segments of rough endoplasmic reticulum. In the second group, the cells were smaller with irregular
nuclei and an electron-dense cytoplasm, which contained short segments of dilated rough endoplasmic
reticulum and numerous mitochondria. The first group
of cells closely resembled osteoblasts, while the
second group had some features of osteoclasts or
their precursors. Phagocytosis occurred within many
of the tumor cells. The branching processes of the
tumor cells were separated by an amorphous ground
substance, which contained collagen-like fibrils and
hydroxyapatite crystals. Crystal deposition was in
some instances in close relation to extracellular
membrane-bound vesicles.

4004   XC-CELL FUSION INDUCED BY MURINE PLASMO-
        CYTOMA CELLS: II. CYTOLOGICAL AND ULTRA-
STRUCTURAL STUDY. (Eng.) Lemay, P. (U 102, INSERM,
2, place de Verdun, 59045 Lille, France); Torpier,
G.; Lefebvre, J. C.; Samaille, J. Biomedicine
23(5):168-175; 1975.

The kinetics and morphologic aspects of XC cell syncytium formation induced by MF_2 cells, a mouse
plasmocytoma-derived cell line, were studied. The
MF_2 cells (3×10^3) and the XC cells (25×10^3) were
cocultured for 18 hr in each well of a Microtest II
tissue culture plate. Only one MF_2 nucleus was
shown to be included in one syncytium by autoradiographic and optical microscopic methods. A partial
inhibition of syncytium formation, either in number
or in size, could be obtained with actinomycin D
(1 µg/ml) added to one or the other cell line, and
to a lesser extent by preincubation of MF_2 cells
with cycloheximide (10 µg/ml). Pretreatment of
either cell line with hydroxyurea (5 mM) did not
inhibit syncytium formation. These results suggested
that a normal RNA synthesis was required to obtain
optimal polykaryon growth. The reduced size of syn-
cytia observed after pretreatment of MF_2 cells with
either actinomycin D or cycloheximide may have resulted from a reduction in the number of C-type particles visible in the electron microscope; these
particles could represent sites for polykaryon induction. Immunoelectron microscopy using a syngeneic
mouse MF_2 cell antiserum and peroxidase labeling revealed a complete mixing and redistribution of the
respective plasma membrane sites of MF_2 and XC cells
on the polykaryon surface. The dilution of the MF_2
cell membrane sites on the syncytium surface is discussed in relation to the possible role of virus
particles in the fusion process. Progressive dilution of budding virus sites would influence the size
of the polykaryon; a minimal density of budding
sites/cell surface unit would therefore be required
to induce or allow the development of a polykaryon.
Experiments to elucidate this point are in progress.

4005   FENESTRAE IN GOLGI AND ENDOPLASMIC
        RETICULUM CISTERNAE OF HUMAN BRAIN TU-
MOURS. (Eng.) Tani, E. (Hyogo Coll. Med., Japan);
Ametani, T.; Nakano, K.; Nishiura, M.; Higashi, N.
Acta Neuropathol. (Berl.) 31(1):13-19; 1975.

tural differential diagnosis between meningiomas
and gliomas.

4007   THE ULTRASTRUCTURAL IDENTIFICATION OF TIS-
        SUE BASOPHILS AND MAST CELLS IN HODGKIN'S
DISEASE. (Eng.) Parmley, R. T. (Med. Univ. South.
Carolina, Charleston); Spicer, S. S.; Wright, N. J.
Lab. Invest. 32(4):469-475; 1975.

Mast cells and basophils, (numbering 2-10 cells per
high power microscopic field) are described in tumor
stroma of patients with untreated Hodgkin's disease.
A tumorous splenic lymph node taken from one to ten
patients contained both cell types. Mast cells alone
were found in large numbers in the pulmonary tumor of
another patient and smaller numbers were found in tumors of five additional patients. Electron microscopy
revealed that the basophils resembled peripheral blood
basophils. They contained caveolae and large round
granules that often contained uniform intragranular
particles and myelin figures. The tumor mast cells
showed a prominent Golgi apparatus and contained numerous granules which varied in density and in particle content. They occasionally showed protruding
granules indicating exocytosis, and caveolae suggesting pinocytosis. Some cells had a subsurface lacunar
system. Mast cells found in reduced numbers in tumor-
free lymph nodes of two other patients contained comparatively more closely packed and structurally uniform granules than tumor mast cells. They contained
more lipid droplets, showed no evidence of pinocytosis
or exocytosis, and they lacked lacunar systems. These
differences suggest that mast cells, and occasionally
basophils infiltrating tumor tissue, are activated
by Hodgkin's disease. Through exocytosis, they evi-
dently play a role in the inflammatory response of
the disease and may mediate immunologic reactions.

4008   THE ULTRASTRUCTURE OF MOUSE NEUROBLASTOMA
        CELLS IN TISSUE CULTURE. (Eng.) Ross, J.
(Dep. Biol., Yale Univ., New Haven, Conn.); Olmstead,
J. B.; Rosenbaum, J. L. *Tissue Cell* 7(1):107-136;
1975.

The different morphological forms of cultured neuroblastoma cells were studied by light and electron
microscopy. A clonal cell line from a murine sym-
pathetic ganglion tumor was grown in suspension or
monolayer cultures to produce round cells without
neurites or extended cells with neurites, resp.
Some monolayer cultures were grown in hypertonic
medium containing 0.18 M sucrose or 0.2 M NaCl to
produce superdifferentiated forms. Cells for light
microscopy were grown on coverslips; those for electron microscopy were fixed in glutaraldehyde, postfixed in OsO_4, and stained with uranyl acetate and
lead citrate. Light microscopy showed the cells
grown in suspension to be spheroid and about 20-
40 μ in diameter with thin undifferentiated cytoplasmic processes about 5-10 μ in length. Those
grown in monolayers were diverse in form with 80-
90% having developed neurites within 24 hr. Some
were round and about 30-60 μ in diameter, while
others were flattened and about 200 μ in diameter.
Neurites varied in number, length, diameter, and
arborization. One to several were present per cell,

and were diversified along their length. When cells reached confluency, they became more homogeneous. The cells grown in hypertonic medium developed longer, more differentiated neurites than normal. Electron microscopy showed the round-type cells had normal organelles, microtubules, and sparse numbers of filaments randomly distributed throughout the cytoplasm without orientation. Virus-like particles were a prominent feature. They were doughnut-shaped or round dense structures about 700-800 Å in diameter. The extended monolayer cells were similar but contained vesicles and neurofilaments typical of normal neurons. Some vesicles were coated with clear centers or dense cores, or had dense cores bounded by smooth membrane. The first type were 1,000-5,000 Å in diameter with radiating T-shaped striae, 150-200 Å long, and were frequently associated with the cell membrane. The last type were about 1,500-2,500 Å in diameter and were frequently clustered. They occasionally had striae. Filaments similar to normal neurofilaments, 90-100 Å in diameter, were found in the neurites and perikaryon. Occasionally they were grouped into bundles. Individual filaments had a periodicity of about 100 Å. Virus-like particles were scattered through the cytoplasm, sometimes occurring in large clumps. They were usually within smooth endoplasmic reticulum cis-ternae. Cell junctions consisted of two apposed membranes with a gap of 200 Å and a dense fuzzy material in the cytoplasm on either side. The hypertonic cells had clusters of spherical vesicles like those of cholinergic neurons; they were seen close to neurite contact points, along the length of the neurite, and in the cell bodies. It is concluded that neuroblastoma lines may form functional junctions with other cells.

4009 ULTRASTRUCTURE OF THE SINUS WALL OF MURINE BONE MARROW IN MYELOGENOUS LEUKEMIA.
(Eng.) Campbell, F. (Health Sci. Cent., Univ. Louisville, Ky.); *Am. J. Anat.* 142(3):319-333; 1975.

The bone marrow ultrastructure of six male RD mice iv injected with 2×10^6 leukemic myelocytes was investigated to determine if alterations occur in the sinus wall. Bone marrow fixative was perfused *via* the vascular system 25 days after inoculation, and sections were embedded in plastic for electron microscopy. Leukemic bone marrow sinuses appeared normal in size, shape and frequency of occurrence. The number of lining cells of the sinus wall appeared normal, and they had organelles similar in distribution and number to those of normal lining cells. The sinus wall was continuous, having openings (0.3-3 μ) only in conjunction with blood cells that were migrating through. These migration pores occurred within, and not between, lining cells. These similarities of leukemic sinuses to normal sinuses indicated that the ability of leukemic cells to pass through lining cells was not due to abnormal migration pores. Adventitial cells were markedly less numerous in leukemic marrow and exhibited marginal degeneration. An increase in thickness and an abnormal distribution of extracellular material near the basal surface of leukemic lining cells were observed. This material was not present near migration pores, and the possibility that it interfered with pore formation is suggested.

Filaments 100 Å in diameter associated with the substance suggested a chemical composition different from normal basal lamina. Virus-like particles were aggregated on the endoplasmic reticulum and in the leukemic cell cytoplasm, and particles 300-500 Å in diameter were seen in the extracellular space. The large number of these particles suggests their involvement in or association with the leukemic state.

4010 COMPARATIVE ELECTRON-MICROSCOPIC FEATURES OF NORMAL, HYPERPLASTIC, AND ADENOMATOUS HUMAN COLONIC EPITHELIUM. II. VARIATIONS IN SURFACE ARCHITECTURE FOUND BY SCANNING ELECTRON MICROSCOPY.
(Eng.) Fenoglio, C. M. (Coll. Physicians and Surgeons of Columbia Univ., 630 West 168 St., New York, N.Y. 10032); Richart, R. M.; Kaye, G. I. *Gastroenterology* 69(1):100-109; 1975.

The high resolution and the great depth of focus possible with the scanning electron microscope aided in defining the surface features of normal human colonic mucosa and the mucosa of human colonic polyps. Thirteen hyperplastic polyps and 15 adenomas were obtained at sigmoidoscopy, clectomy, or polypectomy or from colons resected for carcinoma. Specimens were dehydrated with a graded series of ethanol solutions, and infiltrated with a graded series of ethanol-amyl acetate solutions before critical point drying from liquid CO_2; they were then coated with gold-palladium and examined with the scanning electron microscope. Sections were also examined histologically. This technique demonstrated that the surface of the normal colon has a territorialization that encompasses multiple crypts. The surface of the normal colon was covered with both absorptive cells and goblet cells. Hyperplastic polyps were covered with overdeveloped absorptive cells, and the normal territorialization was present but was distorted. Adenomatous polyps were covered with immature cells and, except for the villous adenomas, goblet cells were absent from the surface. The territorial patterns were totally obliterated in the adenomas. The results confirm the existence of divergent patterns of surface morphology in hyperplastic and adenomatous mucosa corresponding to the divergent cytological patterns that have been previously described from histological, autoradiographic, and transmission electron microscopic studies.

4011 BURKITT'S LYMPHOMA PRESENTING AS ACUTE LEUKAEMIA.
(Eng.) Jaiyesimi, F. (Univ. Coll. Hosp., Ibadan, Nigeria); Oluboyede, O.; Taylor, D.; Familusi, J. B. *Acta Haematol. (Basel)* 54(2):115-119; 1975.

An 8-yr-old Nigerian boy presented with extensive superficial lymphadenopathy, a bleeding diathesis, and numerous tumor cells in blood and bone marrow. Further characterization of the tumor cells confirmed the diagnosis of Burkitt's lymphoma. Peripheral blood film showed numerous blast cells (10-15 μm in diameter) with multiple cytoplasmic vacuolations and 2-3 nucleoli in each nucleus. Iliac bone marrow aspiration revealed complete bone marrow replacement by Burkitt's lymphoma cells morphologically identical

was a high correlation between serum lactic acid
dehydrogenase (LDH) levels and stage of the disease; lowest levels correlated with localized disease, whereas higher LDH levels corresponded to
wide dissemination and/or large abdominal tumors.
Of 22 patients receiving parenteral cyclophosphamide as the primary mode of therapy, complete re-
missions were obtained in 59%; four of these relapsed and three died of the recurrent tumor. The
tumor recurred at the initial tumor site in 3 of
the 4 patients who relapsed. Metabolic complications attributed to chemotherapy, which developed
in 27% of the patients, appeared related to the
mass of tumor present and to rapid and massive
lysis of the malignant cells; such consequences
included primarily hyperkalemia (six cases), hypo-
calcemia, hyperphosphatemia (one case), and lactic
acidosis (one case). There were four sudden deaths
within 48 hr of therapy. Of the seven patients
receiving radiation therapy and/or chemotherapy
other than cyclophosphamide, none survived longer
than six months. Survival was correlated with the
clinical stage of the disease, with a median survival of seven months for all patients in the survey. The site and volume of tumor mass predicted
for prolonged survival. None of the six patients
with bone marrow or CNS involvement remained tumor-
free. The greater frequency of abdominal disease
reported in American Burkitt's lymphoma, as opposed
to the frequent localized jaw tumors of East African
patients was noted and a host reaction against the
lymphoma is suspected. The author also suggests
that the very high growth fraction of Burkitt's
lymphoma may explain its marked sensitivity to
chemotherapy. Sixty percent of patients presenting
only jaw tumor remained alive and free of disease.
The median survival of those presenting abdominal
tumor was seven months; patients experiencing CNS
and bone marrow involvement survived an average of
two months after diagnosis.

4014   BURKITT LYMPHOMA IN THREE AMERICAN CHIL-
        DREN: CLINICAL AND CYTOGENETIC OBSERVA-
TIONS. (Eng.) Hubner, K. F. (Div. Medical and
Health Sciences, Oak Ridge Associated Univ., PO Box
117, Oak Ridge, Tenn. 37830); Littlefield, L. G.
Am. J. Dis. Child. 129(10);1219-1223; 1975.

The clinicopathological observations of three American children with Burkitt's lymphoma are presented
and cytogenetic findings of two of the cases are
described. A 9-yr-old boy developed a small pain-
less lump on the right side of the neck that continued to grow despite penicillin treatment. The lump
was partially excised and appeared microscopically
as lymphoblastic lymphoma. WBC count revealed no
abnormalities except for lymphopenia. Neoplasm scanning showed abnormal uptake of the radioactive nuclide on the right side of the neck and around the
pancreas. Laparotomy revealed a neoplasm involving
the pancreas. The histological diagnosis was malignant lymphoma of the lymphoblastic type, consistent
with Burkitt's lymphoma. A 15-yr-old girl presented
with generalized urticaria, anorexia, nausea, vomit-
ing, weight loss, and dyspnea. Left inguinal lymph
node biopsy and a bone marrow smear disclosed a malignant lymphoma of the poorly differentiated lympho-

cytic type with starry sky pattern, Burkitt type.
The patient had visible nodular swellings in the cervical and supraclavicular regions bilaterally, and
enlarged lymph nodes. The third case, a 4 1/2-yr-
old boy, developed symptoms of partial gastrointestinal obstruction. Exploratory laparotomy revealed
a 475 g retrocecal neoplasm involving the entire ascending colon. Histological diagnosis was malignant
lymphoma, poorly differentiated, lymphocytic type,
involving the colon and adjacent structures. Hematological evaluations of bone marrow smears from
cases 2 and 3 showed a massive infiltration of lymphoma cells into the marrow at various times during
the course of the illness. Because of the variability
of clinical manifestations of presenting symptoms,
early diagnosis and treatment of Burkitt's lymphoma
may be delayed. The prognosis is poor in advanced
cases, especially where there is CNS or bone marrow
involvement.

4015   ABDOMINAL LYMPHOMA WITH α-HEAVY CHAIN
        DISEASE. (Eng.) Plesnicar, S. (Fac.
Medicine, Inst. Oncology, Ljubljana, Yugoslavia);
Sumi-Kriznik, T.; Golouh, R. *Isr. J. Med. Sci.*
11(8):832-837; 1975.

The case of a 27-yr-old man with abdominal lymphoma
is presented. Clinical, immunochemical and pathological studies were performed when the patient presented with abdominal pain, the malabsorption syndrome and marked weight loss. The concentration of
IgA, as measured by a radial diffusion method, was
elevated (860-3,080 mg/100 ml). Immunoelectrophoretic
studies revealed α-heavy chains in the patient's
serum, urine and intestinal juice. Treatment with
cyclophosphamide and corticosteroids produced an
improvement in the patient's condition and he survived for 18 mo after the onset of symptoms. Autopsy
revealed a malignant lymphoma of mixed histiocytic-
lymphocytic type involving mainly the mesenteric
lymph nodes, but also affecting the jejunal mucosa,
liver, spleen and extraabdominal lymph nodes. This
is the first case of abdominal lymphoma with α-heavy
chain disease to be reported from Yugoslavia.

4016   TUMOUR ANGIOGENIC FACTOR ASSOCIATED WITH
        SUBCUTANEOUS LYMPHOMA. (Eng.) Wolf, J.
E. (Baylor Coll. Med., Houston, Tex.); Hubler, W.
R., Jr. *Br. J. Dermatol.* 92(3):273-277; 1975.

In order to test for the presence of a diffusable
Tumor Angiogenic Factor (TAF), 2.0 mm fragments of
a subcutaneous lymphoma from the forehead of a 4-yr-
old Oriental-American girl were implanted in hamsters. The fragments were implanted in direct apposition to the vascular membrane of the cheek pouch membrane of six hamsters or separated from the cheek
pouch vascular bed by 6.0 mm discs of millipore filter (type TA, 0.45 μm pore size, 25 μm thick) in another six hamsters. Implanted materials were observed tor 7-14 days. The vascular response of the
cheek pouch membrane to tumor grafts was contrasted
with that of normal human, keloid, nevocellular nevi,
and inert control materials (Millipore filter and
dialysis membrane). Lymphoma grafts stimulated neo-
vascularization in the hamster cheek pouch directly

tion of bone marrow was taking place. Significantly
greater increases of Hb F levels were observed in
those in remission. The levels of Hb A_2, and carbonic
anhydrases B and C fell to levels found in newborns.
The results indicate that in some leukemias of early
childhood there may be an almost total reversion to
the fetal form of erythropoiesis. They are consistent with the suggestion that childhood leukemias
arise in cell lines which have escaped the normal
processes of differentiation.

4021   MORPHOLOGICAL AND CERTAIN CYTOCHEMICAL
        CHARACTERISTICS OF ACUTE LEUKEMIAS. (Rus.)
Tsyba, N. N. (A. M. Gor'kii Donetsk Medical Inst.,
Donetsk, U.S.S.R.); Novitskaia, V. M. *Lab. Delo*
(5):279-282; 1975.

Morphological and certain cytochemical characteristics of acute leukemias were studied in 51 patients.
The percentages of positive cells in the PAS reaction, myeloperoxidase, phospholipid, and acid
phosphatase tests were 61.6%, 70%, 60%, and 37.8%
in acute myeloblastic leukemia (15 patients). The
leukocyte counts ranged from 4,000-62,000/µl, the
percentage of leukemic myeloblasts was 15-96%, and
the marrow myelokaryocyte count was 150,000-1,060,
000/µl, with prevalence of leukemic myeloblasts
(20-92%). Leukocytopenia (1,500-3,600/µl), thrombocytopenia, and a high percentage of anaplastic
promyelocytes (33-61%) were found in promyelocytic
leukemia (two patients). The PAS reaction was positive in 94.5%, the myeloperoxidase test in 100%,
the phospholipid test in 95%, and the acid phosphatase test in 20%. The PAS reaction was positive
in 61.4% of the cases, the acid phosphatase reaction
in 41.13%, while the myeloperoxidase and phospholipid reactions were negative in acute lymphoblastic
leukemia (19 patients). Leukocyte counts of 1,000-
380,000/µl, 3-91% lymphoblasts were observed. Undifferentiated forms of acute leukemias (five patients) differed only slightly from the lymphoblastic
and myeloblastic forms. Here, all reactions were
negative. A special group of patients with acute
leukemia was identified which was characterized by
43-90% of metaplastic leukemic marrow elements, depression of erythropoiesis and thrombocytopoiesis,
7-24% large leukemia cells, including cells with
vacuolized cytoplasm, and by high acid phosphatase
activity. Morphological changes without cytochemical
shifts were observed in advancing, recurrent, and
terminal cases.

4022   MALIGNANT MELANOMA — A PO-CELL TUMOR?
        (Ger.) Schmidt, H. (Physiolisch-chemi-
sches Institut DDR-402 Halle (Saale), Hollystr. 1,
East Germany); Stintz, A. R. *Arch. Geschwulstforsch.*
45(2):146-152; 1975.

Tissue samples from a primary, recurrent melanoma in
a 57-yr-old patient and from a lymph node metastasis
of a round-cell melanoma in a 53-yr-old patient were
investigated for phenoloxidase-containing cells.
Numerous scattered cells and accumulations of cells
with phenoloxidase were found both in the malignant
melanoma of the skin and in the lymph node metastases. In view of these findings, it is proposed that

phenoloxidase-containing cells are the stem cells of
malignant melanoma, contrary to the common view that
melanoma derives from the melanocytes of the stratum
cylindricum of the epidermis. The occurrence of
phenoloxidase-containing cells in different organs
and tissues gives a good explanation for the high
susceptibility of melanoma to metastases in different
organs and tissues.

4023   BREAST CANCER GENETICS AND CANCER CONTROL:
        TUMOR ASSOCIATION. (Eng.) Lynch. H. T.
(Sch. Medicine, Creighton Univ., Omaha, Nebr. 68178);
Lynch, J.; Lynch, P. *Arch. Surg.* 110(10):1227-1229;
1975.
A study was made of four kindreds in which a familial association existed between carcinoma of the
breast, soft tissue sarcomas, leukemia, and brain
tumors. The mode of transmission was consistent
with that of a Mendelian autosomal dominant genetic
factor. Verified breast cancer was present in a
father, his mother, and his daughter. His son has
a brain tumor (by history) and his grandson, (the
son of the affected daughter), had a histologically
verified rhabdomyosarcoma. This familial aggregation of cancers (except for leukemia, which is absent) is consistent with a newly described familial
breast cancer syndrome. A single pleiotropic, dominantly transmitted gene, possibly interacting with
carcinogenic factors, such as an oncogenic virus,
may be the cause. Histopathology may provide clues
to familial cancer risk in certain families, and
improved cancer control might be achieved through
the identification of familial cancer risk for specific anatomic sites.

4024   RELATIONSHIP BETWEEN EXPERIMENTAL RESULTS
        IN MAMMALS AND MAN. I. CYTOGENETIC ANAL-
YSIS OF BONE MARROW INJURY INDUCED BY A SINGLE DOSE
OF CYCLOPHOSPHAMIDE. (Eng.) Goetz, P. (Res. Inst.
of Child Development, Prague, Czechoslavakia); Sram,
R. J.; Dohnalova, J. *Mutat. Res.* 31(4):247-254; 1975.

The extrapolation of experimental results to man was
studied by cytogenetic bone marrow analysis and micronucleus test in mice, rats and Chinese hamsters.
Furthermore, the frequency of chromosomal aberrations
was compared with the frequencies of polychromatic
RBC containing micronuclei. Cyclophosphamide (CY)
was given at doses of 5, 10, 20, 40, and 80 mg/kg to
ICR mice and Wistar rats; doses of 10, 20, 40, 80,
120, and 160 mg/kg were given to Chinese hamsters.
Five patients with various types of malignancies
until then medically untreated, were iv administered
40 mg CY/kg. Bone marrow cells were examined 24 hr
after the administration. In all rodents, CY induced
a clear-cut dose-effect relationship in the frequency
of breaks and abnormal metaphases, as well as in the
frequency of micronuclei in polychromatic RBC. When
comparing the results in rodents and man at the dose
of 40 mg CY/kg, the sensitivity pattern of species
was mice > rats > Chinese hamsters > man. From this
aspect the possible differences in the metabolism of
CY in analyzed species are discussed. The results
may indicate that micronucleus testing may be a very
suitable method used for screening purposes; however,
the method of classical cytogenetic analysis, espe-

cially the evaluation of breaks, still remains the
most exact and reliable technique.

4025   HETEROCHROMATIN AND CHROMOSOME ABERRATIONS.
        A COMPARATIVE STUDY OF THREE MOUSE CELL
LINES WITH DIFFERENT KARYOTYPE AND HETEROCHROMATIN
DISTRIBUTION. (Eng.) Natarajan, A. T. (Wallenberg
Lab., Lilla Frescati, S-10405 Stockholm 50, Sweden);
Raposa, T. *Hereditas* 80(1):83-90; 1975.

Mitomycin C-induced chromosome aberrations in mouse
cells differing in chromosome number and karyotype
were studied. The cell lines included (1) laboratory
mouse (2n=40), all telocentric chromosomes possessing
centromeric heterochromatin, (2) laboratory mouse x
tobacco mouse hybrid (2n=33) with seven metacentrics
and 26 telocentrics, and (3) ascites tumor cells
(MSWBS; 2n=28-29), with 9-10 biarmed chromosomes,
with heterochromatin localized to the centromeric
regions in all, except two marker chromosomes - one
telocentric with four intercalary blocks, and one biarmed chromosome with two terminal blocks. *In vitro*
cultures of fibroblasts were treated with 0.04-0.5
µg/ml mitomycin. In the case of MSWBS, the animals
carrying tumors were injected with mitomycin C in
isotonic saline (4 µg/ml of ascitic fluid). Both
chromatid **intrachanges** and interchanges were scored.
In laboratory mouse fibroblasts, about 80% of the
aberrations were localized to the heterochromatic
regions. In the tobacco mouse x laboratory mouse
hybrid, the exchanges were between two telocentrics
or a telocentric and a metacentric, but seldom between two metacentrics. This suggests that there
must be a specific control of the association of
chromocenters in the interphase and that there is a
homologous association between the telocentric and
metacentric. In the MSWBS tumor cells, the centromeric heterochromatin was involved more often in
interchanges than the intercalary blocks, without any
restriction between two biarmed chromosomes to form
exchanges. A low frequency of chromatid breaks in
this line may be due to the presence of an extremely
efficient DNA repair system.

4026   ON THOROTRAST LEUKEMIA: EVOLUTION OF
        CLONE OF BONE MARROW CELLS WITH RADIA-
TION-INDUCED CHROMOSOME ABERRATIONS. (Eng.) Vis-
feldt, J. (Univ. Inst. Pathological Anatomy, Juliane
Maries Vej 16, DK-2100 Copenhagen Ø, Denmark); Jen-
sen, G.; Hippe, E. *Acta Pathol. Microbiol. Scand.*
[A] 83(4):373-378; 1975.

Chromosome studies of bone marrow cells from a 62-
yr-old woman, who had been treated with Thorotrast,
were undertaken. The patient had been given 40 ml
of Thorotrast in connection with a neuro-radiologi-
cal examination 34 yr earlier. Clinically, the
patient was considered to be in an incipient myeloid
leukemic phase. One hundred cells were analyzed.
Ninety-seven percent of the cells belonged to a clone
with characteristic marker chromosomes induced by
radiation. The results of the study were compared
with data from chromosome analyses of ^{32}P-treated
patients with polycythemia vera, in whom large
clones occurred in the bone marrow. The results of
chromosome analysis of bone marrow cells from the

maintaining epidermal cells (keratinocytes) derived
from fragments of 50 warts and 16 normal skin specimens in primary tissue culture. Cultures were maintained on Dulbecco's modified Eagle's minimal essential medium with 10% fetal calf serum, L-glutamine
(4.0 mM), penicillin (100 U/ml) and streptomycin
(100 μg/ml) at 35 C in humidified air and 5% CO_2.
Growth occurred in 37% of wart fragments as compared
to 44% of normal skin. The wart keratinocytes continued to proliferate for periods exceeding three
months, but their appearance and growth patterns were
identical to normal skin. Electron microscopic examination of cell sheets from 20 warts and of five parent wart specimens revealed clustered virions in all
nuclei of the parent warts, but no virions were detected in a minimum of 500 nuclei from each of the
cultures. Precipitating antisera to purified whole
human papilloma virus (HPV) virions were prepared in
rabbits. Indirect immunofluorescence studies showed
virions in cryostat sections of parent wart keratin-
izing cell nuclei, but not in wart culture nuclei.
Immunoenzyme studies with horseradish peroxidase-
labeled Fab fragments from anti-HPV and diaminobenzi-
dine similarly revealed virions in parent wart but
not in wart-derived cultures. Although proliferating
wart-derived tissue cultures show no morphologic evidence of virus transformation or virion assembly, the
cultures provide a model for further molecular studies.

4029   FAMILIAL OCCURRENCE OF WILMS'S TUMOR:
        NEPHROBLASTOMA IN ONE OF MONOZYGOUS
TWINS AND IN ANOTHER SIBLING. (Eng.) Juberg, R.
C. (Louisiana State Univ. Sch. Med., Shreveport);
St. Martin, E. C.; Hundley, J. R. *Am. J. Hum.
Genet.* 27(2):155-164; 1975.

A sibship is presented in which a pathologically
documented Wilm's tumor occurred in one member
of a monozygous male twin pair at age two years
and then, nine months later, in his 12-mo-old
sibling. The twins were considered to be monozygotic because of phenotypic similarities, the
probability computed from analysis of blood groups,
and comparison of their dermatoglyphics. There
were no other persons in the kindred with either
the renal tumor or associated malformation, and
the parents were not consanguineous. The tumor in
each of the affected siblings appeared to be single,
contrary to expectation in the hereditary form.
The monozygotic unaffected twin remains without
evidence of tumor at age five years. He therefore
represents an exception to the hypothesis of the
mutagenic origin of Wilm's tumor and provides support for an environmental etiology of nephroblastoma, at least in this family. It can be assumed
that the twins were not always exposed to precisely
the identical qualitative and quantitative environmental factors from conception through the
first 22 mo of postnatal life.

4030   ACUTE MYELOCYTIC LEUKAEMIA AND LEU-
        KAEMIA-ASSOCIATED ANTIGENS IN SISTERS.
(Eng.) Pendergrass, T. W. (Epidemiology Branch,
Natl. Cancer Inst., Bethesda, Md. 20014); Mann,
D. L.; Stoller, R. G.; Halterman, R. H.; Fraumeni,
J. F., Jr. *Lancet* 2(7932):429-431; 1975.

In a sibship of five brothers and seven sisters, three sisters died from acute myelocytic leukemia diagnosed at age 31, 37, and 42 yr, respectively. A fourth sister probably had the same disease; although her death at age 22 was attributed to Hodgkin's disease and aleukemic leukemia, her symptoms and clinical course were similar to those of the older sisters. Laboratory studies on the close relatives revealed that a fifth sister had persistently high RBC sedimentation rates (88 mm/hr) and serum immunoglobulin M levels (232 mg/100 ml). WBC from this sister were lysed by heterologous and allogeneic antisera detecting leukemia-associated antigens. This reaction has been seen previously only in patients with acute leukemia, but it may signify a preleukemic state in a normal member of a leukemia-prone family.

4031   ELEVATED PLASMA TESTOSTERONE AND GONADAL TUMORS IN TWO 46XY "SISTERS". (Eng.) Anderson, C. T., Jr. (Univ. of Wisconsin Hosp., 1300 Univ. Ave., Madison, Wis. 53706); Carlson, I. H. Arch. Pathol. 99(7):360-363; 1975.

The case histories of two sisters, both of the 46XY genotype and demonstrating amenorrhea and virilization, are presented. Both patients had elevated testosterone levels in the peripheral venous blood (250 ng/dl in patient 1). Laparotomy was performed in both sisters who, within a period of two months, were shown to have identical conditions. At laparotomy patient 1 had a 644-g, 17 x 8 x 10-cm, firm nodular mass removed from the left adnexa. Tumor venous effluent showed an elevated testosterone level (6,100 ng/dl). Microscopic examination revealed a dysgerminoma in the right and left ovarian regions, probably arising from a preexisting gonadoblastoma (this was evidenced by the presence of areas of laminar calcification on the tumor). Patient 2, at laparotomy, showed an infantile uterus. Microscopically, the ovaries showed a dysgerminoma and a gonadoblastoma on the left side and a pure gonadoblastoma on the right side. The tumors of both patients had nests of Leydig-like cells and may be the source of the androgen. The author postulates that these sisters were chromosomal males (possibly XO/XY mosaic) who developed internal and external female genitalia. The gonadoblastoma is then thought to have developed in dysgenetic gonads that were the androgen source. Further, some peripheral ovarian stroma may have provided the estrogen, which resulted in breast formation. As the tumor grew and altered to a dysgerminoma, the surrounding stroma may have become compressed, reducing estrogen and increasing androgen production.

4032   FAMILIAL HAEMOPHAGOCYTIC RETICULOSIS. REPORT OF 2 CASES IN SIBS. (Eng.) Botha, J. B. C. (Dept. Pathology, Univ. Cape Town, South Africa); Kahn, L. B.; Kaschula, R. O. C. S. Afr. Med. J. 49(32):1305-1308; 1975.

A four yr old girl and her five mo old brother showed familial hemophagocytic reticulosis. The disease is characterized by a widespread proliferation of histiocytes showing hemophagocytosis and it usually manifests clinically with hematological abnormalities. The initial presentation may be that of a meningo-encephalitic illness, the result of a lympho histiocytic infiltration of the meninges and cerebral tissues. Steroid therapy given to the girl before hospitalization had helped in some clinical improvement of neurological deficit but her condition deteriorated and she developed fever, hepatosplenomegaly, with jaundice and abnormal liver function. Before her death, hemolysis, granulocytopenia and thrombocytopenia were noted. The boy's illness had many features in common with disease process. The significance of abnormalities of his serum lipid pattern is not yet clear.

4033   HEMOPHAGOCYTIC RETICULOSIS: A CASE REPORT WITH INVESTIGATIONS OF IMMUNE AND WHITE CELL FUNCTION. (Eng.) Fullerton, P. (Royal Children's Hosp., Melbourne, Victoria 3052, Australia); Ekert*, H.; Hosking, C.; Tauro, G. P. Cancer 36(2):441-445; 1975.

A five-month-old boy with hemophagocytic reticulosis is described. The diagnosis was made on the basis of hepatomegaly, pancytopenia, and the presence in bone marrow aspirates of many primitive neoplastic histiocytes, some of which were actively phagocytosing RBC. Immunologic investigations revealed a grossly defective phytohemagglutinin (PHA) response of the patient's lymphocytes which improved with chemotherapy (6 mg/m^2 vinblastin and 100 mg/m^2 prednisolone). Low immunoglobulin A levels (16-20 IU/ml) and defective glucose oxidation by phagocytosing cells were demonstrated at diagnosis and have persisted despite chemotherapy. The lack of a normal increase in glucose oxidation after phagocytosis, combined with a normal or greater than normal resting glucose oxidation, suggested that WBC were unable to stimulate the hexose monophosphate shunt following phagocytosis. HL-A typing and chromosome studies did not reveal maternal lymphocytes in the child's circulation; thus, a graft vs host reaction caused by maternofetal transplant of histiocytes was apparently not involved in the pathogenesis of hemophagocytic reticulosis in this patient. The improvement in PHA response which occurred despite treatment with immunosuppressive drugs suggests that immunodepression may be part of the pathogenesis of this disorder. The persistence of the defect in phagocytosing glucose oxidation despite treatment indicates that this may also be related to the pathogenesis of the disease. The patient remains well after 11 mo of treatment with vinblastine and prednisolone.

4034   ENDOCRINE-TYPE GRANULES IN CELLS OF GLOMUS TUMOR OF THE STOMACH. (Eng.) Kim, B. H. (Lutheran Medical Center, Brooklyn, N.Y. 11220); Rosen, Y.; Suen, K. C. Arch. Pathol. 99(10) 544-547; 1975.

The ultrastructure of a glomus tumor removed from the pyloric region of the stomach of a 57-yr-old

the centrolobular zone. In the hepatic parenchyma
of patients afflicted with angiosarcoma of the liver,
the same general alterations were found; of particular interest was the frequency of the focal conspicuous dilatation of sinusoids not related to
lobular topography. Splenic alterations, with or
without hepatic angiosarcoma, included enlarged
organs with grossly visible and conspicuously enlarged organs with grossly visible and conspicuously
enlarged Malpighian follicles. The focal subcapsular
fibrosis was the single most conspicuous lesion in
the precursor stage, while neither the portal
fibrosis nor intralobular fibrosis was considered
pathognomic. The study suggests that the angiosarcoma results from the stimulation of various
hepatic and splenic cells by VC or its metabolites,
presumably formed in the liver. The cause of activation of the hepatocytes may be either a primary
stimulation by VC or a response to a hepatocellular
injury caused by the reagent.

4037   MALIGNANT PERITONEAL MESOTHELIOMA AFTER
        CHOLANGIOGRAPHY WITH THOROTRAST. (Eng.)
Maurer, R. (Pathologisches Institut, Kantonsspital,
Brauerstrasse 15, CH-8401 Winterthur, Switzerland);
Egloff, B. *Cancer* 36(4):1381-1385; 1975.

A report is made of a malignant peritoneal mesothelioma occurring 36 yr after thorotrast administration. A female patient had undergone cholecystectomy after displaying symptoms of gallbladder
disease; during a cholangiogram performed with
thorotrast at operation, the contrast medium spilled
around the bile duct. Twenty-five years later,
hypertension and a nonfunctioning right kidney led
to nephrectomy. Histologically, there was no func-
tioning renal tissue or normal transitional epithelium. Large amounts of thorium dioxide identified
microscopically and by autoradiography led to a
diagnosis of hydronephrosis due to thorotrast-induced
fibrous stenosis of the pyeloureteral junction.
After 36 yr, upper abdominal pains occurred. Abdominal roentgenograms still showed deposits of
contrast medium in the right upper abdominal quadrant, and death followed the development of ascites,
recurrent paralytic ileus, and anuria. At autopsy,
the serosa of the small intestine was found to be
entirely covered by tumor. Microscopically, the
tumor consisted mainly of solid sheets of irregular
round cells with hyperchromatic nuclei and various
amounts of ill-defined cytoplasm. The tumor showed
varying sarcomatous and carcinomatous patterns.
While paravertebral lymph nodes contained thorium
dioxide and tumor tissue, there was no evidence
of distant metastases. The tumor is thus considered
to represent a case of radiation-induced peritoneal
mesothelioma.

4038   MOUSE MAMMARY TUMORS: ALTERATION OF IN-
        CIDENCE AS APPARENT FUNCTION OF STRESS.
(Eng.) Riley, V. (Dept. of Microbiology, Pacific
Northwest Res. Foundation, Seattle, Wash. 98104).
Science 189(4201):465-467; 1975.

The effect of different experimental and environmental conditions on mammary tumor incidence and latency was studied in C3H/He mice carrying the Bittner strain of mammary tumor virus. From 80-100% of these mice usually develop mammary tumors within 8-18 mo after birth when studied under the usual housing and experimental conditions. Four groups of mice were investigated: parous mice housed under conditions of chronic environmental and manipulative stress (group A); nonparous mice housed under the same conditions of chronic environmental and manipulative stress (group B); nonparous mice housed under protective conditions and subjected to moderate intermittent stress (group C); and mice provided with maximum protection from environmental and manipulative stress (group D). The last group was a combination of parous, nonparous, and virgin mice delivered by cesarean section and foster-nursed to deplete the milk-passaged tumor virus. The median tumor latency period for group A was 276 days compared with 358 days for group B. In group C the median latency period was 566 days, a significant extension over both groups A and B. At 400 days of age, the incidence of spontaneous mammary tumors was 92% in group A, 60% in group B, and 7% in group C. Group D had such a small tumor incidence that determination of a median latency period was not possible. At 21 mo of age, only one mammary tumor was detected in 75 surviving mice. The data suggest that moderate chronic or intermittent stress may predispose C3H/He mice to an increased risk of mammary carcinoma, possibly through a resultant compromise of their immunological competence or tumor surveillance system, and that adequate protection from physiological stress may reduce the occurrence of the mammary tumors.

4039   PARTICIPATION OF MYCOPLASMAS IN ONCOGENESIS. (Ger.) Gerlach, F. (Kubeckgasse 12, A-1030 Wien, Austria). *Wien. Med. Wochenschr.* [*Suppl.*] 26 (4):1-12; 1975.

The carcinogenicity of cell-free Mycoplasma preparations isolated from malignant human and animal tumors (carcinomas, sarcomas, hypernephromas, leukemias, lymphogranulomatoses), as well as from normal fetuses, umbilical cords, and placentae was studied in pathogen-free NMRI mice. To achieve long-term exposure, sc depot inoculation was used. Malignant tumors developed in 190 of the 209 inoculated animals (90.9%), while the tumor incidence among 600 control animals was 0.83%. The type of the tumor was independent of the type of the tumor from which the Mycoplasma preparation was obtained. The incidence of multiple tumors was 30.5%. Manifest tumor was preceded by preblastomatous processes, such as polyserositis, hyperplasia of the lymph nodes and of the spleen, and by adenomatous proliferations in inner organs. It is concluded that Mycoplasmas are ubiquitous infective agents specific for tumor induction in general, but not for any particular morphological type or localization. No malignant tumor occurs without the presence of Mycoplasmas, while numerous Mycoplasma infections do not provoke malignant tumors.

4040   COMMON AND UNCOMMON MELANOCYTIC NEVI AND BORDERLINE MELANOMAS. (Eng.) Reed, R. J. (Dermatopathology Lab., 1430 Tulane Ave., New

Orleans, La. 70112); Ichinose, H.; Clark, W. H., Jr.; Mihm, M. C., Jr. *Semin. Oncol.* 2(2):119-147; 1975.

The histologic parameters that distinguish less common, benign melanocytic tumors from common acquired melanocytic nevus are described, and the biologic significance of the deviant patterns is discussed. The histologic diagnosis of evolving malignant melanoma is also reviewed with emphasis on growth phases that are correlated with a good prognosis. Benign tumors discussed include acquired blue nevus, congenital nevus, Spitz tumor, and pigmented spindle cell nevus, as well as acquired melanocytic nevus. Nevus cell patterns in these tumors are compared with those in cytologically borderline lesions including malignant melanomas, halo nevi, and minimal deviation melanomas arising in congenital nevi, spindle cell melanomas, and malignant blue nevus (minimal deviation patterns). Cytologically malignant, biologically indolent melanomas (radial growth phase melanomas) are assessed on the basis of the level of invasion, and it is noted that lesions that have not invaded the dermis or that lack the capacity to survive and actively proliferate in the dermis are biologically borderline lesions with good prognoses. The significance of the host's immune response in regressing melanomas, especially superficial spreading melanomas, is briefly discussed. A brief discussion is also included of two unusual histologic patterns in melanocytic tumors that may be confused with those of evolving melanomas; these are the sheath cell reaction and recurrent nevi.

4041   MULTIFOCAL COLONIC CARCINOMA AND CROHN'S DISEASE. (Eng.) Keighley, M. R. B. (Queen Elizabeth Hosp., Edgbaston, Birmingham, England); Thompson, H.; Alexander-Williams, J. *Surgery* 78(4):534-537; 1975.

A case report of a 65-yr-old man with a 13-yr history of Crohn's disease and multifocal colonic carcinoma is presented and the relation between the two is discussed along with a review of the literature. A barium enema revealed a large polypoid mass in the transverse colon. Also noted were loss of haustral markings and irregular mucosal outline in the proximal colon, and a small polyp in the sigmoid colon. Biopsy was consistent with Crohn's disease. At laparotomy Crohn's disease was found to involve the distal ileum and the entire colon. In addition to the polyps detected by barium enema, two more were found in the ascending colon, one in the transverse and one in the sigmoid. All six polyps were well-differentiated papillary adenocarcinoma of the mucosa only. A lesion in the cecum was also found to be a well-differentiated adenocarcinoma of the mucosa only. Other recent reports and reviews have shown a higher incidence of carcinoma of the bowel in patients with Crohn's disease. A review of reported patients included 34 people with Crohn's disease, four of whom had multiple carcinomas of the bowel; the remainder had only single carcinomas. This incidence of multiple lesions is higher than the incidence of multiple lesions without preexisting inflammatory bowel disease. It is suggested that multiple carcinomas of the bowel are more frequent, and that colonic carcinomas are more common, in patients

after immunosuppression therapy. It is suggested that the triad of thymoma, pure red cell aplasia and immunoglobulin deficiency are manifestations of the interference of an antibody to a common stem-cell component; in this case the inhibition of lymphocyte transformation may have been related to the factor which also inhibited red cell maturation.

4044   METASTATIC THYMOMA WITH MYASTHENIA GRAVIS
        AND PURE RED CELL APLASIA. (Eng.)
DeSevilla, E. (Washington Univ. Sch. Medicine,
510 S. Kingshighway, St. Louis, Mo. 63110); Forrest,
J. V.; Zivnuska, F. R.; Sagel, S. S. *Cancer* 36(3):
1154-1157, 1975.

A 64-yr-old man with a nine-year history of myasthenia gravis presented with left hip pain. Radiography of the pelvis revealed a poorly defined lytic area in the inferior left iliac wing; biopsy showed this to be a metastatic thymoma. A large anterior mediastinal mass was apparent in chest films. Over a two-week period, the left ilium received 3,000 rads through anterior and posterior portals using ^{60}Co. The mediastinal mass was treated with 3,680 rads anteriorly and 1,040 rads posteriorly, also with ^{60}Co. The anterior mediastinal mass slowly decreased in size and the patient remained well except for cholecystitis with cholelithiasis, which necessitated cholecystectomy four months later. Despite the good response to radiotherapy, the patient was readmitted the following year with severe anemia (Hb 6 g; WBC 6,500/mm^3) and pure red cell aplasia. A bone marrow aspirate showed normal megakaryocytic and granulocytic precursors, but no erythrocytic precursors. A film of the pelvis now indicated a lytic area in the right ileum, which was irradiated with ^{60}Co. The red cell aplasia remains unresponsive to anabolic steroids or cyclophosphamide. The pathophysiologic relation of myasthenia gravis and pure red cell aplasia with thymomas is not clear, although an autoimmune disorder is frequently postulated.

4045   DO GENERALIZED METASTASES OCCUR DIRECTLY
        FROM THE PRIMARY? (Eng.) Bross, I. D. J.
(Roswell Park Mem. Inst., Buffalo, N.Y.); Viadana,
E.; Pickren, J. *J. Chronic Dis.* 28(3):149-159;
1975.

Based on 4,728 autopsy reports obtained between 1965-1970, a survey of primary and metastatic cancer sites was made. The data were statistically analyzed to determine whether or not dissemination of metastases from a given primary site is basically a one-step or a multi-step process. Average numbers of metastases were arranged in tables showing the metastases' primary sites and indicating whether or not they had metastasized in certain key sites (lung and liver). In the absence of both lung and liver metastases, the generalized form of cancer was rare (the total number of metastases usually averaged less than three). Whenever metastases were reported in both liver and lung, the generalized form was usually seen at autopsy. This indicates a multi-step metastatic process in which the primary cancer seeds a specific intermediate site from which the

cancer is further disseminated. When neither of
the key sites were reported, the average number of
metastases showed marked differences from one primary to another, suggesting initial differences between primary sites in their metastatic ability.
When both key sites are positive, the average number
of metastases is similar. The distinctions between
primary sites thus diminish once generalization
occurs and the generalized disease can be considered
as a separate entity.

4046   THE RELATIONSHIP OF HISTOLOGY TO THE
          SPREAD OF CANCER. (Eng.) Viadana, E.
(Dept. Biostatistics, Roswell Park Memorial Inst.,
Buffalo, N.Y. 14263); Bross, I. D. J.; Pickren,
J. W. J. Surg. Oncol. 7(3):177-186; 1975.

Autopsy records of 4,728 patients were studied to
determine the relationship between tumor histology
and the metastatic spread of cancer. The various
metastatic sites were grouped into four categories:
the endocrine system, the lymphatic system, the
CNS, and a "remainder" category, which included
parenchymatous organs and the lymphatic system.
The number of metastases within each system was
analyzed by chi-square analysis for males (581)
and females (540). Adenocarcinomas and oat-cell
carcinomas of the lungs were more aggressive in
their metastatic spread than were squamous cell
carcinomas in the lung. Similarly, adenocarcinomas
of the uterine cervix tended to metastasize more
than cervical squamous cell carcinomas. There were
no differences in the metastatic spread of transitional cell carcinomas and adenocarcinomas in the
bladder and kidney; the results were similar for
men and women. Cystadenocarcinomas of the ovaries
appeared to be less widespread than adenocarcinomas
of the same site. The results suggest that adenocarcinomas in general may be more aggressive in
metastatic spread than other histological types.

4047   HISTOGENESIS OF LEIOMYOMATOSIS PERITO-
          NEALIS DISSEMINATA (DISSEMINATED FI-
BROSING DECIDUOSIS). (Eng.) Parmley, T. H. (Johns
Hopkins Hosp., Baltimore, Md. 21205); Woodruff,
J. D.; Winn, K.; Johnson, J. W. C.; Douglas, P. H.
Obstet. Gynecol. 46(5):511-516; 1975.

The clinical course and histopathology in two cases
(both 36-yr-old Negro women) of leiomyomatosis
peritonealis disseminata were studied, and their
case reports are presented. Decidual reactions in
the endocervix, the uterine tube, the pelvic lymph
nodes, and in multiple subperitoneal sites in
the abdominal cavity were seen, and the gradual
replacement of the decidua by fibroblastic proliferation was noted. In addition to routine hematoxylin
and eosin staining; trichrome, reticulum, and PAS
stains were used to define the cellular elements.
In the two cases (one seen immediately after delivery and the other six years after delivery of
normal children) the long fibrillar cells resembled
fibroblasts. The nodule contained dilated capillaries with an associated inflammatory reaction and
the predominant architecture suggested a stromal or
fibrocytic proliferation. The special stains de-

monstrated a single cell rather than a "bundle"
cell arrangement. Within the nodules and peri-
pherally, foci of decidual cells remained. It is
suggested that the stimulus that initiates the
replacement of decidua by fibrous tissue is similar
to that seen at other sites of repair and replacement. The fact that this process occurs in pregnancy suggests that it is the result of continuous
stimulation of the decidua by progesterone and
estrogen. Although the presence of mitotic figures
in what appears to be smooth muscle scattered
throughout the peritoneal cavity suggests the diagnosis of malignancy, leiomyomatosis peritonealis
disseminata is believed by the authors to be the
result of a benign fibrous replacement of decidual
tissue. This condition may be identified by the
following histologic features: (a) cells are quite
long and fibrillar and more closely resemble fibroblasts than muscle cells; (b) their arrangement is
more that of stroma than of the bundles of true
muscle tissue; and (c) there is a dilated capillary
vasculature and an associated inflammatory reaction.
The presence of decidual cells within or at the
periphery of the nodule together with a recent pregnancy confirms the diagnosis.

4048   OSTEOGENIC SARCOMA ASSOCIATED WITH INTERNAL
          FRACTURE FIXATION IN TWO DOGS. (Eng.)
Banks, W. C. (Dept. Veterinary Medicine and Surgery,
Texas A&M Univ., College Station, Tex. 77843);
Morris, E.; Herron, M. R.; Green, R. W. J. Am. Vet.
Med. Assoc. 167(2):166-167; 1975.

4049   MALIGNANT TRANSFORMATION OF BENIGN OSTEO-
          BLASTOMA. A CASE REPORT. (Eng.) Seki, T.
(Sch. Med., Keio Univ., Tokyo, Japan); Fukuda, H.;
Ishii, Y.; Hanaoka, H.; Yatabe, S.; Takano, M.;
Koide, O. J. Bone Joint Surg. [Am.] 57A(3):424-426;
1975.

4050   GIANT INTRA-ARTICULAR OSTEOCHONDROMA OF
          THE KNEE: A CASE REPORT. (Eng.) Sar-
miento, A. (Univ. Miami Sch. Medicine, Miami, Fla.);
Elkins, R. W. J. Bone Joint Surg. [Am.] 57A(4):560-
561; 1975.

4051   SARCOMA IN THE REGION OF ENCEPHALITIC
          ALTERATIONS [abstract]. (Ger.) Boellaard,
J. W. (Tubingen, West Germany). Zentralbl. Allg.
Pathol. 119(3):228; 1975.

4052   ULTRASTRUCTURE OF A TUMOR OF THE TERMINAL
          PLATE [abstract]. (Ger.) Weindl, A.
(Munchen, West Germany); Fahlbusch, R.; Stochdorph,
O. Zentralbl. Allg. Pathol. 119(3):227; 1975.

4053   EXTERNAL TEMPORAL BIOPSY IN A CASE OF
          INFILTRATING LIMBIC GLIOMA: STUDY OF
ASTROCYTIC MEMBRANOUS WRAPPING WHORLS. (Fre.)
Mikol, J. (Faculte de Medicine Cochin-Port Royal,
Centre Psychiatrique Sainte Anne, I Rue Cabanis,
F-75674 Paris Cedex 14, France); Brion*, S.; Thurel,
C. Acta Neuropathol. (Berl.) 32(4):347-352; 1975.

sitat Heidelberg, Heidelberg, West Germany); Schulz,
P. *Schweiz. Med. Wochenschr.* 105(17):543-544; 1975.

4064   PRIMARY MULTIPLE MALIGNANT MELANOMA OF THE
        ORAL CAVITY [abstract]. (Fre.) Mascaro,
J. M. (Valence, Spain); Marques, E.; Botella, R.;
Roman, P.; Aloy, M. *Bull. Soc. Fr. Dermatol. Syphil-
igr.* 82(2):184-185; 1975.

4065   PRIMARY ADENOCARCINOMA OF THE ESOPHAGUS
        ARISING IN A COLUMNAR-LINED ESOPHAGUS.
(Eng.) Stillman, A. E. (2122 North Craycroft Road,
Suite 2G, Tucson, Ariz. 85712); Selwyn, J. I. *Am.
J. Dig. Dis.* 20(6):577-582; 1975.

4066   GLOMUS TUMOR OF THE STOMACH: AN ULTRASTRUC-
        TURAL STUDY. (Eng.) Kanwar, Y. S. (1853
West Polk St., Chicago, Ill. 60612); Manaligod,
J. R.* *Arch. Pathol.* 99(7):392-397; 1975.

4067   DIFFUSE SUBMUCOSAL CYSTS AND CARCINOMA OF
        THE STOMACH. (Eng.) Iwanaga, T. (Center
Adult Diseases, 3 Nakamichi 1-Chome, Higashinari-ku,
Osaka 537, Japan); Koyama, H.; Takahashi, Y.; Tani-
guchi, H.; Wada, A. *Cancer* 36(2):606-614; 1975.

4068   GROWTH PATTERNS OF EARLY GASTRIC CANCER.
        (Eng.) Smirnow, N. (Petrov Res. Inst.
Oncology, Leningrad, U.S.S.R.); Pavlov, K. *Proc.
Int. Cancer Congr. 11th.* Vol. 6 (*Tumors of Specific
Sites*). Florence, Italy, October 20-26, 1974. Edited
by Bucalossi, P.; Veronesi, U.; Cascinelli, N.
New York, American Elsevier, 1975, pp. 248-252.

4069   THE VASO-ACTIVE INTESTINAL POLYPEPTIDE IN
        VERNER-MORRISON SYNDROME. (Ger.) Seif,
F. J. (Medizinische Poliklinik der Universitat, 74
Tubingen 1, Liebermeisterstr. 14, West Germany);
Sadlowski, P.; Heni, F.; Fischer, R.; Bloom, S. R.;
Polak, J. M. *Dtsch. Med. Wochenschr.* 100(9):399-
405; 1975.

4070   PEUTZ-JEGHERS-SYNDROME: COMPLICATIONS AND
        SURGICAL THERAPY. (Ger.) Fabian, W.
(Chirurgischen Klinik des Krankenhauses Nordwest,
6 Frankfurt/M. 90, Steinbacher Hohl 2-26); Becker,
H. *Fortschr. Med.* 93(1):5,8; 1975.

4071   LYMPHOID POLYPOSIS OF THE TERMINAL ILEUM
        AND COLON: A SOURCE OF CLINICAL ERROR.
(Eng.) LoGerfo, F. W. (Univ. Hosp., Boston, Mass.
02118); Mueller, R. F. *Am. Surg.* 41(3):179-180;
1975.

4072   CARCINOMAS OF THE COLON -- HISTOGENESIS.
        (Eng.) Laumonier, R. (Centre Henri Bec-
querel, Rouen, France). *Proc. Int. Cancer Congr.
11th.* Vol. 6 (*Tumors of Specific Sites*). Florence,
Italy, October 20-26, 1974. Edited by Bucalossi, P.;

Veronesi, U.; Cascinelli, N. New York, American
Elsevier, 1975, pp. 308-311.

4073   ENDOCRINE TUMOR OF THE PANCREAS COMPOSED
        OF ARGYROPHIL AND B CELLS. A CORRELATED
LIGHT, IMMUNOFLUORESCENT, AND ULTRASTRUCTURAL STUDY.
(Eng.) Bordi, C. (Istituto di Anatomia ed Istologia
Patologica, Univ. Turin, Italy); Bussolati, G.;
Ballerio, G.; Togni, R. Cancer 35(2):436-444; 1975.

4074   ON PATHOGENESIS OF THE HEMORRHAGIC SYNDROME
        IN CANCER OF THE LIVER COMPLICATED WITH
JAUNDICE. (Rus.) Lys, P. V. (Postgraduate Medical
Inst., Kiev, U.S.S.R.). Vopr. Onkol. 21(6):54-59;
1975.

4075   VERRUCOUS CARCINOMA OF THE LARYNX -- A
        STUDY OF ITS PATHOLOGIC ANATOMY. (Eng.)
Fisher, H. R. (1920 Melwood Drive, Glendale, Calif.
91207). Can. J. Otolaryngol. 4(2):270-277; 1975.

4076   POSSIBLE QUANTIFICATION OF VASCULARIZATION
        AND OF NONPROLIFERATING, HYPOXIC AND
NECROTIC PROPORTIONS OF LUNG CANCER IN MAN. (Ger.)
Winterfeld, G. (Zentralinstitut fur Krebsforschung
der AW der DDR, Robert-Rossle-Klinik, 115 Berlin-
Buch, Lindenberger Weg 80, East Germany); Magdon,
E. Arch. Geschwulstforsch. 45(1):46-55; 1975.

4077   COMPARATIVE ULTRASTRUCTURAL STUDIES: THE
        EFFECTS OF BENZO(A)PYRENE AND CIGARETTE
SMOKE ON THE BRONCHIAL EPITHELIUM OF SYRIAN GOLDEN
HAMSTERS [abstract]. (Eng.) Reznik-Schuller, H.
(Abteilung fur Experimentelle Pathologie, Medizinische
Hochschule Hannover, Karl-Weichert-Allee 9, West
Germany); Mohr, U. Proc. Am. Assoc. Cancer Res.
16:7; 1975.

4078   PRELEUKEMIC CONDITIONS PRESENTING AS
        FAILURE OF BONE MARROW TISSUE. (Pol.)
Krauze-Jaworska, H. (93-513 Lodz, ul. Pabianicka 62,
II Klinika Chorob Wewnetrznych AM, Poland); Polkowska-
Kulesza E.; Krykowski S. Pol. Tyg. Lek. 30(1):23-25;
1975.

4079   A TRANSPLANTABLE MYELOID HAMSTER LEUKEMIA
        WITH HIGH PERIPHERAL LEUKOCYTE COUNTS AND
C-PARTICLES. I. TRANSPLANTATION EXPERIMENTS, PATHO-
GENESIS, AND ELECTRON-MICROSCOPIC OBSERVATIONS.
Bender, E. (Zentralinstitut fur Krebsforschung,
Experimenteller Bereich DDR-1115 Berlin-Buch, Linden-
berger Weg 70, East Germany); Niezabitowski, A.;
Rudolph, M.; Fey, F. Arch. Geschwulstforsch. 45(2):
121-130; 1975.

4080   STUDY OF LEUKOCYTE KINETICS IN ACUTE
        MYELOCYTIC LEUKEMIA UTILIZING CHROMIUM-51.
(Eng.) Rosen, P. J. (Los Angeles Cty. Univ. South.
California Medical Cent.); Eyre, H. J.; Lilien, D. L.;
Perry, S. Med. Pediatr. Oncol. 1(1):51-62; 1975.

4081   CLINICAL AND MORPHOLOGICAL CORRELATIONS IN
        ACUTE PROMYELOCYTIC LEUKEMIA. (Eng.)
Valdivieso, M. (M. D. Anderson Hosp. Tumor Inst.,
Houston, Tex.); Rodriguez, V.; Drewinko, B.; Bodey,
G. P.; Ahearn, M. J.; McCredie, K.; Freireich, E. J.
Med. Pediatr. Oncol. 1(1):37-50; 1975.

4082   CYTOGENETIC EVIDENCE OF IN-VIVO LEUKAEMIC
        TRANSFORMATION OF ENGRAFTED MARROW CELLS.
(Eng.) Goh, K.-o. (Univ. of Rochester and Monroe
Community Hosp., Rochester, N.Y. 14603). Lancet
(7920):1338-1339; 1975.

4083   EOSINOPHILIA AND PLASMACYTOSIS OF THE BONE
        MARROW IN HODGKIN'S DISEASE. (Eng.) Kass,
L. (Dept. Intern. Med., Univ. Michigan, Ann Arbor);
Votaw, M. L. Am. J. Clin. Pathol. 64(2):248-250;
1975.

4084   LYMPHOGRANULOMATOSIS IN AUTOPSIES TODAY.
        (Ger.) Grundmann, E. (Pathologisches
Institut der Universitat, D-4400 Munster i. Westf.,
Westring 17, West Germany); Fritzsche, R. Z.
Krebsforsch. 81(1):7-26; 1975.

4085   ROLE OF CHANGES IN GLUCOCORTICOID ACTIVITY
        OF THE ADRENAL CORTEX IN THE PATHOGENESIS
OF LYMPHOGRANULOMATOSIS. (Eng.) Popov, O. V.
(Hormone Lab., Rostov Inst. Oncol., U.S.S.R.).
Probl. Gematol. Pereliv. Krovi 20(2):8-12; 1975.

4086   ULTRASTRUCTURE OF LYMPH NODE CELLS IN LYM-
        PHOGRANULOMATOSIS. (Rus.) Chernina, L. A.
(N. N. Petrov Res. Inst. Oncology U.S.S.R. Ministry
Health, Leningrad, U.S.S.R.); Savost'ianov, G. A.;
Kolygin, B. A. Vopr. Onkol. 21(2):24-31; 1975.

4087   FAMILY STUDIES IN CASES WITH MALIGNANT
        LYMPHOMAS. (Eng.) Hardmeier, T. (Thur-
gauisches Kantonsspital, CH - 8596 Munsterlingen,
Switzerland); Rellstab, H. Acta Neuropathol.
[Supp.] (Berl.) 6:205-208; 1975.

4088   ACUTE FORM OF MAREK'S DISEASE AND COCCI-
        DIOSIS IN CHICKEN. (Rus.) Kachala, I. I.
(Transcarpathian Oblast Veterinary Lab., U.S.S.R.);
Sabo, A. E.; Korovin, R. N.; Dozorov, B. I. Veter-
inariia (8):52-53; 1975.

4089   INTRAMUSCULAR HEMANGIOMA IN THE HEAD AND
        NECK. (Eng.) Clemis, J. D. (55 E. Wash-
ington St., Chicago, Ill. 60602); Briggs, D. R.;
Changus, G. W. Can. J. Otolaryngol. 4(2):339-347;
1975.

4090   HIGH AMNIOTIC FLUID ALPHA-1-FETOPROTEIN IN
        A CASE OF FETAL SACROCOCCYGEAL TERATOMA.
(Eng.) Schmid, W. (Dept. Pediatrics, Univ. Zurich,
Switzerland); Muhlethaler, J. P. Humangenetik 26(4):
353-354; 1975.

4100   QUANTITATIVE CYTOCHEMICAL AND MORPHOMETRIC
        INVESTIGATIONS ON UTERINE ANGIOMATOSIS
AND STROMAL SARCOMA. (Ger.) Lederer, B. (Patholo-
gisches Institut der Universitat, A-6020 Innsbruck,
Mullerstrasse 44/1, Austria); Weiser, G.; Walter, N.;
Propst, A. Z. Krebsforsch. 83(1):57-64; 1975.

4101   CELL NUCLEUS AND ADENOCARCINOMA OF THE
        ENDOMETRIUM. (Spa.) Ghinelli, C. (Poli-
clinico Pirovano, Division de Patologia, Buenos
Aires, Argentina); De Cesare, F.; Invernati, A. L.;
Terzano, G. Obstet. Ginecol. Lat. Am. 32(11/12):
425-438; 1974.

4102   THE DEVELOPMENT OF THE CORPUS UTERI CANCER
        AGAINST THE BACKGROUND OF ADENOMATOSIS.
(Rus.) Smirnov, E. A. (Cent. Provincial Hosp., Push-
kino, U.S.S.R.). Vopr. Onkol. 21(3):28-35; 1975.

4103   PRECANCEROSIS OF THE UTERINE CERVIX IN
        YOUNG WOMEN. (Ger.) Vacha, K. (Uni-
versitats-Frauenklinik, Hradec Kralove, C.S.S.R.);
Rosol, M.; Kopecny, J. Zentralb. Gynaekol. 97(9):
525-528; 1975.

4104   SCREENING FOR CERVICAL CANCER: CYTOLOGY
        INFLUENCED BY INFLAMMATORY REACTION AND
ITS CAUSES. (Ger.) Naujoks, H. (Abt. fur Klinische
Zytologie, Univ.-Frauenklinik, 6 Frankfurt/M. 70,
Theodor Stern-Kai 7); Granitzka, S. Fortschr. Med.
93(1):21-22; 1975.

4105   GERM-CELL MALIGNANT TUMOURS IN FATHER AND
        SON. (Eng.) Musa, M. B. (Dr. W. W. Cross
Cancer Inst., 11560 University Ave., Edmonton,
Alberta T6G 1Z2, Canada). Can. Med. Assoc. J.
112(10):1201-1202; 1975.

4106   CLASSIFICATION OF MALIGNANT TUMORS OF THE
        THYROID GLAND ACCORDING TO THE 1974 WHO
NOMENCLATURE. HISTOLOGICAL VERIFICATION OF 327 CASES
OF MALIGNANT TUMORS OF THE THYROID GLAND. (CON-
CLUSION). (Ger.) Neracher, H. (Institut fur Patho-
logische Anatomie, Kantonspital, CH-8091 Zurich,
Switzerland); Hedinger*, C. Schweiz. Med. Wochenschr.
105(33):1052-1056; 1975.

4107   A NEW THEORY OF THE CANCER MECHANISM IN
        HUMAN BODY [abstract]. (Eng.) Bajaj, M. M.
(Univ. Delhi, Delhi, India). Bull. Am. Phys. Soc.
20(6):824; 1975.

4108   THE ULTRASTRUCTURE OF NORMAL FIBROBLAST-
        LIKE CELLS AT EARLY STAGES OF SPREADING
IN TISSUE CULTURE. (Rus.) Bragina, E. E. (Dept.
Histology and Cytology, Moscow State Univ., Moscow,
U.S.S.R.). Tsitologiia 17(3):248-253; 1975.

4109   A COMPARATIVE MORPHOMETRICAL STUDY OF THE
        CELL ULTRASTRUCTURE *IN VIVO* AND *IN VITRO*.
.(Rus.) Iagubov, A. S. (Moscow Oncological Res. Inst.,
Ministry of Public Health, U.S.S.R.); Kats, V. A.
Tsitologiia 17(9):1081-1084; 1975.

4110   ULTRASTRUCTURE OF A HUMAN LYMPHOBLASTOID
        CELL LINE: A FREEZE-ETCHING STUDY. (Ita.)
Rebessi, S. (Laboratori di Fisica, Istituto Superiore
di Sanita, Rome, Italy). *Ann. Ist. Super. Sanita*
10(1/2):73-78; 1975.

4111   ULTRASTRUCTURE OF CLEAR CELLS IN HUMAN
        VIRAL WARTS. (Eng.) Laurent, R. (Clinique
Dermatologique Universitaire, Centre Hospitalier,
25030 Besancon, France); Agache, P.; Coume-Marquet, J.
J. Cutaneous Pathol. 2(3):140-148; 1975.

4112   SIX CASES OF FERROALLOY WORKERS' DISEASE.
        (Eng.) Taylor, D. M. (Gwelo, Rhodesia).
Davies, J. C. A. *Cent. Afr. J. Med.* 21(4):67-71;
1975.

See also:

* (Rev):   3601, 3608, 3618, 3620, 3621, 3622,
            3623, 3624, 3625, 3626, 3628, 3632,
            3633, 3639, 3648, 3655, 3656, 3657,
            3658, 3659, 3660, 3666
* (Chem):   3674, 3732, 3757
* (Phys):   3771, 3772
* (Viral): 3794, 3807
* (Immun): 3895, 3903, 3907, 3918, 3929, 3978
* (Epid-Biom): 4129, 4143

4113   ENVIRONMENTAL FACTORS IN CANCER INDUCTION:
APPRAISAL OF EPIDEMIOLOGIC EVIDENCE --
LEUKEMIA, LYMPHOMA, AND RADIATION. (Eng.) Jablon,
S. (Natl. Res. Council, Washington, D. C.). *Proc.
Int. Cancer Congr. 11th.* Vol. 3 *(Cancer Epidemio-
logy, Environmental Factors).* Florence, Italy,
October 20-26, 1974. Edited by Bucalossi, P.; Vero-
nesi, U.; Cascinelli, N. New York, American Elsevier,
1975, pp. 239-243.

The hypothesis that human leukemogenesis follows a
dose-response function which is linear in the neutron
dose and quadratic in the gamma-ray dose was considered on the basis of data on leukemia mortality
among the Japanese A-bomb survivors. Because of the
differences in the construction of the Hiroshima and
Nagasaki bombs, the Hiroshima weapon emitted large
quantities of fast neutrons. For most survivors in
Hiroshima, the estimated kerma (kinetic energy released in materials) from neutrons was about 1/4
that from gamma rays. The Nagasaki weapon emitted
few neutrons and the Nagasaki survivors received
gamma radiation with a very small admixture of neutrons. About 65% of leukemias which occurred in the
survivors have been acute leukemias and 35% chronic
myeloid; there were no cases of chronic lymphatic
leukemia. Leukemias began to occur in the Japanese
survivors about 2 yr after the bombings and reached
a peak in 1950-1954. The standardized mortality
ratio was more than 50 among survivors whose doses
were more than 100 rads. In 1970-1972, the ratio
was 5.6. The leukemia mortality data for the two
cities was fitted, using weighted least squares,
to formulas for the probability of death from leukemia in the period 1950-1972; in one formula the
neutron and gamma doses each entered as first degree
terms; in the second formula the neutron dose entered
as a first degree term, while the gamma-ray dose
entered only as a square. For the equation linear in
both neutrons and gamma rays, there was one discrepancy of large magnitude; among Nagasaki survivors
who received more than 200 rads, the formula predicts
6.5 deaths, whereas 15 actually occurred. For the
second equation, agreement was satisfactory in every
one of the dose cells, including the Nagasaki high-
dose cell. Previously published data on the induction of mammary neoplasms in rats, on opacification
of the murine lens, and on a number of effects arising
from irradiation of dry corn seeds coupled with the
data from this investigation leads to the suggestion
that a derangement of DNA is the mechanism common to
these varied systems. Production of double-stranded
DNA breaks should be a two-hit phenomenon for low
linear energy transfer (LET), but should appear to
be a one-hit phenomenon for high LET radiations such
as neutrons. This evidence points to the double-stranded
DNA break as being the radiation-produced lesion responsible for radiogenic leukemia.

4114   A COHORT ANALYSIS OF U.K. CANCER INCIDENCE
TRENDS FOR CERTAIN SITES. (Eng.) Water-
house, J. A. H. (Birmingham Regional Cancer Regis-
try, Birmingham, England). *Proc. Int. Cancer Congr.
11th.* Vol. 3 *(Cancer Epidemiology, Environmental
Factors).* Florence, Italy, October 20-26, 1974.
Edited by Bucalossi, P.; Veronesi, U.; Cascinelli,
N. New York, American Elsevier, 1975, pp. 180-190.

Evidence for 1960-1971 trends in morbidity rates
of cancer of certain anatomical sites was assessed
on basis of statistics accumulated at the Cancer
Registries of Birmingham in the British midlands
and of the southeastern part of England including
half of the London area. Mean annual incidence
rates by site, sex and age for three time periods
in a span of 12 yr were computed and the points
plotted to form a cohort incidence line. In their
composite pattern, the cohort incidence lines
closely approximated the contour incidence curves
which represented the morbidity risk by age for a
given time period. Similar to the trend world-wi
the rate of incidence of stomach cancer declined
in younger age groups. The rate of incidence of
cancer of the breast in females and ovarian cance
increased in younger populations, whereas in olde
females the rate of endometrial cancer has increa
Rates for cancer of the bladder, pancreas, larynx
and bronchus all showed definite increases. Among
males, the rate of increase of lung and laryngeal
cancer has diminished to zero, however in females
it has been increasing sharply. Evidence incrimi
nating cigarette smoking in lung and laryngeal ca
cer is very strong. Morbidity data from cancer
registries may provide information of predictive
value.

4115   ASSESSING TRENDS OF CANCER INCIDENCE IN
SMALL POPULATIONS. (Eng.) Steinitz, R
(Israel Cancer Registry, Jerusalem, Israel); Cos-
tin, C.; Katz, L. *Proc. Int. Cancer Congr. 11th.*
Vol. 3 *(Cancer Epidemiology, Environmental Factors*
Florence, Italy, October 20-26, 1974. Edited by
Bucalossi, P.; Veronesi, U.; Cascinelli, N. New
York, American Elsevier, 1975, pp. 175-179.

Morbidity data from the time periods 1960-1964 and
1965-1969 are used to compute the range of change
in cancer incidence in the population of Israel.
Several aspects of the population create a unique
situation for epidemiological research: (1) total
population is small, less than three million; (2)
official census statistics divide the population
into Jews and non-Jews; (3) the population can be
analyzed on the basis of continent of origin as
well as country of origin; (4) the age structure
of the Israeli population is bizarre because of
the arrival of immigrants in successive waves.
The population at risk in each of the two time
periods was calculated as the sum of the average
population in each year. It was decided to use
the World population rather than the total Jewish
population as the population at risk and age-speci
fic incidence rates were adjusted to this figure.
Data was tabulated rather than graphed because of
the small numbers involved. Reliability of the re
sults from the two time periods was checked by investigating the incidence of cancer in two sites
with anticipated opposite trends, i.e. cancer of
the stomach and lung cancer. Overall reliability
and comparability of the cancer registry data was
also checked by analyzing the incidence rate of al
malignancies. Significance tests cannot be applie
to cancer morbidity data from Israel at the preser
time, however trends are revealed which warrant
further investigation.

4116   THE CHANGING PATTERN OF CANCER IN SOUTH-
        ERN AFRICA. (Eng.) Keén, P. (Cancer
Res. Unit, South African Inst. for Medical Res.,
Johannesburg, South Africa). *Proc. Int. Cancer
Congr. 11th.* Vol. 3 *(Cancer Epidemiology, Environmental Factors).* Florence, Italy, October 20-26,
1974. Edited by Bucalossi, P.; Veronesi, U.; Cascinelli, N. New York, American Elsevier, 1975, pp.
215-221.

Trends in the incidence of lung, esophageal and
liver cancer in the South African Negro are discussed with data from hospitals in Johannesburg
and the Eastern Transvaal. Lung cancer which represented 8% of cancers in Bantus at the Johannesburg Hospital in 1950 increased to 52% in 1968.
African Negroes in urban areas are taking up smoking cigarettes, and air pollution in native urban
areas is high. The high incidence of antral cancer, 13%, can be traced to the use of a snuff containing large amounts of aromatic hydrocarbons,
chromium, and nickel. Incidence of primary liver
cancer appears to have peaked around 1955 and, for
unknown reasons, is rapidly declining. Esophageal
cancer, extremely rare before 1950, has become the
commonest malignancy in some areas of South Africa.
The cancer first appeared in southernmost areas,
and each year the number of cases diagnosed in the
indigenous population increases in an apparent
chronological order from South to North. No explanation for the reversal of the incidence of lung
and esophageal cancers is given.

4117   MULTIPLE PRIMARY MALIGNANCIES: INDEX
        TUMORS OF THE BREAST AND REPRODUCTIVE
ORGANS. (Eng.). Schoenberg, B. S. (Mayo Graduate
Sch. Medicine, Rochester, Minn.); Christine, B. W.
Proc. Int. Cancer Congr. 11th. Vol. 3 *(Cancer Epi-
demiology, Environmental Factors).* Florence, Italy,
October 20-26, 1974. Edited by Bucalossi, P.;
Veronesi, U.; Cascinelli, N. New York, American
Elsevier, 1975, pp. 294-299.

The occurrence of multiple primary malignancies for
which at least one of the tumors was within the
breast or reproductive organs was investigated using
the Connecticut Tumor Registry (1935-1964) as a data
base. The study followed a group of 18,010 women and
130 men with breast cancer, and a group of 17,405
women and 8,228 men with cancer of the genital organs
for the development of subsequent primary cancer of
any site. In addition, 121,460 patients with a first
primary malignancy of any site were followed for the
subsequent occurrence of primary malignant tumors of
the breast and reproductive organs. The overall survival experience for patients with a first primary
malignancy was converted to person-years of exposure.
The survival experience for all patients with a first
primary cancer of either the breast or reproductive
organs was accumulated. Age-, sex-, and site-specific
incidence rates were applied to the appropriate person-
years of exposure to obtain the expected number of
primary cancers for each group. The coded records of
reported cases were selected and matched with the
first primary cancers to obtain the observed number
of second primaries. Using the same methods, it was
possible to determine the expected number of primary

neoplasms of the breast and genital organs following
primary malignancies of other sites. Excesses of
observed-to-expected second primary cancers occurring
in paired organs, in organs with the same system, or
in structurally or functionally related organs might
be explained on the basis of related tissue being exposed to the same oncogenic factors. Excesses occurring in organs apparently unrelated functionally or
anatomically, in which the association goes only one
way may be caused by something unique about patients
with cancer at site 1 that predisposes them to develo
cancer at site 2, such as a treatment regimen. For
those cases with a two-way association, it may be tha
the same genetic or environmental oncogenic factors
operating in the development of the first cancer were
again operating in the development of the second cancer.

4118   CANCER MORTALITY AMONG MORMONS. (Eng.)
        Enstrom, J. E. (Sch. Public Health, Univ.
California, Los Angeles, Calif. 90024). *Cancer*
36(3):825-841; 1975.

A study was conducted on cancer mortality among Mormons (a religious group in which the use of tobacco,
alcohol, coffee, and tea is forbidden, and the use
of wholesome grains and fruits and moderation in
meat consumption are recommended). Data were taken
from state mortality figures, from church records,
and from a 1965 Alameda County, California study of
health and lifestyles. Preliminary results show
that the 1970-1972 cancer mortality rate among California Mormon adults is about one-half to three-
fourths that of the general California population for
most cancer sites, including many sites with an unclear etiology. Initial indications are that Mormons
as a whole smoke and drink about half as much as the
general population, and that active Mormons abstain
almost completely from tobacco and alcohol. They
appear to be similar to the general white population
in other respects such as socioeconomic status and
urbanization. The average age-adjusted total mortality ratio for nonsmokers at least 35 yr old to United
States whie men in 1960 was 55%, with a range from
46-64% in the eight studies from which these figures
were taken. The 1970-1972 age-adjusted total mortality ratio for California Mormon men at least 35
yr old was 46-69%, depending on the age distribution
assumed. These mortality rates may considerably
overestimate the mortality rates for observant lifetime Mormons, because a large proportion of Mormons
have followed or still follow a lifestyle similar to
the general population; 40% of Mormons in California
are converts, having entered the church as adults, and
another 40% are inactive and tend to ignore Mormon
health practices. Low cancer rates as those seen in
the Mormons have not been observed in any other group
of Americans except Seventh-day Adventists. These
low rates suggest the development and exploration of
new etiologic hypotheses for cancer and may lead to
understanding a lifestyle that produces low risk to
cancer of all types.

4119   EPIDEMIOLOGIC SIMILARITIES AND DIFFEREN-
        CES IN CHILDHOOD LEUKEMIA. (Eng.) Vi-
anna, N. J. (New York State Dept. Health, 84 Holland Ave., Albany, N. Y. 12208); Polan, A. K.

four having a lesser incidence in boys and five in
girls, as compared with that of Greater Bombay.

4121   A GENETIC-EPIDEMIOLOGIC STUDY OF CHRONIC
        OBSTRUCTIVE PULMONARY DISEASE. I. STUDY
DESIGN AND PRELIMINARY OBSERVATIONS. (Eng.) Cohen,
B. H. (Johns Hopkins Univ. Sch. Hygiene and Public
Health, 615 North Wolfe St., Baltimore, Md. 21205);
Ball, W. C., Jr.; Bias, W. B.; Brashears, S.;
Chase, G. A.; Diamond, E. L.; Hsu, S. H.; Kreiss, P.;
Levy, D. A.; Menkes, H. A.; Permutt, S.; Tockman,
M. S. *Johns Hopkins Med. J.* 137(3):95-104; 1975.

A large-scale (3390 subjects) genetic-epidemiologic
investigation of chronic obstructive pulmonary
disease was conducted to examine the role of several
genetic and environmental factors, both singly and
in combination, in the disease. The investigation
consisted of two projects: (a) a modified case-
control study in which biochemical, physiologic,
demographic, personal, family, and sociobiological
data were collected on pulmonary disease patients,
several comparison groups, and selected family
members; and (b) an ancillary project in which
data similar to the first project were collected but
more individuals per kindred were studed to examine
intra- and inter-familial patterns. All subjects,
patients, and relatives received pulmonary function
tests, had blood and saliva samples collected, and
were tested for the ability to taste phenylthiocarbamide. Pulmonary function tests included tests of
pulmonary mechanics, gas distribution, diffusing
capacity, and closing capacity. The alpha$_1$-anti-
trypsin determinations included electrophoretic
phenotyping and serum trypsin inhibitory capacity.
Typing of genetic markers (RBC antigens and enzymes,
plasma proteins, and saliva markers) was used to
determine associations and linkage relationships in
detailed family studies. In all groups studied,
cigarette smokers showed a higher frequency of pulmonary function abnormality than those who never
smoked. There was a higher frequency of Pi variant
phenotypes among patients with chronic obstructive
pulmonary disease than among those without lung
disease. First degree relatives of chronic obstructive pulmonary disease cases showed a significantly
higher frequency of pulmonary function impairment
than first degree relatives of nonpulmonary patients.
The difference was still seen when Pi variant relatives and smokers were excluded, indicating that
there is a familial component in pulmonary dysfunction that is not entirely due to Pi type or smoking.
In the ABO blood group, there seemed to be a low
frequency of blood group B among Caucasian patients
with chronic obstructive pulmonary disease. No unusual patterns in Rh(D) frequency or in the occurrence of phenylthiocarbamide taste blindness were
observed in either whites or blacks. It is suggested
that an interaction of constitutional and extrinsic
factors probably determines both clinical manifestations and disease progression.

4122   SIMILAR GEOGRAPHICAL DISTRIBUTION OF
        MULTIPLE SCLEROSIS AND CANCER OF THE
COLON. (Eng.) Wolfgram, F. (Reed Neurol. Res.

Center, Univ. California, Los Angeles, Calif.
90024). *Acta Neurol. Scand.* 52(4):294-302; 1975.

A search was made of all diseases listed in the
Annual Statistics of the World Health Organization
(WHO) to determine if any disease had the same geographical distribution as multiple sclerosis. A
list was made of the countries in which there was
a significant mortality from multiple sclerosis as
indicated by the WHO Annual Statistics for 1966-
1970. All diseases with significant mortality rates
(83 entries) were compared to the geographical distribution of multiple sclerosis. In the five years
of this survey, 33,849 people died with multiple
sclerosis in the 36 countries selected for study.
The female/male mortality ratio was 1.33 for this
period. Only cancer of the colon had essentially
the same geographic distribution and female/male
ratio as multiple sclerosis. There were 474,927
deaths from cancer of the colon during the survey
years, and the female/male ratio for this was 1.25.
The major exceptions to the parallel distribution
between multiple sclerosis and cancer of the colon
were Poland, Spain, Czechoslovakia and Portugal.
Rectal cancer had a geographical distribution similar
to that of multiple sclerosis and colonic cancer except that it was relatively high in Japan, and the
female/male ratio was 0.85. Coronary heart disease
had a relatively good correlation to the distribution
of multiple sclerosis and cancer of the colon, but
the female/male ratio was 0.74. Epidemiological
evidence indicates that children born in western
societies of Asian and African immigrant parents may
be as susceptible to both multiple sclerosis and
cancer of the colon as are the native-born Caucasian
children. It is hypothesized that both multiple
sclerosis and cancer of the colon have rather long
latent periods in which pathogenic environmental
factors are operative before clinical symptoms are
manifest. It is suggested that a dietary disease
would be plausible in explaining the long latent
period and the susceptibility of immigrant children
to both diseases.

4123   ENVIRONMENTAL FACTORS AND CANCER INCIDENCE
         AND MORTALITY IN DIFFERENT COUNTRIES, WITH
SPECIAL REFERENCE TO DIETARY PRACTICES. (Eng.)
Armstrong, B. (Radcliffe Infirm., Oxford, England);
Doll, R. *Int. J. Cancer* 15(4):617-631; 1975.

Incidence rates for 27 cancers in 23 countries and
mortality rates for 14 cancers in 32 countries were
correlated with a wide range of dietary and other
variables to determine relationships between these
variables and cancer rates. Variables included in
the study were liquid energy during 1955-1957; solid
energy, 1955-1957; population density, 1965; tea
consumption, 1955-1957; coffee, 1955-1957; cigarettes, 1963-1965; total fat; total protein; animal
protein; calories; fats and oils; fish; eggs; meat;
fruit; vegetables; pulses, nuts and seeds; sugar;
potatoes, starchy and other staple foods; cereals;
physician density, 1965; and gross national product.
Product-moment simple and first-order partial correlation coefficients between the cancer rates and
the environmental variables were calculated. The
product-moment correlation method was used in pre-

ference to a rank correlation method as it provides
the most satisfactory partial correlation coefficients. The intercorrelations between all the environmental variables were tabulated in the tri-
angular form. This arrangement permits the identification of environmental variables that are highly
correlated with one another. Dietary variables were
strongly correlated with several types of cancer,
particularly meat consumption with cancer of the
colon and fat consumption with cancers of the breast
and corpus uteri. Examination of a large number of
environmental variables and the calculation of partial coefficients increases the probability that
these associations are not just secondary to an association with some other variable. The data suggest that diet may have an effect upon many cancers,
perhaps by affecting the metabolism of various carcinogens (as has been previously suggested), or by
altering the body's capacity to deal with malignant
cells.

4124   EPIDEMIOLOGICAL STRATEGIES FOR IDENTIFYING
         CARCINOGENS. (Eng.) Janerich, D. T.
(Cancer Control Bureau, New York State Dept. Health,
New York, N.Y.); Lawrence, C. E. *Mutat. Res.* 33(1):
55-63; 1975.

Two situations in which epidemiologic studies were
used to identify carcinogens are reviewed and efforts
to use New York State Cancer Control Bureau data for
systematic surveillance of reported cancer incidence
are discussed. The association between diethylstilbestrol and vaginal cancer in young women and between
vinyl chloride and angiosarcoma of the liver was
confirmed by epidemiological evidence, although in
both cases the association was initially suggested
by clinical observations. In the case of diethylstilbestrol, epidemiologic studies indicated the
limits of the risk associated with intrauterine
exposure; it is suggested that age limits will be
identified for the carcinogenic potential of intrauterine exposure to diethylstilbestrol, just as
intrauterine exposure to X-rays causes an increased
risk of leukemia only during the first decade of
life. In the case of vinyl chloride, its connection
with angiosarcoma of the liver was first based on
clinical observations and the rare nature of the
disorder. Both of these situations indicate that
present methods of detection are crude, and that a
large number of people are exposed to chemicals
before their carcinogenic potential is recognized.
Preliminary efforts to develop a systematic cancer
monitoring methodology in New York State involved
analysis of cancer data from three anatomical sites:
colon, pancreas, and gallbladder. These efforts
yielded one instance in which an unquestionably
high cancer incidence was suggestive of some environmental etiological agent. In two years, a single
county showed unusual rates of pancreatic cancer.
However, preliminary investigation showed no obvious
environmental exposure associated with the cluster,
and in subsequent years the incidence rate returned
to within expected limits. Two alternatives in
devising an epidemiological strategy are considered:
the cohort study and the case history study. The
case history study sometimes runs the risk of
failing to detect a disease risk when one is actually

fect of environmental factors such as air pollution
is considered to be minimal.

4127   OCCUPATIONAL LUNG CANCER AMONG COPPER
        SMELTERS. (Eng.) Kuratsune, M. (Faculty
of Medicine, Kyushu Univ., Fukuoka, Japan). *Proc.
Int. Cancer Congr. 11th.* Vol. 3 *(Cancer Epidemio-
logy, Environmental Factors).* Florence, Italy,
October 20-26, 1974. Edited by Bucalossi, P.;
Veronesi, U.; Cascinelli, N. New York, Americar.
Elsevier, 1975, pp. 98-101.

Two epidemiological studies of occupational lung cancer are reported for: 1) workers in a large copper
refinery and 2) workers in a small arsenic plant and
the inhabitants of the small remote hamlet in which
it was situated. A case control study was carried
out using mortality cards filed in the health center
serving the locality where the copper refinery was
located. The case group consisted of 19 males who
had died from lung cancer during 1967-1969; the control group consisted of 19 males randomly selected
from deaths caused by diseases other than lung cancer.
Cases and controls were matched for age, place and
year of death. Occupational and nonoccupational
history was obtained by family interviews. No significant difference was noted between the groups for
most of the occupational items tested. However, 58%
of the case group had been employees of the copper
refinery compared to 37% of the controls. Of the
case group, 58% had been copper smelters while only
16% of the control group were in this category. No
significant difference was noted for nonoccupational
history. These findings suggest that employment as
smelters in the refinery was the main cause of the
high lung cancer mortality seen among males of the
town. Latent period ranged from 26-48 yr. All 11
cases were officially recognized as being caused by
occupational exposure to arsenic and other compounds
released during the smelting process. In the second
study, all mortality cards kept at the health center
which was responsible for the small hamlet in which
the arsenic plant had been located were examined and
27 deaths from lung cancer were identified from the
entire jurisdictional area during a 10-yr period.
The control group consisted of 27 deaths randomly
selected from deaths caused by diseases other than
lung cancer. Of the lung cancer deaths seen in the
whole jurisdictional area of the health center, 24%
had lived in the town with the arsenic plant which
comprises less than 1% of the population served by
the health center. Similar significant differences
were seen for employment at the arsenic plant but
not for other occupational factors. These facts
suggest that occupational and probably nonoccupational
exposure to arsenic and other compounds released from
the arsenic plant increased the incidence of lung
cancer. It is concluded that heavy exposure to dusts
and fumes released in the smelting of arsenic-
containing ores greatly increases the risk of lung
cancer.

4128   LUNG CANCER IN MEN: A MODEL FOR THE RE-
        LATIONSHIP BETWEEN AGE AT DIAGNOSIS AND
SMOKING HABITS. (Eng.) Weiss, W. (3912 Nether-
field Road, Philadelphia, Pa. 19129). *Am. Rev.
Respir. Dis.* 111(6):883-886; 1975.

A model was constructed to determine the relation between age at diagnosis of lung cancer and smoking habits. A sizeable group of men with a birth spread of ten years who developed lung cancer which was diagnosed within a ten-year period were studied. Thus, the ages of the cancer cases were limited to a 19-yr spread. The multiplicity of birth cohorts and the small segment (only ten years over a 30-40 yr portion of the life span) of the life span distribution of cases observed were confounding and restraining conditions. The curve (age distribution *versus* number of deaths) suggests that a small degree of skewing in the age distribution of lung cancer cases in a ten-year period represents a larger shift in the actual life span age distributions. Thus, the constraint of the conditions causes a damping of the effect of shifts in life span age distribution.

4129   GEOGRAPHIC PATHOLOGY OF OCCULT THYROID CARCINOMAS. *(Eng.)* Fukunaga, F. H. (Kuakini Hosp., Honolulu, Hawaii 96817); Yatani, R. *Cancer* 36(3):1095-1099; 1975.

Because of the considerable variation in the reported prevalence of occult thyroid carcinomas, 1,167 thyroid glands obtained at autopsy from patients in various parts of the world were examined microscopically using identical techniques and diagnostic criteria. Specimens were obtained from patients in southeastern Canada, northeastern Japan, southern Poland, western Colombia, and from Japanese living in Hawaii; they were serially step-sectioned, and a carcinoma was classified as occult if it was less than 1.5 cm in diameter and unsuspected clinically. The criteria for the diagnosis of papillary carcinoma were pale cytoplasm, large vesicular- to ground glass-appearing nuclei with very prominent nuclear membranes that gave etched appearance. The prevalence of occult papillary thyroid carcinoma was significantly higher in Japan (28.4%) and in Hawaiian Japanese (24.2%) when compared with Canada (6%), Poland (9.1%), and Colombia (5.6%). The carcinomas were all papillary except for a single follicular lesion from Colombia. There was no significant sex prevalence. Most of the patients were 40-79 yr of age, but there was no predominant decade. Only the Colombian series had a large number of younger patients, and they showed a slightly lower prevalence of occult carcinomas before age 40. Most papillary thyroid carcinomas grow slowly and probably remain occult for the life of the patient.

4130   CELL KINETICS OF HUMAN SOLID TUMORS FOLLOWING CONTINUOUS INFUSION OF ^{3}H-TDR. (Eng.) Terz, J. J. (Medical Coll. Virginia, Health Sciences Div. Virginia Commonwealth Univ., Richmond, Va.); Curutchet, H. P. *Cell Cycle in Malig. Immun., Proc. Annu. Hanford Biol. Symp., 13th.* Richland, Washington, D.C., U.S. Energy Research and Development Administration, 1975, pp. 323-329.

The cell proliferation kinetics of four human solid tumors was analyzed following the administration of tritiated thymidine (^{3}H-TdR) *in vivo*. Four patients (epidermoid carcinoma of the maxil-

lary antrum, recurrent epidermoid carcinoma of the neck, metastatic carcinoma of the pancreas, and carcinoma of the breast) received a continuous infusion of 2 mCi a day of ^{3}H-TdR for a period ranging from 6-20 days. Biopsies were obtained during and after the infusion period. In each case the mitotic index, labeling index, percent of labeled mitosis, and median grain count were determined. The duration of the cell cyle ranged between 18 hr (epidermoid carcinoma of the neck) to 6 days (adenocarcinoma of the pancreas). The size of the proliferating tumor-cell population rose to 30% at six days and to 60% in 20 days. Ninety percent of the mitoses were labeled at 20 days. In one patient, a simultaneous analysis of the cell kinetics of the normal tissue and tumor counterpart was performed. These studies suggest that a significant segment of the tumor-cell population (40%) does not divide (prolonged G_1 or G_0) at least for 20 days.

4131   THEORETICAL ASPECTS OF GROWTH FRACTION IN A G_0 MODEL. (Eng.) Burns, F. J. (New York Univ. Medical Center, New York, N.Y.). *Cell Cycle in Malig. Immun., Proc. Annu. Hanford Biol. Symp., 13th.* Richland, Washington D.C., U.S. Energy Research and Development Administration, 1975, pp. 315-322.

In a model of the cell cycle previously presented, the proliferating cells (P cells) were subdivided into a G_0 phase from which cells were released randomly at a rate consistent with the proliferation rate of the tissue as a whole and a C phase. The C phase received the cells released from the G_0 phase and was of uniform duration consisting of various subphases including the S phase and the G_2 phase. The model in this paper introduces a rate constant that describes the conversion of P_1 cells into nonproliferating cells (Q cells). The presence of Q cells requires that the growth fraction (GF) be less than 1. Cell loss was assumed to occur either uniformly from all the cells in the population (P cells and Q cells) or solely from the Q population. Expressions for the frequency-of-labeled-mitosis (FLM) curves, percentage of cells in the S phase, and the age distribution functions were derived in terms of GF, the proliferation rate, K_p, the loss rate, K_L, and the durations of the various C subphases. The results indicate that the ratio of the percentage of cells in the S phase to the plateau level of the damped FLM curve is GF/2. The model is fitted to experimental FLM curves of normal tissue (hamster cheek epithelium; intestinal crypt) and 15 tumors with diverse growth and cell-loss characteristics.

4132   SOME ASPECTS OF ERYTHROPOIETIC CELL PROLIFERATION IN ERYTHROLEUKAEMIA. (Eng.) Mitrou, P. S. (Zentrum der Inneren Medizin, Theodor-Stern-Kai 7, D-6000, Frankfurt 70, West Germany); Fischer, M.; Hubner, K. *Acta Haematol (Basel)* 53 (2):65-74; 1975.

Two questions are raised. Are proliferating ery-

Regional Res. Lab., ARS, USDA, Peoria, Ill. 61604); Goulden, M. L.; Lillehoj, E. B.; Kwolek, W. F.; Hesseltine, C. W. *Cereal Foods World* 20(9):442; 1975.

4138   AN EVALUATION OF POTENTIAL MYCOTOXIN-
        PRODUCING MOLDS IN CORN MEAL. (Eng.)
Bullerman, L. B. (Dep. Food Sci. Technol., Univ. Nebraska-Lincoln); Baca, J. M.; Stott, W. T. *Cereal Foods World* 20(5):248-250; 253; 1975.

4139   CYTOKINETIC MEASUREMENTS OF SPONTANEOUS
        C3H MAMMARY TUMORS [abstract]. (Eng.)
Poulakos, L. (Allegheny Gen. Hosp., Pittsburgh, Pa.). *Proc. Am. Assoc. Cancer Res.* 16:25; 1975.

4140   BLOOD-VESSEL NEOPLASMS IN CHILDREN: EPI-
        DEMIOLOGIC ASPECTS. (Eng.) Chabalko,
J. J. (Epidemiology Branch, Natl. Cancer Inst., Bethesda, Md. 20014); Fraumeni*, J. F., Jr. *Med. Pediatr. Oncol.* 1(2):135-141; 1975.

4141   FAMILY HISTORY, PARITY AND ESTROGEN PRE-
        TREATMENT IN PATIENTS WITH BENIGN AND
MALIGNANT BREAST LESIONS [abstract]. (Eng.) Black, M. M. (New York Medical Coll., New York, N.Y. 10029); Leis, H. P., Jr., Kwon, C. S. *Proc. Am. Assoc. Cancer Res.* 16:2, 1975.

4142   AN EPIDEMIOLOGIC STUDY OF OSTEOGENIC SAR-
        COMA IN MALAYSIA. INCIDENCE IN URBAN AS
COMPARED WITH RURAL ENVIRONMENTS AND IN EACH OF THREE SEPARATE RACIAL GROUPS, 1969-1972 [abstract]. (Eng.) Bovill, E. G., Jr. (Univ. California Sch. Medicine, San Francisco, Calif. 94110); Silva, J. F.; Subramanian, N. *J. Bone Joint Surg.* [*Am.*] 57A(7):1026; 1975.

4143   "FAMILIAL TUMORS" IN A MEDICAL ONCOLOGY
        PRACTICE [abstract]. (Eng.) Reimers,
R. (Grand Rapids Oncol. Group, Mich.); Reimers, S.; Borst, J. R.; Moorhead, E. L., II. *Proc. Am. Assoc. Cancer Res.* 16:242; 1975.

4144   EPIDEMIOLOGICAL STUDIES OF EPSTEIN-BARR
        HERPESVIRUS INFECTION IN WESTERN AUSTRALIA.
(Eng.) Lai, P. K. (Div. Microbiol., State Health Lab. Services, Western Australia); Mackay-Scollay, E. M.; Alpers, M. P. *J. Hyg. (Camb.)* 74(3):329-338; 1975.

4145   THE EPIDEMIOLOGY OF LUNG CANCER IN THE
        GERMAN FEDERAL REPUBLIC, WITH SPECIAL
REFERENCE TO REGIONAL DIFFERENCES. (Ger.) Neumann,
G. (7000 Stuttgart 1, Schickhardstr. 35, West Ger-
many). *Prax. Pneumol.* 29(1):32–42; 1975.

4146   TRENDS IN LUNG CANCER MORTALITY IN ITALY.
        (Ita.) D'Alfonso, G. (Clinica Tisiologica,
Universita di Napoli, Naples, Italy); Scarano, L.
Lotta Tuberc. Mal. Polm. Soc. 44(3):409–436; 1974.

4147   REPLICATION KINETICS OF HUMAN BREAST CANCER
        CELLS *IN VIVO* AND *IN VITRO* [abstract].
(Eng.) Post, J. (New York Univ. Res. Serv. Gold-
water Mem. Hosp., Roosevelt Is., New York, N.Y.);
Sklarew, R. J.; Hoffman, J. *Proc. Am. Assoc. Cancer
Res.* 16:34; 1975.

See also:

★	(Rev):	3605,	3606,	3608,	3610,	3613,	3623,
		3627,	3628,	3629,	3630,	3640,	3642,
		3643,	3647,	3648,	3660,	3661,	3662,
		3663,	3664,	3665			
★	(Chem):	3670,	3716				
★	(Phys):	3771,	3773				
★	(Viral):	3807					
'	(Immun):	3899					
★	(Path):	4023,	4032,	4112			

Sell, S. *Cancer Res.* 35(11/Part 1):3021-3026; 1975.

The stability of transplantable hepatocellular carcinomas (THC) derived from animals fed N-2-fluorenylacetamide was studied. The ancuploid THC, 251a and 251c, remained relatively stable through 44 generations. THC 251c, a typical aneuploid tumor, grew rapidly, was very poorly differentiated histologically, showed a hypotriploid chromosomal mode, and produced large quantities of plasma proteins and a_1F. All the diploid THC remained moderately well differentiated histologically and none produce plasma proteins or a_1F. During the initial transplant generations a slow drift to a higher modal number of chromosomes was observed. THC 252 was carried as 3 separate sublines. Each of the sublines showed parallel evolution with a modal range of 42 to 46 chromosomes in the 2nd transplant and a mode at 44-45 by the 10th transplant generation. Little further change through the 23rd generation was seen. A fast growing variant arose spontaneously from the diploid tumor 252 at the 16th generation. The sudden karyotypic evolution in this tumor variant was characterized by increase in tetraploidy and both stable and unstable chromosomal rearrangements. Only the stable rearrangements persisted, and marker chromosomes were common to all of the metaphase cells examined after further transplantations. Despite the striking cytogenetic alterations and a current growth rate of 2.5 weeks, this tumor produces neither plasma proteins nor a_1F. The failure of the variant subline to produce any detectable plasma proteins or a_1F indicates that synthesis of these proteins is strongly associated with the original karyotype of the primary tumor.

4151   INFLUENCE OF UREMIA ON THE PROLIFERATION
        KINETICS OF A SOLID TRANSPLANTABLE
CARCINOMA. (Ger.) Feaux de Lacroix, W. (Pathologisches Institut der Universitat, D-5000 Koln 41, Joseph-Stelzmann-Str. 9 West Germany); Clemens, A.; Klein, P. J.; Castrup, H. J.; Lennartz, K. J. *Beitr. Pathol.* 155(4):410-427; 1975.

The effect of uremia (corresponding to a blood urea level of 133 mg%) on the proliferation kinetics of partially differentiated mammary carcinoma HB was studied in 316 C_3H/Tif adult female mice by autoradiography. The study began on the 21st day after tumor transplantation, or seven days after uremia induction. Uremia was induced by ligation of the right ureter and by infarction of two-thirds of the left kidney. The starvation effect was avoided by additional tube feeding. The growth of the tumor was reduced by uremia. The tumor doubling time was increased from 7.5 days in untreated and sham-operated animals to 15 days in uremic mice. This was due mainly to a reduction of the growth fraction from about 50% to 20%, and to a prolongation of the regeneration time from 15.5 to 19 hr as a result of the prolongation of T_{G1} phase. The cell loss factor was constant at about 81% in all groups. The findings are similar to those previously obtained for the epithelium of the stomach, intestine, and skin, as well as for erythroblasts.

4152   CONTROL OF CELL DIVISION IN THE CORNEA OF
        RATS. III. MITOGENIC EFFECT OF ISOPRO-
TERENOL AND THEOPHYLLINE. *(Eng.)* King, C. D. (Univ.
of Tennessee Center for the Health Sciences, Memphis,
Tenn. 38163); Kauker, M. L.; Cardoso, S. S. *Proc.
Soc. Exp. Biol. Med.* 149(4):840-844, 1975.

The individual and combined effects of isoproterenol
(10 or 25 mg/kg, ip) and theophylline (50 or 100
mg/kg, ip) on the circadian mitotic rhythm and rates
of cell division within the corneal epithelium were
studied in male Holtzman rats. The rats were sacrificed at various intervals after drug treatment, and
the number of mitoses per eye were counted microscopically. Isoproterenol at 25 mg/kg produced an
initial suppression of cell division during the
period when high rates of cell division were observed
in the untreated controls (0700 hr); this suppression was followed by a rise in mitotic activity with
a peak at 2200 hr (corresponding to the control
nadir) and a subsequent decline in activity after
0100 hr. At 10 mg/kg, isoproterenol produced an
initial inhibition of cell division (0700 hr) followed by a peak at 1300 hr, a trough at 1900 hr, and
a second peak at 2300 hr. Theophylline at 100 mg/
kg produced an initial suppression of cell division
(0700 hr), followed by a peak at 1300 hr, a nadir
at 1900 hr, and a second peak at 2300 hr. At 50 mg/
kg, the alterations in circadian rhythm were similar
to those seen with 10 mg/kg isoproterenol. When
isoproterenol (10 mg/kg) and theophylline (50 mg/kg)
were given together, the late evening-early morning
peak of cell division was greatly magnified over the
effects produced by either drug alone. In addition,
the suppression in cell division seen 24 hr after
drug administration with all drug treatments was
greatly magnified after combined drug treatment.
Isoproterenol, but not theophylline, caused a temporary tachycardia and hypotension. It appears unlikely that the cardiovascular responses to isoproterenol play a major role in the activation of
cell division.

4153   SELECTION OF CELL LINES RESISTANT TO CYCLIC
        AMP AND THEOPHYLLINE FROM MOPC 173 MURINE
PLASMOCYTOMA: CROSS-RESISTANCE TO OUABAIN AND CON-
CANAVALIN A. *(Fre.)* Paraf, A. (Institut de Biologie
Moleculaire CNRS, Universite Paris-VII, 2, place
Jussieu, 75005 Paris, France); Faivre, E.; Zilber-
farb, V.; Lelievre, L.; Zachowski, A. *C. R. Acad.
Sci. [D] (Paris)* 281(10):683-685; 1975.

Using cyclic AMP (cAMP) and theophylline as selective
agents, variants resistant to the specific actions
of both these substances were obtained from two
strains (ME_2N sensitive to, and MF_2N insensitive
to contact inhibition) of cells from MOPC 173 murine
plasmacytoma. The strains were also resistant to
other drugs (ouabain and concanavalin A) which are
supposed to act at the level of the plasma membrane.
The cAMP selection developed resistance to cAMP
(50 times for ME_2N, 5 for MF_2N) much more than to
another drug (ouabain: two to four times for ME_2N
and MF_2N; concanavalin A: two for ME_2N). Selection
by a specific inhibitor (ouabain) seems to entail
the appearance of variants comparable to those that
can be obtained with drugs the active site of which

is not well known (cAMP or theophylline) but also
with the selective action of concanavalin A which
acts at the surface of the cell. It is thought that
certain membrane enzymes are associated into complexes with coordinated action; this would explain
the appearance of a crossed resistance. Because of
this coordinated action, there would only be a restricted number of membrane conditions, all others
being lethal. This would imply that, whichever
the effector, one or more enzymatic reactions is
often modified either directly or indirectly. In
simple cases where the specificity of the action is
known (eg, ouabain, Mg^{++}, Na^+, K^+, ATPase) three
hypotheses can explain the pleiotropic modification
observed in the resistant variants: an enzyme mutation which modifies the affinity of the drug for the
enzyme and the interactions of the enzyme with the
lipid environment, a mutation of a membrane protein
which is distinct from the enzyme which acts as a
regulator by coordinating a number of enzymatic
activities, and a modification of the regulator
gene that may interfere with the lipid constituents
of the membrane.

4154   THE FAILURE OF METHOTREXATE TO INHIBIT
        CHICKEN FIBROBLAST PROLIFERATION IN A
SERUM-CONTAINING CULTURE MEDIUM. *(Eng.)* Mitchell,
R. S. (Univ. Colorado Medical Center, Denver, Colo.
80220); Balk*, S. D.; Frank, O.; Baker, H.; Christine, M. J. *Cancer Res.* 35(9):2613-2615; 1975.

Methotrexate (1 μM) abolished the proliferation
of chicken fibroblasts in the plasma-containing
medium with supplemental folic acid but was without
effect when the medium contained serum. The serum
and defibrinogenated plasma (both from chickens)
did not differ significantly in folate content.
The serum contained more than 10 times as much
thymidine as the plasma, suggesting that the release of preformed DNA precursors from the formed
elements of blood during the clotting process is
responsible for the phenomena observed.

4155   CYCLIC AMP AND CELL MORPHOLOGY IN CULTURED
        FIBROBLASTS: EFFECTS ON CELL SHAPE, MICRO-
FILAMENT AND MICROTUBULE DISTRIBUTION, AND ORIENTA-
TION TO SUBSTRATUM. *(Eng.)* Willingham, M. C. (Natl.
Cancer Inst., Bethesda, Md. 20014); Pastan, I.
J. Cell Biol. 67(1):146-159; 1975.

The ultrastructural changes induced by N^6, $O^{2'}$-
dibutyryl cyclic adenosine 3',5' monophosphate
(Bt_2cAMP) in 3T3-4 and L929 cells, and the additional
effects that Bt_2cAMP exerted on the orientation of
3T3 cells to their substratum, were studied. The
change in shape of 3T3 and L929 cells due to 1mM
Bt_2cAMP treatment for 24 hr was accompanied by
altered intracellular distribution of microfilaments
and microtubules. Bt_2cAMP added to cells in low
density culture caused (a) microfilaments to accumulate in bundles near the plasma membrane, mainly at
the cell periphery, and (b) microtubules to accumulate beneath these microfilament bundles. In narrow
cell processes which formed characteristically in
Bt_2cAMP-treated L cells, microtubules accumulated in
parallel arrays near the center of these processes.

in normal and transformed cells was studied with the technique of freeze fracture and electron microscopy. Contact-inhibited 3T3 cells were shown to contain aggregated plasma membrane intramembranous particles, while simian virus 40- transformed 3T3 cells demonstrated a uniform particle distribution. The distribution of intramembranous particles in transformed cells was affected by colchicine or vinblastine ($10^{-7}-10^{-5}$ M) which induced a dose- and time-dependent particle aggregation. These observations suggest that microtubules and other membrane-associated colchicine-sensitive proteins probably influence the distribution of intrinsic membrane proteins and intramembranous particles in nucleated mammalian cells. An aggregated particle distribution was observed in 3T3 cells or colchicine-treated transformed cells frozen in media, phosphate-buffered saline or following brief exposure to glycerol, sucrose or dimethyl sulfoxide containing solutions, independent of whether specimens were rapidly frozen from 37 C, room temperature or 4 C incubations. Cells briefly stabilized in 1% formaldehyde yielded similar patterns of particle distribution as cells rapidly frozen in media or in cryoprotectants. Glutaraldehyde fixation of cells, however, appeared to alter the fracturing process in these cells, as visualized by an altered fracture face appearance, decreased numbers of particles, and no particle aggregates. Differences in membrane organization between normal and transformed cells were therefore demonstrated using a series of preparative methods, and colchicine and vinblastine were shown to modulate intramembranous particle distribution in transformed 3T3 cells.

4158   NONMALIGNANCY OF HYBRIDS DERIVED FROM TWO MOUSE MALIGNANT CELLS. II. ANALYSIS OF MALIGNANCY OF LM(TK$^-$) Cl 1D PARENTAL CELLS. (Eng.) Jami, J. (Institut de Recherches Scientifiques sur le Cancer, Boite Postale 8, 94800 Villejuif, France); Ritz, E. *J. Natl. Cancer Inst.* 54 (1):117-122; 1975.

The possibility that the Cl 1D cell line, an L-cell line derivative, was a mixture of malignant and nonmalignant cells was investigated because previous experiments showed that some hybrids derived from Cl 1D and tumor cell lines grew *in vivo*, whereas others apparently did not have that capacity. More highly malignant Cl 1D cells were not selected by animal passage of tumors obtained from inoculation of cultures. Only 10% of immunologically intact syngeneic C3H and semiallogeneic C3D2F$_1$ and D2C3F$_1$ mice inoculated sc with 1 million or more Cl 1D cells developed tumors. The tumor-producing capacity of this line was increased to 81% in mice X-irradiated (350 R) one day before injection, and 1×10^3 cells were sufficient to produce tumors. These observations suggested that the low capacity of Cl 1D cells to grow *in vivo* did not result from varying degrees of malignancy but mainly from interference of immunogenetic factors between cells and their inbred hosts of origin. Numerous hybrids between tumor and host cells were in all tumors examined.

4159   PROLIFERATION OF MONONUCLEAR PHAGOCYTES
        (KUPFFER CELLS) AND ENDOTHELIAL CELLS IN
REGENERATING RAT LIVER: A LIGHT AND ELECTRON MICRO-
SCOPIC CYTOCHEMICAL STUDY. *(Eng.)* Widmann, J. J.
(Anatomisches Institut der Universitat, Im Neuen-
heimer Feld 307, 69 Heidelberg, West Germany);
Fahimi*, H. D. *Am. J. Pathol.* 80(2):349-366, 1975.

The mitotic potential and possible origin of Kupffer
cells was investigated in regenerating rat liver
after partial hepatectomy using endogenous peroxidase and uptake of large (0.8 μ) latex particles as
markers of Kupffer cells. Female albino rats
(Charles River strain) were subjected to two-thirds
partial hepatectomy and sacrificed 1-11 days after
surgery, using four rats for each time interval.
Thirteen hours prior to sacrifice, each animal received an iv injection of 0.05 ml/100 g of 0.81 μ
latex particles, followed one hour later by 1 mg
vinblastine to arrest the dividing cells in mitosis.
Livers were fixed by perfusion, and pieces of tissue
were processed for cytochemical localization of endogenous peroxidase activity in Kupffer cells.
From each animal, 10-15 1-μ-thick sections stained
with 1% toluidine blue were examined by light microscopy, and dividing sinusoidal cells with and without peroxidase activity were localized. The corresponding tissue blocks were examined by electron
microscopy. The sinusoidal cells exhibited a marked
regenerative response after partial hepatectomy.
Peroxidase activity persisted in endoplasmic reticulum of Kupffer cells during mitosis. Latex particles were exclusively localized in such peroxidase-
positive cells, thus confirming their identity as
Kupffer cells. Quantitative counts revealed that
the peak mitotic activity of Kupffer cells occurred
at 48 hr, whereas that of endothelial cells was at
96 hr after partial hepatectomy. These findings indicate that Kupffer cells are capable of dividing
locally in the liver; no morphologic evidence of
transformation of endothelial cells or monocytes to
Kupffer cells was found. It is concluded that the
Kupffer cells are formed predominantly by local cell
division after partial hepatectomy in the model of
liver regeneration. These observations have established the usefulness of endogenous peroxidase and
latex-labeling as markers of Kupffer cells.

4160   CRITIQUE OF THE "CRITICAL MASS" HYPOTHESIS
        OF THE REGENERATION OF LIVER CELLS. *(Eng.)*
Tongendorff, J. (Max-Planck-Institut fur Biochemie,
8033 Martinsried bei Munchen, West Germany); Trebin,
R.; Ruhenstroth-Bauer*, G. *Am. J. Pathol.* 80(3):
519-524, 1975.

Johnson's hypothesis that the regulation of liver
cell regeneration is triggered by the critical mass
of the cells was examined mathematically and experimentally. The mathematical expression of the "critical mass" theory is intended to describe the rate
sequence of the mitotic response of a regenerating
liver. However, it does not take into account the
increase in cell number and correlative distribution
of cell volume in the regenerating liver. Due to
this defect, the mathematical treatment of Johnson
is untenable as the heuristic demonstration for the
applicability of the "critical masses" theory. To

check the applicability of the total theory to liver
regeneration *in vivo*, four volume distributions of
10,000 individual liver cells from female BDE-inbred
rats were determined at various intervals after
two-thirds hepatectomy. In histologic liver sections, the respective mitotic index was evaluated
by counting 10,000 hepatocytes of each sample. The
analysis of the volume distributions of hepatocytes
and of the rates of mitosis registered after partial
hepatectomy showed markedly variable values for the
alleged "critical volume." The fact that the "critical volume" theory is not sufficient to explain
the mitotic regulation of liver regeneration was
also shown by the experimental results, indicating
that the initial cell hypertrophy that is normally
observed is not a necessary condition for cell division. Thus, both the mathematical considerations
and the experimental results show that Johnson's
"critical mass" theory is not suitable for use in
describing the regulation of liver regeneration.

4161   THE CONTENT OF UNBOUND POLYAMINES IN
        BLOOD PLASMA AND LEUKOCYTES OF PATIENTS
WITH POLYCYTHEMIA VERA. (Eng.) Desser, H. (Ludwig
Boltzmann Institut fur Leukaemieforschung und
Hamatologie, Hanuschkrankenhaus, Heinrich Collin-
strasse 30, A-1140 Vienna, Austria); Hocker, P.;
Weiser, M.; Bohnel, J. *Clin. Chim. Acta* 63(3):
243-247; 1975.

Putrescine, spermidine, and spermine concentrations
were determined in the blood plasma and isolated
WBC of 11 patients with polycythemia vera and in
three healthy subjects. The polyamines were analyzed
on a chromatography column filled with the ion-
exchange resin PA-35. The average values in the
WBC of the healthy volunteers were found to be 1.8
± 1.4 nM putrescine/10^8 cells, 3.0 ± 0.9 nM spermi-
dine/10^8 cells and 12.9 ± 3.8 nM spermine/10^8 cells.
In the plasma of healthy persons the amounts of
the polyamines were below the sensitivity level
of the method employed. In four patients with
polycythemia vera no polyamines were detected.
However, in seven cases 0.1-3 nM polyamines/ml were
found. The level of polyamines in the WBC of six
of these patients was decreased, and in one patient
corresponded to the values found in the normal
range (17.7 ± 6.0 nM polyamines/10^8 cells). Continued blood-letting therapy on three patients led
to values approaching the concentrations found in
normal subjects in both blood plasma and WBC. A
decreased amount of these diamines in the WBC of
the patients was seen to correlate with an elevated
concentration in the plasma.

4162   ACCUMULATION OF POLYAMINES AFTER STIMU-
        LATION OF CELLULAR PROLIFERATION IN
HUMAN DIPLOID FIBROBLASTS. (Eng.) Heby, O. (Univ.
California Medical Center, San Francisco, Calif.);
Marton, L. J.; Zardi, L.; Russell, D. H.; Baserga,
R. *Cell Cycle in Malig. Immun., Proc. Annu. Han-
ford Biol. Symp., 13th.* Richland, Washington,
D.C., U.S. Energy Research and Development Adminis-
tration, 1975, pp. 50-66.

The cellular content of the polyamines putrescine,

spermidine, and spermine was studied at various intervals after confluent monolayers of WI-38 human diploid fibroblasts were stimulated to synthesize DNA by replacing exhausted medium with medium containing 10% fetal calf serum. Spermine was the quantitatively dominating polyamine. In unstimulated cells, spermine content was four times that of spermidine and 18 times that of putrescine. Upon stimulation, a 6-fold increase in putrescine content was observed. The accumulation of putrescine preceded the onset of DNA synthesis and coincided with peak of synthesis of ribosomal RNA. The magnitude of putrescine accumulation depended on the percentage of cells that were stimulated to proliferate. The cellular content of spermidine and spermine increased subsequently to that of putrescine. However, within one hour after stimulation, there was an increase in spermidine content, followed by a gradual decrease. During the remainder of the preprelicative phase, the spermidine content was within the same range as that of unstimulated cells. The activity of putrescine-activated S-adenosyl-L-methionine decarboxylase showed a similar pattern, i.e., an increase within two hours, a depression at about eight hours, and an increase during the period of DNA synthesis. The early stimulation of spermidine synthesis coincides with the increased synthesis of nonhistone chromosomal proteins, the marked rise in chromatin template activity for RNA synthesis, and the increased incorporation of ^{3}H-uridine into the RNA of whole cells, which have been previously shown to occur within the first few hours after stimulation.

4163   DECREASED ^{3}H-URIDINE INCORPORATION AND
        INCREASED ^{3}H-ADENOSINE INCORPORATION BY
HeLa CELLS EXPOSED TO AUTOLOGOUS CULTURE FLUID.
(Eng.) Tilley, R. (Medical Coll. Georgia, Augusta, Ga. 30902); Nair*, C. N. J. Cell. Physiol. 86(2/Suppl. 1/Part II):359-368; 1975.

The ^{3}H-uridine (^{3}H-UR) and ^{3}H-adenosine (^{3}H-AR) incorporation of actively growing HeLa monolayer cultures briefly exposed to the culture fluids (CF) from confluent HeLa cultures was measured. Exposure to CF inhibited the uptake as well as the incorporation of ^{3}H-UR by cultures; less ^{3}H-UR was incorporated in the CF-exposed cultures than in control cultures similarly exposed to fresh medium and labeled. The inhibition of ^{3}H-UR incorporation by CF-exposed cultures could be reduced by increasing the concentration of ^{3}H-UR in the labeling medium. ^{3}H-AR uptake was increased in the CF-exposed cultures. Both the inhibition of ^{3}H-UR incorporation and the stimulation of ^{3}H-AR incorporation were prevented by washing the CF-treated cultures with phosphate-buffered saline before labeling. Similarly, both effects could be produced in HeLa cultures exposed to fresh medium containing 1 x 10^{-5} M uridine instead of to CF. Therefore, the observed effects of CF on label incorporation were probably due to the presence of uridine or a related compound, and the inhibition of ^{3}H-UR incorporation resulted from reduced uptake of ^{3}H-UR rather than from reduced RNA synthesis by exposed cells. The active agent in the CF, formed only when cultures were incubated at physiological temperatures (34-37 C) and not at 45 C, was not a product of medium decay. It was a cellular product formed equally well by cultures incubated in medi containing dialyzed or whole serum, indicating th it was present in calf serum.

4164   AGING OF HUMAN FIBROBLASTS IN VITRO:
        SURFACE FEATURES AND BEHAVIOR OF AGING
WI 38 CELLS. (Eng.) Bowman, P. D. (Div. Natural Sciences, Thimann Lab., Univ. California, Santa Cruz, Calif. 95064); Daniel, C. W. Mech. Ageing Dev. 4(2):147-158; 1975.

WI38 cells at various in vitro ages and a simian virus 40-transformed variant were examined by ^{3}H thymidine autoradiography, time lapse cinematogra and scanning electron microscopy. There was a good correlation between cellular morphology, behavior, and the ability to incorporate ^{3}H thymidi in normal WI38 cells. Small, elongate spindle-shaped cells were usually both active incorporato of ^{3}H thymidine, as determined autoradiographical and they divided actively as determined by time lapse cinematography. Larger, more spread out cells tended to be nondividers. The proportion o large, nondividing cells increased with increasin in vitro age and was correlated with changes in cell surface and behavior. Transformed WI38 cell exhibited no such changes as a function of age. The results support the thesis that in vitro agin reflects the inability of individual cells to und take DNA synthesis and to complete division, but nondividers continue to enlarge and unusual sizes and shapes are attained.

4165   EFFECTS OF WITHDRAWAL OF A SERUM STIMU-
        LUS ON THE PROTEIN-SYNTHESIZING MACHIN
ERY OF CULTURED FIBROBLASTS. (Eng.) Mostafapour M.-K. (Dept. Biology, Massachusetts Inst. Technology, Cambridge, Mass. 02139); Green, H. J. Cell Physiol. 86(2/Suppl. 1/Part II):313-319; 1975.

3T6 cells resting in medium containing 0.5% serum were stimulated to prepare for multiplication by the addition of medium containing 10% serum. Aft a number of hours, when the rate of preribosomal RNA synthesis, total RNA content (mainly ribosoma and the cytoplasmic content of poly A (a measure c poly A(+) mRNA) were considerably elevated, the serum-rich medium was withdrawn, and the original medium replaced. The rate of preribosomal RNA synthesis began to drop within 30 min, but require a much longer time to fall to a new resting level When the serum-rich medium was withdrawn after 12 hr of stimulation, the total RNA content required 12-18 hr to fall to the resting level, whereas cytoplasmic poly A content and the rate of protei synthesis declined more rapidly, reaching a new resting level within eight hr. During the 12 hr following withdrawal of the serum-rich medium an appreciable fraction of the cells initiated DNA synthesis. Presumably, the cellular preparations for DNA synthesis cannot be immediately reversed because of the inertial factors related to the protein-synthesizing machinery.

4166   INDEPENDENCE OF SIALIC ACID LEVELS IN
        NORMAL AND MALIGNANT GROWTH. (Eng.)
Khadapkar, S. V. (Cancer Res. Inst., Tata Memorial
Centre, Parel, Bom bay 400 012, INdia); Sheth, N.
A.; Bhide, S. V. *Cancer Res.* 35(6):1520-1523; 1975.

Sialic acid content in breast or tumor tissue and
in serum of mouse strains that are either susceptible (C3H/J) or resistant (C57BL) to breast cancer was measured at various age periods. Sialic
acid content was also studied in normal lung tissue
and in lung adenoma and hepatoma. Sialic acid levels during nonmalignant growth of a tissue were
measured in breast tissue during pregnancy and
lactation, and in regenerating liver, as well as
in newborn and postnatal liver. The sialic acid
content, when expressed per milligram of protein,
increased in mammary tumor, lung adenoma, and hepatoma. It also increased in nonmalignant growth
of breast tissue during pregnancy and lactation
and of regenerating liver and postnatal liver. Increase in sialic acid per milligram DNA was observed only in lung tumors, regenerating liver,
and postnatal liver. It appears that the changes
in sialic acid level are independent of the normal
or malignant growth of a tissue and that these
changes might be the function of the parameter used
to express the sialic acid values, i.e., either the
DNA content or protein content of a given tissue.

4167   CYTOCHEMICAL AND ULTRASTRUCTURAL STUDIES
        CONCERNING THE CELL COAT GLYCOPROTEINS IN
NORMAL AND TRANSFORMED HUMAN BLOOD LYMPHOCYTES. II.
COMPARISON OF LANTHANUM-RETAINING CELL COAT COMPO-
NENTS IN T AND B LYMPHOCYTES TRANSFORMED BY VARIOUS
KINDS OF STIMULATING AGENTS. (Eng.) Anteunis, A.
(INSERM U. 104, Hopital·Saint-Antoine, Paris 12,
France); Vial, M. *Exp. Cell Res.* 90(1):47-55; 1975.

The lanthanum colloidal technique and electron
microscopy were used to study the modifications occurring during blast formation in the composition
and amounts of the lanthanum-retaining surface components of human blood T and B lymphocytes. Cells
harvested from human blood samples were incubated
with phytohemagglutin (PHA-P) (0.02 μg/ml), concanavalin A (Con A) (25 μg/ml), anti-human-lympho-
cyte horse immunoglobulin (Ig) (ALS, 0.06 mg/ml),
tuberculin (25 μg/ml), or goat antisera directed
against human F(ab')₂ and Fc fragments as well as
against human γ, α, μ, δ, and ε intact heavy chains
(1/20 dilution). In mixed lymphocyte reactions
(MLR), lymphocytes were incubated with mitomycin-
C-treated lymphocytes from a genetically unrelated
donor. Control cell suspensions consisted predominantly of small lymphocytes, most of which
showed a thick continuous electron-opaque layer
external to the membrane. This lanthanum-retaining
material (LRM) was shown to be a fuzzy zone of
varying width surrounding the whole surface of
single cells. Subsequent to mitogen or MLR stimulation, the lymphocytes consisted of two broad categories of blast cells: immunoblasts (T-derived
cells) and plasmablasts (B-derived cells). The PHA-,
Con A-, ALS-, tuberculin-, and MLR-stimulated lymphocytes usually exhibited lanthanum-positive
figures which were thinner than those observed in the

controls. In addition, the outer zone of the cell
coat showed regular cone-shaped deposits of high
electron density. The tuberculin- and MLR-stimulated cultures showed less positive patterns. The
LRM was markedly decreased in cells incubated with
anti-Ig sera directed against human F(ab')₂ and Fc
fragments. Cells incubated with anti-Ig sera directed against the δ, μ, and c heavy chains showed
a thin, more or less continuous, electron-opaque
external layer of varying density and width; treatment with anti-γ-chain and anti-α-chain sera led to
weakly positive or no lanthanum staining. The
physicological significance of the lanthanum alterations remains uncertain.

4168   GLYCOPROTEIN DIFFERENCES BETWEEN THE AS-
        CITES AND CULTURED FORMS OF THE SARCOMA
180 TUMOR. Huggins, J. W. (Dept. Biochemistry,
Oklahoma State Univ., Stillwater, Okla. 74074);
Chestnut, R. W.; Durham. N. N.; Carraway, K. L.
Biochem. Biophys. Res. Commun. 65(2):497-502; 1975.

The cell surface glycoproteins of sarcoma 180 tumor cells grown *in vitro* in cell culture or *in vivo*
as an ascites form in mice were examined by period-
ate-Schiff staining of sodium dodecyl sulfate acrylamide gels and by lactoperoxidase labeling. The
cells were cultured in McCoy's 5a modified medium,
and the ascites cells were grown in the peritoneal
cavity of mice. The cells grown in culture showed
the presence of a set of surface membrane glycoproteins which were absent or markedly decreased
in the ascites form. Conversely, the ascites form
had a glycoprotein that was not evident in the cells
in culture. The other surface membrane polypeptides
and enzyme activities were essentially the same for
the two cell types. The glycoprotein differences
may be related to the ability of the cells to interact with each other or a substratum.

4169   C-REACTIVE PROTEIN IN MALIGNANT TUMOURS.
        (Eng.) Baruah, B. D. (Assam Medical
Coll., Dibrugarh, Assam, India); Gogoi, B. C.
Indian J. Cancer 12(1):39-45; 1975.

To evaluate its potential in the diagnosis and prognosis of cancer, C-reactive protein was quantitatively analyzed in the sera of 100 patients with
different malignant tumors, 30 patients with benign
tumors, and 30 normal individuals. The C-reactive
protein was determined using the quantitative latex
fixation test with anti-C-reactive protein reagent.
The erythrocytic sedimentation rate was estimated
using Westergren's method. The C-reactive protein
test was negative in all of the normal individuals
(16 males and 14 females aged 15-50 yr); the
erythrocytic sedimentation rate in these subjects
ranged from 2-8 mm/hr in the males and from 5-15
mm/hr in the females. In the benign tumor group,
the C-reactive protein test was negative in all
patients except one with fibroleiomyoma of the
uterus, and the erythrocytic sedimentation rate
was within the normal range in all cases except two
with fibroleiomyoma of the uterus. Both the positivity and titer of C-reactive protein increased
according to the extent of the disease in the

synthesis of individual cells were investigated. In addition, a comparative study by the FLM method was presented. Using the diploid cell line B14FAF28 Chinese hamster fibroblasts, the rate of DNA synthesis was measured by means of quantitative autoradiography following a short-term incubation of the cells with 5×10^{-6} M fluorodeoxyuridine (FUdR) and 10^{-5} M [^{14}C]-thymidine (^{14}C-TdR). Grains over individual cells were counted using an incident light microphotometer. The direct measurements of DNA synthesis rate were statistically converted into "flux" parameters for comparison with the FLM method. The rate of ^{14}C-TdR incorporation revealed an inconstant rate of DNA synthesis throughout the S-phase. Seventy-three cells examined had a median ^{14}C-TdR incorporation rate of 21×10^{-18} M/cell/min, with a coefficient of variation of 13.9%. In experiments using exponentially growing cells, a median DNA synthesis time of 6.4-6.6 hr was obtained by the FLM technique, while a value of 6.0-6.1 was derived from the rate of synthesis. The determination of the mean duration of the S-phase in Chinese hamster cells by the two independent methods thus showed good agreement. The correlation of cellular DNA content with thymidine labeling produced no evidence of unlabeled cells having a DNA content characteristic of S-phase cells. While the coefficient of variation of the distribution of DNA synthesis ranged from 28 to 33%, the sources of variations were acknowledged. The major advantages of the new method are its establishment *in vitro* and the requirement of only one tissue sample.

4172   NUCLEOLAR RNA SYNTHESIS IN THE LIVER OF
         PARTIALLY HEPATECTOMIZED AND CORTISOL-
TREATED RATS. (Eng.) Schmid, W. (Institut fur Zellforschung am Deutschen Krebsforschungszentrum Heidelberg, Heidelberg, West Germany); Sekeris, C. E. *Biochim. Biophys. Acta* 402(2):244-252; 1975.

RNA synthesis by isolated nucleoli from male Wistar BR II rat liver was measured 12-14 hr after partial hepatectomy and four hours after cortisol (30 mg/kg, ip) administration. RNA synthesis was increased significantly in these animals over controls. The increased RNA synthetic capacity was also demonstrable in the respective high salt nucleolar extracts and in Biogel A-1.5 filtration fractions of the nucleolar extracts. DNA saturation experiments, using nucleoli and Biogel fractions from control and treated animals as RNA polymerase source, demonstrated that saturation of transcription was reached at the same concentration of exogenous template; this was independent of the extent of RNA synthesis. It is concluded that the activity, and not the amount of nucleolar RNA polymerase, is increased as a result of partial hepatectomy or cortisol administration. Parallel to the effects on RNA polymerase, the activity of RNA-degrading enzymes present in nucleoli was also enhanced by the same treatment.

4173   THE ISOLATION AND CHARACTERIZATION OF
         GLYCOPEPTIDES AND MUCOPOLYSACCHARIDES
FROM PLASMA MEMBRANES OF NORMAL AND REGENERATING LIVERS OF RATS. (Eng.) Akasaki, M. (Faculty Pharmaceutical Sciences, Kyoto Univ., Kyoto 606, Japan);

Kawasaki, T.; Yamashina, I. *FEBS Lett.* 59(1):100-104; 1975.

Glycopeptides and mucopolysaccharides were isolated from plasma membranes of normal and regenerating livers of male Wistar rats; the results were compared with previously published information on hepatoma membranes. Livers from normal rats were homogenized and centrifuged several times before the final pellet was suspended in a small volume of medium, and the suspension was separated through a sucrose gradient. The membranes obtained at the interphase between the cushion and the gradient were collected, washed, and lyophilized, yielding 0.50 mg protein/g liver. Plasma membranes from regenerating livers were obtained from rats that had been 70% hepatectomized. Some of the hepatectomized rats were injected ip three hours prior to death with 80 μCi[^{3}H]glucosamine (to label hexosamines and sialic acid), and with 200 μCi[^{35}S]Na$_2$SO$_4$ (to label O- and N-sulfates of mucopolysaccharides. Plasma membranes were prepared according to the procedure used for normal livers, with a yield of 0.37 mg protein/g liver. Lipids were extracted from the lyophilized plasma membranes, and the membranes were digested with pronase. Mucopolysaccharides were identified using cellulose acetate electrophoresis. Normal and regenerating livers gave similar glycopeptide patterns, all of which were apparently N-glycosidic and acidic in contrast to the production of both N- and O-glycosidic glycopeptides from the hepatoma membranes. The occurrence of heparan sulfate was common to all the plasma membranes from the livers and hepatomas. The presence of O-glycosidic glycopeptides in hepatomas having the capacity to grow in randomly bred rats, and their absence in tissue-specific hepatocytes appears to be consistent with a previously published suggestion that O-glycoproteins (the parent molecules which produce O-glycosidic glycopeptides on proteolytic digestion) may be involved in masking histocompatibility antigens at the cell surface.

4174   A HEPATOMA-ASSOCIATED ALKALINE PHOSPHATASE, THE KASAHARA ISOZYME, COMPARED WITH ONE OF THE ISOZYMES OF FL AMNION CELLS. (Eng.) Higashino, K. (Osaka Univ. Medical Sch., Osaka, Japan); Kudo, S.; Ohtani, R.; Yamamura, Y.; Honda, T.; Sakurai, J. *Ann. N.Y. Acad. Sci.* 259:337-346; 1975.

A comparative study was made among alkaline phosphatase isozymes of hepatoma, FL cells (derived from human amniotic membrane) and a 40-wk-old fresh amniotic membrane. FL cell alkaline phosphatase isozymes were separated by polyacrylamide gel electrophoresis. The dialyzed butanol extract of FL cells contained two main alkaline phosphatase isozymes; the fast moving band corresponded electrophoretically to hepatoma alkaline phosphatase. Amnion alkaline phosphatase differed from hepatoma and FL cell alkaline phosphatase isozymes in electrophoretic mobility. When the effects of L-phenylalanine, L-leucine, and L-homoarginine on the activity of alkaline phosphatase isozymes were compared, the inhibition pattern of the hepatoma alka-

line phosphatase was similar to that of FL cell alkaline phosphatase and different from that of amnion alkaline phosphatase. Hepatoma alkaline phosphatase and the fast-moving FL cell alkaline phosphatase were inhibited only slightly and to the same extent by inorganic phosphate, whereas amnion alkaline phosphatase was very sensitive. The amnion alkaline phosphatase maintained its full activity during heat treatment at 65 C. The slow-moving FL cell alkaline phosphatase was more heat-stable than were the rest of isozymes. The hepatoma alkaline phosphatase and the fast-moving FL cell alkaline phosphatase behaved quite similarly against heat treatment and were less stable. The pH optima and Michaelis constants for phenyl phosphate were not significantly different for these isozymes. Ouchterlony double diffusion tests indicated that hepatoma alkaline phosphatase and the fast-moving FL cell alkaline phosphatase are antigenically indistinguishable. It is not yet known which type of alkaline phosphatase, the hepatoma or the placental type, occurs in the membrane in the early stage of pregnancy but changes during development. It may be regulated by the mechanism that controls the gene expression of the alkaline phosphatase isozymes in the amnion cells. It is suggested that a similar regulatory mechanism may operate in human carcinoma, which decides the preferential occurrence of either the Regan isozyme or the hepatoma alkaline phosphatase. This may explain the clinical finding that the hepatoma alkaline phosphatase occurs exclusively in hepatocellular carcinoma, while the Regan isozyme occurs preferentially in cancers other than hepatoma.

4175   TUMOR-SPECIFIC ALKALINE PHOSPHATASE IN HEPATOMA. (Eng.) Suzuki, H. (First Dept. Internal Medicine, Faculty of Medicine, Univ. Tokyo, Tokyo, Japan); Iino, S.; Endo, Y.; Torii, M.; Miki, K.; Oda, T. *Ann. N.Y. Acad. Sci.* 259:307-320; 1975.

The enzymic and immunological properties, tissue of origin, and clinical significance of human hepatoma alkaline phosphatase were examined. Eleven samples of hepatoma and adjacent uninvolved liver specimens of the mucosa from the intestines of three patients were also obtained at autopsy. Ten fetal livers and intestines (8-24 wk gestation) were obtained; and three placentas, amnions, and chorions were obtained at delivery. Three amnion cell lines were also studied. All tissues were homogenized and extracted in n-butanol; the aqueous extract was precipitated, dissolved in buffer, and the enzyme was separated by disc polyacrylamide gel electrophoresis. α-Fetoprotein and hepatitis B antigen were detected by radioimmunoassay. Eight active bands of serum alkaline phosphatase were observed by polyacrylamide gel disc electrophoresis, with one of these being designated hepatoma alkaline phosphatase. Hepatoma alkaline phosphatase was observed in 16 of 51 cases with hepatocellular carcinoma (31%); it was negative in hepatoblastoma, yolk sac tumor, and congenital bile duct atresia (in which α-fetoprotein was positive), and also in liver diseases such as chronic hepatitis and liver cirrhosis. No correlation was observed between α-fetoprotein and hepatoma alkaline phosphatase. He-

cesses were short when clustered. Thus, the results
show the presence of ATPase associated with plasma
membranes and an increase in this activity when
long neuron-like processes appear.

4177   CHARACTERIZATION OF A NONHEPATIC ALCOHOL
        DEHYDROGENASE FROM RAT HEPATOCELLULAR
CARCINOMA AND STOMACH. (Eng.) Cederbaum, A. I.
(Mount Sinai Sch. Medicine, New York, N.Y. 10029);
Pietrusko, R.; Hempel, J.; Becker, F. F.; Rubin,
E. *Arch. Biochem. Biophys.* 171(1):348-360; 1975.

Ethanol oxidation by the soluble fraction of rat
hepatoma 252 was compared to that of the liver.
The tumor had been induced in a male ACI rat by
administration of *N*-2-fluorenylacetamide. The
hepatoma was transplanted into normal rats by sc
or ip injection; since the rats bearing the tumors
still had a normal liver, hepatoma and liver supernatant fractions were prepared from the same rat,
the latter serving as a control for the former.
Experiments were carried out on tumors that had
undergone 14-18 passages. Ethanol oxidation by
100,000 x g liver and hepatoma supernatants was
NAD-dependent, and the addition of pyrazole reduced
the total steroidal reductase activity by 40-50%,
suggesting the presence of alcohol dehydrogenase.
At low concentrations of ethanol (10.8 mM) the
alcohol dehydrogenase activities of hepatoma and
liver supernatant fractions were comparable. When
the concentration of ethanol was raised to 108 mM,
the activity of the liver enzyme decreased by about
50%, whereas the activity in hepatoma supernatant
fractions was elevated about 4-fold. With 1.3 mM
m-nitrobenzaldehyde, the reducing activity was
32 times higher in hepatoma fractions than that
in normal liver. By contrast the ability to metabolize 5-β-dihydrotestosterone (0.115 mM) and cyclo-
hexanone (12.7 mM) was less than that in supernatant
fractions of the liver. Electrophoresis of the
liver supernatant fractions on ionagar at pH 7.0
revealed only one component that oxidized ethanol.
Hepatoma supernatant fractions contained two components with alcohol dehydrogenase activity; one
with the same electrophoretic mobility as the liver
enzyme, the other showing a slower rate of migration. The latter component, which was absent in
the liver, is designated hepatoma alcohol dehydrogenase. By electrophoresis on starch gels at pH
8.5, it could be demonstrated that the liver and
hepatoma enzymes moved in opposite directions. The
overall substrate specificity characteristics were
similar to those of the liver enzyme in that the
effectiveness of substrates increased with an increase
in chain length and introduction of unsaturation
or an aromatic group. Both liver and hepatoma
alcohol dehydrogenase cross-reacted with antibody
to horse liver alcohol dehydrogenase EE. The
Michaelis constant for ethanol with the hepatoma
enzyme was found to be 223 mM, compared to 0.3 mM
for liver alcohol dehydrogenase; at 1.0 M ethanol
the hepatoma enzyme was not fully saturated with
substrate. The Michaelis constant for 2-hexene-1-
ol was 0.3 mM, indicating that the hepatoma enzyme
is better suited for dehydrogenation of longer
chain alcohols. Stomach alcohol dehydrogenase was
also found to have kinetic properties comparable

to those of the hepatoma enzyme, as well as similar electrophoretic mobility. In addition, the hepatoma enzyme was detected in the serum of rats bearing hepatomas.

4178   STUDIES ON TYROSINE HYDROXYLASE IN NUEROBLASTOMA, IN RELATION TO URINARY LEVELS OF CATECHOLAMINE METABOLITES. (Eng.) Ima-shuku, S. (Dept. Pediatrics, Kyoto Prefectural Univ. Medicine, Kawaramachi, Kamikyoku, Kyoto, Japan); Takada, H.; Sawada, T.; Nakamura, T.; LaBrosse, E. H. *Cancer* 36(2):450-457; 1975.

Tyrosine hydroxylase (TH) activity was determined in 22 neuroblastoma tumors from 15 patients, in one pheochromocytoma, 20 adrenal glands, ten other tumors and organs, and four specimens of sera. The enzyme activity was found only in the neural crest tumors and adrenal glands, but the levels were too low to be detected in the other tumors and in normal liver and kidney tissues. The average specific activity of TH in 23 neural crest tumors was 0.559 ± 0.101; in 20 adrenal glands was 0.418 ± 0.124 nM/mg protein/min. In 13 patients with neuroblastoma, and in the patient with pheochromocytoma, both TH levels in the primary tumors and urinary excretion of vanillyl-mandelic acid (VMA) and homovanillic acid (HVA) were studied. The urinary excretion of VMA by 10 of 13 neuroblastoma patients, and by the patient with pheo-chromocytoma, was significantly to markedly elevated above normal levels; excretion of HVA by 12 of 13 neuroblastoma patients was similarly elevated. These results indicate that tyrosine hydroxylase, an enzyme specifically located in the adrenal medulla and mono-aminergic neurons, is also present in neuroblastoma, a malignant tumor of similar embryologic origin, and in pheochromocytoma. Not only can TH activity in these tumors be demonstrated *in vitro*, but the elevated urinary excretion of VMA and/or HVA by the majority of patients with these tumors also indicates TH activity of the tumors *in vivo*.

4179   THE HISTOCHEMICAL DEMONSTRATION OF ARYL-SULPHATASE IN HUMAN TUMOURS. (Eng.) Koudstaal, J. (Dept. Pathology, "S.S.D.Z.", Sticht-ing Samenwerking Delftse Ziekenhuizen, Reynier de Graefweg 7, Delft, The Netherlands). *Eur. J. Can-cer* 11(11):809-813; 1975.

The activity of arylsulfatase in human tissues and malignant tumors was studied. The enzyme histo-chemical method with 6-bromo-2-naphthyl sulphate gave the best results of the substrates tested. Cells of the stratum granulosum and of the sheath of the hairfollicles of the skin showed a moderate enzyme activity. The strongest activity was seen in the proximal tubules of the kidney. Cells of the ductus epididymis possessed clear activity. Except for a trace in the corpus luteum, the human ovaries showed no enzyme activity. The placenta had activity in the syncytic- and cytotrophoblast cells. The adrenals showed some activity in the medulla. Generally, the enzyme activity of malig-nant tumors was very low, and decreased compared with the activity of normal corresponding cells. A clear enzyme activity was noted in 38 of the 62

after it completes a certain functional life period. Assuming that some messenger RNA molecules in mammalian systems have a high degree of stability, one must conjecture that the regulation of genetic expression by hormonal systems may occur at the level of genetic translation. Various experiments point to the quite probable possibility that hormones exert their effects at two different levels of genetic expression: at the level of transcription; and at the level of the translating unit, the polysome. In neoplastic tissues, especially the liver, it is clear that many if not the majority of hormonal mechanisms regulating genetic expression are abnormal when compared with their normal cell of origin *in vivo*. It remains to be determined whether these mechanisms are predominantly defective at the level of genetic translation or at the level of genetic transcription.

4184   CULTURE OF ZOLLINGER-ELLISON TUMOR CELLS.
(Eng.) Lichtenberger, L. M. (Univ. California Los Angeles Sch. Med.); Lechago, J.; Dockray, G. J.; Passaro, E., Jr. *Gastroenterology* 68 (5/Part 1):1119-1126; 1975.

The molecular form of gastrin synthesized and secreted by six Zollinger-Ellison tumors in culture was studied; the tumors were successfully grown *in vitro* by a monolayer tissue culture technique. The initial gastrin concentration in the medium varied between 0 and 100 ng/ml. Many cytoplasmic secretroy granules were seen in the cells of one culture population. Gastrin secretion was stimulated by the addition of fresh medium to the cultures. Both the culture cells and the medium contained primarily big gastrin (G-34), but smaller amounts of little gastrin (G-17) were also present. Since G-17 was the major molecular form in the original tumor tissue, it is postulated that an enzyme that breaks G-34 down to G-17 is normally present in tumor tissue but has a diminished activity in an *in vitro* environment. Gastrin concentration in the medium decreased with time in culture until no hormone was detected between 2 and 6 wk, possibly because of endocrine cell dedifferentiation and an increased proportion of fibroblasts in the population.

4185   PITUITARY AND SERUM CONCENTRATIONS OF
PROLACTIN AND GH IN SNELL DWARF MICE.
Sinha, Y. N. (Scripps Clinic and Res. Foundation, La Jolla, Calif. 92037); Salocks, C. B.; Vanderlaan, W. P. *Proc. Soc. Exp. Biol. Med.* 150(1): 207-210; 1975.

Secretions of prolactin and growth hormone (GH) in mutant dwarf mice (dw/dw) were studied using homologous radioimmunoassays. Blood samples from 12-13 adult male and female dw/dw mice were collected by orbital puncture and by decapitation. Compared to related normals (+/?), pituitary concentrations of prolactin were very low in dw/dw mice (P<0.01); males had only 35 ng prolactin/mg pituitary tissue and females had only 5 ng/mg in comparison with 1.3 and 4.4 µg/mg in respective controls. GH was 9.6 and 1.8 µg/mg for male and female dw/dw mice, respectively, compared to 67.8 and 37.5 µg/mg in respective controls. Prolactin and GH concentrations were also lower in sera of dw/dw mice, but the relative differences appeared sex-dependent; serum GH was more reduced in males than in females, while serum prolactin was more depressed in females. The data confirm earlier indications of deficiencies in the circulating levels of prolactin and GH in dwarf mice and suggest that the hypoactivity of these hormones may be crucial to some of the ano-.malies found in this mutant.

4186   FIBROBLAST GROWTH FACTOR: ITS LOCALIZA-
        TION, PURIFICATION, MODE OF ACTION, AND
PHYSIOLOGICAL SIGNIFICANCE. (Eng.) Gospodarowicz, D. (The Salk Inst. Biological Studies, San Diego, Calif.); Rudland, P.; Lindstrom, J.; Benirschke, K. *Adv. Metab. Disord.* 8:301-335; 1975.

Cell fibroblast growth factor (FGF), a homogeneous polypeptide hormone, has recently been isolated from bovine pituitary glands and from the brain of vertebrates. The purification, physical properties, and biological activities of the two peptides are reported with emphasis on pituitary FGF. The two peptides have the same apparent molecular wt (13,000) but differ in their N-terminal, which is lysine for pituitary FGF and is blocked for brain FGF. Whether obtained from the brain or pituitary glands, FGF was able to provoke division of cells as diverse as fibroblasts, chondrocytes, adrenal cells, endometrial cells, and glial cells obtained from bovine, murine, and human sources. Studies with 3T3 cells showed that FGF plus hydrocortisone could replace serum for provoking the initial events of cell division and that cells grown to confluency in the presence of 10% serum could be further stimulated to divide by the addition of FGF alone or FGF plus dexamethasone. In polyoma virus-transformed 3T3 cells, the growth-initiating ability of FGF was lost at permissive temperatures. The polypeptide was also shown to increase intracellular cyclic guanosine monophosphate levels in quiescent cultures of BALBc 3T3 cells and to control the pleiotypic responses of the cultures. *In vivo* experiments indicated that FGF could be one of the neurotrophic agents responsible for amphibian limb regeneration. FGF treatment of amputated limb stumps of adult frogs resulted in large deposits of undifferentiated cells resembling a regenerating blastoma. In some cases, differentiation of large masses of chondrocytes and muscle fibers was observed.

4187   PURIFICATION OF A FIBROBLAST GROWTH FAC-
        TOR FROM BOVINE PITUITARY. (Eng.) Gospodarowicz, D. (Salk Inst. Biological Studies, San Diego, Calif. 92112). *J. Biol. Chem.* 260(7):2515-2520; 1975.

The isolation of a fibroblast growth factor (FGF) from bovine pituitary extracts is described. Frozen bovine pituitaries were homogenized (2 kg) in 4l of 0.15 M $(NH_4)_2SO_4$ and centrifuged at 13,000 x g for 20 min. The supernatant was adjusted to pH 3.5 by the addition of 0.5 M HPO_3; the resulting precipitate was then removed by centrifugation. The supernatant was adjusted to pH 6.5 - 7.0 by the addition

of 1 N NaOH; and 290 g of $(NH_4)_2SO_4/l$ was added. The resulting precipitate was discarded and additional $(NH_4)_2SO_4$ (250 g/l) was added to the supernatant. A precipitate with FGF activity was formed and was then collected by centrifugation, distilled with water, dialyzed against water for 24 hr, and lyophilized. Further purification of FGF was achieved by dissolving the FGF fraction to a final concentration of 20 mg/ml in 0.1 M NaPO₄ (pH 6.0) and by chromatography on carboxylmethyl-Sephadex C-50 (CM-Sephadex C-50). The biologically active peak was further purified by chromatography on a series of columns: Sephadex G-75, Sephadex G-50, and CM-Sephadex C-50 (eluted with a linear NaCl gradient). The final product was considered homogeneous because it gave a single peak on Sephadex-75 and CM-Sephadex NaCl gradient elution, and it gave a single band on disc gel electrophoresis. Five milligrams of FGF were obtained per kilogram of pituitary gland. Further analyses revealed that FGF, a polypeptide rich in lysine and arginine, has a molecular weight of 13,300. FGF is a potent mitogenic factor in 3T3 fibroblasts; it was active at concentrations as low as 0.01 ng/ml, and a maximal effect was observed at 1 ng/ml.

4188   PHAGOCYTIC ACTIVITY OF LEUKAEMIC BLASTS.
        (Eng.) Neuwirtova, R. (Medical Faculty, Charles Univ., Prague, Czechoslovakia); Setkova, O.; Houskova, J.; Poch, T.; Dorazilova,V.; Donner, L. *Acta Haematol. (Basel)* 53(2):17-24; 1975.

4189   ISOLATION AND CHARACTERIZATION OF THE
        PLASMA MEMBRANE FROM YOSHIDA HEPATOMA CELLS
(Eng.) Rethy, A. (Inst. General Physiology, Univ. Ferrara, 44100 Ferrara, Italy); Trevisani, A.; Manservigi, R.; Tomasi, V. *J. Membr. Biol.* 20(1/2): 99-110; 1975.

4190   CATION PERMEABILITY AND OUABAIN-INSENSITIVE
        CATION FLUX IN THE EHRLICH ASCITES TUMOR
CELL. (Eng.) Mills, B. (Dept. Biol., Syracuse Univ. N.Y.); Tupper, J. T. *J. Membr. Biol.* 20(1/2):75-97; 1975.

4191   STUDY OF THE MECHANISM OF K+-DEPENDENT
        SWELLING OF EHRLICH ASCITE CARCINOMA (EAC)
CELLS. (Rus.) Polivoda, B. I. (Res. Inst. Medical Radiology, Acad. Med. Sci. U.S.S.R., Obninsk, U.S.S.R *Biofizika* 20(1):160-161; 1975.

4192   THE EFFECT OF ADENOSINE 3',5'-MONOPHOSPHATE
        ON THE CELL CYCLE OF KB CELLS [abstract].
(Eng.) Asnes, C. F. (Boston Univ. Grad. Sch., Mass.) *Diss. Abstr. Int.* B 35(9):4356; 1975.

4193   CYCLIC AMP IN HeLa CELLS STIMULATED WITH
        CHOLERA ENTEROTOXIN AND METHYLXANTHINES
[abstract]. (Eng.) King, J. J. (Worcester Foundation for Experimental Biology, Shrewsbury, Mass.); Maudsley*, D. V. *Br. J. Pharmacol.* 55(2):287P; 1975.

4198   THE GLYCOLYTIC ENZYME ACTIVITY OF THE
        HUMAN CERVIX UTERI. (Eng.) Pedersen,
S. N. (Fibiger Lab., Copenhagen Ø, Dnemark). *Cancer*
35(2):469-474; 1975.

4199   ROLE OF ENZYME TURNOVER AND DNA REPLICATION
        IN REGULATING THE LEVEL OF ASPARTATE
TRANSCARBAMYLASE DURING THE EUCARYOTIC CELL CYCLE.
(Eng.) Wilkins, J. H. (Virginia Polytech. Inst.
State Univ., Blacksburg); Schmidt, R. R. *Fed. Proc.*
34(3):651; 1975.

4200   FORMS OF CYTOSOLIC GLYCEROL-3-PHOSPHATE
        DEHYDROGENASE IN NORMAL AND NEOPLASTIC
TISSUES. (Eng.) Rittmann, L. (Syracuse Univ., N.Y.);
Ostro, M.; Kornbluth, R.; Fondy, T. *Fed. Proc.* 34(3):
676; 1975.

* INDICATES A PLAIN CITATION WITHOUT ACCOMPANYING ABSTRACT

ABE, M.
 -3751*
ABRAMSON, M.
 3998
ACS, G.
 3881*
ADAMS, A.W.
 3709
AGACHE, P.
 4111=
AHEARN, M.J.
 4081*
AINSWORTH, E.J.
 378C*
AKASAKI, M.
 4173
AL-ADNANI, M.S.
 3993
ALBERT, R.E.
 3749*
ALEXANDER, P.
 3900
ALEXANDER-WILLIAMS, J.
 4041
ALLEGRETTI, M.
 3948*
ALLEN, K.H.
 3780*
ALONI, Y.
 3820
ALOY, M.
 4064*
ALPERS, M.P.
 4144*
ALPERT, E.
 3930, 3969*
ALTENAHR, E.
 3622
ALTHOFF, J.
 3701
AL'TSHTEIN, A.D.
 3856*
AMETANI, T.
 4005, 4017
AMLACHER, E.
 3694
AMTRUP, F.
 3935*
ANDERSON, C.T., JR.
 4031
ANTEUNIS, A.
 4167
ANTHONY, H.M.
 3926
ANTHONY, P.P.
 3636*
AOKI, T.
 3615, 3912
ARAI, M.
 3736*
ARAKI, S.
 3751*
ARGYRAKIS, M.
 3729*
ARMSTRONG, B.
 4123
ARMSTRONG, M.Y.K.
 3888

ARNDT, K.A.
 3803
ARNSTEIN, P.
 3705
ARSENEAU, J.C.
 4013
ARVILOMMI, H.
 3985*
ASHLEY, R.L.
 3869*
ASKANAS, V.
 3812
ASNES, C.F.
 4192*
ASOFSKY, R.
 4150
ASPERGREN, K.
 3680
AUFSES, A.H., JR.
 3908
AUGENER, W.
 3984*
AUSSEPE, L.
 3713
AUSTWICK, P.K.C.
 3604
AYOUB, T.
 3785*
AYVAZIAN, J.H.
 3959*
BAARDSEN, A.
 3976*
BABA, T.
 3728*
BABCOCK, V.
 3933*
BACA, J.M.
 4138*
BACHVAROFF, R.
 3959*
BADGER, A.M.
 3950*, 3951*
BAENZIGER, N.
 3863*
BAJAJ, M.M.
 4107*
BAKER, H.
 4154
BAKER, M.S.
 3720*
BALDWIN, J.
 3848*
BALK, S.D.
 4154
BALL, W.C., JR.
 4121
BALLERIO, G.
 4073*
BANERJEE, S.
 3726*
BANKS, P.M.
 4013
BANKS, W.C.
 4048*
BARCLAY, M.
 3632
BARNETT, E.
 3773

BARR, C.C.
 4018
BARTSCH, H.
 3678, 3697
BARUAH, B.C.
 4169
BASERGA, R.
 4162
BATES, R.R.
 3639*
BATTULA, N.
 3839*
BAUM, S.G.
 3824
BEAL, D.D.
 3691
BEAUMARIAGE, M.L.
 3947*
BECKER, F.F.
 4150, 4177
BECKER, F.O.
 3772
BECKER, H.
 4070*
BECKMANN, H.
 3920
BEER, J.Z.
 4170
BENDER, E.
 3811, 4079*
BENEDICT, W.F.
 3720*
BENIRSCHKE, K.
 4186
BENNETT, J.C.
 3962*
BENTLEY, H.P.
 3981*
BENTVELZEN, P.
 3830
BERARD, C.W.
 4013
BERG, J.W.
 4134*
BERKMAN, J.I.
 3974*
BERRY, D.E.
 3734*
BETZ, E.H.
 3947*
BHIDE, S.V.
 4166
BHOOPALAM, N.
 3923
BIAS, W.B.
 4121
BIGI, G.
 3960*
BINZ, H.
 3617
BIRCH, N.
 3674
BIRKELAND, S.A.
 3935*
BIRNIE, G.D.
 3634
BISHOP, D.
 4035

AUTHOR

BITTLINGMAIER, K.
3806
BLACK, M.M.
4141*
BLAMEY, R.W.
3740*
BLOBSTEIN, S.H.
3750*
BLOM, D.H.
3983*
BLOOM, M.L.
3914
BLOOM, S.R.
4069*
BODEY, G.P.
4081*
BOELLAARD, J.W.
4051*
BOHNEL, J.
4161
BOLGAN, A.
4093*
BOLOGNESI, D.P.
3818
BOMHARD, D.
4058*
BONSE, G.
3686
BORDES, M.
3909
BORDI, C.
4073*
BOREK, C.
3783*
BORN, R.
4171
BORST, J.R.
4143*
BOTELLA, R.
4064*
BOTHA, J.B.C.
4032
BOTTOMLEY, R.
3759*
BOUCK, N.P.
3875*
BOULGER, L.R.
4035
BOVILL, E.G., JR.
4142*
BOWDON, B.J.
3765*
BOWMAN, P.D.
4164
BRADA, Z.
3685
BRAGINA, E.E.
4108*
BRASHEARS, S.
4121
BRATTAIN, M.G.
3964*, 3988*
BRAUN, W.H.
3716
BRESNICK, E.
3756*
BRETSCHER, M.S.
3667*

BRIGGS, D.R.
4089*
BRINKHOF, J.
3830
BRINKLEY, B.R.
3620
BRINKMANN, W.
4171
BRION, S.
4053*
BRITTINGER, G.
3984*
BROCKMAN, W.W.
3825, 3826
BRODY, J.I.
3983*
BROITMAN, S.A.
3966*
BROSS, I.D.J.
4045, 4046
BROUET, J.-C.
3918
BROWN, D.T.
3790
BROZNA, J.P.
3886
BRUNETTI, P.
3958*
BRYAN, G.T.
3691, 3737*, 3758*, 3770*
BUBENIK, J.
3902
BUDZICKA, E.
4170
BUFFE, D.
3951*
BULBA, S.
3685
BULLERMAN, L.B.
4138*
BURKLE, G.
3641*
BURLINGHAM, B.T.
3790
BURMICZ, J.S.
3762*
BURNS, F.
3749*
BURNS, F.J.
4131
BURNY, A.
3857*
BURTON, R.M.
3972*
BUSELMAIER, W.
3760*
BUSSOLATI, G.
4073*
BUTLER, J.D.
3610
BYERS, M.J.
3861*
BYRON, P.R.
4043
CAHILL, D.F.
3714
CALLENDER, S.T.
4020

CAM, Y.
4176
CAMPANILE, F.
3975*
CAMPBELL, F.
4009
CANELLOS, G.P.
4013
CANNON, G.
3912
CAP, J.
3953*
CAPPELAERE, P.
3638*
CARDIFF, R.D.
3867*
CARDONA, R.A.
3675
CARDOSO, S.S.
4152
CARLSON, I.H.
4031
CARNAUD, C.
3943*
CARRAWAY, K.L.
4168
CARREL, S.
3910
CARROLL, K.K.
3606
CASE, T.C.
4054*
CASPARY, W.J.
3677
CASSADY, J.R.
3773
CASTRUP, H.J.
4151
CAVALIERI, E.
3742*
CEDERBAUM, A.I.
4177
CHABALKO, J.J.
4140*
CHAN, P.-C.
3741*
CHANGUS, G.W.
4089*
CHASE, G.A.
4121
CHAUVERGNE, J.
3650*, 3932*
CHEN, Y.
3923
CHENG, H.F.
3784*, 3911
CHERNINA, L.A.
4086*
CHESTNUT, R.W.
4168
CHIBA, H.
3938*, 3939*
CHISAKA, N.
3730*
CHIU, C.W.
3770*
CHRISTINE, B.W.
3995, 4117

ABDOMINAL NEOPLASMS
    BURKITT'S LYMPHOMA
        CLINICOPATHOLOGIC STUDY, 4013
    LYMPHOMA
        CASE REPORT, 4015
    SARCOMA, RETICULUM CELL
        CASE REPORT, 4015
        HEAVY CHAIN DISEASE, 4015

ACETAMIDE, N-(ACETYLOXY)-N-9H-FLUOREN-2-YL-
    DNA REPAIR
        FIBROBLASTS, 3721*

ACETAMIDE, N-FLUOREN-2-YL-
    DIETARY FATS
        COCARCINOGENIC EFFECT, 3606
    FIBROBLASTS
        DNA REPAIR, 3674
    HEPATOMA
        ALCOHOL OXIDOREDUCTASES, 4177
        CHROMOSOME ABNORMALITIES, 4150
        FETAL GLOBULINS, 4150
    LIVER NEOPLASMS
        TRANSPLACENTAL CARCINOGENESIS,
          REVIEW, 3601
    LUNG NEOPLASMS
        TRANSPLACENTAL CARCINOGENESIS,
          REVIEW, 3601
    MAMMARY NEOPLASMS, EXPERIMENTAL
        TRANSPLACENTAL CARCINOGENESIS,
          REVIEW, 3601
    STOMACH NEOPLASMS
        ADENOCARCINOMA, 3635*
    XERODERMA PIGMENTOSUM
        DNA REPAIR, 3674

ACETAMIDE, N,N'-9H-FLUOREN-2,7-DIYLBIS-
    HEPATOMA
        LACTATION, 3722*
    LACTATION
        COCARCINOGENIC EFFECT, HEPATOMA,
        MOUSE, 3722*

ACETAMIDE, N-2-PHENANTHRYL-
    FIBROBLASTS
        DNA REPAIR, 3674
    XERODERMA PIGMENTOSUM
        DNA REPAIR, 3674

2-ACETAMIDOFLUORENE
    SEE ACETAMIDE, N-FLUOREN-2-YL-

2-ACETAMINOFLUORENE
    SEE ACETAMIDE, N-FLUOREN-2-YL-

ACETANILIDE, 4'-PHENYL-
    FIBROBLASTS
        DNA REPAIR, 3674
    XERODERMA PIGMENTOSUM
        DNA REPAIR, 3674

ACETANILIDE, 4'-STYRYL-
    FIBROBLASTS
        DNA REPAIR, 3674
    XERODERMA PIGMENTOSUM
        DNA REPAIR, 3674

ACETIC ACID
    SKIN NEOPLASMS
        ENZYMATIC ACTIVITY, 3723*

ACETIC ACID, IMINODI-
    IRON (III) NITRILOTRIACETATO-
        PHOTODECOMPOSITION, 3724*

ACETIC ACID, LEAD (+2) SALT
    ASCORBIC ACID
        TERATOGENIC ACTIVITY, 3727*
    TERATOGENS
        TERATOGENIC ACTIVITY, 3727*

ACETOHYDROXAMIC ACID, N-4-BIPHENYLYL-
    RNA, TRANSFER
        CYTOSOL, 3675

ACETOHYDROXAMIC ACID, N-FLUOREN-2-YL-
    RNA, TRANSFER
        CYTOSOL, 3675

ACETONE
    AFLATOXIN B1
        PEANUTS, 3670

N-ACETOXY-N-2-ACETYLAMINOFLUORENE
    SEE ACETAMIDE, N-(ACETYLOXY)-N-9H-
    FLUOREN-2-YL-

2-ACETYLAMINOFLUORENE
    SEE ACETAMIDE, N-FLUOREN-2-YL

ACID PHOSPHATASE
    LEUKEMIA, LYMPHOBLASTIC
        T-LYMPHOCYTES, 3920
        MORPHOLOGICAL AND CYTOCHEMICAL
          STUDY, 4021
    LEUKEMIA, MYELOBLASTIC
        MORPHOLOGICAL AND CYTOCHEMICAL
          STUDY, 4021

ACRYLIC ACID, 3-CYANO-, ISOBUTYL ESTER
    FIBROBLASTS
        PRECANCEROUS CONDITIONS, 3673

ACTINOMYCIN D
    PLASMACYTOMA
        CELL FUSION, 4004
        ULTRASTRUCTURAL STUDY, 4004

ACUTE DISEASE
    LEUKEMIA
        CELL CYCLE KINETICS, 4131

ADENOACANTHOMA
    SEE ADENOCARCINOMA

ADENOCARCINOMA
    BLADDER NEOPLASMS
        CASE REPORT, 4091*
        NEOPLASM METASTASIS, 4046
        ULTRASTRUCTURAL STUDY, 4091*
    BREAST NEOPLASMS
        CASE REPORT, 3991
        GENETICS, 4023
        NEOPLASMS, MULTIPLE PRIMARY, 3995
        RADIATION, IONIZING, 3991
    BRONCHIAL NEOPLASMS

AUTOANTIBODIES, 3903
CARCINOEMBRYONIC ANTIGEN
   ULTRASTRUCTURAL STUDY, 3912
CERVIX NEOPLASMS
   CASE REPORT, 3690
   ENDOMETRIOSIS, 3690
   ESTROGENS, 3690
   HISTOLOGICAL STUDY, 3997
COLONIC NEOPLASMS
   CARCINOEMBRYONIC ANTIGEN, 3912
ESOPHAGEAL NEOPLASMS
   CASE REPORT, 4065*
GASTROINTESTINAL NEOPLASMS
   CARCINOEMBRYONIC ANTIGEN, 3968*
GYNECOLOGICAL NEOPLASMS
   NEOPLASM METASTASIS, 4046
IMMUNOGLOBULINS
   IMMUNE SERUMS, 3925
INTESTINAL NEOPLASMS
   CARCINOEMBRYONIC ANTIGEN, 3911
   INTESTINAL POLYPS, 4071*
   RADIATION, 3784*
   RADIATION, IONIZING, 3911
INTESTINAL POLYPS
   CASE REPORT, 4071*
   GENETICS, 4071*
KIDNEY NEOPLASMS
   IMIDAZOLE-4-CARBOXAMIDE, 5-AMINO-,
    3691
   NEOPLASM METASTASIS, 4046
LARYNGEAL NEOPLASMS
   EPIDEMIOLOGY, 4075*
LUNG NEOPLASMS
   DIPROPYLAMINE, N-NITROSO-, 3702
   NEOPLASM METASTASIS, 4046
LYMPHOCYTES
   TRANSPLANTATION IMMUNOLOGY, 3890
MACROPHAGES
   TRANSPLANTATION IMMUNOLOGY, 3890
MAMMARY NEOPLASMS, EXPERIMENTAL
   BENZ(A)ANTHRACENE, 7,12-DIMETHYL-,
    3741*
   CYCLOOCTANECARBAMIC ACID, 1,1-BIS
    (P-FLUOROPHENYL)-2-PROPYNYL,
    3684
   DIETARY FATS, 3606
   ESTRADIOL, 3741*, 4180
   IMIDAZOLE-4-CARBOXAMIDE, 5-AMINO-,
    3691
   IMIDAZOLE-4-CARBOXAMIDE, 5-(3,3-
    DIMETHYL-1-TRIAZENO)-, 3691
   IMIDAZOLE-4-CARBOXIMIDE, 5-DIAZO-,
    3691
   4H-IMIDAZOLE-(4,5-D)-V-TRIAZIN-4-
    ONE, 3,7-DIHYDRO-, 3691
   NEOPLASM TRANSPLANTATION, 4149
   PROLACTIN, 3741*
   SUBCELLULAR TUMOR FRACTIONS, 3866*
   VIRUS, MURINE MAMMARY TUMOR, 3829,
    3866*
MYCOBACTERIUM BOVIS
   TRANSPLANTATION IMMUNOLOGY, 3890
MYCOPLASMA
   CARCINOGENIC POTENTIAL, MOUSE,
    4039
NOSE NEOPLASMS
   BUTANOL, 4-(BUTYLNITROSOAMINO)-,
    3701
PANCREATIC NEOPLASMS

REVIEW, 3639*
RADIATION, IONIZING
   TRANSPLANTATION IMMUNOLOGY, 3890
4,4'-STILBENEDIOL, ALPHA,ALPHA'-
   DIETHYL-
   CASE REPORT, 4096*
STOMACH NEOPLASMS
   ACETAMIDE, N-FLOUREN-2-YL-, 3635*
   EPIDEMIOLOGY, DOG, 3635*
   FLUORESCENT ANTIBODY TECHNIQUE,
    3706
   GUANIDINE, 1-METHYL-3-NITRO-1-
    NITROSO-, 3635*, 3706
   IMMUNITY, CELLULAR, 3934*
   MIGRATION INHIBITORY FACTOR, 3934*
   QUINOLINE, 4-NITRO-, 1-OXIDE,
    3635*
THYROID NEOPLASMS
   CASE REPORT, 4002
   EPIDEMIOLOGY, 4106*, 4129
   RADIATION, IONIZING, 3772
   RADIOTHERAPY, 3786*
   ULTRASTRUCTURAL STUDY, 4002
UTERINE NEOPLASMS
   CASE REPORT, 4096*, 4097*
   ULTRASTRUCTURAL STUDY, 4101*
VAGINAL NEOPLASMS
   4,4'-STILBENEDIOL, ALPHA,ALPHA'-
    DIETHYL-, 4124
ZYMOSAN
   TRANSPLANTATION IMMUNOLOGY, 3890

ADENOCARCINOMA, PAPILLARY
   COLONIC NEOPLASMS
    CASE REPORT, 4041

ADENOFIBROMA
   ANTINEOPLASTIC AGENTS
    IMIDAZOLE-4-CARBOXAMIDE, 5-(3-
     METHYL-1-TRIAZENO)-, 3691
   BREAST NEOPLASMS
    CELL NUCLEUS, 4055*
    HISTOLOGICAL STUDY, 4055*
   MAMMARY NEOPLASMS, EXPERIMENTAL
    IMIDAZOLE-4-CARBOXAMIDE, 5-DIAZO-,
     3691
    TRITIUM, 3714

ADENOLYMPHOMA
   CARCINOMA
    CASE REPORT, 4042
   PAROTID NEOPLASMS
    CARCINOMA, 4042

ADENOMA
   ADRENAL GLAND NEOPLASMS
    DIETHYLAMINE, N-NITROSO-, 3696
    4H-IMIDAZOLE-(4,5-D)-V-TRIAZIN-4-
     ONE, 3,7-DIHYDRO-, 3691
    TRANSPLACENTAL CARCINOGENESIS,
     HAMSTER, 3696
   COLONIC NEOPLASMS
    ULTRASTRUCTURAL STUDY, 4010
   KIDNEY NEOPLASMS
    DIPROPYLAMINE, N-NITROSO-, 3702
   LIVER NEOPLASMS
    S-TRIAZINE, HEXAHYDRO-1,3,5-
     TRINITRO-, 3694
   LUNG NEOPLASMS

CARBAMIC ACID, ETHYL ESTER, 3893
CHOLANTHRENE, 3-METHYL-, 3893
DIPROPYLAMINE, N-NITROSO-, 3702
PANCREATIC NEOPLASMS
MONOCROTALINE, 3761*
STOMACH NEOPLASMS
EPIDEMIOLOGY, RAT, 3635*
THYROID NEOPLASMS
EPIDEMIOLOGY, 4106*
RADIATION, IONIZING, 3772
UTERINE NEOPLASMS
ULTRASTRUCTURAL STUDY, 4102*

ADENOMA, CHROMOPHOBE
BRAIN NEOPLASMS
ENDOPLASMIC RETICULUM, 4005
GOLGI APPARATUS, 4005

ADENOSINE CYCLIC 3',5' MONOPHOSPHATE
CELL TRANSFORMATION, NEOPLASTIC
REVIEW, 4156
HELA CELLS
CAFFEINE, 4193*
CHOLERA, 4193*
ENTEROTOXINS, 4193*
THEOPHYLLINE, 4193*
XANTHINE, 3-ISOBUTYL-1-METHYL-,
4193*
NOSE NEOPLASMS
CELL CYCLE KINETICS, 4192*
PLASMACYTOMA
CELL MEMBRANE, 4153
CONCANAVALIN A, 4153
OUABAIN, 4153
VIRUS, KIRSTEN MURINE SARCOMA
CELL MEMBRANE, 4156
VIRUS, MOLONEY MURINE SARCOMA
CELL MEMBRANE, 4156
VIRUS, ROUS SARCOMA
CELL MEMBRANE, 4156

ADENOSINE TRIPHOSPHATASE
NEUROBLASTOMA
5-BETA-CHOLAN-24-OIC ACID, 3-ALPHA,
12-ALPHA-DIHYDROXY-, 4176
DIBUTYRYL CYCLIC AMP, 4176
INSULIN, 4176
SUCROSE, 4176

ADENOSINE TRIPHOSPHATE
VIRUS, HAMSTER SARCOMA
NUCLEOTIDYLTRANSFERASES, 3799
POLY A POLYMERASE, 3799

ADENYL CYCLASE
HEPATOMA
CELL MEMBRANE, 4189*

ADENYLATE KINASE
VIRUS, KIRSTEN MURINE SARCOMA
CELL MEMBRANE, 4156
VIRUS, MOLONEY MURINE SARCOMA
CELL MEMBRANE, 4156
VIRUS, ROUS SARCOMA
CELL MEMBRANE, 4156

ADRENAL GLAND NEOPLASMS
ADENOMA
DIETHYLAMINE, N-NITROSO-, 3696

4H-IMIDAZOLE-(4,5-D)-V-TRIAZIN-4-
ONE, 3,7-DIHYDRO-, 3691
TRANSPLACENTAL CARCINOGENESIS,
HAMSTER, 3696
PHEOCHROMOCYTOMA
DIETHYLAMINE, N-NITROSO-, 3696
TRANSPLACENTAL CARCINOGENESIS,
HAMSTER, 3696

ADRENALECTOMY
HEPATOMA
SERINE DEHYDRATASE, 4183
TYROSINE AMINOTRANSFERASE, 4183

ADRIAMYCIN
VIRUS, ROUS SARCOMA
INFECTIVITY, 3879*

AFLATOXIN B1
ACETONE
PEANUTS, 3670
AFLATOXIN Q1
OXYGENASES, 3669
AMMONIA
PEANUTS, 3670
ASPERGILLUS PARASITICUS
TOXIN DEGRADATION, MYCELIUM, 3671
CARCINOGENIC POTENTIAL
MUTAGENIC ACTIVITY, REVIEW, 3602
FOOD ANALYSIS
FOOD CONTAMINATION, 4137*
FOOD CONTAMINATION
CORN MEAL, 4138*
EXTRACTION, CORN, 4136*
LIVER NEOPLASMS
MORPHOLOGICAL CHANGES, PARENCHYMA,
3719*
MICROSOMES, LIVER
PEPTIDES, 3668
PROTEINS, 3668
POLYRIBOSOMES
PROTEINS, 3668
SODIUM CHLORIDE
PEANUTS, 3670
SODIUM HYDROXIDE
PEANUTS, 3670

AFLATOXIN B2
ASPERGILLUS PARASITICUS
TOXIN DEGRADATION, MYCELIUM, 3671
CARCINOGENIC POTENTIAL
MUTAGENIC ACTIVITY, REVIEW, 3602
FOOD CONTAMINATION
CORN MEAL, 4138*

AFLATOXIN B2A
CARCINOGENIC POTENTIAL
MUTAGENIC ACTIVITY, REVIEW, 3602

AFLATOXIN G1
ASPERGILLUS PARASITICUS
TOXIN DEGRADATION, MYCELIUM, 3671
CARCINOGENIC POTENTIAL
MUTAGENIC ACTIVITY, REVIEW, 3602

AFLATOXIN G2
ASPERGILLUS PARASITICUS
TOXIN DEGRADATION, MYCELIUM, 3671
CARCINOGENIC POTENTIAL

MUTAGENIC ACTIVITY, REVIEW, 3602

AFLATOXIN G2A
    CARCINOGENIC POTENTIAL
        MUTAGENIC ACTIVITY, REVIEW, 3602

AFLATOXIN GM1
    CARCINOGENIC POTENTIAL
        MUTAGENIC ACTIVITY, REVIEW, 3602

AFLATOXIN M1
    CARCINOGENIC POTENTIAL
        MUTAGENIC ACTIVITY, REVIEW, 3602

AFLATOXIN M2
    CARCINOGENIC POTENTIAL
        MUTAGENIC ACTIVITY, REVIEW, 3602

AFLATOXIN P1
    CARCINOGENIC POTENTIAL
        MUTAGENIC ACTIVITY, REVIEW, 3602

AFLATOXIN Q1
    AFLATOXIN B1
        OXYGENASES, 3669
    CARCINOGENIC POTENTIAL
        MUTAGENIC ACTIVITY, REVIEW, 3602

AFLATOXIN RO
    CARCINOGENIC POTENTIAL
        MUTAGENIC ACTIVITY, REVIEW, 3602

AGRANULOCYTOSIS
    GENETICS
        CASE REPORT, 4032
    LEUKEMIA, LYMPHOCYTIC
        T-LYMPHOCYTES, 3918

AIR POLLUTANTS
    SMOKING
        COCARCINOGENIC EFFECT, REVIEW,
        3610

ALANINE, 3-PHENYL-
    HEPATOMA
        ALKALINE PHOSPHATASE, 4174, 4175

ALBUMIN
    CARCINOMA 256, WALKER
        METABOLISM, 4196*
    LEUKEMIA, LYMPHOCYTIC
        ISOLATION AND CHARACTERIZATION,
        4197*

ALCOHOL OXIDOREDUCTASES
    HEPATOMA
        ACETAMIDE, N-FLUOREN-2-YL-, 4177
        ETHYL ALCOHOL, 4177
        PYRAZOLE, 4177

ALKALINE PHOSPHATASE
    HEPATOMA
        ALANINE, 3-PHENYL-, 4174, 4175
        ARGININE, 4174, 4175
        AUSTRALIA ANTIGEN, 4175
        FETAL GLOBULINS, 4175
        IMIDAZOLE, 4175
        ISOENZYMES, 4174
        LEUCINE, 4174

UREA, 4175
LYMPHOMA
    PHOSPHATES, 3862*
RADIATION, IONIZING
    DNA, 3776
TERATOID TUMOR
    SERUM LEVELS, 4175
THYMUS NEOPLASMS
    LYMPHOMA, 3862*

ALKYLATING AGENTS
    CHROMOSOME ABERRATIONS
        ULTRASTRUCTURAL STUDY, 3620
    OCCUPATIONAL HAZARD
        REVIEW, 3662*

ALLOXAN
    VIRUS, ROUS SARCOMA
        CELL TRANSFORMATION, NEOPLASTIC,
        3672

ALPHA 1 ANTITRYPSIN
    BRONCHITIS
        EPIDEMIOLOGY, GENETICS, 4121
    LUNG NEOPLASMS
        EPIDEMIOLOGY, GENETICS, 4121
    PULMONARY EMPHYSEMA
        EPIDEMIOLOGY, GENETICS, 4121

ALPHA-FETOPROTEIN
    SEE FETAL GLOBULINS

ALPHA-AMANITIN
    VIRUS, ROUS SARCOMA
        VIRUS REPLICATION, 3814

AMELOBLASTOMA
    CELL TRANSFORMATION, NEOPLASTIC
        ODONTOGENIC CYSTS, 4063*

AMINO ACIDS
    UREA, NITROSO-
        CARBAMOYLATION, 3765*
    VIRUS, BOVINE ADENO 3
        VIRAL PROTEINS, 3791

AMMONIA
    AFLATOXIN B1
        PEANUTS, 3670

AMOSITE
    SEE ASBESTOS

ANEMIA, APLASTIC
    THYMOMA
        MYASTHENIA GRAVIS, 4044

ANEMIA, FACONI'S
    SEE ANEMIA, APLASTIC

ANGIOBLASTOMA
    SEE ANGIOSARCOMA

ANGIOMATOSIS
    UTERINE NEOPLASMS
        DNA, NEOPLASM, 4100*

ANGIOSARCOMA
    BRAIN NEOPLASMS

ENDOPLASMIC RETICULUM, 4005
GOLGI APPARATUS, 4005
ETHYLENE, CHLORO- POLYMER
OCCUPATIONAL HAZARD, 3663*
LIVER NEOPLASMS
EPIDEMIOLOGY, CHILD, 4140*
ETHYLENE, CHLORO-, 4036, 4124
ETHYLENE, CHLORO- POLYMER, 4036
LUNG NEOPLASMS
ETHYLENE, CHLORO-, 3627
OCCUPATIONAL HAZARD, 3627
MYCOPLASMA
CARCINOGENIC POTENTIAL, MOUSE,
4039
SPLEEN
ETHYLENE, CHLORO-, 4036
ETHYLENE, CHLORO- POLYMER, 4036
VIRUS, LEUKO
CELL TRANSFORMATION, NEOPLASTIC,
3856*
VIRUS REPLICATION, 3856*

ANILINE, N,N-DIMETHYL-P-PHENYLAZO-
FREE RADICALS
LIVER, RAT, 3679

ANTHOPHYLLITE
SEE ASBESTOS

1,8,9-ANTHRACENETRIOL
DNA REPAIR
FIBROBLASTS, 3721*

ANTIBODIES
CARCINOMA, EHRLICH TUMOR
IMMUNOSUPPRESSION, 3960*
LEUKEMIA, LYMPHOCYTIC
T-LYMPHOCYTES, 3961*
MAMMARY NEOPLASMS, EXPERIMENTAL
PLASMINOGEN ACTIVATORS, 3881*
MELANOMA
PLASMINOGEN ACTIVATORS, 3881*
PEPTIDE PEPTIDOHYDROLASES
PLASMINOGEN ACTIVATORS, 3881*
VIRUS, SV40
PLASMINOGEN ACTIVATORS, 3881*

ANTIBODIES, NEOPLASM
T-LYMPHOCYTES
TUBERCULIN TEST, 3938*

ANTIBODIES, VIRAL
BRAIN NEOPLASMS
VIRUS, SIMIAN SARCOMA, 3883*
VIRUS, C-TYPE RNA TUMOR
REVIEW, 3615
VIRUS, EPSTEIN-BARR
INFECTIOUS MONONUCLEOSIS, 4144*
STATISTICAL ANALYSIS, 4144*
VIRUS, POLYOMA, BK
KIDNEY TRANSPLANT, 3873*

ANTIBODY FORMATION
TRANSPLANTATION, HOMOLOGOUS
ANTIGENS, HETEROGENETIC, 3914
VIRUS, MOLONEY MURINE SARCOMA
CONCANAVALIN A, 3808

ANTIGEN-ANTIBODY REACTIONS

CARCINOEMBRYONIC ANTIGEN
PHASE-SPECIFIC ANTIGENS, 3963*
LYMPHOCYTES
ASCITIC FLUID, 3950*
VIRUS, FELINE LEUKEMIA
IMMUNOFLUORESCENCE, 3913
ROSETTE FORMATION, 3913

ANTIGENIC DETERMINANTS
BREAST NEOPLASMS
IMMUNE SERUMS, 3974*
IMMUNOPEROXIDASE STAINING, 3974*
CELL TRANSFORMATION, NEOPLASTIC
VIRUS, ADENO 7, 3975*
COLONIC NEOPLASMS
IMMUNE SERUMS, 3974*
IMMUNOPEROXIDASE STAINING, 3974*
NEURAMINIDASE
OSPOSINS, 3986*
PLASMACYTOMA
B-LYMPHOCYTES, 3919
T-LYMPHOCYTES, 3919
UTERINE NEOPLASMS
IMMUNE SERUMS, 3974*
IMMUNOPEROXIDASE STAINING, 3974*
VIRUS, ADENO 2
VIRAL PROTEINS, 3788
VIRUS, C-TYPE RNA TUMOR
PEPTIDES, 3832
VIRUS, HAMSTER LEUKEMIA
PEPTIDES, 3832
VIRUS, LEUKEMIA
VIRUS, GIBBON LYMPHOMA, 3834
VIRUS, SIMIAN SARCOMA, 3834
VIRUS, MURINE LEUKEMIA
PEPTIDES, 3832
VIRUS, RAUSCHER MURINE LEUKEMIA
PEPTIDES, 3832
VIRUS, SV40
ANTIGENS, NEOPLASM, 3882*

ANTIGENS
CHROMOSOMES
HYBRIDIZATION, 3971*
GENETICS
HYBRIDIZATION, 3971*
OVARIAN NEOPLASMS
CARCINOMA, 3972*

ANTIGENS, HETEROGENETIC
INFECTIOUS MONONUCLEOSIS
IMMUNE SERUMS, 3914
LEUKEMIA, LYMPHOCYTIC
INFECTIOUS MONONUCLEOSIS, 3914
LEUKEMIA, MYELOBLASTIC
INFECTIOUS MONONUCLEOSIS, 3914
LEUKEMIA, MYELOCYTIC
INFECTIOUS MONONUCLEOSIS, 3914
T-LYMPHOCYTES
INFECTIOUS MONONUCLEOSIS, 3914
LYMPHOMA
INFECTIOUS MONONUCLEOSIS, 3914
LYMPHOSARCOMA
INFECTIOUS MONONUCLEOSIS, 3914
SARCOMA, RETICULUM CELL
INFECTIOUS MONONUCLEOSIS, 3914
TRANSPLANTATION, HOMOLOGOUS

ANTIBODY FORMATION, 3914

ANTIGENS, NEOPLASM
   BREAST NEOPLASMS
      REVIEW, 3932*
   CELL TRANSFORMATION, NEOPLASTIC
      VIRUS, ADENO 7, 3975*
   FIBROSARCOMA
      IMMUNITY, CELLULAR, 3943*
   HEPATOMA
      CARBON TETRACHLORIDE, 3942*
   LEUKEMIA
      T-LYMPHOCYTES, 3915
   LEUKEMIA, MYELOBLASTIC
      GENETICS, 4030
   LEUKEMIA, SUBLEUKEMIC
      GENETICS, 4033
   MACROGLOBULINEMIA
      IMMUNE SERUMS, 3973*
   MULTIPLE MYELOMA
      IMMUNE SERUMS, 3973*
   VIRUS, SV40
      ANTIGENIC DETERMINANTS, 3882*

ANTIGENS, VIRAL
   BONE NEOPLASMS
      VIRUS, BOVINE LEUKOSIS, 3830
      VIRUS, C-TYPE, 3830
      VIRUS, RAUSCHER MURINE LEUKEMIA, 3830
      VIRUS, SIMIAN LEUKEMIA, 3830
   BRAIN NEOPLASMS
      VIRUS, SIMIAN SARCOMA, 3883*
   BURKITT'S LYMPHOMA
      VIRUS, EPSTEIN-BARR, 3795, 3844*, 4011
   CHONDROBLASTOMA
      VIRUS, RAUSCHER MURINE LEUKEMIA, 3830
      VIRUS, SIMIAN LEUKEMIA, 3830
   CHONDROSARCOMA
      VIRUS, RAUSCHER MURINE LEUKEMIA, 3830
      VIRUS, SIMIAN LEUKEMIA, 3830
   FIBROSARCOMA
      VIRUS, BOVINE LEUKOSIS, 3830
      VIRUS, RAUSCHER MURINE LEUKEMIA, 3830
      VIRUS, SIMIAN LEUKEMIA, 3830 .
   GIANT CELL TUMORS
      VIRUS, RAUSCHER MURINE LEUKEMIA, 3830
      VIRUS, SIMIAN LEUKEMIA, 3830
   LEUKEMIA, MYELOBLASTIC
      VIRUS, BABOON ENDOGENOUS, 3841*
      VIRUS, C-TYPE, 3841*
      VIRUS, SIMIAN SARCOMA, 3841*
   SARCOIDOSIS
      VIRUS, EPSTEIN-BARR, 3904
      VIRUS, HERPES, 3904
   SARCOMA, OSTEOGENIC
      VIRUS, RAUSCHER MURINE LEUKEMIA, 3830
      VIRUS, SIMIAN LEUKEMIA, 3830
   VIRUS, C-TYPE RNA TUMOR
      CROSS REACTIONS, 3811
      PEPTIDES, 3832
      REVIEW, 3615
   VIRUS, EPSTEIN-BARR

CYTOSINE, 1-BETA-D-ARABINOFURANOSYL-, 3846*
VIRUS, FELINE LEUKEMIA
   CELL MEMBRANE, 3913
   FLUORESCENT ANTIBODY TECHNIC, 3858*
   REVIEW, 3647*
VIRUS, FELINE SARCOMA
   REVIEW, 3647*
VIRUS, HERPES SIMPLEX 1
   ULTRAVIOLET RAYS, 3804
VIRUS, MAREK'S DISEASE HERPES
   IMMUNITY, ACTIVE, 3897
VIRUS, MURINE LEUKEMIA
   GLYCOPROTEINS, 3615
   INTERFERON, 3860*
VIRUS, PAPILLOMA
   COMPLEMENT FIXATION TESTS, 3898
   WART REGRESSION, 3898
VIRUS, SCRIPPS MURINE LEUKEMIA
   ANTINUCLEAR FACTORS, 3861*
VIRUS, SV40
   CELL FUSION, 3823
   TEMPERATURE SENSITIVE MUTANTS, 3822
   ULTRAVIOLET RAYS, 3804
WARTS
   VIRUS, HEPATITIS, 3837
   VIRUS, PAPILLOMA, 3898

ANTILYMPHOCYTE SERUM
   LEUKEMIA, LYMPHOCYTIC
      B-LYMPHOCYTES, 3961*
      T-LYMPHOCYTES, 3961*
   B-LYMPHOCYTES
      LANTHANUM, 4167
   T-LYMPHOCYTES
      LANTHANUM, 4167
   MELANOMA
      NEOPLASM METASTASIS, 4035

ANTIMONY
   CARCINOMA
      DISTRIBUTION STUDY, 3785*
      EXCRETION, 3785*
   FIBROSARCOMA
      DISTRIBUTION STUDY, 3785*
      EXCRETION, 3785*

ANTINEOPLASTIC AGENTS
   ADENOFIBROMA
      IMIDAZOLE-4-CARBOXAMIDE, 5-(3-METHYL-1-TRIAZENO)-, 3691
   IMIDAZOLE-4-CARBOXAMIDE, 5-(3,3-DIMETHYL-1-TRIAZENO)-
      CARCINOGENIC ACTIVITY, 3737*
      CARCINOGENIC ACTIVITY, RAT, 3691
   NEOPLASM TRANSPLANTATION
      ASCITIC CONVERSION, 3941*

ANTISERUM
   SEE IMMUNE SERUMS

ARGININE
   HEPATOMA
      ALKALINE PHOSPHATASE, 4174, 4175

L-ARGININE
   VIRUS, ADENO ASSOCIATED

---VIRUS REPLICATION, 3792

AROCHLOR
    BENZENAMINE, N,N-DIMETHYL-4-((3-
    METHYLPHENYL)AZO)-
        CARCINOGENIC ACTIVITY, 3728*
    HEPATOMA
        BENZENAMINE, N,N-DIMETHYL-4-((3-
        METHYLPHENYL)AZO)-, 3728*

ARSENIC
    LUNG NEOPLASMS
        OCCUPATIONAL HAZARD, 4127

ARTERIOSCLEROSIS
    BENZ(A)ANTHRACENE, 7,12-DIMETHYL-
        CARCINOGENIC ACTIVITY, CHICKEN,
        3749*

ARYL HYDROCARBON HYDROXYLASES
    BARBITURIC ACID, 5-ETHYL-5-PHENYL-
        DNA BINDING, 3742*
    BENZO(A)PYRENE
        ENZYMATIC ACTIVITY, 3676
        GENETIC INTERVENTION, 3676
    CHOLANTHRENE, 3-METHYL
        DNA BINDING, 3742*
        ENZYMATIC ACTIVITY, 3676
        GENETIC INTERVENTION, 3676
    DEXAMETHASONE
        ENZYMATIC ACTIVITY, 3738*
    DNA REPAIR
        SMOKING, 3710
    DNA REPLICATION
        SMOKING, 3710
    ESTRADIOL
        ENZYMATIC ACTIVITY, 3738*
    ESTRIOL
        ENZYMATIC ACTIVITY, 3738*
    ESTRONE
        ENZYMATIC ACTIVITY, 3738*
    PROGESTERONE
        ENZYMATIC ACTIVITY, 3738*
    TESTOSTERONE
        ENZYMATIC ACTIVITY, 3738*

ASBESTOS
    COLONIC NEOPLASMS
        EPIDEMIOLOGY, 4134*
        OCCUPATIONAL HAZARD, 4134*
    LUNG NEOPLASMS
        OCCUPATIONAL HAZARD, 3643*
    RECTAL NEOPLASMS
        EPIDEMIOLOGY, 4134*
    RESPIRATORY TRACT NEOPLASMS
        OCCUPATIONAL HAZARD, 3640*
    SMOKING
        COCARCINOGENIC EFFECT, REVIEW,
        3627
    WATER POLLUTANTS
        REVIEW, 4135*

ASCITIC FLUID
    T-LYMPHOCYTES
        CONCANAVALIN A, 3979*
        LYMPHOCYTE TRAPPING, 3979*

    PLANT AGGLUTININS, 3979*
NEOPLASMS
    T-LYMPHOCYTES, 3979*

ASCORBIC ACID
    ACETIC ACID, LEAD (+2) SALT
        TERATOGENIC ACTIVITY, 3727*

ASPARTATE CARBAMOLYTRANSFERASE
    DNA REPLICATION
        EUCARYOTIC CELL CYCLE, 4199*

ASPERGILLUS FLAVUS
    FOOD CONTAMINATION
        MYCOTOXINS, 4125

ASPERGILLUS PARASITICUS
    AFLATOXIN B1
        TOXIN DEGRADATION, MYCELIUM, 3671
    AFLATOXIN B2
        TOXIN DEGRADATION, MYCELIUM, 3671
    AFLATOXIN G1
        TOXIN DEGRADATION, MYCELIUM, 3671
    AFLATOXIN G2
        TOXIN DEGRADATION, MYCELIUM, 3671
    FOOD CONTAMINATION
        MYCOTOXINS, 4125

ASTROCYTOMA
    BRAIN NEOPLASMS
        CASE REPORT, 4052*
        NEOPLASMS, MULTIPLE PRIMARY, 3995
        ULTRASTRUCTURAL STUDY, 4052*
        VIRUS, SIMIAN SARCOMA, 3883*
    IMMUNOGLOBULINS
        IMMUNE SERUMS, 3925

AUSTRALIA ANTIGEN
    HEPATOMA
        ALKALINE PHOSPHATASE, 4175

AUTOANTIBODIES
    BRONCHIAL NEOPLASMS
        ADENOCARCINOMA, 3903
        CARCINOMA, 3903
        CARCINOMA, EPIDERMOID, 3903
        CARCINOMA, OAT CELL, 3903
    WARTS
        IGM, 3837

AZATHIOPRINE
    SEE PURINE, 6-((1-METHYL-4-
    NITROIMIDAZOL-5-YL)THIO)-

BACTERIA
    DIMETHYLAMINE, N-NITROSO-
        ENZYMATIC ACTIVITY, 3700
    DIPHENYLAMINE, N-NITROSO-
        ENZYMATIC ACTIVITY, 3700
    PYRROLIDINE, 1-NITROSO-
        ENZYMATIC ACTIVITY, 3700

BARBITURIC ACID, 5-ETHYL-5-PHENYL-
    ARYL HYDROCARBON HYDROXYLASES
        DNA BINDING, 3742*
    CYTOCHROME P-450
        DNA BINDING, 3742*
    DIETHYLAMINE, N-NITROSO-
        COCARCINOGENIC ACTIVITY, HAMSTER,

SUBJECT

RAT, 3697
DIMETHYLAMINE, N-NITROSO-
    COCARCINOGENIC ACTIVITY, HAMSTER,
    RAT, 3697
GLUCOSEPHOSPHATASE
    ENZYMATIC ACTIVITY, 3766*
LIVER NEOPLASMS
    PRECANCEROUS CONDITIONS, 3766*
PYRIDINE, 2-(BENZYL(2-(DIMETHYLAMINO)
ETHYL)AMINO)-
    DNA, 3693
PYRIDINE, 2-(BENZYL(2-(METHYL)(NITROSO)
AMINO)ETHYL)AMINO)-
    DNA, 3693

BASOPHILS
HODGKIN'S DISEASE
    ULTRASTRUCTURAL STUDY, 4007

BENZ(A)ANTHRACENE
CYTOCHROME P-450
    DNA BINDING, 3742*
GLYCERALDEHYDEPHOSPHATE DEHYDROGENASE
    ENZYME REPRESSION, LIGHT REACTION,
    3744*
MAMMARY NEOPLASMS, EXPERIMENTAL
    POLYNUCLEOTIDE PHOSPHORYLASE,
    3745*
SALMONELLA TYPHIMURIUM
    MUTAGENIC ACTIVITY, 3678
STRUCTURE-ACTIVITY RELATIONSHIP
    MOLECULAR CONFIRMATION, 3682

BENZ(A)ANTHRACENE, 7,12-DIMETHYL-
ARTERIOSCLEROSIS
    CARCINOGENIC ACTIVITY, CHICKEN,
    3749*
BENZ(A)ANTHRACENE, 8,9-DIOL, 8,9-
    DIHYDRO-7,12-DIMETHYL-
    K-REGION EPOXIDES, 3748*
CHOLANTHRENE, 3-METHYL-
    INCREASE AND SYNTHESIS, 3755*
DIETARY FATS
    COCARCINOGENIC EFFECT, 3606
DNA REPLICATION
    THYMIDINE, 3746*
ESTRADIOL
    PROTEINS, 3747*
GLYCERALDEHYDEPHOSPHATE DEHYDROGENASE
    ENZYME REPRESSION, LIGHT REACTION,
    3744*
LUNG NEOPLASMS
    TRANSPLACENTAL CARCINOGENESIS,
    REVIEW, 3601
MAMMARY NEOPLASMS, EXPERIMENTAL
    ADENOCARCINOMA, 3741*
    BINDING SITES, 3747*
    CYTOSOL, 3740*
    ESTRADIOL, 3680
    OLIGOPEPTIDES, 3752*
    PROGESTERONE, 3680
    TESTOSTERONE, 3680
    TRANSPLACENTAL CARCINOGENESIS,
    REVIEW, 3601
    ULTRASTRUCTURAL STUDY, RAT, 3753*
OVARIAN NEOPLASMS
    FORMATION AND DEVELOPMENT, 3751*
PROLACTIN
    HEMATOPOIETIC STEM CELLS, 3754*

BENZ(A)ANTHRACENE-5,6-DIOL, 5,6-DIHYDRO-
SALMONELLA TYPHIMURIUM
    MUTAGENIC ACTIVITY, 3678

BENZ(A)ANTHRACENE-8,9-DIOL, 8,9-DIHYDRO-
SALMONELLA TYPHIMURIUM
    MUTAGENIC ACTIVITY, 3678

BENZ(A)ANTHRACENE-8,9-DIOL, 8,9-DIHYDRO-7,
12-DIMETHYL-
    BENZ(A)ANTHRACENE, 7,12-DIMETHYL-
    K-REGION EPOXIDES, 3748*
    FIBROBLASTS
    CELL TRANSFORMATION, NEOPLASTIC,
    3748*

BENZ(A)ANTHRACENE-8,9-DIOL, 8,9-DIHYDRO-7-
METHYL-
    BENZ(A)ANTHRACENE, 7-METHYL-
    K-REGION EPOXIDES, 3748*
    FIBROBLASTS
    CELL TRANSFORMATION, NEOPLASTIC,
    3748*
    SALMONELLA TYPHIMURIUM
    MUTAGENIC ACTIVITY, 3678

BENZ(A)ANTHRACENE-8,9-DIOL, 10,11-EPOXY-8,
9-DIHYDRO-
    SALMONELLA TYPHIMURIUM
    MUTAGENIC ACTIVITY, 3678

BENZ(A)ANTHRACENE, 5,6-EPOXY-
SALMONELLA TYPHIMURIUM
    MUTAGENIC ACTIVITY, 3678

BENZ(A)ANTHRACENE, 5,6-EPOXY-7,12-DIMETHYL-
NUCLEIC ACIDS
    BINDING SITES, 3750*

BENZ(A)ANTHRACENE, 7-(METHOXYMETHYL)-12-
METHYL-
    GLYCERALDEHYDEPHOSPHATE DEHYDROGENASE
    ENZYME REPRESSION, LIGHT REACTION,
    3744*

BENZ(A)ANTHRACENE, 7-METHYL-
    BENZ(A)ANTHRACENE-8,9-DIOL, 8,9-
    DIHYDRO-7-METHYL-
    K-REGION EPOXIDES, 3748*

BENZENAMINE, 2-METHYL-4-((2-METHYLPHENYL)
AZO)-
    NERVOUS SYSTEM NEOPLASMS
    TRANSPLACENTAL CARCINOGENESIS,
    REVIEW, 3601

BENZENAMINE, N,N-DIMETHYL-4-((3-
METHYLPHENYL)AZO)-
    AROCHLOR
    CARCINOGENIC ACTIVITY, 3728*
    HEPATOMA
    AROCHLOR, 3728*
    PEPTIDYL TRANSFERASES, 3730*

PEPTIDYL TRANSFERASES
ENZYMATIC ACTIVITY, 3730*

BENZENE, 1-CHLORO-2,4-DINITRO-
T-LYMPHOCYTES
IMMUNE RESPONSE, 3938*

BENZIDINE
ENVIRONMENTAL HAZARD
REVIEW, 3642*

BENZIDINE, 3,3'-DIMETHYL-
BLADDER NEOPLASMS
PRECANCEROUS CONDITIONS, 3731*
HEPATOMA
FETAL GLOBULINS, 3731*

BENZO(A)PYREN-6-OL
BENZO(A)PYRENE
CARCINOGENIC METABOLITE, 3677

BENZO(A)PYREN-5-YLOXY
BENZO(A)PYRENE
CARCINOGENIC METABOLITE, 3677

BENZO(A)PYRENE
ARYL HYDROCARBON HYDROXYLASE
ENZYMATIC ACTIVITY, 3676
GENETIC INTERVENTION, 3676
BENZO(A)PYREN-6-OL
CARCINOGENIC METABOLITE, 3677
BENZO(A)PYREN-5-YLOXY
CARCINOGENIC METABOLITE, 3677
BRONCHIAL NEOPLASMS
HISTOLOGICAL STUDY, HAMSTER, 4077*
PRECANCEROUS CONDITIONS, 4077*
BURKITT'S LYMPHOMA
DNA REPAIR, 3778*
CROTON OIL
COCARCINOGENIC EFFECT, 3733*
DIETARY FATS
COCARCINOGENIC EFFECT, 3606
LUNG NEOPLASMS, 3610
TRANSPLACENTAL CARCINOGENESIS,
REVIEW, 3601
MICROSOMES
BINDING, 3743*
SALMONELLA TYPHIMURIUM
MUTAGENIC ACTIVITY, 3678
SKIN NEOPLASMS
TRANSPLACENTAL CARCINOGENESIS,
REVIEW, 3601
SMOKING
COCARCINOGENIC EFFECT, REVIEW,
3610
STRUCTURE-ACTIVITY RELATIONSHIP
MOLECULAR CONFIRMATION, 3682

BENZO(A)PYRENE-4,5-DIOL, 4,5-DIHYDRO-
SALMONELLA TYPHIMURIUM
MUTAGENIC ACTIVITY, 3678

BENZO(A)PYRENE-7,8-DIOL, 7,8-DIHYDRO-
SALMONELLA TYPHIMURIUM
MUTAGENIC ACTIVITY, 3678

BENZO(A)PYRENE-9,10-DIOL, 9,10-DIHYDRO-
SALMONELLA TYPHIMURIUM

MUTAGENIC ACTIVITY, 3678

BENZO(A)PYRENE-7,8-DIOL, 9,10-EPOXY-7,8-
DIHYDRO-
SALMONELLA TYPHIMURIUM
MUTAGENIC ACTIVITY, 3678

BENZO(A)PYRENE, 4,5-EPOXY-
SALMONELLA TYPHIMURIUM
MUTAGENIC ACTIVITY, 3678

BENZO(G,H,I)PERYLENE
STRUCTURE-ACTIVITY RELATIONSHIP
MOLECULAR CONFIRMATION, 3682

7,8-BENZOFLAVONE
MICROSOMES
BINDING, 3743*

BENZOPYRENE HYDROXYLASE
SEE ALSO HYDROXYLASES/ARYL HYDROCARBON
HYROXYLASES

P-BENZOQUINONE, 2,3,5-TRIS(1-AZIRIDINYL)-
CHROMOSOME ABERRATIONS
BONE MARROW, 3717*
LYMPHOCYTES, 3732*

BENZYL ALCOHOL, 3,4-DIHYDROXY-ALPHA-(
(ISOPROPYLAMINO)METHYL)-
CELL DIVISION
CIRCADIAN RHYTHM, 4152
CORNEA, RAT, 4152

BENZYL ALCOHOL, 3,4-DIHYDROXY-ALPHA-(
(METHYLAMINO)-METHYL)-
SEE EPINEPHRINE

BILE DUCT NEOPLASMS
CHOLANGIOMA
BUTANOL, 4-(BUTYLNITROSOAMINO)-,
3701
DIETHYLAMINE, N-NITROSO-, 3696
TRANSPLACENTAL CARCINOGENESIS,
HAMSTER, 3696
CYSTADENOMA
BUTANOL, 4-(BUTYLNITROSOAMINO)-,
3701

2-BIPHENYLOL
BLADDER NEOPLASMS
OCCUPATIONAL HAZARD, 3660*

BLADDER NEOPLASMS
ADENOCARCINOMA
CASE REPORT, 4091*
NEOPLASM METASTASIS, 4046
ULTRASTRUCTURAL STUDY, 4091*
BENZIDINE, 3,3'-DIMETHYL-
PRECANCEROUS CONDITIONS, 3731*
2-BIPHENYLOL
OCCUPATIONAL HAZARD, 3660*
BUTANOL, 4-(BUTYLNITROSOAMINO)-
DOSE-RESPONSE STUDY, 3701
CARCINOGEN, ENVIRONMENTAL
EPIDEMIOLOGY, SPAIN, REVIEW, 3630

CARCINOMA
  FORMAMIDE, N-(4-(5-NITRO-2-FURYL)-
    2-THIAZOLYL)-, 3736*
CARCINOMA, TRANSITIONAL CELL
  DNA, NEOPLASM, 3989
  NEOPLASM METASTASIS, 4046
CELL TRANSFORMATION, NEOPLASTIC
  CYSTITIS, 4091*
CYCLOHEXANE, 1,2,3,4,5,6-HEXACHLORO-,
  GAMMA ISOMER
  OCCUPATIONAL HAZARD, 3660*
EPIDEMIOLOGY
  AGE FACTORS, GREAT BRITAIN, 4114
FOOD CONTAMINATION
  EPIDEMIOLOGY, SPAIN, REVIEW, 3630
SCHISTOSOMIASIS
  PRECANCEROUS CONDITIONS, 3990
TOBACCO
  EPIDEMIOLOGY, SPAIN, REVIEW, 3630

BLEOMYCIN
  CELL LINE
  CELL TRANSFORMATION, NEOPLASTIC,
    3720*

BLOOD CIRCULATION
  CARCINOMA
  NEOPLASM TRANSPLANTATION, 4059*

BLOOD COAGULATION FACTORS
  NEOPLASM METASTASIS
  REVIEW, 3656*

BLOOD GROUPS
  BRONCHITIS
  EPIDEMIOLOGY, GENETICS, 4121
  CARCINOEMBRYONIC ANTIGEN
  BINDING, 3910
  GLYCOPROTEINS, 3910
  IMMUNE SERUMS, 3910
  LUNG NEOPLASMS
  EPIDEMIOLOGY, GENETICS, 4121
  PULMONARY EMPHYSEMA
  EPIDEMIOLOGY, GENETICS, 4121

BLOOD SEDIMENTATION
  LYMPHOSARCOMA
  DIAGNOSIS AND PROGNOSIS, 4019
  SARCOMA, RETICULUM CELL
  DIAGNOSIS AND PROGNOSIS, 4019

BONE MARROW
  P-BENZOQUININE, 2,3,5-TRIS(1-
    AZIRIDINYL)-
  CHROMOSOME ABERRATIONS, 3717*
  COLCHICINE
  CHROMOSOME ABERRATIONS, 3717*
  CYCLOPHOSPHAMIDE
  CHROMOSOME ABERRATIONS, 3717*,
    4024
  ISONICOTINIC ACID HYDRAZIDE
  CHROMOSOME ABERRATIONS, 3757*
  LEUKEMIA, MYELOBLASTIC
  ULTRASTRUCTURAL STUDY, 4009
  VIRUS-LIKE PARTICLES, 4009
  METHANESULFONIC ACID, ETHYL ESTER
  CHROMOSOME ABERRATIONS, 3717*
  METHANESULFONIC ACID, METHYL ESTER
  CHROMOSOME ABERRATIONS, 3717*

MITOMYCIN C
  CHROMOSOME ABERRATIONS, 3717*
RADIATION
  ROSETTE FORMATION, SHEEP CELLS,
    3947*
  S-TRIAZINE, 2,4,6-TRIS(1-AZIRIDINYL)-
  CHROMOSOME ABERRATIONS, 3717*

BONE MARROW CELLS
  LEUKEMIA
  CHROMOSOME ABERRATIONS, 4082*
  PRECANCEROUS CONDITIONS, 4078*
  LEUKEMIA, LYMPHOBLASTIC
  IMMUNE SERUMS, 3955*
  MORPHOLOGICAL AND CYTOCHEMICAL
    STUDY, 4021
  LEUKEMIA, MYELOBLASTIC
  MORPHOLOGICAL AND CYTOCHEMICAL
    STUDY, 4021
  LEUKEMIA, MYELOCYTIC
  IMMUNE SERUMS, 3955*
  LEUKEMIA, RADIATION-INDUCED
  IMMUNE RESPONSE, 3889
  LYMPHOSARCOMA
  IMMUNE RESPONSE, 3889
  MONOCYTES
  CELL DIVISION, 3899
  PHAGOCYTES
  CELL DIVISION, 3899

BONE NEOPLASMS
  CHONDROMA
  CASE REPORT, 4053*
  OSTEOMA, OSTEOID
  CASE REPORT, 4049*
  RADIATION, IONIZING
  EPIDEMIOLOGY, REVIEW, 3628
  ETIOLOGY, REVIEW, 3628
  SARCOMA, EWING'S
  EPIDEMIOLOGY, CHILD, INDIA, 4120
  SARCOMA, OSTEOGENIC
  CASE REPORT, 4049*
  CASE REPORT, DOG, 4048*
  EPIDEMIOLOGY, CHILD, INDIA, 4120
  EPIDEMIOLOGY, MALAYSIA, 4142*
  VIRUS, POLYOMA, 3874*
  VIRUS, BOVINE LEUKOSIS
  ANTIGENS, VIRAL, 3830
  VIRUS, C-TYPE
  ANTIGENS, VIRAL, 3830
  VIRUS, RAUSCHER MURINE LEUKEMIA
  ANTIGENS, VIRAL, 3830
  VIRUS, MURINE MAMMARY TUMOR, 3830
  VIRUS, SIMIAN LEUKEMIA
  ANTIGENS, VIRAL, 3830

BRACKEN FERN
  TANNIC ACID
  ISOLATION AND CHARACTERIZATION,
    3770*

BRAIN NEOPLASMS
  ADENOMA, CHROMOPHOBE
  ENDOPLASMIC RETICULUM, 4005
  GOLGI APPARATUS, 4005
  ANGIOSARCOMA
  ENDOPLASMIC RETICULUM, 4005
  GOLGI APPARATUS, 4005
  ASTROCYTOMA

CASE REPORT, 4052*
NEOPLASMS, MULTIPLE PRIMARY, 3995
ULTRASTRUCTURAL STUDY, 4052*
VIRUS, SIMIAN SARCOMA, 3883*
EPENDYMOMA
ENDOPLASMIC RETICULUM, 4005
FIBROMA
ULTRASTRUCTURAL STUDY, 3994
GLIOBLASTOMA MULTIFORME
ENDOPLASMIC RETICULUM, 4005
GENETICS, 4023
GOLGI APPARATUS, 4005
NEOPLASMS, MULTIPLE PRIMARY, 3995
GLIOMA
CASE REPORT, 4053*
EPIDEMIOLOGY, CHILD, INDIA, 4120
HODGKIN'S DISEASE
DEOXYRIBONUCLEASE, 4017
PEPSIN, 4017
RIBONUCLEASE, 4017
RNA, 4017
ULTRASTRUCTURAL STUDY, 4017
ISOANTIGENS
TRANSPLACENTAL CARCINOGENESIS,
3892
MEDULLOBLASTOMA
ENDOPLASMIC RETICULUM, 4005
GOLGI APPARATUS, 4005
MENINGIOMA
ENDOPLASMIC RETICULUM, 4005
GOLGI APPARATUS, 4005
NEOPLASMS, MULTIPLE PRIMARY, 3995
NEUROBLASTOMA
EPIDEMIOLOGY, CHILD, INDIA, 4120
NEUROEPITHELIOMA
ENDOPLASMIC RETICULUM, 4005
GOLGI APPARATUS, 4005
NOSE NEOPLASMS
NEOPLASM METASTASIS, 3702
OLIGODENDROGLIOMA
ENDOPLASMIC RETICULUM, 4005
GOLGI APPARATUS, 4005
SARCOMA
ENDOPLASMIC RETICULUM, 4005
GOLGI APPARATUS, 4005
SARCOMA, RETICULUM CELL
CASE REPORT, 4051*
UREA, ETHYL NITROSO-
COPPER (II) SULFATE, 3601
NICKEL (II) SULFATE, 3601
TRANSPLACENTAL CARCINOGENESIS,
REVIEW, 3601
VIRUS, SIMIAN SARCOMA
ANTIBODIES, VIRAL, 3883*
ANTIGENS, VIRAL, 3883*
CELL-TRANSFORMATION, NEOPLASTIC,
3883*

BREAST NEOPLASMS
ADENOCARCINOMA
CASE REPORT, 3991
GENETICS, 4023
NEOPLASMS, MULTIPLE PRIMARY, 3995
RADIATION, IONIZING, 3991
ADENOFIBROMA
CELL NUCLEUS, 4055*
HISTOLOGICAL STUDY, 4055*
AGE FACTORS
EPIDEMIOLOGY, ISRAEL, 4115

FERTILTIY, REVIEW, 3622
ANTIGENIC DETERMINANTS
IMMUNE SERUMS, 3974*
IMMUNOPEROXIDASE STAINING, 3974*
ANTIGENS, NEOPLASM
REVIEW, 3932*
CARCINOEMBRYONIC ANTIGEN
REVIEW, 3932*
CARCINOMA
CELL CYCLE KINETICS, 4130
CELL NUCLEUS, 4055*
COLLAGEN, 3993
HISTOLOGICAL STUDY, 4055*
HYDROXYLASES, 3993
NEOPLASMS, MULTIPLE PRIMARY, 3995
CARCINOMA, DUCTAL
NEOPLASMS, MULTIPLE PRIMARY, 3995
POLYNUCLEOTIDE PHOSPHORYLASE,
3745*
CARCINOMA, MUCINOUS
CELL NUCLEUS, 4055*
HISTOLOGICAL STUDY, 4055*
CARCINOMA, PAPILLARY
CELL NUCLEUS, 4055*
HISTOLOGICAL STUDY, 4055*
CARCINOMA, SCIRRHOUS
CELL NUCLEUS, 4055*
COLLAGEN, 3993
HISTOLOGICAL STUDY, 4055*
HYDROXYLASES, 3993
COLCHICINE
GROWTH SUBSTANCES, 3689
EPIDEMIOLOGY
AGE FACTORS, GREAT BRITAIN, 4114
ESTROGENS
EPIDEMIOLOGY, 4141*
GENETICS
EPIDEMIOLOGY, 4141*
EPIDEMIOLOGY, REVIEW, 4117
GONADOTROPINS
PRECANCEROUS CONDITIONS, MALE,
3992
GROWTH
CELL CYCLE KINETICS, 4147*
GYNECOMASTIA
PRECANCEROUS CONDITIONS, MALE,
3992
HEMANGIOMA
CASE REPORT, 3991
HORMONES
EPIDEMIOLOGY, REVIEW, 4117
IMMUNITY, CELLULAR
REVIEW, 3932*
IMMUNOTHERAPY
REVIEW, 3932*
INSULIN
GROWTH SUBSTANCES, 3689
KLINEFELTER'S SYNDROME
PRECANCEROUS CONDITIONS, MALE,
3992
LYMPHOCYTES
IMMUNITY, CELLULAR, 3936*
LYMPHOSARCOMA
CASE REPORT, 4054*
MONOCYTES
IGG, 3958*
MYOBLASTOMA
NEOPLASM METASTASIS, 4056*
NEOPLASM METASTASIS, 4045

CELL CYCLE KINETICS, 4130
NEOPLASMS
    EPIDEMIOLOGY, REVIEW, 4117
PARITY
    EPIDEMIOLOGY, 4141*
PRECANCEROUS CONDITIONS
    CELL NUCLEUS, 4055*
    COLLAGEN, 3993
    EPIDEMIOLOGY, MALE, 3992
    HISTOLOGICAL STUDY, 4055*
    HYDROXYLASES, 3993
PROLACTIN
    GROWTH SUBSTANCES, 3689
PROTEINS
    ISOLATION AND CHARACTERIZATION, C-
    REACTIVE, 4169
RADIATION, IONIZING
    AGE FACTORS, 4057*
RAUWOLFIA ALKALOIDS
    AGE FACTORS, 4057*
RESERPINE
    STATISTICAL ANALYSIS, REVIEW, 3609
4,4'-STILBENEDIOL, ALPHA,ALPHA'-
    DIETHYL-
    ANTI-ESTROGEN AGENTS, 3740*
SULFATASES
    HISTOCHEMICAL STUDY, 4179
THYMIDINE
    GROWTH SUBSTANCES, 3689
TRANSPLANTATION, HETEROLOGOUS
    NEOPLASM METASTASIS, 4035

BROMODEOXYURIDINE
    SEE URIDINE, 5-BROMO-2'-DEOXY'

5-BROMUDEOXYURIDINE
    SEE URIDINE, 5-BROMO-2'-DEOXY'

BRONCHIAL NEOPLASMS
ADENOCARCINOMA
    AUTOANTIBODIES, 3903
BENZO(A)PYRENE
    HISTOLOGICAL STUDY, HAMSTER, 4077*
    PRECANCEROUS CONDITIONS, 4077*
CAPILLARITY
    HISTOLOGICAL STUDY, 4076*
CARCINOMA
    AUTOANTIBODIES, 3903
    B-LYMPHOCYTES, 3926
    T-LYMPHOCYTES, 3926
CARCINOMA, EPIDERMOID
    AUTOANTIBODIES, 3903
    B-LYMPHOCYTES, 3926
    T-LYMPHOCYTES, 3926
CARCINOMA, OAT CELL
    AUTOANTIBODIES, 3903
    B-LYMPHOCYTES, 3926
    T-LYMPHOCYTES, 3926
B-LYMPHOCYTES
    DIAGNOSIS AND PROGNOSIS, 3926
T-LYMPHOCYTES
    DIAGNOSIS AND PROGNOSIS, 3926
PRECANCEROUS CONDITIONS
    DIAGNOSIS, BRONCHOGRAPHY, 3764*
    DIETHYLAMINE, N-NITRCSO-, 3764*
TOBACCO
    HISTOLOGICAL STUDY, HAMSTER, 4077*
    PRECANCEROUS CONDITIONS, 4077*

BRONCHITIS
ALPHA 1 ANTITRYPSIN
    EPIDEMIOLOGY, GENETICS, 4121
BLOOD GROUPS
    EPIDEMIOLOGY, GENETICS, 4121
PHENOTYPE
    EPIDEMIOLOGY, GENETICS, 4121
SMOKING
    EPIDEMIOLOGY, GENETICS, 4121

BURKITT'S LYMPHOMA
ABDOMINAL NEOPLASMS
    CLINICOPATHOLOGIC STUDY, 4013
BENZO(A)PYRENE
    DNA REPAIR, 3778*
CANDIDA
    IMMUNITY, CELLULAR, 3946*
CELL LINE
    B-LYMPHOCYTES, 3795
CHROMOSOME ABNORMALITIES
    CASE REPORT, CHILD, 4014
CYCLOPHOSPHAMIDE
    CLINICOPATHOLOGIC STUDY, 4013
DIPTHERIA ANTITOXIN
    IMMUNITY, CELLULAR, 3946*
IGM
    IMMUNE SERUMS, 3925
IMMUNITY, CELLULAR
    IMMUNOLOGIC FACTOR, 3946*
INTESTINAL NEOPLASMS
    CLINICOPATHOLOGIC STUDY, 4013
JAW NEOPLASMS
    CLINICOPATHOLOGIC STUDY, 4013
KARYOTYPING
    CASE REPORT, CHILD, 4014
LACTATE DEHYDROGENASE
    CLINICOPATHOLOGIC STUDY, 4013
LEUKEMIA, LYMPHOBLASTIC
    CASE REPORT, 4011
LIP NEOPLASMS
    NEOPLASM METASTASIS, 4012
    VIRUS-LIKE PARTICLES, 4012
METHANESULFONIC ACID, METHYL ESTER
    DNA REPAIR, 3778*
NEOPLASM METASTASIS
    CLINICOPATHOLOGIC STUDY, 4013
NERVOUS SYSTEM NEOPLASMS
    NEOPLASM METASTASIS, 4012
OVARIAN NEOPLASMS
    CLINICOPATHOLOGIC STUDY, 4013
STREPTODORNASE AND STREPTOKINASE
    IMMUNITY, CELLULAR, 3946*
TUBERCULIN
    IMMUNITY, CELLULAR, 3946*
ULTRAVIOLET RAYS
    DNA REPAIR, 3778*
VIRUS, EPSTEIN-BARR
    ANTIGENS, VIRAL, 3795, 3844*, 4011
    DNA, VIRAL, 3845*
    HYBRIDIZATION, 3631

1-BUTANAMINE, N-BUTYL-N-NITROSO-
BUTANOL, 4-(BUTYLNITROSOAMINO)-
    BIOLOGICALLY ACTIVE METABOLITE,
    3701

BUTANOL, 4-(BUTYLNITROSOAMINO)-
BILE DUCT NEOPLASMS
    CHOLANGIOMA, 3701

··CYSTADENOMA, 3701
BLADDER NEOPLASMS
  DOSE-RESPONSE STUDY, 3701
1-BUTANAMINE, N-BUTYL-N-NITROSO-
  BIOLOGICALLY ACTIVE METABOLITE,
  3701
GASTRIC NEOPLASMS
  PAPILLOMA, 3701
LIVER NEOPLASMS
  HEMANGIOENDOTHELIOMA, 3701
NOSE NEOPLASMS
  ADENOCARCINOMA, 3701
RESPIRATORY TRACT NEOPLASMS
  PRECANCEROUS CONDITIONS, 3701
TRACHEAL NEOPLASMS
  POLYPS, 3701

BUTYRIC ACID, 2-ACETAMIDO-4-(ETHYLSULFONYL)-
  BUTYRIC ACID, 2-AMINO-4-(ETHYLTHIO)-
    CARCINOGENIC METABOLITE, RAT, 3685

BUTYRIC ACID, 2-AMINO-4-(ETHYLTHIO)-
  BUTYRIC ACID, 2-ACETAMIDO-4-
    (ETHYLSULFONYL)-
    CARCINOGENIC METABOLITE, RAT, 3685
  BUTYRIC ACID, 4-(ETHYLTHIO)-
    CARCINOGENIC METABOLITE, RAT, 3685
  MICROSOMES
    CELL MEMBRANE, 3735*
  PROTEINS
    SEX FACTORS, 3734*
  RNA REPLICATION
    SEX FACTORS, 3734*

BUTYRIC ACID, 4-(ETHYLTHIO)-
  BUTYRIC ACID, 2-AMINO-4-(ETHYLTHIO)-
    CARCINOGENIC METABOLITE, RAT, 3685

CAFFEINE
  HELA CELLS
    ADENOSINE CYCLIC 3',5'
      MONOPHOSPHATE, 4193*

CANCER
  SEE NEOPLASMS

CANDIDA
  BURKITT'S LYMPHOMA
    IMMUNITY, CELLULAR, 3946*

CANTHARIDIN
  SKIN NEOPLASMS
    ENZYMATIC ACTIVITY, 3723*

CAPILLARITY
  BRONCHIAL NEOPLASMS
    HISTOLOGICAL STUDY, 4076*
  NEOPLASM METASTASIS
    REVIEW, 3657*

CAPSID ANTIGENS
  SEE VIRAL PROTEINS

CARBAMIC ACID, ETHYL ESTER
  EYE NEOPLASMS
    IMMUNOSUPPRESSION, 3893
  HEMANGIOMA
    IMMUNOSUPPRESSION, 3893

HEPATOMA
  ··IMMUNOSUPPRESSION, 3893
LEUKEMIA, MYELOCYTIC
  IMMUNOSUPPRESSION, 3893
LUNG NEOPLASMS
  ADENOMA, 3893
  IMMUNOSUPPRESSION, 3893
  TRANSPLACENTAL CARCINOGENESIS,
    REVIEW, 3601
NEOPLASMS, EXPERIMENTAL
  IMMUNOSUPPRESSION, 3893
RNA POLYMERASE
  CARCINOGENIC ACTIVITY, RAT, 3715

CARBAMIC ACID, METHYL-, 1-NAPHTHYL ESTER
  VIRUS, RNA TUMOR
    CELL TRANSFORMATION, NEOPLASTIC,
      3704

CARBAMIC ACID, N-METHYL-N-NITROSO-, ETHYL
  ESTER
  KIDNEY NEOPLASMS
    TRANSPLACENTAL CARCINOGENESIS,
      REVIEW, 3601
  LEUKEMIA
    TRANSPLACENTAL CARCINOGENESIS,
      REVIEW, 3601
  NERVOUS SYSTEM NEOPLASMS
    TRANSPLACENTAL CARCINOGENESIS,
      REVIEW, 3601
  POLAROGRAPHY
    SPECTROSCOPY, 3762*

CARBAMIC ACID, N-METHYL-N-NITROSO-, 1-
  NAPHTHYL ESTER
  VIRUS, RNA TUMOR
    CELL TRANSFORMATION, NEOPLASTIC,
      3704

CARBON RADIOISOTOPES
  LEUKEMIA, MYELOCYTIC
    CARCINOGENIC EFFECT, MOUSE, 3782*
  LEUKEMIA, RADIATION INDUCED
    CARCINOGENIC EFFECT, MOUSE, 3782*

CARBON TETRACHLORIDE
  HEPATOMA
    ANTIGENS, NEOPLASM, 3942*
    HYPERSENSITIVITY, 3942*

CARBOXYLYASES
  DNA REPLICATION
    FIBROBLASTS, DIPLOID, 4162

CARCINOEMBRYONIC ANTIGEN
  ADENOCARCINOMA
    ULTRASTRUCTURAL STUDY, 3912
  ANTIGEN-ANTIBODY REACTIONS
    PHASE-SPECIFIC ANTIGENS, 3963*
  BLOOD GROUPS
    BINDING, 3910
    GLYCOPROTEINS, 3910
    IMMUNE SERUMS, 3910
  BREAST NEOPLASMS
    REVIEW, 3932*
  COLONIC NEOPLASMS
    ADENOCARCINOMA, 3912
    HYDRAZINE, 1,2-DIMETHYL-, 3909
    ISOLATION AND CHARACTERIZATION,

3968*
  LYMPHOCYTES, 3908
CONCANAVALIN A
  ISOLATION AND CHARACTERIZATION,
  3964*
DIGESTIVE SYSTEM NEOPLASMS
  CARCINOMA, 3969*
  IMMUNE SERUMS, 3969*
DUODENAL NEOPLASMS
  GUANIDINE, 1-METHYL-3-NITRO-1-
  NITROSO-, 3909
FETAL GLOBULINS
  TUMOR CELL SYNTHESIS, 3970*
GASTROINTESTINAL NEOPLASMS
  ADENOCARCINOMA, 3968*
  ISOMERIC VARIANT, 3967*
INTESTINAL NEOPLASMS
  ADENOCARCINOMA, 3911
  IMMUNE SERUMS, 3909
  RADIATION, 3784*
  RADIATION, IONIZING, 3911
LYMPHOCYTES
  PRECANCEROUS CONDITIONS, 3908
NEOPLASM METASTASIS
  ISOELECTRIC PRECIPITATION AND
  ULTRAFILTRATION, 3965*
  ISOLATION AND CHARACTERIZATION,
  3965*
PANCREATIC NEOPLASMS, 3966*
  ISOLATION AND CHARACTERIZATION,
  3966*
SMOKING
  T-LYMPHOCYTES, 3768*
THYROID NEOPLASMS
  CARCINOMA, 3896

CARCINOGEN, CHEMICAL
  COLONIC NEOPLASMS
    EPIDEMIOLOGY, 4124
  ENVIRONMENTAL HAZARD
    REVIEW, 3642*
  ESOPHAGEAL NEOPLASMS
    ANIMAL MODEL, RAT, MOUSE, 3641*
  GALLBLADDER NEOPLASMS
    EPIDEMIOLOGY, 4124
  HEPATOMA
    ANIMAL MODEL, RAT, MOUSE, 3641*
  INTESTINAL NEOPLASMS
    ANIMAL MODEL, RAT, MOUSE, 3641*
  OCCUPATIONAL HAZARD
    REVIEW, 3662*
  PANCREATIC NEOPLASMS
    EPIDEMIOLOGY, 4124
  STOMACH NEOPLASMS
    ANIMAL MODEL, RAT, MOUSE, 3641*

CARCINOGEN, ENVIRONMENTAL
  BLADDER NEOPLASMS
    EPIDEMIOLOGY, SPAIN, REVIEW, 3630
  MYCOTOXINS
    REVIEW, 3604
  NEOPLASMS
    EPIDEMIOLOGY, REVIEW, 4123
    STATISTICAL ANALYSIS, 4123

CARCINOID
  SEE CARCINOID TUMOR

CARCINOMA

ADENOLYMPHOMA
  CASE REPORT, 4042
ANTIMONY
  DISTRIBUTION STUDY, 3785*
  EXCRETION, 3785*
BLADDER NEOPLASMS
  FORMAMIDE, N-(4-(5-NITRO-2-FURYL)-
  2-THIAZOLYL)-, 3736*
BLOOD CIRCULATION
  NEOPLASM TRANSPLANTATION, 4059*
BREAST NEOPLASMS
  CELL CYCLE KINETICS, 4130
  CELL NUCLEUS, 4055*
  COLLAGEN, 3993
  HISTOLOGICAL STUDY, 4055*
  HYDROXYLASES, 3993
  NEOPLASMS, MULTIPLE PRIMARY, 3995
BRONCHIAL NEOPLASMS
  AUTOANTIBODIES, 3903
  B-LYMPHOCYTES, 3926
  T-LYMPHOCYTES, 3926
CARDIAC OUTPUT
  NEOPLASM TRANSPLANTATION, 4059*
CHEMOTAXIS
  ISOLATION AND CHARACTERIZATION,
  INHIBITORY FACTOR, 3886
COLONIC NEOPLASMS
  CASE REPORT, 4041
  PRECANCEROUS CONDITIONS, 4072*
  PREDISPOSING FACTORS, REVIEW, 3623
COMPLEMENT
  CHEMOTAXIS, 3886
DIGESTIVE SYSTEM NEOPLASMS
  CARCINOEMBRYONIC ANTIGEN, 3969*
ESCHERICHIA COLI
  CHEMOTAXIS, 3886
GALLBLADDER NEOPLASMS
  CHONDROITIN, 4194*
  DERMATAN SULFATE, 4194*
  HEPARITIN SULFATE, 4194*
IMMUNOGLOBULINS
  IMMUNE SERUMS, 3925
LEUKOCYTES
  CHEMOTAXIS, 3886
LIPOPROTEINS
  ISOLATION AND CHARACTERIZATION,
  3632
LUNG NEOPLASMS
  EPIDEMIOLOGY, 4146*
  SEX FACTORS, 4146*
MAMMARY NEOPLASMS, EXPERIMENTAL, 3785*
  PLASMINOGEN ACTIVATORS, 3881*
  UREMIA, 4151
  VIRUS, POLYOMA, 3874*
MAXILLARY NEOPLASMS
  CELL CYCLE KINETICS, 4130
MYCOPLASMA
  CARCINOGENIC POTENTIAL, MOUSE,
  4039
NOSE NEOPLASMS
  DIPROPYLAMINE, N-NITROSO-, 3702
OVARIAN NEOPLASMS
  ANTIGENS, 3972*
PANCREATIC NEOPLASMS
  REVIEW, 3639*
PAROTID NEOPLASMS
  ADENOLYMPHOMA, 4042
POLYNUCLEOTIDE PHOSPHORYLASE
  HISTOCHEMISTRY, 3745*

LUNG NEOPLASMS
    DIPROPYLAMINE, N-NITROSO-, 3702
    NEOPLASM METASTASIS, 4046
MOUTH NEOPLASMS
    CLOSTRIDIOPEPTIDASE A, 3998
NOSE NEOPLASMS
    PIPERIDINE, 1-NITROSO-, 3707
PHARYNGEAL NEOPLASMS
    CELL CYCLE KINETICS, 4130
    CLOSTRIDIOPEPTIDASE A, 3998
    PIPERIDINE, 1-NITROSO-, 3707
SKIN NEOPLASMS
    REVIEW, 3618
    ULTRAVIOLET RAYS, 3775
4,4'-STILBENEDIOL, ALPHA,ALPHA'-
  DIETHYL-
    CASE REPORT, 4096*
STOMACH NEOPLASMS
    PIPERIDINE, 3,4-DIBROMO-1-NITROSO-,
      3707
    PIPERIDINE, 3,4-DICHLORO-1-NITROSO-
      , 3707
TRANSPLANTATION, HETEROLOGOUS
    NEOPLASM METASTASIS, 4035
UTERINE NEOPLASMS
    CASE REPORT, 4096*
VIRUS-LIKE PARTICLES
    C-TYPE RNA TUMOR VIRUS, 3835
    ISOLATION AND CHARACTERIZATION,
      3835

CARCINOMA IN SITU
  CERVIX NEOPLASMS
    CELL TRANSFORMATION, NEOPLASTIC,
      3996
    DIAGNOSIS AND THERAPY, 3999
    DNA, NEOPLASM, 3996
    HISTOLOGICAL STUDY, 3997
    PRECANCEROUS CONDITIONS, 3999
    SEX BEHAVIOR, 4094*
  LARYNGEAL NEOPLASMS
    DIAGNOSIS AND THERAPY, 3999
    PRECANCEROUS CONDITIONS, 3999
  STOMACH NEOPLASMS
    PRECANCEROUS CONDITIONS, 4068*

CARCINOMA, MUCINOUS
  BREAST NEOPLASMS
    CELL NUCLEUS, 4055*
    HISTOLOGICAL STUDY, 4055*
  INTESTINAL NEOPLASMS
    CYCLOOCTANECARBAMIC ACID, 1,1-BIS
      (P-FLUOROPHENYL)-2-PROPYNYL,
      3684

CARCINOMA, OAT CELL
  BRONCHIAL NEOPLASMS
    AUTOANTIBODIES, 3903
    B-LYMPHOCYTES, 3926
    T-LYMPHOCYTES, 3926
  LUNG NEOPLASMS
    NEOPLASM METASTASIS, 4046

CARCINOMA, PAPILLARY
  BREAST NEOPLASMS
    CELL NUCLEUS, 4055*
    HISTOLOGICAL STUDY, 4055*
  INTESTINAL NEOPLASMS
    CYCLOOCTANECARBAMIC ACID, 1,1-BIS

(P-FLUOROPHENYL)-2-PROPYNYL,
   3684
LARYNGEAL NEOPLASMS
   EPIDEMIOLOGY, 4075*
THYROID NEOPLASMS
   EPIDEMIOLOGY, 4129
   HISTOLOGIC STUDY, 4129
   NEOPLASM METASTASIS, 4129
   RADIATION, IONIZING, 3772
   RADIOTHERAPY, 3786*

CARCINOMA, RENAL CELL
  SEE ADENOCARCINOMA

CARCINOMA, SCIRRHOUS
  BREAST NEOPLASMS
   CELL NUCLEUS, 4055*
   COLLAGEN, 3993
   HISTOLOGICAL STUDY, 4055*
   HYDROXYLASES, 3993
  INTESTINAL NEOPLASMS
   CYCLOOCTANECARBAMIC ACID, 1,1-BIS
   (P-FLUOROPHENYL)-2-PROPYNYL,
   3684

CARCINOMA, SMALL CELL, LUNG
  SEE CARCINOMA, OAT CELL

CARCINOMA, SQUAMOUS CELL
  SEE CARCINOMA, EPIDERMOID

CARCINOMA, TRANSITIONAL CELL
  BLADDER NEOPLASMS
   DNA, NEOPLASM, 3989
   NEOPLASM METASTASIS, 4046
  KIDNEY NEOPLASMS
   NEOPLASM METASTASIS, 4046

CARCINOMA 256, WALKER
  ALBUMIN
   METABOLISM, 4196*
  IMMUNITY, CELLULAR
   DNA REPLICATION, 3901

CARDIAC OUTPUT
  CARCINOMA
   NEOPLASM TRANSPLANTATION, 4059*

CELL DIFFERENTIATION
  SEE ALSO GROWTH
  ESTROGENS
   DNA, 3607
   RECEPTORS, HORMONE, 3607
   RNA, MESSENGER, 3607
  HODGKIN'S DISEASE
   B-LYMPHOCYTES, 3984*
  LEUKEMIA, LYMPHOCYTIC
   IMMUNOGLOBULINS, SURFACE, 3984*
   B-LYMPHOCYTES, 3984*
  LEUKEMIA, PLASMACYTIC
   IMMUNOGLOBULINS, SURFACE, 3984*
   B-LYMPHOCYTES, 3984*
  PROGESTERONE
   RECEPTORS, HORMONE, 3607
   RNA, MESSENGER, 3607
  VIRUS, FRIEND POLYCYTHEMIA

ERYTHROCYTES, 3849*
CELL DIVISION
  SEE ALSO GROWTH
  BENZYL ALCOHOL, 3,4-DIHYDROXY-ALPHA-(
   (ISOPROPYLAMINO)METHYL)-
   CIRCADIAN RHYTHM, 4152
   CORNEA, RAT, 4152
  BONE MARROW CELLS
   MONOCYTES, 3899
   PHAGOCYTES, 3899
  ENDOTHELIUM
   LIVER REGENERATION, 4159
  KUPFFER CELLS
   LIVER REGENERATION, 4159
  LIVER REGENERATION
   CRITICAL MASS THEORY, 4160
   HEPATOCYTE VOLUME, 4160
  MAMMARY NEOPLASMS, EXPERIMENTAL
   CELL CYCLE KINETICS, 4131
  MELANOMA
   CELL CYCLE KINETICS, 4131
  RADIATION
   CELL CYCLE KINETICS, 3779*
   ENZYMATIC ACTIVITY, 3779*
  SARCOMA
   CELL CYCLE KINETICS, 4131
  THEOPHYLLINE
   CIRCADIAN RHYTHM, 4152
   CORNEA, RAT, 4152
  VIRUS, SV40
   AGE FACTORS, 4164

CELL FUSION
  NASOPHARYNGEAL NEOPLASMS
   VIRUS, EPSTEIN-BARR, 3843*
  PLASMACYTOMA
   ACTINOMYCIN D, 4004
   CYCLOHEXIMIDE, 4004
   ULTRASTRUCTURAL STUDY, 4004
   UREA, HYDROXY-, 4004
  VIRUS, SV40
   ANTIGENS, VIRAL, 3823
   DNA, CIRCULAR, 3823
   RADIATION, IONIZING, 3804

CELL LINE
  BURKITT'S LYMPHOMA
   B-LYMPHOCYTES, 3795
  CELL TRANSFORMATION, NEOPLASTIC
   BLEOMYCIN, 3720*
   CYTOSINE, 1-BETA-D-
   ARABINOFURANOSYL-, 3720*
   METHOTREXATE, 3720*
   URIDINE, 2'-DEOXY-5-FLUORO-, 3720*
  ERYTHROLEUKEMIA
   VIRUS, FRIEND MURINE LEUKEMIA,
   3798

CELL MEMBRANE
  BUTYRIC ACID, 2-AMINO-4-(ETHYLTHIO)-
   MICROSOMES, 3735*
  CARCINOMA, EHRLICH TUMOR
   CATIONS, MONOVALENT, 4190*, 4191*
  ERYTHROCYTES
   REVIEW, 3667*
  GANGLIOSIDES
   COLCHICINE, 3924
   CYTOCHALASIN B, 3924

LIGANDS, 3924
NEURAMINIDASE, 3924
TRYPSIN, 3924
GLYCOPROTEINS
REVIEW, 3667*
HEPATOMA
ADENYL CYCLASE, 4189*
GLYCOPEPTIDES, 4173
HEPARIN, 4173
ISOLATION AND CHARACTERIZATION, 4189*
MUCOPOLYSACCHARIDES, 4173
LEUKEMIA, LYMPHOCYTIC
GANGLIOSIDES, 3924
LYMPHOCYTES
ULTRASTRUCTURAL STUDY, 4110*
MACROPHAGES
ULTRASTRUCTURAL STUDY, 3987*
PLASMACYTOMA
ADENOSINE CYCLIC 3',5'
MONOPHOSPHATE, 4153
THEOPHYLLINE, 4153
RADIATION
ULTRASTRUCTURAL STUDY, 3783*
REVIEW, 3667*
SARCOMA 180, CROCKER
GLYCOPROTEINS, 4168
12-O-TETRADECANOYLPHORBOL-13-ACETATE
SPECTROMETRY, FLUORESCENCE, 3726*
URIDINE, 5-BROMO-2'-DEOXY-
IGM, 3959*
VIRUS, ADENO 2
PEPTIDES, 3788
VIRUS, FELINE LEUKEMIA
ANTIGENS, VIRAL, 3913
VIRUS, KIRSTEN MURINE SARCOMA
ADENOSINE CYCLIC 3',5'
MONOPHOSPHATE, 4156
ADENYLATE KINASE, 4156
VIRUS, MOLONEY MURINE SARCOMA
ADENOSINE CYCLIC 3',5'
MONOPHOSPHATE, 4156
ADENYLATE KINASE, 4156
VIRUS, ROUS SARCOMA
ADENOSINE CYCLIC 3',5'
MONOPHOSPHATE, 4156
ADENYLATE KINASE, 4156
VIRUS, SV40
AGE FACTORS, 4164
COLCHICINE, 4157
UTRASTRUCTURAL STUDY, 4157
VINCALEUCOBLASTINE, 4157

LL NUCLEUS
BREAST NEOPLASMS
ADENOFIBROMA, 4055*
CARCINOMA, 4055*
CARCINOMA, MUCINOUS, 4055*
CARCINOMA, PAPILLARY, 4055*
CARCINOMA, SCIRRHOUS, 4055*
PRECANCEROUS CONDITIONS, 4055*
DNA REPLICATION, 3634
LYMPHOCYTES
ULTRASTRUCTURAL STUDY, 4110*

LL TRANSFORMATION, NEOPLASTIC
ADENOSINE CYCLIC 3',5' MONOPHOSPHATE
REVIEW, 4156
AMELOBLASTOMA

ODONTOGENIC CYSTS, 4063*
ANGIOSARCOMA
VIRUS, LEUKO, 3856*
BLADDER NEOPLASMS
CYSTITIS, 4091*
BRAIN NEOPLASMS
VIRUS, SIMIAN SARCOMA, 3883*
CARCINOMA, EPIDERMOID
ODONTOGENIC CYSTS, 4063*
CELL LINE
BLEOMYCIN, 3720*
CYTOSINE, 1-BETA-D-
ARABINOFURANOSYL-, 3720*
METHOTREXATE, 3720*
URIDINE, 2'-DEOXY-5-FLUORO-, 3720*
CERVIX NEOPLASMS
CARCINOMA IN SITU, 3996
PRECANCEROUS CONDITIONS, 3996
FIBROBLASTS
BENZ(A)ANTHRACENE-8,9-DIOL, 8,9-
DIHYDRO-7,12-DIMETHYL-, 3748*
BENZ(A)ANTHRACENE-8,9-DIOL, 8,9-
DIHYDRO-7-METHYL-, 3748*
ISOENZYMES
REVIEW, 3633
LEUKEMIA L1210
ISOANTIGENS, 3892
LEUKOCYTES
CARCINOGENESIS SCREENING, 3718*
MITOCHONDRIA
ULTRASTRUCTURAL STUDY, 4109*
NEOPLASMS
CELL CYCLE KINETICS, REVIEW, 3653*
ONCOGENIC VIRUSES
CELL CYCLE KINETICS, REVIEW, 3653*
RADIATION, IONIZING
CELL CYCLE KINETICS, REVIEW, 3653*
SARCOMA
ODONTOGENIC CYSTS, 4063*
SARCOMA, OSTEOGENIC
GUANIDINE, 1-METHYL-3-NITRO-1-
NITROSO-, 3705
SKIN NEOPLASMS
MELANOMA, 4040
SPLENIC NEOPLASMS
VIRUS, FRIEND MURINE LEUKEMIA, 3847*
VIRUS, ADENO 7
ANTIGENIC DETERMINANTS, 3975*
ANTIGENS, NEOPLASM, 3975*
VIRUS, AVIAN SARCOMA
ULTRAVIOLET RAYS, 3840*
VIRUS, FRIEND MURINE LEUKEMIA
IMMUNOSUPPRESSION, 3893
VIRUS REPLICATION, 3798
VIRUS, GROSS MURINE LEUKEMIA
ISOANTIGENS, 3892
VIRUS, MOLONEY MURINE SARCOMA
CONCANAVALIN A, 3808
IMMUNOSUPPRESSION, 3893
VIRUS, POLYOMA
ULTRASTRUCTURAL STUDY, 3875*
VIRUS, RNA TUMOR
CARBAMIC ACID, METHYL-, 1-NAPHTHYL
ESTER, 3704
CARBAMIC ACID, N-METHYL-N-NITROSO-,
1-NAPHTHYL ESTER, 3704
VIRUS, ROUS SARCOMA
ALLOXAN, 3672

SUBJECT 17

VIRUS, SV40
    TEMPERATURE SENSITIVE MUTANTS,
    3822
    ULTRASTRUCTURAL STUDY, 4164

CELLS, CULTURED
    DIMETHYLAMINE, N-NITROSO-
    AZAGUANINE, 3699
        MUTAGENIC ACTIVITY, 3699
    SARCOMA 180, CROCKER
        GLYCOPROTEINS, 4168

CELLULAR INCLUSIONS
    GLIOMA
        ULTRASTRUCTURAL STUDY, 4006
    MENINGIOMA
        ULTRASTRUCTURAL STUDY, 4006

CERVIX NEOPLASMS
    ADENOCARCINOMA
        CASE REPORT, 3690
        ENDOMETRIOSIS, 3690
        ESTROGENS, 3690
        HISTOLOGICAL STUDY, 3997
    CARCINOGENESIS SCREENING
        PAP TEST, 4104*
    CARCINOMA, EPIDERMOID
        SEX BEHAVIOR, 4094*
    CARCINOMA IN SITU
        CELL TRANSFORMATION, NEOPLASTIC,
        3996
        DIAGNOSIS AND THERAPY, 3999
        DNA, NEOPLASM, 3996
        HISTOLOGICAL STUDY, 3997
        PRECANCEROUS CONDITIONS, 3999
        SEX BEHAVIOR, 4094*
    GLUCOSEPHOSPHATE
        PRECANCEROUS CONDITIONS, 4198*
    HEXOKINASE
        PRECANCEROUS CONDITIONS, 4198*
    LACTATE DEHYDROGENASE
        PRECANCEROUS CONDITIONS, 4198*
    PHOSPHOFRUCTOKINASE
        PRECANCEROUS CONDITIONS, 4198*
    PRECANCEROUS CONDITIONS
        CELL TRANSFORMATION, NEOPLASTIC,
        3996
        DNA, NEOPLASM, 3996
    PROTEINS
        ISOLATION AND CHARACTERIZATION, C-
        REACTIVE, 4169
    PYRUVATE KINASE
        PRECANCEROUS CONDITIONS, 4198*
    4,4'-STILBENEDIOL, ALPHA,ALPHA'-
    DIETHYL-
        TRANSPLACENTAL CARCINOGENESIS,
        3608
    TRANSPLANTATION, HOMOLOGOUS
        ANTIGENIC STIMULATION, REVIEW,
        3807
    VIRUS, HERPES SIMPLEX 2
        HYBRIDIZATION, 3631

CHEMODECTOMA
    SEE PARAGANGLIOMA, NONCHROMAFFIN

CHEMOTAXIS
    CARCINOMA
        COMPLEMENT, 3886
        ESCHERICHIA COLI, 3886
        ISOLATION AND CHARACTERIZATION,
        INHIBITORY FACTOR, 3886
        LEUKOCYTES, 3886
    HEPATOMA
        COMPLEMENT, 3886
        ESCHERICHIA COLI, 3886
        ISOLATION AND CHARACTERIZATION,
        INHIBITORY FACTOR, 3886
        LEUKOCYTES, 3886

5-BETA-CHOLAN-24-OIC ACID, 3-ALPHA, 12-
ALPHA-DIHYDROXY-
    NEUROBLASTOMA
        ADENOSINE TRIPHOSPHATASE, 4176

CHOLANGIOMA
    BILE DUCT NEOPLASMS
        BUTANOL, 4-(BUTYLNITROSOAMINO)-,
        3701
        DIETHYLAMINE, N-NITROSO-, 3696
        TRANSPLACENTAL CARCINOGENESIS,
        HAMSTER, 3696

CHOLANTHRENE, 3-METHYL-
    ARYL HYDROCARBON HYDROXYLASES
        DNA BINDING, 3742*
        ENZYMATIC ACTIVITY, 3676
        GENETIC INTERVENTION, 3676
    BENZ(A)ANTHRACENE, 7,12-DIMETHYL-
        INCREASE AND SYNTHESIS, 3755*
    CYTOCHROME P-450
        DNA BINDING, 3742*
    DIETARY FATS
        COCARCINOGENIC EFFECT, 3606
    FIBROSARCOMA
        IMMUNITY, CELLULAR, 3945*
        TRANSPLANTATION IMMUNOLOGY, 3945*
    GLUCURONIDASE
        CARCINOGENIC METABOLITE, 3683
        TRANSPLACENTAL CARCINOGENESIS,
        3683
    HYDROLASES
        CARCINOGENIC METABOLITE, 3683
        TRANSPLACENTAL CARCINOGENESIS,
        3683
    LUNG NEOPLASMS
        ADENOMA, 3893
        IMMUNOSUPPRESSION, 3893
    LYMPHOMA
        TRANSPLACENTAL CARCINOGENESIS,
        REVIEW, 3601
    MAMMARY NEOPLASMS, EXPERIMENTAL
        ESTRADIOL, 3680
        PROGESTERONE, 3680
        SARCOMA, 3680
        TESTOSTERONE, 3680
    NEOPLASMS
        IMMUNOSUPPRESSION, 3951*, 3952*
    NEOPLASMS, EXPERIMENTAL

IMMUNOSUPPRESSION, 3893
PROLACTIN
INCREASE AND SYNTHESIS, 3755*
PYRIDINE, 2-(BENZYL(2-(METHYL)(NITROSO)
AMINO)ETHYL)AMINO)-
DNA, 3693
RNA
POLY A, 3756*
SARCOMA
IMMUNITY, CELLULAR, 3934*
IMMUNOGLOBULINS, 3956*
MIGRATION INHIBITORY FACTOR, 3934*
STRUCTURE-ACTIVITY RELATIONSHIP
MOLECULAR CONFIRMATION, 3682
SULFATASES
CARCINOGENIC METABOLITE, 3683
TRANSPLACENTAL CARCINOGENESIS,
3683

CHOLERA
HELA CELLS
ADENOSINE CYCLIC 3',5'
MONOPHOSPHATE, 4193*

CHOLESTEROL
COLONIC NEOPLASMS
DIETARY FATS, 3605
DIETARY FATS
COCARCINOGENIC EFFECT, 3605

CHONDROBLASTOMA
VIRUS, RAUSCHER MURINE LEUKEMIA
ANTIGENS, VIRAL, 3830
VIRUS, SIMIAN LEUKEMIA
ANTIGENS, VIRAL, 3830

CHONDROID SYRINGOMA
SEE NEOPLASMS, EMBRYONAL AND MIXED

CHONDROITIN
GALLBLADDER NEOPLASMS
CARCINOMA, 4194*

CHONDROMA
BONE NEOPLASMS
CASE REPORT, 4050*

CHONDROMYXOID FIBROMA
SEE CHONDROMA

CHONDROSARCOMA
IMMUNOGLOBULINS
IMMUNE SERUMS, 3925
MUCOPOLYSACCHARIDES
ISOLATION AND CHARACTERIZATION,
4195*
VIRUS, RAUSCHER MURINE LEUKEMIA
ANTIGENS, VIRAL, 3830
VIRUS, SIMIAN LEUKEMIA
ANTIGENS, VIRAL, 3830

CHORDOMA
IMMUNOGLOBULINS
IMMUNE SERUMS, 3925

CHORIOCARCINOMA
TESTICULAR NEOPLASMS
CASE REPORT, 4105*
UTERINE NEOPLASMS

ULTRASTRUCTURAL STUDY, RAT, 4098*

CHROMATIN
HELA CELLS
RNA, MESSENGER, 3634

CHROMIUM
OCCUPATIONAL HAZARD
CASE REPORT, 4112*

CHROMOSOME ABERRATIONS
ALKYLATING AGENTS
ULTRASTRUCTURAL STUDY, 3620
P-BENZOQUININE, 2,3,5-TRIS(1-
AZIRIDINYL)-
BONE MARROW, 3717*
P-BENZOQUINONE, 2,3,5-TRIS(1-
AZIRIDINYL)-
LYMPHOCYTES, 3732*
BONE MARROW
CYCLOPHOSPHAMIDE, 4024
COLCHICINE
BONE MARROW, 3717*
CYCLOPHOSPHAMIDE
BONE MARROW, 3717*
GENETICS
REVIEW, 3661*
ISONICOTINIC ACID HYDRAZIDE
BONE MARROW, 3757*
LEUKEMIA
BONE MARROW CELLS, 4082*
CASE REPORT, 4082*
REVIEW, MOLLUSK, 3659*
TRANSPLANTATION, HOMOLOGOUS, 4082*
LEUKEMIA, MYELOBLASTIC
RADIATION, IONIZING, 3771
LEUKEMIA, RADIATION-INDUCED
THORIUM DIOXIDE, 4026
METHANESULFONIC ACID, ETHYL ESTER
BONE MARROW, 3717*
METHANESULFONIC ACID, METHYL ESTER
BONE MARROW, 3717*
MITOMYCIN C
BONE MARROW, 3717*
KARYOTYPING, 4025
RADIATION, IONIZING
DOSE-RESPONSE STUDY, 3777*
ULTRASTRUCTURAL STUDY, 3620
S-TRIAZINE, 2,4,6-TRIS(1-AZIRIDINYL)-
BONE MARROW, 3717*

CHROMOSOME ABNORMALITIES
BURKITT'S LYMPHOMA
CASE REPORT, CHILD, 4014
HEPATOMA
ACETAMIDE, N-FLUOREN-2-YL-, 4150
LYMPHOSARCOMA
CASE REPORT, CHILD, 4014

CHROMOSOMES
ANTIGENS
HYBRIDIZATION, 3971*
DNA
HYBRIDIZATION, REVIEW, 3631
OVARIAN NEOPLASMS
PRECANCEROUS CONDITIONS, 4095*

CHROMOSOMES, HUMAN, 16-18
VIRUS, ADENO 12

PHOSPHOTRANSFERASES, 3793

6-CHRYSENAMINE
LUNG NEOPLASMS
CARCINOGENIC POTENTIAL, 3681
SKIN NEOPLASMS
CARCINOGENIC POTENTIAL, 3681

CHRYSOTILE
SEE ASBESTOS

CIRCADIAN RHYTHM
BENZYL ALCOHOL, 3,4-DIHYCROXY-ALPHA-(
(ISOPROPYLAMINO)METHYL)-
CELL DIVISION, 4152
THEOPHYLLINE
CELL DIVISION, 4152

CLOSTRIDIOPEPTIDASE A
LARYNGEAL NEOPLASMS
CARCINOMA, EPIDERMOIC, 3998
MOUTH NEOPLASMS
CARCINOMA, EPIDERMOIC, 3998
PHARYNGEAL NEOPLASMS
CARCINOMA, EPIDERMOIC, 3998

COAL TAR
DIETARY FATS
COCARCINOGENIC EFFECT, 3606

COCCIDIOSIS
LYMPHOMA
MAREK'S DISEASE, 4088*
MAREK'S DISEASE
CASE REPORT, 4088*

COLCHICINE
BREAST NEOPLASMS
GROWTH SUBSTANCES, 3689
CHROMOSOME ABERRATIONS
BONE MARROW, 3717*
GANGLIOSIDES
CELL MEMBRANE, 3924
VIRUS, SV40
CELL MEMBRANE, 4157

COLLAGEN
BREAST NEOPLASMS
CARCINOMA, 3993
CARCINOMA, SCIRRHOUS, 3993
PRECANCEROUS CONDITIONS, 3993

COLONIC NEOPLASMS
ADENOCARCINOMA
CARCINOEMBRYONIC ANTIGEN, 3912
ADENOCARCINOMA, PAPILLARY
CASE REPORT, 4041
ADENOMA
ULTRASTRUCTURAL STUDY, 4010
ANTIGENIC DETERMINANTS
IMMUNE SERUMS, 3974*
IMMUNOPEROXIDASE STAINING, 3974*
ASBESTOS
EPIDEMIOLOGY, 4134*
OCCUPATIONAL HAZARD, 4134*
CARCINOEMBRYONIC ANTIGEN
ISOLATION AND CHARACTERIZATION,
3968*
LYMPHOCYTES, 3908

CARCINOGEN, CHEMICAL
EPIDEMIOLOGY, 4124
CARCINOMA
CASE REPORT, 4041
PRECANCEROUS CONDITIONS, 4072*
PREDISPOSING FACTORS, REVIEW, 3623
CONCANAVALIN A
BINDING, 3988*
DIETARY FATS
CHOLESTEROL, 3605
DNA REPLICATION
PRECANCEROUS CONDITIONS, 4072*
ENTERITIS, REGIONAL
CASE REPORT, 4041
HYDRAZINE, 1,2-DIMETHYL-
CARCINOEMBRYONIC ANTIGEN, 3909
MULTIPLE SCLEROSIS
EPIDEMIOLOGY, 4122
TRANSIENTS AND MIGRANTS, 4122
POLYPS
CASE REPORT, 4041
ULTRASTRUCTURAL STUDY, 4010

COMPLEMENT
CARCINOMA
CHEMOTAXIS, 3886
HEPATOMA
CHEMOTAXIS, 3886

CONCANAVALIN A
BINDING SITES
SEA URCHIN EMBRYO, 3928
CARCINOEMBRYONIC ANTIGEN
ISOLATION AND CHARACTERIZATION,
3964*
COLONIC NEOPLASMS
BINDING, 3988*
B-LYMPHOCYTES
LANTHANUM, 4167
T-LYMPHOCYTES
ASCITIC FLUID, 3979*
DNA REPLICATION, 3927
LANTHANUM, 4167
PLASMACYTOMA
ADENOSINE CYCLIC 3',5'
MONOPHOSPHATE, 4153
THEOPHYLLINE, 4153
VIRUS, MOLONEY MURINE SARCOMA
ANTIBODY FORMATION, 3808
CELL TRANSFORMATION, NEOPLASTIC,
3808
IMMUNITY, CELLULAR, 3808

CONTACT INHIBITION
VIRUS, SV40
DNA REPLICATION, 3824

CONTRACEPTIVES, ORAL
HEPATOMA
RETINOL, 3637*

COPPER
LUNG NEOPLASMS
OCCUPATIONAL HAZARD, 4127

COPPER (II) SULFATE
BRAIN NEOPLASMS
UREA, ETHYL NITROSO-, 3601

CORONARY DISEASE
    MULTIPLE SCLEROSIS
        EPIDEMIOLOGY, 4122

CORONENE
    LUNG NEOPLASMS, 3610
    SMOKING
        COCARCINOGENIC EFFECT, REVIEW,
        3610

CORTICOTROPIN
    LEUKEMIA, MYELOCYTIC
        IMMUNOSUPPRESSION, 3953*

CORTISOL
    GLUCOSE, 2-DEOXY-
        DNA REPLICATION, 4182
    GROWTH SUBSTANCES
        FIBROBLASTS, 4186
    RNA
        NUCLEOLI, RAT, 4172
    RNA POLYMERASE
        NUCLEOLI, RAT, 4172
    VIRUS, ROUS SARCOMA
        DNA REPLICATION, 4182

CORTISONE
    GLUCOSE, 2-DEOXY-
        DNA REPLICATION, 4182
    HEPATOMA
        SERINE DEHYDRATASE, 4183
        TYROSINE AMINOTRANSFERASE, 4183

CROCIDOLITE
    SEE ASBESTOS

CROTON OIL
    BENZO(A)PYRENE
        COCARCINOGENIC EFFECT, 3733*
    DNA REPAIR
        FIBROBLASTS, 3721*
    LUNG NEOPLASMS
        PAPILLOMA, 3733*

CYCLIC AMP
    SEE ADENOSINE CYCLIC 3',5'
    MONOPHOSPHATE

CYCLIC AMP PHOSPHODIESTERASE
    SEE PHOSPHODIESTERASES

CYCLOHEXANE, 1,2,3,4,5,6-HEXACHLORO-,
GAMMA ISOMER
    BLADDER NEOPLASMS
        OCCUPATIONAL HAZARD, 3660*

CYCLOHEXIMIDE
    PLASMACYTOMA
        CELL FUSION, 4004
        ULTRASTRUCTURAL STUDY, 4004
    RADIATION
        PROTEINS, 3779*
    VIRUS, EPSTEIN-BARR
        VIRUS REPLICATION, 3844*
    VIRUS, HERPES SIMPLEX 2
        VIRAL PROTEINS, 3801

CYCLOOCTANECARBAMIC ACID, 1,1-BIS(P-
FLUOROPHENYL)-2-PROPYNYL

INTESTINAL NEOPLASMS
    CARCINOMA, MUCINOUS, 3684
    CARCINOMA, PAPILLARY, 3684
    CARCINOMA, SCIRRHOUS, 3684
MAMMARY NEOPLASMS, EXPERIMENTAL
    ADENOCARCINOMA, 3684

CYCLOPHOSPHAMIDE
    BONE MARROW
        CHROMOSOME ABERRATIONS, 4024
    BURKITT'S LYMPHOMA
        CLINICOPATHOLOGIC STUDY, 4013
    CHROMOSOME ABERRATIONS
        BONE MARROW, 3717*
    LEUKEMIA, MYELOCYTIC
        IMMUNOSUPPRESSION, 3953*
    MEHTANESULFONIC ACID, METHYL ESTER
        DOMINANT LETHAL TEST, 3760*

CYSTADENOCARCINOMA
    GYNECOLOGICAL NEOPLASMS
        NEOPLASM METASTASIS, 4046
    OVARIAN NEOPLASMS
        PRECANCEROUS CONDITIONS, 4095*

CYSTADENOMA
    BILE DUCT NEOPLASMS
        BUTANOL, 4-(BUTYLNITROSOAMINO)-,
        3701
    OVARIAN NEOPLASMS
        PRECANCEROUS CONDITIONS, 4095*

CYSTITIS
    BLADDER NEOPLASMS
        CELL TRANSFORMATION, NEOPLASTIC,
        4091*

CYTARABINE
    SEE CYTOSINE, 1-BETA-D-
    ARABINOFURANOSYL-

CYTOCHALASIN B
    GANGLIOSIDES
        CELL MEMBRANE, 3924
    VIRUS, SV40
        RNA, MESSENGER, 3820

CYTOCHROME P-450
    BARBITURIC ACID, 5-ETHYL-5-PHENYL-
        DNA BINDING, 3742*
    BENZ(A)ANTHRACENE
        DNA BINDING, 3742*
    CHOLANTHRENE, 3-METHYL
        DNA BINDING, 3742*

CYTOPLASM
    DNA REPLICATION, 3634
    RNA, 3634
    RNA, MESSENGER
        PROTEINS, 3634

CYTOSINE, 1-BETA-D-ARABINOFURANOSYL-
    CELL LINE
        CELL TRANSFORMATION, NEOPLASTIC,
        3720*
    METHANESULFONIC ACID, METHYL ESTER
        DNA REPAIR, 3759*
    VIRUS, EPSTEIN-BARR
        ANTIGENS, VIRAL, 3846*

CYTOSOL
   ACETOHYDROXAMIC ACID, N-4-BIPHENYLYL-
     RNA, TRANSFER, 3675
   ACETOHYDROXAMIC ACID, N-FLUOREN-2-YL-
     RNA, TRANSFER, 3675
   MAMMARY NEOPLASMS, EXPERIMENTAL
     BENZ(A)ANTHRACENE, 7,12-DIMETHYL-,
     3740*

CYTOXAN
   SEE CYCLOPHOSPHAMIDE

DACTINOMYCIN
   SEE ACTINOMYCIN D

DAUNOMYCIN
   VIRUS, ROUS SARCOMA
     INFECTIVITY, 3879*

DEOXYIODOURIDINE
   SEE URIDINE, 2'-DEOXY-5-IODO-

DEOXYRIBONUCLEASE
   BRAIN NEOPLASMS
     HODGKIN'S DISEASE, 4017
   DIETHYLAMINE, N-NITROSO-
     NUCLEIC ACIDS, 3695
   MAMMARY NEOPLASMS, EXPERIMENTAL
     VIRUS, MURINE MAMMARY TUMOR, 3866*

DERMATAN SULFATE
   GALLBLADDER NEOPLASMS
     CARCINOMA, 4194*

DERMATOFIBROMA
   SEE FIBROMA

DERMOID CYST
   UTERINE NEOPLASMS
     HISTOLOGICAL STUDY, CASE REPORT,
     4047

DEXAMETHASONE
   ARYL HYDROCARBON HYDROXYLASES
     ENZYMATIC ACTIVITY, 3738*
   GLUCOSE, 2-DEOXY-
     DNA REPLICATION, 4182
   GROWTH SUBSTANCES
     FIBROBLASTS, 4186

DIABETES MELLITUS
   UREA, METHYL NITROSO-
     ORGANOTROPISM, 3708

DIBENZ(A,H)ACRIDINE
   STRUCTURE-ACTIVITY RELATIONSHIP
     MOLECULAR CONFIRMATION, 3682

5H-DIBENZ(B,F)AZEPNINE, 10,11-DIHYDRO-5-(
   (3-METHYLAMINO)PROPYL)
   NEUROBLASTOMA
     NOREPINEPHRINE, 4181

DIBUTYRYL CYCLIC AMP
   FIBROBLASTS
     ULTRASTRUCTURAL STUDY, 4155
   HEPATOMA
     FETAL GLOBULINS, 3906

     GROWTH, 3906
   NEUROBLASTOMA
     ADENOSINE TRIPHOSPHATASE, 4176

DICYCLOHEXYLAMINE, N-NITROSO-
   POLAROGRAPHY
     SPECTROSCOPY, 3762*

DIET
   NEOPLASMS
     EPIDEMIOLOGY, REVIEW, 4123
     STATISTICAL ANALYSIS, 4123
   NITRIC ACID, SODIUM SALT
     ORGANOTROPISM, CHICKEN EGGS, 3709

DIETARY FATS
   ACETAMIDE, N-FLUOREN-2-YL-
     COCARCINOGENIC EFFECT, 3606
   BENZ(A)ANTHRACENE, 7,12-DIMETHYL-
     COCARCINOGENIC EFFECT, 3606
   BENZO(A)PYRENE
     COCARCINOGENIC EFFECT, 3606
   CHOLANTHRENE, 3-METHYL-
     COCARCINOGENIC EFFECT, 3606
   CHOLESTEROL
     COCARCINOGENIC EFFECT, 3605
   COAL TAR
     COCARCINOGENIC EFFECT, 3606
   COLONIC NEOPLASMS
     CHOLESTEROL, 3605
   HEPATOMA
     M-TOLUIDINE, N,N-DIMETHYL-4-
     (PHENYLAZO)-, 3606
   MAMMARY NEOPLASMS, EXPERIMENTAL
     ADENOCARCINOMA, 3606
   4,4'-STILBENEDIOL, ALPHA,ALPHA'-
     DIETHYL-
     COCARCINOGENIC EFFECT, 3606
   ULTRAVIOLET RAYS
     COCARCINOGENIC EFFECT, 3606

DIETHYLAMINE, N-NITROSO-
   ADRENAL GLAND NEOPLASMS
     ADENOMA, 3696
     PHEOCHROMOCYTOMA, 3696
   BARBITURIC ACID, 5-ETHYL-5-PHENYL-
     COCARCINOGENIC ACTIVITY, HAMSTER,
     RAT, 3697
   BILE DUCT NEOPLASMS
     CHOLANGIOMA, 3696
   BRONCHIAL NEOPLASMS
     PRECANCEROUS CONDITIONS, 3764*
   DNA NUCLEOTIDYTRANSFERASES
     MORPHOLOGICAL EFFECT, LIVER, RAT,
     3703
   HEPATOMA
     FORMATION AND DEVELOPMENT, 3763*
   KIDNEY NEOPLASMS
     TRANSPLACENTAL CARCINOGENESIS,
     REVIEW, 3601
   NEOPLASMS, EXPERIMENTAL
     CARCINOGENIC ACTIVITY, HAMSTER,
     RAT, 3697
   NERVOUS SYSTEM NEOPLASMS
     TRANSPLACENTAL CARCINOGENESIS,
     REVIEW, 3601
   NUCLEIC ACIDS

DEOXYRIBONUCLEASE, 3695
RNA POLYMERASE
    MORPHOLOGICAL EFFECT, LIVER, RAT,
    3703
SALMONELLA TYPHIMURIUM
    MUTAGENIC ACTIVITY, 3697
STOMACH NEOPLASMS
    PAPILLOMA, 3696
TRACHEAL NEOPLASMS
    PAPILLOMA, 3696

DIETHYLNITROSAMINE
SEE DIETHYLAMINE, N-NITROSO-

DIETHYLSTILBESTROL
SEE 4,4'-STILBENEDIOL, ALPHA, ALPHA'-
    DIETHYL-

DIGESTIVE SYSTEM NEOPLASMS
CARCINOEMBRYONIC ANTIGEN
    IMMUNE SERUMS, 3969*
CARCINOMA
    CARCINOEMBRYONIC ANTIGEN, 3969*
EPIDEMIOLOGY
    MORMONS, 4118
NEOPLASMS, MULTIPLE PRIMARY, 3995

DIHYDROAFLATOXIN B1
SEE AFLATOXIN B2

DIMETHYL BENZANTHRACENE
SEE BENZ(A)ANTHRACENE, 7,12-DIMETHYL-

DIMETHYL SULFOXIDE
SEE METHANE, SULFINYLBIS-
    :

DIMETHYLAMINE, N-NITROSO-
BACTERIA
    ENZYMATIC ACTIVITY, 3700
BARBITURIC ACID, 5-ETHYL-5-PHENYL-
    COCARCINOGENIC ACTIVITY, HAMSTER,
    RAT, 3697
CELLS, CULTURED
    AZAGUANINE, 3699
    MUTAGENIC ACTIVITY, 3699
DNA
    ALKYLATION, CELL-CYCLE DEPENDENCY,
    3698
    BINDING, 3698
MAMMARY NEOPLASMS, EXPERIMENTAL
    MUTAGENIC ACTIVITY, 3699
MICROSOMES, LIVER
    MUTAGENIC ACTIVITY, 3699
NADP
    MUTAGENIC ACTIVITY, 3699
NEOPLASMS, EXPERIMENTAL
    CARCINOGENIC ACTIVITY, HAMSTER,
    RAT, 3697
POLAROGRAPHY
    SPECTROSCOPY, 3762*
SALMONELLA TYPHIMURIUM
    MUTAGENIC ACTIVITY, 3697

DIMETHYLAMINOAZOBENZENE
SEE ANILINE, N,N-DIMETHYL-P-PHENYLAZO-

1,2-DIMETHYLHYDRAZINE   .

SEE HYDRAZINE, 1,2-DIMETHYL-

DIMETHYLNITROSAMINE
SEE DIMETHYLAMINE, N-NITROSO-

DIPHENYLAMINE, N-NITROSO-
BACTERIA
    ENZYMATIC ACTIVITY, 3700

DIPROPYLAMINE, N-NITROSO-
ESOPHAGEAL NEOPLASMS
    PAPILLOMA, 3702
KIDNEY NEOPLASMS
    ADENOMA, 3702
LIVER NEOPLASMS
    HEMANGIOENDOTHELIOMA, 3702
LUNG NEOPLASMS
    ADENOCARCINOMA, 3702
    ADENOMA, 3702
    CARCINOMA, EPIDERMOID, 3702
NOSE NEOPLASMS
    CARCINOMA, 3702

DIPTHERIA ANTITOXIN
BURKITT'S LYMPHOMA
    IMMUNITY, CELLULAR, 3946*

DISGERMINOMA
FETAL GLOBULINS
    VITELLINE MEMBRANE, 3907
OVARIAN NEOPLASMS
    CASE REPORT, 4031
    SEX CHROMOSOMES, 4031
    TESTOSTERONE, 4031

DNA
BARBITURIC ACID, 5-ETHYL-5-PHENYL-
    PYRIDINE, 2-(BENZYL(2-
    (DIMETHYLAMINO)ETHYL)AMINO)-,
    3693
    PYRIDINE, 2-(BENZYL(2-(METHYL)
    (NITROSO)AMINO)ETHYL)AMINO)-,
    3693
CHOLANTHRENE, 3-METHYL-
    PYRIDINE, 2-(BENZYL(2-(METHYL)
    (NITROSO)AMINO)ETHYL)AMINO)-,
    3693
CHROMOSOMES
    HYBRIDIZATION, REVIEW, 3631
DIMETHYLAMINE, N-NITROSO-
    ALKYLATION, CELL-CYCLE DEPENDENCY,
    3698
    BINDING, 3698
ERYTHROLEUKEMIA
    ERYTHROPOIESIS, 4132
ESTRADIOL
    THYMIDINE INCORPORATION, 3680
ESTROGENS
    CELL DIFFERENTIATION, 3607
FIBROSARCOMA
    SIALIC ACIDS, 4166
GROWTH SUBSTANCES
    FIBROBLASTS, 4186
HEPATOMA
    SIALIC ACIDS, 4166
LEUKEMIA, LYMPHOBLASTIC
    RADIATION, IONIZING, 4170
LUNG NEOPLASMS
    SIALIC ACIDS, 4166

MAMMARY NEOPLASMS, EXPERIMENTAL
    RADIATION INJURIES, EXPERIMENTAL,
      4113
    SIALIC ACIDS, 4166
MICROSOMES, LIVER
    PYRIDINE, 2-(BENZYL(2-
      (DIMETHYLAMINO)ETHYL)AMINO)-,
      3693
    PYRIDINE, 2-(BENZYL(2-(METHYL)
      (NITROSO)AMINO)ETHYL)AMINO)-,
      3693
PROGESTERONE
    THYMIDINE INCORPORATION, 3680
RADIATION, IONIZING
    ALKALINE PHOSPHATASE, 3776
    DNA POLYMERASE, 3776
    ENDONUCLEASES, 3776
    RADIATION DOSAGE, 4113
SODIUM AZIDE
    BASE-SUBSTITUTION MUTAGENESIS,
      3729*
TESTOSTERONE
    THYMIDINE INCORPORATION, 3680
VIRUS, FROG 3
    SYNTHESIS INHIBITION, HOST CELL,
      3851*
VIRUS, SV40
    STRAND SEPARATION, TEMPERATURE-
      DEPENDENT, 3821

DNA, CIRCULAR
    VIRUS, SV40
      CELL FUSION, 3823

DNA, NEOPLASM
    BLADDER NEOPLASMS
      CARCINOMA, TRANSITIONAL CELL, 3989
    CERVIX NEOPLASMS
      CARCINOMA IN SITU, 3996
      PRECANCEROUS CONDITIONS, 3996
    UTERINE NEOPLASMS
      ANGIOMATOSIS, 4100*
      MESENCHYMOMA, 4100*
      SARCOMA, 4100*

DNA NUCLEOTIDYLTRANSFERASES
    SEE ALSO DNA POLYMERASE
    DIETHYLAMINE, N-NITROSO-
      MORPHOLOGICAL EFFECT, LIVER, RAT,
      3703
    MORPHOLINE, N-NITROSO-
      MORPHOLOGICAL EFFECT, LIVER, RAT,
      3703
    VIRUS, AVIAN MYELOBLASTOSIS
      BASE-PAIRING ERRORS, 3839*
      POLYNUCLEOTIDES, 3839*
    VIRUS, FELINE LEUKEMIA
      ISOLATION AND CHARACTERIZATION,
      3836
    VIRUS, KILHAM
      ISOLATION AND CHARACTERIZATION,
      3855*
    VIRUS, RAUSCHER MURINE LEUKEMIA
      ISOLATION AND CHARACTERIZATION,
      3836
    VIRUS, RD114
      ISOLATION AND CHARACTERIZATION,
      3836

DNA POLYMERASE
    SEE ALSO DNA NUCLEOTIDYLTRANSFERASES
    RADIATION, IONIZING
      DNA, 3776

DNA REPAIR
    ACETAMIDE, N-(ACETYLOXY)-N-9H-FLUOREN-
      2-YL-
      FIBROBLASTS, 3721*
    1,8,9-ANTHRACENETRIOL
      FIBROBLASTS, 3721*
    ARYL HYDROCARBON HYDROXYLASES
      SMOKING, 3710
    BURKITT'S LYMPHOMA
      BENZO(A)PYRENE, 3778*
      METHANESULFONIC ACID, METHYL ESTER,
      3778*
      ULTRAVIOLET RAYS, 3778*
    CROTON OIL
      FIBROBLASTS, 3721*
    CYTOSINE, 1-BETA-D-ARABINOFURANOSYL-
      METHANESULFONIC ACID, METHYL ESTER,
      3759*
    FIBROBLASTS
      ACETAMIDE, N-FLUOREN-2-YL-, 3674
      ACETAMIDE, N-2-PHENANTHRYL-, 3674
      ACETANILIDE, 4'-PHENYL-, 3674
      ACETANILIDE, 4'-STYRYL-, 3674
      SMOKING, 3710
      ULTRAVIOLET RAYS, 3674
    SMOKING
      DNA REPLICATION, 3769*
    12-O-TETRADECANOYLPHORBOL-13-ACETATE
      FIBROBLASTS, 3721*
    VIRUS, ADENO 2
      NITROUS ACID, SODIUM SALT, 3789
    XERODERMA PIGMENTOSUM
      ACETAMIDE, N-FLUOREN-2-YL-, 3674
      ACETAMIDE, N-2-PHENANTHRYL-, 3674
      ACETANILIDE, 4'-PHENYL-, 3674
      ACETANILIDE, 4'-STYRYL-, 3674
      ULTRAVIOLET RAYS, 3674

DNA REPLICATION
    ARYL HYDROCARBON HYDROXYLASES
      SMOKING, 3710
    ASPARTATE CARBAMOYLTRANSFERASE
      EUCARYOTIC CELL CYCLE, 4199*
    BENZ(A)ANTHRACENE, 7,12-DIMETHYL-
      THYMIDINE, 3746*
    CARBOXYLYASES
      FIBROBLASTS, DIPLOID, 4162
    CARCINOMA, EHRLICH TUMOR
      IMMUNITY, CELLULAR, 3901
    CARCINOMA 256, WALKER
      IMMUNITY, CELLULAR, 3901
    CELL CYCLE KINETICS, 4171
    CELL NUCLEUS, 3634
    COLONIC NEOPLASMS
      PRECANCEROUS CONDITIONS, 4072*
    CORTISOL
      GLUCOSE, 2-DEOXY-, 4182
    CORTISONE
      GLUCOSE, 2-DEOXY-, 4182
    CYTOPLASM, 3634
    DEXAMETHASONE
      GLUCOSE, 2-DEOXY-, 4182

FETAL GLOBULINS
  PLANT AGGLUTININS, 3954*
FIBROBLASTS
  GROWTH SUBSTANCES, 4165
  SMOKING, 3710
HEPATOMA
  IMMUNITY, CELLULAR, 3901
HODGKIN'S DISEASE
  T-LYMPHOCYTES, 3921
INFECTIOUS MONONUCLEOSIS
  T-LYMPHOCYTES, 3921
LYMPHOCYTES
  LYMPHOKINES, 3922
T-LYMPHOCYTES
  CONCANAVALIN A, 3927
  MONOCYTES, 3927
NICOTINE
  LYMPHOCYTE TRANSFORMATION, 3712
PUTRESCINE
  FIBROBLASTS, DIPLOID, 4162
SMOKING
  DNA REPAIR, 3769*
SPERMIDINE
  FIBROBLASTS, DIPLOID, 4162
SPERMINE
  FIBROBLASTS, DIPLOID, 4162
VIRUS, BOVINE ADENO 3
  THYMIDINE INCORPORATION, 3791
VIRUS, HEPRES SIMPLEX 2
  MAGNESIUM, 3802
VIRUS, HERPES SIMPLEX
  THYMIDINE, 3806
VIRUS, HERPES SIMPLEX 1
  ULTRAVIOLET RAYS, 3804
VIRUS, ROUS SARCOMA
  CORTISOL, 4182
  REVERSE TRANSCRIPTASE, 3880*
VIRUS, SV40
  ALPHA COMPONENT, 3824
  CONTACT INHIBITION, 3824
  EVOLUTIONARY VARIANTS, 3826
  INITIATION SITES, CLONED VARIANTS,
    3825
  ULTRAVIOLET RAYS, 3804

DNA, VIRAL
  BURKITT'S LYMPHOMA
    VIRUS, EPSTEIN-BARR, 3845*
  LEUKEMIA
    REVIEW, 3614
  NEOPLASMS
    HYBRIDIZATION, REVIEW, 3631
  VIRUS, ADENO 2
    ISOLATION AND CHARACTERIZATION,
      3790
  VIRUS, BOVINE ADENO 3
    ISOLATION AND CHARACTERIZATION,
      3791
  VIRUS, CELO
    RNA, 3827
  VIRUS, HERPES SIMPLEX 1
    ENDONUCLEASES, 3805
  VIRUS, HERPES SIMPLEX 2
    ENDONUCLEASES, 3805
    TEMPERATURE-SENSITIVE MUTANTS,
      3853*
  VIRUS, MURINE MAMMARY TUMOR
    REVERSE TRANSCRIPTASE, 3869*
  VIRUS, MURINE SARCOMA-LEUKEMIA

VIRUS REPLICATION, 3864*
VIRUS, SV40
  TEMPERATURE SENSITIVE MUTANTS,
    3822

DUODENAL NEOPLASMS
  GUANIDINE, 1-METHYL-3-NITRO-1-NITROSO-
    CARCINOEMBRYONIC ANTIGEN, 3909

EHRLICH ASCITES CARCINOMA
  SEE CARCINOMA, EHRLICH TUMOR

EHRLICH ASCITES TUMOR (CELL)
  SEE CARCINOMA, EHRLICH TUMOR

EHTYLENE, CHLORO-
  ENVIRONMENTAL HAZARD
    CONTROL AND MANAGEMENT, 3664*

ENDOMETRIOSIS
  CERVIX NEOPLASMS
    ADENOCARCINOMA, 3690
  UTERINE NEOPLASMS
    ULTRASTRUCTURAL STUDY, RABBIT,
      4099*

ENDONUCLEASES
  RADIATION, IONIZING
    DNA, 3776
  VIRUS, HERPES SIMPLEX 1
    DNA, VIRAL, 3805
  VIRUS, HERPES SIMPLEX 2
    DNA, VIRAL, 3805
  VIRUS, SV40
    RNA, MESSENGER, 3819
    RNA, RIBOSOMAL, 3819

ENDOPLASMIC RETICULUM
  BRAIN NEOPLASMS
    ADENOMA, CHROMOPHOBE, 4005
    ANGIOSARCOMA, 4005
    EPENDYMOMA, 4005
    GLIOBLASTOMA MULTIFORME, 4005
    MEDULLOBLASTOMA, 4005
    MENINGIOMA, 4005
    NEUROEPITHELIOMA, 4005
    OLIGODENDROGLIOMA, 4005
    SARCOMA, 4005

ENDOTHELIUM
  CELL DIVISION
    LIVER REGENERATION, 4159

ENTERITIS, REGIONAL
  COLONIC NEOPLASMS
    CASE REPORT, 4041

ENTEROTOXINS
  HELA CELLS
    ADENOSINE CYCLIC 3',5'
      MONOPHOSPHATE, 4193*

ENVIRONMENTAL HAZARD
  BENZIDINE
    REVIEW, 3642*
  CARCINOGEN, CHEMICAL
    REVIEW, 3642*
  EHTYLENE, CHLORO-
    CONTROL AND MANAGEMENT, 3664*

ETHYLENE, CHLORO-
      HYGIENIC ASPECTS, REVIEW, 3665*
ETHYLENE, CHLORO- POLYMER
      HYGIENIC ASPECTS, REVIEW, 3665*
RADIATION
      PLUTONIUM, 3645*

EOSINOPHILIA
   HODGKIN'S DISEASE
      PRECANCEROUS CONDITIONS, 4083*

EPENDYMOMA
   BRAIN NEOPLASMS
      ENDOPLASMIC RETICULUM, 4005

ERYTHROCYTES
   CELL MEMBRANE
      REVIEW, 3667*
   VIRUS, FRIEND POLYCYTHEMIA
      CELL DIFFERENTIATION, 3849*
   VIRUS, HERPES SIMPLEX 1
      ULTRAVIOLET RAYS, 3804
   VIRUS, SV40
      ULTRAVIOLET RAYS, 3804

ERYTHROLEUKEMIA
   ERYTHROPOIESIS
      CHILD, 4020
      DNA, 4132
   SPLENIC NEOPLASMS
      VIRUS, FRIEND MURINE LEUKEMIA,
         3847*
   VIRUS, FRIEND MURINE LEUKEMIA
      CELL LINE, 3798

ERYTHROPOIESIS
   ERYTHROLEUKEMIA
      CHILD, 4020
      DNA, 4132
   LEUKEMIA, LYMPHOBLASTIC
      CHILD, 4020
   LEUKEMIA, LYMPHOCYTIC
      CHILD, 4020
   LEUKEMIA, MYELOCYTIC
      CHILD, 4020

ESCHERICHIA COLI
   CARCINOMA
      CHEMOTAXIS, 3886
   HEPATOMA
      CHEMOTAXIS, 3886

ESOPHAGEAL NEOPLASMS
   ADENOCARCINOMA
      CASE REPORT, 4065*
   AGE FACTORS
      EPIDEMIOLOGY, ISRAEL, 4115
   CARCINOGEN, CHEMICAL
      ANIMAL MODEL, RAT, MOUSE, 3641*
   CARCINOMA, EPIDERMOID
      PIPERIDINE, 3,4-DIBROMO-1-NITROSO-,
         3707
      PIPERIDINE, 3,4-DICHLORO-1-NITROSO-,
         3707
   EPIDEMIOLOGY
      AFRICA, 4116
      MORMONS, 4118
   LEIOMYOMA
      EPIDEMIOLOGY, 4000

SEE CARBAMIC ACID, ETHYL ESTER

ETHYL METHANE SULFONATE
  SEE METHANESULFONIC ACID, ETHYL ESTER

N-ETHYL-N-NITROSOUREA
  SEE UREA, ETHYL NITROSO-

1-ETHYL-1-NITROSOUREA
  SEE UREA, ETHYL NITROSO-

ETHYLENE, CHLORO-
  ENVIRONMENTAL HAZARD
    HYGIENIC ASPECTS, REVIEW, 3665*
  HEPATOMA
    OCCUPATIONAL HAZARD, 3663*
  LIVER NEOPLASMS
    ANGIOSARCOMA, 4036, 4124
    EPIDEMIOLOGY, 4124
  LUNG NEOPLASMS
    ANGIOSARCOMA, 3627
    MESOTHELIOMA, 3627
  MICROSOMES, LIVER
    CARCINOGENIC POTENTIAL, 3686
  OCCUPATIONAL HAZARD
    REVIEW, 3663*
  SPLEEN
    ANGIOSARCOMA, 4036
  TAGGED
    SYNTHESIS, 3716

ETHYLENE, CHLORO- POLYMER
  ANGIOSARCOMA
    OCCUPATIONAL HAZARD, 3663*
  ENVIRONMENTAL HAZARD
    HYGIENIC ASPECTS, REVIEW, 3665*
  HEPATOMA
    OCCUPATIONAL HAZARD, 3663*
  LIVER NEOPLASMS
    ANGIOSARCOMA, 4036
  OCCUPATIONAL HAZARD
    REVIEW, 3663*
  SPLEEN
    ANGIOSARCOMA, 4036

ETHYLENE, 1,1-DICHLORO-
  MICROSOMES, LIVER
    CARCINOGENIC POTENTIAL, 3686

ETHYLENE, 1,2-DICHLORO-
  MICROSOMES, LIVER
    CARCINOGENIC POTENTIAL, 3686

ETHYLENE, TETRACHLORO-
  MICROSOMES, LIVER
    CARCINOGENIC POTENTIAL, 3686

ETHYLENE, TRICHLORO-
  MICROSOMES, LIVER
    CARCINOGENIC POTENTIAL, 3686

ETHYLNITROSOUREA
  SEE UREA, ETHYL NITROSO-

EXCRETION
  CARCINOMA
    ANTIMONY, 3785*
  FIBROSARCOMA
    ANTIMONY, 3785*

EYE NEOPLASMS
  CARBAMIC ACID, ETHYL ESTER
    IMMUNOSUPPRESSION, 3893
  RETINOBLASTOMA
    EPIDEMIOLOGY, CHILD, INDIA, 4120
  SARCOMA, RETICULUM CELL
    CASE REPORT, 4018
    UVEITIS, 4018

FATTY ACIDS
  VIRUS, KIRSTEN MURINE SARCOMA
    CELL TRANSFORMATION, NEOPLASTIC,
      3877*

FERRITIN
  HELA CELLS
    ISOLATION AND CHARACTERIZATION,
      3930
  HEPATOMA
    ISOLATION AND CHARACTERIZATION,
      3930

FETAL GLOBULINS
  CARCINOEMBRYONIC ANTIGEN
    TUMOR CELL SYNTHESIS, 3970*
  DISGERMINOMA
    VITELLINE MEMBRANE, 3907
  HELA CELLS
    ISOLATION AND CHARACTERIZATION,
      3930
  HEPATOMA
    ACETAMIDE, N-FLUOREN-2-YL-, 4150
    ALKALINE PHOSPHATASE, 4175
    BENZIDINE, 3,3'-DIMETHYL-, 3731*
    DIBUTYRYL CYCLIC AMP, 3906
    ISOLATION AND CHARACTERIZATION,
      3930
    POLYRIBOSOMES, 3906
    STEROIDS, 3636*
  LYMPHOSARCOMA
    DIAGNOSOS AND PROGNOSIS, 4019
  PLANT AGGLUTININS
    DNA REPLICATION, 3954*
    IMMUNOSUPPRESSION, 3954*
  SARCOMA, RETICULUM CELL
    DIAGNOSOS AND PROGNOSIS, 4019
  TERATOID TUMOR
    CASE REPORT, FETUS, 4090*
    VITELLINE MEMBRANE, 3907
  UTERINE NEOPLASMS
    VIRUS, MURINE SARCOMA, 3865*

ALPHA-FETOPROTEIN
  SEE FETAL GLOBULINS

FIBRINOGEN
  LYMPHOSARCOMA
    DIAGNOSOS AND PROGNOSIS, 4019
  SARCOMA, RETICULUM CELL
    DIAGNOSOS AND PROGNOSIS, 4019

FIBROBLASTS
  ACETAMIDE, N-(ACETYLOXY)-N-9H-FLUOREN-
    2-YL-
    DNA REPAIR, 3721*
  ACRYLIC ACID, 3-CYANO-, ISOBUTYL ESTER
    PRECANCEROUS CONDITIONS, 3673
  1,8,9-ANTHRACENETRIOL
    DNA REPAIR, 3721*

BENZ(A)ANTHRACENE-8,9-DICL, 8,9-
DIHYDRO-7,12-DIMETHYL-
CELL TRANSFORMATION, NEOPLASTIC,
3748*
BENZ(A)ANTHRACENE-8,9-DICL, 8,9-
DIHYDRO-7-METHYL-
CELL TRANSFORMATION, NEOPLASTIC,
3748*
CROTON OIL
DNA REPAIR, 3721*
DIBUTYRYL CYCLIC AMP
ULTRASTRUCTURAL STUDY, 4155
DNA REPAIR
ACETAMIDE, N-FLUOREN-2-YL-, 3674
ACETAMIDE, N-2-PHENANTHRYL-, 3674
ACETANILIDE, 4'-PHENYL-, 3674
ACETANILIDE, 4'-STYRYL-, 3674
SMOKING, 3710
ULTRAVIOLET RAYS, 3674
DNA REPLICATION
GROWTH SUBSTANCES, 4165
SMOKING, 3710
GROWTH
ULTRASTRUCTURAL STUDY, 4108*
GROWTH SUBSTANCES
CORTISOL, 4186
DEXAMETHASONE, 4186
DNA, 4186
ISOLATION AND CHARACTERIZATION,
4186, 4187
REVIEW, 4186
POLY A
GROWTH SUBSTANCES, 4165
RNA REPLICATION
GROWTH SUBSTANCES, 4165
12-O-TETRADECANOYLPHORBOL-13-ACETATE
DNA REPAIR, 3721*

FIBROMA
BRAIN NEOPLASMS
ULTRASTRUCTURAL STUDY, 3994

FIBROSARCOMA
ANTIMONY
DISTRIBUTION STUDY, 3785*
EXCRETION, 3785*
CHOLANTHRENE, 3-METHYL-
IMMUNITY, CELLULAR, 3945*
TRANSPLANTATION IMMUNOLOGY, 3945*
EPIDEMIOLOGY
CHILD INDIA, 4120
IMMUNITY, CELLULAR
ANTIGENS, NEOPLASM, 3943*
HISTOCOMPATIBILITY ANTIGENS, 3943*
MYCOPLASMA
CARCINOGENIC POTENTIAL, MOUSE,
4039
NEOPLASMS, CONNECTIVE TISSUE
VIRUS, CELO, 3794
SIALIC ACIDS
DNA, 4166
NEOPLASM TRANSPLANTATION, 4166
TRANSPLANTATION IMMUNOLOGY
HYBRIDIZATION, 4158
IMMUNOGENETICS, 4158
VIRUS, BOVINE LEUKOSIS
ANTIGENS, VIRAL, 3830
VIRUS, CELO
INTERCELLULAR ATTACHMENTS, 3794

ULTRASTRUCTURAL STUDY, PLASMA
MEMBRANE, 3794
VIRUS, RAUSCHER MURINE LEUKEMIA
ANTIGENS, VIRAL, 3830
VIRUS, MURINE MAMMARY TUMOR, 3830
VIRUS, SIMIAN LEUKEMIA
ANTIGENS, VIRAL, 3830

FLUOCINOLONE ACETONIDE
SKIN NEOPLASMS
ANTINEOPLASTIC AGENTS, 3738*

FLUORANTHENE
VIRUS, FRIEND MURINE LEUKEMIA
SPLENOMEGALY, 3797

FLUORENYL ACETAMIDE
SEE ACETAMIDE, N-FLUOREN-2-YL-

FOLIC ACID
METHOTREXATE
FIBROBLASTS, CHICKEN, 4154

FOLLICLE-STIMULATING HORMONE
SEE FSH

FOOD ANALYSIS
AFLATOXIN B1
FOOD CONTAMINATION, 4137*

FOOD CONTAMINATION
AFLATOXIN B1
CORN MEAL, 4138*
EXTRACTION, CORN, 4136*
FOOD ANALYSIS, 4137*
AFLATOXIN B2
CORN MEAL, 4138*
ASPERGILLUS FLAVUS
MYCOTOXINS, 4125
ASPERGILLUS PARASITICUS
MYCOTOXINS, 4125
BLADDER NEOPLASMS
EPIDEMIOLOGY, SPAIN, REVIEW, 3630
HEPATOMA
REVIEW, 3603
MYCOTOXINS
CORN MEAL, 4138*

FORMAMIDE, N-(4-(5-NITRO-2-FURYL)-2-
THIAZOLYL)-
BLADDER NEOPLASMS
CARCINOMA, 3736*

FURYLFURAMIDE
SEE ACRYLAMIDE, 2-(2-FURYL)-3-(5-NITRO-
2-FURYL)-

GALLBLADDER NEOPLASMS
CARCINOGEN, CHEMICAL
EPIDEMIOLOGY, 4124
CARCINOMA
CHONDROITIN, 4194*
DERMATAN SULFATE, 4194*
HEPARITIN SULFATE, 4194*

GANGLIOSIDES
COLCHICINE
CELL MEMBRANE, 3924
CYTOCHALASIN B

CELL MEMBRANE, 3924
LEUKEMIA, LYMPHOCYTIC
    CELL MEMBRANE, 3924
LIGANDS
    CELL MEMBRANE, 3924
NEURAMINIDASE
    CELL MEMBRANE, 3924
TRYPSIN
    CELL MEMBRANE, 3924

GASTRIC NEOPLASMS
PAPILLOMA
    BUTANOL, 4-(BUTYLNITROSOAMINO)-,
    3701

GASTRIN
STOMACH NEOPLASMS
    ENZYMATIC ACTIVITY, 4184
    TISSUE CULTURE, 4184

GASTROINTESTINAL NEOPLASMS
ADENOCARCINOMA
    CARCINOEMBRYONIC ANTIGEN, 3968*
CARCINOEMBRYONIC ANTIGEN
    ISOMERIC VARIANT, 3967*
EPIDEMIOLOGY
    AGE FACTORS, GREAT BRITAIN, 4114
NEOPLASM METASTASIS, 4045
TRANSPLANTATION, HETEROLOGOUS
    NEOPLASM METASTASIS, 4035

GENETICS
ADENOCARCINOMA
    INTESTINAL POLYPS, 4071*
AGRANULOCYTOSIS
    CASE REPORT, 4032
ANTIGENS
    HYBRIDIZATION, 3971*
BRAIN NEOPLASMS
    GLIOBLASTOMA MULTIFORME, 4023
BREAST NEOPLASMS
    ADENOCARCINOMA, 4023
    EPIDEMIOLOGY, 4141*
    EPIDEMIOLOGY, REVIEW, 4117
CHROMOSOME ABERRATIONS
    REVIEW, 3661*
HEPATOMEGALY
    CASE REPORT, 4032
HISTOCOMPATIBILITY ANTIGENS
    REVIEW, 3654*
HODGKIN'S DISEASE, 4087*
    REVIEW, 4087*
LEUKEMIA, LYMPHOBLASTIC
    IMMUNE RESPONSE, 3929
LEUKEMIA, MYELOBLASTIC
    ANTIGENS, NEOPLASM, 4030
LEUKEMIA, MYELOCYTIC, 4087*
LEUKEMIA, SUBLEUKEMIC
    ANTIGENS, NEOPLASM, 4030
MELANOMA, 4087*
MUTATION
    REVIEW, 3661*
NEOPLASMS
    EPIDEMIOLOGY, 4143*
NEPHROBLASTOMA
    CASE REPORT, 4029
NEUROFIBROMATOSIS
    MUTATION, 4027
RETICULOSIS, HEMOPHAGOCYTIC

CASE REPORT, 4033
IMMUNE RESPONSE, 4033
PREDNISOLONE, 4033
SARCOMA, RETICULUM CELL, 4087*
    REVIEW, 3650*
SPLENOMEGALY
    CASE REPORT, 4032
TESTICULAR NEOPLASMS
    CASE REPORT, 4105*
THROMBOPENIA
    CASE REPORT, 4032
UROGENITAL NEOPLASMS
    EPIDEMIOLOGY, REVIEW, 4117
VIRUS, MURINE LEUKEMIA
    ISOLATION AND CHARACTERIZATION,
    3859*

GERM-FREE LIFE
LEUKEMIA
    IMIDAZOLE-4-CARBOXAMIDE, 5-(3,3-
    DIMETHYL-1-TRIAZENO)-, 3758*
RHABDOMYOSARCOMA
    IMIDAZOLE-4-CARBOXAMIDE, 5-(3,3-
    DIMETHYL-1-TRIAZENO)-, 3758*

GIANT CELL TUMORS
IMMUNOGLOBULINS
    IMMUNE SERUMS, 3925

VIRUS, RAUSCHER MURINE LEUKEMIA
    ANTIGENS, VIRAL, 3830
VIRUS, SIMIAN LEUKEMIA
    ANTIGENS, VIRAL, 3830

GLIOBLASTOMA MULTIFORME
BRAIN NEOPLASMS
    ENDOPLASMIC RETICULUM, 4005
    GENETICS, 4023
    GOLGI APPARATUS, 4005
    NEOPLASMS, MULTIPLE PRIMARY, 3995

GLIOMA
BRAIN NEOPLASMS
    CASE REPORT, 4053*
    EPIDEMIOLOGY, CHILD, INDIA, 4120
CELLULAR INCLUSIONS
    ULTRASTRUCTURAL STUDY, 4006

GLOMANGIOMA
STOMACH NEOPLASMS
    CASE REPORT, 4034
    CASE REPORT, REVIEW, 4066*
    ENDOCRINE-TYPE GRANULES, 4034
    ULTRASTRUCTURAL STUDY, 4034, 4066*

GLUCOCORTICOIDS
HODGKIN'S DISEASE
    METABOLISM, 4085*

GLUCOSAMINE
VIRUS, AVIAN SARCOMA
    GLYCOPROTEINS, 3818
    VIRAL PROTEINS, 3818
    VIRUS REPLICATION, 3818
VIRUS, B77
    GLYCOPROTEINS, 3818
    VIRAL PROTEINS, 3818
    VIRUS REPLICATION, 3818

GLUCOSE, 2-DEOXY-
    CORTISOL
        DNA REPLICATION, 4182
    CORTISONE
        DNA REPLICATION, 4182
    DEXAMETHASONE
        DNA REPLICATION, 4182

GLUCOSEPHOSPHATASE
    BARBITURIC ACID, 5-ETHYL-5-PHENYL-
        ENZYMATIC ACTIVITY, 3766*

GLUCOSEPHOSPHATE
    CERVIX NEOPLASMS
        PRECANCEROUS CONDITIONS, 4198*

GLUCOSEPHOSPHATE DEHYDROGENASE
    MAMMARY NEOPLASMS, EXPERIMENTAL
        ESTRADIOL, 4180

GLUCURONIDASE
    CHOLANTHRENE, 3-METHYL-
        CARCINOGENIC METABOLITE, 3683
        TRANSPLACENTAL CARCINOGENESIS,
        3683

GLYCERALDEHYDEPHOSPHATE DEHYDROGENASE
    BENZ(A)ANTHRACENE
        ENZYME REPRESSION, LIGHT REACTION,
        3744*
    BENZ(A)ANTHRACENE, 7,12-DIMETHYL-
        ENZYME REPRESSION, LIGHT REACTION,
        3744*
    BENZ(A)ANTHRACENE, 7-(METHOXYMETHYL)-
    12-METHYL-
        ENZYME REPRESSION, LIGHT REACTION,
        3744*

GLYCEROLPHOSPHATE DEHYDROGENASE
    LEUKEMIA
        ISOLATION AND CHARACTERIZATION,
        4200*

GLYCOGEN
    LEUKEMIA, MYELOBLASTIC
        IMMUNITY, CELLULAR, 3900

GLYCOPEPTIDES
    HEPATOMA
        CELL MEMBRANE, 4173
        HISTOCOMPATIBILITY ANTIGENS, 4173

GLYCOPROTEINS
    CARCINOEMBRYONIC ANTIGEN
        BLOOD GROUPS, 3910
    CELL MEMBRANE
        REVIEW, 3667*
    SARCOMA 180, CROCKER
        ASCITES FORM, 4168
        CELL MEMBRANE, 4168
        CELLS, CULTURED, 4168
    VIRUS, AVIAN SARCOMA
        GLUCOSAMINE, 3818
    VIRUS, B77
        GLUCOSAMINE, 3818
    VIRUS, MURINE LEUKEMIA
        ANTIGENS, VIRAL, 3615
    VIRUS, MURINE MAMMARY TUMOR
        ISOLATION AND CHARACTERIZATION,

        3829
    VIRUS, RNA TUMOR
        REVIEW, 3614

GOLGI APPARATUS
    BRAIN NEOPLASMS
        ADENOMA, CHROMOPHOBE, 4005
        ANGIOSARCOMA, 4005
        GLIOBLASTOMA MULTIFORME, 4005
        MEDULLOBLASTOMA, 4005
        MENINGIOMA, 4005
        NEUROEPITHELIOMA, 4005
        OLIGODENDROGLIOMA, 4005
        SARCOMA, 4005

GONADOTROPINS
    BREAST NEOPLASMS
        PRECANCEROUS CONDITIONS, MALE,
        3992

GRAFT REJECTION
    SARCOMA
        IMMUNE SERUMS, 3902
        IMMUNITY, PASSIVE, 3902

GRAFT VS HOST REACTION
    LEUKEMIA
        TRANSPLANTATION, HOMOLOGOUS, 3940*
    LYMPHOMA
        GRAFT VS HOST REACTION, 3878*
        HISTOCOMPATIBILITY TESTING, 3888
        VIRUS, C-TYPE RNA TUMOR, 3878*
    SARCOMA
        TRANSPLANTATION, HOMOLOGOUS, 3940*

GROWTH
    SEE ALSO CELL DIVISION/CELL
        DIFFERENTIATION
    BREAST NEOPLASMS
        CELL CYCLE KINETICS, 4147*
    FIBROBLASTS
        ULTRASTRUCTURAL STUDY, 4108*
    HEPATOMA
        DIBUTYRYL CYCLIC AMP, 3906
        THEOPHYLLINE, 3906
    LYMPHOMA
        MACROPHAGES, 3949*
    MAMMARY NEOPLASMS, EXPERIMENTAL
        CELL CYCLE KINETICS, 4139*
        NEOPLASM TRANSPLANTATION, 4149
    VIRUS, ADENO 7
        TEMPERATURE DEPENDENCE, 3884*
    VIRUS, SV40
        TEMPERATURE DEPENDENCE, 3884*

GROWTH SUBSTANCES
    BREAST NEOPLASMS
        COLCHICINE, 3689
        INSULIN, 3689
        PROLACTIN, 3689
        THYMIDINE, 3689
    DNA REPLICATION
        FIBROBLASTS, 4165
    FIBROBLASTS
        CORTISOL, 4186
        DEXAMETHASONE, 4186
        DNA, 4186
        ISOLATION AND CHARACTERIZATION,
        4186, 4187

REVIEW, 4186
HEAD-AND NECK NEOPLASMS
LYMPHOMA, 4016
POLY A
FIBROBLASTS, 4165
RNA REPLICATION
FIBROBLASTS, 4165
VIRUS, KIRSTEN MURINE SARCOMA
CELL TRANSFORMATION, NEOPLASTIC,
3877*
VIRUS, POLYOMA
TEMPERATURE-SENSITIVE MUTANTS,
4186

GUANIDINE, 1-METHYL-3-NITRO-1-NITROSO-
DUODENAL NEOPLASMS
CARCINOEMBRYONIC ANTIGEN, 3909
SARCOMA, OSTEOGENIC
CELL TRANSFORMATION, NEOPLASTIC,
3705
STOMACH NEOPLASMS
ADENOCARCINOMA, 3635*, 3706
IMMUNE RESPONSE, 3939*

GYNECOLOGIC NEOPLASMS
ADENOCARCINOMA
NEOPLASM METASTASIS, 4046
AGE FACTORS
EPIDEMIOLOGY, ISRAEL, 4115
CYSTADENOCARCINOMA
NEOPLASM METASTASIS, 4046
EPIDEMIOLOGY
AGE FACTORS, GREAT BRITAIN, 4114
MORMONS, 4118
NEOPLASMS, MULTIPLE PRIMARY, 3995
SULFATASES
HISTOCHEMICAL STUDY, 4179

GYNECOMASTIA
BREAST NEOPLASMS
PRECANCEROUS CONDITIONS, MALE,
3992

HAPTOGLOBINS
LYMPHOSARCOMA
DIAGNOSIS AND PROGNOSIS, 4019
SARCOMA, RETICULUM CELL
DIAGNOSIS AND PROGNOSIS, 4019

HEAD AND NECK NEOPLASMS
HEMANGIOMA
CASE REPORT, 4089*
ULTRASTRUCTURAL STUDY, 4089*
LYMPHOMA
GROWTH SUBSTANCES, 4016
TRANSPLANTATION, HETEROLOGOUS,
4016

HEAVY CHAIN DISEASE
ABDOMINAL NEOPLASMS
SARCOMA, RETICULUM CELL, 4015

HELA CELLS
CAFFEINE
ADENOSINE CYCLIC 3',5'
MONOPHOSPHATE, 4193*
CHOLERA
ADENOSINE CYCLIC 3',5'
MONOPHOSPHATE, 4193*
CHROMATIN
RNA, MESSENGER, 3634
CULTURE MEDIA
ADENOSINE INCORPORATION, 4163
URIDINE INCORPORATION, 4163
ENTEROTOXINS
ADENOSINE CYCLIC 3',5'
MONOPHOSPHATE, 4193*
FERRITIN
ISOLATION AND CHARACTERIZATION,
3930
FETAL GLOBULINS
ISOLATION AND CHARACTERIZATION,
3930
THEOPHYLLINE
ADENOSINE CYCLIC 3',5'
MONOPHOSPHATE, 4193*
TRANSPLANTATION, HETEROLOGOUS
NEOPLASM METASTASIS, 4035
URIDINE
NUCLEOSIDE EXCRETION, 4163
XANTHINE, 3-ISOBUTYL-1-METHYL-
ADENOSINE CYCLIC 3',5'
MONOPHOSPHATE, 4193*

HEMANGIOBLASTOMA
SEE ANGIOSARCOMA

HEMANGIOENDOTHELIOMA
LIVER NEOPLASMS
BUTANOL, 4-(BUTYLNITROSOAMINO)-,
3701
DIPROPYLAMINE, N-NITROSO-, 3702
EPIDEMIOLOGY, CHILD, 4140*
PROPYLAMINE, N-METHYL-N-NITROSO-,
3702

HEMANGIOMA
BREAST NEOPLASMS
CASE REPORT, 3991
CARBAMIC ACID, ETHYL ESTER
IMMUNOSUPPRESSION, 3893
HEAD AND NECK NEOPLASMS
CASE REPORT, 4089*
ULTRASTRUCTURAL STUDY, 4089*
LIP NEOPLASMS
EPIDEMIOLOGY, 4000
LIVER NEOPLASMS
S-TRIAZINE, HEXAHYDRO-1,3,5-
TRINITRO-, 3694
LYMPHOSARCOMA
RADIATION, IONIZING, 3773
MOUTH NEOPLASMS
EPIDEMIOLOGY, 4000
RADIATION, IONIZING
EPIDEMIOLOGY, 3773
SALIVARY GLAND NEOPLASMS
EPIDEMIOLOGY, 4000
SARCOMA, EWING'S
RADIATION, IONIZING, 3773
SKIN NEOPLASMS
EPIDEMIOLOGY, 4000

EPIDEMIOLOGY, CHILD, 4140*
STOMACH NEOPLASMS
EPIDEMIOLOGY, 4000
TERATOID TUMOR
RADIATION, IONIZING, 3773

HEMANGIOPERICYTOMA
LIVER NEOPLASMS
EPIDEMIOLOGY, CHILD, 4140*

HEMATOPOIETIC STEM CELLS
BENZ(A)ANTHRACENE, 7,12-CIMETHYL-
PROLACTIN, 3754*
VIRUS, FRIEND-SPLEEN FOCUS FORMING
VIRUS REPLICATION, 3796

HEMATOSARCOMA
SEE ANGIOSARCOMA

HEME
VIRUS, FRIEND MURINE LEUKEMIA
PROTEIN BIOSYNTHESIS, 3850*

HEMOGLOBULINS
VIRUS, FRIEND POLYCYTHEMIA
SYNTHESIS, 3849*

HEPARIN
HEPATOMA
CELL MEMBRANE, 4173

HEPARITIN SULFATE
GALLBLADDER NEOPLASMS
CARCINOMA, 4194*

HEPATOCARCINOMA
SEE HEPATOMA

HEPATOCELLULAR CARCINOMA
SEE HEPATOMA

HEPATOMA
ACETAMIDE, N-FLUOREN-2-YL-
ALCOHOL OXIDOREDUCTASES, 4177
CHROMOSOME ABNORMALITIES, 4150
FETAL GLOBULINS, 4150
ACETAMIDE, N,N'-9H-FLUOREN-2,7-DIYLBIS-

LACTATION, 3722*
ADRENALECTOMY
SERINE DEHYDRATASE, 4183
TYROSINE AMINOTRANSFERASE, 4183
ALKALINE PHOSPHATASE
ALANINE, 3-PHENYL-, 4174, 4175
ARGININE, 4174, 4175
AUSTRALIA ANTIGEN, 4175
FETAL GLOBULINS, 4175
IMIDAZOLE, 4175
ISOENZYMES, 4174
LEUCINE, 4174
UREA, 4175
BENZENAMINE, N,N-DIMETHYL-4-((3-
METHYLPHENYL)AZO)-
AROCHLOR, 3728*
PEPTIDYL TRANSFERASES, 3730*
BENZIDINE, 3,3'-DIMETHYL-
FETAL GLOBULINS, 3731*
CARBAMIC ACID, ETHYL ESTER
IMMUNOSUPPRESSION, 3893

CARBON TETRACHLORIDE
ANTIGENS, NEOPLASM, 3942*
HYPERSENSITIVITY, 3942*
CARCINOGEN, CHEMICAL
ANIMAL MODEL, RAT, MOUSE, 3641*
CELL MEMBRANE
ADENYL CYCLASE, 4189*
ISOLATION AND CHARACTERIZATION,
4189*
CHEMOTAXIS
ISOLATION AND CHARACTERIZATION,
INHIBITORY FACTOR, 3886
COMPLEMENT
CHEMOTAXIS, 3886
CONTRACEPTIVES, ORAL
RETINOL, 3637*
CORTISONE
SERINE DEHYDRATASE, 4183
TYROSINE AMINOTRANSFERASE, 4183
DIBUTYRYL CYCLIC AMP
FETAL GLOBULINS, 3906
GROWTH, 3906
DIETARY FATS
M-TOLUIDINE, N,N-DIMETHYL-4-
(PHENYLAZO)-, 3606
DIETHYLAMINE, N-NITROSO-
FORMATION AND DEVELOPMENT, 3763*
ESCHERICHIA COLI
CHEMOTAXIS, 3886
ETHYL ALCOHOL
ALCOHOL OXIDOREDUCTASES, 4177
ETHYLENE, CHLORO-
OCCUPATIONAL HAZARD, 3663*
ETHYLENE, CHLORO- POLYMER
OCCUPATIONAL HAZARD, 3663*
FERRITIN
ISOLATION AND CHARACTERIZATION,
3930
FETAL GLOBULINS
ISOLATION AND CHARACTERIZATION,
3930
FOOD CONTAMINATION
REVIEW, 3603
GLYCOPEPTIDES
CELL MEMBRANE, 4173
HISTOCOMPATIBILITY ANTIGENS, 4173
HEPARIN
CELL MEMBRANE, 4173
IMMUNITY, CELLULAR
DNA REPLICATION, 3901
LEUKEMIA L1210
IMMUNOTHERAPY, 3937*
LEUKOCYTES
CHEMOTAXIS, 3886
MUCOPOLYSACCHARIDES
CELL MEMBRANE, 4173
MYCOTOXINS
REVIEW, 3603
POLYRIBOSOMES
FETAL GLOBULINS, 3906
PYRAZOLE
ALCOHOL OXIDOREDUCTASES, 4177
SIALIC ACIDS
DNA, 4166
STEROIDS
FETAL GLOBULINS, 3636*
SULFATASES
HISTOCHEMICAL STUDY, 4179
THEOPHYLLINE

GROWTH, 3906

HEPATOMEGALY
    GENETICS
        CASE REPORT, 4032

HEXOKINASE
    CERVIX NEOPLASMS
        PRECANCEROUS CONDITICNS, 4198*

HISTOCOMPATIBILITY ANTIGENS
    FIBROSARCOMA
        IMMUNITY, CELLULAR, 3943*
    GENETICS
        REVIEW, 3654*
    HEPATOMA
        GLYCOPEPTIDES, 4173
    IMMUNE SERUMS
        REVIEW, 3654*
    LEUKEMIA
        ISOLATION AND CHARACTERIZATICN,
        3915
    LEUKEMIA, LYMPHOBLASTIC
        IMMUNE RESPONSE, 3929
    B-LYMPHOCYTES
        REVIEW, 3654*
    T-LYMPHOCYTES
        REVIEW, 3654*
    LYMPHOMA
        VIRUS, MAREK'S DISEASE HERPES,
        3980*
    VIRUS, BOVINE ADENO 3
        VIRUS, BOVINE ADENO 1, 3916
        VIRUS, BOVINE ADENO 2, 3916
    VIRUS, FRIEND MURINE LEUKEMIA
        VIRUS REPLICATION, 3798
    VIRUS, MAREK'S DISEASE HERPES
        B-LYMPHOCYTES, 3980*
        T-LYMPHOCYTES, 3980*

HISTONES
    LEUKEMIA, LYMPHOBLASTIC
        RADIATION, IONIZING, 4170

HODGKIN'S DISEASE
    BASOPHILS
        ULTRASTRUCTURAL STUDY, 4007
    BRAIN NEOPLASMS
        DEOXYRIBONUCLEASE, 4017
        PEPSIN, 4017
        RIBONUCLEASE, 4017
        RNA, 4017
        ULTRASTRUCTURAL STUDY, 4017
    EOSINOPHILIA
        PRECANCEROUS CONDITICNS, 4083*
    EPIDEMIOLOGY
        CHILD INDIA, 4120
        HISTOLOGICAL STUDY, 4084*
    GENETICS, 4087*
        REVIEW, 4087*
    GLUCOCORTICOIDS
        METABOLISM, 4085*
    IMMUNOGLOBULINS, SURFACE, 3984*
    LIPOPROTEINS
        ISOLATION AND CHARACTERIZATION,
        3632
    LYMPHOCYTES
        ULTRASTRUCTURAL STUDY, 4086*
    B-LYMPHOCYTES

CELL DIFFERENTIATION, 3984*
IMMUNE SERUMS, 3921
T-LYMPHOCYTES
    DNA REPLICATION, 3921
    ROSETTE FORMATION, 3921
LYMPHOMA
    CLASSIFICATION, 3625
MAST CELLS
    ULTRASTRUCTURAL STUDY, 4007
PROTEINS
    ISOLATION AND CHARACTERIZATION, C-
    REACTIVE, 4169

HORMONES
    BREAST NEOPLASMS
        EPIDEMIOLOGY, REVIEW, 4117
    RNA, MESSENGER
        REVIEW, 4183
    UROGENITAL NEOPLASMS
        EPIDEMIOLOGY, REVIEW, 4117

HYBRID CELLS
    VIRUS, ADENO 12
        PHOSPHOTRANSFERASES, 3793
        THYMIDINE KINASE, 3793

HYDRAZINE, 1,2-DIETHYL-
    NERVOUS SYSTEM NEOPLASMS
        TRANSPLACENTAL CARCINOGENESIS,
        REVIEW, 3601

HYDRAZINE, 1,2-DIMETHYL-
    COLONIC NEOPLASMS
        CARCINOEMBRYONIC ANTIGEN, 3909

HYDROCORTISONE
    SEE CORTISOL

HYDROLASES
    CHOLANTHRENE, 3-METHYL-
        CARCINOGENIC METABOLITE, 3683
        TRANSPLACENTAL CARCINOGENESIS,
        3683

N-HYDROXY-2-ACETYLAMINOFLUORENE
    SEE ACETOHYDROXAMIC ACID, N-FLUOREN-2-
    YL-

HYDROXYLASES
    BREAST NEOPLASMS
        CARCINOMA, 3993
        CARCINOMA, SCIRRHOUS, 3993
        PRECANCEROUS CONDITIONS, 3993

HYPERGAMMAGLOBULINEMIA
    THYMOMA
        IMMUNITY, CELLULAR, 4043

HYPERNEPHROID CARCINOMA
    SEE ADENOCARCINOMA

HYPERNEPHROMA
    SEE ADENOCARCINOMA

HYPERSENSITIVITY
    HEPATOMA
        CARBON TETRACHLORIDE, 3942*
    MAMMARY NEOPLASMS, EXPERIMENTAL
        IMMUNIZATION, 3942*

IGG
    BREAST NEOPLASMS
        MONOCYTES, 3958*
    LUNG NEOPLASMS
        MONOCYTES, 3958*
    LYMPHOMA
        MONOCYTES, 3958*
    MELANOMA
        MONOCYTES, 3958*
    SARCOMA, OSTEOGENIC
        IMMUNE SERUMS, 3925
    THYMOMA
        MONOCYTES, 3958*

IGM
    BURKITT'S LYMPHOMA
        IMMUNE SERUMS, 3925
    SARCOMA, OSTEOGENIC
        IMMUNE SERUMS, 3925
    URIDINE, 5-BROMO-2'-DEOXY-
        CELL MEMBRANE, 3959*
    WARTS
        AUTOANTIBODIES, 3837

IMIDAZOLE
    HEPATOMA
        ALKALINE PHOSPHATASE, 4175

IMIDAZOLE-4-CARBOXAMIDE, 5-AMINO-
    KIDNEY NEOPLASMS
        ADENOCARCINOMA, 3691
    MAMMARY NEOPLASMS, EXPERIMENTAL
        ADENOCARCINOMA, 3691
    THYROID NEOPLASMS
        CARCINOMA, 3691

IMIDAZOLE-4-CARBOXAMIDE, 5-DIAZO-
    MAMMARY NEOPLASMS, EXPERIMENTAL
        ADENOCARCINOMA, 3691
        ADENOFIBROMA, 3691

IMIDAZOLE-4-CARBOXAMIDE, 5-(3,3-DIMETHYL-1-
    TRIAZENO)-
    ANTINEOPLASTIC AGENTS
        CARCINOGENIC ACTIVITY, 3737*
        CARCINOGENIC ACTIVITY, RAT, 3691
    LEUKEMIA
        GERM-FREE LIFE, 3758*
    MAMMARY NEOPLASMS, EXPERIMENTAL
        ADENOCARCINOMA, 3691
    RHABDOMYOSARCOMA
        GERM-FREE LIFE, 3758*
    THYMUS NEOPLASMS
        LYMPHOSARCOMA, 3691, 3758*

IMIDAZOLE-4-CARBOXAMIDE, 5-(3-METHYL-1-
    TRIAZENO)-
    ANTINEOPLASTIC AGENTS
        ADENOFIBROMA, 3691
    UTERINE NEOPLASMS
        LEIOMYOSARCOMA, 3691

4H-IMIDAZOLE-(4,5-D)-V-TRIAZIN-4-ONE, 3,7-
    DIHYDRO-
    ADRENAL GLAND NEOPLASMS
        ADENOMA, 3691
    MAMMARY NEOPLASMS, EXPERIMENTAL
        ADENOCARCINOMA, 3691

2-IMIDAZOLIDINETHIONE
    THYROID NEOPLASMS
        CARCINOGENIC ACTIVITY, RAT, 3687

IMIPRAMINE
    NEUROBLASTOMA
        NOREPINEPHRINE, 4181

IMMUNE SERUMS
    ADENOCARCINOMA
        IMMUNOGLOBULINS, 3925
    ASTROCYTOMA
        IMMUNOGLOBULINS, 3925
    BREAST NEOPLASMS
        ANTIGENIC DETERMINANTS, 3974*
    BURKITT'S LYMPHOMA
        IGM, 3925
    CARCINOEMBRYONIC ANTIGEN
        BLOOD GROUPS, 3910
    CARCINOMA
        IMMUNOGLOBULINS, 3925
    CHONDROSARCOMA
        IMMUNOGLOBULINS, 3925
    CHORDOMA
        IMMUNOGLOBULINS, 3925
    COLONIC NEOPLASMS
        ANTIGENIC DETERMINANTS, 3974*
    DIGESTIVE SYSTEM NEOPLASMS
        CARCINOEMBRYONIC ANTIGEN, 3969*
    GIANT CELL TUMOR
        IMMUNOGLOBULINS, 3925
    HISTOCOMPATIBILITY ANTIGENS
        REVIEW, 3654*
    HODGKIN'S DISEASE
        B-LYMPHOCYTES, 3921
    INFECTIOUS MONONUCLEOSIS
        ANTIGENS, HETEROGENETIC, 3914
        B-LYMPHOCYTES, 3921
    INTESTINAL NEOPLASMS
        CARCINOEMBRYONIC ANTIGEN, 3909
    LEUKEMIA, LYMPHOBLASTIC
        BONE MARROW CELLS, 3955*
    LEUKEMIA, LYMPHOCYTIC
        B-LYMPHOCYTES, 3921
        T-LYMPHOCYTES, 3918, 3961*
    LEUKEMIA, MYELOCYTIC
        BONE MARROW CELLS, 3955*
    B-LYMPHOCYTES
        LANTHANUM, 4167
    LYMPHOMA
        B-LYMPHOCYTES, 3921
    MACROGLOBULINEMIA
        ANTIGENS, NEOPLASM, 3973*
    MULTIPLE MYELOMA
        ANTIGENS, NEOPLASM, 3973*
        IMMUNOGLOBULINS, SURFACE, 3923
    SARCOMA
        GRAFT REJECTION, 3902
        IMMUNITY, CELLULAR, 3902

TRANSPLANTATION IMMUNOLOGY, 3902
SARCOMA, OSTEOGENIC
   IGG, 3925
   IGM, 3925
   IMMUNOGLOBULINS, 3925
UTERINE NEOPLASMS
   ANTIGENIC DETERMINANTS, 3974*

IMMUNITY, ACTIVE
  VIRUS, MAREK'S DISEASE HERPES
   ANTIGENS, VIRAL, 3897

IMMUNITY, CELLULAR
  BREAST NEOPLASMS
   LYMPHOCYTES, 3936*
   REVIEW, 3932*
  BURKITT'S LYMPHOMA
   CANDIDA, 3946*
   DIPTHERIA ANTITOXIN, 3946*
   IMMUNOLOGIC FACTOR, 3946*
   STREPTODORNASE AND STREPTOKINASE,
    3946*
   TUBERCULIN, 3946*
  CARCINOMA, EHRLICH TUMOR
   DNA REPLICATION, 3901
   TRANSPLANTATION IMMUNOLOGY, 3934*
  CARCINOMA 256, WALKER
   DNA REPLICATION, 3901
  FIBROSARCOMA
   ANTIGENS, NEOPLASM, 3943*
   CHOLANTHRENE, 3-METHYL-, 3945*
   HISTOCOMPATIBILITY ANTIGENS, 3943*
  HEPATOMA
   DNA REPLICATION, 3901
  IMMUNOSUPPRESSION
   REVIEW, 3652*
  LEUKEMIA
   REVIEW, 3626
  LEUKEMIA, MYELOBLASTIC
   GLYCOGEN, 3900
   MONOCYTES, 3900
  LUNG NEOPLASMS
   LEUKEMIA, MONOCYTIC, 3933*
   REVIEW, 3640*
  LYMPHOCYTES
   MACROPHAGES, 3931*
  LYMPHOSARCOMA
   ESTRADIOL, 3945*
  MAMMARY NEOPLASMS, EXPERIMENTAL
   VACCINES, ATTENUATED, 3619
   VIRUS, LACTATE DEHYDROGENASE, 3619
  MULTIPLE MYELOMA
   VIRUS, HERPES SIMPLEX, 3894
  NEOPLASMS
   REVIEW, 3626, 3651*
  NEOPLASMS, EXPERIMENTAL
   VIRUS, SV40, 3891
  SARCOMA
   CHOLANTHRENE, 3-METHYL-, 3934*
   IMMUNE SERUMS, 3902
  SARCOMA, RETICULUM CELL
   REVIEW, 3650*
  STOMACH NEOPLASMS
   ADENOCARCINOMA, 3934*
  THYMOMA
   HYPERGAMMAGLOBULINEMIA, 4043
  UTERINE NEOPLASMS
   VIRUS, MURINE SARCOMA, 3865*
  VIRUS, HERPES

REVIEW, 3616
VIRUS, MOLONEY MURINE SARCOMA
  CONCANAVALIN A, 3808

IMMUNITY, PASSIVE
  SARCOMA
   GRAFT REJECTION, 3902

IMMUNIZATION
  MAMMARY NEOPLASMS, EXPERIMENTAL
   HYPERSENSITIVITY, 3942*

IMMUNOGENETICS
  FIBROSARCOMA
   TRANSPLANTATION IMMUNOLOGY, 4158
  L CELLS
   TRANSPLANTATION IMMUNOLOGY, 4158

IMMUNOGLOBULIN FRAGMENTS
  MYELOMA PROTEINS
   ISOLATION AND CHARACTERIZATION,
    3957*

IMMUNOGLOBULINS
  SEE ALSO IGA/IGD/IGE/IGG/IGM
  ADENOCARCINOMA
   IMMUNE SERUMS, 3925
  ASTROCYTOMA
   IMMUNE SERUMS, 3925
  CARCINOMA
   IMMUNE SERUMS, 3925
  CHONDROSARCOMA
   IMMUNE SERUMS, 3925
  CHORDOMA
   IMMUNE SERUMS, 3925
  GIANT CELL TUMOR
   IMMUNE SERUMS, 3925
  LEUKEMIA, LYMPHOBLASTIC
   IMMUNE RESPONSE, 3929
  LEUKEMIA, LYMPHOCYTIC
   ISOLATION AND CHARACTERIZATION,
    4197*
  LEUKEMIA, MYELOBLASTIC
   VIRUS, C-TYPE, 3841*
  SARCOMA
   CHOLANTHRENE, 3-METHYL-, 3956*
  SARCOMA, OSTEOGENIC
   IMMUNE SERUMS, 3925

IMMUNOGLOBULINS, FAB
  MYELOMA PROTEINS
   ISOLATION AND CHARACTERIZATION,
    3957*

IMMUNOGLOBULINS, SURFACE
  HODGKIN'S DISEASE, 3984*
  LEUKEMIA, LYMPHOCYTIC
   CELL DIFFERENTIATION, 3984*
   B-LYMPHOCYTES, 3918
  LEUKEMIA, PLASMACYTIC
   CELL DIFFERENTIATION, 3984*
  MULTIPLE MYELOMA
   IMMUNE SERUMS, 3923
   B-LYMPHOCYTES, 3923
   RNA, NEOPLASM, 3923

IMMUNOSUPPRESSION
  CARCINOMA, EHRLICH TUMOR
   ANTIBODIES, 3960*

EYE NEOPLASMS
  CARBAMIC ACID, ETHYL ESTER, 3893
FETAL GLOBULINS
  PLANT AGGLUTININS, 3954*
HEMANGIOMA
  CARBAMIC ACID, ETHYL ESTER, 3893
HEPATOMA
  CARBAMIC ACID, ETHYL ESTER, 3893
IMMUNITY, CELLULAR
  REVIEW, 3652*
LEUKEMIA
  REVIEW, 3652*
LEUKEMIA, MYELOCYTIC
  CARBAMIC ACID, ETHYL ESTER, 3893
  CORTICOTROPIN, 3953*
  CYCLOPHOSPHAMIDE, 3953*
LUNG NEOPLASMS
  CARBAMIC ACID, ETHYL ESTER, 3893
  CHOLANTHRENE, 3-METHYL-, 3893
LYMPHATIC NEOPLASMS
  REVIEW, 3652*
LYMPHOCYTES
  ASCITIC FLUID, 3950*
T-LYMPHOCYTES
  MACROPHAGES, 3949*
NEOPLASM TRANSPLANTATION
  CARCINOGENIC ACTIVITY, 3944*
NEOPLASMS
  ASCITIC FLUID, 3950*
  CHOLANTHRENE, 3-METHYL-, 3951*,
    3952*
  T-LYMPHOCYTES, 3983*
NEOPLASMS, EXPERIMENTAL
  CARBAMIC ACID, ETHYL ESTER, 3893
  CHOLANTHRENE, 3-METHYL-, 3893
  VIRUS, SV40, 3891
RHABDOMYOSARCOMA
  LYMPHOCYTES, 3952*
SARCOMA, MAST CELL
  B-LYMPHOCYTES, 3887
  T-LYMPHOCYTES, 3887
SARCOMA, RETICULUM CELL
  REVIEW, 3650*
SARCOMA, YOSHIDA
  TRANSPLANTATION IMMUNOLOGY, 3948*
TRANSPLANTATION, HOMOLOGOUS
  ANTIGENIC STIMULATION, REVIEW,
    3807
VIRUS, FRIEND MURINE LEUKEMIA
  CELL TRANSFORMATION, NEOPLASTIC,
    3893
  ULTRASTRUCTURAL STUDY, SPLEEN,
    MOUSE, 3895
VIRUS, MOLONEY MURINE SARCOMA
  CELL TRANSFORMATION, NEOPLASTIC,
    3893

IMMUNOTHERAPY
  BREAST NEOPLASMS
    REVIEW, 3932*
  NEOPLASMS
    REVIEW, 3651*

INCLUSION BODIES, VIRAL
  LEUKEMIA
    REVIEW, 3614
  PLASMACYTOMA
    A-TYPE PARTICLES, 3870*
  VIRUS, GUINEA PIG LEUKEMIA

VIRUS, B-TYPE, 3852*
VIRUS, RNA TUMOR
  REVIEW, 3614

INFECTIOUS MONONUCLEOSIS
  IMMUNE SERUMS
    ANTIGENS, HETEROGENETIC, 3914
  LEUKEMIA, LYMPHOCYTIC
    ANTIGENS, HETEROGENETIC, 3914
  LEUKEMIA, MYELOBLASTIC
    ANTIGENS, HETEROGENETIC, 3914
  LEUKEMIA, MYELOCYTIC
    ANTIGENS, HETEROGENETIC, 3914
  B-LYMPHOCYTES
    IMMUNE SERUMS, 3921
  T-LYMPHOCYTES
    ANTIGENS, HETEROGENETIC, 3914
    DNA REPLICATION, 3921
    ROSETTE FORMATION, 3921
  LYMPHOMA
    ANTIGENS, HETEROGENETIC, 3914
  LYMPHOSARCOMA
    ANTIGENS, HETEROGENETIC, 3914
  SARCOMA, RETICULUM CELL
    ANTIGENS, HETEROGENETIC, 3914
  VIRUS, EPSTEIN-BARR
    ANTIBODIES, VIRAL, 4144*

INSULIN
  BREAST NEOPLASMS
    GROWTH SUBSTANCES, 3689
  NEUROBLASTOMA
    ADENOSINE TRIPHOSPHATASE, 4176

INTERFERON
  VIRUS, FRIEND MURINE LEUKEMIA
    REGRESSION, 3905
  VIRUS, MURINE LEUKEMIA
    ANTIGENS, VIRAL, 3860*
    VIRUS REPLICATION, 3860*

INTESTINAL NEOPLASMS
  ADENOCARCINOMA
    CARCINOEMBRYONIC ANTIGEN, 3911
    INTESTINAL POLYPS, 4071*
    RADIATION, 3784*
    RADIATION, IONIZING, 3911
  AGE FACTORS
    EPIDEMIOLOGY, ISRAEL, 4115
  BURKITT'S LYMPHOMA
    CLINICOPATHOLOGIC STUDY, 4013
  CARCINOEMBRYONIC ANTIGEN
    IMMUNE SERUMS, 3909
  CARCINOGEN, CHEMICAL
    ANIMAL MODEL, RAT, MOUSE, 3641*
  CARCINOMA, MUCINOUS
    CYCLOOCTANECARBAMIC ACID, 1,1-BIS
      (P-FLUOROPHENYL)-2-PROPYNYL,
      3684
  CARCINOMA, PAPILLARY
    CYCLOOCTANECARBAMIC ACID, 1,1-BIS
      (P-FLUOROPHENYL)-2-PROPYNYL,
      3684
  CARCINOMA, SCIRRHOUS
    CYCLOOCTANECARBAMIC ACID, 1,1-BIS
      (P-FLUOROPHENYL)-2-PROPYNYL,
      3684
  EPIDEMIOLOGY
    MORMONS, 4118

LEIOMYOMA
EPIDEMIOLOGY, 4000
PEUTZ-JEGHERS SYNDROME
DIAGNOSIS AND PROGNOSIS, 4070*
RADIATION
CARCINOEMBRYONIC ANTIGEN, 3784*
RADIATION, IONIZING
CARCINOEMBRYONIC ANTIGEN, 3911
SULFATASES
HISTOCHEMICAL STUDY, 4179

INTESTINAL POLYPS
ADENOCARCINOMA
CASE REPORT, 4071*
GENETICS, 4071*
INTESTINAL NEOPLASMS
ADENOCARCINOMA, 4071*

IODINE ISOTOPES
THYROID GLAND
CARCINOGENIC ACTIVITY, 3611

IRON (III) NITRILOTRIACETATO-
ACETIC ACID, IMINODI-
PHOTODECOMPOSITION, 3724*

ISLET CELL TUMOR
PANCREATIC NEOPLASMS
CASE REPORT, 4069*, 4073*
PEPTIDES, 4069*
ULTRASTRUCTURAL STUDY, 4073*

ISOANTIBODIES
THYMOMA
IMMUNE RESPONSE, 4043

ISOANTIGENS
BRAIN NEOPLASMS
TRANSPLACENTAL CARCINOGENESIS,
3892
LEUKEMIA
IODINATION, 3962*
ISOLATION AND CHARACTERIZATION,
3962*
LEUKEMIA L1210
CELL TRANSFORMATION, NEOPLASTIC,
3892
IMMUNE RESPONSE, 3892
LUNG NEOPLASMS
TRANSPLACENTAL CARCINOGENESIS,
3892
LYMPHOMA
IMMUNE RESPONSE, 3892
MAMMARY NEOPLASMS, EXPERIMENTAL, 3892
IMMUNE RESPONSE, 3892
MULTIPLE MYELOMA
IMMUNE RESPONSE, 3892
NEUROBLASTOMA
IMMUNE RESPONSE, 3892
TERATOID TUMOR
IMMUNE RESPONSE, 3892
VIRUS, FRIEND MURINE LEUKEMIA
IMMUNE RESPONSE, 3892
VIRUS, GROSS MURINE LEUKEMIA
CELL TRANSFORMATION, NEOPLASTIC,
3892
IMMUNE RESPONSE, 3892
VIRUS, MOLONEY MURINE SARCOMA
IMMUNE RESPONSE, 3892

VIRUS, RAUSCHER MURINE LEUKEMIA
IMMUNE RESPONSE, 3892

ISOENZYMES
CELL TRANSFORMATION, NEOPLASTIC
REVIEW, 3633
HEPATOMA
ALKALINE PHOSPHATASE, 4174
NEOPLASMS
REVIEW, 3633

ISOGENIC TRANSPLANTATION
SEE TRANSPLANTATION, HOMOLOGOUS

ISONICOTINIC ACID HYDRAZIDE
CHROMOSOME ABERRATIONS
BONE MARROW, 3757*

JAW NEOPLASMS
BURKITT'S LYMPHOMA
CLINICOPATHOLOGIC STUDY, 4013

KARYOTYPING
BURKITT'S LYMPHOMA
CASE REPORT, CHILD, 4014
CHROMOSOME ABERRATIONS
MITOMYCIN C, 4025
L CELLS
HYBRIDIZATION, 4158
LYMPHOSARCOMA
CASE REPORT, CHILD, 4014

KERATOCANTHOMA
FETAL MEMBRANES
ULTRASTRUCTURAL STUDY, 4062*

KIDNEY
VIRUS, POLYOMA
TRANSPLANTATION, HOMOLOGOUS, 3872*

KIDNEY NEOPLASMS
ADENOCARCINOMA
IMIDAZOLE-4-CARBOXAMIDE, 5-AMINO-,
3691
NEOPLASM METASTASIS, 4046
ADENOMA
DIPROPYLAMINE, N-NITROSO-, 3702
CARBAMIC ACID, N-METHYL-N-NITROSO-,
ETHYL ESTER
TRANSPLACENTAL CARCINOGENESIS,
REVIEW, 3601
CARCINOMA, TRANSITIONAL CELL
NEOPLASM METASTASIS, 4046
DIETHYLAMINE, N-NITROSO
TRANSPLACENTAL CARCINOGENESIS,
REVIEW, 3601
LEIOMYOMA
EPIDEMIOLOGY, 4000
NEOPLASM METASTASIS, 4045
NEPHROBLASTOMA
DIAGNOSIS AND PROGNOSIS, 4133*
EPIDEMIOLOGY, CHILD, 4133*
EPIDEMIOLOGY, CHILD, INDIA, 4120
4,4'-STILBENEDIOL, ALPHA,ALPHA'-
DIETHYL-
ANTINEOPLASTIC AGENTS, 3739*
SULFATASES
HISTOCHEMICAL STUDY, 4179
VIRUS, C-TYPE RNA TUMOR

ANTIVIRAL AGENTS, 3739*
VIRUS-LIKE PARTICLES, 3739*

KLINEFELTER'S SYNDROME
BREAST NEOPLASMS
PRECANCEROUS CONDITIONS, MALE,
3992

KUPFFER CELLS
CELL DIVISION
LIVER REGENERATION, 4159

L CELLS
KARYOTYPING
HYBRIDIZATION, 4158
LEUKEMIA
TERATOID TUMOR, 4158
TRANSPLANTATION IMMUNOLOGY
HYBRIDIZATION, 4158
IMMUNOGENETICS, 4158

LACTATE DEHYDROGENASE
BURKITT'S LYMPHOMA
CLINICOPATHOLOGIC STUDY, 4013
CERVIX NEOPLASMS
PRECANCEROUS CONDITIONS, 4198*

LACTATION
ACETAMIDE, N,N'-9H-FLUOREN-2,7-DIYLBIS-

COCARCINOGENIC EFFECT, HEPATOMA,
MOUSE, 3722*
HEPATOMA
ACETAMIDE, N,N'-9H-FLUOREN-2,7-
DIYLBIS-, 3722*
MAMMARY NEOPLASMS, EXPERIMENTAL
SULPIRIDE, 3713

LANTHANUM
B-LYMPHOCYTES
ANTILYMPHOCYTE SERUM, 4167
CONCANAVALIN A, 4167
IMMUNE SERUMS, 4167
MITOMYCIN C, 4167
PLANT AGGLUTININS, 4167
TUBERCULIN, 4167
T-LYMPHOCYTES
ANTILYMPHOCYTE SERUM, 4167
CONCANAVALIN A, 4167
MITOMYCIN C, 4167
PLANT AGGLUTININS, 4167
TUBERCULIN, 4167

LARYNGEAL NEOPLASMS
ADENOCARCINOMA
EPIDEMIOLOGY, 4075*
CARCINOMA, EPIDERMOID
CLOSTRIDIOPEPTIDASE A, 3998
EPIDEMIOLOGY, 4075*
CARCINOMA IN SITU
DIAGNOSIS AND THERAPY, 3999
PRECANCEROUS CONDITIONS, 3999
CARCINOMA, PAPILLARY
EPIDEMIOLOGY, 4075*
EPIDEMIOLOGY
AGE FACTORS, GREAT BRITAIN, 4114
MORMOMS, 4118

LEIOMYOMA
EPIDEMIOLOGY, 4000
NEOPLASM METASTASIS, 4045
SMOKING
AGE FACTORS, GREAT BRITAIN, 4114

LEIOMYOMA
ESOPHAGEAL NEOPLASMS
EPIDEMIOLOGY, 4000
INTESTINAL NEOPLASMS
EPIDEMIOLOGY, 4000
KIDNEY NEOPLASMS
EPIDEMIOLOGY, 4000
LARYNGEAL NEOPLASMS
EPIDEMIOLOGY, 4000
LIVER NEOPLASMS
EPIDEMIOLOGY, 4000
MOUTH NEOPLASMS
EPIDEMIOLOGY, 4000
SALIVARY GLAND NEOPLASMS
EPIDEMIOLOGY, 4000
SKIN NEOPLASMS
EPIDEMIOLOGY, 4000
STOMACH NEOPLASMS
EPIDEMIOLOGY, 4000
UROGENITAL NEOPLASMS
EPIDEMIOLOGY, 4000
UTERINE NEOPLASMS
C-REACTIVE PROTEIN, 4169
HISTOLOGICAL STUDY, CASE REPORT,
4047

LEIOMYOSARCOMA
LIPOPROTEINS
ISOLATION AND CHARACTERIZATION,
3632
PROSTATIC NEOPLASMS
CASE REPORT, 4092*
UTERINE NEOPLASMS
IMIDAZOLE-4-CARBOXAMIDE, 5-(3-
METHYL-1-TRIAZENO)-, 3691
VIRUS, POLYOMA, 3874*

LEUCINE
HEPATOMA
ALKALINE PHOSPHATASE, 4174

LEUKEMIA
ACUTE DISEASE
CELL CYCLE KINETICS, 4131
AGE FACTORS
EPIDEMIOLOGY, 4119
BONE-MARROW CELLS
CHROMOSOME ABERRATIONS, 4082*
PRECANCEROUS CONDITIONS, 4078*
CARBAMIC ACID, N-METHYL-N-NITROSO-,
ETHYL ESTER
TRANSPLACENTAL CARCINOGENESIS,
REVIEW, 3601
CELL CYCLE KINETICS
REVIEW, MOLLUSK, 3659*
CHROMOSOME ABERRATIONS
CASE REPORT, 4082*
REVIEW, MOLLUSK, 3659*
TRANSPLANTATION, HOMOLOGOUS, 4082*
DNA-RNA HYBRIDIZATION
REVIEW, 3614
DNA, VIRAL
REVIEW, 3614

EPIDEMIOLOGY
  MORMOMS, 4118
GLYCEROLPHOSPHATE DEHYDROGENASE
  ISOLATION AND CHARACTERIZATION,
  4200*
HISTOCOMPATIBILITY ANTIGENS
  ISOLATION AND CHARACTERIZATION,
  3915
IMIDAZOLE-4-CARBOXAMIDE, 5-(3,3-
  DIMETHYL-1-TRIAZENOI-
  GERM-FREE LIFE, 3758*
IMMUNITY, CELLULAR
  REVIEW, 3626
IMMUNOSUPPRESSION
  REVIEW, 3652*
INCLUSION BODIES, VIRAL
  REVIEW, 3614
ISOANTIGENS
  IODINATION, 3962*
  ISOLATION AND CHARACTERIZATION,
  3962*
L CELLS
  TERATOID TUMOR, 4158
LEUKOCYTES
  BIOCHEMISTRY AND METABOLISM,
  REVIEW, 3666*
LYMPHOCYTES
  PRECANCEROUS CONDITIONS, 3978*
T-LYMPHOCYTES
  ANTIGENS, NEOPLASM, 3915
MEGAKARYOCYTES
  EPIDEMIOLOGY, 4119
NEOPLASM METASTASIS
  STATISTICAL ANALYSIS, 4107*
PLUTONIUM
  DISTRIBUTION, RAT, 3774
PROTEINS
  ISOLATION AND CHARACTERIZATION, C-
  REACTIVE, 4169
RADIATION, IONIZING
  EPIDEMIOLOGY, HIROSHIMA, NAGASAKI,
  4113
  RADIATION DOSAGE, 4113
REVERSE TRANSCRIPTASE
  REVIEW, 3614
TRANSPLANTATION, HOMOLOGOUS
  GRAFT VS HOST REACTION, 3940*
TRANSPLANTATION IMMUNOLOGY
  REVIEW, 3626
VIRUS, C-TYPE RNA TUMOR
  VIRUS-LIKE PARTICLES, 3649*
VIRUS, RAUSCHER MURINE LEUKEMIA
  REVIEW, 3614
VIRUS, SIMIAN SARCOMA
  REVIEW, 3614

LEUKEMIA, ACUTE GRANULOCYTIC
  SEE LEUKEMIA, MYELOBLASTIC

LEUKEMIA, ACUTE LYMPHOCYTIC
  SEE LEUKEMIA, LYMPHOBLASTIC

LEUKEMIA, CHRONIC GRANULOCYTIC
  SEE LEUKEMIA, MYELOCYTIC

LEUKEMIA L1210
  HEPATOMA
    IMMUNOTHERAPY, 3937*
  ISOANTIGENS

CELL TRANSFORMATION, NEOPLASTIC,
  3892
IMMUNE RESPONSE, 3892

LEUKEMIA, LYMPHOBLASTIC
  ACID PHOSPHATASE
    T-LYMPHOCYTES, 3920
    MORPHOLOGICAL AND CYTOCHEMICAL
    STUDY, 4021
  AGE FACTORS
    EPIDEMIOLOGY, 4119
  BONE MARROW CELLS
    IMMUNE SERUMS, 3955*
    MORPHOLOGICAL AND CYTOCHEMICAL
    STUDY, 4021
  BURKITT'S LYMPHOMA
    CASE REPORT, 4011
  EPIDEMIOLOGY
    CHILD INDIA, 4120
  ERYTHROPOIESIS
    CHILD, 4020
  GENETICS
    IMMUNE RESPONSE, 3929
  HISTOCOMPATIBILITY ANTIGENS
    IMMUNE RESPONSE, 3929
  IMMUNOGLOBULINS
    IMMUNE RESPONSE, 3929
  LYMPHOCYTES
    IMMUNE RESPONSE, 3929
  T-LYMPHOCYTES
    NEURAMINIDASE, 3920
    ROSETTE FORMATION, 3920
  NEUTROPHILS
    PHAGOCYTOSIS, 4188*
  PEROXIDASES
    MORPHOLOGICAL AND CYTOCHEMICAL
    STUDY, 4021
  RADIATION, IONIZING
    DNA, 4170
    HISTONES, 4170

LEUKEMIA, LYMPHOCYTIC
  AGE FACTORS
    EPIDEMIOLOGY, 4119
  ALBUMIN
    ISOLATION AND CHARACTERIZATION,
    4197*
  EPIDEMIOLOGY
    CHILD INDIA, 4120
  ERYTHROPOIESIS
    CHILD, 4020
  GANGLIOSIDES
    CELL MEMBRANE, 3924
  IMMUNOGLOBULINS
    ISOLATION AND CHARACTERIZATION,
    4197*
  IMMUNOGLOBULINS, SURFACE
    CELL DIFFERENTIATION, 3984*
  INFECTIOUS MONONUCLEOSIS
    ANTIGENS, HETEROGENETIC, 3914
  B-LYMPHOCYTES
    ANTILYMPHOCYTE SERUM, 3961*
    CELL DIFFERENTIATION, 3984*
    IMMUNE SERUMS, 3921
    IMMUNOGLOBULINS, SURFACE, 3918
    ROSETTE FORMATION, 3917
  T-LYMPHOCYTES
    AGRANULOCYTOSIS, 3918
    ANTIBODIES, 3961*

ANTILYMPHOCYTE SERUM, 3961*
IMMUNE RESPONSE, 3917
IMMUNE SERUMS, 3918, 3961*
PITUITARY GLAND, 3981*
ROSETTE FORMATION, 3917, 3918,
3921
SKIN NEOPLASMS, 3918
SPLENOMEGALY, 3918
PLANT AGGLUTININS, 3917
PREGNA-1,4-DIENE-3,11,20-TRIONE, 17,21-
DIHYDROXY-, 3917
TRANSFERRIN
ISOLATION AND CHARACTERIZATION,
4197*
VIRUS, GROSS MURINE LEUKEMIA
T-LYMPHOCYTES, 3981*
PITUITARY GLAND, 3981*

LEUKEMIA, MONOCYTIC
LUNG NEOPLASMS
IMMUNITY, CELLULAR, 3933*
TRANSPLANTATION IMMUNOLOGY, 3933*

LEUKEMIA, MYELOBLASTIC
ACID PHOSPHATASE
MORPHOLOGICAL AND CYTOCHEMICAL
STUDY, 4021
AGE FACTORS
EPIDEMIOLOGY, 4119
BONE MARROW
ULTRASTRUCTURAL STUDY, 4009
VIRUS-LIKE PARTICLES, 4009
BONE MARROW CELLS
MORPHOLOGICAL AND CYTOCHEMICAL
STUDY, 4021
EPIDEMIOLOGY
CHILD INDIA, 4120
GENETICS
ANTIGENS, NEOPLASM, 4030
GLYCOGEN
IMMUNITY, CELLULAR, 3900
INFECTIOUS MONONUCLEOSIS
ANTIGENS, HETEROGENETIC, 3914
MONOCYTES
IMMUNE RESPONSE, 3900
IMMUNITY, CELLULAR, 3900
NEUTROPHILS
PHAGOCYTOSIS, 4188*
PEROXIDASES
MORPHOLOGICAL AND CYTOCHEMICAL
STUDY, 4021
PHOSPHOLIPIDS
MORPHOLOGICAL AND CYTOCHEMICAL
STUDY, 4021
RADIATION, IONIZING
CASE REPORT, 3771
CHROMOSOME ABERRATIONS, 3771
VIRUS, BABOON ENDOGENOUS
ANTIGENS, VIRAL, 3841*
REVERSE TRANSCRIPTASE, 3841*
VIRUS, C-TYPE
ANTIGENS, VIRAL, 3841*
IMMUNOGLOBULINS, 3841*
REVERSE TRANSCRIPTASE, 3841*
VIRUS, C-TYPE RNA TUMOR
ISOLATION AND CHARACTERIZATION,
3834
VIRUS REPLICATION, 3834
VIRUS, SIMIAN SARCOMA

ANTIGENS, VIRAL, 3841*
REVERSE TRANSCRIPTASE, 3841*

LEUKEMIA, MYELOCYTIC
AGE FACTORS
EPIDEMIOLOGY, 4119
BONE MARROW CELLS
IMMUNE SERUMS, 3955*
CARBAMIC ACID, ETHYL ESTER
IMMUNOSUPPRESSION, 3893
CARBON RADIOISOTOPES
CARCINOGENIC EFFECT, MOUSE, 3782*
CORTICOTROPIN
IMMUNOSUPPRESSION, 3953*
CYCLOPHOSPHAMIDE
IMMUNOSUPPRESSION, 3953*
ERYTHROPOIESIS
CHILD, 4020
GENETICS, 4087*
INFECTIOUS MONONUCLEOSIS
ANTIGENS, HETEROGENETIC, 3914
LEUKOCYTES
KINETICS, CHROMIUM LABELING, 4080*
ORGANOTROPISM, 4080*
LIPOPROTEINS
ISOLATION AND CHARACTERIZATION,
3632
MYOBLASTOMA
NEOPLASM METASTASIS, 4056*
NEOPLASM TRANSPLANTATION
ULTRASTRUCTURAL STUDY, HAMSTER,
4079*
NEUTROPHILS
PHAGOCYTOSIS, 4188*
PRECANCEROUS CONDITIONS
CASE REPORT, 4081*
ULTRASTRUCTURAL STUDY, 4081*
PROTEINS
ISOLATION AND CHARACTERIZATION, C-
REACTIVE, 4169
RADIATION, IONIZING
CARCINOGENIC EFFECT, MOUSE, 3782*
VIRUS, C-TYPE MURINE MYELOMA
ULTRASTRUCTURAL STUDY, HAMSTER,
4079*
VIRUS, C-TYPE RNA TUMOR
REVERSE TRANSCRIPTASE, 3842*
VIRUS-LIKE PARTICLES
ULTRASTRUCTURAL STUDY, HAMSTER,
4079*

LEUKEMIA, MYELOGENOUS
SEE LEUKEMIA, MYELOCYTIC

LEUKEMIA, MYELOID
SEE LEUKEMIA, MYELOCYTIC

LEUKEMIA, PLASMACYTIC
IMMUNOGLOBULINS, SURFACE
CELL DIFFERENTIATION, 3984*
B-LYMPHOCYTES
CELL DIFFERENTIATION, 3984*

LEUKEMIA, PROMYELOCYTIC
SEE LEUKEMIA, MYELOBLASTIC

LEUKEMIA, RADIATION-INDUCED
  BONE MARROW CELLS
    IMMUNE RESPONSE, 3889
  CARBON RADIOISOTOPES
    CARCINOGENIC EFFECT, MOUSE, 3782*
  T-LYMPHOCYTES
    IMMUNE RESPONSE, 3889
  RADIATION CHIMERA
    IMMUNE RESPONSE, 3889
  RADIATION, IONIZING
    CARCINOGENIC EFFECT, MOUSE, 3782*
  THORIUM DIOXIDE
    CHROMOSOME ABERRATIONS, 4026

LEUKEMIA, SUBLEUKEMIC
  GENETICS
    ANTIGENS, NEOPLASM, 4030

LEUKOCYTES
  CARCINOMA
    CHEMOTAXIS, 3886
  CELL TRANSFORMATION, NEOPLASTIC
    CARCINOGENESIS SCREENING, 3718*
  HEPATOMA
    CHEMOTAXIS, 3886
  LEUKEMIA
    BIOCHEMISTRY AND METABOLISM,
      REVIEW, 3666*
  LEUKEMIA, MYELOCYTIC
    KINETICS, CHROMIUM LABELING, 4080*
    ORGANOTROPISM, 4080*
  METHYLPREDNISOLONE
    POLYMORPHONUCLEAR NEUTROPHILS,
      3976*
  POLYCYTHEMIA VERA
    PUTRESCINE, 4161
    SPERMINE, 4161
  PURINE, 6-((1-METHYL-4-NITROIMIDAZOL-5-
    YL)THIO)-
    POLYMORPHONUCLEAR NEUTROPHILS,
      3976*
  VIRUS, HERPES TURKEY
    DETECTION, 3854*
  VIRUS, MAREK'S DISEASE HERPES
    DETECTION, 3854*

LIGANDS
  GANGLIOSIDES
    CELL MEMBRANE, 3924

LIP NEOPLASMS
  BURKITT'S LYMPHOMA
    NEOPLASM METASTASIS, 4012
    VIRUS-LIKE PARTICLES, 4012
  HEMANGIOMA
    EPIDEMIOLOGY, 4000

LIPOPROTEINS
  CARCINOMA
    ISOLATION AND CHARACTERIZATION,
      3632
  HODGKIN'S DISEASE
    ISOLATION AND CHARACTERIZATION,
      3632
  LEIOMYOSARCOMA
    ISOLATION AND CHARACTERIZATION,
      3632
  LEUKEMIA, MYELOCYTIC
    ISOLATION AND CHARACTERIZATION,
      3632
  MULTIPLE MYELOMA
    ISOLATION AND CHARACTERIZATION,
      3632
  NEOPLASM METASTASIS
    ISOLATION AND CHARACTERIZATION,
      3632

LIVER
  RADIATION INJURIES, EXPERIMENTAL
    PATHOPHYSIOLOGICAL STUDY, RAT,
      3787*

LIVER CELL CARCINOMA
  SEE HEPATOMA

LIVER NEOPLASMS
  ACETAMIDE, N-FLUOREN-2-YL-
    TRANSPLACENTAL CARCINOGENESIS,
      REVIEW, 3601
  ADENOMA
    S-TRIAZINE, HEXAHYDRO-1,3,5-
      TRINITRO-, 3694
  AFLATOXIN B1
    MORPHOLOGICAL CHANGES, PARENCHYMA,
      3719*
  ANGIOSARCOMA
    EPIDEMIOLOGY, CHILD, 4140*
    ETHYLENE, CHLORO-, 4036, 4124
    ETHYLENE, CHLORO- POLYMER, 4036
  BARBITURIC ACID, 5-ETHYL-5-PHENYL-
    PRECANCEROUS CONDITIONS, 3766*
  EPIDEMIOLOGY
    AFRICA, 4116
  ETHYLENE, CHLORO-
    EPIDEMIOLOGY, 4124
  HEMANGIOENDOTHELIOMA
    BUTANOL, 4-(BUTYLNITROSOAMINO)-,
      3701
    DIPROPYLAMINE, N-NITROSO-, 3702
    EPIDEMIOLOGY, CHILD, 4140*
    PROPYLAMINE, N-METHYL-N-NITROSO-,
      3702
  HEMANGIOMA
    S-TRIAZINE, HEXAHYDRO-1,3,5-
      TRINITRO-, 3694
  HEMANGIOPERICYTOMA
    EPIDEMIOLOGY, CHILD, 4140*
  HEMOSTASIS, ALTERATIONS
    JAUNDICE, COMPLICATIONS, 4074*
  LEIOMYOMA
    EPIDEMIOLOGY, 4000

LUCANTHONE
  RADIATION
    RNA REPLICATION, 3779*

LUNG NEOPLASMS
  ACETAMIDE, N-FLUOREN-2-YL-
    TRANSPLACENTAL CARCINOGENESIS,
      REVIEW, 3601
  ADENOCARCINOMA
    DIPROPYLAMINE, N-NITROSO-, 3702
    NEOPLASM METASTASIS, 4046
  ADENOMA

CARBAMIC ACID, ETHYL ESTER, 3893
CHOLANTHRENE, 3-METHYL-, 3893
DIPROPYLAMINE, N-NITROSO-, 3702
ALPHA 1 ANTITRYPSIN
EPIDEMIOLOGY, GENETICS, 4121
ANGIOSARCOMA
ETHYLENE, CHLORO-, 3627
OCCUPATIONAL HAZARD, 3627
ARSENIC
OCCUPATIONAL HAZARD, 4127
ASBESTOS
OCCUPATIONAL HAZARD, 3643*
BENZ(A)ANTHRACENE, 7,12-DIMETHYL-
TRANSPLACENTAL CARCINOGENESIS,
REVIEW, 3601
BENZO(A)PYRENE, 3610
TRANSPLACENTAL CARCINOGENESIS,
REVIEW, 3601
BLOOD GROUPS
EPIDEMIOLOGY, GENETICS, 4121
CARBAMIC ACID, ETHYL ESTER
IMMUNOSUPPRESSION, 3893
TRANSPLACENTAL CARCINOGENESIS,
REVIEW, 3601
CARCINOMA
EPIDEMIOLOGY, 4146*
SEX FACTORS, 4146*
CARCINOMA, EPIDERMOID
DIPROPYLAMINE, N-NITROSO-, 3702
NEOPLASM METASTASIS, 4046
CARCINOMA, OAT CELL
NEOPLASM METASTASIS, 4046
CELL CYCLE KINETICS
TISSUE CULTURE SUBPOPULATIONS,
HAMSTER, 4148
CHOLANTHRENE, 3-METHYL-
IMMUNOSUPPRESSION, 3893
6-CHRYSENAMINE
CARCINOGENIC POTENTIAL, 3681
COPPER
OCCUPATIONAL HAZARD, 4127
CORONENE, 3610
EPIDEMIOLOGY
AGE FACTORS, GREAT BRITAIN, 4114
MORMONS, 4118
IMMUNITY, CELLULAR
REVIEW, 3640*
ISOANTIGENS
TRANSPLACENTAL CARCINOGENESIS,
3892
LEUKEMIA, MONOCYTIC
IMMUNITY, CELLULAR, 3933*
TRANSPLANTATION IMMUNOLOGY, 3933*
MESOTHELIOMA
ETHYLENE, CHLORO-, 3627
OCCUPATIONAL HAZARD, 3627, 3643*
METALS
CARCINOGENIC POTENTIAL, REVIEW,
3610
MONOCYTES
IGG, 3958*
NEOPLASM METASTASIS, 4045
OCCUPATIONAL HAZARD
EPIDEMIOLOGY, JAPAN, 4127
PAPILLOMA
CROTON OIL, 3733*
PHENOTYPE
EPIDEMIOLOGY, GENETICS, 4121
PLUTONIUM

AGE FACTORS, 3781*
PRECANCEROUS CONDITIONS, 3781*
RADIATION INJURIES, EXPERIMENTAL,
3781*
SIALIC ACIDS
DNA, 4166
SMOKING, 3610
AGE FACTORS, GREAT BRITAIN, 4114
AGE FACTORS, STATISTICAL ANALYSIS,
4128
EPIDEMIOLOGY, AFRICA, 4116
EPIDEMIOLOGY, GENETICS, 4121
EPIDEMIOLOGY, IMMIGRANTS, 4126
UREA, ETHYL NITROSO-
TRANSPLACENTAL CARCINOGENESIS,
3892

LYMPHATIC NEOPLASMS
IMMUNOSUPPRESSION
REVIEW, 3652*
SARCOMA
VIRUS, POLYOMA, BK, 3810
TRANSPLANTATION, HOMOLOGOUS
ANTIGENIC STIMULATION, REVIEW,
3807

LYMPHOCYTE TRANSFORMATION
B-LYMPHOCYTES
AUTOLOGOUS STIMULATION, 3885
THYMIDINE INCORPORATION, 3885
T-LYMPHOCYTES
THYMIDINE INCORPORATION, 3885
NICOTINE
DNA REPLICATION, 3712

LYMPHOCYTES
ADENOCARCINOMA
TRANSPLANTATION IMMUNOLOGY, 3890
ASCITIC FLUID
ANTIGEN-ANTIBODY REACTIONS, 3950*
IMMUNOSUPPRESSION, 3950*
P-BENZOQUINONE, 2,3,5-TRIS(1-
AZIRIDINYL)-
CHROMOSOME ABERRATIONS, 3732*
BREAST NEOPLASMS
IMMUNITY, CELLULAR, 3936*
CARCINOEMBRYONIC ANTIGEN
PRECANCEROUS CONDITIONS, 3908
CELL MEMBRANE
ULTRASTRUCTURAL STUDY, 4110*
CELL NUCLEUS
ULTRASTRUCTURAL STUDY, 4110*
COLONIC NEOPLASMS
CARCINOEMBRYONIC ANTIGEN, 3908
HODGKIN'S DISEASE
ULTRASTRUCTURAL STUDY, 4086*
IMMUNITY, CELLULAR
MACROPHAGES, 3931*
LEUKEMIA
PRECANCEROUS CONDITIONS, 3978*
LEUKEMIA, LYMPHOBLASTIC
IMMUNE RESPONSE, 3929
LYMPHOKINES
DNA REPLICATION, 3922
PROTEINS, 3922
RNA REPLICATION, 3922
METHYLPREDNISOLONE
POLYMORPHONUCLEAR NEUTROPHILS,
3976*

PLANT AGGLUTININS
  BINDING SITES, AGE FACTORS, 3977*
PURINE, 6-((1-METHYL-4-NITROIMIDAZOL-5-
  YL)THIO)-
    POLYMORPHONUCLEAR NEUTROPHILS,
    3976*
RHABDOMYOSARCOMA
  IMMUNOSUPPRESSION, 3952*
VIRUS, BOVINE LEUKEMIA
  MORPHOLOGY, 3857*
VIRUS, HERPES TURKEY
  DETECTION, 3854*
VIRUS, MAREK'S DISEASE HERPES
  DETECTION, 3854*

-LYMPHOCYTES
ANTILYMPHOCYTE SERUM
  LANTHANUM, 4167
BINDING SITES
  GENOTYPE, 3617
BRONCHIAL NEOPLASMS
  CARCINOMA, 3926
  CARCINOMA, EPIDERMOID, 3926
  CARCINOMA, OAT CELL, 3926
  DIAGNOSIS AND PROGNOSIS, 3926
BURKITT'S LYMPHOMA
  CELL LINE, 3795
CONCANAVALIN A
  LANTHANUM, 4167
HISTOCOMPATIBILITY ANTIGENS
  REVIEW, 3654*
HODGKIN'S DISEASE
  CELL DIFFERENTIATION, 3984*
  IMMUNE SERUMS, 3921
IMMUNE SERUMS
  LANTHANUM, 4167
INFECTIOUS MONONUCLEOSIS
  IMMUNE SERUMS, 3921
LEUKEMIA, LYMPHOCYTIC
  ANTILYMPHOCYTE SERUM, 3961*
  CELL DIFFERENTIATION, 3984*
  IMMUNE SERUMS, 3921
  IMMUNOGLOBULINS, SURFACE, 3918
  ROSETTE FORMATION, 3917
LEUKEMIA, PLASMACYTIC
  CELL DIFFERENTIATION, 3984*
LYMPHOCYTE TRANSFORMATION
  AUTOLOGOUS STIMULATION, 3885
  THYMIDINE INCORPORATION, 3885
LYMPHOMA
  IMMUNE SERUMS, 3921
MITOMYCIN C
  LANTHANUM, 4167
MULTIPLE MYELOMA
  IMMUNOGLOBULINS, SURFACE, 3923
PLANT AGGLUTININS
  LANTHANUM, 4167
PLASMACYTOMA
  ANTIGENIC DETERMINANTS, 3919
SARCOMA, MAST CELL
  IMMUNOSUPPRESSION, 3887
TUBERCULIN
  LANTHANUM, 4167
VIRUS, MAREK'S DISEASE HERPES
  HISTOCOMPATIBILITY ANTIGENS, 3980*

LYMPHOCYTES
ANTIBODIES, NEOPLASM
  TUBERCULIN TEST, 3938*

ANTILYMPHOCYTE SERUM
  LANTHANUM, 4167
ASCITIC FLUID
  CONCANAVALIN A, 3979*
  LYMPHOCYTE TRAPPING, 3979*
  PLANT AGGLUTININS, 3979*
BENZENE, 1-CHLORO-2,4-DINITRO-
  IMMUNE RESPONSE, 3938*
BINDING SITES
  GENOTYPE, 3617
BRONCHIAL NEOPLASMS
  CARCINOMA, 3926
  CARCINOMA, EPIDERMOID, 3926
  CARCINOMA, OAT CELL, 3926
  DIAGNOSIS AND PROGNOSIS, 3926
CONCANAVALIN A
  DNA REPLICATION, 3927
  LANTHANUM, 4167
HISTOCOMPATIBILITY ANTIGENS
  REVIEW, 3654*
HODGKIN'S DISEASE
  DNA REPLICATION, 3921
  ROSETTE FORMATION, 3921
IMMUNOSUPPRESSION
  MACROPHAGES, 3949*
INFECTIOUS MONONUCLEOSIS
  ANTIGENS, HETEROGENETIC, 3914
  DNA REPLICATION, 3921
  ROSETTE FORMATION, 3921
LEUKEMIA
  ANTIGENS, NEOPLASM, 3915
LEUKEMIA, LYMPHOBLASTIC
  ACID PHOSPHATASE, 3920
  NEURAMINIDASE, 3920
  ROSETTE FORMATION, 3920
LEUKEMIA, LYMPHOCYTIC
  AGRANULOCYTOSIS, 3918
  ANTIBODIES, 3961*
  ANTILYMPHOCYTE SERUM, 3961*
  IMMUNE RESPONSE, 3917
  IMMUNE SERUMS, 3918, 3961*
  PITUITARY GLAND, 3981*
  ROSETTE FORMATION, 3917, 3918,
    3921
  SKIN NEOPLASMS, 3918
  SPLENOMEGALY, 3918
  VIRUS, GROSS MURINE LEUKEMIA,
    3981*
LEUKEMIA, RADIATION-INDUCED
  IMMUNE RESPONSE, 3889
LYMPHOCYTE TRANSFORMATION
  THYMIDINE INCORPORATION, 3885
LYMPHOMA
  ROSETTE FORMATION, 3921
LYMPHOSARCOMA
  IMMUNE RESPONSE, 3889
MAMMARY NEOPLASMS, EXPERIMENTAL
  VIRUS, MURINE MAMMARY TUMOR, 3868*
MITOMYCIN C
  LANTHANUM, 4167
MONOCYTES
  DNA REPLICATION, 3927
NEOPLASMS
  ASCITIC FLUID, 3979*
  IMMUNOSUPPRESSION, 3983*
PLANT AGGLUTININS
  LANTHANUM, 4167
PLASMACYTOMA
  ANTIGENIC DETERMINANTS, 3919

RADIATION, IONIZING
    IMMUNE RESPONSE, 3982*
SARCOMA, MAST CELL
    IMMUNOSUPPRESSION, 3887
SMOKING
    CARCINOEMBRYONIC ANTIGEN, 3768*
TUBERCULIN
    LANTHANUM, 4167
VIRUS, MAREK'S DISEASE HERPES
    HISTOCOMPATIBILITY ANTIGENS, 3980*

LYMPHOKINES
    LYMPHOCYTES
        DNA REPLICATION, 3922
        PROTEINS, 3922
        RNA REPLICATION, 3922
    MONOCYTES
        SYNTHESIS, SPONTANEOUS, 3985*
    VIRUS, HERPES SAIMIRI
        VIRUS REPLICATION, 3922
    VIRUS, HERPES SIMPLEX 1
        VIRUS REPLICATION, 3922

LYMPHOMA (GENERAL AND UNSPECIFIED)
    SEE ALSO HODGKIN'S DISEASE/BURKITT'S
        LYMPHOMA/ SARCOMA, RETICULUM CELL/
        LYMPHOSARCOMA
    ABDOMINAL NEOPLASMS
        CASE REPORT, 4015
    ALKALINE PHOSPHATASE
        PHOSPHATES, 3862*
    CHOLANTHRENE, 3-METHYL-
        TRANSPLACENTAL CARCINOGENESIS,
        REVIEW, 3601
    EPIDEMIOLOGY
        MORMONS, 4118
        MORMONS, 4118
    GRAFT VS HOST REACTION
        GRAFT VS HOST REACTICN, 3878*
        HISTOCOMPATIBILITY TESTING, 3888
    GROWTH
        MACROPHAGES, 3949*
    HEAD AND NECK NEOPLASMS
        GROWTH SUBSTANCES, 4016
        TRANSPLANTATION, HETEROLOGOUS,
        4016
    HODGKIN'S DISEASE
        CLASSIFICATION, 3625
    INFECTIOUS MONONUCLEOSIS
        ANTIGENS, HETEROGENETIC, 3914
    ISOANTIGENS
        IMMUNE RESPONSE, 3892
    B-LYMPHOCYTES
        IMMUNE SERUMS, 3921
    T-LYMPHOCYTES
        ROSETTE FORMATION, 3921
    MAREK'S DISEASE
        COCCIDIOSIS, 4088*
    MONOCYTES
        IGG, 3958*
    RADIATION, IONIZING
        EPIDEMIOLOGY, HIROSHIMA, NAGASAKI,
        4113
        MORTALITY, 3780*
        RADIATION DOSAGE, 4113
    THYMUS NEOPLASMS
        ALKALINE PHOSPHATASE, 3862*

VIRUS, C-TYPE RNA TUMOR
    GRAFT VS HOST REACTICN, 3878*
VIRUS, HERPES SAIMIRI
    ISOLATION AND CHARACTERIZATION,
    3922
VIRUS, MAREK'S DISEASE HERPES
    HISTOCOMPATIBILITY ANTIGENS, 3980
VIRUS, SCRIPPS MURINE LEUKEMIA
    IMMUNOLOGY, 3861*
    VERTICAL TRANSMISSION, 3861*

LYMPHOSARCOMA
    BLOOD SEDIMENTATION
        DIAGNOSOS AND PROGNOSIS, 4019
    BONE MARROW CELLS
        IMMUNE RESPONSE, 3889
    BREAST NEOPLASMS
        CASE REPORT, 4054*
    CHROMOSOME ABNORMALITIES
        CASE REPORT, CHILD, 4014
    EPIDEMIOLOGY
        CHILD INDIA, 4120
    ESTRADIOL
        IMMUNITY, CELLULAR, 3945*
        TRANSPLANTATION IMMUNOLOGY, 3945*
    FETAL GLOBULINS
        DIAGNOSOS AND PROGNOSIS, 4019
    FIBRINOGEN
        DIAGNOSOS AND PROGNOSIS, 4019
    HAPTOGLOBINS
        DIAGNOSOS AND PROGNOSIS, 4019
    HEMANGIOMA
        RADIATION, IONIZING, 3773
    INFECTIOUS MONONUCLEOSIS
        ANTIGENS, HETEROGENETIC, 3914
    KARYOTYPING
        CASE REPORT, CHILD, 4014
    T-LYMPHOCYTES
        IMMUNE RESPONSE, 3889
    MYCOPLASMA
        CARCINOGENIC POTENTIAL, MOUSE,
        4039
    PROTEINS
        ISOLATION AND CHARACTERIZATION, C-
        REACTIVE, 4169
    RADIATION CHIMERA
        IMMUNE RESPONSE, 3889
    RADIATION, IONIZING
        IMMUNE RESPONSE, 3889
    THYMUS NEOPLASMS
        IMIDAZOLE-4-CARBOXAMIDE, 5-(3,3-
        DIMETHYL-1-TRIAZENO)-, 3691,
        3758*

MACROGLOBULINEMIA
    ANTIGENS, NEOPLASM
        IMMUNE SERUMS, 3973*

MACROPHAGES
    ADENOCARCINOMA
        TRANSPLANTATION IMMUNOLOGY, 3890
    CELL MEMBRANE
        ULTRASTRUCTURAL STUDY, 3987*
    IMMUNITY, CELLULAR
        LYMPHOCYTES, 3931*
    T-LYMPHOCYTES

IMMUNOSUPPRESSION, 3949*
LYMPHOMA
GROWTH, 3949*
NEOPLASMS
VIRUS, MOLONEY MURINE SARCOMA,
3949*

MAGNESIUM
VIRUS, HAMSTER SARCOMA
NUCLEOTIDYLTRANSFERASES, 3799
POLY A POLYMERASE, 3799
VIRUS, HEPRES SIMPLEX 2
DNA REPLICATION, 3832

MAMMARY NEOPLASMS, EXPERIMENTAL
ACETAMIDE, N-FLUOREN-2-YL-
TRANSPLACENTAL CARCINOGENESIS,
REVIEW, 3601
ADENOCARCINOMA
BENZ(A)ANTHRACENE, 7,12-DIMETHYL-,
3741*
CYCLOOCTANECARBAMIC ACID, 1,1-BIS
(P-FLUOROPHENYL)-2-PROPYNYL,
3684
DIETARY FATS, 3606
ESTRADIOL, 3741*, 4180
IMIDAZOLE-4-CARBOXAMIDE, 5-AMINO-,
3691
IMIDAZOLE-4-CARBOXAMIDE, 5-(3,3-
DIMETHYL-1-TRIAZENO)-, 3691
IMIDAZOLE-4-CARBOXIMIDE, 5-DIAZO-,
3691
4H-IMIDAZOLE-(4,5-D)-V-TRIAZIN-4-
ONE, 3,7-DIHYDRO-, 3691
NEOPLASM TRANSPLANTATION, 4149
PROLACTIN, 3741*
SUBCELLULAR TUMOR FRACTIONS, 3866*
VIRUS, MURINE MAMMARY TUMOR, 3829,
3866*
ADENOFIBROMA
IMIDAZOLE-4-CARBOXAMIDE, 5-DIAZO-,
3691
TRITIUM, 3714
ANTIBODIES
PLASMINOGEN ACTIVATORS, 3881*
BENZ(A)ANTHRACENE
POLYNUCLEOTIDE PHOSPHORYLASE,
3745*
BENZ(A)ANTHRACENE, 7,12-DIMETHYL-
BINDING SITES, 3747*
CYTOSOL, 3740*
ESTRADIOL, 3680
OLIGOPEPTIDES, 3752*
PROGESTERONE, 3680
TESTOSTERONE, 3680
TRANSPLACENTAL CARCINOGENESIS,
REVIEW, 3601
ULTRASTRUCTURAL STUDY, RAT, 3753*
CARCINOMA, 3785*
PLASMINOGEN ACTIVATORS, 3881*
UREMIA, 4151
VIRUS, POLYOMA, 3874*
CELL DIVISION
CELL CYCLE KINETICS, 4131
CHOLANTHRENE, 3-METHYL-
ESTRADIOL, 3680
PROGESTERONE, 3680
TESTOSTERONE, 3680
DEOXYRIBONUCLEASE

VIRUS, MURINE MAMMARY TUMOR, 3866*
DIMETHYLAMINE, N-NITROSO-
MUTAGENIC ACTIVITY, 3699
ESTRADIOL
GLUCOSEPHOSPHATE DEHYDROGENASE,
4180
GROWTH
CELL CYCLE KINETICS, 4139*
IMMUNIZATION
HYPERSENSITIVITY, 3942*
ISOANTIGENS, 3892
IMMUNE RESPONSE, 3892
MYCOPLASMA
CARCINOGENIC POTENTIAL, MOUSE,
4039
MYXOMA
ULTRASTRUCTURAL STUDY, DOG, 4058*
NEOPLASM TRANSPLANTATION
GROWTH, 4149
PORNASE
VIRUS, MURINE MAMMARY TUMOR, 3866*
PROLACTIN
ESTROGENS, 3638*
1-PIPERAZINEETHANOL, 4-(3-(2-
CHLOROPHENOTHIAZIN-10-YL)PROPYL)-
, 3688
PITUITARY GLAND NEOPLASMS, 3638*
RADIATION INJURIES, EXPERIMENTAL
DNA, 4113
RADIATION, IONIZING
TRANSPLANTATION IMMUNOLOGY, 4149
ULTRASTRUCTURAL STUDY, 3866*
VIRUS, MURINE MAMMARY TUMOR, 3866*
RIBONUCLEASE
VIRUS, MURINE MAMMARY TUMOR, 3866*
SARCOMA
CHOLANTHRENE, 3-METHYL-, 3680
SIALIC ACIDS
DNA, 4166
SOMATOTROPIN
1-PIPERAZINEETHANOL, 4-(3-(2-
CHLOROPHENOTHIAZIN-10-YL)PROPYL)-
, 3688
SULPIRIDE
LACTAION, 3713
PROLACTIN, 3713
TRANSPLANTATION, HOMOLOGOUS
SPLEEN, 3940*
UREMIA
CELL CYCLE KINETICS, 4151
VACCINES, ATTENUATED
IMMUNITY, CELLULAR, 3619
VIRUS, BITTNER MURINE MAMMARY TUMOR
STRESS, 4038
VIRUS, LACTATE DEHYDROGENASE
IMMUNITY, CELLULAR, 3619
VIRUS, MURINE MAMMARY TUMOR
T-LYMPHOCYTES, 3868*
STRESS, 3868*, 4038

MANGANESE
VIRUS, HAMSTER SARCOMA
NUCLEOTIDYLTRANSFERASES, 3799
POLY A POLYMERASE, 3799

MAREK'S DISEASE
COCCIDIOSIS
CASE REPORT, 4088*
LYMPHOMA

COCCIDIOSIS, 4088*

MAST CELLS
  HODGKIN'S DISEASE
    ULTRASTRUCTURAL STUDY, 4007

MAXILLARY NEOPLASMS
  CARCINOMA
    CELL CYCLE KINETICS, 4130

MEDIASTINAL NEOPLASMS
  NEOPLASM METASTASIS
    CASE REPORT, 4044

MEDULLOBLASTOMA
  BRAIN NEOPLASMS
    ENDOPLASMIC RETICULUM, 4005
    GOLGI APPARATUS, 4005

MEDULLOEPITHELIOMA
  SEE NEUROEPITHELIOMA

MEGAKARYOCYTES
  LEUKEMIA
    EPIDEMIOLOGY, 4119

MEHTANESULFONIC ACID, METHYL ESTER
  CYCLOPHOSPHAMIDE
    DOMINANT LETHAL TEST, 3760*
  MUTAGENS
    DOMINANT LETHAL TEST, 3760*

MELANOMA
  ANTIBODIES
    PLASMINOGEN ACTIVATORS, 3881*
  ANTILYMPHOCYTE SERUM
    NEOPLASM METASTASIS, 4035
  CELL DIVISION
    CELL CYCLE KINETICS, 4131
  GENETICS, 4087*
  MONOCYTES
    IGG, 3958*
  MOUTH NEOPLASMS
    CASE REPORT, 4064*
  NEOPLASM METASTASIS
    EPIDEMIOLOGY, CHILD, 3655*
  NEVUS, PIGMENTED
    EPIDEMIOLOGY, CHILD, 3655*
  SKIN NEOPLASMS
    CASE REPORT, 4022
    CELL TRANSFORMATION, NEOPLASTIC,
      4040
    HISTOLOGICAL STUDY, 4040
    IMMUNE RESPONSE, 4040
    PRECANCEROUS CONDITIONS, 4040
    UTLRASTRUCTURAL STUDY, 4061*
  SULFATASES
    HISTOCHEMICAL STUDY, 4179
  TRANSPLACENTAL CARCINOGENESIS
    EPIDEMIOLOGY, CHILD, 3655*
  TRANSPLANTATION, HETEROLCGOUS
    NEOPLASM METASTASIS, 4035

MENINGIOMA
  BRAIN NEOPLASMS
    ENDOPLASMIC RETICULUM, 4005
    GOLGI APPARATUS, 4005
    NEOPLASMS, MULTIPLE PRIMARY, 3995
  CELLULAR INCLUSIONS

ULTRASTRUCTURAL STUDY, 4006

MESENCHYMOMA
  TESTICULAR NEOPLASMS
    CASE REPORT, 4093*
  UTERINE NEOPLASMS
    DNA, NEOPLASM, 4100*

MESOTHELIOMA
  LUNG NEOPLASMS
    ETHYLENE, CHLORO-, 3627
    OCCUPATIONAL HAZARD, 3627, 3643*
  PERITONEAL NEOPLASMS
    THORIUM DIOXIDE, 4037

METALS
  LUNG NEOPLASMS
    CARCINOGENIC POTENTIAL, REVIEW,
      3610
  RESPIRATORY TRACT NEOPLASMS
    OCCUPATIONAL HAZARD, 3640*
  SMOKING
    COCARCINOGENIC EFFECT, REVIEW,
      3610

METHANE, SULFINYLBIS-
  PLASMACYTOMA
    A-TYPE PARTICLES, 3870*

METHANESULFONIC ACID, ETHYL ESTER
  CHROMOSOME ABERRATIONS
    BONE MARROW, 3717*

METHANESULFONIC ACID, METHYL ESTER
  BURKITT'S LYMPHOMA
    DNA REPAIR, 3778*
  CHROMOSOME ABERRATIONS
    BONE MARROW, 3717*
  CYTOSINE, 1-BETA-D-ARABINOFURANOSYL-
    DNA REPAIR, 3759*

METHANOL, (METHYL-ONN-AZOXY)-
  LIVER REGENERATION
    MEGALOCYTOSIS, 3692

METHOTREXATE
  CELL LINE
    CELL TRANSFORMATION, NEOPLASTIC,
      3720*
  FOLIC ACID
    FIBROBLASTS, CHICKEN, 4154
  VIRUS, FRIEND MURINE LEUKEMIA
    SPLENOMEGALY, 3797

N-METHYL-N'-NITRO-N-NITROSOGUANIDINE
  SEE GUANIDINE, 1-METHYL-3-NITRO-1-
    NITROSO-

N-METHYL-N-NITROSOUREA
  SEE UREA, METHYL NITROSO-

1-METHYL-1-NITROSOUREA
  SEE UREA, METHYL NITROSO-

METHYL SULFOXIDE
  SEE METHANE, SULFINYLBIS-

3-METHYLCHOLANTHRENE
  SEE CHOLANTHRENE, 3-METHYL-

```
MONOCROTALINE
    PANCREATIC NEOPLASMS
        ADENOMA, 3761*

MONOCYTES
    BONE MARROW CELLS
        CELL DIVISION, 3899
    BREAST NEOPLASMS
        IGG, 3958*
    LEUKEMIA, MYELOBLASTIC
        IMMUNE RESPONSE, 3900
        IMMUNITY, CELLULAR, 3900
    LUNG NEOPLASMS
        IGG, 3958*
    T-LYMPHOCYTES
        DNA REPLICATION, 3927
    LYMPHOKINES
        SYNTHESIS, SPONTANEOUS, 3985*
    LYMPHOMA
        IGG, 3958*
    MELANOMA
        IGG, 3958*
    THYMOMA
        IGG, 3958*

MORPHOLINE, N-NITROSO-
    DNA NUCLEOTIDYTRANSFERASES
        MORPHOLOGICAL EFFECT, LIVER, RAT,
            3703
    RNA POLYMERASE
        MORPHOLOGICAL EFFECT, LIVER, RAT,
            3703

MOUTH NEOPLASMS
    CARCINOMA, EPIDERMOID
        CLOSTRIDIOPEPTIDASE A, 3998
    HEMANGIOMA
        EPIDEMIOLOGY, 4000
    LEIOMYOMA
        EPIDEMIOLOGY, 4000
    MELANOMA
        CASE REPORT, 4064*
    NEOPLASM METASTASIS, 4045

MUCOPOLYSACCHARIDES
    CHONDROSARCOMA
        ISOLATION AND CHARACTERIZATION,
            4195*
    HEPATOMA
        CELL MEMBRANE, 4173

MULTIPLE MYELOMA
    ANTIGENS, NEOPLASM
        IMMUNE SERUMS, 3973*
    IMMUNE SERUMS
        IMMUNOGLOBULINS, SURFACE, 3923
    ISOANTIGENS
        IMMUNE RESPONSE, 3892
    LIPOPROTEINS
        ISOLATION AND CHARACTERIZATION,
            3632
    B-LYMPHOCYTES
        IMMUNOGLOBULINS, SURFACE, 3923
    PROTEINS
        ISOLATION AND CHARACTERIZATION, C-
            REACTIVE, 4169
    RADIATION, IONIZING
        EPIDEMIOLOGY, HIROSHIMA, NAGASAKI,
            4113
```

01

RADIATION DOSAGE, 4113
RNA, NEOPLASM
 IMMUNOGLOBULINS, SURFACE, 3923
VIRUS, HERPES SIMPLEX
 CASE REPORT, 3894
 IMMUNITY, CELLULAR, 3894

MULTIPLE SCLEROSIS
 COLONIC NEOPLASMS
  EPIDEMIOLOGY, 4122
  TRANSIENTS AND MIGRANTS, 4122
 CORONARY DISEASE
  EPIDEMIOLOGY, 4122
 RHEUMATIC HEART DISEASE
  EPIDEMIOLOGY, 4122
 STOMACH NEOPLASMS
  EPIDEMIOLOGY, 4122
 TUBERCULOSIS, PULMONARY
  EPIDEMIOLOGY, 4122

MUTAGENS
 MEHTANESULFONIC ACID, METHYL ESTER
  DOMINANT LETHAL TEST, 3760*

MUTATION
 GENETICS
  REVIEW, 3661*
 NEUROFIBROMATOSIS
  GENETICS, 4027
 RADIATION, IONIZING
  DOSE-RESPONSE STUDY, 3777*

MYASTHENIA GRAVIS
 THYMOMA
  ANEMIA, APLASTIC, 4044

MYCOBACTERIUM BOVIS
 ADENOCARCINOMA
  TRANSPLANTATION IMMUNOLOGY, 3890

MYCOBACTERIUM BUTYRICUM
 VIRUS, FRIEND MURINE LEUKEMIA
  SPLENOMEGALY, 3797

MYCOPLASMA
 ADENOCARCINOMA
  CARCINOGENIC POTENTIAL, MOUSE,
  4039
 ANGIOSARCOMA
  CARCINOGENIC POTENTIAL, MOUSE,
  4039
 CARCINOMA
  CARCINOGENIC POTENTIAL, MOUSE,
  4039
 FIBROSARCOMA
  CARCINOGENIC POTENTIAL, MOUSE,
  4039
 LYMPHOSARCOMA
  CARCINOGENIC POTENTIAL, MOUSE,
  4039
 MAMMARY NEOPLASMS, EXPERIMENTAL
  CARCINOGENIC POTENTIAL, MOUSE,
  4039
 MYOSARCOMA
  CARCINOGENIC POTENTIAL, MOUSE,
  4039
 SARCOMA
  CARCINOGENIC POTENTIAL, MOUSE,
  4039

SARCOMA, RETICULUM CELL
 CARCINOGENIC POTENTIAL, MOUSE,
 4039

MYCOTOXINS
 ASPERGILLUS FLAVUS
  FOOD CONTAMINATION, 4125
 ASPERGILLUS PARASITICUS
  FOOD CONTAMINATION, 4125
 CARCINOGEN, ENVIRONMENTAL
  REVIEW, 3604
 FOOD CONTAMINATION
  CORN MEAL, 4138*
 HEPATOMA
  REVIEW, 3603

MYELOMA
 SEE MULTIPLE MYELOMA

MYELOMA PROTEINS
 IMMUNOGLOBULIN FRAGMENTS
  ISOLATION AND CHARACTERIZATION,
  3957*
 IMMUNOGLOBULINS, FAB
  ISOLATION AND CHARACTERIZATION,
  3957*

MYOBLASTOMA
 BREAST NEOPLASMS
  NEOPLASM METASTASIS, 4056*
 LEUKEMIA, MYELOCYTIC
  NEOPLASM METASTASIS, 4056*

MYOSARCOMA
 MYCOPLASMA
  CARCINOGENIC POTENTIAL, MOUSE,
  4039

MYXOMA
 MAMMARY NEOPLASMS, EXPERIMENTAL
  ULTRASTRUCTURAL STUDY, DOG, 4058*

NADP
 DIMETHYLAMINE, N-NITROSO-
  MUTAGENIC ACTIVITY, 3699

ALPHA-NAPHTHOFLAVONE-
 SEE 7,8-BENZOFLAVONE

BETA-NAPHTHOFLAVONE
 SEE 5,6-BENZOFLAVONE

2-NAPHTHYLAMINE
 OCCUPATIONAL HAZARD
  WORKER PROTECTION, REVIEW, 3629

NASOPHARYNGEAL NEOPLASMS
 NEOPLASM METASTASIS, 4045
 PROTEINS
  ISOLATION AND CHARACTERIZATION, C
  REACTIVE, 4169
 VIRUS, C-TYPE MURINE MYELOMA
  FLUORESCENT ANTIBODY TECHNIC,
  3843*
 VIRUS, EPSTEIN-BARR
  CELL FUSION, 3843*
  HYBRIDIZATION, 3631

NEOPLASM METASTASIS

BLADDER NEOPLASMS
   ADENOCARCINOMA, 4J46
   CARCINOMA, TRANSITIONAL CELL, 4046
BLOOD COAGULATION FACTORS
   REVIEW, 3656*
BREAST NEOPLASMS, 4045
   CELL CYCLE KINETICS, 413J
   MYOBLASTOMA, 4C56*
   TRANSPLANTATION, HETEROLOGOUS,
    4035
BURKITT'S LYMPHOMA
   CLINICOPATHOLOGIC STUDY, 4013
CAPILLARITY
   REVIEW, 3657*
CARCINOEMBRYONIC ANTIGEN
   ISOELECTRIC PRECIPITATION AND
    ULTRAFILTRATION, 3965*
   ISOLATION AND CHARACTERIZATION,
    3965*
CARCINOMA, BASAL CELL
   TRANSPLANTATION, HETEROLOGOUS,
    4035
CARCINOMA, EPIDERMOID
   TRANSPLANTATION, HETEROLOGOUS,
    4035
GASTROINTESTINAL NEOPLASMS, 4045
   TRANSPLANTATION, HETEROLOGOUS,
    4035
GYNECOLOGICAL NEOPLASMS
   ADENOCARCINOMA, 4046
   CYSTADENOCARCINOMA, 4046
HELA CELLS
   TRANSPLANTATION, HETEROLOGOUS,
    4035
KIDNEY NEOPLASMS, 4J45
   ADENOCARCINOMA, 4046
   CARCINOMA, TRANSITIONAL CELL, 4046
LARYNGEAL NEOPLASMS, 4045
LEUKEMIA
   STATISTICAL ANALYSIS, 4107*
LEUKEMIA, MYELOCYTIC
   MYOBLASTOMA, 4056*
LIP NEOPLASMS
   BURKITT'S LYMPHOMA, 4012
LIPOPROTEINS
   ISOLATION AND CHARACTERIZATION,
    3632
LUNG NEOPLASMS, 4045
   ADENOCARCINOMA, 4J46
   CARCINOMA, EPIDERMOID, 4046
   CARCINOMA, OAT CELL, 4046
MEDIASTINAL NEOPLASMS
   CASE REPORT, 4044
MELANOMA
   ANTILYMPHOCYTE SERUM, 4035
   EPIDEMIOLOGY, CHILD, 3655*
   TRANSPLANTATION, HETEROLOGOUS,
    4035
MOUTH NEOPLASMS, 4045
NASOPHARYNGEAL NEOPLASMS, 4045
NEOPLASM CIRCULATING CELLS
   REVIEW, 3658*
NEOPLASM SEEDING
   STATISTICAL ANALYSIS, 4107*
NERVOUS SYSTEM NEOPLASMS
   BURKITT'S LYMPHOMA, 4012
NOSE NEOPLASMS
   BRAIN NEOPLASMS, 3702
PANCREATIC NEOPLASMS, 4045

   TRANSPLANTATION, HETEROLOGOUS, 4035
RADIATION, IONIZING
   NEOPLASM TRANSPLANTATION, 4035
THYMOMA
   CASE REPORT, 4044
THYROID NEOPLASMS
   CARCINOMA, PAPILLARY, 4129
UROGENITAL NEOPLASMS, 4045

NEOPLASM SEEDING
   NEOPLASM METASTASIS
    STATISTICAL ANALYSIS, 4107*

NEOPLASM TRANSPLANTATION
   ANTINEOPLASTIC AGENTS
    ASCITIC CONVERSION, 3941*
   FIBROSARCOMA
    SIALIC ACIDS, 4166
   IMMUNOSUPPRESSION
    CARCINOGENIC ACTIVITY, 3944*
   LEUKEMIA, MYELOCYTIC
    ULTRASTRUCTURAL STUDY, HAMSTER,
     4079*
   MAMMARY NEOPLASMS, EXPERIMENTAL
    ADENOCARCINOMA, 4149
    GROWTH, 4149
   RADIATION, IONIZING
    NEOPLASM METASTASIS, 4035
   TESTICULAR NEOPLASMS
    TERATOID TUMOR, 3621

NEOPLASMS (GENERAL AND UNSPECIFIED)
   SEE ALSO UNDER PARTICULAR SITE
   ASCITIC FLUID
    IMMUNOSUPPRESSION, 3950*
    T-LYMPHOCYTES, 3979*
   BREAST NEOPLASMS
    EPIDEMIOLOGY, REVIEW, 4117
   CARCINOGEN, ENVIRONMENTAL
    EPIDEMIOLOGY, REVIEW, 4123
    STATISTICAL ANALYSIS, 4123
   CELL TRANSFORMATION, NEOPLASTIC
    CELL CYCLE KINETICS, REVIEW, 3653*
   CHOLANTHRENE, 3-METHYL-
    IMMUNOSUPPRESSION, 3951*, 3952*
   DIET
    EPIDEMIOLOGY, REVIEW, 4123
    STATISTICAL ANALYSIS, 4123
   DNA, VIRAL
    HYBRIDIZATION, REVIEW, 3631
   GENETICS
    EPIDEMIOLOGY, 4143*
   IMMUNITY, CELLULAR
    REVIEW, 3626, 3651*
   IMMUNOTHERAPY
    REVIEW, 3651*
   ISOENZYMES
    REVIEW, 3633
   T-LYMPHOCYTES
    IMMUNOSUPPRESSION, 3983*
   RADIATION, IONIZING
    RISK-BENEFIT CALCULATION, 3613
   TRANSPLANTATION IMMUNOLOGY
    REVIEW, 3626
   UROGENITAL NEOPLASMS
    EPIDEMIOLOGY, REVIEW, 4117
   VIRUS, ADENO
    HYBRIDIZATION, 3631
   VIRUS, HERPES TURKEY

CHORIOALLANTOIC MEMBRANE, 3854*
VIRUS, MAREK'S DISEASE HERPES
    CHORIOALLANTOIC MEMBRANE, 3854*
VIRUS, MOLONEY MURINE SARCOMA
    MACROPHAGES, 3949*
VIRUS, SHOPE PAPILLOMA
    HYBRIDIZATION, 3631

NEOPLASMS, CONNECTIVE TISSUE
    FIBROSARCOMA
        VIRUS, CELO, 3794
    SARCOMA
        VIRUS, CELO, 3794

NEOPLASMS, EMBRYONAL AND MIXED
    PHARYNGEAL NEOPLASMS
        CASE REPORT, 4001
        TONSILITIS, 4001

NEOPLASMS, EXPERIMENTAL
    CARBAMIC ACID, ETHYL ESTER
        IMMUNOSUPPRESSION, 3893
    CHOLANTHRENE, 3-METHYL-
        IMMUNOSUPPRESSION, 3893
    DIETHYLAMINE, N-NITROSO-
        CARCINOGENIC ACTIVITY, HAMSTER,
        RAT, 3697
    DIMETHYLAMINE, N-NITROSO-
        CARCINOGENIC ACTIVITY, HAMSTER,
        RAT, 3697
    VIRUS, FRIEND MURINE LEUKEMIA
        TRANSPLANTATION, HOMOLOGOUS, 3798
    VIRUS, SV40
        IMMUNITY, CELLULAR, 3891
        IMMUNOSUPPRESSION, 3891
        TRANSPLANTATION IMMUNOLOGY, 3891

NEOPLASMS, MULTIPLE PRIMARY
    BRAIN NEOPLASMS
        ASTROCYTOMA, 3995
        GLIOBLASTOMA MULTIFORME, 3995
        MENINGIOMA, 3995
    BREAST NEOPLASMS
        ADENOCARCINOMA, 3995
        CARCINOMA, 3995
        CARCINOMA, DUCTAL, 3995
    DIGESTIVE SYSTEM NEOPLASMS, 3995
    GYNECOLOGIC NEOPLASMS, 3995
    NERVOUS SYSTEM NEOPLASMS, 3995
    PINEALOMA, 3995
    PROSTATIC NEOPLASMS, 3995
    SPINAL NEOPLASMS
        CARCINOMA, 3995

NEOPLASMS, MUSCLE TISSUE
    SARCOMA
        ULTRASTRUCTURAL STUDY, REVIEW,
        3624

NEPHROBLASTOMA
    GENETICS
        CASE REPORT, 4029
    KIDNEY NEOPLASMS
        DIAGNOSIS AND PROGNOSIS, 4133*
        EPIDEMIOLOGY, CHILD, 4133*
        EPIDEMIOLOGY, CHILD, INDIA, 4120

PROTEINS
    ISOLATION AND CHARACTERIZATION, C-
    REACTIVE, 4169

NERVOUS SYSTEM NEOPLASMS
    BENZENAMINE, 2-METHYL-4-((2-
    METHYLPHENYL)AZO)-
        TRANSPLACENTAL CARCINOGENESIS,
        REVIEW, 3601
    BURKITT'S LYMPHOMA
        NEOPLASM METASTASIS, 4012
    CARBAMIC ACID, N-METHYL-N-NITROSO-,
    ETHYL ESTER
        TRANSPLACENTAL CARCINOGENESIS,
        REVIEW, 3601
    DIETHYLAMINE, N-NITROSO
        TRANSPLACENTAL CARCINOGENESIS,
        REVIEW, 3601
    EPIDEMIOLOGY
        MORMONS, 4118
    ETHANE, AZOXY-
        TRANSPLACENTAL CARCINOGENESIS,
        REVIEW, 3601
    HYDRAZINE, 1,2-DIETHYL-
        TRANSPLACENTAL CARCINOGENESIS,
        REVIEW, 3601
    NEOPLASMS, MULTIPLE PRIMARY, 3995
    P-TOLUAMIDE, N-ISOPROPYL-ALPHA-(2-
    METHYLHYDRAZINO)-,
        TRANSPLACENTAL CARCINOGENESIS,
        REVIEW, 3601
    UREA, ETHYL NITROSO-
        TRANSPLACENTAL CARCINOGENESIS,
        REVIEW, 3601
    UREA, N-NITROSO-N-PROPYL-
        TRANSPLACENTAL CARCINOGENESIS,
        REVIEW, 3601

NEURAMINIDASE
    ANTIGENIC DETERMINANTS
        OSPOSINS, 3986*
    GANGLIOSIDES
        CELL MEMBRANE, 3924
    LEUKEMIA, LYMPHOBLASTIC
        T-LYMPHOCYTES, 3920

NEUROBLASTOMA
    BRAIN NEOPLASMS
        EPIDEMIOLOGY, CHILD, INDIA, 4120
    5-BETA-CHOLAN-24-OIC ACID, 3-ALPHA, 12
    ALPHA-DIHYDROXY-
        ADENOSINE TRIPHOSPHATASE, 4176
    DIBUTYRYL CYCLIC AMP
        ADENOSINE TRIPHOSPHATASE, 4176
    5H-DIBENZ(B,F)AZEPINE, 10,11-DIHYDRO-
    5-((3-METHYLAMINO)PROPYL)
        NOREPINEPHRINE, 4181
    IMIPRAMINE
        NOREPINEPHRINE, 4181
    INSULIN
        ADENOSINE TRIPHOSPHATASE, 4176
    ISOANTIGENS
        IMMUNE RESPONSE, 3892
    OUABAIN
        NOREPINEPHRINE, 4181
    PHENETHYLAMINE, ALPHA-METHYL-
        NOREPINEPHRINE, 4181
    PHENOL, 2,4-DINITRO-
        NOREPINEPHRINE, 4181

NOREPINEPHRINE
    NEUROBLASTOMA
        5H-DIBENZ(B,F)AZEPNINE, 10,11-
           DIHYDRO-5-((3-METHYLAMINO)PROPYL)
           , 4181
        IMIPRAMINE, 4181
        OUABAIN, 4181
        PHENETHYLAMINE, ALPHA-METHYL-,
           4181
        PHENOL, 2,4-DINITRO-, 4181

NOSE NEOPLASMS
    ADENOCARCINOMA
        BUTANOL, 4-(BUTYLNITROSOAMINO)-,
           3701
    ADENOSINE CYCLIC 3',5' MONOPHOSPHATE
        CELL CYCLE KINETICS, 4192*
    BRAIN NEOPLASMS
        NEOPLASM METASTASIS, 3702
    CARCINOMA
        DIPROPYLAMINE, N-NITROSO-, 3702
    CARCINOMA, EPIDERMOID
        PIPERIDINE, 1-NITROSO-, 3707
    PAPILLOMA
        2-PROPANOL, 1-NITROSO(PROPYL)AMINO-
           , 3702

NUCLEIC ACIDS
    SEE ALSO DNA/RNA
    BENZ(A)ANTHRACENE, 5,6-EPOXY-7,12-
        DIMETHYL-
-    ··   BINDING SITES, 3750*
    DIETHYLAMINE, N-NITROSO-
        DEOXYRIBONUCLEASE, 3695

NUCLEOTIDES
    VIRUS, HAMSTER SARCOMA
        POLY A POLYMERASE, 3799

NUCLEOTIDYLTRANSFERASES
    VIRUS, HAMSTER SARCOMA
        ADENOSINE TRIPHOSPHATE, 3799
        MAGNESIUM, 3799
        MANGANESE, 3799

NUTRITION
    SEE DIET

OCCUPATIONAL HAZARD
    ALKYLATING AGENTS
        REVIEW, 3662*
    ANGIOSARCOMA
        ETHYLENE, CHLORO- POLYMER, 3663*
    BLADDER NEOPLASMS
        2-BIPHENYLOL, 3660*
        CYCLOHEXANE, 1,2,3,4,5,6-
           HEXACHLORO-, GAMMA ISOMER, 3660*
    CARCINOGEN, CHEMICAL
        REVIEW, 3662*
    CHROMIUM
        CASE REPORT, 4112*
    COLONIC NEOPLASMS
        ASBESTOS, 4134*
    ETHYLENE, CHLORO-
        REVIEW, 3663*
    ETHYLENE, CHLORO- POLYMER
        REVIEW, 3663*
    HEPATOMA
        ETHYLENE, CHLORO-, 3663*

ETHYLENE, CHLORO- POLYMER, 3663*
LUNG NEOPLASMS
    ANGIOSARCOMA, 3627
    ARSENIC, 4127
    ASBESTOS, 3643*
    COPPER, 4127
    EPIDEMIOLOGY, JAPAN, 4127
    MESOTHELIOMA, 3627, 3643*
2-NAPHTHYLAMINE
    WORKER PROTECTION, REVIEW, 3629
RESPIRATORY TRACT NEOPLASMS
    ASBESTOS, 3640*
    CARCINOGEN, CHEMICAL, 3640*
    CARCINOMA, 4145*
    METALS, 3640*
    REVIEW, 3640*

ODONTOGENIC CYSTS
    AMELOBLASTOMA
        CELL TRANSFORMATION, NEOPLASTIC,
        4063*
    CARCINOMA, EPIDERMOID
        CELL TRANSFORMATION, NEOPLASTIC,
        4063*
    SARCOMA
        CELL TRANSFORMATION, NEOPLASTIC,
        4063*

OLIGODENDROGLIOMA
    BRAIN NEOPLASMS
        ENDOPLASMIC RETICULUM, 4005
        GOLGI APPARATUS, 4005

OLIGOPEPTIDES
    MAMMARY NEOPLASMS, EXPERIMENTAL
        BENZ(A)ANTHRACENE, 7,12-DIMETHYL-,
        3752*

ONCOGENIC VIRUSES
    CELL TRANSFORMATION, NEOPLASTIC
        CELL CYCLE KINETICS, REVIEW, 3653*
    REVIEW, 3646*

ORBITAL NEOPLASMS
    RETINOBLASTOMA
        CASE REPORT, CHILD, 4003
    SARCOMA, OSTEOGENIC
        RADIATION, 4003
        ULTRASTRUCTURAL STUDY, 4003

OSPOSINS
    NEURAMINIDASE
        ANTIGENIC DETERMINANTS, 3986*

OSTEOMA, OSTEOID
    BONE NEOPLASMS
        CASE REPORT, 4049*
    SARCOMA, OSTEOGENIC
        CASE REPORT, 4049*

OUABAIN
    CARCINOMA, EHRLICH TUMOR
        CATIONS, MONOVALENT, 4190*
    NEUROBLASTOMA
        NOREPINEPHRINE, 4181
    PLASMACYTOMA
        ADENOSINE CYCLIC 3',5'
        MONOPHOSPHATE, 4153
        THEOPHYLLINE, 4153

OVARIAN NEOPLASMS
    BENZ(A)ANTHRACENE, 7,12-DIMETHYL-
        FORMATION AND DEVELOPMENT, 3751*
    BURKITT'S LYMPHOMA
        CLINICOPATHOLOGIC STUDY, 4013
    CARCINOMA
        ANTIGENS, 3972*
    CHROMOSOMES
        PRECANCEROUS CONDITIONS, 4095*
    CYSTADENOCARCINOMA
        PRECANCEROUS CONDITIONS, 4095*
    CYSTADENOMA
        PRECANCEROUS CONDITIONS, 4095*
    DISGERMINOMA
        CASE REPORT, 4031
        SEX CHROMOSOMES, 4031
        TESTOSTERONE, 4031
    THECA CELL TUMOR
        VIRUS, POLYOMA, 3874*

OXYGENASES
    AFLATOXIN Q1
        AFLATOXIN B1, 3669

PANCREATIC NEOPLASMS
    ADENOCARCINOMA
        REVIEW, 3639*
    ADENOMA
        MONOCROTALINE, 3761*
    AGE FACTORS
        EPIDEMIOLOGY, ISRAEL, 4115
    CARCINOEMBRYONIC ANTIGEN, 3966*
        ISOLATION AND CHARACTERIZATION,
        3966*
    CARCINOGEN, CHEMICAL
        EPIDEMIOLOGY, 4124
        REVIEW, 3639*
    CARCINOMA
        REVIEW, 3639*
    EPIDEMIOLOGY
        MORMONS, 4118
    ISLET CELL TUMOR
        CASE REPORT, 4069*, 4073*
        PEPTIDES, 4069*
        ULTRASTRUCTURAL STUDY, 4073*
    NEOPLASM METASTASIS, 4045
    SULFATASES
        HISTOCHEMICAL STUDY, 4179
    TRANSPLANTATION, HETEROLOGOUS
        NEOPLASM METASTASIS, 4035

PAPILLOMA
    ESOPHAGEAL NEOPLASMS
        DIPROPYLAMINE, N-NITROSO-, 3702
        PIPERIDINE, 3,4-DIBROMO-1-NITROS
        3707
        PIPERIDINE, 3,4-DICHLORO-1-NITRO
        , 3707
        PROPYLAMINE, N-METHYL-N-NITROSO-,
        3702
    GASTRIC NEOPLASMS
        BUTANOL, 4-(BUTYLNITROSOAMINO)-,
        3701
    LUNG NEOPLASMS
        CROTON OIL, 3733*
    NOSE NEOPLASMS
        2-PROPANOL, 1-NITROSO(PROPYL)AMI
        , 3702
    STOMACH NEOPLASMS

50-,
    MORPHOLOGICAL AND CYTOCHEMICAL
     STUDY, 4021
  LEUKEMIA, MYELOBLASTIC
   MORPHOLOGICAL AND CYTOCHEMICAL
    STUDY, 4021

PERTUSSIS VACCINE
 VIRUS, FRIEND MURINE LEUKEMIA
  SPLENOMEGALY, 3797

PEUTZ-JEGHERS SYNDROME
 INTESTINAL NEOPLASMS
  DIAGNOSIS AND PROGNOSIS, 4070*

PHAGOCYTES
 BONE MARROW CELLS
  CELL DIVISION, 3899

PHAGOCYTOSIS
 LEUKEMIA, LYMPHOBLASTIC
  NEUTROPHILS, 4188*
 LEUKEMIA, MYELOBLASTIC
  NEUTROPHILS, 4188*
 LEUKEMIA, MYELOCYTIC
  NEUTROPHILS, 4188*

PHARYNGEAL NEOPLASMS
 CARCINOMA, EPIDERMOID
  CELL CYCLE KINETICS, 4130
  CLOSTRIDIOPEPTIDASE A, 3998
  PIPERIDINE, 1-NITROSO-, 3707
 EPIDEMIOLOGY
  MORMONS, 4118
 NEOPLASMS, EMBRYON AND MIXED
  CASE REPORT, 4001
  TONSILITIS, 4001

PHENETHYLAMINE, ALPHA-METHYL-
 NEUROBLASTOMA
  NOREPINEPHRINE, 4181

PHENOBARBITAL
 SEE BARBITURIC ACID, 5-ETHYL-5-PHENYL-

PHENOBARBITONE
 SEE BARBITURIC ACID, 5-ETHYL-5-PHENYL-

PHENOL, 2,4-DINITRO-
 NEUROBLASTOMA
  NOREPINEPHRINE, 4181

PHENOLSULFONPHTHALEIN
 VIRUS, HERPES SIMPLEX
  PHOTOINACTIVATION, 3803

PHENOTYPE
 BRONCHITIS
  EPIDEMIOLOGY, GENETICS, 4121
 LUNG NEOPLASMS
  EPIDEMIOLOGY, GENETICS, 4121
 PULMONARY EMPHYSEMA
  EPIDEMIOLOGY, GENETICS, 4121

PHENYLACETATES
 NEUROBLASTOMA
  TYROSINE HYDROXYLASE, 4178
 PHEOCHROMOCYTOMA
  TYROSINE HYDROXYLASE, 4178

PHEOCHROMOCYTOMA
    ADRENAL GLAND NEOPLASMS
        DIETHYLAMINE, N-NITROSO-, 3696
        TRANSPLACENTAL CARCINOGENESIS,
          HAMSTER, 3696
    TYROSINE HYDROXYLASE
        PHENYLACETATES, 4178
        VANILMANDELIC ACID, 4178

PHORBOL MYRISTATE ACETATE
    SEE 12-O-TETRADECANOYLPHORBOL-13-
      ACETATE

PHOSPHATASES
    VIRUS, ROUS SARCOMA
        REVERSE TRANSCRIPTASE, 3880*

PHOSPHATES
    LYMPHOMA
        ALKALINE PHOSPHATASE, 3862*

PHOSPHOFRUCTOKINASE
    CERVIX NEOPLASMS
        PRECANCEROUS CONDITIONS, 4198*

PHOSPHOLIPIDS
    LEUKEMIA, MYELOBLASTIC
        MORPHOLOGICAL AND CYTOCHEMICAL
         STUDY, 4021

PHOSPHORIBOSYL PYROPHOSPHATE SYNTHETASE
    SEE PHOSPHOTRANSFERASES, ATP

PHOSPHORYLCHOLINE
    SEE CHOLINE

PHOSPHOTRANSFERASES
    VIRUS, ADENO 12
        CHROMOSOMES, HUMAN, 16-18, 3793
        HYBRID CELLS, 3793

PHYTOHEMAGGLUTININ
    SEE PLANT AGGLUTININS

PINEALOMA
    NEOPLASMS, MULTIPLE PRIMARY, 3995

1-PIPERAZINEETHANOL, 4-(3-(2-
    CHLOROPHENOTHIAZIN-10-YL)PROPYL)-
    MAMMARY NEOPLASMS, EXPERIMENTAL
        PROLACTIN, 3688
        SOMATOTROPIN, 3688

PIPERIDINE, 3,4-DIBROMO-1-NITROSO-
    ESOPHAGEAL NEOPLASMS
        CARCINOMA, EPIDERMOIC, 3707
        PAPILLOMA, 3707
    STOMACH NEOPLASMS
        CARCINOMA, EPIDERMOIC, 3707
        PAPILLOMA, 3707

PIPERIDINE, 3,4-DICHLORO-1-NITROSO-
    ESOPHAGEAL NEOPLASMS
        CARCINOMA, EPIDERMOID, 3707
        PAPILLOMA, 3707
    STOMACH NEOPLASMS
        CARCINOMA, EPIDERMOID, 3707

PIPERIDINE, 1-NITROSO-

NOSE NEOPLASMS
    CARCINOMA, EPIDERMOID, 3707
PHARYNGEAL NEOPLASMS
    CARCINOMA, EPIDERMOID, 3707
POLAROGRAPHY
    SPECTROSCOPY, 3762*

PITUITARY GLAND NEOPLASMS
    MAMMARY NEOPLASMS, EXPERIMENTAL
        PROLACTIN, 3638*

PITUITARY GROWTH HORMONE
    SEE SOMATOTROPIN

PLANT AGGLUTININS
    FETAL GLOBULINS
        DNA REPLICATION, 3954*
        IMMUNOSUPPRESSION, 3954*
    LEUKEMIA, LYMPHOCYTIC, 3917
    LYMPHOCYTES
        BINDING SITES, AGE FACTORS, 3977*
    B-LYMPHOCYTES
        LANTHANUM, 4167
    T-LYMPHOCYTES
        ASCITIC FLUID, 3979*
        LANTHANUM, 4167
    THYMOMA
        IMMUNE RESPONSE, 4043

PLASMACYTOMA
    ACTINOMYCIN D
        CELL FUSION, 4004
        ULTRASTRUCTURAL STUDY, 4004
    ADENOSINE CYCLIC 3',5' MONOPHOSPHATE
        CELL MEMBRANE, 4153
        CONCANAVALIN A, 4153
        OUABAIN, 4153
    CELL FUSION
        ULTRASTRUCTURAL STUDY, 4004
    CYCLOHEXIMIDE
        CELL FUSION, 4004
        ULTRASTRUCTURAL STUDY, 4004
    INCLUSION BODIES, VIRAL
        A-TYPE PARTICLES, 3870*
    B-LYMPHOCYTES
        ANTIGENIC DETERMINANTS, 3919
    T-LYMPHOCYTES
        ANTIGENIC DETERMINANTS, 3919
    METHANE, SULFINYLBIS-
        A-TYPE PARTICLES, 3870*
    RNA, NEOPLASM
        HYBRIDIZATION,, 3863*
    THEOPHYLLINE
        CELL MEMBRANE, 4153
        CONCANAVALIN A, 4153
        OUABAIN, 4153
    UREA, HYDROXY-
        CELL FUSION, 4004
        ULTRASTRUCTURAL STUDY, 4004
    URIDINE, 2'-DEOXY-5-IODO-
        A-TYPE PARTICLES, 3870*
    VIRUS-LIKE PARTICLES
        ISOLATION AND CHARACTERIZATION,
         3863*

PLASMS
    POLYCYTHEMIA VERA
        PUTRESCINE, 4161
        SPERMIDINE, 4161

AFLATOXIN B1
    PROTEINS, 3668
HEPATOMA
    FETAL GLOBULINS, 3906

774

POLYVINYL CHLORIDE
    SEE ETHYLENE, CHLORO- POLYMER

CNS, 3781*
EXPERIMENTAL,   PORNASE
                       MAMMARY NEOPLASMS, EXPERIMENTAL
                           VIRUS, MURINE MAMMARY TUMOR, 3866*
, 3645*
                       PRECANCEROUS CONDITIONS
44*   ACRYLIC ACID, 3-CYANO-, ISOBUTYL ESTER
                               FIBROBLASTS, 3673
VIEW, 3612   BLADDER NEOPLASMS
                               BENZIDINE, 3,3'-DIMETHYL-, 3731*
                               SCHISTOSOMIASIS, 3990
                           BREAST NEOPLASMS
                               CELL NUCLEUS, 4055*
                               COLLAGEN, 3993
165   EPIDEMIOLOGY, MALE, 3992
                               HISTOLOGICAL STUDY, 4055*
                               HYDROXYLASES, 3993
                           BRONCHIAL NEOPLASMS
TE, 3799   BENZO(A)PYRENE, 4077*
                               DIAGNOSIS, BRONCHOGRAPHY, 3764*
                               DIETHYLAMINE, N-NITROSO-, 3764*
                               TOBACCO, 4077*
                           CARCINOEMBRYONIC ANTIGEN
                               LYMPHOCYTES, 3908
                           CERVIX NEOPLASMS
KEMIA   CARCINOMA IN SITU, 3999
                               CELL TRANSFORMATION, NEOPLASTIC,
                                   3996
                               DNA, NEOPLASM, 3996
US FORMING   GLUCOSEPHOSPHATE, 4198*
 3796   HEXOKINASE, 4198*
                               LACTATE DEHYDROGENASE, 4198*
                               PHOSPHOFRUCTOKINASE, 4198*
                               PYRUVATE KINASE, 4198*
                           COLONIC NEOPLASMS
                               CARCINOMA, 4072*
                               DNA REPLICATION, 4072*
                           HODGKIN'S DISEASE
                               EOSINOPHILIA, 4083*
                           LARYNGEAL NEOPLASMS
                               CARCINOMA IN SITU, 3999
                           LEUKEMIA
                               BONE MARROW CELLS, 4078*
745*   LYMPHOCYTES, 3978*
                           LEUKEMIA, MYELOCYTIC
*   CASE REPORT, 4081*
IMENTAL   ULTRASTRUCTURAL STUDY, 4081*
745*   LIVER NEOPLASMS
                               BARBITURIC ACID, 5-ETHYL-5-PHENYL-,
                                   3766*
SIS   LUNG NEOPLASMS
FERASES, 3839*   PLUTONIUM, 3781*
                           OVARIAN NEOPLASMS
                               CHROMOSOMES, 4095*
                               CYSTADENOCARCINOMA, 4095*
                               CYSTADENOMA, 4095*
4, 4010   RECTAL NEOPLASMS
                               CARCINOMA, 4072*
ROSOAMINO)-,   RESPIRATORY TRACT NEOPLASMS
                               BUTANOL, 4-(BUTYLNITROSOAMINO)-,
                                   3701
                           SARCOMA, OSTEOGENIC

        SUBJECT 55

CASE REPORT, 4049*
SKIN NEOPLASMS
  MELANOMA, 4040
  NEVUS, PIGMENTED, 4040
  ULTRAVIOLET RAYS, 3775
STOMACH NEOPLASMS
  CARCINOMA, 4067*, 4068*
  CARCINOMA IN SITU, 4068*
UTERINE NEOPLASMS
  AGE FACTORS, 4103*
  EPIDEMIOLOGY, 4103*

PREDNISOLONE
  RETICULOSIS, HEMOPHAGOCYTIC
    GENETICS, 4033.

PREDNISONE
  SEE PREGNA-1,4-DIENE-3,11,20-TRIONE,
    17,21-DIHYDROXY-

PREGNA-1,4-DIENE-3,11,20-TRIONE, 17,21-
  DIHYDROXY-
    LEUKEMIA, LYMPHOCYTIC, 3917

PROGESTERONE
  ARYL HYDROCARBON HYDROXYLASES
    ENZYMATIC ACTIVITY, 3738*
  DNA
    THYMIDINE INCORPORATION, 3680
  MAMMARY NEOPLASMS, EXPERIMENTAL
    BENZ(A)ANTHRACENE, 7,12-DIMETHYL-,
      3680
    CHOLANTHRENE, 3-METHYL-, 3680
  RECEPTORS, HORMONE
    CELL DIFFERENTIATION, 3607
  RNA, MESSENGER
    CELL DIFFERENTIATION, 3607
  UTERINE NEOPLASMS
    CASE REPORT, 4097*

PROLACTIN
  BENZ(A)ANTHRACENE, 7,12-DIMETHYL-
    HEMATOPOIETIC STEM CELLS, 3754*
  BREAST NEOPLASMS
    GROWTH SUBSTANCES, 3689
  CHOLANTHRENE, 3-METHYL-
    INCREASE AND SYNTHESIS, 3755*
  MAMMARY NEOPLASMS, EXPERIMENTAL
    ADENOCARCINOMA, 3741*
    ESTROGENS, 3638*
    1-PIPERAZINEETHANOL, 4-(3-(2-
      CHLOROPHENOTHIAZIN-10-YL)PROPYL)-
      , 3688
    PITUITARY GLAND NEOPLASMS, 3638*
    SULPIRIDE, 3713
  SOMATOTROPIN
    RADIOIMMUNOASSAY, MOUSE, 4185

2-PROPANOL, 1-NITROSO(PROPYL)AMINO-
  NOSE NEOPLASMS
    PAPILLOMA, 3702

PROPYLAMINE, N-METHYL-N-NITROSO-
  ESOPHAGEAL NEOPLASMS
    PAPILLOMA, 3702
  LIVER NEOPLASMS
    HEMANGIOENDOTHELIOMA, 3702

PROSTATIC NEOPLASMS

LEIOMYOSARCOMA
  CASE REPORT, 4092*
  NEOPLASMS, MULTIPLE PRIMARY, 3995
SARCOMA
  CASE REPORT, 4092*
VIRUS, HERPES SIMPLEX 1
  REVIEW, 3648*
VIRUS, HERPES SIMPLEX 2
  REVIEW, 3648*

PROTEIN KINASE
  VIRUS, ROUS SARCOMA
    REVERSE TRANSCRIPTASE, 3880*

PROTEINS
  BREAST NEOPLASMS
    ISOLATION AND CHARACTERIZATION, C-
      REACTIVE, 4169
  BUTYRIC ACID, 2-AMINO-4-(ETHYLTHIO)-
    SEX FACTORS, 3734*
  CERVIX NEOPLASMS
    ISOLATION AND CHARACTERIZATION, C-
      REACTIVE, 4169
  CYCLOHEXIMIDE
    RADIATION, 3779*
  ESOPHAGEAL NEOPLASMS
    ISOLATION AND CHARACTERIZATION, C-
      REACTIVE, 4169
  ESTRADIOL
    BENZ(A)ANTHRACENE, 7,12-DIMETHYL-,
      3747*
  HODGKIN'S DISEASE
    ISOLATION AND CHARACTERIZATION, C-
      REACTIVE, 4169
  LEUKEMIA
    ISOLATION AND CHARACTERIZATION, C-
      REACTIVE, 4169
  LEUKEMIA, MYELOCYTIC
    ISOLATION AND CHARACTERIZATION, C-
      REACTIVE, 4169
  LYMPHOCYTES
    LYMPHOKINES, 3922
  LYMPHOSARCOMA
    ISOLATION AND CHARACTERIZATION, C-
      REACTIVE, 4169
  MICROSOMES, LIVER
    AFLATOXIN B1, 3668
  MULTIPLE MYELOMA
    ISOLATION AND CHARACTERIZATION, C-
      REACTIVE, 4169
  NASOPHARYNGEAL NEOPLASMS
    ISOLATION AND CHARACTERIZATION, C-
      REACTIVE, 4169
  NEPHROBLASTOMA
    ISOLATION AND CHARACTERIZATION, C-
      REACTIVE, 4169
  POLYRIBOSOMES
    AFLATOXIN B1, 3668
  RNA, MESSENGER
    CYTOPLASM, 3634
  SARCOMA
    ISOLATION AND CHARACTERIZATION, C-
      REACTIVE, 4169
  TONSILLAR NEOPLASMS
    ISOLATION AND CHARACTERIZATION, C-
      REACTIVE, 4169
  UREA, NITROSO-
    CARBAMOYLATION, 3765*
  VIRUS, FRIEND MURINE LEUKEMIA

BIOSYNTHESIS, 3850*

EUDOTUMORS, INFLAMMATORY
SEE FIBROMA

LMONARY EMPHYSEMA
ALPHA 1 ANTITRYPSIN
EPIDEMIOLOGY, GENETICS, 4121
BLOOD GROUPS
EPIDEMIOLOGY, GENETICS, 4121
PHENOTYPE
EPIDEMIOLOGY, GENETICS, 4121
SMOKING
EPIDEMIOLOGY, GENETICS, 4121

LMONARY NEOPLASMS
RADIATION, IONIZING
MORTALITY, 3780*

RINE, 6-((1-METHYL-4-NITRO IMIDAZOL-5-YL)
THIO)-
LEUKOCYTES
POLYMORPHONUCLEAR NEUTROPHILS,
3976*
LYMPHOCYTES
POLYMORPHONUCLEAR NEUTROPHILS,
3976*

RINE-6-THIOL
THYMOMA
IMMUNE RESPONSE, 4043

TRESCINE
DNA REPLICATION
FIBROBLASTS, DIPLOID, 4162
POLYCYTHEMIA VERA
LEUKOCYTES, 4161
PLASMS, 4161

RAZOLE
HEPATOMA
ALCOHOL OXIDOREDUCTASES, 4177

RIDINE, 2-(BENZYL(2-(DIMETHYLAMINO)ETHYL)
AMINO)-
BARBITURIC ACID, 5-ETHYL-5-PHENYL-
DNA, 3693
MICROSOMES, LIVER
DNA, 3693

RIDINE, 2-(BENZYL(2-(METHYL)(NITROSO)
AMINO)ETHYL)AMINO)-
BARBITURIC ACID, 5-ETHYL-5-PHENYL-
DNA, 3693
CHOLANTHRENE, 3-METHYL-
DNA, 3693
MICROSOMES, LIVER
DNA, 3693

RROLIDINE, 1-NITROSO-
BACTERIA
ENZYMATIC ACTIVITY, 3700

RUVATE KINASE
CERVIX NEOPLASMS
PRECANCEROUS CONDITIONS, 4198*

INOLINE
QUINOLINE, 2-CHLORO-

CARCINOGENIC ACTIVITY, 3767*

QUINOLINE, 2-CHLORO-
QUINOLINE
CARCINOGENIC ACTIVITY, 3767*

QUINOLINE, 4-NITRO-, 1-OXIDE
STOMACH NEOPLASMS
ADENOCARCINOMA, 3635*

RADIATION
BONE MARROW
ROSETTE FORMATION, SHEEP CELLS,
3947*
CELL DIVISION
CELL CYCLE KINETICS, 3779*
ENZYMATIC ACTIVITY, 3779*
CELL MEMBRANE
ULTRASTRUCTURAL STUDY, 3783*
CYCLOHEXIMIDE
PROTEINS, 3779*
INTESTINAL NEOPLASMS
ADENOCARCINOMA, 3784*
CARCINOEMBRYONIC ANTIGEN, 3784*
LUCANTHONE
RNA REPLICATION, 3779*
ORBITAL NEOPLASMS
SARCOMA, OSTEOGENIC, 4003
PLUTONIUM
ENVIRONMENTAL HAZARD, 3645*
TRITIUM
HORMONE POTENTIATION, 3714
PREGNANCY EFFECTS, 3714

RADIATION CHIMERA
LEUKEMIA, RADIATION-INDUCED
IMMUNE RESPONSE, 3889
LYMPHOSARCOMA
IMMUNE RESPONSE, 3889

RADIATION INJURIES
PLUTONIUM
RADIATION DOSAGE, 3644*

RADIATION INJURIES, EXPERIMENTAL
LIVER
PATHOPHYSIOLOGICAL STUDY, RAT,
3787*
LUNG NEOPLASMS
PLUTONIUM, 3781*
MAMMARY NEOPLASMS, EXPERIMENTAL
DNA, 4113

RADIATION, IONIZING
ADENOCARCINOMA
TRANSPLANTATION IMMUNOLOGY, 3890
ALKALINE PHOSPHATASE
DNA, 3776
BONE NEOPLASMS
EPIDEMIOLOGY, REVIEW, 3628
ETIOLOGY, REVIEW, 3628
BREAST NEOPLASMS
ADENOCARCINOMA, 3991
AGE FACTORS, 4057*
CELL TRANSFORMATION, NEOPLASTIC
CELL CYCLE KINETICS, REVIEW, 3653
CHROMOSOME ABERRATIONS
DOSE-RESPONSE STUDY, 3777*
ULTRASTRUCTURAL STUDY, 3620

DNA
    RADIATION DOSAGE, 4113
DNA POLYMERASE
    DNA, 3776
ENDONUCLEASES
    DNA, 3776
HEMANGIOMA
    EPIDEMIOLOGY, 3773
INTESTINAL NEOPLASMS
    ADENOCARCINOMA, 3911
    CARCINOEMBRYONIC ANTIGEN, 3911
LEUKEMIA
    EPIDEMIOLOGY, HIROSHIMA, NAGASAKI,
       4113
    RADIATION DOSAGE, 4113
LEUKEMIA, LYMPHOBLASTIC
    DNA, 4170
    HISTONES, 4170
LEUKEMIA, MYELOBLASTIC
    CASE REPORT, 3771
    CHROMOSOME ABERRATIONS, 3771
LEUKEMIA, MYELOCYTIC
    CARCINOGENIC EFFECT, MOUSE, 3782*
LEUKEMIA, RADIATION INDUCED
    CARCINOGENIC EFFECT, MOUSE, 3782*
LYMPHOMA
    EPIDEMIOLOGY, HIROSHIMA, NAGASAKI,
       4113
    MORTALITY, 3780*
    RADIATION DOSAGE, 4113
LYMPHOSARCOMA
    HEMANGIOMA, 3773
    IMMUNE RESPONSE, 3889
MAMMARY NEOPLASMS, EXPERIMENTAL
    TRANSPLANTATION IMMUNOLOGY, 4149
    ULTRASTRUCTURAL STUDY, 3866*
    VIRUS, MURINE MAMMARY TUMOR, 3866*
MULTIPLE MYELOMA
    EPIDEMIOLOGY, HIROSHIMA, NAGASAKI,
       4113
    RADIATION DOSAGE, 4113
MUTATION
    DOSE-RESPONSE STUDY, 3777*
NEOPLASM TRANSPLANTATION
    NEOPLASM METASTASIS, 4035
NEOPLASMS
    RISK-BENEFIT CALCULATION, 3613
PLUTONIUM
    RADIATION DOSAGE, REVIEW, 3612
POLYCYTHEMIA
    VIRUS, FRIEND-SPLEEN FOCUS FORMING,
       3796
PULMONARY NEOPLASMS
    MORTALITY, 3780*
SARCOMA, EWING'S
    HEMANGIOMA, 3773
SARCOMA, RETICULUM CELL
    MORTALITY, 3780*
    REVIEW, 3650*
T-LYMPHOCYTES
    IMMUNE RESPONSE, 3982*
TERATOID TUMOR
    HEMANGIOMA, 3773
THYROID GLAND
    CARCINOGENIC ACTIVITY, 3611
THYROID NEOPLASMS
    ADENOCARCINOMA, 3772
    ADENOMA, 3772
    CARCINOMA, PAPILLARY, 3772

UREMIA
    IMMUNOLOGICAL POTENTIAL, 3935*
VIRUS, FRIEND-SPLEEN FOCUS FORMING
    SPLEEN COLONY HISTOLOGY, 3796
VIRUS, SV40
    CELL FUSION, 3804

RAUWOLFIA ALKALOIDS
    BREAST NEOPLASMS
       AGE FACTORS, 4057*

RECEPTORS, HORMONE
    ESTROGENS
       CELL DIFFERENTIATION, 3607
    PROGESTERONE
       CELL DIFFERENTIATION, 3607

RECTAL NEOPLASMS
    ASBESTOS
       EPIDEMIOLOGY, 4134*
    CARCINOMA
       PRECANCEROUS CONDITIONS, 4072*

RESERPINE
    BREAST NEOPLASMS
       STATISTICAL ANALYSIS, REVIEW, 3609

RESPIRATORY TRACT NEOPLASMS
    ASBESTOS
       OCCUPATIONAL HAZARD, 3640*
    CARCINOMA
       AGE FACTORS, 4145*
       EPIDEMIOLOGY, 4145*
       OCCUPATIONAL HAZARD, 4145*
       SEX FACTORS, 4145*
    METALS
       OCCUPATIONAL HAZARD, 3640*
    OCCUPATIONAL HAZARD
       CARCINOGEN, CHEMICAL, 3640*
       REVIEW, 3640*
    PRECANCEROUS CONDITIONS
       BUTANOL, 4-(BUTYLNITROSOAMINO)-,
        3701
    SMOKING
       ANIMAL MODEL, HAMSTER, 3711

RETICULOSARCOMA
    SEE SARCOMA, RETICULUM CELL

RETICULOSIS, HEMOPHAGOCYTIC
    GENETICS
       CASE REPORT, 4033
       IMMUNE RESPONSE, 4033
       PREDNISOLONE, 4033

RETINOBLASTOMA
    EYE NEOPLASMS
       EPIDEMIOLOGY, CHILD, INDIA, 4120
    ORBITAL NEOPLASMS
       CASE REPORT, CHILD, 4003

RETINOL
    HEPATOMA
       CONTRACEPTIVES, ORAL, 3637*

REVERSE TRANSCRIPTASE
    LEUKEMIA

REVIEW, 3614
LEUKEMIA, MYELOBLASTIC
    VIRUS, BABOON ENDOGENOUS, 3841*
    VIRUS, C-TYPE, 3841*
    VIRUS, SIMIAN SARCOMA, 3841*
LEUKEMIA, MYELOCYTIC
    VIRUS, C-TYPE RNA TUMOR, 3842*
VIRUS, C-TYPE
    DNA-RNA HYBRIDIZATION, 3833
VIRUS, GUINEA PIG LEUKEMIA
    VIRUS, B-TYPE, 3852*
VIRUS, MURINE MAMMARY TUMOR
    DNA, VIRAL, 3869*
    RNA, VIRAL, 3867*
VIRUS, ROUS SARCOMA
    DNA REPLICATION, 3880*
    ISOLATION AND CHARACTERIZATION,
        3816
    PHOSPHATASES, 3880*
    PROTEIN KINASE, 3880*

RHABDOMYOSARCOMA
    IMIDAZOLE-4-CARBOXAMIDE, 5-[3,3-
        DIMETHYL-1-TRIAZENO]-
        GERM-FREE LIFE, 3758*
    LYMPHOCYTES
        IMMUNOSUPPRESSION, 3952*

RHEUMATIC HEART DISEASE
    MULTIPLE SCLEROSIS
        EPIDEMIOLOGY, 4122

RIBONUCLEASE
    BRAIN NEOPLASMS
        HODGKIN'S DISEASE, 4017
    MAMMARY NEOPLASMS, EXPERIMENTAL
        VIRUS, MURINE MAMMARY TUMOR, 3866*
    VIRUS, CELO
        RNA, 3827

RIFAMYCIN
    VIRUS, FRIEND MURINE LEUKEMIA
        SPLENOMEGALY, 3797

RNA
    BRAIN NEOPLASMS
        HODGKIN'S DISEASE, 4017
    CHOLANTHRENE, 3-METHYL-
        POLY A, 3756*
    CORTISOL
        NUCLEOLI, RAT, 4172
    CYTOPLASM, 3634
    VIRUS, CELO
        DNA, VIRAL, 3827
        RIBONUCLEASE, 3827
    VIRUS, FROG 3
        SYNTHESIS INHIBITION, HOST CELL,
            3851*

RNA, MESSENGER
    CYTOPLASM
        PROTEINS, 3634
    ESTROGENS
        CELL DIFFERENTIATION, 3607
    HELA CELLS
        CHROMATIN, 3634
    HORMONES
        REVIEW, 4183
    PROGESTERONE

CELL DIFFERENTIATION, 3607
VIRUS, HAMSTER SARCOMA
    POLY A POLYMERASE, 3799
VIRUS, SV40
    CYTOCHALASIN B, 3820
    ENDONUCLEASES, 3819
    HALF-LIFE STUDY, 3820

RNA, NEOPLASM
    MULTIPLE MYELOMA
        IMMUNOGLOBULINS, SURFACE, 3923
    PLASMACYTOMA
        HYBRIDIZATION,, 3863*
    VIRUS, HARVEY MURINE SARCOMA
        HYBRIDIZATION,, 3863*
    VIRUS, MOLONEY MURINE SARCOMA
        HYBRIDIZATION,, 3863*

RNA POLYMERASE
    CARBAMIC ACID, ETHYL ESTER
        CARCINOGENIC ACTIVITY, RAT, 3715
    CORTISOL
        NUCLEOLI, RAT, 4172
    DIETHYLAMINE, N-NITROSO-
        MORPHOLOGICAL EFFECT, LIVER, RAT,
            3703
    MORPHOLINE, N-NITROSO-
        MORPHOLOGICAL EFFECT, LIVER, RAT,
            3703
    VIRUS, ROUS SARCOMA
        VIRUS REPLICATION, 3814

RNA REPLICATION
    BUTYRIC ACID, 2-AMINO-4-[ETHYLTHIO]-
        SEX FACTORS, 3734*
    FIBROBLASTS
        GROWTH SUBSTANCES, 4165
    LUCANTHONE
        RADIATION, 3779*
    LYMPHOCYTES
        LYMPHOKINES, 3922

RNA, RIBOSOMAL
    VIRUS, SV40
        DNA-RNA HYBRIDIZATION, 3819
        ENDONUCLEASES, 3819

RNA, TRANSFER
    ACETOHYDROXAMIC ACID, N-4-BIPHENYLYL-
        CYTOSOL, 3675
    ACETOHYDROXAMIC ACID, N-FLUOREN-2-YL-
        CYTOSOL, 3675

RNA, VIRAL
    VIRUS, C-TYPE
        DNA-RNA HYBRIDIZATION, 3833
    VIRUS, HARVEY MURINE SARCOMA
        DNA-RNA HYBRIDIZATION, 3800
    VIRUS, MOLONEY MURINE LEUKEMIA
        DNA-RNA HYBRIDIZATION, 3800
    VIRUS, MURINE MAMMARY TUMOR
        REVERSE TRANSCRIPTASE, 3867*

SALIVARY GLAND NEOPLASMS
    HEMANGIOMA
        EPIDEMIOLOGY, 4000
    LEIOMYOMA
        EPIDEMIOLOGY, 4000

SALMONELLA TYPHIMURIUM
  BENZ(A)ANTHRACENE,
    MUTAGENIC ACTIVITY, 3678
  BENZ(A)ANTHRACENE-5,6-DICL, 5,6-
    DIHYDRO-
    MUTAGENIC ACTIVITY, 3678
  BENZ(A)ANTHRACENE-8,9-DICL, 8,9-
    DIHYDRO-
    MUTAGENIC ACTIVITY, 3678
  BENZ(A)ANTHRACENE-8,9-DICL, 8,9-
    DIHYDRO-7-METHYL-
    MUTAGENIC ACTIVITY, 3678
  BENZ(A)ANTHRACENE-8,9-DICL, 10,11-
    EPOXY-8,9-DIHYDRO-
    MUTAGENIC ACTIVITY, 3678
  BENZ(A)ANTHRACENE, 5,6-EPOXY-
    MUTAGENIC ACTIVITY, 3678
  BENZO(A)PYRENE
    MUTAGENIC ACTIVITY, 3678
  BENZO(A)PYRENE-4,5-DIOL, 4,5-DIHYDRO-
    MUTAGENIC ACTIVITY, 3678
  BENZO(A)PYRENE-7,8-DIOL, 7,8-DIHYDRO-
    MUTAGENIC ACTIVITY, 3678
  BENZO(A)PYRENE-9,10-DIOL, 9,10-DIHYDRO-

    MUTAGENIC ACTIVITY, 3678
  BENZO(A)PYRENE-7,8-DIOL, 9,10-EPOXY-7,
    8-DIHYDRO-
    MUTAGENIC ACTIVITY, 3678
  BENZO(A)PYRENE, 4,5-EPOXY-
    MUTAGENIC ACTIVITY, 3678
  DIETHYLAMINE, N-NITROSO-
    MUTAGENIC ACTIVITY, 3697
  DIMETHYLAMINE, N-NITROSO-
    MUTAGENIC ACTIVITY, 3697

SARCOIDOSIS
  VIRUS, EPSTEIN-BARR
    ANTIGENS, VIRAL, 3904
  VIRUS, HERPES
    ANTIGENS, VIRAL, 3904

SARCOMA
  BRAIN NEOPLASMS
    ENDOPLASMIC RETICULUM, 4005
    GOLGI APPARATUS, 4005
  CELL DIVISION
    CELL CYCLE KINETICS, 4131
  CELL TRANSFORMATION, NEOPLASTIC
    ODONTOGENIC CYSTS, 4063*
  CHOLANTHRENE, 3-METHYL-
    IMMUNITY, CELLULAR, 3934*
    IMMUNOGLOBULINS, 3956*
    MIGRATION INHIBITORY FACTOR, 3934*
  IMMUNE SERUMS
    GRAFT REJECTION, 3902
    IMMUNITY, CELLULAR, 3902
    TRANSPLANTATION IMMUNOLOGY, 3902
  IMMUNITY, PASSIVE
    GRAFT REJECTION, 3902
  LYMPHATIC NEOPLASMS
    VIRUS, POLYOMA, BK, 3810
  MAMMARY NEOPLASMS, EXPERIMENTAL
    CHOLANTHRENE, 3-METHYL-, 3680
  MYCOPLASMA
    CARCINOGENIC POTENTIAL, MOUSE,
    4039
  NEOPLASMS, CONNECTIVE TISSUE
    VIRUS, CELO, 3794

NEOPLASMS, MUSCLE TISSUE
  ULTRASTRUCTURAL STUDY, REVIEW,
  3624
PROSTATIC NEOPLASMS
  CASE REPORT, 4092*
PROTEINS
  ISOLATION AND CHARACTERIZATION, C-
  REACTIVE, 4169
THYROID NEOPLASMS
  EPIDEMIOLOGY, 4106*
TRANSPLANTATION, HOMOLOGOUS
  GRAFT VS HOST REACTION, 3940*
UTERINE NEOPLASMS
  DNA, NEOPLASM, 4100*
VIRUS, CELO
  INTERCELLULAR ATTACHMENTS, 3794
  ULTRASTRUCTURAL STUDY, PLASMA
  MEMBRANE, 3794
VIRUS, ROUS SARCOMA
  TRANSPLANTATION, HOMOLOGOUS, 3902

SARCOMA 180, CROCKER
  GLYCOPROTEINS
    ASCITIES FORM, 4168
    CELL MEMBRANE, 4168
    CELLS, CULTURED, 4168

SARCOMA, EWING'S
  BONE NEOPLASMS
    EPIDEMIOLOGY, CHILD, INDIA, 4120
  HEMANGIOMA
    RADIATION, IONIZING, 3773

SARCOMA, MAST CELL
  B-LYMPHOCYTES
    IMMUNOSUPPRESSION, 3887
  T-LYMPHOCYTES
    IMMUNOSUPPRESSION, 3887

SARCOMA, OSTEOGENIC
  BONE NEOPLASMS
    CASE REPORT, 4049*
    CASE REPORT, DOG, 4048*
    EPIDEMIOLOGY, CHILD, INDIA, 4120
    EPIDEMIOLOGY, MALAYSIA, 4142*
    VIRUS, POLYOMA, 3874*
  GUANIDINE, 1-METHYL-3-NITRO-1-NITROSO-
    CELL TRANSFORMATION, NEOPLASTIC,
    3705
  IGG
    IMMUNE SERUMS, 3925
  IGM
    IMMUNE SERUMS, 3925
  IMMUNOGLOBULINS
    IMMUNE SERUMS, 3925
  ORBITAL NEOPLASMS
    RADIATION, 4003
    ULTRASTRUCTURAL STUDY, 4003
  OSTEOMA, OSTEOID
    CASE REPORT, 4049*
  PRECANCEROUS CONDITIONS
    CASE REPORT, 4049*
  VIRUS, RAUSCHER MURINE LEUKEMIA
    ANTIGENS, VIRAL, 3830
  VIRUS, SIMIAN LEUKEMIA
    ANTIGENS, VIRAL, 3830

SARCOMA, RETICULUM CELL
  ABDOMINAL NEOPLASMS

CASE REPORT, 4015
HEAVY CHAIN DISEASE, 4015
BLOOD SEDIMENTATION
DIAGNOSOS AND PROGNOSIS, 4019
BRAIN NEOPLASMS
CASE REPORT, 4051*
EYE NEOPLASMS
CASE REPORT, 4018
UVEITIS, 4018
FETAL GLOBULINS
DIAGNOSOS AND PROGNOSIS, 4019
FIBRINOGEN
DIAGNOSOS AND PROGNOSIS, 4019
GENETICS, 4087*
REVIEW, 3650*
HAPTOGLOBINS
DIAGNOSOS AND PROGNOSIS, 4019
IMMUNITY, CELLULAR
REVIEW, 3650*
IMMUNOSUPPRESSION
REVIEW, 3650*
INFECTIOUS MONONUCLEOSIS
ANTIGENS, HETEROGENETIC, 3914
MYCOPLASMA
CARCINOGENIC POTENTIAL, MOUSE,
4039
RADIATION, IONIZING
MORTALITY, 3780*
REVIEW, 3650*

SARCOMA, YOSHIDA
TRANSPLANTATION IMMUNOLOGY
IMMUNOSUPPRESSION, 3948*

SCHISTOSOMIASIS
BLADDER NEOPLASMS
PRECANCEROUS CONDITICNS, 3990

SEMINOMA
SEE DISGERMINOMA

SERINE DEHYDRATASE
HEPATOMA
ADRENALECTOMY, 4183
CORTISONE, 4183

SEX CHROMOSOMES
OVARIAN NEOPLASMS
DISGERMINOMA, 4031

SIALIC ACIDS
FIBROSARCOMA
DNA, 4166
NEOPLASM TRANSPLANTATION, 4166
HEPATOMA
DNA, 4166
LUNG NEOPLASMS
DNA, 4166
MAMMARY NEOPLASMS, EXPERIMENTAL
DNA, 4166

SKIN NEOPLASMS
ACETIC ACID
ENZYMATIC ACTIVITY, 3723*
BENZO(A)PYRENE
TRANSPLACENTAL CARCINOGENESIS,
REVIEW, 3601
CANTHARIDIN
ENZYMATIC ACTIVITY, 3723*

CARCINOMA, BASAL CELL
REVIEW, 3618
CARCINOMA, EPIDERMOID
REVIEW, 3618
ULTRAVIOLET RAYS, 3775
6-CHRYSENAMINE
CARCINOGENIC POTENTIAL, 3681
FLUOCINOLONE ACETONIDE
ANTINEOPLASTIC AGENTS, 3738*
HEMANGIOMA
EPIDEMIOLOGY, 4030
EPIDEMIOLOGY, CHILD, 4140*
LEIOMYOMA
EPIDEMIOLOGY, 4030
LEUKEMIA, LYMPHOCYTIC
T-LYMPHOCYTES, 3918
MELANOMA
CASE REPORT, 4022
CELL TRANSFORMATION, NEOPLASTIC,
4040
HISTOLOGICAL STUDY, 4040
IMMUNE RESPONSE, 4040
PRECANCEROUS CONDITIONS, 4040
UTLRASTRUCTURAL STUDY, 4061*
NEVUS
UTLRASTRUCTURAL STUDY, 4061*
NEVUS, PIGMENTED
HISTOLOGICAL STUDY, 4040
PRECANCEROUS CONDITIONS, 4040
TRANSPLANTATION, HOMOLOGOUS
ANTIGENIC STIMULATION, REVIEW,
3807
ULTRAVIOLET RAYS
PRECANCEROUS CONDITIONS, 3775

SMOKING
AIR POLLUTANTS
COCARCINOGENIC EFFECT, REVIEW,
3610
ASBESTOS
COCARCINOGENIC EFFECT, REVIEW,
3627
BENZO(A)PYRENE
COCARCINOGENIC EFFECT, REVIEW,
3610
BRONCHITIS
EPIDEMIOLOGY, GENETICS, 4121
CARCINOEMBRYONIC ANTIGEN
T-LYMPHOCYTES, 3768*
CORONENE
COCARCINOGENIC EFFECT, REVIEW,
3610
DNA REPAIR
ARYL HYDROCARBON HYDROXYLASES,
3710
FIBROBLASTS, 3710
DNA REPLICATION
ARYL HYDROCARBON HYDROXYLASES,
3710
DNA REPAIR, 3769*
FIBROBLASTS, 3710
LARYNGEAL NEOPLASMS
AGE FACTORS, GREAT BRITAIN, 4114
LUNG NEOPLASMS, 3610
AGE FACTORS, GREAT BRITAIN, 4114
AGE FACTORS, STATISTICAL ANALYSIS,
4128
EPIDEMIOLOGY, AFRICA, 4116
EPIDEMIOLOGY, GENETICS, 4121

EPIDEMIOLOGY, IMMIGRANTS, 4126
METALS
    COCARCINOGENIC EFFECT, REVIEW,
        3613
PULMONARY EMPHYSEMA
    EPIDEMIOLOGY, GENETICS, 4121
RESPIRATORY SYSTEM NEOPLASMS
    ANIMAL MODEL, HAMSTER, 3711

SODIUM AZIDE
    DNA
        BASE-SUBSTITUTION MUTAGENESIS,
            3729*

SODIUM CHLORIDE
    AFLATOXIN B1
        PEANUTS, 3670

SODIUM HYDROXIDE
    AFLATOXIN B1
        PEANUTS, 3670

SOMATOTROPIN
    MAMMARY NEOPLASMS, EXPERIMENTAL
        1-PIPERAZINEETHANOL, 4-(3-(2-
            CHLOROPHENOTHIAZIN-10-YL)PROPYL)-
            , 3688
    PROLACTIN
        RADIOIMMUNOASSAY, MOUSE, 4185

SPERMIDINE
    DNA REPLICATION
        FIBROBLASTS, DIPLOID, 4162
    POLYCYTHEMIA VERA
        PLASMS, 4161

SPERMINE
    DNA REPLICATION
        FIBROBLASTS, DIPLOID, 4162
    POLYCYTHEMIA VERA
        LEUKOCYTES, 4161
        PLASMS, 4161

SPINAL NEOPLASMS
    CARCINOMA
        NEOPLASMS, MULTIPLE PRIMARY, 3995
    TERATOID TUMOR
        CASE REPORT, FETUS, 4090*

SPLEEN
    ANGIOSARCOMA
        ETHYLENE, CHLORO-, 4036
        ETHYLENE, CHLORO- POLYMER, 4036

SPLENIC NEOPLASMS
    ERYTHROLEUKEMIA
        VIRUS, FRIEND MURINE LEUKEMIA,
            3847*
    VIRUS, FRIEND MURINE LEUKEMIA
        CELL TRANSFORMATION, NEOPLASTIC,
            3847*
        TRANSPLANTATION IMMUNOLOGY, 3847*

SPLENOMEGALY
    GENETICS
        CASE REPORT, 4032

STATOLON
    VIRUS, FRIEND MURINE LEUKEMIA

REGRESSION, 3905

STEROIDS
    HEPATOMA
        FETAL GLOBULINS, 3636*

4,4'-STILBENEDIOL, ALPHA,ALPHA'-DIETHYL-
    ADENOCARCINOMA
        CASE REPORT, 4096*
    BREAST NEOPLASMS
        ANTI-ESTROGEN AGENTS, 3740*
    CARCINOMA, EPIDERMOID
        CASE REPORT, 4096*
    CERVIX NEOPLASMS
        TRANSPLACENTAL CARCINOGENESIS,
            3608
    DIETARY FATS
        COCARCINOGENIC EFFECT, 3606
    KIDNEY NEOPLASMS
        ANTINEOPLASTIC AGENTS, 3739*
    UTERINE NEOPLASMS
        CASE REPORT, 4097*
    VAGINAL NEOPLASMS
        ADENOCARCINOMA, 4124
        EPIDEMIOLOGY, 4124
        TRANSPLACENTAL CARCINOGENESIS,
            3608

STILBESTROL
    SEE 4,4'-STILBENEDIOL, ALPHA, ALPHA'-
        DIETHYL-

STOMACH NEOPLASMS
    ADENOCARCINOMA
        ACETAMIDE, N-FLOUREN-2-YL-, 3635*
        EPIDEMIOLOGY, DOG, 3635*
        FLUORESCENT ANTIBODY TECHNIQUE,
            3706
        GUANIDINE, 1-METHYL-3-NITRO-1-
            NITROSO-, 3635*, 3706
        IMMUNITY, CELLULAR, 3934*
        MIGRATION INHIBITORY FACTOR, 3934*
        QUINOLINE, 4-NITRO-, 1-OXIDE,
            3635*
    ADENOMA
        EPIDEMIOLOGY, RAT, 3635*
    AGE FACTORS
        EPIDEMIOLOGY, ISRAEL, 4115
    CARCINOGEN, CHEMICAL
        ANIMAL MODEL, RAT, MOUSE, 3641*
    CARCINOMA
        PRECANCEROUS CONDITIONS, 4067*,
            4068*
    CARCINOMA, EPIDERMOID
        PIPERIDINE, 3,4-DIBROMO-1-NITROSO-,
            3707
        PIPERIDINE, 3,4-DICHLORO-1-NITROSO-
            , 3707
    CARCINOMA IN SITU
        PRECANCEROUS CONDITIONS, 4068*
    GASTRIN
        ENZYMATIC ACTIVITY, 4184
        TISSUE CULTURE, 4184
    GLOMANGIOMA
        CASE REPORT, 4034
        CASE REPORT, REVIEW, 4066*
        ENDOCRINE-TYPE GRANULES, 4034
        ULTRASTRUCTURAL STUDY, 4034, 4066*
    GUANIDINE, 1-METHYL-3-NITRO-1-NITROSO-

IMMUNE RESPONSE, 3939*
HEMANGIOMA
EPIDEMIOLOGY, 4000
LEIOMYOMA
EPIDEMIOLOGY, 4000
MULTIPLE SCLEROSIS
EPIDEMIOLOGY, 4122
PAPILLOMA
DIETHYLAMINE, N-NITROSO-, 3696
PIPERIDINE, 3,4-DIBROMO-1-NITROSO-, 3707
SEX FACTORS, 3696
TRANSPLACENTAL CARCINOGENESIS, HAMSTER, 3696
SULFATASES
HISTOCHEMICAL STUDY, 4179
TOBACCO
EPIDEMIOLOGY, AFRICA, 4116

STREPTODORNASE AND STREPTOKINASE
BURKITT'S LYMPHOMA
IMMUNITY, CELLULAR, 3946*

STRESS
MAMMARY NEOPLASMS, EXPERIMENTAL
VIRUS, BITTNER MURINE MAMMARY TUMOR, 4038
VIRUS, MURINE MAMMARY TUMOR, 3868*, 4038

SUCROSE
NEUROBLASTOMA
ADENOSINE TRIPHOSPHATASE, 4176

SULFATASES
BREAST NEOPLASMS
HISTOCHEMICAL STUDY, 4179
CHOLANTHRENE, 3-METHYL-
CARCINOGENIC METABOLITE, 3683
TRANSPLACENTAL CARCINOGENESIS, 3683
GYNECOLOGIC NEOPLASMS
HISTOCHEMICAL STUDY, 4179
HEPATOMA
HISTOCHEMICAL STUDY, 4179
INTESTINAL NEOPLASMS
HISTOCHEMICAL STUDY, 4179
KIDNEY NEOPLASMS
HISTOCHEMICAL STUDY, 4179
MELANOMA
HISTOCHEMICAL STUDY, 4179
PANCREATIC NEOPLASMS
HISTOCHEMICAL STUDY, 4179
STOMACH NEOPLASMS
HISTOCHEMICAL STUDY, 4179
TESTICULAR NEOPLASMS
HISTOCHEMICAL STUDY, 4179

SULPIRIDE
MAMMARY NEOPLASMS, EXPERIMENTAL
LACTAION, 3713
PROLACTIN, 3713

TANNIC ACID
BRACKEN FERN

ISOLATION AND CHARACTERIZATION, 3770*

TERATOGENS
ACETIC ACID, LEAD (+2) SALT
TERATOGENIC ACTIVITY, 3727*

TERATOID TUMOR
ALKALINE PHOSPHATASE
SERUM LEVELS, 4175
FETAL GLOBULINS
CASE REPORT, FETUS, 4090*
VITELLINE MEMBRANE, 3907
HEMANGIOMA
RADIATION, IONIZING, 3773
ISOANTIGENS
IMMUNE RESPONSE, 3892
L CELLS
LEUKEMIA, 4158
SPINAL NEOPLASMS
CASE REPORT, FETUS, 4090*
TESTICULAR NEOPLASMS
CASE REPORT, 4105*
NEOPLASM TRANSPLANTATION, 3621

TERATOMA, EMBRYONAL
SEE TERATOID TUMOR

TESTICULAR NEOPLASMS
CHORIOCARCINOMA
CASE REPORT, 4105*
GENETICS
CASE REPORT, 4105*
MESENCHYMOMA
CASE REPORT, 4093*
SULFATASES
HISTOCHEMICAL STUDY, 4179
TERATOID TUMOR
CASE REPORT, 4105*
NEOPLASM TRANSPLANTATION, 3621

TESTOSTERONE
ARYL HYDROCARBON HYDROXYLASES
ENZYMATIC ACTIVITY, 3738*
DNA
THYMIDINE INCORPORATION, 3680
MAMMARY NEOPLASMS, EXPERIMENTAL
BENZ(A)ANTHRACENE, 7,12-DIMETHYL-, 3680
CHOLANTHRENE, 3-METHYL-, 3680
OVARIAN NEOPLASMS
DISGERMINOMA, 4031

12-O-TETRADECANOYLPHORBOL-13-ACETATE
CELL MEMBRANE
SPECTROMETRY, FLUORESCENCE, 3726*
DNA REPAIR
FIBROBLASTS, 3721*
12-O-TETRADECANOYLPHORBOLOL-13-ACETATE
ACTIVE INTERMEDIATE, 3725*

12-O-TETRADECANOYLPHORBOLOL-13-ACETATE
12-O-TETRADECANOYLPHORBOL-13-ACETATE
ACTIVE INTERMEDIATE, 3725*

THECA CELL TUMOR
OVARIAN NEOPLASMS
VIRUS, POLYOMA, 3874*

THEOPHYLLINE
    CELL DIVISION
        CIRCADIAN RHYTHM, 4152
        CORNEA, RAT, 4152
    HELA CELLS
        ADENOSINE CYCLIC 3',5'
          MONOPHOSPHATE, 4193*
    HEPATOMA
        GROWTH, 3906
    PLASMACYTOMA
        CELL MEMBRANE, 4153
        CONCANAVALIN A, 4153
        OUABAIN, 4153

THORIUM DIOXIDE
    LEUKEMIA, RADIATION-INDUCED
        CHROMOSOME ABERRATIONS, 4026
    PERITONEAL NEOPLASMS
        MESOTHELIOMA, 4037

THOROTRAST
    SEE THORIUM DIOXIDE

THROMBOPENIA
    GENETICS
        CASE REPORT, 4032

THYMIDINE
    BENZ(A)ANTHRACENE, 7,12-DIMETHYL-
        DNA REPLICATION, 3746*
    BREAST NEOPLASMS
        GROWTH SUBSTANCES, 3689
    VIRUS, HERPES SIMPLEX
        DNA REPLICATION, 3806

THYMIDINE KINASE
    VIRUS, ADENO 12
        HYBRID CELLS, 3793

THYMOMA
    HYPERGAMMAGLOBULINEMIA
        IMMUNITY, CELLULAR, 4043
    ISOANTIBODIES
        IMMUNE RESPONSE, 4043
    MONOCYTES
        IGG, 3958*
    MYASTHENIA GRAVIS
        ANEMIA, APLASTIC, 4044
    NEOPLASM METASTASIS
        CASE REPORT, 4044
    PLANT AGGLUTININS
        IMMUNE RESPONSE, 4043
    PURINE-6-THIOL
        IMMUNE RESPONSE, 4043
    TUBERCULIN
        IMMUNE RESPONSE, 4043

THYMUS NEOPLASMS
    LYMPHOMA
        ALKALINE PHOSPHATASE, 3862*
    LYMPHOSARCOMA
        IMIDAZOLE-4-CARBOXAMIDE, 5-(3,3-
          DIMETHYL-1-TRIAZENO)-, 3691,
          3758*

THYROID NEOPLASMS
    ADENOCARCINOMA
        CASE REPORT, 4002
        EPIDEMIOLOGY, 4106*, 4129

        RADIATION, IONIZING, 3772
        RADIOTHERAPY, 3786*
        ULTRASTRUCTURAL STUDY, 4002
    ADENOMA
        EPIDEMIOLOGY, 4106*
        RADIATION, IONIZING, 3772
    CARCINOMA
        CARCINOEMBRYONIC ANTIGEN, 3896
        CASE REPORT, 4002
        EPIDEMIOLOGY, 4106*
        IMIDAZOLE-4-CARBOXAMIDE, 5-AMINO-,
          3691
        IMMUNE RESPONSE, 3896
        ULTRASTRUCTURAL STUDY, 4002
    CARCINOMA, PAPILLARY
        EPIDEMIOLOGY, 4129
        HISTOLOGIC STUDY, 4129
        NEOPLASM METASTASIS, 4129
        RADIATION, IONIZING, 3772
        RADIOTHERAPY, 3786*
    2-IMIDAZOLIDINETHIONE
        CARCINOGENIC ACTIVITY, RAT, 3687
    PAPILLOMA
        EPIDEMIOLOGY, 4106*
    SARCOMA
        EPIDEMIOLOGY, 4106*

TILORONE
    VIRUS, FRIEND MURINE LEUKEMIA
        SPLENOMEGALY, 3797

TISSUE CULTURE
    STOMACH NEOPLASMS
        GASTRIN, 4184

TOBACCO
    BLADDER NEOPLASMS
        EPIDEMIOLOGY, SPAIN, REVIEW, 3630
    BRONCHIAL NEOPLASMS
        HISTOLOGICAL STUDY, HAMSTER, 4077*
        PRECANCEROUS CONDITIONS, 4077*
    STOMACH NEOPLASMS
        EPIDEMIOLOGY, AFRICA, 4116

M-TOLUIDINE, N,N-DIMETHYL-4-(PHENYLAZO)-
    HEPATOMA
        DIETARY FATS, 3606

P-TOLUAMIDE, N-ISOPROPYL-ALPHA-(2-
    METHYLHYDRAZINO)-,
    NERVOUS SYSTEM NEOPLASMS
        TRANSPLACENTAL CARCINOGENESIS,
          REVIEW, 3601

TONSILITIS
    PHARYNGEAL NEOPLASMS
        NEOPLASMS, EMBRYON AND MIXED, 4001

TONSILLAR NEOPLASMS
    PROTEINS
        ISOLATION AND CHARACTERIZATION, C-
          REACTIVE, 4169

TRACHEAL NEOPLASMS
    PAPILLOMA
        DIETHYLAMINE, N-NITROSO-, 3696
    POLYPS
        BUTANOL, 4-(BUTYLNITROSOAMINO)-,
          3701

TRANSFERRIN
  LEUKEMIA, LYMPHOCYTIC
    ISOLATION AND CHARACTERIZATION,
    4197*

TRANSIENTS AND MIGRANTS
  COLONIC NEOPLASMS
    MULTIPLE SCLEROSIS, 4122

TRANSPLACENTAL CARCINOGENESIS
  BRAIN NEOPLASMS
    ISOANTIGENS, 3892

TRANSPLANTATION ANTIGENS
  SEE HISTOCOMPATIBILITY ANTIGENS

TRANSPLANTATION, HETEROLOGOUS
  BREAST NEOPLASMS
    NEOPLASM METASTASIS, 4035
  CARCINOMA, BASAL CELL
    NEOPLASM METASTASIS, 4035
  CARCINOMA, EPIDERMOID
    NEOPLASM METASTASIS, 4035
  GASTROINTESTINAL NEOPLASMS
    NEOPLASM METASTASIS, 4035
  HEAD AND NECK NEOPLASMS
    LYMPHOMA, 4016
  HELA CELLS
    NEOPLASM METASTASIS, 4035
  MELANOMA
    NEOPLASM METASTASIS, 4035
  PANCREATIC NEOPLASMS
    NEOPLASM METASTASIS, 4335

TRANSPLANTATION, HOMOLOGOUS
  ANTIBODY FORMATION
    ANTIGENS, HETEROGENETIC, 3914
  CERVIX NEOPLASMS
    ANTIGENIC STIMULATION, REVIEW,
    3807
  IMMUNOSUPPRESSION
    ANTIGENIC STIMULATION, REVIEW,
    3807
  LEUKEMIA
    CHROMOSOME ABERRATIONS, 4082*
    GRAFT VS HOST REACTION, 3940*
  LYMPHATIC NEOPLASMS
    ANTIGENIC STIMULATION, REVIEW,
    3807
  MAMMARY NEOPLASMS, EXPERIMENTAL
    SPLEEN, 3940*
  NEOPLASMS, EXPERIMENTAL
    VIRUS, FRIEND MURINE LEUKEMIA,
    3798
  SARCOMA
    GRAFT VS HOST REACTION, 3940*
    VIRUS, ROUS SARCOMA, 3902
  SKIN NEOPLASMS
    ANTIGENIC STIMULATION, REVIEW,
    3807
  VIRUS, CYTOMEGALO
    ANTIGENIC STIMULATION, REVIEW,
    3807
  VIRUS, HERPES SIMPLEX 1
    ANTIGENIC STIMULATION, REVIEW,
    3807
  VIRUS, HERPES SIMPLEX 2
    ANTIGENIC STIMULATION, REVIEW,
    3807

VIRUS, HERPES ZOSTER
    ANTIGENIC STIMULATION, REVIEW,
    3807
VIRUS, POLYOMA
    KIDNEY, 3872*

TRANSPLANTATION IMMUNOLOGY
  ADENOCARCINOMA
    LYMPHOCYTES, 3890
    MACROPHAGES, 3890
    MYCOBACTERIUM BOVIS, 3890
    RADIATION, IONIZING, 3890
    ZYMOSAN, 3890
  CARCINOMA, EHRLICH TUMOR
    IMMUNITY, CELLULAR, 3934*
    MIGRATION INHIBITORY FACTOR, 3934*
  FIBROSARCOMA
    CHOLANTHRENE, 3-METHYL-, 3945*
    HYBRIDIZATION, 4158
    IMMUNOGENETICS, 4158
  IMMUNOTHERAPY
    REVIEW, 3653*
  L CELLS
    HYBRIDIZATION, 4158
    IMMUNOGENETICS, 4158
  LEUKEMIA
    REVIEW, 3626
  LUNG NEOPLASMS
    LEUKEMIA, MONOCYTIC, 3933*
  LYMPHOSARCOMA
    ESTRADIOL, 3945*
  MAMMARY NEOPLASMS, EXPERIMENTAL
    RADIATION, IONIZING, 4149
  NEOPLASMS
    REVIEW, 3626
  NEOPLASMS, EXPERIMENTAL
    VIRUS, SV40, 3891
  SARCOMA
    IMMUNE SERUMS, 3902
  SARCOMA, YOSHIDA
    IMMUNOSUPPRESSION, 3948*
  SPLENIC NEOPLASMS
    VIRUS, FRIEND MURINE LEUKEMIA,
    3847*
  VIRUS, BOVINE ADENO 3
    VIRUS, ADENO 12, 3916
    VIRUS, BOVINE ADENO 1, 3916
    VIRUS, BOVINE ADENO 2, 3916
    VIRUS, CELO, 3916
    VIRUS, SIMIAN ADENO 7, 3916

S-TRIAZINE, HEXAHYDRO-1,3,5-TRINITRO-
  LIVER NEOPLASMS
    ADENOMA, 3694
    HEMANGIOMA, 3694

S-TRIAZINE, 2,4,6-TRIS(1-AZIRIDINYL)-
  CHROMOSOME ABERRATIONS
    BONE MARROW, 3717*

TRITIUM
  MAMMARY NEOPLASMS, EXPERIMENTAL
    ADENOFIBROMA, 3714
  RADIATION
    HORMONE POTENTIATION, 3714
    PREGNANCY EFFECTS, 3714

TRYPSIN
  GANGLIOSIDES

CELL MEMBRANE, 3924

TUBERCULIN
    BURKITT'S LYMPHOMA
        IMMUNITY, CELLULAR, 3946*
    B-LYMPHOCYTES
        LANTHANUM, 4167
    T-LYMPHOCYTES
        LANTHANUM, 4167
    THYMOMA
        IMMUNE RESPONSE, 4043

TUBERCULOSIS, PULMONARY
    MULTIPLE SCLEROSIS
        EPIDEMIOLOGY, 4122

TYROSINE
    VIRUS, ADENO 2
        PEPTIDES, 3788

TYROSINE AMINOTRANSFERASE
    HEPATOMA
        ADRENALECTOMY, 4183
        CORTISONE, 4183

TYROSINE HYDROXYLASE
    NEUROBLASTOMA
        PHENYLACETATES, 4178
        VANILMANDELIC ACID, 4178
    PHEOCHROMOCYTOMA
        PHENYLACETATES, 4178
        VANILMANDELIC ACID, 4178

ULTRAVIOLET RAYS
    BURKITT'S LYMPHOMA
        DNA REPAIR, 3778*
    DIETARY FATS
        COCARCINOGENIC EFFECT, 3606
    FIBROBLASTS
        DNA REPAIR, 3674
    SKIN NEOPLASMS
        CARCINOMA, EPIDERMOID, 3775
        PRECANCEROUS CONDITIONS, 3775
    VIRUS, AVIAN SARCOMA
        CELL TRANSFORMATION, NEOPLASTIC,
        3840*
    VIRUS, HERPES SIMPLEX 1
        ANTIGENS, VIRAL, 3804
        DNA REPLICATION, 3804
        ERYTHROCYTES, 3804
    VIRUS, SV40
        ANTIGENS, VIRAL, 3804
        DNA REPLICATION, 3804
        ERYTHROCYTES, 3804
    XERODERMA PIGMENTOSUM
        DNA REPAIR, 3674

UREA
    HEPATOMA
        ALKALINE PHOSPHATASE, 4175

UREA, ETHYL NITROSO-
    BRAIN NEOPLASMS
        COPPER (II) SULFATE, 3601
        NICKEL (II) SULFATE, 3601
        TRANSPLACENTAL CARCINOGENESIS,
        REVIEW, 3601
    LUNG NEOPLASMS
        TRANSPLACENTAL CARCINOGENESIS,
        3892
    NERVOUS SYSTEM NEOPLASMS
        TRANSPLACENTAL CARCINOGENESIS,
        REVIEW, 3601

UREA, HYDROXY-
    PLASMACYTOMA
        CELL FUSION, 4004
        ULTRASTRUCTURAL STUDY, 4004

UREA, METHYL NITROSO-
    DIABETES MELLITUS
        ORGANOTROPISM, 3708

UREA, N-NITROSO-N-PROPYL-
    NERVOUS SYSTEM NEOPLASMS
        TRANSPLACENTAL CARCINOGENESIS,
        REVIEW, 3601

UREA, NITROSO-
    AMINO ACIDS
        CARBAMOYLATION, 3765*
    PEPTIDES
        CARBAMOYLATION, 3765*
    PROTEINS
        CARBAMOYLATION, 3765*

UREMIA
    MAMMARY NEOPLASMS, EXPERIMENTAL
        CARCINOMA, 4151
        CELL CYCLE KINETICS, 4151
    RADIATION, IONIZING
        IMMUNOLOGICAL POTENTIAL, 3935*

URETHANE
    SEE CARBAMIC ACID, ETHYL ESTER

URIDINE
    HELA CELLS
        NUCLEOSIDE EXCRETION, 4163

URIDINE, 5-BROMO-2'-DEOXY-
    CELL MEMBRANE
        IGM, 3959*

URIDINE, 2'-DEOXY-5-FLUORO-
    CELL LINE
        CELL TRANSFORMATION, NEOPLASTIC,
        3720*

URIDINE, 2'-DEOXY-5-IODO-
    CARCINOMA
        VIRUS, SHOPE PAPILLOMA, 3871*
    PLASMACYTOMA
        A-TYPE PARTICLES, 3870*
    VIRUS, SHOPE PAPILLOMA
        GENOME ACTIVATION, 3871*

URIDINE KINASE
    SEE PHOSPHOTRANSFERASES, ATP

UROGENITAL NEOPLASMS
    EPIDEMIOLOGY
        HORMONS, 4118
    GENETICS
        EPIDEMIOLOGY, REVIEW, 4117
    HORMONES
        EPIDEMIOLOGY, REVIEW, 4117
    LEIOMYOMA

EPIDEMIOLOGY, 4000
NEOPLASM METASTASIS, 4045
NEOPLASMS
EPIDEMIOLOGY, REVIEW, 4117

UTERINE NEOPLASMS
ADENOCARCINOMA
CASE REPORT, 4096*, 4097*
ULTRASTRUCTURAL STUDY, 4101*
ADENOMA
ULTRASTRUCTURAL STUDY, 4102*
ANGIOMATOSIS
DNA, NEOPLASM, 4100*
ANTIGENIC DETERMINANTS
IMMUNE SERUMS, 3974*
IMMUNOPEROXIDASE STAINING, 3974*
CARCINOMA, EPIDERMOID
CASE REPORT, 4096*
CHORIOCARCINOMA
ULTRASTRUCTURAL STUDY, RAT, 4098*
DERMOID CYST
HISTOLOGICAL STUDY, CASE REPORT,
4047
ENDOMETRIOSIS
ULTRASTRUCTURAL STUDY, RABBIT,
4099*
LEIOMYOMA
C-REACTIVE PROTEIN, 4169
HISTOLOGICAL STUDY, CASE REPORT,
4047
LEIOMYOSARCOMA
IMIDAZOLE-4-CARBOXAMIDE, 5-(3-
METHYL-1-TRIAZENO)-, 3691
VIRUS, POLYOMA, 3874*
MESENCHYMOMA
DNA, NEOPLASM, 4100*
PRECANCEROUS CONDITIONS
AGE FACTORS, 4103*
EPIDEMIOLOGY, 4103*
PROGESTERONE
CASE REPORT, 4097*
SARCOMA
DNA, NEOPLASM, 4100*
4,4'-STILBENEDIOL, ALPHA,ALPHA'-
DIETHYL-
CASE REPORT, 4097*
VIRUS, MURINE SARCOMA
FETAL GLOBULINS, 3865*
IMMUNITY, CELLULAR, 3865*

UVEITIS
EYE NEOPLASMS
SARCOMA, RETICULUM CELL, 4018

VACCINES
VIRUS, CYTOMEGALO
REVIEW, 3616
VIRUS, HERPES
REVIEW, 3616
VIRUS, VARICELLA-ZOSTER
REVIEW, 3616

VACCINES, ATTENUATED
MAMMARY NEOPLASMS, EXPERIMENTAL
IMMUNITY, CELLULAR, 3619

VAGINAL NEOPLASMS
ADENOCARCINOMA
4,4'-STILBENEDIOL, ALPHA,ALPHA'-

DIETHYL-, 4124
4,4'-STILBENEDIOL, ALPHA,ALPHA'-
DIETHYL-
EPIDEMIOLOGY, 4124
TRANSPLACENTAL CARCINOGENESIS,
3608

VANILMANDELIC ACID
NEUROBLASTOMA
TYROSINE HYDROXYLASE, 4178
PHEOCHROMOCYTOMA
TYROSINE HYDROXYLASE, 4178

VINCALEUCOBLASTINE
VIRUS, SV40
CELL MEMBRANE, 4157

VINCRISTINE
SEE LEUROCRISTINE

VINYL CHLORIDE
SEE ETHYLENE, CHLORO-

VIRAL PROTEINS
VIRUS, ADENO 2
ANTIGENIC DETERMINANTS, 3788
ISOLATION AND CHARACTERIZATION,
3788, 3790
VIRUS, AVIAN MYELOBLASTOSIS
SYNTHESIS IN VITRO, 3813
VIRUS, AVIAN SARCOMA
GLUCOSAMINE, 3818
VIRUS, B77
GLUCOSAMINE, 3818
VIRUS, BOVINE ADENO 3
AMINO ACIDS, 3791
ISOLATION AND CHARACTERIZATION,
3791
VIRUS, HERPES SIMPLEX 2
CYCLOHEXIMIDE, 3801
VIRUS, ROUS SARCOMA
ULTRASTRUCTURAL STUDY, BYRAN
STRAIN, 3815

VIRUS, ADENO
NEOPLASMS
HYBRIDIZATION, 3631
VIRUS, SV40
TEMPERATURE SENSITIVE MUTANTS,
3822

VIRUS, ADENO 2
DNA REPAIR
NITROUS ACID, SODIUM SALT, 3789
DNA, VIRAL
ISOLATION AND CHARACTERIZATION,
3790
PEPTIDES
CELL MEMBRANE, 3788
ISOLATION AND CHARACTERIZATION,
3790
TYROSINE, 3788
VIRAL PROTEINS
ANTIGENIC DETERMINANTS, 3788
ISOLATION AND CHARACTERIZATION,
3788, 3790

VIRUS, ADENO 5
VIRUS, ADENO 7

HOMOLOGY, NUCLEOTIDE SEQUENCE,
3838*

VIRUS, ADENO 7
CELL TRANSFORMATION, NEOPLASTIC
ANTIGENIC DETERMINANTS, 3975*
ANTIGENS, NEOPLASM, 3975*
GROWTH
TEMPERATURE DEPENDENCE, 3884*
VIRUS, ADENO 5
HOMOLOGY, NUCLEOTIDE SEQUENCE,
3838*

VIRUS, ADENO 12
HYBRID CELLS
PHOSPHOTRANSFERASES, 3793
THYMIDINE KINASE, 3793
PHOSPHOTRANSFERASES
CHROMOSOMES, HUMAN, 16-18, 3793
VIRUS, BOVINE ADENO 3
TRANSPLANTATION IMMUNOLOGY, 3916

VIRUS, ADENO ASSOCIATED
VIRUS, HELPER
VIRUS REPLICATION, 3792
VIRUS REPLICATION
L-ARGININE, 3792
ULTRASTRUCTURAL STUDY, 3792
VIRUS, SV15
VIRUS REPLICATION, 3792

VIRUS, AVIAN MYELOBLASTOSIS
DNA NUCLEOTIDYLTRANSFERASES
BASE-PAIRING ERRORS, 3839*
POLYNUCLEOTIDES, 3839*
ELECTROPHORESIS
ISOLATION AND CHARACTERIZATION,
3828
VIRAL PROTEINS
SYNTHESIS IN VITRO, 3813

VIRUS, AVIAN SARCOMA
GENES, RECESSIVE
RESISTANCE AND SUSCEPTIBILITY,
3817
GLYCOPROTEINS
GLUCOSAMINE, 3818
ULTRAVIOLET RAYS
CELL TRANSFORMATION, NEOPLASTIC,
3840*
VIRAL PROTEINS
GLUCOSAMINE, 3818
VIRUS REPLICATION
GLUCOSAMINE, 3818

VIRUS, B77
GLYCOPROTEINS
GLUCOSAMINE, 3818
VIRAL PROTEINS
GLUCOSAMINE, 3818
VIRUS REPLICATION
GLUCOSAMINE, 3818

VIRUS, B-TYPE
VIRUS, GUINEA PIG LEUKEMIA
INCLUSION BODIES, VIRAL, 3852*
REVERSE TRANSCRIPTASE, 3852*

VIRUS, BABOON ENDOGENOUS

LEUKEMIA, MYELOBLASTIC
ANTIGENS, VIRAL, 3841*
REVERSE TRANSCRIPTASE, 3841*

VIRUS, BITTNER MURINE MAMMARY TUMOR
MAMMARY NEOPLASMS, EXPERIMENTAL
STRESS, 4038

VIRUS, BOVINE ADENO 1
VIRUS, BOVINE ADENO 3
HISTOCOMPATIBILITY ANTIGENS, 3916
TRANSPLANTATION IMMUNOLOGY, 3916

VIRUS, BOVINE ADENO 2
VIRUS, BOVINE ADENO 3
HISTOCOMPATIBILITY ANTIGENS, 3916
TRANSPLANTATION IMMUNOLOGY, 3916

VIRUS, BOVINE ADENO 3
DNA REPLICATION
THYMIDINE INCORPORATION, 3791
DNA, VIRAL
ISOLATION AND CHARACTERIZATION,
3791
PEPTIDES
ISOLATION AND CHARACTERIZATION,
3791
VIRAL PROTEINS
AMINO ACIDS, 3791
ISOLATION AND CHARACTERIZATION,
3791
VIRUS, ADENO 12
TRANSPLANTATION IMMUNOLOGY, 3916
VIRUS, BOVINE ADENO 1
HISTOCOMPATIBILITY ANTIGENS, 3916
TRANSPLANTATION IMMUNOLOGY, 3916
VIRUS, BOVINE ADENO 2
HISTOCOMPATIBILITY ANTIGENS, 3916
TRANSPLANTATION IMMUNOLOGY, 3916
VIRUS, CELO
TRANSPLANTATION IMMUNOLOGY, 3916
VIRUS, SIMIAN ADENO 7
TRANSPLANTATION IMMUNOLOGY, 3916

VIRUS, BOVINE LEUKEMIA
LYMPHOCYTES
MORPHOLOGY, 3857*

VIRUS, BOVINE LEUKOSIS
BONE NEOPLASMS
ANTIGENS, VIRAL, 3830
FIBROSARCOMA
ANTIGENS, VIRAL, 3830

VIRUS, C-TYPE
BONE NEOPLASMS
ANTIGENS, VIRAL, 3830
LEUKEMIA, MYELOBLASTIC
ANTIGENS, VIRAL, 3841*
IMMUNOGLOBULINS, 3841*
REVERSE TRANSCRIPTASE, 3841*
REVERSE TRANSCRIPTASE
DNA-RNA HYBRIDIZATION, 3833
RNA, VIRAL
DNA-RNA HYBRIDIZATION, 3833

VIRUS, C-TYPE MURINE MYELOMA
LEUKEMIA, MYELOCYTIC
ULTRASTRUCTURAL STUDY, HAMSTER,

4079*
NASOPHARYNGEAL NEOPLASMS
    FLUORESCENT ANTIBODY TECHNIC,
    3843*

VIRUS, C-TYPE RNA TUMOR
    ANTIBODIES, VIRAL
        REVIEW, 3615
    ANTIGENIC DETERIMANTS
        PEPTIDES, 3832
    ANTIGENS, VIRAL
        CROSS REACTIONS, 3811
        PEPTIDES, 3832
        REVIEW, 3615
    KIDNEY NEOPLASMS
        ANTIVIRAL AGENTS, 3739*
        VIRUS-LIKE PARTICLES, 3739*
    LEUKEMIA
        VIRUS-LIKE PARTICLES, 3649*
    LEUKEMIA, MYELOBLASTIC
        ISOLATION AND CHARACTERIZATION,
        3834
        VIRUS REPLICATION, 3834
    LEUKEMIA, MYELOCYTIC
        REVERSE TRANSCRIPTASE, 3842*
    LYMPHOMA
        GRAFT VS HOST REACTION, 3878*
    VIRUS-LIKE PARTICLES
        EMBRYO MUSCLE, CHICKEN, 3812
        ISOLATION AND CHARACTERIZATION,
        3812
        PLACENTA, RAT, 3831
    VIRUS, MASON-PFIZER MONKEY
        CROSS REACTIONS, 3811

VIRUS, CELO
    FIBROSARCOMA
        INTERCELLULAR ATTACHMENTS, 3794
        ULTRASTRUCTURAL STUDY, PLASMA
        MEMBRANE, 3794
    NEOPLASMS, CONNECTIVE TISSUE
        FIBROSARCOMA, 3794
        SARCOMA, 3794
    RNA
        DNA, VIRAL, 3827
        RIBONUCLEASE, 3827
    SARCOMA
        INTERCELLULAR ATTACHMENTS, 3794
        ULTRASTRUCTURAL STUDY, PLASMA
        MEMBRANE, 3794
    VIRUS, BOVINE ADENO 3
        TRANSPLANTATION IMMUNOLOGY, 3916

VIRUS, CYTOMEGALO
    TRANSPLANTATION, HOMOLOGOUS
        ANTIGENIC STIMULATION, REVIEW,
        3807
    VACCINES
        REVIEW, 3616

VIRUS, DNA
    VIRUS, EPSTEIN-BARR
        REVIEW, 3614
    VIRUS, SV40
        REVIEW, 3614

VIRUS, EPSTEIN-BARR
    ANTIBODIES, VIRAL
        INFECTIOUS MONONUCLEOSIS, 4144*

STATISTICAL ANALYSIS, 4144*
ANTIGENS, VIRAL
    CYTOSINE, 1-BETA-D-
        ARABINOFURANOSYL-, 3846*
BURKITT'S LYMPHOMA
    ANTIGENS, VIRAL, 3795, 3844**, 4011
    DNA, VIRAL, 3845*
    HYBRIDIZATION, 3631
CYCLOHEXIMIDE
    VIRUS REPLICATION, 3844*
NASOPHARYNGEAL NEOPLASMS
    CELL FUSION, 3843*
    HYBRIDIZATION, 3631
SARCOIDOSIS
    ANTIGENS, VIRAL, 3904
VIRUS, DNA
    REVIEW, 3614

VIRUS, FELINE LEUKEMIA
    ANTIGEN-ANTIBODY REACTIONS
        IMMUNOFLUORESCENCE, 3913
        ROSETTE FORMATION, 3913
    ANTIGENS, VIRAL
        CELL MEMBRANE, 3913
        FLUORESCENT ANTIBODY TECHNIC,
        3858*
        REVIEW, 3647*
    DNA NUCLEOTIDYTRANSFERASES
        ISOLATION AND CHARACTERIZATION,
        3836
    ELECTROPHORESIS
        ISOLATION AND CHARACTERIZATION,
        3828
    VIRUS, REPLICATION
        REVIEW, 3647*

VIRUS, FELINE SARCOMA
    ANTIGENS, VIRAL
        REVIEW, 3647*
    VIRUS, REPLICATION
        REVIEW, 3647*

VIRUS, FRIEND MURINE LEUKEMIA
    CELL TRANSFORMATION, NEOPLASTIC
        IMMUNOSUPPRESSION, 3893
    ERYTHROLEUKEMIA
        CELL LINE, 3798
    FLUORANTHENE
        SPLENOMEGALY, 3797
    HEME
        PROTEIN BIOSYNTHESIS, 3850*
    IMMUNOSUPPRESSION
        ULTRASTRUCTURAL STUDY, SPLEEN,
        MOUSE, 3895
    INTERFERON
        REGRESSION, 3905
    ISOANTIGENS
        IMMUNE RESPONSE, 3892
    METHOTREXATE
        SPLENOMEGALY, 3797
    MYCOBACTERIUM BUTYRICUM
        SPLENOMEGALY, 3797
    NEOPLASMS, EXPERIMENTAL
        TRANSPLANTATION, HOMOLOGOUS, 3798
    PERTUSSIS VACCINE
        SPLENOMEGALY, 3797
    POLY I-C
        REGRESSION, 3905
    PROTEINS

BIOSYNTHESIS, 3850*
REGRESSION
    HISTOPATHOLOGIC STUDY, 3848*
RIFAMYCIN
    SPLENOMEGALY, 3797
SPLENIC NEOPLASMS
    CELL TRANSFORMATION, NEOPLASTIC,
        3847*
    ERYTHROLEUKEMIA, 3847*
    TRANSPLANTATION IMMUNOLOGY, 3847*
STATOLON
    REGRESSION, 3905
TILORONE
    SPLENOMEGALY, 3797
VIRUS, MOLONEY MURINE LEUKEMIA
    VIRUS RECOVERY, 3798
VIRUS REPLICATION
    CELL TRANSFORMATION, NEOPLASTIC,
        3798
    HISTOCOMPATIBILITY ANTIGENS, 3798

VIRUS, FRIEND POLYCYTHEMIA
ERYTHROCYTES
    CELL DIFFERENTIATION, 3849*
HEMOGLOBULINS
    SYNTHESIS, 3849*

VIRUS, FRIEND-SPLEEN FOCUS FORMING
POLYCYTHEMIA
    RADIATION, IONIZING, 3796
RADIATION, IONIZING
    SPLEEN COLONY HISTOLOGY, 3796
VIRUS REPLICATION
    HEMATOPOIETIC STEM CELLS, 3796

VIRUS, FROG 3
DNA
    SYNTHESIS INHIBITION, HOST CELL,
        3851*
RNA
    SYNTHESIS INHIBITION, HOST CELL,
        3851*

VIRUS, GIBBON LYMPHOMA
VIRUS, LEUKEMIA
    ANTIGENIC DETERMINANTS, 3834

VIRUS, GROSS MURINE LEUKEMIA
ISOANTIGENS
    CELL TRANSFORMATION, NEOPLASTIC,
        3892
    IMMUNE RESPONSE, 3892
LEUKEMIA, LYMPHOCYTIC
    T-LYMPHOCYTES, 3981*
    PITUITARY GLAND, 3981*

VIRUS, GUINEA PIG LEUKEMIA
VIRUS, B-TYPE
    INCLUSION BODIES, VIRAL, 3852*
    REVERSE TRANSCRIPTASE, 3852*

VIRUS, HAMSTER LEUKEMIA
ANTIGENIC DETERMINANTS
    PEPTIDES, 3832

VIRUS, HAMSTER SARCOMA
NUCLEOTIDYLTRANSFERASES
    ADENOSINE TRIPHOSPHATE, 3799
    MAGNESIUM, 3799

MANGANESE, 3799
POLY A POLYMERASE
    ADENOSINE TRIPHOSPHATE, 3799
    MAGNESIUM, 3799
    MANGANESE, 3799
    NUCLEOTIDES, 3799
    RNA, MESSENGER, 3799

VIRUS, HARVEY MURINE SARCOMA
RNA, NEOPLASM
    HYBRIDIZATION,, 3863*
RNA, VIRAL
    DNA-RNA HYBRIDIZATION, 3800

VIRUS, HELPER
VIRUS, ADENO ASSOCIATED
    VIRUS REPLICATION, 3792

VIRUS, HEPATITIS
WARTS
    ANTIGENS, VIRAL, 3837

VIRUS, HEPRES SIMPLEX 2
DNA REPLICATION
    MAGNESIUM, 3802

VIRUS, HERPES
IMMUNITY, CELLULAR
    REVIEW, 3616
SARCOIDOSIS
    ANTIGENS, VIRAL, 3904
VACCINES
    REVIEW, 3616

VIRUS, HERPES SAIMIRI
LYMPHOKINES
    VIRUS REPLICATION, 3922
LYMPHOMA
    ISOLATION AND CHARACTERIZATION,
        3922

VIRUS, HERPES SIMPLEX
DNA REPLICATION
    THYMIDINE, 3806
MULTIPLE MYELOMA
    CASE REPORT, 3894
    IMMUNITY, CELLULAR, 3894
NEUTRAL RED
    PHOTOINACTIVATION, 3803
PHENOLSULFONPHTHALEIN
    PHOTOINACTIVATION, 3803

VIRUS, HERPES SIMPLEX 1
DNA, VIRAL
    ENDONUCLEASES, 3805
LYMPHOKINES
    VIRUS REPLICATION, 3922
PROSTATIC NEOPLASMS
    REVIEW, 3648*
TRANSPLANTATION, HOMOLOGOUS
    ANTIGENIC STIMULATION, REVIEW,
        3807
ULTRAVIOLET RAYS
    ANTIGENS, VIRAL, 3804
    DNA REPLICATION, 3804
    ERYTHROCYTES, 3804

VIRUS, HERPES SIMPLEX 2
CERVIX NEOPLASMS

HYBRIDIZATION, 3631
CYCLOHEXIMIDE
    VIRAL PROTEINS, 3801
DNA, VIRAL
    ENDONUCLEASES, 3805
    TEMPERATURE-SENSITIVE MUTANTS,
        3853*
PEPTIDES
    ISOLATION AND CHARACTERIZATION,
        HEP-2 CELLS, 3801
PROSTATIC NEOPLASMS
    REVIEW, 3648*
TRANSPLANTATION, HOMOLOGOUS
    ANTIGENIC STIMULATION, REVIEW,
        3807
VIRUS REPLICATION
    TEMPERATURE-SENSITIVE MUTANTS,
        3853*

VIRUS, HERPES TURKEY
    LEUKOCYTES
        DETECTION, 3854*
    LYMPHOCYTES
        DETECTION, 3854*
    NEOPLASMS
        CHORIOALLANTOIC MEMBRANE, 3854*

VIRUS, HERPES ZOSTER
    TRANSPLANTATION, HOMOLOGOUS
        ANTIGENIC STIMULATION, REVIEW,
            3807

VIRUS, KILHAM
    DNA NUCLEOTIDYTRANSFERASES
        ISOLATION AND CHARACTERIZATION,
            3855*

VIRUS, KIRSTEN MURINE SARCOMA
    ADENOSINE CYCLIC 3',5' MONOPHOSPHATE
        CELL MEMBRANE, 4156
    ADENYLATE KINASE
        CELL MEMBRANE, 4156
    FATTY ACIDS
        CELL TRANSFORMATION, NEOPLASTIC,
            3877*
    GROWTH SUBSTANCES
        CELL TRANSFORMATION, NEOPLASTIC,
            3877*
    VITAMIN E
        CELL TRANSFORMATION, NEOPLASTIC,
            3877*

VIRUS, LACTATE DEHYDROGENASE
    MAMMARY NEOPLASMS, EXPERIMENTAL
        IMMUNITY, CELLULAR, 3619

VIRUS, LEUKEMIA
    VIRUS, GIBBON LYMPHOMA
        ANTIGENIC DETERMINANTS, 3834
    VIRUS, MURINE SARCOMA
        VIRUS RESCUE, 3834
    VIRUS, SIMIAN SARCOMA
        ANTIGENIC DETERMINANTS, 3834

VIRUS, LEUKO
    ANGIOSARCOMA
        CELL TRANSFORMATION, NEOPLASTIC,
            3856*
        VIRUS REPLICATION, 3856*

VIRUS-LIKE PARTICLES
    CARCINOMA, EPIDERMOID
        C-TYPE RNA TUMOR VIRUS, 3835
        ISOLATION AND CHARACTERIZATION,
            3835
    KIDNEY NEOPLASMS
        VIRUS, C-TYPE RNA TUMOR, 3739*
    LEUKEMIA
        VIRUS, C-TYPE RNA TUMOR, 3649*
    LEUKEMIA, MYELOBLASTIC
        BONE MARROW, 4009
    LEUKEMIA, MYELOCYTIC
        ULTRASTRUCTURAL STUDY, HAMSTER,
            4079*
    LIP NEOPLASMS
        BURKITT'S LYMPHOMA, 4012
    NEUROBLASTOMA
        ULTRASTRUCTURAL STUDY, 4008
    PLASMACYTOMA
        ISOLATION AND CHARACTERIZATION,
            3863*
    VIRUS, C-TYPE RNA TUMOR
        EMBRYO MUSCLE, CHICKEN, 3812
        ISOLATION AND CHARACTERIZATION,
            3812
        PLACENTA, RAT, 3831
    VIRUS, MURINE MAMMARY TUMOR
        ISOLATION AND CHARACTERIZATION,
            3867*
    WARTS
        ULTRASTRUCTURAL STUDY, 4111*

VIRUS, MAREK'S DISEASE HERPES
    ANTIGENS, VIRAL
        IMMUNITY, ACTIVE, 3897
    LEUKOCYTES
        DETECTION, 3854*
    LYMPHOCYTES
        DETECTION, 3854*
    B-LYMPHOCYTES
        HISTOCOMPATIBILITY ANTIGENS, 3980*
    T-LYMPHOCYTES
        HISTOCOMPATIBILITY ANTIGENS, 3980*
    LYMPHOMA
        HISTOCOMPATIBILITY ANTIGENS, 3980*
    NEOPLASMS
        CHORIOALLANTOIC MEMBRANE, 3854*

VIRUS, MASON-PFIZER MONKEY
    VIRUS, C-TYPE RNA TUMOR
        CROSS REACTIONS, 3811

VIRUS, MOLONEY MURINE LEUKEMIA
    RNA, VIRAL
        DNA-RNA HYBRIDIZATION, 3800
    VIRUS, FRIEND MURINE LEUKEMIA
        VIRUS RECOVERY, 3798
    VIRUS REPLICATION
        TEMPERATURE SENSITIVE MUTANTS,
            3809
        ULTRASTRUCTURAL STUDY, TB CELLS,
            3809

VIRUS, MOLONEY MURINE SARCOMA
    ADENOSINE CYCLIC 3',5' MONOPHOSPHATE
        CELL MEMBRANE, 4156
    ADENYLATE KINASE
        CELL MEMBRANE, 4156
    CELL TRANSFORMATION, NEOPLASTIC

IMMUNOSUPPRESSION, 3893
CONCANAVALIN A
    ANTIBODY FORMATION, 3808
    CELL TRANSFORMATION, NEOPLASTIC,
        3808
    IMMUNITY, CELLULAR, 3808
ISOANTIGENS
    IMMUNE RESPONSE, 3892
NEOPLASMS
    MACROPHAGES, 3949*
RNA, NEOPLASM
    HYBRIDIZATION,, 3863*

VIRUS, MURINE LEUKEMIA
    ANTIGENIC DETERIMANTS
        PEPTIDES, 3832
    ANTIGENS, VIRAL
        GLYCOPROTEINS, 3615
    ELECTROPHORESIS
        ISOLATION AND CHARACTERIZATION,
            3828
    GENETICS
        ISOLATION AND CHARACTERIZATION,
            3859*
    INTERFERON
        ANTIGENS, VIRAL, 3860*
        VIRUS REPLICATION, 3860*

VIRUS, MURINE MAMMARY TUMOR
    BONE NEOPLASMS
        VIRUS, RAUSCHER MURINE LEUKEMIA,
            3830
    DNA, VIRAL
        REVERSE TRANSCRIPTASE, 3869*
    ELECTROPHORESIS
        ISOLATION AND CHARACTERIZATION,
            3828
    FIBROSARCOMA
        VIRUS, RAUSCHER MURINE LEUKEMIA,
            3830
    GLYCOPROTEINS
        ISOLATION AND CHARACTERIZATION,
            3829
    MAMMARY NEOPLASMS, EXPERIMENTAL
        ADENOCARCINOMA, 3829, 3866*
        DEOXYRIBONUCLEASE, 3866*
        T-LYMPHOCYTES, 3868*
        PORNASE, 3866*
        RADIATION, IONIZING, 3866*
        RIBONUCLEASE, 3866*
        STRESS, 3868*, 4038
    RNA, VIRAL
        REVERSE TRANSCRIPTASE, 3867*
    VIRUS-LIKE PARTICLES
        ISOLATION AND CHARACTERIZATION,
            3867*

VIRUS, MURINE SARCOMA
    UTERINE NEOPLASMS
        FETAL GLOBULINS, 3865*
        IMMUNITY, CELLULAR, 3865*
    VIRUS, LEUKEMIA
        VIRUS RESCUE, 3834

VIRUS, MURINE SARCOMA-LEUKEMIA
    DNA, VIRAL
        VIRUS REPLICATION, 3864*

VIRUS, PAPILLOMA

ANTIGENS, VIRAL
    COMPLEMENT FIXATION TESTS, 3898
    WART REGRESSION, 3898
WARTS
    ANTIGENS, VIRAL, 3898
    ULTRASTRUCTURAL STUDY, 4028

VIRUS, POLYOMA
    BONE NEOPLASMS
        SARCOMA, OSTEOGENIC, 3874*
    CELL TRANSFORMATION, NEOPLASTIC
        ULTRASTRUCTURAL STUDY, 3875*
    GROWTH SUBSTANCES
        TEMPERATURE-SENSITIVE MUTANTS,
            4186
    ISOLATION AND CHARACTERIZATION
        ULTRASTRUCTURAL STUDY, 3872*
    MAMMARY NEOPLASMS, EXPERIMENTAL
        CARCINOMA, 3874*
    OVARIAN NEOPLASMS
        THECA CELL TUMOR, 3874*
    TRANSPLANTATION, HOMOLOGOUS
        KIDNEY, 3872*
    UTERINE NEOPLASMS
        LEIOMYOSARCOMA, 3874*
    VIRUS, SENDAI
        ULTRASTRUCTURAL STUDY, 3875*

VIRUS, POLYOMA, BK
    ANTIBODIES, VIRAL
        KIDNEY TRANSPLANT, 3873*
    LYMPHATIC NEOPLASMS
        SARCOMA, 3810
    VIRUS, SV40
        CROSS REACTIONS, 3810

VIRUS, RAUSCHER MURINE LEUKEMIA
    ANTIGENIC DETERIMANTS
        PEPTIDES, 3832
    BONE NEOPLASMS
        ANTIGENS, VIRAL, 3830
        VIRUS, MURINE MAMMARY TUMOR, 3830
    CHONDROBLASTOMA
        ANTIGENS, VIRAL, 3830
    CHONDROSARCOMA
        ANTIGENS, VIRAL, 3830
    DNA NUCLEOTIDYTRANSFERASES
        ISOLATION AND CHARACTERIZATION,
            3836
    FIBROSARCOMA
        ANTIGENS, VIRAL, 3830
        VIRUS, MURINE MAMMARY TUMOR, 3830
    GIANT CELL TUMORS
        ANTIGENS, VIRAL, 3830
    ISOANTIGENS
        IMMUNE RESPONSE, 3892
    LEUKEMIA
        REVIEW, 3614
    SARCOMA, OSTEOGENIC
        ANTIGENS, VIRAL, 3830

VIRUS, RD114
    DNA NUCLEOTIDYTRANSFERASES
        ISOLATION AND CHARACTERIZATION,
            3836

VIRUS REPLICATION
    ANGIOSARCOMA
        VIRUS, LEUKO, 3856*

LEUKEMIA, MYELOBLASTIC
    VIRUS, C.-TYPE RNA TUMOR, 3834
VIRUS, ADENO ASSOCIATED
    L-ARGININE, 3792
    ULTRASTRUCTURAL STUDY, 3792
    VIRUS, HELPER, 3792
    VIRUS, SV15, 3792
VIRUS, AVIAN SARCOMA
    GLUCOSAMINE, 3818
VIRUS, B77
    GLUCOSAMINE, 3818
VIRUS, EPSTEIN-BARR
    CYCLOHEXIMIDE, 3844*
VIRUS, FELINE LEUKEMIA
    REVIEW, 3647*
VIRUS, FELINE SARCOMA
    REVIEW, 3647*
VIRUS, FRIEND MURINE LEUKEMIA
    CELL TRANSFORMATION, NEOPLASTIC,
        3798
    HISTOCOMPATIBILITY ANTIGENS, 3798
VIRUS, FRIEND-SPLEEN FOCUS FORMING
    HEMATOPOIETIC STEM CELLS, 3796
VIRUS, HERPES SAIMIRI
    LYMPHOKINES, 3922
VIRUS, HERPES SIMPLEX 1
    LYMPHOKINES, 3922
VIRUS, HERPES SIMPLEX 2
    TEMPERATURE-SENSITIVE MUTANTS,
        3853*
VIRUS, MOLONEY MURINE LEUKEMIA
    TEMPERATURE SENSITIVE MUTANTS,
        3809
    ULTRASTRUCTURAL STUDY, TB CELLS,
        3809
VIRUS, MURINE LEUKEMIA
    INTERFERON, 3860*
VIRUS, MURINE SARCOMA-LEUKEMIA
    DNA, VIRAL, 3864*
VIRUS, ROUS SARCOMA
    ALPHA-AMANITIN, 3814
    RNA POLYMERASE, 3814

VIRUS, RNA TUMOR
    CARBAMIC ACID, METHYL-, 1-NAPHTHYL
    ESTER
        CELL TRANSFORMATION, NEOPLASTIC,
            3704
    CARBAMIC ACID, N-METHYL-N-NITROSO-, 1-
    NAPHTHYL ESTER
        CELL TRANSFORMATION, NEOPLASTIC,
            3704
    GLYCOPROTEINS
        REVIEW, 3614
    INCLUSION BODIES, VIRAL
        REVIEW, 3614

VIRUS, ROUS SARCOMA
    ADENOSINE CYCLIC 3',5' MCNOPHOSPHATE
        CELL MEMBRANE, 4156
    ADENYLATE KINASE
        CELL MEMBRANE, 4156
    ADRIAMYCIN
        INFECTIVITY, 3879*
    ALLOXAN

CELL TRANSFORMATION, NEOPLASTIC,
    3672
ALPHA-AMANITIN
    VIRUS REPLICATION, 3814
CORTISOL
    DNA REPLICATION, 4182
DAUNOMYCIN
    INFECTIVITY, 3879*
DNA REPLICATION
    REVERSE TRANSCRIPTASE, 3880*
PHOSPHATASES
    REVERSE TRANSCRIPTASE, 3880*
PROTEIN KINASE
    REVERSE TRANSCRIPTASE, 3880*
REVERSE TRANSCRIPTASE
    ISOLATION AND CHARACTERIZATION,
        3816
RNA POLYMERASE
    VIRUS REPLICATION, 3814
SARCOMA
    TRANSPLANTATION, HOMOLOGOUS, 3902
SURFACE PROPERTIES
    ULTRASTRUCTURAL STUDY, BYRAN
        STRAIN, 3815
VIRAL PROTEINS
    ULTRASTRUCTURAL STUDY, BYRAN
        STRAIN, 3815

VIRUS, SCRIPPS MURINE LEUKEMIA
    ANTINUCLEAR FACTORS, 3861*
        ANTIGENS, VIRAL, 3861*
    LYMPHOMA
        IMMUNOLOGY, 3861*
        VERTICAL TRANSMISSION, 3861*

VIRUS, SENDAI
    VIRUS, POLYOMA
        ULTRASTRUCTURAL STUDY, 3875*

VIRUS, SHOPE PAPILLOMA
    CARCINOMA
        URIDINE, 2'-DEOXY-5-IODO-, 3871*
    NEOPLASMS
        HYBRIDIZATION, 3631
    URIDINE, 2'-DEOXY-5-IODO-
        GENOME ACTIVATION, 3871*

VIRUS, SIMIAN ADENO 7
    VIRUS, BOVINE ADENO 3
        TRANSPLANTATION IMMUNOLOGY, 3916

VIRUS, SIMIAN LEUKEMIA
    BONE NEOPLASMS
        ANTIGENS, VIRAL, 3830
    CHONDROBLASTOMA
        ANTIGENS, VIRAL, 3830
    CHONDROSARCOMA
        ANTIGENS, VIRAL, 3830
    FIBROSARCOMA
        ANTIGENS, VIRAL, 3830
    GIANT CELL TUMORS
        ANTIGENS, VIRAL, 3830
    SARCOMA, OSTEOGENIC
        ANTIGENS, VIRAL, 3830

VIRUS, SIMIAN SARCOMA
    BRAIN NEOPLASMS
        ANTIBODIES, VIRAL, 3883*
        ANTIGENS, VIRAL, 3883*

ASTROCYTOMA, 3883*
CELL TRANSFORMATION, NEOPLASTIC,
3883*
LEUKEMIA
REVIEW, 3614
LEUKEMIA, MYELOBLASTIC
ANTIGENS, VIRAL, 3841*
REVERSE TRANSCRIPTASE, 3841*
VIRUS, LEUKEMIA
ANTIGENIC DETERMINANTS, 3834

VIRUS, SV15
VIRUS, ADENO ASSOCIATED
VIRUS REPLICATION, 3792

VIRUS, SV40
ANTIBODIES
PLASMINOGEN ACTIVATORS, 3881*
ANTIGENS, NEOPLASM
ANTIGENIC DETERMINANTS, 3882*
ANTIGENS, VIRAL
TEMPERATURE SENSITIVE MUTANTS,
3822
CELL DIVISION
AGE FACTORS, 4164
CELL FUSION
ANTIGENS, VIRAL, 3823
DNA, CIRCULAR, 3823
CELL MEMBRANE
AGE FACTORS, 4164
UTRASTRUCTURAL STUDY, 4157
CELL TRANSFORMATION, NEOPLASTIC
TEMPERATURE SENSITIVE MUTANTS,
3822
ULTRASTRUCTURAL STUDY, 4164
COLCHICINE
CELL MEMBRANE, 4157
DNA
STRAND SEPARATION, TEMPERATURE-
DEPENDENT, 3821
DNA REPLICATION
ALPHA COMPONENT, 3824
CONTACT INHIBITION, 3824
EVOLUTIONARY VARIANTS, 3826
INITIATION SITES, CLONED VARIANTS,
3825
DNA, VIRAL
TEMPERATURE SENSITIVE MUTANTS,
3822
GROWTH
TEMPERATURE DEPENDENCE, 3884*
NEOPLASMS, EXPERIMENTAL
IMMUNITY, CELLULAR, 3891
IMMUNOSUPPRESSION, 3891
TRANSPLANTATION IMMUNOLOGY, 3891
RADIATION, IONIZING
CELL FUSION, 3804
RNA, MESSENGER
CYTOCHALASIN B, 3820
ENDONUCLEASES, 3819
HALF-LIFE STUDY, 3820
RNA, RIBOSOMAL
DNA-RNA HYBRIDIZATION, 3819
ENDONUCLEASES, 3819
ULTRAVIOLET RAYS
ANTIGENS, VIRAL, 3804
DNA REPLICATION, 3804
ERYTHROCYTES, 3804
VINCALEUCOBLASTINE

CELL MEMBRANE, 4157
VIRUS, ADENO
TEMPERATURE SENSITIVE MUTANTS,
3822
VIRUS, DNA
REVIEW, 3614
VIRUS, POLYOMA, BK
CROSS REACTIONS, 3810

VIRUS, VARICELLA-ZOSTER
VACCINES
REVIEW, 3616

VITAMIN A
SEE RETINOL

VITAMIN C
SEE ASCORBIC ACID

VITAMIN E
VIRUS, KIRSTEN MURINE SARCOMA
CELL TRANSFORMATION, NEOPLASTIC,
3877*

VITELLINE MEMBRANE
DISGERMINOMA
FETAL GLOBULINS, 3907
TERATOID TUMOR
FETAL GLOBULINS, 3907

WARTS
AUTOANTIBODIES
IGM, 3837
VIRUS, HEPATITIS
ANTIGENS, VIRAL, 3837
VIRUS-LIKE PARTICLES
ULTRASTRUCTURAL STUDY, 4111*
VIRUS, PAPILLOMA
ANTIGENS, VIRAL, 3898
ULTRASTRUCTURAL STUDY, 4028

WATER POLLUTANTS
ASBESTOS
REVIEW, 4135*

WILM'S TUMOR
SEE NEPHROBLASTOMA

XANTHINE, 3-ISOBUTYL-1-METHYL-
HELA CELLS
ADENOSINE CYCLIC 3',5'.
MONOPHOSPHATE, 4193*

XERODERMA PIGMENTOSUM
DNA REPAIR
ACETAMIDE, N-FLUOREN-2-YL-, 3674
ACETAMIDE, N-2-PHENANTHRYL-, 3674
ACETANILIDE, 4'-PHENYL-, 3674
ACETANILIDE, 4'-STYRYL-, 3674
ULTRAVIOLET RAYS, 3674

ZYMOSAN
ADENOCARCINOMA
TRANSPLANTATION IMMUNOLOGY, 3890

CHEMICAL ABSTRACT SERVICES REGISTRY NUMBER INDEX

CAS REGISTRY NUMBER	ABSTRACT NUMBER	CAS REGISTRY NUMBER	ABSTRACT NUMBER
36066	4153, 4181	56495	3601, 3606, 3618, 3676, 3680, 3682, 3683, 3693, 3843
50022	4182, 4186	56531	3606, 3608, 4124
50066	3693, 3697	56553	3678, 3682
50180	4013, 4024	57136	4175
50237	4172, 4182, 4186	57830	3607, 3680
50248	4033	57885	3605
50282	3680, 4180	57976	3601, 3606, 3680
50328	3601, 3606, 3610, 3676, 3677, 3678, 3682	58220	3680, 4031
50442	4043	58399	3688
50555	3609	58968	4163
50715	3672	59052	3797, 4154
50760	4004	59303	4154
50895	3689, 3806	59892	3703
51796	3601, 3715, 3893	60117	3679
53032	3917	60151	4181
53065	4182	62759	3697, 3698, 3699, 3700
53952	3675	63252	3704
53963	3601, 3606, 3674, 4150, 4177	64175	4177
54115	3712	64868	3689, 3924, 4151
55174	3685	66819	3801, 4004
55185	3695, 3696, 3697, 3703	67210	3685

CHEMICAL ABSTRACT SERVICES REGISTRY NUMBER INDEX

CAS REGISTRY NUMBER	ABSTRACT NUMBER	CAS REGISTRY NUMBER	ABSTRACT NUMBER
67641	3670	226368	3682
70257	3705, 3706, 3909	288324	4175
71443	4161, 4162	304289	3723
75014	3627, 3686, 3716, 4036, 4124	366701	3601
75354	3686	540590	3684
79016	3686	540738	3909
83443	4176	548862	3606
86306	3700	553242	3803
91598	3629	573568	4181
91816	3693	586174	4181
96457	3687	590965	3692
97563	3601	615532	3601
100754	3707	621647	3702
110601	4161, 4162	684935	3708
121824	3694	759739	3601, 3892
124209	4161, 4162	816579	3601
127071	4004	865214	4157
127184	3686	924163	3701
154176	4182	924469	3702
191071	3610	930552	3700
191242	3682	1162658	3602, 3668, 3669, 3670, 3671
206440	3797		

CAS REGISTRY NUMBER	ABSTRACT NUMBER	CAS REGISTRY NUMBER	ABSTRACT N
1310732......................	3670	7664417......................	3670
1314201......................	4026, 4037	7758987......................	3601
1332214......................	3627	7786814......................	3601
1385951......................	3602, 3671	8001589......................	3606
1615801......................	3601	8063943......................	4036
3817116......................	3701	9002077......................	3924
4075790......................	3674	9002624......................	3688, 3689, 371
4120778......................	3674	9002862......................	4036
4463223......................	3675	9004108......................	3689
6795239......................	3602	9005496......................	4173
6885570......................	3602	9007732......................	3930
7220817......................	3602, 3671	9010724......................	3890
7241987......................	3602, 3671	10028178......................	3714
7439910......................	4167	12001284......................	3627
7439954......................	3799	12001295......................	3627
7439965......................	3799, 3802	12172735......................	3627
7440075......................	3612, 3774	13929356......................	3797
7440382......................	4127	14056101......................	3673
7553562......................	3611	16301261......................	3601
7631994......................	3709	17068789......................	3627
7632000......................	3789	17878545......................	3602
7647145......................	3670	18559972......................	3674

CHEMICAL ABSTRACT SERVICES REGISTRY NUMBER INDEX

CAS REGISTRY NUMBER	ABSTRACT NUMBER	CAS REGISTRY NUMBER	ABSTRACT NUMBER
20241107...............	3602		
21150209...............	3814		
32215024...............	3602		
33953730...............	3677		
39603537...............	3702		

ER	WLN	ABSTRACT NUMBER

L B666J EMV1 3674

L B677 MV&T&J CO1 DO1 EO1 JMV1 NO1 .... 3689, 3924, 4151

L C6566 1A PJ 3797

L D6 B666J 3678, 3682

L D6 B666J C J 3601, 3606, 3680

L D6 B6666 2AB TJ 3601, 3606, 3610, 3676, 3677, 3678, 3682

L E5 B666 LUTJ A E FY&3Y OQ -B&AEFO ... 3605

L E5 B666 OV MUTJ A CQ E FV1Q FQ -B&ACEF 4172, 4182, 4186

L E5 B666 OV MUTJ A E FQ -B&AEF 3680, 4031

L E5 B666 OV MUTJ A E FV1 -B&AEF 3607, 3680

L E5 B666TJ A DQ E FY2VR OQ 4176

L E5 B666TJ A1Q CQ E IQ MQ OQ F- QT5OV EHJ& OO- BT6OTJ CQ DQ EQ F........... 4153, 4181

L E5 B666TTT&J E FQ OQ 3680, 4180

L E6 B6656 1A T&&&T&J R 3601, 3606, 3618, 3676, 3680, 3682, 3683, 3693, 3843

716,

L66J BOVM1 3704

L66J CZ 3629

L8TJ AMVOXR DF&R DF&1UU1 3684

674,

ONNR&R 3700

ONN1&VO2 3601

ONN1&1 .... 3697, 3698, 3699, 3700

WLN 1

WLNI	ABSTRACT NUMBER	WLN	ABSTRACT NUMBER
ONN2&2	3695, 3696, 3697, 3703	T5NTJ ANO	3700
ONN3&3	3702	T56 BM DN FN HNJ ISH	4043
ONN4&4	3701	T6MPOTJ BO BN2G2G	4013, 4024
QR DY2& 2U	3606, 3608, 4124	T6MVMVVVJ	3672
QVYZ2S2	3685	T6N CN ENTJ ANW CNW ENW	3694
QVYZ2S2 -DL	3685	T6N DOTJ ANO	3703
Q1NUNO&1	3692	T6NJ BN1R&2N1&1	3693
Q2	4177	T6NJ C- BT5NTJ A	3712
Q4N4&NO	3701	T6NTJ ANO	3707
T C6 B5665 2AB S BX IN QN NU JH&&TTTTJ FO1 I KVO1 KQ LOV1 M2 E- NT F6 E596 A BN LM&&TTJ NVO1 RQ R2	4157	T6VMVMV FHJ F2 FR	3693, 3697
T C666 BN INJ E FZ LN1&1 &GH	3803	T6VMVTJ E1YQ- BL6VTJ D F	3801, 4004
T C666 BN ISJ EG B3- AT6N DNTJ D2Q	3688	T66 BN DN GN JNJ CZ EQ H1MR DVMYVQ2VQ .	4154
T C666 BO EV INJ D FZ N G- K-/VM- OT5-16 AN FVN IVN LVO PVM SVTJ G J KY N RY 2	4004	T66 BN DN GN JNJ CZ EZ H1N1&R DVMYVQ2VQ	3797, 4154
T F5 C6 B655 DOV GV OO QO RUT&&TTJ LO1	3602, 3668, 3669, 3670, 3671	VH1YQYQYQ1Q -BAA -D	4182
T F5 C6 B655 DOV GV OO QOT&&TTJ LO1 ...	3602, 3671	WNMYUM&N1&NO	3705, 3706, 390
T F6 C6 B655 DOV GVQ PO ROT&&TTJ MO1 ..	3602, 3671	WNR BQ CNN	4181
T F6 D5 C666 EM ON&&TTTJ HOT SOVR CO1 DO1 DO1& TO1	3609	ZH	3670
T G6 D6 B666 CNJ	3682	ZVN1&NO	3708
TMYMTJ BUS	3687	ZVN2&NO	3601, 3892
T5M CNJ	4175	ZVO2	3601, 3715, 389
		ZVZ	4175

WISWESSER LINE NOTATION INDEX

IACT NUMBER

, 4162

CARCINOGENESIS ABSTRACTS

A monthly publication of the

National Cancer Institute

Editor

George P. Studzinski, M.D.

College of Medicine and Dentistry

of New Jersey, Newark

DEPOSITORY

DEC 1 5 1976

UNIV. OF ILL. LIBRARY
AT URBANA-CHAMPAIGN

Associate Editor

Jussi J. Saukkonen, M.D.

Jefferson Medical College, Philadelphia

NCI Staff Consultants

Elizabeth Weisburger, Ph.D.

Joan W. Chase, M.S.

Literature Selected, Abstracted, and Indexed

by
The Franklin Institute Research Laboratories
Science Information Services
Biomedical Section

Bruce H. Kleinstein, Ph.D., J.D., Group Manager,
Biomedical Projects

Ruthann E. Auchinleck
Production Editor

Contract Number NO1-CP-43293

Public Health Service, USDHEW

DHEW Publication No. (NIH) 76-301

CONTENTS

	Cross Reference Abbreviations	Abstracts, Citations	Page
REVIEW	(Rev)	4201-4265	751
CHEMICAL CARCINOGENESIS	(Chem)	4266-4361	763
PHYSICAL CARCINOGENESIS	(Phys)	4362-4372	780
VIRAL CARCINOGENESIS	(Viral)	4373-4434	782
IMMUNOLOGY	(Immun)	4435-4555	794
PATHOGENESIS	(Path)	4556-4643	818
EPIDEMIOLOGY AND BIOMETRY	(Epid-Biom)	4644-4682	832
MISCELLANEOUS	(Misc)	4683-4800	840
AUTHOR INDEX			
SUBJECT INDEX			
CHEMICAL ABSTRACT SERVICES REGISTRY NUMBER INDEX			
WISWESSER LINE NOTATION INDEX			

and the accumulation of injected estrogen in their
cell nuclei *in vivo,* apparently still bound to a
4-5S form of the receptor. Experiments employing
the R3230AC transplantable mammary carcinoma have
indicated that the chromatin of the autonomous
breast tumor cells possesses the capacity for extensive receptor-estrogen interaction. Cytoplasmic
receptor-estrogen also binds equally well to nuclei
prepared from target and nontarget tissues, in a
manner strictly proportional to the available receptor-estrogen. The quantitation of estrogen receptor in human breast cancer specimens is described. Receptors in 154 primary tumors and 72 metastatic tumors have been assayed. The estrogen receptor values in primary tumors range from zero to
almost 1,000 femtoM/mg of cytosol protein. Data
from 436 treatment trials in 380 patients are also
reviewed, and clinical correlations are presented.
Thirty-three percent of 211 ablative therapy treatment trials yielded objective tumor regression; 8%
of 94 trials in patients with negative tumor receptor values were successful, while 55% of 107
trials in patients with positive tumor receptor
values succeeded. Of 170 additive therapy trials,
8% of 82 trials in patients with negative tumor
receptor values succeeded, compared to 60% of 85
trials in patients with positive tumor receptor
values. In addition, 27% of 55 trials yielded
responses to a variety of endocrine therapies, including antiestrogens and aminoglutethimide. Thus
the data reveal that estrogen receptor assays can be
helpful in predicting the results of endocrine therapy for metastatic breast cancer. (32 references)

g

4203   VINYL CHLORIDE AND PUBLIC HEALTH. (Eng.)
        Anonymous. *Med. J. Aust.* 1(15):455-456;
1975.

Current knowledge about occupational and consumer
exposure to vinyl chloride (VC) is reviewed. The
manufacture of polyvinyl chloride (PVC) is described;
the reaction of chlorine and ethylene, the suspension method, the addition of modifying agents, and
the exothermic polymerization is noted. The possibility of transcutaneous absorption to worker's hands
is acknowledged, as industrial atmospheric VC concentrations of 600-1,000 ppm are found. The characteristic clinical picture occurring in workers
chronically exposed to the gas is entitled "vinyl-
chloride-krankheit": symptoms may include sclerotic
changes in the skin of the hands, cold sensitivity,
Raynaud's phenomenon, and band-like osteolysis of
the terminal phalanges. In addition, thrombocytopenia and splenomegaly are frequently found, and
pulmonary fibrosis, restrictive ventilation, and
diffusion defects also occur. However, greatest
concern and publicity arises from the discovery of
angiosarcoma of the liver in five VC workers. The
association of VC with the otherwise rare angiosarcoma is highly suspicious, and the carcinogenic
effect of VC is demonstrated in animals. However,
the translation of such experimental findings to
man must consider species differences, the possibility of co-carcinogens, and differing latent
intervals. The present upper limits of VC exposure

in the U.K. are set at 200 ppm, while a limit of
50 ppm is recommended in the U.S. In addition, the
limit of VC monomer is 10 ppm for rigid PVC containers and one ppm for flexible PVC films in contact with foods, while one part per 10 million is
allowed in foods packaged in PVC containers. The
newly acquired intolerance to pollution and the
conflicting employer *vs* employee forces, are discussed. However, it is stated that the potential
dangers of VC relate to the worker with the basic
material, and not to the user of the finished product. (9 references)

4204   VINYL CHLORIDE: ITS SAFETY HAZARDS.
        (Dut.) Zielhuis, R. L. (No affiliation
given). *Ned. Tijdschr. Geneeskd.* 119(2):63-66;
1975.

Recent literature on the carcinogenicity of vinyl
chloride (VC), an intermediary in PVC production,
is reviewed. Angiosarcoma of the liver, a rare
tumor, was found in workers exposed to VC during
production and processing in the USA, Great Britain,
Sweden, and West Germany. A retrospective survey
revealed increasing incidence of tumors of the digestive and respiratory tracts and of lymphomas with
increasing intensity and duration of occupational
VC exposure among 7,000 workers. Skin and lung
carcinomas were found in rats exposed to 30,000 ppm
VC. Malignant tumors were found in rats, hamsters
and mice exposed to 50-10,000 ppm VC for 12 mo to
130 wk. The discovery of the carcinogenic properties of VC has resulted in a drastic lowering of the
maximum allowable conc. of VC in workplace air from
200-500 ppm to 50-25 ppm and even to as low as 1 ppm
in manufacturing plants in the USA. (15 references)

4205   RESEARCH FINDS AROMATIC AMINES EXPOSURE
        CAUSE OF INCREASED NUMBER OF TUMORS.
(Eng.) Orjelick, R. (Div. Occupational Health and
Radiation Control, Health Commission of New South
Wales, Australia). *Int. J. Occup. Health Saf.* 44(5):
46-47; 1975.

The relationship of occupational exposure to aromatic amines and an increased incidence of urinary
tract tumors is reviewed. A statistically significant association between exposure and subsequent
tumor development has been found for 4-aminodiphenyl,
benzidine, 6-naphthylamine, and 2-naphthylamine;
a strong association is suspected for 2-acetylamino-
fluorene, 3,3'-dichlorobenzidine, 4'-dimethylamino-
azobenzene, N'nitrosodimethylamine, diphenylamine,
O-tolidine, O-dianisidine, α-naphthylamine thiourea,
auramine, and magenta. Human absorption occurs *via*
inhalation of the dust or vapor, absorption through
intact skin, and by ingestion; individual hygiene
is considered a most important factor. The proportion of bladder tumors estimated to be related to
industrial carcinogen exposure varies from 1 to 33%.
Occupations considered to have a definite causative
association with bladder cancer include rubber and

4209   THE EFFECTS OF DIAGNOSTIC X-RAY EXPOSURE
ON THE HUMAN FETUS: AN EXAMINATION OF
THE EVIDENCE. (Eng.) Oppenheim, B. E. (Dept.
Radiology, Univ. Chicago, 950 E. 59th St., Chicago,
Ill. 60637); Griem, M. L.; Meier, P. *Radiology*
114(3):529-534; 1975.

The effects of diagnostic x-ray exposure on the
human fetus are reviewed. Prenatal irradiation
and the gross death rate reveals that among 6000
caucasian and 9000 black children exposed *in
utero* to diagnostic x-rays, the death rate is mul-
tiplied by two for the white children and does not
change for the black children. Another report shows
that the rate of prenatal exposure to x-rays is
about 50% higher for children with malignancies
than for the controls, corresponding to a relative
risk of malignancy following prenatal irradiation
of 1.6. A statistically significant alteration of
the sex ratio is found for offspring of black women
who themselves have been exposed *in utero* to diagnostic x-rays before 30 wk of fetal life. The
relation between preconception irradiation and leukemia with regard only to maternal exposure reveals
the risk to be between 1.55 and 1.73. Preconception irradiation is also linked to mongolism.
Each of these studies is subject to criticism and
for each study proving one point there are one or
more studies proving the exact reverse. Comparisons with studies in which radiation exposure occurred on a nonselective basis demonstrate a significant discrepancy in each instance. This casts
some doubt on the validity of all these studies,
and suggests that exposure to diagnostic x-rays
may be less harmful than is sometimes claimed.
(51 references)

4210   MALIGNANT NEOPLASMS IN CHILDREN. (Rus.)
Durnov, L. A. (Inst. Experimental and
Clinical Oncology, Moscow, USSR); Kiselev, A. V.
Pediatriia (6):85-89; 1975.

Epidemiological, clinical and therapeutic aspects
of malignant tumors in childhood are reviewed.
Burkitt's tumor is believed to be of viral etiology.
Recently, cats and dogs are suspected to be the
primary hosts of oncogenic viruses causing leukemia
in humans. Metastases of malignant tumors into the
placenta and fetus were observed in 24 cases during
the 1866-1966 period. These cases include melanoma
as the primary tumor in 11 cases, breast cancer in
four cases, cancer of the lungs and stomach in two
cases each, and lymphosarcoma, sarcoma of the hip,
of the ethmoid bone, adrenal gland and ovary in one
case each. The increased tumor morbidity among
children irradiated during their intrauterine lives
indicates that irradiation is a risk factor in genetically predisposed children. Irradiation is most
hazardous during the first three months of intrauterine life. Inborn developmental defects and
neuroblastoma are suspected to have a common etiology. The familial cancer syndrome is believed to
be due to an interaction of genetic and exogenous
(viral) factors. (46 references)

4211   STATISTICAL REVIEW OF CANCER IN ENGLAND
        AND WALES. (Eng.) Anonymous. *Br. Med. J.*
4(5991):245-246; 1975.

The most recent survey of the voluntary cancer registration scheme in England and Wales is discussed.
While noted to have potential epidemiological significance, the inconsistent reporting and level of
inaccuracy is suggested to mask the subtle clues
emerging from time-and-space cluster analysis.
Cancer rates of 332.2 and 300.7/100,000 are reported
for males and females, respectively. Lung cancer
accounts for 29% of all cancer in men, while breast
cancer is responsible for 24% of all cancers reported
in women. The registration of gastric cancer is
falling slowly, but not as fast as the decline in
mortality might suggest, while the rates of other
alimentary tract tumors are increasing. There is
also an apparent rise of Hodgkin's disease and myeloma cases. However, registrations of borderline
conditions, such as myelofibrosis and polycythemia,
are assumed to indicate gross underestimates of
their frequency. There is also considerable variation in the apparent incidences of carcinoma in
situ of the cervix; regional reports range from 1.1
to 15.9 per 100,000. Cancers of the lip, salivary
gland, larynx, uterus, testis, and eye have five
year-age corrected survival rates above 50%, while
tumors of the large bowel, breast, prostate, and
bladder, and the lymphomas have corrected five year
survival rates of 25-50%. Although the five-year
survival is considered good reason to assume cure
for some cancers, cancers of the breast, head and
neck, cervix, prostate, and the lymphomas may account for deaths up to 15 yr after initial treatment.
No conclusions on the effectiveness of new types of
treatment can yet be made, but the accurate recording
of cases is stressed. (1 reference)

4212   DIFFERENCES IN SURFACE MEMBRANE COMPONENTS
        BETWEEN NORMAL, VIRAL-TRANSFORMED AND
REVERTANT CELLS. (Eng.) Black, P. H. (Massachusetts
General Hosp., Boston, Mass.); Chou, I. N.; Roblin,
R. *Proc. Int. Cancer Congr. 11th.* Vol. 1 *(Cell
Biology and Tumor Immunology).* Florence, Italy,
October 20-26, 1974. Edited by Bucalossi, P.;
Veronesi, U.; Cascinelli, N. New York, American
Elsevier, 1975, pp. 119-129.

Experiments investigating intracellular and extracellular membrane stabilization mechanisms are
reviewed, and changes found in the plasma membrane
of transformed cells are described. Electron microscopic examination of thin sections of untransformed
3T3 mouse cells, simian virus 40 (SV40)-transformed
mouse cells (SV-3T3), and revertant cells (Rev SV-
3T3) has revealed a population of 70 A alpha filaments in the peripheral cytoplasm of confluent 3T3
and revertant cells. Abundant alpha filaments are
correlated with the presence of contact inhibition
of growth. While the precise mechanism is unknown,
the data suggest a role in modulating cytoplasmic
viscosity and in motility control of the cell. An
interaction of the filament with the plasma membrane acting to rigidify the plasma membrane is
also suggested, as is the involvement of the micro-

fer factor is considered a soluble indicator of
cellular immunity, capable of inducing nonsensitized
lymphocytes to produce lymphokines. Clinically,
transfer factor activates uncommitted and nonsensitized recipient lymphocytes, confers specific
skin sensitivity, is exceedingly potent, but does not
transfer the capacity for antibody formation.
Transfer factor is clinically used to correct certain primary immunodeficiency states; favorable
responses are reported in patients with generalized
vaccinia, herpes zoster, neonatal herpes virus
infections, measles, "giant-cell" pneumonia, subacute scloerosing panencephalitis, veruca vulgaris,
plus numerous other infectious diseases. However,
despite previous reports, a randomized double-blind
study of 30 patients failed to confirm a casual
relationship between therapy with transfer factor
and wart regression. Much attention is directed to
the treatment of hepatitis B virus infection with
transfer factor; temporary falls in the levels of
hepatitis B antigen are reported after the use of
transfer factor in patients with antigen-positive
chronic active hepatitis. Administration of "specific" transfer factor prepared from subjects recently recovered from hepatitis B infection increases
the number of T lymphocytes and suggests stimulation of cell-mediated immunity, while treatment
with lymphocytes of a recovered patient results in
liver function tests implying liver damage. While
generally therapeutically effective, the risk of
transfer factor exciting generalized hypersensitivity reactions to dessimated microbial or other
antigens is also recognized. (no references)

4216   A SPECULATIVE COMPARISON OF ALLOTYPIC
        SUPPRESSION AND ALLOTYPIC BLAST TRANS-
FORMATION: CONTROL OF EXPRESSION OF T AND B CELL
IMMUNOGLOBULIN ALLOTYPIC MARKERS BY A B CELL
PRODUCT. (Eng.) Sell, S. (Univ. California, San
Diego, Medical Sch., La Jolla, Calif. 92037).
Transplant. Rev. 27:135-156; 1975.

A comparison of the similarities and differences
between allotype suppression and allotype transformation is presented; special attention is given
to the control of expression of T and B cell allotypes of rabbit immunoglobulin (Ig) molecules.
Eight shared characteristics of allotype suppression
and allotype transformation are described, while
it is noted that only rabbit allotypic sera will
produce suppression. A discussion of allelic
expression of lymphoid cells notes demonstrations
of allelic exclusion and concludes that the Ig-
bearing cells upon which antiallotypic sera act
have the capacity to express and synthesize the
products of both Ig alleles. The phenomenon of
mixed allotypic sequential stimulation is discussed,
the restimulation of blocked lymphocyte cultures is
described, and antiallotypic sequential stimulation
experiments are diagramatically summarized. Studies
on the expression of lymphocyte surface allotypic
markers are correlated with events occurring during
blast transformation, and the mixed allotype
sequential stimulation and endocytosis findings are
compared to the *in vitro* allotype suppression
reversal. In view of data supporting allelic in-

clusion in heterozygous lymphoid cells reactive to antiallotypic sera, and the lack of evidence for T-cell mediated suppression in the rabbit, a direct cell model for allotypic suppression based on double allelic expression is proposed. The B-cell product suppression of T and B-cell allotypic markers is noted, and the single cell model for allotypic suppression is symbolically diagramed. It is postulated that antiallotype antibody induces suppression of expression of that allotypic marker in uncommitted T and B cells. Committment to a given allotype is suggested to occur only in differentiated cells, reversibility is allowed to less differentiated T and/or B cells; allotypic suppression thus represents control of T and B-cell expression by a B-cell product. (108 references)

4217   REDUCED INCIDENCE OF SPONTANEOUS TUMORS: ANOTHER STATISTICAL ANALYSIS. (Eng.) Casagrande, J. (Sch. Medicine, Univ. Southern California, Los Angeles, Calif. 90033); Pike, M. C. *Science* 190(4216):808-809; 1975.

Another statistical analysis is given to previously reported data on reduced spontaneous mammary tumor incidence in C3H/He mice treated with polyadenylate-polyuridylate. The recalculation, correcting for the intercurrent deaths, revealed a 49% rate of mammary tumors in controls and a 44% tumor rate for treated animals, a statistically insignificant difference. The proper statistical methodology requires knowledge of the actual times of intercurrent deaths and times of diagnosis of mammary tumors. The tentative conclusion is that such treatment prevents tumors in a proportion of animals, but fails to delay the appearance of the tumor in those mice it fails to protect completely. (3 references)

4218   THE RESPONSE TO PHOSPHORYLCHOLINE: DISSECTING AN IMMUNE RESPONSE. (Eng.) Kohler, H. (La Rabida Inst. and Dept. Pathology and Biochemistry, Univ. Chicago, Chicago, Ill. 60649). *Transplant. Rev.* 27:24-56; 1975.

Studies on the antibody response to phosphorylcholine (PC), as characterized by restrictions in the epitope specificity, idiotype composition, immunoglobulin (Ig) class, and involvement of cell types, are presented and discussed. The inhibition of PC-specific plaques elicited in BALB/c mice is noted; the similarity between PC myeloma proteins and anti-PC antibodies is investigated by comparing their affinity constants in equilibrium dialysis, the fingerprints of their light chains, and their banding profiles in isoelectric focusing. The antibodies to PC in BALB/c mice are thus determined to be of oligoclonal origin. Additional studies note that anti-idiotypic antibodies are antireceptor antibodies. The presence of TEPC-15 idiotype on the surface of immunocompetent cells for PC is demonstrated by morphological, biochemical, and functional assays, while other experiments have isolated and characterized the receptor Ig for PC with biochemical methods; specific suppression by heterologous

steroid hormone receptor protein moves into the nucleus, associates with the chromatin, and affects a regulatory gene locus. Assuming only one such locus, femaleness represents the constitutive or noninduced state, while an XY male carrying such a mutant locus is the noninducible, e.g. the testicular feminization (Tfm) mutation. Studies on affected Tfm/Y mice have shown that every conceivable androgen-target organ of mutants becomes totally nonresponsive to even massive doses of effective androgens. Consequently, most androgen target organs are missing altogether in Tfm/Y mutants; in the few remaining target organs, androgen-non-responsiveness is correlated with virtual absence of the high affinity nuclear-cytosol androgen-receptor protein. It is concluded that the Tfm locus regulates the entire process of mammalian sexual differentiation by specifying the high affinity nuclear-cytosol androgen-receptor protein. It is further hypothesized that mammalian sex determination and differentiation may be under the control of merely two regulatory genes, both residing on the sex chromosomes. (9 references)

4222   MALIGNANT LYMPHOMA. (Eng.) Stuart, A. E. (Dept. Pathology, Univ. Edinburgh, Edinburgh, Scotland). *J. R. Coll. Surg. Edinb.* 20(5): 332-347; 1975.

The presentation, treatment, and pathology of malignant lymphoma are described; special attention is given to Hodgkin's disease. Following a brief outline of lymphoma group-treatment of patients, a description of the clinical and pathological staging of Hodgkin's disease is presented. Four clinical stages, noting the features of the original biopsy, history and physical examination, laboratory studies, and all radiographic procedures, are described. Pathological staging is determined *via* involvement found at laparotomy or by any tissue removal; procedures of the staging laparotomy are described. The Lukes histological classification of Hodgkin's disease is advocated, and consists of the following groups: (1) diffuse lesions, (2) nodular lesions, (3) nodular sclerosis, (4) mixed cellularity, (5) diffuse fibrosis, and (6) reticular type. The simpler Rye classification is also noted. The predictable spread and stability of Hodgkin's disease is discussed; nodular sclerosis is found the most stable, while lymphocyte predominance is the least stable. The diverse classifications and interpretations of non-Hodgkin's lymphoma are acknowledged and evaluated; the Rappaport classification is supported. Factors affecting the prognosis in non-Hodgkin lymphoma are discussed; survival is related to the histology. Positive bone marrow biopsies are often associated with splenic involvement; an association is also found between involved cervical or supraclavicular nodes and upper para-aortic lymphadenopathy. The histology and epidemiology and Burkitt's lymphoma, mediastinal lymphoma, angioblastic-immunoblastic lymphadenopathy, and extra-nodal lymphoma - including the thyroid, primary gastric, and mammary lymphomas - are also discussed. Conditions confused with malignant lymphoma are described, and novel cell receptor studies are summarized. (21 references)

4223   CERVICAL DYSPLASIA -- DIAGNOSIS, PROGNOSIS,
        AND MANAGEMENT. (Eng.) Sedlis, A. (No
affiliation given). *Prog. Gynecol.* 6:559-581; 1975.

The diagnosis, prognosis, and management of cervical
dysplasia is reviewed, and the role of dysplasia as
a cancer precursor is investigated. Microscopic
features distinguishing dysplasia from the normal
epithelium include inhibited maturation, increased
proliferation, pleomorphism, and variable positions
of epithelial cells. The degree of neoplastic transformation of epithelium varies with different forms
of dysplasia; based on the percentage of cells undergoing neoplastic transformation, dysplasia is classified as mild, moderate, or severe. Conditions found
to simulate dysplasia include epithelial atypia,
tissue repair, inflammation, and condylomata acumin-
ata or papillomata. Histological diagnosis of dysplasia *via* conization, random punch biopsies, colpos-
copy, and colpomicroscopy is discussed. Various
cytological features are noted; the differences between the various degrees of dysplasia are also reflected in the desquamated cells. Indirect statistical data supporting the concept of the premalignant
potential of dysplasia reveal that dysplasia usually
preceeds carcinoma *in situ*, that prevalence and
incidence of dysplasia become smaller in successive
age groups, and that the majority of cervical cancer
cases originate in patients with dysplasia. Indirect
evidence for the neoplastic character of dysplasia
is also obtained from results of tissue culture,
electron microscopy studies, and microspectrophoto-
metric determination of nuclear DNA contents. While
prospective studies provide various figures for progression and regression, a statistical model indicates
all forms of dysplasia show progression eventually
culminating in carcinoma unless the lesion is treated
or disturbed by biopsy. Factors influencing the
management of dysplasia, and the objectives of treatment are discussed. Treatment depends on the histological grade, patient's age, and desire for reproduction, and may involve hysterectomy, cryosurgery,
or electrocauterization. It is noted that the recognition and treatment of dysplasia is a significant
achievement in the control of cervical cancer.
(22 references)

4224   CHROMOSOMAL LOCALIZATION OF ENZYME
        INFORMATION IN MAN. (Eng.) Khan, P. M.
(Dept. Human Genetics, State Univ. Leiden, Wassenaarseweg 62, Leiden, Netherlands); Bootsma, D.
Proc. Int. Cancer Congr. 11th. Vol. 1 *(Cell Biology
and Tumor Immunology).* Florence, Italy, October
20-26, 1974. Edited by Bucalossi, P.; Veronesi,
U.; Cascinelli, N. New York, American Elsevier,
1975, pp. 56-63.

Current knowledge on the gene map of man, with
particular reference to the loci involved in the
production of enzymes, is presented. Cell fusion
techniques are useful in showing that the loss of
certain chromosomes from hybrids between malignant
and other malignant or nonmalignant cells does
influence the tumorigenic capability of the cell
hybrid; somatic cell hybrids have demonstrated the
occurrence of a specific integration site for the

4227   GENOME COMPLEXITY AND *IN VIVO* TRANSCRIPTION
        IN HUMAN LEUKEMIC LEUKOCYTES. (Eng.)
Saunders, G. F. (M. D. Anderson Hospital and Tumor
Inst., Houston, Tex. 77025); Chuang, C. R.; Sawada,
H. *Acta Haematol. (Basel)* 54(4):227-233; 1975.

The composition of satellite DNAs and the transcriptional activity of the different kinetic classes of
DNA are examined. At least eight well-defined satel-
lites are isolated from the human genome. A tabulated summary of the physical properties of these
human satellite DNAs, including buoyant densities in
CsCl, chromosome location, and alkaline CsCl strand
separation, is presented. In addition, studies reveal highly repetitive nonsatellite sequences arranged in blocks 1100 nucleotides long and interspersed with intermediate repetitive and nonrepetitive sequences greater than 2500 nucleotides long.
Analyses of several classes of normal and leukemic
human leukocytes find similar levels of unique DNA
sequence transcription in the total cellular RNA;
approximately 2.5-3.9% of the nonrepetitive sequences
are thus represented. Assessment of repetitive sequence transcription *via in vitro* chromatin transcription and hybridization competition experiments
with total cellular RNA shows chromatins from the
most metabolically active cells are better templates.
Hybridization of chromatin transcripts estimates
4.2-5.7% of the repetitive sequences are available
for transcription. The results suggest greater
differences among classes of human lymphocytes in
the repetitive transcripts than in the nonrepetitive
transcripts. It is further suggested that leukemia
results from an impairment of the cell differentiation
process, and that such impairment involves transcription of DNA sequences in malignant cells which are
not transcribed in normally differentiated cells.
(16 references)

4228   EXCISION REPAIR IN HUMAN CELLS. (Eng.)
        Bootsma, D. (Dept. Cell Biology and Gene-
tics, Erasmus Univ., Rotterdam, Netherlands); de
Weerd-Kastelein, E. A. *Proc. Int. Cancer Congr.
11th.* Vol. 1 *(Cell Biology and Tumor Immunology).*
Florence, Italy, October 20-26, 1974. Edited by
Bucalossi, P.; Veronesi, U.; Cascinelli, N. New
York, American Elsevier, 1975, pp. 164-169.

General DNA repair mechanisms and specifically
defective DNA repair in xeroderma pigmentosum (XP)
are discussed; in addition, evidence suggesting
a relationship between DNA repair and carcinogenesis
is presented. Three cellular recovery systems are
demonstrated in UV-exposed bacteria; photoreacti-
vation, excision repair, and post-replication repair.
Since "unscheduled DNA synthesis" (UDS) and "repair
replication" have been demonstrated in human cells,
it is assumed that an excision repair process also
exists in human cells. The removal of UV-induced
thymine dimers from the DNA of human cells has been
demonstrated *via* chromatography, an immunological
assay, and also through use of a dimer-specific
endonuclease. The biological significance of such
UDS and repair replication has been revealed through
the discovery of their inhibition in XP. Following
a discussion of the clinical characteristics of

XP, it is suggested that the phenotypic differences of the De Sanctis-Cacchione and the Hebra-Kaposi forms of the disease result from the variable expression of the same genetic defect. Exposure of XP cells to UV radiation results in low or negligible levels of UDS and repair replication, indicating a defect in the excision repair process; however, XP cells exposed to X-rays retain normal levels of UDS and repair replication. While it is thus inferred that the XP cells are defective in an early enzymatic step, several XP patients have not shown such a defect. Experimental evidence for genetic heterogeneity in patients is discussed. The combined data from biochemical analysis of XP cells and complementation results from cell fusion studies have indicated the involvement of at least five different genes. Such complementation groups are tabulated, and the involvement of a coordinated enzyme complex in the enzymatic repair mechanism is hypothesized. Such discoveries also suggest a causal relationship between defective DNA repair and carcinogenesis, and indicate that DNA repair processes may play a role in the cell transformation. (29 references)

4229   REGULATORS OF CELL DIVISION: ENDOGENOUS
        MITOTIC INHIBITORS OF MAMMALIAN CELLS.
(Eng.) Lozzio, B. B. (Univ. Tennessee Memorial Res. Center and Hosp., Knoxville, Tenn.); Lozzio, C. B.; Bamberger, E. G.; Lair, S. V. *Int. Rev. Cytol.* 42: 1-45; 1975.

Antimitotic substances found in a variety of tissues and sera of humans and animals, and mitosis inhibitors produced by cultured cells are reviewed. The limitation of mitotic activity in high-density cultures cannot be explained solely on the basis of available nutrients. Specific and nonspecific inhibitors of cultured cell growth are reported. Crude extracts from murine embryo skin and placenta have retarded the growth of some carcinomas, spontaneous and transplanted tumors. Antimitotic substances were also found in extracts of human amniotic and chorionic membranes, and in molluscs and amphibian ovaries. Specific and nonspecific mitotic inhibitors partially purified from mammalian and amphibian kidneys have a lipoid composition, while the existence of a chalone is reported in amphibian kidneys. A variety of inhibitors have been isolated from liver tissue of various species; some fit the criteria for a liver chalone, while others are distinctly different in the biological properties. Such inhibitors have been intensively investigated for their effect on DNA, RNA, and protein synthesis. Some tissues of tumor-bearing animals and/or the tumor itself are found to contain mitotic inhibitors. Specific inhibitors of malignant cell multiplication extracted from tumors of rats, mice, and hamsters have included epidermal carcinoma, melanoma, lymphoma, and mylogenous leukemia. They are partially characterized as lipids, proteins, and polypeptides. Striated muscle extracts also contain components cytotoxic to malignant cells *in vitro* and *in vivo*. Tissue-specific epidermal chalones have been found to depress mitotic activity and DNA synthesis. Homologous antimitotic spleen

4249   SEROLOGICAL APPROACH TO TUMOUR-ASSOCIATED
        ANTIGENS. (Eng.) Moore, M. (Christie
Hosp. and Holt Radium Inst., Manchester M20 9BX,
England). *Proc. Int. Cancer Congr. 11th.* Vol. 1
(*Cell Biology and Tumor Immunology*). Florence,
Italy, October 20-26, 1974. Edited by Bucalossi, P.;
Veronesi, U.; Coscinelli, N. New York, American
Elsevier, 1975, pp. 238-243. (57 references)

4250   TUMOR ANTIGENS IN THE TUMOR-HOST RELATION-
        SHIP. (Eng.) Smith, R. T. (Univ. Florida
Coll. Medicine, Gainesville, Fla. 32610). *Interaction of Radiation and Host Immune Defense Mechanisms
in Malignancy*, Conference, 5th, The Greenbrier.
White Sulphur Springs, West Virginia, March 23-27,
1974. Chaired by Bond, V. P.; Hellman, S.; Order,
S. E.; Suit, H. D.; Withers, H. R. Brookhaven National
Laboratory Associated Universities, Inc., 1974,
pp. 17-22. (4 references)

4251   TUMOR IMMUNITY: SOME IMPLICATIONS OF *IN
        VITRO* STUDIES ON LYMPHOCYTE-MEDIATED REAC-
TIVITY TO ANIMAL AND HUMAN NEOPLASMS. (Eng.) Hell-
strom, I. (Univ. Washington Sch. Medicine, Seattle,
Wash. 98195); Hellstrom, K. E. *Interaction of
Radiation and Host Immune Defense Mechanisms in
Malignancy*, Conference 5th, The Greenbrier. White
Sulfur Springs, West Virginia, March 23-27, 1974.
Chaired by Bond, V. P.; Hellman, S.; Order, S. E.;
Suit, H. D.; Withers, H. R. Brookhaven National
Laboratory Associated Universities, Inc., 1974, pp.
23-29. (28 references)

4252   IMMUNOLOGICAL DEFICIENCY STATES AND
        MALIGNANCY. (Eng.) Cottier, H. (Inst.
Pathology, Univ. Bern, Bern, Switzerland); Hess, M.
W.; Keller, H. U.; Luscieti, P.; Sordat, B. *Inter-
action of Radiation and Host Immune Defense Mechanisms
in Malignancy*, Conference, 5th, The Greenbrier.
White Sulphur Springs, West Virginia, March 23-27,
1974. Chaired by Bond, V. P.; Hellman, S.; Order,
S. E.; Suit, H. D.; Withers, H. R. Brookhaven
National Laboratory Associated Universities, Inc.,
1974, pp. 30-51. (125 references)

4253   IMMUNOLOGICAL ASPECTS OF GASTROINTESTINAL
        NEOPLASIAS. (Ger.) Hess, M. W. (Patho-
logisches Institut der Universitat, Freiburgstrasse
30, CH-3010 Bern, Switzerland); Zimmermann, A.;
Brun del Re, G.; Cottier, H. *Schweiz. Med. Wochenschr.*
105(18):570-575; 1975. (37 references)

4254   IMMUNOLOGY, TUMOR MARKERS, AND PANCREATIC
        CANCER. (Eng.) Zamcheck, N. (Boston City
Hosp., Boston, Mass.). *J. Surg. Oncol.* 7(2):155-165;
1975. (43 references)

4255   PROSPECTIVES FOR CELL AND ORGAN CULTURE
        SYSTEMS IN THE STUDY OF PANCREATIC CARCINOMA.
(Eng.) Jamieson, J. D. (Yale Univ. Sch. Medicine,
New Haven, Conn. 06510). *J. Surg. Oncol.* 7(2):139-
141; 1975. (12 references)

4256   PREVENTION OF CANCERS OF THE RECTUM AND
        COLON [editorial]. (Fre.) Soullard, J.
(Hopital Beaujon, 100, boulevard du General-Leclerc,
F 92110 Clichy, France); Potet*, F. *Arch. Fr.
Mal. App. Dig.* 64(3):197-200; 1975. (15 references)

4257   EPITHELIOMAS OF THE RHINOPHARYNX: RECENT
        CASES. INTRODUCTION. (Fre.) Cachin, Y.
(French Association for Cancer Study, France). *Bull.
Cancer (Paris)* 62(3):249-250; 1975. (No references)

4258   MALIGNANT MELANOMA OF THE SKIN: CLINICAL
        ASPECTS AND CURRENT VIEWS OF PROGNOSIS.
(Fre.) Barriere, H. (Clinique Dermatologique, Hotel-
Dieu, 44 Nantes, France); Litoux*, P. *Ouest Med.*
28(4):179-188; 1975. (No references)

4259   CURRENT TRENDS IN THE CARCINOMA *IN SITU*
        OF THE VULVA. (Ita.) Carenza, L. (Univer-
sita di Roma - II Cattedra di Patologia Ostetrica
e Ginecologica, Italy); D'Alessandro, P.; Cardillo,
M. R. *Patol. Clin. Ostet. Ginecol.* 2(6):269-282;
1974. (69 references)

4260   PRELEUKEMIA. (Eng.) Rothstein, G. (Univ.
        Utah Coll. Med., Salt Lake City); Wintrobe,
M. M. *Adv. Intern. Med.* 20:363-378; 1975. (62
references)

4261   TERATOCARCINOMAS AS A MODEL SYSTEM FOR THE
        STUDY OF EMBRYOGENESIS AND NEOPLASIA.
(Eng.) Martin, G. R. (Dept. Pediatrics, Univ. California, San Francisco, Calif. 94143). *Cell* 5(3):
229-243; 1975. (79 references)

tones were applied to the shaved backs of male
C57BL mice, injected sc into female CFLP mice, and
incubated with hamster embryo cells *in vitro*.
Chemical studies indicated that the primary products
of the acid-catalyzed dehydration of *cis*-9,10-di-
hydro-9,10-dihydroxyphenanthrene, *cis*-4,5-dihydro-
4,5-dihydroxybenzo(*a*)pyrene, and *cis*-5,6-dihydro-5,
6-dihydroxy-7,12-dimethylbenz(*a*)anthracene were, in
each case, the K-region phenols. The phenols of
DMBA rearranged into the keto forms described above.
The keto forms of the K-region phenols of DMBA were
more stable than the enol forms. The K-region ketones of DMBA showed no carcinogenic activity *in
vivo* and no transforming capacity *in vitro*. The
data indicate that the K-region ketones of DMBA
probably do not mediate the carcinogenic activity of
DMBA.

4268   LOCALIZATION OF POLYCYCLIC HYDROCARBON
         CARCINOGENS IN THE LUNG FOLLOWING INTRA-
TRACHEAL INSTILLATION IN GELATIN SOLUTION. (Eng.)
Kennedy, A. R. (Harvard Sch. Public Health, Boston,
Mass.); Little, J. B. *Cancer Res.* 35(6):1563-1567;
1975.

The deposition and localization of 1.6% benzo(*a*)-
pyrene (BP) suspended in a 1.5% gelatin-0.9% NaCl
solution was studied in male Syrian hamsters by
UV fluorescence microscopy. The majority of BP
particles were in the 0.5-1.0 μm size range. Changes
in BP deposition were evaluated immediately after
intratracheal administration of the BP suspension
and at intervals of 1-2 hr, 24-48 hr, 5-7 days,
and 2-4 wk to 3 mo. The BP particles were initially seen as intense white-yellow fluorescent particles distributed in the alveolar region as dispersed particles and occasional aggregates. At
1-2 hr, most particles were evenly distributed
throughout the peripheral lung; some *in situ* metabolism of BP was suggested by the development in
the peripheral lung of pale yellow fluorescent areas
without distinct particles. Although large numbers
of BP-filled macrophages were seen in the upper
airways by 24 hr after instillation, little BP
appeared to penetrate into the bronchial epithelium.
After five days, few BP particles were left in the
alveolar region; occasional clumps of particles
were found in the peripheral lung macrophages and
in the larger airways. Some of the pale yellow
fluorescence that persisted in the peripheral lung
appeared to be extracellular. After 2-4 wk, essentially no BP particles remained in the peripheral
lung, macrophages, or larger airways. The pale
yellow areas persisted in the peripheral lung,
but their color was unstable on exposure to UV or
tungsten light. These results differ from those
previously obtained when BP was administered on
small ferric oxide particles. Retention of BP
was longer with the ferric oxide particles and the
particles were deposited primarily on the epithelium
of the larger airways. The BP-gelatin-0.9% NaCl
solution technique appears to be a useful model
for delivering the carcinogen dose to the alveolar
region. As previously suggested by tumor induction
experiments with 6.6-7.8% BP, this region is the
major site of deposition and retention of the carcinogen.

4269   TEMPERATURE-SENSITIVE MUTANTS OF CHEMIC-
        ALLY TRANSFORMED EPITHELIAL CELLS. (Eng.)
Yamaguchi, N. (Columbia Univ. Coll. Physicians Surg.,
New York, N.Y.); Weinstein, I. B. *Proc. Natl. Acad.
Sci. USA* 72(1):214-218; 1975.

The first isolation of temperature-sensitive (TS)
mutants of epithelial cells transformed with chemical carcinogens was studied. The TS mutant, W-8,
was obtained from K-16 normal epithelial rat liver
cells exposed to 0.5 µg/ml *N*-acetoxy-2-acetylamino-
fluorene. When 10^5 or 10^6 W-8 cells were injected
sc into newborn Sprague-Dawley rats, 1-2-cm locally
invasive tumors composed of epitheloid cells appeared within 3-4 wk. The wild type W-8 clone was
obtained by isolating a colony at 41 C in 1.2%
methyl cellulose. Like the wild type W-8, the TS
mutants readily grew in agar suspension at 36 C,
but (in contrast to the wild type) they did not
grow at 40 C. Detailed studies of one of the mutants (TS-223), which displayed the most extreme
response to temperature, indicated that it also had
reduced cloning efficiency at high temperature in
monolayer culture and a lower saturation density.
Scanning electron microscopy revealed that at 40 C,
confluent cultures of TS 223 consisted of a monolayer of generally flat polygonal cells; at 36 C,
the cultures contained many patches of piled-up
cells that were spherical and had rougher surface
membranes. All of these cellular changes were reversible with higher or lower temperatures. TS
lesions appear to reside in a host cell gene that
modulates expression of the transformed phenotype.
When the 36 C and 40 C cultures were at their pla-
teaus in saturation density, there was a very low
level, of thymidine incorporation. A change to
fresh medium at both 36 C and 40 C led to a burst
of thymidine incorporation. While the TS mutants
display minimal features of transformation complicating characterization, they may be useful in out-
lining minimal changes required for the expression
of the transformed phenotype in epithelial cells.

4270   LACK OF INDUCTION OF DOMINANT LETHALS
        IN MICE BY ORALLY ADMINISTERED AF-2.
(Eng.) Soares, E. R. (Natl. Inst. Environmental
Health Science, Res. Triangle Park, N.C. 27709);
Sheridan, W. *Mutat. Res.* 31(4):235-240; 1975.

Toxic and potential mutagenic effects of oral
trans-2-(2-furyl)-3-(5-nitro-2-furyl) acrylamide
(AF-2) were studied in C3H/HeJ, DBA/2J, and CBA/J
mice. In the toxicity tests, AF-2, suspended in
distilled water or 40% ethanol or dissolved in
pure dimethylsulfoxide, was administered by gavage
in doses of 200-900 mg/kg. There was no significant increase in deaths of AF-2-treated animals
compared to controls unless mice were deprived of
food and water for 24 hr prior to treatment.
Starved mice died within 48 hr of receiving AF-2,
and postmortem examination revealed a bright or-
ange precipitate (probably AF-2) in the stomachs.
In the dominant lethal tests, each of the total of
150 DBA/2J mice was treated, in groups of 30, with
300-450 mg/kg of AF-2 and paired with three C57BL/
6J female mice for seven days. Pairing with other
sets of three females was repeated for six conse-

fractions from different tissues on precarcinogens
(2) *in vitro vs in vivo* sensitivity of the target
cells used in assay. The relevancy of the DNA
repair test for carcinogenic potential is improved
by applying multiple low doses of the carcinogens
to the cell system. This approach more closely
approximates the usual exposure in humans than the
single large dose. It is suggested that several
endpoints be used to determine carcinogenic potential of chemicals and/or viruses, i.e. DNA repair
synthesis, chromosome aberration, incidence of
neoplastic transformation. Based on tests with
80 chemicals, all carcinogens can be said to produce DNA lesions followed by a measurable level
of DNA repair synthesis.

4274   ISOZYME PATTERN OF FRUCTOSE DIPHOSPHATE
         ALDOLASE DURING HEPATOCARCINOGENESIS
INDUCED BY 2-ACETYLAMINOFLUORENE IN RAT LIVER.
(Eng.) Silber, D. (Sch. Medicine, Pasteura 4,
50367 Wroclaw, Poland); Checinska, E.; Rabczynski,
J.; Kochman, M. *Int. J. Cancer* 16(4):675-681;
1975.

Changes in the isoenzyme pattern of aldolase, during
hepatocarcinogenesis induced by 2-acetylaminofluo-
rene (AAF) in male Wistar rats, were studied. The
rats were fed bread and, once/wk, carrots. In the
group given 11.25 mg AAF/rat/day, the fructose-1,6-
diphosphatase:fructose-1-phosphatase activity ratio
increased from 1.47 after 4 mo, to 8.0 at 9.5 mo,
compared to 1.2 for the control rats. After 9 mo
of AAF administration, cholangioadenomas changing
into malignant tumors were found. All of the A-B
hybrid enzymes were demonstrated by electrophoresis
at this time. These results support the hypothesis
of dedifferentiation during neoplastic transformation.

4275   EVIDENCE FOR A SECOND ARYLHYDROXAMIC
         ACID ACYLTRANSFERASE SPECIES IN THE SMALL
INTESTINE OF THE RAT. (Eng.) Olive, C. W. (Univ.
Chicago Pritzker Sch. Medicine, Chicago, Ill. 60616);
King, C. M. *Chem. Biol. Interact.* 11(6):599-604;
1975.

The possibility of existence of tissue-specific
acyltransferase species was studied with antiserum
to the enzyme. The small intestine of the Sprague-
Dawley rat contains two species of arylhydroxamic
acid acyltransferases. These enzymes were separated
by gel filtration on Sephadex G-100. The smaller
species had the mobility of rat liver acyltransferase and was precipitated with antiserum directed
against the liver enzyme. The larger species was
not precipitated with this antiserum and was not
detected in the soluble fraction from liver. Both
species showed greater activity with *N*-hydroxy-*N*-2-
acetylaminofluorene than with *N*-hydroxy-*N*-4-acetyl-
aminobiphenyl (*N*-hydroxy-AABP) as substrate. The
smaller species showed more activity toward *N*-
hydroxy-AABP than the larger species did.

4276   CONDITIONS MODIFYING DEVELOPMENT OF TUMORS
        IN MICE AT VARIOUS SITES BY BENZO(a)PYRENE.
(Eng.) Vesselinovitch, S. D. (Franklin McLean Memor-
ial Res. Inst., Univ. Chicago, Chicago, Ill. 60637);
Kyriazis, A. P.; Mihailovich, N.; Rao, K. V. N.
Cancer Res. 35(11/Part 1):2948-2953; 1975.

The modifying roles of age, sex, and strain of mice
on the incidence, multiplicity, and spectrum of tumors induced by benzo(a)pyrene were investigated.
The first-generation (F₁) hybrids of C57BL/6J x
C3HeB/FeJ and C3HeB/FeJ x A/J mice of both sexes
were given single ip injections (75 or 150 μg/g)
of benzo(a)pyrene at 1, 15, or 42 days of age. Experimental animals were allowed to live their life-
spans, while animals in control groups were killed
at 52, 90, 142, or 170 wk of age. Animals treated
with benzo(a)pyrene died, in general, by the 100th
wk of age due to development of liver, lung, stomach, and lymphoreticular tumors. Few of the control
animals died during that same observation period.
The age of mice at the time of exposure to the carcinogen modified development of tumors at all the
sites. The sex of animals influenced the development of liver and lymphoreticular tumors. The C3HeB/
FeJ x A/J F₁ hybrids, developed lung tumors more
readily than did the C57BL/6J x C3HeB/FeJ F₁ mice,
which had significantly more liver tumors and neoplasms of the lymphoreticular system than the former strain. No strain difference was observed in
regard to tumors at other sites. Higher doses of
benzo(a)pyrene were more effective in inducing lung,
liver, and stomach tumors. In addition, five cases
of pancreatic ductal adenoma and adenocarcinoma
were observed in carcinogen-treated mice. The data
indicate that the preweaning animals are more sensitive to carcinogenic response than are the adults;
the authors suggest that the relative cell immatur-
ity, high rate of macromolecular synthesis, and a
consequently high rate of cell replication, as well
as low immunological competence, are probably causally related to the high susceptibility of these
tissues to the inception of carcinogenesis.

4277   INHIBITION OF HEPATIC ARYL HYDROCARBON
        HYDROXYLASE BY 3-METHYLCHOLANTHRENE,
7,8-BENZOFLAVONE AND OTHER INDUCERS ADDED *IN VITRO*.
(Eng.) Jellinck, P. H. (Dept. Biochemistry, Queen's
Univ., Kingston, Ontario, Canada); Smith, G.; New-
combe, A.-M. *Chem. Biol. Interact.* 11(5):459-468;
1975.

The inhibition of aryl hydrocarbon hydroxylase by
3-methylcholanthrene, 7,8-benzoflavone, estrogens,
and testosterone *in vitro* was investigated in liver
microsomes from female Holtzman rats treated with
3-methylcholanthrene or 7,8-benzo-flavone (5 mg in
5 ml sesame oil, ip) 18 hr before sacrifice. A
supernatant fraction of the liver (microsomes plus
cytosol) was obtained at 8,000 x g. Microsomes from
50 mg liver were incubated with [12-¹⁴C]-dimethyl-
benz(a)anthracene or [3,6-¹⁴C]-benzopyrene (2 μg)
and NADPH (0.3 mM) and one of the following: 3-
methylcholanthrene and 7,8-benzoflavone at 2.5, 0.5,
and 0.25 μg/ml; and estradiol-17β, diethylstilbestrol, and testosterone, at 2.5 μg/ml. Both 3-methyl-

administration of CPA was followed by a midnight
administration of VBL, a strong antitumoral effect was
observed, the shrinkage of tumors being more marked
after sequential CPA-VBL treatment than after CPA
alone. However, the percentage of definite cures
effected by CPA alone was not significantly increased
by the sequential treatment. The results suggest
that the relative synchrony normally induced in
tumors by the circadian activity can be replaced by a
much better synchrony, induced by an alkylating
agent.

4281   INCREASED THYMIDINE UPTAKE BY METHYLCHOL-
        ANTHRENE-TREATED C3H/10T1/2 CELLS. (Eng.)
Bairstow, F. (New York Univ. Medical Center, New
York, N.Y. 10016); Heidelberger, C. *Int. J. Cancer*
16(3):370-375; 1975.

The uptake of ^{3}H-thymidine by postconfluent cultures of methylcholanthrene (MCA)-transformed C3H/
10T1/2 mouse embryo cells was measured at time intervals up to 110 days following MCA treatment.
MCA (1-15 μg/ml) was added to the cultures for 24
hr one day following seeding. In other experiments,
^{3}H-thymidine uptake was measured in mixtures of
MCA-transformed and nontransformed 10T1/2 cells.
In comparison with nontransformed cultures, the
loss of tritium from the incubation medium was
much greater with the MCA-transformed cells. The
loss of tritium from the medium was well correlated
with the uptake of ^{3}H-thymidine by the cells as
determined by autoradiography. Tritium uptake was
also considerably elevated above control (nontransformed) levels in admixtures containing 2% or 10%
transformed cells; with more than 20% transformed
cells in the mixture, there was no appreciable
change in thymidine uptake. Measurements made at
various intervals up to 110 days revealed that after 25 days the MCA-treated cultures incorporated
significantly more thymidine than acetone-treated
controls. The increased uptake corresponded with
the appearance of Type III transformed foci in the
MCA-treated cultures. The results suggest that
thymidine uptake may eventually be used as an additional criterion of cell transformation.

A

4282   CHEMICAL CARCINOGENESIS IN DIFFUSION
        CHAMBERS. (Eng.) Parmiani, G. (Istituto
Nazionale per lo Studio e la Cura dei Tumori, Via
G. Venezian 1, Milan, Italy). *Proc. Int. Cancer
Congr. 11th*. Vol. 2 *(Chemical and Viral Oncogenesis.*
Florence, Italy, October 20-26, 1974. Edited by
Bucalossi, P.; Veronesi, U.; Cascinelli, N. New
York, American Elsevier, 1975, pp. 113-117.

d

Suitability of the diffusion chamber (DC) for the
study of host-mediated chemical carcinogenesis and
the immunogenicity of methylcholanthrene-induced
tumors is investigated. Lung tissue of fetal mice
was placed on the internal filters of the DCs,
subsequently implanted ip in adult BALB/c mice.
Five doses of urethane (1 mg/kg, ip) were administered to mice over a ten-day period. DCs were

transferred sc to immunodepressed (thymectomy plus
400 rads) BALB/c mice killed 8, 18, and 50 wk later
In a similar experiment, fetal liver tissue from
C3HF mice was placed in DCs, and a single dose of
N-nitrosodimethylamine (4 µg/g, sc) was given to
mice one week after ip implantation. DCs were
transferred to immunodepressed C3HF mice which
were killed at 8 and 16 wk. Histological examination of the lung and liver tissue in the DCs revealed
no neoplastic transformation. Connective tissue
predominated in most excised transplants. DCs were
used to study the theory that the immunogenicity
of tumors produced by carcinogenic transformed
target cells depends on the stage of the cell cycle
in which the target cell comes in contact with the
chemical. BALB/c mice fibroblasts were placed on
filters of DCs, then implanted ip in syngeneic mice.
A filter with 30 µg 3-methylcholanthrene (MCA) was
implanted (1) simultaneously and removed after one
week (2) added to DCs after fourth week of implantation. At the end of fifth week, filters were transferred sc to BALB/c x C3HF mice. If MCA was in contact with fibroblasts during the first week of growth
mice were more likely to develop sarcoma tumor
nodules (16/62) than if the MCA was added at later
stages (7/41). Early MCA contact resulted in 6/16
immunogenic tumors, whereas all tumors induced by
fibroblasts subjected to late treatment with MCA
were immunogenic, as tested by the growth excision
method. The DC *in vitro-in vivo* technique can be
used to study hydrocarbon carcinogenesis in immunologic experiments, but is not suitable for carcinogenic chemicals such as urethan or N-nitrosodimethylamine.

4283   INFLUENCE OF A CHRONIC ADMINISTRATION OF
        DIETHYLNITROSAMINE ON THE RELATION BE-
TWEEN SPECIFIC TISSULAR AND DIVISION FUNCTIONS IN
THE RAT LIVER. (Eng.) Van Cantfort, J. (Service
de Chimie medicale, Universite de Liege, 151, Bd.
de la Constitution, 4000 Liege, Belgium); Barbason,
H. *Eur. J. Cancer* 11(8):531-536; 1975.

The relation between the division function and
specific tissular function in the rat liver (which
normally follow a circadian rhythm and are mutually exclusive, each predominating at a different
point in the life cycle of the animal) was studied
in diethylnitrosamine (DEN)-treated (10 mg/kg/day
in the drinking water for 25 or 50 days) male
Sprague-Dawley rats. At the end of DEN treatment,
the animals were sacrificed, at 10 a.m. or 10 p.m.,
or were partially hepatectomized at 10 a.m. and
then sacrificed 12-72 hr later. The mitotic index
and the activities of cholesterol-7α-hydroxylase,
benzopyrene-hydroxylase, and aminopyrine-N-de-
methylase were measured in the livers after sacrifice. In the DEN-treated animals, the mitotic index was 1% higher than in the normal controls, but
no circadian rhythm of mitoses was detected. However, the circadian rhythm of cholesterol-7α-hy-
droxylase was either partially (25 days) or totally
(55 days) abolished by the DEN treatment, the enzymatic activity reaching a steady state equilibrium intermediate between the highest and lowest

Sprague-Dawley rat, and Syrian golden hamster respiratory tissues, as well as in primary lung neoplasia in humans. Prior to tissue removal, the animals were subjected for at least ten days to a light-dark regimen (start of 12-hr light period at 7 a.m.; tissues removed between 10:30 and 12:00 hr). [^{14}C]DMN was purified by Dowex-1-bisulfite column chromatography to remove a contaminant (probably [^{14}C]formaldehyde) interfering with the enzyme assay. Since formaldehyde and methyl carbonium ions (which yield methanol with water) are considered to be the primary products of DMN metabolism, tissue slices were assayed for the production of [^{14}C]CO$_2$ from ^{14}C-labeled methanol, formaldehyde, formate, and DMN. Oxidation of formate to CO$_2$ was very much rate-limiting, although the oxidation of formaldehyde to formate was not. This rate-limiting step was circumvented by introducing quantitative chemical oxidation of formate to CO$_2$ by mercury(II)chloride following the enzymic reaction. Since oxidation of methanol to CO$_2$ proved to be insignificant, production of CO$_2$ from DMN by lung enzymes and HgCl$_2$ may serve as a parameter for N-demethylating activity and the production of the suspected carcinogenically active methyl carbonium ions. The DMN-N-demethylating activities of lung tissue slices of the two mouse strains with widely different susceptibilities to formation of lung adenomas by DMN differed significantly, but the difference seemed too small to explain the divergence in tumorigenic response. The enzymatic activities decreased in hamster bronchus, hamster trachea, hamster lung, GRS/A mouse lung, C3HF/A mouse lung, human lung, and Sprague-Dawley rat lung, in that order. The reported resistance of the hamster respiratory system to tumor induction by DMN may therefore not be due to poor DMN-N-demethylating capacity.

4288   THE INACTIVATION OF BACTERIOPHAGE R17 BY ETHYLATING AGENTS: THE LETHAL LESIONS. (Eng.) Shooter, K. V. (Royal Cancer Hosp., London SW3 6JB Great Britain); Howse, R. *Chem. Biol. Interact.* 11(6):563-573; 1975.

The toxic action of ethyl methanesulphonate (EMS) and N-ethyl-N-nitrosourea (ENUA) on single-strand RNA was studied. Labeled ethylating agents [1-^{14}C]EMS (5.3 mCi/mM) and [^{3}H]ENUA (116 mCi/mM) were introduced into suspensions of bacteriophage R17 for periods of 16 and 2 hr, respectively. Unreacted reagents were separated from the bacteriophage on a Sephadex G-100 column. 3-Alkylcytosine, 7-alkylguanine, O^6-alkylguanine, 1-alkyladenine and 3-alkyladenine as well as various unidentified reactants were quantified as a fraction of the total bound radioactivity. At the mean LD for EMS, 8 M ethyl were bound per mole RNA, and for ENUA 3.5 M ethyl were bound. There was very little difference in the amount of alkylated products produced in experiments with methylating agents and EMS. However, ENUA was responsible for a greater production of 3-alkyladenine and unidentified reactants as well as a lowered production of 7-alkylguanine. Assuming that 7-alkylguanine is not a lethal lesion, the methylating

agents and EMS produce lethal lesions in the form
of 3-alkylcytosine, O^6-alkylguanine, and chain
breaks as a result of hydrolysis of ethyl phosphotriesters. These agents account for only .65
lethal lesions per molecule for ENUA. The possibility that 3-alkyladenine constitutes a lethal
lesion for ENUA unlike the others is suggested.

4289   A MAMMALIAN SPOT TEST: INDUCTION OF GE-
        NETIC ALTERATIONS IN PIGMENT CELLS OF
MOUSE EMBRYOS WITH X-RAYS AND CHEMICAL MUTAGENS.
(Eng.) Fahrig, R. (Zentrallaboratorium fur Muta-
genitatsprufung der Deutschen Forschungsgemein-
schaft, D-7800 Freiburg Breisacherstr. 33, West
Germany). *Mol. Gen. Genet.* 138(4):309-314; 1975.

The effects of X-irradation or of chemical mutagens
on the expression of recessive color genes were
studied. Embryos heterozygous for five recessive
coat-color genes from the cross C 57 BL/6 J Han x
T-stock were X-irradiated with 100 rads or treated
in utero with 50 mg/kg methyl methanesulfonate (MMS)
and ethyl methanesulfonate (EMS), respectively. Controls consisted of irradiated embryos of C 57 BL x
C 57 BL matings homozygous wild-type for the genes
under study, and nontreated offspring of both types
of mating. The colors of the spots observed in the
adult fur were either due to expression of the recessive coat genes or were white. Irradiated and
mutagen-treated offspring of C 57 BL x T-stock matings had almost exclusively nonwhite spots, distributed randomly over the mouse surface. Irradiated offspring of C 57 BL x C 57 BL matings had only
white spots which were always midventral. In nontreated offspring of both types of mating, no spot
could be observed. After correcting for white midventral spots observed in the one type of control,
the frequency of expression of one or the other of
the recessive color genes was calculated to be a-
bout 11% for embryos irradiated with 100 rads at
11 days postconception, about 8% for embryos irradiated with 100 rads at 10 1/2 days postconception,
about 1% for embryos irradiated with 100 rads at
nine days postconception, about 7% for embryos
treated with 50 mg MMS/kg at 10.5 days postconception, and about 8% for embryos treated with 50 mg
EMS/kg at 10.5 days postconception. It is assumed
that the white midventral spots are preferentially
the result of pigment cell killing, while the nonwhite spots are preferentially the result of gene
mutations or recombinational processes like mitotic
crossing over and mitotic gene conversion. Of nu-
merical and structural chromosome aberrations, only
those come into question which are able to pass the
filter of several mitoses. Therefore, the test
system described is supposed to cover only heritable
DNA-alterations, but the whole spectrum of them.

4290   HISTOLOGICAL CHANGES IN THE SUBMANDIBULAR
        GLANDS OF RATS AFTER INTRADUCTAL INJEC-
TION OF CHEMICAL CARCINOGENS. (Eng.) Takeuchi, J.
(Sch. Dent. Aichi-Gakuin Univ., Nagoya, Japan);
Miura, K.; Katoh, Y.; Usizima, H. *Acta Pathol. Jpn.*
25(1):1-13; 1975.

Early histologic changes in the duct, acini, and inter-

1978, New York, N.Y. 10029); Gildin, J. *J. Natl. Cancer Inst.* 55(2):385-391; 1975.

The effect of BCG injection into the colon wall (6.7 x 10^6 organisms) on the induction of colon tumors by 1,2-dimethylhydrazine dihydrochloride (DMH, 30 mg/kg intragastrically once weekly for five weeks) was examined in male Sprague-Dawley and Fischer Rats. Twenty-five rats received BCG alone; 21 were treated with BCG following the last dose of DMH; and five received BCG prior to DMH treatment. In 14 animals with DMH-induced colon tumors, BCG was injected directly into the tumors; in three rats with resected tumors, BCG was injected into the colon wall adjacent to the resection site. Animals treated with BCG alone developed a generalized infection that persisted for 21 wk. There was no mucosal ulceration or evidence of severe necrosis in either Sprague-Dawley or Fischer rats. Rats infected with BCG after DMH treatment developed colon tumors at the same rate and in the same incidence as in DMH-treated controls given Tween culture medium (7H9). However, 37% of the tumors in BCG-infected rats were mucinous adenocarcinomas, compared with 9% in rats given 7H9. Moreover, 47% of BCG-treated animals with colon tumors had metastases to the abdominal lymph nodes, the serosal surface of other gut segments, or (in one rat), the lung; metastases were rare in 7H9-treated rats. In many BCG-infected animals, primary or metastatic tumors were intimately associated with granulomas and acid-fast bacilli. The severity and persistence of BCG infection were enhanced in rats given DMH, compared with rats given BCG alone. Prior BCG infection had no effect on the induction of colon tumors by DMH, and no tumor metastasized. Only one of the five pretreated animals had evidence of infection when examined 45-59 wk after BCG injection. Direct injection of BCG into tumors had no effect on tumor size, although active infection and granuloma formation occurred; infection persisted in the adjacent colon and the abdominal lymph nodes. In animals subjected to tumor resection before intracolonic BCG injection, infection with BCG did not interfere with the healing of the colon or skin lesions.

4294　　TUMOR INDUCTION STUDIES WITH *n*-BUTYL-
　　　　AND *n*-PROPYLHYDRAZINE HYDROCHLORIDES IN
SWISS MICE. (Eng.) Nagel, D. (Univ. Nebraska Medical Center, 42nd Street and Dewey Avenue, Omaha, Nebr. 68105); Shimizu, H.; Toth, B. *Eur. J. Cancer* 11(7):473-478; 1975.

Long-term studies on the tumor-inducing capabilities of *n*-butylhydrazine hydrochloride (BH) and *n*-propylhydrazine hydrochloride (PH) were undertaken. Solutions of 0.0125% BH and 0.025% PH were given separately and continuously in the drinking water of 6-wk-old, randomly-bred Swiss albino mice for the remainder of their lives. The lung tumor incidence was 53% in BH-treated and 49% in PH-treated animals, while the corresponding incidence in the untreated controls was 22%. Histopathological examinations of lung tumors revealed the characteristic appearance of adenomas and adenocarcinomas. The treatments had no apparent effect on the development of other types of neoplasms. The study demonstrates for the first time the tumorigenicity of

these chemicals. Because of widespread use of sub-
stituted hydrazines in the environment, their
possible significance is discussed.

4295   *IN VIVO* CHROMOSOME-DAMAGING EFFECT OF
        CYCLOHEXYLAMINE IN THE CHINESE HAMSTER.
(Eng.) van Went-de Vries, G. F. (Lab. Pharmacology
and Toxicology, Natl. Inst. Public Health, P. O.
Box 1, Bilthoven, The Netherlands); Freudenthal, J.;
Hogendoorn, A. M.; Kragten, M. C. T.; Gramberg, L.
G. *Food Cosmet. Toxicol.* 13(4):415-418; 1975.

The chromosome-breaking activity of cyclohexylamine
(CHA) was tested by oral administration of the
chemical in 20 Chinese hamsters. The hamsters were
treated po with CHA of high purity given in three
daily doses of 200 mg/kg/day. The purity of the CHA
was checked by mass spectrometry and in view of the
instability of this compound, it was found advisable
to handle it at a low pH (2.2-2.4) and in a nitrogen
atmosphere. Chromosome analysis was carried out on
phytohemagglutinin-stimulated lymphocytes of the
peripheral blood before and after the administration.
A significant increase (P < 0.0005) in the number of
structural aberrations of the chromosomes was found
after the treatment. A ring chromosome, several
exchange figures and numerous fragments and breaks
were registered. Because the lymphocyte cultures
necessitate removal of a large volume of blood from
the hamsters, the influence of hematopoiesis on the
number of chromosome aberrations is questioned.

4296   PATHOLOGIC ANALYSIS OF CHEMICAL NEPH-
        RITIS IN RATS INDUCED BY *N*-(3,5-
DICHLOROPHENYL)SUCCINIMIDE. (Eng.) Sugihara, S.
(Nagoya City Univ. Medical Sch., Mizuho-ku, Nagoya
467, Japan); Shinohara, Y.; Miyata, Y.; Inoue, K.;
Ito*, N. *Lab. Invest.* 33(3):219-230; 1975.

The effects of N-(3,5-dichlorophenyl)succinimide
(NDPS) administration to 78 male Wistar rats were
studied by blood chemical, histopathologic, histochemical, and ultrastructural examinations. NDPS
was given to 72 rats in drinking water for 4, 8,
12, 16, 20, and 24 wk; the remaining rats served
as untreated controls. Administration of 5,000 ppm
NDPS induced severe damage of the kidneys, but did
not affect other organs. Administration of 2,500
ppm NDPS had little effect, even on the kidneys.
Chemical analysis of the blood showed that the urea
nitrogen level increased with the increase in the
period of NDPS administration. Histopathologic
examination showed intensive cell infiltration,
mainly of lymphocytes, into the renal interstitium
after NDPS administration for only four weeks.
After NDPS administration for 12 wk or more, typical
interstitial nephritis was observed. Histochemical
analysis showed that after only one week various
enzymatic activities in the proximal convoluted
tubules were markedly decreased whereas their activities in the distal convoluted tubules were relatively well preserved for at least four weeks.
These results indicate that NDPS predominantly at-
tacked the proximal convoluted tubules. After
four weeks, the most prominent ultrastructural
changes were seen in the proximal convoluted tubules

plated at an inoculum such that 1×10^4 cells/5-cm
dish survived for the transformation assay and 30-
100 cells/dish survived for the assay of plating
efficiency. Caffeine was present in the culture
medium for the first 48 hr after plating. The medium was then replaced with fresh, caffeine-free
medium, and was changed twice a week. Dishes were
fixed with methanol and stained with Giemsa either
10 or 30 days after plating and scored for plating
efficiency and transformation frequency, respectively. In the first 48 hr after treatment, caffeine
reduced the survival of treated cells but not that
of untreated cells. Transformation frequency per
surviving cell was also reduced by caffeine. Both
survival and transformation frequency of treated
cells were inversely proportional to the concentration of caffeine. The inhibitory effect of caffeine
on transformation frequency was also dependent on
the concentration of 4-nitroquinoline-1-oxide. The
exposure of cells to caffeine either before or 48
hr after treatment with 4-nitroquinoline-1-oxide
had no marked effect on survival or on transformation
frequency. It is suggested that the reduction of
survival and transformation frequency caused by caffeine in 4-nitroquinoline-1-oxide-treated A31-714
cells is related to the observation that caffeine
inhibits postreplication repair in mouse cells. The
results support the somatic mutation theory of cell
transformation.

4301   PANCREATIC ISLET-CELL AND OTHER TUMORS IN
        RATS GIVEN HELIOTRINE, A MONOESTER PYRROLI-
ZIDINE ALKALOID, AND NICOTINAMIDE. (Eng.) Schoental,
R. (Royal Veterinary Coll., Royal Coll. St., London
NW1 England); *Cancer Res.* 35(8):2020-2024; 1975.

The long term effects of the monoester pyrrolizidine
alkaloid, heliotrine were studied in 34 weanling Por-
ton-Wistar rats pretreated with nicotinamide. Three
of six male rats, surviving 22-27.5 mo after one or
two intragastric doses of heliotrine (230 mg/kg body
wt), and nicotinamide (350 mg/kg body wt, developed
pancreatic islet-cell tumors, accompanied in one of
the rats by transitory cell papillomas of the urinary
bladder and interstitial testicular tumors and in another by a hepatoma. The lesions in the livers showed progression from megalocytosis, to microscopic hepatocellular hyperplasia, to increasingly larger nodules and hepatoma. One rat, given heliotrine but no
nicotinamide, also developed adenoma of the pancreatic islet cells. Heliotrine, in which the crucial
double bond in the pyrrolizidine moiety is sterically
hindered, appears to be less readily sequestered by
the liver and also to affect other organs. Alkylation of nicotinamide at the N-1 position prevents its
being reused for coenzyme biosynthesis. Hence, pretreatment of rats with large doses of nicotinamide
prevents the depletion of nicotinamide adenine di-
nucleotide coenzymes and liver necrosis in rats given heliotrine (230 mg/kg).

4302   DEHYDRORETRONECINE-INDUCED RHABDOMYOSAR-
        COMAS IN RATS. (Eng.) Allen, J. R. (Univ.
Wisconsin Medical Sch., Madison, Wis. 53706); Hsu,
I. C.; Carstens, L. A. *Cancer Res.* 35(4):997-1002;
1975.

Two groups of 75 male Sprague-Dawley rats were given sc injections of either monocrotaline (5 mg/kg) or its major detectable metabolite, dehydroretronecine (20 and 10 mg/kg), biweekly for one yr. Tissues obtained from partial hepatectomies performed at four mo on a portion of these animals showed that both compounds caused a decided inhibition of mitotic division in regnerating liver. Rhabdomyosarcomas developed at the site of dehydroretronecine injection in 51.6% of the rats and in 3.3% of the monocrotaline-treated rats. Metastatic lesions were recorded in 8.3% of these animals. In addition, 10% of the monocrotaline-treated rats developed other tumors that included myelogenous leukemias, hepatocellular carcinomas, and pulmonary adenomas. These data indicate that either monocrotaline or its metabolite dehydroretronecine are capable of causing neoplastic transformation in the tissues of experimental animals.

4303   BIOCHEMICAL CHANGES IN GR MOUSE MAMMARY
        TISSUE DURING HORMONAL TUMOR INDUCTION.
(Eng.) Daehnfeldt, J. L. (Fibiger Laboratory, Ndr. Frihavnsgade 70, 2100 Kobenhavn, Denmark); Schulein, M.; Briand, P. *Eur. J. Cancer* 11(7):509-515; 1975.

Changes in glucose-6-phosphate dehydrogenase (G-6-PDH) and lactose synthetase (LS) activity, and in hormone receptor capacity, occurring during hormonal induction of GR mouse mammary tumors were studied; these parameters were compared in tumorous and nontumorous tissue from the same animals. Treatment of spayed 10-12 wk-old virgin GR mice with estrone (0.5 µg/ml in drinking water) and progesterone (5-10 mg sc pellets weekly) led to mammary tumor development after about 12 wk. Pronounced tissue changes, however, were observed during the first three weeks after the start of hormonal treatment. The total weight of the glands more than doubled during this period, while the protein/DNA ratio decreased. G-6-PDH activity rose by 25% during the first two weeks of hormonal treatment calculated on a protein basis, but fell again to the initial value during the remainder of the induction period. LS A-protein activity and B-protein content rose to a steadily high level after two weeks of treatment. The non-blocked cytoplasmic estradiol binding capacity decreased after hormonal treatment of spayed virgins. In the induced tumors, the RNA/DNA ratio was unchanged, and the protein/DNA ratio decreased by 25% compared to nontumorous mammary tissue from the same mice. The activity of G-6-PDH increased by 50% as calculated on a protein basis. The increased LS A-protein activity remained unchanged, while LS B-protein disappeared. There was ten times as much non-blocked cytoplasmic estradiol receptor protein as in the nontumorous tissue from the same animals. These experiments show that high estradiol receptor content and undetectable levels of LS B-protein are characteristic changes in the horomone-induced GR mouse mammary tumors.

4304   COVALENT BINDING OF *TRANS*-STILBENE TO
        RAT LIVER MICROSOMES. (Eng.) Docks,
E. L. (U.S. Borax Res., Anaheim, Calif.); Krishna, G. *Biochem. Pharmacol.* 24(21):1965-1969; 1975.

sulated and there was marked increase in vascularity.
A 29-yr-old woman who had been taking BCP for six
years, noted a mobile, nontender mass in her abdomen.
At exploration, a large tumor mass was found in the
left lobe of the liver and a partial left hepatectomy
was performed. Microscopically, the hepatocytes were
well differentiated and no increase in bile ducts
or blood vessels was noted. A 29-yr-old woman who
had been on BCP for six years presented with sudden
onset of right upper quadrant pain, fever and pre-
cipitous anemia. A large hematoma was found in the
right lobe of the liver. This syndrome must be considered in young women on BCP who present with abdominal pain as this may represent massive intrahepatic or intraperitoneal hemorrhage. Diagnosis is
established by palpation of a mass, an abnormal liver
scan, angiography or peritoneoscopy. Large tumors
must be resected and continued use of BCP is contra-
indicated.

4308   LIVER TUMORS IN WOMEN ON CONTRACEPTIVE
        STEROIDS. (Eng.) Christopherson, W. M.
(Health Sciences Center, Louisville, Ky. 40201);
Mays, E. T.; Barrows, G. H. *Obstet. Gynecol.* 46(2):
221-223; 1975.

Of 13 cases of liver tumors in young women who had
taken oral contraceptives for 1-8 yr, eight were
focal nodular hyperplasia. One of the patients died
of hemorrhage from a ruptured benign hepatoma during
pregnancy. One case was a liver cell carcinoma which
also bled. Physicians involved in the care of persons taking oral contraceptives should be aware of
this possible complication, which usually presents
with evidence of acute abdomen.

4309   SPUTUM CYTOLOGY AMONG FREQUENT USERS OF
        PRESSURIZED SPRAY CANS. (Eng.) Good, W. ·
O. (Montrose Mem. Hosp., Colo.): Ellison, C.; Archer,
V. E. *Cancer Res.* 35(2):316-321; 1975.

Sputum cultures were taken among frequent users of
aerosol spray cans and among other persons to determine if aerosol preparations alter the flora of
the bronchial tree or contain carcinogenic agents.
Sputum samples were taken from 50 white adult patients, "frequent users", from 200 white adult patients in the same geographical area, and from 50
white nonpatients from another geographical area.
An attempt was made to match each of the individuals
in the "frequent users" group with a similar person
from the other two groups on the basis of age, sex
and cigarette-smoking habits. All three groups of
samples were collected in 50% ethanol.solution containing 2% polyethylene glycol 1540, sent to the
same cytology laboratory, and treated identically.
Each sputum sample was classified on the basis of
the most atypical cell seen. Atypical metaplastic
changes in exfoliated cells were compared between
the groups. Combined moderate and mild atypical
sputum classifications were significantly more frequent among the "frequent users" than among the non-
patient controls($p < 0.005$) and among the "frequent
users" than among the patient controls ($p < 0.10$).
The absence of any normal sputum samples among the
"frequent users" is quite striking and suggests that

some agents used in some pressurized aerosol cans
may be carcinogenic and that further exploration
be done on aerosol ingredients. Until such studies
exonerate pressurized aerosol products, they should
be used in ways which minimize aerosol inhalation.

4310   MEASUREMENTS OF VINYL CHLORIDE FROM
          AEROSOL SPRAYS. (Eng.) Gay, B. W., Jr.
(Environmental Protection Agency, Natl. Environmental
Res. Center, Res. Triangle Park, N.C. 27711);
Lonneman, W. A.; Bridbord, K.; Moran, J. B. *Ann.
N.Y. Acad. Sci.* 246:286-295; 1975.

Human exposure levels and the decay of vinyl chloride in home and office environments were studied.
Aerosol hair and insect sprays known to contain
vinyl chloride were released under conditions simu-
lating actual use conditions in typical home and
office environments. Cryogenic trapping concentration procedures were used to collect samples and
introduce them into a gas chromatographic system for
analysis. Detection of vinyl chloride was by flame
ionization. In all experiments, high concentrations
of vinyl chloride were detected when the vinyl
chloride-containing aerosol sprays were used. A 30-
sec release of hairspray in a 56,000-1 room produced
vinyl chloride in concentrations of 13.10-13.70 ppm
two minutes after spray; the concentrations were
0.06-0.10 ppm 360 min after spray. A 60-sec release
of hairspray in a 29,300-1 room produced concentrations of 122.7-124.4 ppm during the spray, with
0.12-o.13 ppm remaining 120 min later. A 30-sec
release of insect spray in a 21,400-1 room produced
a vinyl chloride concentration of 380.1-383.6 ppm
one minute after spray, with a concentration of
0.65-0.83 ppm remaining in an adjacent hall 151 min
later. A 120-sec release of insect spray in a
133,000-1 room produced an initial concentration of
41.64-41.9 ppm vinyl chloride, with concentration at
120 min of 0.01-0.012 ppm. In all cases, the user
was exposed to high peak concentrations, even with
good ventilation in the room. When the aerosol spray
was released in smaller-sized rooms, higher concentrations of vinyl chloride persisted for some time,
the decrease in concentration with time apparently
being a dilution effect of room ventilation. Based
on these data, predicted past exposures to vinyl
chloride resulting from the use of household products
were determined; these exposures were substantially
above current proposed occupational standards for this
substance, especially among women. On this basis,
the frequency of angiosarcoma of the liver among the
general population may be increased in coming years.

4311   THE VALUE OF PREDICTIVE EXPERIMENTAL
          BIOASSAYS IN OCCUPATIONAL AND ENVIRON-
MENTAL CARCINOGENESES. AN EXAMPLE: VINYL CHLORIDE.
(Eng.) Maltoni, C. (Istituto di Oncologia "Felice
Addarii", Ospedali di Bologna, Ente Ospedaliero
Regionale. Viale Ercolani 4/2, 401 38 Bologna,
Italy). *Ambio* 4(1):18-23; 1975.

The first completed experiment of a series on the
effects of vinyl chloride (VC) indicates that inhalation of VC by Sprague Dawley rats produced: Zymbal
gland carcinomas, nephroblastomas, angiosarcomas

4319   INDUCTION OF CHROMOSOME DAMAGE IN FIBRO-
        BLASTS FROM GENETIC INSTABILITY SYNDROMES
[abstract]. (Eng.) Wolman, S. R. (New York Univ.
Sch. Medicine, New York, N.Y. 10016); Auerbach,
A. D. *Proc. Am. Assoc. Cancer Res.* 16:69; 1975.

4320   ALTERED REGULATION OF α-AMINOLEVULINIC ACID
        SYNTHETASE (ALAS) IN ETHIONINE CARCINO-
GENESIS [abstract]. (Eng.) Woods, J. S. (Natl. Inst.
Environ. Health Sci., Res. Triangle Park, N.C. 27709).
Proc. Am. Assoc. Cancer Res. 16:40; 1975.

4321   THE REACTION OF ^{14}C-LABELLED BIS-(CHLORO-
        METHYL) ETHER WITH DNA [abstract]. (Eng.)
Goldschmidt, B. M. (New York Univ. Medical Center,
New York, N.Y. 10016); Van Duuren, B. L.; Frenkel,
K. *Proc. Am. Assoc. Cancer Res.* 16:66; 1975.

4322   A SYNERGISTIC EFFECT OF ULTRAVIOLET LIGHT
        AND ETHIDIUM BROMIDE ON PETITE INDUCTION IN
YEAST. (Eng.) Hixon, S. C. (Fondation Curie, Inst.
du Radium, Batiment 110, F-91405 Orsay, France);
Yielding, K. L. *Mutat. Res.* 29(1):159-163; 1975.

4323   INHIBITION OF DNA DEGRADATION BY BOUND 2-
        ACETAMIDOFLUORENE [abstract]. (Eng.)
Scribner, J. D. (Fred Hutchinson Cancer Res. Center,
Seattle, Wash. 98104); Naimy, N. K. *Proc. Am. Assoc.
Cancer Res.* 16:65; 1975.

4324   ENZYMES INVOLVED IN MUTAGEN FORMATION FROM
        2-ACETYLAMINOFLUORENE [abstract]. (Eng.)
Shaw, C. R. (M.D. Anderson Hosp. and Tumor Inst.,
Houston, Tex. 77025); Stout, D. L.; Baptist, J. N.
Proc. Am. Assoc. Cancer Res. 16:199; 1975.

4325   ALTERATION OF DEOXYRIBONUCLEIC ACID TRANS-
        FORMING ACTIVITY IN *BACILLUS SUBTILIS* BY
N(OH)-AMINOACETYLFLUORENE [abstract]. (Eng.) Felkner,
I. C. (Univ. Texas System Cancer Center and Health
Science Center, Houston, Tex. 77025); Matney, T. S.;
Shaw, C. R.; Hickey, R. C. *Proc. Am. Assoc. Cancer
Res.* 16:200; 1975.

4326   CARCINOGENESIS BY LOCAL APPLICATION OF
        FLUORENYLHYDROXAMIC ACIDS TO THE MAMMARY
GLAND OF THE RAT [abstract]. (Eng.) Malejka-Giganti,
D. (VA Hosp., Minneapolis, Minn. 55417); Gutmann,
H. R.; Rydell, R. E. *Proc. Am. Assoc. Cancer Res.*
16:62; 1975.

4327   CATECHOLAMINES IN URINE AND TUMOR TISSUE IN
        CASES OF UNI- AND BILATERAL PHEOCHROMO-
CYTOMA. (Ger.) Poch, G. (Institut fur experimentelle
und klinische Pharmakologie, Graz, Austria); Kukovetz,
W. R.; Becker, H.; Cesnik, H. *Med. Klin.* 70(5):190-
196; 1975.

4328   TUMORIGENICITY OF *n* PROPYL-, *n* AMYL- AND
        n ALLYL-HYDRAZINES. TOXICITY OF AGARITINE
[abstract]. (Eng.) Toth, B. (Univ. Nebraska Medical
Center, Omaha, Nebr. 68105); Nagel, D.; Shimizu, H.;
Sornson, H.; Issenberg, P.; Erickson, J. *Proc. Am.
Assoc. Cancer Res.* 16:61; 1975.

4329   ALTERATIONS IN HEPATIC POLYAMINE LEVELS IN
        RATS FOLLOWING HYDRAZINE TREATMENT [ab-
stract]. (Eng.) Banks, W. L., Jr. (Medical Coll.
Virginia, Virginia Commonwealth Univ., Richmond,
Va. 23298); Hubbard, V. S. *Proc. Am. Assoc.
Cancer Res.* 16:110; 1975.

4330   GROWTH RATE OF 1,2 DIMETHYL-HYDRAZINE
        INDUCED COLON ADENOCARCINOMA IN RAT [ab-
stract]. (Eng.) Maskens, A. P. (Univ. Louvain,
Louvain Sch. Medicine, Louvain, Belgium); Meersseman,
F. P.; Rahier, J. *Proc. Am. Assoc. Cancer Res.*
16:17; 1975.

4331   CHROMOSOME ANALYSES AFTER PROPHYLACTIC
        AND THERAPEUTIC APPLICATION OF INH IN MAN.
[abstract]. Obe, G. (Genetisches Inst. der FU, Berlin,
West Germany); Beek, B.; Gebhart, E.; Fonatsch, C.;
Bauchinger, M.; Schmid, E. *Mutat. Res.* 29(2):257-
258; 1975.

4332   CHROMOSOME ANALYSIS IN SPERMATOGONIA OF
        MAMMALS [abstract]. (Eng.) Miltenburger,
H. G. (Zool. Inst., Tech. Univ., Darmstadt, West
Germany); Korte, A.; Muller, D.; Rathenberg, R.
Mutat. Res. 29(2):255-256; 1975.

4333   ARYL AND HETEROCYCLIC DIAZO COMPOUNDS AS
        POTENTIAL ENVIRONMENTAL CARCINOGENS [ab-
stract]. (Eng.) Lower, G. M., Jr. (Univ. Wisconsin
Medical Sch., Madison, Wis. 53706); Bryan, G. T.
Proc. Am. Assoc. Cancer Res. 16:42; 1975.

4334   THE EFFECT OF UNSATURATED FATTY ACIDS ON
        CARCINOGENIC ACTIVITY OF OVERHEATED SUN-
FLOWER OIL. (Rus.) Vysheslavova, M. Ia. (Inst.
Nutr. U.S.S.R. Acad. Med. Sci., Moscow); Kozlova, I.
N. *Vopr. Onkol.* 21(4):40-44; 1975.

4335   A NEW CARCINOGEN IS PRESENT IN BEEF AND OTHER
        MEATS [abstract]. (Eng.) Shamberger, R. J.
(Cleveland Clinic Foundation, Cleveland, Ohio 44106);
Willis, C. E. *Proc. Am. Assoc. Cancer Res.* 16:68;
1975.

4336   COMPUTERIZED GAS CHROMATOGRAPHIC-MASS
        SPECTROMETRIC ANALYSIS OF POLYCYCLIC AROMATIC
HYDROCARBONS IN ENVIRONMENTAL SAMPLES. (Eng.) Lao,
R. C. (Chemistry Div., Air Pollution Control Direc-
torate, Dept. of the Environment, Ottawa, Canada);
Thomas, R. S.; Monkman, J. L. *J. Chromatogr.* 112:
681-700; 1975.

Japan); Teratani, M.; Takahashi, A.; Uchino, H. *Proc. Am. Assoc. Cancer Res.* 16:203; 1975.

4355   EARLY NEOPLASTIC EVENTS IN TRANSPLACENTAL
         CARCINOGENESIS [abstract]. (Eng.) Swen-
berg, J. A. (Ohio State Univ., Columbus, Ohio 43210);
Clendenon, N.; Denlinger, R.; Gordon, W. *Proc. Am.
Assoc. Cancer Res.* 16:63; 1975.

h,

4356   CARCINOGENIC ACTION OF QUINOXALINE 1,4-
         DIOXIDE IN RATS. (Eng.) Tucker, M. J.
(Pharm. Div., Imp. Chem. Ind. Ltd., Macclesfield,
England). *J. Natl. Cancer Inst.* 55(1):137-146; 1975.

4357   INDUCTION OF ORNITHINE DECARBOXYLASE AND
         S-ADENOSYL-L-METHIONINE DECARBOXYLASE IN
MOUSE EPIDERMIS BY TUMOR PROMOTERS [abstract].
(Eng.) O'Brien, T. G. (McArdle Lab., Univ. Wis-
consin, Madison, Wis. 53706); Boutwell, R. K.
Proc. Am. Assoc. Cancer Res. 16:62; 1975.

4358   STIMULATED DNA SYNTHESIS IN MOUSE EPIDERMAL
         CELL CULTURES TREATED WITH 12-O-TETRA-
DECANOYL PHORBOL-13-ACETATE [abstract]. (Eng.)
Yuspa, S. H. (Natl. Cancer Inst., Bethesda, Md.
20014); Hennings, H.; Ben, T. *Proc. Am. Assoc.
Cancer Res.* 16:69; 1975.

4359   MUTAGENIC EFFECT OF THIO-TEPA ON LABORATORY
         MICE. IV. EFFECT OF THE GENOTYPE AND SEX
ON THE RATE OF CHROMOSOME ABERRATIONS IN BONE MARROW
CELLS. (Rus.) Surkova, N. I. (Res. Lab. of Experi-
mental-Biological Models, Acad. Medical Sciences of
U.S.S.R., Moscow Region, U.S.S.R.); Malashenko, A. M.
Genetika 11(1):61-72; 1975.

4360   MUTAGENICITY OF VINYL CHLORIDE AND DERIVA-
         TIVES IN *S. TYPHIMURIUM* STRAINS [abstract].
(Eng.) Bartsch, H. (International Agency for Re-
search on Cancer, 69008 Lyon, France); Malaveille, C.;
Montesano, R.; Croisy, A.; Jacquignon, P. *Proc.
Am. Assoc. Cancer Res.* 16:22; 1975.

4361   PROPERTIES OF ACIDIC SACCHARIDES PRODUCED
         BY B16 MELANOMA CELLS TREATED WITH 1-
METHYL-3-ISOBUTYLXANTHINE. (Eng.) Banks, J. (Mil-
ton S. Hershey Medical Center, Pennsylvania State
Univ., Hershey, Pa. 17033); Kreider, J. W.; Satob,
C.; Davidson*, E. A. *Cancer Res.* 35(9):2383-2389;
1975.

See also:

* (Rev):   4201, 4204, 4203, 4204, 4205, 4206
            4230, 4231, 4232, 4233, 4234, 4236,
            4237, 4262
* (Phys): 4363, 4367, 4368, 4370
* (Immun): 4435, 4437, 4493, 4495, 4501, 4512,
            4532, 4533, 4544, 4546
* (Path): 4556, 4567, 4572, 4599, 4607, 4624
* (Epid-Biom): 4657, 4663, 4664, 4665, 4667,
               4673, 4676, 4677, 4678

4362   CELLULAR INJURY AND CELL PROLIFERATION
        IN SKIN CARCINOGENESIS BY UV LIGHT.
(Eng.) Stenback, F. (Univ. Nebraska Medical Center, Omaha, Nebr. 68105). *Oncology* 31(2):61-75;
1975.

The effects on mouse skin of varying doses of light
from a high-voltage mercury lamp were compared to
the alterations occurring after repeated light applications from a carbon-arc xenon sunlamp. Skin
samples were taken immediately before, immediately
after, and at intervals ranging from 30 min to 30
days after irradiation. A single dose of UV light
at a carcinogenic wavelength (5.5×10^7 erg/cm^2)
caused necrosis and ulceration, followed by scar-
ring, but failed to produce tumors. In the case
of three small doses (1.5×10^8 erg/cm^2), the neoplastic response was predominantly epithelial, resulting in formation of papillomas and squamous
cell carcinomas. Increasing the dose (2.75×10^8
erg/cm^2, five doses; or 3.9×10^9 erg/cm^2, 20 doses)
caused formation of fibromas and fibrosarcomas of
the ear. However, repeated application of UV light
having a spectrum similar to sunlight failed to induce tumors. In these studies, only animals showing
initially destructive lesions ultimately presented
tumor formation, while the hyperplasia induced by
UV irradiation showed no relationship to neoplastic
transformation. The results emphasize the significance of dose and wavelength in evaluating effects
of UV-light skin carcinogenesis.

4363   ULTRAVIOLET LIGHT IRRADIATION AS INITIATING
        AGENT IN SKIN TUMOR FORMATION BY THE TWO-
STAGE METHOD. (Eng.) Stenback, F. (Dept. Pathol.,
Univ. Oulu, Finland). *Eur. J. Cancer* 11(4):241-246;
1975.

The biological and morphological characteristics of
neoplastic progression in Swiss mouse skin induced by
the two-stage method, using UV light as an initiator
and croton oil as a promoter, were studied by biological and histological methods. UV light (5.5×10^7
erg/cm^2) proved to be an effective initiator in skin
tumor formation, when followed by repeated applications of croton oil. In comparison with experiments
using 7,12-dimethylbenz(a)anthracene (50 µg) as an
initiator, the average latent period was longer and
the number of tumor-bearing animals and of tumors was
lower. Morphologically, an orderly epidermal hyperplasia, progressing through papillomatous hyperplasia
and ending in extroverted squamous cell papillomas,
occurred in only a few animals. The proliferation of
epithelial cells induced by multiple UV irradiation
was disorderly with distinct cytological and histological abnormalities. A few squamous cell carcinomas
developed from the borders of ulcers induced by repeated UV light applications. Efforts to increase
the neoplastic response, by giving either UV light
ten times prior to promotion or croton oil immediately
after each of 16 UV irradiations, did not increase
the total number of tumors but did induce some malignant ones. These studies emphasize the specificity
of initiation and promotion in skin tumor formation
and the different actions of physical and chemical
carcinogens on animal skin, depending upon the com-

bination of progression and persistence of destructi-
and proliferative actions specific to each agent.

4364   THE RISK OF TUMOR INDUCTION IN MAN FOL-
        LOWING MEDICAL IRRADIATION FOR MALIGNANT
NEOPLASM. (Eng.) Seydel, H. G. (American Oncologic
Hosp. of the Fox Chase Cancer Center, Central and
Shelmire Aves., Philadelphia, Pa. 19111). *Cancer*
35(6):1641-1645;]975.

The possible carcinogenic effects of therapeutic
irradiation for malignancy were assessed. A review
of second malignant tumors (SMT) from the joint
tumor regristry of the Fox Chase center was carried
out. Cases entered between 1929 and 1973 for primary sites in the cervix, oral cavity, and oropharynx were analyzed. The number of five-year
survivors was determined and the incidence of SMT
was expressed as a percentage. Three patients developed adenocarcinoma of the endometrium in a
uterus which had been previously irradiated for
squamous cell carcinoma of the cervix. The average
dose was 8100 rads. There were 354 five-year survivors among the 933 patients treated for cancer.
SMT in the form of adenocarcinoma of the endometrium
occurred in 0.85% of the five-year survivors. Nine
SMT of the oropharynx and the oral cavity occurred
in the irradiated mucosa of 611 five-year survivors
among 1464 patients treated by radiotherapy. The
average dose was 6800 rads. SMT occurred in 1.14%
of the five-year survivors. Review of the tumor
registry revealed 407 patients who were treated surgically for cancers of the oral cavity or oropharynx
SMT in the upper respiratory and digestive tract
occurred in 8 (2%). SMT of this region following
radiotherapeutic treatment, including lesions outside the irradiated area, was 23 among 1464 patients
(1.6%).

4365   FUNCTION OF THE UVR MARKER IN DARK RE-
        PAIR OF DNA MOLECULES. (Eng.)
Sedliakova, M. (Cancer Res. Inst., Slovak Acad.
Sciences, 880 32 Bratislava, Czechoslovakia);
Brozmanova, J.; Slezarikova, V.; Masek, F.; Fandlova
E. *Neoplasma* 22(4):361-384; 1975.

The function of the *uvr* marker in dark repair of
DNA molecules was investigated in cultures of
Escherichia coli K12 and *E. coli* 15T⁻, both strains
of the *uvr⁺ rec⁺* genotype, and in an *E. coli* K12
rec⁺ uvr⁻ culture. Cell suspensions were irradiated
(254 nm radiation, 6.3 ergs/mm/sec) in a 1-2 mm
thick layer with continuous stirring. DNA was
labeled with [³H]thymidine for various periods after
irradiation. In some experiments, the cultures were
pretreated to produce amino acid and thymine starvation, amino acid starvation only, or thymine
starvation only. Thymine dimers were determined
radiochromatographically in the fraction precipitated in trichloroacetic acid. Molecular weight of
NDA was estimated from alkaline sucrose gradient
sedimentation profiles. Combined amino acid and
thymine prestarvation resulted in depression of
dimer excision in both *E. coli* strains of the *uvr⁺*
genotype. Depression of excision, however, was

covery from division delay in a way that resembled
the recovery from sublethal or potentially lethal
radiation damage. This provides further evidence
that the respective radiosensitive targets are
distinct.

4367   EXPERIMENTAL MODIFICATION OF PHOTOCARCINO-
        GENESIS. II. FLUORESCENT WHITENING AGENTS
AND SIMULATED SOLAR UVR. (Eng.) Forbes, P. D.
(Temple Univ. Health Sci. Cent., Philadelphia, Pa.);
Urbach, F. *Food Cosmet. Toxicol.* 13(3):339-342;
1975.

4368   EXPERIMENTAL MODIFICATION OF PHOTOCARCINO-
        GENESIS. III. SIMULATION OF EXPOSURE TO
SUNLIGHT AND FLUORESCENT WHITENING AGENTS. (Eng.)
Forbes, P. D. (Temple Univ. Health Sci. Cent.,
Philadelphia, Pa.); Urbach, F. *Food Cosmet. Toxicol.*
13(3):343-345; 1975.

4369   POTENTIAL GENETIC EFFECTS OF PHOTOTHERAPY
        FOR NEONATAL JAUNDICE [abstract]. (Eng.)
Speck, W. T. (Dept. Pediatr., Columbia Univ., New
York, N.Y.); Rosenkranz, H. S. *Proc. Am. Assoc.
Cancer Res.* 16:16; 1975.

4370   PHOTOTOXICITY AND PHOTOCARCINOGENESIS:
        COMPARATIVE EFFECTS OF ANTHRACENE AND 8-
METHOXYPSORALEN IN THE SKIN OF MICE [abstract].
(Eng.) Forbes, P. D. (Temple Univ. Health Sci.
Cent., Philadelphia, Pa.); Davies, R. E.; Urbach,
F. *Proc. Am. Assoc. Cancer Res.* 16:120; 1975.

4371   MULTIPLE FLUOROSCOPY OF THE CHEST:
        CARCINOGENICITY FOR THE FEMALE BREAST AND
IMPLICATIONS FOR BREAST CANCER SCREENING PROGRAMS.
(Eng.) Delarue, N. C. (Suite 530, 170 St. George St.,
Toronto, Ont. M5R 2M8, Canada); Gale, G.; Ronald,
A. *Can. Med. Assoc. J.* 112(12):1405-1412; 1975.

4372   CANCER OCCURRENCE AFTER REPEATED FLUORO-
        SCOPY DURING PNEUMOTHORAX THERAPY [ab-
stract]. (Eng.) Uzman, B. G. (Veterans Administration Hosp., Shreveport, La.). *IRCS Med. Sci.* 3(10/
Suppl.):23; 1975.

See also:

* (Rev):   4207, 4208, 4209, 4235, 4237, 4260
* (Chem):   4322, 4339
* (Immun): 4438, 4439, 4511
* (Path):   4610, 4635
* (Epid-Biom): 4653, 4662, 4675, 4680

4373   GENERATION OF AVIAN MYELOBLASTOSIS VIRUS
        STRUCTURAL PROTEINS BY PROTEOLYTIC CLEAV-
AGE OF A PRECURSOR POLYPEPTIDE. (Eng.) Vogt, V. M.
(Inst. General Microbiology, Univ. Bern, Altenber-
grain 21, 3013 Bern, Switzerland); Eisenman, R ,;
Diggelmann, H. *J. Mol. Biol.* 96(3):471-493; 1975.

A cleavage pathway for the generation of the structural proteins of avian myeloblastosis virus (AMV)
was studied. AMV-infected primary chick fibroblasts
were pulse-labeled with [^{35}S]methionine, incubated,
and then lysed with cold lysis buffer. Cytoplasmic
fractions were prepared by centrifugation of the lysates; to the supernatant from each plate of cells,
2-5 μg of AMV and 0.1 ml of rabbit anti-AMV serum
were added. The immune precipitate was collected by
centrifugation and was analyzed by sodium dodecyl
sulfate-polyacrylamide gel electrophoresis and agarose
gel filtration. Gel electrophoresis of the immune
precipitates of the labeled cells revealed five metabolically unstable radioactive proteins (pr12,
pr32, pr60, pr66 and pr76). Double-labeled ion-
exchange chromatography of the tryptic digests of
these polypeptides demonstrated that the largest precursor, pr76, contained the amino acid sequences of
all four of the internal structural proteins that
bear the group-specific antigenic determinants of the
virus (p28, p17, p10, and p14). Pr76 appeared not
to contain the sequence of the fifth and smallest internal virion protein (p7). The four smaller precursors were intermediate cleavage products of pr76.
The arrangement of the virion proteins in pr76 was
determined by labeling the cells shortly after inhibiting polypeptide chain initiation. The relative
amounts of radioactivity in both the completed virion
proteins and the tryptic peptides of pr76 implied
the same order for 3 of the 4 proteins, although the
exact position of one protein remains uncertain. The
cleavage of percursors proceeded in crude extracts of
AMV-infected cells. Thus, proteolysis was resistant
to several protease inhibitors, but was completely
blocked by the addition of agents that disrupt membranes. It is suggested that the processing of pr76
begins by removal of a C-terminal fragment, which in
turn is trimmed to yield virion p14. The larger polypeptide remaining after the first cleavage ultimately
forms virion proteins p28, p17, and p10. It is likely
that the precursors are attached to and cleaved on the
cell membrane.

4374   SYNTHESIS AND FUNCTION OF AVIAN SARCOMA
        VIRUS-SPECIFIC DNA IN PERMISSIVE AND NON-
PERMISSIVE HOST CELLS. *(Eng.)* Varmus, H. E.
(Dept. Microbiology, Univ. California, San Francisco,
Calif.); Guntaka, R. V.; Deng, C. T.; Domenik, C.;
Bishop, J. M. *Proc. Int. Cancer Congr. 11th.* Vol.
2 *(Chemical and Viral Oncogenesis)*. Florence,
Italy, October 20-26, 1974. Edited by Bucalossi,
P.; Veronesi, U.; Cascinelli, N. New York, American
Elsevier, 1975, pp. 272-277.

Recent attempts to describe the mechanism(s) by
which RNA tumor virus genes are synthesized, inserted into the DNA of the host cell, and subsequently transcribed into viral DNA are summarized.
The experiments were performed with several strains
of avian sarcoma virus, principally B77 virus, and

4377   QUANTITATION OF AVIAN RNA TUMOR VIRUS RE-
VERSE TRANSCRIPTASE BY RADIOIMMUNOASSAY.
(Eng.) Panet, A. (Hadassah Medical Sch., Jerusalem,
Israel); Baltimore, D.; Hanafusa, T. *J. Virol.*
16(1):146-152; 1975.

A radioimmunoassay (RIA) that can detect and quantitate 3-5 ng. of the avian RNA tumor virus reverse
transcriptase is described. Avian myeloblastosis
virus (AMV, strain A) reverse transcriptase was purified by chromatography on DEAE-Sephadex and phosphocellulose columns; the preparation consisted of two
subunits, α and β. The purified enzyme was treated
with $Na^{125}I$, and the RIA was carried out using rat
anti-AMW reverse transcriptase serum. DNA polymerase
activity was also assayed. With high antiserum concentrations, 87% of the iodinated reverse transcriptase was precipitated. The assay detected no antigenic sites in Rous sarcoma virus α virions or in
virions of a murine RNA tumor virus (Moloney murine
leukemia virus). About 70 molecules of reverse
transcriptase were found per AMV virion using either
this assay or an assay based on antibody inhibition
of enzymatic activity; the RIA indicated that about
3.5% of the virion proteins were reverse transcriptase. The RIA detected about 270 ng of enzyme per
milligram of cell protein in virus-producing cells;
whereas uninfected cells contained much less antigenic materials, they contained some determinants
able to displace radiolabeled antigen. No additional antigenic determinants on reverse transcriptase could be detected that were not found on the
α subunit. The RIA was 7-fold less sensitive than
the enzymatic assay of DNA polymerase for purified
enzyme and 9-fold less sensitive for disrupted
virions. However, the RIA is useful in that it can
detect antigen using small amounts of protein and
in the presence of inhibitors.

4378   ANIMAL DEPENDENT RNA POLYMERASES: PARTIAL
PURIFICATION AND PROPERTIES OF THREE CLAS-
SES OF RNA POLYMERASES FROM UNINFECTED AND ADENO-
VIRUS-INFECTED HeLa CELLS. (Eng.) Hossenlopp, P.
(Institut de Chimie Biologique, Faculte de Medecine
de Strasbourg, 11 Rue Humann, F-67085 Strasbourg-
Cedex, France); Wells, D.; Chambon, P. *Eur. J. Bio-
chem.* 58(1):237-251; 1975.

The class A, B, and C DNA-dependent RNA polymerases
from uninfected and adenovirus 2-infected HeLa cells
were partially purified and characterized. The enzymes were solubilized, and multiple peaks of enzyme
activity were separated by DEAE-Sephadex chromatography. Aliquots of the DEAE-Sephadex peaks were
then precipitated with ammonium sulfate and subjected to glycerol density gradient ultracentrifugation.
In addition to the class A and B enzyme activities
(insensitive and sensitive to inhibition by 10 mM
α-amanitin, respectively), three peaks of class C
enzyme activity were found that were sensitive to
inhibition by 0.1 mM α-amanitin. Rechromatography
of these class C peaks indicated that they were not
chromatographic artifacts. The class C enzymes differed from the class A and B enzymes not only in
their sensitivity to α-amanitin, but in the immunological properties and their ability to transcribe
native calf thymus DNA at high ionic strength. How-

ever, the ionic strength optimum and the divalent cation requirements of the class C enzymes were markedly dependent on the nature and the amount of template in the reaction. The multiple enzymes isolated from the infected and uninfected cells showed neither qualitative nor quantitative differences. All of the partially purified HeLa cell RNA polymerases were able to transcribe an intact double-stranded adenovirus 2 DNA under conditions where no transcription occurred with purified calf thymus A1 and B RNA polymerases. Since the multiple enzymes were devoid of endonuclease and exonuclease activities, the ability of the partially purified enzymes to transcribe adenovirus 2 DNA could not be ascribed to the initiation of RNA synthesis at nicks of single-stranded regions of the DNA. No differences in transcriptional ability were found between corresponding enzyme classes from normal and infected cells, but a comparison of the ability of the various enzyme classes to transcribe intact viral DNA revealed large differences. Although partially purified HeLa class A and B enzymes were able to initiate on the intact viral DNA only to a limited extent, the class C enzymes transcribed intact duplex DNA more efficiently than any other class of eukaryotic polymerase yet reported.

4379   ISOLATION AND CHARACTERIZATION OF ADENO-
        VIRUS TYPE 12 DNA BINDING PROTEINS. ·(Eng.)
Rosenwirth, B. (Inst. Virology, Univ. Cologne, Cologne, Germany); Shiroki, K.; Levine, A. J.; Shimojo, H. *Virology* 67(1):14-23; 1975.

The isolation of a set of single strand-specific DNA binding proteins from adenovirus type 12-infected African green monkey kidney (AGMK) cells is described, and the binding proteins are characterized. The proteins were labeled with [^{3}H]leucine, [^{14}C]leucine, or [^{35}S]methionine, after which the cell extract was centrifuged and the supernatant containing the DNA binding proteins was applied to a DNA-cellulose column containing single-stranded calf thymus DNA. The binding proteins thus eluted were dialyzed against buffer and added to single- or double-stranded type 12 adenovirus DNA. The DNA binding proteins were also subjected to sodium dodecyl sulfate-gel electrophoresis and autoradiography. Infection of the AGMK cells with adenovirus type 12 resulted in the production of two single-stranded specific DNA binding proteins of molecular weight 60,000 and 48,000, respectively. Both proteins were synthesized in the absence of viral DNA replication and neither protein appeared to correspond to any polypeptide detected in mature adenovirus virions. Temperature-sensitive mutants from three different early, DNA-negative complementation groups (tsA, tsB, and tsC) were also tested for the production of these proteins at the permissive and nonpermissive temperatures. Mutants of the tsB and tsC classes produced both proteins at 32 and 40 C, while the tsA mutants produced both proteins at 32 C but neither protein at 40 C. The results suggest that the adenovirus type 12 tsA gene induces or codes for the DNA binding proteins, and that adenoviruses in general produce a single strand-specific DNA binding protein that is required for viral DNA replication.

4380   MURINE XENOTROPIC TYPE C VIRUSES: I. DIS-
        TRIBUTION AND FURTHER CHARACTERIZATION OF
THE VIRUS IN NZB MICE. (Eng.) Levy, J. A. (Dept. Medicine, Univ. California, San Francisco, Calif. 94143); Kazan, P.; Varnier, O.; Kleiman, H. *J. Virol.* 16(4):844-853; 1975.

The distribution and properties of NZB-MuLV, an X-tropic murine leukemia virus (MuLV) found in all embryonic and adult New Zealand Black (NZB) mouse cells were studied. Cultures of NZB embryo (NZB-ME) and other mammalian cells were checked for the presence of the group-specific (gs), complement-fixing (CF) antigen. Cultures were also examined for the induction of plaques after cocultivation with XC cells. Virus genome rescue was also attempted by cocultivation of HT-1, NRK-Harvey, or MSV-8 cells; with the tissue being assayed fo for NZB-MuLV. The distribution of NZB-MuLV in NZB mice was determined by focus formation assays, and the ability of the NZB-MuLV present in NZB-ME cells to 'help" in focus formation by MSV was also determined. NZB-MuLV was recovered from over 50 adult NZB mice and 15 NZB embryos. Its presence was best detected by measuring its ability to rescue MSV genome from a non-virus-producing MSV-transformed rat cell. The virus was able to serve as a helper for the replication of MSV. In addition, it was found to have a distinct type-specific coat and was a prototype for a third serotype of mouse type C viruses, NZB. The xenotropic virus showed a wide host range, being able to infect even avian cells. It was produced spontaneously by all cells cultivated from NZB tissues and accounted for the high concentration of viral antigens associated with NZB tissues. The extent of virus production was similar in male and female mice, and all cell clones established from embryos also produced it. A variability in the intracellular regulation of virus replication was suggested because tissue cells from the same animal differed quantitatively in their ability to produce xenotropic viruses. The xenotropic virus may have an evolutionary role, and since enhanced spontaneous virus production is associated with cells from NZB mice, the virus may play a role in the autoimmune disease observed in this mouse strain.

4381   MACROMOLECULAR SYNTHESIS IN CELLS IN-
        FECTED WITH FROG VIRUS 3. III. VIRUS-
SPECIFIC PROTEIN SYNTHESIS BY TEMPERATURE-SENSITIVE MUTANTS. (Eng.) Goorha, R. (St. Jude Children's Res. Hosp., P.O. Box 318, Memphis, Tenn. 38101); Naegele, R. F.; Purifoy, D.; Granoff, A. *Virology* 66(2):428-439; 1975.

Six temperature-sensitive (ts) mutants of frog virus 3 (FV 3), five DNA$^+$ and one DNA$^-$, representing six separate complementation groups, were examined for their intracellular patterns of virus-specific protein synthesis at both the permissive (23 C) and nonpermissive (30 C) temperatures. To unmask the viral proteins, protein synthesis in the host fathead minnow (FHM) cells was inhibited by exposing the cells to heat-inactivated Δ mutant. The cells were then exposed to infectious virus and incubated at 23 or 30 C. At 2, 4, and 8 hr after viral infec-

tion of 4.03 mM. 2'-Deoxythymidine-5'-monophos-
phate (dThd-5'-P) competitively inhibited TK2 at
all concentrations tested, but its effect on TK1
was concentration-dependent; at 500 μM, it stimulated TK1 activity, while at 8 mM it inhibited it.
During enzyme kinetic studies, TK1 displayed substrate inhibition which was reversed by dThd-5'-P,
but after polyacrylamide gel electrophoresis, the
stimulation by dThd-5'-P disappeared. The results
suggest the existence on the TK1 molecule of a second binding site for 2'-deoxythymidine which medi-
ates substrate inhibition and which can also be occupied by dTdh-5'-P; the second binding site appears
to be separated from the catalytic center.

4383   MAREK'S DISEASE VIRUS (KEKAVA STRAIN)
        REPLICATION IN CHICKENS, CHICK EMBRYOS AND
CELL CULTURES. (Eng.) Yakovleva, L. S. (Inst. of
Experimental and Clinical Oncology, U.S.S.R. Acad. of
Medical Sciences, 115478 Moscow, U.S.S.R.); Mazurenko,
N. P.; Gunenkova, N. K.; Vinogradov, V. N.; Pavlov-
skaya, A. I.; Chernyakhovskaya, I. *Acta Virol.*
19(4):293-298; 1975.

The Kekava strain of Marek's disease virus (MDV-
Kekara) was passaged *in vivo*, *in ovo*, and *in vitro*.
In the course of 12 passages of MDV-Kekava in chickens,
the morbidity varied greatly (from 23-50%). MDV-
Kekava produced plaques in cultures of chick embryo
kidney and adult chicken kidney cells and chick embryo
fibroblasts (CEF). The virus adaptation to the cultures was very slow. MDV-Kekava induced the formation
of pocks on the chorioallantoic membranes (CAM) of
chick embryos, but the proportion of embryos with CAM
lesions did not exceed 24%. Serial passaging of the
virus in chick embryos beyond the fifth passage was
unsuccessful. The results of virus isolation in
chickens, cell cultures and chick embryos indicate
the possibility of a long-term latent virus carrier
state in chickens without development of tumors.

4384   MURINE LEUKEMIA VIRUS: DETECTION OF UN-
        INTEGRATED DOUBLE-STRANDED DNA FORMS OF
THE PROVIRUS. *(Eng.)* Gianni, A. M. (Istituto di
Patologia Generale I, Facolta di Medicina, Universita Statale di Milano, Italy); Smotkin, D.; Weinberg, R. A. *Proc. Natl. Acad. Sci. U.S.A.* 72(2):447-
451, 1975.

Two forms of double-stranded viral DNA were found
which proved to be putative intermediates in the
flow of viral genetic information between the virion
RNA and the integrated viral DNA genome. Roller
bottle cultures of mouse bone marrow (JLS V-9)
cells were infected with 10^7 PFU/ml of Moloney
murine leukemia virus, and the viral DNA was extracted from the infected cells. Viral DNA was
detected in hybridization tests using either [³H]-
cDNA or [^{125}I]-RNA, and proviral DNA was assayed by
incubation with either of these probes. Both
strands of viral DNA appeared to reach a plateau
within six hours of infection, although the DNA
complementary to [^{125}I]-labeled RNA reached its
maximum prior to the DNA complementary to the [³H]-
DNA probe. The degree of hybridization of both
probes was comparable. Centrifugation of the pro-

viral DNA in CsCl under conditions of isopycnic banding resulted in one band of detectable proviral DNA at a density of 1.7133. Additional fractionation of the proviral DNA on an ethidium bromide CsCl (EtBr-CsCl) gradient showed that a substantial proportion of the DNA banded at a density indicative of a closed double-stranded circular DNA. In a neutral sucrose gradient, the DNA from the dense band in the EtBr-CsCl gradients sedimented at a rate of about 30 S, with a smaller, variable shoulder at 19-20 S. After storage for one month at -20 C, the prominent peak was observed at 19 S. In an alkaline sucrose gradient, half of the viral DNA sedimented with a coefficient of 70 S. The DNA found in the less-dense band of the EtBr-CsCl gradient contained two components: the rapidly sedimenting component behaved like double-stranded linear or nicked-circular DNA of 6×10^6 daltons, and the slowly sedimenting material appeared to contain DNA fragments of 5×10^5 daltons. It was concluded that the DNA found in the denser band of the EtBr-CsCl gradient consisted of double-stranded, closed-circular DNA molecules containing no alkali-labile linkages and of molecular wt 6×10^6 daltons.

4385   DEPENDENCE OF MOLONEY MURINE LEUKEMIA
        VIRUS PRODUCTION ON CELL GROWTH. (Eng.)
Paskind, M. P. (Center for Cancer Res., Massachusetts Inst. Technology, 77 Mass. Ave., Cambridge, Mass. 02139); Weinberg, R. A.; Baltimore, D. *Virology* 67(1):242-248; 1975.

The effects of release from growth arrest (G_0) on the production of Moloney murine leukemia virus (M-MuLV) was studied, using a virus-producing clone derived from cells chronically infected with M-MuLV (JLS V-11). The cultures became stationary at subconfluent density three days after seeding and were released from the stationary phase one day later. DNA synthesis was measured after incubation for 1-hr periods with 0.25 µCi/ml of [^{3}H]thymidine. Reverse transcriptase was also assayed using poly(A) x oligo(dT) or poly(C) x oligo(dG) as template primer, and XC-cell assays were used to determine the effects of cycloheximide (10 µg/ml) and actinomycin D (0.1 µg/ml) on virus release. Finally, a poly(U)·filter assay was used to determine whether the first wave of virus particles released after the resumption of cell growth utilized preexisting viral RNA. The production of M-MuLV ceased in the stationary JLS V-11 cells. Stimulation of the cells to reenter the growth cycle synchronously was followed by two waves of virus release. The first wave, occurring within four hours after the cells were released from G_0, was unaffected by actinomycin D; the second wave, coinciding roughly with the mitotic period of the following cell cycle, was prevented by actinomycin D. Both waves of virus release were inhibited by cycloheximide. Thus, the first wave of released virus appeared to contain viral RNA which was already present in the resting cells, while the second wave of released virus particles carried newly formed viral RNA. The results further suggest that some form of translational, rather than transcriptional, control is responsible for the reinitiation of virus production during the transition from quiescence to growth.

4390   STRUCTURE OF B77 SARCOMA VIRUS RNA: STA-
BILIZATION OF RNA AFTER PACKAGING.
(Eng.) Stoltzfus, C. M. (Vanderbilt Univ. Sch. Medi-
cine, Nashville, Tenn. 37232); Snyder, P. N. *J.*
Virol. 16(5):1161-1170; 1975.

The RNA structure and maturation of Bratislava 77
(B77) sarcoma virus was investigated. A cloned
stock of virus was isolated from a single focus of
transformed chick embryo fibroblast cells. Plates
of cells transformed and producing B77 virus were
labeled with [5-^{3}H]uridine and [^{32}P]phosphate; the
virus was then purified and RNA extracted. Denaturation of 65S RNA into subunits was prevented by
increasing the extraction salt concentration from
0.1 to 0.5 mM NaCl. Pronase digestion of immature
RNA resulted in little, if any, nicking. However the
proportion of 60-70S RNA to 35S RNA was strongly
dependent on the extraction temperature; the rate
of stabilization at 45 C was increased four-fold
over the rate at 37 C. The incubation temperature
also affected the RNA dissociation. A change in
T_m, from 56 C to 67.5 C, with increasing virus age
was also noted. Evidence suggests that the inter-
subunit linkages in immature B77 60-70S RNA are
unstable and spontaneously dissociate when the
virus is extracted with phenol at room temperature
in the presence of 0.1 M NaCl. It is further suggested that the observed increase in T_m with virus
age results from an increase in stability of base-
paired regions.

4391   EFFECT OF SERUM CONCENTRATION ON ADENYL-
ATE CYCLASE ACTIVITIES OF UNTRANSFORMED
BHK CELLS AND POLYOMA VIRUS-TRANSFORMED DERIVA-
TIVES. (Eng.) Matsumoto, T. (Sapporo Medical Col-
lege, Minami-1-jo, Nishi-17-chome, Chuo-ku, Sapporo
060, Japan); Uchida, T. *Gann.* 66(4):393-397; 1975.

Adenylate cyclase activities of untransformed hamster
BHK 21/13 cells and their polyoma virus-transformed
derivatives were compared using cells cultured in low
and high serum concentrations. Untransformed cells
cultured in a medium containing 10% serum had an enzyme activity 1.5 times higher than that of transformed cells cultured under the same conditions.
When the cells were cultured in a medium containing
0.5% serum that stopped proliferation of untransformed
cells, the untransformed cells had a two-fold higher
adenylate cyclase activity than the transformed cells.
The full activity measured in the presence of
10 mM NaF was two-fold higher in the untransformed
than in the transformed cells. The low serum concentration increased the enzyme activity of both untransformed cells, but the enzyme activity of transformed
cells in the presence of 0.5% serum did not exceed the
activity of untransformed cells in 10% serum. Intracellular cyclic AMP levels paralleled the activity of
adenylate cyclase, but no correlation was found between cyclic AMP levels and cyclic AMP phosphodiesterase activities.

4392   HOMOLOGY AND RELATIONSHIP BETWEEN THE
GENOMES OF PAPOVAVIRUSES, BK VIRUS AND
SIMIAN VIRUS 40. (Eng.) Khoury, G. (Natl. Inst.
of Allergy and Infectious Diseases, Bethesda, Md.

20014); Howley, P. M.; Garon, C.; Mullarkey, M. F.;
Takemoto, K. K.; Martin, M. A. *Proc. Natl. Acad.
Sci. USA* 72(7):2563-2567; 1975.

The polynucleotide sequence homology between Papovavirus BK virus (BKV) DNA and the simian virus 40
(SV40) DNAs was studied, and the common regions
were localized on the physical map of SV40 DNA.
The respective DNAs were digested with restriction
enzymes, and DNA-DNA hybridization experiments
were carried out using the minus DNA strain of
the 11 ^{32}P-labeled *Hind* SV40 fragments and a 10- to
20-fold molar excess of ^{3}H-labeled sheared BKV or
polyoma DNA. DNA-DNA filter hybridization experiments and DNA-complementary RNA annealing reactions
were also performed. *Eco*RI-cleaved linear BKV and
SV40 DNAs were denatured and studied by electron
microscopy, or the DNAs were added to adenovirus
2+ ND$_2$ DNA prior to denaturing. The DNA-DNA hybridization studies and analysis of the hetero-
duplex molecules between *Eco*RI linear molecules of
SV40 and BKV DNAs established the portions of the
SV40 genome that are homologous to BKV. The shared
sequences were located principally in the late gene
region of the SV40 DNA, corresponding to portions
of *Hind* fragments C, D, G, J, and K, and to a lesser
extent *Hind* fragments E and F. Little or no homology was detected between BKV DNA and the early
region of the SV40 genome. The general location of
homologous regions at the termini of the RI-linear
molecules and the constant central region of hetero-
ology suggested that the BKV and SV40 genomes have
the same basic genome. The studies with SV40 complementary RNA further indicated that the strand
orientation of the shared nucleotide sequences are
the same for both SV40 and BKV. The precise evolutionary relationship of these two agents, is,
however, unknown.

4393   THE ALTERATION OF RIBOSOMES FOR mRNA
        SELECTION CONCERNED WITH ADENOVIRUS GROWTH
IN SV40-INFECTED SIMIAN CELLS. (Eng.) Nakajima, K.
(Inst. Medical Science, P.O. Takanawa, Tokyo, Japan);
Oda, K. *Virology* 67(1)85-93; 1975.

The simian virus 40 (SV40) helper function to support human adenovirus (Ad) growth in African green
monkey kidney (AGMK) cells previously infected with
SV40 was studied. Confluent monolayer cultures were
infected with SV40 (0.2-20 plaque-forming units
(PFU)/cell), and 24 hr later were infected with Ad2
(10 PFU/cell). The amounts of DNA and messenger
RNA (mRNA) synthesis in these cells were estimated
by hybridization of pulse-labeled DNA or RNA with
an excess of SV40 or cellular DNA. The amount of
Ad2 virion antigens synthesized in the doubly infected cells was estimated by a microcomplement-fix-
ation test. Protein synthesis with preincubation
S30 (prepared from SV40-infected cells) was studied
to determine the alteration of the protein synthesizing system to be able to translate late Ad2 mRNA for
hexon. Formation of the 80S initiation complex for
protein synthesis was carried out in the presence of
high salt-washed ribosomes from preincubation AGMK
cell S30, Met-transferRNA, and ^{3}H-labeled late Ad2
mRNA (24S). The effect of input multiplicities on
the alteration of ribosomes was also investigated.

monstrate that the SV40-specific proteins induced by Ad2$^+$ ND2, Ad2$^+$ ND4, and Ad2$^+$ ND5 are metabolically unstable, disappearing almost completely in 12 hr. These proteins are not present in purified virions. Two nonstructural Ad2-specific proteins have been demonstrated in Ad2 and hybrid virus-infected cells which have a smaller apparent molecular wt after a short pulse than after a pulse followed by a chase. The molecular wt increase during the chase may be cuased by the addition of carbyhydrate to a polypeptide backbone.

4397   ELECTRON MICROSCOPE LOCALIZATION OF A PROTEIN BOUND NEAR THE ORIGIN OF SIMIAN VIRUS 40 DNA REPLICATION. (Eng.) Griffith, J. (Stanford Univ. Med. Cent., Calif.); Dieckmann, M.; Berg, P. *J. Virol.* 15(1):167-172; 1975.

Electron microscopy was utilized in investigating the location and possible function of a stably bound protein "knob" on simian virus 40 (SV40) DNA. Monkey kidney cell cultures (CU-1) infected with SV40 virus (strain Rh9 11) were used. Throughout infection of the cultures by SV40 virus, most of the viral DNA existed as a non-encapsulated, nucleoprotein complex. After Triton X-100 extraction and incubation with 1 M NaCl, however, a new, high salt-stable protein-DNA complex was discovered. Examination by high resolution electron microscopy, revealed the tightly twisted rods characteristic of SV40 form 1 DNA. Nicking by brief X-ray treatment yielded circular structures containing one "knob" per DNA molecule as the predominant molecular species. Disruption of purified virus particles and CsCl gradient banding revealed structures closely resembling the salt-stable protein-DNA complexes recovered directly from the infected cells. Cleavage of the salt-stable protein-DNA with *eco* R1 endonuclease yielded unit length linear molecules, with the "knobs" occurring predominantly at 0.64-0.75 SV40 fractional lengths from the farther end. Cleavage with *hpa*II endonuclease set the location of the "knobs" at 0.70$\pm$0.05 on the SV40 DNA map. Thus, high resolution electron microscopy has located the site of the 2 or 3 globular protein unit "knob". Speculations on the function and genetic origin of the stably bound protein on the SV40 DNA as related to location suggest a role in initiating DNA replication or promoting some essential step in expression.

4398 ^{1}H AND ^{13}C NUCLEAR MAGNETIC RESONANCE SPECTRA OF THE LIPIDS IN NORMAL AND SV 40 VIRUS-TRANSFORMED HAMSTER EMBRYO FIBROBLAST MEMBRANES. (Eng.) Nicolau, C. (Institut fur Strahlenchemie im Max-Planck-Institut fur Kohlenforschung, 433 Mulheim a. d. Ruhr West Germany); Dietrich, W.; Steiner, M. R.; Steiner, S.; Melnick, J. L. *Biochim. Biophys. Acta* 382(3):311-321; 1975.

Well-resolved ^{1}H and ^{13}C nuclear magnetic resonance spectra were obtained with normal and simian virus 40-transformed membranes of hamster embryo fibroblasts. Estimation of the ratio of $^{13}CT_2$ values of the normal to transformed cell membranes showed an increased intermolecular motion in the transformed cell membranes. The temperature dependence of the

$(CH_2)n$ line in the ^1H spectra in the range 298-343 K showed an activation energy for the lateral diffusion of the fluid phospholipid regions in the normal cell membranes while the transformed ones showed practically no temperature dependence in this temperature range. The fluidity of the phospholipid region in the transformed cell membrane seems to be significantly higher than that observed in the normal cell material. These data support and extend the findings concerning the mobility of the concanavalin A binding/agglutinizing sites on the surface of normal and virus-transformed cells and suggest further approaches to the study of the membrane alterations in tumor cells.

4399   DNA INFECTIVITY AND THE INDUCTION OF HOST
        DNA SYNTHESIS WITH TEMPERATURE-SENSITIVE
MUTANTS OF SIMIAN VIRUS 40. (Eng.) Chou, J. Y.
(Natl. Inst. Arthritis Metab. Dig. Dis., Bethesda,
Md.); Martin, R. G. *J. Virol.* 15(1):145-150; 1975.

The induction of host DNA synthesis and the initiation of viral DNA synthesis (infectivity) were studied in mutant classes of Simian virus 40 (SV40). DNA extracts were prepared from SV40-infected CV-1 cell monolayers at a concentration of five plaque-forming units (PFU)/cell. Infectivity was studied *via* experiments: equal parts of a DNA solution prepared from the extracts and virus suspension containing 2×10^6 PFU/ml were added to confluent monolayers of CV-1 cells. For induction experiments, cells were labeled by incubation for one hour with 1.5 ml of modified Eagle medium without serum supplement containing 100 μCi/ml ^3H-thymidine. DNA from D-mutant class was infective at 40 C (D101 yielded 2×10^5 PFU/ml), while A, B, C, and BC were significantly less infective (A207, 20 PFU/ml; B201, 2.4×10^2 PFU/ml; C260, 3×10^2 PFU/ml; BC230, 2 PFU/ml). Coinfection of A virions with wild-type or D-mutant DNA did not enhance viral replication (A209 virion + D202 DNA, 17.1 counts/min thymidine uptake; A209 + wild type DNA, 31.5 counts/min; D202 DNA, 20.9 counts/min; wild type DNA, 29.1 counts/min). A-mutants were defective in inducing host DNA synthesis at 42.5 C (A241, 42.5 C, 14.8×10^3 counts/min; 33 C, 25.2×10^3 counts/min). It is suggested that the D function is not required for the production of infectious virions. It is also concluded that the A function is pleiotropic, being required for both the initiation of viral DNA synthesis and the induction of host DNA synthesis.

4400   DETECTION OF AN ANTIGEN RELATED TO MASON-
        PFIZER VIRUS IN MALIGNANT HUMAN BREAST
TUMORS. (Eng.) Yeh, J. (John L. Smith Memorial Inst.
for Cancer Res., Pfizer Inc., Maywood, N. J. 07607);
Ahmed, M.; Mayyasi, S. A.; Alessi, A. A. *Science*
190(4214):583-584; 1975.

A protein antigenically related to the Mason-Pfizer monkey virus (M-PMV) p27 was detected in human malignant breast tumors. Frozen malignant breast tumors selected after gross and microscopic pathological examination were homogenized and extracted with ether. The soluble proteins were futher purified by diethylaminoethyl-cellulose ion exchange column chromatography, and the fractions expected

to elude p27 collected and concentrated. Of 18 breast tumor specimens tested, eight induced the release of 25% of the maximum precipitable ^{125}I-labeled M-PMV p27; the range of ^{125}I release was 27-78%. Each of the tumor specimens showed competition curves of similar slope to that developed with purified M-PMV as the competing antigen, suggesting similar reacting proteins. The amount of p27-like proteins detected ranged from 0.3 to 5.7 ng/mg of tissue extract. No detectable antigen related to M-PMV was found in 12 human tumors of other origin, or two normal placentas tested. Furthermore, nonspecific competition by the malignant breast tissue extracts was eliminated *via* their lack of competition against SSV-1 p30. Partially purified tissue material quantitatively tested for p27 by competition radioimmunoassay revealed a 50% recovery of the p27 after purification procedures, and suggested a very low level of viruslike antigens in malignant breast tissue.

4401   INDUCTION OF ENDOGENOUS AND OF SPLEEN FOCUS
        FORMING VIRUSES DURING DIMETHYLSULFOXIDE-IN-
DUCED DIFFERENTIATION OF MOUSE ERYTHROLEUKEMIA CELLS
TRANSFORMED BY SPLEEN FOCUS-FORMING VIRUS. (Eng.)
Dube, S. K. (Max-Planck Institut fur Experimentelle
Medizin, Gottingen, West Germany); Pragnell, I. B.;
Kluge, N.; Gaedicke, G.; Steinheider, G.; Ostertag,
W. *Proc. Natl. Acad. Sci. USA* 72(5):1863-1867; 1975.

Mouse erythroleukemia cells in culture provided an opportunity to study the role of the erythroid cell-transforming spleen focus-forming virus (SFFV) during dimethylsulfoxide (Me$_2$SO)-induced erythroid differentiation of these cells. Me$_2$SO (1 or 1.5%) induced erythroid differentiation and a 10- to 100-fold increase in spleen focus formation in an erythroleukemie cell clone established by infection of N type DBA/2 mice with NB tropic SFFV. This induction was correlated with the induction of globin messenger RNA synthesis. No increase in virus induction occurred in a clone of SFFV-transformed cells that do not differentiate in the presence of Me$_2$SO, indicating that induction of virus release is dependent on erythroid differentiation. The Friend virus (FV) complex released by uninduced cells was NB tropic, since it resulted in a comparable increase in spleen focus formation in N type DBA/2 and B type BALB/c mice. The virus released by Me$_2$SO-induced cells was N tropic, as it induced a 20-fold increase in spleen focus formation in DBA/2 mice but no detectable increase in BALB/c mice. To determine if the induced virus was exclusively N tropic endogenous virus, the effects of externally added NB tropic LLV-F helper virus, the second component of the FV complex, were studied. Added NB tropic LLV-F did not increase the number of spleen focus-forming units released by either uninduced or induced cells, but did induce identical increases in SFFV titers in B and N type mice. Me$_2$SO therefore induces SFFV as well as an endogenous virus. This joint induction could mean that transformed cells release a nondefective endogenous virus with the ability to transform erythroid cells. The data indicate that the endogenous virus and the erythroid focus-forming virus interact during Me$_2$SO-induced erythroid differentiation.

4407 EFFECTS OF 5-TUNGSTO-2-ANTIMONIATE (HPA-
23) IN ONCOGENIC DNA AND RNA VIRUS-CELL
SYSTEMS [abstract]. (Eng.) Ablashi, D. V. (Natl.
Cancer Inst., Bethesda, Md.); Twardzik, D. R.;
Easton, J. M.; Armstrong, G. R.; Luetzeler, J.;
Jasmin, C.; Chermann, J. C. *Proc. Am. Assoc.
Cancer Res.* 16:31; 1975.

4408 DEFECTIVE TRANSFORMING VIRUSES CAUSE
ERYTHROLEUKEMIA [abstract]. (Eng.)
Scher, C. D. (Child. Hosp., Boston, Mass.); Scolnick,
E. M. *Proc. Am. Assoc. Cancer Res.* 16:43; 1975.

4409 TRANSFORMATION OF CULTURED RAT ADRENOCORTI-
CAL CELLS BY KIRSTEN MURINE SARCOMA VIRUS
[abstract]. (Eng.) Auersperg, N. (Cancer Res. Centre,
Univ. British Columbia, Vancouver, B.C. V6T 1W5,
Canada); Jull, J. W.; Hudson, J. B.; Goddard, E. J.
Proc. Am. Assoc. Cancer Res. 16:67; 1975.

4410 DIFFERENCES IN DETECTION OF SIMIAN SARCOMA
VIRUS 70S RNA AND REVERSE TRANSCRIPTASE IN
CELLS CULTURED *IN VITRO* AND *IN VIVO* [abstract]. Sax-
ton, R. E. (Div. Surg. Oncol., Univ. California Los
Angeles); Rangel, D.; Morton, D. L. *Proc. Am. Assoc.
Cancer Res.* 16:171; 1975.

4411 IDENTIFICATION OF SIMIAN ONCORNA VIRUS
(SSV_1) RNAs. (Fre.) Ravicovitch, R. E.
(Laboratoire d'Hematologie Experimentale, UER d'Hema-
tologie Experimentale, UER d'Hematologie, Hopital
Saint-Louis, Paris, France); Salle, M.; Robert-Robin,
J.; d'Auriol, L.; Peries, J. *C. R. Acad. Sci. [D]
(Paris)* 280(5):681-684; 1975.

4412 A COMPARATIVE EVALUATION OF ISOLATION
METHODS FOR INTRACELLULAR VIRAL REVERSE
TRANSCRIPTASE USING A MODEL SYSTEM [abstract]. (Eng.)
Allaudeen, H. S. (Natl. Cancer Inst., Bethesda, Md.);
Sarngadharan, M. G.; Gallo, R. C. *Proc. Am. Assoc.
Cancer Res.* 16:263; 1975.

4413 COMPARATIVE ANALYSIS OF PLASMA MEMBRANES
OF NORMAL AND TRANSFORMED CELL PHENOTYPES
[abstract]. (Eng.) Nigam, V. N. (Cell Biology,
University of Sherbrooke, Quebec, Canada); Brailovsky,
C. *Proc. Am. Assoc. Cancer Res.* 16:196; 1975.

4414 IN VITRO TRANSCRIPTION AND TRANSLATION OF
SV40 DNA [abstract]. (Eng.) Sharma, P. R.
(Dept. Molecular Biology, Univ. Geneva,. CH-1211
Geneve 4, Switzerland); Weil, R. *Experientia* 31(6)
748; 1975.

4415 INTEGRATION OF SV40 [abstract]. (Eng.)
Prasad, I. (New York Univ. Sch. Med.,
N.Y.); Basilico, C. *Proc. Am. Assoc. Cancer Res.*
16:20; 1975.

4416   SUPERCOILING OF SV40 DNA AND CHROMATIN
        STRUCTURE [abstract]. (Eng.) Germond,
J. E. (Inst. Suisse de Recherche Experimentale sur le
Cancer, CH-1011 Lausanne, Switzerland); Hirt, B.;
Oudet, P.; Gross-Bellard, M.; Chambon, P. *Experientia*
31(6):739; 1975.

4417   COMPARATIVE PEPTIDE ANALYSIS OF STRUCTURAL
        PROTEINS FROM SIMIAN VIRUS 40 AND SOME
TEMPERATURE SENSITIVE MUTANTS [abstract]. (Eng.)
Fey, G. (Inst. Suisse de Recherches Experimentales
sur le Cancer, CH-1011 Lausanne, Switzerland); Hirt,
B. *Experientia* 31(6):738; 1975.

4418   PROPERTIES OF VIRUS RESCUED FROM A LINE
        OF SV40-TRANSFORMED HUMAN CELLS AND ITS
REVERTANT VARIANTS [abstract]. (Eng.) Kashmiri,
S. V. S. (Wistar Inst., Philadelphia, Pa.); Tan,
K. B.; Diamond, L. *Proc. Am. Assoc. Cancer Res.*
16:189; 1975.

4419   TUMORIGENICITY IN *NUDE* MICE OF VIRUS
        TRANSFORMED CELLS IS CORRELATED SPECIFICALLY
WITH ANCHORAGE-INDEPENDENT GROWTH *IN VITRO* [abstract].
(Eng.) Freedman, V. H. (Albert Einstein Coll. Medi-
cine, Bronx, N.Y.); Shin, S. *Genetics* 80(3/Part 1/
Suppl.):s32; 1975.

4420   EPSTEIN-BARR VIRUS IN INFECTIOUS MONONU-
        CLEOSIS AND BURKITT LYMPHOMA [abstract].
(Eng.) Gerber, P. (Food and Drug Adm., Bethesda,
Md.); Nkrumah, F. K.; Prichett, R.; Kieff, E.
Proc. Am. Assoc. Cancer Res. 16:23; 1975.

4421   CHARACTERISTICS OF CELLULAR TRANSFORMATION
        ASSOCIATED WITH THE EPSTEIN-BARR VIRUS IN
NON-LYMPHOBLASTOID CELLS [abstract]. (Eng.) Glaser,
R. (M. S. Hershey Med. Cent. Pennsylvania State
Univ., Hershey, Pa.); Decker, B.; Ladda, R.;
Shows, T. B. *Proc. Am. Assoc. Cancer Res.* 16:70;
1975.

4422   OPTIMIZATION OF CULTURE CONDITIONS FOR
        PRODUCTION AND ASSAY OF HIGH TITERED
EPSTEIN-BARR VIRUS [abstract]. (Eng.) Charamella,
L. J. (Natl. Cancer Inst., Frederick, Md.); Brown,
B. L.; Nagle, S. C. *Proc. Am. Assoc. Cancer Res.*
16:157; 1975.

4423   VARIATIONS IN PHENOTYPIC EXPRESSION OF
        MMTV IN MOUSE MAMMARY TUMOR-DERIVED CELL
CLONES [abstract]. (Eng.) Arthur, L. O. (Natl.
Cancer Inst., Frederick, Md.); Fine, D. L.; Plowman,
J. K.; Hillman, E. A. *Proc. Am. Assoc. Cancer Res.*
16:168; 1975,

4424   THE H-2 COMPLEX AS DETERMINANT FOR SUS-
        CEPTIBILITY TO MAMMARY TUMOR VIRUS [ab-
stract]. (Eng.) Muhlbock, O. (No affiliation
given); Dux, A. *Z. Immunitaetsforsch.* 149(5):
359-360; 1975.

4425   ANALYSIS OF EXTRINSICALLY LABELED MuMTV
        [abstract]. Sheffield, J. B. (Inst. Med. Res.,
Camden, N.J.); Daly, T. M.; Taraschi, N.; Moore,
D. H. *Proc. Am. Assoc. Cancer Res.* 16:94; 1975.

4426   EFFECT OF DEXAMETHASONE ON THE MOUSE MAM-
        MARY TUMOR VIRUS GROWN IN CAT KIDNEY CELLS
[abstract]. (Eng.) Lasfargues, E. Y. (Inst. Med.
Res., Camden, N.J.); Dion, A.; Lasfargues, J. C.;
Moore, D. H. *Proc. Am. Assoc. Cancer Res.* 16:28;
1975.

4427   IDENTIFICATION OF CYTOPLASMICALLY SYNTHE-
        SIZED MTV SPECIFIC POLYPEPTIDES [abstract]
(Eng.) Dickson, C. (Cancer Res. Lab., Univ. California, Berkeley). *Proc. Am. Assoc. Cancer Res.*
16:26; 1975.

4428   SYNTHESIS AND METHYLATION OF tRNA IN
        ADENOVIRUS-INFECTED KB CELLS [abstract].
Sehulster, L. M. (Rutgers Med. Sch., Piscataway,
N.J.); Varricchio, F.; Raska, K., Jr. *Proc. Am.
Assoc. Cancer Res.* 16:71; 1975.

4429   CHARACTERIZATION OF POLYOMA m-RNAs [ab-
        stract]. (Eng.) Salomon, C. (Dept.
Biologie Moleculaire, Univ. Geneve, CH-1211 Geneve
4, Switzerland); Turler, H. *Experientia* 31(6):746;
1975.

4430   POLYOMA VIRUS-SPECIFIC RNA IN PRODUCTIVELY
        INFECTED AND TRANSFORMED CELLS [abstract].
(Eng.) Beard, P. (Swiss Inst. for Experimental Can-
cer Res., CH-1011 Lausanne, Switzerland); Acheson,
N. H. *Experientia* 31(6):734; 1975.

4431   STRAND ORIGIN OF POLYOMA VIRUS-SPECIFIC
        'GIANT' NUCLEAR RNA AND 16S AND 19S
MESSENGER RNA [abstract]. (Eng.) Acheson, N. H.
(ISREC, CH-1011 Lausanne, Switzerland); Beard, P.
Experientia 31(6):733; 1975.

4432   TRANSLATION *IN VITRO* OF POLYOMA VIRUS
        SPECIFIC (16 AND 19S) m-RNA [abstract].
(Eng.) von der Helm, K. (ISREC, CH-1011 Lausanne,
Switzerland); Acheson, N. *Experientia* 31(6):740;
1975.

4433   ELECTRON MICROSCOPIC STUDIES ON THE DEVELOP
        MENT AND CYTOPATHOLOGY OF A HUMAN POLYOMA
VIRUS (BK) [abstract]. (Eng.) Lecatsas, G. (Dept.
Microbiology, Univ. Pretoria, Pretoria, South Africa)
Prozesky, O. W. *S. Afr. Med. J.* 49(26):1054; 1975.

A AND LEUKEMIA INCIDENCE IN PASSIVELY
TED NEONATES [abstract]. Buffett,
ark Mem. Inst., Buffalo, N.Y.).
Cancer Res. 16:97; 1975.

  4213, 4214, 4215, 4219, 4237,
  4239, 4240, 4241, 4242, 4243,
  4245, 4257, 4263
0, 4478, 4480, 4506, 4519, 4529,
6
, 4595
 4659, 4669

4435   A STUDY OF IMMUNOLOGICAL SPECIFICITY OF
        TRANSPLACENTALLY INDUCED NEURINOMAS IN RATS.
(Rus.) Okulov, V. B. (N. N. Petrov Inst. Oncol.,
Min. Publ. Health U.S.S.R., Leningrad, U.S.S.R.);
Kolodin, V. I. *Vestn. Akad. Med. Nauk SSSR* (2):60-
63; 1975.

Tumors of the peripheral nervous system (neurinomas),
induced transplacentally in rats with N-nitroso-N-
ethyl urea were studied by precipitation in agar and
immunoelectrophoresis for their immunological specificity. Females were administered a 40 mg/kg-dose
of N-nitroso-N-ethyl urea ip between the 17th and
19th days of gestation. Neurinomas of the peripheral
nervous system were found in 17 of 23 animals in
the offspring, but no tumor was detected in six
animals macroscopically. The tumor was localized
in the supramaxillary branch of the trigeminal
nerve in 11 cases, in the lumbosacral plexus in
three cases, in the nerves of the extremities in
two cases, and in the mesenteric plexus of the small
intestine in one case. A protein type antigen with
the mobility of beta-globulins was detected both in
the primary tumor tissue and in transplanted tumors
over up to 50 passages. No such antigen was found
in extracts of the brain and peripheral nerve tissues in control animals. In 12 out of 14 animals
with transplacentally induced primary neurinomas,
and in all animals with tumor transplants, the
antigen was present in the blood serum. In rats with
tumors of other tissues, including tumors induced by
nitroso compounds, no neurinoma antigen was found in
the tumor or in the serum. The neurinoma antigen
was found to differ from other specific proteins in
the nerve tissue.

4436   REDUCED CAPACITY TO PRODUCE SPECIFIC
        'EFFECTOR' CELLS AFTER INJECTION OF CBA
MICE WITH C3H CELLS. (Eng.) Lilliehook, B. (Dept.
Tumor Biology, Karolinska Institutet, S-104 01
Stockholm, Sweden); Jacobsson, H.; Blomgren, H.
Scand. J. Immunol. 4(5/6):463-469; 1975.

A study was made to characterize further the
previously demonstrated immunological reactivity
of CBA mice to infusion with H-2 compatible, M-
antigen-incompatible C3H cells, whereby thymus
or peripheral lymphocytes of the infused CBA mice
exhibit a decrease in mixed lymphocyte culture
(MLC) reactivity with C3H lymphocytes. In the
present tests, 1- to 2-month-old CBA mice were
injected iv with 10^7 spleen cells from syngeneic
mice or from C3H x CBA hybrids. Three weeks later,
host-*versus*-graft (HVG) tests were carried out by
injecting the host animals with C3H spleen cells
into the left footpad and C3H x CBA spleen cells
into the right footpad. Enlargement of the local
lymph node of the right footpad relative to that
of the left occurred in the mice preinjected with
syngeneic cells, whereas there was no detectable
response in the animals preinjected with C3H x CBA
cells. The specificity of C3H x CBA cells for reducing the responsiveness was confirmed by results
of further control tests in which CBA x C57B1 cells
were used instead of C3H x CBA cells for injection
into the right footpad. Lymphocytes from CBA mice
showing a decreased HVG response to C3H x CBA

toire d'Anatomie pathologique, Universite de
Liege, 1 rue des Bonnes Villes, B 4000 Liege,
Belgium); Hiesche, K.; Möller, E.; Revesz, L.;
Haot, J. *C. R. Soc. Biol. (Paris)* 169(3):717-
723; 1975.

Three hundred and seventy-three male CBA mice,
aged two months, were exposed to a 400 R dose
of x-rays and injected with SRBC to determine
the influence of sublethal irradiation on the
immunological response of the spleen to the
foreign blood cells. In 185 cases, one of the
posterior limbs was shielded from irradiation.
Nonirradiated controls numbered 87. The SRBC
were injected (2/10 ml of a solution containing
10% RBC) into a lateral vein of the tail of 69
controls and of the irradiated mice, on days
8, 12, 16, 20, and 30 following irradiation. In
the hematopoietic marrow of the normal mice, the
number of natural rosettes doubled following injection, and remained above the normal value until
the end of observations (18 days). After irradiation, the rosette formation capacity (RFC) decreased, and antigenic stimulation had very little effect. After immunization, the RFC rose
rapidly in normal mice, with a maximum on day
5 (16 times its normal value). In the irradiated
mice, the immunocyte adherence of the splenic
cells was significantly reduced and immunological
response was more spread out (maximum on days 8
and 12). Response improved if the interval between irradiation and immunization was increased.
However, recovery was still incomplete by day 30.
The difference between mice irradiated *in toto*
and those whose limbs were shielded was insignificant. In nonirradiated mice, cells which secrete immunoglobulin M (IgM) and immunoglobulin
G (IgG) (hemolysin-secreting cells:HSC) increased
significantly following immunization. Irradiation decreased both types of hemolytic secretions.
Results indicate that type B HSCs seem to recover
faster than RFCs of B and T origin. The small
rise of RFCs in the marrow of nonirradiated mice
after immunization is explained by an incapacity
of the bone marrow cells to become immunocompetent in the absence of a differentiation process
which takes place in a peripheral structure.
Bone marrow cells, irradiated or not, respond
very little to antigenic stimulation.

4440   ANTIGENIC CHANGES IN CULTURED MURINE LYM-
        PHOMAS AFTER RETRANSPLANTATION INTO SYN-
GENEIC HOSTS. (Eng.) Cikes, M. (Swiss Inst. Exp.
Cancer Res., Lausanne). *J. Natl. Cancer Inst.* 54
(4):903-906; 1975.

Changes of murine leukemia virus (MLV)-determined
cell-surface antigen (MV-CSA) and H-2 antigen on
reimplantation of cultured murine lymphomas into
syngeneic hosts were described. YAC and YCAB lymphomas were induced in an A/Sn(H-2^a/H-2^a) and an
(A/Sn x A.CA)F$_1$ (H-2^a/H-2^f) mouse by inoculation
with MLV during the neonatal period. Cell cultures
were derived from ascites forms of the lymphomas.
Anti-H-2 sera were provided by A.CA anti-A/Sn serum and A/Sn anti-A.SW serum. Anti-MV-CSA serum was
produced in (A/Sn x C57BL)F$_1$ mice by immunization

with ascites YAC lymphoma cells. The reciprocal
of the number of cells capable of reducing the cytotoxic potency of the antiserum to 50% was taken
as an index of the number of available antigen
sites on the cell surface. After retransplantation
of cultured YAC lymphoma cells into syngeneic hosts,
the antigenic patterns of the cells tended to re-
vert to that characteristic of the *in vivo*-propo-
gated lymphoma lines, as demonstrated by a fourfold
increase in H-2 antigen and a 92% decrease of MV-
CSA antigen. Similar results were obtained with
the YCAB lymphoma line. In addition, a single passage of WLH-1 cultured spontaneous lymphoma cells
resulted in almost complete restoration of the H-2
antigen content, and a drastically decreased capacity to absorb anti-MV-CSA serum. Thus reversion
to the antigenic pattern characteristic of the corresponding lines propogated *in vivo* was observed,
although the mechanisms remain unclear.

4441   SPECIFIC PARTIAL DEPLETION OF GRAFT-*VS*-HOST
        ACTIVITY BY INCUBATION AND CENTRIFUGATION OF
MOUSE SPLEEN CELLS ON ALLOGENEIC SPLEEN CELL MONO-
LAYERS. (Eng.) Mage, M. G. (Lab. Biochemistry,
Natl. Cancer Inst., Bethesda, Md. 20014); McHugh,
L. L. *J. Immunol.* 115(4):911-913; 1975.

Spleen cells (from BALB/c mice immunized with·the
C57BL/6 lymphoma EL4, or from non-immune BALB/c) were
incubated on monolayers of [C57BL/6 x BALB/cF_1
(B_6CF_1) spleen cells to investigate the graft-*vs*-
host (GVH) reaction. The incubation was carried out
for ½ hr or for one hr at 37 C followed by centrifugation for five min at 70 x g to 110 x g at 34 to 37
C. Control monolayers were BALB/c spleen cells. As
measured by the Simonsen spleen weight assay in
neonatal mice, GVH activity was partially depleted in
cell populations nonadherent to B_6CF_1 monolayers.
Residual GVH activity of these nonadhe:nt cells was
about half that of cells incubated on t·e control
syngeneic monolayers (the mean of eigh experiments
was 49% ± 11% S.D.). Two or three consecutive cycles
of incubation and centrifugation did not significant·
ly diminished the residual GVH activity. It is suggested that spleen cells with GVH activity are heterogenous with respect to binding to allogeneic target
cells under the above conditions. Cell populations
non-adherent to third-part [A x AL]F_1 monolayers retained full activity, and cell populations partially
depleted of GVH activity in B_6CF_1 neonates had full
activity in third-party [BALB/c x AL]F_1 neonates.

4442   SPLENIC REGULATION OF LYMPHOCYTE TRAPPING
        IN LYMPH NODES DRAINING TUMOR GRAFTS.
(Eng.) Fightlin, R. S. (Yale Univ. Sch. Medicine,
New Haven, Conn.); Lytton, B.; Gershon, R. K. *J.
Immunol.* 115(2):345-350; 1975.

The trapping of chromium-labeled splenic lymphocytes
in the lymph nodes and spleens of tumor-inoculated
mice was measured. Both histocompatible and incompatible tumors were studied. Methylcholanthrene
(MCA)-induced lymphoma, L1210, was inoculated sc
into DBA/2 (syngeneic) and C57BL/6 (allogeneic) mice.
A BALB/c MCA sarcoma was also injected into histo-
incompatible C57BL/6 animals. Tumor resections were

was absorbed with homogenized liver, spleen cells
and L2C leukemic blood cells until all cytotoxicity
for L2C cells was removed. Macrophages for use in
tests of macrophage migration inhibition factor in
supernatants of MC-D and LBC cultures were obtained
from normal guinea pigs administered paraffin oil,
ip. It was found in exploratory experiments that
LBC were induced by co-cultivation of MC-D with
normal lymphoid cells from all sources tested. Foci
of LBC were observed on days 4-6 of co-cultivation,
which was followed by rapid proliferation of LBC
and by the destruction of the tumor cells. Treatment of lymphoid cell preparations with the rabbit
anti-T serum prior to co-cultivation with MC-D
cells reduced or totally eliminated their ability
to develop cytotoxicity. Supernatants of LBC cultures, freed of viable cells by freezing and thaw-
ing, inhibited the migration of peritoneal exudate
cells, but did not contain lymphotoxic or a skin
reactive factor. The absence of the lymphotoxin
and skin reactive factor did not, however, militate
against the T-cell nature of the LBC since the evidence presently available for liberation of lymphotoxin from guinea-pig lymphocytes is controversial
and there is not definite proof of the T-cell origin of these substances. It was concluded that
the cytotoxic LBC cells were of T-cell origin.

4446   SELECTIVE SUPPRESSION OF THE IMMUNE RE-
        SPONSE EMPLOYING THE GRAFT-VS.-HOST RE-
ACTION. (Eng.) Lapp, W. S. (Dept. Physiology,
McGill Univ., Montreal, Quebec, Canada); Treiber,
W.; Elie, R. *Transplant. Proc.* 7(1/Suppl. 1):393-
397; 1975.

Experiments were performed to determine whether
graft-*vs*-host (GvH)-induced immunosuppression is
due to suppressed T-cell helper function and if
GvH mice can develop cell-mediated immunity (CMI)
and humoral immunity (HI) after antigenic challenge.
Inbred A, CBA, C57BL/6, and various F_1 hybrid mice
were used. Normal spleen and thymus cells separated from GvH spleen cells (GvHSC) by a cell-imper-
meable membrane restored the plaque-forming cell
(PFC) response of the GVHSC to SRBC. Treatment of
the restoring cells with anti-theta serum prevented
the restoration of the PFC response. When GvHSC
were used as restoring cells and their adherent
cell population removed, they were able to restore
the PFC response of responding GvHSC. These results indicated that T-cell helper function is inhibited in GvH mice and that T-cell activity is regulated by an adherent cell population. The antigenic stimulation of CMI and HI was studied in
(CBA x A)F_1 mice grafted with C57BL/6 skin 20 days
after GvH induction with A strain lymphoid cells.
The animals could be induced to reject the skin
allografts if challenged with C57BL/6 kidney cells
(B6KC) at least once before allografting. However,
GvH mice treated with B6KC or with B6KC plus a skin
graft followed by a second treatment with B6KC did
not produce hemagglutinating antibodies to C57BL/6
antigens. Regardless of their previous treatment,
all GvH mice failed to produce PFC in response to
SRBC. The selective stimulation of CMI observed in
this experiment could serve as a model to study the
effect of CMI on tumor growth in the absence of HI.

Preliminary experiments demonstrated that preimmunization of GvH mice with nonviable sarcoma cells markedly retarded tumor growth after subsequent challenge with viable sarcoma cells.

4447 . IMMUNOSUPPRESSION BY MURINE LEUKEMIA VIRUSES. (Eng.) Friedman, H. (No affiliation given). In: *Viruses and Immunity: Toward Understanding Viral Immunology and Immunopathology* edited by Koprowski, C., Koprowski, H. New York, Academic Press, Inc., 1975, pp. 17-47.

Recent studies have shown that murine RNA leukemogenic viruses are immunosuppressive. A review of previous work in this area is presented, and some new studies are described. Deficiencies in a variety of antibody responses are consistently observed in leukemia virus-infected mice. The mechanism of such virus induced immunosuppression is, however, not fully understood. Immunosuppression may be due to direct effects of tumor virus infection on different cell types involved in various pathways of the immune response. Studies with a Friend leukemia virus (FLV) model have permitted extensive analyses of various effects of leukemia virus infection on different parameters of antibody formation. Infection of mice with FLV suppresses the primary immune response to SRBC and *Escherichia coli* lipopolysaccharide. Both 19S IgM and 7S IgG hemolysin-forming cells are markedly suppressed in FLV-infected mice challenged with SRBC, with greater depression of the 7S response than 19S response. The time interval between virus infection and challenge immunization, as well as the dose of virus, affected the response. Antigen reactive cells or antibody precursor cells in the B lymphocyte population were the main target for immunosuppression. The numbers of antigen-binding cells for SRBC, both in nonimmune and RBC-immunized mice, were also markedly suppressed. *In vitro* studies indicated that virus-infected leukemic cells, but not cell-free virus, could prevent *in vitro* immunization of normal spleen cells. There was no inhibitory effect on antibody-forming cells directly. Electron microscopy showed the presence of typical leukemia virus particles budding from the surface of many lymphoid cells in the process of secreting antibody to RBC. Thus, leukemia virus infection and antibody formation are not mutually exclusive events at the level of individual immuno-competent cells. FLV infection did not significantly affect the antigen processing mechanism and/or phagocytic cells. A true primary immune response to *Vibrio cholerae* somatic antigen was not affected by FLV infection. In contrast, the secondary immune response to *Vibrio* was readily suppressed, suggesting that antigen-stimulated or antigen-activated immuno-competent cells may be more susceptible to virus infection. The sum total of these studies of the effects of FLV infection on antibody formation indicates that immunosuppression is due mainly to virus-induced alteration of antibody precursor cells *per se*, or alternately, to the preferential infection of antigen-activated lymphoid cells, with a concomitant "autoimmune" response to those cells after they acquired leukemia virus-associated antigens on their surfaces.

11-day-old mouse embryos. The transplantability of ascitic hepatoma was reduced by antisera to 9- and 11-day-old mouse embryos more than by antitumor sera and by antisera to 13-day-old mouse embryos and adult mouse tissues. The duration of the survival of mice inoculated with tumor cells incubated with antisera to 5- and 7-day-old chicken embryos and to 9- and 13-day-old mouse embryos was longer than that of mice which received tumor cells incubated with antisera to 11- and 13-day-old chicken embryos, to 18-day-old mouse embryos, and to tissues of adult mice and chickens. The survival time was the same after incubation of the tumor cells with antisera to 5-day-old chicken embryos, 9- and 13-day-old mouse embryos, or to the tumor. The antiserum to 11-day-old mouse embryo showed a more powerful antitumor effect than the antitumor antiserum. The findings indicate that antisera to mouse and chicken embryos at the early stages of embryogenesis have the highest tumor inhibiting activity, and that these tumor cells acquire certain traits of embryonal cells and tissues of early stages of the embryogenesis.

4452   SERUM ANTI-IMMUNOGLOBULINS IN MULTIPLE
         MYELOMA AND BENIGN MONOCLONAL GAMMOPATHY.
(Eng.) Lindstrom, F. D. (Univ. New Mexico Sch. Med., Albuquerque); Williams, R. C., Jr. *Clin. Immunol. Immunopathol.* 3(4):503-513; 1975.

Sera from 111 patients with immunoglobulin G (IgG) myeloma (MM) and 76 patients with IgG monoclonal gammopathy (BMG) were examined for the presence of anti-IgG antibodies by the method of agglutination of Rh-positive cells coated with various human incomplete Rh antibodies. Definite positive agglutinating reactions were seen in 49.5% of the MM patients compared with 26.3% of the patients with BMG and 28% of 100 age-matched controls. In general, the specificity of the anti-IgG agglutinating activities in most sera appeared to be directed against determinants present in whole human Cohn Fraction II and in some instances Fc, but not present in Fab or F(ab)$_2$. No clear evidence for anti-idiotypic reactivity with autologous M components could be obtained. In most sera, anti-IgG factors appeared to be present in low-molecular weight fractions (3.5-5.0S). The apparent increase in anti-Ig factors in MM as opposed to BMG may reflect a basic difference in control mechanisms. It is possible that tumor-specific antigens produce enhancing antibodies capable of abrogating a cell-mediated immune response of benign plasma cells and encouraging the proliferation of malignant plasma cells.

4453   TRIGGERING MECHANISM OF B LYMPHOCYTES.
         II. INDUCTION OF ORNITHINE DECARBOXYLASE
IN B CELLS BY ANTI-IMMUNOGLOBULIN AND ENHANCING SOLUBLE FACTOR. (Eng.) Watanabe, T. (Osaka Univ. Medical Sch., Fukushima-ku, Osaka, 553, Japan); Kishimoto*, T.; Miyake, T.; Nishizawa, Y.; Inoue, H.; Takeda, Y.; Yamamura, Y. *J. Immunol.* 115(5):1185-1190; 1975.

In vitro stimulation of rabbit lymphocytes with anti-immunoglobulin antibody (anti-Ig) induced ornithine decarboxylase (ODC) in the cells. The enzyme activity reached the maximum at eight hr after stimulation of the cell with antibody and returned to the unstimulated level after 20 hr. The optimum concentration of antibody to produce the maximum activity of ODC was 25 µg/ml, which was in the same range as that required for the maximum production of IgG and the highest rate of DNA synthesis. ODC was again induced with soluble factor (SF) in the cells that had been previously treated with anti-Ig antibody for 24 hr. The treatment of the cells with anti-thymocyte serum and complement did not affect the activity of the enzyme induced by anti-Ig or by SF, indicating that enzyme induction was an event occurring in B cells and not in T cells. The addition of mitotic inhibitors, such as hydroxyurea (HU) or cytosine arabinoside (Ara-C), to the culture with anti-Ig did not inhibit the production of ODC. The presence of cytochalasin-B with anti-Ig did not inbibit, but rather enhanced, the synthesis of ODC. On the other hand, the presence of HU or cytochalasin-B with soluble factor completely suppressed the induction of ODC. These results suggested that intracellular events that trigger the induction of ODC by the interaction of Ig receptor with anti-Ig may be different from the events that induce the same enzyme triggered by SF.

4454　　MULTIPLE INDIVIDUAL AND CROSS-SPECIFIC
　　　　 IDIOTYPES ON 13 LEVAN-BINDING MYELOMA
PROTEINS OF BALB/c MICE. (Eng.) Lieberman, R. (Natl. Inst. Allergy and Infectious Diseases, Bethesda, Md. 20014); Potter, M.; Humphrey, W.; Jr.; Mushinski, E. B.; Vrana, M. *J. Exp. Med.* 142(1): 106-119; 1975.

Thirteen levan-binding myeloma proteins (LBMP) of BALB/c origin were classified into two groups with different binding specificities; one group of 11 proteins bound β2→1 fructosans, and a second group of two proteins bound fructosans probably of β2→6 linkage. Anti-idiotypic sera prepared to ten of the proteins in the appropriate strains of mice identified numerous idiotypic determinants. Each protein used for immunization had its own unique individual idiotypic specificities (IdI) and in addition most of the proteins carried 2-9 cross-specific or shared idiotypes (IdX) that were found only among LBMP, and not found in 106 non-LBMP. Most of the IdX determinants and only four of the IdI determinants of the β2→1 fructosan binding group were located in the antigen-binding site. The multiplicity of antigenic differences in this functionally related group of immunoglobulins reveals an unexpected degree of heterogeneity in V-regions that appears to be unrelated to binding.

4455　　DIFFERENCES IN SUSCEPTIBILITY OF MATURE
　　　　 AND IMMATURE MOUSE B LYMPHOCYTES TO ANTI-
IMMUNOGLOBULIN-INDUCED IMMUNOGLOBULIN SUPPRESSION
IN VITRO: POSSIBLE IMPLICATIONS FOR B-CELL TOLERANCE TO SELF. (Eng.) Raff, M. C. (Dept. Zoology, University Coll. London, WC1E 6BT, England); Owen, J. J. T.; Cooper, M. D.; Lawton, A. R., III; Megson, M.; Gathings, W. E. *J. Exp. Med.* 142(5):1052-1064; 1975.

Organ explants and dissociated cells from fetal and adult BALB/c mouse lymphoid tissues were used to study the *in vitro* effects of anti-Ig antibodies on developing and mature immunoglobulin M (IgM)-bearing B lymphocytes. Purified goat antibodies against mouse µ-chains and rabbit antibodies against mouse immunoglobulin determinants, and their Fab fragments, inhibited the development of IgM-bearing B cells in explant cultures of 14-day mouse fetal liver, and caused the disappearance of cell surface IgM in explant and dissociated cell cultures of more develope lymphoid tissues. While treatment of cultures of fetal or newborn liver, or adult bone marrow, with low concentrations (<10 µg/ml) of anti-Ig for 24 hr or less caused the complete, but reversible, disappearance (modulation) of cell surface IgM, treatment for 48 hr or more produced irreversible IgM suppression. In contrast, anti-Ig-induced suppression of cell surface IgM in cultures of adult spleen or lymph nodes required much higher concentrations of antibody (≥ 100 µg/ml) and was always reversible. These differences between immature and mature IgM-bearing cells could not be related to differences in the amount of surface IgM on the cells. The remarkable sensitivity of newly formed B cells to IgM modulation and irreversible IgM suppression when ligands bind to their Ig receptors, may have important implications for B-cell tolerance to self antigens.

4456　　DEVELOPMENT AND INHIBITION OF CYTOTOXIC
　　　　 ANTIBODY AGAINST SPONTANEOUS MURINE BREAST
CANCER. (Eng.) Stolfi, R. L. (Catholic Med. Cent. Brooklyn Queens, Inc., Woodhaven, N.Y.); Fugmann, R. A.; Stolfi, L. M.; Martin, D. S. *J. Immunol.* 114 (6):1824-1830; 1975.

An investigation was conducted to elucidate the various components and to determine the pattern of humoral immunity in a spontaneous, autochthonous murine breast tumor system. Female (BALB/cfC3H x DBA/8)F_1 hybrid mice (CD$_8$F$_1$), infected maternally with mammary tumor virus, or CD$_8$F$_1$ male mice bearing a first-generation tumor transplant were used. For the detection of tumor-specific, complement-dependent cytotoxic antibody, reaction conditions were modified to maximize cytolytic complementation with mouse serum. When modified conditions were used, tumor-specific cytotoxic antibody was detected in high dilutions of sera (1:100-1:1000), but not at low dilutions (1:10). Since such a pattern of reactivity suggested the presence of an inhibitor of cytotoxic antibody in high concentrations of serum, further tests were conducted to determine its nature. Sequential measurements subsequent to primary tumor implantation revealed that, although the time of appearance of this inhibitor varied from tumor to tumor, it always became detectable simultaneously with significant levels of lymphocyte-blocking activity, the inhibitor of cytotoxic antibody activity was shown to operate through a competitive antigen-binding and blocking mechanism. These data suggest the possibility that a single serum moiety may be responsible for the efferent blocking of both the cellular and the humoral components of the immune tumor rejection response.

netics of antibody formation in tumor-bearing hosts
and transplant recipients.

4459   DETECTION OF ANTIBODIES TO Ia ANTIGENS IN
        H-2 ALLOANTISERA. (Eng.) Sachs, D. H.
(Natl. Cancer Inst., Bethesda, Md.); Fathman, C. G.;
Cone, J. L.; Dickler, H. B. *Transplant. Proc.* 7(1/
Suppl. 1):123-126; 1975.

The presence of Ir-associated (Ia) murine cell-surface antigen specificities in H-2 antisera was studied by absorption and testing of antisera on lymphoid cells from H-2-recombinant strains differing
genetically only in the Ir region. An H-2 allo-
antiserum [(B6 x A)F$_1$ anti-B10.D2], for which recombinants capable of mapping component Ia specificities are not yet available, was examined for its
ability to show differential absorptive properties
and to block the binding of aggregates to the B-cell
Fc receptor. In a fashion analogous to that previously observed with H-2 antisera containing known
Ia specificities, absorption of (B6 x A)F$_1$ anti-
B10.D2 antiserum with an Ia-negative ascites tumor
removed the reactivity with the tumor but not the
reactivity with a subpopulation of normal splenic
lymphocytes. As shown by testing of the absorbed
antiserum on nylon column-separated spleen cells,
the residual cytotoxic activity was directed predominantly against B cells. Preincubation of splenic lymphocytes with the antiserum increased the
percentage of cells staining with fluorescent rabbit antimouse immunoglobulin reagent (which detects
B-cell surface immunoglobulin), but markedly inhibited the binding of immunoglobulin aggregates to
Fc receptors. Lack of complete absorption by the
Ia-negative tumor and the ability of the serum to
block binding to the Fc receptor both indicate the
probable presence of anti-Ia activity with this H-2
antiserum.

4460   SENSITIVITY OF THE EPSTEIN-BARR VIRUS
        TRANSFORMED HUMAN LYMPHOID CELL LINES TO
INTERFERON. (Eng.) Adams, A. (Dept. Tumor Biology,
Karolinska Inst., Stockholm 60, Sweden); Strander,
H.; Cantell, K. *J. Gen. Virol.* 28(2):207-217; 1975.

The effect of interferon on expression of Epstein-
Barr virus (EBV) early gene functions in 14 human
EBV-transformed lymphoid cell lines was investigated.
The 'early antigen' synthesis which follows either
EBV superinfection of established lymphoid cell
lines or 5'-iododeoxyuridine activation of the intrinsic EBV genomes harbored by these cells was suppressed with interferon (1, 10, 100, 1000, and
10,000 U/ml, 48 hr). In contrast, the spontaneous
early antigen expression that occurs in a few cells
in the producer cell lines could not be blocked with
interferon. The lymphoid cell lines tested differed
in their ability to acquire an antiviral state after
exposure to interferon. Several cell lines were
also growth inhibited by the interferon preparations.
The antiviral and growth inhibitory activities of
different interferon preparations could not be separated by a number of criteria. The growth inhibitory
activity was nondialysable, nonsedimentable (two
hours, 100,000 x g), and was destroyed by trypsin

digestion. It also precipitated with interferon
when crude WBC interferon preparations were heated
at 70 C for one hour, and the antiviral and growth
inhibitory activities had the same kinetics of inactivation both at 60 and 75 C. Final proof that the
anti-growth substance is interferon itself will require access to interferon purified to homogeneity.

4461   CLINICAL SIGNIFICANCE OF α_2H-GLOBULIN.
        (Eng.) Yachi, A. (Sapporo Medical Coll.,
Sapporo, Japan); Akahonai, Y.; Takahashi, A.; Wada,
T. *Ann. N.Y. Acad. Sci.* 259:435-445; 1975.

Incidences of α_2H-globulin (α_2H), an iron-containing
glycoprotein, in sera of patients with neoplastic
and nonneoplastic diseases are reported. Serum α_2H
was determined by a counterimmunoelectrophoresis
procedure in which an agarose gel plate containing
4% specific anti-α_2H serum was punched out with wells
3 mm in diameter into which 9 μl samples of test
serum were introduced. The plate was then subjected
to constant current of 2 mA/cm plate width for a
running time of about 5 hr. Concentration of α_2H was
expressed as an arbitrary unit defined in terms of a
17 mm peak height of precipitate corresponding to
1 U/ml. In the sera of 13 of 37 cases (35%) with
histologically proven hepatocellular carcinoma, α_2H
was positive. In sera of patients with gastric, pancreatic, colonic, and bronchogenic carcinoma, incidences of α_2H were 8.3-38.5%. The incidence was
93% for patients with acute myelogenous leukemia
(AML), and 60% for patients with multiple myeloma.
The changes in serum α_2H levels in cases with AML
showed increases in stages of relapse, whereas
levels decreased in remission following chemotherapy. In chronic myelogenous leukemia, however,
only 1 of 7 cases were α_2H-positive, this case
being in a stage of acute blastic crisis. In contrast, healthy adult subjects were all α_2H-negative.
With regard to cord blood serum, α_2H was detectable in 4 of 42 cases. In nonneoplastic disorders,
however, α_2H was positive in sera of 3 out of 5
cases with aplastic anemia. All other cases with
various diseases including hepatic, renal, and
connective tissue disorders were α_2H-negative. It
was not known whether or not serum α_2H is identical with tumor-associated and acidic isoferritin
of carcinofetal or carcinoplacental origin.

4462   ANTIGENIC IDENTITY OF ALPHA-MACROFETO-
        PROTEIN AND ACUTE PHASE PROTEIN IN THE
RAT. (Eng.) Boss, J. H. (Hebrew Univ.-Hadassah
Medical Sch., Jerusalem, Israel); Dishon, T.; Rosen-
mann, E. *Biomedicine* 23(6):196-197; 1975.

α-Macrofetoprotein (AMFP) is present in the serum
of rat embryos and neonates, pregnant and postpar-
tum females, and hepatoma-bearing animals. Acute
phase protein (APP), an alpha-2-AP globulin, occurs in the serum of rats with acute toxic liver
injury. The possibility that these proteins might
be identical was investigated by cross reacting
antisera raised in rabbits against embryonic serum
(ES) and against inflammatory serum (IS) from adult
male rats injected with turpentine (1 ml, sc and
im) 24 hr previously. In Ouchtelony's double im-

treatment was confirmed by reelectrophoresis of separated and similarly treated AFP_a and AFP_b. Two bands of sialized or desialized AFP were also observed on isoelectric focusing. Both AFP_a and AFP_b treated with and without neuraminidase gave single fused precipitin lines against the antiserum in Ouchterlony double-diffusion analysis. On the basis of the changes in electrophoresis mobilities of the intermediates following neuraminidase treatment, AFP_a and AFP_b were estimated to have at least 2.5 and 4.5 molecules of sialic acid per molecule, respectively.

4466   EVOLUTION OF α-FETOPROTEIN SERUM LEVELS
             THROUGHOUT LIFE IN HUMANS AND RATS, AND
DURING PREGNANCY IN THE RAT. (Eng.) Masseyeff, R. (Faculte de Medicine, Nice, France); Gilli, J.; Krebs, B.; Calluaud, A.; Bonet, C. *Ann. N.Y. Acad. Sci.* 259:17-28; 1975.

The evolution of serum α-fetoprotein (AFP) levels in man and rat was studied throughout the life span, using a radioimmunoassay method (RIA) with specific antisera for human and rat AFP, respectively. Sera of 201 clinically normal children, 192 healthy adult blood donors, and 16 individuals aged 70-98 were assayed. Sera of 66 Sprague-Dawley rats aged 3-70 wk were also assayed. In man, AFP levels decreased steeply during the first year and reached a low basal range by the end of the second year. This low basal range was maintained throughout adult life. The rat showed a similar pattern, although at a higher concentration at every stage. A difference between these two species was that puberty occurred while AFP was still decreasing in the rat, whereas in many, the adult basal level was achieved long before the onset of puberty. In both species the normal range was more scattered during the beginning of life than it was thereafter. The evolution of maternal serum AFP was also studied in 29 Sprague-Dawley pregnant rats. AFP levels began to increase sharply on the 10th-13th day of pregnancy. From the 13th to the 19th day a deceleration of AFP increase was observed that led to a decrease between the 17th and 19th day. From the 19th to the 21st day AFP levels again increased steeply. A significant correlation was observed on the 21st day of pregnancy between the number of fetuses of the progeny and the maternal level of AFP. A reduction of the AFP level occurred following hemihysterectomy that was grossly proportional to the number of remaining fetuses. The persistance of a stable level during adult life can be explained solely by the stability of the synthesis of this protein. Conversely, in pregnancy the evolution of the serum AFP curve is dependent of the fetal synthesis, maternal and fetal catabolism, and permeability of the placental barrier.

4467   α-FETOPROTEIN LEVELS IN NORMAL MALES
             FROM SEVEN ETHNIC GROUPS WITH DIFFERENT
HEPATOCELLULAR CARCINOMA RISKS. (Eng.) Sizaret, P. (International Agency for Res. on Cancer, 69008 Lyon, France); Tuyns, A.; Martel, N.; Jouvenceaux, A.; Levin, A.; Ong, Y. W.; Rive, J. *Ann. NY Acad. Sci.* 259:136-155; 1975.

α-Fetoprotein (AFP) levels of 1,335 men (15 yr and older) of seven ethnic groups (Chinese, Indians, and Malays from Singapore, Caucasians from Lyon, and Negroes from Nairobi, forest, and the savanna region of the Ivory Coast) were determined by radioimmunoassay. These levels were obtained in an attempt to correlate AFP with hepatocellular carcinoma (HCC) risk. A few elevated levels (up to 30 nU/ml) were detected in some normal individuals, especially in the older age-groups. There was a systematic age-dependency of AFP levels particularly evident in the groups from Singapore-Lyon, in which there was a 50% AFP increase between the ages of 20 and 40. Comparison between Africans on the one hand and people from Singapore-Lyon on the other hand revealed highly significant differences (p < 0.001), especially in the younger groups, whereas Chinese, Malays, and Indians from Singapore had very similar AFP patterns; this suggests an important role for environmental factors in the regulation of AFP levels. The age dependency of the presumed effect of environmental factors is in keeping with previous data showing that young animals respond more vigorously to AFP-stimulating factors. Although the incidence of HCC differs in the three Singapore groups (the highest in Chinese and the lowest in Indians), no relationship was observed here between mean AFP level and HCC incidence in Singapore.

4468   EARLY DETECTION OF HEPATOMA: PROSPECTIVE
          STUDY IN LIVER CIRRHOSIS USING PASSIVE
HEMAGGLUTINATION AND THE RADIOIMMUNOASSAY. (Eng.)
Lehmann, F.-G. (Dept. Medicine, Univ. Marburg, Marburg/Lahn, West Germany). Ann. NY Acad. Sci. 259: 196-210; 1975.

A prospective study of serum α-fetoprotein (AFP) elevation in patients with liver cirrhosis was carried out, and the results were compared with clinical conditions and biochemical parameters (SGOT, SGPT, bilirubin, prothrombin complex time, and γ-globulin). The possibility was also investigated of a general screening of patients with precancerous disease. The passive hemagglutination method involved the use of SRBC to which rabbit IgG antibody prepared against crystalline human AFP was bound with bis-diazotized benzidine. The radioimmunoassay was carried out with ^{125}I-labeled AFP in a double antibody technique using anti-AFP serum from the rabbit as the first antibody and anti-rabbit-γ-globulin serum from the goat as second antibody. Double diffusion and electroimmunoosmophoresis tests were also carried out. The study included 215 patients with liver cirrhosis, of which 165 showed negative results in all four tests. Thirty-one patients of this group died during the study, and were found on autopsy to have liver cirrhosis without carcinoma. Three out of the 215 patients showed positive results in all four tests. The three patients revealed primary liver cell carcinomas. Forty-seven out of the 215 patients yielded negative results by the less sensitive double diffusion and electroimmunoosmophoresis tests, but positive results in the sensitive passive hemagglutination and radioimmunoassay tests. In approximately 90% of the cases, a temporary rise of AFP up to 2,000 ng/ml was not associated with tumor growth at the time of

measurement. There was no correlation of AFP level with actual liver dystrophy as determined by the various biochemical parameters. Primary liver cell carcinoma was found to be more frequent in post-hepatitic than in postalcoholic, cryptogenic, and other cirrhosis, and to be more frequent in Australia-antigen positive cases than in Australia-antigen negative cases. Routine serological tumor antigen screening of patients with a precancerous disease is considered to be useful.

4469   SERUM ALPHA-FETOPROTEIN AND BETA-HUMAN
          CHORIONIC GONADOTROPIN LEVELS IN
PATIENTS WITH NON-SEMINOMATOUS GERM CELL TESTICULAR
CANCER. (Eng.) Lange, P. H. (Univ. Minnesota, Coll. Health Sciences, Minneapolis, Minn.); Hakala, T. R.; Fraley, E. E. Minn. Med. 58(11):813-815; 1975.

Sera from 146 patients with genito-urinary disease were assayed for the presence of alpha fetoprotein (AFP) and beta human chorionic gonadotropin (B-HCG) by recently developed radioimmune assays. Elevated AFP values were found only in patients with non-seminomatous germ cell tumors of the testicle (12/22). B-HCG elevation was present in hypernephroma (3/50), transitional cell carcinoma (3/46), and in testicular tumor patients (12/22), but the levels were greater than 10 mIU/ml in the latter group only. Of 15 sera tested both by gel-agar precipitation and radioimmune assay, ten had elevated AFP values by radioimmune assay, whereas only three were positive by the conventional method. The use of these new serum assays for AFP and B-HCG may have broad applications, both to the diagnosis and management of certain testicular neoplasms.

4470   HUMAN α-FETOPROTEIN IMMUNOCHEMICAL ANAL-
          YSIS OF ISOPROTEINS. (Eng.) Alpert, E.
(Harvard Medical Sch., Boston, Mass. 02114); Perencevich, R. C. Ann. N.Y. Acad. Sci. 259:131-135; 1975.

The basis for separation of human α-fetoprotein (AFP) into two bands during isoelectric focusing was studied. Resolution of the isoproteins into a more uniform band after neuraminidase treatment suggested that the presence of terminal sialic acids contributed to heterogeneity. Staining of the gels after immunofixation with specific antibody suggests that there may be quantitative or qualitative differences in the antigenic determinants of one of the isoproteins. The existence of AFP isoproteins may explain the wide variation found in AFP standards and serum concentrations.

4471   PLASMA LEVELS OF CARCINOEMBRYONIC ANTIGEN
          IN BRONCHIAL CARCINOMA AND CHRONIC BRON-
CHITIS. (Eng.) Pauwels, R. (Academic Hosp., State Univ., Ghent, Belgium); van der Straeten, M. Thorax 30(5):560-562; 1975.

To determine the clinical value of plasma carcinoembryonic antigen (CEA) in bronchial carcinoma and also whether plasma CEA levels could be used to

of three components of carcinoembryonic antigen (CEA),
the antigen was extracted from liver metastases of
colonic tumor tissue with perchloric acid and purified
by chromatography on Sepharose 4B and Sephadex G-200.
It was then fractionated on a column of concanavalin A
Sepharose to give unbound fraction 1, 2% methyl glu-
eoside elutable fraction 2, and 10% methyl glucoside
elutable fraction 3. The recovery of protein in each
fraction was 10, 50 and 20%, respectively. To test
the effect of removal of sialic acid on the fractionation, CEA was also treated with neuraminidase. Rabbit
antisera were prepared against unfractionated CEA
and fraction 2 and were absorbed with normal colon,
spleen, and human plasma before use in double diffusion studies. Established antisera which were mono-
specific for CEA-unique antigen A were obtained by
donation. Circular dichroism spectra were measured
on a Cary 61 recording spectropolarimeter. Results
of the double diffusion tests showed that each fraction reacted with the established anti-A antisera
and also with antisera specific for a determinant,
C, common to both CEA and a glycoprotein of normal
tissues. Removal of sialic acid from CEA appeared
to have no effect on its fractionation on concanavalin A Sepharose or on its binding to antibody.
Further, the presence of concanavalin A up to a 40-
fold excess did not inhibit the binding of CEA to
antibody. These results suggest that con A binding
sites are unlikely to be exposed by removal of sialic
acid. Results of carbohydrate analyses showed differences in the three subfractions with respect to
content of neutral sugar, galactose, fucose, hexosa-
mine, and sialic acid. Results of circular dichroism
studies also indicated the existence of chemical differences.

4474   IMMUNOPATHOLOGICAL STUDIES ON CEA AND CEA-
        ASSOCIATED ANTIGENS WITH REEVALUATION OF
THE CANCER SPECIFICITIES OF CEA. (Eng.) Mori, T.
(Osaka Univ. Medical Sch.. Osaka, Japan); Wakumoto,
H.; Shimano, T.; Lee, P.-K.; Higashi, H. *Ann. N.Y.
Acad. Sci.* 259:412-416; 1975.

The distribution of targets to anti-carcinoembryonic
antigen (CEA) produced in guinea pigs, was studied
in 228 surgical specimens. Four antibodies, anti-
$CEAf_1$ (fecal antigen), anti-CEAgc (reacting with
gastric and colonic cancer), anti-CEAg (gastric cancer only) and anti-$CEAf_2$ (from a patient with advanced gastric cancer) were used. CEAgc was detected
in 52 of 84 gastric cancers, all of 14 colo-rectal
cancers, 1 of 4 pancreatic cancers, as well as in
atypical glands of the digestive organs. None of
the examined tissues from other organs fluoresced
with anti-CEAgc. Atypical glands in the stomach
with positive CEAgc showed intestinal metaplasia,
but not *vice versa*. Differentiated or mucinous
cancers of the stomach were mostly CEA-positive.
Atypical glands with CEAgc in the colon were often
observed in ulcerative and granulomatous colitis,
in polyps, and in areas bordering on cancers. Anti-
CEAg showed fairly good reactivity to gastric cancer. The CEA in adenocarcinoma of organs other than
digestive system was a CEA-associated fecal antigen.
The CEA in squamous cell carcinoma of the endodermal
organs was a different CEA-associated fecal antigen.
CEA was detected exclusively in adenocarcinoma and

atypical glands of the digestive organs. Minor
antigenic differences of CEA among gastric cancer,
colonic cancer, and atypical glands were demonstrated.

4475   EMBRYONAL ANTIGEN OF HAMSTER ADENOVIRUS
        SARCOMA. (Rus.) Bashkaev, I. S. (The
P. A. Herzen Res. Inst. of Oncology, Moscow, USSR);
Ageenko, A. I. *Vopr. Onkol.* 21(6):93-96; 1975.

Sarcomas induced in hamster by human adenovirus
type 12 were investigated for the presence of embryonal antigen. Sarcoma type A-12, induced by the
introduction of human adenovirus type 12 into newborn hamsters, sarcoma type A-12, transplanted into
adult hamsters, and sarcomas induced in hamsters
by the introduction of cell suspensions from lungs
without macroscopically visible metastases of sarcoma A-12-bearing animals were used for the experiments. The embryonal tissue was taken from pregnant
hamsters between the 10th and 13th days of pregnancy.
Embryonal antigen was detected in 50% of both primary and transplanted A-12 sarcomas. This antigen
was identical with one of the embryonal antigens
detected in the embryonal extract with antiembryonal
serum. This embryonal antigen was not present in
tumors induced with lung cell suspensions from
tumor-bearing animals. The embryonal antigen was
found to be serologically identical with the second
type of lung tissue antigen that is detectable by
sarcoma A-12-specific serum. Consequently, this
antigen is also present in the hamster lung. The
tissue protein that is serologically identical with
the embryonal antigen is apparently increased in
the adenovirus-induced tumor. Since this antigen
is specific for the embryon, and partly also for the
tumor, it may be regarded as a carcinofetal antigen.
The negative reaction with the antiembryonal serum
in 50% of the A-12 sarcomas tested is apparently
due to the low concentration of the embryonal antigen in these tumors.

4476   INVESTIGATION OF SPECIFIC ANTIGENS IN
        PROSTATIC CANCER. (Eng.) Moncure, C. W.
(Medical Coll. Virginia, Virginia Commonwealth
Univ., Richmond, Va. 23219); Johnston, C. L., Jr.;
Koontz, W. W., Jr.; Smith, M. J. V. *Cancer Chemo-
ther. Rep. (Part 1)* 59(1):105-110; 1975.

New Zealand white rabbits immunized with normal
human prostatic homogenates (0.5 ml/week for 16 wk,
sc) have been shown to develop antibodies specific
for one form of prostatic acid phosphatase and a
protein antigen which does not appear to possess
enzymatic activity. Antibodies were also developed
against several other proteins, esterase, two aminopeptidases, and a second form of acid phosphatase.
Extensive absorption of the antisera with gluteral-
dehydepolymerized serum and kidney homogenate removed all antibodies except those which react with
the specific acid phosphatase and the specific protein. These two antigens have been separated from
each other by ion-exchange chromatography. The
partially purified acid phosphatase has been used
to produce antiserum employed for the detection of
prostatic cancer in bone marrow aspirates. Four
patients with known stage IV prostatic cancer were

4481   THE OCCURRENCE OF THE HL-B ALLOANTIGENS
        ON THE CELLS OF UNCLASSIFIED ACUTE LYMPHO-
BLASTIC LEUKEMIAS. (Eng.) Fu, S. M. (Rockefeller
Univ. New York, N.Y. 10021); Winchester, R. J.;
Kunkel, H. G. *J. Exp. Med.* 142(5):1334-1338; 1975.

The application of B-cell markers from pregnancy
sera to the classification of leukemia is described.
Six cases of acute lymphoblastic leukemia (ALL) were
studied by T- and B-cell membrane markers. Cases
with a high percentage of peripheral lymphoblasts
were selected. In cases A and B, 98-99% of the
lymphoblasts formed rosettes with sheep RBC; these
lymphoblasts had neither surface immunoglobulin
(Ig) nor the Fc receptors. In the remaining four
cases (C-F), the lymphoblasts had no detectable
surface Ig of classes δ, μ, γ, α, κ, and λ. In
cases E and F, only 60% and 41% of the mononucleated
cells were blasts. These blasts did not form E
rosettes, whereas a substantial percentage of the
small lymphocytes did. Indirect fluorescent studies with HL-B-specific pregnancy sera were performed. In cases A and B, no significant numbers
of the lymphoblasts stained with any of the sera.
In case C, two of the four sera used stained practically all the lymphoblasts while one stained only
20% and the other stained less than 2%. A similar
staining pattern was shown in case D. In cases E
and F, the 57% and 36% of cells stained by a pregnancy serum represented the staining of the majority
of lymphoblasts in the peripheral blood of these
cases. Cases A and B appear to be T-cell leukemias.
The detection of Ia-related HL-B antigens on the
cells from cases C-F indicate that these cases may
be of B-cell lineage.

4482   HL-A IN HODGKIN'S DISEASE. III. A PROS-
        PECTIVE STUDY. *(Eng.)* Kissmeyer-Nielsen,
F. (Univ. Hosp., Copenhagen, Denmark), Kjerbye, K.
E.; Lamm, L. U. *Transplant Rev.* 22:168-174, 1975.

A prospective study was performed by HL-A typing of
all new patients included in the Danish Hodgkin
project, LYGRA, and the results were compared to
those of previous studies. All 201 patients (80
women and 121 men) followed the same protocol concerning diagnosis and treatment. The control material consisted of 562 normal caucasian blood
donors which were randomly HL-A typed. The frequency of the majority of HL-A antigens was similar to the control sample; the most deviating of
the HL-A antigens was HL-A1, involving a 10% increase in patients as compared to the controls.
The results did not confirm previous findings in
two independent series of patients and did not support the increase of other HL-A antigens. One explanation for the discrepancies is that the prospective nature of this study accounts for the differences from the previously published mixtures of
prospective and retrospective cases. Twenty-six
patients dying from malignant diseases showed a
decreased frequency of antigens as compared to an
increase in these antigens found in previous retrospective studies. The authors conclude that HL-A1
is increased in patients acquiring Hodgkin's disease, although the relative risk is only slightly
increased (1.53) and that there seems to be an as-

sociation between HL-A1 and HL-A8 and sex, age of
onset, and histology.

**4483   LYMPHOCYTE-ASSOCIATED ANTIGENS IN PATIENTS
WITH ALVEOLAR CELL CARCINOMA.** (Eng.) Mohr,
J. A. (Veterans Admin. Hosp., 921 N.E. 13 St., Okla-
homa City, Okla.); Nordquist, R. E.; Coalson, R. E.;
Rhoades, E. R.; Coalson, J. J. J. Lab. Clin. Med.
86(3):491-493; 1975.

Antiserum produced in rabbits against a virus-like
particle pellet recovered from human alveolar cell
carcinoma (ACC), was used for the detection of lymphocyte-associated antigen in patients with ACC. The
method used was direct immunofluorescence. Five of
11 patients with ACC demonstrated bright fluorescence,
with each having 2-12% total fluorescing lymphocytes.
All five of these patients had widespread pulmonary
ACC and/or metastatic disease, while the six ACC
patients without fluorescence had localized pulmonary
ACC. None of the other test subjects (55 healthy
subjects, 14 patients with malignancies other than
ACC, 20 patients with nonmalignant disease, and
seven patients with lupus nephritis) had any cells
that fluoresced. The results demonstrate that lymphocytes from some patients with ACC possess antigen
and the presence of lymphocyte fluorescence in
patients with ACC suggests widespread or metastatic
disease. The presence of antigen may be of both
diagnostic and prognostic value.

**4484   EXTRACTION OF TUMOR-SPECIFIC ANTIGEN FROM
CELLS AND PLASMA MEMBRANES OF LINE-10 HEPA-
TOMA.** (Eng.) Leonard, E. J. (Natl. Cancer Inst.,
Bethesda, Md. 20014); Richardson, A. L.; Hardy, A.
S.; Rapp, H. J. J. Natl. Cancer Inst. 55(1):73-79;
1975.

The properties of tumor-specific antigen from line-
10 ascites variants of diethylnitrosamine-induced
hepatoma were studied. Tumor-specific antigen was
extracted with 3 M KCl from line-10 guinea pig hepatoma cells. The yield of antigenic activity, estimated by production of delayed cutaneous hypersensitivity reactions in line-10 immune guinea pig,
was 10-30% of the antigen present in intact cells.
By ultracentrifugation criteria, the extracted antigen was soluble. Gel filtration, ion exchange chromatography, and salting-out studies showed that the
antigen was heterogeneous in size and net charge.
The possibility that 3 M KCl extracted a homogeneous population of molecules associating into poly-
mers of various sizes at low ionic strength was
ruled out by heterogeneity on Sephadex G-200 chromatography at high ionic strength. After osmotic
lysis of sucrose-loaded line-10 cells, whole plasma
membranes or large membrane fragments were obtained
in a yield of about 20%. The isolation procedure
did not cause detectable loss of membrane antigenic
activity. The membranes had 33 skin test U/mg membrane protein, compared to the intact cell value of
1.7 skin test U/mg cell protein. Extracts of plasma
membranes had 10-30% of the antigenic activity of
the starting membrane material. In contrast to the
wide variety of proteins liberated from intact cells,
much of the protein extracted from the membranes was
in the molecular wt range above 250,000.

**4485   TUMOR-SPECIFIC ANTIGEN RELATED TO RAT
HISTOCOMPATIBILITY ANTIGENS.** (Eng.) Bo-
wen, J. G. (Lab. for Experimental Surgery, Erasmus
Univ., Rotterdam, The Netherlands); Baldwin, R. W.
Nature 258(5530):75-76; 1975.

A tumor-specific antigen preparation isolated from
the serum of Wistar rats bearing transplants of a
4-dimethylaminoazobenzene-induced hepatoma (D23)
was used to test the hypothesis that tumor-associated antigens (TAAs) of chemically induced tumors
are modified histocompatibility antigens. Antisera
against highly purified hepatoma D23 antigen were
raised in New Zealand white rabbits and Sloaker
rats and were tested by the indirect membrane immunofluorescence test for the presence of antibody directed against the tumor-specific antigen on viable
D23 cells. The antisera reacted not only against
D23 cells but also against two immunologically distinct hepatomas, D30 and D33, indicating that the
sera recognized antigenic similarities between the
three tumors. After absorption in vivo and in vitro
with normal Wistar liver membranes, the antisera
retained their reactivity against D23 but lost
their reactivity against D30 and D33. The possibility that hepatoma D23 antigen is a modified histo-
compatability antigen was further explored by ex-
amining the binding of ^{125}I-labeled D23 antigen to
columns of immunoadsorbents prepared from Sloaker
antinormal Wistar liver antiserum, normal Sloaker
serum, Wistar antihepatoma D30 antiserum, and Wistar
antihepatoma D23 serum. Results of the binding assay showed that D23 antigen has points of identity
with antigens recognized by Sloaker rats on normal
Wistar liver membranes. This work confirms that the
hepatoma D23 antigen is a modified histocompatibility antigen.

**4486   LYMPHOCYTE RESPONSE TO AUTOLOGOUS TUMOR
ANTIGEN(S) AND PHYTOHEMAGGLUTININ IN
OVARIAN CANCER PATIENTS.** (Eng.) Chatterjee, M.
(Gynecology Dept., Roswell Park Memorial Inst.,
666 Elm St., Buffalo, N.Y. 14203); Barlow, J. J.;
Allen, E. J.; Chung, W. S.; Piver, M. S. Cancer
36(3):956-962; 1975.

The response of lymphocytes from 11 patients with
ovarian carcinoma to autologous tumor antigens
and phytohemagglutinin (PHA) was studied. Lymphocytes from age matched normal women served as controls. Tumor antigen was prepared by homogenizing
tumor tissue, and immunodiffusion was performed to
determine antigen concentration (5 mg/ml). Lymphocyte suspensions were prepared on each patient and
control and were incubated with 2 µCi of tritiated
thimidine in the presence of tumor antigen or PHA.
Blastogenic responses were calculated as dpm of
tritiated thymidine incorporated. Tumor extracts
were tested for potential cytotoxic effect and
were found to be negative with both autologous and
homologous lymphocytes, and with 10% homologous
serum tumor extract was shown to have no inhibitory
effect. It was found that four tumor extracts
shared the major precipitin band of pooled tumor
extracts indicating the presence of cystadeno
carcinoma associated antigens in these extracts and
only these extracts which had tumor associated

reactive against an antigenic determinant(s) was
determined using anti-idiotypic antibodies as a probe.
Anti-idiotypic antibodies made against the antigen-
binding receptors of T lymphocytes for a given antigen (Ag-B locus antigens in Lewis and DA rats) were
shown to react with IgG antibodies of the same antigen-binding reactivity. Using such anti-idiotypic
antibodies, normal Lewis T lymphocytes of B and T
type could be visualized by the use of anti-(Lewis-
anti-DA) antibodies. Visualization was possible
using direct fluorescent antibody tests or autoradiography. Using the first technique and naked eye
observations, 6.2% of normal Lewis T lymphocytes
expressed idiotypic markers signifying anti-DA
reactivity, whereas anti-DA-reactive B lymphocytes
as measured by this approach was about 1.1%. Autoradiography on purified normal Lewis T lymphocytes
gave similar figures. When comparing the intensity
of fluorescence at the single cell level using quantitative cytofluorometry anti-idiotypic antibodies
reactive with T lymphocytes gave a similar degree of
intensity as was obtained using anti-Ig antibodies
against B lymphocytes.

4489   SHARED IDIOTYPIC DETERMINANTS ON B AND
        T LYMPHOCYTES REACTIVE AGAINST THE SAME
ANTIGENIC DETERMINANTS: IV. ISOLATION OF TWO
GROUPS OF NATURALLY OCCURRING, IDIOTYPIC MOLECULES
WITH SPECIFIC ANTIGEN-BINDING ACTIVITY IN THE SERUM
AND URINE OF NORMAL RATS. (Eng.) Binz, H. (Uppsala
Univ. Medical Sch., Box 562, 751 22 Uppsala,
Sweden); Wigzell, H. *Scand. J. Immunol.* 4(5/6):
591-600; 1975.

To provide further information on the T cell antigen-binding receptor that shares idiotype with IgG
antibodies against alloantigen, an investigation
was carried out on the isolation and properties of
naturally-occurring idiotypic molecules carrying
the desired antigen-binding specificity. Rats used
in the study were inbred Lewis, DN, and BN strains,
possessing alloantigens $Ag-B^1$, $Ag-B^4$, and $Ag-B^3$,
respectively, together with certain F_1 hybrids. T
cells were prepared by fractionation of spleen or
lymph node cell suspensions on columns of Degalan
beads coated with rat Ig and rabbit anti-rat Ig.
Anti-idiotypic antisera were prepared by immunizing
(Lewis x DA)F_1 rats with Lewis T lymphocytes.
Materials tested for presence of idiotypic molecules
carrying the desired antigen-binding specificity
were serum and urine, collected from normal Lewis
rats. Assays for detection of anti-idiotypic
molecules were based on inhibition of binding of
known anti-idiotypic antisera to purified Lewis
peripheral T cells in a previously developed procedure. For assays of antigen-binding capacity of
the naturally occurring idiotypic factors to
(L x DA)F_1 cells, the idiotypic preparations were
first isolated from normal serum and urine by immunoabsorption onto anti-idiotypic immunoabsorbants,
followed by labeling of the purified fractions with
^{125}I. It was found that sera and urine of normal
rats contain naturally-occurring antibody-like
molecules with reactivity to allogeneic Ag-B antigens. Such molecules contained both antigen-binding
capacity for the relevant antigens and the idiotypic markers signifying such specific reactivity.

The preparations contained two groups of molecules, one around 7S in size and the other around 35,000 in molecular weight. Only the smaller molecules were found in the urine. Purified natural anti-Ag-B factors, when inoculated into rabbits or chickens, led to the production of anti-idiotypic antibodies that selectively inactivated rat T cells and that had the capacity to react against the relevant Ag-B antigens while leaving other reactivity intact. It is concluded that the system that was developed allows the purification of naturally-oc-curring idiotypic B- and T-cell products possessing antigen-binding specificity and is useful for further biochemical and functional analysis.

4490   ROLE OF B CELLS IN THE EXPRESSION OF THE
         PPD RESPONSE OF HUMAN LYMPHOCYTES *IN
VITRO. (Eng.) Blomgren, H. (Radiumhemmet, Karo-
linska Sjukhuset, S-104 01 Stockholm 60, Sweden).
Scand. J. Immunol. 4(5/6):499-510; 1975.

An investigation was carried out to study the types of lymphocytes that are involved in the *in vitro* response to purified protein derivative (PPD) of tuberculin. A crude lymphocyte preparation, separated from blood of healthy human donors by centrifugation on a Ficoll-Isopaque gradient, was designated as nonpurified. Following removal of phagocytic cells by magnetic action on cells which had phagocytosed iron and by adherence to plastic Falcon flasks, T and B cells were separated from one another by incubation with SRBC, sedimentation of rosetted T cells in a Ficoll-Isopaque gradient, and collection of B cells from the fluid inter-phase. T cells were identified by their ability to form rosettes with SRBC; B cells were identified by their membrane-bound immunoglobulin molecules, using a fluorescent staining method. PPD stimula-tion of lymphocytes was assayed by a microculture technique involving uptake of ^{3}H-thymidine. The concentration of PPD used was 10 µg/ml. The results of kinetic studies on PPD stimulation of the dif-ferent lymphocyte preparations showed that the non-purified lymphocytes exhibited a peak thymidine up-take after seven days of culture, while the greatest proliferation of T cells occurred later, and B cells showed minimal isotope incorporation during the entire culture period. Although T and B cells cultured separately showed very weak responses during the first seven days of culture, mixtures of T and B cells were strongly stimulated during this period. Further, the response of the mixtures appeared optimal at a ratio of T cells to B cells of 1:1. Mitomycin treatment almost completely blocked the capacity the both T and B cells to incorporate thymidine when stimulated with PPD, and mixtures of mitomycin-treated T cells and normal cells yielded higher stimulation than mixtures of normal T cells or mitomycin-treated B cells. This indicated that T cells exposed to PPD release fac-tors that stimulate B cells. Results of experiments on stimulatory effects of supernatant culture media from PPD-treated T or B cells confirmed the indica-tion that PPD-stimulated T lymphocytes, but not B lymphocytes, released the stimulatory factors. The results did not show, however, whether PPD-stimu-lated lymphocytes release factors that are mitogenic

principle showed positive Coomassie brilliant blue
staining for protein but no Alcian blue staining
for glycoprotein. The amino acid composition was
found to be four residues of aspartic acid, one of
threonine, five of serine, eight of glutamic acid,
two of proline, five of glycine, two of alanine,
one of leucine, one of lysine, and two of arginine.
Based on leucine as a unit, the molecular wt was
3,220.

4495   IMMUNOCOMPETENT LYMPHOCYTES IN RATS HOST-
        ING DMBA-INDUCED MAMMARY TUMOURS. (Eng.)
Kellen, J. A. (Sunnybrook Medical Centre, 2075 Bay-
view Ave., Toronto, Ontario, M4N 3M5, Canada);
Mirakian, A. *Wadley Med. Bull.* 5(1):1-7; 1975.

The binding of tumor-associated proteins of 7,12-
dimethylbenz(a)anthracene (DMBA)-induced mammary
tumors to peripheral lymphocytes and spleen cells
of tumor-bearing animals was investigated. The
binding was visualized by labeling the tumor proteins with fluorescein. The tumors were induced
in female virgin Wistar rats. The tumors were removed by surgery, homogenized, and centrifuged at
150,000 x g for 60 min. The soluble protein of the
supernatant was then labeled with fluorescein-iso-
thiocyanate. Periperal lymphocytes were isolated
from heparinized blood of tumor-bearing animals by
centrifugation on Lymphoprep; spleen cells, by a
similar procedure. The cells were incubated for
60 min at room temperature with the labeled tumor
protein preparation, washed with saline, and prepared as a smear for UV microscopic observation.
At least 1,000 cells were counted, and the incidence
of fluorescence was assessed. The labeled proteins
appeared to be bound to the membranes of the mononuclear cell preparations. All lymphocyte preparations from spleens of tumor-bearing hosts showed
a frequency of fluorescent cells of 1-3%. This
indicates that DMBA-induced mammary tumors contain
specific soluble proteins that are antigenic and
are recognized by the cellular immune system of the
host. The percentage of fluorescent cells was lower
in DMBA-injected rats without tumors, but not to a
statistically significant degree. This suggests
that even subclinical stages without apparent and
palpable nodules already manifest themselves through
the formation of specific tumor-associated proteins.
No uptake of labeled tumor proteins was seen in lymphocytes of healthy controls. The results with preparations of lymphocytes from peripheral blood
proved to be inconclusive.

4496   SPONTANEOUS ROSETTE AND ROSETTE-INHIBITION
        TESTS ON FRESH AND CRYOPRESERVED LYMPHO-
CYTES. (Eng.) Gross, R. L. (New England Deaconess
Hosp., Boston, Mass.); Levin, A. G.; Peers, F. G.;
Steel, C. M.; Tsu, T. *Cryobiology* 12(5):455-462;
1975.

Peripheral blood lymphocytes from 13 healthy adults
(Africans, Asians, and Europeans, 24-53-yr-old) were
compared in the fresh state and after controlled
freezing in liquid nitogen. The mean percentage of
cells forming spontaneous rosettes with sheep red
blood cells was 52% in the fresh samples and 57%
in frozen-thawed material. These figures are not

significantly different. Fresh and frozen cells
proved equally sensitive to the action of an antilymphocyte globulin preparation in inhibiting rosette formation. It is clear that valid results
may be obtained through the use of frozen preserved
lymphocytes in rosette and rosette-inhibition tests.

4497   CLASSIFICATION OF LYMPHOCYTES IN NASOPHAR-
        YNGEAL CARCINOMA (NPC) BIOPSIES. (Eng.)
Jondal, M. (Dept. Tumor Biology, Karolinska Inst.,
10401 Stockholm 60, Sweden); Klein G. *Biomedicine*
23(5):163-165; 1975.

Seven biopsies taken from nasopharyngeal carcinoma
(NPC) tissues were analyzed to establish the relative proportion of T and B cells. The vast majority
of these cells had the capacity to form sheep RBC rosettes (mean 87%) and were thus classified as T cells.
It is likely that the normal progenitor cell of NPC
is susceptible to EBV infection. The regular genome-
positive state of the anaplastic NPC-carcinoma cell
suggests that this tumor preferentially, if not exclusively, arises from a genome carrying cell. Varying proportions of the infiltrating T-lymphocytes
might represent a host cell sensitization against
tumor-associated antigens.

4498   PHYTOHEMAGGLUTININ-INDUCED LYMPHOCYTE
        TRANSFORMATION IN NEWLY PRESENTING PATIENTS
WITH PRIMARY CARCINOMA OF THE LUNG. (Eng.) Barnes,
E. W. (Royal Infirmary, Edinburgh, Scotland); Farmer,
A.; Penhale, W. J.; Irvine, W. J.; Roscoe, P.; Horne,
N. W. *Cancer* 36(1):187-193; 1975.

The transformation of peripheral blood lymphocytes
in response to phytohemagglutinin (PHA) was studied
in 37 newly presenting patients who were subsequently
proven to have carcinoma of the lung. When compared
with healthy age- and sex-matched normal controls,
no differences were noted in the thymidine uptake
of the unstimulated lymphocytes in culture or in the
response to any of the three dose levels of PHA used
(0.32, 0.63, and 1.25 µl/ml). Neither the extent of
the spread of the carcinoma nor the type of its histology showed any correlation with the PHA response,
but the response was significantly depressed within
14 days of death in seven patients. The transformation indicies of normal lymphocytes in the presence
of sera from 20 lung cancer patients at various
stages in the disease and during treatment with BCG,
compared with those obtained from sera of controls
indicated no evidence for serum inhibitory factors
in the lymphocytes of the cancer group. The results
suggest that a primary depression in general lymphocyte responsiveness that could have accounted for
the initiation of the disease does not occur. However, since depression does appear to follow the
development of the disease, this suggests that the
depression must be of a more selective nature, rather than a general loss.

4499   HETEROGENEITY OF T CELLS: DIFFERENT RE-
        PRESENTATIONS OF Ly ALLOANTIGENS ON KILLER
CELLS AND HELPER CELLS. (Eng.) Oettgen, H. F.
(Memorial Sloan-Kettering Cancer Center, New York,
N.Y.); Kisielow, P.; Shiku, H.; Bean, M. A.; Hirst,

from mice treated ip with the different test preparations. Activation of the macrophages was determined by adding P 815 target cells to the monolayers, incubating for 24 hr, adding ^{3}H-thymidine
four hours before the end of the incubation period,
and measuring uptake of radioactivity in the cells.
Results were expressed as percentage growth inhibition. The preparations tested for activity
included pheno-killed *Mycobacterium smegmatis*
cells; interphase material (IPM), a fraction extracted from *M. smegmatis;* Neo-WSA and WSA, two
water-soluble adjuvants also extracted from *M.
smegmatis;* muramyl dipeptide, a synthetic adjuvant
component of the mycobacterial cell wall; muramyl
L-alanine, a nonadjuvant cell wall component; and
a lipopolysaccharide (LPS) prepared from *Salmonella
enteritidis*. In the *in vitro* tests, WSA and the
non-adjuvant muramyl peptide elicited weak responses, while IPM and the adjuvant muramyl peptide elicited strong responses. In the *in vivo*
tests, macrophages from mice treated with IPM,
New-WSA, WSA, or LPS strongly inhibited mastocytoma cell growth, while macrophages from mice
treated with the muramyl peptides did not differ
from those from normal mice. Macrophages of LPS-
or WSA-treated mice were highly activated if they
were harvested 4 or 8 days after *in vivo* stimulation, but this response decreased and disappeared
later. In contrast, a very strong macrophage activation could still be demonstrated 64 days after
injection of IPM. The antitumor effect of the IPM
could not be related to a direct cytotoxic effect
on the mastocytoma cells. Since of all the preparations tested, previous work had shown that
only IPM protected mice against a syngeneic leukemia, it could be concluded that macrophage activation was correlated with adjuvant activity
rather than with *in vivo* tumor activity.

4503   TOLERANCE INDUCTION WITH BOVINE γ GLOB-
        ULIN IN MOUSE RADIATION CHIMAERAS DEPENDS
ON MACROPHAGES. (Eng.) Lukic, M. L. (Immunology
Res. Center, Univ. Belgrade, Vojvode Stepe 458,
11000 Belgrade, Yugoslavia); Leskowitz, S. *Nature*
252(5484):605-607; 1974.

Carrageenin, a polysaccharide isolated from seaweed
and known to be toxic for macrophages, was tested
for its effects on tolerance induced in BALB/c mice
by ultracentrifuged bovine γ globulin (sBGG, 2 mg,
ip). sBGG was administered 24 hr after the last
injection of carrageenin; one week later the mice
with immunized with 100 μg-BGG in complete Freund's
adjuvant (CFA) and then challenged with 2 μg ^{125}I-
BGG. Pretreatment with 0.5 mg or 4 x 10.5 mg favored tolerance induction in the mice, whereas the
immune response of mice pretreated with BGG-CFA
only was not affected. In spleen cell transfer experiments, lethally irradiated (850 rads) DBA/2
mice were tested for tolerance induction after reconstitution with BALB/c bone marrow cells (2 x 10^7).
The mice were injected with 2 mg sBGG 4, 6, or 12
wk after reconstitution, followed by immunization
with BGG-CFA and then challenge with ^{125}I-BGG. At
4 and 6 wk after reconstitution, the mice showed
nonimmune clearance typical or normal DBA/2 mice
made tolerant. After 12 wk, 10 of 11 mice showed

the characteristic resistance to tolerance of the
BALB/c donors. These results provide strong evidence that strain differences in the induction of
tolerance by antigens such as BGG are related to
an event occurring at the macrophage level.

4504   IMMUNOLOGICAL SIGNIFICANCE OF CERTAIN MOR-
        PHOLOGICAL ASPECTS OF SATELLITE LYMPH NODES
OF LARYNGEAL CARCINOMAS. , (Ita.) Cortesina, G.
(Clinica Otorinolaringologica dell'Universita,
10100 Torino - Via Genova, 3, Italy); Pia, F.;
Mancini, P.; Boido, C.; Giordano, C. *Canoro* 27(6):
373-388; 1974.

The behavior of lymphocytes and macrophages in tumor
metastases of larynx carcinomas were studied in 14
men and 1 woman. A relatively low number of germinal
centers were found in the early stages of disease,
with an increase of these in the advanced stages.
A strong hyperplasia of macrophages and T lymphocytes in cortical and paracortical sinuses was also
observed in the early phases of tumoral diffusion.
The results seem in accord with the response of body
defense to tumoral invasion.

4505   OUABAIN BINDING TO INTACT LYMPHOCYTES: EN-
        HANCEMENT BY PHYTOHEMAGGLUTININ AND LEUCO-
AGGLUTININ. (Eng.) Quastel, M. R. (Isotope Dept.,
Soroka Medical Center, Beer Sheva, Israel); Kaplan,
J. G. *Exp. Cell Res.* 94(2):351-362; 1975.

A study was made of the kinetics of binding of ouabain as an indicator of membrane changes associated
with lymphocyte blastoid activation by phytohemagglutinin (PHA), leucoagglutinin (LA), concanavalin A (Con A), and pokeweed mitogen (PWM). Cultures of lymphocytes derived from human peripheral
blood were used throughout. ^{3}H-ouabain, dissolved
in benzene-alcohol, was added to the cultures to
give final concentrations varying from 10^{-8} to 10^{-6}
M. The cell cultures were incubated at 37 C in the
presence of the ^{3}H-ouabain for 1-33 hr; at different times, cells were removed and tested for radioactivity. The results were expressed in terms of
the number of ouabain glycoside molecules taken up
by the cell. Both PHA and LA increased the rate
of binding of the glycoside within minutes. The
saturation number of binding sites was estimated
to be 1.25×10^5/cell for nonstimulated lymphocytes. PHA caused the saturation level to rise
to values of about 2.3×10^5/cell. The rate of
^{3}H-ouabain binding was very sensitive to potassium
concentration, and the specificity of ^{3}H-ouabain
binding to K^+ receptors on the lymphocytes was
demonstrated by the inhibitory effect of K^+ at
various concentrations on the degree of binding.
The very early increase in both ouabain binding
and ^{42}K transport for mitogen-stimulated cells,
the increase in V_{max} though not K_m found for both
^{3}H-ouabain and ^{42}K transport, and the similar increase of ^{3}H-ouabain binding compared with that of
^{42}K transport following PHA administration all support the view that ^{3}H-ouabain serves as a label
for the K^+ site. Both ouabain and digoxin dis-
placed the label. Estimates were provided for the
affinity constants for uptake and turnover of oua-

4515   INCREASED CANCER IN UREMIC PATIENTS [abstract]. (Eng.) Matas, A. J. (Univ. Minnesota, Minneapolis); Najarian, J. S.; Simmons, R. L. *Proc. Am. Assoc. Cancer Res.* 16:48; 1975.

4516   MICROCYTOTOXICITY ASSAY OF THE IMMUNOGENICITY OF HYPERPLASTIC ALVEOLAR NODULE (HAN) LINES IN STRAIN BALB/C MICE [abstract]. (Eng.) Heppner, G. (Roger Williams Gen. Hosp., Brown Univ., Providence, R.I.); Kopp, J.; Medina, D. *Proc. Am. Assoc. Cancer Res.* 16:55; 1975.

4517   CORRELATION OF *IN VITRO* IMMUNOLOGICAL REACTIONS WITH *IN VIVO* TUMOR IMMUNITY [abstract]. (Eng.) Rodrigues, D. (Litton Bionetics Res. Lab., Kensington, Md.); Ting, C. C. *Proc. Am. Assoc. Cancer Res.* 16:99; 1975.

4518   T CELL-DEPENDENT HELPER AND SUPPRESSIVE INFLUENCES IN AN ADOPTIVE IgG ANTIBODY RESPONSE. (Eng.) Arrenbrecht, S. (Weizmann Inst. Sci., Rehovot, Israel); Mitchell, G. F. *Immunology* 28(3):485-495; 1975.

4519   IMMUNOSUPPRESSION BY LEUKEMIA VIRUSES. XI. EFFECT OF FRIEND LEUKEMIA VIRUS ON HUMORAL IMMUNE COMPETENCE OF LEUKEMIA-RESISTANT C57BL/6 MICE. (Eng.) Ceglowski, W. S. (Dept. Microbiol., Pennsylvania State Univ., University Park, Pa. 19141); Campbell, B. P.; Friedman, H. *J. Immunol.* 114(1/Part 1):231-236; 1975.

4520   QUANTITATION OF CELL-MEDIATED IMMUNITY: RESPONSE TO DINITROCHLOROBENZENE AND UBIQUITOUS ANTIGENS. (Eng.) Thompson, C. D. (Royal Melbourne Hosp., Victoria 3050, Australia); Whittingham, S.; Mackay*, I. R.; Khoo, S. K.; Toh, B. H.; Stagg, R. J. *Can. Med. Assoc. J.* 112(9):1078-1081; 1975.

4521   IDENTIFICATION OF FAMILY MEMBERS WITH CELL MEDIATED IMMUNITY TO SARCOMA ANTIGENS [abstract]. (Eng.) Neidhart, J. A. (Ohio State Univ., Columbus, Ohio); Ertel, I. J.; LoBuglio, A. F. *Proc. Am. Assoc. Cancer Res.* 16:114; 1975.

4522   CELLULAR IMMUNITY TO TUMOR-ASSOCIATED ANTIGENS IN SARCOMA PATIENTS AND THEIR RELATIVES [abstract]. (Eng.) Boddie, A. W., Jr. (Sepulveda Veterans Adm. Hosp., Calif.); Urist, M. M.; Holmes, E. C.; Sparks, F. C. *Proc. Am. Assoc. Cancer Res.* 16:261; 1975.

4523   ROSETTE-FORMING ABILITY OF THYMUS-DERIVED LYMPHOCYTES IN CELL-MEDIATED IMMUNITY. I. DELAYED HYPERSENSITIVITY AND *IN VITRO* CYTOTOXICITY. (Eng.) Elliott, B. E. (Basel Inst. Immunology, 487 Grenzacherstrasse, CH-4058 Basel, Switzerland); Haskill, J. S.; Axelrad, M. A. *J. Exp. Med.* 141(3):584-599; 1975.

4524   LYMPHOCYTE DYSFUNCTION AND IMPAIRED CELLU-
        LAR IMMUNITY IN HODGKIN'S DISEASE [ab-
stract]. (Eng.) Graze, P. (Bureau of Biologics and
Natl. Navy Medical Center, Bethesda, Md. 20014);
Pitts, R. B. *Proc. Am. Assoc. Cancer Res.* 16:24;
1975.

4525   AN IMMUNO-ULTRASTRUCTURAL STUDY OF SUR-
        FACE IMMUNOGLOBULINS IN METASTASIZING AND
NON-METASTASIZING HAMSTER LYMPHOMAS. (Eng.) Machi-
nami, R. (Faculty Medicine, Univ. Tokyo, Hongo,
Tokyo 113, Japan); Carter, R. L.; Birbeck, M. S. C.
Eur. J. Cancer 11(2):87-90; 1975.

4526   SURFACE PROTEINS OF HUMAN ACUTE MYELO-
        GENOUS LEUKEMIA (AML) BLASTS [abstract].
(Eng.) Cotropia, J. P. (M. D. Anderson Hosp. and
Tumor Inst., Houston, Tex.). *Proc. Am. Assoc. Cancer
Res.* 16:42; 1975.

4527   CHANGES IN SURFACE IMMUNOGLOBULINS (SIg)
        AND SURFACE ULTRASTRUCTURE IN CHRONIC
LYMPHOCYTIC LEUKEMIA (CLL) [abstract]. Brynes, R. K.
(Univ. Chicago, Franklin McLean Memorial Res..Inst.,
Chicago, Ill.); Golomb, H. M.; Reese, C.; Ultmann,
J. E. *Proc. Am. Assoc. Cancer Res.* 16:115; 1975.

4528   AUTO-ANTIBODIES IN MALIGNANT DISEASE.
        (Eng.) Whitehouse, J. M. A. (Royal Hosp.
St. Bartholomew, West Smithfield, London EC1A 7BE,
England); Fairley, G. *Bull. Soc. Sci. Med. Grand
Duche Luxemb.* 110(2):197-202; 1973.

4529   FLUORESCENT ANTIBODY AND FERRITIN-LABELED
        ANTIBODY STUDIES OF ANTI-MURINE LEUKEMIA
VIRUS FRACTIONS GP-69-71 AND P30 ON ONCORNA VIRUSES
[abstract]. (Eng.) Oshiro, L. S. (California State
Dept. Health, Berkeley, Calif.); Riggs, J. L.;
Lennette, E. H. *Proc. Am. Assoc. Cancer Res.* 16:104;
1975.

4530   SHEEP RED BLOOD CELL ANTIBODIES IN CANCER
        [abstract]. (Eng.) Hirshaut, Y. (Mem.
Sloan-Kettering Cancer Cent., New York, N.Y.); Fass,
B.; Sethi, J.; Teitelbaum, H. *Proc. Am. Assoc. Cancer
Res.* 16:180; 1975.

4531   STUDIES WITH TUMOR ASSOCIATED ANTIBODIES IN
        ISOGENEIC AND ALLOGENEIC RAT TUMOR SYSTEMS
[abstract]. (Eng.) Della Penta, D. W. (Univ. Rochester, N.Y.). *Diss. Abstr. Int. B* 35(7):3181; 1975.

4532   IMMUNOPROTECTION WITH SOLUBLE TUMOR ANTIGEN
        FROM CULTURED CELLS [abstract]. (Eng.)
Pellis, N. R. (Northwest. Univ., Chicago, Ill.);
Kahan, B. D. *Proc. Am. Assoc. Cancer Res.* 16:47;
1975.

4533   EVIDENCE FOR EXISTENCE OF SOLUBLE TUMOR
        ANTIGENS *IN VIVO* [abstract]. (Eng.)
Detrick-Hooks, B. (Natl. Inst. Health, Bethesda,
Md.); Smith, H. G. *Proc. Am. Assoc. Cancer Res.*
16:36; 1975.

4534   CARCINOEMBRYONIC ANTIGEN AND CARCINOID
        TUMORS. (Eng.) Feldman, J. M. (Veterans
Admin. Hosp., Durham, N.C. 27710); Plonk, J. W.
Ann. Intern. Med. 83(1):82-83; 1975.

4535   IMMUNOTHERAPY OF L1210 LEUKEMIA WITH
        ISOLATED, ANTIGENIC SUBPOPULATIONS OF
L1210 CELLS [abstract]. (Eng.) Killion, J. J.
(Oklahoma Med. Res. Found., Oklahoma City, Okla.)
LeFever, A.; Cantrell, J. L.; Bierig, K.; Koll-
morgen, G. M. *Proc. Am. Assoc. Cancer Res.* 16:22;
1975.

4536   AUGMENTATION OF LYMPHOCYTE CYTOTOXICITY
        BY ANTIBODY TO *HERPESVIRUS SAIMIRI* ANTIGE
[abstract]. (Eng.) Prevost, J.-M. (Natl. Inst.
Health, Bethesda, Md.); Pearson, G. R. *Proc. Am.
Assoc. Cancer Res.* 16:64; 1975.

4537   SPLENIC LYMPHOCYTES IN CHRONIC GRANULOCYT
        LEUKAEMIA. (Eng.) Kaur, J. (Royal Post-
graduate Medical Sch., London W12, England); Spiers,
A. S. D.; Galton, D. A. G. *Lancet* 2(7923):40; 1975.

4538   VALUE OF LYMPHOCYTE STUDY BY SURFACE MARKE
        IN NON-HODGKIN'S MALIGNANT LYMPHOMA [ab-
stract]. (Eng.) Gajl-Peczalska, K. J. (Univ.
Minnesota, Minneapolis); Bloomfield, C. D.; Sosin, H
Kersey, J. H. *Proc. Am. Assoc. Cancer Res.* 16:60;
1975.

4539   INHIBITION OF MITOGEN INDUCED LYMPHOCYTE
        BLASTOGENESIS BY SERA OF MELANOMA
PATIENTS [abstract]. (Eng.) Rangel, D. M. (Div.
Surg. Oncol., Univ. California Los Angeles). *Proc.
Am. Assoc. Cancer Res.* 16:261; 1975.

4540   *IN VITRO* STIMULATION OF T AND B CELLS BY
        TUMOR CELLS [abstract]. (Eng.) Burk,
M. W. (Univ. Minnesota, Minneapolis); Yu, S.; Mc-
Khann, C. F. *Proc. Am. Assoc. Cancer Res.* 16:164;
1975.

4541   ELECTRON MICROSCOPIC LOCALIZATION OF ACID
        PHOSPHATASE ACTIVITY WITHIN HODGKIN'S
DISEASE LYMPH NODES [abstract]. (Eng.) Rowan, R. A
(Stanford Univ. Sch. Med., Calif.); Masek, M. A.;
Thompson, J. M.; Frenster, J. H. *Proc. Am. Assoc.
Cancer Res.* 16:10; 1975.

4542   LYMPHOCYTE MEDIATED IMMUNITY AGAINST
        OSTEOSARCOMA CELLS IN TISSUE CULTURE [ab-
stract]. (Eng.) Singh, I. (Med. Coll. Ohio, Toledc

4550   A USEFUL MODEL FOR THE STUDY OF HOST
        IMMUNITY AND METASTASES FOLLOWING TUMOUR
EXCISION [abstract]. (Eng.) Bray, A. E. (Perth
Med. Cent., Western Australia); Keast, D. *Clin.*
Exp. Pharmacol. Physiol. 2(1):83; 1975.

4551   IMMUNOLOGICAL STUDIES ON HUMAN CHRONIC
        MYELOGENOUS LEUKEMIA CELLS [abstract].
(Eng.) Whitson, M. E. (Univ. Tennessee). *Diss.*
Abstr. Int. B 35(11):5504-5505; 1975.

4552   MEASUREMENT OF SPECIFIC IMMUNE RESPONSES
        IN ACUTE LYMPHOCYTIC LEUKAEMIA OF CHILD-
HOOD. (Eng.) O'Keefe, D. E. (Royal Children's
Hosp. Res. Found., Melbourne, Australia); Wilson,
F. C.; Ekert, H.; Jose, D. G. *Clin. Exp. Pharmacol.*
Physiol. 2(1):80-81; 1975.

4553   THE APPLICATION OF ROSETTE-FORMATION TEST
        IN THE EVALUATION OF IMMUNOLOGICAL REAC-
TIVITY IN PATIENTS WITH MELANOMA. (Rus.) Trapezni-
kov, N. N. (Inst. Exp. Clin. Oncol , Acad. Med. Sci.
U.S.S.R., Moscow, U.S.S.R.); Kupin, V. I.; Radzik-
hovskaia, R. M.; Kadagidze, Z. G. *Dokl. Akad. Nauk*
S.S.S.R. 222(2):468-471; 1975.

4554   EVIDENCE FOR A MEMBRANE CARRIER MOLECULE
        COMMON TO EMBRYONAL AND TUMOUR-SPECIFIC
ANTIGENIC DETERMINANTS EXPRESSED BY A MOUSE TRANS-
PLANTABLE TUMOUR. (Eng.) Comoglio, P. M. (Istituto
di Anatomia Umana, Corso Massimo d'Azeglio 52,
Lav. Ist. Anat. Istol. Patol. Perugia 35(1):29-
logy 29(2):353-364; 1975.

4555   IMMUNOLOGIC INDUCTION OF RETICULUM CELL
        SARCOMA: DONOR-TYPE LYMPHOMAS IN THE
GRAFT-VERSUS-HOST MODEL. (Eng.) Gleichmann, E.
(Medizinische Hochschule Hannover, D-3000 Hannover,
Karl-Wiechert-Allee 9, West Germany); Peters, K.;
Lattmann, E.; Gleichmann, H. *Eur. J. Immunol.*
5(6):406-412; 1975.

See also:

*   (Rev):   4214, 4215, 4216, 4218, 4237, 4240,
            4246, 4247, 4248, 4249, 4250, 4251,
            4252, 4253, 4261, 4263
*   (Chem): 4293
*   (Viral): 4403, 4404, 4424, 4434
*   (Path): 4557, 4566, 4582, 4587, 4594, 4601
*   (Epid-Biom): 4650

4556   MULTIPLE NEOPLASMS IN TWO SIBLINGS WITH
        A VARIANT FORM OF FANCONI'S ANEMIA.
(Eng.) Sarna, G. (S. Perry, Natl. Inst. Health,
Building 1, Room 111, Bethesda, Md. 20014); Toma-
sulo, P.; Lotz, M. J.; Bubinak, J. F.; Shulman,
N. R. *Cancer* 36(3):1029-1033; 1975.

The case reports of two brothers with a variant form
of Fanconi's anemia who developed multiple neoplasms
after prolonged survival and treatment with androgens
are described. The patients were diagnosed at ages
7 and 10 yr as being Fanconi's anemia variants on
the basis of marrow hypoplasia, family history,
multiple chromosomal aberrations, and, in one who
was tested, abnormal fibroblast transformation with
Simian virus 40. The first brother received total
cumulative doses of testosterone cyclopentyl proprio-
nate (16.8 g), norethandrolone (31.6 g), oxandrolone
(2.2 g), and oxymethalone (9.0 g) over 75 mo. The
second brother received norethandrolone (56.7 g),
oxandrolone (2.1 g), and oxymethalone (1.2 g) over
88 mo. The first brother developed two separate
oral squamous cell carcinomata, and the other
developed acute leukemia and hepatoma. Androgens
may have had a carcinogenic role in the appearance
of the hepatic neoplasm. There is an increased
incidence of neoplasm associated with Fanconi's
anemia. This may be related to frequent spontaneous
chromosomal aberrations and/or to increased cellular
susceptibility to viral transformation.

4557   CUTANEOUS T-CELL LYMPHOMAS: THE SEZARY
        SYNDROME, MYCOSIS FUNGOIDES, AND RELATED
DISORDERS. (Eng.) Lutzner, M. (Natl. Cancer Inst.,
Bldg. 10, Room 12N-238, Bethesda, Md. 20014);
Edelson, R.; Schein, P.; Green, I.; Kirkpatrick,
C.; Ahmed, A. *Ann. Intern. Med.* 83(4):534-552;
1975.

Neoplastic T cells from four patients with the
Sezary syndrome were unable to translate various
membrane stimuli, which included responses to
phytohemagglutinin, concanavalin A, mitomycin C,
allogeneic cells, blastogenic factor, and the mitogenic activity of anti-T-cell serums. This lack
of reactivity was not secondary to the production
of inhibitors. Cells from these patients also
lacked the ability to kill antibody-coated target
cells. Cells from these patients did not produce
migratory inhibitory factor in response to various
stimuli; plasma from these patients did show migra-
tory-inhibitory-factory-like activity. The Sezary
cells seem to be T cells with abnormalities intrinsic to their neoplastic nature. It seems appropriate to classify the large and small cell variants
of the Sezary syndrome, mycosis fungoides, and related lymphoproliferative disorders under the broad
heading "cutaneous T-cell lymphomas." In all cases,
the processes seem to originate in the skin, and
quite remarkably the thymus remains uninvolved.

4558   ON A MALIGNANT VARIANT OF ABRIKOSOV'S
        TUMOURS. (Rus.) Apatenko, A. K. (No
affiliation); Sementsov, P. N. *Arkh. Patol.* 37(2):
16-21; 1975.

A malignant variant of Abrikosoff's tumor was found

in two men and one woman. The tumor was localized
in the soft tissues of a finger, under the skin of
the upper lip and in the soft tissues of the lumbar
region, respectively. Locally destructive infiltrative growth, polymorphism of the cells and nucle
irregular patterns of amitotic cell division, and
hyperchromatism of the cell nuclei were the characteristic features of all tumors. Isolated polynuclear cells were also found.

4559   GIANT MELANOSOMES IN MOLES AND IN NORMAL
        HUMAN EPIDERMIS. (Ger.) Konrad, K.
(Abteilung fur Experimentelle Dermatologie, I.
Universitats-Hautklinik, Vienna, Austria); Honigsmann, H. *Wien. Klin. Wochenschr.* 87(5):173-177;
1975.

Giant melanosomes were found in normal human epidermis (two cases), in nevi (four cases), and in
melanoderma (one case). This finding indicates
that giant melanosomes are not diagnostic for
von Recklinghausen's disease (neurofibromatosis).

4560   NEVUS SPILUS--A PIGMENT NEVUS WITH GIANT
        MELANOSOMES. CLINICAL, HISTOLOGICAL AND
FINE STRUCTURAL ASPECTS. (Ger.) Konrad, K. (I.
Univ-Hautklinik, Alserstrasse 4, A-1090 Wien,
Austria); Honigsmann, H.; Wolff, K. *Hautarzt* 25
(12):585-593; 1975.

Clinical, histological and fine structural aspects
of nevus spilus in ten patients (six men and four
women, aged 18-65 yr) are described. Nevus spilus
had been present since childhood in six patients,
but it was impossible to determine whether it was
inborn. Neither the anamneses nor the clinical
examinations furnished any indication of Recklinghausen's disease or other systemic diseases involving pigmentation disturbances. Despite the presence of nevus cells, the chance of malignant trans
formation is slim. The giant melanosomes detected
in part of the melanocytes and in nevus cells are
interpreted as an expression of disturbances in
the morphogenesis of individual melanosomes rather
than as disturbances in melanogenesis or in pigment
formation. The giant melanosomes contain active
tryosinase. It was possible to distinguish these
melanosomes from granules found in Chediak-Higashi
syndrome. Giant melanosomes are assumed to occur
also in pigment disturbances other than nevus spilu

4561   SUBCUTANEOUS ANGIOLYMPHOID HYPERPLASIA
        (KIMURA DISEASE). REPORT OF A CASE. (En)
Kim, B. H. (Lutheran Medical Center, Brooklyn, N.Y.
11220); Sithian, N.; Cucolo*, G. F. *Arch. Surg.*
110(10):1246-1248, 1975.

A case of subcutaneous angiolymphoid hyperplasia
(Kimura disease) examined at its early stage is
reported. A ten-year-old boy presented with a
nontender, movable, and firm mass growing in his
left cheek. Results of laboratory tests were within
normal limits, and no blood eosinophilia was reported. Under general anesthesia, a transverse
incision was made over the buccal aspect of the

Electron microscopy revealed large histiocytic
epithelioid cells characterized by lobulated nuclei
with marginal chromatin, irregularly ruffled plasma
membranes with interdigitating processes, and numerous small intracytoplasmic vacuoles with pleomorphic
contents. The fibrocytic cells were fusiform in
configuration with ellipsoid nuclei and extended
cytoplasmic processes. The cell membrane was often
closely associated with the randomly arrayed colla-
genous fibers of uniform diameter. The lack of
systemic disease and clinical behavior of the
enlarging tumor suggested a benign fibrous histiocytoma.

4564   THE NATURAL HISTORY OF MALIGNANT MELANOMA
        OF THE CHOROID: SMALL VS LARGE TUMORS.
(Eng.) Davidorf, F. H. (Dep. Ophthalmol., Ohio
State Univ., Columbus); Lang, J. R. *Trans. Am.
Acad. Ophthalmol. Otolaryngol.* 79(2):310-320; 1975.

The growth characteristics and natural history of
264 malignant melanomas of the choroid were studied.
Small melanomas were compared with larger melanomas.
Of the 264 melanomas, 50 were designated as small
melanomas (less than 10 mm in diameter and 3 mm in
elevation). A large percentage (22.0%) of mixed
and epithelioid tumors (considered to be more malignant) were found in the group of small tumors.
In 33 patients with a possible five-year survival,
two (6.1%) died of metastases. Of 22 possible ten-
year follow-ups, five patients died of metastatic
melanoma (22.7%). These deaths occurred in one
patient with a spindle cell A melanoma, three with
spindle cell B melanomas, and one with a mixed melanoma. All five of these patients were women,
and all had posterior pole tumors. No deaths occurred in patients with tumors less than 7 mm in
diameter and 2 mm in elevation. The relatively
good five-year survival in the patients with smaller melanomas supports the views held by previous
authors that small melanomas carry a better prognosis than larger tumors. However, mixed and epithelioid tumors in this series do not seem to have
a worse prognosis than spindle cell tumors. Cell
type may be of less prognostic significance that
tumor size in the case of small melanomas.

4565   GASTRIC MUCOSAL DYSPLASIAS: WHAT IS
        THEIR CLINICAL SIGNIFICANCE? (Ger.)
Oehlert, W. (Pathologisches Institut der Universitat, 78 Freiburg, Albertstr. 19 West Germany); Keller, P.; Henke, M.; Strauch, M. *Deutsch. Med. Wochenschr.* 100(39):1950-1956; 1975.

Using criteria of proliferation kinetics, karyo-
metry, histology and histochemistry, definition
of normal and dysplastic surface epithelium of the
gastric mucosa was attempted. Instead of such
terms as surface carcinoma, intra-epithelial carcinoma and carcinoma-in-situ, the term dysplasia
III of the surface epithelium is suggested. Serial
biopsies (2-5) during 1 mo to 2 yr, of gastric mucosa from 285 patients with dysplasia II/III, demonstrated reversibility or persistence of such
changes without transition into early carcinoma.
In this series, 116 (40.7%) of the lesions im-

819

proved, 142 (49.8%) were unchanged, and 27 (9.5%)
progressed. Of the 8,262 (3,539 women) patients
biopsied in this 2 yr study, 5,639 had pathologic
findings.

4566   PRIMARY BILIARY CIRRHOSIS WITH CANCER OF
        THE AMPULLA OF VATER. (Eng.) Makipour,
H. (Cleveland Clinic, Cleveland, Ohio); Winkelman,
E. I. *Cleve. Clin. Q.* 42(2):209-213; 1975.

A case history is presented of carcinoma of the
ampulla of Vater occurring with primary biliary
cirrhosis and the value of the antimitochondrial
antibody and the potential role of endoscopic re-
tograde cholangiopancreatography in establishing
a correct diagnosis is discussed. A 57-yr-old
woman presented with jaundice and diarrhea of 8-9
mo duration. Other complaints were mild right upper quadrant abdominal discomfort, generalized bone
pain, and a 3.6-kg wt loss. Antimitochondrial
antibody was positive at greater than 1:600 serum
dilution; smooth muscle antibody and antinuclear
factor were negative. Other clinical findings were
interpreted as compatible with primary biliary cirrhosis. She was discharged and a high protein diet
and diphenoxylate hydrochloride were prescribed
for control of the diarrhea. During the next six
months pruritus appeared and the diarrhea and weight
loss continued. The patient was readmitted to the
hospital; the results of physical examination were
unchanged, except for the presence of bilateral
periobital xanthelasma. At laparotomy a 3.5-cm
firm, rubbery mass was found in the head of the
pancreas. Operative cholangiogram showed complete
obstruction of the distal common duct with proximal
dilatation. Transduodenal biopsy of the ampulla
of Vater showed well-differentiated adenocarcinoma.
The patient tolerated a Whipple's procedure well
but for the next 30 mo continued to have severe
diarrhea and generalized bone pain. The patient
died on the fourth day after being readmitted to
the hospital. Autopsy showed the capsular surface
of the liver was finely nodular and the entire liver
parenchyma was deep green. There was no evidence
of recurrence of metastatic tumor. A review of the
literature indicated that the antimitochondrial
antibody is an accurate, sensitive test that is
positive in 80-85% of all patients with primary
biliary cirrhosis and falsely positive in only 2-
4%. Lesions of the biliary tract, however, can
produce a false positive test result, and these
lesions can coexist with primary biliary cirrhosis.
Since the iv cholangiogram does not visualize the
biliary tree in most cases, diagnostic studies are
limited to the transhepatic cholangiogram or endoscopic retrograde cholangiopancreatography. Some
physicians are reluctant to use percutaneous transhepatic cholangiography as a routine procedure because of reported complications. Endoscopic retrograde cholangiopancreatography has been used successfully to exclude obstruction in the biliary
tree and is probably the safest procedure at
present.

4567   RUPTURED BENIGN HEPATOMA ASSOCIATED WITH
        AN ORAL CONTRACEPTIVE: A CASE REPORT.
(Eng.) Stenwig, A. E. (Norwegian Radium Hosp.,

in the cancer patients and follicular phase controls.
The urinary 17-ketosteroid levels were lower in the
cancer patients than in the controls. However, a-
mong the 17-ketosteroid fractions, the level of dehydroepiandrosterone was higher in the cancer patients than in the ovulatory controls; the levels
of etiocholanolone and androsterone were similar in
the two groups.

4571   CANCER OF THE VAGINA. (Eng.) Davis,
        P. C. (Doctors' Building, NW Annex,
Asheville, N. C. 28801); Franklin, E. W., III.
South. Med. J. 68(10):1239-1242; 1975.

Experience in the diagnosis and management of 37
cases of cancer of the vagina is reported. Twenty-
five of the 37 patients had squamous cell carcinoma
of the vagina, and seven of these cases were intra-
epithelial. The remaining 12 patients had sarcoma,
melanoma, or adenocarcinoma. All but four of the
patients with squamous cell carcinoma were over
50 yr old; among the patients with sarcomas of
diverse histologic types, the age range was 2.5-
77 yr. Bleeding or bloody vaginal discharge was
the most common presenting symptom with all cell
types. Exophytic lesions accounted for 46% of the
30 invasive cancers, including 50% of the epidermoid carcinomas. In 60% of the cases, the lesion
was located in the upper portion of the vagina.
Seven of the 37 patients·had a second primary malignancy, including two lesions of the skin beyond
the anogenital epithelium. One patient had a stage
I adenocarcinoma of the endometrium and another
had a large squamous cell carcinoma of the lung.
Three patients had previously been treated for
carcinoma *in situ* of the cervix by hysterectomy.
Over half the patients received radiation therapy,
usually external cobalt radiation. Primary surgical treatment in one-third of the patients ranged
from local resection or vaginectomy with skin graft
to total pelvic exenteration. Nineteen of the 37
patients are living and well, 13 died as a result
of their disease, and another five are living with
persistent disease. Among the 18 treatment failures,
ten had local recurrence only, while six had distant as well as local recurrence and two had distant
metastases without local recurrence. As in all
malignancies, early diagnosis and prompt, adequate
treatment offer the best opportunity for cure in
vaginal cancer.

4572   PROMPT REGRESSION OF CYSTIC VAGINAL
        ADENOSIS FOLLOWING CESSATION OF ORAL
CONTRACEPTIVE THERAPY. (Eng.) Strand, C. L.
(Jersey City Medical Center, Jersey City, N.J.
07304); Windhager*, H. A.; Kim, E. H.; Chiranand,
N. *Am. J. Clin. Pathol.* 64(4):483-487; 1975.

The cessation of an oral contraceptive (Ovulen-21)
was found to result in prompt regression of cystic
vaginal adenosis. A 31-yr-old Caucasian woman,
gravida 5, para 4, presented with a three-month
history of progressive dyspareunia, dysuria, and
profuse vaginal discharge. On pelvic examination,
the exocervix and entire vagina showed marked
reddish discoloration; numerous discrete and ag-

gregated cystic nodules were palpated in the upper
two-thirds of the vagina. Laboratory data were
within normal values, while culture of the material
from the cervix and vagina grew *Escherichia coli*.
An electroconization biopsy of the cervix showed
numerous nabothian cysts and proliferative endometrium. A vaginal biopsy showed cystic vaginal
adenosis. Although no therapy was received for
vaginal adenosis, the vaginal discharge diminished
and the dysuria disappeared within two months after
cessation of oral contraceptive therapy. A repeated biopsy showed only a few pea-sized nodules
in the left lateral formix. The regression is
considered suggestive of a causal-relationship
between long-lasting Ovulen-21 treatment and appearance of a symptomatic type of vaginal adenosis.

4573   PAPILLARY CYSTADENOFIBROMA OF ENDOMETRIUM:
         A HISTOCHEMICAL AND ULTRASTRUCTURAL STUDY.
(Eng.) Grimalt, M. (*Jewish General Hosp., 3755 Cote
Ste. Catherine Road, Montreal H3T 1E2 Canada); Ar-
guelles, M.; Ferenczy,* A. *Cancer* 36(1):137-144;
1975.

The light and electron microscopic as well as the
histochemical characteristics of a papillary cystadenofibroma of the endometrium are described. The neoplasm arose in the lower uterine segment and was composed of a florid fibroblastic growth arranged in
club-shaped papillae projecting into clefts and cystic
spaces. The epithelium lining the plicae, recesses,
and cysts was exclusively of the mucous-secreting
type and bore identical histochemical and subcellular
characteristics to that of the normal endocervical
epithelium. Essentially similar papillary lesions
were previously reported in the endocervix, endometrium, and fallopian tube. The pathogenesis and differential diagnosis of this distinct neoplasm are discussed in the light of the available morphological
data. Absence of mitosis is not a reliable criterion
for differentiating papillary adenofibroma from low-
grade endometrial stromal sarcoma. However, the
presence of abundant mitoses in a well-differentiated
endometrial stromal cell growth undoubtedly favors
sarcoma.

4574   THE INCIDENCE OF YOLK SAC TUMOR (ENDODERMAL
         SINUS TUMOR) ELEMENTS IN GERM CELL TUMORS
OF THE TESTIS IN ADULTS. (Eng.) Talerman, A. (Inst.
of Radiotherapy, Postbus 5201, Rotterdam, Holland).
Cancer 36(1):211-215; 1975.

The incidence of yolk sac tumor (endodermal sinus tumor) elements was studied in 147 germ cell neoplasms
of the testis in adults observed over a $4\frac{1}{2}$-yr period.
Excluding 79 cases of pure seminoma, yolk sac tumor
elements were found in 26 (38%) of 68 tumors; in
eight tumors the yolk sac tumor was the predominant
element. Yolk sac tumor elements were found admixed
with all other germ cell tumor elements. Tumors composed entirely of yolk sac tumor were not encountered.
Fifteen (57%) of the 26 patients with tumors containing yolk sac tumor elements have died of their disease
during the period under study, compared with 13(31%)
of the remaining 42 patients, suggesting that the
prognosis of adults with testicular tumors containing

ropneumonia and thoracic trauma with pleural hemorrhage) were found in the anamneses of 12 patients.

4579   ENDOBRACHYESOPHAGUS AND ADENOCARCINOMA.
        (Fre.) Savary, M. (Av. Haldimand 19b,
CH-1400 Yverdon, Switzerland); Naef, A.-P.; Ozzello,
L.; Roethlisberger, B. *Schweiz. Med. Wochenschr.*
105(18):575-579; 1975.

During 6168 esophagoscopies, 164 cylindrical epithelial patches were detected endoscopically with
a regular progressive incidence from year to year.
This increased detection may be due to improved
optics, and the progressive training of endoscopists. The cylindrical epithelial islets of the
upper (19%) or lower (4%) esophagus can be distinguished endoscopically from the widespread cylindrical
epithelial lining of the lower esophagus (77%). The
first type was not associated in any significant
manner with any organic pathology. It is of congenital origin usually without pathological significance. Statistically, however, the second type
seems to be in the category of peptic esophagitis
in its chronic form, and to constitute a form of
cicatrization. The incidence (8.8%) of adenocarcinoma of the lower esophagus in association with
this second form of cylindrical epithelial lining
is significant. It is advisable for peptic esophagitis to be treated before it reaches its chronic
stage and, in particular, before the appearance of
cylindrical epithelial scars.

4580   ASSEMBLY OF MICROTUBULES ONTO KINETO-
        CHORES OF ISOLATED MITOTIC CHROMOSOMES
OF HeLa CELLS. (Eng.) Telzer, B. R. (Dept. Biology, Yale Univ., New Haven, Conn. 06520); Moses,
M. J.; Rosenbaum, J. L. *Proc. Natl. Acad. Sci.
U.S.A.* 72(10):4023-4027; 1975.

The capacity of HeLa kinetochores to act as preferential sites for the *in vitro* initiation and assembly of tubular subunits into microtubules was
investigated. Electron microscopy showed kinetochores to be distinct from the chromatin with which
they are associated. When unfixed chromosomes were
immobilized by attachment to grids and incubated
with chick brain tubulin, microtubules assembled
onto the kinetochores. This demonstrated the
competence of kinetochores in isolated chromosomes
to act *in vitro* as microtubule assembly sites, and
suggested that they also possess this capacity *in
vivo*. In addition, the results provided a possible
means for isolating and characterizing kinetochores.

4581   FAMILIAL NEUROBLASTOMA. (Eng.) Pegelow,
        C. H. (Los Angeles County--Univ. Southern
California Medical center, 1129 N. State St., Los
Angeles, Calif. 90033); Ebbin, A. J.; Powars, D.;
Towner, J. W. *J. Pediatr.* 87(5):763-765; 1975.

A neuroblastoma was diagnosed in the infant daughter
of parents each of whom had a child with neuroblastoma by a previous marriage. Routine and trypsin
Giemsa-banded karyotype analysis demonstrated that
the patient had cytogenetic abnormalities derived

from each parental line. A number 21p-q- chromosome was found in the proposita, her mother, and maternal grandfather. Examination of this chromosome in prophases that had long chromosomes did not reveal a ring configuration. This atypical number 21 chromosome had a slightly narrower band 21q11. An atypical number 11 chromosome was found in the proposita and her father. Band 11q21 was wide and band 11q23 was narrow. This is apparently the first reported instance of familial neuroblastoma in which trypsin Giemsa-banding was used in chromosomal analysis. The atypical chromosomes and other unusual banding patterns should be explored in instances of neuroblastoma.

4582   VARIED MANIFESTATIONS OF A FAMILIAL LYMPHOPROLIFERATIVE DISORDER. (Eng.)
Fraumeni, J. F., Jr. (Natl. Cancer Inst., A521 Landow Building, Bethesda, Md. 20014); Wertelecki, W.; Blattner, W. A.; Jensen, R. D.; Leventhal, B. G. *Am. J. Med.* 59(1):145-151; 1975.

Immunologic studies were done on a family with diverse forms of lymphoproliferative neoplasia. In a sibship of nine adults, four died of lymphocytic or histiocytic lymphomas, and one of Waldenström's macroglobulinemia (immunoglobulin M [IgM], kappa type) complicated by adenocarcinoma of the lung. In the next generation, one member died of Hodgkin's disease; four of nine healthy persons had impaired lymphocyte transformation *in vitro* in response to phytohemagglutinin-P (PHA-P). Three of these had polyclonal elevations in IgM levels. One had depressed responses to both PHA-P and streptolysin O. Adenocarcinoma of the lung developed in one woman with immune defects, and lymphocytic leukemia developed in her 3-yr-old grandson. This family of Jewish origin migrated from Russia to Massachusetts in 1900. The sibs in generation II had different occupations and lived apart as adults. These findings point to a genetically regulated defect of immunity expressed as diverse lymphoproliferative disorders, including polyclonal and monoclonal IgM gammopathies. The occurrence of pulmonary adenocarcinoma in two members suggests genetic and immunologic determinants in these instances.

4583   CYTOGENETIC STUDIES OF THE SPLEEN IN CHRONIC GRANULOCYTIC LEUKAEMIA. (Eng.)
Spiers, A. S. D. (Royal Hobart Hospital Clinical Sch., Collins St., Hobart, Tasmania 7000, Australia); Baikie*, A. G.; Dartnall, J. A.; Cox, J. I. *Aust. N.Z. J. Med.* 5(4):295-305; 1975.

Cytogenetic studies were performed on splenic tissue from 12 patients with chronic granulocytic leukemia (CGL). The diagnosis had been established by examination of films of peripheral blood and bone marrow aspirate; the stage of the disease varied widely. Six elective splenectomies were performed, as were three for pancytopenia. Splenic tissue was cultured by three methods, metaphases were examined by light microscopy, scored for the presence of Philadelphia chromosome (Ph^1), and the chromosome numbers counted. The disease was in its chronic phase in eight pati-

increase in the B clone with reduced surface immunoglobulin level, by increased rosette-forming
T-lymphocyte count, and by reduced blast transformation of T-lymphocytes. When the blast transformation capacity of the proliferating T-clone
was normal, and intact B-lymphocytes were absent,
proliferation of B-lymphocytes with reduced surface globulin level, absence of B cells with normal surface immunoglobulin concentration, increased rosette-forming T-cell count, and unchanged blast transformation of T-cells were observed.

4588   STUDY OF ACUTE TRANSFORMATION IN 45 CASES
        OF CHRONIC MYELOID LEUKEMIA. (Fre.)
Duhamel, G. (Hopital Saint-Antoine, F 75571 Paris
Cedex 12, France); Najman, A.; Gorin, N.; Deloux,
J.; Elghouzi, M. *Nouv. Rev. Fr. Hematol.* 15(2):
301-302; 1975.

Clinical, hematological, and biological symptoms
of acute transformation were studied in 45 cases
of chronic myeloid leukemia. Clinical signs preceded hematological symptoms in two of three cases;
changes in general condition were observed in 58%,
splenomegaly in 71%, fever in 48%, signs of hemorrhage in 18%, hepatomegaly in 18%, and bone pain
in 15%. A hyperleukocytosis above 30,000 was observed in 79%; anemia and thrombocytopenia, however, appeared late and unobtrusively. Leukoblas-
tosis is the main diagnostic criterion; the 20%
rate of blood blasts chosen for diagnosis of an
acute state was found in 50%. Bone marrow leuko-
blastosis was found in 80% of the patients. The
rise in leukocytic alkaline phosphatase was significant in 80%; chromosomal changes could not be
observed. Acute transformation occurred following
a chronic phase (average length 38 mo). The acute
phase was brief; 20% died one month after its discovery, and 50% died three months later. The therapeutic association of purinethol, methotrexate,
and prednisone was compared to that of cytosine
and prednisone; complete remission was obtained
in only 8% with the first treatment, and did not
last over two months. The second association was
slightly more successful. Therapy can do very
little once the acute transformation occurs.

4589   A SCANNING ELECTRON MICROSCOPIC STUDY
        OF 34 CASES OF ACUTE GRANULOCYTIC,
MYELOMONOCYTIC, MONOBLASTIC AND HISTIOCYTIC
LEUKEMIA. (Eng.) Polliack, A. (Sloan-Kettering
Inst. for Cancer Res., 410 East 68th St., New York,
N.Y. 10021); McKenzie, S.; Gee, T.; Lampen, N.;
de Harven*, E.; Clarkson, B. D. *Am. J. Med.*
59:308-315; 1975.

The surface architecture of leukemic cells in 34
patients with acute nonlymphoblastic leukemia was
examined. Six patients with myeloblastic, four
with promyelocytic, ten with myelomonocytic, eight
with monocytic, four with histiocytic, and two with
undifferentiated leukemia were studied. Most leukemic histiocytes and monocytes appeared similar
and were characterized by the presence of large,
well developed broad-based ruffled membranes or

prominent raised ridge-like profiles, resembling
in this respect normal monocytes. Most cells from
patients with acute promyelocytic or myeloblastic
leukemia exhibited narrower ridge-like profiles,
whereas some showed ruffles or microvilli. Patients
with myelomonocytic leukemia showed mixed populations of cells with ridge-like profiles and ruffled
membranes, whereas cells from two patients with
undifferentiated leukemia had smooth surfaces,
similar to those encountered in cells from patients
with acute lymphoblastic leukemia. It appears
from the literature that nonlymphoblastic and lymphoblastic leukemia cells (particularly histiocytes
and monocytes) can frequently be distinguished on
the basis of their surface architecture. The surface features of leukemic histiocytes and monocytes
are similar, suggesting that they may belong to the
same cell series. The monocytes seem to have
characteristic surface features recognizable with
the scanning electron microscope and differ from
most cells from patients with acute granulocytic
leukemia. Although overlap of surface features
and misidentification can occur, scanning electron
microscopy is a useful adjunct to other modes of
microscopy in the study and diagnosis of acute
leukemia.

4590   SOME PARAMETERS OF THE PROLIFERATIVE
         ACTIVITY OF BONE MARROW BLAST CELLS IN
PATIENTS WITH ACUTE LEUKEMIA AT DIFFERENT STAGES
OF THE DISEASE. (Rus.) Lapotnikov, V. A. (I. P.
Pavlov First Leningrad Medical Inst., U.S.S.R.);
Nemirovskii, V. S.; Burd, I. E.; Dombrovskaia, N.
V. *Probl. Gematol. Pereliv. Krovi* 20(3):24-27;
1975.

The mitotic index and the *in vitro* incorporation
of tritiated thymidine (labeling index) in bone
marrow blast cells were studied in 36 patients
with acute leukemia at different stages of the
disease. The labeling index was relatively high
before therapy was indicated, but the labeling
index of large blasts was significantly lower
than that of normal myeloblasts. Cytostatic drug
therapy failed to cause any decrease in the labeling index either of the entire blast pool or
of blasts with diameters larger than 23 μ. However, the therapy reduced the mitotic index of
the leukocytes to 1% against 2.9% as measured
before therapy. There was little difference in
the labeling index and the mitotic index between
untreated patients and patients who were receiving
maintenance therapy in remission. However, the
total number of thymidine-incorporating leukocytes
in the mitotic group (blasts, promyelocytes, myelocytes) was significantly higher than in untreated patients. The increased mitotic index
in remission may be due to a blockade of mitoses
in the metaphase by vinblastine, or to a prolongation of the mitosis. The labeling index was
highest during recurrence (14.2% against 5.2%
before therapy and 6.6% in remission), which was
primarily due to thymidine incorporation by large
blasts. The percentage of DNA-synthesizing small
blasts increased to 6.8% aginst 2.5% in untreated
patients. These small blasts probably originate
from dividing large blasts. The persistence of

4595   HISTOPATHOLOGICAL OBSERVATIONS OF SPON-
        TANEOUS REGRESSION OF ROUS SARCOMAS IN
JAPANESE QUAILS. (Eng.) Yoshikawa, Y. (Dept.
Measles Virus, Natl. Inst. Health, Musashi-Murayama,
Tokyo 190-12, Japan); Yamanouchi, K.; Takahashi,
R.; Fujiwara, K. *Jpn. J. Med. Sci. Biol.* 28(3):
189-200; 1975.

Tumors induced in Japanese quails by the Schmidt-
Ruppin strain of Rous sarcoma virus, subgroup A,
were examined histopathologically. Stock virus
was inoculated sc at both wingwebs of 32 quails.
At various intervals post inoculation, the quails
were necropsied and tumor masses collected and
examined. The tumors had reached a maximum size
of 11 x 11 mm at day 14; tumor masses were then
almost globular, solid, and elastic, reaching to
the muscle layers without showing clear boundaries.
Most tumors began to regress at 18 days, while
some continued to progress. These included five
progressing with hematoma, and four progressing
without hematoma. Microscopically, three phases
of tumor development were noted in regressors.
Phase I, days 4-7, involved predominant infiltration
of heterophils. Phase II, days 10-14, was characterized by necrosis of tumor cells and focal accumulation of lymphoid cells. In Phase III, days 18-
24, tumor cells and heterophils disappeared, small
foci of hemorrhage were demonstrated, and plasma
cells frequently infiltrated. Microscopic changes
in progressors followed a similar pattern. Progressors without hematoma experienced marked growth
of tumor cells with characteristic features of
typical fibrosarcoma. Hemorrhagic areas surrounded
by connective tissues occupied most of the lesions
in progressors with hematoma. No marked variation
in histologic changes differentiating regressors
and progressors were noted in samples taken prior
to day 18. In addition, the cellular reaction of
the progressors without hematoma had features similar
to phase II as late as 24 days, suggesting a lack
of cell-mediated destruction of tumor cells suppos-
edly occurring in regressors. It is suggested
that progressors with hematoma represent a regressing process which failed to follow the regular
course.

4596   WILMS' TUMOR: HISTOLOGIC VARIATION AND
        PROGNOSIS. (Eng.) Lawler, W. (Dept.
Pathology, Stopford Building, The University Man-
chester M13 9PT, England); Marsden, H. B.; Palmer,
M. K. *Cancer* 36(3):1122-1126; 1975.

A histologic analysis of 75 consecutive cases of
Wilms' tumor in the Manchester (England) hospital
region was carried out to determine whether histologic classification at the time of initial surgery
would be helpful in assessing the prognosis. Eight
of the tumors could not be histologically classified; the remaining 67 were classified on the presence of tubules, smooth and/or striated muscle,
glomerular structures, well-defined nests of typical nephric blastemal cells among a background of
mesenchyme, and larger, more open cells with prominent nucleoli. The degree of tubule formation
was classified as 0, +, ++, or +++. The 0 group

had no tubules; the + group and ++ groups had definite but scanty tubules; and the +++ group had a very large number of tubules, most tumor cells being involved in tubule formation. Survival time increased with increasing tubule formation; 83% of the +++ group were alive with no sign of recurrence three years after diagnosis, and none of the group 0 patients were alive at this time. With regard to the other features, survival was better when glomerular structures were present than when they were absent, with a small as opposed to a large amount of smooth muscle, and with no cell nests. The presence or absence of larger cells showed no relationship with survival. There was no correlation between age and histological type, and no one histological feature was a significant predictor of prognosis in individual cases. Histological classification was correlated with clinical stage as well as survival, suggesting that the histological classification presented may be very useful in predicting the prognosis of children with Wilms' tumor.

4597   THE GROWTH CHARACTERISTICS OF THE METAS-
         TATIC WISTAR/FURTH WILMS' TUMOR MODEL.
(Eng.) Murphy, G. P. (Dept. of Experimental Surgery, Roswell Park Memorial Inst., 666 Elm St., Buffalo, N.Y. 14263); Williams, P. D.; Klein, R. Res. Commun. Chem. Pathol. Pharmacol. 12(2):397-404; 1975.

Growth characteristics, metastatic spread, and survival times were evaluated in a transplantable Wistar/Furth rat Wilms' tumor model. The tumor doses (1 x 10^5, 1 x 10^4, and 1 x 10^3 live tumor cells/cc per animal) were injected in an attempt to find the optimum level of tumor take. Lung metastases were frequent following tumor injection by all routes. However, tumor spread to the lungs was the least frequent following sc injection of the tumor. Survival time for the im group was statistically longer at all tumor dose levels. For the 1 x 10^5 tumor dosage, survival time ranged from a mean of 27 days for the intrarenal group to a mean 42 days for the im group. For the 1 x 10^4 dosage, survival time ranged from a mean of 37 days for the ip group to a mean of 51 days for the im group. It is concluded that this animal tumor closely resembles the human Wilms' tumor, and that the point at which the transplanted tumor fails to be successfully transplanted is below the dosage level of 1 x 10^3 tumor cells.

4598   GROWTH OF A PLASMA CELL MYELOMA IN LATHY-
         RITIC MICE. (Eng.) Martino, L. J. (St.
Louis Univ. Sch. of Medicine, 1402 S Grand Blvd., St. Louis, Mo.); Yeager, V. L.; Taylor, J. J. Arch. Pathol. 99(10):536-539; 1975.

The growth of a plasma cell myeloma (Adj PC-5) was studied in 4- to 6-wk-old male and female BALB/c mice made lathyritic by the administration of 1.6% ß-aminoproprionitrile (BAPN) in powdered food. The mice received an iv injection of a 0.1-ml cell suspension containing between 10^5 and 10^6 Adj PC-5 cells. Histologic sections of bones from these mice

were compared with those from controls fed a normal diet, from mice treated with BAPN alone, and from mice given only tumor cells. Bones from control mice appeared normal. Bones from mice treated with tumor cells and BAPN exhibited histological signs of lathyrism similar to those in mice treated with BAPN only, and of tumor invasion similar to that in the mice treated only with tumor. However, the number of bones in which the medullary cavities were filled with tumor cells was greatly decreased, compared wit the number in mice treated with tumor cells only. Additionally, the lathyritic response to BAPN was decreased in tumor-bearing animals, as evidenced by the absence of bone spicules in thickened periosteum and a 50% reduction in the number of epiphyseal rents. The percentage of bones with osteolytic or osteoplastic lesions was the same in the two groups of tumor-bearing animals, as was the number of animals that developed ascites or became paralyzed. The percentage of bones in which tumor cells lifted the periosteum or spread into the muscles through the fibrous periosteum was also the same. In addition to retarding tumor growth in the medullary cavity, BAPN caused osteoporosis and reduced the tensile strength of collagen, both of which allowed extramedullary tumor growth to proceed as usual.

4599   IN VITRO CULTURE OF CELLS ISOLATED FROM
         DIMETHYLNITROSAMINE-INDUCED RENAL MESEN-
CHYMAL TUMORS OF THE RAT. I. QUALITATIVE MORPHO-
LOGY. (Eng.) Hard, G. C. (Baker Med. Res. Inst., Prahran, Australia); Borland, R. J. Natl. Cancer Inst. 54(5):1085-1095; 1975.

The in vitro morphology of cells isolated from renal mesenchymal tumors induced in outbred Porton Wistar or inbred Wistar rats by a single ip injection of dimethylnitrosamine (60 mg/kg) was followed in short- and long-term culture. The cells were examined daily by phase-contrast microscopy. The morphology of cells isolated from normal rat-kidney cortex was studied for comparison. Primary tumor-cell isolates consisted of a range of mesenchymal cell forms interspersed between discrete, homogeneous islands of cohesive epithelium-like cells. The latter were presumably derived from renal parenchyma that had been engulfed within the in vivo neoplasm by proliferating tumor cells, since this epithelial component did not survive serial subculture. Established tumor cell lines consisted of a pleomorphic range of mesenchymal forms resembling the descriptions of various cell types in vitro, including fibroblasts, smooth muscle cells, and endothelial cells. A range of mesenchymal cell forms in culture therefore correlated with the heterogeneous histologic nature of the in vivo neoplasm in which fibroblasts, smooth muscle fibers, and endothelium-like cells were also represented. Multinucleate or polymorphonuclear giant cells were characteristic of tumor cell cultures, whereas large, expanded polygonal cells with longitudinal striations were characteristic of normal kidney cell cultures. Some tumor cell cultures were typified by cells with cytoplasmic vacuolation and some by the acquisition of an epithelioid form at high cell

4603   SCANNING ELECTRON MICROSCOPY OF NORMAL AND
        ABNORMAL EXFOLIATED CERVICAL SQUAMOUS CELLS.
(Eng.) Murphy, F. (Dept. Obstet. Gynaecol., Univ.
Birmingham, England); Allen, J. M.; Jordan, J. A.;
Williams, A. E. *Br. J. Obstet. Gynaecol.* 82(1):44-
51; 1975.

4604   BRONCHIAL CARCINOID TUMOR WITH LIVER METAS-
        TASES AND FLUSH SYNDROME. ULTRASTRUCTURAL
STUDY. (Fre.) Patricot, L.-M. (Hopital de la Croix-
Rousse, F. 69317 Lyon, Cedex 1, France); Vauzelle,
J.-L.; Leung, T.-K; Feroldi, J. *Arch. Anat. Pathol.*
23(1):5-13; 1975.

4605   CARCINOGENESIS STUDIES ON HUMAN BRONCHIAL
        EPITHELIUM (BE) [abstract]. (Eng.) Trump,
B. F. (Univ. Maryland, Baltimore, Md. 21201); Barrett,
L. A.; McDowell, E. M. *Proc. Am. Assoc. Cancer Res.*
16:59; 1975.

4606   FAMILIAL BRONCHIAL CARCINOID AND POLYENDO-
        CRINE ADENOMATOSIS. (Fre.) Dry, J. (Hopi-
tal Rothschild, 43, bd de Picpus, F 75571 Paris Cedex
11, France); Lebrigand, H.; Pradalier, A.; Leynadier,
F.; Huguier, M. *Ann. Med. Interne (Paris)* 126(6/7):
491-496; 1975.

4607   DEVELOPMENT OF MALIGNANCY DURING METHO-
        TREXATE TREATMENT. (Heb.) Schewach-
Millet, M. (Haim Sheba Medical Center, Tel Hashomer,
Israel); Feinstein, A.; Rafael, D. *Harefuah* 88(8):
366-367, 400; 1975.

4608   LEIOMYOSARCOMA OF SUPERFICIAL SOFT TISSUE
        WITH UNUSUAL METASTASIS. (Eng.) Kaplan,
G. (Veterans Administration Hosp., New York, N.Y.);
Varga, P. *N.Y. State J. Med.* 75(12):2240-2241; 1975.

4609   ELIMINATION OF INDUCED CHROMOSOME ABERRA-
        TIONS IN EMBRYOGENESIS OF MICE [abstract].
(Eng.) Buselmaier, B. (Institut fur Anthropologie
und Humangenetik Universitat Heidelberg, Neuenheimer-
feld 328, West Germany); Hansmann, I. *Mutat. Res.*
29(2):215-216; 1975.

4610   CHROMOSOME OBSERVATIONS IN C57B1 STRAIN
        LEUKEMIC MICE. (Fre.) Leonard, A.
(Departement de Radiobiologie, Centre d'Etude de
l'Energie nucleaire, B-2400 Mol, Belgique); Maisin,
J. R.; Mattelin, G. *C. R. Soc. Biol. (Paris)* 169(2):
464-467; 1975.

4611   MITOTIC ANOMALIES AS THE SOURCE OF MUTATION
        OF THE GENOME: POLYPLOIDIZATION CYCLES--
SEGREGATION IN MAMMALIAN CELL POPULATION. (Eng.)
Rizzoni, M. (Centro di Studio di Genetica Evolu-
zionistica del C.N.R. and Istituto di Genetica,
Universita di Roma, Rome, Italy); Spirito, F.;
Perticone, P.; De Salvia, R.; Ricordy, R.; Tanzarella,
C.; Palitti, F. *Mutat. Res.* 29(2):200-201; 1975.

4612   IDENTIFICATION OF CHROMOSOMAL ABERRATIONS
        OBSERVED IN HUMAN CANCER CELLS BY R BANDING.
(Fre.) Ayraud, N. (Laboratoire de Cytogenetique,
U.E.R. de Medecine et Centre Antoine Lacassagne,
06000 Nice, France). *C. R. Soc. Biol. (Paris)* 169(2):
365-373; 1975.

4613   SEX-CHROMOSOME LOSS IN HUMAN TUMOURS. (Eng.)
        Zankl, H. (Dept. Human Genetics, Univ.
Saarland, 6650 Homburg/Saar, West Germany); Seidel,
H.; Zang, K. D. *Lancet* 1(7900):221; 1975.

4614   SIMULTANEOUS OCCURRENCE OF HIATUS HERNIA AND
        CARCINOMATA IN THE CARDIA AND THE ESOPHAGUS.
(Dan.) Nilsson, T. (laege, Plantagevej 14, DK-3460
Birkerod, Denmark); Halkier, E. *Ugeskr. Laeger*
137(4):199-201; 1975.

4615   INTESTINAL LYMPHANGIECTASIA AND TUMOR SPREAD
        BY REVERSAL OF LYMPHATIC FLOW IN ASSOCIATION
WITH A CARCINOMA OF THE COLON. (Fre.) Debray, C.
(Hopital Bichat, 170, bd Ney, F 75877 Paris Cedex
18, France); Leymarios*, J.; Marche, C.; Malmonte,
A. M.; Le Canuet, R. *Nouv. Presse Med.* 4(33):2391-
2392; 1975.

4616   NEOPLASTIC PANETH CELLS: OCCURRENCE IN AN
        ADENOCARCINOMA OF A MECKEL'S DIVERTICULUM.
(Eng.) Scharfenberg, J. C. (Alton Ochsner Med.
Found. Clin., New Orleans, La.); DeCamp, P. T.
Am. J. Clin. Pathol. 64(2):204-208; 1975.

4617   CARCINOID SYNDROME. (Fle.) Carpentier,
        J. (Evergemse Steenweg, 154, 9030 Wondel-
gem, Belgium); Brys, R.; De Keyser, R. *Tijdschr.*
Gastroenterol. 18(2):100-105; 1975.

4618   CANCER IN CROHN'S DISEASE -- DANGER OF A
        BY-PASSED LOOP. (Eng.) Greenstein, A. J.
(Mt. Sinai Sch. Med., City Univ. of N.Y., Fifth
Ave. & 100th St., New York, N.Y. 10029); Janowitz,
H. D. *Am. J. Gastroenterol.* 64(2):122-124; 1975.

4619   PRIMARY HEPATIC MALIGNANCY IN PREGNANT
        WOMEN. (Eng.) Purtilo, D. T. (Univ.
Massachusetts Medical Sch., 55 N. Lake Ave., Wor-
cester, Mass. 01604); Clark, J. V.; Williams, R.
Am. J. Obstet. Gynecol. 121(1):41-44; 1975.

4620   MULTIPLE LYMPHANGIOENDOTHELIOMA OF THE
        SPLEEN IN A 13-YEAR-OLD GIRL. (Eng.)
Hamoudi, A. B. (Children's Hosp., Columbus, Ohio
43205); Vassy, L. E.; Morse, T. S. *Arch. Pathol.*
99(11):605-606; 1975.

4621   PRELEUKEMIAS. (Ger.) Schott, G. (I. Med.
        Klinik, 95 Zwickau, Karl-Keil-Strasse 35,
East Germany). *Z. Gesamte Inn. Med.* 30(4):159-161;
1975.

4622   HISTAMINE METABOLISM IN POLYCYTHAEMIA
        VERA. (Eng.) Westin, J. (Sahlgren's
Hosp., S-413 45 Goteborg, Sweden); Granerus, G.;
Weinfeld, A.; Wetterquist, H. *Scand. J. Haematol.*
15(1):45-57; 1975.

4623   A CASE OF ATYPICAL ACUTE MYELOSIS: RNA
        LEUKEMIA. (Fre.) Schneiberg, K. (Insti-
tut d'Oncologie, ul. Wybrzeze Armi Czerwonej 15,
44-101 Gliwice, Poland); Wieczorkiewicz, A.; Rytwin-
ski, K.; Bartnikowa, W. *Acta Haematol.* 53(2):49-55;
1975.

4624   ACUTE MYELOID LEUKEMIA FOLLOWING MELPHALAN
        THERAPY FOR IgA PARAPROTEINEMIA. (Ger.)
Vogtli, W. (1. Medizinische Universitatsklinik,
Kantonsspital, CH-4004 Basel, Switzerland); Nagel*,
G. A.; Bianchi, L.; Truog, P. *Schweiz. Med. Wochen-*
schr. 105(1):27-28; 1975.

4625   LEUKEMIC MITOCHONDRIA. III. ACUTE LYMPHO
        BLASTIC LEUKEMIA. (Eng.) Schumacher, H.
R. (Harrisburg Hosp., Pa.); Szekely, I. E.; Fisher,
D. R. *Am. J. Pathol.* 78(1):49-58; 1975.

4626   EVOLUTION OF CHRONIC MYELOID LEUKEMIAS
        TURNED "ACUTE". (Fre.) Lederlin, P.
(la Clinique Medicale A de Monsieur le Professeur
R. Herbeuval, C.H.U.-Nancy, France); Puchelle,
J. C.; Aymard, J. P.; Thibaut, G.; Guerci, O.;
Herbeuval, R. *Ann. Med. Nancy* 14:191-193; 1975.

4627   CULTURE OF LEUKEMIC CELLS. II. NEW
        ACHIEVEMENTS IN CYTOGENETICS OF CHRONIC
MYELOID LEUKEMIA. (Pol.) Rozynkowa, D. (Inst.
Clinical Pathology, Medical Acad., Lublin, Poland);
Stepien, J. *Pol. Arch. Med. Wewn.* 54(1):39-45; 1975.

4628   CONTRIBUTION TO THE STUDY OF NEUROLOGICAL
        MANIFESTATIONS OF CHRONIC LYMPHATIC
LEUKEMIA. ROLE OF ASSOCIATED GLOBULIN ABNORMALITIES.
A DESCRIPTION OF 8 CASES. (Fre.) Masson, R. (Ser-
vice d'Hematologie, Hopital Edouard-Herriot, Lyon,
France); Boucher, M.; Joyeux, O.; Chazot, G.; Mar-
tin, C.; Favre-Gilly, J.; Revol, L. *Rev. Neurol.*
(Paris) 131(6):373-385; 1975.

4629   MARKER CHROMOSOME 14q+ IN MULTIPLE MYELOMA.
        (Eng.) Philip, P. (Dept. Medicine A Rigs-
hospitalet, DK-2100 Copenhagen O, Denmark). *Here-*
ditas 80(1):155-156; 1975.

4630   GENETIC HETEROGENEITY IN FAMILIAL BREAST
        CANCER [abstract]. (Eng.) Lynch, H. T.
(Creighton Univ., Omaha, Nebr.); Guirgis, H. A.;
Mulcahy, G.; Lynch, J.; King, M. C.; Maloney, K.;
Thul, P.; Thomas, R. *Proc. Am. Assoc. Cancer Res.*
16:190; 1975.

4638   ON TWO TYPES OF MALIGNANT NEOPLASM OF
        GENITALS AND MAMMARY GLANDS. (Ukr.)
Voitenko, V. P. (Inst. Gerontology, Acad. Sciences,
Ukrainian S.S.R.). *Dopov. Akad. Nauk 'Ukr. RSR Ser.
B* (1):79-80; 1975.

4639   ENDOMETRIAL STROMAL TUMOR IN A CHIMPANZEE.
        (Eng.) Toft, J. D., II (USAF Sch. Aero-
space Medicine, Brooks Air Force Base, Tex. 78235);
Mac Kenzie, W. F. *Vet. Pathol.* 12(1):32-36; 1975.

4640   LYMPHATIC OBSTRUCTION IN CARCINOMATOUS
        ASCITES. (Eng.) Feldman, G. B. (Natl.
Cancer Inst., Bethesda, Md.). *Cancer Res.* 35(2):
325-332; 1975.

4641   ADRENAL MEDULLARY HYPERPLASIA IN FAMILIAL
        MEDULLARY THYROID CARCINOMA [abstract].
(Eng.) DeLellis, R. A. (New England Med. Cent.
Hosp., Boston, Mass.); Wolfe, H. J.; Gang, D. L.;
Gagel, R. F. *Am. J. Pathol.* 78(1):40a; 1975.

4642   CLASSIFICATION OF MALIGNANT TUMORS OF THE
        THYROID GLAND ACCORDING TO THE WHO 1974
NOMENCLATURE. HISTOLOGICAL VERIFICATION OF 327
CASES OF MALIGNANT TUMORS OF THE THYROID GLAND.
(Ger.) Neracher, H. (Institut fur Pathologische
Anatomie, Kantonsspital, CH-8091 Zurich, Switzerland);
Hedinger*, C. *Schweiz. Med. Wochenschr.* 105(32):
1000-1006; 1975.

4643   PATHOLOGY OF THYMOMAS -- SELECTED TOPICS.
        (Pol.) Mioduszewska, O. (Instytut Onko-
logii, Wawelska 15, 02-034 Warszawa, Poland).
Nowotwory 25(1):7-14; 1975.

See also:

* (Rev):   4208, 4219, 4220, 4221, 4222, 4223,
            4224, 4235, 4240, 4254, 4255, 4256,
            4257, 4258, 4259, 4260, 4261
* (Chem): 4270, 4295, 4296, 4305, 4306, 4308,
            4312, 4331, 4332, 4339, 4355
* (Phys): 4362
* (Viral): 4400
* (Immun): 4468, 4504, 4514, 4515, 4525, 4537,
            4544, 4547
* (Epid-Biom): 4646, 4649, 4650, 4652, 4655,
                 4660, 4661, 4672

4644   MORTALITY DUE TO MALIGNANT NEOPLASMS,
        AUSTRALIA, 1950-1972 -- A SUMMARY OF
TRENDS AT THE MAJOR SITES. (Eng.) Scott, G. (Dept.
Preventive and Social Medicine, Univ. Sydney, Syd-
ney, Australia). *Cancer Forum* (5):250-255; 1975.

The trends in mortality in Australia from the leading sites of malignant neoplasms were studied for
1950-1972 by means of Standardized Mortality Ratios.
Until the establishment of the NSW Central Cancer
Registry in 1972, no population-based cancer incidence data was available in Australia. Hence it is
difficult to determine whether the trends in mortality are related to changes in incidence. Sites
included in the tables of Standardized Mortality
Ratios are: digestive organs plus peritoneum;
stomach; large intestine; rectum; pancreas; and
lung. In these tables data for males and females
are presented separately. Tables for females only
include breast, cervix, corpus, and ovary; and for
males, prostate. There has been a significant decline in mortality from gastric cancers in both male
and female accompanied by a rise in mortality from
malignant tumors of the pancreas. Mortality from
cervcial cancer has apparently declined despite
problems of classification and coding of cancers
of the female genital organs. Incidence and mortality from lung cancer have risen sharply. Mortality and incidence appear to be slowly increasing
for ovarian cancer, Hodgkin's disease, and leukemia.

4645   CANCER IN FINLAND 1953-1970: INCIDENCE,
        MORTALITY, PREVALENCE. (Eng.) Teppo, L.
(Finnish Cancer Registry, The Institute for Statis-
tical and Epidemiological Cancer Research, Finland);
Hakama, M.; Hakulinen, T.; Lehtonen, M.; Saxen, E.
Acta Pathol. Microbiol. Scand. [A] *Supplement 252*
:1-66; 1975.

Data relating to the incidence, mortality, and prevalence of cancer and of tumors of selected sites in
Finland from 1953 to 1970 is presented. The information was compiled by the Finnish Cancer Registry --
The Institute for Statistical and Epidemiological
Cancer Research and the Childhood Cancer Registry.
Some facts are presented about Finland, the organization of the Registry, and of the material and defini-
tions used in this study. The specific sites for
which data are presented are: lip, esophagus, stomach,
colon, rectum, liver, gallbladder, pancreas, larynx,
lung, breast, cervix, uterus, ovary, testis, prostate,
kidney, urinary bladder, skin, thyroid gland, and
brain. In addition, information concerning Hodgkin's
disease, leukemia and childhood cancers is presented.
There were 28,414 new cases of cancer diagnosed in
males and 26,375 cases in females between 1966 and
1970. Lung, stomach, and prostate (30.9, 14.6, and
8.3%, respectively) were the leading types of cancer
in males, while breast, uterus, and stomach (18.8,
13.0, and 12.2% respectively) predominated in females.
The incidence rates of cancer of the lung, larynx, and
stomach were much higher among Finnish men than for
men of Denmark, Sweden, and Norway. However, the
incidence rates of breast, prostate, colon, and rectum were comparatively low in Finland. Cancer of the

4650   THE INCIDENCE OF MALIGNANCIES IN TRANS-
        PLANT RECIPIENTS. (Eng.) Penn, I. (Univ.
Colorado Sch. Medicine, Denver, Colo.). *Transplant.*
Proc. 7(2):323-326; 1975.

De novo cancers were encountered in 27 (5.6%) of
483 renal-homograft recipients followed for up to
11.5 yr. Clinical features of *de novo* malignancies were studied in a series of 234 patients, of
whom 231 received kidney transplants, 2 received
hearts, and 1 a liver transplant. Among the 241
malignancies that developed, there were 95 cancers
of the skin and lips, 66 solid lymphomas, 18 carcinomas of the uterine cervix, 12 carcinomas of
the lung, and 50 miscellaneous malignancies. The
patients with lymphomas differed from those with
other cancers in that they were slightly younger
(34.9 *vs* 39.6 yr), the tumors occurred earlier after transplantation (20.6 *vs* 31.6 mo), were more
frequently treated with antilymphocyte globulin
(40% *vs* 25.4%), and more often received organs
from unrelated donors (7.6% *vs* 65.4%). Only 11%
of the lymphoma patients are alive at present compared with 82% of those with carcinomas of the skin,
lips, or uterine cervix and 25% of patients with
other types of cancers. The development of cancer
could not be related to any individual immunosuppressive measure but appeared to be a result of
immunosuppression in general. Preexisting neoplasms may have been present in two donors and in
22 recipients. Presumably the malignancies present
at time of transplantation were small but grew rapidly under the influence of immunosuppressive
therapy. The incidence of *de novo* cancer in organ-
transplant recipients is not high enough to contra-
indicate the performance of transplantations, particularly as many of the neoplasms are of low-grade
malignancy and are easily controlled with conventional cancer therapy. In the case of high-grade
neoplasms, consideration should be given to reducing, or even stopping, immunosuppressive therapy.
Recovery of the lymphoreticular system has been observed in several patients treated in this manner.

4651   INCIDENCE OF MALIGNANT TUMORS IN U.S.
        CHILDREN. (Eng.) Young, J. L., Jr.
(Natl. Cancer Inst., Bethesda, Md.); Miller, R. W.
J. Pediatr. 86(2):254-258; 1975.

Incidence rates of childhood cancer in the United
States from 1969 through 1971 were determined in
the Third National Cancer Survey, designed by the
Biometry Branch of the National Cancer Institute.
The survey was conducted in seven metropolitan
areas (Detroit, Minneapolis-St. Paul, Pittsburgh,
Atlanta, Birmingham, Dallas-Fort Worth, and San
Francisco-Oakland), as well as in the states of
Iowa and Colorado. In these areas, 1,925 malignant
tumors were diagnosed in white children under age
15 yr, 225 in black children, 16 in other nonwhites,
and 7 with race not specified. The annual incidence
rates for all malignant tumors were 124.5/million
whites and 97.8/million blacks. These rates were,
respectively, 63% and 84% higher than the corresponding annual mortality rates, 1960-1969. The

rank order of the seven most common malignat tumors
in childhood was unchanged when measured by incidence or by mortality rates: (1) leukemia, (2)
tumors of the CNS, (3) lymphoma, (4) neuroblastoma,
(5) Wilms' tumor, (6) bone cancer, and (7) rhabdomyosarcoma. The data for whites were generally in
accord with past experience, except for the seemingly high frequency of malignant carcinoid. Five car-
cinoid-type bronchial adenomas were diagnosed in
children aged 10-13 yr. Among blacks, there was an
unexplained deficiency of CNS tumors in males. Only
14 males were affected compared with 41 black females, although 1960-1969 mortality data for children gave CNS tumors as the cause of death in 509
black males *versus* 499 females. Since this survey
was population based and had a nearly complete as-
certainment of cases, it provides the best estimate
yet of rates of childhood cancer in the United States.

4652   CLINICAL ASPECTS OF CARCINOMA OF THE LIVER.
        (Eng.) Frey, P. (Departement fur Innere
Medizin der Universitat, Kantonsspital, CH-8091
Zurich, Ramistr. 100, Switzerland); Schmid, M.;
Knoblauch, M. *Dtsch. Med. Wochenschr.* 100(32):1625-
1629; 1975.

The incidence and diagnosis of epi-cirrhotic liver
carcinoma was studied. Between 1969 and 1973 there
were 84 cases of primary liver carcinoma and 467 of
cirrhosis among 10,211 autopsies. In 74 the carcinoma developed in a cirrhotic liver. Clinically
the diagnosis of carcinoma was made in 14 cases.
The mean survival time after the diagnosis was seven
wk; after onset of the terminal symptoms 13 wk.
Typical clinical features were decompensated, hyper-
trophic, often coarsely granular, liver cirrhosis.
The best diagnostic method was laparoscopy and determination of α_1-fetoprotein and cholestatic enzymes.

4653 ^{99m}Tc-PERTECHNETATE THYROID SCINTIGRAPHY
         IN PATIENTS PREDISPOSED TO THYROID NEO-
PLASMS BY PRIOR RADIOTHERAPY TO THE HEAD AND NECK.
(Eng.) Arnold, J. (Michael Reese Hosp. & Medical
Center, 2900 S. Ellis Ave., Chicago, Ill. 60616);
Pinsky, S.; Ryo, U. Y.; Frohman, L.; Schneider, A.;
Favus, M.; Stachura, M.; Arnold, M.; Colman, M.
Radiology 115(2):653-657; 1975.

The possibility that radiotherapy to the thymus of
infants might predispose to the development of thyroid neoplasms was investigated. Histories, physical examinations and ^{99m}Tc-pertechnetate imaging
were used to evaluate 1,452 persons who had had x-
ray therapy to the neck region for benign conditions
18-35 yr previously. Thyroid imaging was begin 20-
30 min after iv administration of 5 mCi ^{99m}Tc sodium
pertechnetate. Thyroid abnormalities were found in
301 patients (21%) and thyroid malignancy was found
in 29% of 193 patients with abnormal studies who
were explored surgically. Thyroid imaging detected
96% of these abnormalities, 40% of which would not
have been unequivocally established by physical examination alone. Nuclear imaging could not distinguish benign from malignant lesions.

gist. Various localization and surgical techniques
were proposed, and modified radical mastectomy with
removal of the breast and axillary nodes is supported
as the necessary present approach.

4659   THE AMERICAN BURKITT LYMPHOMA REGISTRY:
          A PROGRESS REPORT. (Eng.) Levine, P.
H. (Natl. Cancer Inst., Bethesda, Md. 20014);
Cho, B. R.; Connelly, R. R.; Berard, C. W.; O'
Conor, G. T.; Dorfman, R. F.; Easton, J. M.;
DeVita, V. T. *Ann. Intern. Med.* 83(1):31-36; 1975.

Demographic and clinical features of 114 patients
confirmed by the American Burkitt Lymphoma Registry in its first three years of operation are
presented and compared with known features of
African Burkitt's lymphoma. In addition, the survival data up to December, 1973 are reviewed.
The registry includes information on incidence,
presenting features, clinical course, epidemiological aspects, and laboratory parameters. The
tumor shows a general predilection for incidence
in the jaw, gonads, gastrointestinal tract, and
CNS in both American and African patients, but
incidence of bone marrow and peripheral lymphoma
is higher in American patients. Response to
chemotherapy was found to be similar in both
groups. Epidemiologic features that appear to
link American and African Burkitt's lymphoma,
such as fewer cases in high-altitude areas and
clustering of patients in time and space, require
more study. The absence of the Epstein-Barr virus
genome in African and American cases raises the
possibility that agents other than this virus may
be responsible for inducing some cases of the   .
disease on both continents.

4660   CLINICO-EPIDEMIOLOGICAL CHARACTERISTICS
          OF LYMPHOGRANULOMATOSIS. (Rus.) Plot-
nikov, Iu. K. (1st Dept. of Hosp. Therapy, D. I. Ul'
ianov Kuibyshev Medical Inst., USSR). *Ter. Arkh.*
42(6):56-60; 1975.

Clinico-epidemiological observations of 239 patients,
with lymphogranulomatosis diagnosed during the
1965-1973 period were compared with observations of
790 persons in a control group. There were no
statistically significant differences in the occupational distribution, even though the percentage
of workers employed in the petrochemical and lacquer
industries and of those exposed to engine exhaust
gases was somewhat but insignificantly higher among
the patients than in the control group. Angina with
subsequent tonsillectomy was 1.8 times more frequent
among patients than in the control, however, the difference was significant statistically. The incidence
of penumonia, cholecystitis, otitis, dysentery, etc.,
was considerably higher among the patients (15.7%
against 9.6%). The incidences of gastric and duodenal ulcer, chronic gastritis, uterine fibromyoma,
adenoma of the prostate gland, intestinal polyps,
chronic eczema, psoriasis and bronchial asthma were
largely the same in the two groups during the last
four years preceding the diagnosis of lymphogranulo-

matosis of the epidemiological survey. During the same period, the incidence of chronic inflammatory diseases, such as cholecystitis, otitis, pneumonia of stage II and III, tonsillitis, pyelonephritis and colitis was 12.9% among patients against 7.6% in the control. The incidence of malignant tumors was 1.8 times more frequent among closest relatives (parents, brothers and sisters) of the patients than among those of the control subjects. The findings indicate the possible role of earlier and concurrent inflammatory and pyegenic processes in the development of lymphogranulomatosis, i.e., that the latter may be preceded by immunological disturbances.

4661   ACUTE NONLYMPHOCYTIC LEUKEMIA. AN ADULT
         CLUSTER. (Eng.) Bartsch, D. C. (942
Lowry Medical Arts Bldg., 350 St. Peter St., St.
Paul, Minn. 55102); Springher, F.; Falk, H.
J.A.M.A. 232(13):1333-1336; 1975.

Between 1970 and 1973, seven cases of acute nonlymphocytic leukemia (five cases of acute myelomonoeytic leukemia and two cases of acute myelocytic leukemia) were diagnosed in the village of Elmwood, Wisconsin, and the surrounding area. The patients ranged in age from 14-97 yr. Interpersonal relationships were noted among all patients in the group. This cluster of cases represents a greater than 20-fold increase in the expected incidence of this disease for this population. Morphologic similarities in the blast cells were seen in all cases; these included abundant cytoplasm, with small amounts of azure granulation, and nuclei that often showed soft folding and convolutions. This cluster of cases may bear a relationship to bovine leukemia because the town was located in dairy farming country and because a number of the patients had either worked at or lived near the town creamery.

4662   PLUTONIUM FALLOUT IN TOKYO. (Eng.)
         Miyake, Y. (Geochemistry Res. Assoc.,
Tokyo, Japan); Katsuragi, Y.; Sugimura, Y. *Pap.
Materol. Geophys.* 26(1):1-8; 1975.

The monthly fall rates of plutonium 239 and 238 in Tokyo during the period from Jan. 1967 to the end of 1973 are reported. The cumulative amounts of plutonium fallout in Tokyo from the beginning of nuclear explosions to the end of 1973 are estimated to be $1.2 mCi/km^2$ and $55 \mu Ci/km^2$, respectively, for plutonium 239 and 238. Owing to the accidental release of 238 due to a satellite failure, the ratio of 238 to 239 increased abruptly from 1.8% in 1966 to 16% in 1967. The maximum value of the ratio, 31%, was observed in 1970, and then the value decreased gradually to 6.3% in 1973. The ratio of 239 to Sr-90 during the period of 1968 to 1973, in which the Chinese bomb tests prevailed, was about 1%, which was lower than the former value of 1.6% in the tests conducted by the USA and the USSR. The total amounts of deposition in the northern hemisphere of 239 and 238 were calculated to be about 200 kCi and 9 kCi, respectively. Of the 9kCi of plutonium, 2 kCi was of satellite origin.

area, most nitrosamine concentration values were less than 3 ppb. In a village in the high incidence area, high values of 15.5 ppb dimethylnitrosamine and 1.5 ppb diethylnitrosamine were found in a dish containing rice, spinach and yogurt. These findings of nitrosamines in food samples from the high incidence area were not, however, confirmed by mass spectrometry. The nitrate level in the water samples was within international standards. Ninety-five percent of the water samples had nitrite concentrations within normal limits, and in 5% no nitrite was found. It is concluded that no recognized exposure to any exogenous carcinogens was found and no single factor responsible for the high incidence of esophageal cancer in the Caspian littoral of Iran was identified. It is suggested that a combination of factors connected with climate, rate of precipitation, soil, vegetation and agriculture plus socio-economic level and cultural habits might be responsible for the differences in incidence of esophageal cancer in the two regions studied.

4666   INFERENCES FROM DATA: DEVELOPMENT OF
        MATHEMATICAL MODELS. (Eng.) Wagner,
G. (German Cancer Res. Center, Heidelberg, West
Germany); Tautu, P. *Proc. Int. Cancer Congr. 11th.*
Vol. 3 *(Cancer Epidemiology, Environmental Factors).*
Florence, Bucalossi, P.; Veronesi, U.; Cascinelli,
N. New York, American Elsevier, 1975, pp. 11-17.

Mathematical models previously developed for the epidemiology of non-infectious diseases are examined, and new models utilizing the latest theoretical, observational and experimental knowledge are suggested. Relationship between cancer incidence and age has been studied frequently, but the voluminous incidence data collected in the last 15 yr does not reveal the nature of the cancer process. There are no mathematical models providing etiologic or causal factors in cancer. Even the existence of a causal relationship between smoking and lung cancer is still open to discussion. Special statistical models have been developed to determine if leukemia cases cluster in time and space which would be termed structural models. A single mathematical model for the genetic contribution in cancer is not possible because of the multiple aspects involved in the cell division process. An illness-death stochastic model describing the relapse, recovery and death of cancer patients has been devised. All the previously mentioned models are "data-reduction" models or mathematical forms for an observed distribution of data capable of submission to significance tests. "Data-generating" models would incorporate the following characteristics: (1) knowledge of the natural history of the cancers; (2) the "taxonomic" adjacency relation based on a particular index of similarity, e.g. genetic or immunogenetic and the "biotic" adjacency relation which assumes common environmental or socio-cultural factors; (3) mathematical description of the whole carcinogenisis process with formulation of a system of events contributing to the process. The age-incidence curve would actually be the age-specific failure rate or age at which tumor developed as result of failure of individual to overcome contributing pathogenic events; (4) distribution of etiological factors and end results of their interaction into two sets and establishment of many-one correspondences between the two sets. This system is static and for continous process relationships, the two sets of elements must be represented as two parallel systems of multiple events. The "data-generating" mathematical model attempts to explain the process and suggest new experiments and observations.

4667   AN IMPROVED SEMI-QUANTITATIVE METHOD FOR THE ESTIMATION OF AFLATOXIN M_1 IN LIQUID MILK. (Eng.) Patterson, D. S. P. (Central Veterinary Lab., Weybridge, Surrey, KT15 3NB, England); Roberts, B. A. *Food Cosmet. Toxicol.* 13(5):541-542; 1975.

An improved method for the estimation of aflatoxin M_1 in milk eliminates the risk of emulsion formation, reduces the working time for analysis prior to thin-layer chromatography, and increases the sensitivity of analysis to 0.1 µg/l of milk. The method involves mixing liquid or reconstituted powdered milk with acetone (100 ml) and then transferring it to a dialysis sac. The latter is allowed to equilibrate overnight in a Kilner or similar jar containing 66% aqueous acetone. The dialysis sac is then drained into the jar and the toxin is extracted from the diffusate with chloroform (at least 250 ml). The extract is passed through a bed of anhydrous Na_2SO_4 as are two further extracts obtained with 50 ml chloroform. Following evaporation of the combined chloroform extracts, the residue is redissolved in 50-300 ml chloroform, depending on the likely level of aflatoxin M_1 contamination. The semiquantitative estimation of aflatoxin M_1 in the extract is carried out by visual matching under UV light. Silica-gel-coated thin-layer plates are used and chromatograms are developed with chloroform-acetone-propan-2-ol. The detection limit of 0.1 µg/l compares favorably with that of the more elaborate procedure of Schuller et al and is the level of sensitivity currently required for prospective milk surveys.

4668   THE INCREASING INCIDENCE OF BREAST CANCER IN SASKATCHEWAN. (Eng.) Barclay, T. H. C. (Allan Blair Memorial Clinic, Saskatchewan Cancer Commission, Regina, Saskatechewan, Canada); Black, M. M.; Hankey, B. F.; Cutler, S. J. *Proc. Int. Cancer Congr. 11th.* Vol. 3 *(Cancer Epidemiology, Environmental Factors).* Florence, Italy, October 20-26, 1974. Edited by Bucalossi, P.; Veronesi, U.; Cascinelli, N. New York, American Elsevier, 1975, pp. 282-288.

A study was conducted to determine if the reported increase in breast cancer in Saskatchewan was real or only apparent. Factors which may be responsible for an apparent increase in incidence rates and which were investigated in this study were an improvement in the case finding program, changes in the age

distribution within the population, and shifts in diagnostic criteria. In Saskatchewan, there is a case finding program in the form of a comprehensive cancer diagnostic and treatment service which is available to all residents. In addition, all tissue removed at all surgical procedures, must be examined by an experienced pathologist. Pathology reports with a cancer annotation and copies of all death certificates are made available to the Cancer Commission. These sources are used to identify patients with malignant disease, who are not referred to the clinic. A retrospective study, using all the available sources of information concerning the diagnosis of breast cancer in Saskatchewan indicated that the observed increase in incidence rate between 1946-1972 was not due to improved case finding in the later years. During this period of time the age distribution of the female population in Saskatchewan did change. An increase was evident in all 5-yr age groups, except 35-39 yr (in which no material change occurred). The annual number of patients with breast cancer rose over this span of 27 yr from 132 to 331. After the influence of age change was eliminated, there was still an increase of 57% in the incidence rate. It was considered possible that the changes in incidence rate might be explained by differences in interpretation of the histology of tumors by the pathologists. Available original histology slides from women who had been diagnosed in 1946, 1951, 1960, and 1968 as having mammary cancer and from women with benign breast tumor seen in the same years were re-evaluated in a blind study. For the entire series of 469 malignant tumors, there was considerable agreement between the original and the revised diagnoses. The small differences which did exist indicated that cancer of the breast was overdiagnosed by the pathologists in the earlier three selected years. In the re-examination of 423 benign lesions, only one invasive cancer was found and four tumors diagnosed as benign were seen to exhibit *in situ* carcinomatous change. It is concluded that the increase in the incidence of breast cancer over the past 27 yr is real.

4669   HEPATOCELLULAR CARCINOMA AND THE HEPATITIS B VIRUS: DETECTION OF ANTIBODY TO HEPATITIS B CORE ANTIGEN. (Fre.) Maupas, P. (Lab. de Virologie, Faculte de Medecine et de Pharmacie, 2 bis, boulevard Tonnelle, 37000 Tours, France); Larouze, B.; Saimot, G.; Payet, M.; Werner, B.; O'Connell, A.; Millman, I.; London, W. T.; Blumberg, B. S. *Nouv. Presse Med.* 4(27):1962-1964; 1975.

4670   PROSTATIC TUMORS IN IRAN. (Eng.) Habibi, A. (Churchill Ave., Tehran, Iran); Manouchehri, A.; Diba, M. H.; Sajadi, M.; Ghavam, M. *Int. Surg.* 60(8):405-407; 1975.

4671   ANALYSIS OF SURVIVAL AND RECURRENCE VS. PATIENT AND DOCTOR DELAY IN TREATMENT OF BREAST CANCER. (Eng.) Dennis, C. R. (Downstate Med. Cent., Brooklyn, N.Y.); Gardner*, B.; Lim, B. *Cancer* 35(3):714-720; 1975.

4682   CANCER PATTERNS IN AUSTRALIA: 1950-1973.
(Eng.) Gray, N. (Anti-Cancer Council of
Victoria, Melbourne, Australia). *Proc. Int. Cancer
Congr. 11th.* Vol. 3 *(Cancer Epidemiology, Environ-
mental Factors)*. Florence, Italy, October 20-26,
1974. Edited by Bucalossi, P.; Veronesi, U.;
Cascinelli, N. New York, American Elsevier, 1975,
pp. 191-200.

See also:

* (Rev): 4203, 4205, 4206, 4208, 4209, 4210,
          4211, 4217, 4237, 4238, 4240, 4257,
          4262, 4263, 4264, 4265
* (Chem): 4291, 4305, 4309, 4310
* (Phys): 4364, 4366, 4371
* (Viral): 4434
* (Immun): 4467, 4515
* (Path): 4570

4683   ESTABLISHMENT AND SOME BIOLOGICAL
        CHARACTERISTICS OF HUMAN HEPATOMA CELL
LINES. (Eng.) Doi, I. (Okayama Univ. Medical
Sch., Shikatacho 2-5-1, Okayama 700, Japan); Namba,
M.; Sato, J. *Gann* 66(4):385-392; 1975.

Two cell lines of human hepatoma, HLE and HLF lines,
were established *in vitro* from the hepatocellular
carcinoma of a 68-yr-old man. One clone (HLEC)
was obtained from a single HLE cell. The cells of
HLE and HLEC were epithelial-like and both of these
cells demonstrated glycogen granules in the cytoplasm when stained with periodic acid and Schiff
reagent. Although HLF cells resembled fibroblasts
in morphology, they appeared to have originated
from hepatoma cells, judging from epithelial characteristics in aggregates reconstituted by rotation
culture and heterotransplantability. HLE cells
produced α-fetoprotein until day 187 of culture,
but HLF cells did not produce α-fetoprotein at any
period examined. The chromosome number of both
cell lines was distributed near the triploid range.
HLF cells were transplantable into the cheek pouch
of adult Syrian hamsters treated with cortisone
acetate (5 mg ip, one day before and three days
after implantation), but HLE cells were not. The
absence of α-fetoprotein production in HLF cells
suggests either that they were derived from hepatoma cells not producing α-fetoprotein, or that they
may have lost the capacity to produce α-fetoprotein
in vitro.

4684   NON-MENDELIAN SEGREGATION IN HYBRIDS
        BETWEEN CHINESE HAMSTER CELLS. (Eng.)
Harris, M. (Dept. Zoology, Univ. California,
Berkeley, Calif. 94720). *J. Cell. Physiol.* 86(2/-
Suppl. 1/Part II):413-429; 1975.

Mechanisms of segregation in hybrids between Chinese
hamster cells (272 and 599) were studied using
clonal analysis. Hybrid cells were grown in HAT
medium and subjected to back selection with bromodeoxyuridine (BUDR) or azaguanine (AZG). In AZG
or BUDR at 30 μg/ml, segregation began with a random,
high-frequency event that gave rise to cells capable
of growth in both HAT and back selection medium,
unlike the precursor hybrid or original parental
cell types. BUDR-resistant segregants were propagated serially in the presence of BUDR, and were
examined by clonal analysis for changes in plating
properties during long term culture. During 240
days the HAT/BUDR plating ratio for segregant cells
declined continuously, from approximately ten down
to 10^{-6}. A parallel decrease was observed in the
rate of H^3-thymidine incorporation, along with a
drop in thymidine kinase activity. These shifts
took place only in the presence of BUDR, and were
reversed by altered selection in HAT medium. Clonal
studies showed that the evolution of segregant
properties occurred in most cells, and did not
arise from variation and selection of minority cell
types. These properties of the segregating system
are not consistent with models based on gene mutation, chromosome rearrangements, or chromosome
loss. The evolution of segregants resembles more
closely a sorting-out progress, taking place by

trophoresis. Both variant clones contained a unique protein which double labeling experiments indicated was a nonhistone protein. Cells labeled with ^{14}C-methionine and ^{3}H-tryptophan showed a high rate of incorporation of tryptophan, which is incorporated selectively into nonhistone proteins. The results indicate that temperature-sensitive cells could be a useful tool for study of the control of the cell cycle in mammalian cell systems, particularly in the identification of endogenous components necessary for regulation of cell growth.

4689   DIFFERENCES IN THE UREA-EXTRACTED OR
         FIBROUS PROTEINS OF MOUSE EPIDERMIS AT
VARIOUS STAGES OF MALIGNANT TRANSFORMATION. (Eng.)
Carruthers, C. (Dept. of Immunology Res., Roswell
Park Memorial Inst., Buffalo, N.Y. 14203); *Oncology* 30:125-133; 1974.

The differences in the urea proteins or mouse epidermis at various stages of neoplastic transformation are described. The urea-extractable proteins of normal and hyperplastic epidermis, papilloma, and squamous-cell carcinoma were fractionated by gel filtration on Sepharose 4B columns to yield excluded and retarded urea protein fractions. There was a considerable difference in the elution patterns of the urea protein of normal epidermis and carcinoma. The level of excluded proteins decreased from normal epidermis through hyperplastic epidermis, papilloma, and carcinoma; retarded protein fractions increased proportionately. The urea proteins of normal and hyperplastic epidermis had consistent elution patterns, whereas papilloma and carcinoma were less consistent, but still comparable. Normal epidermis showed a large sharp peak of proteins excluded by Sepharose 4B and a broad peak for the retarded fractions; carcinoma urea proteins showed a small peak for excluded proteins and a large, sharp peak for retarded proteins. When the urea proteins of all four tissue types were ultracentrifuged 100,000 x g for 2, 6, or 16 hr, gel and supernatant fractions were obtained. The amounts of urea protein obtained from normal epidermis by centrifugation and filtration were in good agreement. Similar results were observed with hyperplastic epidermis; however, the amount of protein in the excluded peak was less than that of normal epidermis. By gel filtration of papilloma and carcinoma urea proteins, 15-16% was found in the excluded peak, and 84-85% was found in the retarded protein peak. In contrast, centrifugation demonstrated nearly equal amounts of protein in the gel and supernatant fractions. Immunodiffusion of the fractions obtained by gel filtration and centrifugation in agar against antisera or IgG fractions of antisera, raised against the various urea proteins resulted in precipitin bands of identity, suggesting that the gel and supernatant fractions are closely related and differ only in their degree of polymerization. The results show that antigenic differences exist between the fibrous proteins of normal epidermis and papilloma or carcinoma, but the latter two tissues are antigenically similar with respect to these proteins.

4690   FIBROBLAST TUBULIN. (Eng.) Ostlund, R.
            (Natl. Cancer Inst., Bethesda, Md. 20014);
Pastan*, I. *Biochemistry* 14(18):4064-4068; 1975.

Tubulin prepared from fibroblastic cells was characterized, and the content and properties of tubulin
from normal and transformed cells were compared.
Tubulin purified from L cells by Sepharose 4B and
DEAE-cellulose chromatography had properties similar
to brain tubulin. The specific activity of colchicine binding was 1.1 M of colchicine bound per
100,000 g of protein. Between 0.07 and 0.45% of the
total protein of L cells was active tubulin. Proteins differing biochemically from tubulin but co-
electrophoresing with it in sodium dodecyl sulfate
gels were encountered. Colchicine-binding activity
of virus transformed normal rat kidney (NRK) fibroblasts measured by a filter assay using a whole cell
homogenate was reduced when compared to the parent
NRK cells. L cells growing on plastic or glass surfaces had five times as much colchicine-binding activity as the same cells growing in spinner culture.
Apparently, growth on a surface increases the tubulin
content of L cells.

4691   THE HALF-LIFE OF POLYADENYLATED POLYSOMAL
            RNA FROM NORMAL AND TRANSFORMED CELLS IN
MONOLAYER CULTURE. (Eng.) Sensky, T. E. (Faculty
Medical Sciences, Univ. Coll. London, Gower St.,
London, WC1E 6BT, U.K.); Haines, M. E.; Rees, K. R.
Biochim. Biophys. Acta 407(4):430-438; 1975.

RNA half-lives were estimated in five monolayer cell
lines, including two pairs of normal cells and their
transformed counterparts to ascertain whether alterations in control of gene expression involve changes
in messenger (mRNA) stability. A line of transformed Chinese hamster kidney cells (transformed CHK)
and their untransformed counterparts (CHK), CV-1
African Green Monkey kidney cells, BALB/c 3T3 mouse
embryo fibroblasts, SV40-transformed 3T3 fibroblasts
and a strain of 3T3 cells (XYT) were all labeled for
at least two generations with 0.1 µCi/ml $^{32}P_i$. Medium, containing 0.1 µCi/ml $^{32}P_i$, 5 µCi/ml [^{3}H]uri-
dine, and 5 µg/ml unlabeled uridine, was prepared in
bulk so that labeling conditions would be uniform
throughout the experiments. Experiments were started
with the cells within two days after passage and continued for up to 40 hr. Cell growth was still exponential at the end of each experiment. The extracted
polysomal RNA was fractionated on oligo(dT)-cellu-
lose, the high and low salt fractions were precipitated with trichloroacetic acid, and acid-insoluble
specific activity was determined as the ^{3}H/^{32}P ratio.
The results for the polyadenylated fractions in all
cases fit with those expected from a model in which
the whole fraction has a single half-life of less
than one generation time. From both the transformed/
untransformed cell pairs, there was evidence that a
relationship exists between cell generation time and
the half-life of the polyadenylated polysomal RNA
fraction; the relationship persists even through the
process of transformation. It is concluded that considerable alteration in the pattern of RNA stability
is unlikely to be obligatory in *in vitro* transformation.

that the particles are componenets of the RNP net-
work whose fragmentation occurs as a consequence of
two processes: (a) activation of nuclear nucleases
and (b) shearing forces.

4696   BIOCHEMICAL ESTIMATION OF THE BASIC DYE-
        BINDING CAPACITY OF RNA FROM RAT HEPA-
TOMA. (Eng.) Lepage, R. (Institut du Cancer de
Montreal, Montreal, Quebec, Canada); de Lamirande,
G.; Daoust, R. *Cancer Res.* 35(1):45-48; 1975.

The RNA content of cell fractions was determined
in normal male Sprague-Dawley rat liver and in
solid Novikoff hepatoma to verify the possibility
of an increased cytoplasmic RNA content in tumors.
The possibility that cytoplasmic fractions of Novikoff hepatoma show greater affinity for basic dyes
than corresponding normal fractions was examined
by means of a test-tube toluidine blue-binding assay. The dye-binding capacity of total cytoplasmic
fractions from tumors was 75% higher than normal
after Carnoy fixation, which retains mostly ribosomal RNA. Assays on fresh ribosomes indicated
that tumor ribosomes bound 71% more toluidine blue
per milligram of RNA than the ribosomal preparation from normal liver. This study thus demonstrates
a greater affinity of tumor RNA for basic dyes, and
a comparison of biochemical and cytophotometric
analyses suggests that an increase in basophilia
by a factor of about 2 would be due to a qualitative alteration in ribosomal RNA molecules and/or
ribosome structure in cancer cells.

4697   THE MESSENGER RNA SEQUENCES IN GROWING AND
        RESTING MOUSE FIBROBLASTS. (Eng.) Wil-
liams, J. G. (Dept. Biology, Massachusetts Inst.
Technology, 77 Massachusetts Ave., Cambridge, Mass.
02139); Penman, S. *Cell* 6(2):197-206; 1975.

The sequences present in cytoplasmic messenger RNA
(mRNA) in resting and growing cells of the mouse
fibroblast line 3T6 were examined under controlled
cellular proliferation by hybridization of complementary DNA (cDNA) to its RNA template. RNA was
extracted at room temperature with phenol chloroform. The abundance and complexity classes of mRNA
in growing 3T6 cells were compared to those in other
established cell lines. The overall complexity measured for mRNA from HeLa cells and from mouse fibroblast lines 3T6, SV-PY-3T3, and L were quantitatively similar; they corresponded to approximately 10,000
and 8,400 sequences, respectively. The relative
amount of the two major abundance classes and their
complexities was identical in the three mouse fibroblast lines despite their different histories. HeLa
cell mRNA was significantly different in the amount
and complexity of the two major classes. Despite a
reduced mRNA content, resting and growing 3T6 cells
had a similar distribution with respect to its two
mRNA complexity classes. Cross hybridization of
cDNA and mRNA was employed to determine sequence
homology between mRNA from resting and growing cell
populations. The majority of mRNA sequences were
the same in the two states. When the common se-

quences were removed, cross hybridization revealed
that about 3% of the mRNA in resting cells was not
present in the growing state; the opposite cross
demonstrated that 3% of the mRNA in growing cells
was not present in the resting cells. These differences may result from alterations in gene expression that are related to the growing state of the
cell. The results suggest that mRNA populations of
established lines of cultured mammalian cells resemble one another with respect to complexity and
abundance classes. The similarity between the mouse
fibroblast lines suggests either a stable pattern
originating in the tissue or animal of origin, or a
common selection pressure leading to quantitatively
similar patterns of expression.

4698   RIBOSOMAL RNA METABOLISM IN SYNCHRONIZED
        PLASMACYTOMA CELLS. (Eng.) Eikhom, T.
S. (Dep. Biochem., Univ. Bergen, Norway); Abraham,
K. A.; Dowhen, R. M. *Exp. Cell Res.* 91(2):301-309;
1975.

Ribosome synthesis and metabolism were studied in
a mouse plasmacytoma cell line synchronized by iso-
leucine deprivation. Ribosomal RNA (rRNA) was
characterized by gel electrophoresis. The rate
of ribosome synthesis (as measured by the appearance of labeled rRNA in the cytoplasm) was low
during the G1 phase, increased rapidly during
the S phase, remained high during part of the G2
phase, and dropped to a minimum during mitosis.
A slowdown in the increasing rate of RNA synthesis was observed during the middle of the S phase.
No significant decrease in the total nucleotide
pool per cell could be observed during the S phase.
The accumulation of RNA (as determined by absorb-
ance measurements) was highest during the S and
G2 phases. Pulse labeling of rRNA and pulse chase
experiments demonstrated that newly synthesized
ribosomal subunits entered into free polysomes to
the highest extent during the S phase. The percentage of membrane-bound polysomes of total polysomes increased during the G1 phase, as did the
percentage of labeled rRNA in the membrane-bound
fraction. The results suggest that newly synthesized ribosomal subunits enter into free or membrane-bound polysomes depending on the requirements of the cell. During the S phase they enter
preferentially into free polysomes to produce
mainly internal proteins; during the G1 phase they
enter to a greater extent into membrane-bound polysomes, perhaps to produce more proteins destined
for secretion.

4699   MEMBRANE-BOUND RIBOSOMES OF MYELOMA
        CELLS. II. KINETIC STUDIES ON THE
ENTRY OF NEWLY MADE RIBOSOMAL SUBUNITS INTO THE
FREE AND THE MEMBRANE-BOUND RIBOSOMAL PARTICLES.
(Eng.) Mechler, B. (Dept. Biochemistry, Univ.
Cambridge, England); Vassalli, P. *J. Cell Biol.*
67(1):16-24; 1975.

The kinetics of appearance of newly made 60S and
40S ribosomal subunits in the free and membrane-
bound ribosomal particles of MOPC 21 (P3K) mouse

of butyric acid and inhibitors of DNA synthesis was
determined by assaying cells for the percent of
benzidine-positive (Hb-containing cells). At 1 mM,
butyric acid readily induced erythroid differentiation in both log phase and stationary phase
cells. Cytosine arabinoside (1 μg/ml) and hydroxyurea (0.1 mM) quickly inhibited both DNA synthesis
and cell division in butyric acid-treated cells
but had no effect on cell differentiation. These
results appear to rule out more complex models for
globin gene expression which require gene replication or cell division (or both).

4703   DEMONSTRATION THAT A MOUSE IMMUNOGLOBULIN
        LIGHT CHAIN MESSENGER RNA HYBRIDIZES EXCLU-
SIVELY WITH UNIQUE DNA. (Eng.) Rabbitts, T.H. (Med-
ical Res. Council, Lab. Molecular Biology, Hills
Road, Cambridge CB2 2QH, England); Jarvis, J. M.;
Milstein, C. *Cell* 6(1):5-12; 1975.

^{32}P-labeled light chain messenger RNA was prepared
from mouse MOPC 21 myeloma cells. The messenger
RNA was hybridized to purifed repetitive nuclear DNA
and both the hybridized (repetitive ^{32}P-RNA) and nonhybridized (nonrepetitive ^{32}P-RNA) fractions were
isolated. A Tl ribonuclease fingerprint was given
only by the nonhybridized RNA which showed that oligonucleotides derived from the variable and constant
regions of the light chain messenger RNA. The fingerprint also showed that oligonucleotides were derived
from the untranslated regions of the light chain messenger RNA. The nonrepetitive ^{32}P-RNA rehybridized
only with the unique fraction of total nuclear DNA.
The rapidly hybridizing part of the unfractionated
^{32}P-RNA preparation, therefore, is not a component of
the light chain messenger RNA itself. Complementary
DNA was prepared with reverse transcriptase using unlabeled light chain messenger RNA as template, and
the transcripts were fractionated into various size
classes. Complementary DNA molecules greater than
900 bases in length hybridized with both the initial
messenger RNA and with the nonrepetitive ^{32}P-RNA but
failed by hybridize with excess purified repetitive
^{32}P-RNA. The rapidly hybridizing component of the
messenger RNA fraction, therefore, does not appear
to be transcribed by reverse transcriptase. It is
concluded that, under the experimental conditions
used, the light chain messenger RNA hybridizes exclusively with unique DNA.

4704   CHARACTERIZATION OF NOVIKOFF HEPATOMA
        mRNA METHYLATION AND HETEROGENEITY IN
THE METHYLATED 5' TERMINUS. (Eng.) Desrosiers,
R. C. (Dept. Molecular Biophysics and Biochemistry,
Yale Univ., New Haven, Conn. 06520); Friderici,
K. H.; Rottman, F. M. *Biochemistry* 14(20):4367-
4374; 1975.

KOH digestion of methyl-labeled poly(A)$^+$ Novikoff
hepatoma messenger RNA (mRNA) purified by (dT)-
cellulose chromatography produced mononucleotide
and multiple peaks of a large oligonucleotide
(-6 to -8 charge) when separated on the basis of
charge by Pellionex-WAX high-speed liquid chroma-

tography in 7 M urea. Heat denaturation of the
RNA before application to (dT)-cellulose was required to release contaminants (mostly 18S ribosomal RNA) that persisted even after repeated
binding to (dT)-cellulose at room temperature.
Analysis of the purified poly(A)$^+$ mRNA by enzyme
digestion, acid hydrolysis, and a variety of
chromatographic techniques demonstrated that the
mononucleotide (53%) was due entirely to N^6-methyl-
adenosine. The large oligonucleotides (47%) were
found to contain 7-methylguanosine and the 2'-O-
methyl derivatives of all four nucleosides. No
radioactivity was found associated with the poly(A)
segment. Periodate oxidation of the mRNA followed
by β elimination released only labeled 7-methyl-
guanine consistent with a blocked 5' terminus containing an unusual 5'-5' bond. Alkaline phosphatase treatment of intact mRNA had no effect on the
migration of the KOH produced oligonucleotides on
Pellionex-WAX. When RNA from which 7-methylguanine
was removed by β elimination was used for the phosphatase treatment, distinct dinucleotides (NmpNp)
and trinucleotides (NmpNmpNp) occurred after KOH
hydrolysis and Pellionex-WAX chromatography. Thus,
Novikoff hepatoma poly(A)$^+$ mRNA molecules can contain either one or two 2'-O-methylnucleotides
linked by a 5'-5' bond to a terminal 7-methyl-
guanosine, and the 2'-O-methylation can occur with
any of the four nucleotides. The 5' terminus may
be represented by m^{7}G$^{5'}$ppp$^{5'}$(Nmp)$_{1or2}$Np, a general
structure proposed earlier as a possible 5' terminus
for all eucaryotic mRNA molecules. The composition
analyses indicated that there are three N^6-methyl-
adenosine residues, one 7-methylguanosine residue,
and 1.7 2'-O-methylnucleoside residues per average
mRNA molecule.

4705   TRANSCRIPTION OF MOUSE DNA. (Eng.)
         Ivanov, I. G. (Inst. Biochemistry, Bul-
garian Acad. Sciences; Sofia 13, Bulgaria); Zlateva,
B. S.; Markov, G. G. *Dokl. Bolg. Akad. Nauk* 28(6):
813-816; 1975.

The transcription of the three basic families of
mouse DNA (highly repeated, intermediate, and
unique) was studied. DNA was isolated from mouse
liver and Ehrlich ascites tumor cells (EAT), and
[^{3}H]-DNA was obtained from EAT cells by injecting
the animals with [^{3}H]thymidine on the seventh day
after transplantation. Labeled and unlabeled
pre-mRNA (HnRNA) was isolated from mouse liver and
EAT, and hybridization experiments were performed
in solution and on membrane filters. Saturation of
highly repeated and intermediate DNA with the 63 C
fraction of heterogeneous nuclear RNA (63-RNA) was
reached at RNA concentrations of 25-50 µg/ml. With
85-RNA, the saturation curves of the same DNAs
tended toward a biphasic character. Both heterogeneous RNA fractions hybridized with the total DNA
to about the same degree (6%), although the intermediate DNA had a greater ability to hybridize with
63- and 85-RNA. Under the conditions used for the
isolation of highly repeated DNA, the renaturation
of satellite DNA was completed. Under these conditions, about 15% of the DNA behaved as highly
repeated, and it was transcribed to a considerable

4710   THE SYNTHESIS OF FORSSMAN GLYCOLIPID IN
        CLONES OF NIL 2 HAMSTER FIBROBLASTS
GROWN IN MONOLAYER OR SPINNER CULTURE. (Eng.)
Sakiyama, H. (Div. Physiology and Pathology, Natl.
Inst. Radiological Sciences, 9-1, 4 chome, Ana-
gawa Chiba-shi 280, Japan); Terasima, T. *Cancer
Res.* 35(7):1723-1726; 1975.

The amount of Forssman glycolipid (GL-5) was investigated in two clones of Nil hamster cells
grown either in monolayer or in spinner culture.
GL-5 assayed by the complement-fixation inhibition
test increased with increasing cell density in
the monolayer. However, cells grown in spinner
cultures failed to show the density-dependent response in both clones examined. Cells transformed
by hamster sarcoma virus did not show the density-
dependent increase of GL-5, even when grown in
monolayer. The effect of transfer from confluent
to sparse cultures on the amount and the synthesis
of GL-5 was also examined. The amount of GL-5
did not change significantly before the first cell
division and then decreased until the third day
after transfer; in contrast, the synthesis of GL-5
dropped markedly before the first cell division
and continued to decrease until three days after
transfer. It is suggested that the presence of a
high level of density-sensitive glycolipid does
not by itself inhibit cell division and that intercellular contact is probably not the sole factor
responsible for activation of GL-5 synthesis.

4711   ACYLTRANSFERASES AND THE BIOSYNTHESIS OF
        PULMONARY SURFACTANT LIPID IN ADENOMA
ALVEOLAR TYPE II CELLS. (Eng.) Snyder, F. (Medical
and Health Sciences Div., Oak Ridge Associated Univ.,
Oak Ridge, Tenn. 37830); Malone, B. *Biochem. Biophys.
Res. Commun.* 66(3):914-919; 1975.

Alveolar type II adenoma cells were used to investigate the nature of the acyltransferases associated
with the endoplasmic reticulum. The adenomas were
harvested from the lungs of female BALB/c mice that
had been injected ip with urethan (1 mg/g for three
consecutive days) approximately six months earlier.
Acyltransferase activities were determined in a
system containing an acyl acceptor, adenoma microsomes (0.8 mg protein), CoA (200 μM), ATP (10 mM),
Mg^{2+} (4 mM), and $[1-^{14}C]$palmitic acid (33.3 μM).
All samples were incubated for 20 min at 37 C. The
microsomal preparations were found to contain acyltransferases that transferred palmitate to *rac*-
glycerol-3-P to form phosphatidic acid, to dihydroxy-
acetone-P to form acyldihydroxyacetone-P, and to
1-acyl-*sn*-glycero-3-phosphocholine to form 3-*sn*-
phosphatidylcholine. The data demonstrate that the
microsomal preparations can catalyze significant
incorporation of palmitic acid into the 2-position
of the disaturated species of 3-sn-phosphatidylcho-
line independently of phosphatidic acid formation
as evidenced by the fact that *sn*-glycerol-3-P and
Ca^{2+} (which inhibits choline phorphotransferase)
did not influence the incorporation of palmitic
acid into the main surfactant lipid. Thus, a de-
acylation-acylation reaction involving 2-lysophos-
phatidylcholine appears to be an important pathway

for the synthesis of surfactant lipid in alveolar
type II cells; the control of acyl specificity at
the 2-position is determined by the relative concentrations of the coparticipating substrates,
1-palmitoyl-sn-glycero-3-phosphocholine and palmi-
toyl-CoA.

4712   MARKED REDUCTION OF CYCLIC GMP PHOSPHO-
        DIESTERASE ACTIVITY IN VIRALLY TRANS-
FORMED MOUSE FIBROBLASTS. (Eng.) Lynch, T. J.
(St. Jude Children's Res. Hosp., Memphis, Tenn.
38101); Tallant, E. A.; Cheung, W. Y. Biochem.
Biophys. Res. Commun. 65(3):1115-1122; 1975.

Cyclic AMP phosphodiesterase and cyclic GMP phosphodiesterase activities were determined in normal 3T3
and transformed mouse fibroblasts and the effect of
dibutyryl cyclic AMP on these enzymes was investigated. Normal mouse fibroblasts and fibroblasts
transformed by simian virus 40 were plated in culture dishes to a final density of 150,000-400,000
cells/plate in modified Eagle's medium and harvested
before confluency. In some cultures, the medium was
changed 48 hr after plating and replaced with medium
containing 1 mM dibutyryl cyclic AMP, 1 mM theophylline, and sometimes actinomycin D (1 µg/ml).
The cells were incubated for an additional ten hours
before being harvested and assayed. Phosphodiesterase was assayed with radioactive cyclic AMP or
cyclic GMP using an anionic exchange resin. In the
normal cells, the specific activities of cyclic AMP
phosphodiesterase and cyclic GMP phosphodiesterase
were 62 and 41, respectively; in the transformed
cells, the two activities were 47 and 4, respectively. In the transformed fibroblasts, therefore,
both enzyme activities were significantly lower
but the reduction was much more pronounced with
cyclic GMP phosphodiesterase. Incubation with
dibutyryl cyclic AMP increased both cyclic AMP and
cyclic GMP phosphodiesterase activities, but the
increase was larger with cyclic AMP than with cyclic
GMP phosphodiesterase. The elevation of both activities was blocked by actinomycin D, indicating that
the increase of enzyme activity involved the synthesis of new messenger RNA. The differential increase in cyclic AMP phosphodiesterase by dibutyryl
cyclic AMP and the disparate reduction in cyclic
GMP phosphodiesterase in the transformed fibroblasts
suggest that the two enzymes are regulated separately.

4713   PURIFICATION OF ARGINASES FROM HUMAN-
        LEUKEMIC LYMPHOCYTES AND GRANULOCYTES:
STUDY OF THEIR PHYSICOCHEMICAL AND KINETIC PROPERTIES.
(Eng.) Reyero, C. (Institut fur Biochemie, Tier-
arztliche Hochschule, Linke Bahngasse 11, A-1030
Wien, Austria); Dorner, F. Eur. J. Biochem. 56(1):
137-147; 1975.

Because purification studies on normal human WBC
is impeded by lack of sufficient starting material, the possible chemical differences of the arginases of leukemic lymphocytes and leukemic granulocytes were investigated. Arginase was isolated
from granulocytes of a patient with chronic myelocytic leukemia and from lymphocytes of a patient

and 1b synthesized ribosomal RNA-like products, whereas forms 11a, 11b, and the parent enzyme mixture synthesized compounds that were more similar to DNA. No species specificity was found for DNA templates, and denatured DNA was consistently preferred to the native template by 11a and 11b; the two kinds of template were about equally efficient for forms 1a and 1b. Although forms 1a, 11b, and 11a were all inhibited by daunomycin (5.0, 10.0, and 20.0 µg/ml), form 11a was preferentially affected. $3',5'$-Cyclic (2×10^{-6} M), $3',5'$-cyclic guanosine monophosphate (2×10^{-6} M) and gibberellic acid (2.8 - 277 µg/ml) implicated as RNA polymerase regulators in other systems, were generally ineffective. The levels of nuclear RNA polymerase activity, per milligram of DNA, in three mouse ascites tumors (EAC, 6C3HED lymphosarcoma, and TA3 adenocarcinoma) were compared with those from three normal mouse tissues (kidney, liver, and spleen). On the average, the tumor nuclei contained 8.9, 1.5, 2.7, 20.0, and 3.8 times the activities of RNA polymerases 1a, 1b, 11a, 11b, and 111, respectively, as the normal cells, but the difference was significant only for 11b.

4717 CHARACTERIZATION OF TERMINAL DEOXYNUCLEO-TIDYLTRANSFERASE IN A CELL LINE (8402) DERIVED FROM A PATIENT WITH ACUTE LYMPHOBLASTIC LEUKEMIA. (Eng.) Sarin, P. S. (Natl. Cancer Inst., NIH, Bethesda, Md.); Gallo, R. C. *Biochem. Biophys. Res. Commun.* 65(2):673-682; 1975.

Terminal deoxynucleotidyltransferase in a cell line (8402) derived from the blood cells of a patient with acute lymphoblastic leukemia was characterized. The activity is comparable to the levels present in thymus gland. The terminal transferase from these cells was purified by successive chromatography on DEAE cellulose, phosphocellulose, and hydroxyapatite columns to approximately 900-fold. The purified enzyme had a sedimentation value of 3.4S, a pH optimum of 7.8, a Mn^{2+} optimum of 0.1 mM, and a Mg^{2+} optimum of 7 mM. The purified enzyme efficiently utilized a number of oligo or poly deoxynucleotide initiators and was similar in its properties to the terminal deoxynucleotidyltransferase purified from calf thymus and from the blast cells of some patients with chronic myelogenous leukemia.

4718 ISOLATION AND SOME PROPERTIES OF EXORI-BONUCLEASE FROM RAT LIVER CELL NUCLEI. (Rus.) Skridonenko, A. D. (Res. Inst. Health Resort Sci., Odessa, USSR). *Biokhimiia* 40(1):13-20; 1975.

Exoribonuclease was first obtained and purified from Ehrlich's mouse ascites cell nuclei, mouse liver cells and other tissues and its properties were described in the late 60's. Working on the ribonuclease action of rat liver cell nuclei, the author succeeded in isolating and purifying an exoribonuclease analogous to that of Ehrlich's ascites cells. Lazarus and Sporn's method for the isolation of rat liver cell nuclei and determination of exoribonuclease activity was used. The enzyme was obtained by extraction from a rat liver homogenate, first at

pH 6.0, then at pH 8.0, followed by a dual fractionation with ammonium sulfate. Further purification
was attained by columnar chromatography (DEAE-Sephadex A-50). Its maximal activity was reached at
pH 7.4-7.6 in an acid buffer (tris-HCl, tris-ace-
tate). In order to be optimally effective, the
presence of Mg ions was required. Chromatographic
data showed that exoribonuclease was essentially
localized in the chromatin of the cell nucleus
(80-90%), the balance (10-20%) in acid and residual
proteins. The investigations indicated that the
ribonuclease split synthetic polynucleotides and
RNA forming ribonucleoside-5'-phosphates. The enzyme preferably reacted with single-stranded polyribonucleotides and required much more time to hydro-
lyze multi-chain-containing helical structures. The
author concluded that the enzyme studied broke down
the "non-informative" (unstable, fast disintegrating
within the nucleus) portion of the newly formed high-
polymer cellular RNA which, however, remains to be
corroborated.

4719   THE STEROID ALCOHOL AND ESTROGEN SULFO-
        TRANSFERASES IN RODENT AND HUMAN MAMMARY
TUMORS. (Eng.) Godefroi, V. C. (Michigan Cancer
Foundation, Detroit, Mich. 48201); Locke, E. R.;
Singh, D. V.; Brooks, S. C. *Cancer Res.* 35(7):
1791-1798; 1975.

Rodent and human mammary tumor systems were investigated to relate the steroid alcohol and estrogen
sulfotransferase activities to the hormonal dependency of the tumor as determined by estrogen receptor content. Unlike the normal mammary gland
or the hyperplastic alveolar nodule, rodent mammary neoplasms displayed significant levels of
these two sulfotransferases. In the hormone-independent mouse tumors produced from outgrowth
lines D_1, D_2, and D_8, high dehydroepiandrosterone
sulfotransferase activity was characteristic of
the rapidity with which hyperplastic alveolar nodules developed into a neoplasm (V_{max} = 52.8 vs 1.8
fM/min/mg protein) while estrone sulfotransferase
activity was either not detectable or low (V_{max} =
5.5 fM). After oophorectomy of BALB/c mice bearing
slowly developing tumors, both sulfotransferases
in the nonregressing neoplasms showed marked increases in activity (V_{max} dehydroepiandrosterone =
30.0 fM; V_{max} estrone = 18.5 fM). Strain differences were found to influence the steroid-sulfating
capability but not the estrogen receptor content of
hormone-dependent rat mammary tumors. In Wistar-
Lewis rats the steroid alcohol sulfotransferase
activity was at least 35 times higher than in the
Sprague-Dawley strain. As was observed in the
mouse mammary tumor, Sprague-Dawley rat neoplasms
that grew in the absence of ovarian hormones contained significantly greater levels of the steroid
alcohol sulfotransferase. A possible correlation
between presence of the steroid alcohol sulfotransferase and the estrogen receptor protein was observed in a limited number of human breast carcinomas. A positive clinical response of four patients
to hormonal manipulation was matched in all four
cases with the presence of estrogen receptor protein
and in three with the presence of the steroid alco-

glands. Some degree of amyloidosis was seen in half the mice of both sexes, beginning at eight months in males and 12 months in females. Hyperplastic endometrium and myometrial changes seen in the old females suggested that this mouse stock might be useful in the investigation of spontaneous reproductive pathology. The many spontaneous tumors found in the mice (most of them occurring after 18 mo) made the detection of weak carcinogens difficult in a two-year experiment. Variability in tumor incidence among small groups of mice emphasized the need for adequate samples.

4724 PLACENTAL TRANSFER OF TRITIUM-LABELED PROSTAGLANDIN A_1 IN THE RAT. (Eng.) Persaud, T. V. N. (Dept. Anatomy, Univ. Manitoba, Winnipeg, Manitoba, Canada. R 3E OW3); Jackson, C. W. *Exp. Pathol. (Jena)* 10(5/6):353-358; 1975.

The placental transfer and tissue distribution of [^{3}H]-prostaglandin A_1 (^{3}H-PGA$_1$) were studied in pregnant Long Evans rats. When ^{3}H-PGA, (25 µCi iv) was given on the 21st gestational day, significant levels were found in the fetal gut, kidney, and brain, within five min. Subsequently these levels increased or remained constant. In rats injected on day 18, the dpm/min/mg wet wt in the amniotic fluid increased from 8,520 at five min, to 58,600 at 24 hr. It appears that PGA, acts directly on the conceptus.

4725 PROPERTIES OF THE GENOME IN NORMAL AND SV40 TRANSFORMED WI38 HUMAN DIPLOID FIBROBLASTS. II. METABOLISM AND BINDING OF HISTONES. (Eng.) Krause, M. O. (Dept. Biochemistry, Univ. Florida, Gainesville, Fla. 32610); Stein, G. S.* *Exp. Cell Res.* 92(1):175-190; 1975.

The composition, metabolism and binding of the histone fractions from normal and simian virus 40 (SV40)-transformed WI38 human diploid fibroblasts were compared. The fibroblasts were grown in monolayer culture in Eagle's basal medium containing 10% fetal calf serum. Metabolism of histones was determined by labeling cells either with a mixture of ^{3}H-arginine and ^{3}H-lysine and ^{32}P or with ^{3}H-acetate. Two alternate procedures were used for nuclear isolation. The first method involved lysis of cells by treatment with 80 mM NaCl, 20 mM EDTA, and 1% Triton X-100 at pH 7.2. Under these conditions, (F1, F3, F2b, and F2a1) all five histone fractions were retained in the isolated nuclei. The second method involved lysis of cells at pH 2.4 with 10 mM Mg^{2+}, 10 mM Ca^{2+}, 25 mM citric acid, 0.1 M sucrose and 0.1% Triton X-100. This method resulted in optimal recovery of arginine-rich histones F3 and F2a1. Histones were extracted from the nuclei of both normal and transformed cells at 4C with 0.25 M HCl and were fractionated by polyacrylamide gel electrophoresis. While the relative amount of each histone fraction was similar in normal and SV40-transformed cells, substantial increases in the levels of F3 acetylation and F1 (lysine rich) and F2a2 phosphorylation were observed for the histones of the transformed cells. Differences in extractability of arginine-rich histones with 0.25 M HCl were also observed. F3 was extracted more rapidly than F2a1 from nuclei of normal WI38 fibroblasts, but the reverse was true in the transformed cells. The extractability of histones may reflect the tightness with which the histone molecules are associated with one another and with the other components of the chromatin complex. Differences in histone binding may be mediated at several levels; modifications in the histones themselves may result in conformation changes which are likely to alter intermolecular associations. It is also possible that the nonhistone chromosomal proteins associated with the genome of SV40-transformed cells may be responsible for viral-induced variations in histone binding. It is not clear whether modifications in histone binding are functionally relevant to the altered biological properties associated with viral transformation, but previously published evidence suggests that the binding of histones to DNA may be instrumental in dictating transcriptional properties of the genome.

4726　　INCREASED AMIDOPHOSPHORIBOSYLTRANSFERASE
　　　　(AT) ACTIVITY IN HEPATOMAS [abstract].
(Eng.) Weber, G. (Indiana Univ. Sch. Med., Indiana-polis, Ind.); Prajda, N.　*Proc. Am. Assoc. Cancer Res.* 16:50; 1975.

4727　　NEUTROPHIL ALKALINE PHOSPHATASE: COMPARI-
　　　　SON OF ENZYMES FROM NORMAL SUBJECTS AND
PATIENTS WITH POLYCYTHEMIA VERA AND CHRONIC MYELO-GENOUS LEUKEMIA. (Eng.) Rosenblum, D. (Washington Univ. Sch. Med., St. Louis, Mo.); Petzold, S. J. *Blood* 45(3):335-343; 1975.

4728　　HUMAN TUMOR NON-REGAN ALKALINE PHOSPHATASE
　　　　ISOENZYME [abstract]. (Eng.) Fishman,
W. H. (Tufts Cancer Res. Cent., Boston, Mass.); Driscoll, S.; Miyayama, H.; Fishman, L. *Proc. Am. Assoc. Cancer Res.* 16:195; 1975.

4729　　CELL SURFACE MARKERS IN LYMPHOID MALIGNANCY
　　　　[abstract]. (Eng.) Castleberry, R. P.
(Univ. Alabama Med. Cent., Birmingham); Keightley, R. G.; Lawton, A. R.; Moreno, H. *Clin. Res.* 23(1): 73A; 1975.

4730　　CHARACTERIZATION OF 200 CELL LINES FROM
　　　　HUMAN TUMORS [abstract]. (Eng.) Fogh,
J. (Mem. Sloan-Kettering Cancer Cent., New York, N.Y.). *Proc. Am. Assoc. Cancer Res.* 16:31; 1975.

4731　　THE HEPATIC CHALONE: II. CHEMICAL AND
　　　　BIOLOGICAL PROPERTIES OF THE RABBIT LIVER
CHALONE. (Eng.) Deschamps, Y. (Departement de Bio-chimie, Universite de Montreal, 6128, Montreal 101, P.Q., Canada); Verly*, W. G. *Biomedicine* 22(3):195-208; 1975.

4732　　AN EXPERIMENTAL MOUSE TESTICULAR TERATOMA
　　　　AS A MODEL FOR NEUROEPITHELIAL NEOPLASIA
AND DIFFERENTIATION [abstract]. (Eng.) VandenBerg, S. R. (Stanford Univ. Sch. Medicine, Stanford, Calif Herman, M. M.; Spence, A. M.; Sipe, J. C.; Ludwin, S Bignami, A. *J. Neuropathol. Exp. Neurol.* 34(1):94-9 1975.

4733　　SPECIFICITY AND ACTION MECHANISM OF GRANUL
　　　　CYTE CHALONE AND ITS ROLE IN THE REGULATOR
SYSTEM OF GRANULOPOIESIS. (Ger.) Paukovits, W. R. (Institut fur Krebsforschung der Universitat Wien, Wien, Austria). *Oesterr. Z. Onkol.* 2(2/3):51-56; 1975.

4734　　THE REGULATION OF CELL PROLIFERATION BY
　　　　CHALONES: EXPERIMENTAL INVESTIGATIONS ON
EPIDERMAL HYPERPLASIA. (Ger.) Rohrbach, R. (Patho-logisches Institut der Universitat Freiburg im Breisgau, West Germany). *Veroeff. Morphol. Pathol.* 99:1-67; 1975.

4735　　THE BINDING OF TRITIATED ELONGATION FACTOR
　　　　1 AND 2 TO RIBOSOMES FROM KREBS II MOUSE
ASCITES TUMOR CELLS. (Eng.) Nolan, R. D. (Sandoz Forschungsinstitut, Wien, Austria); Grasmuk, H.; Drews, J. *Eur. J. Biochem.* 50(2):391-402; 1975.

4736　　CLONAL DEVELOPMENT IN TUMOROUS CULTURES
　　　　OF *CREPIS CAPILLARIS*. (Eng.) Sacristan, M. D. (Max-Planck-Institut fur Biologie, Abt. Mel-chers, Tubingen, West Germany). *Naturwissenschaften* 62(3):139-140; 1975.

4737　　ISOLATION OF CELL SUBPOPULATIONS FROM *IN*
　　　　VITRO TUMOR MODELS BY SEDIMENTATION AT
UNIT GRAVITY [abstract]. (Eng.) Durand, R. E. (Univ. Wisconsin Med. Sch., Madison, Wis.). *Bio-phys. J.* 15(2):248a; 1975.

4738　　CHARACTERIZATION OF A MULTIPLE MYELOMA
　　　　CELL LINE OF HUMAN ORIGIN [abstract].
(Eng.) Lozzio, C. B. (Univ. Tennessee Memorial Res. Center, Knoxville, Tenn. 37920); Solomon, A.; Lozzio, B. B.; Klepper, M. B. *IRCS Med. Sci.* 3(10/ Supp.):14; 1975.

4739　　GROWTH OF CANINE MAMMARY TUMORS IN NUDE
　　　　MICE [abstract]. (Eng.) Faulkin, L. J.
(Univ. of California, Davis, Ca.); Mitchell, D. J. *Proc. Am. Assoc. Cancer Res* 16:183; 1975.

4740　　GROWTH OF MALIGNANT TUMORS TRANSPLANTED
　　　　IN ATHYMIC MICE: THE VASCULAR FACTOR
[abstract]. (Eng.) Machado, E. A. (Univ. Tennessee Memorial Res. Center, Knoxville, Tenn. 37920); Lozzio, C. B.; Lozzio, B. B. *IRCS Med. Sci.* 3(10/ Suppl):17; 1975.

4750   A RETINOIC ACID BINDING PROTEIN OF CHICK
        EMBRYO METATARSAL SKIN [abstract]. (Eng.)
Sani, B. P. (Southern Res. Inst., Birmingham, Ala.).
Proc. Am. Assoc. Cancer Res. 16:23; 1975.

4751   EFFECT OF ACTINOMYCIN D ON THE PROTEIN
        SYNTHESIS CHARACTERISTIC OF HEPATOMA PC-1.
(Ukr.) Polishchuk, A. S. (Inst. Biochemistry, Acad.
Sciences, Kiev, Ukranian S.S.R.); Korotkoruchko,
V. P.; Danko, I. M. *Dopov. Akad. Nauk Ukr. R.S.R.*
Ser. B (2):158-160; 1975.

4752   IMMUNOFLUORESCENCE DETECTION OF CASEIN IN
        HUMAN MAMMARY DYSPLASTIC AND NEOPLASTIC
TISSUES. (Eng.) Bussolati, G. (Istituto Anatomia
Istologia Patologica II, Universita Torino, Italy);
Pich, A.; Alfani, V. *Virchows Arch. [Pathol. Anat.]*
365(1):15-21; 1975.

4753   RELATIVE RATES OF ALTERNATIVE PATHWAYS OF
        PURINE NUCLEOTIDE BIOSYNTHESIS IN EHRLICH
ASCITES TUMOR CELLS *IN VIVO* [abstract]. (Eng.)
Smith, C. M. (Univ. Alberta Cancer Res. Unit,
Edmonton, Alberta, Canada, T6G 2E1); Henderson, J. F.
Proc. Am. Assoc. Cancer Res. 16:56; 1975.

4754   DNA CONTENT IN THE BLAST CELLS OF PATIENTS
        SUFFERING FROM ACUTE LEUKEMIA. (Rus.)
Reuk, V. D. (Central Inst. Hematology and Blood Trans-
fusion U.S.S.R. Acad. Medical Sci., Moscow, U.S.S.R.);
Gerasimova, N. A.; Dultsina, S. M. *Probl. Gematol.*
Pereliv. Krovi 20(5):53-56; 1975.

4755   ON THE ABSENCE OF SPECIFIC mRNA SPECIES IN
        HEPATOMA CELLS [abstract]. (Eng.) Feigel-
son, P. (Inst. Cancer Res., Coll. Phys. Surg.,
Columbia Univ., New York, N.Y.); Murthy, L. R.; Sippel,
A.; Morris, H. P. *Proc. Am. Assoc. Cancer Res.* 16:26;
1975.

4756   SITE-DIRECTED MUTAGENESIS: EFFECT OF AN
        EXTRACISTRONIC MUTATION IN PHAGE Qβ RNA
[abstract]. (Eng.) Sabo, D. L. (Inst. Molekularbiologie I, Univ. Zurich, CH-8049, Zurich, Switzerland); Bandle, E.; Weissmann, C. *Experientia* 31(6):
746; 1975.

4757   THE CAUSE OF CANCER AND CANCER ITSELF--
        LIPOPEROXIDE AND DNA RADICAL [abstract].
(Eng.) Fukuzumi, K. (Dept. Applied Chemistry,
Nagoya Univ., Nagoya, Japan). *Gann, Proc. Jpn.*
Cancer Assoc., 34th Annual Meeting, October 1975.
p. 26.

4758   ENZYME REGULATION IN HYPERPLASTIC NODULES
        OF RAT LIVER DURING HEPATOCARCINOGENESIS
[abstract]. (Eng.) Kitagawa, T. (Cancer Inst.,
Tokyo, Japan). *Gann, Proc. Jpn. Cancer Assoc.,*
34th Annual Meeting, October 1865. p. 16.

4759   HISTOCHEMICAL STUDY ON THE ALKALINE PHOS-
        PHATASE ACTIVITY DURING HEPATOCARCINO-
GENESIS [abstract]. (Jpn.) Yoshida, Y. (Sapporo
Medical Coll., Sapporo, Japan); Kaneko, A.; Chisaka,
N.; Dempo, K.; Onoe, T. *Gann, Proc. Jpn. Cancer
Assoc., 34th Annual Meeting, October 1975.* p. 34.

4760   CFA-TRANSFERASE OF MOUSE TUMOR CELLS [ab-
        stract]. (Eng.) Okada, H. (Natl. Cancer
Center Res. Inst., Tokyo, Japan); Cyong, J. *Gann,
Proc. Jpn. Cancer Assoc., 34th Annual Meeting,
October 1975.* p. 57.

4761   EARLY CELLULAR RESPONSES TO DIVERSE
        GROWTH STIMULI INDEPENDENT OF PROTEIN AND
RNA SYNTHESIS. (Eng.) Rubin, H. (Dept. Molecular
Biology and Virus Lab., Univ. California, Berkeley,
Calif. 94720); Koide, T. *J. Cell. Physiol.* 86(1):
47-58; 1975.

4762   COMPARATIVE STUDIES ON THE NUCLEOTIDE
        SEQUENCE OF RIBOSOMAL 5.8S RNA FROM NORMAL
AND MALIGNANT TISSUES [abstract]. (Eng.) Nazar,
R. N. (Baylor Coll. Med., Houston, Tex.); Sitz, T.
O.; Busch, H. *Proc. Am. Assoc. Cancer Res.* 16:53;
1975.

4763   THE PRIMARY STRUCTURE OF THE MAJOR CYTO-
        PLASMIC VALINE tRNA OF MOUSE MYELOMA
CELLS. (Eng.) Piper, P. W. (Medical Res. Council,
Lab. Molecular Biology, Cambridge, England). *Eur.
J. Biochem.* 51(1):295-304; 1975.

4764   STUDIES ON LOW MOLECULAR WEIGHT NUCLEAR
        RNA SPECIES OF NOVIKOFF HEPATOMA AND
RAT LIVER [abstract]. (Eng.) Ro-Choi, T. S. (Baylor Coll. Med., Houston, Tex.); Henning, D.; Raj,
N. B.; Busch, H. *Proc. Am. Assoc. Cancer Res.* 16:
72; 1975.

4765   FATE OF TUMOR CELLS IN BRAIN IS RELATED
        TO GLUTAMINE OR ASPARAGINE LEVELS IN
BRAIN DURING L-GLUTAMINASE AND L-ASPARAGINASE
THERAPY [abstract]. (Eng.) Dolowy, W. C. (Chicago
Med. Sch., Ill.). *Proc. Am. Assoc. Cancer Res.*
16:4; 1975.

4766   IMMUNOCHEMICAL PATTERN OF ASPARTATE AMINO-
        TRANSFERASE ISOZYMES IN SEVERAL RODENTS
AND IN EHRLICH ASCITES CELLS. (Eng.) Kopelovich,
L. (Memorial Sloan-Kettering Cancer Center, New
York, N.Y. 10012). *Proc. Soc. Exp. Biol. Med.*
148(2):410-413; 1975.

4767   PHYSIOLOGICAL GENETICS OF MELANOTIC TUMOURS
        IN *DROSOPHILA MELANOGASTER*. VII. THE ROLE
OF CHOLINE IN THE EXPRESSION OF THE TUMOUR GENE *tu bw*
AND OF ITS SUPPRESSOR, *su-tu*. (Eng.) Sparrow, J. C.
(Dept. Biology, Univ. York, Heslington, York, England);
Sang, J. H. *Genet. Res.* 24(2):215-227; 1975.

4785   EXTRACTION OF A FACTOR FROM EHRLICH ASCITES
        TUMOR CELLS THAT INCREASES THE ACTIVITY
OF THE FETAL ISOZYME OF PYRUVATE KINASE IN MOUSE
LIVER. (Eng.) Ibsen, K. H. (California Coll. Med.,
Univ. California, Irvine, Calif.); Basabe, J. R.;
Lopez, T. P. *Cancer Res.* 35(1):180-188; 1975.

4786   CHANGES IN THE LEVELS OF RIBONUCLEOTIDE
        REDUCTASE, THYMIDINE KINASE AND DNA SYN-
THESIS IN EHRLICH TUMOR CELLS *IN VITRO* [abstract].
(Eng.) Cory, J. G. (Univ. South Florida Coll.
Med., Tampa, Fla.); Mansell, M. M.; Whitford, T. W.,
Jr. *Proc. Am. Assoc. Cancer Res.* 16:27; 1975.

4787   ALTERATIONS IN GLYCOLIPID COMPOSITIONS OF
        HUMAN COLONIC ADENOCARCINOMA [abstract].
(Eng.) Siddiqui, B. (V.A. Hosp., San Francisco,
Calif. 94121); Kim, Y. S. *Proc. Am. Assoc. Cancer
Res.* 16:195; 1975.

4788   TRANSGLUTAMINASE IN TRANSFORMATION AND
        MALIGNANCY [abstract]. (Eng.) Birck-
bichler, P. J. (Samuel Roberts Noble Found., Inc.,
Ardmore, Okla.); Orr, G. R.; Conway, E.; Carter,
H. A.; Patterson, M. K., Jr. *Fed. Proc.* 34(3):614;
1975.

4789   ACTIVITY OF ASPARAGINASE AND GLUTAMINASE
        OF CHICKENS WITH EXPERIMENTAL SARCOMATOSIS.
(Ukr.) Antonov, V. S. (Ukrainian Res. Inst. Experi-
mental Veterinary, Kharkiv, U.S.S.R.); Klenina, N. V.
Ukr. Biochim. Zh. 47(3):335-338; 1975.

4790   POLYSOME RIBONUCLEASES OF LEUKEMIC CELLS.
        (Ukr.) Shliakhovenko, V. O. (Inst. Onco-
logy Problems, Acad. Sciences, Ukrainian S.S.R.,
Kiev, U.S.S.R.); Negrii, H. Z.; Kozak, V. V. *Ukr.
Biokhim. Zh.* 47(3):327-331; 1975.

4791   FACTORS INFLUENCING THE EFFECT OF HORMONES
        ON THE ACCUMULATION OF CYCLIC AMP IN
CULTURED HUMAN ASTROCYTOMA CELLS. (Eng.) Clark,
R. B. (Univ. Colorado Med. Sch., Denver); Su, Y.-F.;
Ortmann, R.; Cubeddu X, L.; Johnson, G. L.; Perkins,
J. P. *Metabolism* 24(3):343-358; 1975.

4792   ESTROGEN RECEPTORS IN MALE PATIENTS WITH
        GYNECOMASTIA AND BREAST CANCER [abstract].
(Eng.) Leclercq, G. (Institut Jules Bordet, 1000
Brussels, Belgium); Verhest, A.; Deboel, M. C.; Matt-
heiem, W. H.; Lejour, M.; Heuson, J. C. *Acta Endocrinol.* [*Suppl.*] *(Kbh.)* 199:281; 1975.

4793   THE *IN VIVO* UPTAKE OF TRITIATED ESTRADIOL
        IN CARCINOMA OF THE BREAST. (Eng.)
Viladiu, P. (Hospital de la Santa Cruz y San Pablo,
Univ. Barcelona, Spain). *Surg. Gynecol. Obstet.*
140(4):544-546; 1975.

4794   EPIDERMAL GROWTH FACTOR: ALTERED BINDING
        EFFICIENCY IN AGING HUMAN FIBROBLASTS [ab-
stract]. (Eng.) Ladda, R. (Coll. Med., Pennsylvania
state Univ., Hershey). *Fed. Proc.* 34(3):276; 1975.

MISCELLANEOUS

4795   HORMONE PRODUCTION BY NON-ENDOCRINE TUMORS
   ·   [abstract]. (Eng.) Rosen, S. W. (Natl.
Inst. Health, Bethesda, Md. 20014). *Acta Endocrinol. [Suppl.] (Kbh.)* 199:75; 1975.

4796   REGULATION OF CYCLIC AMP IN MAMMARY TISSUES:
      CELLULAR LOCALIZATION OF PROLACTIN EFFECT
[abstract]. (Eng.) Scott, D. F. (Med. Coll. Georgia,
Augusta); Howard, E. F. *Proc. Am. Assoc. Cancer Res.*
16:51; 1975.

4797   CALCITONIN PRODUCTION BY RAT THYROID TUMOURS.
   ·   (Eng.) Triggs, S. M. (Welsh Natl. Sch.
Medicine, Cardiff, CF4 4XN, Wales); Hesch, R. D.;
Woodhead, J. S.; Williams, E. D. *J. Endocrinol.*
66(1):37-43; 1975.

* INDICATES A PLAIN CITATION WITHOUT ACCOMPANYING ABSTRACT

.M.

i.L.

. J.

, A.

,S.

A.J.

, H.

INTESTINAL NEOPLASMS
    HISTOCHEMICAL STUDY, 4770*
PROSTATIC NEOPLASMS
    ANTIBODIES, 4476
    ANTIGENS, NEOPLASM, 4476
    IMMUNE SERUMS, 4476

ACRYLAMIDE, 2-(2-FURYL)-3-(5-NITRO-2-FURYL)-
    FOOD PRESERVATIVES
        MUTAGENIC EFFECT, MOUSE, 4270
    RETINOBLASTOMA
        MUTANTS, ENVIRONMENTAL, JAPAN,
            4664

ACTINOMYCIN D
    HEPATOMA
        DNA, 4707
        PROTEINS, 4751*
        RNA, 4751*
    VIRUS, MOLONEY MURINE LEUKEMIA
        VIRUS REPLICATION, 4385
    VIRUS, SV40
        PHOSPHODIESTERASES, 4712

ACUTE DISEASE
    LEUKEMIA
        CELL DIVISION, 4590
        CELLULAR INCLUSIONS, 4601*
        THYMIDINE INCORPORATION,
            LEUKOCYTES, 4590
    LEUKEMIA, MYELOCYTIC
        CHROMOSOMES, HUMAN, 21-22, 4626*

ACYLTRANSFERASES
    ACETOHYDROXAMIC ACID, N-FLUOREN-2-YL-
        SMALL INTESTINE, RAT, 4275
        TISSUE-SPECIFICITY, 4275
    LUNG NEOPLASMS
        ADENOMA, 4711

ADENINE NUCLEOTIDES
    LEUKEMIA
        BLOOD PLATELETS, 4591

ADENOACANTHOMA
    SEE ADENOCARCINOMA

ADENOCARCINOMA
    BENZO(A)PYRENE
        COMMON STATISTICAL FACTORS, SITE-
            SPECIFICITY, 4276
    BREAST NEOPLASMS
        CHROMOSOME ABNORMALITIES, 4612*
        MULTIFOCAL PRIMARY CARCINOMA, 4569
    CERVIX NEOPLASMS
        MULTIFOCAL PRIMARY CARCINOMA, 4569
    COLONIC NEOPLASMS
        ADENOSINE CYCLIC 3',5'
            MONOPHOSPHATE, 4744*
        ADENOSINE TRIPHOSPHATE, 4744*
        ADENYL CYCLASE, 4744*
        BACILLUS CALMETTE-GUERIN, 4293
        CARCINOEMBRYONIC ANTIGEN, 4474
        GLYCOLIPIDS, 4787*
        HYDRAZINE, 1,2-DIMETHYL-, 4330*
        MECKEL'S DIVERTICULUM, 4616*
        PROTEIN-LOSING ENTEROPATHIES,
            4615*

RADIATION, IONIZING, 4364
ESOPHAGEAL NEOPLASMS
    DIAPHRAGMATIC HERNIA, 4614*
    EARYL DIAGNOSIS, ENDOSCOPY, 4579
GASTROINTESTINAL NEOPLASMS
    CASE REPORT, 4618*
    ENTERITIS, REGIONAL, 4618*
GLYCOPROTEINS
    GASTRIC JUICE, 4748*
GYNECOLOGIC NEOPLASMS
    4,4'-STILBENEDIOL, ALPHA,ALPHA'-
      DIETHYL-, 4305
    TRANSPLACENTAL CARCINOGENESIS,
      4305
LIVER NEOPLASMS
    LIVER CIRRHOSIS, OBSTRUCTIVE, 4566
LUNG NEOPLASMS
    EPIDEMIOLOGY, MOUSE, 4723
    HYDRAZINE, BUTYL-,
      MONOHYDROCHLORIDE, 4294
    HYDRAZINE, PROPYL-,
      MONOHYDROCHLORIDE, 4294
    IMMUNE RESPONSE, 4582
    LYMPHOCYTE TRANSFORMATION, 4498
MAMMARY NEOPLASMS, EXPERIMENTAL
    ADRIAMYCIN, 4292
    CHROMATIN, 4743*
    EPIDEMIOLOGY, MOUSE, 4723
NOSE NEOPLASMS
    QUINOXALINE, 1,4-DIOXIDE, 4356*
PANCREATIC NEOPLASMS
    BENZO(A)PYRENE, 4276.
    CARCINOEMBRYONIC ANTIGEN, 4474
    STOMACH NEOPLASMS, 4474
    TISSUE CULTURE, 4255*
PROSTATIC NEOPLASMS
    CASTRATION, 4575
    EPIDEMIOLOGY, IRAN, 4670*
    NEOPLASM METASTASIS, 4575
REVERSE TRANSCRIPTASE
    DNA, 4716
SALIVARY GLAND NEOPLASMS
    BENZ(A)ANTHRACENE, 7,12-DIMETHYL-,
      4290
STOMACH NEOPLASMS
    CARCINOEMBRYONIC ANTIGEN, 4474
    DIAPHRAGMATIC HERNIA, 4614*
    GLYCOPROTEINS, 4748*
    INCIDENCE, MOUSE, 4723
THYROID NEOPLASMS
    CLASSIFICATION, 4642*
    NEOPLASM METASTASIS, 4653
    RADIATION, IONIZING, 4653
UTERINE NEOPLASMS
    INCIDENCE, MOUSE, 4723
    MULTIFOCAL PRIMARY CARCINOMA, 4569
    RADIATION, IONIZING, 4364
VAGINAL NEOPLASMS
    DIAGNOSIS AND PROGNOSIS, 4571

ADENOFIBROMA
    MAMMARY NEOPLASMS, EXPERIMENTAL
      ADRIAMYCIN, 4292
      ETHYLENE, CHLORO-, 4311, 4312
    UTERINE NEOPLASMS
      ULTRASTRUCTURAL STUDY, 4573

ADENOMA
    ADRENAL GLAND NEOPLASMS

        INCIDENCE, MOUSE, 4723
COLONIC NEOPLASMS
    REVIEW, 4256*
LUNG NEOPLASMS
    ACYLTRANSFERASES, 4711
    CARBAMIC ACID, ETHYL ESTER,
    EPIDEMIOLOGY, MOUSE, 4723
    HYDRAZINE, BUTYL-,
      MONOHYDROCHLORIDE, 4294
    HYDRAZINE, PROPYL-,
      MONOHYDROCHLORIDE, 4294
    MONOCROTALINE, 4302
    UREA, METHYL NITROSO-, 4437
PITUITARY NEOPLASMS
    ETHYLENE, CHLORO-, 4311
PULMONARY NEOPLASMS
    ETHYLENE, CHLORO-, 4312
RECTAL NEOPLASMS
    REVIEW, 4256*
SALIVARY GLAND NEOPLASMS
    EPIDEMIOLOGY, 4672*

ADENOMA, CHROMOPHOBE
-   PITUITARY NEOPLASMS
      HYPOTHROIDISM, 4576
      SECRETORY GRANULES, DIAMETER

ADENOMATOSIS, FAMILIAL ENDOCRINE
    BRONCHIAL NEOPLASMS
      CARCINOID TUMOR, 4606*

ADENOSINE
    ASTROCYTOMA
      ADENOSINE CYCLIC 3,5 MONOPHO
        4791*
    L CELLS
      RESISTANT PHENOTYPE, 4685

ADENOSINE CYCLIC 3',5' MONOPHOSPHATE
    ASTROCYTOMA
      ADENOSINE, 4791*
      CATECHOLAMINES, 4791*
      PROSTAGLANDINS, 4791*
    CARCINOMA, EHRLICH TUMOR
      REVERSE TRANSCRIPTASE, 4716
    COLONIC NEOPLASMS
      ADENOCARCINOMA, 4744*
    INSULIN
      GROWTH SUBSTANCES, 4225
    MAMMARY NEOPLASMS, EXPERIMENTAL
      EPINEPHRINE, 4796*
      PROLACTIN, 4796*
      PROSTAGLANDINS E, 4796*
    MELANOMA
      MSH, 4721
    VIRUS, POLYOMA
      GROWTH SUBSTANCES, 4391

ADENOSINE TRIPHOSPHATE

FOOD CONTAMINATION
    CORN, 4677*

AGE FACTORS
    CERVIX NEOPLASMS
        EPIDEMIOLOGY, 4674*
    HEPATOMA
        EPIDEMIOLOGY, MOUSE, 4723
    LEUKEMIA
        EPIDEMIOLOGY, MOUSE, 4723
    LYMPHATIC NEOPLASMS
        EPIDEMIOLOGY, MOUSE, 4723
    NEOPLASMS, EXPERIMENTAL
        EPIDEMIOLOGY, MOUSE, 4723

AGRANULOCYTOSIS
    LYMPHOCYTES
        CELLULAR INCLUSIONS, 4601*

AIR POLLUTANTS
    LUNG NEOPLASMS
        BIOASSAY, 4237*

ALANINE, 3-(P-(BIS(2-CHLOROETHYL)AMINO)
PHENYL)-, L-
    LEUKEMIA, MYELOBLASTIC
        CASE REPORT, 4624*
        PARAPROTEINEMIA, 4624*

ALBUMINS
    HEPATOMA
        CELL CYCLE KINETICS, 4463

ALCOHOLIC BEVERAGES
    1-BUTANOL, 3-METHYL-
        CARCINOGENIC ACTIVITY, 4318*
    ISOBUTYL ALCOHOL
        CARCINOGENIC ACTIVITY, 4318*
    PROPYL ALCOHOL
        CARCINOGENIC ACTIVITY, 4318*

ALFATOXIN M1
    MILK
        QUANTITATION METHOD, 4667

ALKALINE PHOSPHATASE
    COLITIS, ULCERATIVE
        LEUKOCYTES, 4715
    HEPATOMA
        ENZYMATIC ACTIVITY, 4759*
    LEUKEMIA, MYELOCYTIC
        LEUKOCYTES, 4588
        NEUTROPHILS, 4727*
    NEOPLASMS
        ISOENZYMES, 4728*
    PLACENTA
        ISOENZYMES, 4728*
    POLYCYTHEMIA VERA
        NEUTROPHILS, 4727*

ALKYLATING AGENTS
    ETHER, BIS(CHLOROMETHYL)-
        DNA, 4321*
    2-OXETANONE
        DNA, 4321*

ALPHA 1 ANTITRYPSIN
    GASTROINTESTINAL NEOPLASMS
        SMOKING, 4657

    LEUKEMIA
        SMOKING, 4657
    LUNG NEOPLASMS
        SMOKING, 4657
    LYMPHOMA
        SMOKING, 4657
    NEOPLASMS
        SMOKING, 4657
    SMOKING
        SERUM LEVELS, HOSPITALIZED
            PATIENTS, 4657

ALPHA-FETOPROTEIN
    SEE FETAL GLOBULINS

ALPHA GLOBULINS
    CHOLANGIOMA
        SERUM LEVEL, 4461
    FETAL GLOBULINS
        ISOLATION AND CHARACTERIZATION,
            4462
    GASTROINTESTINAL NEOPLASMS
        SERUM LEVEL, 4461
    HEPATOMA
        FETAL GLOBULINS, 4461
        ISOLATION AND CHARACTERIZATION,
            4464
    HODGKIN'S DISEASE
        SERUM LEVEL, 4461
    LEUKEMIA, MYELOBLASTIC
        SERUM LEVEL, 4461
    LEUKEMIA, MYELOCYTIC
        SERUM LEVEL, 4461
    LUNG NEOPLASMS
        SERUM LEVEL, 4461
    MULTIPLE MYELOMA
        SERUM LEVEL, 4461
    THYROID NEOPLASMS
        SERUM LEVEL, 4461

ALPHA-AMANITINE
    CARCINOMA, EHRLICH TUMOR
        REVERSE TRANSCRIPTASE, 4716
    HELA CELLS
        DNA REPLICATION, 4706
        VIRUS, ADENO 2, 4378

AMINO ACIDS
    ULTRAVIOLET RAYS
        DNA REPAIR, 4365

AMINO ACYL T RNA SYNTHETASES
    HEPATOMA
        ISOLATION AND CHARACTERIZATION,
            4773*

AMINOLEVULINIC ACID SYNTHETASE
    BUTYRIC ACID, 2-AMINO-4-(ETHYLTHIO)-
        MITOCHONDRIA, LIVER, 4320*

AMINOPEPTIDASES
    PROSTATIC NEOPLASMS
        ANTIBODIES, 4476

AMINOPYRINE N-DEMETHYLASE
    DIETHYLAMINE, N-NITROSO-
        LIVER, RAT, 4283

AMOSITE

LYMPHOSARCOMA
    HISTOCOMPATIBILITY ANTIGENS, 4459
NEOPLASMS, EXPERIMENTAL
    VIRUS, POLYOMA, 4458
SARCOMA
    CHOLANTHRENE, 3-METHYL-, 4458
    T-LYMPHOCYTES, 4499
VIRUS, HERPES SAIMIRI
    ANTIGENS, VIRAL, 4536*
VIRUS, TURKEY HERPES
    ANTIGENS, VIRAL, 4404*

ANTIGEN-ANTIBODY REACTIONS
COLONIC NEOPLASMS
    CARCINOEMBRYONIC ANTIGEN, 4473
IMMUNITY, CELLULAR
    LYMPHOCYTES, 4251*
    NEGATIVE ASSOCIATION, REVIEW,
        4251*
LEUKEMIA
    IMMUNE SERUMS, 4449
B-LYMPHOCYTES
    IMMUNOGLOBULINS, SURFACE, 4455
T-LYMPHOCYTES
    HISTOCOMPATIBILITY ANTIGENS, 4489
SARCOMA
    IMMUNE SERUMS, 4449
VIRUS, HERPES SAIMIRI
    CELL TRANSFORMATION, NEOPLASTIC,
        4536*

ANTIGENIC DETERMINANTS
COLONIC NEOPLASMS
    CARCINOEMBRYONIC ANTIGEN, 4473
FETAL GLOBULINS
    PROTEINS, 4470
HEPATOMA
    HISTOCOMPATIBILITY ANTIGENS, 4485
LYMPHATIC NEOPLASMS
    DNA, 4545*
B-LYMPHOCYTES
    CELL MEMBRANE, 4488
T-LYMPHOCYTES
    CELL MEMBRANE, 4488
NEOPLASM TRANSPLANTATION
    MEMBRANE CARRIER MODEL, 4554*
SARCOMA
    VIRUS, HERPES SIMPLEX, 4213
VIRUS, MURINE LEUKEMIA
    ISOLATION AND CHARACTERIZATION,
        4380

ANTIGENS
AUTOIMMUNE DISEASES
    IMMUNITY, CELLULAR, 4520*
    T-LYMPHOCYTES, 4520*
BENZENE, 1-CHLORO-2,4-DINITRO-
    IMMUNITY, CELLULAR, 4520*
    T-LYMPHOCYTES, 4520*
CELL MEMBRANE
    CELL DIFFERENTIATION, 4477
    FIBROBLASTS, 4477
LYMPHOMA
    T-LYMPHOCYTES, 4491
LYMPHOSARCOMA
    HYPERSENSITIVITY, DELAYED, 4538*
NEOPLASM
    IMMUNITY, CELLULAR, 4520*
    T-LYMPHOCYTES, 4520*

5

* INDICATES A PLAIN CITATION WITHOUT ACCOMPANYING ABSTRACT

ANTIGENS, BACTERIAL
    T-LYMPHOCYTES
        HELPER CELL GENERATION, 4500
    PHOSPHORYLCHOLINE
        IMMUNE RESPONSE, REVIEW, 4218

ANTIGENS, NEOPLASM
    HEPATOMA
        CELL MEMBRANE, 4484
        DIETHYLAMINE, N-NITROSO-, 4484,
            4533*
        IMMUNE SERUMS, 4485
        ISOLATION AND CHARACTERIZATION,
            4484
    LEUKEMIA L1210
        IMMUNOTHERAPY, 4535*
    LEUKEMIA, LYMPHOBLASTIC
        ANTIBODIES, NEOPLASM, 4552*
        IMMUNITY, CELLULAR, 4552*
    MAMMARY NEOPLASMS, EXPERIMENTAL
        LYMPHOCYTE BINDING, 4495
    MELANOMA
        HYPERSENSITIVITY, DELAYED, 4479
        IMMUNITY, CELLULAR, 4479
    MITOMYCIN C
        B-LYMPHOCYTES, 4540*
        T-LYMPHOCYTES, 4540*
    NEOPLASMS
        REVIEW, 4246*
    NEOPLASMS, EXPERIMENTAL
        VIRUS, POLYOMA, 4458
    NEURILEMMOMA
        ISOLATION AND CHARACTERIZATION,
            4435
    OVARIAN NEOPLASMS
        LYMPHOCYTE TRANSFORMATION, 4487
    PROSTATIC NEOPLASMS
        ACID PHOSPHATASE, 4476
    SARCOMA
        CHOLANTHRENE, 3-METHYL-, 4250*,
            4458, 4532*
        IMMUNITY, CELLULAR, 4521*, 4522*
    TERATOID TUMOR
        REVIEW, 4261*
    TUMOR-HOST RELATIONSHIP
        IMMUNOLOGY, REVIEW, 4250*

ANTIGENS, VIRAL
    BREAST NEOPLASMS
        DNA-RNA HYBRIDIZATION, 4478
        VIRUS, D-TYPE RNA TUMOR, 4478
        VIRUS, HERPES SIMPLEX 2, 4478
        VIRUS, MASON-PFIZER MONKEY, 4400
    BURKITT'S LYMPHOMA
        VIRUS, EPSTEIN-BARR, 4420*
    CARCINOMA, BRONCHIOLAR
        VIRUS-LIKE PARTICLES, 4483
    INFECTIOUS MONONUCLEOSIS
        VIRUS, EPSTEIN-BARR, 4420*
    LYMPHOMA
        CONCANAVALIN A, 4506
        TRANSPLANTATION IMMUNOLOGY, 4440
        VIRUS, MURINE LEUKEMIA, 4440
    VIRUS, EPSTEIN-BARR
        INTERFERON, 4460

        URIDINE, 5-BROMO-2'-DEOXY-, 4460
    VIRUS, FELINE LEUKEMIA
        REVIEW, 4240*
    VIRUS, FELINE SARCOMA
        REVIEW, 4240*
    VIRUS, FRIEND MURINE LEUKEMIA
        IMMUNOSUPPRESSION, 4447
    VIRUS, HERPES SAIMIRI
        ANTIBODY FORMATION, 4536*
    VIRUS, HERPES SIMPLEX
        IMMUNOGLOBULINS, 4214
    VIRUS, HERPES SIMPLEX 2
        CARCINOGENIC POTENTIAL, REVIEW,
            4213
    VIRUS, KIRSTEN MURINE SARCOMA
        VIRUS, KIRSTEN MURINE LEUKEMIA,
            4387
        VIRUS, SIMIAN SARCOMA, 4387
    VIRUS, MAREK'S DISEASE HERPES
        VIRUS REPLICATION, 4383
    VIRUS, MURINE LEUKEMIA
        X-TROPIC VIRUS, 4380
    VIRUS, RADIATION MURINE LEUKEMIA
        VIRUS REPLICATION, 4386
    VIRUS, TURKEY HERPES
        ANTIBODY FORMATION, 4404*

ANTIHYPERTENSIVE AGENTS
    NEOPLASMS
        EPIDEMIOLOGY, CHILD, 4654

ANTILYMPHOCYTE SERUM
    NEOPLASMS
        TRANSPLANTATION, HOMOLOGOUS, 4650
    SARCOMA, OSTEOGENIC
        RADIATION, 4235*
    VIRUS, HERPES SAIMIRI
        ISOLATION AND CHARACTERIZATION,
            4403*

ANTISERUM
    SEE IMMUNE SERUMS

ARGINASE
    LEUKEMIA, LYMPHOCYTIC
        LYMPHOCYTES, 4713
    LEUKEMIA, MYELOCYTIC
        LYMPHOCYTES, 4713

ARYL HYDROCARBON HYDROXYLASES
    BARBITURIC ACID, 5-ETHYL-5-PHENYL,
        SODIUM SALT
        CYTOCHROME P-450, 4278
    BENZO(A)PYRENE
        CYTOCHROME P-450, 4278
    5,6-BENZOFLAVONE
        CYTOCHROME P-450, 4278
    7,8-BENZOFLAVONE
        CYTOCHROME P-450, 4278
        INHIBITION, LIVER, 4277
    BENZOTHIAZOLE, 2-(P-CHLOROPHENYL)-
        CYTOCHROME P-450, 4278
    CHOLANTHRENE, 3-METHYL-
        CYTOCHROME P-450, 4278
        INHIBITION, LIVER, 4277
    ESTROGENS
        INHIBITION, LIVER, 4277
    ISOPROTERENOL
        CYTOCHROME P-450, 4278

EPIDEMIOLOGY, REVIEW, 4263*
GENETICS, 4263*

AUTOANTIBODIES
    CARCINOMA
        FLUORESCENT ANTIBODY TECHNIC,
            4528*
    MELANOMA
        FLUORESCENT ANTIBODY TECHNIC,
            4528*
    NEUROBLASTOMA
        FLUORESCENT ANTIBODY TECHNIC,
            4528*
    PHOSPHORYLCHOLINE
        IMMUNOGLOBULINS, 4218

AUTOIMMUNE DISEASES
    ANTIGENS
        IMMUNITY, CELLULAR, 4520*
        T-LYMPHOCYTES, 4520*
    BENZENE, 1-CHLORO-2,4-DINITRO-
        IMMUNITY, CELLULAR, 4520*
        T-LYMPHOCYTES, 4520*

AVIAN LEUKOSIS
    VIRUS, ROUS SARCOMA
        SUBGROUP A, 4376
        SUBGROUP E, 4376

AZATHIOPRINE
    SEE PURINE, 6-((1-METHYL-4-
        NITROIMIDAZOL-5-YL)THIO)-

BACILLUS CALMETTE-GUERIN
    COLONIC NEOPLASMS
        ADENOCARCINOMA, 4293
        HYDRAZINE, 1,2-DIMETHYL-, 4293
    HYDRAZINE, 1,2-DIMETHYL
        CARCINOGENIC ACTIVITY, RAT, 4293
    LUNG NEOPLASMS
        NEOPLASM METASTASIS, 4293
    LYMPHATIC NEOPLASMS
        NEOPLASM METASTASIS, 4293

BACTERIOPHAGES
    METHANESULFONIC ACID, ETHYL ESTER
        RNA, 4288
    UREA, ETHYL NITROSO-
        RNA, 4288
    RNA
        EXTRACISTRONIC MUTATION, 4756*

BARBITURIC ACID, 5-ETHYL-5-PHENYL, SODIUM
  SALT
    ARYL HYDROCARBON HYDROXYLASES
        CYTOCHROME P-450, 4278

BASOPHILS
    HEPATOMA
        RNA, NEOPLASM, 4696

BCG VACCINATION
    LEUKEMIA, LYMPHOBLASTIC
        IMMUNITY, CELLULAR, 4552*
    LYMPHOMA
        VIRUS, HERPES SAIMIRI, 4448

BENCE JONES PROTEIN
  MULTIPLE MYELOMA
    CHROMOSOME ABERRATIONS, 4738*

BENZ(A)ANTHRACEN-5-ONE, 5,6-DIHYDRO-7,12-
DIMETHYL-
  BENZ(A)ANTHRACENE, 7,12-DIMETHYL-
    CARCINOGENIC ACTIVITY, 4267
    K-REGION KETONES, 4267

BENZ(A)ANTHRACEN-6-ONE, 5,6-DIHYDRO-7,12-
DIMETHYL-
  BENZ(A)ANTHRACENE, 7,12-DIMETHYL-
    CARCINOGENIC ACTIVITY, 4267
    K-REGION KETONES, 4267

BENZ(A)ANTHRACENE, 7,12-DIMETHYL-
  BENZ(A)ANTHRACEN-5-ONE, 5,6-DIHYDRO-7,
  12-DIMETHYL-
    CARCINOGENIC ACTIVITY, 4267
    K-REGION KETONES, 4267
  BENZ(A)ANTHRACEN-6-ONE, 5,6-DIHYDRO-7,
  12-DIMETHYL-
    CARCINOGENIC ACTIVITY, 4267
    K-REGION KETONES, 4267
  CELL TRANSFORMATION, NEOPLASTIC
    ACETYLTRANSFERASES, 4741*
    CELL FUSION, 4339*
  DIETHYLAMINE, N-NITROSO-
    NUCLEIC ACIDS, 4271
    SITE-SPECIFICITY, LIVER, 4271
  KERATOACANTHOMA
    CROTON OIL, 4363
  LEUKEMIA
    IMMUNE SERUMS, 4444
  MAMMARY NEOPLASMS, EXPERIMENTAL
    ESTRADIOL, 4202
    IMMUNITY, CELLULAR, 4495
    TRANSFERASES, 4719
  NEOPLASMS, EXPERIMENTAL
    ANTIBODIES, NEOPLASM, 4531*
    IMMUNE ADHERENCE REACTION, 4531*
  PAPILLOMA
    CROTON OIL, 4363
  SALIVARY GLAND NEOPLASMS
    ADENOCARCINOMA, 4290
    PRECANCEROUS CONDITIONS, 4290
    RHABDOMYOSARCOMA, 4290

BENZENAMINE, N,N-DIMETHYL-4-((3-
METHYLPHENYL)AZO)-
  HEPATOMA
    CHROMATIN, 4746*
    GLUTAMYL TRANSPEPTIDASE, 4788*
    PROTEIN KINASE, 4746*

BENZENE
  ANEMIA, APLASTIC
    EPIDEMIOLOGY, OCCUPATIONAL HAZARD,
    4663
  ERYTHROLEUKEMIA
    EPIDEMIOLOGY, OCCUPATIONAL HAZARD,
    4663
  LEUKEMIA
    EPIDEMIOLOGY, OCCUPATIONAL HAZARD,
    4663
  LEUKEMIA, LYMPHOCYTIC
    EPIDEMIOLOGY, OCCUPATIONAL HAZARD,
    4663

LEUKEMIA, MYELOBLASTIC
    EPIDEMIOLOGY, OCCUPATIONAL
    4663
LEUKEMIA, MYELOCYTIC
    EPIDEMIOLOGY, OCCUPATIONAL
    4663

BENZENE, 1-CHLORO-2,4-DINITRO-
  ANTIGENS
    IMMUNITY, CELLULAR, 4520*
    T-LYMPHOCYTES, 4520*
  AUTOIMMUNE DISEASES
    IMMUNITY, CELLULAR, 4520*
    T-LYMPHOCYTES, 4520*
  NEOPLASM
    IMMUNITY, CELLULAR, 4520*
    T-LYMPHOCYTES, 4520*

BENZIDINE
  OCCUPATIONAL HAZARD
    EPIDEMIOLOGY, 4205

BENZIDINE, 3,3'-DICHLORO-
  OCCUPATIONAL HAZARD
    EPIDEMIOLOGY, 4205

BENZIDINE, 3,3'-DIMETHOXY-
  OCCUPATIONAL HAZARD
    EPIDEMIOLOGY, 4205

BENZO(A)PYREN-3-OL
  BENZO(A)PYREN-6-OL
    SPECTROMETRY, FLUORESCENCE,

BENZO(A)PYREN-6-OL
  BENZO(A)PYREN-3-OL
    SPECTROMETRY, FLUORESCENCE,

BENZO(A)PYRENE
  ADENOCARCINOMA
    COMMON STATISTICAL FACTORS,
      SPECIFICITY, 4276
  ARYL HYDROCARBON HYDROXYLASES
    CYTOCHROME P-450, 4278
  ENDOPLASMIC RETICULUM
    RIBOSOMES, 4266
  HEPATOMA
    COMMON STATISTICAL FACTORS,
      SPECIFICITY, 4276
  LUNG NEOPLASMS
    COMMON STATISTICAL FACTORS,
      SPECIFICITY, 4276
  LYMPHATIC NEOPLASMS
    COMMON STATISTICAL FACTORS,
      SPECIFICITY, 4276
    SARCOMA, 4544*
  NEOPLASMS, EXPERIMENTAL
    ANTIBODIES, NEOPLASM, 4531*
    IMMUNE ADHERENCE REACTION, 4
  PANCREATIC NEOPLASMS
    ADENOCARCINOMA, 4276
    PAPILLOMA, 4276
  PAPILLOMA
    COMMON STATISTICAL FACTORS,
      SPECIFICITY, 4276

TRYPTOPHAN, 4769*
LEIOMYOMA
CASE REPORT, 4637*
PHEOCHROMOCYTOMA
CASE REPORT, 4636*

BLOOD PLATELETS
LEUKEMIA
ADENINE NUCLEOTIDES, 4591
ADENOSINE TRIPHOSPHATE, 4591
PLATELET AGGREGATION, 4591
PRECANCEROUS CONDITIONS, 4591
ULTRASTRUCTURAL STUDY, 4591

BONE MARROW CELLS
CHALONES
CELL MEMBRANE, 4733*
CHROMOSOME ABERRATIONS
PHOSPHINE SULFIDE, TRIS(1-
AZIRIDINYL)-, 4359*
LEUKEMIA
DNA, 4754*
LEUKEMIA, LYMPHOBLASTIC
REVIEW, 4220
LEUKEMIA, MONOBLASTIC
PRECANCEROUS CONDITIONS, 4220
REVIEW, 4220
LEUKEMIA, MONOCYTIC
PRECANCEROUS CONDITIONS, 4220
REVIEW, 4220
LEUKEMIA, MYELOCYTIC
PRECANCEROUS CONDITIONS, 4220
REVIEW, 4220
TRANSPLANTATION IMMUNOLOGY
GRAFT VS HOST REACTION, 4436

BONE NEOPLASMS
BURKITT'S LYMPHOMA
EPIDEMIOLOGY, 4659
EPIDEMIOLOGY
CHILD, 4651
LEIOMYOSARCOMA
CASE REPORT, 4608*
RADIATION, IONIZING
DOSE-RESPONSE STUDY, 4680*

BRAIN NEOPLASMS
CARCINOMA, EHRLICH TUMOR
ASPARINE, 4765*
GLUTAMINE, 4765*
CHORIOCARCINOMA
NEOPLASM METASTASIS, 4720
FIBROSARCOMA
CASE REPORT, 4631*
GENETICS, 4631*
GLIOMA
IMMUNITY, CELLULAR, 4549*
MENINGIOMA
IMMUNITY, CELLULAR, 4549*
NEURILEMMOMA
IMMUNITY, CELLULAR, 4549*
NEUROBLASTOMA
ETHYLENE, CHLORO-, 4311, 4312
SEX FACTORS
EPIDEMIOLOGY, FINLAND, 4645

BREAST NEOPLASMS
ACID PHOSPHATASE
HISTOCHEMICAL STUDY, 4770*

ADENOCARCINOMA
    CHROMOSOME ABNORMALITIES, 4612*
    MULTIFOCAL PRIMARY CARCINOMA, 4569
ANTIGENS, VIRAL
    DNA-RNA HYBRIDIZATION, 4478
CARCINOEMBRYONIC ANTIGEN
    PROTEINS, 4248*
CARCINOMA IN SITU
    EPIDEMIOLOGY, SASKATCHEWAN, 4668
CELL TRANSFORMATION, NEOPLASTIC
    MATHEMATICAL MODELS, EVALUATION,
    4666
EPIDEMIOLOGY
    AUSTRALIA, 4644
    FLUOROSCOPY, 4371*
    TREATMENT DELAY AND SURVIVAL,
    4671*
ESTRADIOL
    ADIPOSE TISSUE, 4793*
    NEOPLASM METASTASIS, 4793*
    URINE, 4570
ESTRIOL
    URINE, 4570
ESTROGENS
    CHROMOSOMES, 4792*
    RECEPTORS, HORMONE, 4792*
ESTRONE
    URINE, 4570
GALACTOSIDASES
    HISTOCHEMICAL STUDY, 4770*
GENETICS
    EPIDEMIOLOGY, 4630*
GLUCURONIDASE
    HISTOCHEMICAL STUDY, 4770*
HODGKIN'S DISEASE
    RADIATION, IONIZING, 4364
HORMONES
    THERAPY, GUIDELINES, 4776*
17-HYDROXYCORTICOSTEROIDS
    URINE, 4570
17-KETOSTEROIDS
    URINE, 4570
MELANOMA
    GENETICS, 4799*
NEOPLASM METASTASIS
    DIAGNOSIS AND PROGNOSIS, 4658
    EPIDEMIOLOGY, 4658
    EPIDEMIOLOGY, SASKATCHEWAN, 4668
PRECANCEROUS CONDITIONS
    EPIDEMIOLOGY, SASKATCHEWAN, 4668
RADIATION
    FLUOROSCOPY, 4371*
    REPEATED FLUOROSCOPY HISTORY,
    4372*
RADIATION, IONIZING
    DOSE-RESPONSE STUDY, 4680*
RECEPTORS, HORMONE
    REVIEW, 4202
RESERPINE
    EPIDEMIOLOGY, REVIEW, 4232*
SEX CHROMATIN
    DIAGNOSIS AND PROGNOSIS, 4638*
    NEOPLASM METASTASIS, 4638*
SEX FACTORS
    EPIDEMIOLOGY, FINLAND, 4645
SULFATASES
    HISTOCHEMICAL STUDY, 4770*
TRANSFERASES
    ESTROGENS, 4719

RECEPTORS, HORMONE, 4719
UREMIA
    IMMUNOSUPPRESSION, 4515*
VIRUS, D-TYPE RNA TUMOR
    ANTIGENS, VIRAL, 4478
VIRUS, HERPES SIMPLEX 2
    ANTIGENS, VIRAL, 4478
VIRUS, MASON-PFIZER MONKEY
    ANTIGENS, VIRAL, 4400

BROMODEOXYURIDINE
    SEE URIDINE, 5-BROMO-2'-DEOXY'

5-BROMODEOXYURIDINE
    SEE URIDINE, 5-BROMO-2'-DEOXY'

BRONCHIAL NEOPLASMS
    CARCINOEMBRYONIC ANTIGEN
        SERUM LEVELS, 4775*
    CARCINOID TUMOR
        ADENOMATOSIS, FAMILIAL ENDOCRINE
        4606*
        MALIGNANT CARCINOID SYNDROME,
        4604*
        NEOPLASM METASTASIS, 4604*
    CARCINOMA
        CARCINOEMBRYONIC ANTIGEN, 4471
    GONADOTROPINS, CHORIONIC
        SERUM LEVELS, 4775*
    MALIGNANT CARCINOID SYNDROME
        CASE REPORT, 4604*
    PRECANCEROUS CONDITIONS
        DNA, 4605*
        RNA, 4605*

BRONCHITIS
    CHRONIC DISEASE
        CARCINOEMBRYONIC ANTIGEN, 4471

BURKITT'S LYMPHOMA
    ADRENAL GLAND NEOPLASMS
        NEOPLASM METASTASIS, 4647
    BONE NEOPLASMS
        EPIDEMIOLOGY, 4659
    CONCANAVALIN A
        CELL TRANSFORMATION, NEOPLASTIC,
        4506
        NEURAMINIDASE, 4506
        TRYPSIN, 4506
    GASTROINTESTINAL NEOPLASMS
        EPIDEMIOLOGY, 4659
    HISTOCOMPATIBILITY ANTIGENS
        STATISTICAL ANALYSIS, 4798*
    NERVOUS SYSTEM NEOPLASMS
        EPIDEMIOLOGY, 4659
    OVARIAN NEOPLASMS
        EPIDEMIOLOGY, 4659
    PLANT AGGLUTININS
        CELL TRANSFORMATION, NEOPLASTIC,
        4506
    TESTICULAR NEOPLASMS
        EPIDEMIOLOGY, 4659
    VIRUS, EPSTEIN-BARR
        ANTIGENS, VIRAL, 4420*

BUTANE, 1,2:3,4-DIEPOXY-
    ANEMIA, APLASTIC
        CHROMOSOME ABERRATIONS, 4319*
    FIBROBLASTS

CARBOXY-LYASES
   CROTON OIL
      ENZYMATIC ACTIVITY, 4357*
   HEPATOMA
      DIURNAL OSCILLATION, 4772*
   HYDRAZINE
      ENZYMATIC ACTIVITY, 4329*
   12-O-TETRADECANOYLPHORBOL-13-ACETATE
      ENZYMATIC ACTIVITY, 4357*

CARCINOEMBRYONIC ANTIGEN
   BREAST NEOPLASMS
      PROTEINS, 4248*
   BRONCHIAL NEOPLASMS
      CARCINOMA, 4471
      SERUM LEVELS, 4775*
   BRONCHITIS
      CHRONIC DISEASE, 4471
   COLONIC NEOPLASMS
      ADENOCARCINOMA, 4474
      ANTIGEN-ANTIBODY REACTIONS, 4473
      ANTIGENIC DETERMINANTS, 4473
      BINDING SITES, CONCANAVALIN A,
         4473
      ISOLATION AND CHARACTERIZATION,
         4472, 4473
      NEOPLASM METASTASIS, 4472, 4473
      SPECTRUM ANALYSIS, 4473
   ESOPHAGEAL NEOPLASMS
      CARCINOMA, EPIDERMOID, 4474
   GASTROINTESTINAL NEOPLASMS
      CARCINOID TUMOR, 4534*
      PROTEINS, 4248*
      SERUM LEVELS, 4775*
   LIVER NEOPLASMS
      PROTEINS, 4248*
   LUNG NEOPLASMS
      CARCINOMA, EPIDERMOID, 4474
      PROTEINS, 4248*
   OVARIAN NEOPLASMS
      LYMPHOCYTE TRANSFORMATION, 4487
   PANCREATIC NEOPLASMS
      ADENOCARCINOMA, 4474
      THERAPY MONITORING, 4254*
   RECTAL NEOPLASMS
      CARCINOMA, EPIDERMOID, 4474
   SARCOMA
      VIRUS, ADENO 12, 4475
   STOMACH NEOPLASMS
      ADENOCARCINOMA, 4474
      METAPLASIA, 4474
      PROTEINS, 4248*
   UROGENITAL NEOPLASMS
      PROTEINS, 4248*
   UTERINE NEOPLASMS
      CARCINOMA, EPIDERMOID, 4474

CARCINOGEN, CHEMICAL
   DNA
      DOSE-RESPONSE STUDY, 4272
   NEOPLASMS
      EPIDEMIOLOGY, REVIEW, 4230*

CARCINOGEN, ENVIRONMENTAL
   ESOPHAGEAL NEOPLASMS
      EPIDEMIOLOGY, IRAN, 4665

CARCINOID
    SEE CARCINOID TUMOR

CARCINOID TUMOR
    BRONCHIAL NEOPLASMS
        ADENOMATOSIS, FAMILIAL ENDOCRINE,
        4606*
        MALIGNANT CARCINOID SYNDROME,
        4604*
        NEOPLASM METASTASIS, 4604*
    EPIDEMIOLOGY
        CHILD, 4651
    GASTROINTESTINAL NEOPLASMS
        CARCINOEMBRYONIC ANTIGEN, 4534*
    SKIN NEOPLASMS
        CASE REPORT, 4634*

CARCINOMA
    AUTOANTIBODIES
        FLUORESCENT ANTIBODY TECHNIC,
        4528*
    BRONCHIAL NEOPLASMS
        CARCINOEMBRYONIC ANTIGEN, 4471
    HEAD AND NECK NEOPLASMS
        VIRUS, HERPES SIMPLEX, 4214
    INTESTINAL NEOPLASMS
        METHANOL, (METHYLNITROSOAMINO)-,
        ACETATE (ESTER), 4350*
    LIP NEOPLASMS
        ULTRASTRUCTURAL STUDY, 4558
        VIRUS, HERPES SIMPLEX, 4214
    LIVER NEOPLASMS
        MONOCROTALINE, 4302
    MAMMARY NEOPLASMS, EXPERIMENTAL
        CASEIN, 4752*
        ETHYLENE, CHLORO-, 4312
    NASOPHARYNGEAL NEOPLASMS
        REVIEW, 4257*
    NEOPLASMS, SOFT TISSUE
        CASE REPORT, 4558
        ULTRASTRUCTURAL STUDY, 4558
    SKIN NEOPLASMS
        ETHYLENE, CHLORO-, 4312
    STOMACH NEOPLASMS
        CLASSIFICATION, 4565
        EPIDEMIOLOGY, EUROPE, 4649
        GUANIDINE, 1-METHYL-3-NITRO-1-
        NITROSO-, 4347*
        QUINOLINE, 4-NITRO-, 1-OXIDE,
        4347*
    THYROID NEOPLASMS
        ADRENAL MEDULLA, 4641*
        CLASSIFICATION, 4642*

CARCINOMA, ANAPLASTIC
    PROSTATIC NEOPLASMS
        EPIDEMIOLOGY, IRAN, 4670*
    THYROID NEOPLASMS
        CLASSIFICATION, 4642*

CARCINOMA, BASAL CELL
    NOSE NEOPLASMS
        QUINOXALINE, 1,4-DIOXIDE, 4356*

CARCINOMA, BRONCHIOLAR
    VIRUS-LIKE PARTICLES
        ANTIGENS, VIRAL, 4483
        IMMUNE SERUMS, 4483

CARCINOMA, PAPILLARY
    PROSTATIC NEOPLASMS
        EPIDEMIOLOGY, IRAN, 4670*
    THYROID NEOPLASMS
        CLASSIFICATION, 4642*
        NEOPLASM METASTASIS, 4653
        RADIATION, IONIZING, 4653

CARCINOMA, RENAL CELL
    SEE ADENOCARCINOMA

CARCINOMA, SMALL CELL, LUNG
    SEE CARCINOMA, OAT CELL

CARCINOMA, SQUAMOUS CELL
    SEE CARCINOMA, EPIDERMOID

CARCINOMA, TRANSITIONAL CELL
    BLADDER NEOPLASMS
        CELL MEMBRANE, 4747*
        RADIATION, IONIZING, 4364

CARRAGEEN
    GAMMA GLOBULINS
        IMMUNE RESPONSE, 4503

CARRAGEENEN
    SEE CARRAGEEN

CARRIER PROTEINS
    RETINOIC ACID
        CHICK EMBRYO, 4750*
        ISOLATION AND CHARACTERIZATION,
          4750*
    VIRUS, ADENO 12
        DNA, 4379

CASEIN
    MAMMARY NEOPLASMS, EXPERIMENTAL
        CARCINOMA, 4752*
        CARCINOMA, DUCTAL, 4752*
        PRECANCEROUS CONDITIONS, 4752*

CATECHOLAMINES
    ADRENAL GLAND NEOPLASMS
        PHEOCHROMOCYTOMA, 4327*
    ASTROCYTOMA
        ADENOSINE CYCLIC 3,5 MONOPHOSPHATE,
          4791*
        ADENYL CYCLASE, 4791*

CELL ADHESION
    VIRUS, SV40
        CELL MEMBRANE, 4212
        MUCOPOLYSACCHARIDES, 4212

CELL AGGREGATION
    FIBROBLASTS
        CELL TRANSFORMATION, NEOPLASTIC,
          4800*
    VIRUS, HAMSTER SARCOMA
        GLYCOLIPIDS, 4710

CELL DIFFERENTIATION
    SEE ALSO GROWTH

ANTIGENS
    CELL MEMBRANE, 4477
ERYTHROLEUKEMIA
    BUTYRIC ACID, 4702
LEUKEMIA, MYELOCYTIC
    GRANULOCYTES, 4686
    GROWTH SUBSTANCES, 4686
T-LYMPHOCYTES
    GRAFT VS HOST REACTION, 4494
    THYMUS HORMONE, 4494
LYMPHOMA
    B-LYMPHOCYTES, 4491
    T-LYMPHOCYTES, 4491
NEPHROBLASTOMA
    DIAGNOSIS AND PROGNOSIS, 4596
    HISTOLOGICAL STUDY, 4596
NERVOUS SYSTEM NEOPLASMS
    ANIMAL MODEL, MOUSE, 4732*
TERATOID TUMOR
    EMBRYO, 4600
    REVIEW, 4261*
TESTICULAR NEOPLASMS
    TERATOID TUMOR, 4732*

CELL DIVISION
    SEE ALSO GROWTH
    DIETHYLAMINE, N-NITROSO-
        LIVER, RAT, 4283
    DNA REPLICATION
        CHALONES, 4229
        MITOTIC INHIBITORS, ENDOGENOUS,
            4229
    FIBROBLASTS
        RADIATION, IONIZING, 4366
        REVIEW, 4226
    HELA CELLS
        REVIEW, 4226
    LEUKEMIA
        ACUTE DISEASE, 4590
    LEUKEMIA, MYELOCYTIC
        CHROMOSOMES, HUMAN, 21-22, 4583

CELL FUSION
    CELL TRANSFORMATION, NEOPLASTIC
        BENZ(A)ANTHRACENE, 7,12-DIMETHYL-,
            4339*

CELL LINE
    NEOPLASMS
        ISOLATION AND CHARACTERIZATION,
            4730*
    TRANSFERASES
        ISOLATION AND CHARACTERIZATION,
            4760*
        TUMOR-CELL PROTECTION, 4760*

CELL MEMBRANE
    ANTIGENS
        CELL DIFFERENTIATION, 4477
        FIBROBLASTS, 4477
    BLADDER NEOPLASMS
        CARCINOMA, TRANSITIONAL CELL,
            4747*
    BONE MARROW CELLS
        CHALONES, 4733*
    CELL TRANSFORMATION, NEOPLASTIC
        GLUTAMYL TRANSPEPTIDASE, 4788*
    HEPATOMA
        ANTIGENS, NEOPLASM, 4484

LYMPHOCYTES
    BLASTOID ACTIVATION, 4505
    CONCANAVALIN A, 4505
    PLANT AGGLUTININS, 4505
B-LYMPHOCYTES
    ANTIGENIC DETERMINANTS, 4488
T-LYMPHOCYTES
    ANTIGENIC DETERMINANTS, 4488
LYMPHOSARCOMA
    HISTOCOMPATIBILITY ANTIGENS, 4459
MELANOMA
    FREEZE-ETCH STUDY, NUCLEAR PORES,
        4593
PLASMACYTOMA
    RIBOSOMES, 4699
SARCOMA
    FREEZE-ETCH STUDY, NUCLEAR PORES,
        4593
VIRUS, HAMSTER SARCOMA
    GLYCOLIPIDS, 4710
VIRUS, SV40
    CELL ADHESION, 4212
    CELL TRANSFORMATION, NEOPLASTIC,
        4413*
    CONCANAVALIN A, 4212, 4398
    GLYCOLIPIDS, 4212
    GLYCOPROTEINS, 4212
    HEPARIN, 4212
    MUCOPOLYSACCHARIDES, 4212
    PEPTIDE HYDROLASES, 4212
    PHOSPHOLIPIDS, 4212, 4398
    REVIEW, MOUSE, 4212
    ULTRASTRUCTURAL STUDY, 4212

CELL MIGRATION INHIBITION
    SARCOMA
        LEUKOCYTES, 4522*

CELL NUCLEUS
    LEUCOSARCOMA
        CASE REPORT, 4592
        ULTRASTRUCTURAL STUDY, 4592
    METHIONINE
        CELL CYCLE KINETICS, 4688
        TEMPERATURE-SENSITIVE CLONES, 4688
    TRYPTOPHAN
        CELL CYCLE KINETICS, 4688
        TEMPERATURE-SENSITIVE CLONES, 4688

CELL SEPARATION
    CELL TRANSFORMATION, NEOPLASTIC
        CYTOTOXICITY TESTS, 4737*
        SEDIMENTATION VELOCITY, 4737*

CELL TRANSFORMATION, NEOPLASTIC
    ADRENAL GLAND NEOPLASMS
        VIRUS, KIRSTEN MURINE SARCOMA,
            4409*
    BENZ(A)ANTHRACENE, 7,12-DIMETHYL-
        ACETYLTRANSFERASES, 4741*
        CELL FUSION, 4339*
    BREAST NEOPLASMS
        MATHEMATICAL MODELS, EVALUATION,
            4666
    BURKITT'S LYMPHOMA
        CONCANAVALIN A, 4506
        PLANT AGGLUTININS, 4506
    CARBAMIC ACID, ETHYL ESTER
        CELL CYCLE DEPENDENCE, 4282

ONCOGENIC VIRUSES, REVIEW, 4239*
5-TUNGSTO-2-ANTIMONIATE, 4407*
VIRUS, FELINE LEUKEMIA
REVIEW, 4240*
VIRUS, FELINE SARCOMA
REVIEW, 4240*
VIRUS, HERPES SAIMIRI
ANTIGEN-ANTIBODY REACTIONS, 4536*
LYMPHOCYTES, 4403*
ONCOGENIC VIRUSES, REVIEW, 4239*
5-TUNGSTO-2-ANTIMONIATE, 4407*
VIRUS INHIBITORS, 4403*
VIRUS, HERPES SIMPLEX
PHOSPHONOACETIC ACID, 4406*
REVIEW, 4214
VIRUS, HERPES SIMPLEX 1
ONCOGENIC VIRUSES, REVIEW, 4239*
VIRUS, HERPES SIMPLEX 2
ONCOGENIC VIRUSES, REVIEW, 4239*
VIRUS, POLYOMA SV40 HYBRID
RNA, MESSENGER, 4697
VIRUS, ROUS SARCOMA
ADENYL CYCLASE, 4375
PHOSPHODIESTERASES, 4375
VIRUS, SIMIAN SARCOMA
REVERSE TRANSCRIPTASE, 4410*
RNA POLYMERASE, 4410*
VIRUS, SV40
CELL MEMBRANE, 4413*
IMMUNOSUPPRESSION, 4419*
ISOLATION AND CHARACTERIZATION,
4418*
PHOSPHODIESTERASES, 4712
RNA, MESSENGER, 4691
VIRUS REPLICATION, 4419*

CELLULAR INCLUSIONS
AGRANULOCYTOSIS
LYMPHOCYTES, 4601*
INFECTIOUS MONONUCLEOSIS
LYMPHOCYTES, 4601*
LEUKEMIA
ACUTE DISEASE, 4601*
LEUKEMIA, LYMPHOCYTIC
LYMPHOCYTES, 4601*
MENINGIOMA
HYALINE, 4602*
ULTRASTRUCTURAL STUDY, 4602*
MUCOPOLYSACCHARIDOSIS
LYMPHOCYTES, 4601*

CEREBRELLAR NEOPLASMS
ASTROCYTOMA
EPIDEMIOLOGY, RHODESIA, 4648
CRANIOPHARYNGIOMA
EPIDEMIOLOGY, RHODESIA, 4648
EPENDYMOMA
EPIDEMIOLOGY, RHODESIA, 4648
GLIOBASTOMA MULTIFORME
EPIDEMIOLOGY, RHODESIA, 4648
MEDULLOBLASTOMA
EPIDEMIOLOGY, RHODESIA, 4648
MENINGIOMA
EPIDEMIOLOGY, RHODESIA, 4648
NEURILEMMOMA
EPIDEMIOLOGY, RHODESIA, 4648

CEREBROSPINAL FLUID
LEUKEMIA, LYMPHOCYTIC

IMMUNOGLOBULINS, 4628*

CERVIX NEOPLASMS
    ADENOCARCINOMA
        MULTIFOCAL PRIMARY CARCINOMA, 4569
    AGE FACTORS
        EPIDEMIOLOGY, 4674*
    CARCINOMA, EPIDERMOID
        CONTRACEPTIVES, ORAL, 4306
        MULTIFOCAL PRIMARY CARCINOMA, 4569
        RADIATION, IONIZING, 4364
    CARCINOMA IN SITU, 4569
        EPIDEMIOLOGY, 4674*
        ULTRASTRUCTURAL STUDY, 4603*
    CARCINOMA, MUCINOUS
        MULTIFOCAL PRIMARY CARCINOMA, 4569
    CELL TRANSFORMATION, NEOPLASTIC
        MATHEMATICAL MODELS, EVALUATION,
            4666
    NEOPLASM METASTASIS
        EPIDEMIOLOGY, 4674*
    PRECANCEROUS CONDITIONS
        CONTRACEPTIVES, ORAL, 4306
        DIAGNOSIS AND PROGNOSIS, REVIEW,
            4223
        EPIDEMIOLOGY, 4674*
        ULTRASTRUCTURAL STUDY, 4603*
    PREGNANCY, MULTIPLE
        EPIDEMIOLOGY, 4674*
    TRANSPLANTATION, HOMOLOGOUS
        IMMUNOSUPPRESSION, 4515*
    UREMIA
        IMMUNOSUPPRESSION, 4515*
    VIRUS, HERPES SIMPLEX 2
        SEROEPIDEMIOLOGIC STUDIES, REVIEW,
            4213

STOMACH NEOPLASMS
    PANCREATIC NEOPLASMS
        ADENOCARCINOMA, 4474

CHALONES
    BONE MARROW CELLS
        CELL MEMBRANE, 4733*
    CELL DIVISION
        DNA REPLICATION, 4229
    DNA REPLICATION
        CELL CYCLE KINETICS, 4734*
        EPIDERMAL CELLS, 4734*
        LIVER REGENERATION, 4731*
        REVIEW, 4734*
    HEPATOMA
        ANILINE, N,N-DIMETHYL-P-PHENYLAZO-,
            4731*
        DNA REPLICATION, 4731*
    LIVER
        ISOLATION AND CHARACTERIZATION,
            RABBIT, 4731*
    PROTEINS
        LIVER REGENERATION, 4731*
    RNA
        LIVER REGENERATION, 4731*

CHEMODECTOMA
    SEE PARAGANGLIOMA, NONCHROMAFFIN

CHLORINE
    WATER POLLUTATNTS
        REVIEW, 4206

```
                         CLONE CELLS, 4736*
                    POLYCYTHEMIA VERA
                         1,4-BUTANEDIOL, DIMETHANESULFONATE,
                         4584
(3-                      CYCLOPHOSPHAMIDE, 4584
                         KARYOTYPING, 4584
                         RADIATION, IONIZING, 4584
                    UTERINE NEOPLASMS
                         CARCINOMA, EPIDERMOID, 4612*


               CHROMOSOME MAPPING
                    ENZYMES
                         HYBRIDIZATION, 4224
                         REVIEW, 4224


               CHROMOSOMES
9*                  BREAST NEOPLASMS
                         ESTROGENS, 4792*
                    HELA CELLS
                         ULTRASTRUCTURAL STUDY, 4580
                    LEUKEMIA, MYELOCYTIC
                         ACUTE TRANSFORMATION, 4588
                    VIRUS, SV40
                         SITE SPECIFICITY, 4415*
9*

               CHROMOSOMES, HUMAN, 13-15
                    MULTIPLE MYELOMA
                         CHROMOSOME ABNORMALITIES, 4629*


               CHROMOSOMES, HUMAN, 21-22
                    LEUKEMIA, MYELOCYTIC
27*                      ACUTE DISEASE, 4626*
                         1,4-BUTANEDIOL DIMETHYLSULFONATE,
                         4583
                         CELL DIVISION, 4583
                         CHROMOSOME ABERRATIONS, 4627*
                         MOSAICISM, INFANT, 4586
                         SPLEEN, 4583
R                   NEUROBLASTOMA
                         GENETICS, 4581


               CHRONIC DISEASE
                    BRONCHITIS
                         CARCINOEMBRYONIC ANTIGEN, 4471


               CHRYSOTILE
                    SEE ASBESTOS
9*

               CIRCADIAN RHYTHM
                    LEUKEMIA, LYMPHOCYTIC
                         ENZYMES, 4778*
                    LEUKEMIA, MYELOCYTIC
                         ENZYMES, 4778*


               CITRATES
                    HEPATOMA
                         DIET, 4768*


               CLONE CELLS
29*                 HEPATOMA
                         FETAL GLOBULINS, 4683
                    PLANT TUMORS
                         CHROMOSOME ABNORMALITIES, 4736*
```

COLCHICINE
    L CELLS
        NERVE TISSUE PROTEINS, 4690
    VIRUS, KIRSTEN MURINE SARCOMA
        NERVE TISSUE PROTEINS, 4690
    VIRUS, MOLONEY MURINE SARCOMA
        NERVE TISSUE PROTEINS, 4690
    VIRUS, ROUS SARCOMA
        NERVE TISSUE PROTEINS, 4690

COLITIS
    HODGKIN'S DISEASE
        EPIDEMIOLOGY, 4660

COLITIS, ULCERATIVE
    LEUKOCYTES
        ALKALINE PHOSPHATASE, 4715

COLONIC NEOPLASMS
    ADENOCARCINOMA
        ADENOSINE CYCLIC 3',5'
          MONOPHOSPHATE, 4744*
        ADENOSINE TRIPHOSPHATE, 4744*
        ADENYL CYCLASE, 4744*
        BACILLUS CALMETTE-GUERIN, 4293
        CARCINOEMBRYONIC ANTIGEN, 4474
        GLYCOLIPIDS, 4787*
        HYDRAZINE, 1,2-DIMETHYL-, 4330*
        MECKEL'S DIVERTICULUM, 4616*
        PROTEIN-LOSING ENTEROPATHIES,
          4615*
        RADIATION, IONIZING, 4364
    ADENOMA
        REVIEW, 4256*
    CARCINOEMBRYONIC ANTIGEN
        ANTIGEN-ANTIBODY REACTIONS, 4473
        ANTIGENIC DETERMINANTS, 4473
        BINDING SITES, CONCANAVALIN A,
          4473
        ISOLATION AND CHARACTERIZATION,
          4472, 4473
        NEOPLASM METASTASIS, 4472, 4473
        SPECTRUM ANALYSIS, 4473
    HYDRAZINE, 1,2-DIMETHYL-
        BACILLUS CALMETTE-GUERIN, 4293
    METHANOL, (METHYL-ONN-AZOXY)-, ACETATE
      (ESTER)
        SITE SPECIFICITY, 4341*
    POLYPS
        METHANOL, (METHYLNITROSOAMINO)-,
          ACETATE (ESTER), 4350*
    PRECANCEROUS CONDITIONS
        REVIEW, 4256*
    PROTEIN-LOSING ENTEROPATHIES
        CASE REPORT, 4615*
    TRANSPLANTATION, HETEROLOGOUS
        GROWTH, 4740*
        IMMUNOSUPPRESSION, 4740*
    TRANSPLANTATION, HOMOLOGOUS
        IMMUNOSUPPRESSION, 4515*

COMPLEMENT
    12-O-TETRADECANOYLPHORBOL-13-ACETATE
        LYSOSOMES, 4714

CONCANAVALIN A
    BURKITT'S LYMPHOMA
        CELL TRANSFORMATION, NEOPLASTIC,
          4506

ARYL HYDROCARBON HYDROXYLASES,
4278
BENZO(A)PYRENE
ARYL HYDROCARBON HYDROXYLASES,
4278
5,6-BENZOFLAVONE
ARYL HYDROCARBON HYDROXYLASES,
4278
7,8-BENZOFLAVONE
ARYL HYDROCARBON HYDROXYLASES,
4278
BENZOTHIAZOLE, 2-(P-CHLOROPHENYL)-
ARYL HYDROCARBON HYDROXYLASES,
4278
CHOLANTHRENE, 3-METHYL-
ARYL HYDROCARBON HYDROXYLASES,
4278
ISOPROTERENOL
ARYL HYDROCARBON HYDROXYLASES,
4278
OXAZOLE, 2,5-DIPHENYL-
ARYL HYDROCARBON HYDROXYLASES,
4278

CYTOSINE
LEUKEMIA, MYELOCYTIC
ACUTE TRANSFORMATION, 4588

CYTOSINE, 1-BETA-D-ARABINOFURANOSYL-
ERYTHROLEUKEMIA
BUTYRIC ACID, 4702
B-LYMPHOCYTES
ORNITHINE DECARBOXYLASE, 4453

CYTOSINE NUCLEOTIDES
HELA CELLS
DNA REPLICATION, 4706

CYTOXAN
SEE CYCLOPHOSPHAMIDE

DACTINOMYCIN
SEE ACTINOMYCIN D

DAUNOMYCIN
CARCINOMA, EHRLICH TUMOR
REVERSE TRANSCRIPTASE, 4716

DEOXYIODOURIDINE
SEE URIDINE, 2'-DEOXY-5-IODO-

DEOXYRIBONUCLEOSIDES
VIRUS, HERPES SIMPLEX 1
THYMIDINE KINASE, 4382
VIRUS, HERPES SIMPLEX 2
THYMIDINE KINASE, 4382

DEOXYRIBONUCLEOTIDES
HELA CELLS
DNA REPLICATION, 4706
VIRUS, HERPES SIMPLEX 1
THYMIDINE KINASE, 4382
VIRUS, HERPES SIMPLEX 2
THYMIDINE KINASE, 4382

DERMATOFIBROMA
SEE FIBROMA

DEXAMETHASONE

19

VIRUS, MURINE MAMMARY TUMOR
    REVERSE TRANSCRIPTASE, 4423*,
        4426*

DIAPHRAGMATIC HERNIA
    ESOPHAGEAL NEOPLASMS
        ADENOCARCINOMA, 4614*
    STOMACH NEOPLASMS
        ADENOCARCINOMA, 4614*

DIBUTYRYL CYCLIC AMP
    VIRUS, SV40
        PHOSPHODIESTERASES, 4712

DIELDRIN
    ENVIRONMENTAL HAZARD
        CARCINOGENIC ACTIVITY, REVIEW,
            4230*
    PESTICIDES
        CARCINOGENIC ACTIVITY, REVIEW,
            4230*

DIET
    HEPATOMA
        CITRATES, 4768*
        FATTY ACIDS, 4768*
        LIGASES, 4768*
        OXIDOREDUCTASES, 4768*
    PROSTATIC NEOPLASMS
        EPIDEMIOLOGY, JAPAN, 4265*

DIETHYLAMINE
    NITROUS ACID, SODIUM SALT
        COCARCINOGENIC EFFECT, GUINEA PIG,
            4284
        DIMETHYLAMINE, N-NITROSO-, 4284

DIETHYLAMINE, N-NITROSO-
    AMINOPYRINE N-DEMETHYLASE
        LIVER, RAT, 4283
    BENZ(A)ANTHRACENE, 7,12-DIMETHYL-
        NUCLEIC ACIDS, 4271
        SITE-SPECIFICITY, LIVER, 4271
    CELL DIVISION
        LIVER, RAT, 4283
    CELL TRANSFORMATION, NEOPLASTIC
        CELL CYCLE DEPENDENCE, 4282
    ESOPHAGEAL NEOPLASMS
        EPIDEMIOLOGY, IRAN, 4665
    HEPATOMA
        ANTIGENS, NEOPLASM, 4484, 4533*
        TUMOR-TETANUS ASSAY, 4315*
    HYDROXYLASES
        LIVER, RAT, 4283
    METHANOL, (METHYLNITROSOAMINO)-
        CARCINOGENIC METABOLITE, 4350*
    STEROID HYDROXYLASES
        LIVER, RAT, 4283

DIETHYLNITROSAMINE
    SEE DIETHYLAMINE, N-NITROSO-

DIETHYLSTILBESTROL
    SEE 4,4'-STILBENEDIOL, ALPHA, ALPHA'-
        DIETHYL-

DIGESTIVE SYSTEM NEOPLASMS
    CARCINOMA, MUCINOUS
        CHROMOSOME ABNORMALITIES, 4612*

RETICS
NEOPLASMS
EPIDEMIOLOGY, CHILD, 4654

ACETAMIDE, N-FLUOREN-2-YL-
ACETOHYDROXAMIC ACID, N-FLUOREN-2-
YL-, 4325*
BINDING, 4323*
ADENOCARCINOMA
REVERSE TRANSCRIPTASE, 4716
BENZO(A)PYRENE-4,5-DIOL
BINDING, 4337*
BENZO(A)PYRENE-7,8-DIOL, 7,8-DIHYDRO-
BINDING, 4337*
BRONCHIAL NEOPLASMS
PRECANCEROUS CONDITIONS, 4605*
CARCINOGEN, CHEMICAL
DOSE-RESPONSE STUDY, 4272
CARCINOMA, EHRLICH TUMOR
DNA-RNA HYBRIDIZATION, 4705
REVERSE TRANSCRIPTASE, 4716
ETHER, BIS(CHLOROMETHYL)-
ALKYLATING AGENTS, 4321*
HEPATOMA
ACTINOMYCIN D, 4707
LEUKEMIA
BONE MARROW CELLS, 4754*
LEUKOCYTES, 4708
LYMPHATIC NEOPLASMS
ANTIGENIC DETERMINANTS, 4545*
LYMPHOSARCOMA
REVERSE TRANSCRIPTASE, 4716
MUTAGENS
DOSE-RESPONSE STUDY, 4272
MUTATION
PHOTOTHERAPY, 4369*
NEOPLASMS
LIPOPEROXIDE, 4757*
2-OXETANONE
ALKYLATING AGENTS, 4321*
VIRUS, ADENO 12
CARRIER PROTEINS, 4379
VIRUS, EPSTEIN-BARR
CELL TRANSFORMATION, NEOPLASTIC,
4226
VIRUS, PAPOVA, BK
ISOLATION AND CHARACTERIZATION,
4392
POLYNUCLEOTIDES, 4392
VIRUS, SV40
ISOLATION AND CHARACTERIZATION,
4392
POLYNUCLEOTIDES, 4392

, CIRCULAR
VIRUS, MOLONEY MURINE LEUKEMIA
ISOLATION AND CHARACTERIZATION,
4384

NUCLEOTIDYLTRANSFERASES
SEE ALSO DNA POLYMERASE
LEUKEMIA, LYMPHOBLASTIC
ISOLATION AND CHARACTERIZATION,
4717
LEUKEMIA, MYELOCYTIC
ISOLATION AND CHARACTERIZATION,
4717
LYMPHATIC NEOPLASMS

SERUM INHIBITORY FACTOR, 4774*
MULTIPLE MYELOMA
SERUM INHIBITORY FACTOR, 4774*
PHOSPHONOACETIC ACID
ENZYMATIC ACTIVITY, 4406*
5-TUNGSTO-2-ANTIMONIATE
ENZYMATIC ACTIVITY, 4407*

DNA POLYMERASE
SEE ALSO DNA NUCLEOTIDYLTRANSFERASES
VIRUS, AVIAN MYELOBLASTOSIS
ISOLATION AND CHARACTERIZATION,
4377

DNA REPAIR
CARCINOMA, EHRLICH TUMOR
CELL CYCLE KINETICS, 4342*
METHOTREXATE, 4342*
DNA REPLICATION
CELL TRANSFORMATION, NEOPLASTIC,
4228
ESCHERICHIA COLI
ULTRAVIOLET RAYS, 4365
FIBROBLASTS
CARCINOGENIC POTENTIAL, 4273
GUANIDINE, 1-METHYL-3-NITRO-1-NITROSO-
GUANINE, 4346*
ULTRAVIOLET RAYS
AMINO ACIDS, 4365
REVIEW, 4228
THYMINE, 4365
XERODERMA PIGMENTOSUM
GENETICS, 4228
HYBRIDIZATION, 4228
RADIATION, 4228
REVIEW, 4228
ULTRAVIOLET RAYS, 4228

DNA REPLICATION
CARCINOMA, EHRLICH TUMOR
SUBCELLULAR FRACTIONS, 4786*
CELL DIVISION
CHALONES, 4229
MITOTIC INHIBITORS, ENDOGENOUS,
4229
CHALONES
CELL CYCLE KINETICS, 4734*
EPIDERMAL CELLS, 4734*
LIVER REGENERATION, 4731*
REVIEW, 4734*
DNA REPAIR
CELL TRANSFORMATION, NEOPLASTIC,
4228
GLUCOSE, 2-DEOXY-
CHICK EMBRYO CELLS, 4761*
HELA CELLS
ADENOSINE TRIPHOSPHATE, 4706
ALPHA-AMANITINE, 4706
CELL CYCLE KINETICS, 4679*
CHROMATIN, 4693
CYTOSINE NUCLEOTIDES, 4706
DEOXYRIBONUCLEOTIDES, 4706
GUANOSINE TRIPHOSPHATE, 4706
MAGNESIUM, 4706
RIBONUCLEASE, 4706
SPERMIDINE, 4706
SPERMINE, 4706
VIRUS, ADENO 2, 4378
VIRUS, VACCINIA, 4402

HEPATOMA
    CHALONES, 4731*
LEUKEMIA
    PROTEIN CONFORMATION, 4227
B-LYMPHOCYTES
    ANTI-ANTIBODIES, 4453
PROPIONIC ACID, 2-AMINO-2-METHYL-
    CHICK EMBRYO CELLS, 4761*
REVERSE TRANSCRIPTASE
    REVIEW, 4243*
SKIN NEOPLASMS
    12-O-TETRADECANOYLPHORBOL-13-
        ACETATE, 4358*
ULTRAVIOLET RAYS
    REVIEW, 4228
VIRUS, ADENO 2
    VIRUS, HELPER, 4393
VIRUS, AVIAN RNA SARCOMA
    CELL TRANSFORMATION, NEOPLASTIC,
        4374
    ETHIDIUM BROMIDE, 4374
VIRUS, B77
    CELL TRANSFORMATION, NEOPLASTIC,
        4374
    ETHIDIUM BROMIDE, 4374
VIRUS, FROG 3
    VIRAL PROTEINS, 4381
VIRUS, MOLONEY MURINE LEUKEMIA
    BONE MARROW CELLS, MOUSE, 4384
VIRUS, SV40
    TEMPERATURE SENSITIVE MUTANTS,
        4399
    TRANSLATION, GENETIC, 4414*
ZINC
    CHICK EMBRYO CELLS, 4761*

DNA, SATELLITE
LEUKEMIA
    ISOLATION AND CHARACTERIZATION,
        4227

DNA, VIRAL
LEUKEMIA, MYELOBLASTIC
    VIRUS, C-TYPE, 4245*
VIRIS, PAPOVA
    REVIEW, 4241*
VIRUS, ADENO 7
    URIDINE, 2'-DEOXY-5-IODO-, 4395
VIRUS, MOLONEY MURINE LEUKEMIA
    ISOLATION AND CHARACTERIZATION,
        4384
VIRUS, POLYOMA, BK
    REVIEW, 4241*
VIRUS, SV40
    ENDONUCLEASES, 4397
    ISOLATION AND CHARACTERIZATION,
        4418*
    RADIATION, 4397
    TEMPERATURE SENSITIVE MUTANTS,
        4399
    ULTRASTRUCTURAL STUDY, 4397, 4416*
    VIRUS REPLICATION, 4399
VIRUS, VACCINIA
    MITOMYCIN C, 4402

DOPAMINE-B-HYDROXYLASE
NEUROBLASTOMA
    CELL CYCLE KINETICS, 4780*

MUTAGENIC ACTIVITY, 4344*
ULTRAVIOLET RAYS
DNA REPAIR, 4365
UREA, METHYL NITROSO-
MUTAGENIC ACTIVITY, 4344*

ESOPHAGEAL NEOPLASMS
ADENOCARCINOMA
DIAPHRAGMATIC HERNIA, 4614*
EARYL DIAGNOSIS, ENDOSCOPY, 4579
ANILINE, N-METHYL-N-NITROSO-
ANIMAL MODEL, RAT, 4550*
CARCINOGEN, ENVIRONMENTAL
EPIDEMIOLOGY, IRAN, 4665
CARCINOMA, EPIDERMOID
CARCINOEMBRYONIC ANTIGEN, 4474
DIETHYLAMINE, N-NITROSO-
EPIDEMIOLOGY, IRAN, 4665
DIMETHYLAMINE, N-NITROSO-
EPIDEMIOLOGY, IRAN, 4665
ESOPHAGITIS
PRECANCEROUS CONDITIONS, 4579

ESOPHAGITIS
ESOPHAGEAL NEOPLASMS
PRECANCEROUS CONDITIONS, 4579.

ESTERASES
INTESTINAL NEOPLASMS
HISTOCHEMICAL STUDY, 4770*
PROSTATIC NEOPLASMS
ANTIBODIES, 4476

ESTRADIOL
BREAST NEOPLASMS
ADIPOSE TISSUE, 4793*
NEOPLASM METASTASIS, 4793*
URINE, 4570
MAMMARY NEOPLASMS, EXPERIMENTAL
BENZ(A)ANTHRACENE, 7,12-DIMETHYL-,
4202
CHROMATIN, 4202
ESTROGEN RECEPTORS, REVIEW, 4202
RNA, TRANSFER, METHYLTRANSFERASES
UTERUS, RAT, 4701

17-BETA ESTRADIOL
SEE ESTRADIOL

ESTRIOL
BREAST NEOPLASMS
URINE, 4570

ESTROGENS
ARYL HYDROCARBON HYDROXYLASES
INHIBITION, LIVER, 4277
BREAST NEOPLASMS
CHROMOSOMES, 4792*
RECEPTORS, HORMONE, 4792*
TRANSFERASES, 4719
MAMMARY NEOPLASMS, EXPERIMENTAL
TRANSFERASES, 4719
UTERINE NEOPLASMS
EPIDEMIOLOGY, 4673*

ESTRONE
BREAST NEOPLASMS
URINE, 4570
MAMMARY NEOPLASMS, EXPERIMENTAL

GLUCOSEPHOSPHATE DEHYDROGENASE,
    4303
LACTOSE SYNTHETASE, 4303

ETHANOL, 2-MERCAPTO-
    FETAL GLOBULINS
        IMMUNOLOGICAL AND ELECTROPHORETIC
            PROPERTIES, 4465

ETHER, BIS(CHLOROMETHYL)-
    DNA
        ALKYLATING AGENTS, 4321*

ETHIDIUM BROMIDE
    SACCHAROMYCES CEREVISIAE
        MUTAGENIC ACTIVITY, 4322*
    ULTRAVIOLET RAYS
        MUTAGENIC ACTIVITY, 4322*
    VIRUS, AVIAN RNA SARCOMA
        DNA REPLICATION, 4374
    VIRUS, B77
        DNA REPLICATION, 4374

ETHIONINE
    SEE BUTYRIC ACID, 2-AMINO-4-(ETHYLTHIO)-

ETHYL CARBAMATE
    SEE CARBAMIC ACID, ETHYL ESTER

ETHYL METHANE SULFONATE
    SEE METHANESULFONIC ACID, ETHYL ESTER

N-ETHYL-N-NITROSOUREA
    SEE UREA, ETHYL NITROSO-

1-ETHYL-1-NITROSOUREA
    SEE UREA, ETHYL NITROSO-

ETHYLENE, CHLORO-
    AEROSOLS
        EXPOSURE LEVELS, 4310
    ANGIOSARCOMA
        OCCUPATIONAL HAZARD, REVIEW, 4204
    BRAIN NEOPLASMS
        NEUROBLASTOMA, 4311, 4312
    ETHYLENE, 1,1-DICHLORO-
        COCARCINOGENIC EFFECT, 4313
    HEAD AND NECK NEOPLASMS
        ANGIOSARCOMA, 4311
    KIDNEY NEOPLASMS
        NEPHROBLASTOMA, 4311, 4312
    LIP NEOPLASMS
        ANGIOSARCOMA, 4311
    LIVER NEOPLASMS
        ANGIOMA, 4311
        ANGIOSARCOMA, 4204, 4311, 4312,
            4313
        HEMANGIOMA, 4312
    LUNG NEOPLASMS
        CARCINOGENIC ACTIVITY, RAT, 4204
    MAMMARY NEOPLASMS, EXPERIMENTAL
        ADENOFIBROMA, 4311, 4312
        CARCINOMA, 4312
    PITUITARY NEOPLASMS
        ADENOMA, 4311
    PULMONARY NEOPLASMS
        ADENOMA, 4312
    SALMONELLA TYPHIMURIUM

ULTRAVIOLET RAYS, 4362
MOUTH NEOPLASMS
RADIATION, IONIZING, 4364
NEOPLASM TRANSPLANTATION
CHOLANTHRENE, 3-METHYL-, 4509*
IMMUNE SERUMS, 4509*
NOSE NEOPLASMS
QUINOXALINE, 1,4-DIOXIDE, 4356*
PHARYNGEAL NEOPLASMS
RADIATION, IONIZING, 4364
SKIN NEOPLASMS
EPIDEMIOLOGY, MOUSE, 4723
ULTRAVIOLET RAYS, 4363

FLUORENYL ACETAMIDE
SEE ACETAMIDE, N-FLUOREN-2-YL-

FOLLICLE-STIMULATING HORMONE
SEE FSH

FOOD ADDITIVES
NITROUS ACID, SODIUM SALT
CARCINOGENIC ACTIVITY, 4349*
MICROBIOLOGY, 4352*
REVIEW, 4353*

FOOD CONTAMINATION
AFLATOXIN M1
CORN, 4677*
MALONALDEHYDE
MEAT, 4335*

FOOD PRESERVATIVES
ACRYLAMIDE, 2-(2-FURYL)-3-(5-NITRO-2-
FURYL)-
MUTAGENIC EFFECT, MOUSE, 4270

FORMALDEHYDE
MOUTH NEOPLASMS
CARCINOMA, EPIDERMOID, 4556

FORSSMAN ANTIGEN
LEUKEMIA
ANTIBODIES, 4530*
LYMPHOMA
ANTIBODIES, 4530*
LYMPHOSARCOMA
ANTIBODIES, 4530*
SARCOMA
ANTIBODIES, 4530*

FRUCTOSEDIPHOSPHATE ALDOLASE
ACETAMIDE, N-FLUOREN-2-YL-
ISOENZYMES, 4274
LIVER NEOPLASMS
ACETAMIDE, N-FLUOREN-2-YL-, 4274.

FUNGICIDES, INDUSTRIAL
HORDEUM VULGARE
CHROMOSOME ABERRATIONS, 4291
PLANTS
CHROMOSOME ABERRATIONS, 4291
VICIA FABA
CHROMOSOME ABERRATIONS, 4291

7H-FURO(3,2-G)(1)BENZOPYRAN-7-ONE, 9-
METHOXY-
ANTHRACENE
COCARCINOGENIC EFFECT, 4370*

* INDICATES A PLAIN CITATION

SKIN NEOPLASMS
    ANTHRACENE, 4370*

FURYLFURAMIDE
    SEE ACRYLAMIDE, 2-(2-FURYL)-3-(5-NITRO-
    2-FURYL)-

GALACTOSIDASES
    BREAST NEOPLASMS
        HISTOCHEMICAL STUDY, 4770*
    INTESTINAL NEOPLASMS
        HISTOCHEMICAL STUDY, 4770*

GAMMA GLOBULINS
    CARRAGEEN
        IMMUNE RESPONSE, 4503
    RADIATION, IONIZING
        IMMUNE RESPONSE, 4503

GASTRIC MUCOSA
    STOMACH NEOPLASMS
        PRECANCEROUS CONDITIONS, 4565

GASTRIN
    STOMACH NEOPLASMS
        ISOLATION AND CHARACTERIZATION,
        4722
    ZOLLINGER-ELLISON SYNDROME
        ISOLATION AND CHARACTERIZATION,
        4722

GASTROINTESTINAL NEOPLASMS
    ADENOCARCINOMA
        CASE REPORT, 4618*
        ENTERITIS, REGIONAL, 4618*
    ALPHA GLOBULINS
        SERUM LEVEL, 4461
    ASBESTOS
        OCCUPATIONAL HAZARD, 4262*
    BURKITT'S LYMPHOMA
        EPIDEMIOLOGY, 4659
    CARCINOEMBRYONIC ANTIGEN
        PROTEINS, 4248*
        SERUM LEVELS, 4775*
    CARCINOID TUMOR
        CARCINOEMBRYONIC ANTIGEN, 4534*
    EPIDEMIOLOGY
        AUSTRALIA, 4644
    GENETICS
        GENETICS, 4799*
    GONADOTROPINS, CHORIONIC
        SERUM LEVELS, 4775*
    LEIOMYOMA
        CASE REPORT, 4637*
    SEX FACTORS
        EPIDEMIOLOGY, FINLAND, 4645
    SMOKING
        ALPHA 1 ANTITRYPSIN, 4657

GENETICS
    ANEMIA, APLASTIC
        PRECANCEROUS CONDITIONS, 4556
    BLADDER NEOPLASMS
        TRYPTOPHAN, 4769*
    BRAIN NEOPLASMS
        FIBROSARCOMA, 4631*
    BREAST NEOPLASMS
        EPIDEMIOLOGY, 4630*
    CAFFEINE

GONADOTROPINS, CHORIONIC
    BRONCHIAL NEOPLASMS
        SERUM LEVELS, 4775*
    GASTROINTESTINAL NEOPLASMS
        SERUM LEVELS, 4775*
    NEOPLASMS
        NONENDOCRINE TUMOR, 4795*
    TESTICULAR NEOPLASMS
        RADIOIMMUNOASSAY, 4469
    UROGENITAL NEOPLASMS
        RADIOIMMUNOASSAY, 4469

GRAFT VS HOST REACTION
    BONE MARROW CELLS
        TRANSPLANTATION IMMUNOLOGY, 4436
    ISOANTIGENS
        ANTIBODY FORMATION, 4446
    B-LYMPHOCYTES
        TRANSPLANTATION IMMUNOLOGY, 4436
    T-LYMPHOCYTES
        CELL DIFFERENTIATION, 4494
        IMMUNE SERUMS, 4446
        IMMUNITY, CELLULAR, 4446
        ISOANTIGENS, 4446
        TRANSPLANTATION IMMUNOLOGY, 4436
    LYMPHOMA
        TRANSPLANTATION IMMUNOLOGY, 4441
    SARCOMA
        IMMUNOSUPPRESSION, 4446
    SARCOMA, RETICULUM CELL
        LYMPHOMA, 4555*

GRANULOCYTES
    LEUKEMIA, MYELOCYTIC
        CELL DIFFERENTIATION, 4686
        ULTRASTRUCTURAL STUDY, 4686

GROWTH
    SEE ALSO CELL DIVISION/CELL
      DIFFERENTIATION
    COLONIC NEOPLASMS
        TRANSPLANTATION, HETEROLOGOUS,
          4740*
    HEPATOMA
        IMMUNE SERUMS, 4451
    LEUKEMIA, MYELOCYTIC
        TRANSPLANTATION, HETEROLOGOUS,
          4740*
    MAMMARY NEOPLASMS, EXPERIMENTAL
        TRANSPLANTATION, HOMOLOGOUS, 4740*
    MELANOMA
        MSH, 4721
    NEPHROBLASTOMA
        NEOPLASM TRANSPLANTATION, 4597
    TERATOID TUMOR
        MOSAICISM, 4600

GROWTH SUBSTANCES
    ADENOSINE CYCLIC 3',5'MONOPHOSPHATE
        INSULIN, 4225
    FIBROBLASTS
        BINDING, 4794*
        MEMBRANE RECEPTORS, 4794*
    LEUKEMIA, MYELOCYTIC
        CELL DIFFERENTIATION, 4686
    VIRUS, POLYOMA
        ADENOSINE CYCLIC 3,5'
          MONOPHOSPHATE, 4391
        ADENYL CYCLASE, 4391

PHOSPHODIESTERASES, 4391

GUANIDINE, 1-METHYL-3-NITRO-1-NITROSO-
    DNA REPAIR
        GUANINE, 4346*
    ESCHERICHIA COLI
        MUTAGENIC ACTIVITY, 4344*
    SALIVARY GLAND NEOPLASMS
        PRECANCEROUS CONDITIONS, 4290
    SODIUM CHLORIDE
        COCARCINOGENIC EFFECT, 4347*
    STOMACH NEOPLASMS
        CARCINOMA, 4347*

GUANIDINE, MONOHYDROCHLORIDE
    FETAL GLOBULINS
        IMMUNOLOGICAL AND ELECTROPHORETIC
            PROPERTIES, 4465

GUANINE
    GUANIDINE, 1-METHYL-3-NITRO-1-NITROSO-
        DNA REPAIR, 4346*

GUANINE, 1-METHYL-, 3-OXIDE
    URIC ACID, 1-METHYL-
        CARCINOGENIC METABOLITE, 4345*

GUANOSINE CYCLIC 3',5' MONOPHOSPHATE
    CARCINOMA, EHRLICH TUMOR
        REVERSE TRANSCRIPTASE, 4716

GUANOSINE TRIPHOSPHATE
    HELA CELLS
        DNA REPLICATION, 4706

GYNECOLOGIC NEOPLASMS
    ADENOCARCINOMA
        4,4'-STILBENEDIOL, ALPHA,ALPHA'-
            DIETHYL-, 4305
        TRANSPLACENTAL CARCINOGENESIS,
            4305
    EPIDEMIOLOGY
        AUSTRALIA, 4644
    4,4'-STILBENEDIOL, ALPHA,ALPHA'-
        DIETHYL-
        DIAGNOSIS AND PROGNOSIS, 4305
        EPIDEMIOLOGY, 4305
        PRECANCEROUS CONDITIONS, 4305

HAMANGIOENDOETHELIOMA
    THYROID NEOPLASMS
        CLASSIFICATION, 4642*

HEAD AND NECK NEOPLASMS
    ANGIOSARCOMA
        ETHYLENE, CHLORO-, 4311
    CARCINOMA
        VIRUS, HERPES SIMPLEX, 4214

HEART NEOPLASMS
    FIBROMA
        HISTOLOGICAL STUDY, 4562

HELA CELLS
    ADENOSINE TRIPHOSPHATE
        DNA REPLICATION, 4706
    ALPHA-AMANITINE
        DNA REPLICATION, 4706
    CELL DIVISION

REVIEW, 4226
CHROMATIN
    DNA REPLICATION, 4693
CHROMOSOMES
    ULTRASTRUCTURAL STUDY, 4580
CYTOSINE NUCLEOTIDES
    DNA REPLICATION, 4706
DEOXYRIBONUCLEOTIDES
    DNA REPLICATION, 4706
DNA REPLICATION
    CELL.CYCLE KINETICS, 4679*
GUANOSINE TRIPHOSPHATE
    DNA REPLICATION, 4706
MAGNESIUM
    DNA REPLICATION, 4706
MICROTUBULES
    ULTRASTRUCTURAL STUDY, 4580
RIBONUCLEASE
    DNA REPLICATION, 4706
RNA, MESSENGER
    ISOLATION AND CHARACTERIZATION
        MOUSE, 4697
RNA, RIBOSOMAL
    ISOLATION AND CHARACTERIZATION
        4762*
SPERMIDINE
    DNA REPLICATION, 4706
SPERMINE
    DNA REPLICATION, 4706
VIRUS, ADENO 2
    ALPHA-AMANITINE, 4378
    DNA REPLICATION, 4378
    RNA POLYMERASE, 4378
    RNA REPLICATION, 4378
VIRUS, ADENO 2 - SV40 HYBRID
    VIRAL PROTEINS, 4396
VIRUS, VACCINIA
    DNA REPLICATION, 4402

HELIOTRINE
    PANCREATIC NEOPLASMS
        ISLET. CELL TUMOR, 4301

HEMANGIOBLASTOMA
    SEE ANGIOSARCOMA

HEMANGIOMA
    LIVER NEOPLASMS
        ETHYLENE, CHLORO-, 4312
    SPLENIC NEOPLASMS
        INCIDENCE, MOUSE, 4723
    VIRUS, ROUS SARCOMA
        SUBGROUP E, 4376

HEMANGIOPERICYTOMA
    NASOPHARYNGEAL NEOPLASMS
        RADIATION, IONIZING, 4364

HEMATOMA
    HEMORRHAGE
        CONTRACEPTIVES, ORAL, 4307
    VIRUS, ROUS SARCOMA
        NEOPLASM REGRESSION, SPONTANEOL
            4595

HEMATOPOIESIS
    CYCLOHEXYLAMINE
        LYMPHOCYTES, 4295
    LEUKOCYTES

COPPER, 4501

HEMATOSARCOMA
  SEE ANGIOSARCOMA

HEMOGLOBINS
  ERYTHROLEUKEMIA
    FIBROBLASTS, 4692

HEMORRHAGE
  HEMATOMA
    CONTRACEPTIVES, ORAL, 4307

HEPARIN
  OVARIAN NEOPLASMS
    ASCITES, 4640*
  VIRUS, SV40
    CELL MEMBRANE, 4212

HEPATITIS
  IMMUNITY, CELLULAR, 4215
  TRANSFER FACTOR
    HYPERSENSITIVITY, 4215

HEPATOCARCINOMA
  SEE HEPATOMA

HEPATOCELLULAR CARCINOMA
  SEE HEPATOMA

HEPATOMA
  ACTINOMYCIN D
    DNA, 4707
    PROTEINS, 4751*
    RNA, 4751*
  AGE FACTORS
    EPIDEMIOLOGY, MOUSE, 4723
  ALBUMINS
    CELL CYCLE KINETICS, 4463
  ALKALINE PHOSPHATASE
    ENZYMATIC ACTIVITY, 4759*
  ALPHA GLOBULINS
    FETAL GLOBULINS, 4461
    ISOLATION AND CHARACTERIZATION,
      4464
  AMINO ACYL T RNA SYNTHETASES
    ISOLATION AND CHARACTERIZATION,
      4773*
  ANILINE, N,N-DIMETHYL-P-PHENYLAZO-
    CARCINOGENIC ACTIVITY, RAT, 4317*
    CHALONES, 4731*
    HISTOCOMPATIBILITY ANTIGENS, 4485
    TRANSPLACENTAL CARCINOGENESIS,
      4317*
    TUMOR-TETANUS ASSAY, 4315*
  ANTIGENS, NEOPLASM
    IMMUNE SERUMS, 4485
    ISOLATION AND CHARACTERIZATION,
      4484
  AUSTRALIA ANTIGEN
    EPIDEMIOLOGY, SENEGAL, FRENCH
      SUDAN, 4669*
  BENZENAMINE, N,N-DIMETHYL-4-((3-
    METHYLPHENYL)AZO)-
    CHROMATIN, 4746*
    PROTEIN KINASE, 4746*
  BENZENZMINE, N,N-DIMETHYL-4-((3-
    METHYLPHENYL)AZO)-
    GLUTAMYL TRANSPEPTIDASE, 4788*

BENZO(A)PYRENE
  COMMON STATISTICAL FACTORS, SITE-
    SPECIFICITY, 4276
CARBOXY-LYASES
  DIURNAL OSCILLATION, 4772*
CELL MEMBRANE
  ANTIGENS, NEOPLASM, 4484
CITRATES
  DIET, 4768*
CONTRACEPTIVES, ORAL
  CASE REPORT, TUMOR RUPTURE, 4567
  PRECANCEROUS CONDITIONS, 4308
CORTISOL
  RNA REPLICATION, 4749*
  TRYPTOPHAN OXYGENASE, 4755*
DIETHYLAMINE, N-NITROSO-
  ANTIGENS, NEOPLASM, 4484, 4533*
  TUMOR-TETANUS ASSAY, 4315*
DNA REPLICATION
  CHALONES, 4731*
FATTY ACIDS
  DIET, 4768*
FETAL GLOBULINS
  CELL CYCLE KINETICS, 4463
  CLONE CELLS, 4683
  PRECANCEROUS CONDITIONS, 4468
  RISK FACTORS, 4467
HISTOCOMPATIBILITY ANTIGENS
  ANTIGENIC DETERMINANTS, 4485
IMMUNE SERUMS
  GROWTH, 4451
  NEOPLASM TRANSPLANTATION, 4451
LIGASES
  DIET, 4768*
LIVER CIRRHOSIS
  AUSTRALIA ANTIGEN, 4468
  EPIDEMIOLOGY, SWITZERLAND, 4652
  FETAL GLOBULINS, 4468
NEOPLASM TRANSPLANTATION
  IMMUNE RESPONSE, 4510*
NORETHANDROLONE
  ANEMIA, APLASTIC, 4556
NUCLEOPROTEINS
  COMPLEXES, U1 AND U2 RNA, 4694
OLIGOPEPTIDES
  NEOPLASM METASTASIS, 4777*
OXANDROLONE
  ANEMIA, APLASTIC, 4556
OXIDOREDUCTASES
  DIET, 4768*
OXYMETHOLONE
  ANEMIA, APLASTIC, 4556
PENTOSYLTRANSFERASES
  GROWTH RATE, 4726*
PEPTIDE HYDROLASES
  NEOPLASM METASTASIS, 4777*
PRECANCEROUS CONDITIONS
  ENZYMATIC ACTIVITY, RAT, 4758*
PROTEIN KINASE
  CHROMATIN, 4746*
RNA
  ISOLATION AND CHARACTERIZATION,
    4764*
RNA, MESSENGER
  POLY A, 4704
  TRYPTOPHAN OXYGENASE, 4755*
RNA, NEOPLASM
  BASOPHILS, 4696
  BINDING, TOLUIDINE BLUE, 4696

RNA, RIBOSOMAL
    ISOLATION AND CHARACTERIZATION,
        4762*
VIRUS, HEPATITIS B
    AUSTRALIA ANTIGEN, 4669*

HISTAMINE
    LEUKEMIA, MYELOCYTIC
        METABOLISM, 4622*
    POLYCYTHEMIA VERA
        METABOLISM, 4622*

HISTIOCYTOMA
    EYE NEOPLASMS
        CASE REPORT, 4563

HISTOCOMPATIBILITY ANTIGENS
    BURKITT'S LYMPHOMA
        STATISTICAL ANALYSIS, 4798*
    HEPATOMA
        ANILINE, N,N-DIMETHYL-P-PHENYLAZO-,
            4485
        ANTIGENIC DETERMINANTS, 4485
    HODGKIN'S DISEASE
        DIAGNOSIS AND PROGNOSIS, 4482
    LEUKEMIA
        ANTIBODY FORMATION, 4444
        IMMUNITY, CELLULAR, 4444
    T-LYMPHOCYTES
        ANTI-ANTIBODIES, 4489
        ANTIGEN-ANTIBODY REACTIONS, 4489
        IGG, 4489
    LYMPHOMA
        TRANSPLANTATION IMMUNOLOGY, 4440
        VIRUS, MURINE LEUKEMIA, 4440
    LYMPHOSARCOMA
        ANTIBODY FORMATION, 4459
        CELL MEMBRANE, 4459
    MAMMARY NEOPLASMS, EXPERIMENTAL
        SARCOMA, RETICULUM CELL, 4543*
    RADIATION
        T-LYMPHOCYTES, 4438
    SARCOMA, RETICULUM CELL
        TRANSPLANTATION IMMUNOLOGY, 4543*
    VIRUS, MURINE MAMMARY TUMOR
        VIRUS SUSCEPTIBILITY, 4424*
    VIRUS, RAUSCHER MURINE LEUKEMIA
        VIRUS REPLICATION, 4480

HISTONES
    VIRUS, SV40
        FIBROBLASTS, 4725

HODGKIN'S DISEASE
    ACID PHOSPHATASE
        LOCALIZATION, LYMPH NODES, 4541*
    ALPHA GLOBULINS
        SERUM LEVEL, 4461
    BREAST NEOPLASMS
        RADIATION, IONIZING, 4364
    CHOLECYSTITIS
        EPIDEMIOLOGY, 4660
    CLASSIFICATION
        REVIEW, 4222
    COLITIS
        EPIDEMIOLOGY, 4660
    CONCANAVALIN A
        CELL TRANSFORMATION, NEOPLASTIC,
            4506

HYDRAZINE, PROPYL-
    LUNG NEOPLASMS
        CARCINOGENIC ACTIVITY, 4328*

HYDRAZINE, PROPYL-, MONOHYDROCHLORIDE
    LUNG NEOPLASMS
        ADENOCARCINOMA, 4294
        ADENOMA, 4294

HYDROCORTISONE
    SEE CORTISOL

HYDROLASES
    MELANOMA
        MELANOSOMES, BIOCHEMICAL STUDY,
        4782*

N-HYDROXY-2-ACETYLAMINOFLUORENE
    SEE ACETOHYDROXAMIC ACID, N-FLUOREN-2-
        YL-

17-HYDROXYCORTICOSTEROIDS
    BREAST NEOPLASMS
        URINE, 4570

HYDROXYLASES
    DIETHYLAMINE, N-NITROSO-
        LIVER, RAT, 4283

HYPERNEPHROID CARCINOMA
    SEE ADENOCARCINOMA

HYPERNEPHROMA
    SEE ADENOCARCINOMA

HYPERPLASIA
    LARYNGEAL NEOPLASMS
        MACROPHAGES, 4504
    SKIN NEOPLASMS
        CASE REPORT, ANGIOLYMPHOID, 4561

HYPERSENSITIVITY
    TRANSFER FACTOR
        HEPATITIS, 4215

HYPERSENSITIVITY, DELAYED
    T-LYMPHOCYTES
        EFFECTOR CELLS, ISOLATION AND
            CHARACTERIZATION, 4523*
    LYMPHOSARCOMA
        ANTIGENS, 4538*
    MELANOMA
        ANTIGENS, NEOPLASM, 4479

HYPOTHROIDISM
    PITUITARY NEOPLASMS
        ADENOMA, CHROMOPHOBE, 4576

IGA
    PHOSPHORYLCHOLINE
        IMMUNE RESPONSE, REVIEW, 4218
    PLASMACYTOMA
        ULTRASTRUCTURAL STUDY, 4594
    VIRUS, HERPES SIMPLEX
        REVIEW, 4214

IGG

LEUKEMIA
    ANTIBODIES, 4449
LEUKEMIA, MYELOCYTIC
    IMMUNOGLOBULINS, FAB, 4526*
    IMMUNOGLOBULINS, FC, 4526*
T-LYMPHOCYTES
    ANTIBODY FORMATION, 4518*
    HISTOCOMPATIBILITY ANTIGENS, 4489
MULTIPLE MYELOMA
    ANTI-ANTIBODIES, 4452
SARCOMA
    ANTIBODIES, 4449
VIRUS, HERPES SIMPLEX
    REVIEW, 4214

IGM
    B-LYMPHOCYTES
        IMMUNOGLOBULINS, SURFACE, 4455
    LYMPHOMA
        B-LYMPHOCYTES, 4491
    VIRUS, HERPES SIMPLEX
        REVIEW, 4214

IMIDAZOLE-4-CARBOXAMIDE, 5-AMINO-
    IMIDAZOLE-4-CARBOXAMIDE, 4-DIAZO-
        CARCINOGENIC ACTIVITY, 4333*

IMIDAZOLE-4-CARBOXAMIDE, 5-(3,3-BIS-
(CHLOROETHYL)-1-TRIAZENO)-
    MELANOMA
        IMMUNITY, CELLULAR, 4546*
        IMMUNOSUPPRESSION, 4546*

IMIDAZOLE-4-CARBOXAMIDE, 4-DIAZO-
    IMIDAZOLE-4-CARBOXAMIDE, 5-AMINO-
        CARCINOGENIC ACTIVITY, 4333*

IMMUNE ADHERENCE REACTION
    NEOPLASMS, EXPERIMENTAL
        BENZ(A)ANTHRACENE, 7,12-DIMETHYL-,
        4531*
        BENZO(A)PYRENE, 4531*
        CHOLANTHRENE, 3-METHYL-, 4531*

IMMUNE SERUMS
    CARCINOMA, BRONCHIOLAR
        VIRUS-LIKE PARTICLES, 4483
    FIBROSARCOMA
        NEOPLASM TRANSPLANTATION, 4509*
    HEPATOMA
        ANTIGENS, NEOPLASM, 4485
        GROWTH, 4451
        NEOPLASM TRANSPLANTATION, 4451
    IMMUNOGLOBULINS
        MYELOMA PROTEINS, 4454
    LEUKEMIA
        ANTIGEN-ANTIBODY REACTIONS, 4449
        BENZ(A)ANTHRACENE, 7,12-DIMETHYL-,
        4444
        TRANSPLANTATION, HOMOLOGOUS, 4449
    LEUKEMIA, LYMPHOBLASTIC
        LYMPHOCYTE-DEPENDENT ANTIBODY,
        4457
    LEUKEMIA, MYELOCYTIC
        DIAGNOSIS AND PROGNOSIS, 4551*
        LYMPHOCYTE-DEPENDENT ANTIBODY,
        4457
    B-LYMPHOCYTES
        T-LYMPHOCYTES, 4453

T-LYMPHOCYTES
    ANTIBODY FORMATION, 4518*
    GRAFT VS HOST REACTION, 4446
LYMPHOMA
    TRANSPLANTATION IMMUNOLOGY, 4440
    VIRUS, MURINE LEUKEMIA, 4440
MAMMARY NEOPLASMS, EXPERIMENTAL
    ANTIBODIES, NEOPLASM, 4456
MELANOMA
    LYMPHOCYTE TRANSFORMATION, 4539*
NEOPLASMS
    TRANSPLANTATION, HOMOLOGOUS, 4650
NEOPLASMS, EXPERIMENTAL
    VIRUS, ADENO 7 - SV40 HYBRID, 4450
PHOSPHORYLCHOLINE
    IMMUNOGLOBULINS, 4218
PROSTATIC NEOPLASMS
    ACID PHOSPHATASE, 4476
SARCOMA
    ANTIGEN-ANTIBODY REACTIONS, 4449
    LYMPHOCYTE TRANSFORMATION, 4493
    T-LYMPHOCYTES, 4499
    MIXED LYMPHOCYTE CULTURE REACTION,
    4493
    TRANSPLANTATION, HOMOLOGOUS, 4449
SARCOMA, OSTEOGENIC
    RADIATION, 4235*
VIRUS, ADENO 7 - SV40 HYBRID
    IMMUNITY, CELLULAR, 4450
    VIRUS, SV40, 4450
VIRUS, FRIEND MURINE LEUKEMIA
    IMMUNE RESPONSE, 4517*
VIRUS, MURINE LEUKEMIA
    FERRITIN, 4529*
    FLUORESCENT ANTIBODY TECHNIC,
    4529*

IMMUNITY, CELLULAR
    ANTIGEN-ANTIBODY REACTIONS
        LYMPHOCYTES, 4251*
        NEGATIVE ASSOCIATION, REVIEW,
        4251*
    AUTOIMMUNE DISEASES
        ANTIGENS, 4520*
        BENZENE, 1-CHLORO-2,4-DINITRO-,
        4520*
    BENZENE, 1-CHLORO-2,4-DINITRO-
        ANTIGENS, 4520*
    BRAIN NEOPLASMS
        GLIOMA, 4549*
        MENINGIOMA, 4549*
        NEURILEMMOMA, 4549*
    HEPATITIS, 4215
    HODGKIN'S DISEASE
        T-LYMPHOCYTES, 4524*
    IMMUNOTHERAPY
        REVIEW, 4247*
    LEUKEMIA
        CYCLOPHOSPHAMIDE, 4444
        HISTOCOMPATIBILITY ANTIGENS, 4444
        TRANSPLANTATION, HOMOLOGOUS, 4444
    LEUKEMIA, LYMPHOBLASTIC
        ANTIGENS, NEOPLASM, 4552*
        BCG VACCINATION, 4552*
    T-LYMPHOCYTES
        EFFECTOR CELLS, ISOLATION AND
        CHARACTERIZATION, 4523*
        GRAFT VS HOST REACTION, 4446
    MAMMARY NEOPLASMS, EXPERIMENTAL

BENZ(A)ANTHRACENE, 7,12-DIMETHYL
    4495
    VIRUS, MURINE MAMMARY TUMOR, 445
MELANOMA
    ANTIGENS, NEOPLASM, 4479
    IMIDAZOLE-4-CARBOXAMIDE, 5-(3,3-
    BIS-(CHLOROETHYL)-1-TRIAZENO)-
    4546*
    ROSETTE FORMATION, 4553*
NEOPLASM
    ANTIGENS, 4520*
    BENZENE, 1-CHLORO-2,4-DINITRO-,
    4520*
NEOPLASMS
    REVIEW, 4246*
OVARIAN NEOPLASMS
    LYMPHOCYTES, 4486
SARCOMA
    ANTIGENS, NEOPLASM, 4522*
    ANTIGENS, NEOPLASMS, 4521*
    CHOLANTHRENE, 3-METHYL-, 4499
    GENETICS, 4521*
    LYMPHOCYTES, 4521*
    T-LYMPHOCYTES, 4499
SARCOMA, OSTEOGENIC
    LYMPHOCYTES, 4542*
THYROID NEOPLASMS
    GENETICS, 4547*
VIRUS, ADENO 7 - SV40 HYBRID
    IMMUNE SERUMS, 4450
    T-LYMPHOCYTES, 4450
VIRUS, HERPES SIMPLEX
    T-LYMPHOCYTES, 4214
    REVIEW, 4214
VIRUS, MOLONEY MURINE SARCOMA
    CORYNEBACTERIUM PARVUM, 4548*

IMMUNITY, PASSIVE
    T-LYMPHOCYTES
        ANTIBODY FORMATION, 4518*

IMMUNIZATION
    LEUKEMIA
        EPIDEMIOLOGY, MICE, 4434*

IMMUNOGLOBULINS
    SEE ALSO IGA/IGD/IGE/IGG/IGM
    LEUKEMIA, LYMPHOCYTIC
        CEREBROSPINAL FLUID, 4628*
        NEOPLASM METASTASIS, 4628*
    B-LYMPHOCYTES
        PHOSPHORYLCHOLINE, 4218
        SEQUENTIAL STIMULATION, 4216
    T-LYMPHOCYTES
        SEQUENTIAL STIMULATION, 4216
    LYMPHOMA
        ULTRASTRUCTURAL STUDY, HAMSTER,
        4525*
    LYMPHOSARCOMA
        ABNORMALITIES, 4538*
    MYELOMA PROTEINS
        BINDING, 4454
        IMMUNE SERUMS, 4454
    PHOSPHORYLCHOLINE
        AUTOANTIBODIES, 4218
        IMMUNE RESPONSE, REVIEW, 4218
        IMMUNE SERUMS, 4218
        IMMUNOSUPPRESSION, 4218
    RADIATION, IONIZING

B-LYMPHOCYTES, 4439
VIRUS, HERPES SIMPLEX
    ANTIGENS, VIRAL, 4214
    REVIEW, 4214

IMMUNOGLOBULINS, FAB
    LEUKEMIA, MYELOCYTIC
        IGG, 4526*

IMMUNOGLOBULINS, FC
    LEUKEMIA, MYELOCYTIC
        IGG, 4526*
    LYMPHOSARCOMA
        B-LYMPHOCYTES, 4459
        T-LYMPHOCYTES, 4459

IMMUNOGLOBULINS, SURFACE
    LEUKEMIA, LYMPHOBLASTIC
        B-LYMPHOCYTES, 4481
    LEUKEMIA, LYMPHOCYTIC
        IMMUNE SERUMS, 4527*
        B-LYMPHOCYTES, 4587
        T-LYMPHOCYTES, 4587
    B-LYMPHOCYTES
        ANTIGEN-ANTIBODY REACTIONS, 4455
        IGM, 4455
        IMMUNOSUPPRESSION, 4455

IMMUNOLOGIC DEFICIENCY SYNDROMES
    NEOPLASMS
        RELATIONSHIP, REVIEW, 4252*

IMMUNOSUPPRESSION
    BREAST NEOPLASMS
        UREMIA, 4515*
    CERVIX NEOPLASMS
        TRANSPLANTATION, HOMOLOGOUS, 4515*
        UREMIA, 4515*
    COLONIC NEOPLASMS
        TRANSPLANTATION, HETEROLOGOUS,
        4740*
        TRANSPLANTATION, HOMOLOGOUS, 4515*
    DISGERMINOMA
        TRANSPLANTATION, HOMOLOGOUS, 4515*
    ISLET CELL TUMOR
        UREMIA, 4515*
    KIDNEY NEOPLASMS
        UREMIA, 4515*
    LEUKEMIA
        UREMIA, 4515*
    LEUKEMIA, MYELOCYTIC
        TRANSPLANTATION, HETEROLOGOUS,
        4740*
    LUNG NEOPLASMS
        UREMIA, 4515*
    B-LYMPHOCYTES
        IMMUNOGLOBULINS, SURFACE, 4455
    LYMPHOMA
        LEPROSY, 4514*
    MAMMARY NEOPLASMS, EXPERIMENTAL
        TRANSPLANTATION, HOMOLOGOUS, 4740*
        TRANSPLANTATION IMMUNOLOGY, 4739*
    MELANOMA
        IMIDAZOLE-4-CARBOXAMIDE, 5-(3,3-
        BIS-(CHLOROETHYL)-1-TRIAZENO)-,
        4546*
        LYMPHOCYTES, 4539*
    NEOPLASM TRANSPLANTATION
        ULTRAVIOLET RAYS, 4512*

NEOPLASMS
    TRANSPLANTATION, HOMOLOGOUS, 4650
PHOSPHORYLCHOLINE
    IMMUNOGLOBULINS, 4218
SARCOMA
    GRAFT VS HOST REACTION, 4446
SARCOMA, OSTEOGENIC
    RADIATION, 4235*
SARCOMA, RETICULUM CELL
    TRANSPLANTATION, HOMOLOGOUS, 4515*
SKIN NEOPLASMS
    TRANSPLANTATION, HOMOLOGOUS, 4515*
    WARTS, 4219
THYROID NEOPLASMS
    UREMIA, 4515*
VIRUS, FRIEND MURINE LEUKEMIA
    ANTIGENS, VIRAL, 4447
    IMMUNE RESPONSE, 4519*
    B-LYMPHOCYTES, 4447
    T-LYMPHOCYTES, 4447
VIRUS, MOLONEY MURINE SARCOMA
    CORYNEBACTERIUM PARVUM, 4548*
VIRUS, MURINE LEUKEMIA
    REVIEW, 4447
VIRUS, SV40
    CELL TRANSFORMATION, NEOPLASTIC,
    4419*

IMMUNOTHERAPY
    IMMUNITY, CELLULAR
        REVIEW, 4247*
    NEOPLASMS
        REVIEW, 4246*

INCLUSION BODIES, VIRAL
    SKIN NEOPLASMS
        WARTS, 4219

INFECTIOUS MONONUCLEOSIS
    LYMPHOCYTES
        CELLULAR INCLUSIONS, 4601*
    VIRUS, EPSTEIN-BARR
        ANTIGENS, VIRAL, 4420*

INSULIN
    ADENOSINE CYCLIC 3',5'MONOPHOSPHATE
        GROWTH SUBSTANCES, 4225

INTERFERON
    VIRUS, EPSTEIN-BARR
        ANTIGENS, VIRAL, 4460
        URIDINE, 5-BROMO-2'-DEOXY-, 4460

INTESTINAL NEOPLASMS
    ACID PHOSPHATASE
        HISTOCHEMICAL STUDY, 4770*
    CARCINOMA
        METHANOL, (METHYLNITROSOAMINO)-,
        ACETATE (ESTER), 4350*
    ESTERASES
        HISTOCHEMICAL STUDY, 4770*
    GALACTOSIDASES
        HISTOCHEMICAL STUDY, 4770*
    GLUCURONIDASE
        HISTOCHEMICAL STUDY, 4770*
    LYMPHOSARCOMA
        INTESTINAL MICROFLORA, 4253*
        PASSAGE RATE, CONTENTS, 4253*
    MALIGNANT CARCINOID SYNDROME

CASE REPORT, 4617*
METHANOL, (METHYL-ONN-AZOXY)-, ACETATE
(ESTER)
SITE SPECIFICITY, 4341*
SULFATASES
HISTOCHEMICAL STUDY, 4770*

IRON
NEOPLASMS
EPIDEMIOLOGY, CHILD, 4654

ISLET CELL TUMOR
PANCREATIC NEOPLASMS
HELIOTRINE, 4301
NICOTINAMIDE, 4301
UREMIA
IMMUNOSUPPRESSION, 4515*

ISOANTIGENS
ANTIBODY FORMATION
GRAFT VS HOST REACTION, 4446
LEUKEMIA, LYMPHOBLASTIC
HL-B-SPECIFIC PREGNANCY SERA, 4481
B-LYMPHOCYTES, 4481
T-LYMPHOCYTES
GRAFT VS HOST REACTION, 4446
SARCOMA
T-LYMPHOCYTES, 4499

ISOBUTYL ALCOHOL
ALCOHOLIC BEVERAGES
CARCINOGENIC ACTIVITY, 4318*

ISOENZYMES
ACETAMIDE, N-FLUOREN-2-YL-
FRUCTOSEDIPHOSPHATE ALDOLASE, 4274
MAMMARY NEOPLASMS, EXPERIMENTAL
GLUCOSEPHOSPHATE DEHYDROGENASE,
4779*
NEOPLASMS
ALKALINE PHOSPHATASE, 4728*
PLACENTA
ALKALINE PHOSPHATASE, 4728*

ISOGENIC TRANSPLANTATION
SEE TRANSPLANTATION, HOMOLOGOUS

ISONICOTINIC ACID HYDRAZIDE
CHROMOSOME ABERRATIONS
EVALUATION, 4331*
SPERMATOZOA, 4332*

ISOPROTERENOL
ARYL HYDROCARBON HYDROXYLASES
CYTOCHROME P-450, 4278

JAUNDICE, NEONATAL
GENETICS
PHOTOTHERAPY, 4369*

KARYOTYPING
LEUKEMIA
MEGAKARYOCYTES, 4585
POLYCYTHEMIA VERA
CHROMOSOME ABNORMALITIES, 4584
TERATOID TUMOR
REVIEW, 4261*

KERATOACANTHOMA

CROTON OIL
BENZ(A)ANTHRACENE, 7,12-DIMETHYL
4363

17-KETOSTEROIDS
BREAST NEOPLASMS
URINE, 4570

KIDNEY NEOPLASMS
ACETIC ACID, LEAD (4+) SALT
DIMETHYLAMINE, N-NITROSO-, 4351*
NEPHRECTOMY, 4351*
DIMETHYLAMINE, N-NITROSO-
HISTOLOGICAL STUDY, 4599
NEPHRECTOMY, 4351*
MESENCHYMOMA
DIMETHYLAMINE, N-NITROSO-, 4599
NEPHROBLASTOMA
ETHYLENE, CHLORO-, 4311, 4312
PENTOSYLTRANSFERASES
RAT, 4726*
RADIATION, IONIZING
DOSE-RESPONSE STUDY, 4680*
THORIUM DIOXIDE
CASE REPORT, 4635*
UREMIA
IMMUNOSUPPRESSION, 4515*

L CELLS
ADENOSINE
RESISTANT PHENOTYPE, 4685
NERVE TISSUE PROTEINS
COLCHICINE, 4690
RNA, MESSENGER
ISOLATION AND CHARACTERIZATION,
MOUSE, 4697
TRIAZINE-3,5(2H,4H)-DIONE, 2-BETA-D-
RIBOFURANOSYL-
RESISTANT PHENOTYPE, 4685

LACTATE DEHYDROGENASE
12-O-TETRADECANOYLPHORBOL-13-ACETATE
LEUKOCYTES, 4714

LACTOSE SYNTHETASE
MAMMARY NEOPLASMS, EXPERIMENTAL
ESTRONE, 4303
PROGESTERONE, 4303

LARYNGEAL NEOPLASMS
ASBESTOS
OCCUPATIONAL HAZARD, 4262*
LYMPHOCYTES
NEOPLASM METASTASIS, 4504
T-LYMPHOCYTES
NEOPLASM METASTASIS, 4504
MACROPHAGES
HYPERPLASIA, 4504
SEX FACTORS
EPIDEMIOLOGY, FINLAND, 4645

LEIOMYOMA
BLADDER NEOPLASMS
CASE REPORT, 4637*
GASTROINTESTINAL NEOPLASMS

CHILD, 4651
FORSSMAN ANTIGEN
ANTIBODIES, 4530*
HISTOCOMPATIBILITY ANTIGENS
ANTIBODY FORMATION, 4444
IMMUNITY, CELLULAR, 4444
IMMUNE SERUMS
ANTIGEN-ANTIBODY REACTIONS, 4449
IMMUNIZATION
EPIDEMIOLOGY, MICE, 4434*
LEUKOCYTES
DNA, 4708
MEGAKARYOCYTES
CASE REPORT, 4585
KARYOTYPING, 4585
NORETHANDROLONE
ANEMIA, APLASTIC, 4556
OXANDROLONE
ANEMIA, APLASTIC, 4556
OXYMETHOLONE
ANEMIA, APLASTIC, 4556
PLANT AGGLUTININS
CELL TRANSFORMATION, NEOPLASTIC,
4506
POLYRIBOSOMES
PROTEINS, 4790*
RIBONUCLEASE, 4790*
PRECANCEROUS CONDITIONS
CASE REPORT, 4556
REVIEW, 4260*
RADIATION
OCCUPATIONAL HAZARD, 4208
RADIATION, IONIZING
DOSE-RESPONSE STUDY, 4680*
SMOKING
ALPHA 1 ANTITRYPSIN, 4657
TRANSPLANTATION, HOMOLOGOUS
IMMUNE SERUMS, 4449
IMMUNITY, CELLULAR, 4444
UREMIA
IMMUNOSUPPRESSION, 4515*
VIRUS, SV40
ANEMIA, APLASTIC, 4556

LEUKEMIA, ACUTE GRANULOCYTIC
SEE LEUKEMIA, MYELOBLASTIC

LEUKEMIA, ACUTE LYMPHOCYTIC
SEE LEUKEMIA, LYMPHOBLASTIC

LEUKEMIA, CHRONIC GRANULOCYTIC
SEE LEUKEMIA, MYELOCYTIC

LEUKEMIA, EXPERIMENTAL
MYCOBACTERIUM BOVIS
TRANSPLANTATION IMMUNOLOGY, 4511*

LEUKEMIA L1210
ANTIGENS, NEOPLASM
IMMUNOTHERAPY, 4535*
CORYNEBACTERIUM PARVUM
IMMUNOLOGIC ENHANCEMENT, 4508*
LYMPHOMA
TRANSPLANTATION, HOMOLOGOUS, 4442

LEUKEMIA, LYMPHOBLASTIC
ANTIGENS, NEOPLASM
ANTIBODIES, NEOPLASM, 4552*
IMMUNITY, CELLULAR, 4552*

BCG VACCINATION
　IMMUNITY, CELLULAR, 4552*
BONE MARROW CELLS
　REVIEW, 4220
CYTIDINE CYCLIC 3',5' MONOPHOSPHATE
　EXCRETION, 4745*
DNA NUCLEOTIDYLTRANSFERASES
　ISOLATION AND CHARACTERIZATION,
　　4717
GENETICS
　IMMUNE RESPONSE, 4582
IMMUNE SERUMS
　LYMPHOCYTE-DEPENDENT ANTIBODY,
　　4457
ISOANTIGENS
　HL-B-SPECIFIC PREGNANCY SERA, 4481
LEUKEMIA, MYELOCYTIC
　CROSS REACTIONS, 4457
B-LYMPHOCYTES
　IMMUNOGLOBULINS, SURFACE, 4481
　ISOANTIGENS, 4481
T-LYMPHOCYTES
　ISOLATION AND CHARACTERIZATION,
　　4729*
　ROSETTE FORMATION, 4481
MITOCHONDRIA
　ULTRASTRUCTURAL STUDY, 4625*

LEUKEMIA, LYMPHOCYTIC
BENZENE
　EPIDEMIOLOGY, OCCUPATIONAL HAZARD,
　　4663
ENZYMES
　CIRCADIAN RHYTHM, 4778*
ERYTHRODERMA
　IMMUNE RESPONSE, 4557
　T-LYMPHOCYTES, 4557
IMMUNOGLOBULINS
　CEREBROSPINAL FLUID, 4628*
　NEOPLASM METASTASIS, 4628*
IMMUNOGLOBULINS, SURFACE
　IMMUNE SERUMS, 4527*
LEUKOCYTES
　ENZYMATIC ACTIVITY, 4778*
LYMPHOCYTES
　ARGINASE, 4713
　CELLULAR INCLUSIONS, 4601*
B-LYMPHOCYTES
　IMMUNOGLOBULINS, SURFACE, 4587
　ISOLATION AND CHARACTERIZATION,
　　4587
　LYMPHOCYTE TRANSFORMATION, 4587
T-LYMPHOCYTES
　IMMUNOGLOBULINS, SURFACE, 4587
　ISOLATION AND CHARACTERIZATION,
　　4587
　LYMPHOCYTE TRANSFORMATION, 4587
NERVOUS SYSTEM
　CASE REPORT, 4628*
OCCUPATIONAL HAZARD
　EPIDEMIOLOGY, 4661
SUDDEN INFANT DEATH
　EPIDEMIOLOGY, GREAT BRITAIN, 4655
SURFACE PROPERTIES
　ULTRASTRUCTURAL STUDY, 4527*

LEUKEMIA, MONOBLASTIC
BLOOD CELLS
　ULTRASTRUCTURAL STUDY, 4589

BONE MARROW CELLS
　PRECANCEROUS CONDITIONS, 4220
　REVIEW, 4220

LEUKEMIA, MONOCYTIC
BLOOD CELLS
　ULTRASTRUCTURAL STUDY, 4589
BONE MARROW CELLS
　PRECANCEROUS CONDITIONS, 4220
　REVIEW, 4220

LEUKEMIA, MYELOBLASTIC
ALANINE, 3-(P-(BIS(2-CHLOROETHYL)AMINO
　PHENYL)-, L-
　　CASE REPORT, 4624*
　　PARAPROTEINEMIA, 4624*
ALPHA GLOBULINS
　SERUM LEVEL, 4461
BENZENE
　EPIDEMIOLOGY, OCCUPATIONAL HAZARD,
　　4663
BLOOD CELLS
　ULTRASTRUCTURAL STUDY, 4589
CYTIDINE CYCLIC 3',5' MONOPHOSPHATE
　EXCRETION, 4745*
MONOCROTALINE
　HEPATOECTOMY, 4302
NASOPHARYNGEAL NEOPLASMS
　CASE REPORT, 4623*
　RNA, RIBOSOMAL, 4623*
　ULTRASTRUCTURAL STUDY, 4623*
OCCUPATIONAL HAZARD
　BOVINE LEUKEMIA, RELATIONSHIP,
　　4661
　EPIDEMIOLOGY, 4661
VIRUS, C-TYPE
　DNA, VIRAL, 4245*
　REVIEW, 4245*
　RNA, VIRAL, 4245*

LEUKEMIA, MYELOCYTIC
ACUTE DISEASE
　CHROMOSOMES, HUMAN, 21-22, 4626*
ALPHA GLOBULINS
　SERUM LEVEL, 4461
BENZENE
　EPIDEMIOLOGY, OCCUPATIONAL HAZARD,
　　4663
BLOOD CELLS
　ULTRASTRUCTURAL STUDY, 4589
BONE MARROW CELLS
　PRECANCEROUS CONDITIONS, 4220
　REVIEW, 4220
CHROMOSOME ABERRATIONS
　MOSAICISM, INFANT, 4586
CHROMOSOMES
　ACUTE TRANSFORMATION, 4588
CHROMOSOMES, HUMAN, 21-22
　1,4-BUTANEDIOL DIMETHYLSULFONATE,
　　4583
　CELL DIVISION, 4583
　CHROMOSOME ABERRATIONS, 4627*
　MOSAICISM, INFANT, 4586
　SPLEEN, 4583
CYTOSINE
　ACUTE TRANSFORMATION, 4508
DNA NUCLEOTIDYLTRANSFERASES
　ISOLATION AND CHARACTERIZATION,
　　4717

DRUG THERAPY
   CASE REPORT, 4626*
ENZYMES
   CIRCADIAN RHYTHM, 4778*
GRANULOCYTES
   CELL DIFFERENTIATION, 4686
   ULTRASTRUCTURAL STUDY, 4686
GROWTH SUBSTANCES
   CELL DIFFERENTIATION, 4686
HISTAMINE
   METABOLISM, 4622*
IGG
   IMMUNOGLOBULINS, FAB, 4526*
   IMMUNOGLOBULINS, FC, 4526*
IMMUNE SERUMS
   DIAGNOSIS AND PROGNOSIS, 4551*
   LYMPHOCYTE-DEPENDENT ANTIBODY,
     4457
LEUKEMIA, LYMPHOBLASTIC
   CROSS REACTIONS, 4457
LEUKOCYTES
   ACUTE TRANSFORMATION, 4588
   ALKALINE PHOSPHATASE, 4588
   ENZYMATIC ACTIVITY, 4778*
LYMPHOCYTES
   ARGINASE, 4713
T-LYMPHOCYTES
   ISOLATION AND CHARACTERIZATION,
     SPLEEN, 4537*
MEGAKARYOCYTES
   CASE REPORT, 4585
METHOTREXATE
   ACUTE TRANSFORMATION, 4588
NEUTROPHILS
   ALKALINE PHOSPHATASE, 4727*
OCCUPATIONAL HAZARD
   EPIDEMIOLOGY, 4661
PREDNISONE
   ACUTE TRANSFORMATION, 4588
PURINE-6-THIOL
   ACUTE TRANSFORMATION, 4588
RADIATION, IONIZING
   CHROMOSOME ABERRATIONS, 4583
SUDDEN INFANT DEATH
   EPIDEMIOLOGY, GREAT BRITAIN, 4655
TRANSPLANTATION, HETEROLOGOUS
   GROWTH, 4740*
   IMMUNOSUPPRESSION, 4740*
UREA, 1-BUTYL-1-NITROSO-
   MODEL TEST SYSTEM, 4354*

EUKEMIA, MYELOGENOUS
   SEE LEUKEMIA, MYELOCYTIC

EUKEMIA, MYELOID
   SEE LEUKEMIA, MYELOCYTIC

EUKEMIA, PROMYELOCYTIC
   SEE LEUKEMIA, MYELOBLASTIC

EUKOCYTES
   COLITIS, ULCERATIVE
     ALKALINE PHOSPHATASE, 4715
   HEMATOPOIESIS
     COPPER, 4501
   LEUKEMIA
     DNA, 4708
   LEUKEMIA, LYMPHOCYTIC
     ENZYMATIC ACTIVITY, 4778*

LEUKEMIA, MYELOCYTIC
   ACUTE TRANSFORMATION, 4588
   ALKALINE PHOSPHATASE, 4588
   ENZYMATIC ACTIVITY, 4778*
SARCOMA
   CELL MIGRATION INHIBITION, 4522*
12-O-TETRADECANOYLPHORBOL-13-ACETATE
   GLUCURONIDASE, 4714
   LACTATE DEHYDROGENASE, 4714
   MURAMIDASE, 4714

LIGASES
HEPATOMA
   DIET, 4768*

LIP NEOPLASMS, 4558
ANGIOSARCOMA
   ETHYLENE, CHLORO-, 4311
CARCINOMA
   ULTRASTRUCTURAL STUDY, 4558
   VIRUS, HERPES SIMPLEX, 4214

LIPIDS
XANTHOMASTOSIS
   SERUM LEVELS, 4784*

LIPOPROTEINS
XANTHOMASTOSIS
   SERUM LEVELS, 4784*

LIVER
CHALONES
   ISOLATION AND CHARACTERIZATION,
     RABBIT, 4731*
MITOSIS
   MONOCROTALINE, 4302
   RETRONECINE, 3,8-DIDEHYDRO-, 4302

LIVER CELL CARCINOMA
   SEE HEPATOMA

LIVER CIRRHOSIS
HEPATOMA
   AUSTRALIA ANTIGEN, 4468
   EPIDEMIOLOGY, SWITZERLAND, 4652
   FETAL GLOBULINS, 4468

LIVER CIRRHOSIS, OBSTRUCTIVE
LIVER NEOPLASMS
   ADENOCARCINOMA, 4566
   CASE REPORT, 4566

LIVER NEOPLASMS
ACETAMIDE, N-FLUOREN-2-YL-
   FRUCTOSEDIPHOSPHATE ALDOLASE, 4274
ADENOCARCINOMA
   LIVER CIRRHOSIS, OBSTRUCTIVE, 4566
ANGIOMA
   ETHYLENE, CHLORO-, 4311
ANGIOSARCOMA
   ETHYLENE, CHLORO-, 4204, 4311,
     4312, 4313
   ETHYLENE, 1,1-DICHLORO-, 4313
AUSTRALIA ANTIGEN
   EPIDEMIOLOGY, REVIEW, 4263*
   GENETICS, 4263*
CARCINOEMBRYONIC ANTIGEN

        PROTEINS, 4248*
    CARCINOMA
        MONOCROTALINE, 4302
    HEMANGIOMA
        ETHYLENE, CHLORO-, 4312
    LIVER CIRRHOSIS, OBSTRUCTIVE
        CASE REPORT, 4566
    PREGNANCY
        CASE REPORT, 4619*

LIVER REGENERATION
    CHALONES
        DNA REPLICATION, 4731*
        PROTEINS, 4731*
        RNA, 4731*

LUNG NEOPLASMS
    ADENOCARCINOMA
        EPIDEMIOLOGY, MOUSE, 4723
        HYDRAZINE, BUTYL-,
          MONOHYDROCHLORIDE, 4294
        HYDRAZINE, PROPYL-,
          MONOHYDROCHLORIDE, 4294
        IMMUNE RESPONSE, 4582
        LYMPHOCYTE TRANSFORMATION, 4498
    ADENOMA
        ACYLTRANSFERASES, 4711
        CARBAMIC ACID, ETHYL ESTER, 4711
        EPIDEMIOLOGY, MOUSE, 4723
        HYDRAZINE, BUTYL-,
          MONOHYDROCHLORIDE, 4294
        HYDRAZINE, PROPYL-,
          MONOHYDROCHLORIDE, 4294
        MONOCROTALINE, 4302
        UREA, METHYL NITROSO-, 4437
    AIR POLLUTANTS
        BIOASSAY, 4237*
    ALPHA GLOBULINS
        SERUM LEVEL, 4461
    ASBESTOS
        OCCUPATIONAL HAZARD, 4262*
    BENZO(A)PYRENE
        COMMON STATISTICAL FACTORS, SITE-
          SPECIFICITY, 4276
    CARCINOEMBRYONIC ANTIGEN
        PROTEINS, 4248*
    CARCINOMA, EPIDERMOID
        CARCINOEMBRYONIC ANTIGEN, 4474
        CHROMOSOME ABNORMALITIES, 4612*
        LYMPHOCYTE TRANSFORMATION, 4498
    CARCINOMA, OAT CELL
        LYMPHOCYTE TRANSFORMATION, 4498
    EPIDEMIOLOGY
        AUSTRALIA, 4644, 4682*
    ETHYLENE, CHLORO-
        CARCINOGENIC ACTIVITY, RAT, 4204
    GENETICS
        GENETICS, 4582
        NEOPLASMS, MULTIPLE PRIMARY, 4799*
    HYDRAZINE, ALLYL-
        CARCINOGENIC ACTIVITY, 4328*
    HYDRAZINE, PENTYL-
        CARCINOGENIC ACTIVITY, 4328*
    HYDRAZINE, PROPYL-
        CARCINOGENIC ACTIVITY, 4328*
    LEIOMYOSARCOMA
        CASE REPORT, 4608*
        NEOPLASM METASTASIS, 4608*
    T-LYMPHOCYTES

IMMUNOGLOBULINS, SURFACE, 4587
ISOLATION AND CHARACTERIZATION,
4587
LYMPHOCYTE TRANSFORMATION, 4587
T-LYMPHOCYTES
IMMUNE SERUMS, 4453
LYMPHOMA
CELL DIFFERENTIATION, 4491
IGM, 4491
ISOLATION AND CHARACTERIZATION,
4538*
ISOLATION AND CHARACTERIZATION,
MOUSE, 4491
LYMPHOSARCOMA
IMMUNOGLOBULINS, FC, 4459
ISOLATION AND CHARACTERIZATION,
4538*
ROSETTE FORMATION, 4492
MITOMYCIN C
ANTIGENS, NEOPLASM, 4540*
NASOPHARYNGEAL NEOPLASMS
IMMUNOGLOBULINS, SURFACE, 4497
PHOSPHORYLCHOLINE
IMMUNE RESPONSE, REVIEW, 4218
IMMUNOGLOBULINS, 4218
RADIATION, IONIZING
IMMUNE RESPONSE, 4439
IMMUNOGLOBULINS, 4439
TRANSPLANTATION IMMUNOLOGY
GRAFT VS HOST REACTION, 4436
TUBERCULIN
IMMUNE RESPONSE, 4490
MITOMYCIN C, 4490
UREA, HYDROXY-
ORNITHINE DECARBOXYLASE, 4453
VIRUS, FRIEND MURINE LEUKEMIA
IMMUNOSUPPRESSION, 4447

T-LYMPHOCYTES
ANTIBODY FORMATION
IGG, 4518*
IMMUNITY, PASSIVE, 4518*
ANTIGENS, BACTERIAL
HELPER CELL GENERATION, 4500
AUTOIMMUNE DISEASES
ANTIGENS, 4520*
BENZENE, 1-CHLORO-2,4-DINITRO-,
4520*
BENZENE, 1-CHLORO-2,4-DINITRO-
ANTIGENS, 4520*
CELL DIFFERENTIATION
GRAFT VS HOST REACTION, 4494
THYMUS HORMONE, 4494
CELL MEMBRANE
ANTIGENIC DETERMINANTS, 4488
HISTOCOMPATIBILITY ANTIGENS
ANTI-ANTIBODIES, 4489
ANTIGEN-ANTIBODY REACTIONS, 4489
IGG, 4489
HODGKIN'S DISEASE
IMMUNITY, CELLULAR, 4524*
HYPERSENSITIVITY, DELAYED
EFFECTOR CELLS, ISOLATION AND
CHARACTERIZATION, 4523*
IMMUNE SERUMS
ANTIBODY FORMATION, 4518*
GRAFT VS HOST REACTION, 4446
IMMUNITY, CELLULAR
EFFECTOR CELLS, ISOLATION AND

CHARACTERIZATION, 4523*
GRAFT VS HOST REACTION, 4446
IMMUNOGLOBULINS
SEQUENTIAL STIMULATION, 4216
ISOANTIGENS
GRAFT VS HOST REACTION, 4446
LARYNGEAL NEOPLASMS
NEOPLASM METASTASIS, 4504
LEUKEMIA, LYMPHOBLASTIC
ISOLATION AND CHARACTERIZATION,
4729*
ROSETTE FORMATION, 4481
LEUKEMIA, LYMPHOCYTIC
ERYTHRODERMA, 4557
IMMUNOGLOBULINS, SURFACE, 4587
ISOLATION AND CHARACTERIZATION,
4587
LYMPHOCYTE TRANSFORMATION, 4587
LEUKEMIA, MYELOCYTIC
ISOLATION AND CHARACTERIZATION,
SPLEEN, 4537*
LUNG NEOPLASMS
LYMPHOCYTE TRANSFORMATION, 4498
B-LYMPHOCYTES
IMMUNE SERUMS, 4453
LYMPHOMA
ANTIGENS, 4491
CELL DIFFERENTIATION, 4491
ISOLATION AND CHARACTERIZATION,
4538*
ISOLATION AND CHARACTERIZATION,
MOUSE, 4491
TRANSPLANTATION, HOMOLOGOUS, 4442
LYMPHOSARCOMA
IMMUNOGLOBULINS, FC, 4459
ISOLATION AND CHARACTERIZATION,
4538*, 4729*
ROSETTE FORMATION, 4492
MITOMYCIN C
ANTIGENS, NEOPLASM, 4540*
MYCOSIS FUNGOIDES
IMMUNE RESPONSE, 4557
ISOLATION AND CHARACTERIZATION,
4557
NASOPHARYNGEAL NEOPLASMS
ROSETTE FORMATION, 4497
NEOPLASM
ANTIGENS, 4520*
BENZENE, 1-CHLORO-2,4-DINITRO-,
4520*
PHOSPHORYLCHOLINE
IMMUNE RESPONSE, REVIEW, 4218
PLASMACYTOMA
CYCLOPHOSPHAMIDE, 4443
MITOMYCIN C, 4443
RADIATION
HISTOCOMPATIBILITY ANTIGENS, 4438
RADIATION, IONIZING
IMMUNE RESPONSE, 4439
ROSETTE FORMATION, 4496
SARCOMA
ANTIBODY FORMATION, 4499
IMMUNE SERUMS, 4499
IMMUNITY, CELLULAR, 4499
ISOANTIGENS, 4499
SKIN NEOPLASMS
LYMPHOMA, 4557
TRANSPLANTATION IMMUNOLOGY
GRAFT VS HOST REACTION, 4436

MALIGNANT CARCINOID SYNDROME
   BRONCHIAL NEOPLASMS
      CARCINOID TUMOR, 4604*
      CASE REPORT, 4604*
   INTESTINAL NEOPLASMS
      CASE REPORT, 4617*

MALONALDEHYDE
   FOOD CONTAMINATION
      MEAT, 4335*

MAMMARY NEOPLASMS, EXPERIMENTAL
   ACETOHYDROXAMIC ACID, N-FLUOREN-2-YL-
      CARCINOGENIC ACTIVITY, RAT, 4326*
   ADENOCARCINOMA
      ADRIAMYCIN, 4292
      CHROMATIN, 4743*
      EPIDEMIOLOGY, MOUSE, 4723
   ADENOFIBROMA
      ADRIAMYCIN, 4292
      ETHYLENE, CHLORO-, 4311, 4312
   ANTIBODIES, NEOPLASM
      IMMUNE SERUMS, 4456
   ANTIGENS, NEOPLASM
      LYMPHOCYTE BINDING, 4495
   BENZ(A)ANTHRACENE, 7,12-DIMETHYL-
      ESTRADIOL, 4202
      IMMUNITY, CELLULAR, 4495
      TRANSFERASES, 4719
   2,3-BUTANEDIOL, 1,4-DIMERCAPTO-
      GLUCOSEPHOSPHATE DEHYDROGENASE,
         4771*
   CARCINOMA
      CASEIN, 4752*
      ETHYLENE, CHLORO-, 4312
   CARCINOMA, DUCTAL
      CASEIN, 4752*
   CHROMATIN
      ESTRADIOL, 4202
      PROTEINS, 4743*
   EPINEPHRINE
      ADENOSINE CYCLIC 3',5'
         MONOPHOSPHATE, 4796*
   ESTRADIOL
      ESTROGEN RECEPTORS, REVIEW, 4202
   GLUCOSEPHOSPHATE DEHYDROGENASE
      ESTRONE, 4303
      ISOENZYMES, 4779*
      ISOLATION AND CHARACTERIZATION,
         4771*
      PRECANCEROUS CONDITIONS, 4771*
      PROGESTERONE, 4303
   LACTOSE SYNTHETASE
      ESTRONE, 4303
      PROGESTERONE, 4303
   LYMPHOCYTES
      IMMUNE RESPONSE, 4516*
   POLY A-U
      STATISTICAL ANALYSIS, 4217
   PRECANCEROUS CONDITIONS
      CASEIN, 4752*
      EPIDEMIOLOGY, MOUSE, 4723
      IMMUNE RESPONSE, 4516*
   PROLACTIN
      ADENOSINE CYCLIC 3',5'
         MONOPHOSPHATE, 4796*
      ESTROGEN RECEPTORS, REVIEW, 4202
   PROSTAGLANDINS E
      ADENOSINE CYCLIC 3',5'

S, 4440

S, 4440

4538*

3*
53*

TION,

TION,

TION,

CETATE

MONOPHOSPHATE, 4796*
SARCOMA, RETICULUM CELL
    HISTOCOMPATIBILITY ANTIGENS, 4543*
    TRANSPLANTATION IMMUNOLOGY, 4543*
TRANSFERASES
    ESTROGENS, 4719
    RECEPTORS, HORMONE, 4719
TRANSPLANTATION, HOMOLOGOUS
    ANTIBODIES, NEOPLASM, 4456
    GROWTH, 4740*
    IMMUNOSUPPRESSION, 4740*
TRANSPLANTATION IMMUNOLOGY
    IMMUNOSUPPRESSION, 4739*
VIRUS, MURINE MAMMARY TUMOR
    ANTIBODIES, NEOPLASM, 4456
    IMMUNITY, CELLULAR, 4456

MANDIBULAR NEOPLASMS
ODONTOGENIC TUMOR
    NEOPLASM SEEDING, 4568

MEDIASTINAL NEOPLASMS
LYMPHOMA
    UREA, METHYL NITROSO-, 4437

MEDULLOBLASTOMA
CEREBRELLAR NEOPLASMS
    EPIDEMIOLOGY, RHODESIA, 4648

MEDULLOEPITHELIOMA
SEE NEUROEPITHELIOMA

MEGAKARYOCYTES
LEUKEMIA
    CASE REPORT, 4585
    KARYOTYPING, 4585
LEUKEMIA, MYELOCYTIC
    CASE REPORT, 4585

MEHTANESULFONIC ACID, ETHYL ESTER
SACCHAROMYCES CEREVISIAE
    MUTAGENIC ACTIVITY, 4343*
SACCHAROMYCES POMBE
    MUTAGENIC ACTIVITY, 4343*

MELANIN
NEOPLASMS, EXPERIMENTAL
    DROSOPHILA MELANOGASTER, 4767*
    GENETICS, 4767*

MELANOCYTES
NEUROFIBROMATOSIS
    DIFFERENTIAL DIAGNOSITCS, 4559
NEVUS
    DIFFERENTIAL DIAGNOSITCS, 4559
NEVUS, PIGMENTED
    ULTRASTRUCTURAL STUDY, 4560

MELANOMA
ADENOSINE, CYCLIC 3',5' MONOPHOSPHATE
    MSH, 4721
ANTIGENS, NEOPLASM
    HYPERSENSITIVITY, DELAYED, 4479
    IMMUNITY, CELLULAR, 4479
AUTOANTIBODIES
    FLUORESCENT ANTIBODY TECHNIC, 4528*
BREAST NEOPLASMS
    GENETICS, 4799*

CELL MEMBRANE
    FREEZE-ETCH STUDY, NUCLEAR PORES 4593
CHOROID NEOPLASMS
    EPIDEMIOLOGY, 4564
    NEOPLASM METASTASIS, 4564
    PROGNOSIS, TUMOR SIZE, 4564
DIPHENOL OXIDASES
    ISOLATION AND CHARACTERIZATION, 4781*
    MELANOSOMES, BIOCHEMICAL STUDY, 4782*
EAR NEOPLASMS
    MELANOSOMES, BIOCHEMICAL STUDY, 4782*
EPIDEMIOLOGY
    AUSTRALIA, 4644
GENETICS
    NEOPLASMS, MULTIPLE PRIMARY, 479
GROWTH
    MSH, 4721
HYDROLASES
    MELANOSOMES, BIOCHEMICAL STUDY, 4782*
IMIDAZOLE-4-CARBOXAMIDE, 5-(3,3-BIS-(CHLOROETHYL)-1-TRIAZENO)-
    IMMUNITY, CELLULAR, 4546*
    IMMUNOSUPPRESSION, 4546*
IMMUNE SERUMS
    LYMPHOCYTE TRANSFORMATION, 4539*
IMMUNITY, CELLULAR
    ROSETTE FORMATION, 4553*
LYMPHOCYTES
    IMMUNOSUPPRESSION, 4539*
NEOPLASM TRANSPLANTATION
    ATHYMIC MICE, 4438
PEROXIDASES
    MELANOSOMES, BIOCHEMICAL STUDY, 4782*
PHARYNGEAL NEOPLASMS
    RADIATION, IONIZING, 4235*
PRECANCEROUS CONDITIONS
    DIAGNOSIS AND PROGNOSIS, REVIEW, 4258*
SKIN NEOPLASMS
    DIAGNOSIS AND PROGNOSIS, REVIEW, 4258*
    EPIDEMIOLOGY, AFRICA, 4646
VAGINAL NEOPLASMS
    DIAGNOSIS AND PROGNOSIS, 4571
XANTHINE, 3-ISOBUTYL-1-METHYL-
    GLYCOPEPTIDES, 4361*

MELOCHIA TOMENTOSA
MELOCHINONE
    ISOLATION AND CHARACTERIZATION, 4299

MELOCHINONE
MELOCHIA TOMENTOSA
    ISOLATION AND CHARACTERIZATION, 4299

MENINGIOMA
BRAIN NEOPLASMS
    IMMUNITY, CELLULAR, 4549*
CELLULAR INCLUSIONS
    HYALINE, 4602*
    ULTRASTRUCTURAL STUDY, 4602*

METHIONINE
    CELL NUCLEUS
        CELL CYCLE KINETICS, 4688
        TEMPERATURE-SENSITIVE CLONES, 4688

METHOTREXATE
    CARCINOMA, BRONCHOGENIC
        CASE REPORT, 4607*
        NEOPLASM METASTASIS, 4607*
        PSORIASIS, 4607*
    CARCINOMA, EHRLICH TUMOR
        DNA REPAIR, 4342*
    LEUKEMIA, MYELOCYTIC
        ACUTE TRANSFORMATION, 4588

N-METHYL-N'-NITRO-N-NITROSOGUANIDINE
    SEE GUANIDINE, 1-METHYL-3-NITRO-1-
        NITROSO-

N-METHYL-N-NITROSOUREA
    SEE UREA, METHYL NITROSO-

1-METHYL-1-NITROSOUREA
    SEE UREA, METHYL NITROSO-

METHYL SULFOXIDE
    SEE METHANE, SULFINYLBIS-

METHYLATION
    VIRUS, ADENO 2
        RNA, TRANSFER, METHYLTRANSFERASES,
          4428*

3-METHYLCHOLANTHRENE
    SEE CHOLANTHRENE, 3-METHYL-

20-METHYLCHOLANTHRENE
    SEE CHOLANTHRENE, 3-METHYL-

METHYLNITROSOUREA
    SEE UREA, METHYL NITROSO-

LEUKEMIA, MYELOCYTIC, 4583

MICROSOMES, LIVER
    STILBENE
        BINDING SITES, RAT, 4304

MICROTUBULES
    HELA CELLS
        ULTRASTRUCTURAL STUDY, 4580

MILK
    ALFATOXIN M1
        QUANTITATION METHOD, 4667

MITOCHONDRIA
    LEUKEMIA, LYMPHOBLASTIC
        ULTRASTRUCTURAL STUDY, 4625*

MITOCHONDRIA, LIVER
    BUTYRIC ACID, 2-AMINO-4-(ETHYLTHIO)-
        AMINOLEVULINIC ACID SYNTHETASE,
          4320*

MITOMYCIN C
    B-LYMPHOCYTES
        ANTIGENS, NEOPLASM, 4540*
        TUBERCULIN, 4490

T-LYMPHOCYTES
    ANTIGENS, NEOPLASM, 4540*
    TUBERCULIN, 4490
PLASMACYTOMA
    T-LYMPHOCYTES, 4443
    TRANSPLANTATION IMMUNOLOGY, 4443
VIRUS, VACCINIA
    DNA, VIRAL, 4402

MITOSIS
    LIVER
        MONOCROTALINE, 4302
        RETRONECINE, 3,8-DIDEHYDRO-, 4302
    MUTATION
        POLYPLOIDIZATION-SEGREGATION,
          4611*
    SARCOMA
        CYCLOPHOSPHAMIDE, 4280
    TRACHEAL NEOPLASMS
        BENZO(A)PYRENE, 4338*

MIXED FUNCTION OXIDASE
    SEE OXIDOREDUCTASES

MONOCRCTALINE
    LEUKEMIA, MYELOBLASTIC
        HEPATOECTOMY, 4302
    LIVER
        MITOSIS, 4302
    LIVER NEOPLASMS
        CARCINOMA, 4302
    LUNG NEOPLASMS
        ADENOMA, 4302

MOSAICISM
    TERATOID TUMOR
        GROWTH, 4600

MOUTH NEOPLASMS
    CARCINOMA, EPIDERMOID
        ANEMIA, APLASTIC, 4556
        FORMALDEHYDE, 4556
        NORETHANDROLONE, 4556
        OXANDROLONE, 4556
        RADIATION, IONIZING, 4364
    FIBROSARCOMA
        RADIATION, IONIZING, 4364
    SMOKING
        EPIDEMIOLOGY, INDIA, 4656

MSH
    MELANOMA
        ADENOSINE, CYCLIC 3',5'
          MONOPHOSPHATE, 4721
        GROWTH, 4721

MUCOPOLYSACCHARIDES
    VIRUS, SV40
        CELL ADHESION, 4212
        CELL MEMBRANE, 4212

MUCOPOLYSACCHARIDOSIS
    LYMPHOCYTES
        CELLULAR INCLUSIONS, 4601*

MULTIPLE MYELOMA
    ALPHA GLOBULINS
        SERUM LEVEL, 4461
    CHROMOSOME ABERRATIONS

HEPATOMA
    OLIGOPEPTIDES, 4777*
    PEPTIDE HYDROLASES, 4777*
LARYNGEAL NEOPLASMS
    LYMPHOCYTES, 4504
    T-LYMPHOCYTES, 4504
LEUKEMIA, LYMPHOCYTIC
    IMMUNOGLOBULINS, 4628*
LUNG NEOPLASMS
    BACILLUS CALMETTE-GUERIN, 4293
    RADIATION, 4235*
LYMPHATIC NEOPLASMS
    BACILLUS CALMETTE-GUERIN, 4293
    SARCOMA, 4544*
NEPHROBLASTOMA
    NEOPLASM TRANSPLANTATION, 4597
PROSTATIC NEOPLASMS
    ADENOCARCINOMA, 4575
PULMONARY NEOPLASMS
    SARCOMA, 4544*
SARCOMA
    VIRUS, HERPES SIMPLEX, 4213
SARCOMA, OSTEOGENIC
    RADIATION, 4235*
THYMOMA
    HISTOLOGICAL STUDY, 4643*
THYROID NEOPLASMS
    ADENOCARCINOMA, 4653
    CARCINOMA, PAPILLARY, 4653
UROGENITAL NEOPLASMS
    SEX CHROMATIN, 4638*

NEOPLASM REGRESSION, SPONTANEOUS
HEMATOMA
    VIRUS, ROUS SARCOMA, 4595
SARCOMA
    VIRUS, ROUS SARCOMA, 4595

NEOPLASM SEEDING
MANDIBULAR NEOPLASMS
    ODONTOGENIC TUMOR, 4568

NEOPLASM TRANSPLANTATION
ANTIGENIC DETERMINANTS
    MEMBRANE CARRIER MODEL, 4554*
CARCINOMA, EHRLICH TUMOR
    BRAIN, 4765*
FIBROSARCOMA
    CHOLANTHRENE, 3-METHYL-, 4509*
    IMMUNE SERUMS, 4509*
GLIOMA
    ATHYMIC MICE, 4438
HEPATOMA
    IMMUNE RESPONSE, 4510*
    IMMUNE SERUMS, 4451
LYMPHOMA
    ATHYMIC MICE, 4438
    VIRUS, MOLONEY MURINE LEUKEMIA,
        4438
MELANOMA
    ATHYMIC MICE, 4438
MULTIPLE MYELOMA
    PROPIONITRILE, 3-AMINO-, 4598
NEPHROBLASTOMA
    GROWTH, 4597
    NEOPLASM METASTASIS, 4597
NERVOUS SYSTEM NEOPLASMS
    EPENDYMOMA, 4632*
    GLIOMA, 4632*

NEURILEMMOMA
    UREA, ETHYL NITROSO-, 4435
OVARIAN NEOPLASMS
    ASCITES, 4640*
    LYMPHATIC OBSTRUCTION, 4640*
PLASMACYTOMA
    ENDOTOXINS, 4513*
TERATOID TUMOR
    REVIEW, 4261*
ULTRAVIOLET RAYS
    IMMUNOSUPPRESSION, 4512*

NEOPLASMS (GENERAL AND UNSPECIFIED)
    SEE ALSO UNDER PARTICULAR SITE
    ALKALINE PHOSPHATASE
        ISOENZYMES, 4728*
    ANTIGENS
        IMMUNITY, CELLULAR, 4520*
        T-LYMPHOCYTES, 4520*
    ANTIGENS, NEOPLASMS
        REVIEW, 4246*
    ANTIHYPERTENSIVE AGENTS
        EPIDEMIOLOGY, CHILD, 4654
    BENZENE, 1-CHLORO-2,4-DINITRO-
        IMMUNITY, CELLULAR, 4520*
        T-LYMPHOCYTES, 4520*
    CARCINOGEN, CHEMICAL
        EPIDEMIOLOGY, REVIEW, 4230*
    CELL LINE
        ISOLATION AND CHARACTERIZATION,
            4730*
    CHROMOSOME ABNORMALITIES
        R-BANDING, 4612*
    DIAGNOSIS AND PROGNOSIS
        REVIEW, 4264*
    DIURETICS
        EPIDEMIOLOGY, CHILD, 4654
    DNA
        LIPOPEROXIDE, 4757*
    EPIDEMIOLOGY
        CHILD, 4651
        CONSIDERATIONS, REVIEW, 4681*
        REVIEW, 4264*
        STATISTICAL ANALYSIS, REVIEW, 4211
    EVOLUTION
        CARCINOGEN, NON-INDUSTRIAL, 4233*
        REVIEW, 4233*
    GENETICS
        MATHEMATICAL MODELS, EVALUATION,
            4666
    GONADOTROPINS, CHORIONIC
        NONENDOCRINE TUMOR, 4795*
    HORMONES, ECTOPIC
        NONENDOCRINE TUMOR, 4795*
    IMMUNITY, CELLULAR
        REVIEW, 4246*
    IMMUNOLOGIC DEFICIENCY SYNDROMES
        RELATIONSHIP, REVIEW, 4252*
    IMMUNOTHERAPY
        REVIEW, 4246*
    IRON
        EPIDEMIOLOGY, CHILD, 4654
    ONCOGENIC VIRUSES
        EPIDEMIOLOGY, CHILD, REVIEW, 4210
    RADIATION, IONIZING
        DOSE-RESPONSE STUDY, 4680*
        EPIDEMIOLOGY, CHILD, REVIEW, 4210
        EPIDEMIOLOGY, FETUS, 4209
        POINT SOURCE, 4680*

DOPAMINE-B-HYDROXYLASE
   CELL CYCLE KINETICS, 4780*
EPIDEMIOLOGY
   CHILD, 4651
GENETICS
   CHROMOSOMES, HUMAN, 21-22, 4581

NEUROEPITHELIOMA
  NOSE NEOPLASMS
    QUINOXALINE, 1,4-DIOXIDE, 4356*

NEUROFIBROMATOSIS
  MELANOCYTES
    DIFFERENTIAL DIAGNOSITCS, 4559

NEUTROPHILS
  LEUKEMIA, MYELOCYTIC
    ALKALINE PHOSPHATASE, 4727*
  POLYCYTHEMIA VERA
    ALKALINE PHOSPHATASE, 4727*

NEVUS
  MELANOCYTES
    DIFFERENTIAL DIAGNOSITCS, 4559

NEVUS, PIGMENTED
  MELANOCYTES
    ULTRASTRUCTURAL STUDY, 4560

NICOTINAMIDE
  PANCREATIC NEOPLASMS
    ISLET CELL TUMOR, 4301

4-NITROQUINOLINE-1-OXIDE
  SEE QUINOLINE, 4-NITRO-, 1-OXIDE

N-NITROSO-N-METHYL UREA
  SEE UREA, METHYL NITROSO-

N-NITROSOMETHYLUREA
  SEE UREA, METHYL NITROSO-

NITROUS ACID, SODIUM SALT
  DIETHYLAMINE
    COCARCINOGENIC EFFECT, GUINEA PIG,
      4284
    DIMETHYLAMINE, N-NITROSO-, 4284
  FOOD ADDITIVES
    CARCINOGENIC ACTIVITY, 4349*
    MICROBIOLOGY, 4352*
    REVIEW, 4353*

NITROXYL RADICAL
  CARCINOMA, EHRLICH TUMOR
    ISOLATION AND CHARACTERIZATION,
      4742*
  LYMPHOSARCOMA
    ISOLATION AND CHARACTERIZATION,
      4742*

NOREPINEPHRINE
  ADRENAL GLAND NEOPLASMS
    PHEOCHROMOCYTOMA, 4327*

NORETHANDROLONE
  HEPATOMA
    ANEMIA, APLASTIC, 4556
  LEUKEMIA
    ANEMIA, APLASTIC, 4556

MOUTH NEOPLASMS
    CARCINOMA, EPIDERMOID, 4556

NOSE NEOPLASMS
    ADENOCARCINOMA
        QUINOXALINE, 1,4-DIOXIDE, 4356*
    CARCINOMA, BASAL CELL
        QUINOXALINE, 1,4-DIOXIDE, 4356*
    CARCINOMA, EPIDERMOID
        QUINOXALINE, 1,4-DIOXIDE, 4356*
    FIBROSARCOMA
        QUINOXALINE, 1,4-DIOXIDE, 4356*
    LEIOMYOSARCOMA
        CASE REPORT, 4608*
    NEUROEPITHELIOMA
        QUINOXALINE, 1,4-DIOXIDE, 4356*

NUCLEIC ACIDS
    SEE ALSO DNA/RNA
    DIETHYLAMINE, N-NITROSO-
        BENZ(A)ANTHRACENE, 7,12-DIMETHYL-,
        4271
    NERVOUS SYSTEM NEOPLASMS
        ASTROCYTOMA, 4783*
        GLIOBASTOMA MULTIFORME, 4783*
        OLIGODENDROGLIOMA, 4783*
        1-TRIAZINE, 3,3-DIMETHYL-1-PHENYL-,
        4783*
    VIRUS, AVIAN LEUKOSIS
        ISOLATION AND CHARACTERIZATION,
        4244*
    VIRUS, C-TYPE RNA TUMOR
        ISOLATION AND CHARACTERIZATION,
        4244*
    VIRUS, EPSTEIN-BARR
        ISOLATION AND CHARACTERIZATION,
        4244*
    VIRUS, HERPES SIMPLEX
        ISOLATION AND CHARACTERIZATION,
        4244*
    VIRUS, MURINE MAMMARY TUMOR
        ISOLATION AND CHARACTERIZATION,
        4244*
    VIRUS, SV40
        ISOLATION AND CHARACTERIZATION,
        4244*

NUCLEOPROTEINS
    HEPATOMA
        COMPLEXES, UI AND U2 RNA, 4694
    RNA
        ISOLATION AND CHARACTERIZATION,
        4695

NUTRITION
    SEE DIET

OBESITY
    UTERINE NEOPLASMS
        EPIDEMIOLOGY, 4673*

OCCUPATIONAL HAZARD
    ACETAMIDE, N-FLUOREN-2-YL-
        EPIDEMIOLOGY, 4205
    ANILINE, N,N-DIMETHYL-P-PHENYLAZO-
        EPIDEMIOLOGY, 4205
    AURAMINE
        EPIDEMIOLOGY, 4205
    BENZIDINE

        EPIDEMIOLOGY, 4205
    BENZIDINE, 3,3'-DICHLORO-
        EPIDEMIOLOGY, 4205
    BENZIDINE, 3,3'-DIMETHOXY-
        EPIDEMIOLOGY, 4205
    4-BIPHENYLAMINE
        EPIDEMIOLOGY, 4205
    DIMETHYLAMINE, N-NITROSO-
        EPIDEMIOLOGY, 4205
    DIPHENYLAMINE
        EPIDEMIOLOGY, 4205
    GASTROINTESTINAL NEOPLASMS
        ASBESTOS, 4262*
    HODGKIN'S DISEASE
        EPIDEMIOLOGY, 4660, 4661
    LARYNGEAL NEOPLASMS
        ASBESTOS, 4262*
    LEUKEMIA
        RADIATION, 4208
    LEUKEMIA, LYMPHOCYTIC
        EPIDEMIOLOGY, 4661
    LEUKEMIA, MYELOBLASTIC
        BOVINE LEUKEMIA, RELATIONSHIP,
        4661
        EPIDEMIOLOGY, 4661
    LEUKEMIA, MYELOCYTIC
        EPIDEMIOLOGY, 4661
    LUNG NEOPLASMS
        ASBESTOS, 4262*
    2-NAPHTHYLAMINE
        EPIDEMIOLOGY, 4205
    PERITONEAL NEOPLASMS
        ASBESTOS, 4262*
    PLEURAL NEOPLASMS
        ASBESTOS, 4262*
    PLUTONIUM
        PRECANCEROUS CONDITIONS, 4207
    O-TOLUIDINE
        EPIDEMIOLOGY, 4205
    UREA, 1-(1-NAPHTHYL)-2-THIO-
        EPIDEMIOLOGY, 4205

ODONTOGENIC TUMOR
    MANDIBULAR NEOPLASMS
        NEOPLASM SEEDING, 4568

OLIGODENDROGLIOMA
    NERVOUS SYSTEM NEOPLASMS
        NUCLEIC ACIDS, 4783*

OLIGONUCLEOTIDES
    RNA, MESSENGER
        ISOLATION AND CHARACTERIZATION,
        4704

OLIGOPEPTIDES
    HEPATOMA
        NEOPLASM METASTASIS, 4777*

ONCOGENIC VIRUSES
    NEOPLASMS
        EPIDEMIOLOGY, CHILD, REVIEW, 4210

ORNITHINE DECARBOXYLASE
    B-LYMPHOCYTES
        ANTI-ANTIBODIES, 4453
        CYTOCHALASIN B, 4453

CYTOSINE, 1-BETA-D-
ARABINOFURANOSYL-, 4453
UREA, HYDROXY-, 4453

OUABAIN
LYMPHOCYTES
BINDING SITES, 4505

OVARIAN NEOPLASMS
ANTIGENS, NEOPLASM
LYMPHOCYTE TRANSFORMATION, 4487
BURKITT'S LYMPHOMA
EPIDEMIOLOGY, 4659
CARCINOEMBRYONIC ANTIGEN
LYMPHOCYTE TRANSFORMATION, 4487
CARCINOMA, MUCINOUS
MULTIFOCAL PRIMARY CARCINOMA, 4569
CORTISOL
ASCITES, 4640*
CYSTADENOMA, PAPILLARY
LYMPHOCYTE TRANSFORMATION, 4487
HEPARIN
ASCITES, 4640*
IMMUNITY, CELLULAR
LYMPHOCYTES, 4486
NEOPLASM TRANSPLANTATION
ASCITES, 4640*
LYMPHATIC OBSTRUCTION, 4640*
WARFARIN
ASCITES, 4640*

OXANDROLONE
HEPATOMA
ANEMIA, APLASTIC, 4556
LEUKEMIA
ANEMIA, APLASTIC, 4556
MOUTH NEOPLASMS
CARCINOMA, EPIDERMOID, 4556

OXAZOLE, 2,5-DIPHENYL-
ARYL HYDROCARBON HYDROXYLASES
CYTOCHROME P-450, 4278

2-OXETANONE
DNA
ALKYLATING AGENTS, 4321*

OXIDOREDUCTASES
ANILINE, N,N-DIMETHYL-P-PHENYLAZO-
ENZYMATIC ACTIVITY, 4298
HEPATOMA
DIET, 4768*

OXIDOREDUCTASES, N-DEMETHYLATING
ANILINE, N,N-DIMETHYL-P-PHENYLAZO-
ENZYMATIC ACTIVITY, 4298
LUNG NEOPLASMS
DIMETHYLAMINE, N-NITROSO-, 4287

OXYMETHOLONE
HEPATOMA
ANEMIA, APLASTIC, 4556
LEUKEMIA
ANEMIA, APLASTIC, 4556

PALATAL NEOPLASMS
SMOKING
EPIDEMIOLOGY, INDIA, 4656

PANCREATIC NEOPLASMS
ADENOCARCINOMA
BENZO(A)PYRENE, 4276
CARCINOEMBRYONIC ANTIGEN, 4474
STOMACH NEOPLASMS, 4474
TISSUE CULTURE, 4255*
CARCINOEMBRYONIC ANTIGEN
THERAPY MONITORING, 4254*
ISLET CELL TUMOR
HELIOTRINE, 4301
NICOTINAMIDE, 4301
PAPILLOMA
BENZO(A)PYRENE, 4276

PAPILLOMA
BENZO(A)PYRENE
COMMON STATISTICAL FACTORS, SITE-
SPECIFICITY, 4276
CROTON OIL
BENZ(A)ANTHRACENE, 7,12-DIMETHYL-,
4363
ULTRAVIOLET RAYS, 4363
PANCREATIC NEOPLASMS
BENZO(A)PYRENE, 4276
PROTEINS
CELL TRANSFORMATION, NEOPLASTIC,
4689
ISOLATION AND CHARACTERIZATION,
4689
PRECANCEROUS CONDITIONS, 4689
SKIN NEOPLASMS
ULTRAVIOLET RAYS, 4362, 4363

PARAFFIN OIL
PLASMACYTOMA
ENDOTOXINS, 4513*

PARAPROTEINEMIA
LEUKEMIA, MYELOBLASTIC
ALANINE, 3-(P-(BIS(2-CHLOROETHYL)
AMINO)PHENYL)-, L-, 4624*

PEANUTS
MYCOTOXINS
ENVIRONMENTAL HAZARD, 4676*

PENTOSYLTRANSFERASES
HEPATOMA
GROWTH RATE, 4726*
KIDNEY NEOPLASMS
RAT, 4726*

PEPTIDE HYDROLASES
HEPATOMA
NEOPLASM METASTASIS, 4777*
VIRUS, SV40
CELL MEMBRANE, 4212
CHOLESTEROL, 4212

PEPTIDES
VIRUS, AVIAN MYELOBLASTOSIS
VIRAL PROTEINS, 4373
VIRUS, SV40
TEMPERATURE SENSITIVE MUTANTS,
4417*

PERITONEAL NEOPLASMS
  ASBESTOS
    OCCUPATIONAL HAZARD, 4262*
  EPIDEMIOLOGY
    AUSTRALIA, 4644

PEROXIDASES
  MELANOMA
    MELANOSOMES, BIOCHEMICAL STUDY,
    4782*

PESTICIDES
  DIELDRIN
    CARCINOGENIC ACTIVITY, REVIEW,
    4230*

PHARYNGEAL NEOPLASMS
  CARCINOMA, EPIDERMOID
    RADIATION, IONIZING, 4364
  EPIDEMIOLOGY
    AUSTRALIA, 4644

  FIBROSARCOMA
    RADIATION, IONIZING, 4364
  MELANOMA
    RADIATION, IONIZING, 4235*
  SMOKING
    EPIDEMIOLOGY, INDIA, 4656

PHENOBARBITAL
  SEE BARBITURIC ACID, 5-ETHYL-5-PHENYL-

PHENOBARBITONE
  SEE BARBITURIC ACID, 5-ETHYL-5-PHENYL-

PHEOCHROMOCYTOMA
  ADRENAL GLAND NEOPLASMS
    CATECHOLAMINES, 4327*
    INCIDENCE, MOUSE, 4723
    NOREPINEPHRINE, 4327*
  BLADDER NEOPLASMS
    CASE REPORT, 4636*

PHORBOL MYRISTATE ACETATE
  SEE 12-O-TETRADECANOYLPHORBOL-13-
  ACETATE

PHOSPHINE SULFIDE, TRIS(1-AZIRIDINYL)-
  BONE MARROW CELLS
    CHROMOSOME ABERRATIONS, 4359*

PHOSPHODIESTERASES
  VIRUS, B77
    CELL TRANSFORMATION, NEOPLASTIC,
    4375
    TEMPERATURE SENSITIVE MUTANTS,
    4375
  VIRUS, POLYOMA
    GROWTH SUBSTANCES, 4391
  VIRUS, ROUS SARCOMA
    CELL TRANSFORMATION, NEOPLASTIC,
    4375
    TEMPERATURE SENSITIVE MUTANTS,
    4375
  VIRUS, SV40
    ACTINOMYCIN D, 4712
    CELL TRANSFORMATION, NEOPLASTIC,
    4712

    DIBUTYRYL CYCLIC AMP, 4712

PHOSPHOLIPIDS
  VIRUS, SV40
    CELL MEMBRANE, 4212, 4398

PHOSPHONOACETIC ACID
  CELL TRANSFORMATION, NEOPLASTIC
    VIRUS, HERPES SIMPLEX, 4406*
  DNA NUCLEOTIDYLTRANSFERASES
    ENZYMATIC ACTIVITY, 4406*

PHOSPHORIBOSYL PYROPHOSPHATE SYNTHETASE
  SEE PHOSPHOTRANSFERASES, ATP

PHOSPHORYLCHOLINE
  SEE CHOLINE
  ANTIGENS, BACTERIAL
    IMMUNE RESPONSE, REVIEW, 4218
  IGA
    IMMUNE RESPONSE, REVIEW, 4218
  IMMUNOGLOBULINS
    AUTOANTIBODIES, 4218
    IMMUNE RESPONSE, REVIEW, 4218
    IMMUNE SERUMS, 4218
    IMMUNOSUPPRESSION, 4218
  B-LYMPHOCYTES
    IMMUNE RESPONSE, REVIEW, 4218
    IMMUNOGLOBULINS, 4218
  T-LYMPHOCYTES
    IMMUNE RESPONSE, REVIEW, 4218
  MYELOMA PROTEINS
    IMMUNE RESPONSE, REVIEW, 4218

PHOSPHOTRANSFERASES
  VIRUS, ADENO
    CELL CYCLE KINETICS, 4394
  VIRUS, SV40
    CELL CYCLE KINETICS, 4394

PHYTOHEMAGGLUTININ
  SEE PLANT AGGLUTININS

PIPERIDINE, 1-NITROSO-
  SUBSTITUENT GROUP
    CARCINOGENIC ACTIVITY, 4348*

PITUITARY GROWTH HORMONE
  SEE SOMATOTROPIN

PITUITARY NEOPLASMS
  ADENOMA
    ETHYLENE, CHLORO-, 4311
  ADENOMA, CHROMOPHOBE
    HYPOTHROIDISM, 4576
    SECRETORY GRANULES, DIAMETER, 457

PLACENTA
  ALKALINE PHOSPHATASE
    ISOENZYMES, 4728*

PLANT AGGLUTININS
  BURKITT'S LYMPHOMA
    CELL TRANSFORMATION, NEOPLASTIC,
    4506
  GLIOMA
    LYMPHOCYTES, 4549*
  HODGKIN'S DISEASE
    CELL TRANSFORMATION, NEOPLASTIC,

4506
LEUKEMIA
  CELL TRANSFORMATION, NEOPLASTIC,
    4506
LYMPHOCYTES
  CELL MEMBRANE, 4505
LYMPHOMA
  CELL TRANSFORMATION, NEOPLASTIC,
    4506
MULTIPLE MYELOMA
  CELL TRANSFORMATION, NEOPLASTIC,
    4506
SARCOMA
  TRANSPLANTATION, HOMOLOGOUS, 4493

PLANT TUMORS
  CHROMOSOME ABNORMALITIES
    CLONE CELLS, 4736*

PLANTS
  CHROMOSOME ABERRATIONS
    FUNGICIDES, INDUSTRIAL, 4291
  METHANE, DIAZO-
    MUTATION, 4340*

PLASMA CELLS
  MULTIPLE MYELOMA
    ULTRASTRUCTURAL STUDY, 4738*

PLASMACYTOMA
  CYCLOPHOSPHAMIDE
    T-LYMPHOCYTES, 4443
    TRANSPLANTATION IMMUNOLOGY, 4443
  ENDOTOXINS
    DOSE-RESPONSE STUDY, 4513*
    IMMUNE RESPONSE, 4513*
    NEOPLASM TRANSPLANTATION, 4513*
    PARAFFIN OIL, 4513*
  IGA
    ULTRASTRUCTURAL STUDY, 4594
  MITOMYCIN C
    T-LYMPHOCYTES, 4443
    TRANSPLANTATION IMMUNOLOGY, 4443
  RIBOSOMES
    CELL MEMBRANE, 4699
    KINETICS, 4699
  RNA, MESSENGER
    DNA-RNA HYBRIDIZATION, 4703
    ISOLATION AND CHARACTERIZATION,
      MOUSE, 4703
  RNA, RIBOSOMAL
    CELL CYCLE KINETICS, 4698
    POLYSOMES, 4698

PLASMINOGEN
  MULTIPLE MYELOMA
    URIDINE, 5-BROMO-2-'-DEOXY-, 4687

PLATELET AGGREGATION
  LEUKEMIA
    BLOOD PLATELETS, 4591

PLEURAL NEOPLASMS
  ASBESTOS
    OCCUPATIONAL HAZARD, 4262*

PLUTONIUM
  PRECANCEROUS CONDITIONS
    OCCUPATIONAL HAZARD, 4207

RADIOACTIVE FALLOUT
  TOKYO, RADIOISOTOPES 239 AND 238,
    4662

PNEUMONIA
  HODGKIN'S DISEASE
    EPIDEMIOLOGY, 4660
  LUNG NEOPLASMS
    MESOTHELIOMA, 4578

POLY A
  HEPATOMA
    RNA, MESSENGER, 4704
  SARCOMA 180, CROCKER
    ISOLATION AND CHARACTERIZATION,
      4700
    RNA, MESSENGER, 4700

POLY A-U
  MAMMARY NEOPLASMS, EXPERIMENTAL
    STATISTICAL ANALYSIS, 4217

POLYCYCLIC HYDROCARBONS
  ENVIRONMENTAL HAZARD
    ISOLATION AND CHARACTERIZATION,
      4336*

POLYCYTHEMIA VERA
  1,4-BUTANEDIOL, DIMETHANESULFONATE
    CHROMOSOME ABNORMALITIES, 4584
  CYCLOPHOSPHAMIDE
    CHROMOSOME ABNORMALITIES, 4584
  HISTAMINE
    METABOLISM, 4622*
  KARYOTYPING
    CHROMOSOME ABNORMALITIES, 4584
  NEUTROPHILS
    ALKALINE PHOSPHATASE, 4727*
  RADIATION, IONIZING
    CHROMOSOME ABNORMALITIES, 4584

POLYNUCLEOTIDES
  VIRUS, PAPOVA, BK
    DNA, 4392
  VIRUS, SV40
    DNA, 4392

POLYPS
  COLONIC NEOPLASMS
    METHANOL, (METHYLNITROSOAMINO)-,
      ACETATE (ESTER), 4350*

POLYRIBOSOMES
  LEUKEMIA
    PROTEINS, 4790*
    RIBONUCLEASE, 4790*

POLYSOMES
  PLASMACYTOMA
    RNA, RIBOSOMAL, 4698

POLYVINYL CHLORIDE
  SEE ETHYLENE, CHLORO- POLYMER

PRECANCEROUS CONDITIONS
  ANEMIA, APLASTIC
    CHROMOSOME ABNORMALITIES, 4556
    GENETICS, 4556
  BREAST NEOPLASMS

EPIDEMIOLOGY, SASKATCHEWAN, 4668
BRONCHIAL NEOPLASMS
    DNA, 4605*
    RNA, 4605*
CARCINOMA, EPIDERMOID
    PROTEINS, 4689
CERVIX NEOPLASMS
    CONTRACEPTIVES, ORAL, 4306
    DIAGNOSIS AND PROGNOSIS, REVIEW,
    4223
    EPIDEMIOLOGY, 4674*
    ULTRASTRUCTURAL STUDY, 4603*
COLONIC NEOPLASMS
    REVIEW, 4256*
ESOPHAGEAL NEOPLASMS
    ESOPHAGITIS, 4579
GYNECOLOGIC NEOPLASMS
    4,4'-STILBENEDIOL, ALPHA,ALPHA'-
    DIETHYL-, 4305
HEPATOMA
    CONTRACEPTIVES, ORAL, 4308
    ENZYMATIC ACTIVITY, RAT, 4758*
    FETAL GLOBULINS, 4468
LEUKEMIA
    BLOOD PLATELETS, 4591
    CASE REPORT, 4556
    REVIEW, 4260*
LEUKEMIA, MONOBLASTIC
    BONE MARROW CELLS, 4220
LEUKEMIA, MONOCYTIC
    BONE MARROW CELLS, 4220
LEUKEMIA, MYELOCYTIC
    BONE MARROW CELLS, 4220
LUNG NEOPLASMS
    AEROSOLS, 4309
MAMMARY NEOPLASMS, EXPERIMENTAL
    CASEIN, 4752*
    EPIDEMIOLOGY, MOUSE, 4723
    IMMUNE RESPONSE, 4516*
MELANOMA
    DIAGNOSIS AND PROGNOSIS, REVIEW,
    4258*
PAPILLOMA
    PROTEINS, 4689
PLUTONIUM
    OCCUPATIONAL HAZARD, 4207
RECTAL NEOPLASMS
    REVIEW, 4256*
SALIVARY GLAND NEOPLASMS
    BENZ(A)ANTHRACENE, 7,12-DIMETHYL-,
    4290
    C.I. ACID RED 26, DISODIUM SALT,
    4290
    CARBAMIC ACID, N-METHYL-N-NITROSO-,
    ETHYL ESTER, 4290
    CHOLANTHRENE, 3-METHYL-, 4290
    GUANIDINE, 1-METHYL-3-NITRO-1-
    NITROSO-, 4290
    QUINOLINE, 4-NITRO-, 1-OXIDE, 4290
SKIN NEOPLASMS
    ULTRAVIOLET RAYS, 4363
STOMACH NEOPLASMS
    EPIDEMIOLOGY, EUROPE, 4649
    GASTRIC MUCOSA, 4565
    HISTOLOGICAL STUDY, 4565
TRACHEAL NEOPLASMS
    ASBESTOS, 4314*
    BENZO(A)PYRENE, 4338*
VAGINAL NEOPLASMS

CONTRACEPTIVES, ORAL, 4572

PREDNISONE
    SEE PREGNA-1,4-DIENE-3,11,20-TRIONE,
    17,21-DIHYDROXY-
    LEUKEMIA, MYELOCYTIC
    ACUTE TRANSFORMATION, 4588

PREGN-5-EN-20-ONE, 3-HYDROXY-, (3BETA)-
    CHORIOCARCINOMA
    METABOLITES, 4720

PREGNANCY
    FETAL GLOBULINS
    RADIOIMMUNOASSAY, RAT, 4466
    LIVER NEOPLASMS
    CASE REPORT, 4619*

PREGNANCY, MULTIPLE
    CERVIX NEOPLASMS
    EPIDEMIOLOGY, 4674*

PROGESTERONE
    MAMMARY NEOPLASMS, EXPERIMENTAL
    GLUCOSEPHOSPHATE DEHYDROGENASE,
    4303
    LACTOSE SYNTHETASE, 4303

PROLACTIN
    MAMMARY NEOPLASMS, EXPERIMENTAL
    ADENOSINE CYCLIC 3',5'
    MONOPHOSPHATE, 4796*
    ESTROGEN RECEPTORS, REVIEW, 4202

PROPIONIC ACID, 2-AMINO-2-METHYL-
    DNA REPLICATION
    CHICK EMBRYO CELLS, 4761*

PROPIONITRILE, 3-AMINO-
    MULTIPLE MYELOMA
    ANTINEOPLASTIC EFFECT, 4598
    NEOPLASM TRANSPLANTATION, 4598

PROPYL ALCOHOL
    ALCOHOLIC BEVERAGES
    CARCINOGENIC ACTIVITY, 4318*

PROSTAGLANDINS
    ASTROCYTOMA
    ADENOSINE CYCLIC 3,5 MONOPHOSPHATE,
    4791*
    ADENYL CYCLASE, 4791*
    TRANSPLACENTAL ACTIVITY
    METABOLISM, RAT, 4724

PROSTAGLANDINS E
    MAMMARY NEOPLASMS, EXPERIMENTAL
    ADENOSINE CYCLIC 3',5'
    MONOPHOSPHATE, 4796*

PROSTATIC NEOPLASMS
    ACID PHOSPHATASE
    ANTIBODIES, 4476
    ANTIGENS, NEOPLASM, 4476
    IMMUNE SERUMS, 4476
    ADENOCARCINOMA
    CASTRATION, 4575
    EPIDEMIOLOGY, IRAN, 4670*
    NEOPLASM METASTASIS, 4575

4689
    PRECANCEROUS CONDITIONS, 4689
STOMACH NEOPLASMS
    CARCINOEMBRYONIC ANTIGEN, 4248*
UROGENITAL NEOPLASMS
    CARCINOEMBRYONIC ANTIGEN, 4248*
VIRUS, SV40
    FIBROBLASTS, 4725

PSEUDOTUMORS, INFLAMMATORY
    SEE FIBROMA

PSORIASIS
    CARCINOMA, BRONCHOGENIC
        METHOTREXATE, 4607*

PULMONARY NEOPLASMS
    ADENOMA
        ETHYLENE, CHLORO-, 4312
    SARCOMA
        BENZO(A)PYRENE, 4544*
        NEOPLASM METASTASIS, 4544*

PURINE NUCLEOTIDES
    CARCINOMA, EHRLICH TUMOR
        BIOSYNTHESIS, 4753*

PURINE-6-THIOL
    LEUKEMIA, MYELOCYTIC
        ACUTE TRANSFORMATION, 4588

IH-PYRROLE-2,5-DIONE, 1-ETHYL-
    FETAL GLOBULINS
        IMMUNOLOGICAL AND ELECTROPHORETIC
        PROPERTIES, 4465

PYRUVATE KINASE
    CARCINOMA, EHRLICH TUMOR
        ENZYMATIC ACTIVITY, 4785*

QUINOLINE, 4-NITRO-, 1-OXIDE
    CELL TRANSFORMATION, NEOPLASTIC
        CAFFEINE, 4300
    SALIVARY GLAND NEOPLASMS
        PRECANCEROUS CONDITIONS, 4290
    SODIUM CHLORIDE
        COCARCINOGENIC EFFECT, 4347*
    STOMACH NEOPLASMS
        CARCINOMA, 4347*

QUINOXALINE, 1,4-DIOXIDE
    NOSE NEOPLASMS
        ADENOCARCINOMA, 4356*
        CARCINOMA, BASAL CELL, 4356*
        CARCINOMA, EPIDERMOID, 4356*
        FIBROSARCOMA, 4356*
        NEUROEPITHELIOMA, 4356*

RADIATION
    BREAST NEOPLASMS
        FLUOROSCOPY, 4371*
        REPEATED FLUOROSCOPY HISTORY,
        4372*
    HISTOCOMPATIBILITY ANTIGENS
        T-LYMPHOCYTES, 4438
    LEUKEMIA
        OCCUPATIONAL HAZARD, 4208
    LUNG NEOPLASMS
        NEOPLASM METASTASIS, 4235*

SARCOMA, OSTEOGENIC
   ANTILYMPHOCYTE SERUM, 4235*
   IMMUNE SERUMS, 4235*
   IMMUNOSUPPRESSION, 4235*
   NEOPLASM METASTASIS, 4235*
SPLENIC NEOPLASMS
   LYMPHOMA, 4438
VIRUS, SV40
   DNA, VIRAL, 4397
XERODERMA PIGMENTOSUM
   DNA REPAIR, 4228

RADIATION, IONIZING
   BLADDER NEOPLASMS
      CARCINOMA, TRANSITIONAL CELL, 4364
   BONE NEOPLASMS
      DOSE-RESPONSE STUDY, 4680*
   BREAST NEOPLASMS
      DOSE-RESPONSE STUDY, 4680*
      HODGKIN'S DISEASE, 4364
   CELL TRANSFORMATION, NEOPLASTIC
      DOSE-RESPONSE STUDY, 4680*
   CERVIX NEOPLASMS
      CARCINOMA, EPIDERMOID, 4364
   CHROMOSOME ABERRATIONS
      GENES, RECESSIVE, 4289
   COLONIC NEOPLASMS
      ADENOCARCINOMA, 4364
   DOWN'S SYNDROME
      EPIDEMIOLOGY, FETUS, 4209
   FIBROBLASTS
      CELL DIVISION, 4366
   GAMMA GLOBULINS
      IMMUNE RESPONSE, 4503
   KIDNEY NEOPLASMS
      DOSE-RESPONSE STUDY, 4680*
   LEUKEMIA
      DOSE-RESPONSE STUDY, 4680*
   LEUKEMIA, MYELOCYTIC
      CHROMOSOME ABERRATIONS, 4583
   LUNG NEOPLASMS
      DOSE-RESPONSE STUDY, 4680*
   B-LYMPHOCYTES
      IMMUNE RESPONSE, 4439
      IMMUNOGLOBULINS, 4439
   T-LYMPHOCYTES
      IMMUNE RESPONSE, 4439
   MOUTH NEOPLASMS
      CARCINOMA, EPIDERMOID, 4364
      FIBROSARCOMA, 4364
   NASOPHARYNGEAL NEOPLASMS
      HEMANGIOPERICYTOMA, 4364
   NEOPLASMS
      DOSE-RESPONSE STUDY, 4680*
      EPIDEMIOLOGY, CHILD, REVIEW, 4210
      EPIDEMIOLOGY, FETUS, 4209
      POINT SOURCE, 4680*
   NEOPLASMS, EXPERIMENTAL
      DOSE-RESPONSE STUDY, 4680*
   PHARYNGEAL NEOPLASMS
      CARCINOMA, EPIDERMOID, 4364
      FIBROSARCOMA, 4364
      MELANOMA, 4235*
   POLYCYTHEMIA VERA
      CHROMOSOME ABNORMALITIES, 4584
   RADIOISOTOPES
      TISSUE DISTRIBUTION, 4680*
   SARCOMA
      LYMPHOCYTE TRANSFORMATION, 4493

SKIN NEOPLASMS
   CARCINOMA, EPIDERMOID, 4364
   DOSE-RESPONSE STUDY, 4680*
THYROID NEOPLASMS
   ADENOCARCINOMA, 4653
   CARCINOMA, PAPILLARY, 4653
   DOSE-RESPONSE STUDY, 4680*
TONSILLAR NEOPLASMS
   CARCINOMA, EPIDERMOID, 4235*
URANIUM
   ENVIRONMENTAL HAZARD, 4675*
UTERINE NEOPLASMS
   ADENOCARCINOMA, 4364

RADIOACTIVE FALLOUT
   PLUTONIUM
      TOKYO, RADIOISOTOPES 239 AND 238,
      4662

RADIOISOTOPES
   RADIATION, IONIZING
      TISSUE DISTRIBUTION, 4680*

RAUWOLFIA ALKALOIDS
   CARCINOGENIC POTENTIAL
      REVIEW, 4231*

RECEPTORS, HORMONE
   BREAST NEOPLASMS
      ESTROGENS, 4792*
      REVIEW, 4202
      TRANSFERASES, 4719
   MAMMARY NEOPLASMS, EXPERIMENTAL
      TRANSFERASES, 4719
   SEX DIFFERENTIATION
      GENETICS, 4221
      SEX CHROMOSOMES, 4221
      TESTOSTERONE, 4221

RECTAL NEOPLASMS
   ADENOMA
      REVIEW, 4256*
   CARCINOMA, EPIDERMOID
      CARCINOEMBRYONIC ANTIGEN, 4474
   PRECANCEROUS CONDITIONS
      REVIEW, 4256*

RESERPINE
   BREAST NEOPLASMS
      EPIDEMIOLOGY, REVIEW, 4232*

RETICULOSARCOMA
   SEE SARCOMA, RETICULUM CELL

RETINOBLASTOMA
   ACRYLAMIDE, 2-(2-FURYL)-3-(5-NITRO-2-
      FURYL)-
      MUTANTS, ENVIRONMENTAL, JAPAN,
      4664
   GENETICS
      STATISTICAL ANALYSIS, JAPAN, 4664
   MUTAGENS
      STATISTICAL ANALYSIS, JAPAN, 4664

RETINOIC ACID
   CARRIER PROTEINS
      CHICK EMBRYO, 4750*
      ISOLATION AND CHARACTERIZATION,
      4750*

SUBCELLULAR FRACTIONS, 4786*

RIBOSOMES
    ACETAMIDE, N-FLUOREN-2-YL-
        ENDOPLASMIC RETICULUM, 4266
    BENZO(A)PYRENE
        ENDOPLASMIC RETICULUM, 4266
    CARCINOMA, EHRLICH TUMOR
        ELONGATION FACTORS, 4735*
    PLASMACYTOMA
        CELL MEMBRANE, 4699
        KINETICS, 4699
    VALERIC ACID, 2,2-DIPHENYL-, 2-
        (DIETHYLAMINO)ETHYL ESTER
        ENDOPLASMIC RETICULUM, 4266
    VIRUS, ADENO 2
        VIRUS, HELPER, 4393

RNA
    BACTERIOPHAGES
        METHANESULFONIC ACID, ETHYL ESTER,
            4288
        UREA, ETHYL NITROSO-, 4288
    BACTERIPHAGES
        EXTRACISTRONIC MUTATION, 4756*
    BRONCHIAL NEOPLASMS
        PRECANCEROUS CONDITIONS, 4605*
    CHALONES
        LIVER REGENERATION, 4731*
    HEPATOMA
        ACTINOMYCIN D, 4751*
        ISOLATION AND CHARACTERIZATION,
            4764*
    NUCLEOPROTEINS
        ISOLATION AND CHARACTERIZATION,
            4695
    TRANSFER FACTOR
        MECHANISM OF ACTION, 4215
    VIRUS, POLYOMA
        CELL LINE, HAMSTER, 4430*
        GIANT MOLECULES, 4431*

RNA, MESSENGER
    CELL TRANSFORMATION, NEOPLASTIC
        ISOLATION AND CHARACTERIZATION,
            4691
    ERYTHROLEUKEMIA
        VIRUS, FRIEND SPLEEN FOCUS-FORMING,
            4401
    HELA CELLS
        ISOLATION AND CHARACTERIZATION,
            MOUSE, 4697
    HEPATOMA
        POLY A, 4704
        TRYPTOPHAN OXYGENASE, 4755*
    L CELLS
        ISOLATION AND CHARACTERIZATION,
            MOUSE, 4697
    OLIGONUCLEOTIDES
        ISOLATION AND CHARACTERIZATION,
            4704
    PLASMACYTOMA
        DNA-RNA HYBRIDIZATION, 4703
        ISOLATION AND CHARACTERIZATION,
            MOUSE, 4703
    SARCOMA 180, CROCKER
        POLY A, 4700
    VIRUS, ADENO 2
        VIRUS, HELPER, 4393

VIRUS, MURINE MAMMARY TUMOR
 VIRAL REPLICATION, 4427*
VIRUS, POLYOMA
 GIANT MOLECULES, 4431*
 ISOLATION AND CHARACTERIZATION,
  4429*
 TRANSLATION, GENETIC, 4432*
VIRUS, POLYOMA SV40 HYBRID
 CELL TRANSFORMATION, NEOPLASTIC,
  4697
VIRUS, SV40
 CELL TRANSFORMATION, NEOPLASTIC,
  4691
 ISOLATION AND CHARACTERIZATION,
  4691

RNA, NEOPLASM
 HEPATOMA
  BASOPHILS, 4696
  BINDING, TOLUIDINE BLUE, 4696

RNA POLYMERASE
 ACETAMIDE, N-(ACETYLOXY)-N-9H-FLUOREN-
  2-YL-
  BINDING, 4323*
 ANILINE, N,N-DIMETHYL-P-PHENYLAZO-
  ENZYMATIC ACTIVITY, 4316*
 HELA CELLS
  VIRUS, ADENO 2, 4378
 VIRUS, SIMIAN SARCOMA
  CELL TRANSFORMATION, NEOPLASTIC,
   4410*
  VIRUS REPLICATION, 4410*

RNA REPLICATION
 HELA CELLS
  VIRUS, ADENO 2, 4378
 HEPATOMA
  CORTISOL, 4749*

RNA, RIBOSOMAL
 HELA CELLS
  ISOLATION AND CHARACTERIZATION,
   4762*
 HEPATOMA
  ISOLATION AND CHARACTERIZATION,
   4762*
 NASOPHARYNGEAL NEOPLASMS
  LEUKEMIA, MYELOBLASTIC, 4623*
 PLASMACYTOMA
  CELL CYCLE KINETICS, 4698
  POLYSOMES, 4698

RNA, TRANSFER
 MULTIPLE MYELOMA
  VALINE, 4763*
 TRYPTOPHAN
  ISOLATION AND CHARACTERIZATION,
   4389
 VIRUS, ROUS SARCOMA
  ISOLATION AND CHARACTERIZATION,
   4389

RNA, TRANSFER, METHYLTRANSFERASES
 ESTRADIOL
  UTERUS, RAT, 4701
 VIRUS, ADENO 2
  KB CELLS, 4428*
  METHYLATION, 4428*

SARCOMA 180, CROCKER
POLY A
ISOLATION AND CHARACTERIZATION,
4700
RNA, MESSENGER, 4700

SARCOMA, OSTEOGENIC
LYMPHOCYTES
IMMUNITY, CELLULAR, 4542*
RADIATION
ANTILYMPHOCYTE SERUM, 4235*
IMMUNE SERUMS, 4235*
IMMUNOSUPPRESSION, 4235*
NEOPLASM METASTASIS, 4235*
VIRUS, KIRSTEN MURINE LEUKEMIA
VIRUS REPLICATION, 4387
VIRUS, SIMIAN SARCOMA
VIRUS REPLICATION, 4387

SARCOMA, RETICULUM CELL
GENETICS
IMMUNE RESPONSE, 4582
LYMPHOMA
GRAFT VS HOST REACTION, 4555*
MAMMARY NEOPLASMS, EXPERIMENTAL
HISTOCOMPATIBILITY ANTIGENS, 4543*
TRANSPLANTATION IMMUNOLOGY, 4543*
TRANSPLANTATION, HOMOLOGOUS
IMMUNOSUPPRESSION, 4515*
TRANSPLANTATION IMMUNOLOGY
HISTOCOMPATIBILITY ANTIGENS, 4543*

SEMINOMA
SEE DISGERMINOMA

SEX BEHAVIOR
PROSTATIC NEOPLASMS
EPIDEMIOLOGY, JAPAN, 4265*

SEX CHROMATIN
BREAST NEOPLASMS
DIAGNOSIS AND PROGNOSIS, 4638*
NEOPLASM METASTASIS, 4638*
UROGENITAL NEOPLASMS
DIAGNOSIS AND PROGNOSIS, 4638*
NEOPLASM METASTASIS, 4638*

SEX CHROMOSOMES
MENINGIOMA
AGE FACTORS, 4613*
SEX DIFFERENTIATION
RECEPTORS, HORMONE, 4221

SEX DIFFERENTIATION
RECEPTORS, HORMONE
GENETICS, 4221
SEX CHROMOSOMES, 4221
TESTOSTERONE, 4221

SKIN NEOPLASMS
ANTHRACENE
7H-FURO(3,2-G)(1)BENZOPYRAN-7-ONE,
9-METHOXY-, 4370*
CARCINOID TUMOR
CASE REPORT, 4634*
CARCINOMA
ETHYLENE, CHLORO-, 4312
CARCINOMA, EPIDERMOID
RADIATION, IONIZING, 4-

ULTRAVIOLET RAYS, 4362, 4363
ETHYLENE, CHLORO-
    CARCINOGENIC ACTIVITY, RAT, 4204
FIBROMA
    ULTRAVIOLET RAYS, 4363
FIBROSARCOMA
    EPIDEMIOLOGY, MOUSE, 4723
    ULTRAVIOLET RAYS, 4363
HYPERPLASIA
    CASE REPORT, ANGIOLYMPHOID, 4561
LYMPHOMA
    T-LYMPHOCYTES, 4557
MELANOMA
    DIAGNOSIS AND PROGNOSIS, REVIEW,
       4258*
    EPIDEMIOLOGY, AFRICA, 4646
PAPILLOMA
    ULTRAVIOLET RAYS, 4362, 4363
PRECANCEROUS CONDITIONS
    ULTRAVIOLET RAYS, 4363
RADIATION, IONIZING
    DOSE-RESPONSE STUDY, 4680*
RHABDOMYOSARCOMA
    RETRONECINE, 3,8-DIDEHYDRO-, 4302
SEX FACTORS
    EPIDEMIOLOGY, FINLAND, 4645
12-O-TETRADECANOYLPHORBOL-13-ACETATE
    DNA REPLICATION, 4358*
TRANSPLANTATION, HOMOLOGOUS
    IMMUNOSUPPRESSION, 4515*
ULTRAVIOLET RAYS
    WHITENING AGENTS, 4367*, 4368*
WARTS
    IMMUNOSUPPRESSION, 4219
    INCLUSION BODIES, VIRAL, 4219
    REVIEW, 4219
XERODERMA PIGMENTOSUM
    CELL REPAIR DEFICIENCY, 4633*

SMOKING
    ALPHA 1 ANTITRYPSIN
        SERUM LEVELS, HOSPITALIZED
           PATIENTS, 4657
    GASTROINTESTINAL NEOPLASMS
        ALPHA 1 ANTITRYPSIN, 4657
    LEUKEMIA
        ALPHA 1 ANTITRYPSIN, 4657
    LUNG NEOPLASMS
        ALPHA 1 ANTITRYPSIN, 4657
        EPIDEMIOLOGY, 4682*
        MATHEMATICAL MODELS, EVALUATION,
           4666
    LYMPHOMA
        ALPHA 1 ANTITRYPSIN, 4657
    MOUTH NEOPLASMS
        EPIDEMIOLOGY, INDIA, 4656
    NEOPLASMS
        ALPHA 1 ANTITRYPSIN, 4657
    PALATAL NEOPLASMS
        EPIDEMIOLOGY, INDIA, 4656
    PHARYNGEAL NEOPLASMS
        EPIDEMIOLOGY, INDIA, 4656
    PROSTATIC NEOPLASMS
        EPIDEMIOLOGY, JAPAN, 4265*
    TONGUE NEOPLASMS
        EPIDEMIOLOGY, INDIA, 4656

SODIUM CHLORIDE
    GUANIDINE, 1-METHYL-3-NITRO-1-NITROSO-

GASTRIN
   ISOLATION AND CHARACTERIZATION,
    - 4722
   METAPLASIA
    CARCINOEMBRYONIC ANTIGEN, 4474
   PRECANCEROUS CONDITIONS
    EPIDEMIOLOGY, EUROPE, 4649
    GASTRIC MUCOSA, 4565
    HISTOLOGICAL STUDY, 4565

SUBCELLULAR FRACTIONS .
   CARCINOMA, EHRLICH TUMOR
    DNA REPLICATION, 4786*
    RIBONUCLEOTIDE REDUCTASES, 4786*
    THYMIDINE KINASE, 4786*

SUCCINIMIDE, N-(3,5-DICHLOROPHENYL)-
   NEPHRITIS, INTERSTITIAL, 4296

SUDDEN INFANT DEATH
   LEUKEMIA, LYMPHOCYTIC
    EPIDEMIOLOGY, GREAT BRITAIN, 4655
   LEUKEMIA, MYELOCYTIC
    EPIDEMIOLOGY, GREAT BRITAIN, 4655

SULFATASES
   ANILINE, N,N-DIMETHYL-P-PHENYLAZO-
    ENZYMATIC ACTIVITY, 4298
   BREAST NEOPLASMS
    HISTOCHEMICAL STUDY, 4770*
   INTESTINAL NEOPLASMS
    HISTOCHEMICAL STUDY, 4770*

SUNFLOWER SEED OIL
   FATTY ACIDS
    CARCINOGENIC ACTIVITY, RAT, 4334*

SURFACE PROPERTIES
   LEUKEMIA, LYMPHOCYTIC
    ULTRASTRUCTURAL STUDY, 4527*

TALC
   ASBESTOS
    RISK FACTOR, REVIEW, 4234*

TERATOID TUMOR
   ANTIGENS, NEOPLASM
    REVIEW, 4261*
   CELL DIFFERENTIATION
    EMBRYO, 4600
    REVIEW, 4261*
   GROWTH
    MOSAICISM, 4600
   KARYOTYPING
    REVIEW, 4261*
   LUNG NEOPLASMS
    HISTOLOGICAL STUDY, 4577
    THYMIC TISSUE, 4577
   NEOPLASM TRANSPLANTATION
    REVIEW, 4261*
   TESTICULAR NEOPLASMS
    CELL DIFFERENTIATION, 4732*
    FETAL GLOBULINS, 4574

TERATOMA, EMBRYONAL
   SEE TERATOID TUMOR

TESTICULAR NEOPLASMS
   BURKITT'S LYMPHOMA

EPIDEMIOLOGY, 4659
CHORIOCARCINOMA
   FETAL GLOBULINS, 4574
DISGERMINOMA
   FETAL GLOBULINS, 4574
FETAL GLOBULINS
   RADIOIMMUNOASSAY, 4469
GONADOTROPINS, CHORIONIC
   RADIOIMMUNOASSAY, 4469
TERATOID TUMOR
   CELL DIFFERENTIATION, 4732*
   FETAL GLOBULINS, 4574

TESTOSTERONE
   ARYL HYDROCARBON HYDROXYLASES
    INHIBITION, LIVER, 4277
   PROSTATIC NEOPLASMS
    SERUM LEVELS, 4575
   SEX DIFFERENTIATION
    RECEPTORS, HORMONE, 4221

12-O-TETRADECANOYLPHORBOL-13-ACETATE
   CARBOXY-LASES
    ENZYMATIC ACTIVITY, 4357*
   COMPLEMENT
    LYSOSOMES, 4714
   CYTOCHALASIN B
    LYSOSOMES, 4714
   LEUKOCYTES
    GLUCURONIDASE, 4714
    LACTATE DEHYDROGENASE, 4714
    MURAMIDASE, 4714
   SKIN NEOPLASMS
    DNA REPLICATION, 4358*
   ZYMOSAN
    LYSOSOMES, 4714

THORIUM DIOXIDE
   KIDNEY NEOPLASMS
    CASE REPORT, 4635*

THOROTRAST
   SEE THORIUM DIOXIDE

THYMECTOMY
   LYMPHOMA
    UREA, METHYL NITROSO-, 4437

THYMIDINE KINASE
   CARCINOMA, EHRLICH TUMOR
    SUBCELLULAR FRACTIONS, 4786*
   VIRUS, HERPES SIMPLEX 1
    DEOXYRIBONUCLEOSIDES, 4382
    DEOXYRIBONUCLEOTIDES, 4382
   VIRUS, HERPES SIMPLEX 2
    DEOXYRIBONUCLEOSIDES, 4382
    DEOXYRIBONUCLEOTIDES, 4382

THYMINE
   ULTRAVIOLET RAYS
    DNA REPAIR, 4365

THYMOMA
   NEOPLASM METASTASIS
    HISTOLOGICAL STUDY, 4643*

THYMUS EXTRACTS
   HORMONES
    ISOLATION AND CHARACTERIZATION,

4494

THYMUS NEOPLASMS
  LYMPHOMA
    UREA, METHYL NITROSO-, 4437

THYROID NEOPLASMS
  ADENOCARCINOMA
    CLASSIFICATION, 4642*
    NEOPLASM METASTASIS, 4653
    RADIATION, IONIZING, 4653
  ALPHA GLOBULINS
    SERUM LEVEL, 4461
  CALCITONIN
    C CELL, RAT, 4797*
  CARCINOMA
    ADRENAL MEDULLA, 4641*
    CLASSIFICATION, 4642*
  CARCINOMA, ANAPLASTIC
    CLASSIFICATION, 4642*
  CARCINOMA, PAPILLARY
    CLASSIFICATION, 4642*
    NEOPLASM METASTASIS, 4653
    RADIATION, IONIZING, 4653
  HAMANGIOENDOETHELIOMA
    CLASSIFICATION, 4642*
  IMMUNITY, CELLULAR
    GENETICS, 4547*
  RADIATION, IONIZING
    DOSE-RESPONSE STUDY, 4680*
  SEX FACTORS
    EPIDEMIOLOGY, FINLAND, 4645
  UREMIA
    IMMUNOSUPPRESSION, 4515*

TISSUE CULTURE
  PANCREATIC NEOPLASMS
    ADENOCARCINOMA, 4255*

TOBACCO SMOKE
  SEE SMOKING

O-TOLUIDINE
  OCCUPATIONAL HAZARD
    EPIDEMIOLOGY, 4205

TONGUE NEOPLASMS
  SMOKING
    EPIDEMIOLOGY, INDIA, 4656

TONSILLAR NEOPLASMS
  CARCINOMA, EPIDERMOID
    RADIATION, IONIZING, 4235*

TONSILLITIS
  HODGKIN'S DISEASE
    EPIDEMIOLOGY, 4660

TRACHEAL NEOPLASMS
  ASBESTOS
    PRECANCEROUS CONDITIONS, 4314*
  BENZO(A)PYRENE
    MITOSIS, 4338*
    PRECANCEROUS CONDITIONS, 4338*

TRANSFER FACTOR
  HEPATITIS
    HYPERSENSITIVITY, 4215
  RNA

MECHANISM OF ACTION, 4215

TRANSFERASES
  BREAST NEOPLASMS
    ESTROGENS, 4719
    RECEPTORS, HORMONE, 4719
  CELL LINE
    ISOLATION AND CHARACTERIZATION,
    4760*
    TUMOR-CELL PROTECTION, 4760*
  DIMETHYLAMINE, N-NITROSO-
    CELL TRANSFORMATION, NEOPLASTIC,
    4709
    TEMPERATURE SENSITIVE CELLS,
    HAMSTER, 4709
  MAMMARY NEOPLASMS, EXPERIMENTAL
    BENZ(A)ANTHRACENE, 7,12-DIMETHYL-,
    4719
    ESTROGENS, 4719
    RECEPTORS, HORMONE, 4719

TRANSLATION, GENETIC
  VIRUS, SV40
    DNA REPLICATION, 4414*

TRANSPLANATION, HOMOLOGOUS
  NEOPLASMS, EXPERIMENTAL
    VIRUS, ADENO 7 - SV40 HYBRID, 445

TRANSPLANTATION ANTIGENS
  SEE HISTOCOMPATIBILITY ANTIGENS

TRANSPLANTATION, HETEROLOGOUS
  COLONIC NEOPLASMS
    GROWTH, 4740*
    IMMUNOSUPPRESSION, 4740*
  LEUKEMIA, MYELOCYTIC
    GROWTH, 4740*
    IMMUNOSUPPRESSION, 4740*

TRANSPLANTATION, HOMOLOGOUS
  CERVIX NEOPLASMS
    IMMUNOSUPPRESSION, 4515*
  COLONIC NEOPLASMS
    IMMUNOSUPPRESSION, 4515*
  DISGERMINOMA
    IMMUNOSUPPRESSION, 4515*
  LEUKEMIA
    IMMUNE SERUMS, 4449
    IMMUNITY, CELLULAR, 4444
  LYMPHOMA
    CHOLANTHRENE, 3-METHYL-, 4442
    IMMUNE RESPONSE, 4442
    LEUKEMIA L1210, 4442
    LYMPHOCYTE TRAPPING, 4442
    T-LYMPHOCYTES, 4442
  MAMMARY NEOPLASMS, EXPERIMENTAL
    ANTIBODIES, NEOPLASM, 4456
    GROWTH, 4740*
    IMMUNOSUPPRESSION, 4740*
  NEOPLASMS
    ANTILYMPHOCYTE SERUM, 4650
    EPIDEMIOLOGY, 4650
    IMMUNE SERUMS, 4650
    IMMUNOSUPPRESSION, 4650
  SARCOMA
    BENZO(A)PYRENE, 4493
    CHOLANTHRENE, 3-METHYL-, 4493
    IMMUNE SERUMS, 4449

PLANT AGGLUTININS, 4493
SARCOMA, RETICULUM CELL
  IMMUNOSUPPRESSION, 4515*
SKIN NEOPLASMS
  IMMUNOSUPPRESSION, 4515*

TRANSPLANTATION IMMUNOLOGY
  BONE MARROW CELLS
    GRAFT VS HOST REACTION, 4436
  LEUKEMIA, EXPERIMENTAL
    MYCOBACTERIUM BOVIS, 4511*
  B-LYMPHOCYTES
    GRAFT VS HOST REACTION, 4436
  T-LYMPHOCYTES
    GRAFT VS HOST REACTION, 4436
  LYMPHOMA
    ANTIGENS, VIRAL, 4440
    GRAFT VS HOST REACTION, 4441
    HISTOCOMPATIBILITY ANTIGENS, 4440
    IMMUNE SERUMS, 4440
  MAMMARY NEOPLASMS, EXPERIMENTAL
    IMMUNOSUPPRESSION, 4739*
    SARCOMA, RETICULUM CELL, 4543*
  NEOPLASMS, EXPERIMENTAL
    CHOLANTHRENE, 3-METHYL-, 4511*
  PLASMACYTOMA
    CYCLOPHOSPHAMIDE, 4443
    MITOMYCIN C, 4443
  SARCOMA
    MYCOBACTERIUM BOVIS, 4511*
  SARCOMA, RETICULUM CELL
    HISTOCOMPATIBILITY ANTIGENS, 4543*

1-TRIAZINE, 3,3-DIMETHYL-1-PHENYL-
  NERVOUS SYSTEM NEOPLASMS
    NUCLEIC ACIDS, 4783*

TRIAZINE-3,5(2H,4H)-DIONE, 2-BETA-D-
  RIBOFURANOSYL-
  L CELLS
    RESISTANT PHENOTYPE, 4685

TRYPSIN
  BURKITT'S LYMPHOMA
    CONCANAVALIN A, 4506

TRYPTOPHAN
  BLADDER NEOPLASMS
    GENETICS, 4769*
  CELL NUCLEUS
    CELL CYCLE KINETICS, 4688
    TEMPERATURE-SENSITIVE CLONES, 4688
  RNA, TRANSFER
    ISOLATION AND CHARACTERIZATION,
      4389

TRYPTOPHAN OXYGENASE
  HEPATOMA
    CORTISOL, 4755*
    RNA, MESSENGER, 4755*

TUBERCULIN
  B-LYMPHOCYTES
    IMMUNE RESPONSE, 4490
    MITOMYCIN C, 4490
  T-LYMPHOCYTES
    IMMUNE RESPONSE, 4490
    MITOMYCIN C, 4490

TUBERCULOSIS
  UTERINE NEOPLASMS
    CASE REPORT, MONKEY, 4639*

5-TUNGSTO-2-ANTIMONIATE
  CELL TRANSFORMATION, NEOPLASTIC
    VIRUS, EPSTEIN-BARR, 4407*
    VIRUS, HERPES SAIMIRI, 4407*
  DNA NUCLEOTIDYLTRANSFERASES
    ENZYMATIC ACTIVITY, 4407*
  REVERSE TRANSCRIPTASE
    ENZYMATIC ACTIVITY, 4407*

ULTRAVIOLET RAYS
  AMINO ACIDS
    DNA REPAIR, 4365
  CARCINOMA, EPIDERMOID
    CROTON OIL, 4363
  DNA REPAIR
    REVIEW, 4228
  DNA REPLICATION
    REVIEW, 4228
  EAR NEOPLASMS
    FIBROMA, 4362
    FIBROSARCOMA, 4362
  ESCHERICHIA COLI
    DNA REPAIR, 4365
  ETHIDIUM BROMIDE
    MUTAGENIC ACTIVITY, 4322*
  NEOPLASM TRANSPLANTATION
    IMMUNOSUPPRESSION, 4512*
  PAPILLOMA
    CROTON OIL, 4363
  SKIN NEOPLASMS
    CARCINOMA, EPIDERMOID, 4362, 4363
    FIBROMA, 4363
    FIBROSARCOMA, 4363
    PAPILLOMA, 4362, 4363
    PRECANCEROUS CONDITIONS, 4363
    WHITENING AGENTS, 4367*, 4368*
  THYMINE
    DNA REPAIR, 4365
  XERODERMA PIGMENTOSUM
    DNA REPAIR, 4228

URANIUM
  RADIATION, IONIZING
    ENVIRONMENTAL HAZARD, 4675*

UREA
  FETAL GLOBULINS
    IMMUNOLOGICAL AND ELECTROPHORETIC
      PROPERTIES, 4465

UREA, 1-BUTYL-1-NITROSO-
  LEUKEMIA, MYELOCYTIC
    MODEL TEST SYSTEM, 4354*

UREA, ETHYL NITROSO-
  BACTERIOPHAGES
    RNA, 4288
  NERVOUS SYSTEM NEOPLASMS
    NEURILEMMOMA, 4355*
  NEURILEMMOMA
    NEOPLASM TRANSPLANTATION, 4435
    TRANSPLACENTAL CARCINOGENESIS,
      4355*, 4435

UREA, HYDROXY-

* INDICATES A PLAIN CITATION WITHOUT ACCOMPANYING ABSTRACT

ERYTHROLEUKEMIA
    BUTYRIC ACID, 4702
    B-LYMPHOCYTES
        ORNITHINE DECARBOXYLASE, 4453

UREA, METHYL NITROSO-
    ESCHERICHIA COLI
        MUTAGENIC ACTIVITY, 4344*
    LUNG NEOPLASMS
        ADENOMA, 4437
    LYMPHOMA
        THYMECTOMY, 4437
    MEDIASTINAL NEOPLASMS
        LYMPHOMA, 4437
    THYMUS NEOPLASMS
        LYMPHOMA, 4437

UREA, 1-(1-NAPHTHYL)-2-THIO-
    OCCUPATIONAL HAZARD
        EPIDEMIOLOGY, 4205

UREMIA
    BREAST NEOPLASMS
        IMMUNOSUPPRESSION, 4515*
    CERVIX NEOPLASMS
        IMMUNOSUPPRESSION, 4515*
    ISLET CELL TUMOR
        IMMUNOSUPPRESSION, 4515*
    KIDNEY NEOPLASMS
        IMMUNOSUPPRESSION, 4515*
    LEUKEMIA
        IMMUNOSUPPRESSION, 4515*
    LUNG NEOPLASMS
        IMMUNOSUPPRESSION, 4515*
    NEOPLASMS
        POST-TRANSPLANTATION, 4515*
        PRETRANSPLANTATION, 4515*
    THYROID NEOPLASMS
        IMMUNOSUPPRESSION, 4515*

URETHANE
    SEE CARBAMIC ACID, ETHYL ESTER

URIC ACID, 1-METHYL-
    GUANINE, 1-METHYL-, 3-OXIDE
        CARCINOGENIC METABOLITE, 4345*

URIDINE, 5-BROMO-2'-DEOXY-
    HYBRID CELLS
        RESISTANT SEGREGANTS, 4684
    MULTIPLE MYELOMA
        CELL TRANSFORMATION, NEOPLASTIC,
        4687
        FIBRINOLYSIS, 4687
        PLASMINOGEN, 4687
    VIRUS, EPSTEIN-BARR
        ANTIGENS, VIRAL, 4460
        INTERFERON, 4460

URIDINE, 2'-DEOXY-5-IODO-
    VIRUS, ADENO 7
        DNA, VIRAL, 4395
        VIRUS REPLICATION, 4395

URIDINE KINASE

SEE PHOSPHOTRANSFERASES, ATP

URINE
    BREAST NEOPLASMS
        ESTRADIOL, 4570
        ESTRIOL, 4570
        ESTRONE, 4570
        17-HYDROXYCORTICOSTEROIDS, 4570
        17-KETOSTEROIDS, 4570

UROGENITAL NEOPLASMS
    CARCINOEMBRYONIC ANTIGEN
        PROTEINS, 4248*
    FETAL GLOBULINS
        RADIOIMMUNOASSAY, 4469
    GONADOTROPINS, CHORIONIC
        RADIOIMMUNOASSAY, 4469
    MALE, 4670*
    SEX CHROMATIN
        DIAGNOSIS AND PROGNOSIS, 4638*
        NEOPLASM METASTASIS, 4638*
    SEX FACTORS
        EPIDEMIOLOGY, FINLAND, 4645

UTERINE NEOPLASMS
    ADENOCARCINOMA
        INCIDENCE, MOUSE, 4723
        MULTIFOCAL PRIMARY CARCINOMA, 4569
        RADIATION, IONIZING, 4364
    ADENOFIBROMA
        ULTRASTRUCTURAL STUDY, 4573
    ANGIOSARCOMA
        ETHYLENE, CHLORO-, 4311
    CARCINOMA, EPIDERMOID
        CARCINOEMBRYONIC ANTIGEN, 4474
        CHROMOSOME ABNORMALITIES, 4612*
    CYSTADENOMA, PAPILLARY
        ULTRASTRUCTURAL STUDY, 4573
    ESTROGENS
        EPIDEMIOLOGY, 4673*
    HODGKIN'S DISEASE
        EPIDEMIOLOGY, 4660
    LEIOMYOMA
        CASE REPORT, 4637*
        CASE REPORT, MONKEY, 4639*
        INCIDENCE, MOUSE, 4723
    MENOPAUSE
        EPIDEMIOLOGY, 4673*
    OBESITY
        EPIDEMIOLOGY, 4673*
    SARCOMA
        ULTRASTRUCTURAL STUDY, 4573
    TUBERCULOSIS
        CASE REPORT, MONKEY, 4639*

VAGINAL NEOPLASMS
    ADENOCARCINOMA
        DIAGNOSIS AND PROGNOSIS, 4571
    CARCINOMA, EPIDERMOID
        DIAGNOSIS AND PROGNOSIS, 4571
        MULTIFOCAL PRIMARY CARCINOMA, 4569
        REVIEW, 4259*
    CARCINOMA IN SITU
        MULTIFOCAL PRIMARY CARCINOMA, 4569
        REVIEW, 4259*
    CONTRACEPTIVES, ORAL
        CASE REPORT, 4572
    LEIOMYOMA
        CASE REPORT, 4637*

RNA REPLICATION, 4378
RIBOSOMES
    VIRUS, HELPER, 4393
RNA, MESSENGER
    VIRUS, HELPER, 4393
RNA, TRANSFER, METHYLTRANSFERASES
    KB CELLS, 4428*
    METHYLATION, 4428*
VIRAL PROTEINS
    VIRUS, HELPER, 4393
VIRUS, SV40
    VIRUS, HELPER, 4393

VIRUS, ADENO 7
    URIDINE, 2'-DEOXY-5-IODO-
        DNA, VIRAL, 4395
        VIRUS REPLICATION, 4395

VIRUS, ADENO 12
    DNA
        CARRIER PROTEINS, 4379
    SARCOMA
        CARCINOEMBRYONIC ANTIGEN, 4475

VIRUS, ADENO 2 - SV40 HYBRID
    HELA CELLS
        VIRAL PROTEINS, 4396

VIRUS, ADENO 7 - SV40 HYBRID
    IMMUNE SERUMS
        IMMUNITY, CELLULAR, 4450
        VIRUS, SV40, 4450
    T-LYMPHOCYTES
        IMMUNITY, CELLULAR, 4450
    NEOPLASMS, EXPERIMENTAL
        IMMUNE SERUMS, 4450
        TRANSPLANATION, HOMOLOGOUS, 4450

VIRUS, AVIAN LEUKOSIS
    NUCLEIC ACIDS
        ISOLATION AND CHARACTERIZATION,
        4244*

VIRUS, AVIAN MYELOBLASTOSIS
    DNA POLYMERASE
        ISOLATION AND CHARACTERIZATION,
        4377
    REVERSE TRANSCRIPTASE
        ISOLATION AND CHARACTERIZATION,
        4377
    VIRAL PROTEINS
        ISOLATION AND CHARACTERIZATION,
        4373
        PEPTIDES, 4373
        PROTEOLYTIC CLEAVAGE, 4373
    VIRUS, MOLONEY MURINE LEUKEMIA
        REVERSE TRANSCRIPTASE, 4377
    VIRUS, ROUS SARCOMA
        REVERSE TRANSCRIPTASE, 4377

VIRUS, AVIAN RNA SARCOMA
    DNA REPLICATION
        CELL TRANSFORMATION, NEOPLASTIC,
        4374
    ETHIDIUM BROMIDE
        DNA REPLICATION, 4374

VIRUS, AVIAN RNA TUMOR
    VIRUS, HERPES SIMPLEX 2

GROWTH EFFECTS, CHICK EMBRYO,
4405*

VIRUS, B77
ADENYL CYCLASE
CELL TRANSFORMATION, NEOPLASTIC,
4375
TEMPERATURE SENSITIVE MUTANTS,
4375
DNA REPLICATION
CELL TRANSFORMATION, NEOPLASTIC,
4374
ETHIDIUM BROMIDE
DNA REPLICATION, 4374
PHOSPHODIESTERASES
CELL TRANSFORMATION, NEOPLASTIC,
4375
TEMPERATURE SENSITIVE MUTANTS,
4375
RNA, VIRAL
ISOLATION AND CHARACTERIZATION, 60-
70S SUBUNITS, 4390

VIRUS, C-TYPE
LEUKEMIA, MYELOBLASTIC
DNA, VIRAL, 4245*
REVIEW, 4245*
RNA, VIRAL, 4245*

VIRUS, C-TYPE RNA TUMOR
NUCLEIC ACIDS
ISOLATION AND CHARACTERIZATION,
4244*

VIRUS CULTIVATION
VIRUS, EPSTEIN-BARR
INFECTIOUS VIRUS, 4422*
TRANSFORMING VIRUS, 4422*

VIRUS, CYTOMEGALO
CELL TRANSFORMATION, NEOPLASTIC
ONCOGENIC VIRUSES, REVIEW, 4239*

VIRUS, D-TYPE RNA TUMOR
BREAST NEOPLASMS
ANTIGENS, VIRAL, 4478

VIRUS, EPSTEIN-BARR
BURKITT'S LYMPHOMA
ANTIGENS, VIRAL, 4420*
CELL TRANSFORMATION, NEOPLASTIC
FIBROBLASTS, 4421*
ONCOGENIC VIRUSES, REVIEW, 4239*
5-TUNGSTO-2-ANTIMONIATE, 4407*
DNA
CELL TRANSFORMATION, NEOPLASTIC,
4226
INFECTIOUS MONONUCLEOSIS
ANTIGENS, VIRAL, 4420*
INTERFERON
ANTIGENS, VIRAL, 4460
NASOPHARYNGEAL NEOPLASMS
EPIDEMIOLOGY, REVIEW, 4238*
LYMPHOCYTE TRANSFORMATION, 4497
NUCLEIC ACIDS
ISOLATION AND CHARACTERIZATION,
4244*
URIDINE, 5-BROMO-2'-DEOXY-
ANTIGENS, VIRAL, 4460

INTERFERON, 4460
VIRUS CULTIVATION
INFECTIOUS VIRUS, 4422*
TRANSFORMING VIRUS, 4422*

VIRUS, FELINE LEUKEMIA
ANTIBODIES, NEOPLASM
REVIEW, 4240*
ANTIGENS, VIRAL
REVIEW, 4240*
CELL TRANSFORMATION, NEOPLASTIC
REVIEW, 4240*

VIRUS, FELINE SARCOMA
ANTIBODIES, NEOPLASM
REVIEW, 4240*
ANTIGENS, VIRAL
REVIEW, 4240*
CELL TRANSFORMATION, NEOPLASTIC
REVIEW, 4240*

VIRUS, FRIEND MURINE LEUKEMIA
IMMUNE SERUMS
IMMUNE RESPONSE, 4517*
IMMUNOSUPPRESSION
ANTIGENS, VIRAL, 4447
IMMUNE RESPONSE, 4519*
B-LYMPHOCYTES, 4447
T-LYMPHOCYTES, 4447

VIRUS, FRIEND SPLEEN FOCUS-FORMING
ERYTHROLEUKEMIA
METHANE, SULFINYLBIS-, 4401
RNA, MESSENGER, 4401
RNA, VIRAL, 4401

VIRUS, FROG 3
VIRAL PROTEINS
DNA REPLICATION, 4381
TEMPERATURE SENSITIVE MUTANTS
4381

VIRUS, HAMSTER SARCOMA
GLYCOLIPIDS
CELL AGGREGATION, 4710
CELL MEMBRANE, 4710

VIRUS, HARVEY MURINE SARCOMA
ERYTHROLEUKEMIA
VIRUS REPLICATION, 4408*

VIRUS, HELPER
VIRUS, ADENO 2
DNA REPLICATION, 4393
RIBOSOMES, 4393
RNA, MESSENGER, 4393
VIRAL PROTEINS, 4393
VIRUS, SV40
VIRUS, ADENO 2, 4393

VIRUS, HEPATITIS B
HEPATOMA
AUSTRALIA ANTIGEN, 4669*

VIRUS, HERPES SAIMIRI
ANTIGEN-ANTIBODY REACTIONS
CELL TRANSFORMATION, NEOPLASTI
4536*
ANTIGENS, VIRAL

ANTIBODY FORMATION, 4536*
ANTILYMPHOCYTE SERUM
    ISOLATION AND CHARACTERIZATION,
    4403*
CELL TRANSFORMATION, NEOPLASTIC
    ONCOGENIC VIRUSES, REVIEW, 4239*
    5-TUNGSTO-2-ANTIMONIATE, 4407*
LYMPHOCYTES
    CELL TRANSFORMATION, NEOPLASTIC,
    4403*
    IMMUNE RESPONSE, 4536*
LYMPHOMA
    BCG VACCINATION, 4448
VIRUS INHIBITORS
    CELL TRANSFORMATION, NEOPLASTIC,
    4403*

IRUS, HERPES SIMPLEX
ANTIGENS, VIRAL
    IMMUNOGLOBULINS, 4214
CELL TRANSFORMATION, NEOPLASTIC
    PHOSPHONOACETIC ACID, 4406*
    REVIEW, 4214
HEAD AND NECK NEOPLASMS
    CARCINOMA, 4214
IGA
    REVIEW, 4214
IGG
    REVIEW, 4214
IGM
    REVIEW, 4214
IMMUNITY, CELLULAR
    REVIEW, 4214
IMMUNOGLOBULINS
    REVIEW, 4214
LIP NEOPLASMS
    CARCINOMA, 4214
T-LYMPHOCYTES
    IMMUNITY, CELLULAR, 4214
NUCLEIC ACIDS
    ISOLATION AND CHARACTERIZATION,
    4244*
SARCOMA
    ANTIGENIC DETERMINANTS, 4213
    NEOPLASM METASTASIS, 4213

IRUS, HERPES SIMPLEX 1
CELL TRANSFORMATION, NEOPLASTIC
    ONCOGENIC VIRUSES, REVIEW, 4239*
DEOXYRIBONUCLEOSIDES
    THYMIDINE KINASE, 4382
DEOXYRIBONUCLEOTIDES
    THYMIDINE KINASE, 4382

IRUS, HERPES SIMPLEX 2
ANTIGENS, VIRAL
    CARCINOGENIC POTENTIAL, REVIEW,
    4213
BREAST NEOPLASMS
    ANTIGENS, VIRAL, 4478
CELL TRANSFORMATION, NEOPLASTIC
    ONCOGENIC VIRUSES, REVIEW, 4239*
CERVIX NEOPLASMS
    SEROEPIDEMIOLOGIC STUDIES, REVIEW,
    4213
DEOXYRIBONUCLEOSIDES
    THYMIDINE KINASE, 4382
DEOXYRIBONUCLEOTIDES
    THYMIDINE KINASE, 4382

VIRUS, AVIAN RNA TUMOR
    GROWTH EFFECTS, CHICK EMBRYO, 4405*

VIRUS INHIBITORS
VIRUS, HERPES SAIMIRI
    CELL TRANSFORMATION, NEOPLASTIC,
    4403*

VIRUS, KIRSTEN MURINE LEUKEMIA
SARCOMA, OSTEOGENIC
    VIRUS REPLICATION, 4387
VIRUS, KIRSTEN MURINE SARCOMA
    ANTIGENS, VIRAL, 4387
    VIRUS REPLICATION, 4387

VIRUS, KIRSTEN MURINE SARCOMA
ADRENAL GLAND NEOPLASMS
    CELL TRANSFORMATION, NEOPLASTIC,
    4409*
ERYTHROLEUKEMIA
    VIRUS REPLICATION, 4408*
NERVE TISSUE PROTEINS
    COLCHICINE, 4690
VIRUS, KIRSTEN MURINE LEUKEMIA
    ANTIGENS, VIRAL, 4387
    VIRUS REPLICATION, 4387
VIRUS, MOLONEY MURINE SARCOMA
    NERVE TISSUE PROTEINS, 4690
VIRUS, SIMIAN SARCOMA
    ANTIGENS, VIRAL, 4387
    VIRUS REPLICATION, 4387

VIRUS-LIKE PARTICLES
CARCINOMA, BRONCHIOLAR
    ANTIGENS, VIRAL, 4483
    IMMUNE SERUMS, 4483
REVERSE TRANSCRIPTASE
    CLASSIFICATION, REVIEW, 4242*

VIRUS, MAREK'S DISEASE HERPES
ANTIGENS, VIRAL
    VIRUS REPLICATION, 4383
VIRUS REPLICATION
    ISOLATION AND CHARACTERIZATION,
    CHICK EMBRYO, 4383

VIRUS, MASON-PFIZER MONKEY
BREAST NEOPLASMS
    ANTIGENS, VIRAL, 4400

VIRUS, MOLONEY MURINE LEUKEMIA
ACTINOMYCIN D
    VIRUS REPLICATION, 4385
CYCLOHEXIMIDE
    VIRUS REPLICATION, 4385
DNA, CIRCULAR
    ISOLATION AND CHARACTERIZATION,
    4384
DNA REPLICATION
    BONE MARROW CELLS, MOUSE, 4384
DNA, VIRAL
    ISOLATION AND CHARACTERIZATION,
    4384
NEOPLASM TRANSPLANTATION
    LYMPHOMA, 4438
REVERSE TRANSCRIPTASE
    RADIOIMMUNOASSAY, 4377
VIRUS, AVIAN MYELOBLASTOSIS
    REVERSE TRANSCRIPTASE, 4377
VIRUS REPLICATION

CELL CYCLE KINETICS, 4385

VIRUS, MOLONEY MURINE SARCOMA
    IMMUNITY, CELLULAR
        CORYNEBACTERIUM PARVUM, 4548*
    IMMUNOSUPPRESSION
        CORYNEBACTERIUM PARVUM, 4548*
    NERVE TISSUE PROTEINS
        COLCHICINE, 4690
        VIRUS, KIRSTEN MURINE SARCOMA,
        4690

VIRUS, MURINE LEUKEMIA
    ANTIGENIC DETERMINANTS
        ISOLATION AND CHARACTERIZATION,
        4380
    ANTIGENS, VIRAL
        X-TROPIC VIRUS, 4380
    IMMUNE SERUMS
        FERRITIN, 4529*
        FLUORESCENT ANTIBODY TECHNIC, 4529*
    IMMUNOSUPPRESSION
        REVIEW, 4447
    LYMPHOMA
        ANTIGENS, VIRAL, 4440
        HISTOCOMPATIBILITY ANTIGENS, 4440
        IMMUNE SERUMS, 4440
    VIRUS REPLICATION
        AUTOIMMUNE DISEASES, MICE, 4380
        X-TROPIC VIRUS, 4380

VIRUS, MURINE MAMMARY TUMOR
    DEXAMETHASONE
        REVERSE TRANSCRIPTASE, 4423*
    GENETICS
        REVERSE TRANSCRIPTASE, 4423*
    GLYCOPROTEINS
        ISOLATION AND CHARACTERIZATION,
        4425*
    HISTOCOMPATIBILITY ANTIGENS
        VIRUS SUSCEPTIBILITY, 4424*
    MAMMARY NEOPLASMS, EXPERIMENTAL
        ANTIBODIES, NEOPLASM, 4456
        IMMUNITY, CELLULAR, 4456
    NUCLEIC ACIDS
        ISOLATION AND CHARACTERIZATION,
        4244*
    REVERSE TRANSCRIPTASE
        DEXAMETHASONE, 4426*
    RNA, MESSENGER
        VIRAL REPLICATION, 4427*

VIRUS, MURINE SARCOMA
    VIRUS, RADIATION MURINE LEUKEMIA
        VIRUS REPLICATION, 4386

VIRUS, PAPOVA, BK
    DNA
        ISOLATION AND CHARACTERIZATION, 4392
        POLYNUCLEOTIDES, 4392
    VIRUS, POLYOMA, BK
        VIRUS, SV40, 4241*

VIRUS, POLYOMA
    ADENOSINE CYCLIC 3,5' MONOPHOSPHATE
        GROWTH SUBSTANCES, 4391
    ADENYL CYCLASE
        GROWTH SUBSTANCES, 4391
    NEOPLASMS, EXPERIMENTAL

ULTRASTRUCTURAL STUDY, 4212
CELL TRANSFORMATION, NEOPLASTIC
    CELL MEMBRANE, 4413*
    ISOLATION AND CHARACTERIZATION,
    4418*
CHROMATIN
    ULTRASTRUCTURAL STUDY, 4416*
CHROMOSOMES
    SITE SPECIFICITY, 4415*
CONCANAVALIN A
    CELL MEMBRANE, 4212, 4398
DIBUTYRYL CYCLIC AMP
    PHOSPHODIESTERASES, 4712
DNA
    ISOLATION AND CHARACTERIZATION,
    4392
    POLYNUCLEOTIDES, 4392
DNA REPLICATION
    TEMPERATURE SENSITIVE MUTANTS,
    4399
    TRANSLATION, GENETIC, 4414*
DNA, VIRAL
    ENDONUCLEASES, 4397
    ISOLATION AND CHARACTERIZATION,
    4418*
    RADIATION, 4397
    TEMPERATURE SENSITIVE MUTANTS,
    4399
    ULTRASTRUCTURAL STUDY, 4397, 4416*
    VIRUS REPLICATION, 4399
FIBROBLASTS
    HISTONES, 4725
    PROTEINS, 4725
GLYCOLIPIDS
    CELL MEMBRANE, 4212
GLYCOPROTEINS
    CELL MEMBRANE, 4212
HEPARIN
    CELL MEMBRANE, 4212
IMMUNOSUPPRESSION
    CELL TRANSFORMATION, NEOPLASTIC,
    4419*
LEUKEMIA
    ANEMIA, APLASTIC, 4556
MUCOPOLYSACCHARIDES
    CELL ADHESION, 4212
    CELL MEMBRANE, 4212
NUCLEIC ACIDS
    ISOLATION AND CHARACTERIZATION,
    4244*
PEPTIDE HYDROLASES
    CELL MEMBRANE, 4212
    CHOLESTEROL, 4212
PEPTIDES
    TEMPERATURE SENSITIVE MUTANTS,
    4417*
PHOSPHODIESTERASES
    CELL TRANSFORMATION, NEOPLASTIC,
    4712
PHOSPHOLIPIDS
    CELL MEMBRANE, 4212, 4398
PHOSPHOTRANSFERASES
    CELL CYCLE KINETICS, 4394
RNA, MESSENGER
    CELL TRANSFORMATION, NEOPLASTIC,
    4691
    ISOLATION AND CHARACTERIZATION,
    4691
VIRAL PROTEINS

ISOLATION AND CHARACTERIZATION,
4418*
VIRUS, ADENO 2
VIRUS, HELPER, 4393
VIRUS, ADENO 7 - SV40 HYBRID
IMMUNE SERUMS, 4450
VIRUS, POLYOMA, BK
VIRUS, PAPOVA, 4241*
REVIEW, 4241*
VIRUS REPLICATION
CELL TRANSFORMATION, NEOPLASTIC,
4419*

VIRUS, TURKEY HERPES
ANTIGENS, VIRAL
ANTIBODY FORMATION, 4404*

VIRUS, VACCINIA
DNA, VIRAL
MITOMYCIN C, 4402
HELA CELLS
DNA REPLICATION, 4402

VITAMIN A
SEE RETINOL

VITAMIN C
SEE ASCORBIC ACID

VITAMINS
NEOPLASMS
EPIDEMIOLOGY, CHILD, 4654

VULVAR NEOPLASMS
CARCINOMA IN SITU
MULTIFOCAL PRIMARY CARCINOMA, 4569

WARFARIN
OVARIAN NEOPLASMS
ASCITES, 4640*

WARTS
SKIN NEOPLASMS
IMMUNOSUPPRESSION, 4219
INCLUSION BODIES, VIRAL, 4219
REVIEW, 4219

WATER POLLUTANTS
BENZO(A)PYRENE
DETECTION METHOD, 4678*

WATER POLLUTATNTS
CHLORINE
REVIEW, 4206

WILM'S TUMOR
SEE NEPHROBLASTOMA

WOUNDS AND INJURIES
LUNG NEOPLASMS
MESOTHELIOMA, 4578

XANTHINE, 3-ISOBUTYL-1-METHYL-
MELANOMA
GLYCOPEPTIDES, 4361*

XANTHOMASTOSIS
LIPIDS
SERUM LEVELS, 4784*

LIPOPROTEINS
SERUM LEVELS, 4784*

XERODERMA PIGMENTOSUM
BUTANE, 1,2:3,4-DIEPOXY-
CHROMOSOME ABERRATIONS, 4319*
DNA REPAIR
HYBRIDIZATION, 4228
REVIEW, 4228
GENETICS
DNA REPAIR, 4228
RADIATION
DNA REPAIR, 4228
SKIN NEOPLASMS
CELL REPAIR DEFICIENCY, 4633*
ULTRAVIOLET RAYS
DNA REPAIR, 4228

ZINC
DNA REPLICATION
CHICK EMBRYO CELLS, 4761*

ZOLLINGER-ELLISON SYNDROME
GASTRIN
ISOLATION AND CHARACTERIZATION,
4722

ZYMOSAN
12-O-TETRADECANOYLPHORBOL-13-ACETAT
LYSOSOMES, 4714

~MBER	ABSTRACT NUMBER	CAS REGISTRY NUMBER	ABSTRACT NUMBER
....	4505	55185........................ (Cont.)	4665
....	4556	55981........................	4583
....	4465	56495........................	4277, 4278, 4281, 4282, 4290, 4442, 4445, 4458, 4493, 4499
....	4402, 4443, 4490	56531........................	4305
....	4280, 4443, 4444, 4584	56575........................	4290, 4300
....	4510	57136........................	4465
....	4202, 4510, 4701	57307........................	4278
....	4266, 4268, 4276, 4278, 4493	57830........................	4303
....	4588	57885........................	4212
....	4385, 4706, 4707, 4712	58082........................	4300
....	4278	57976........................	4202
....	4282, 4711	58220........................	4221, 4271, 4575
....	4556	58617........................	4685
....	4588	59052........................	4588
....	4303, 4570	59143........................	4460, 4684, 4687
....	4556	60117........................	4205, 4298, 4485
....	4275	60242........................	4465
....	4205, 4266, 4274	62500........................	4288, 4289
....	4389, 4688	62680........................	4266
....	4685	62759........................	4205, 4282, 4284, 4286, 4287, 4599, 4665, 4709
....	4395	63683........................	4688
....	4271, 4282, 4283, 4484,	64868........................	4690

CHEMICAL ABSTRACT SERVICES REGISTRY NUMBER INDEX

CAS REGISTRY NUMBER	ABSTRACT NUMBER	CAS REGISTRY NUMBER	ABSTRACT NUMBER
65714	4365	109897	4284
66273	4289	119904	4205
66819	4385	122394	4205
67685	4401	124209	4706
70257	4290	127071	4453, 4702
71432	4663	128530	4465
71443	4706	145131	4720
73233	4267, 4271, 4290, 4363, 4389, 4444, 4495, 4688, 4719	147944	4453, 4702
		151188	4598
75014	4203, 4204, 4310, 4311, 4312, 4313	303333	4301
75354	4313	315220	4302
77065	4716	492808	4205
86884	4205	540738	4293
91598	4205	604591	4277, 4278
91941	4205	615532	4290
92671	4205	645498	4304
92717	4278	684935	4437
92875	4205	759739	4288, 4435
95534	4205	1239458	4374
98920	4301	1407154	4716
107926	4702	3688537	4270, 4664
108918	4295	3761533	4290

CHEMICAL ABSTRACT SERVICES REGISTRY NUMBER INDEX

CAS REGISTRY NUMBER	ABSTRACT NUMBER	CAS REGISTRY NUMBER	ABSTRACT NUMBER
6051872	4278	35693993	4297
6098448	4269		
6795239	4667		
7439896	4654		
7439954	4706		
7440075	4207, 4662		
7632000	4284		
7782505	4206		
8001283	4363		
8015950	4503		
9002077	4506		
9002624	4202		
9004108	4225		
9005496	4212		
9008111	4460		
13345216	4279		
16561298	4717		
21150209	4378, 4706, 4716		
23107122	4302		
23214928	4292		
24096535	4296		
33953730	4279		

ABSTRACT NUMBER	WLN	ABSTRACT NUMBER
.... 4206	L66J BMYZUS 4205	
.... 4706	L66J CZ 4205	
.N-O-Q 4284	ONN1&VO2 4290	
I 4313	ONN1&1 .... 4205, 4282, 428 4286, 4287, 459 4665, 4709	
.... 4203, 4204, 4310, 4311, 4312, 4313	ONN2&2 .... 4271, 4282, 428 4484, 4665	
56 HHJ EMV1 .... 4205, 4266, 4274	OS1&1 4401	
56 HHJ ENQV1 4275	QR BQ DYQ1MY -L 4278	
77 MV&T&J CO1 DO1 EO1 JMV1 NO1 4690	QR DY2& 2U 4305	
B666J C J 4202, 4267, 4271, 4290, 4363, 4444, 4495, 4719	QV3 4702	
B6666 2AB TJ 4266, 4268, 4276, 4278, 4493	R 4663	
.B666 FVTTT&J E OQ 4303, 4570	SH2Q 4465	
˙666 LUTJ A E FY&3Y OQ -B&AEFO ... 4212		
B666 OV MUTJ A E FQ -B&AEF 4221, 4271, 4575	T C666 BO EV INJ D FZ N G- K-/VM-OT5-16 AN FVN IVN LVO PVM SVTJ G J KY N RY 2 .... 4385, 4706, 470 4712	
B666 OV MUTJ A E FV1 -B&AEF 4303	T D3 B556 DN EM JV MVTTT&J GO1 H1OVZ KZ L 4402, 4443, 449	
B666TJ A1Q CQ E IQ MQ QQ F- DT5OV HJ& OO- BT6OTJ CQ DQ EQ F 4505	T5N COJ BR& DR 4278	
B666TTT&J E FQ GQ OQ 4510	T5OJ BYVZU1- BT5OJ ENW 4270, 4664	
B666TTT&J E FQ OQ 4202, 4510, 4701	T5VNVJ B2 4465	
B6656 1A T&&&T&J R 4277, 4278, 4281, 4282, 4290, 4442, 4445, 4458, 4493, 4499	T55 AN CUTJ D1OVXQY&&YO1 FQ 4301	
	T55 AN CUTJ FQ D1OV1- ET5OVTJ C D EQ .. 4302	
AZ 4295	T56 BM DN FN HNJ ISH 4588	

WLN	ABSTRACT NUMBER	WLN	ABSTRACT NUMBE
T56 BN DN FN HNJ IZ D- BT5OTJ CQ DQ E1Q -A&CD 4685		ZR DR DZ 4205	
T56 BN DN FNVNVJ B F H 4300		ZVN1&NO 4437	
T6MPOTJ BO BN2G2G 4280, 4443, 4444, 4584		ZVN2&NO 4288, 4435	
T6MVMVJ E 4365		ZVO2 .... 4282, 4711	
T6NJ CVZ 4301		ZVZ 4465	
T6NNVMVJ B- BT5OTJ CQ DQ E1Q -B&CD 4685		ZYZUM &GH 4465	
T6NVMVJ EE A- I.T5OTJ B1Q CQ -A&C 4460, 4684, 4687		Z2CN 4598	
T6NVMVJ EI A- ET5OTJ B1Q CQ -A&C 4395		Z3M4M3Z 4706	
T6NVNJ DZ A- BT5OTJ CQ DQ E1Q 4453, 4702		1MM1 4293	
T6VMVMV FHJ F2 FR 4278		1N1&R D- 2YUM 4205	
T6VMVTJ E1YQ- .L6VTJ D F 4385		1N1&R DNUNR 4205, 4298,	
T66 BN DN JNJ CZ EZ H1N1&R DVMYVQ2VQ .. 4588		1OR BZ E- 2 4205	
T66 BNJ.BO ENW 4290, 4300		2M2 4284	
VHH 4556		3XR&R&VO2N2&2 4266	
WNMYUM&N1&NO 4290			
WS1&O1 4289			
WS1&O2 .... 4288, 4298			
WS1&O2 2U -C 4583			
ZR B 4205			
ZR BG DR DZ CG 4205			
ZR DR 4205			

Lightning Source UK Ltd.
Milton Keynes UK
UKHW022201140219
337291UK00006B/612/P